Selected Titles in This Series

Volume

10 **Maurice Auslander:** Selected Works of Maurice Auslander, Parts 1 and 2 (Idun Reiten, Sverre O. Smalø, and Øyvind Solberg, Editors), 1999

9 **Lipman Bers:** Selected Works of Lipman Bers: Papers on Complex Analysis, Parts 1 and 2 (Irwin Kra and Bernard Maskit, Editors), 1998

8 **Walter E. Thirring:** Selected Papers of Walter E. Thirring with Commentaries, 1998

7 **Robert Steinberg:** Robert Steinberg Collected Papers, 1997

6 **Julia Robinson:** The Collected Works of Julia Robinson (Solomon Feferman, Editor), 1996

5 **Freeman Dyson:** Selected Papers of Freeman Dyson with Commentary, 1996

4 **Witold Hurewicz:** Collected Works of Witold Hurewicz (Krystyna Kuperberg, Editor), 1995

3.2 **A. Adrian Albert:** A. Adrian Albert Collected Mathematical Papers: Nonassociative Algebras and Miscellany, Part 2 (Richard E. Block, Nathan Jacobson, J. Marshall Osborn, David J. Saltman, and Daniel Zelinsky, Editors), 1993

3.1 **A. Adrian Albert:** A. Adrian Albert Collected Mathematical Papers: Associative Algebras and Riemann Matrices, Part 1 (Richard E. Block, Nathan Jacobson, J. Marshall Osborn, David J. Saltman, and Daniel Zelinsky, Editors), 1993

2 **Salomon Bochner:** Collected Papers of Salomon Bochner, Parts 1–4 (Robert C. Gunning, Editor), 1992

1 **R. H. Bing:** The Collected Papers of R. H. Bing, Parts 1 and 2 (Sukhjit Singh, Steve Armentrout, and Robert J. Daverman, Editors), 1988

SELECTED WORKS OF

MAURICE AUSLANDER

(Photograph courtesy of Gordana Todorov.)

MAURICE AUSLANDER
August 3, 1926–November 18, 1994

SELECTED WORKS OF MAURICE AUSLANDER

Part 2

Idun Reiten
Sverre O. Smalø
Øyvind Solberg
Editors

American Mathematical Society
Providence, Rhode Island

1991 *Mathematics Subject Classification.* Primary 13-xx, 14-xx, 16-xx, 18-xx, 19-xx, 20-xx.

Library of Congress Cataloging-in-Publication Data

Auslander, Maurice.

[Essays. Selections]

Selected works of Maurice Auslander / Idun Reiten, Sverre O. Smalø, Øyvind Solberg, editors.

p. cm.

Includes bibliographical references.

ISBN 0-8218-0679-3 (set). — ISBN 0-8218-0998-9 (pt. 1). — ISBN 0-8218-1000-6 (pt. 2)

1. Algebra. I. Reiten, Idun, 1942– . II. Smalø, Sverre O. III. Solberg, Øyvind, 1961– . IV. Title.

QA155.2.A882 1998

512—dc21 98-2926

CIP

A complete list of acknowledgments can be found at the back of this publication.

∞ The paper used in this book is acid-free and falls within the guidelines established to ensure permanence and durability.

Visit the AMS home page at URL: `http://www.ams.org/`

10 9 8 7 6 5 4 3 2 1 04 03 02 01 00 99

Contents

Foreword ix

Publication list for Maurice Auslander xi

Chapter VI. Lattices over general orders 1
- Functors and morphisms determined by objects 3
- Applications of morphisms determined by modules 115
- A survey of existence theorems for almost split sequences 157

Chapter VII. Tilting theory and homologically finite subcategories 163
- Coxeter functors without diagrams 165
- Preprojective modules over artin algebras 211
- Almost split sequences in subcategories 273
- Applications of contravariantly finite subcategories 303
- Homological theory of idempotent ideals 345

Chapter VIII. Almost split sequences and commutative rings 371
- Isolated singularities and existence of almost split sequences 373
- Rational singularities and almost split sequences 401
- Almost split sequences for rational double points 423
- The Cohen-Macaulay type of Cohen-Macaulay rings 435
- Almost split sequences for Cohen-Macaulay-modules 459
- The what, where and why of almost split sequences 465
- Cohen-Macaulay modules for graded Cohen-Macaulay rings and their completions 473
- Graded modules and their completions 485

Chapter IX. Grothendieck groups and Cohen-Macaulay approximations 497
- Grothendieck groups of algebras and orders 499
- Grothendieck groups of algebras with nilpotent annihilators 551
- The homological theory of maximal Cohen-Macaulay approximations 555
- Liftings and weak liftings of modules 589

Chapter X. Relative theory and syzygy modules 635
- Relative homology and representation theory I, Relative homology and homologically finite subcategories 637
- Relative homology and representation theory II, Relative cotilting theory 661
- k-Gorenstein algebras and syzygy modules 691
- Syzygy modules for noetherian rings 719

Acknowledgments 739

Foreword

In view of Maurice Auslander's important contributions to many parts of algebra, there is great interest in the gathering of several of his papers into the present volumes: "Selected Works of Maurice Auslander". In these we are pleased to present to the mathematical public the core of Auslander's work. Experts also, acquainted perhaps with only parts of his scientific production, will enjoy the wide range of his contributions and the impact of his work on the development of modern algebra.

Because of Auslander's long and close connection with the algebra group in Trondheim, through numerous visits and joint papers, we found it natural to undertake the task of editing such volumes. The American Mathematical Society, of which Auslander was a long time member, was the appropriate choice of publisher.

The selected works start with an article from the Notices of the AMS, written shortly after his death, which in addition to a short summary of his mathematical accomplishments also deals with aspects of his personality.

Even though Auslander worked in many different areas, there are characteristic features common to most of his work. The use of homological methods, including functor categories, should especially be mentioned. While his early work is mostly concerned with commutative rings and the later work mainly with artin algebras, he was always interested in finding common features and common settings for these situations. A synthesis of the topics was given through his theory of general orders, with applications to singularity theory. Typical for his work is also the large number of papers initiating new directions, and his name is attached to several concepts in algebra, including algebras, rings, modules, quivers, sequences, translates, conditions. In addition he is responsible for influential conjectures.

Despite these interconnections, we feel that it is easier to get an impression of Auslander's contribution by grouping related papers together. According to this principle we have organized the selected papers in ten chapters, of which five appear in each volume. For each chapter we provide some background material, interrelationship between the papers, and an indication of further developments. For this we give no specific references. Even though a large number of mathematicians would have deserved to be cited, we have limited mentioning names to a minimum.

Space requirements did not allow to include all of Auslander's papers. With the exception of the Queen Mary College notes which are not readily available, monographs are not included. A main criterion in the difficult selection process was to reflect the broad range and impact of his contributions. While the inclusion of some of the papers was unquestionable, some arbitrariness is involved in other cases.

Due to low quality of typing, several papers have been retyped. To be faithful to the original, we have made changes only when there were obvious typographical errors or inconsistencies in notation. We take the full responsibility for any new typos which may have been introduced.

We are grateful to several colleagues who offered their opinion in the selection process, and suggested improvements in what we have written. We also thank our doctoral students Aslak Buan, Ole Enge, Inger Heidi Slungård and Stig Venås for their help with proof reading the retyped manuscripts.

We express our thanks to the American Mathematical Society, and especially to Chris Thivierge, for fruitful collaboration. We also appreciate financial support from the Norwegian University of Science and Technology and from the Maurice Auslander Award of the American-Scandinavian Foundation.

Trondheim, November, 1997.

Idun Reiten Sverre O. Smalø Øyvind Solberg

Publication list for Maurice Auslander

I* [1] *On the dimension of modules and algebras* (III), *Global dimension*, Nagoya Math. J. 9 (1955) 67-77.

I [2] (with R. C. Lyndon) *Commutator subgroups of free groups*, Amer. J. Math. 77 (1955) 929–931.

[3] (with D. Buchsbaum) *Homological dimension in noetherian rings*, Proc. Nat. Acac. Sci. 42 (1956) 36–38.

I [4] *On the dimension of modules and algebras* (VI), *Comparison of global and algebra dimension*, Nagoya Math. J. 11 (1957) 61–65.

I [5] *On regular group rings*, Proc. Amer. Math. Soc. 8 (1957) 658–665.

I [6] (with D. Buchsbaum) *Homological dimension in local rings*, Trans. Amer. Math. Soc. 85 (1957) 390–405.

I [7] (with D. Buchsbaum) *Homological dimension in noetherian rings* II, Trans. Amer. Math. Soc. 88 (1958) 194–206.

[8] (with L. Auslander) *Solvable Lie groups and locally Euclidean Riemann spaces*, Proc. Amer. Math. Soc. 9 (1958) 933–941.

[9] (with A. Rosenberg) *Dimension of ideals in polynomial rings*, Can. J. Math. 10 (1958) 287–293.

I [10] (with D. Buchsbaum) *Codimension and multiplicity*, Ann. of Math. 68 (1958) 625–657. Correction, Ann. of Math. (2) 70 (1959) 395–397.

[11] *Remark on automorphisms of groups*, Proc. Amer. Math. Soc. 9 (1958) 229–230.

I [12] (with D. Buchsbaum) *Unique factorization in regular local rings*, Proc. Nat. Acad. Sci. USA 45 (1959) 733–734.

II [13] (with D. Buchsbaum) *On ramification theory in noetherian rings*, Amer. J. Math. 81 (1959) 749–765.

II [14] (with O. Goldman) *Maximal orders*, Trans. Amer. Math. Soc. 97 (1960) 1–24.

II [15] (with O. Goldman) *The Brauer group of a commutative ring*, Trans. Amer. Math. Soc. 97 (1960) 367–409.

II [16] *Modules over unramified regular local rings*, Illinois J. Math. 5 (1961) 631–647.

[17] (with D. Buchsbaum) *Invariant factors and two criteria for projectivity of modules*, Trans. Amer. Math. Soc. 104 (1962) 516–522.

II [18] *On the purity of the branch locus*, Amer. J. Math. 84 (1962) 116–125.

I [19] *A remark on a paper of M. Hironaka*, Amer. J. Math. 84 (1962) 8–10.

II [20] (with D. S. Rim) *Ramification index and multiplicity*, Illinois J. Math. 7 (1963) 566–581.

II [21] *Modules over unramified regular local rings*, Proc. International Congr. Mathematicians, Stockholm (1962) 230–233, Inst. Mittag-Leffler, Djursholm, 1963.

III [22] *Coherent functors*, Proc. Conf. Categorial Algebra (La Jolla, Calif. 1965) 189–231, Springer, New York, 1966.

*The Roman numerals indicate the chapter in which the paper appears.

[23] *Remarks on a theorem of Bourbaki*, Nagoya Math. J. 27 (1966) 361–369.

[24] *Anneaux de Gorenstein, et torsion en algèbre commutative*, Seminar Pierre Samuel, 1966/67 (Notes by M. Mangeney, C. Peskine and L. Szpiro), École Normale Supérieure de Jeunes Filles Secretariat mathematique, Paris 1967, 69 pp.

[25] *Comments on the functor* $\mathrm{Ext}^1(C,\ )$, Topology 8 (1969) 151–166.

II [26] (with A. Brumer) *Brauer groups of discrete valuation rings*, Nederl. Akad. Wetensch. Proc. Ser. A, 71, no. 3 (1968) 286–296.

[27] (with M. Bridger) *Stable module theory*, Memoirs of AMS. 94 (1969), American Mathematical Society, Providence, R.I. 1969, 146 pp.

IV [28] *Representation dimension of artin algebras*, Queen Mary College, Mathematics Notes, University of London (1971).

IV [29] (with K. W. Roggenkamp) *A characterization of orders of finite lattice type*, Invent. Math. 17 (1972) 79–84.

[30] *Representation theory of artin algebras*, Brandeis Univ. 1972 (Notes by I. Reiten).

III [31] (with I. Reiten) *Stable equivalence of artin algebras*, Proc. Conf. on Orders, Group rings and related topics (Ohio 1972) 8–71, Lecture Notes in Math. 353, Springer, Berlin 1973.

[32] (with D. Buchsbaum) Groups, Rings, Modules, Harper and Row. (1974) 470 pp.

IV [33] *Representation theory of artin algebras* I, Comm. Algebra 1 (1974) 177–268.

IV [34] *Representation theory of artin algebras* II, Comm. in Algebra 1 (1974) 269–310.

III [35] (with I. Reiten) *Stable equivalence of dualizing R-varieties*, Adv. in Math. 12 (1974) No. 3, 306–366.

IV [36] (with I. Reiten) *Representation theory of artin algebras* III*: Almost split sequences*, Comm. Algebra 3 (1975) 239–294.

[37] (with I. Reiten) *Stable equivalence of dualizing R-varieties* II*: Hereditary dualizing R-varieties*, Adv. in Math. 17, (1975) No. 2, 93–121.

[38] (with I. Reiten) *Stable equivalence of dualizing R-varieties* III*: Dualizing R-varieties stably equivalent to hereditary dualizing R-varieties*, Adv. in Math. 17 (1975) No. 2, 122–142.

[39] (with I. Reiten) *Stable equivalence of dualizing R-varieties* IV*: Higher global dimension*, Adv. in Math. 17, (1975) No. 2, 143–166.

[40] (with I. Reiten) *Stable equivalence of dualizing R-varieties* V*: Artin algebras stably equivalent to hereditary algebras*, Adv. in Math. 17, (1975) No. 2, 167–195.

V [41] (with I. Reiten) *On a generalized version of the Nakayama conjecture*, Proc. Amer. Math. Soc. 52 (1975) 69–74.

[42] *Almost split sequences* I, Proc. Int. Conf. on representations of algebras, Ottawa 1974, Lecture Notes in Math. 488 Springer-Verlag (1975) 1–8.

[43] (with I. Reiten) *Almost split sequences* II, Proc. Int. Conf. on representations of algebras, Carleton Univ., (Ottawa 1974), Lecture Notes in Math. 488, Springer Verlag (1975) 9–19.

[44] (with E. L. Green and I. Reiten) *Modules having waists*, Proc. Int. Conf. on representations of algebras, Ottawa 1974, Lecture Notes in Math. 488, Springer Verlag (1975) 20–28.

V [45] (with E. L. Green and I. Reiten) *Modules with waists*, Illinois J. Math. 19 (1975) 467–477.

[46] *Categorical methods in the representation theory of artin rings*, Univ. of Trondheim 1975, (Notes by S. O. Smalø).

IV [47] *Large modules over artin algebras*, Algebra, Topology and Category theory, (A collection of papers in honor of Samuel Eilenberg) 1–17 Academic Press, New York, 1976.

[48] (with I. Reiten) *On the representation type of triangular matrix rings*, J. London Math. Soc. (2), 12 (1976) No. 3, 371–382.

IV [49] (with I. Reiten) *Representation theory of artin algebras* IV*: Invariants given by almost split sequences*, Comm. Algebra 5 (1977) 443–518.

IV [50] (with I. Reiten) *Representation theory of artin algebras* V*: Methods for computing almost split sequences and irreducible morphisms*, Comm. Algebra 5 (1977) No. 5, 519–554.

[51] *Existence theorems for almost split sequences*, Ring theory II (Proc. Second Conf., Univ. Oklahoma, 1975) 1–44. Lecture Notes in Pure Appl. Math. 26, Dekker, New York, 1977.

IV [52] (with I. Reiten) *Representation theory of artin algebras* VI*: A functorial approach to almost split sequences*, Comm. Algebra 6 (1978) No. 3, 257–300.

[53] (with M. I. Platzeck and I. Reiten) *Periodic modules over weakly symmetric algebras*, J. Pure Appl. Algebra 11 (1977) 279–291.

IV [54] (with M. I. Platzeck) *Representation theory of hereditary artin algebras*, Representation theory of algebras, (Proc. Conf. Temple Univ., 1976) 389–424. Lecture Notes in Pure Appl. Math. 37, Dekker, New York, 1978.

VI [55] *Functors and morphisms determined by objects*, Representation theory of algebras (Proc. Conf. Temple Univ., 1976), 1–244. Lecture Notes in Pure Appl. Math. 37, Dekker, New York 1978.

VI [56] *Applications of morphisms determined by modules*, Representation theory of algebras (Proc. Conf. Temple Univ. 1976), 245–327. Lecture Notes in Pure Appl. Math. 37, Dekker, New York 1978.

VII [57] (with M. I. Platzeck and I. Reiten) *Coxeter functors without diagrams*, Trans. Amer. Math. Soc. 250 (1979) 1–46.

IV [58] (with R. Bautista, M. I. Platzeck, I. Reiten and S. O. Smalø) *Almost split sequences whose middle term has at most two indecomposable summands*, Canad. J. Math. 31, (1979) No 5, 942–960.

[59] *Preprojective modules over artin algebras*, Ring theory (Proc. Antwerp Conf. (NATO Adv. Study Inst.), Antwerp, 1978) 361–384. Lecture Notes in Pure appl. Math. 51, Dekker, New York 1979.

[60] (with S. O. Smalø) *Preprojective modules: An introduction and some applications*, Representation theory II (Proc. ICRA II, Carleton Univ., Ottawa 1979) 48–73. Lecture Notes in Math. 832, Springer, Berlin, 1980.

[61] (with I. Reiten) *Uniserial functors*, Representation theory II (Proc. ICRA II, Carleton Univ., Ottawa 1979) 1–47, Lecture Notes in Math. 832, Springer, Berlin 1980.

VII [62] (with S. O. Smalø) *Preprojective modules over artin algebras*, J. Algebra 66, (1980) No. 1, 61–122.

VII [63] (with S. O. Smalø) *Almost split sequences in subcategories*, J. Algebra 69, (1981) No. 2, 426–454, with Addendum J. Algebra, 71 (1981) No. 2 ,592–594.

[64] (with S. O. Smalø) *Preprojective lattices over classical orders.* Integral representations and applications. (Oberwolfach 1980) 326–345. Lecture Notes in Math. 882, Springer, Berlin-New York 1981.

[65] *Representation theory of finite-dimensional algebras*, Algebraists' homage: papers in ring theory and related topics (New Haven, Conn,. 1981), 27–39, Contemp. Math. 13, Amer. Math. Soc., Providence, R.I., 1982.

[66] (with S. O. Smalø) *Lattices over orders: Finitely presented functors and preprojective partitions*, Trans. Amer. Math. Soc. 273 (1982) No. 2, 433–445.

III [67] *A functorial approach to representation theory*, Representations of algebras. (ICRA III, Puebla 1980) 105–179. Lecture Notes in Math. 944, Springer, Berlin-New York, 1982.

IV [68] *Relations for Grothendieck groups of artin algebras*, Proc. Amer. Math. Soc. 91 (1984) No. 3, 336–340.

V [69] (with I. Reiten) *Modules determined by their composition factors*, Illinois J. Math. 29 (1985) No. 2, 280–301.

[70] *Finite type implies isolated singularity*, Orders and their applications (Oberwolfach, 1984), 1–4, Lecture Notes in Math. 1142, Springer, Berlin-New York, 1985.

VIII [71] *Isolated singularities and existence of almost split sequences*, Representation theory II (ICRA IV, Ottawa, 1984) 194–241, Lecture Notes in Math. 1178, Springer, Berlin-New York, 1986.

V [72] (with J. F. Carlson) *Almost split sequences and group rings*, J. Algebra 103 (1986) No. 1, 122–140.

IX [73] (with I. Reiten) *Grothendieck groups of algebras and orders*, J. Pure Appl. Algebra 39 (1986) No. 1–2, 1–51.

VIII [74] *Rational singularties and almost split sequences*, Trans. Amer. Math. Soc. 293 (1986) No. 2, 511–531.

[75] (with I. Reiten) *McKay quivers and extended Dynkin diagrams*, Trans. Amer. Math. Soc. 293 (1986) No. 1, 293–301.

VI [76] *A survey of existence theorems for almost split sequences*, Representations of algebras (Durham, 1985), 81–89, London Math. Soc. Lecture Note Ser. 116, Cambridge Univ. Press, Cambridge-New York, 1986.

[77] *Almost split sequences and algebraic geometry*, Representations of algebras (Durham, 1985), 165–179, London Math. Soc. Lecture Note Ser. 116, Cambridge Univ. Press, Cambridge-New York, 1986.

VIII [78] (with I. Reiten) *Almost split sequences for rational double points*, Trans. Amer. Math. Soc. 302 (1987) No. 1, 87–97.

VIII [79] (with I. Reiten) *The Cohen-Macaulay type of Cohen-Macaulay rings*, Adv. in Math., 73 (1989), No. 1, 1–23.

[80] (with I. Reiten) *Almost split sequences in dimension two*, Adv. in Math. 66 (1987) No. 1, 88–118.

[81] (with I. Reiten) *Almost split sequences for $\mathbb{Z}$-graded rings*, Singularities, representation of algebras and vector bundles (Lambrecht 1985), 232–243, Lecture Notes in Math. 1273, Springer, Berlin-New York(1987).

[82] (with I. Reiten) *Almost split sequences for abelian group graded rings*, J. Algebra,114 (1988) No. 1, 29–39.

VIII [83] (with I. Reiten) *Almost split sequences for Cohen-Macaulay modules*, Math. Ann. 277 (1987) No. 2, 345–349.

VIII [84] *The what, where and why of almost split sequences*, Proceedings of the International Congress of Mathematicians, vol. 1–2, (Berkeley, CA, 1986) 338–345, Amer. Math. Soc., Providence, RI, 1987.

VIII [85] (with I. Reiten) *Cohen-Macaulay modules for graded Cohen-Macaulay rings and their completions*, Commutative Algebra, (Berkeley, CA, 1987) 21–31, Math. Sci. Res. Inst. Publ. 15 Springer, Berlin-New York, 1989.

IX [86] (with I. Reiten) *Grothendieck groups of algebras with nilpotent annihilators*, Proc. Amer. Math. Soc. 103 (1988), No. 4, 1022–1024.

[87] (with E. L. Green) *Trace quotient modules*, Illinois J. Math. 32 (1988) No. 3, 534–556.

II [88] (with I. Reiten and S. O. Smalø) *Galois actions on rings and finite Galois coverings*, Math. Scand. 65 (1989) No. 1, 5–32.

IX [89] (with R. O. Buchweitz) *The homological theory of maximal Cohen-Macaulay approximations*, Colloque en l'honneur de Pierre Samuel (Orsay 1987), Mém. Soc. Math. France, (N.S.) No. 38 (1989) 5–37.

VIII [90] (with I. Reiten) *Graded modules and their completions*, Topics in Algebra, Part 1 (Warsaw 1989) 181–192, Banach Center Publ. 26, Part 1, PWN, Warsaw 1990.

[91] (with E. L. Green) *Modules over endomorphism rings*, Comm. Algebra 20 (5) (1992) 1259–1278.

VII [92] (with I. Reiten) *Applications of contravariantly finite subcategories*, Adv. in Math. 86, No. 1 (1991) 111–152.

[93] (with I. Reiten) *Cohen-Macaulay and Gorenstein artin algebras*, Representation theory of finite groups and finite-dimensional algebras, (Bielefeld 1991) 221–245, Progr. Math. 95, Birkhauser, Basel (1991).

V [94] (with I. Reiten) *On a theorem of E. Green on the dual of the transpose*, Representations of finite-dimensional algebras (Tsukuba 1990) 53–65, CMS Conf. Proc., 11, Amer. Math. Soc., Providence RI (1991) 53–65.

[95] (with I. Reiten) *Homologically finite subcategories*, Representations of algebras and related topics (ICRA V, Tsukuba 1990). London Math. Soc. Lecture Note Ser. 168, 1–42 (199).

VII [96] (with M. I. Platzeck and G. Todorov) *Homological theory of idempotent ideals*, Trans. Amer. Math. Soc. 332 (1992) No. 2, 667–692.

IX [97] (with S. Ding and Ø. Solberg) *Liftings and weak liftings of modules*, J. Algebra 156 (1993) No. 2, 273–317.

X [98] (with Ø. Solberg) *Relative homology and representation theory* I, *Relative homology and homologically finite subcategories*, Comm. Algebra 21 (9) (1993) 2995–3031.

X [99] (with Ø. Solberg) *Relative homology and representation theory* II, *Relative cotilting theory*, Comm. Algebra 21 (9) (1993) 3033–3079.

[100] (with Ø. Solberg) *Relative homology and representation theory* III, *Cotilting modules and Wedderburn correspondence*, Comm. Algebra 21 (9) (1993) 3081–3097.

[101] (with Ø. Solberg) *Gorenstein algebras and algebras with dominant dimension at least 2*, Comm. Algebra 21 (11) (1993) 3897–3934.

X [102] (with I. Reiten) *k-Gorenstein algebras and syzygy modules*, J. Pure Appl. Algebra 92 (1994) 1–27.

[103] (with M. Kleiner) *The Green correspondence for adjoint functors*, Representations of algebras (ICRA VI, Ottawa 1992) 69–81, CMS Conf. Proc. 14, Amer. Math. Soc., Providence, RI, 1993.

[104] (with Ø. Solberg) *Relative homology*, Finite-dimensional algebras and related topics (ICRA VI, Ottawa 1992) Kluwer Academic Publishers, Nato Adv. Sci. Inst. Ser. C Math. Phys. Sci. 424 (1994) 347–359.

III [105] (with M. Kleiner) *Adjoint functors and an extension of the Green correspondance for group representations*, Adv. in Math. 104, (1994) No. 2, 297–314.

[106] (with I. Reiten and S. O. Smalø) *Representation theory of artin algebras*, Cambridge Univ. Press, Cambridge 1995.

III [107] (with I. Reiten) $D\,\mathrm{Tr}$*-periodic modules and functors*, Representation theory of algebras (ICRA VII, Cocoyoc 1994) 39–50, CMS Conf. Proc. 18, Amer. Math. Soc., Providence, RI, 1996.

X [108] (with I. Reiten) *Syzygy modules for noetherian rings*, J. Algebra 183 (1996) No. 1, 167–185.

CHAPTER VI

Lattices over general orders

[55] Functors and morphisms determined by objects (1978).
[56] Applications of morphisms determined by modules (1978).
[76] A survey of existence theorems for almost split sequences (1986).

While Auslander's main topic of research since the early seventies was the representation theory of artin algebras, he kept also during this period of time his previous interest in the more general theory of noetherian rings, in particular orders and commutative rings. The papers in this chapter deal with a general theory containing these classes of rings as special cases.

After proving a result about artin algebras, Auslander would usually try to see if there were generalizations to arbitrary noetherian rings, or at least analogous results for commutative rings or for orders. In particular, he applied this point of view to the theory of almost split sequences and irreducible maps. A key problem was whether there was an analogue of the important operation $D\operatorname{Tr}$ (the dual of the transpose) in other settings than for artin algebras. He was encouraged about such a development after first proving existence of almost split sequences for lattices over classical orders.

The main focus of [55] is a generalization of almost split sequences in a different direction. Auslander introduced a new type of morphisms, called morphisms right (or left) determined by a module. Like for almost split sequences the definition make sense in very general contexts. He investigated thoroughly the existence of such morphisms for modules over arbitrary rings, and the related concept of a subfunctor determined by a module. He also described the kernels of morphisms right determined by a module, and like for right almost split maps, the transpose played again a central role, by appearing in a crucial formula relating Ext and Hom. Analogous results were given in the general setup of lattices over orders, where the orders are certain noetherian algebras over commutative Gorenstein rings and the lattices certain modules which are (maximal) Cohen-Macaulay over the Gorenstein ring. The transpose is replaced by a modified version, namely composed with a syzygy. This amazing theory contains both modules over artin algebras, lattices over classical orders and (maximal) Cohen-Macaulay modules over certain commutative rings.

A theory of almost split sequences for lattices over general orders was already started in [51]. In [55] it is obtained through specialization of the general theory of morphisms (and subfunctors) determined by modules. Also quite general existence theorems for almost split sequences over arbitrary rings are obtained by specializing the results on morphisms determined by modules. For example, if C is a finitely presented module with local endomorphism ring, there is an almost split sequence with right hand term C.

Specializations of the general theory to artin algebras are given in [56]. This paper also contains a discussion of generalizations of the first Brauer-Thrall conjecture. A very important and influential result from this paper is that if an indecomposable algebra has a finite component of irreducible maps, then the algebra is of finite representation type.

This result, and its later extension to more general settings, has been a useful criterion for proving finite representation type.

The main application of the existence theorems for almost split sequences over general orders has been to commutative ring theory and singularities. This topic will be discussed in Chapter VIII. The existence theorems for arbitrary rings have been useful for work on artin rings and in model theory.

Auslander himself was very excited about the general theory of morphisms determined by modules and was convinced that there were important applications still to come. In order to make this theory more accessible, in the setting of artin algebras, it is discussed in the last chapter of the book [106].

Functors and morphisms determined by objects[1]

Maurice Auslander[2]

Introduction

In the course of our study of the representation theory of artin algebras, I. Reiten and I introduced the notions of right or left almost split morphisms, as well as almost split sequences, of finitely generated modules over artin algebras. In view of the important role these notions have played in elucidating some questions in the representation theory of artin algebras (see [1] for an account of some of the results that have been obtained along these lines), it is natural to wonder if these ideas are special to artin algebras or are of broader interest. This is an especially tempting question since nowhere in the definitions of these notions is the fact that one is dealing with finitely generated modules over artin algebras used. In fact, the formulations given for artin algebras of all these notions make sense in arbitrary abelian categories and some even in arbitrary categories. Moreover, many of the proofs given for artin algebras carry over verbatim to these categorical settings. Thus the problem of generalization boils down basically to the question of when these types of morphisms and exact sequences exist.

This point of view was pursued in [3] where it was shown that right or left almost split morphisms and almost split sequences do indeed exist in more general contexts than artin algebras. That paper dealt with various categories of modules over rings Λ which are algebras over complete noetherian local rings R such that Λ is a finitely generated R-module. Amongst other things, it was shown that if R is a complete discrete valuation ring and Λ is an R-order in the classical sense, then the category of Λ-lattices has a theory of right and left almost split morphisms and almost split sequences entirely analogous to that for the category of finitely generated modules over artin algebras.

In this paper we generalize our previous work in a somewhat different direction. Instead of finding various new contexts in which right or left almost split morphisms exist, we introduce a new type of morphisms, those which are either right or left determined by objects. We will presently see that right and left almost split morphisms are special cases of morphisms which are right or left determined by objects. While the notion of a morphism being determined by an object is a general categorical one, as in the case of almost split morphisms, the problem of their existence is not at all clear, and it is with this question that this paper is mainly concerned.

In an effort to make what appears to be a somewhat long and technically involved paper easier to follow, the rest of this introduction is devoted to giving a brief survey of the basic notions and results of the paper. For the most part, the general organization

[1]Lecture Notes in Pure Appl. Math. 37, 1–244 (1978)
[2]Written with partial support of NSF Grant MCS 72-04584.

of the survey follows that of the paper, but certain liberties have been taken when this seemed appropriate to aid the exposition of the basic ideas. To simplify the presentation, we assume that all categories are additive, even though much of what is said of a purely categorical nature can be carried out in arbitrary categories.

Let $f\colon B \to C$ be a morphism in the category $\mathcal{C}$. We recall (see [7], [11]) that f is said to be right almost split if a) f is not a splittable epimorphism (f is said to be a splittable epimorphism if there is a morphism $s\colon C \to B$ such that $fs = 1_C$) and b) given a morphism $g\colon X \to C$ in $\mathcal{C}$ which is not a splittable epimorphism, then there is a morphism $h\colon X \to B$ such that $fh = g$. It is easily checked that if $f\colon B \to C$ is right almost split, then a) the endomorphism ring $\operatorname{End} C$ of C is a local ring, b) $\operatorname{Im}(\operatorname{Hom}_{\mathcal{C}}(C,B) \xrightarrow{(C,f)} \operatorname{End} C)$ is the unique maximal right ideal of $\operatorname{End} C$ and c) if $g\colon Y \to C$ is a morphism in $\mathcal{C}$ such that $\operatorname{Im}(\operatorname{Hom}_{\mathcal{C}}(C,Y) \xrightarrow{(C,g)} \operatorname{End} C)$ is contained in the unique maximal right ideal of $\operatorname{End} C$, then there is a morphism $h\colon Y \to B$ such that $g = fh$. In fact, a little thought suffices to show that a morphism $f\colon B \to C$ in $\mathcal{C}$ is right almost split if and only if it satisfies conditions a), b) and c). This formulation suggests the following definition.

A morphism $f\colon B \to C$ in $\mathcal{C}$ is said to be right determined by an object X in $\mathcal{C}$ (or more simply f is right X-determined) if a morphism $g\colon Y \to C$ has the property $\operatorname{Im}(\operatorname{Hom}_{\mathcal{C}}(X,Y) \xrightarrow{(X,g)} \operatorname{Hom}_{\mathcal{C}}(X,C))$ is contained in $\operatorname{Im}(\operatorname{Hom}_{\mathcal{C}}(X,B) \xrightarrow{(X,f)} \operatorname{Hom}_{\mathcal{C}}(X,C))$ if and only if there is a morphism $h\colon Y \to B$ such that $fh = g$. Before discussing the question of the existence of morphisms which are right determined by objects in $\mathcal{C}$, we take up the question of the uniqueness of such morphisms.

Suppose $f\colon B \to C$ is a morphism in $\mathcal{C}$ and X an object in $\mathcal{C}$. Then it is easily seen that $H = \operatorname{Im}(\operatorname{Hom}_{\mathcal{C}}(X,B) \xrightarrow{(X,f)} \operatorname{Hom}_{\mathcal{C}}(X,C))$ is an $(\operatorname{End} X)^{\mathrm{op}}$-submodule of the $(\operatorname{End} X)^{\mathrm{op}}$-module $\operatorname{Hom}_{\mathcal{C}}(X,C)$, where $(\operatorname{End} X)^{\mathrm{op}}$ denotes the opposite ring of $\operatorname{End} X$ and the abelian group $\operatorname{Hom}_{\mathcal{C}}(X,C)$ is considered an $(\operatorname{End} X)^{\mathrm{op}}$-module (all modules are left modules) by means of the usual operation $t \cdot f$ where $t \cdot f$ is the composition ft for all t in $\operatorname{End} X$ and f in $\operatorname{Hom}_{\mathcal{C}}(X,C)$. Now if $f\colon B \to C$ is right X-determined then a morphism $g\colon Y \to C$ has the property that there is a morphism $h\colon Y \to B$ such that $g = fh$ if and only if the $(\operatorname{End} X)^{\mathrm{op}}$-submodule $\operatorname{Im}(\operatorname{Hom}_{\mathcal{C}}(X,Y) \xrightarrow{(X,g)} \operatorname{Hom}_{\mathcal{C}}(X,C))$ of $\operatorname{Hom}_{\mathcal{C}}(X,C)$ is contained in the $(\operatorname{End} X)^{\mathrm{op}}$-submodule $H = \operatorname{Im}(\operatorname{Hom}_{\mathcal{C}}(X,B) \xrightarrow{(X,f)} \operatorname{Hom}_{\mathcal{C}}(X,C))$ of $\operatorname{Hom}_{\mathcal{C}}(X,C)$. Thus two right X-determined morphisms $f\colon B \to C$ and $f'\colon B' \to C$ have the property that the $(\operatorname{End} X)^{\mathrm{op}}$-submodules $\operatorname{Im}(X,f)$ and $\operatorname{Im}(X,f')$ of $\operatorname{Hom}_{\mathcal{C}}(X,C)$ are the same, if and only if there is a commutative diagram

$$\begin{array}{ccc} B & \xrightarrow{f} & C \\ {\scriptstyle h}\downarrow & & \| \\ B' & \xrightarrow{f'} & C \\ {\scriptstyle h'}\downarrow & & \| \\ B & \xrightarrow{f} & C. \end{array}$$

This observation easily shows that a right X-determined morphism $f\colon B \to C$ need not necessarily be determined even up to isomorphism by the $(\operatorname{End} X)^{\mathrm{op}}$-submodule $\operatorname{Im}(X,f)$ of $\operatorname{Hom}_{\mathcal{C}}(X,C)$, where by an isomorphism from $f\colon B \to C$ to $f'\colon B' \to C$ we mean an isomorphism $u\colon B \to B'$ such that $f = f'u$. However, as we now point out, there is a simple condition that can be put on a morphism $f\colon B \to C$ which guarantees that if it is right X-determined, then it is uniquely determined, up to isomorphism, by the $(\operatorname{End} X)^{\text{op}}$-submodule $\operatorname{Im}(X, f)$ of $\operatorname{Hom}_{\mathcal{C}}(X, C)$.

We say that a morphism $f\colon B \to C$ is right minimal provided an endomorphism $u\colon B \to B$ is an isomorphism whenever $fu = f$. Our previous remarks show that two right X-determined morphisms which are also right minimal $f\colon B \to C$ and $f'\colon B' \to C$ are isomorphic if and only if $\operatorname{Im}(X, f) = \operatorname{Im}(X, f')$. With these preliminary general remarks in mind we now turn our attention to the question of the existence of right determined morphisms. All of the existence theorems we know are based on the following theorem which is proven in Section 3 of Chapter I.

Let $\mathcal{C}$ be the category $\operatorname{Mod} \Lambda$ of all Λ-modules for some arbitrary ring Λ. Suppose X is a finitely presented Λ-module and C is an arbitrary Λ-module. Let H be an $(\operatorname{End} X)^{\text{op}}$-submodule of $\operatorname{Hom}_\Lambda(X, C)$ containing the $(\operatorname{End} X)^{\text{op}}$-submodule $P(X, C)$ consisting of all morphisms $h\colon X \to C$ which are a composition $X \to P \to C$ with P a projective Λ-module. Then there is an epimorphism $f\colon B \to C$ of Λ-modules which is right minimal and right X-determined such that $\operatorname{Im}(X, f) = H$. Furthermore, this right minimal and right X-determined morphism $f\colon B \to C$ has the following property. Suppose $f'\colon B' \to C$ is any right X-determined morphism such that $\operatorname{Im}(X, f') = H$, then $f'\colon B' \to C$ is isomorphic to $g\colon B \amalg B'' \to C$ where $B \amalg B''$ is the direct sum of B and some other module B'' and $g|_B = f$ while $g|_{B''} = 0$.

An obvious question to ask concerning the right minimal and right X-determined epimorphism $f\colon B \to C$ is what is $\operatorname{Ker} f$? While a complete answer to this question can be found in Section 3, we briefly indicate the form of the answer.

Let $P_1 \xrightarrow{v} P_0 \xrightarrow{\epsilon} X \to 0$ be a projective presentation of X, i.e. $P_1 \xrightarrow{v} P_0 \xrightarrow{\epsilon} X \to 0$ is an exact sequence with the P_i finitely generated projective Λ-modules. Then $\operatorname{Hom}_\Lambda(P_i, \Lambda)$ is a finitely generated projective Λ^{op}-module and so $\operatorname{Tr} X = \operatorname{Coker}(\operatorname{Hom}_\Lambda(P_0, \Lambda) \to \operatorname{Hom}_\Lambda(P_1, \Lambda))$ is a finitely presented Λ^{op}-module. Let $\Gamma = \operatorname{End}_{\Lambda^{\text{op}}}(\operatorname{Tr} X)$. Then there is an injective Γ-module I such that the abelian group $\operatorname{Hom}_\Gamma(\operatorname{Tr} X, I)$ together with the Λ-module structure induced from the Λ^{op}-module structure of Λ^{op} on $\operatorname{Tr} X$ is isomorphic to the kernel of the right minimal and right X-determined epimorphism $f\colon B \to C$. The particular injective Γ-module I used in this construction is the injective envelope of $\operatorname{Hom}_\Lambda(X, C)/H$, which is not only an $(\operatorname{End} X)^{\text{op}}$-module, but also a Γ-module in a natural way, as explained in Section 3.

Before giving some of the applications of this main existence theorem, we point out that no comparable theorem exists for the dual notion of morphisms left determined by an object. More precisely, a morphism $f\colon A \to B$ in a category $\mathcal{C}$ is said to be left determined by Y in $\mathcal{C}$ (or is left Y-determined) if a morphism $g\colon A \to X$ has the property $\operatorname{Im}(\operatorname{Hom}_{\mathcal{C}}(X, Y) \xrightarrow{(g,Y)} \operatorname{Hom}_{\mathcal{C}}(A, Y))$ is contained in $\operatorname{Im}(\operatorname{Hom}_{\mathcal{C}}(B, Y) \xrightarrow{(f,Y)} \operatorname{Hom}_{\mathcal{C}}(A, Y))$ if and only if there is a morphism $h\colon B \to X$ such that $hf = g$. All the known existence theorems for morphisms left determined by objects are derived from existence theorems for morphisms right determined by objects by means of some sort of duality. This and other uses of dualities is part of the reason why most of the theory developed in this paper is for algebras rather than rings.

Rather than getting involved in the technicalities needed to give the applications of the main existence theorem to the derivation of other existence theorems in the full generality given in the latter part of Chapter I, I simply give a few sample results in the classical cases of finite dimensional algebras over a field and classical orders over Dedekind rings.

Let Λ be a finite dimensional algebra over a field k. Suppose $0 \to A \xrightarrow{g} B \xrightarrow{f} C \to 0$ is an exact sequence of finitely generated Λ-modules. Then f is right determined by $\operatorname{Tr}(\operatorname{Hom}_k(A, k))$ and g is left determined by $\operatorname{Hom}_k(\operatorname{Tr} C, k)$. More generally, any morphism $f\colon B \to C$ of finitely generated Λ-modules is right determined by $\operatorname{Tr}(\operatorname{Hom}_k(\operatorname{Ker} f, k)) \amalg \Lambda$ and is left determined by $\operatorname{Hom}_k(\operatorname{Tr}(\operatorname{Coker} f), k) \amalg \operatorname{Hom}_k(\Lambda, k)$. Thus we see that as far as finite dimensional algebras are concerned, the property of morphisms being left and right determined by modules is a universal property of morphisms between finitely generated modules.

Suppose now that R is a Dedekind domain (not a field) and Λ an R-order in the classical sense. Let $L(\Lambda)$ be the category of Λ-lattices, i.e. Λ-modules which are finitely generated projective R-modules. Suppose X is a Λ-lattice. Then we define the Λ^{op}-lattice $\operatorname{Tr}_L X$ as follows. Let $P_1 \xrightarrow{b} P_0 \xrightarrow{\epsilon} X \to 0$ be a projective presentation of X. Then $\operatorname{Tr}_L X = \operatorname{Im}(\operatorname{Hom}_\Lambda(P_0, \Lambda) \to \operatorname{Hom}_\Lambda(P_1, \Lambda))$ by definition. Now suppose $0 \to A \xrightarrow{g} B \xrightarrow{f} C \to 0$ is an exact sequence of Λ-modules each of which is a Λ-lattice. Viewing g and f as morphisms in $L(\Lambda)$, we have that f is right $\operatorname{Tr}_L(\operatorname{Hom}_R(A, B)$-determined in $L(\Lambda)$ and g is left $\operatorname{Hom}_R(\operatorname{Tr}_L C, R)$-determined in $L(\Lambda)$. More generally, a morphism $f\colon B \to C$ in $L(\Lambda)$ is both left and right determined by objects in $L(\Lambda)$.

Chapter II is devoted to applying some of the results of Chapter I to showing that simple functors on suitable categories of modules are finitely presented. In this connection the notion of a subfunctor of a functor being determined by an object in a category is introduced. Suppose $F\colon \mathcal{C}^{\mathrm{op}} \to \mathrm{Ab}$ is a functor (all functors are additive) where $\mathcal{C}^{\mathrm{op}}$ is the opposite category of $\mathcal{C}$. A subfunctor G of F is said to be determined by an object X in $\mathcal{C}$ if a subfunctor G' of F is contained in G whenever $G'(X) \subseteq G(X)$. The connection between this notion and that of a morphism $f\colon B \to C$ in $\mathcal{C}$ being right determined by X is given by the following easily verified result. A morphism $f\colon B \to C$ is right X-determined if and only if the induced morphism $(\ ,f)\colon (\ ,B) \to (\ ,C)$ of representable functors has the property that the subfunctor $\operatorname{Im}(\ ,f)$ of $(\ ,C)$ is determined by X.

Suppose now that G and G' are two X-determined subfunctors of a functor $F\colon \mathcal{C}^{\mathrm{op}} \to \mathrm{Ab}$. Then it is obvious that $G = G'$ if and only if $G(X) = G'(X)$. Thus an X-determined subfunctor G of F is uniquely determined by the $(\operatorname{End} X)^{\mathrm{op}}$-submodule $G(X)$ of $F(X)$. This suggests the question: If H is an $(\operatorname{End} X)^{\mathrm{op}}$-submodule of $F(X)$ is there an X-determined subfunctor G of F such that $G(X) = H$? It is easy to see that this question has an affirmative answer for each $(\operatorname{End} X)^{\mathrm{op}}$-submodule H of F. We denote by F_H the unique X-determined subfunctor of F such that $F_H(X) = H$.

In particular, for each C in $\mathcal{C}$ and each right ideal $\underline{a}$ of $\operatorname{End} C$, there is associated the C-determined subfunctor $(\ ,C)_{\underline{a}}$ of $(\ ,C)$. It is not very difficult to see that $\underline{a}$ is a maximal right ideal of $\operatorname{End} C$ if and only if $(\ ,C)_{\underline{a}}$ is a maximal subfunctor of $(\ ,C)$. Moreover a subfunctor G of $(\ ,C)$ is maximal if and only if $G(C)$ is a maximal right ideal of $\operatorname{End} C$ and $G = (\ ,C)_{G(C)}$. Thus there is a one-to-one correspondence between the maximal right ideals $\underline{m}$ of $\operatorname{End} C$ and the maximal subfunctors of $(\ ,C)$, the correspondence being given by $\underline{m} \mapsto (\ ,C)_{\underline{m}}$ for all maximal right ideals of $\operatorname{End} C$. From this it follows that a simple functor $S\colon \mathcal{C}^{\mathrm{op}} \to \mathrm{Ab}$ such that $S(C) \neq 0$ is isomorphic to $(\ ,C)/(\ ,C)_{\underline{m}}$ for some maximal right ideal $\underline{m}$ of $\operatorname{End} C$. Moreover, since each simple functor $S \neq 0$, there is some C in $\mathcal{C}$ such that $S(C) \neq 0$. Thus all simple functors from $\mathcal{C}^{\mathrm{op}}$ to Ab are isomorphic to $(\ ,C)/(\ ,C)_{\underline{m}}$ for some C in $\mathcal{C}$ and maximal right ideal $\underline{m}$ of $\operatorname{End} C$.

Next we recall that a functor $F\colon \mathcal{C}^{\mathrm{op}} \to \mathrm{Ab}$ is said to be finitely presented if there is an exact sequence of functors $(\ ,A) \to (\ ,B) \to F \to 0$. Therefore the simple functors are finitely presented if and only if for each C in $\mathcal{C}$ and maximal ideal $\underline{m}$ of $\operatorname{End} C$ there is a right C-determined morphism $f\colon B \to C$ such that $\operatorname{Im}((C, B) \xrightarrow{(C,f)} (C, C)) = \underline{m}$.

Chapter II is mainly devoted to showing how the results of Chapter I concerning morphisms which are right determined by objects in certain categories $\mathcal{C}$ of modules over particular rings can be used to prove that the simple functors from $\mathcal{C}^{\mathrm{op}}$ to Ab are finitely presented. In particular, if Λ is a finite dimensional algebra over a field k, then the simple functors from $(\operatorname{mod} \Lambda)^{\mathrm{op}}$ to Ab are finitely presented where $\operatorname{mod} \Lambda$ is the category of all finitely generated Λ-modules. Also if Λ is an order over a Dedekind domain R, then all the simple functors from $(L(\Lambda))^{\mathrm{op}}$ to Ab are finitely presented. Similar results for simple functors from a category $\mathcal{C}$ to Ab are also established. These results lead us back to almost split sequences, the starting point of this whole discussion.

While the major portion of this paper is devoted to exploring the notions we just finished discussing, there are sprinkled throughout the paper various applications not directly connected with the development of the general theory. It is hoped that some of these notions will prove useful in studying other such questions, as has already happened in the case of finite dimensional algebras over fields as indicated by some of the other papers in this volume (see [1] and [9]).

Finally, I would like to thank Professor Robert Gordon for his skill and diligence in organizing the conference, the NFS and Temple University for their financial backing, and the participants in the conference for their spirit and good humor.

Chapter I: Existence theorems

This chapter is devoted to proving the basic existence theorems concerning functors and morphisms determined by objects in various categories of modules over particular rings. We begin with the definition and some of the basic properties of functors from a category to abelian groups determined by objects in the category.

1. Subfunctors determined by objects

We recall that a category $\mathcal{C}$ is said to be a *preadditive category* if a) for each pair of objects X, Y in $\mathcal{C}$ the set (X, Y) of morphisms from X to Y has an abelian group structure and b) the maps $(X, Y) \times (Y, Z) \to (X, Z)$ given by $(f, g) \to gf$, the composition of f and g in $\mathcal{C}$, is bilinear for all X, Y, Z in $\mathcal{C}$. Also if $\mathcal{C}$ and $\mathcal{D}$ are preadditive categories, then a functor $F\colon \mathcal{C} \to \mathcal{D}$ is said to be *additive* if the morphisms $F\colon (X, Y) \to (F(X), F(Y))$ are group morphisms for all X, Y in $\mathcal{C}$.

From now on we assume, unless stated to the contrary, that all categories are preadditive and all functors are additive. The collection of all functors from the category $\mathcal{C}$ to the category $\mathcal{D}$ will be denoted by $(\mathcal{C}, \mathcal{D})$. Finally, we denote the opposite category of $\mathcal{C}$ by $\mathcal{C}^{\mathrm{op}}$. Clearly $\mathcal{C}^{\mathrm{op}}$ is a preadditive category. We will usually follow the convention of considering the contravariant functors from $\mathcal{C}$ to $\mathcal{D}$ as the functors (covariant) from $\mathcal{C}^{\mathrm{op}}$ to $\mathcal{D}$.

Suppose C is an object in $\mathcal{C}$. Then the abelian group (C, C) together with the multiplication given by the composition of morphisms in $\mathcal{C}$ is a ring called the *endomorphism ring* of C and is denoted by $\operatorname{End} C$. Assume now that $G\colon \mathcal{C}^{\mathrm{op}} \to \mathrm{Ab}$ is a functor, where Ab is the category of abelian groups. Then for each C in $\mathcal{C}$, the abelian group $G(C)$ is an $(\operatorname{End} C)^{\mathrm{op}}$-module where $(\operatorname{End} C)^{\mathrm{op}}$ is the opposite ring of $\operatorname{End} C$, the operation of f on $G(C)$ being given by $f \cdot x = G(f)(x)$ for all f in $\operatorname{End} C$ and x in $G(C)$. Clearly if

$\alpha\colon F \to G$ is a morphism of functors from $\mathcal{C}^{\mathrm{op}}$ to Ab, then for each C in $\mathcal{C}$ the morphism $\alpha_C\colon F(C) \to G(C)$ is a morphism of $(\operatorname{End} C)^{\mathrm{op}}$-modules. The morphism $\alpha\colon F \to G$ is said to be an *inclusion morphism* if $\alpha_C\colon F(C) \to G(C)$ is an inclusion of $(\operatorname{End} C)^{\mathrm{op}}$-modules for each C in $\mathcal{C}$. A functor F is said to be a *subfunctor* of G, if there is an inclusion morphism $\alpha\colon F \to G$. We now point out a way of describing the subfunctors of a functor.

LEMMA 1.1. *Suppose G is in $(\mathcal{C}^{\mathrm{op}}, \mathrm{Ab})$. If for each X in $\mathcal{C}$ we are given a subgroup A_X of $G(X)$ such that $G(f)(A_Y) \subset A_X$ for all morphisms $f\colon X \to Y$ in $\mathcal{C}$, then there is a unique subfunctor F of G such that $F(X) = A_X$ for all X in $\mathcal{C}$.*

Of particular concern to us are the subfunctors of a functor in $(\mathcal{C}^{\mathrm{op}}, \mathrm{Ab})$ determined by objects in $\mathcal{C}$, a notion we now define.

Let X be in $\mathcal{C}$. A subfunctor F of a functor $G\colon \mathcal{C}^{\mathrm{op}} \to \mathrm{Ab}$ is said to be *determined* by X if a subfunctor F' of G is contained in F whenever $F'(X) \subset F(X)$. Obviously if F and F' are two subfunctors of G determined by X, then $F = F'$ if and only if $F(X) = F'(X)$. A natural question to ask at this point is whether for each $(\operatorname{End} X)^{\mathrm{op}}$-submodule H of $G(X)$ there is a subfunctor F of G determined by X such that $F(X) = H$. We now show that this question has an affirmative answer.

Suppose G is in $(\mathcal{C}^{\mathrm{op}}, \mathrm{Ab})$, X is an object of $\mathcal{C}$ and H is an $(\operatorname{End} X)^{\mathrm{op}}$-submodule of $G(X)$. For each C in $\mathcal{C}$ define A_C to be the subgroup of $G(C)$ consisting of all x in $G(C)$ such that for each $f\colon X \to C$ the element $G(f)(x)$ in $G(X)$ is contained in H. It is easily checked that for each morphism $g\colon U \to V$ in $\mathcal{C}$ we have $G(g)(A_V) \subset A_U$. Thus, by Lemma 1.1, there is a unique subfunctor, which we denote by G_H, of G having the property $G_H(C) = A_C$ for all C in $\mathcal{C}$. Furthermore, the fact that H is an $(\operatorname{End} X)^{\mathrm{op}}$-submodule of $G(X)$ implies that $G_H(X) = H$.

It is a straightforward matter to deduce the following facts concerning subfunctors determined by objects from our discussion so far.

PROPOSITION 1.2. *Let X be an object in $\mathcal{C}$ and G a functor in $(\mathcal{C}^{\mathrm{op}}, \mathrm{Ab})$.*

a) *If H is an $(\operatorname{End} X)^{\mathrm{op}}$-submodule of $G(X)$, then the subfunctor G_H of G is determined by X and has the property $G_H(X) = H$.*

b) *A subfunctor F of G is determined by X if and only if $F = G_H$ where $H = F(X)$.*

c) *The map $H \mapsto G_H$ gives a bijection between the $(\operatorname{End} X)^{\mathrm{op}}$-submodules H of $G(X)$ and the subfunctors of G determined by X.*

As an immediate consequence of the definition of a subfunctor being determined by an object we have the following.

PROPOSITION 1.3. *Suppose X is in $\mathcal{C}$ and $F' \subset F \subset G$ are in $(\mathcal{C}^{\mathrm{op}}, \mathrm{Ab})$. Then $F \subset G$ is determined by X if and only if $F/F' \subset G/F'$ is determined by X.*

Since $\mathcal{C} = (\mathcal{C}^{\mathrm{op}})^{\mathrm{op}}$, we have that $(\mathcal{C}, \mathrm{Ab}) = ((\mathcal{C}^{\mathrm{op}})^{\mathrm{op}}, \mathrm{Ab})$. This observation shows how to give analogous definitions and results for functors in $(\mathcal{C}, \mathrm{Ab})$ to those given for functors in $(\mathcal{C}^{\mathrm{op}}, \mathrm{Ab})$. These analogues will be used freely in the rest of the paper.

I would like to thank D. Eisenbud for suggesting the terminology "a subfunctor determined by an object" and the construction of the subfunctors G_H of G.

2. Morphisms determined by objects

This section is mainly devoted to describing when a subfunctor of a functor determined by an object is finitely generated. For simplicity of exposition we assume throughout this section that $\mathcal{C}$ is an additive and not just a preadditive category. We will be using constantly the well known result (Yoneda's lemma) that if G is in $(\mathcal{C}^{\mathrm{op}}, \mathrm{Ab})$ and x is in $G(X)$, then there is a unique morphism $\beta\colon (\ ,X) \to G$ such that $\beta_X(1_X) = x$.

Recall (see [6] for more details) that a functor F in $(\mathcal{C}^{\mathrm{op}}, \mathrm{Ab})$ is said to be *finitely generated* if there is an epimorphism $(\ ,C) \to F$ for some C in $\mathcal{C}$. This definition is based on the fact that the representable functors $(\ ,C)$ are projective objects in $(\mathcal{C}, \mathrm{Ab})$ which have many of the same categorical properties as finitely generated projective modules. We now develop a criterion for when a subfunctor F of G determined by an object X in $\mathcal{C}$ is finitely generated.

PROPOSITION 2.1. *Let B be in $\mathcal{C}$ and G in $(\mathcal{C}, \mathrm{Ab})$. A morphism $(\ ,B) \xrightarrow{\alpha} G$ has the property that $\mathrm{Im}\,\alpha$ is determined by an object X in $\mathcal{C}$ if and only if given any morphism $\beta\colon (\ ,L) \to G$ with the property $\beta(\ ,L)(X) \subset \alpha(\ ,B)(X)$, then there is a morphism $f\colon L \to B$ such that $\beta = \alpha(\ ,f)$.*

PROOF. Suppose $\mathrm{Im}\,\alpha$ is a subfunctor of G determined by X. Let $\beta\colon (\ ,L) \to G$ be such that $\beta(\ ,L)(X) \subset \alpha(\ ,B)(X)$. Then the subfunctor $\beta(\ ,L)$ of G is contained in $\alpha(\ ,B)$ since $\alpha(\ ,B)$ is determined by X. Therefore there is a morphism $\gamma\colon (\ ,L) \to (\ ,B)$ such that $\alpha\gamma = \beta$. But $\gamma\colon (\ ,L) \to (\ ,B)$ is of the form $(\ ,f)\colon (\ ,L) \to (\ ,B)$ for some $f\colon L \to B$. Thus we have shown that if $\mathrm{Im}\,\alpha$ is determined by X in G, and $\beta\colon (\ ,L) \to G$ is such that $\beta(\ ,L)(X) \subset \alpha(\ ,B)(X)$, then there is a morphism $f\colon L \to B$ such that $\alpha(\ ,f) = \beta$.

The rest of the proof is trivial. □

An easily verified consequence of this result is the following.

COROLLARY 2.2. *Let X be in $\mathcal{C}$ and F a subfunctor of G in $(\mathcal{C}^{\mathrm{op}}, \mathrm{Ab})$ determined by X.*

a) *A morphism $\alpha\colon (\ ,B) \to G$ has the property $\mathrm{Im}\,\alpha = F$ if and only if*

i) $\mathrm{Im}\,\alpha(X) = F(X)$ *and*

ii) *Given $\beta\colon (\ ,L) \to G$ such that $\mathrm{Im}\,\beta(X) \subset F(X)$, then there is a morphism $f\colon L \to B$ such that $\alpha(\ ,f) = \beta$.*

b) *F is finitely generated if and only if there is a morphism $\alpha\colon (\ ,B) \to G$ satisfying* i) *and* ii) *of* a).

We now apply these observations to the case $G = (\ ,C)$ for some C in $\mathcal{C}$. We begin with some terminology which will be used throughout the rest of this paper.

Let $f\colon B \to C$ be a morphism in $\mathcal{C}$ and X an object in $\mathcal{C}$.

a) f is said to be *right X-determined* if the subfunctor $\mathrm{Im}(\ ,f)$ of $(\ ,C)$ is determined by X.

b) f is said to be *left X-determined* if the subfunctor $\mathrm{Im}(f,\ )$ of $(B,\ )$ is determined by X.

The following characterization of right (left) X-determined morphisms can easily be deduced from Proposition 2.1 and its analogue for covariant functors.

PROPOSITION 2.3. *Let $f\colon B \to C$ be a morphism in $\mathcal{C}$ and X an object in $\mathcal{C}$.*

a) *$f\colon B \to C$ is right X-determined if and only if given any morphism $g\colon L \to C$ in $\mathcal{C}$ such that $\mathrm{Im}(\ ,g)(X) \subset \mathrm{Im}(\ ,f)(X)$, there is an $h\colon L \to B$ with the property $fh = g$.*

b) *$f\colon B \to C$ is left X-determined if and only if given any morphism $j\colon B \to L$ such that $\mathrm{Im}(j,)(X) \subset \mathrm{Im}(f,\ )(X)$, there is a $k\colon C \to L$ such that $kf = j$.*

The following criteria for when a morphism is right or left determined by an object are very useful in practice.

PROPOSITION 2.4. *Let* $f\colon B \to C$ *be a morphism in* $\mathcal{C}$ *and* X *an object in* $\mathcal{C}$. *Suppose* H *is an* $(\operatorname{End} X)^{\mathrm{op}}$*-submodule of* (X, C). *Then the following are equivalent.*

a) *A morphism* $h\colon L \to C$ *in* $\mathcal{C}$ *satisfies* $\operatorname{Im}(\ ,h)(X) \subset H$ *if and only if there is a morphism* $g\colon L \to B$ *such that* $fg = h$.

b) f *is a right* X*-determined morphism with* $\operatorname{Im}(\ ,f)(X) = H$.

PROOF. a) implies b). It follows from the hypothesis of a) that in order to show that f is right X-determined, it suffices to show that $\operatorname{Im}(\ ,f)(X) = H$. Since for the morphism $f\colon B \to C$ there is a morphism $h\colon B \to B$ satisfying $fh = f$, namely $h = \mathrm{id}_B$, we know by a) that $\operatorname{Im}(\ ,f)(X) = H$. We now show that $\operatorname{Im}(\ ,f)(X) \supset H$. For let $h\colon X \to C$ be in H. Then $\operatorname{Im}(\ ,h)(X)$ is simply the $(\operatorname{End} X)^{\mathrm{op}}$-submodule of (X, C) generated by h and so h is in $\operatorname{Im}(\ ,h)(X) \subset H$. Therefore there is a $g\colon X \to B$ such that $h = fg$. This implies that $\operatorname{Im}(\ ,h)(X) \subset \operatorname{Im}(\ ,f)(X)$ and so h is in $\operatorname{Im}(\ ,f)(X)$. Since this is true for all h in H we have that $\operatorname{Im}(\ ,f)(X) \supset H$. This completes the proof that a) implies b).

b) implies a) is trivial. □

Of course, we also have the companion result concerning morphisms which are left determined by objects which we now state without proof.

PROPOSITION 2.5. *Let* $f\colon B \to C$ *be a morphism in* $\mathcal{C}$ *and* X *an object in* $\mathcal{C}$. *Suppose* H *is an* $\operatorname{End} X$*-submodule of* (B, X). *Then the following statements are equivalent.*

a) *A morphism* $h\colon B \to L$ *in* $\mathcal{C}$ *satisfies* $\operatorname{Im}(h,\ )(X) \subset H$ *if and only if there is a morphism* $g\colon C \to L$ *such that* $gf = h$.

b) f *is a left* X*-determined morphism with* $\operatorname{Im}(f,\ )(X) = H$.

The following useful result is worth noting.

PROPOSITION 2.6. *Let* $f\colon B \to C$ *be a morphism in* $\mathcal{C}$ *and suppose* $X = X_1 \amalg X_2$.

a) *If* $f\colon B \to C$ *is right* X_1*-determined then* f *is right* X*-determined.*

b) *If* $f\colon B \to C$ *is left* X_1*-determined then* f *is left* X*-determined.*

PROOF. a) Suppose $f\colon B \to C$ is right X_1-determined. Let $h\colon L \to C$ be a morphism such that $\operatorname{Im}(\ ,h)(X)$ is contained in $\operatorname{Im}(\ ,f)(X)$. Since $\operatorname{Im}(\ ,h)$ and $\operatorname{Im}(\ ,f)$ are additive subfunctors of $(\ ,C)$, it follows that $\operatorname{Im}(\ ,h)(X_1) \subset \operatorname{Im}(\ ,f)(X_1)$. Thus there is a morphism $t\colon L \to B$ such that $ft = h$ since $f\colon B \to C$ is right X_1-determined.

b) Dual of a). □

As a consequence of this result, we have the following.

PROPOSITION 2.7. *Suppose* $h\colon A \to C$ *is the composition* $A \xrightarrow{g} B \xrightarrow{f} C$.

a) *If* f *is a monomorphism which is right* X_1*-determined and* g *is right* X_2*-determined, then* h *is right* $X = X_1 \amalg X_2$*-determined.*

b) *If* g *is an epimorphism which is left* X_1*-determined and* f *is left* X_2*-determined, then* h *is left* $X = X_1 \amalg X_2$*-determined.*

PROOF. Suppose $v\colon L \to C$ is such that $\operatorname{Im}(\ ,v)(X) \subset \operatorname{Im}(\ ,h)(X)$. Then $\operatorname{Im}(\ ,v)(X) \subset \operatorname{Im}(\ ,f)(X)$. Since f is right X_1-determined we know by Proposition 2.6 that f is right X-determined. Therefore, there is a morphism $t\colon L \to B$ such that $ft = v$. Thus $\operatorname{Im}(X,f)(\operatorname{Im}(X,t)(X)) \subset \operatorname{Im}(X,f)(\operatorname{Im}(X,g)(X))$. This implies $\operatorname{Im}(\ ,t)(X) \subset \operatorname{Im}(\ ,g)(X)$ since f and hence (X, f) is a monomorphism. Again by Proposition 2.6, we know that g is right X-determined since $X = X_1 \amalg X_2$ and g is right X_2-determined. Thus there is a morphism $u\colon L \to A$ such that $gu = t$. Therefore $fgu = ft = v$ and so $u\colon L \to A$ has the property that $hu = v$. Consequently, h is right $X_1 \amalg X_2$-determined as claimed.

b) Dual of a). □

Usually we are not only interested in when a functor G in $(\mathcal{C}^{\mathrm{op}}, \mathrm{Ab})$ is finitely generated, but also in when it has a projective cover which is finitely generated (see [6] for definition and properties of projective covers in $(\mathcal{C}^{\mathrm{op}}, \mathrm{Ab})$). The following description of finitely generated projective covers is quite useful in showing when finitely generated functors have projective covers.

PROPOSITION 2.8. *Let B be in $\mathcal{C}$ and G in $(\mathcal{C}^{\mathrm{op}}, \mathrm{Ab})$.*

a) *A morphism $\alpha\colon (\ ,B) \to G$ is a projective cover of its image if and only if an endomorphism $f\colon B \to B$ is an isomorphism whenever $\alpha = \alpha(\ ,f)$.*

b) *In particular, a morphism $g\colon B \to C$ in $\mathcal{C}$ has the property that $(\ ,g)\colon (\ ,B) \to (\ ,C)$ is a projective cover of* $\mathrm{Im}(\ ,g)$ *if and only if an endomorphism $f\colon B \to B$ is an isomorphism whenever $gf = g$.*

PROOF. a) Follows easily from the definition of projective covers as can be seen in [6].

b) Trivial consequence of a). □

An easily verified consequence of Proposition 2.8 is the following result.

COROLLARY 2.9. *Let $g\colon B \to C$ and $h\colon A \to C$ be morphisms in $\mathcal{C}$ such that* $\mathrm{Im}(\ ,g) = \mathrm{Im}(\ ,h)$. *Suppose $(\ ,g)\colon (\ ,B) \to (\ ,C)$ is a projective cover of* $\mathrm{Im}(\ ,g)$.

a) *There are morphisms $s\colon B \to A$ and $t\colon A \to B$ satisfying $hs = g$ and $gt = s$ such that $ts = 1_B$.*

b) *If $(\ ,h)\colon (\ ,A) \to (\ ,C)$ is a projective cover of* $\mathrm{Im}(\ ,h)$ *and $t\colon B \to A$ is a morphism such that $ht = g$, then t is an isomorphism.*

c) *$(\ ,h)\colon (\ ,A) \to (\ ,C)$ is a projective cover for* $\mathrm{Im}(\ ,h)$ *if and only if there is an isomorphism $t\colon B \to A$ such that $ht = g$.*

The reader should develop the analogues of Proposition 2.6 and Corollary 2.7 for covariant functors. This result suggests the following terminology.

Let $f\colon B \to C$ be a morphism in $\mathcal{C}$.

a) f is said to be *right minimal* if an endomorphism $h\colon B \to B$ is an isomorphism whenever $fh = f$.

b) f is said to be *left minimal* if an endomorphism $u\colon C \to C$ is an isomorphism whenever $uf = f$.

c) f is said to be a *minimal right X-determined morphism* if it is both right X-determined and right minimal.

d) f is said to be a *minimal left X-determined morphism* if it is both left X-determined and left minimal.

The following uniqueness theorem follows easily from the definitions involved.

PROPOSITION 2.10. *Let X be in $\mathcal{C}$.*

a) *Suppose $f\colon B \to C$ and $f'\colon B' \to C$ are two minimal right X-determined morphisms. Then there is an isomorphism $u\colon B \to B'$ such that $f = f'u$ if and only if* $\mathrm{Im}(\ ,f)(X) = \mathrm{Im}(\ ,f')(X)$.

b) *Suppose $h\colon A \to B$ and $h'\colon A \to B'$ are two minimal left X-determined morphisms. Then there is an isomorphism $v\colon B \to B'$ such that $vh = h'$ if and only if* $\mathrm{Im}(h, \ )(X) = \mathrm{Im}(h', \ )(X)$.

Before giving our next result it is convenient to introduce the following definitions.

Let $f\colon X \to Y$ be a morphism in $\mathcal{C}$. Then f is said to be a *splittable monomorphism* if there is a morphism $g\colon Y \to X$ such that $gf = 1_X$. Dually, f is said to be a *splittable epimorphism* if there is an $h\colon Y \to X$ such that $fh = 1_Y$.

PROPOSITION 2.11. *Assume idempotents in $\mathcal{C}$ splits. Let $f\colon B \to C$ be a right minimal X-determined morphism with $H = \operatorname{Im}(\ ,f)(X)$. Then the following statements are equivalent for a morphism $g\colon A \to C$ with $\operatorname{Im}(\ ,g)(X) \subset H$.*

a) *If g is the composition $A \xrightarrow{h} L \xrightarrow{j} C$ where j has the property $\operatorname{Im}(\ ,j)(X) \subset H$, then h is a splittable monomorphism.*

b) *If $h\colon A \to B$ is such that $fh = g$, then h is a splittable monomorphism.*

c) *There is a splittable monomorphism $h\colon A \to B$ such that $fh = g$.*

PROOF. a) implies b). Trivial.

b) implies c). Since $f\colon B \to C$ is right X-determined and $\operatorname{Im}(\ ,g)(X) \subset H = \operatorname{Im}(\ ,f)(X)$, it follows that there is an $h\colon A \to B$ such that $fh = g$.

c) implies a). Since idempotents in $\mathcal{C}$ split and $h\colon A \to B$ is a splittable monomorphism, it follows that there is an isomorphism $u\colon A \amalg A' \to B$ such that the composition $A \to A \amalg A' \xrightarrow{u} B$ is the morphism h. Suppose g is the composition $A \xrightarrow{h} L \xrightarrow{j} C$ where j has the property $\operatorname{Im}(\ ,j)(X) \subset H$. Then letting $v\colon L \amalg A' \to C$ be the morphism induced by j and the composition $A' \to B \xrightarrow{f} C$, we have i) $\operatorname{Im}(\ ,v)(X) \subset H$ and ii) there is a morphism $w\colon B \to L \amalg A'$ such that the diagram

$$\begin{array}{ccc} A \coprod A' & \longrightarrow & L \coprod A' \\ \downarrow{\scriptstyle u} & & \| \\ B & \xrightarrow{w} & L \coprod A', \end{array}$$

where $A \amalg A' \to L \amalg A'$ is induced by $1_A\colon A' \to A'$ and $h\colon A \to L$, commutes. Hence we obtain the commutative diagram

$$\begin{array}{ccc} A & \xrightarrow{h} & L \\ {\scriptstyle g}\downarrow & & \downarrow{\scriptstyle \text{inc}} \\ B & \xrightarrow{w} & L \coprod A' \\ \downarrow{\scriptstyle f} & & \downarrow{\scriptstyle v} \\ C & = & C. \end{array}$$

Since $\operatorname{Im}(\ ,v)(X) \subset \operatorname{Im}(\ ,f)(X)$ and $f\colon B \to C$ is right X-determined, there is a morphism $z\colon L \amalg A' \to B$ such that $v = fz$. The fact that $f\colon B \to C$ is also right minimal implies that zw is an isomorphism. Thus we obtain the commutative diagram

$$\begin{array}{ccccc} A & \xrightarrow{h} & L & & \\ {\scriptstyle g}\downarrow & & \downarrow{\scriptstyle \text{inc}} & & \\ B & \xrightarrow{w} & L \coprod A' & \xrightarrow{(zw)^{-1}z} & B \end{array}$$

with $((zw)^{-1}z)w = 1_B$. Hence if we let $p\colon B \to A$ be such that $ph = 1_A$ (remember $h\colon A \to B$ is a splittable monomorphism) then the composition $A \xrightarrow{h} L \xrightarrow{\text{inc}}$

$L \amalg A' \xrightarrow{(zw)^{-1}z} B \xrightarrow{p} A$ is the identity, which shows that $h\colon A \to L$ is a splittable monomorphism. This finishes the proof that c) implies a). □

As usual we obtain the dual result for left minimal morphisms determined by an object in $\mathcal{C}$ which we state without proof.

PROPOSITION 2.12. *Assume idempotents in $\mathcal{C}$ split. Let $f\colon C \to B$ be a left minimal X-determined morphism with $H = \operatorname{Im}(f, \ )(X)$. Then the following are equivalent for a morphism $g\colon C \to A$ with $\operatorname{Im}(g, \ )(X) \subset H$.*

a) *If g is the composition $C \xrightarrow{j} L \xrightarrow{h} A$ where j has the property $\operatorname{Im}(j, \ )(X) \subset H$, then h is a splittable epimorphism.*

b) *If $h\colon B \to A$ is such that $hf = g$, then h is a splittable epimorphism.*

c) *There is a splittable epimorphism $h\colon B \to A$ such that $hf = g$.*

The rest of this chapter is mainly devoted to developing existence theorems for morphisms which are either left or right determined by objects in certain full subcategories $\mathcal{C}$ of the category of modules of various types of rings. Our usual procedure will be to first establish existence theorems for morphisms right determined by some module and then use duality considerations to derive existence theorems for left determined morphisms. In this connection we point out now how left and right determined morphisms are connected by a duality between categories.

Let $D\colon \mathcal{C} \to \mathcal{C}'$ be a duality between the additive categories $\mathcal{C}$ and $\mathcal{C}'$. Then for each C in $\mathcal{C}$ the duality D induces a ring isomorphism $D\colon \operatorname{End} C \to (\operatorname{End} D(C))^{\mathrm{op}}$ which we will usually consider an identification. With this convention in mind we have the following easily verified properties of the duality.

LEMMA 2.13. *Let B and C be in $\mathcal{C}$ and $D\colon \mathcal{C} \to \mathcal{C}'$ a duality.*

a) *$D\colon (B,C) \to (D(C), D(B))$ is an $(\operatorname{End} B)^{\mathrm{op}}$-$\operatorname{End} C$ bimodule isomorphism.*

b) *$H \subset (B,C)$ is an $\operatorname{End} C$-submodule of (B,C) if and only if $D(H) \subset (D(C), D(B))$ is an $\operatorname{End} C$-submodule of $(D(C), D(B))$.*

c) *If $H \subset (B,C)$ is an $\operatorname{End} C$-submodule of (B,C), then $D\colon (B,C) \to (D(C), D(B))$ induces an $\operatorname{End} C$-isomorphism $D\colon (B,C)/H \to (D(C), D(B))/D(H)$.*

Applying these observations one easily obtains the following.

PROPOSITION 2.14. *Suppose $D\colon \mathcal{C} \to \mathcal{C}'$ is a duality. Let $g\colon A \to B$ be a morphism in $\mathcal{C}$ and X an object in $\mathcal{C}$.*

a) *If $t\colon A \to L$ is a morphism in $\mathcal{C}$, then $D(\operatorname{Im}(t, \ )(X)) = \operatorname{Im}(\ , D(t))(D(X))$.*

b) *$D(\operatorname{Im}(g, \ )(X)) = \operatorname{Im}(\ , D(g))(D(X))$.*

c) *g is left X-determined in $\mathcal{C}$ if and only if $D(g)\colon D(B) \to D(A)$ is right $D(X)$-determined in $\mathcal{C}'$.*

d) *g is left minimal if and only if $D(g)\colon D(B) \to D(A)$ is right minimal.*

e) *g is minimal left $D(X)$-determined in $\mathcal{C}$ if and only if $D(g)\colon D(B) \to D(A)$ is minimal right $D(X)$-determined in $\mathcal{C}'$.*

3. General existence theorems

Throughout this section Λ is a fixed ring. We denote the category of all Λ-modules by $\operatorname{Mod} \Lambda$ and the category of all finitely presented Λ-modules by $\operatorname{mod} \Lambda$. Our primary purpose in this section is to investigate when an epimorphism $f\colon B \to C$ in $\operatorname{Mod} \Lambda$ is right X-determined with X in $\operatorname{mod} \Lambda$. The results of this section form the basis of all existence theorems that we know for morphisms determined by modules. In order to state more explicitly the main result of this section we need the following definitions and notations.

We designate the opposite ring of Λ by Λ^{op}. Following the convention of considering right Λ-modules as left Λ^{op}-modules, we denote the category of right Λ-modules by $\operatorname{Mod}\Lambda^{\mathrm{op}}$ and the category of right finitely presented Λ-modules by $\operatorname{mod}\Lambda^{\mathrm{op}}$. In accordance with our previous notational practice in this paper, we will often denote $\operatorname{Hom}_\Lambda(A,B)$, the group of Λ-modules from A to B in $\operatorname{Mod}\Lambda$, by (A,B). As usual we consider (A,B) an $(\operatorname{End}A)^{\mathrm{op}}$-$\operatorname{End}B$ bimodule by means of the obvious operations of $\operatorname{End}A$ on A and $\operatorname{End}B$ on B. Finally for each pair A, B in $\operatorname{Mod}\Lambda$ we denote by $P(A,B)$ the abelian subgroup of (A,B) consisting of all morphisms $f\colon A\to B$ which can be factored as $A\xrightarrow{g}P\xrightarrow{h}B$ with P projective. It is easily checked that $P(A,B)$ is an $(\operatorname{End}A)^{\mathrm{op}}-\operatorname{End}B$ submodule of (A,B). The importance of this submodule is indicated by the following result.

PROPOSITION 3.1. *Suppose $f\colon B\to C$ is a morphism in* $\operatorname{Mod}\Lambda$ *and X is in* $\operatorname{Mod}\Lambda$.

a) *If f is an epimorphism, then the* $(\operatorname{End}X)^{\mathrm{op}}$*-submodule* $\operatorname{Im}(\ ,f)(X)$ *of (X,C) contains $P(X,C)$.*

b) *If f is right X-determined, then f is an epimorphism if and only if* $\operatorname{Im}(\ ,f)(X)$ *contains $P(X,C)$.*

PROOF. Suppose $f\colon B\to C$ is an epimorphism. Let $t\colon X\to C$ be in $P(X,C)$. Then t can be written as a composition $X\xrightarrow{u}P\xrightarrow{v}C$ with P projective. Since $f\colon B\to C$ is an epimorphism, there is a morphism $w\colon P\to B$ such that $fw=v$. Therefore $t\colon X\to C$ can be written as the composition fwu and so t is in $\operatorname{Im}(\ ,f)(X)$. Therefore $P(X,C)\subset\operatorname{Im}(\ ,f)(X)$.

b) Since we know by a) that if $f\colon B\to C$ is an epimorphism, then $\operatorname{Im}(\ ,f)(X)\supset P(X,C)$ we only have to show that if $f\colon B\to C$ is right X-determined and $\operatorname{Im}(\ ,f)(X)\supset P(X,C)$, then $f\colon B\to C$ is an epimorphism. Let $g\colon P\to C\to 0$ be exact with P projective. Then $\operatorname{Im}(\ ,g)(X)\subset P(X,C)\subset\operatorname{Im}(\ ,f)(X)$. Since by hypothesis $f\colon B\to C$ is right X-determined, it follows that there is a morphism $t\colon P\to B$ such that $g=ft$. The fact that g is an epimorphism implies that f is an epimorphism, our desired result. □

We can now state the main result of this section.

Let X be in $\operatorname{mod}\Lambda$ and C in $\operatorname{Mod}\Lambda$. Suppose H is an $(\operatorname{End}X)^{\mathrm{op}}$-submodule of (X,C) containing $P(X,C)$. Then there is a minimal right X-determined epimorphism $f\colon B\to C$ such that $(\operatorname{Im}(\ ,f))(X)=H$.

Before proceeding to the proof of this theorem, we recall some definitions and results we will need in the course of the proof (see [12] for more details).

We begin by recalling some easily verified properties of the subgroup $P(A,B)$ of (A,B).

a) If $f\in P(A,B)$ and $g\in(B,C)$, then $gf\in P(A,C)$.

b) If $f\in(A,B)$ and $g\in P(B,C)$, then $gf\in P(A,C)$.

c) If $f\colon P\to C$ is an epimorphism with P projective, then $\operatorname{Im}(X,P)\xrightarrow{(X,f)}(X,C)=P(X,C)$ for all X in $\operatorname{Mod}\Lambda$.

Next we recall the definition of the additive category $\underline{\operatorname{Mod}}\Lambda$, the category of Λ-modules modulo projectives. The objects of $\underline{\operatorname{Mod}}\Lambda$ are the same as those of $\operatorname{Mod}\Lambda$. For each pair of objects A, B in $\underline{\operatorname{Mod}}\Lambda$, the group of morphisms $\underline{\operatorname{Hom}}(A,B)$ from A to B is the group $\operatorname{Hom}_\Lambda(A,B)/P(A,B)$. For each f in $\operatorname{Hom}_\Lambda(A,B)$ we denote by $\underline{f}$ the image of f in $\operatorname{Hom}_\Lambda(A,B)/P(A,B)$. Further the composition of morphisms in $\underline{\operatorname{Mod}}\Lambda$ is defined by the formula $\underline{g}\ \underline{f} = \underline{gf}$ for $f\colon A \to B$ and $g\colon B \to C$. The fact that this composition is well defined follows from properties a) and b) cited above.

We denote by $\underline{\mathrm{mod}}\Lambda$ the full subcategory of $\underline{\mathrm{Mod}}\Lambda$ whose objects are in $\mathrm{mod}\,\Lambda$, the category of finitely presented Λ-modules. There is a well known duality (see [8] and [10]) $\mathrm{Tr}\colon \underline{\mathrm{mod}}\Lambda \to \underline{\mathrm{mod}}\Lambda^{\mathrm{op}}$, called the *transpose*, which plays a central role throughout this paper and is now briefly described.

Let $\mathcal{P}(\Lambda)$ and $\mathcal{P}(\Lambda^{\mathrm{op}})$ denote the categories of finitely generated projective Λ and Λ^{op}-modules respectively. It is well known that if P is in $\mathcal{P}(\Lambda)$, then the Λ^{op}-module $\mathrm{Hom}_\Lambda(P,\Lambda)$ is in $\mathcal{P}(\Lambda^{\mathrm{op}})$ and that the functor $\mathrm{Hom}_\Lambda(\ ,\Lambda)\colon \mathcal{P}(\Lambda) \to \mathcal{P}(\Lambda^{\mathrm{op}})$ is a duality. While this duality cannot be extended to a duality $\mathrm{mod}\,\Lambda \to \mathrm{mod}\,\Lambda^{\mathrm{op}}$, it can be extended to a duality $\mathrm{Tr}\colon \underline{\mathrm{mod}}\Lambda \to \underline{\mathrm{mod}}\Lambda^{\mathrm{op}}$ as follows.

For each A in $\mathrm{mod}\,\Lambda$ choose a fixed projective presentation $P_1(A) \xrightarrow{u} P_0(A) \to A \to 0$ with the $P_i(A)$ in $\mathcal{P}(\Lambda)$. Then $\mathrm{Hom}_\Lambda(u,\Lambda)\colon \mathrm{Hom}_\Lambda(P_0(A),\Lambda) \to \mathrm{Hom}_\Lambda(P_1(A),\Lambda)$ is a morphism in $\mathcal{P}(\Lambda^{\mathrm{op}})$ and so $\mathrm{Coker}\,\mathrm{Hom}_\Lambda(u,\Lambda)$ is in $\mathrm{mod}\,\Lambda^{\mathrm{op}}$. Define $\mathrm{Tr}\,A = \mathrm{Coker}(\mathrm{Hom}_\Lambda(u,\Lambda))$. Suppose $f\colon A \to B$ is a morphism in $\mathrm{mod}\,\Lambda$. There is then an exact commutative diagram

$$\begin{array}{ccccccc}
P_1(A) & \longrightarrow & P_0(A) & \longrightarrow & A & \longrightarrow & 0\\
\downarrow{\scriptstyle f_1} & & \downarrow{\scriptstyle f_0} & & \downarrow{\scriptstyle f} & & \\
P_1(B) & \longrightarrow & P_0(B) & \longrightarrow & B & \longrightarrow & 0
\end{array}$$

which induces an exact commutative diagram

$$\begin{array}{ccccccc}
\mathrm{Hom}_\Lambda(P_0(B),\Lambda) & \longrightarrow & \mathrm{Hom}_\Lambda(P_1(B),\Lambda) & \longrightarrow & \mathrm{Tr}\,B & \longrightarrow & 0\\
\downarrow{\scriptstyle \mathrm{Hom}_\Lambda(f_0,\Lambda)} & & \downarrow{\scriptstyle \mathrm{Hom}_\Lambda(f_1,\Lambda)} & & \downarrow{\scriptstyle h} & & \\
\mathrm{Hom}_\Lambda(P_0(A),\Lambda) & \longrightarrow & \mathrm{Hom}_\Lambda(P_1(A),\Lambda) & \longrightarrow & \mathrm{Tr}\,A & \longrightarrow & 0.
\end{array}$$

While the morphism h depends on the choice of f_0 and f_1, the image $\underline{h}$ of h in $\underline{\mathrm{Hom}}_\Lambda(\mathrm{Tr}\,B,\mathrm{Tr}\,A)$ does not depend on these choices and so we define $\mathrm{Tr}(f) = \underline{h}$. This then gives rise to a contravariant functor $\mathrm{Tr}\colon \underline{\mathrm{mod}}\Lambda \to \underline{\mathrm{mod}}\Lambda^{\mathrm{op}}$ which can be seen to be a duality using the fact that $\mathrm{Hom}_\Lambda(\ ,\Lambda)\colon \mathcal{P}(\Lambda) \to \mathcal{P}(\Lambda^{\mathrm{op}})$ is a duality.

Obviously the construction of the duality Tr depends on the initial choices of the projective presentations $P_1(A) \to P_0(A) \to A \to 0$. However, standard homological arguments show that changing these presentations changes the associated dualities $\underline{\mathrm{mod}}\Lambda \to \underline{\mathrm{mod}}\Lambda^{\mathrm{op}}$ by prescribed isomorphisms and so the duality $\mathrm{Tr}\colon \underline{\mathrm{mod}}\Lambda \to \underline{\mathrm{mod}}\Lambda^{\mathrm{op}}$ is essentially unique.

Before going on to show the role the dualities $\mathrm{Tr}\colon \underline{\mathrm{mod}}\Lambda \to \underline{\mathrm{mod}}\Lambda^{\mathrm{op}}$ play in the study of epimorphisms in $\mathrm{Mod}\,\Lambda$ which are right determined by modules in $\mathrm{mod}\,\Lambda$, it is important to make a few observations concerning these dualities.

First it is important to note that if Tr_1 and Tr_2 are two dualities gotten by using different projective presentations for the modules in $\mathrm{mod}\,\Lambda$, then $\mathrm{Tr}_1(X)$ and $\mathrm{Tr}_2(X)$ need not be isomorphic in $\mathrm{mod}\,\Lambda$ even though they are isomorphic in $\underline{\mathrm{mod}}\Lambda$. What can be shown is that there are projective modules P and Q in $\mathrm{mod}\,\Lambda$ such that $\mathrm{Tr}_1(X) \amalg P \cong \mathrm{Tr}_2(X) \amalg Q$ in $\mathrm{mod}\,\Lambda$.

Second, it is not difficult to show that given two dualities $\mathrm{Tr}_1 : \underline{\mathrm{mod}}\Lambda \to \underline{\mathrm{mod}}\Lambda^{\mathrm{op}}$ and $\mathrm{Tr}_2 : \underline{\mathrm{mod}}\Lambda^{\mathrm{op}} \to \underline{\mathrm{mod}}\Lambda$, then $\mathrm{Tr}_2\,\mathrm{Tr}_1 \cong 1_{\underline{\mathrm{mod}}\Lambda}$ and $\mathrm{Tr}_1\,\mathrm{Tr}_2 \cong 1_{\underline{\mathrm{mod}}\Lambda}$. However, it is not necessarily the case that $\mathrm{Tr}_2\,\mathrm{Tr}_1(X) \cong X$ in $\mathrm{mod}\,\Lambda$ for all X in $\mathrm{mod}\,\Lambda$. However, it can be shown without a great deal of difficulty that given $\mathrm{Tr}_1 : \underline{\mathrm{mod}}\Lambda \to \underline{\mathrm{mod}}\Lambda^{\mathrm{op}}$ there is a $\mathrm{Tr}_2\colon \underline{\mathrm{mod}}\Lambda^{\mathrm{op}} \to \underline{\mathrm{mod}}\Lambda$ such that $\mathrm{Tr}_2\,\mathrm{Tr}_1(X) \cong X$ in $\mathrm{mod}\,\Lambda$ for all X in $\mathrm{mod}\,\Lambda$. The reader should be cautioned that it is not necessarily the case that just because the dualities $\mathrm{Tr}_1\colon \underline{\mathrm{mod}}\Lambda \to \underline{\mathrm{mod}}\Lambda^{\mathrm{op}}$ and $\mathrm{Tr}_2\colon \underline{\mathrm{mod}}\Lambda^{\mathrm{op}} \to \underline{\mathrm{mod}}\Lambda$ have the property that $\mathrm{Tr}_2\,\mathrm{Tr}_1(X) \cong X$ in $\mathrm{mod}\,\Lambda$ for all nonprojective X in $\mathrm{mod}\,\Lambda$, then $\mathrm{Tr}_1\,\mathrm{Tr}_2(Y) \cong Y$ in $\mathrm{mod}\,\Lambda^{\mathrm{op}}$ for all nonprojective Y in $\mathrm{mod}\,\Lambda^{\mathrm{op}}$. In order to simplify notation, we assume from now on that we have chosen two dualities $\mathrm{Tr}_1\colon \underline{\mathrm{mod}}\Lambda \to \underline{\mathrm{mod}}\Lambda^{\mathrm{op}}$ and $\mathrm{Tr}_2\colon \underline{\mathrm{mod}}\Lambda^{\mathrm{op}} \to \underline{\mathrm{mod}}\Lambda$ with the property $\mathrm{Tr}_2\,\mathrm{Tr}_1(X) \cong X$ for all X in $\mathrm{mod}\,\Lambda$ both of which we denote simply by Tr.

With this convention in mind we turn our attention to showing how the duality $\mathrm{Tr}\colon \underline{\mathrm{mod}}\Lambda \to \underline{\mathrm{mod}}\Lambda^{\mathrm{op}}$ is connected with epimorphisms which are right determined by modules in $\mathrm{mod}\,\Lambda$.

Our use of the transpose in this section is based on the following.

PROPOSITION 3.2. *For each X in $\mathrm{mod}\,\Lambda$ the functors $\underline{\mathrm{Hom}}_\Lambda(X,\ )\colon \mathrm{Mod}\,\Lambda \to \mathrm{Ab}$ and $\mathrm{Tor}_1^\Lambda(\mathrm{Tr}\,X,\ )\colon \mathrm{Mod}\,\Lambda \to \mathrm{Ab}$ are isomorphic in a way which is functorial in X. Specifically, given $f\colon X \to Y$ in $\mathrm{mod}\,\Lambda$ the diagram*

$$\begin{array}{ccc} \underline{\mathrm{Hom}}_\Lambda(Y,\ ) & \xrightarrow{\alpha_Y} & \mathrm{Tor}_1^\Lambda(\mathrm{Tr}\,Y,\ ) \\ \downarrow{\scriptstyle (\underline{f},\ )} & & \downarrow{\scriptstyle \mathrm{Tor}_1^\Lambda(\mathrm{Tr}\,f,\ )} \\ \underline{\mathrm{Hom}}_\Lambda(X,\ ) & \xrightarrow{\alpha_X} & \mathrm{Tor}_1^\Lambda(\mathrm{Tr}\,X,\ ) \end{array}$$

commutes.

PROOF. The proof is an obvious generalization of the one given in [10 Proposition 2.2] for artin algebras Λ. □

With these preliminaries in mind, we turn our attention to proving the result cited in the beginning of this section. We first prove a somewhat weaker result, which is very useful in its own right, from which we deduce our desired result.

Let X be in $\mathrm{Mod}\,\Lambda$. Then X is also an $\mathrm{End}\,X$-module. Since the operations of Λ and $\mathrm{End}\,X$ on X commute, X is a Λ-$\mathrm{End}\,X$ bimodule. More generally given any ring morphism $\Sigma \to \mathrm{End}\,X$, the induced operation of Σ on X gives rise to a Λ-Σ bimodule structure on X. All bimodule structures Λ-Σ on X are obtained in this way.

Let X be in $\mathrm{mod}\,\Lambda$. Then $\mathrm{Tr}\,X$ is in $\mathrm{mod}\,\Lambda^{\mathrm{op}}$. Suppose $\Gamma = \mathrm{End}\,\mathrm{Tr}\,X$ and $\Sigma \to \Gamma$ is a ring morphism so that $\mathrm{Tr}\,X$ is a Λ^{op}-Γ bimodule. Then the operation of Λ^{op} on $\mathrm{Tr}\,X$ makes $\mathrm{Hom}_\Sigma(\mathrm{Tr}\,X, Y)$ a Λ-module for each Σ-module Y. Moreover, it is important to observe that the usual isomorphism $\mathrm{Hom}_\Gamma(\mathrm{Tr}\,X, \mathrm{Hom}_\Sigma(\Gamma, Y)) \to \mathrm{Hom}_\Sigma(\mathrm{Tr}\,X, Y)$ is an isomorphism of Λ-modules for each Σ-module Y. We recall the following well known result which plays a critical role in much of this paper.

PROPOSITION 3.3. *Let X and Σ be as above. Suppose C is in $\mathrm{Mod}\,\Lambda$ and Y is in $\mathrm{Mod}\,\Sigma$. Then there are morphisms*

$$\mathrm{Ext}^1_\Lambda(C, \mathrm{Hom}_\Sigma(\mathrm{Tr}\,X, Y)) \to \mathrm{Hom}_\Sigma(\mathrm{Tor}_1^\Lambda(\mathrm{Tr}\,X, C), Y)$$

which are functorial in C and Y and which are isomorphisms whenever Y is an injective Σ-module. Moreover the diagram of isomorphisms

$$\begin{array}{ccc} \mathrm{Ext}^1_\Lambda(C, \mathrm{Hom}_\Sigma(\mathrm{Tr}\,X, Y)) & \longrightarrow & \mathrm{Hom}_\Sigma(\mathrm{Tor}_1^\Lambda(\mathrm{Tr}\,X, C), Y) \\ \downarrow & & \downarrow \\ \mathrm{Ext}^1_\Lambda(C, \mathrm{Hom}_\Gamma(\mathrm{Tr}\,X, \mathrm{Hom}_\Sigma(\Gamma, Y))) & \longrightarrow & \mathrm{Hom}^1_\Gamma(\mathrm{Tor}_1^\Lambda(\mathrm{Tr}\,X, C), \mathrm{Hom}_\Sigma(\Gamma, Y)) \end{array}$$

commutes.

PROOF. See [14 VI, Proposition 5.1]. □

Our use of Proposition 3.3 depends on the following observation. The duality $\mathrm{Tr}\colon \underline{\mathrm{mod}}\Lambda^{\mathrm{op}} \to \underline{\mathrm{mod}}\Lambda$ gives rise to an isomorphism $\underline{\mathrm{End}}\,\mathrm{Tr}\,X \to (\underline{\mathrm{End}}X)^{\mathrm{op}}$ where $\underline{\mathrm{End}}\,\mathrm{Tr}\,X = \mathrm{End}\,X/P(X,X)$. Thus for each C in $\mathrm{Mod}\,\Lambda$ the $(\underline{\mathrm{End}}X)^{\mathrm{op}}$-module $\underline{\mathrm{Hom}}_\Lambda(X,C)$ is also an $\underline{\mathrm{End}}\,\mathrm{Tr}\,X$-module and thus an $\mathrm{End}\,\mathrm{Tr}\,X$-module. Hence if $\mathrm{Tr}\,X$ is a Λ^{op}-Σ bimodule, then $\underline{\mathrm{Hom}}_\Lambda(X,C)$ is a Σ-module for all C in $\mathrm{Mod}\,\Lambda$ by means of the morphism $\Sigma \to \mathrm{End}\,\mathrm{Tr}\,X$. Moreover the functorial isomorphisms $\mathrm{Tor}_1^\Lambda(\mathrm{Tr}\,X,C) \to \underline{\mathrm{Hom}}_\Lambda(X,C)$ referred to in Proposition 3.2 are Σ-isomorphisms. Thus we obtain the following version of Proposition 3.3.

PROPOSITION 3.4. *Let C and X be Λ-modules with X in* $\mathrm{mod}\,\Lambda$. *Suppose* $\mathrm{Tr}\,X$ *is a Λ^{op}-Σ bimodule and I is an injective Σ-module. Then there are isomorphisms*

$$\mathrm{Ext}^1_\Lambda(C,\mathrm{Hom}_\Sigma(\mathrm{Tr}\,X,I)) \to \mathrm{Hom}_\Sigma(\underline{\mathrm{Hom}}_\Lambda(X,C),I)$$

functorial in C and I.

Moreover the diagram of isomorphisms

$$\begin{array}{ccc}
\mathrm{Ext}^1_\Lambda(C,\mathrm{Hom}_\Sigma(\mathrm{Tr}\,X,I)) & \longrightarrow & \mathrm{Hom}_\Sigma(\underline{\mathrm{Hom}}_\Lambda(X,C),I) \\
\downarrow & & \downarrow \\
\mathrm{Ext}^1_\Lambda(C,\mathrm{Hom}_\Gamma(\mathrm{Tr}\,X,\mathrm{Hom}_\Sigma(\Gamma,I))) & \longrightarrow & \mathrm{Hom}_\Gamma(\underline{\mathrm{Hom}}_\Lambda(X,C),\mathrm{Hom}_\Sigma(\Gamma,I))
\end{array}$$

commutes.

PROOF. Immediate once one observes that since I is an injective Σ-module, $\mathrm{Hom}_\Sigma(\Gamma,I)$ is an injective Γ-module. □

We now investigate some of the connections between an exact sequence $x\colon 0 \to \mathrm{Hom}_\Sigma(\mathrm{Tr}\,X,I) \xrightarrow{g} B \xrightarrow{f} C \to 0$ in $\mathrm{Ext}^1_\Lambda(C,\mathrm{Hom}_\Sigma(\mathrm{Tr}\,X,I))$ and the Σ-morphism $v\colon \underline{\mathrm{Hom}}_\Lambda(X,C) \to I$ in $\mathrm{Hom}_\Sigma(\underline{\mathrm{Hom}}_\Lambda(X,C),I)$ corresponding to x under the isomorphism $\mathrm{Ext}^1_\Lambda(C,\mathrm{Hom}_\Sigma(\mathrm{Tr}\,X,I)) \to \mathrm{Hom}_\Sigma(\underline{\mathrm{Hom}}_\Lambda(X,C),I)$. Under these circumstances we have the following result.

THEOREM 3.5. a) *A morphism $h\colon L \to C$ in* $\mathrm{Mod}\,\Lambda$ *can be written as $ft = h$ for some morphism $t\colon L \to B$ if and only if* $\mathrm{Im}(\ ,h)(X) \subset H$, *where H is the preimage of* $\mathrm{Ker}\,v$ *under the epimorphism* $\mathrm{Hom}_\Lambda(X,C) \to \underline{\mathrm{Hom}}_\Lambda(X,C)$.

b) *f is right X-determined with* $\mathrm{Im}(\ ,f)(X)$ *contained in H.*

c) $\mathrm{Im}(\ ,f)(X)$ *is the maximal* $(\mathrm{End}\,X)^{\mathrm{op}}$*-submodule of H, i.e.* $\mathrm{Im}(\ ,f)(X)$ *is the* $(\mathrm{End}\,X)^{\mathrm{op}}$*-submodule of H generated by all* $(\mathrm{End}\,X)^{\mathrm{op}}$*-submodules.*

d) $\mathrm{Im}(\ ,f)(X) = H$ *if and only if H is an* $(\mathrm{End}\,X)^{\mathrm{op}}$*-submodule of* $\mathrm{Hom}_\Lambda(X,C)$.

PROOF. a) The assertion that there is a $t\colon L \to B$ such that $ft = h$ is equivalent to the top line splitting in the pull back diagram

$$\begin{array}{ccccccccc}
0 & \longrightarrow & \mathrm{Hom}_\Sigma(\mathrm{Tr}\,X,I) & \longrightarrow & V & \longrightarrow & L & \longrightarrow & 0 \\
 & & \downarrow & & \downarrow & & \downarrow & & \\
0 & \longrightarrow & \mathrm{Hom}_\Sigma(\mathrm{Tr}\,X,I) & \xrightarrow{g} & B & \xrightarrow{f} & C & \longrightarrow & 0,
\end{array}$$

or equivalently, that $\operatorname{Ext}^1_\Lambda(h, \operatorname{Hom}_\Sigma(\operatorname{Tr} X, I))(x) = 0$ where x is the element of $\operatorname{Ext}^1(C, \operatorname{Hom}_\Sigma(\operatorname{Tr} X, I))$ represented by $0 \to \operatorname{Hom}_\Sigma(\operatorname{Tr} X, I) \xrightarrow{g} B \xrightarrow{f} C \to 0$. From the commutative diagram

$$\begin{array}{ccc} \operatorname{Ext}^1_\Lambda(C, \operatorname{Hom}_\Sigma(\operatorname{Tr} X, I)) & \approx & \operatorname{Hom}_\Sigma(\underline{\operatorname{Hom}}_\Lambda(X, C), I) \\ \downarrow {\scriptstyle \operatorname{Ext}^1_\Lambda(h, \operatorname{Hom}_\Sigma(\operatorname{Tr} X, I))} & & \downarrow {\scriptstyle (\underline{\operatorname{Hom}}_\Lambda(X,h), I)} \\ \operatorname{Ext}^1_\Lambda(L, \operatorname{Hom}_\Sigma(\operatorname{Tr} X, I)) & \approx & \operatorname{Hom}_\Sigma(\underline{\operatorname{Hom}}_\Lambda(X, L), I) \end{array}$$

we deduce that $\operatorname{Ext}^1_\Lambda(h, \operatorname{Hom}_\Sigma(\operatorname{Tr} X, I))(x) = 0$ if and only if $(\underline{\operatorname{Hom}}_\Lambda(X, h), I)(v) = 0$. Since $(\underline{\operatorname{Hom}}_\Lambda(X, h), I)(v)$ is the composition $\underline{\operatorname{Hom}}_\Lambda(X, L) \xrightarrow{\underline{\operatorname{Hom}}_\Lambda(X,h)} \underline{\operatorname{Hom}}_\Lambda(X, C) \xrightarrow{v} I$, we have that $(\underline{\operatorname{Hom}}_\Lambda(X, h), I)(v) = 0$ if and only if $\operatorname{Im}(\underline{\operatorname{Hom}}_\Lambda(X, h)) \subset \operatorname{Ker} v$. From this it follows that $(\underline{\operatorname{Hom}}_\Lambda(X, h), I)(v) = 0$ if and only if $\operatorname{Im}(\operatorname{Hom}_\Lambda(X, h)) \subset H$, the preimage of $\operatorname{Ker} v$ under the epimorphism $(X, C) \to \underline{\operatorname{Hom}}_\Lambda(X, C)$. Since $\operatorname{Im}(\operatorname{Hom}_\Lambda(X, h)) = \operatorname{Im}(\ , h)(X)$, the proof of a) is complete.

b) Since $f 1_B = f$, it follows from a) that $\operatorname{Im}(\ , f)(X) \subset H$. Now suppose $h\colon L \to C$ is such that $\operatorname{Im}(\ , h)(X) \subset \operatorname{Im}(\ , f)(X)$. Then $\operatorname{Im}(\ , h)(X) \subset H$ and so by part a) there is $t\colon L \to B$ such that $h = ft$. Therefore f is right X-determined.

c) Suppose J is the maximal $(\operatorname{End} X)^{\mathrm{op}}$-submodule of H. Then for each $h\colon L \to C$ such that $\operatorname{Im}(\ , h)(X) \subset H$ we have that $\operatorname{Im}(\ , h)(X) \subset J$. Thus J is an $(\operatorname{End} X)^{\mathrm{op}}$-submodule of $\operatorname{Hom}_\Lambda(X, C)$ with the property that a morphism $h\colon L \to C$ has the property that there is a morphism $t\colon L \to B$ such that $ft = h$ if and only if $\operatorname{Im}(\ , h)(X)$ is contained in J. It then follows from Proposition 2.4 that $\operatorname{Im}(\ , f)(X) = J$.

d) Trivial consequence of c). □

Before going on to give various applications of this result, we point out that the proof of Theorem 3.5 can be used to establish a somewhat more general result which we state now for future reference.

THEOREM 3.6. *Let $\mathcal{C}$ be a full additive subcategory of* $\operatorname{Mod}\Lambda$. *Suppose X is a Λ-Σ^{op} bimodule with X in $\mathcal{C}$ and Y is a Σ-module such that there exists an isomorphism of functors from $\mathcal{C}$ to* Ab, $\operatorname{Ext}^1_\Lambda(\ , A) \cong \operatorname{Hom}_\Sigma(\underline{\operatorname{Hom}}_\Lambda(X,\ ), Y)$ *for some A in* $\operatorname{Mod}\Lambda$. *Finally suppose $x\colon 0 \to A \xrightarrow{g} B \xrightarrow{f} C \to 0$ is in $\operatorname{Ext}^1_\Lambda(C, A)$ with C in $\mathcal{C}$ and $v\colon \underline{\operatorname{Hom}}_\Lambda(X, C) \to Y$ the element of $\operatorname{Hom}_\Sigma(\underline{\operatorname{Hom}}_\Lambda(X, C), Y)$ corresponding to x under the isomorphism $\operatorname{Ext}^1_\Lambda(C, A) \to \operatorname{Hom}_\Sigma(\underline{\operatorname{Hom}}_\Lambda(X, C), Y)$. Then*

a) *A morphism $h\colon L \to C$ in $\mathcal{C}$ can be written as $ft = h$ for some $t\colon L \to B$ if and only if $\operatorname{Im}(\ , h)(X) \subset H$ where H is the preimage of $\operatorname{Ker} v$ under the epimorphism $\operatorname{Hom}_\Lambda(X, C) \to \underline{\operatorname{Hom}}_\Lambda(X, C)$.*

Further, suppose A is in $\mathcal{C}$ and $\mathcal{C}$ is closed under extensions. Then we have the following.

b) *f is right X-determined in $\mathcal{C}$ with $\operatorname{Im}(\ , f)(X) \subset H$.*

c) *$\operatorname{Im}(\ , f)(X)$ is the maximal $(\operatorname{End} X)^{\mathrm{op}}$-submodule of H.*

d) *$\operatorname{Im}(\ , f)(X) = H$ if and only if H is an $(\operatorname{End} X)^{\mathrm{op}}$-submodule of (X, C).*

We now apply Theorem 3.5 to obtain part of the existence theorem for right X-determined morphisms given at the beginning of this section.

THEOREM 3.7. *Let X be in* $\operatorname{mod}\Lambda$, *$\Sigma$ a ring such that $\operatorname{Tr} X$ is a Λ^{op}-Σ bimodule. Suppose C is in* $\operatorname{Mod}\Lambda$ *and H is an $(\operatorname{End} X)^{\mathrm{op}}$-submodule of (X, C) containing $P(X, C)$. Finally, let I be an injective Σ-module such that there is a Σ-morphism $v\colon \underline{\operatorname{Hom}}_\Lambda(X, C) \to$*

I with $\operatorname{Ker} v = H/P(X, C)$. *Then there is an exact sequence of Λ-modules* $x\colon 0 \to \operatorname{Hom}_\Sigma(\operatorname{Tr} X, I) \xrightarrow{g} B \xrightarrow{f} C \to 0$ *satisfying*

a) $\operatorname{Im}(\ , f)(X) = H$ *and*

b) f *is right X-determined.*

Furthermore the element x in $\operatorname{Ext}^1_\Lambda(C, \operatorname{Hom}_\Sigma(\operatorname{Tr} X, I))$ *is the element corresponding to* $v\colon \underline{\operatorname{Hom}}(X, C) \to I$ *under the isomorphism* $\operatorname{Ext}^1_\Lambda(C, \operatorname{Hom}_\Sigma(\operatorname{Tr} X, I)) \to \operatorname{Hom}_\Sigma(\underline{\operatorname{Hom}}_\Lambda(X, C), I)$.

Proof. Let $x\colon 0 \to \operatorname{Hom}_\Sigma(\operatorname{Tr} X, I) \to B \xrightarrow{f} C \to 0$ in $\operatorname{Ext}^1_\Lambda(C, \operatorname{Hom}_\Sigma(\operatorname{Tr} X, I))$ be the element corresponding to $v\colon \underline{\operatorname{Hom}}_\Lambda(X, C) \to I$ under the isomorphism $\operatorname{Ext}^1_\Lambda(C, \operatorname{Hom}_\Sigma(\operatorname{Tr} X, I)) \to \operatorname{Hom}_\Sigma(\underline{\operatorname{Hom}}_\Lambda(X, C), I)$. It then follows trivially from Theorem 3.5 that $0 \to \operatorname{Hom}_\Sigma(\operatorname{Tr} X, I) \to B \xrightarrow{f} C \to 0$ has the desired properties. □

Having shown that for X in $\operatorname{mod} \Lambda$ and C in $\operatorname{Mod} \Lambda$, and H an $(\operatorname{End} X)^{\mathrm{op}}$-submodule of (X, C) containing $P(X, C)$, there is an epimorphism $f\colon B \to C$ which is right X-determined with $\operatorname{Im}(\ , f)(X) = H$, we now show that there is a minimal right X-determined epimorphism $f\colon B \to C$ such that $\operatorname{Im}(\ , f)(X) = H$. To do this we will need the following property of finitely presented modules which will be proven in the appendix of this paper.

Proposition 3.8. *Let Y be a finitely presented Λ^{op}-module and let* $\Gamma = \operatorname{End} Y$. *Further, let $\mathcal{I}$ be the full subcategory of* $\operatorname{Mod} \Gamma$ *whose objects are the injective Γ-modules. Then the functor* $\operatorname{Hom}_\Gamma(Y, \ )\colon \mathcal{I} \to \operatorname{Mod} \Lambda$ *given by* $I \mapsto \operatorname{Hom}_\Gamma(Y, I)$ *for each injective Γ-module I in $\mathcal{I}$, is fully faithful.*

We now apply this result to obtain the following consequence of Theorem 3.5.

Theorem 3.9. *Let X be in* $\operatorname{mod} \Lambda$ *and* $\Gamma = \operatorname{End} \operatorname{Tr} X$. *Suppose C is in* $\operatorname{Mod} \Lambda$ *and H is an* $(\operatorname{End} X)^{\mathrm{op}}$*-submodule of* (X, C) *containing* $P(X, C)$. *Further, let* $v\colon \underline{\operatorname{Hom}}_\Lambda(X, C) \to I$ *be a Γ-morphism with I an injective Γ-module such that* $\operatorname{Ker} v = H/P(X, C)$ *and the induced morphism* $\underline{\operatorname{Hom}}_\Lambda(X, C)/H/P(X, C) \to I$ *is an injective envelope. Then the short exact sequence* $x\colon 0 \to \operatorname{Hom}_\Gamma(\operatorname{Tr} X, I) \xrightarrow{g} B \xrightarrow{f} C \to 0$ *in* $\operatorname{Ext}^1_\Lambda(C, \operatorname{Hom}_\Gamma(\operatorname{Tr} X, I))$ *corresponding to the morphism* $v\colon \underline{\operatorname{Hom}}_\Lambda(X, C) \to I$ *under the isomorphism* $\operatorname{Ext}^1_\Lambda(C, \operatorname{Hom}_\Gamma(\operatorname{Tr} X, I)) \to \operatorname{Hom}_\Gamma(\underline{\operatorname{Hom}}_\Lambda(X, C), I)$ *has the property that* $f\colon B \to C$ *is a minimal right X-determined morphism such that* $\operatorname{Im}(\ , f)(X) = H$.

Proof. Letting $\Sigma = \Gamma$ in Theorem 3.5, we know that $f\colon B \to C$ is a right X-determined morphism such that $(\operatorname{Im}(\ , f))(X) = H$. Therefore we only have to show that $f\colon B \to C$ is right minimal.

Let $h\colon B \to B$ be such that $fh = f$. We want to show that h is an isomorphism. Since $0 \to \operatorname{Hom}_\Gamma(\operatorname{Tr} X, I) \xrightarrow{g} B \xrightarrow{f} C \to 0$ is exact, there is a commutative diagram

$$\begin{array}{ccccccccc}
0 & \longrightarrow & \operatorname{Hom}_\Gamma(\operatorname{Tr} X, I) & \xrightarrow{g} & B & \xrightarrow{f} & C & \longrightarrow & 0 \\
 & & \downarrow{\scriptstyle j} & & \downarrow{\scriptstyle h} & & \| & & \\
0 & \longrightarrow & \operatorname{Hom}_\Gamma(\operatorname{Tr} X, I) & \xrightarrow{g} & B & \xrightarrow{f} & C & \longrightarrow & 0.
\end{array}$$

Therefore to show that h is an isomorphism, it suffices to show that j is an isomorphism. By Proposition 3.8 we know that there is a Γ-morphism $t\colon I \to I$ such that $\operatorname{Hom}_\Gamma(\operatorname{Tr} X, t)\colon \operatorname{Hom}_\Gamma(\operatorname{Tr} X, I) \to \operatorname{Hom}_\Gamma(\operatorname{Tr} X, I)$ is the morphism $j\colon \operatorname{Hom}_\Gamma(\operatorname{Tr} X, I) \to$

$\operatorname{Hom}_\Gamma(\operatorname{Tr} X, I)$. We now show that j is an isomorphism by showing that $t\colon I \to I$ is an isomorphism, thus finishing the proof of the theorem.

By Proposition 3.4 we have the commutative diagram

$$\begin{array}{ccc} \operatorname{Ext}^1_\Lambda(C, \operatorname{Hom}_\Gamma(\operatorname{Tr} X, I)) & \longrightarrow & \operatorname{Hom}_\Gamma(\underline{\operatorname{Hom}}_\Lambda(X,C), I) \\ {\scriptstyle \operatorname{Ext}^1_\Lambda(C,j)}\downarrow & & \downarrow{\scriptstyle \operatorname{Hom}_\Gamma(\underline{\operatorname{Hom}}_\Lambda(X,C),t)} \\ \operatorname{Ext}^1_\Lambda(C, \operatorname{Hom}_\Gamma(\operatorname{Tr} X, I)) & \longrightarrow & \operatorname{Hom}_\Gamma(\underline{\operatorname{Hom}}_\Lambda(X,C), I) \end{array}$$

Since $\operatorname{Ext}^1_\Lambda(C,j)(x) = x$, we have that $\operatorname{Hom}_\Gamma(\underline{\operatorname{Hom}}_\Lambda(X,C),t)(v) = v$. Hence

$$\begin{array}{ccc} \underline{\operatorname{Hom}}_\Lambda(X,C) & \xrightarrow{v} & I \\ \| & & \downarrow{\scriptstyle t} \\ \underline{\operatorname{Hom}}_\Lambda(X,C) & \xrightarrow{v} & I \end{array}$$

commutes. This gives rise to the exact commutative diagram

$$(*) \qquad \begin{array}{ccccc} 0 & \longrightarrow & \underline{\operatorname{Hom}}_\Lambda(X,C)/\operatorname{Ker} v & \longrightarrow & I \\ & & \| & & \downarrow{\scriptstyle t} \\ 0 & \longrightarrow & \underline{\operatorname{Hom}}_\Lambda(X,C)/\operatorname{Ker} v & \longrightarrow & I. \end{array}$$

The fact that $0 \to \underline{\operatorname{Hom}}_\Lambda(X,C)/\operatorname{Ker} v \to I$ is an injective envelope implies that $t\colon I \to I$ is an isomorphism.

To see this observe that $\operatorname{Ker} t = 0$ since from the commutative diagram $(*)$ it follows that $\operatorname{Ker} t \cap \underline{\operatorname{Hom}}_\Lambda(X,C)/\operatorname{Ker} v = (0)$. Hence $\operatorname{Im} t$ is an injective submodule of I containing $\underline{\operatorname{Hom}}_\Lambda(X,C)/\operatorname{Ker} v$. Therefore $\operatorname{Im} t = I$ since $0 \to \underline{\operatorname{Hom}}_\Lambda(X,C)/\operatorname{Ker} v \to I$ is an injective envelope. Consequently $t\colon I \to I$ is an isomorphism and the proof of the theorem is complete. □

Before going on with our discussion of epimorphisms which are right determined by a module, we point out the following well known and easily verified facts about extensions which we will use freely throughout the rest of the paper.

PROPOSITION 3.10. *Suppose $A = A_1 \amalg A_2$ and $x\colon 0 \to A \xrightarrow{g} B \xrightarrow{f} C \to 0$ is an exact sequence. Let $x_i\colon 0 \to A_i \xrightarrow{g_i} B_i \xrightarrow{f_i} C \to 0$ for $i = 1,2$ be the components of x under the natural isomorphism $\operatorname{Ext}^1_\Lambda(C, A_1 \amalg A_2) \to \operatorname{Ext}^1_\Lambda(C, A_1) \amalg \operatorname{Ext}^1_\Lambda(C, A_2)$. Then the following are equivalent.*

a) $x_1 = 0$.

b) *there exists a commutative diagram*

$$\begin{array}{ccccccccc} 0 & \longrightarrow & A_2 & \xrightarrow{g_2} & B_2 & \xrightarrow{f_2} & C & \longrightarrow & 0 \\ & & \downarrow{\scriptstyle \text{inc}} & & \downarrow{\scriptstyle u} & & \| & & \\ 0 & \longrightarrow & A & \xrightarrow{g} & B & \xrightarrow{f} & C & & \\ & & \downarrow{\scriptstyle p} & & \downarrow{\scriptstyle v} & & & & \\ 0 & \longrightarrow & A_2 & \longrightarrow & B_2 & \longrightarrow & C & & \end{array}$$

with inc: $A_2 \to A_1 \amalg A_2$ *the natural inclusion,* $p\colon A_1 \amalg A_2 \to A_2$ *the natural projection and* vu *an isomorphism.*

c) $g(A_1)$ *has a complement in* B *containing* $g(A_2)$.

As a consequence of Theorem 3.9 we have the following.

PROPOSITION 3.11. *Suppose* X *is in* $\operatorname{mod}\Lambda$ *and* $x\colon 0 \to A \xrightarrow{g} B \xrightarrow{f} C \to 0$ *is an exact sequence in* $\operatorname{Mod}\Lambda$ *with* f *right* X*-determined. Let* $0 \to A' \xrightarrow{f'} B' \xrightarrow{g'} C \to 0$ *be an exact sequence in* $\operatorname{Mod}\Lambda$ *satisfying* $\operatorname{Im}(\ ,f')(X) = \operatorname{Im}(\ ,f)(X)$ *and* f' *is a minimal right* X*-determined morphism.*

a) *There exists a commutative diagram*

$$\begin{array}{ccccccccc}
0 & \longrightarrow & A' & \xrightarrow{g'} & B' & \xrightarrow{f'} & C & \longrightarrow & 0 \\
 & & \downarrow{\scriptstyle u} & & \downarrow{\scriptstyle v} & & \| & & \\
0 & \longrightarrow & A & \xrightarrow{g} & B & \xrightarrow{f} & C & \longrightarrow & 0 \\
 & & \downarrow{\scriptstyle w} & & \downarrow{\scriptstyle z} & & \| & & \\
0 & \longrightarrow & A' & \xrightarrow{g'} & B' & \xrightarrow{f'} & C & &
\end{array}$$

such that $zv = 1_{B'}$, *and* $wu = 1_{A'}$.

b) x *splits if and only if* $\operatorname{Im}(\ ,f)(X) = (X,C)$.

c) *Suppose* x *does not split. Then* f *is a right minimal morphism if and only if the element* x *in* $\operatorname{Ext}^1_\Lambda(C,A)$ *has the property that if* $A \cong A_1 \amalg A_2$ *with* A_1 *and* A_2 *not zero, then the component* x_i *in* $\operatorname{Ext}^1_\Lambda(C,A_i)$ *for* $i = 1,2$ *of* x *under the canonical isomorphism* $\operatorname{Ext}^1_\Lambda(C,A) \to \amalg_{i=1}^2 \operatorname{Ext}^1_\Lambda(C,A_i)$ *are both not zero.*

PROOF. a) Since f and f' are right X-determined morphisms with $\operatorname{Im}(\ ,f)(X) = \operatorname{Im}(\ ,f')(X)$, there is a commutative exact diagram

$$\begin{array}{ccccccccc}
0 & \longrightarrow & A' & \xrightarrow{g'} & B' & \xrightarrow{f'} & C & \longrightarrow & 0 \\
 & & \downarrow{\scriptstyle u} & & \downarrow{\scriptstyle v} & & \| & & \\
0 & \longrightarrow & A & \xrightarrow{g} & B & \xrightarrow{f} & C & \longrightarrow & 0 \\
 & & \downarrow{\scriptstyle u'} & & \downarrow{\scriptstyle v'} & & \| & & \\
0 & \longrightarrow & A' & \xrightarrow{g'} & B' & \xrightarrow{f'} & C. & &
\end{array}$$

Since $f'\colon B' \to C$ is right minimal, the fact that $f'v'v = f'$ implies that $v'v$ is an isomorphism. Letting $z = (v'v)^{-1}v'$ and w the morphism such that $g'w = zg$, we have that $zv = 1_{B'}$ and $wu = 1_{A'}$.

b) Clearly if x splits then $\operatorname{Im}(\ ,f)(X) = (X,C)$. Suppose $\operatorname{Im}(\ ,f)(X) = (X,C)$. Then $1_C\colon C \to C$ is a minimal right X-determined morphism such that $\operatorname{Im}(\ ,1_C)(X) = (X,C) = \operatorname{Im}(\ ,f)(X)$. It then follows from a) that $A' = 0$. Applying a) again we have that x splits.

c) Since x does not split, $\operatorname{Im}(\ ,f)(X) \neq (X,C)$ and so $\operatorname{Im}(\ ,f')(X) \neq (X,C)$. Hence $A' \neq 0$ (see part a). Suppose f is not right minimal. Then the splittable monomorphism $u\colon A' \to A$ is not an isomorphism. Hence letting $\operatorname{Coker} u = A''$, we have that $A \cong A' \amalg A''$

with $A' \neq (0) \neq A''$. It then follows from a) that we have a commutative exact diagram

$$\begin{array}{ccccccccc}
 & & A'' & = & A'' & & & & \\
 & & \amalg & & \amalg & & & & \\
0 & \longrightarrow & A' & \xrightarrow{g'} & B' & \xrightarrow{f'} & C & \longrightarrow & 0 \\
 & & \Vert\wr & & \Vert\wr & & \Vert & & \\
0 & \longrightarrow & A & \longrightarrow & B & \xrightarrow{f} & C & \longrightarrow & 0.
\end{array}$$

From this it follows that the component of x in $\operatorname{Ext}^1_\Lambda(C, A'')$ is zero under the isomorphism $\operatorname{Ext}^1_\Lambda(C, A) \cong (\operatorname{Ext}^1_\Lambda(C, A') \amalg \operatorname{Ext}^1_\Lambda(C, A'')$. Thus f is a minimal right morphism if given any representation of $A \cong A_1 \amalg A_2$ with $A_i \neq 0$ for $i = 1, 2$, both of the components of x in $\operatorname{Ext}^1_\Lambda(C, A_i)$ are nonzero.

The rest of c) is easily established. □

The following uniqueness theorem is worth noting.

PROPOSITION 3.12. *Let X be in $\operatorname{mod}\Lambda$ and suppose $0 \to A \xrightarrow{g} B \xrightarrow{f} C \to 0$ and $0 \to A' \xrightarrow{g'} B' \xrightarrow{f'} C \to 0$ are two exact sequences with f and f' minimal right X-determined morphisms.*

a) *If $\operatorname{Im}(\ ,f)(X) = \operatorname{Im}(\ ,f')(X)$, then there is a commutative exact diagram*

$$\begin{array}{ccccccccc}
0 & \longrightarrow & A & \xrightarrow{g} & B & \xrightarrow{f} & C & \longrightarrow & 0 \\
 & & \downarrow{\scriptstyle u} & & \downarrow{\scriptstyle v} & & \Vert & & \\
0 & \longrightarrow & A' & \xrightarrow{g'} & B & \xrightarrow{f'} & C & \longrightarrow & 0
\end{array}$$

and any such commutative diagram has the property that v, and hence u, is an isomorphism.

b) $\operatorname{Im}(\ ,f)(X) = \operatorname{Im}(\ ,f')(X)$ *if and only if there is a commutative diagram*

$$\begin{array}{ccccccccc}
0 & \longrightarrow & A & \xrightarrow{g} & B & \xrightarrow{f} & C & \longrightarrow & 0 \\
 & & \downarrow{\scriptstyle u} & & \downarrow{\scriptstyle v} & & \Vert & & \\
0 & \longrightarrow & A' & \xrightarrow{g'} & B' & \xrightarrow{f'} & C & \longrightarrow & 0
\end{array}$$

with v, and hence u, is isomorphism.

PROOF. Is a trivial consequence of Proposition 3.10. See also Proposition 2.8. □

Let $\mathcal{U}$ be the full subcategory of $\operatorname{Mod}\Lambda$ consisting of finite sums of Λ-modules isomorphic to modules of the form $\operatorname{Hom}_\Sigma(X, I)$, where X is a Λ^{op}-Σ bimodule which is a finitely presented Λ^{op}-module and I is an injective Σ-module. The rest of this section is devoted to showing that if $0 \to A \xrightarrow{g} B \xrightarrow{f} C$ is an exact sequence of Λ-modules with A in $\mathcal{U}$, then f is right X-determined for some X in $\operatorname{mod}\Lambda$. To this end it is convenient to have the following description of when an exact sequence $0 \to A \xrightarrow{g} B \xrightarrow{f} C \to 0$ of Λ-modules has the property f is right X-determined for some X in $\operatorname{Mod}\Lambda$.

LEMMA 3.13. *Let $x\colon 0 \to A \xrightarrow{g} B \xrightarrow{f} C \to 0$ be an exact sequence in $\operatorname{Mod}\Lambda$ and X a Λ-module. Then f is right X-determined if and only if the element x in $\operatorname{Ext}^1_\Lambda(C, A)$ has the following property.*

Let $h\colon L \to C$ be a morphism in $\operatorname{Mod}\Lambda$. *Then* $\operatorname{Ext}^1_\Lambda(h, A)(x) = 0$ *whenever for each morphism* $j\colon X \to L$ *we have that* $\operatorname{Ext}^1_\Lambda(hj, A)(x) = 0$.

PROOF. Suppose $t\colon Y \to C$ is an arbitrary morphism in $\operatorname{Mod}\Lambda$. It is well known that $\operatorname{Ext}^1_\Lambda(t, A)(x) = 0$ if and only if there exists $u\colon Y \to B$ such that $fu = t$. In view of this observation it is easy to show that a morphism $h\colon L \to C$ has the property that $\operatorname{Ext}^1_\Lambda(hj, A)(x) = 0$ for all j in (X, L) if and only if $\operatorname{Im}(\ ,h)(X) \subset \operatorname{Im}(\ ,f)(X)$. Our desired result now follows trivially. □

As a consequence of this description of when an epimorphism is determined by a module, we have the following general result.

PROPOSITION 3.14. *Let* $\{A_i\}_{i\in I}$ *be a family of* Λ*-modules and* $x\colon 0 \to \prod_{i\in I} A_i \xrightarrow{g} B \xrightarrow{f} C \to 0$ *an exact sequence in* $\operatorname{Mod}\Lambda$. *Let* x_i *in* $\operatorname{Ext}^1_\Lambda(C, A_i)$ *be the* i^{th}*-component of* x *under the canonical isomorphism* $\operatorname{Ext}^1_\Lambda(C, \prod_{i\in I} A_i) \to \prod \operatorname{Ext}^1_\Lambda(C, A_i)$ *and let* $x_i\colon 0 \to A_i \xrightarrow{g_i} B_i \xrightarrow{f_i} C \to 0$ *be the exact sequence corresponding to* x_i. *Then a* Λ*-module* X *has the property that* $f\colon B \to C$ *is right* X*-determined if each* $f_i\colon B_i \to C$ *is right* X*-determined.*

PROOF. Recall that the projectives $\prod_{i\in I} A_i \to A_j$ induce an isomorphism of functors $\operatorname{Ext}^1_\Lambda(\ ,\prod_{i\in I} A_i) \to \prod_{i\in I}\operatorname{Ext}^1_\Lambda(\ ,A_i)$. In particular, given any morphism $t\colon Y \to C$ we have that $\operatorname{Ext}^1_\Lambda(t, C)(x) = (\operatorname{Ext}^1_\Lambda(t, C)(x_i))_{i\in I}$. Thus we have that if $h\colon L \to C$ is a morphism, then a morphism $j\colon X \to L$ has the property $\operatorname{Ext}^1_\Lambda(jh, \prod_{i\in I} A_i)(x_i)_{i\in I} = 0$ if and only if $\operatorname{Ext}^1_\Lambda(jh, A_i)(x_i) = 0$ for each i in I.

Assume that each f_i is right X-determined and let $h\colon L \to C$ be a morphism such that for each $j\colon X \to L$ we have $\operatorname{Ext}^1_\Lambda(hj, \prod_{i\in I} A_i)(x_i) = 0$. Then by our preliminary remarks we have that $\operatorname{Ext}^1_\Lambda(hj, A_i)(x_i) = 0$ for all i in I. Hence $\operatorname{Ext}^1_\Lambda(h, A_i)(x) = 0$ for all i in I by Lemma 3.12 since each f_i is right X-determined. Consequently $\operatorname{Ext}^1_\Lambda(h, \prod_{i\in I} A_i)(x) = 0$. Thus we have by Lemma 3.13 that f is right X-determined. □

As a consequence of this proposition we have the following result.

COROLLARY 3.15. *Let* $x\colon 0 \to \prod_{i\in I} A_i \xrightarrow{g} B \xrightarrow{f} C \to 0$ *be in* $\operatorname{Ext}^1_\Lambda(C, \prod A_i)$ *and let* $x_i\colon 0 \to A_i \xrightarrow{g_i} B \xrightarrow{f_i} C \to 0$ *in* $\operatorname{Ext}^1_\Lambda(C, A_i)$ *be the* i^{th} *projection of* x. *If each* f_i *is right* X_i*-determined, then* f *is right* $X = \coprod_{i\in I} X_i$*-determined.*

PROOF. The fact that each f_i is right X_i-determined implies that each f_i is right X-determined since X contains each X_i as a summand (see Proposition 2.6). Thus f is right X-determined by Proposition 3.14. □

The following is a trivial consequence of this corollary.

PROPOSITION 3.16. *Let* $\{X_j\}_{j\in J}$ *be a family of* Λ^{op}-Σ_j *bimodules with each* X_j *a finitely presented* Λ^{op}*-module and let* I_j *be an injective* Σ_j*-module. Finally let* A_j *be the* Λ*-module* $\operatorname{Hom}_{\Sigma_j}(X_j, I_j)$ *for each* j *in* A_j. *If* $x\colon 0 \to \prod_{j\in J} A_j \xrightarrow{g} B \xrightarrow{f} C \to 0$ *is exact, then* f *is right* $\coprod_{j\in J} \operatorname{Tr} X_j$*-determined.*

PROOF. Let $x_j\colon 0 \to A_j \xrightarrow{g_j} B_j \xrightarrow{f_j} C \to 0$ be the jth projection of x. Since $\operatorname{Tr}(\operatorname{Tr} X_j) \cong X_j$, and so $A_j =\cong \operatorname{Hom}_{\Sigma_j}(\operatorname{Tr}(\operatorname{Tr} X_j), I_j)$ it follows from Theorem 3.5 that f_j is right $\operatorname{Tr} X_j$-determined. Since this is true for each j in J, it follows from Corollary 3.15 that f is right $\coprod \operatorname{Tr} X_j$-determined. □

We are now in a position to prove the result cited earlier.

THEOREM 3.17. *Let* $0 \to A \xrightarrow{g} B \xrightarrow{f} C$ *be an exact sequence of* Λ*-modules. Suppose* $A = \prod_{j\in J} \operatorname{Hom}_{\Sigma_j}(X_j, I_j)$ *where each* X_j *is a* Λ^{op}-Σ_j *bimodule which is a finitely presented* Λ^{op}*-module and each* I_j *is* Σ_j*-injective. Let* $X = \coprod_{j\in J} \operatorname{Tr} X_j$*. Then*

a) *If* f *is an epimorphism, then* f *is right* X*-determined.*

b) f *is right* $X \coprod \Lambda$*-determined in general.*

c) *If* J *is finite, then* $X \coprod \Lambda$ *is in* $\operatorname{mod} \Lambda$.

PROOF. a) This is simply Proposition 3.15.

b) By a) we know that the induced epimorphism $h\colon B \to \operatorname{Im} f$ is right X-determined. Further it is trivial to show that the inclusion $\operatorname{inc}\colon \operatorname{Im} f \to C$ is right Λ-determined. Thus the composition $B \xrightarrow{h} \operatorname{Im} f \xrightarrow{\operatorname{inc}} C$ is $X \coprod \Lambda$-determined by Proposition 2.7.

c) Obvious. □

Added in proof: The main existence theorem of this section shows that if X is a finitely presented Λ-module and C is an arbitrary Λ-module, then for each $(\operatorname{End} X)^{\text{op}}$-submodule H of (X, C) containing $P(X, C)$ there is a minimal right X-determined morphism $f\colon B \to C$ such that $\operatorname{Im}(\ , f)(X) = H$. In this addendum we show that the same result is valid for arbitrary $(\operatorname{End} X)^{\text{op}}$-submodules H of (X, C), not only for those containing $P(X, C)$. The method of proof is to reduce the general case to the special one we have studied in this section. This reduction is based on the following general observations. Throughout this discussion we consider the canonical isomorphism $(\Lambda, C) \to C$ given by $f \mapsto f(1)$ an identification for all Λ-modules C.

PROPOSITION 3.18. *Suppose* F *is a subfunctor of* $(\ , C)$ *for some* Λ*-module* C*.*

a) *Let* L *be a* Λ*-module. A morphism* $f\colon L \to C$ *has the property* $\operatorname{Im}(\ , f) \subset F$ *if and only if the element* f *in* (L, C) *is contained in* $F(L)$*.*

b) *Suppose* $F(\Lambda) = (\Lambda, C)$*. Then* $F(X) \supset P(X, C)$ *for all finitely generated* Λ*-modules* X*.*

c) *If* C' *is a submodule of* C*, then* $(\ , C')$ *is a* Λ*-determined subfunctor of* $(\ , C)$*.*

d) *Letting* C' *be the submodule* $F(\Lambda)$ *of* C*, we have that* $F \subset (\ , C')$ *and* $F(X) \supset P(X, C')$ *for all finitely generated* Λ*-modules* X*.*

PROOF. a) Trivial.

b) Since $F(\Lambda) = (\Lambda, C)$, we have that $F(\Lambda^n)(\Lambda^n, C)$ for all positive integers n. Hence by a), we know that if f is in (Λ^n, C) then $\operatorname{Im}(\ , f)(\ , \Lambda^n) \subset F$ for all positive integers n. Suppose $h\colon \coprod_{i\in I} \Lambda_i \to C$ is an epimorphism with each $\Lambda_i = \Lambda$. Then for each finite subset J of L the restriction $h_J\colon \coprod_{i\in J} \Lambda_i \to C$ of h to $\coprod_{i\in J} \Lambda_i$ has the property $\operatorname{Im}(\ , h_J) \subset F$.

Let f be in $P(X, C)$ where X is a finitely generated Λ-module. Then there is a morphism $g\colon X \to \coprod_{i\in I} \Lambda_i$ such that $f = hg$. Since X is finitely generated, there is a finite subset J of I such that $\operatorname{Im} g \subset \coprod_{i\in J} \Lambda_i$. Thus $f = h_J g$. Since $\operatorname{Im}(\ , h_J) \subset F$, it follows that $\operatorname{Im}(\ , f) \subset F$. This means by a), that f is in $F(X)$. Hence $F(X) \supset P(X, C)$.

c) Suppose C' is a submodule of C. In order to show that $(\ , C')$ is a Λ-determined subfunctor of $(\ , C)$ we have to show that a subfunctor F of $(\ , C)$ is contained in $(\ , C')$ if $F(\Lambda) \subset (\Lambda, C')$. Suppose $F(\Lambda) \subset (\Lambda, C')$. Let L be a Λ-module. Then $F(L) \subset (L, C)$. In order to show that $F(L) \subset (L, C')$, we have to show that each morphism $f\colon L \to C$ in $F(L)$ has the property $\operatorname{Im} f \subset C'$, or, equivalently, $\operatorname{Im}((\Lambda, L) \xrightarrow{(\Lambda, f)} (\Lambda, C)) \subset (\Lambda, C')$. Since $f\colon L \to C$ is in $F(L)$, we know by a) that $\operatorname{Im}(\ , f) \subset F$. In particular, $\operatorname{Im}((\Lambda, L) \xrightarrow{(\Lambda, f)} (\Lambda, C)) \subset F(\Lambda) \subset C'$, which finishes the proof of c).

d) Trivial consequence of c) and b). □

We now apply these observations to obtain our promised result.

THEOREM 3.19. *Let X and C be Λ-modules with X finitely generated. Suppose H is an $(\operatorname{End} X)^{\mathrm{op}}$-submodule of (X, C). Then there is an exact sequence $0 \to A \xrightarrow{g} B \xrightarrow{f} C$ of Λ-modules with f a minimal right X-determined morphism such that $\operatorname{Im}(\ ,f)(X) = H$. Moreover, if we let $C' = \operatorname{Im} f$ and let $0 \to A \xrightarrow{g} B \xrightarrow{f'} C' \to 0$ be the induced exact sequence, then*

a) H is an $(\operatorname{End} X)^{\mathrm{op}}$-submodule of (X, C') containing $P(X, C')$.

b) f' is minimal right X-determined with $\operatorname{Im}(\ ,f')(X) = H$.

PROOF. Let H be an $(\operatorname{End} X)^{\mathrm{op}}$-submodule of (X, C) and suppose $(\ ,C)_H$ is the X-determined subfunctor of $(\ ,C)$ such that $(\ ,C)_H(X) = H$. Let C' be the submodule $(\ ,C)_H(\Lambda)$ of C. Then by Proposition 3.18 we know that $(\ ,C)_H \subset (\ ,C')$ and $H = (\ ,C)_H(X)$ is an $(\operatorname{End} X)^{\mathrm{op}}$-submodule of (X, C') containing $P(X, C')$. Therefore we know there is an exact sequence $0 \to A \xrightarrow{g} B \xrightarrow{f'} C' \to 0$ with f' a minimal right X-determined morphism such that $\operatorname{Im}(\ ,f')(X) = H$. But the fact that $(\ ,C)_H$ is an X-determined subfunctor of $(\ ,C)$ implies that $(\ ,C)_H$ is an X-determined subfunctor of $(\ ,C')$. Hence $\operatorname{Im}(\ ,f') = (\ ,C)_H$ since they are both X-determined subfunctors of $(\ ,C')$ such that $\operatorname{Im}(\ ,f')(X) = (\ ,C)_H(X)$. Letting $f\colon B \to C$ be the composition $B \xrightarrow{f'} C' \xrightarrow{\text{inc}} C$, we conclude from the above remarks that $\operatorname{Im}(\ ,f) = (\ ,C)_H$. Hence f is a right X-determined morphism with $\operatorname{Im}(\ ,f)(X) = H$. The fact that f is right minimal is an immediate consequence of the fact that f' is right minimal. Hence the proof of the theorem is complete. □

4. Noetherian R-algebras

This section is devoted to specializing the results of Section 3 to the case Λ is a noetherian algebra, a notion we now define.

By a *R-algebra* Λ we mean a commutative ring R together with a ring morphism $f\colon R \to \Lambda$ with $f(R)$ contained in the center of Λ. If $f\colon R \to \Lambda$ is an R-algebra, then the morphism $g\colon R \to \Lambda^{\mathrm{op}}$ given by $g(r) = f(r)$ for all r in R is also an R-algebra which we denote by Λ^{op}. We say that an R-algebra Λ is a *noetherian R-algebra* if R is a noetherian ring and Λ is a finitely generated R-module. Clearly an R-algebra Λ is a noetherian R-algebra if and only if the R-algebra Λ^{op} is a noetherian R-algebra. Thus if Λ is a noetherian R-algebra, then Λ is a noetherian ring, i.e. Λ is left and right noetherian.

It is not difficult to see that a ring Λ has the property that Λ is a noetherian R-algebra for some R if and only if the center of Λ is a noetherian ring and Λ is a finitely generated module over its center. We say that a ring Λ is a *noetherian algebra* if its center is noetherian and it is a finitely generated module over its center. Throughout the rest of this section we assume that R is a commutative noetherian ring (unless stated explicitly to the contrary).

Before applying the results of Section 3 to noetherian algebras we remind the reader of some well known facts concerning modules over noetherian algebras.

Let R be a commutative noetherian ring. We denote by $I(M)$ an injective envelope of an R-module M. The injective R-module $\coprod_{\underline{m}} I(R/\underline{m})$ where $\underline{m}$ runs over all maximal ideals of R, which we denote by I_R, plays an important role in our considerations of noetherian R-algebras primarily because of the following properties which we state without proof. The reader is referred to [13], [16], and [17] for the basic facts concerning commutative noetherian rings we need.

PROPOSITION 4.1. *Let R be a commutative noetherian ring. Then the R-module $I_R = \coprod_{\underline{m}} I(R/\underline{m})$, where $\underline{m}$ ranges over all maximal ideals of R, has the following properties.*

a) *I_R is an injective cogenerator of* $\operatorname{Mod} R$.

b) *Let A be a submodule of I_R. Then A is of finite length if and only if it is finitely generated.*

c) *A finitely generated R-module is of finite length if and only if there is a monomorphism $A \to I_R^n$, where I_R^n means the sum of n copies of I_R with n a nonnegative integer ($I^0 = (0)$ by convention).*

d) *An R-module A has finite length if and only if $\operatorname{Hom}_R(A, I_R)$ has finite length.*

e) *Denoting the full subcategory of $\operatorname{Mod} \Lambda$ whose objects are the R-modules of finite length by $\mathrm{f.l.}\,R$, the functor $\operatorname{Hom}_R(\ , I_R)\colon \mathrm{f.l.}\,R \to \mathrm{f.l.}\,R$ is a duality with inverse $\operatorname{Hom}_R(\ , I_R)$. The isomorphism $I_{\mathrm{f.l.}\,R} \to \operatorname{Hom}_R(\operatorname{Hom}_R(\ , I_R), I_R)$ is given by the functorial isomorphism $\phi\colon A \to \operatorname{Hom}_R(\operatorname{Hom}_R(A, I_R), I_R)$ where $\phi_M(a)(f) = f(a)$ for all a in A and f in $\operatorname{Hom}_R(M, I_R)$.*

Suppose now that Λ is a noetherian R-algebra. Then every Λ-module A is also an R-module. Moreover for each pair of Λ-modules A and B, $\operatorname{Hom}_\Lambda(A, B)$ is also an R-submodule of $\operatorname{Hom}_R(A, B)$. In particular, $\operatorname{End}_\Lambda(A)$ is an R-subalgebra of $\operatorname{End}_R(A)$. We now list some well known connections between viewing a Λ-module as a Λ-module and as an R-module.

PROPOSITION 4.2. *Let Λ be a noetherian R-algebra.*

a) *A Λ-module A is a finitely generated Λ-module if and only if A is a finitely generated R-module.*

b) *If A and B are finitely generated Λ-modules, then $\operatorname{Hom}_\Lambda(A, B)$ is a finitely generated R-module.*

c) *If A is a finitely generated Λ-module, then $\operatorname{End}_\Lambda(A)$ is a noetherian R-algebra.*

d) *The following statements are equivalent for a Λ-module A.*

i) *A is of finite length over Λ.*
ii) *A is of finite length over R.*
iii) *$\operatorname{Hom}_R(A, I_R)$ is a Λ^{op}-module of finite length.*
iv) *$A_{\underline{p}} = 0$ for all nonmaximal prime ideals $\underline{p}$ of R.*

e) *The functor $\operatorname{Hom}_R(\ , I_R)\colon \operatorname{Mod}\Lambda \to \operatorname{Mod}\Lambda^{\mathrm{op}}$ induces a duality $\operatorname{Hom}_R(\ , I_R)\colon \mathrm{f.l.}.\Lambda \to \mathrm{f.l.}\,\Lambda^{\mathrm{op}}$ with inverse $\operatorname{Hom}_R(\ , I_R)\colon \mathrm{f.l.}\,\Lambda^{\mathrm{op}} \to \mathrm{f.l.}\,\Lambda$ and the functorial isomorphisms $\phi_A\colon A \to \operatorname{Hom}_R(\operatorname{Hom}_R(A, I_R), I_R)$ the usual ones.*

f) *If A and B are Λ-modules of finite length, then $\operatorname{Hom}_\Lambda(A, B)$ is an R-module of finite length.*

As a consequence of these observations we have the following.

PROPOSITION 4.3. *Let Λ be a noetherian R-algebra.*

a) *$\operatorname{Hom}_R(\Lambda^{\mathrm{op}}, I_R)$ is an injective cogenerator for $\operatorname{Mod}\Lambda$.*

b) *A finitely generated Λ-module A is of finite length if and only if there is a monomorphism $A \to \operatorname{Hom}_R(\Lambda^{\mathrm{op}}, I_R)^n$ for some nonnegative integer n.*

c) *Suppose A is a Λ-module of finite length. Then $\operatorname{Soc} A$, the socle of A, is isomorphic to $\coprod_{i=1}^{k} S_i^{k_i}$ with the S_i a finite set of nonisomorphic simple Λ-modules. Thus there is a monomorphism $A \to \operatorname{Hom}_R(\Lambda^{\mathrm{op}}, I_R)^n$ where $n = \max$ of $n_1, \ldots, n_k$.*

PROOF. a) Since Λ^{op} is a projective Λ^{op}-module and I_R is an injective R-module, it is well known that $\mathrm{Hom}_R(\Lambda^{\mathrm{op}}, I_R)$ is an injective Λ-module. Therefore to finish showing that $\mathrm{Hom}_R(\Lambda^{\mathrm{op}}, I_R)$ is an injective cogenerator for $\mathrm{Mod}\,\Lambda$, we have to show that every simple Λ-module is isomorphic to a submodule of $\mathrm{Hom}_R(\Lambda^{\mathrm{op}}, I_R)$. Let S be a simple Λ-module. The fact that $\mathrm{Hom}_R(\ , I_R)\colon \mathrm{f.l.}\,\Lambda \to \mathrm{f.l.}\,\Lambda^{\mathrm{op}}$ is a duality shows that $T = \mathrm{Hom}_R(S, I_R)$ is a simple Λ^{op}-module. Thus there is an epimorphism $\Lambda^{\mathrm{op}} \to T$ which induces a monomorphism $\mathrm{Hom}_R(T, I_R) \to \mathrm{Hom}_R(\Lambda^{\mathrm{op}}, I_R)$. Since $S \cong \mathrm{Hom}_R(T, I_R)$, we are done.

The rest of the proof of the proposition follows in a straightforward way from a) and some of the previous observations. □

As our final preliminary remark we point out the following.

PROPOSITION 4.4. *Let X be a Λ-module. Then the standard isomorphisms* $\mathrm{Hom}_\Lambda(X, \mathrm{Hom}_R(\Lambda^{\mathrm{op}}, I_R)) \cong \mathrm{Hom}_R(X, I_R)$ *which are functorial in X, give an isomorphism of functors* $\mathrm{Hom}_\Lambda(\ , \mathrm{Hom}_R(\Lambda^{\mathrm{op}}, I_R)) \cong \mathrm{Hom}_R(\ , I_R)$ *which we will sometimes view as an identification.*

With these preliminary remarks in mind, we turn to applying some of the results of Section 3 to noetherian algebras. Throughout the rest of this section we will denote the full subcategory of $\mathrm{Mod}\,\Lambda$ whose objects are the finitely generated Λ-modules by $\mathrm{noeth}\,\Lambda$ and the full subcategory of $\mathrm{Mod}\,\Lambda$ whose objects are the artinian Λ-modules by $\mathrm{art}\,\Lambda$.

We begin with the following analogue of Theorem 3.5. However before stating this result, it is convenient to introduce the following terminology.

Let $0 \to A \xrightarrow{g} B \xrightarrow{f} C \to 0$ be an exact sequence. An exact sequence $0 \to A' \xrightarrow{g'} B' \xrightarrow{f'} C \to 0$ is said to be a *summand* over C of $0 \to A \xrightarrow{g} B \xrightarrow{f} C \to 0$ if there is a commutative diagram

$$\begin{array}{ccccccccc}
0 & \longrightarrow & A' & \xrightarrow{g'} & B' & \xrightarrow{f'} & C & \longrightarrow & 0 \\
 & & \downarrow{\scriptstyle u} & & \downarrow{\scriptstyle v} & & \| & & \\
0 & \longrightarrow & A & \xrightarrow{g} & B & \xrightarrow{f} & C & \longrightarrow & 0 \\
 & & \downarrow{\scriptstyle u'} & & \downarrow{\scriptstyle v'} & & \| & & \\
0 & \longrightarrow & A' & \xrightarrow{g'} & B' & \xrightarrow{f'} & C & &
\end{array}$$

with $v'v = 1_B$, and $u'u = 1_A$.

An exact sequence $0 \to A \xrightarrow{g'} B' \to C' \to 0$ is said to be a *summand over* A of $0 \to A \xrightarrow{g} B \xrightarrow{f} C \to 0$ if there is a commutative diagram

$$\begin{array}{ccccccccc}
0 & \longrightarrow & A & \xrightarrow{g'} & B' & \xrightarrow{f'} & C' & \longrightarrow & 0 \\
 & & \| & & \downarrow{\scriptstyle u} & & \downarrow{\scriptstyle v} & & \\
0 & \longrightarrow & A & \xrightarrow{g} & B & \xrightarrow{f} & C & \longrightarrow & 0 \\
 & & \| & & \downarrow{\scriptstyle u'} & & \downarrow{\scriptstyle v'} & & \\
0 & \longrightarrow & A & \xrightarrow{g'} & B' & \xrightarrow{f'} & C' & \longrightarrow & 0
\end{array}$$

with $u'u = 1_B$, and $v'v = 1_{C'}$.

THEOREM 4.5. *Let Λ be a noetherian R-algebra, X and C in* $\operatorname{noeth}\Lambda$ *and A the Λ-module* $\operatorname{Hom}_R(\operatorname{Tr} X, I_R^n)$ *with n a positive integer. Further let* $\Gamma = \operatorname{End}\operatorname{Tr} X$. *Suppose* $0 \to A \xrightarrow{g} B \xrightarrow{f} C \to 0$ *is an exact sequence in* $\operatorname{Mod}\Lambda$.

a) *f is right X-determined.*

b) $(X,C)/\operatorname{Im}(\ ,f)(X)$ *is a Γ-module of finite length which is isomorphic to a submodule of* $\operatorname{Hom}_R(\Gamma^{\mathrm{op}}, I)^n$.

c) *There is an exact sequence* $0 \to A' \xrightarrow{g'} B' \xrightarrow{f'} C \to 0$ *which is a summand over C of* $0 \to A \xrightarrow{g} B \xrightarrow{f} C \to 0$ *with f' a minimal right X-determined morphism such that* $\operatorname{Im}(\ ,f)(X) = \operatorname{Im}(\ ,f')(X)$.

d) *The following are equivalent for an exact sequence* $0 \to A'' \xrightarrow{g''} B'' \xrightarrow{f''} C \to 0$ *with f'' right X-determined which is a summand of C over* $0 \to A \xrightarrow{g} B \xrightarrow{f} C \to 0$:

i) *f'' is minimal right X-determined.*

ii) *If A''' is a submodule of A'' such that $g''(A''')$ is a summand of B'', then $A''' = 0$.*

PROOF. a) An immediate consequence of Theorem 3.5.

b) Using the identification $\operatorname{Hom}_R(\operatorname{Tr} X, I_R^n) = \operatorname{Hom}_\Gamma(\operatorname{Tr} X, \operatorname{Hom}_R(\Gamma, I_R^n))$, we have by Theorem 3.5 that the Γ-morphism $v\colon (X,C)/P(X,C) \to \operatorname{Hom}_R(\Gamma^{\mathrm{op}}, I_R^n)$ corresponding to the extension $0 \to A \xrightarrow{g} B \xrightarrow{f} C \to 0$ under the isomorphism $\operatorname{Ext}^1_\Lambda(C, \operatorname{Hom}_\Gamma(\operatorname{Tr} X, \operatorname{Hom}_R(\Gamma, I_R^n))) \to \operatorname{Hom}_\Gamma(\underline{\operatorname{Hom}}_\Lambda(X,C), \operatorname{Hom}_R(\Gamma^{\mathrm{op}}, I_R^n))$ has the property $\operatorname{Ker} v = \operatorname{Im}(\ ,f)(X)/P(X,C)$. Hence the Γ-module $(X,C)/\operatorname{Im}(\ ,f)(X)$ is isomorphic to a submodule of $\operatorname{Hom}_R(\Gamma^{\mathrm{op}}, I_R^n)$. Since X and C are noetherian Λ-modules, (X,C), and hence $(X,C)/\operatorname{Im}(\ ,f)(X)$, is a finitely generated R-module. Therefore $(X,C)/\operatorname{Im}(\ ,f)(X)$ is a finitely generated Γ-submodule of $\operatorname{Hom}_R(\Gamma^{\mathrm{op}}, I_R^n)$ and is thus of finite length over Γ.

c) An immediate consequence of Proposition 3.11.

d) The fact that i) and ii) are equivalent follows from Proposition 3.10 and 3.11. □

Theorem 4.5 suggests the following existence theorem for morphisms which are right determined by objects in $\operatorname{noeth}\Lambda$.

THEOREM 4.6. *Let X and C be in* $\operatorname{noeth}\Lambda$. *Suppose H is an* $(\operatorname{End} X)^{\mathrm{op}}$*-submodule of (X,C) containing $P(X,C)$ such that $(X,C)/H$ is an* $(\operatorname{End} X)^{\mathrm{op}}$*-module of finite length. Then* $\operatorname{Soc}((X,C)/H) = \coprod_{i=1}^k S_i^{n_i}$ *where $S_1, \dots, S_k$ is a finite number of nonisomorphic simple* $(\operatorname{End} X)^{\mathrm{op}}$*-modules. Letting $n = \max$ of $n_1, \dots, n_k$, we have that there is an exact sequence of Λ-modules*

$$0 \to \operatorname{Hom}_R(\operatorname{Tr} X, I_R)^n \to B \xrightarrow{f} C \to 0$$

such that f is right X-determined and $\operatorname{Im}(\ ,f)(X) = H$.

PROOF. Since H contains $P(X,C)$ we have that $(X,C)/H$ is an $(\underline{\operatorname{End}} X)^{\mathrm{op}}$-module. Let $\Gamma = \operatorname{End}\operatorname{Tr} X$. Since the duality $\operatorname{Tr}\colon \underline{\operatorname{noeth}}\Lambda^{\mathrm{op}} \to \underline{\operatorname{noeth}}\Lambda$ gives an R-algebra isomorphism $\underline{\operatorname{End}}\operatorname{Tr} X \to (\underline{\operatorname{End}} X)^{\mathrm{op}}$, $(X,C)/H$ can be viewed as Γ-module. Moreover the simple $(\operatorname{End} X)^{\mathrm{op}}$-modules $S_1, \dots, S_k$ are $(\underline{\operatorname{End}} X)^{\mathrm{op}}$-modules and thus simple Γ-modules with the property that $\operatorname{Soc}((X,C)/H) = \coprod_{i=1}^k S_i^{n_i}$. By our preliminary remarks, we know that $\operatorname{Hom}_R(\Gamma^{\mathrm{op}}, I_R)$ is an injective Γ-module which is a cogenerator. Thus each $S_i \subset \operatorname{Hom}_R(\Gamma^{\mathrm{op}}, I_R)^n$. Hence there is a Γ-morphism $v\colon (X,C)/P(X,C) \to \operatorname{Hom}_R(\Gamma^{\mathrm{op}}, I_R)^n$ with $\operatorname{Ker} v = H/P(X,C)$.

By Theorem 3.5 we know that the extension $0 \to \operatorname{Hom}_\Gamma(\operatorname{Tr} X, \operatorname{Hom}_R(\Gamma^{\mathrm{op}}, I_R)^n) \to B \xrightarrow{f} C \to 0$ in $\operatorname{Ext}^1_\Lambda(C, \operatorname{Hom}_\Gamma(\operatorname{Tr} X, \operatorname{Hom}_R(\Gamma^{\mathrm{op}}, I_R)^n))$ corresponding to the morphism

$v\colon (X,C)/P(X,C) \to \operatorname{Hom}_R(\Gamma^{\mathrm{op}}, I_R)^n$ has the property that f is a right X-determined morphism such that $\operatorname{Im}(\ ,f)(X) = H$. Since $\operatorname{Hom}_\Gamma(\operatorname{Tr} X, \operatorname{Hom}_R(\Gamma^{\mathrm{op}}, R))$ is isomorphic to $\operatorname{Hom}_R(\operatorname{Tr} X, I_R)$, see Proposition 4.4, we have that $\operatorname{Hom}_\Gamma(\operatorname{Tr} X, \operatorname{Hom}_R(\Gamma^{\mathrm{op}}, I_R)^n) \cong \operatorname{Hom}_\Gamma(\operatorname{Tr} X, \operatorname{Hom}_R(\Gamma^{\mathrm{op}}, I_R))^n \cong \operatorname{Hom}_R(\operatorname{Tr} X, I_R)^n$. This finishes the proof of the theorem. □

Added in proof: Using arguments similar to those given in the proof of Theorem 3.19, we can generalize Theorem 4.6 as follows.

THEOREM 4.7. *Let X and C be in* noeth Λ. *Suppose H is an* $(\operatorname{End} X)^{\mathrm{op}}$*-submodule of (X,C) such that $(X,C)/H$ is an* $(\operatorname{End} X)^{\mathrm{op}}$*-module of finite length. Then there is an exact sequence* $0 \to \operatorname{Hom}_R(\operatorname{Tr} X, I_R)^n \xrightarrow{g} B \xrightarrow{f} C$ *such that f is right X-determined and* $\operatorname{Im}(\ ,f)(X) = H$.

PROOF. Let $(\ ,C)_H$ be the X-determined subfunctor of $(\ ,C)$ such that $(\ ,C)_H(X) = H$. Letting C' be the submodule $(\ ,C)_H(\Lambda)$ of C, we know by Proposition 3.18 that $(\ ,C)_H \subset (\ ,C')$ and $H \supset P(X,C')$. Since $(X,C')/H \subset (X,C)/H$ and $(X,C)/H$ is of finite length over $(\operatorname{End} X)^{\mathrm{op}}$, we know that $(X,C')/H$ is of finite length over $(\operatorname{End} X)^{\mathrm{op}}$. Therefore by Theorem 4.6, we know there is an exact sequence $0 \to \operatorname{Hom}_R(\operatorname{Tr} X, I)^n \xrightarrow{g} B \xrightarrow{f'} C' \to 0$ such that f' is right X-determined with $\operatorname{Im}(\ ,f')(X) = H$. Using the same arguments as those given in the proof of Theorem 3.19, we see that the exact sequence $0 \to \operatorname{Hom}_R(\operatorname{Tr} X, I)^n \xrightarrow{g} B \xrightarrow{f} C$, where f is the composition $B \xrightarrow{f'} C' \xrightarrow{\mathrm{inc}} C$, has our desired properties. □

5. Noetherian algebras over complete local rings

Our purpose in this section is to show how some of the special properties of the category of Λ-modules in the case Λ is a noetherian R-algebra with R a complete local ring are reflected when our previous results are considered in this particular context. We assume throughout this section, unless stated to the contrary, that R is a complete, noetherian local ring and Λ is a noetherian R-algebra. Our discussion begins with a review of the properties of Mod Λ pertinent to our purposes.

Every noetherian R-algebra Λ is semiperfect, i.e. every finitely generated Λ-module has a projective cover or, equivalently, $\Lambda/\operatorname{rad}\Lambda$ is semi-simple and every idempotent in $\Lambda/\operatorname{rad}\Lambda$ is the image of an idempotent in Λ. In particular if A is in noeth Λ, then End A is a noetherian R-algebra by means of the morphism $g\colon R \to \operatorname{End} A$ given by $g(r)(a) = ra$ for all a in A (remember A is also an R-module). Hence End A is semiperfect. From this the following properties of the category noeth Λ can be deduced.

PROPOSITION 5.1. a) *A in* noeth Λ *is indecomposable if and only if* End A *is a local ring.*

b) *Each nonzero A in* mod Λ *can be written as a finite sum $\coprod_{i=1}^n A_i$ of indecomposable Λ-modules A_i and this decomposition is unique up to isomorphism (i.e. if $\coprod_{i=1}^n A_i = A = \coprod_{j=1}^m B_j$ with the A_i and B_j indecomposable modules, then $m = n$ and, after suitable relabeling, $A_i \cong B_i$ for all i).*

Since R is a local ring, rad R is the unique maximal ideal in R and $I_R = I(R/\operatorname{rad} R)$. Because R is also complete the following properties of the functors $D\colon \operatorname{Mod}\Lambda \to \operatorname{Mod}\Lambda^{\mathrm{op}}$ and $D\colon \operatorname{Mod}\Lambda^{\mathrm{op}} \to \operatorname{Mod}\Lambda$ defined by $D(A) = \operatorname{Hom}_R(A, R)$ can be deduced from the corresponding well known properties of the functor $D\colon \operatorname{Mod} R \to \operatorname{Mod} R$. For each A in

$\operatorname{Mod}\Lambda$ or in $\operatorname{Mod}\Lambda^{\text{op}}$ we define $\phi_A\colon A \to D^2A$ by $\phi_A(a)(f) = f(a)$ for all a in A. Then we have the following.

a) The ϕ_A are monomorphisms which are functorial in A.

b) If A is in $\operatorname{noeth}\Lambda$, then $D(A)$ is in $\operatorname{art}\Lambda^{\text{op}}$, the category of artinian Λ-modules, and ϕ_A is an isomorphism.

c) If A is in $\operatorname{art}\Lambda$, then $D(A)$ is in $\operatorname{noeth}\Lambda$ and ϕ_A is an isomorphism.

d) The functors $D\colon \operatorname{Mod}\Lambda \to \operatorname{Mod}\Lambda^{\text{op}}$ and $D\colon \operatorname{Mod}\Lambda^{\text{op}} \to \operatorname{Mod}\Lambda$ induce inverse dualities $D\colon \operatorname{noeth}\Lambda \to \operatorname{art}\Lambda^{\text{op}}$ and $D\colon \operatorname{art}\Lambda^{\text{op}} \to \operatorname{noeth}\Lambda$ with the isomorphisms $A \to D^2A$ being the morphisms ϕ_A for all A in $\operatorname{noeth}\Lambda$ or in $\operatorname{art}\Lambda^{\text{op}}$.

The duality $D\colon \operatorname{noeth}\Lambda \to \operatorname{art}\Lambda^{\text{op}}$ shows that the category $\operatorname{art}\Lambda$ has the following properties analogous to those given for $\operatorname{noeth}\Lambda$ in Proposition 5.1.

PROPOSITION 5.2. a) *$\operatorname{End}A$ is a noetherian R-algebra and so is semi-perfect for all A in* $\operatorname{art}\Lambda$.

b) *If A is in* $\operatorname{art}A$ *then A is indecomposable if and only if* $\operatorname{End}A$ *is local.*

c) *Every A in* $\operatorname{art}\Lambda$ *can be written as a finite sum $\coprod_{i=1}^n A_i$ of indecomposable Λ-modules A_i and this representation is unique up to isomorphism.*

d) *If $A \to I(A)$ is an injective envelope of A in* $\operatorname{art}\Lambda$, *then $I(A)$ is in* $\operatorname{art}\Lambda$.

Not only does the functor $\operatorname{Hom}_R(\ ,I_R)$ have special properties when R is a complete local ring, but also the duality $\operatorname{Tr}\colon \operatorname{noeth}\Lambda \to \operatorname{noeth}\Lambda^{\text{op}}$, as we now point out. This is due to the fact that the modules in $\operatorname{noeth}\Lambda$ have minimal projective presentations.

The duality $\operatorname{Tr}\colon \underline{\operatorname{mod}}\Lambda \to \underline{\operatorname{mod}}\Lambda^{\text{op}}$ described in Section 3 for arbitrary rings Λ has the property that for each M in $\operatorname{mod}\Lambda$ the object $\operatorname{Tr}M$ in $\operatorname{mod}\Lambda$ depends on the particular projective presentation $P_1 \to P_0 \to M \to 0$ used in the construction of the functor Tr. It is only when $\operatorname{Tr}M$ is viewed as an object in $\underline{\operatorname{mod}}\Lambda$ that it is independent (up to a canonical isomorphism) of the particular projective presentation for M used in constructing Tr.

Suppose now that each M in $\operatorname{mod}\Lambda$ has a minimal projective presentation $P_1 \xrightarrow{f} P_0 \to M \to 0$. Then the exact sequence $(P_0,\Lambda) \to (P_1,\Lambda) \to \operatorname{Coker}(f,\Lambda) \to 0$ is easily seen to be a minimal projective presentation for the Λ^{op}-module $\operatorname{Coker}(f,\Lambda)$. From this it follows that each Λ^{op}-module in $\operatorname{mod}\Lambda^{\text{op}}$ has a minimal projective presentation. Hence in this case, it is natural to use only minimal projective presentations to define the functors $\operatorname{Tr}\colon \underline{\operatorname{mod}}\Lambda \to \underline{\operatorname{mod}}\Lambda^{\text{op}}$ and $\operatorname{Tr}\colon \underline{\operatorname{mod}}\Lambda^{\text{op}} \to \underline{\operatorname{mod}}\Lambda$. This convention has the advantage that since minimal projective presentations are unique up to isomorphism, $\operatorname{Tr}M$ is uniquely determined (up to canonical isomorphism) not only in $\underline{\operatorname{mod}}\Lambda$ but in $\operatorname{mod}\Lambda$ as well. So from now on whenever the ring Λ has the property that each M in $\operatorname{mod}\Lambda$ has a minimal projective presentation, we will use only minimal projective presentations whenever projective presentations are needed. In particular, we will only use minimal projective presentations in constructing the duality $\operatorname{Tr}\colon \underline{\operatorname{mod}}\Lambda \to \underline{\operatorname{mod}}\Lambda^{\text{op}}$.

Assume now that not only does each M in $\operatorname{mod}\Lambda$ and $\operatorname{mod}\Lambda^{\text{op}}$ have a minimal projective presentation, but also that $\operatorname{End}M$ is semiperfect for all M in $\operatorname{mod}\Lambda$ and $\operatorname{mod}\Lambda^{\text{op}}$. Let $\operatorname{mod}_P\Lambda$ denote the full subcategory of $\operatorname{mod}\Lambda$ consisting of those M with no nonzero projective summands. Then it is not difficult to verify that the map $\operatorname{Tr}\colon \operatorname{Ob}(\operatorname{mod}\Lambda) \to \operatorname{Ob}(\operatorname{mod}\Lambda^{\text{op}})$ has the following properties.

a) $\operatorname{Tr}M = 0$ if and only if M is projective.

b) $\operatorname{Tr}M$ is in $\operatorname{mod}_P\Lambda^{\text{op}}$ for all M in $\operatorname{mod}\Lambda$. Hence Tr induces $\operatorname{Tr}\colon \operatorname{mod}_P\Lambda \to \operatorname{mod}_P\Lambda^{\text{op}}$.

c) $M \cong \operatorname{Tr}(\operatorname{Tr}M) \amalg P$ with P a projective Λ-module.

d) $M \cong \operatorname{Tr}(\operatorname{Tr}M)$ if and only if M is in $\operatorname{mod}_P\Lambda$.

e) If M_1 and M_2 are in $\operatorname{mod}_P\Lambda$, then $M_1 \cong M_2$ if and only if $\operatorname{Tr}M_1 \cong \operatorname{Tr}M_2$.

f) If $M = M_1 \amalg M_2$ in $\operatorname{mod}\Lambda$, then $\operatorname{Tr} M \cong \operatorname{Tr} M_1 \amalg \operatorname{Tr} M_2$.

g) If M is in $\operatorname{mod}_P \Lambda$, then M is indecomposable if and only if $\operatorname{Tr} M$ is indecomposable.

h) Let M be in $\operatorname{mod}_P \Lambda$, then $M \cong M_1 \amalg \cdots \amalg M_n$ is a decomposition of M into indecomposable modules if and only if $\operatorname{Tr} M \cong \operatorname{Tr} M_1 \amalg \cdots \amalg \operatorname{Tr} M_n$ is a decomposition of $\operatorname{Tr} M$ into indecomposable Λ^{op}-modules.

In connection with these observations we recall the notion of a representation equivalence. Let $\mathcal{C}$ and $\mathcal{D}$ be additive categories. A map $F\colon \operatorname{Ob}\mathcal{C} \to \operatorname{Ob}\mathcal{D}$ is said to be a *representation equivalence* (see [2] for a discussion of this notion when F is given by a functor from $\mathcal{C}$ to $\mathcal{D}$) if it has the following properties:

a) C_1 and C_2 in $\mathcal{C}$ are isomorphic if and only if $F(C_1) \cong F(C_2)$ in $\mathcal{D}$;

b) If $C = C_1 \amalg C_2$, then $F(C) \cong F(C_1) \amalg F(C_2)$;

c) Given D in $\mathcal{D}$, there is a C in $\mathcal{C}$ such that $F(C) \cong D$.

d) C in $\mathcal{C}$ is indecomposable if and only if $F(C)$ is indecomposable in $\mathcal{D}$ and

e) If $C \cong C_1 \amalg \cdots \amalg C_n$ with the C_i indecomposable, then $F(C) \cong F(C_1) \amalg \cdots \amalg F(C_n)$ with the $F(C_i)$ indecomposable.

Suppose the objects in $\operatorname{mod}\Lambda$ and $\operatorname{mod}\Lambda^{\text{op}}$ have minimal projective resolution and that $\operatorname{End} M$ is semiperfect for all M in $\operatorname{mod}\Lambda$ and $\operatorname{mod}\Lambda^{\text{op}}$. Suppose now that the duality $\operatorname{Tr}\colon \underline{\operatorname{mod}}\Lambda \to \underline{\operatorname{mod}}\Lambda^{\text{op}}$ is defined using minimal projective presentations. Then our previous remarks show that the induced duality $\operatorname{Tr}\colon \underline{\operatorname{mod}}_P\Lambda \to \underline{\operatorname{mod}}_P\Lambda^{\text{op}}$ has the property that the associated map $\operatorname{Tr}\colon \operatorname{mod}_P \Lambda \to \operatorname{mod}_P \Lambda^{\text{op}}$ is a representation equivalences. We will see other examples of representation equivalences as we go along.

In particular, if Λ is a noetherian R-algebra with R a complete local ring, then the above remarks apply to $\operatorname{noeth}\Lambda = \operatorname{mod}\Lambda$ since the necessary hypothesis is satisfied by $\operatorname{noeth}\Lambda$ and $\operatorname{noeth}\Lambda^{\text{op}}$. It should also be observed that the inclusion $\underline{\operatorname{noeth}}_P\Lambda \to \underline{\operatorname{noeth}}\Lambda$ is an equivalence of categories where $\operatorname{noeth}_P \Lambda$ is the full subcategory of $\operatorname{noeth}\Lambda$ whose objects are those of $\operatorname{noeth}_P \Lambda$. This equivalence $\underline{\operatorname{noeth}}_P\Lambda \to \underline{\operatorname{noeth}}\Lambda$ will often be considered an identification.

Suppose now that Λ is a noetherian R-algebra with R a complete local ring. Let $D\colon \operatorname{noeth}\Lambda \to \operatorname{art}\Lambda^{\text{op}}$ be the usual duality. From the fact that each module in $\operatorname{noeth}\Lambda$ has a projective cover in $\operatorname{noeth}\Lambda$, it follows that each module in $\operatorname{art}\Lambda$ has an injective envelope in $\operatorname{art}\Lambda^{\text{op}}$. As we have already seen, each X in $\operatorname{art}\Lambda^{\text{op}}$ can be written uniquely (up to isomorphism) as $X_1 \amalg \cdots \amalg X_n$ with the X_i indecomposable. Thus each X in $\operatorname{art}\Lambda^{\text{op}}$ can be written uniquely up to isomorphism as $U \amalg I$ where U has no nonzero injective summands and I is injective.

Let $\operatorname{art}_I \Lambda^{\text{op}}$ denote the full subcategory of $\operatorname{art}\Lambda^{\text{op}}$ whose objects are those with no nonzero injective summands. Then the duality $D\colon \operatorname{noeth}\Lambda \to \operatorname{art}\Lambda^{\text{op}}$ induces a duality $D\colon \operatorname{noeth}_P \Lambda \to \operatorname{art}_I \Lambda^{\text{op}}$. We thus obtain maps $D\operatorname{Tr}\colon \operatorname{Ob}(\operatorname{noeth}_P \Lambda) \to \operatorname{Ob}(\operatorname{art}_I \Lambda)$ and $\operatorname{Tr} D\colon \operatorname{Ob}(\operatorname{art}_I \Lambda) \to \operatorname{Ob}(\operatorname{noeth}_P \Lambda)$ such that $\operatorname{Tr} D(D\operatorname{Tr}(X)) \cong X$ for all X in $\operatorname{noeth}_P \Lambda$ and $D\operatorname{Tr}(\operatorname{Tr} D(Y)) \cong Y$ for all Y in $\operatorname{art}_I \Lambda$. In addition the maps $D\operatorname{Tr}\colon \operatorname{Ob}(\operatorname{noeth}_P \Lambda) \to \operatorname{Ob}(\operatorname{art}_I \Lambda)$ and $\operatorname{Tr} D\colon \operatorname{Ob}(\operatorname{art}_I \Lambda) \to \operatorname{Ob}(\operatorname{noeth}_P \Lambda)$ are representation equivalences. This follows from the fact that $\operatorname{Tr}\colon \operatorname{Ob}(\operatorname{noeth}_P \Lambda) \to \operatorname{Ob}(\operatorname{noeth}_P \Lambda^{\text{op}})$ and $D\colon \operatorname{Ob}(\operatorname{noeth}_P \Lambda) \to \operatorname{Ob}(\operatorname{art}_I \Lambda^{\text{op}})$ are representation equivalences.

In order to give a categorical interpretation of these maps on the $\operatorname{Ob}(\operatorname{noeth}_P \Lambda)$ and $\operatorname{Ob}(\operatorname{art}_I \Lambda)$ we need to introduce the following general definition and notations.

In analogy with the definition of $\underline{\operatorname{Mod}}\Lambda$, we define for an arbitrary ring Λ the category $\operatorname{Mod}\Lambda$ modulo injectives, which we denote by $\overline{\operatorname{Mod}}\Lambda$, in the following way. The objects of $\overline{\operatorname{Mod}}\Lambda$ are those of $\operatorname{Mod}\Lambda$ and the morphisms $\overline{\operatorname{Hom}}_\Lambda(X,Y)$ in $\overline{\operatorname{Mod}}\Lambda$ is defined to be $\operatorname{Hom}_\Lambda(X,Y)/I(X,Y)$ where $I(X,Y)$ is the subgroup consisting of all those morphisms $f\colon X \to Y$ which can be written as $X \to I \to Y$ with I injective. For each f in

$\mathrm{Hom}_\Lambda(X,Y)$, denote its image in $\overline{\mathrm{Hom}}_\Lambda(X,Y)$ by $\overline{f}$. The the composition of morphisms $\overline{f}, \overline{g}$ in $\overline{\mathrm{Mod}}\Lambda$ is given by $\overline{f}\overline{g} = \overline{(fg)}$. It is worth noting that if $X \to I$ is a monomorphism with I injective, then the sequence $(I,Y) \to (X,Y) \to \overline{\mathrm{Hom}}_\Lambda(X,Y) \to 0$ is exact.

Returning to the case where Λ is a noetherian R-algebra with R a complete local ring, we denote by $\overline{\mathrm{art}}\Lambda$ the full subcategory of $\overline{\mathrm{Mod}}\Lambda$ whose objects are the objects of $\mathrm{art}\,\Lambda$. Denoting by $\overline{\mathrm{art}}_I\Lambda$ the full subcategory of $\overline{\mathrm{art}}\Lambda$ whose objects are in $\mathrm{art}_I\,\Lambda$, we see that the inclusion $\overline{\mathrm{art}}_I\Lambda \to \overline{\mathrm{art}}\Lambda$ is an equivalence of categories which we often consider an identification. Also it is not difficult to see that the duality $D\colon \mathrm{noeth}\,\Lambda \to \mathrm{art}\,\Lambda$ induces a duality $D\colon \underline{\mathrm{noeth}}\Lambda \to \overline{\mathrm{art}}\Lambda$. Combining this duality with the dualities $\mathrm{Tr}\colon \underline{\mathrm{noeth}}\Lambda \to \underline{\mathrm{noeth}}\Lambda^{\mathrm{op}}$ and $\mathrm{Tr}\colon \underline{\mathrm{noeth}}\Lambda^{\mathrm{op}} \to \underline{\mathrm{noeth}}\Lambda$, we obtain the equivalences $D\,\mathrm{Tr}\colon \underline{\mathrm{noeth}}\Lambda \to \overline{\mathrm{art}}\Lambda^{\mathrm{op}}$ and $\mathrm{Tr}\,D\colon \overline{\mathrm{art}}\Lambda^{\mathrm{op}} \to \underline{\mathrm{noeth}}\Lambda$ which are inverses of each other. These play a fundamental role in much of what follows.

With these preliminaries in mind, we now begin our discussion of morphisms determined by modules for Λ a noetherian R-algebra with R a complete local ring. We start with the following analogue of the first part of Theorem 4.5.

PROPOSITION 5.3. *Suppose Λ is a noetherian R-algebra with R a complete local ring. Let $0 \to A \xrightarrow{g} B \xrightarrow{f} C \to 0$ be an exact sequence in* $\mathrm{Mod}\,\Lambda$ *with A in* $\mathrm{art}_I\,\Lambda$ *and C in* $\mathrm{noeth}\,\Lambda$ *and let X be the Λ-module* $\mathrm{Tr}\,DA$ *in* $\mathrm{noeth}\,\Lambda$. *Finally, let* $\Gamma = \mathrm{End}\,D(A)$.

a) *f is right X-determined.*

b) $(X,C)/\mathrm{Im}(\ ,f)(X)$ *is a Γ-module of finite length isomorphic to a submodule of* $\mathrm{Hom}_R(\Gamma^{\mathrm{op}}, I_R)$.

c) *There is a summand $0 \to A' \xrightarrow{g'} B' \xrightarrow{f'} C \to 0$ over C of $0 \to A \xrightarrow{g} B \xrightarrow{f} C \to 0$ with f' a minimal right X-determined morphism such that* $\mathrm{Im}(\ ,f')(X) = \mathrm{Im}(\ ,f)(X)$.

PROOF. This result is essentially a restatement of a) – c) of Theorem 4.5 with $n = 1$ once one observes that since A is in $\mathrm{art}_I\,\Lambda$, $A = D\,\mathrm{Tr}\,X$ where $X = \mathrm{Tr}\,DA$. This fact is a trivial consequence of our preliminary comments. □

In order to simplify the statements of our next result which is a strengthened version of the remainder of Theorem 4.5, we introduce the following terminology.

Suppose $0 \to A \xrightarrow{g} B \xrightarrow{f} C \to 0$ is an exact sequence. An exact sequence $0 \to A' \xrightarrow{g'} B' \xrightarrow{f'} C \to 0$ is said to be *isomorphic over C* to $0 \to A \xrightarrow{g} B \xrightarrow{f} C \to 0$ if there is a commutative diagram

$$\begin{array}{ccccccccc} 0 & \longrightarrow & A' & \xrightarrow{g'} & B' & \xrightarrow{f'} & C & \longrightarrow & 0 \\ & & \downarrow{\scriptstyle u} & & \downarrow{\scriptstyle v} & & \| & & \\ 0 & \longrightarrow & A & \xrightarrow{g} & B & \xrightarrow{f} & C & \longrightarrow & 0 \end{array}$$

with u and v isomorphisms. An exact sequence $0 \to A \xrightarrow{g'} B' \xrightarrow{f'} C' \to 0$ is said to be *isomorphic over A* to $0 \to A \xrightarrow{g} B \xrightarrow{f} C \to 0$ if there is a commutative diagram

$$\begin{array}{ccccccccc} 0 & \longrightarrow & A & \xrightarrow{g'} & B' & \xrightarrow{f'} & C' & \longrightarrow & 0 \\ & & \| & & \downarrow{\scriptstyle u} & & \downarrow{\scriptstyle v} & & \\ 0 & \longrightarrow & A & \xrightarrow{g} & B & \xrightarrow{f} & C & \longrightarrow & 0 \end{array}$$

with u and v isomorphisms.

PROPOSITION 5.4. *Suppose* $0 \to A \xrightarrow{g} B \xrightarrow{f} C \to 0$ *is an exact sequence in* $\operatorname{Mod}\Lambda$ *with* A *in* $\operatorname{art}_I \Lambda$ *and* C *in* $\operatorname{noeth}\Lambda$.

a) f *is right minimal if and only if the zero module is the only submodule* A' *of* A *such that* $g(A')$ *is a summand of* B.

b) *There are summands* A' *of* A *maximal with respect to* $g(A')$ *being a summand of* B. *Suppose* A_1 *and* A_2 *are two such summands of* A.

i) *The induced exact sequences* $0 \to A/A_i \xrightarrow{g_i} B/g(A_i) \xrightarrow{f_i} C \to 0$ *have the property that* f_i *is minimal right* $\operatorname{Tr} DA$*-determined and* $\operatorname{Im}(\ ,f_i)(X) = \operatorname{Im}(\ ,f)(X)$ *for* $i = 1, 2$.

ii) *The exact sequences* $0 \to A/A_i \xrightarrow{g_i} B/g(A_i) \xrightarrow{f_i} C \to 0$ *are isomorphic over* C.

iii) A_1 *and* A_2 *are isomorphic summands of* A *and* A/A_1 *and* A/A_2 *are isomorphic summands of* A.

PROOF. Since f is right $\operatorname{Tr} DA$-determined by Proposition 5.3, a) follows from a) of Theorem 4.5.

b) Since A is artinian, the set of summands of A has both the ascending and the descending chain condition. Thus there are summands A' of A maximal with respect to $g(A')$ being a summand of B.

i) First we observe that because of the way the A_i were chosen, it is not hard to show that (0) is the only submodule A' of A/A_i such that $g_i(A')$ is a summand of $B/g(A_i)$ for $i = 1, 2$. Hence, i) follows immediately from a)

ii) This follows from the fact that the f_i are minimal right $\operatorname{Tr} DA$-determined morphisms such that $\operatorname{Im}(\ ,f_1)(X) = \operatorname{Im}(\ ,f)(X) = \operatorname{Im}(\ ,f_2)(X)$ (see Proposition 3.12).

iii) That A_1 and A_2 are isomorphic summands of A is a consequence of ii). That $A/A_1 \cong A/A_2$ follows from the fact that $\operatorname{art}\Lambda$ is a Krull-Schmidt category. □

As our next result we have the following restatement of Theorem 4.6.

THEOREM 5.5. *Let* X *and* C *be in* $\operatorname{noeth}_P \Lambda$. *Suppose* H *is an* $(\operatorname{End} X)^{\mathrm{op}}$*-submodule of* (X, C) *containing* $P(X, C)$ *such that* $(X, C)/H$ *is an* $(\operatorname{End} X)^{\mathrm{op}}$*-module of finite length. Then* $\operatorname{Soc}((X, C)/H) = \coprod_{i=1}^{k} S_i^{n_i}$ *where* $S_1, \dots, S_k$ *is a finite number of nonisomorphic simple* $(\operatorname{End} X)^{\mathrm{op}}$*-modules. Letting* $n = \max$ *of* $n_1, \dots, n_k$, *we have that there is an exact sequence of* Λ*-modules* $0 \to (D \operatorname{Tr} X)^n \to B \xrightarrow{f} C \to 0$ *such that* f *is right* X*-determined and* $\operatorname{Im}(\ ,f)(X) = H$.

We now establish some results which are not for the most part merely rephrasing of general results for arbitrary noetherian algebras, but are particular to noetherian R-algebras where R is a complete local ring. We begin with the following uniqueness theorem.

PROPOSITION 5.6. *Let* $0 \to A \xrightarrow{g} B \xrightarrow{f} C \to 0$ *be an exact sequence in* $\operatorname{Mod}\Lambda$ *with* A *in* $\operatorname{art}_I \Lambda$ *and* C *in* $\operatorname{noeth}\Lambda$ *such that* f *is right minimal. Suppose* $A = \coprod_{i=1}^{k} V_i^{n_i}$ *with the* $n_i > 0$ *and the* $V_1, \dots, V_k$ *nonisomorphic indecomposable* Λ*-modules. Letting* $V = \coprod_{i=1}^{k} V_i$, *we have the following.*

a) f *is right* $\operatorname{Tr} DV$*-determined.*

b) *If* X *is in* $\operatorname{noeth}\Lambda$ *such that* f *is right* X*-determined but not right* X'*-determined for any proper summand* X' *of* X, *then* $X \cong \operatorname{Tr} DV$.

PROOF. a) Suppose $Y = \coprod_{j=1}^{m} Y_j$ where some of the Y_j may be isomorphic. Let $h\colon L \to C$ be a morphism in $\operatorname{Mod}\Lambda$. If for every morphism $t_j\colon Y_j \to L$ there is a morphism

$s_j\colon Y_j \to B$ such that $ht_j = fs_j$, then clearly for each morphism $t\colon \coprod_{j=1}^m Y_j \to L$ there is a morphism $s\colon \coprod_{j=1}^m Y_j \to B$ such that $ht = fs$. On the other hand, suppose for every $t\colon \coprod_{j=1}^m Y_j \to L$ there is an $s\colon \coprod_{j=1}^m Y_j \to B$ such that $ht = fs$. Then given $t_k\colon Y_k \to L$ for some $k = 1, \dots, m$, there is a morphism $t\colon \coprod_{j=1}^m Y_j \to L$ such that t_k is the composition $Y_k \xrightarrow{i_k} \coprod_{j=1}^m Y_j \xrightarrow{t} L$, where i_k is the natural inclusion $Y_k \to \coprod_{j=1}^m Y_j$. Let $s\colon \amalg Y_j \to B$ be such that $ht = fs$. Then $h(t_k) = h(ti_k) = f(si_k)$ and so the morphism $s_k = si_k\colon Y_k \to B$ has the property $ht_k = f(s_k)$. Hence we have shown that for each morphism $t\colon \coprod_{j=1}^m Y_j \to L$ there is a morphism $s\colon \coprod_{j=1}^m Y_j \to B$ such that $ht = fs$ if and only if for each $t_j\colon Y_j \to L$ there is an $s_j\colon Y \to B$ such that $ht_j = fs_j$.

Now suppose that the indexing is such that $Y_1, \dots, Y_n$ are all the nonisomorphic Y_j where $j = 1, \dots, m$. It now follows from what we have shown that a morphism $h\colon L \to C$ has the property that for each $t\colon \coprod_{i=1}^m Y_i \to L$ there is a morphism $s\colon \coprod_{i=1}^m Y_i \to B$ such that $ht = fs$ if and only if for each $u\colon \coprod_{i=1}^n Y_i \to L$ there is a $v\colon \coprod_{i=1}^n Y_i \to B$ such that $hu = fv$. From this it follows that f is right $\coprod_{j=1}^m Y_j$-determined if and only if f is right $\coprod_{i=1}^n Y_i$-determined.

Now we know by Proposition 5.3 that f is right $\operatorname{Tr} DA$-determined, hence by what we have just shown we know that f is right $\operatorname{Tr} DV$-determined. This finishes the proof of a).

b) We first show that X is in $\operatorname{noeth}_P \Lambda$. For suppose $\coprod_{i=1}^m X_i$ is a representation of X as a sum of indecomposable modules and one X_i projective, say X_1. Since $f\colon B \to C$ is an epimorphism, given any morphism $h\colon L \to C$ in $\operatorname{Mod} \Lambda$ and any morphism $t\colon X_1 \to L$ there is an $s\colon X_1 \to B$ such that $ht = fs$. Thus from the proof of a) it follows that if $h\colon L \to C$ is a morphism in $\operatorname{Mod}\Lambda$ then for each $t\colon \coprod_{i=1}^m X_i \to L$ there is an $s\colon \coprod_{i=1}^m X_i \to B$ such that $ht = fs$ if and only if for each $u\colon \coprod_{i=2}^m X_i \to L$ there is a morphism $v\colon \coprod_{i=2}^m X_i \to B$ such that $hu = fv$. Therefore the fact that f is right X-determined implies that f is right $\coprod_{i=2}^m X_i$-determined. But this is impossible since $\coprod_{i=2}^m X_i$ is a proper summand of X. Hence X is in $\operatorname{noeth}_P \Lambda$.

Suppose f is right X-determined. Then there is by Theorem 5.5 an exact sequence $0 \to (D \operatorname{Tr} X)^n \to B' \xrightarrow{f'} C \to 0$ such that f' is right X-determined and $\operatorname{Im}(\ ,f')(X) = \operatorname{Im}(\ ,f)(X)$. The fact that f is right minimal means that f is a minimal right X-determined morphism. Hence $0 \to A \to B \to C \to 0$ is a summand over C of $0 \to (D \operatorname{Tr} X)^n \to B' \xrightarrow{f'} C \to 0$. Therefore A is isomorphic to a summand of $(D \operatorname{Tr} X)^n$ which implies that V is isomorphic to a summand of $D \operatorname{Tr} X$. From this it follows that $\operatorname{Tr} DV$ is isomorphic to a summand of $\operatorname{Tr} D(D \operatorname{Tr} X)$ which is isomorphic to X since X is in $\operatorname{noeth}_P \Lambda$. But by a) we know that f is right $\operatorname{Tr} DV$-determined. Therefore $\operatorname{Tr} DV \cong X$ since no proper summand X' of X has the property f is right X'-determined. This finishes the proof of part b) as well as the entire proposition. □

In our next result we apply our methods to derive a general theorem concerning extensions of noetherian by artinian modules.

THEOREM 5.7. *Let A and C be in $\operatorname{art}_I \Lambda$ and $\operatorname{noeth} \Lambda$ respectively. Suppose $(X,C)/P(X,C)$ is an $(\operatorname{End} X)^{\mathrm{op}}$-module of finite length where $X = \operatorname{Tr} DA$. Let $S_1, \dots, S_d$ be a complete set of nonisomorphic simple $(\operatorname{End} X)^{\mathrm{op}}$-modules and for each $(\operatorname{End} X)^{\mathrm{op}}$-submodule H of (X,C) containing $P(X,C)$ let $n_1(H), \dots, n_d(H)$ be the uniquely determined nonnegative integers such that the $(\operatorname{End} X)^{\mathrm{op}}$-socle of $(X,C)/H$ is isomorphic to $\coprod_{i=1}^d S_i^{n_i(H)}$. Finally let $n = \max\{n_i(H)\}$ as i runs through $1, \dots, d$ and H runs through all $(\operatorname{End} X)^{\mathrm{op}}$-submodules of (X,C) containing $P(X,C)$.*

a) *n is finite.*

b) *If $k > n$ and $0 \to A^k \xrightarrow{g} B \xrightarrow{f} C \to 0$ is exact, then there is a submodule A' of A^k isomorphic to A^{k-n} such that $g(A')$ is a summand of B.*

PROOF. a) Since each $n_i(H) \leq t$, the length of $(X, C)/P(X, C)$, it follows that $n \leq t$ and is thus finite.

b) Let $\Gamma = \operatorname{End} \operatorname{Tr} X$. Since $X = \operatorname{Tr} DA$ is in $\operatorname{noeth}_P \Lambda$ we know that $P(\operatorname{Tr} X, \operatorname{Tr} X)$ is contained in the radical of Γ and so $\underline{\Gamma} = \underline{\operatorname{End}} \operatorname{Tr} X$ has the same simple modules as Γ. Similarly the simple $(\underline{\operatorname{End}} X)^{\mathrm{op}}$-modules are the same as the simple $(\operatorname{End} X)^{\mathrm{op}}$-modules. Thus $S_1, \dots, S_d$ is a complete set of nonisomorphic Γ-modules by means of the isomorphism $\operatorname{Tr}\colon (\underline{\operatorname{End}} X)^{\mathrm{op}} \to \underline{\operatorname{End}} \operatorname{Tr} X$. Finally, letting P_i be a Γ^{op}-projective cover of the simple Γ^{op}-module $D(S_i)$ we have that the induced morphisms $S_i \to D(P_i)$ are injective envelopes for the S_i and $\Gamma^{\mathrm{op}} = \coprod_{i=1}^{d} P_i^{m_i}$ for some positive integers m_i.

Let $k > n$ and let $x\colon 0 \to A^k \xrightarrow{g} B \xrightarrow{f} C \to 0$ be an exact sequence. Then since A is in $\operatorname{art}_I \Lambda$ we know that $A = D \operatorname{Tr} X = \operatorname{Hom}_R(\operatorname{Tr} X, I_R) = \operatorname{Hom}_\Gamma(\operatorname{Tr} X, \operatorname{Hom}_R(\Gamma^{\mathrm{op}}, I_R))$, thus $A^k = \operatorname{Hom}_\Gamma(\operatorname{Tr} X, D(\Gamma^{\mathrm{op}})^k)$. Let $v\colon (X, C)/P(X, C) \to D(\Gamma^{\mathrm{op}})^k$ be the Γ-morphism corresponding to the element x in $\operatorname{Ext}^1_\Lambda(C, \operatorname{Hom}_\Gamma(\operatorname{Tr} X, D(\Gamma^{\mathrm{op}})^k)$. Denoting by H the submodule of (X, C) containing $P(X, C)$ such that $H/P(X, C) = \operatorname{Ker} v$, we have that the socle of $(X, C)/H$ is $\coprod_{i=1}^{d} S_i^{n_i(H)}$ and therefore $I = D(\coprod_{i=1}^{d} P_i^{n_i(H)})$ is an injective envelope of $(X, C)/H$. Since $D(\Gamma^{\mathrm{op}})^k = \amalg D(P_i^{km_i})$ we can write $D(\Gamma^{\mathrm{op}})^k$ as a sum $I \amalg \coprod_{i=1}^{d} D(P_i^{km_i - n_i(H)}) = I \amalg \coprod_{i=1}^{d} D(P_i^{nm_i - n_i(H)}) \amalg D(\Gamma^{\mathrm{op}})^{k-n}$ since $nm_i \geq n_i(H)$ for all $i = 1, \dots, d$. This gives a decomposition $U \amalg D(\Gamma^{\mathrm{op}})^{k-n}$ of $D(\Gamma^{\mathrm{op}})^k$ such that $\operatorname{Im} v \subset U$ and $U \cong D(\Gamma^{\mathrm{op}})^n$. Therefore the projection $p\colon U \amalg D(\Gamma^{\mathrm{op}})^{k-n} \to D(\Gamma^{\mathrm{op}})^{k-n}$ has the property $pv = 0$. Since the diagram

$$\begin{array}{ccc}
\operatorname{Ext}^1_\Lambda(C \operatorname{Hom}_\Gamma(\operatorname{Tr} X, D(\Gamma^{\mathrm{op}})^k) & \xrightarrow{\cong} & \operatorname{Hom}_\Gamma((X, C)/P(X, C), D(\Gamma^{\mathrm{op}})^k) \\
\downarrow{\scriptstyle \operatorname{Ext}^1(C, \operatorname{Hom}_\Gamma(\operatorname{Tr} X, p))} & & \downarrow{\scriptstyle \operatorname{Hom}_\Gamma((X,C)/P(X,C), p)} \\
\operatorname{Ext}^1_\Lambda(C, \operatorname{Hom}_\Gamma(\operatorname{Tr} X, D(\Gamma^{\mathrm{op}})^{k-n}) & \xrightarrow{\cong} & \operatorname{Hom}_\Gamma((X, C)/P(X, C), D(\Gamma^{\mathrm{op}})^{k-n})
\end{array}$$

commutes, it follows that $\operatorname{Ext}^1_\Lambda(C, \operatorname{Hom}_\Gamma(\operatorname{Tr} X, p))(x) = 0$. Hence there is a commutative diagram

$$\begin{array}{ccccccc}
0 \longrightarrow & \operatorname{Hom}_\Gamma(\operatorname{Tr} X, D(\Gamma^{\mathrm{op}})^k) & \xrightarrow{g} & B & \longrightarrow & C & \longrightarrow 0 \\
& \downarrow{\scriptstyle \operatorname{Hom}_\Gamma(\operatorname{Tr} X, p)} & \swarrow & & & & \\
& \operatorname{Hom}_\Gamma(\operatorname{Tr} X, D(\Gamma^{\mathrm{op}})^{k-n}). & & & & &
\end{array}$$

Therefore the submodule $A' = \operatorname{Hom}_\Gamma(\operatorname{Tr} X, D(\Gamma^{\mathrm{op}})^k) = A^k$ has the property that $g(A')$ is a summand of B. Since $A' = \operatorname{Hom}_\Gamma(\operatorname{Tr} X, D(\Gamma^{\mathrm{op}})^{k-n})$ is isomorphic to $D(\operatorname{Tr} X)^{k-n} \cong A^{k-n}$, the proposition has been established. □

Added to the proof: Using arguments similar to those already used in the addenda to Sections 3 and 4, we can generalize Theorem 5.5 as follows.

THEOREM 5.8. *Let X and C be in $\operatorname{noeth}_P \Lambda$. Suppose H is an $(\operatorname{End} X)^{\mathrm{op}}$-submodule of (X, C) such that $(X, C)/H$ is an $(\operatorname{End} X)^{\mathrm{op}}$-module of finite length. Then there is an exact sequence of Λ-modules*

$$0 \to (D \operatorname{Tr} X)^n \xrightarrow{g} B \xrightarrow{f} C$$

such that f is right X-determined and $\mathrm{Im}(\ ,f)(X) = H$. *Moreover there is a submodule A of B such that $g(A)$ is a summand of B such that the exact sequence* $0 \to (D\,\mathrm{Tr}\,X)^n/A \xrightarrow{g'} B/g(A) \xrightarrow{f'} C$ *has the property that f' is minimal right X-determined with* $\mathrm{Im}(\ ,f')(X) = H$.

6. Morphisms left determined by modules

As in Section 5, we assume throughout this section, unless stated explicitly to the contrary, that Λ is a noetherian R-algebra with R a complete local ring. We showed in Section 5 that if $0 \to A \xrightarrow{g} B \xrightarrow{f} C \to 0$ is an exact sequence with A in $\mathrm{art}_I\,\Lambda$ and C in $\mathrm{noeth}\,\Lambda$, then f is right $\mathrm{Tr}\,DA$-determined. In view of this result it is natural to ask if g is left X-determined for some module X. Eventually (see Section 10) we will show that g is left $D\,\mathrm{Tr}\,C$-determined. However, we now only show that g is left $D\,\mathrm{Tr}\,C$-determined relative to a certain full subcategory $\mathrm{arno}\,\Lambda$ of Λ, which we now describe.

Let $0 \to A \to B \to C \to 0$ be an exact sequence of Λ-modules with A in $\mathrm{art}\,\Lambda$ and C in $\mathrm{noeth}\,\Lambda$. In general the module B is neither artinian nor noetherian. This suggests considering the full subcategory $\mathrm{arno}\,\Lambda$ of $\mathrm{Mod}\,\Lambda$ whose objects consists of those Λ-modules B such that there is an exact sequence $0 \to B' \to B \to B'' \to 0$ with B' in $\mathrm{art}\,\Lambda$ and B'' in $\mathrm{noeth}\,\Lambda$. The following result concerning the category $\mathrm{arno}\,\Lambda$ is easily verified.

PROPOSITION 6.1. *The full subcategory* $\mathrm{arno}\,\Lambda$ *of* $\mathrm{Mod}\,\Lambda$ *has the following properties.*

a) *Let $0 \to A' \to A \to A'' \to 0$ be an exact sequence in* $\mathrm{Mod}\,\Lambda$. *Then A is in* $\mathrm{arno}\,\Lambda$ *if and only if A' and A'' are in* $\mathrm{arno}\,\Lambda$.

b) *A Λ-module A is in* $\mathrm{arno}\,\Lambda$ *if and only if $D(A)$ is in* $\mathrm{arno}\,\Lambda^{\mathrm{op}}$.

c) *If A is in* $\mathrm{arno}\,\Lambda$, *then $\phi_A\colon A \to D^2A$ is an isomorphism.*

d) *The functors $D\colon$* $\mathrm{Mod}\,\Lambda \to \mathrm{Mod}\,\Lambda^{\mathrm{op}}$ *and $D\colon$* $\mathrm{Mod}\,\Lambda^{\mathrm{op}} \to \mathrm{Mod}\,\Lambda$ *induce inverse dualities $D\colon$* $\mathrm{arno}\,\Lambda \to \mathrm{arno}\,\Lambda^{\mathrm{op}}$ *and $D\colon$* $\mathrm{arno}\,\Lambda^{\mathrm{op}} \to \mathrm{arno}\,\Lambda$.

Thus we see that $\mathrm{arno}\,\Lambda$ is a full abelian subcategory of $\mathrm{Mod}\,\Lambda$ which is closed under extensions such that the inclusion $\mathrm{arno}\,\Lambda \to \mathrm{Mod}\,\Lambda$ is exact. Viewing $\mathrm{arno}\,\Lambda$ as an additive category in its own right we can ask if a morphism $f\colon B \to C$ in $\mathrm{arno}\,\Lambda$ is right X-determined for some X in $\mathrm{arno}\,\Lambda$, i.e. if f and X have the property that for a morphism $h\colon L \to C$ in $\mathrm{arno}\,\Lambda$ there is an $s\colon L \to B$ such that $h = sf$ whenever $\mathrm{Im}(\ ,f)(X) \supset \mathrm{Im}(\ ,h)(X)$. Clearly if f is right X-determined in $\mathrm{Mod}\,\Lambda$, then f is right X-determined in $\mathrm{arno}\,\Lambda$. However the converse is not so clear. Analogous definitions and comments can be made for morphisms being left determined in $\mathrm{arno}\,\Lambda$. The rest of this section is devoted to showing how the results in Section 5 concerning right determined morphisms in $\mathrm{Mod}\,\Lambda$ can be translated to results concerning left determined morphisms in $\mathrm{arno}\,\Lambda$ using the duality $D\colon \mathrm{arno}\,\Lambda \to \mathrm{arno}\,\Lambda^{\mathrm{op}}$. These results are based on the observations made at the end of Section 2 concerning dual additive categories which will be used freely.

For each X in $\mathrm{art}\,\Lambda$ the duality $D\colon \mathrm{art}\,\Lambda \to \mathrm{noeth}\,\Lambda^{\mathrm{op}}$ induces a ring isomorphism $D\colon \mathrm{End}\,X \to (\mathrm{End}\,D(X))^{\mathrm{op}}$ which we will often view as an identification. With this convention in mind we have the following easily verified properties of the duality D.

PROPOSITION 6.2. *Let X and Y be in* $\mathrm{art}\,\Lambda$ *and $D\colon$* $\mathrm{art}\,\Lambda \to \mathrm{noeth}\,\Lambda^{\mathrm{op}}$ *the usual duality.*

a) *$D\colon (X,Y) \to (D(Y),D(X))$ is an* $(\mathrm{End}\,X)^{\mathrm{op}}$-$\mathrm{End}\,Y$ *bimodule isomorphism.*

b) *If* $H \subset (X,Y)$ *is an* $\operatorname{End} Y$*-submodule of* (X,Y), *then* $D(H) \subset (D(Y), D(X))$ *is an* $(\operatorname{End} D(Y))^{\mathrm{op}}$*-submodule of* $(D(Y), D(X))$.

c) *Suppose* H *is an* $\operatorname{End} Y$*-submodule of* (X,Y).

i) $H = I(X,Y)$ *if and only if* $D(H) = P(D(Y), D(X))$.

ii) $H \supset I(X,Y)$ *if and only if* $D(H) \supset P(D(Y), D(X))$.

iii) $D\colon (X,Y) \to (D(Y), D(X))$ *induces an* $\operatorname{End} Y$*-isomorphism* $D\colon (X,Y)/H \to (D(Y), D(X))/D(H)$.

PROOF. Left to the reader. □

We say that a sequence $0 \to A \xrightarrow{g} B \xrightarrow{f} C \to 0$ of Λ-modules is exact in $\operatorname{arno}\Lambda$ if it is an exact sequence of Λ-modules and A, B, and C are in $\operatorname{arno}\Lambda$.

It is the following easily verified result which enables us to transfer what has been shown in Section 5 to morphisms which are left determined in $\operatorname{arno}\Lambda$.

PROPOSITION 6.3. *Let* $0 \to A \xrightarrow{g} B \xrightarrow{f} C \to 0$ *be an exact sequence of* Λ*-modules with* A *in* $\operatorname{art}\Lambda$ *and* C *in* $\operatorname{noeth}\Lambda$. *Let* X *be in* $\operatorname{art}\Lambda$ *and let* H *be the* $\operatorname{End} X$*-submodule* $\operatorname{Im}(g,\ )(X)$ *of* (A, X).

a) $H \supset I(A,X)$.

b) *The exact sequence of* Λ^{op}*-modules* $0 \to D(C) \xrightarrow{D(f)} D(B) \xrightarrow{D(g)} D(A) \to 0$ *has the property that* $D(H)$ *is the* $(\operatorname{End} D(X))^{\mathrm{op}}$*-submodule* $\operatorname{Im}(\ , D(g))(D(X))$ *of* $(D(X), D(A))$.

c) *More generally, if* $t\colon A \to L$ *is a morphism in* $\operatorname{arno}\Lambda$, *then* $D(\operatorname{Im}(t,\ )(X)) = \operatorname{Im}(\ , D(t))(D(X))$.

d) g *is left* X*-determined in* $\operatorname{arno}\Lambda$ *if and only if* $D(g)$ *is right* $D(X)$*-determined in* $\operatorname{arno}\Lambda$.

e) g *is minimal left* X*-determined in* $\operatorname{arno}\Lambda$ *if and only if* $D(g)$ *is minimal right* $D(X)$*-determined in* $\operatorname{arno}\Lambda$.

A straightforward application of Proposition 6.3 to Proposition 5.3 yields the following result.

PROPOSITION 6.4. *Let* $0 \to A \xrightarrow{g} B \xrightarrow{f} C \to 0$ *be an exact sequence in* $\operatorname{arno}\Lambda$ *with* A *in* $\operatorname{art}\Lambda$ *and* C *in* $\operatorname{noeth}_P\Lambda$ *and let* $X = D\operatorname{Tr} C$.

a) g *is left* X*-determined in* $\operatorname{arno}\Lambda$.

b) $(A,X)/\operatorname{Im}(g,\ )(X)$ *is an* $\operatorname{End} X$*-module of finite length which is isomorphic to a submodule of* $\operatorname{Hom}_R(\operatorname{End}(X)^{\mathrm{op}}, I_R)$.

c) *There is a summand* $0 \to A \xrightarrow{g'} B' \xrightarrow{f'} C' \to 0$ *over* A *of* $0 \to A \xrightarrow{g} B \xrightarrow{f} C \to 0$ *with* g' *a minimal left* X*-determined morphism such that* $\operatorname{Im}(g',\ )(X) = \operatorname{Im}(g,\ )(X)$.

As a consequence of Proposition 6.4 or of applying Proposition 6.3 to Proposition 5.4, we obtain the following analogue of Proposition 5.4.

PROPOSITION 6.5. *Suppose* $0 \to A \xrightarrow{g} B \xrightarrow{f} C \to 0$ *is an exact sequence in* $\operatorname{arno}\Lambda$ *with* A *in* $\operatorname{art}\Lambda$ *and* C *in* $\operatorname{noeth}\Lambda$.

a) g *is left minimal if and only if there are no proper submodules* C' *of* C *such that the composition* $B \to C \to C/C'$ *is a splittable epimorphism.*

b) *Obviously there are submodules* C' *minimal with respect to the composition* $B \to C \to C/C'$ *being a splittable epimorphism. Suppose* C_1 *and* C_2 *are two such submodules.*

i) C_1 *and* C_2 *are summands of* C *and there are summands* B_1 *and* B_2 *of* B *such that* $f(B_i) = C_i$ *and* $f|_{B_i}\colon B_i \to C_i$ *are isomorphisms for* $i = 1, 2$. *Moreover the exact sequences* $0 \to A \xrightarrow{g_i} B/B_i \xrightarrow{f_i} C/C_i \to 0$ *have the property that* g_i *is minimal left* X*-determined and* $\operatorname{Im}(g_i,\ )(X) = \operatorname{Im}(g,\ )(X)$ *for* $i = 1, 2$.

ii) *The exact sequences* $0 \to A \to B/B_i \to C/C_i \to 0$ *are isomorphic over* A.

iii) C/C_1 *and* C/C_2 *are isomorphic summands of* C *and* C_1 *and* C_2 *are isomorphic summands of* C.

We now prove the following existence theorem which is an analogue of Theorem 5.5.

THEOREM 6.6. *Let* A *and* X *be in* $\operatorname{art}\Lambda$. *Let* H *be an* $\operatorname{End} X$*-submodule of* (A, X) *such that* $H \supset I(A, X)$ *and the* $\operatorname{End} X$*-module* $(A, X)/H$ *is of finite length with socle* $\coprod_{i=1}^{k} S_i^{n_i}$ *where* $S_1, \dots, S_k$ *is a complete set of nonisomorphic simple* $\operatorname{End} X$*-modules. Finally let* $n = \max\{n_1, \dots, n_k\}$. *Then*

a) *There exists an exact sequence* $0 \to A \xrightarrow{g} B \xrightarrow{f} \operatorname{Tr} D(X)^n \to 0$ *with* g *a left* X*-determined morphism in* $\operatorname{arno}\Lambda$ *such that* $(\operatorname{Im}(g,\ ))(X) = H$.

b) *Letting* $C = (\operatorname{Tr} D(X))^n$, *the exact sequence* $0 \to A \xrightarrow{g} B \xrightarrow{f} C \to 0$ *can be written as a sum*

$$\begin{array}{ccccccccc} 0 & \longrightarrow & A & \xrightarrow{g'} & B' & \xrightarrow{f'} & C_1 & \longrightarrow & 0 \\ & & & & \amalg & & \amalg & & \\ & & & & C_2 & = & C_2 & & \end{array}$$

where g' *is a minimal left* X*-determined morphism in* $\operatorname{arno}\Lambda$ *such that* $(\operatorname{Im}(g',\ ))(X) = H$.

PROOF. a) Since A and X are in $\operatorname{art}\Lambda$, we know that $D(A)$ and $D(X)$ are in $\operatorname{noeth}\Lambda^{\mathrm{op}}$. Also we know by Proposition 6.2 that the $\operatorname{End} D(X)^{\mathrm{op}}$-submodule $D(H)$ of $(D(X), D(A))$ contains $P(D(X), D(A))$ and that the $(\operatorname{End} D(X))^{\mathrm{op}}$-module $(D(X), D(A))/D(H)$ is an essential extension of its socle which is of the form $\coprod_{i=1}^{k} T_i^{n_i}$ where $T_1, \dots, T_k$ is a complete set of nonisomorphic simple $\operatorname{End}(D(X))^{\mathrm{op}}$-modules. Thus by Theorem 5.5 we know there is an exact sequence $0 \to (D \operatorname{Tr} D(X))^n \xrightarrow{h} B' \xrightarrow{j} D(A) \to 0$ of Λ^{op}-modules with j a right $D(X)$-determined morphism such that $(\operatorname{Im}(\ , j))(D(X)) = D(H)$. Applying the functor D to this exact sequence we obtain the exact sequence $0 \to A \xrightarrow{D(j)} D(B') \to (\operatorname{Tr} D(X))^n \to 0$. By Proposition 6.3 we know that $D(j)$ is a left X-determined morphism with $(\operatorname{Im}(D(j),\ ))(X) = H$. This completes the proof of a).

b) This follows readily from Theorem 5.3. □

The following translation of Proposition 5.6 can be obtained by applying Proposition 6.3 to Proposition 5.6.

PROPOSITION 6.7. *Let* $0 \to A \xrightarrow{g} B \xrightarrow{f} C \to 0$ *be an exact sequence in* $\operatorname{arno}\Lambda$ *with* A *in* $\operatorname{art}\Lambda$ *and* C *in* $\operatorname{noeth}_P \Lambda$ *such that* g *is left minimal. Suppose* $C = \coprod_{i=1}^{k} W_i^{n_i}$ *with the* $n_i > 0$ *and the* W_i *nonisomorphic indecomposable* Λ*-modules. Letting* $W = \coprod_{i=1}^{k} W_i$, *we have the following.*

a) g *is left* $D \operatorname{Tr} W$*-determined.*

b) *If* X *is in* $\operatorname{art}\Lambda$ *such that* g *is left* X*-determined but not left* X'*-determined for any proper summand* X' *of* X, *then* $X \cong D \operatorname{Tr} W$.

Our next result is the following analogue of Proposition 5.7.

PROPOSITION 6.8. *Let A and C be in* $\operatorname{art}\Lambda$ *and* $\operatorname{noeth}_P \Lambda$ *respectively. Suppose* $(A,X)/I(A,X)$ *is an* $\operatorname{End} X$*-module of finite length t where $X = D\operatorname{Tr} C$. Let $S_1, \dots, S_d$ be a complete set of nonisomorphic simple* $\operatorname{End} X$*-modules and for each* $\operatorname{End} X$*-submodule H of (A,X) containing $I(A,X)$, let $n_1(H), \dots, n_d(H)$ be the uniquely determined non-negative integers such that the* $(\operatorname{End} X)$*-socle of $(A,X)/H$ is isomorphic to $\coprod_{i=1}^{d} S_i^{n_i(H)}$. Finally let $n = \max\{n_i(H)\}$ as i runs through $1, \dots, d$ and H runs through all* $\operatorname{End} X$*-submodules of (A,X) containing $I(A,X)$.*

a) *n is finite.*

b) *If $k > n$ and $0 \to A \to B \to C^k \to 0$ is exact, then there is an exact sequence $0 \to A \to B \to C^n \to 0$ which is a summand over A of $0 \to A \to B \to C^k \to 0$.*

Using the duality $D\colon \operatorname{noeth}\Lambda \to \operatorname{art}\Lambda^{\mathrm{op}}$, the following result can be derived which gives a different starting point for the results of this section. It is included here without proof for the sake of completeness.

PROPOSITION 6.9. *Let C be in* $\operatorname{noeth}\Lambda$ *and A in* $\operatorname{art}\Lambda$*. Further, let $\Gamma = \operatorname{End} D\operatorname{Tr} C$. Then there is an isomorphism of R-modules*

$$\operatorname{Ext}^1_\Lambda(C, A) \to \operatorname{Hom}_\Gamma(\overline{\operatorname{Hom}_\Lambda}(A, D\operatorname{Tr} C), \operatorname{Hom}_R(\Gamma^{\mathrm{op}}, I_R))$$

which is functorial in A in $\operatorname{art}\Lambda$*. Moreover if $0 \to A \xrightarrow{g} B \xrightarrow{f} C \to 0$ in $\operatorname{Ext}^1_\Lambda(C, A)$ and the Γ-morphism $v\colon \overline{\operatorname{Hom}_\Lambda}(A, D\operatorname{Tr} C) \to \operatorname{Hom}_R(\Gamma^{\mathrm{op}}, I_R)$ correspond under this isomorphism, then*

a) *g is left $D\operatorname{Tr} C$-determined.* $\operatorname{Im}(g, \ )$ *is a Γ-submodule of $(A, D\operatorname{Tr} C)$ such that*

b) $\operatorname{Im}(g, \ )(D\operatorname{Tr} C) \supset I(A, D\operatorname{Tr} C)$ *and* $\operatorname{Im}(g, \ )(D\operatorname{Tr} C)/I(A, D\operatorname{Tr} C) = \operatorname{Ker} u$.

Added in proof: Applying the duality $D\colon \operatorname{noeth}\Lambda \to \operatorname{art}\Lambda^{\mathrm{op}}$ to Theorem 5.8 we obtain the following generalization of Theorem 6.6.

THEOREM 6.10. *Let A and X be in* $\operatorname{art}\Lambda$*. Let H be an* $\operatorname{End} X$*-submodule of (A,X) such that $(A,X)/H$ is of finite length over* $\operatorname{End} X$*. Then there is an exact sequence of Λ-modules*

$$A \xrightarrow{g} B \xrightarrow{f} \operatorname{Tr} D(X)^n \to 0$$

with g left X-determined morphism in $\operatorname{arno}\Lambda$ *such that* $\operatorname{Im}(g, \ )(X) = H$*. Moreover, letting $C = (\operatorname{Tr} D(X))^n$, the exact sequence $A \xrightarrow{g'} B \xrightarrow{f'} C \to 0$ can be written as a sum*

$$\begin{array}{ccccccc} A & \xrightarrow{g'} & B' & \xrightarrow{f'} & C' & \longrightarrow & 0 \\ & & \amalg & & \amalg & & \\ & & C_1 & \xrightarrow{\cong} & C_2 & & \end{array}$$

where g' is a minimal left X-determined morphism in $\operatorname{arno}\Lambda$ *such that* $(\operatorname{Im}(g', \ ))(X) = H$.

7. Orders and lattices

The two classical situations where the representation theory of rings have been studied most extensively are a) finitely generated modules over a finite dimensional algebra over a field and b) lattices over orders over Dedekind domains. One of the aims of the rest of this chapter is to see what our earlier results tell us about representation theory in these classical situations. Rather than dealing directly with these two seemingly different situations, we develop in this section a more general notion of orders and lattices which includes amongst a great many other examples, these classical situations as special cases. We will then interpret our previous results for these generalized orders and lattices.

We say that a commutative noetherian ring X is *equidimensional* if $\dim R_{\underline{m}} = \dim R$ for all maximal ideals $\underline{m}$ of R. In particular if R is equidimensional, then $\dim R$ is finite. It is easily seen that if R is equidimensional then $R/(x)$ is equidimensional for each nonzero divisor x in R.

We recall that a local ring R is said to be Gorenstein if $\text{inj. dim. } R < \infty$ (see [13] and [16] for basic properties of Gorenstein rings). If R is a local Gorenstein ring of dimension d, then $\text{inj. dim. } R = d$ and a minimal injective resolution $0 \to R \to I_0(R) \to \cdots \to I_d(R) \to 0$ of R over R has the following properties: a) $I_d(R) = I_R$ the injective envelope of $R/\operatorname{rad} R$ and b) $\operatorname{Hom}_R(M, I_j(R)) = 0$ for all R-modules M of finite length and $j < d$. An arbitrary commutative noetherian ring R is said to be a *Gorenstein* ring if $R_{\underline{p}}$ is *Gorenstein* for all prime ideals $\underline{p}$ of R. The following properties of equidimensional rings are well known.

PROPOSITION 7.1. *Let R be an equidimensional ring of dimension d.*

a) *R is Gorenstein if and only if* $\text{inj. dim. } R < \infty$.

b) *If R is Gorenstein, then* $\text{inj. dim. .} R = d$.

c) *If R is Gorenstein and $0 \to R \to I_0(R) \to \cdots \to I_d(R) \to 0$ is a minimal injective resolution of R, then $I_d(R) \cong I_R$ and a module M in* $\operatorname{noeth} R$ *is of finite length if and only if* $\operatorname{Hom}_R(M, I_j(R)) = 0$ *for all $j < d$.*

d) *If R is Gorenstein and x in R is regular (i.e. x is a nonunit which is not a zero divisor in R), then $R/(x)$ is an equidimensional Gorenstein ring.*

e) *If R is Gorenstein, then every localization of R is a Gorenstein ring.*

Before discussing equidimensional Gorenstein rings further we point out some examples: a) fields, b) Dedekind domains, e) equidimensional regular rings such as affine rings of nonsingular affine varieties, d) 0-dimensional local rings with a simple socle, e) rings derived from examples a) through c) by taking factor rings by ideals generated by a regular element.

Let M be a nonzero finitely generated R-module with R a local ring. A sequence $x_1, \dots, x_r$ of elements in $\operatorname{rad} R$ is said to be a regular M-sequence if x_j is $M/(x_1, \dots, x_{j-1})M$-regular for all $j = 1, \dots, d$, i.e. x_j is not a zero divisor in $M/(x_1, \dots, x_{j-1})M$ where $(x_1, \dots, x_{j-1})$ is the ideal generated by $x_1, \dots, x_{j-1}$. A regular M-sequence $x_1, \dots, x_r$ is said to be maximal if every element of $\operatorname{rad} R$ is not regular in $M/(x_1, \dots, x_r)M$. It is known that every regular M-sequence can be extended to a maximal regular M-sequence and that all maximal regular M-sequences have the same length (i.e. same number of elements). The length of the maximal regular M-sequence is denoted by $\operatorname{depth} M$. In general $\operatorname{depth} M \leq \dim R$. We recall that $\operatorname{depth} M$ is the smallest integer i such that $\operatorname{Ext}^i_R(R/\operatorname{rad} R, M) \neq 0$. A module M over R is said to be a *Cohen-Macaulay module* if $\operatorname{depth} M = \dim R$.

Suppose R is an arbitrary commutative noetherian ring. A finitely generated R-module M is called a *Cohen-Macaulay module* if $M_{\underline{p}}$ is a Cohen-Macaulay $R_{\underline{p}}$-module for all prime ideals $\underline{p}$ of R. Moreover R is said to be a *Cohen-Macaulay ring* if it is a Cohen-Macaulay module over itself. Of particular importance to us is the fact that if R is a Gorenstein ring then R is a Cohen-Macaulay ring. Thus all our previous examples of Gorenstein rings also give examples of Cohen-Macaulay rings. The following are examples of Cohen-Macaulay modules: a) if R is a field, then all nonzero finite dimensional vector spaces are Cohen-Macaulay; b) if R is an artin ring, then all finitely generated nonzero modules are Cohen-Macaulay modules; c) if R is a Dedekind domain, then a nonzero finitely generated module M is Cohen-Macaulay if and only if M is projective, or equivalently, if and only if M is torsion free; d) if R is a regular ring, then a finitely generated R-module M is Cohen-Macaulay if and only if M is projective; e) if R is a two-dimensional integrally closed domain, then a nonzero module M is Cohen-Macaulay if and only if M is reflexive, i.e. the usual morphism $M \to \operatorname{Hom}_R(\operatorname{Hom}(M,R),R)$ is an isomorphism. In particular R is Cohen-Macaulay as are all the nonzero finitely generated projective R-modules.

We now summarize the main properties of Cohen-Macaulay modules over a Gorenstein ring R which we will need. All of these properties are obvious in case R is a field or a Dedekind ring.

PROPOSITION 7.2. *Assume that R is a Gorenstein ring of dimension d (not necessarily equidimensional).*

a) *Suppose $0 \to M_1 \to M_2 \to M_3 \to 0$ is an exact sequence in* noeth R.

i) *If M_1, M_3 are Cohen-Macaulay then M_2 is Cohen-Macaulay.*

ii) *If M_2 and M_3 are Cohen-Macaulay then M_1 is Cohen-Macaulay.*

iii) *If the sequence splits, then M_2 is Cohen-Macaulay if and only if M_1 and M_3 are Cohen-Macaulay.*

b) *If $0 \to M_d \xrightarrow{f_d} M_{d-1} \to \cdots \xrightarrow{f_1} M_0$ is an exact sequence in* noeth R *with the M_i Cohen-Macaulay for $i = 0, \dots, d-1$, then* $\operatorname{depth} \operatorname{Im} f_i \geq i$ *for all $i \geq 1$ and so M_d is Cohen-Macaulay.*

c) *If M is Cohen-Macaulay, then $\operatorname{Ext}^i_R(M,R) = 0$ for all $i > 0$.*

d) *If M is Cohen-Macaulay, then $\operatorname{Hom}_R(M,R) = M^*$ is Cohen-Macaulay.*

e) *Each Cohen-Macaulay module is reflexive, i.e. the natural morphism $M \to M^{**}$ is an isomorphism.*

f) *Let $\mathcal{CM}(R)$ denote the full subcategory of* noeth R *whose objects are the Cohen-Macaulay modules. The functor $\operatorname{Hom}_R(\ ,R)\colon \mathcal{CM}(R) \to \mathcal{CM}(R)$ is a duality.*

g) *If $0 \to M_1 \to M_2 \to M_3 \to 0$ is exact with the M_i in $\mathcal{CM}(R)$, then $0 \to M_3^* \to M_2^* \to M_1^* \to 0$ is exact with the M_i^* in $\mathcal{CM}(R)$.*

h) *If M is in $\mathcal{CM}(R)$, then $M_{\underline{p}}$ is in $\mathcal{CM}(R)$ for all prime ideals $\underline{p}$ in R.*

Let Λ be a noetherian R-algebra with R an equidimensional Gorenstein ring. We say that a Λ-module is a Cohen-Macaulay module if it is finitely generated and is Cohen-Macaulay when viewed as an R-module. We say that a Λ-module M is a *lattice* if it is a Cohen-Macaulay Λ-module and $M_{\underline{p}}$ and $\operatorname{Hom}_R(M,R)_{\underline{p}}$ are $\Lambda_{\underline{p}}$-projective and $\Lambda^{\mathrm{op}}_{\underline{p}}$-projective modules respectively for all nonmaximal prime ideals $\underline{p}$ in R. We denote the full subcategory of noeth Λ whose objects are the Λ-lattices by $L(\Lambda)$. Finally we say a noetherian R-algebra Λ is an *R-order* if R is an equidimensional Gorenstein ring and Λ is in $L(\Lambda)$. Before discussing lattices over R-orders we point out some examples.

a) If R is a field or, more generally, a 0-dimensional Gorenstein ring, then every noetherian R-algebra Λ is an R-order and $L(\Lambda) = \operatorname{noeth}\Lambda = \operatorname{art}\Lambda$.

b) Suppose Λ is a noetherian R-algebra with R a Dedekind ring with field of quotients K. Then Λ is an R-order if and only if Λ is a projective R-module and $\Sigma = K \otimes_R \Lambda$ is a selfinjective ring. If Λ is an R-order, then $L(\Lambda)$ consists of the Λ-modules M which are projective R-modules such that $K \otimes_R M$ is a projective Σ-module. It should be noted in particular that if Λ is a projective R-module such that $\Sigma = K \otimes \Lambda$ is semisimple, then Λ is an R-order and the Λ-lattices are precisely the Λ-modules which are projective R-modules. Thus we see that the classical orders and lattices fit into this more general scheme.

c) Suppose R is an equidimensional Gorenstein ring. Then R is an R-order and the R-lattices consist of those Cohen-Macaulay R-modules M such that $M_{\underline{p}}$ is $R_{\underline{p}}$-free for all nonmaximal primes $\underline{p}$ of R.

d) Suppose R is a two-dimensional integrally closed Gorenstein ring. Then R is an R-order and the R-lattices are precisely the reflexive R-modules.

We now give some of the basic properties of Λ-lattices over an R-order Λ. These follow readily from the definitions and Proposition 7.2. Moreover they are obvious for R a field or a Dedekind domain.

PROPOSITION 7.3. *Let Λ be an R-order with $d = \dim R$.*

a) *Λ^{op} is an R-order.*

b) *For M in* $\text{noeth}\,\Lambda$ *the following are equivalent:*

i) *M is in $L(\Lambda)$.*

ii) *M is Cohen-Macaulay and $M_{\underline{p}}$ is $\Lambda_{\underline{p}}$-projective for all nonmaximal primes $\underline{p}$ in R.*

iii) *M is Cohen-Macaulay and $M_{\underline{m}}$ is $\Lambda_{\underline{m}}$-projective for all but a finite number of maximal ideals $\underline{m}$ of R.*

iv) *M is Cohen-Macaulay and* $\text{Hom}_R(M, R)$ *is in $L(\Lambda^{\text{op}})$.*

c) *The functor* $\text{Hom}_R(\ ,R)\colon L(\Lambda) \to L(\Lambda^{\text{op}})$ *is a duality.*

d) *Suppose $0 \to M_1 \to M_2 \to M_3 \to 0$ is an exact sequence in* $\text{noeth}\,\Lambda$.

i) *If the sequence splits, then M_2 is in $L(\Lambda)$ if and only if M_1 and M_3 are in $L(\Lambda)$.*

ii) *If M_1 and M_3 are in $L(\Lambda)$, then M_2 is in $L(\Lambda)$.*

iii) *If M_2 and M_3 are in $L(\Lambda)$, then M_1 is in $L(\Lambda)$.*

iv) *If the M_i are in $L(\Lambda)$, then $0 \to M_3^* \to M_2^* \to M_1^* \to 0$ is an exact sequence with the M_i^* in $L(\Lambda^{\text{op}})$ where $M_i^* = \text{Hom}(M_i, R)$.*

e) *If $\cdots \to P_{i+1} \xrightarrow{d_{i+1}} P_i \to \cdots \to P_1 \xrightarrow{d_1} P_0 \xrightarrow{d_0} M \to 0$ is a projective resolution of M and M has the property $M_{\underline{p}}$ is $\Lambda_{\underline{p}}$ projective for all nonmaximal prime ideals $\underline{p}$, then* $\text{Im}\, d_i$ *is a lattice for all $i \geq \dim R$.*

We now use these basic facts concerning lattices to define various functors which will enable us to translate most of our results in earlier sections into results concerning lattices.

First we recall the functor $\Omega\colon \underline{\text{noeth}}\Lambda \to \underline{\text{noeth}}\Lambda$ defined as follows. For each M in $\text{noeth}\,\Lambda$ choose an epimorphism $P(M) \to M \to 0$ with $P(M)$ projective and define $\Omega M = \text{Ker}(P(M) \to M)$. As usual if M has a projective cover we choose $P(M) \to M \to 0$ to be a projective cover. Clearly if $f\colon M \to N$ is a morphism in $\text{noeth}\,\Lambda$, then there exists a commutative exact diagram

$$\begin{array}{ccccccccc}
0 & \longrightarrow & \Omega M & \longrightarrow & P(M) & \longrightarrow & M & \longrightarrow & 0 \\
 & & \downarrow{\scriptstyle f_1} & & \downarrow{\scriptstyle f_0} & & \downarrow{\scriptstyle f} & & \\
0 & \longrightarrow & \Omega N & \longrightarrow & P(N) & \longrightarrow & N & \longrightarrow & 0.
\end{array}$$

While the morphism $f_1\colon \Omega(M) \to \Omega(N)$ depends on the particular choice of the morphism $f_0\colon P(M) \to P(N)$ which makes the diagram commute, different choices of the f_0 result in morphisms $\Omega M \to \Omega N$ which differ by a morphism factoring through a projective module. Thus if we define $\Omega(f)\colon \Omega M \to \Omega N$ to be $\underline{f}_1$ in $\underline{\text{Hom}}_\Lambda(\Omega M, \Omega N)$ where $f_1\colon \Omega M \to \Omega N$

is any morphism for which there is a commutative diagram

$$\begin{array}{ccccccccc} 0 & \longrightarrow & \Omega M & \longrightarrow & P(M) & \longrightarrow & M & \longrightarrow & 0 \\ & & \downarrow{\scriptstyle f_1} & & \downarrow{\scriptstyle f_0} & & \downarrow & & \\ 0 & \longrightarrow & \Omega N & \longrightarrow & P(N) & \longrightarrow & N & \longrightarrow & 0, \end{array}$$

we obtain the functor $\Omega\colon \underline{\text{noeth}}\Lambda \to \underline{\text{noeth}}\Lambda$. More generally, we define $\Omega^i\colon \underline{\text{noeth}}\Lambda \to \underline{\text{noeth}}\Lambda$ for all nonnegative integers by induction on i as follows: Ω^0 is the identity and $\Omega^{i+1} = \Omega(\Omega^i)$ for all $i \geq 0$.

Next we observe that if $f_1, f_2\colon M \to N$ are morphisms in noeth Λ then the morphisms $\text{Ext}^1_\Lambda(f_1, \ ), \text{Ext}^1_\Lambda(f_2, \ )\colon \text{Ext}^1_\Lambda(N, \ ) \to \text{Ext}^1_\Lambda(M, \ )$ are the same if and only if $\underline{f}_1$ and $\underline{f}_2$ are equal in $\underline{\text{Hom}}_\Lambda(M, N)$ (see [15]). Thus for each $\underline{f}$ in $\underline{\text{Hom}}_\Lambda(M, N)$ we can define $\text{Ext}^1_\Lambda(\underline{f}, \ )\colon \text{Ext}^1_\Lambda(N, \ ) \to \text{Ext}^1_\Lambda(M, \ )$ to be $\text{Ext}^1_\Lambda(f, \ )\colon \text{Ext}^1_\Lambda(N, \ ) \to \text{Ext}^1_\Lambda(M, \ )$ without introducing ambiguities. We now list some of the properties of the functor $\Omega\colon \underline{\text{noeth}}\Lambda \to \underline{\text{noeth}}\Lambda$ which are well known and easily checked.

PROPOSITION 7.4. *Let Λ be an arbitrary noetherian ring. Then the functors $\Omega^i\colon \underline{\text{noeth}}\Lambda \to \underline{\text{noeth}}\Lambda$ have the following properties for each $i > 0$.*

a) *For each M in* noeth Λ *there is an isomorphism $\phi_M\colon \text{Ext}^1_\Lambda(\Omega^i M, \ ) \to \text{Ext}^{i+1}(M, \ )$ which is functorial in M, i.e. given a morphism $\underline{f}\colon M \to N$ in $\underline{\text{noeth}}\Lambda$, the diagram*

$$\begin{array}{ccc} \text{Ext}^1_\Lambda(\Omega^i N, \ ) & \xrightarrow{\phi_N} & \text{Ext}^{i+1}_\Lambda(N, \ ) \\ \downarrow{\scriptstyle \text{Ext}^1_\Lambda(\Omega^i \underline{f}, \ )} & & \downarrow{\scriptstyle \text{Ext}^{i+1}_\Lambda(\underline{f}, \ )} \\ \text{Ext}^1_\Lambda(\Omega^i M, \ ) & \xrightarrow[\phi_M]{} & \text{Ext}^{i+1}_\Lambda(M, \ ) \end{array}$$

commutes.

b) *If $\text{Ext}^j_\Lambda(M, \Lambda) = 0$ for $j = 1, \dots, i$, then $\Omega^i\colon \underline{\text{Hom}}_\Lambda(\ , M) \to \underline{\text{Hom}}_\Lambda(\ , \Omega^i M)$ is an isomorphism.*

PROOF. a) Since $0 \to \Omega^i(M) \to P_{i-1} \to \cdots \to P_0 \to M \to 0$ is exact with the P_j projective, we have the usual dimension shift isomorphism $\phi_M\colon \text{Ext}^1_\Lambda(\Omega^i M, \ ) \to \text{Ext}^{i+1}_\Lambda(M, \ )$. It is not difficult to check that the ϕ_M have the desired properties.

b) Follows from [8, Proposition 1.39]. □

Suppose now that Λ is an R-order with $d = \dim R$. We denote by $J(\Lambda)$ the full subcategory of noeth Λ consisting of all M in noeth Λ satisfying both of the following conditions: a) $M_{\underline{p}}$ is $\Lambda_{\underline{p}}$-projective for all nonmaximal prime ideals $\underline{p}$ of R and b) $\text{Ext}^i_\Lambda(M, \Lambda) = 0$ for all $i = 1, \dots, d$. Let $\underline{J}(\Lambda)$ be the full subcategory of $\underline{\text{noeth}}\Lambda$ whose objects are in $J(\Lambda)$. By Proposition 7.3, condition a) implies that if M is in $J(\Lambda)$, then $\Omega^d M$ is in $L(\Lambda)$. Thus Ω^d induces a functor $\Omega^d\colon \underline{J}(\Lambda) \to \underline{L}(\Lambda)$. Condition b) together with Proposition 7.4 shows that $\Omega^d\colon \underline{J}(\Lambda) \to \underline{L}(\Lambda)$ is fully faithful. It is our aim now to show that Ω^d is dense and hence an equivalence of categories. That is, we want to show that if X is a nonprojective lattice, then there is an M in $J(\Lambda)$ such that $\Omega^d(M) \cong X$ in $\underline{L}(\Lambda)$. The proof of this fact depends on the following result.

PROPOSITION 7.5. *Suppose Λ is an R-order with* $\dim R = d$. *For each M in $L(\Lambda)$ we have that* $\text{Tr}\, M$ *is in $J(\Lambda^{\text{op}})$.*

PROOF. The proof takes several steps. We begin with the following lemmas.

LEMMA 7.6. *Let Λ be an arbitrary noetherian R-algebra (R an arbitrary commutative noetherian ring) and let M be in* $\operatorname{noeth}\Lambda$.

a) *M is of finite length over Λ if and only if $M_{\underline{p}} = 0$ for all nonmaximal prime ideals $\underline{p}$ of R.*

b) *The natural morphism $\operatorname{Ext}^i_\Lambda(M,N)_{\underline{p}} \to \operatorname{Ext}^i_{\Lambda_{\underline{p}}}(M_{\underline{p}},N_{\underline{p}})$ is an isomorphism for all N in $\operatorname{Mod}\Lambda$ and all $i \geq 0$.*

c) $\underline{\operatorname{Hom}}_\Lambda(M,N)_{\underline{p}} \cong \underline{\operatorname{Hom}}_{\Lambda_{\underline{p}}}(M_{\underline{p}},N_{\underline{p}})$.

d) *The following are equivalent:*

i) *$M_{\underline{p}}$ is $\Lambda_{\underline{p}}$-projective for all nonmaximal prime ideals $\underline{p}$ of R.*

ii) *$\underline{\operatorname{Hom}}_\Lambda(M,N)$ and $\underline{\operatorname{Hom}}_\Lambda(N,M)$ are of finite length over R for all N in* $\operatorname{noeth}\Lambda$.

iii) *$\underline{\operatorname{End}}(M)$ is of finite length over R.*

e) *$M_{\underline{p}}$ is $\Lambda_{\underline{p}}$-projective for all nonmaximal prime ideals $\underline{p}$ of R if and only if $(\operatorname{Tr} M)_{\underline{p}}$ is $\Lambda_{\underline{p}}$-projective for all nonmaximal prime ideals $\underline{p}$ of R.*

PROOF. a) and b) Standard facts in commutative ring theory.

c) Let $P \to N \to 0$ be exact with P projective. Then $\operatorname{Hom}_\Lambda(M,P) \to \operatorname{Hom}_\Lambda(M,N) \to \underline{\operatorname{Hom}}_\Lambda(M,N) \to 0$ is exact. Hence for each prime ideal $\underline{p}$ of R we have the commutative exact diagram

$$\begin{array}{ccccccc}
\operatorname{Hom}_\Lambda(M,P_{\underline{p}}) & \longrightarrow & \operatorname{Hom}_\Lambda(M,N)_{\underline{p}} & \longrightarrow & \underline{\operatorname{Hom}}_\Lambda(M,N)_{\underline{p}} & \longrightarrow & 0 \\
\| & & \| & & \downarrow & & \\
\operatorname{Hom}_{\Lambda_{\underline{p}}}(M_{\underline{p}},P_{\underline{p}}) & \longrightarrow & \operatorname{Hom}_{\Lambda_{\underline{p}}}(M_{\underline{p}},N_{\underline{p}}) & \longrightarrow & \underline{\operatorname{Hom}}_{\Lambda_{\underline{p}}}(M_{\underline{p}},N_{\underline{p}}) & \longrightarrow & 0.
\end{array}$$

From this it follows that $\underline{\operatorname{Hom}}_\Lambda(M,N)_{\underline{p}} \to \underline{\operatorname{Hom}}_{\Lambda_{\underline{p}}}(M_{\underline{p}},N_{\underline{p}})$ is an isomorphism.

d) i) implies ii). Since $M_{\underline{p}}$ is $\Lambda_{\underline{p}}$-projective for all nonmaximal prime ideals, we have that $\underline{\operatorname{Hom}}_{\Lambda_{\underline{p}}}(N_{\underline{p}},M_{\underline{p}}) = 0 = \underline{\operatorname{Hom}}_\Lambda(M_{\underline{p}},N_{\underline{p}})$ for all N in $\operatorname{noeth}\Lambda$ and all nonmaximal prime ideals $\underline{p}$ of R. The fact that $\underline{\operatorname{Hom}}_{\Lambda_{\underline{p}}}(N_{\underline{p}},M_{\underline{p}}) = (\underline{\operatorname{Hom}}_\Lambda(M,N))_{\underline{p}}$ implies that $\underline{\operatorname{Hom}}_\Lambda(N,M)_{\underline{p}} = 0 = \underline{\operatorname{Hom}}_{\Lambda_{\underline{p}}}(M,N)_{\underline{p}}$ for all nonmaximal prime ideals $\underline{p}$ of R. Hence $\underline{\operatorname{Hom}}_\Lambda(N,M)$ and $\underline{\operatorname{Hom}}_\Lambda(M,N)$ are of finite length over R by part a).

ii) implies iii). Trivial.

iii) implies i). By a) we know that $\operatorname{End}_\Lambda(M)$ is of finite length if and only if $\operatorname{End}_{\Lambda_{\underline{p}}}(M_{\underline{p}}) = 0$ for all nonmaximal primes $\underline{p}$ of R. Then by c) we have that $\underline{\operatorname{End}}_\Lambda(M)$ is of finite length if and only if $\underline{\operatorname{End}}_{\Lambda_{\underline{p}}}(M_{\underline{p}}) = 0$ for all nonmaximal prime ideals $\underline{p}$ of R. Since $\underline{\operatorname{End}}_{\Lambda_{\underline{p}}}(M_{\underline{p}}) = 0$ if and only if $M_{\underline{p}}$ is $\Lambda_{\underline{p}}$-projective, we have our desired result.

e) The duality $\operatorname{Tr}\colon \underline{\operatorname{noeth}\Lambda} \to \underline{\operatorname{noeth}\Lambda^{\mathrm{op}}}$ gives an isomorphism $\operatorname{Tr}\colon \underline{\operatorname{End}}_\Lambda(M) \to \underline{\operatorname{End}}_{\Lambda^{\mathrm{op}}}(\operatorname{Tr} M)$ of R-modules. Thus $\underline{\operatorname{End}}_\Lambda(M)$ has finite length if and only if $\underline{\operatorname{End}}_{\Lambda^{\mathrm{op}}}(\operatorname{Tr} M)$ is of finite length. □

REMARK. The connection between this lemma and Proposition 7.5 is that the lemma shows that if M is in $L(\Lambda)$, then $(\operatorname{Tr} M)_{\underline{p}}$ is $\Lambda_{\underline{p}}$-projective for all nonmaximal prime ideals $\underline{p}$, a necessary condition for $\operatorname{Tr} M$ to be in $J(\Lambda^{\mathrm{op}})$.

The proof of the next lemma can be found in [8 Proposition 2.6]

LEMMA 7.7. *Let Λ be a noetherian ring and M in* $\operatorname{noeth}\Lambda$. *There is an exact sequence $0 \to \operatorname{Ext}^1_\Lambda(\operatorname{Tr} M,\Lambda) \to M \to \operatorname{Hom}_\Lambda(\operatorname{Hom}_\Lambda(M,\Lambda),\Lambda) \to \operatorname{Ext}^2(\operatorname{Tr} M,\Lambda) \to 0$ where $M \to \operatorname{Hom}_\Lambda(\operatorname{Hom}_\Lambda(M,\Lambda),\Lambda)$ is the usual morphism.*

We now apply this to prove the following which establishes Proposition 7.5 for $\dim R = 0, 1, 2$.

LEMMA 7.8 *Let* Λ *be an* R*-order with* $\dim R = d$ *and* M *a module in* $L(\Lambda)$.

a) *If* $d \geq 1$, *then* $\operatorname{Ext}^1_\Lambda(\operatorname{Tr} M, \Lambda) = 0$ *or, equivalently,* $0 \to M \to \operatorname{Hom}_\Lambda(\operatorname{Hom}_\Lambda(M, \Lambda), \Lambda)$ *is exact.*

b) *If* $d \geq 2$, *then* $\operatorname{Ext}^i_\Lambda(\operatorname{Tr} M, \Lambda) = 0$ *for* $i = 1, 2$ *or, equivalently* $M \to \operatorname{Hom}_\Lambda(\operatorname{Hom}_\Lambda(M, \Lambda), \Lambda)$ *is an isomorphism.*

PROOF. The usual localization arguments show that it suffices to prove the lemma when R is a local ring. So we assume R is a local ring.

a) Since $(\operatorname{Tr} M)_{\underline{p}}$ is $\Lambda_{\underline{p}}$-projective for all nonmaximal prime ideals $\underline{p}$ of R, we have by Lemma 7.6 that $\operatorname{Ext}^1_\Lambda(\operatorname{Tr} M, \Lambda)_{\underline{p}} = \operatorname{Ext}^1_{\Lambda_{\underline{p}}}((\operatorname{Tr} M)_{\underline{p}}, \Lambda_{\underline{p}}) = 0$ for all nonmaximal prime ideals $\underline{p}$ of R. Hence $\operatorname{Ext}^1_\Lambda(\operatorname{Tr} M, \Lambda)$ has finite length. By Lemma 7.7 we have that $0 \to \operatorname{Ext}^1_\Lambda(\operatorname{Tr} M, \Lambda) \to M \to \operatorname{Hom}_\Lambda(\operatorname{Hom}_\Lambda(M, \Lambda), \Lambda)$ is exact which means that $\operatorname{Ext}^1_\Lambda(\operatorname{Tr} M, \Lambda) = 0$ since $\operatorname{depth} M \geq 1$.

b) Suppose $d \geq 2$. Then by a) and Lemma 7.7 we have the exact sequence

$$(*) \quad 0 \to M \to \operatorname{Hom}_\Lambda(\operatorname{Hom}_\Lambda(M, \Lambda), \Lambda) \to \operatorname{Ext}^2_\Lambda(\operatorname{Tr} M, \Lambda) \to 0.$$

Now the fact that $d \geq 2$ implies that $\operatorname{depth} M$ and $\operatorname{depth} \Lambda$ is at least 2. From the fact that $\operatorname{depth} \Lambda \geq 2$ it follows trivially that $\operatorname{depth} \operatorname{Hom}_\Lambda(\operatorname{Hom}_\Lambda(M, \Lambda), \Lambda) \geq 1$. But $\operatorname{Ext}^2_\Lambda(\operatorname{Tr} M, \Lambda)$ has finite length since $(\operatorname{Tr} M)_{\underline{p}}$ is $\Lambda_{\underline{p}}$-projective for all nonmaximal prime ideals $\underline{p}$ of R. Hence if $\operatorname{Ext}^2_\Lambda(\operatorname{Tr} M, \Lambda) \neq (0)$, then $\operatorname{depth} \operatorname{Ext}^2_\Lambda(\operatorname{Tr} M, \Lambda) = 0$. Since $\operatorname{depth} X$ is the smallest integer i such that $\operatorname{Ext}^i_R(R/\operatorname{rad} R, X) \neq 0$, it follows from the exact sequence $(*)$ that $\operatorname{depth} M = 1$. This contradicts the fact that $\operatorname{depth} M = \dim R \geq 2$. Therefore if $\dim R \geq 2$ and M is a lattice, then $\operatorname{Ext}^i_\Lambda(\operatorname{Tr} M, \Lambda) = 0$ for $i = 1, 2$. □

We now return to the proof of Proposition 7.5.

If $d = 0$ there is nothing to prove. The cases $d = 1, 2$ have been established in the last lemma. We now show that if $\dim R \geq 3$, then $\operatorname{Ext}^i_\Lambda(\operatorname{Tr} M, \Lambda) = 0$ for $i = 1, 2, 3$.

Let $P_1 \to P_0 \to M \to 0$ be a projective presentation for M in $L(\Lambda)$. Then we have the exact sequence

$$0 \to \operatorname{Hom}_\Lambda(M, \Lambda) \to \operatorname{Hom}_\Lambda(P_0, \Lambda) \to \operatorname{Hom}_\Lambda(P_1, \Lambda) \to \operatorname{Tr} M \to 0$$

so $\operatorname{Ext}^{i+2}_{\Lambda^{\mathrm{op}}}(\operatorname{Tr} M, \Lambda^{\mathrm{op}}) \cong \operatorname{Ext}^i_{\Lambda^{\mathrm{op}}}(\operatorname{Hom}_\Lambda(M, \Lambda), \Lambda^{\mathrm{op}})$ for all $i \geq 1$. Since $\dim R \geq 3$, we know that $\operatorname{Ext}^i_\Lambda(\operatorname{Tr} M, \Lambda) = 0$ for $i = 1, 2$ (see Lemma 7.8). Hence in order to show that $\operatorname{Ext}^i_\Lambda(\operatorname{Tr} M, \Lambda) = 0$ for $i = 3$, it suffices to prove that $\operatorname{Ext}^1_{\Lambda^{\mathrm{op}}}(\operatorname{Hom}_\Lambda(M, \Lambda), \Lambda^{\mathrm{op}}) = 0$.

The fact that $\operatorname{depth} R = \operatorname{depth} \Lambda = \operatorname{depth} M \geq 1$ implies that there is an element x in $\operatorname{rad} R$ which is simultaneously R, Λ, and M regular.

From the exact sequence $0 \to \Lambda^{\mathrm{op}} \xrightarrow{x} \Lambda^{\mathrm{op}} \to \Lambda^{\mathrm{op}}/x\Lambda^{\mathrm{op}} \to 0$ we deduce the exact sequence $0 \to \operatorname{Hom}_{\Lambda^{\mathrm{op}}}(\operatorname{Hom}_\Lambda(M, \Lambda), \Lambda^{\mathrm{op}}) \xrightarrow{x} \operatorname{Hom}_{\Lambda^{\mathrm{op}}}(\operatorname{Hom}_\Lambda(M, \Lambda), \Lambda^{\mathrm{op}}) \to \operatorname{Hom}_{\Lambda^{\mathrm{op}}}(\operatorname{Hom}_\Lambda(M, \Lambda), \Lambda^{\mathrm{op}}/x\Lambda^{\mathrm{op}}) \to \operatorname{Ext}^1_{\Lambda^{\mathrm{op}}}(\operatorname{Hom}_\Lambda(M, \Lambda), \Lambda^{\mathrm{op}}) \xrightarrow{x} \operatorname{Ext}^1_{\Lambda^{\mathrm{op}}}(\operatorname{Hom}_\Lambda(M, \Lambda), \Lambda^{\mathrm{op}})$. Since $M_{\underline{p}}$ is $\Lambda_{\underline{p}}$-projective for $\underline{p}$ nonmaximal, we see that $\operatorname{Hom}_\Lambda(M, \Lambda)_{\underline{p}}$ is $\Lambda^{\mathrm{op}}_{\underline{p}}$-projective for all $\underline{p}$ nonmaximal. Hence $\operatorname{Ext}^1_{\Lambda^{\mathrm{op}}}(\operatorname{Hom}_\Lambda(M, \Lambda), \Lambda^{\mathrm{op}})$ is of finite length. Suppose it is not zero. Then multiplication by x has a non-trivial kernel K of finite length. Also since $\dim R \geq 3$, we know that $M = \operatorname{Hom}_{\Lambda^{\mathrm{op}}}(\operatorname{Hom}_\Lambda(M, \Lambda), \Lambda^{\mathrm{op}})$. So we deduce the exact sequence $0 \to M/xM \to \operatorname{Hom}_{\Lambda^{\mathrm{op}}}(\operatorname{Hom}_\Lambda(M, \Lambda), \Lambda^{\mathrm{op}}/x\Lambda^{\mathrm{op}}) \to K \to 0$ whose first 2 terms have depth ≥ 2 while $\operatorname{depth} K = 0$. Since this is impossible (use the fact that depth X is the smallest integer i such that $\operatorname{Ext}^i_R(R/\operatorname{rad} R, X) \neq 0$), this shows that $K = 0 = \operatorname{Ext}^1_{\Lambda^{\mathrm{op}}}(\operatorname{Hom}_\Lambda(M,\Lambda),\Lambda)$ if $\dim R \geq 3$. Hence we know that Proposition 7.5 is true for $\dim R \geq 3$. We now finish the proof of Proposition 7.5 by induction on $\dim R = d$.

Suppose Proposition 7.5 is true for $\dim R = k \geq 2$ and let $\dim R = k+1$. Since $\operatorname{Ext}^{i+2}_{\Lambda^{\mathrm{op}}}(\operatorname{Tr} M, \Lambda^{\mathrm{op}}) \cong \operatorname{Ext}^i_\Lambda(\operatorname{Hom}_\Lambda(M,\Lambda),\Lambda^{\mathrm{op}})$ for all $i \geq 1$ it suffices to show that $\operatorname{Ext}^i_\Lambda(\operatorname{Hom}_\Lambda(M,\Lambda),\Lambda^{\mathrm{op}}) = 0$ for $i = 1,\dots,k-1$. As before we can find an x in $\operatorname{rad} R$ such that x is R, Λ, and M regular at the same time. Hence M/xM is a lattice over the R/xR-order $\Lambda/x\Lambda$ with $\dim R/xR = k$. Thus $\operatorname{Ext}^i_{\Lambda/x\Lambda}(\operatorname{Hom}_{\Lambda/x\Lambda}(M/xM,\Lambda/x\Lambda))$ is zero for $i = 1,\dots,k-2$ by the induction hypothesis. Since $k \geq 3$, we know that $\operatorname{Ext}^1_{\Lambda^{\mathrm{op}}}(\operatorname{Hom}_\Lambda(M,\Lambda),\Lambda^{\mathrm{op}}) = 0$ so $\operatorname{Hom}_\Lambda(M,\Lambda)/x\operatorname{Hom}_\Lambda(M,\Lambda) \cong \operatorname{Hom}_\Lambda(M,\Lambda/x\Lambda) \cong \operatorname{Hom}_{\Lambda/x\Lambda}(M/xM,\Lambda/x\Lambda)$. From this it follows that $\operatorname{Ext}^i_{\Lambda^{\mathrm{op}}/x\Lambda^{\mathrm{op}}}(\operatorname{Hom}_{\Lambda/x\Lambda}(M/x\Lambda,\Lambda/x\Lambda),\Lambda^{\mathrm{op}}/x\Lambda^{\mathrm{op}}) \cong \operatorname{Ext}^i_{\Lambda^{\mathrm{op}}}(\operatorname{Hom}_\Lambda(M,\Lambda),\Lambda^{\mathrm{op}}/x\Lambda^{\mathrm{op}})$ for all i. Thus we have that $\operatorname{Ext}^i_{\Lambda^{\mathrm{op}}}(\operatorname{Hom}_\Lambda(M,\Lambda),\Lambda^{\mathrm{op}}/x\Lambda^{\mathrm{op}}) = 0$ for $i = 1,\dots,k-2$.

From the exact sequence $0 \to \Lambda^{\mathrm{op}} \xrightarrow{x} \Lambda^{\mathrm{op}} \to \Lambda^{\mathrm{op}}/x\Lambda^{\mathrm{op}}$ we deduce the long exact sequence $\operatorname{Ext}^1_{\Lambda^{\mathrm{op}}}(\operatorname{Hom}_\Lambda(M,\Lambda),\Lambda^{\mathrm{op}}/x\Lambda^{\mathrm{op}}) \to \operatorname{Ext}^2_{\Lambda^{\mathrm{op}}}(\operatorname{Hom}_\Lambda(M,\Lambda),\Lambda^{\mathrm{op}} \xrightarrow{x} \operatorname{Ext}^2_{\Lambda^{\mathrm{op}}}(\operatorname{Hom}_\Lambda(M,\Lambda),\Lambda^{\mathrm{op}}) \to \cdots \to \operatorname{Ext}^{k-2}_{\Lambda^{\mathrm{op}}}(\operatorname{Hom}_\Lambda(M,\Lambda),\Lambda^{\mathrm{op}}/x\Lambda^{\mathrm{op}}) \to \operatorname{Ext}^{k-1}_{\Lambda^{\mathrm{op}}}(\operatorname{Hom}_\Lambda(M,\Lambda),\Lambda^{\mathrm{op}}) \xrightarrow{x} \operatorname{Ext}^{k-1}_{\Lambda^{\mathrm{op}}}(\operatorname{Hom}_\Lambda(M,\Lambda),\Lambda^{\mathrm{op}})$. Since $\operatorname{Ext}^i_{\Lambda^{\mathrm{op}}}(\operatorname{Hom}_\Lambda(M,\Lambda),\Lambda^{\mathrm{op}}/x\Lambda^{\mathrm{op}}) = 0$ for $i = 1,\dots,k-2$, it follows that multiplication by x in $\operatorname{rad} R$ is a monomorphism on $\operatorname{Ext}^i_{\Lambda^{\mathrm{op}}}(\operatorname{Hom}_\Lambda(M,\Lambda),\Lambda^{\mathrm{op}})$ for $i = 2,\dots,k-1$. Therefore $\operatorname{Ext}^i_{\Lambda^{\mathrm{op}}}(\operatorname{Hom}_\Lambda(M,\Lambda),\Lambda^{\mathrm{op}}) = 0$ for $i = 2,\dots,k-1$ since each $\operatorname{Ext}^i_{\Lambda^{\mathrm{op}}}(\operatorname{Hom}_\Lambda(M,\Lambda),\Lambda^{\mathrm{op}})$ is of finite length for $i \geq 1$ (remember $\operatorname{Hom}_\Lambda(M,\Lambda)_{\underline{p}}$ is $\Lambda_{\underline{p}}$-projective for all nonmaximal $\underline{p}$). Because $\dim R = k+1$ with $k \geq 3$, we also know that $\operatorname{Ext}^1_{\Lambda^{\mathrm{op}}}(\operatorname{Hom}_\Lambda(M,\Lambda),\Lambda^{\mathrm{op}}) = 0$. Thus we have shown that if $\dim R = k+1$ with $k \geq 3$, then $\operatorname{Ext}^i_{\Lambda^{\mathrm{op}}}(\operatorname{Hom}_\Lambda(M,\Lambda),\Lambda^{\mathrm{op}}) = 0$ for $i = 1,\dots,(k+1)-2$. This finishes the proof that $\operatorname{Ext}^1_{\Lambda^{\mathrm{op}}}(\operatorname{Tr} M,\Lambda^{\mathrm{op}}) = 0$ for $i = 1,\dots,\dim R$ for M in $L(\Lambda)$. Hence we have established Proposition 7.5, that $\operatorname{Tr} M$ is in $J(\Lambda^{\mathrm{op}})$ if M is in $L(\Lambda)$. □

We now point out several consequences of Proposition 7.5 of particular interest to us in this paper.

THEOREM 7.8. *Let Λ be an R-order with $\dim R = d$. The functor $\Omega^d\colon \underline{J}(\Lambda) \to \underline{L}(\Lambda)$ is an equivalence of categories.*

PROOF. As noted before, we already know that $\Omega^d\colon \underline{J}(\Lambda) \to \underline{L}(\Lambda)$ is fully faithful. Therefore we only have to show that Ω^d is dense.

Let M be in $L(\Lambda)$. Since $\operatorname{Tr} M$ is in $J(\Lambda^{\mathrm{op}})$ by Proposition 7.5, we know that $\operatorname{Ext}^i_{\Lambda^{\mathrm{op}}}(\operatorname{Tr} M,\Lambda^{\mathrm{op}}) = 0$ for $i = 1,\dots,d$. Hence in the terminology of [8, page 59], we know that M is d-torsion free. Thus by [8, Theorem 2.17], there is an exact sequence of Λ-modules

$$(*)\quad 0 \to M \to P_1 \to \cdots \to P_d \to U \to 0$$

with the P_i projective Λ-modules such that

$$(**)\ 0 \to \operatorname{Hom}_\Lambda(U,\Lambda) \to \operatorname{Hom}_\Lambda(P_d,\Lambda) \to \cdots \to \operatorname{Hom}_\Lambda(P_1,\Lambda) \to \operatorname{Hom}_\Lambda(M,\Lambda) \to 0$$

is exact. We now show that U is in $J(\Lambda)$.

Since for each nonmaximal prime ideal $\underline{p}$ in R, we know that $M_{\underline{p}}$ is $\Lambda_{\underline{p}}$-projective, it follows from $(*)$ that $\operatorname{pd}_{\Lambda_{\underline{p}}} U_{\underline{p}} \leq d$ for all nonmaximal prime ideals $\underline{p}$ in R. On the other hand we also have by $(**)$ that $\operatorname{Ext}^i_\Lambda(U,\Lambda) = 0$ for $i = 1,\dots,d$ so that $\operatorname{Ext}^i_{\Lambda_{\underline{p}}}(U_{\underline{p}},\Lambda_{\underline{p}}) = 0$

for $i = 1, \dots, d$ for all nonmaximal $\underline{p}$ in R. Since $\Lambda_{\underline{p}}$ is noetherian, the fact that $\operatorname{pd}_{\Lambda_{\underline{p}}} U_{\underline{p}} \leq d$ and $\operatorname{Ext}^i_{\Lambda_{\underline{p}}}(U_{\underline{p}}, \Lambda_{\underline{p}}) = 0$ for $i = 1, \dots, d$ for all nonmaximal prime ideals $\underline{p}$ of R, implies that $U_{\underline{p}}$ is $\Lambda_{\underline{p}}$-projective for all nonmaximal prime ideals $\underline{p}$ of R. Combining this with the fact that $\operatorname{Ext}^i_\Lambda(U, \Lambda) = 0$ for $i = 1, \dots, d$ we have that U is in $J(\Lambda)$.

Since $0 \to M \to P_1 \to \cdots \to P_d \to U \to 0$ is exact with the P_i projective Λ-modules for $i = 1, \dots, d$, we know that $\Omega^d(U) \cong M$ in $\underline{L}(\Lambda)$. Therefore we have established that $M \cong \Omega^d(U)$ in $\underline{L}(\Lambda)$ with U in $\underline{J}(\Lambda)$. This completes the proof that $\Omega^d \colon \underline{J}(\Lambda) \to \underline{L}(\Lambda)$ is an equivalence of categories. □

Next we strengthen Proposition 7.5 as follows.

THEOREM 7.9. *The duality* $\operatorname{Tr} \colon \underline{\operatorname{noeth}}\Lambda \to \underline{\operatorname{noeth}}\Lambda^{\mathrm{op}}$ *induces a duality* $\operatorname{Tr} \colon \underline{L}(\Lambda) \to \underline{J}(\Lambda^{\mathrm{op}})$.

PROOF. By Proposition 7.5, we know that if M is in $L(\Lambda)$, then $\operatorname{Tr} M$ is in $J(\Lambda^{\mathrm{op}})$. Hence the duality $\operatorname{Tr} \colon \underline{\operatorname{noeth}}\Lambda \to \underline{\operatorname{noeth}}\Lambda^{\mathrm{op}}$ induces a contravariant functor $\operatorname{Tr} \colon \underline{L}(\Lambda) \to \underline{J}(\Lambda^{\mathrm{op}})$ which is fully faithful. Hence to show that this fully faithful functor $\operatorname{Tr} \colon \underline{L}(\Lambda) \to \underline{J}(\Lambda^{\mathrm{op}})$ is a duality it suffices to show that is is dense. Since $\operatorname{Tr}\operatorname{Tr} \colon \underline{\operatorname{noeth}}\Lambda \to \underline{\operatorname{noeth}}\Lambda$ is isomorphic to the identity, this amounts to showing that if M is in $J(\Lambda^{\mathrm{op}})$, then $\operatorname{Tr} M$ is in $L(\Lambda)$.

Suppose M is in $J(\Lambda^{\mathrm{op}})$. Then $M_{\underline{p}}$ is $\Lambda^{\mathrm{op}}_{\underline{p}}$-projective for all nonmaximal prime ideals $\underline{p}$ of R, or equivalently, $\underline{\operatorname{End}} M$ is an artin ring. Since $\operatorname{Tr} \colon \underline{\operatorname{End}} M \to (\underline{\operatorname{End}} \operatorname{Tr} M)^{\mathrm{op}}$ is a ring isomorphism, we have that $\underline{\operatorname{End}} \operatorname{Tr} M$ is an artin ring and so $(\operatorname{Tr} M)_{\underline{p}}$ is $\Lambda_{\underline{p}}$-projective for all nonmaximal prime ideals $\underline{p}$ of R. Therefore to show that $\operatorname{Tr} M$ is in $L(\Lambda)$ it suffices to show that $\operatorname{Tr} M$ is Cohen-Macaulay.

Let $d = \dim R$ and let $\cdots \to P_{d+1} \to P_d \to \cdots \to P_0 \to M \to 0$ be a projective Λ^{op}-resolution of M. Since M is in $J(\Lambda^{\mathrm{op}})$ we know that $\operatorname{Ext}^i_{\Lambda^{\mathrm{op}}}(M, \Lambda^{\mathrm{op}}) = 0$ for $i = 1, \dots, d$. Therefore

$$0 \to \operatorname{Hom}_{\Lambda^{\mathrm{op}}}(M, \Lambda^{\mathrm{op}}) \to \operatorname{Hom}_{\Lambda^{\mathrm{op}}}(P_0, \Lambda^{\mathrm{op}}) \to \cdots \to \operatorname{Hom}_{\Lambda^{\mathrm{op}}}(P_{d+1}, \Lambda^{\mathrm{op}})$$

is an exact sequence of Λ-modules with the $\operatorname{Hom}_{\Lambda^{\mathrm{op}}}(P_i, \Lambda^{\mathrm{op}})$ projective modules. Thus we have the exact sequence

$$0 \to \operatorname{Tr} M \to \operatorname{Hom}_{\Lambda^{\mathrm{op}}}(P_2, \Lambda^{\mathrm{op}}) \to \cdots \to \operatorname{Hom}_{\Lambda^{\mathrm{op}}}(P_{d+1}, \Lambda^{\mathrm{op}})$$

Since the $\operatorname{Hom}_{\Lambda^{\mathrm{op}}}(P_i, \Lambda^{\mathrm{op}})$ are Cohen-Macaulay modules for $i = 2, \dots, d+1$ because they are projective Λ-modules, it follows that $\operatorname{Tr} M$ is a Cohen-Macaulay module (see Proposition 7.2). □

As a consequence of Theorems 7.8 and 7.9 we have that the composition of functors $\underline{L}(\Lambda) \xrightarrow{\operatorname{Tr}} \underline{J}(\Lambda^{\mathrm{op}}) \xrightarrow{\Omega^d} \underline{L}(\Lambda^{\mathrm{op}})$ is a duality since Tr is a duality and Ω^d is an equivalence of categories. We will denote this duality by $\operatorname{Tr}_L \colon \underline{L}(\Lambda) \to \underline{L}(\Lambda^{\mathrm{op}})$. As we shall see, the duality Tr_L plays the role for lattices that the duality Tr plays for noetherian modules.

Using the results of this section, it is a matter of carrying out obvious calculations to verify the following.

PROPOSITION 7.10. *The contravariant functors* $\operatorname{Tr}_L \colon \underline{L}(\Lambda) \to \underline{L}(\Lambda^{\mathrm{op}})$ *and* $\operatorname{Tr}_L \colon \underline{L}(\Lambda^{\mathrm{op}}) \to \underline{L}(\Lambda)$ *are inverse dualities.*

As in the case of the ordinary transpose, the functors Tr_L depend on choices of projective resolutions of modules in $\operatorname{noeth}\Lambda$. So while we always have that $\operatorname{Tr}_L(\operatorname{Tr}_L X) \cong X$ in $\underline{L}(\Lambda)$ for each X in $L(\Lambda)$, it need not be the case that $\operatorname{Tr}_L \operatorname{Tr}_L X \cong X$ in $L(\Lambda)$ for all

X in $L(\Lambda)$ for arbitrary choices of transpose functors $\underline{L}(\Lambda) \to \underline{L}(\Lambda^{\mathrm{op}})$ and $\underline{L}(\Lambda^{\mathrm{op}}) \to \underline{L}(\Lambda)$. However there are transpose functors $\mathrm{Tr}_{L'}\colon \underline{L}(\Lambda) \to \underline{L}(\Lambda^{\mathrm{op}})$ and $\mathrm{Tr}_{L''}\colon \underline{L}(\Lambda^{\mathrm{op}}) \to \underline{L}(\Lambda)$ such that $\mathrm{Tr}_{L''}\,\mathrm{Tr}_{L'}(X) \cong X$ for all X in $L(\Lambda)$. Throughout this section we denote by $\mathrm{Tr}_L\colon \underline{L}(\Lambda) \to \underline{L}(\Lambda^{\mathrm{op}})$ and $\mathrm{Tr}_L\colon \underline{L}(\Lambda^{\mathrm{op}}) \to \underline{L}(\Lambda)$ a fixed pair of transpose functors with the property $\mathrm{Tr}_L\,\mathrm{Tr}_L(X) \cong X$ for all X in $L(\Lambda)$.

We end this section by pointing out the following easily verified facts concerning the categories $L(\Lambda)$, $J(\Lambda)$ and the duality $\mathrm{Tr}_L\colon \underline{L}(\Lambda) \to \underline{L}(\Lambda^{\mathrm{op}})$ for low values of $d = \dim R$.

Case 1: $d = 0$. Then $\operatorname{noeth}\Lambda = L(\Lambda) = J(\Lambda)$ and $\mathrm{Tr}_L\colon \underline{L}(\Lambda) \to \underline{L}(\Lambda^{\mathrm{op}})$ is the ordinary transpose $\mathrm{Tr}\colon \underline{\operatorname{noeth}}\Lambda \to \underline{\operatorname{noeth}}\Lambda^{\mathrm{op}}$.

Case 2: $d = 1$. Let M be in $L(\Lambda)$ and $P_1 \to P_0 \to M \to 0$ a projective presentation of M. If $0 \to K \to \operatorname{Hom}_\Lambda(P_1, \Lambda) \to \operatorname{Tr} M \to 0$ is exact, then $K \cong \mathrm{Tr}_L(M)$ in $\underline{L}(\Lambda^{\mathrm{op}})$.

Case 3: $d = 2$. $\mathrm{Tr}_L(M) \cong \operatorname{Hom}_\Lambda(M, \Lambda)$ in $\underline{L}(\Lambda^{\mathrm{op}})$ for all M in $L(\Lambda)$.

8. Lattice determined morphisms of lattices

Throughout this section we assume that Λ is an R-order with $d = \dim R$. Our aim is to study morphisms in the additive category $L(\Lambda)$ of Λ-lattices. In particular, we derive results for $L(\Lambda)$ analogous to those derived for $\operatorname{Mod}\Lambda$ in Section 3 and $\operatorname{noeth}\Lambda$ in Section 6. In order to accomplish this we need some preliminary definitions and results. We begin with some analogues of definitions and results which are well known for classical orders.

We say that a sequence $\to M_1 \xrightarrow{f_1} M_2 \xrightarrow{f_2} M_3 \to M_4 \to \cdots$ of morphisms in $L(\Lambda)$ is exact if it is exact as a sequence of Λ-modules and each nonzero $\operatorname{Im} f_i$ is a lattice. We say that a lattice C is projective in $L(\Lambda)$ if every exact sequence $0 \to A \to B \to C \to 0$ in $L(\Lambda)$ splits. We say that a lattice A is injective in $L(\Lambda)$ if every exact sequence $0 \to A \to B \to C \to 0$ in $L(\Lambda)$ splits.

PROPOSITION 8.1. *Let X be in $L(\Lambda)$.*

a) *X is projective in $L(\Lambda)$ if and only if X is a projective Λ-module.*

b) *X is injective in $L(\Lambda)$ if and only if there is a projective lattice P in $L(\Lambda^{\mathrm{op}})$ such that $X \cong P^*$ where $P^* = \operatorname{Hom}_R(P, R)$.*

PROOF. a) Suppose X is projective in $L(\Lambda)$. Then there is an exact sequence $0 \to K \to P \to X \to 0$ in $\operatorname{noeth}\Lambda$ with P projective. Since X and P are in $L(\Lambda)$, it follows that K is in $L(\Lambda)$ (see Proposition 7.3). Thus $0 \to K \to P \to X \to 0$ is an exact sequence in $L(\Lambda)$ and therefore splits, which means that X is a projective Λ-module. The fact that X being a projective Λ-module implies that X is projective in $L(\Lambda)$ is trivial.

b) Suppose X is injective in $L(\Lambda)$. Let $0 \to A \to B \to X^* \to 0$ be an exact sequence in $L(\Lambda^{\mathrm{op}})$. Then $0 \to X^{**} \to B^* \to A^* \to 0$ is exact in $L(\Lambda)$ (see Proposition 7.3). Since, by Proposition 7.3, we know that $X \cong X^{**}$, the fact that X is injective in $L(\Lambda)$ implies that the sequence $0 \to X^{**} \to B^* \to A^* \to 0$ splits. Using the fact that the functor $\operatorname{Hom}_R(\ , R)\colon L(\Lambda) \to L(\Lambda^{\mathrm{op}})$ is a duality, (see Proposition 7.3) it follows that the sequence $0 \to A \to B \to X^* \to 0$ splits, which implies that X^* is a projective lattice. Thus $X \cong (X^*)^*$ with X^* a projective Λ^{op}-lattice, which gives our desired result. A similar argument shows that if $X \cong P^*$ with P a projective Λ^{op}-lattice, then X is injective in $L(\Lambda)$. Thus b) is established. □

As a consequence of this result we have the following.

PROPOSITION 8.2. *Let X be in $L(\Lambda)$.*

a) *There is an injective resolution of* X *in* $L(\Lambda)$, *i.e. there is an exact sequence in* $L(\Lambda)$

$$0 \to X \to I_0 \to I_1 \to \cdots \to I_n \to \cdots$$

with the I_j *injective lattices for all* $j \geq 0$.

b) *If* A *is a lattice and* $0 \to X \to I_0 \to I_1 \to \cdots \to I_n \to \cdots$ *is an injective resolution of* X *in* $L(\Lambda)$, *then* $H^i(\mathrm{Hom}_\Lambda(A, I)) \cong \mathrm{Ext}^i_\Lambda(A, X)$ *for all* $i \geq 0$ *where* I *is the complex* $0 \to I_0 \to I_1 \to \cdots \to I_n \to \cdots$.

PROOF. a) Let $\cdots \to P_n \to \cdots \to P_1 \to P_0 \to X^* \to 0$ be a projective resolution for X^* in $L(\Lambda^{\mathrm{op}})$. Then $\cdots \to P_n \to \cdots \to P_1 \to P_0 \to X^* \to 0$ is an exact sequence in $L(\Lambda^{\mathrm{op}})$ and so $0 \to X^{**} \to P_0^* \to P_1^* \to \cdots \to P_n^* \to \cdots$ is an exact sequence in $L(\Lambda)$ (see Proposition 7.3). Since, by Proposition 8.1, each of the P_i^* is an injective Λ-lattice, part a) has been established.

b) We first observe that if A and C are in $L(\Lambda)$ and $0 \to A \to B \to C \to 0$ is an exact sequence in noeth Λ, then B is in $L(\Lambda)$ and so the sequence $0 \to A \to B \to C \to 0$ is exact in $L(\Lambda)$. Part b) follows easily from this observation using standard arguments from homological algebra. □

Let $I(L(\Lambda))$ be the full subcategory of $L(\Lambda)$ whose objects are the injective Λ-lattices. For each pair of Λ-lattices A and C define $I(A, C)$ to be the R-submodule of $\mathrm{Hom}_\Lambda(A, C)$ consisting of all morphisms $A \to C$ which can be written as a composition $A \to I \to C$ with I in $I(L(\Lambda))$. Using the duality $\mathrm{Hom}_\Lambda(\ , R)\colon L(\Lambda) \to L(\Lambda^{\mathrm{op}})$ it is not difficult to see that if $0 \to A \to I \to L \to 0$ is exact in $L(\Lambda)$ with I an injective Λ-lattice, then $\mathrm{Im}(\mathrm{Hom}_\Lambda(I, C) \to \mathrm{Hom}_\Lambda(A, C)) = I(A, C)$.

Define the category $\overline{L}(\Lambda)$ to be the preadditive category whose objects are the Λ-lattices and whose morphisms are $\overline{\mathrm{Hom}_\Lambda}(A, C) = \mathrm{Hom}_\Lambda(A, C)/I(A, C)$ for all A and C in $L(\Lambda)$. As an immediate consequence of the definitions involved we have the following.

PROPOSITION 8.3. a) *The duality* $\mathrm{Hom}_R(\ , R)\colon L(\Lambda) \to L(\Lambda^{\mathrm{op}})$ *induces a duality* $\underline{L}(\Lambda) \to \overline{L}(\Lambda^{\mathrm{op}})$.

b) *The composition* $\underline{L}(\Lambda) \xrightarrow{\mathrm{Tr}_L} \underline{L}(\Lambda^{\mathrm{op}}) \xrightarrow{\mathrm{Hom}_R(\ ,R)} \overline{L}(\Lambda)$ *is an equivalence of categories with inverse the composition* $\overline{L}(\Lambda) \xrightarrow{\mathrm{Hom}_R(\ ,R)} \underline{L}(\Lambda^{\mathrm{op}}) \xrightarrow{\mathrm{Tr}_L} \underline{L}(\Lambda)$.

The following is another useful preliminary result.

LEMMA 8.4. *Let* $0 \to R \to I_0 \to \cdots \to I_d \to 0$ *be a minimal injective resolution of* R *over* R *and let* M *be in* $L(\Lambda)$.

a) $\mathrm{Hom}_R(M, I_j)$ *is an injective* Λ^{op}*-module for* $j = 0, \ldots, d-1$.

b) $0 \to \mathrm{Hom}_R(M, R) \to \mathrm{Hom}_R(M, I_0) \to \cdots \to \mathrm{Hom}_R(M, I_d) \to 0$ *is an exact sequence of* Λ^{op}*-modules with the* $\mathrm{Hom}_R(M, I_j)$ *injective for* $j = 0, \ldots, d-1$.

c) *There is an isomorphism of functors* $\mathrm{Ext}^1_{\Lambda^{\mathrm{op}}}(\ , \mathrm{Hom}_R(M, I_d)) \cong \mathrm{Ext}^{d+1}_{\Lambda^{\mathrm{op}}}(\ , \mathrm{Hom}_R(M, R))$ *which is functorial in* M *in* $L(\Lambda)$.

PROOF. a) Suppose $0 \leq j \leq d-1$. If $d = 0$, there is nothing to prove so we may as well assume that $d > 0$. Since I_j is an injective R-module, we know that for each Λ^{op}-module A we have an isomorphism $\mathrm{Ext}^1_{\Lambda^{\mathrm{op}}}(A, \mathrm{Hom}_R(M, I_j)) \cong \mathrm{Hom}_R(\mathrm{Tor}_1^\Lambda(A, M), I_j)$. Because M is in $L(\Lambda)$, we know that $M_{\underline{p}}$ is $\Lambda_{\underline{p}}$-projective for all nonmaximal prime ideals $\underline{p}$ in R. Hence $\mathrm{Tor}_1^\Lambda(A, M)_{\underline{p}} = \mathrm{Tor}_1^{\Lambda_{\underline{p}}}(A_{\underline{p}}, M_{\underline{p}}) = 0$ for all nonmaximal prime ideals $\underline{p}$ in R. Thus $\mathrm{Tor}_1^\Lambda(A, M)$ is an R-module which is locally of finite length, i.e. every finitely generated submodule of $\operatorname{Tor}_1^\Lambda(A,M)$ is of finite length. Since R is a Gorenstein ring and $0 \le j \le d-1$, we know that $\operatorname{Hom}_R(X,I_j) = 0$ for all X of finite length. Therefore $\operatorname{Hom}_R(\operatorname{Tor}_1^\Lambda(A,M),I_j) = 0 = \operatorname{Ext}^1_{\Lambda^{\mathrm{op}}}(A,\operatorname{Hom}_R(M,I_j))$. Since this is true for all Λ^{op}-modules A, we have that $\operatorname{Hom}_R(M,I_j)$ is Λ^{op}-injective for $0 \le j < d-1$.

b) The fact that M is in $L(\Lambda)$ implies that M is a Cohen-Macaulay module over R. Therefore we know that $\operatorname{Ext}^i_R(M,R) = 0$ for $i > 0$ since R is a Gorenstein ring. Hence the sequence of Λ^{op}-modules $0 \to \operatorname{Hom}_R(M,R) \to \operatorname{Hom}_R(M,I_0) \to \cdots \to \operatorname{Hom}_R(M,I_d) \to 0$ is exact. The fact that $\operatorname{Hom}_R(M,I_j)$ is Λ^{op}-injective for $j = 0,\dots,d-1$ was shown in a).

c) Suppose M and N are in $L(\Lambda)$ and $f\colon M \to N$ is a morphism. We then have the commutative diagram of Λ^{op}-modules

$$\begin{array}{ccccccccc}
0 & \longrightarrow & \operatorname{Hom}_R(N,R) & \longrightarrow & \operatorname{Hom}_R(N,I_0) & \longrightarrow \cdots \longrightarrow & \operatorname{Hom}_R(N,I_d) & \longrightarrow & 0\\
 & & \Big\downarrow{\scriptstyle \operatorname{Hom}_R(f,R)} & & \Big\downarrow{\scriptstyle \operatorname{Hom}_R(f,I_0)} & & \Big\downarrow{\scriptstyle \operatorname{Hom}_R(f,I_d)} & & \\
0 & \longrightarrow & \operatorname{Hom}_R(M,R) & \longrightarrow & \operatorname{Hom}_R(M,I_0) & \longrightarrow \cdots \longrightarrow & \operatorname{Hom}_R(M,I_d) & \longrightarrow & 0.
\end{array}$$

By b) we know that this diagram is exact and the $\operatorname{Hom}_R(M,I_j)$ are injective Λ^{op}-modules for $j = 0,\dots,d-1$. Therefore by standard dimension shift arguments we obtain the following exact commutative diagram

$$\begin{array}{ccccccc}
0 & \longrightarrow & \operatorname{Ext}^1_{\Lambda^{\mathrm{op}}}(\ ,\operatorname{Hom}_R(N,I_d)) & \longrightarrow & \operatorname{Ext}^{d+1}_{\Lambda^{\mathrm{op}}}(\ ,\operatorname{Hom}_R(N,R)) & \longrightarrow & 0\\
 & & \Big\downarrow{\scriptstyle \operatorname{Hom}_R(f,I_d)} & & \Big\downarrow{\scriptstyle \operatorname{Hom}_R(f,R)} & & \\
0 & \longrightarrow & \operatorname{Ext}^1_{\Lambda^{\mathrm{op}}}(\ ,\operatorname{Hom}_R(M,I_d)) & \longrightarrow & \operatorname{Ext}^{d+1}_{\Lambda^{\mathrm{op}}}(\ ,\operatorname{Hom}_R(M,R)) & \longrightarrow & 0
\end{array}$$

which is our desired result. □

Next we recall from Section 7 that the functor $\Omega^d\colon \underline{\operatorname{noeth}}\Lambda \to \underline{\operatorname{noeth}}\Lambda$ induces an equivalence of categories $\Omega^d\colon \underline{J}(\Lambda) \to \underline{L}(\Lambda)$ where $J(\Lambda)$ is the full subcategory of $\operatorname{noeth}\Lambda$ consisting of those M in $\operatorname{noeth}\Lambda$ satisfying $M_{\underline{p}}$ is $\Lambda_{\underline{p}}$-projective for nonmaximal prime ideals $\underline{p}$ in R and $\operatorname{Ext}^i_\Lambda(M,\Lambda) = 0$ for $i = 1,\dots,d$. We also agreed to denote the inverse of $\Omega^d\colon \underline{J}(\Lambda) \to \underline{L}(\Lambda)$ by $\Omega^{-d}\colon \underline{L}(\Lambda) \to \underline{J}(\Lambda)$.

Let C be in $\underline{L}(\Lambda)$. Since $\Omega^d(\Omega^{-d}C) \cong C$ in $\underline{L}(\Lambda)$, we have isomorphisms $\operatorname{Ext}^1_\Lambda(C,\ ) \cong \operatorname{Ext}^{d+1}(\Omega^{-d}C,\ )$ which are functorial in C in $\underline{L}(\Lambda)$. Further let X be in $L(\Lambda)$. Then combining the various isomorphisms given above we have $\operatorname{Ext}^1_\Lambda(C,X) \cong \operatorname{Ext}^1_\Lambda(C,\operatorname{Hom}_R(X^*,R)) \cong \operatorname{Ext}^{d+1}_\Lambda(C,\operatorname{Hom}_R(X^*,I_d)) \cong \operatorname{Ext}^1_\Lambda(\Omega^d C,\operatorname{Hom}_R(X^*,I_d))$ which are functorial in C and X in $\underline{L}(\Lambda)$. By Proposition 3.3 we have the isomorphism $\operatorname{Ext}^1_\Lambda(\Omega^{-d}C,\operatorname{Hom}_R(X^*,I_d)) \cong \operatorname{Hom}_R(\underline{\operatorname{Hom}}_\Lambda(\operatorname{Tr}(X),\Omega^{-d}C),I_d)$ which is functorial in C and X in $\underline{L}(\Lambda)$. Since X^* is in $L(\Lambda^{\mathrm{op}})$, we have that $\operatorname{Tr}(X^*)$ is in $J(\Lambda)$ (see Proposition 7.5) as is $\Omega^{-d}C$. Hence the equivalence of categories $\Omega^d\colon \underline{J}(\Lambda) \to \underline{L}(\Lambda)$ gives an isomorphism $\Omega^d\colon \underline{\operatorname{Hom}}_\Lambda(\operatorname{Tr}X^*,\Omega^{-d}C) \to \underline{\operatorname{Hom}}_\Lambda(\Omega^d\operatorname{Tr}(X^*),C)$ functorial in X and C in $\underline{L}(\Lambda)$. But by definition $\Omega^d\operatorname{Tr}(X^*) = \operatorname{Tr}_L(X^*)$. Therefore we have the isomorphism $\Omega^d\colon \underline{\operatorname{Hom}}_\Lambda(\operatorname{Tr}X^*,\Omega^{-d}C) \to \underline{\operatorname{Hom}}_\Lambda(\operatorname{Tr}_L X^*,C)$ which is functorial in C and X in $\underline{L}(\Lambda)$. Hence we have the isomorphism $\operatorname{Ext}^1_\Lambda(\Omega^{-d}C,\operatorname{Hom}_R(X^*,I_d)) \cong \operatorname{Hom}_R(\underline{\operatorname{Hom}}_\Lambda(\operatorname{Tr}_L X^*,C),I_d)$ which is functorial in C and X in $\underline{L}(\Lambda)$. Therefore combining these isomorphisms we obtain the following result.

PROPOSITION 8.7. *Let C and X be in $L(\Lambda)$. Then we have an isomorphism $\operatorname{Ext}^1_\Lambda(C,X) \cong \operatorname{Hom}_R(\underline{\operatorname{Hom}}_\Lambda(\operatorname{Tr}_L X^*,C),I_d)$, functorial in C and X in $\underline{L}(\Lambda)$.*

Using the equivalence of categories $\operatorname{Hom}_R(\ ,R)\operatorname{Tr}_L\colon \underline{L}(\Lambda) \to \overline{L}(\Lambda)$, this result can be restated as follows.

PROPOSITION 8.8. *Let C and X be in $L(\Lambda)$. Then there is an isomorphism $\operatorname{Ext}^1_\Lambda(C, \operatorname{Hom}_R(\operatorname{Tr}_L X, R)) \cong \operatorname{Hom}_R(\underline{\operatorname{Hom}}_\Lambda(X,C), I_d)$ which is functorial in C and X in $\underline{L}(\Lambda)$.*

The following result is an easy consequence of Proposition 8.8.

PROPOSITION 8.9. *Let C and X be in $L(\Lambda)$ and n a positive integer. Then there is an isomorphism $\operatorname{Ext}^1_\Lambda(C, \operatorname{Hom}_R(\operatorname{Tr}_L X, R^n)) \cong \operatorname{Hom}_R(\underline{\operatorname{Hom}}_\Lambda(X,C), I^n_d)$ which is functorial in C and X in $\underline{L}(\Lambda)$.*

We now investigate some of the connections between an exact sequence $x\colon 0 \to \operatorname{Hom}_R(\operatorname{Tr}_L X, R^n) \xrightarrow{g} B \xrightarrow{f} C \to 0$ in $L(\Lambda)$ and the R-morphism $v\colon \underline{\operatorname{Hom}}_\Lambda(X,C) \to I^n_d$ corresponding to x under the isomorphism $\operatorname{Ext}^1_\Lambda(C, \operatorname{Hom}_R(\operatorname{Tr}_L X, R^n)) \cong \operatorname{Hom}_R(\underline{\operatorname{Hom}}_\Lambda(X,C), I^n_d)$. Under this hypothesis we have the following consequence of Theorem 3.6.

THEOREM 8.10. *Let H be the R-submodule of (X,C) containing $P(X,C)$ such that $H/P(X,C) = \operatorname{Ker} v$.*

a) *A morphism $h\colon L \to C$ in $L(\Lambda)$ can be written as $ft = h$ for some morphism $t\colon L \to B$ if and only if $\operatorname{Im}(\ ,h)(X) \subset H$.*

b) *f is right X-determined in $L(\Lambda)$ and $\operatorname{Im}(\ ,f)(X)$ is the maximal $(\operatorname{End} X)^{\mathrm{op}}$-submodule of H.*

c) *$\operatorname{Im}(\ ,f)(X) = H$ if and only if H is an $(\operatorname{End} X)^{\mathrm{op}}$-submodule of (X,C).*

PROOF. All one has to do is check that the hypothesis of Theorem 3.6 is satisfied in this case. □

We now restate Proposition 8.9 and Theorem 8.10 in a form particularly useful for our immediate purposes.

THEOREM 8.11. *Let X and C be in $L(\Lambda)$ and n a positive integer and let $\Sigma = (\operatorname{End} X)^{\mathrm{op}}$.*

a) *There is an isomorphism $\operatorname{Ext}^1_\Lambda(C, \operatorname{Hom}_R(\operatorname{Tr}_L X, R^n)) \cong \operatorname{Hom}_\Sigma(\underline{\operatorname{Hom}}_\Lambda(X,C), \operatorname{Hom}_R(\Sigma^{\mathrm{op}}, I^n_d))$ which is functorial in C in $\underline{L}(\Lambda)$.*

b) *Let $v\colon \underline{\operatorname{Hom}}_\Lambda(X,C) \to \operatorname{Hom}_R(\Sigma^{\mathrm{op}}, I^n_d)$ be the Σ-morphism corresponding to an extension $0 \to \operatorname{Hom}_R(\operatorname{Tr}_L X, R^n) \xrightarrow{g} B \xrightarrow{f} C \to 0$ under the isomorphism given in a) and let H be the Σ-submodule of (X,C) containing $P(X,C)$ such that $H/P(X,C) = \operatorname{Ker} v$.*

i) *f is right X-determined.*

ii) *$\operatorname{Im}(\ ,f)(X) = H$.*

PROOF. a) We know that there is an isomorphism $\operatorname{Hom}_R(\underline{\operatorname{Hom}}_\Lambda(X,C), I^n_d) \cong \operatorname{Hom}_\Sigma(\underline{\operatorname{Hom}}_\Lambda(X,C), \operatorname{Hom}_R(\Sigma^{\mathrm{op}}, I^n_d)$ which is functorial in C in $\underline{L}(\Lambda)$. Part a) now follows trivially from Proposition 8.9.

b) Follows from a) and Theorem 3.6. □

As an easy consequence of Theorem 8.11 we have the following.

PROPOSITION 8.12. *Let $0 \to A \xrightarrow{g} B \xrightarrow{f} C \to 0$ be an exact sequence in $L(\Lambda)$. Then f is right $\operatorname{Tr}_L A^*$-determined in $L(\Lambda)$.*

PROOF. We know that $A \cong \operatorname{Hom}_R(\operatorname{Tr}_L(\operatorname{Tr}_L A^*), R)$. Using this observation, the proposition is a trivial consequence of Theorem 8.11. □

In order to give further applications of Theorem 8.11 it is convenient to have the following facts.

LEMMA 8.13. *Let X and C be in $L(\Lambda)$ and let $\Sigma = (\operatorname{End} X)^{\mathrm{op}}$.*

a) *$\underline{\operatorname{Hom}}_\Lambda(X, C)$ is a Σ-module of finite length since it is an R-module of finite length.*

b) *Suppose H is a Σ-submodule of (X, C) containing $P(X, C)$. Then $(X, C)/H$ is a Σ-module of finite length. Hence the socle of $(X, C)/H$ is of the form $\coprod_{i=1}^k S_i^{n_i}$ with the S_i nonisomorphic Σ-modules and the n_i positive integers. Let $n = \max\{n_1, \dots, n_k\}$. Then there is a Σ-morphism $v\colon \underline{\operatorname{Hom}}_\Lambda(X, C) \to \operatorname{Hom}_R(\Sigma^{\mathrm{op}}, I_d^n)$ with $H/P(X, C) = \operatorname{Ker} v$.*

PROOF. a) Since X and C are in noeth Λ and $X_{\underline{p}}$ is $\Lambda_{\underline{p}}$-projective for all nonmaximal prime ideals $\underline{p}$ we have by Lemma 7.6 that $\underline{\operatorname{Hom}}_\Lambda(X, C)$ is an R-module of finite length and so a Σ-module of finite length.

b) We know that $\operatorname{Hom}_R(\Sigma^{\mathrm{op}}, I_d)$ is an injective cogenerator for Σ and so each simple Σ-module is contained in $\operatorname{Hom}_R(\Sigma^{\mathrm{op}}, I_d)$. From this observation it follows trivially that $\coprod S_i^{n_i}$ is contained in $\operatorname{Hom}_R(\Sigma^{\mathrm{op}}, I_d^n) = \operatorname{Hom}_R(\Sigma^{\mathrm{op}}, I_d)^n$. Thus there is a Σ-monomorphism $(X, C)/H \to \operatorname{Hom}_R(\Sigma^{\mathrm{op}}, I_d^n)$ and hence the composition $\underline{\operatorname{Hom}}_\Lambda(X, C) \to (X, C)/H \to \operatorname{Hom}_R(\Sigma^{\mathrm{op}}, I_d^n)$ is our desired Σ-morphism. □

As a consequence of Theorem 8.11 and Lemma 8.13 we obtain the following analogue of Theorem 3.7.

THEOREM 8.14. *Let C and X be in $L(\Lambda)$ and let $\Sigma = (\operatorname{End} X)^{\mathrm{op}}$. Suppose H is a Σ-submodule of (X, C) containing $P(X, C)$ and suppose $\coprod_{i=1}^k S_i^{n_i}$ is isomorphic to the socle of $(X, C)/H$ where the S_i are nonisomorphic simple Σ-modules and the n_i are positive integers. Let $n = \max\{n_1, \dots, n_k\}$. Then there is an exact sequence $x\colon 0 \to \operatorname{Hom}_R(\operatorname{Tr}_L X, R^n) \xrightarrow{g} B \xrightarrow{f} C \to 0$ satisfying:*

a) $\operatorname{Im}(\ , f)(X) = H$.

b) *f is right X-determined in $L(\Lambda)$.*

So far we have been concentrating on how epimorphisms in $L(\Lambda)$ are right determined in $L(\Lambda)$. The duality $\operatorname{Hom}_R(\ , R)\colon L(\Lambda) \to L(\Lambda^{\mathrm{op}})$ enables us to translate these results to results about how monomorphisms in $L(\Lambda)$ are left determined in $L(\Lambda)$. The proofs, which are not given, follow easily from the remarks concerning dual categories at the end of Section 2. The reader should also consult the analogous results to Section 6 for noetherian algebras over complete local rings.

PROPOSITION 8.15. *Let $0 \to A \xrightarrow{g} B \xrightarrow{f} C \to 0$ be an exact sequence in $L(\Lambda)$. Then g is left $(\operatorname{Tr}_L C)^*$-determined.*

PROOF. Dual of Proposition 8.12. □

As the dual of Theorem 8.14 we obtain the following.

THEOREM 8.16. *Let A and X be in $L(\Lambda)$. Let $\Sigma = \operatorname{End} X$ and suppose H is a Σ-submodule of (A, X) containing $I(A, X)$. Then $(A, X)/I$ is a Σ-module of finite length whose socle is of the form $\coprod_{i=1}^k S_i^{n_i}$ with the S_i nonisomorphic Σ-modules and the n_i positive integers. Let $n = \max\{n_1, \dots, n_k\}$. Then there is an exact sequence $0 \to A \xrightarrow{g} B \xrightarrow{f} \operatorname{Tr}_L(\operatorname{Hom}_R(X, R^n) \to 0$ in $L(\Lambda)$ such that*

a) *g is left X-determined.*
b) $\operatorname{Im}(g,\ )(X) = H$.

Finally we point out the following version of Proposition 8.7 which could have been used as the starting point for deriving the last two results.

PROPOSITION 8.17. *Let X and A be in $L(\Lambda)$. Then there is an isomorphism $\operatorname{Ext}^1_\Lambda(X, A) \cong \operatorname{Hom}_R(\overline{\operatorname{Hom}_\Lambda}(A, (\operatorname{Tr}_L X)^*), I_d)$ which is functorial in X and A.*

PROOF. The equivalence of categories $\underline{L}(\Lambda) \to \overline{L}(\Lambda)$ given by $X \mapsto \operatorname{Hom}_R(\operatorname{Tr}_L X, R)$ gives isomorphisms $\underline{\operatorname{Hom}}_\Lambda(U, V) \to \overline{\operatorname{Hom}_\Lambda}(\operatorname{Hom}_R(\operatorname{Tr}_L U, R), \operatorname{Hom}_R(\operatorname{Tr}_L V, R))$ functorial in U and V. Letting $U = \operatorname{Tr}_L A^*$ and $V = X$ we obtain isomorphisms $\underline{\operatorname{Hom}}_\Lambda(\operatorname{Tr}_L A^*, X) \to \overline{\operatorname{Hom}_\Lambda}(A, (\operatorname{Tr}_L X)^*)$ functorial in A and X. Thus the functorial isomorphism given in Proposition 8.7

$$\operatorname{Ext}^1_\Lambda(X, A) \to \operatorname{Hom}_R(\underline{\operatorname{Hom}}_\Lambda(\operatorname{Tr}_L A^*, X), I_d)$$

becomes the functorial isomorphism

$$\operatorname{Ext}^1_\Lambda(X, A) \to \operatorname{Hom}_R(\overline{\operatorname{Hom}_\Lambda}(A, (\operatorname{Tr}_L X)^*), I_d),$$

which is our desired result. □

9. Orders over complete local rings

Throughout this section we assume unless stated to the contrary, that Λ is an R-order where R is a complete local ring of dimension d. Our purpose is to explore what some of the special features of the category $\operatorname{noeth}\Lambda$ already explored in Section 5 enable us to say about Λ-lattices. The results here parallel those of Section 5 and 6.

Let $L(\Lambda)$ be the category of lattices. Then we know that $L(\Lambda)$ is a Krull-Schmidt category, i.e. every lattice M has a unique (up to isomorphism) representation as a sum $M_1 \amalg \cdots \amalg M_n$ with the M_i indecomposable lattices. Thus each lattice M has a unique (up to isomorphism) representation as a sum $M_1 \amalg P$ where M_1 has no nonzero projective summands and P is projective. Similarly every lattice M has a unique (up to isomorphism) representation as $N \amalg I$ where N has no nonzero injective lattice summands and I is an injective lattice. We denote by $L_P(\Lambda)$ the full subcategory of $L(\Lambda)$ whose objects are those lattices with no nonzero projective summands. Similarly we denote by $L_I(\Lambda)$ the full subcategory of $L(\Lambda)$ whose objects are those lattices with no nonzero summands which are injective lattices. Obviously the duality $D\colon L(\Lambda) \to L(\Lambda^{\mathrm{op}})$ given by $D(M) = \operatorname{Hom}_R(M, R)$ induces a duality $D\colon L_P(\Lambda) \to L_I(\Lambda^{\mathrm{op}})$. Also if we denote by $\underline{L}_P(\Lambda)$ the full subcategory of $\underline{L}(\Lambda)$ whose objects are in $L_P(\Lambda)$, then the inclusion $\underline{L}_P(\Lambda) \to \underline{L}(\Lambda)$ is an equivalence of categories. Similarly, letting $\overline{L}_I(\Lambda)$ be the full subcategory of $\overline{L}(\Lambda)$ whose objects are in $L_I(\Lambda)$, then the inclusion $\overline{L}_I(\Lambda) \to \overline{L}(\Lambda)$ is an equivalence of categories. Finally the duality $D\colon L_P(\Lambda) \to L_I(\Lambda^{\mathrm{op}})$ induces a duality $D\colon \underline{L}_P(\Lambda) \to \overline{L}_I(\Lambda^{\mathrm{op}})$.

Following our usual convention for noetherian algebras over complete local rings when talking about a projective resolution of a lattice M we will mean, unless stated to the contrary, that the resolution is minimal. The fact that every object in $L(\Lambda)$ has a minimal projective resolution implies that every object in $L(\Lambda)$ has a minimal lattice injective resolution. For let M be in $L(\Lambda)$. Then $D(M)$ is in $L(\Lambda^{\mathrm{op}})$ and so has a minimal projective resolution $\cdots \to P_n \to \cdots \to P_0 \to D(M) \to 0$ in $L(\Lambda)$. Then $M = D^2(M) \to D(P_0) \to \cdots \to D(P_n) \to \cdots$ is a minimal lattice injective resolution of M.

We have already seen in Section 5 that the transpose $\operatorname{Tr}\colon \underline{\operatorname{noeth}\Lambda} \to \underline{\operatorname{noeth}\Lambda^{\mathrm{op}}}$ has the property that $\operatorname{Tr} M$ is in $\operatorname{noeth}_P \Lambda^{\mathrm{op}}$. Thus if M is in $L(\Lambda)$, then $\operatorname{Tr} M$ is in $J_P(\Lambda^{\mathrm{op}})$

where $J_P(\Lambda^{\mathrm{op}})$ is the full subcategory of $J(\Lambda^{\mathrm{op}})$ whose objects have no nonzero projective summands. Clearly the inclusion $\underline{J}_P(\Lambda^{\mathrm{op}}) \to \underline{J}(\Lambda^{\mathrm{op}})$ is an equivalence of categories. Thus the duality $\mathrm{Tr}\colon \underline{\mathrm{noeth}\Lambda} \to \underline{\mathrm{noeth}\Lambda^{\mathrm{op}}}$ induces the duality $\mathrm{Tr}\colon \underline{L}_P(\Lambda) \to \underline{J}_P(\Lambda^{\mathrm{op}})$. Also the fact that $\mathrm{Tr}\colon \mathrm{Ob}\,\mathrm{noeth}_P\,\Lambda \to \mathrm{Ob}\,\mathrm{noeth}_P\,\Lambda^{\mathrm{op}}$ is a representation equivalence implies that $\mathrm{Tr}\colon \mathrm{Ob}\,L_P(\Lambda) \to \mathrm{Ob}\,J_P(\Lambda^{\mathrm{op}})$ is also a representation equivalence.

We now want to show that the duality $\mathrm{Tr}_L\colon \underline{L}(\Lambda) \to \underline{L}(\Lambda^{\mathrm{op}})$ has the property that if M is in $L_P(\Lambda)$, then $\mathrm{Tr}_L(M)$ is in $L_P(\Lambda^{\mathrm{op}})$ and the map $\mathrm{Tr}_L\colon \mathrm{Ob}\,L_P(\Lambda) \to \mathrm{Ob}\,L_P(\Lambda^{\mathrm{op}})$ is a representation equivalence. We begin with the following.

LEMMA 9.1. *Let $\mathcal{V}$ be the full subcategory of* $\mathrm{noeth}_P\,\Lambda$ *consisting of all objects M such that* $\mathrm{Ext}^1_\Lambda(M,\Lambda) = 0$.

a) *If M is in $\mathcal{V}$, then $\Omega^1(M)$ is in* $\mathrm{noeth}_P\,\Lambda$.

b) $\Omega^1\colon \underline{\mathrm{Hom}}_\Lambda(M,N) \to \underline{\mathrm{Hom}}_\Lambda(\Omega^1(M),\Omega^1(N))$ *is an isomorphism for all M and N in $\mathcal{V}$.*

c) *If M and N are in $\mathcal{V}$, then $M \cong N$ if and only if $\Omega^1(M) \cong \Omega^1(N)$.*

d) *If $M \cong M_1 \amalg M_2$ in $\mathcal{V}$, then $\Omega^1(M) \cong \Omega^1(M_1) \amalg \Omega^1(M_2)$.*

e) *M in $\mathcal{V}$ is indecomposable if and only if $\Omega^1(M)$ is indecomposable.*

f) *If $M \cong M_1 \amalg \cdots \amalg M_k$ with the M_i indecomposable, then $\Omega^1(M) \cong \Omega^1(M_1) \amalg \cdots \amalg \Omega^1(M_k)$ with the $\Omega^1(M_i)$ indecomposable modules in* $\mathrm{noeth}_P\,\Lambda$.

PROOF. a) Let $P \to M \to 0$ be a projective cover. Then $0 \to \Omega^1(M) \xrightarrow{f} P \xrightarrow{g} M \to 0$ is exact. Suppose Q is a projective summand of $\Omega^1(M)$, i.e. there is a morphism $u\colon Q \to \Omega^1(M)$ which is a splittable monomorphism. Since $\mathrm{Ext}^1_\Lambda(M,\Lambda) = 0$, we have the following exact commutative diagram

$$\begin{array}{ccccccccc} 0 & \longrightarrow & \mathrm{Hom}_\Lambda(M,\Lambda) & \longrightarrow & \mathrm{Hom}_\Lambda(P,\Lambda) & \xrightarrow{\mathrm{Hom}(f,\Lambda)} & \mathrm{Hom}_\Lambda(\Omega^1(M),\Lambda) & \longrightarrow & 0 \\ & & & & & & \downarrow {\scriptstyle \mathrm{Hom}_\Lambda(u,\Lambda)} & & \\ & & & & & & \mathrm{Hom}_\Lambda(Q,\Lambda) & & \\ & & & & & & \downarrow & & \\ & & & & & & 0 & & \end{array}$$

Because $\mathrm{Hom}_\Lambda(Q,\Lambda)$ is Λ^{op}-projective and the composition $\mathrm{Hom}_\Lambda(fu,\Lambda)\colon \mathrm{Hom}_\Lambda(P,\Lambda) \to \mathrm{Hom}_\Lambda(Q,\Lambda)$ is an epimorphism, there is a morphism $s\colon \mathrm{Hom}_\Lambda(Q,\Lambda) \to \mathrm{Hom}_\Lambda(P,\Lambda)$ such that $\mathrm{Hom}_\Lambda(fu,\Lambda)s = 1_{\mathrm{Hom}_\Lambda(Q,\Lambda)}$. Hence $1_{\mathrm{Hom}_\Lambda(\mathrm{Hom}_\Lambda(Q,\Lambda),\Lambda)} = \mathrm{Hom}_\Lambda(s,\Lambda)\,\mathrm{Hom}_\Lambda(\mathrm{Hom}_\Lambda(fu,\Lambda),\Lambda)$. From this it follows that the composition

$$Q \xrightarrow{fu} P \to \mathrm{Hom}_\Lambda(\mathrm{Hom}_\Lambda(P,\Lambda),\Lambda) \xrightarrow{\mathrm{Hom}_\Lambda(s,\Lambda)} \mathrm{Hom}_\Lambda(\mathrm{Hom}_\Lambda(Q,\Lambda),\Lambda) \to 0$$

is the identity, where $P \to \mathrm{Hom}_\Lambda(\mathrm{Hom}_\Lambda(P,\Lambda),\Lambda)$ and $\mathrm{Hom}_\Lambda(\mathrm{Hom}_\Lambda(Q,\Lambda),\Lambda) \to Q$ are the standard isomorphisms. Thus $Q \xrightarrow{fu} P$ is a splittable monomorphism. Since $gfu = 0$, a complement P' of $\mathrm{Im}\,fu$ has the property $g|_{P'}\colon P' \to M$ is an epimorphism. The fact that $g\colon P \to M \to 0$ is a projective cover implies that $P' = P$ or equivalently $Q = 0$. Hence $\Omega^1(M)$ is in $\mathrm{noeth}_P\,\Lambda$.

b) Follows from the fact that $\mathrm{Ext}^1_\Lambda(N,\Lambda) = 0$ (see Proposition 7.4).

c) It is obvious that if $M \cong N$, then $\Omega^1(M) \cong \Omega^1(N)$. Suppose $f\colon \Omega^1(M) \to \Omega^1(N)$ and $g\colon \Omega^1(N) \to \Omega^1(M)$ have the property $gf = 1_{\Omega^1(M)}$ and $fg = 1_{\Omega^1(N)}$. Then by b) there are morphisms $u\colon M \to N$ such that $\Omega^1(\underline{u}) = \underline{f}$ and $v\colon N \to M$ such that $\Omega^1(\underline{v}) = \underline{g}$. Since $\Omega^1(\underline{v}\ \underline{u}) = \Omega^1(\underline{v})\Omega^1(\underline{u}) = \underline{g}\ \underline{f} = \underline{gf} = \underline{1}_{\Omega^1(M)}$, it follows from b) that

$\underline{vu} = \underline{1}_M$. Because R is complete and M is in $\operatorname{noeth}_P \Lambda$, the fact that $\underline{vu} = \underline{1}_M$ implies that $vu\colon M \to M$ is an isomorphism. A similar argument shows that $uv\colon N \to N$ is an isomorphism. Hence M and N are isomorphic.

d) Follows from the fact that the projective cover of a sum of two modules is the sum of the projective covers of each module.

e) Since R is complete, we know that X in $\operatorname{noeth}\Lambda$ is indecomposable if and only if $\operatorname{End} X$ is local. Further, if X is in $\operatorname{noeth}_P \Lambda$, we have that $P(X, X) \subset \operatorname{rad}\operatorname{End} X$ and so X is indecomposable if and only if $\underline{\operatorname{End}} X$ is local. But by b) $\Omega^1\colon \underline{\operatorname{End}} M \to \underline{\operatorname{End}}\Omega^1(M)$ is an isomorphism. Hence $\underline{\operatorname{End}} M$ is local if and only if $\underline{\operatorname{End}}\Omega^1(M)$ is local. Since M and $\Omega^1(M)$ are both in $\operatorname{noeth}_P \Lambda$, it follows that M is indecomposable if and only if $\Omega^1(M)$ is indecomposable.

f) Follows from previous results. □

As a consequence of this lemma we have

PROPOSITION 9.2. *Let $d = \dim R$. Then the equivalence of categories $\Omega^d\colon \underline{J}(\Lambda) \to \underline{L}(\Lambda)$ has the following properties.*

a) *If M is in $J_P(\Lambda)$, then $\Omega^d(M)$ is in $L_P(\Lambda)$.*

b) *The induced map $\Omega^d\colon \operatorname{Ob}(\underline{J}_P(\Lambda)) \to \operatorname{Ob}(\underline{L}_P(\Lambda))$ is a representation equivalence.*

PROOF. If $d = 0$, then $J(\Lambda) = L(\Lambda)$ and $\Omega^0\colon J(\Lambda) \to L(\Lambda)$ is the identity so there is nothing to prove. Hence we can assume that $d > 0$.

a) If M is in $J(\Lambda)$, then we know that $\operatorname{Ext}^i_\Lambda(M, \Lambda) = 0$ for $i = 1, \dots, d$. Suppose M is in $J_P(\Lambda)$. Then arguing by induction on i, we have by Lemma 9.1 that $\Omega^i(M)$ is in $\operatorname{noeth}_P \Lambda$ for $i = 1, \dots, d$. Thus a) has been established.

b) Suppose M and N are in $J_P(\Lambda)$. Then arguing by induction on i, we have by Lemma 9.1 that $M \cong N$ if and only if $\Omega^i(M) \cong \Omega^i(N)$ for $i = 1, \dots, d$. Thus $M \cong N$ if and only if $\Omega^d(M) \cong \Omega^d(N)$. A similar argument shows that if M is in $J_P(\Lambda)$ and $M \cong M_1 \amalg M_2$, then $\Omega^d(M) \cong \Omega^d(M_1) \amalg \Omega^d(M_2)$. Finally, the fact that $\Omega^d\colon \underline{J}(\Lambda) \to \underline{L}(\Lambda)$ is an equivalence of categories implies that the induced functor $\Omega^d\colon \underline{J}_P(\Lambda) \to \underline{L}_P(\Lambda)$ is an equivalence of categories. Hence given N in $L_P(\Lambda)$, there is an M in $J_P(\Lambda)$ such that $\Omega^d(M) \cong N$ in $\underline{L}_P(\Lambda)$. Since R is complete and $\Omega^d(M)$ and N are in $L_P(\Lambda)$, the fact that $\Omega^d(M) \cong N$ in $\underline{L}_P(\Lambda)$ implies that $\Omega^d(M) \cong N$ in $L_P(\Lambda)$. This completes the proof that $\Omega^d\colon \operatorname{Ob}(\underline{J}_P(\Lambda)) \to \operatorname{Ob}(\underline{L}_P(\Lambda))$ is a representation equivalence. □

Combining these results we obtain the following.

COROLLARY 9.3. *Let $d = \dim R$. Then we have the following.*

a) *The duality $\operatorname{Tr}\colon \underline{L}(\Lambda) \to \underline{J}(\Lambda^{\mathrm{op}})$ induces a representation equivalence $\operatorname{Tr}\colon \operatorname{Ob}(\underline{L}_P(\Lambda)) \to \operatorname{Ob}(\underline{J}_P(\Lambda^{\mathrm{op}}))$.*

b) *The equivalence $\Omega^d\colon \underline{J}(\Lambda) \to \underline{L}(\Lambda)$ of categories induces a representation equivalence $\Omega^d\colon \operatorname{Ob}(\underline{J}_P(\Lambda)) \to \operatorname{Ob}(\underline{L}_P(\Lambda))$.*

c) *The duality $\operatorname{Tr}_L\colon \underline{L}(\Lambda) \to \underline{L}(\Lambda^{\mathrm{op}})$ induces a representation equivalence $\operatorname{Tr}_L\colon \operatorname{Ob}(\underline{L}_P(\Lambda)) \to \operatorname{Ob}(\underline{L}_P(\Lambda^{\mathrm{op}}))$ with the property $\operatorname{Tr}_L \operatorname{Tr}_L(X) \cong X$ for all X in $\underline{L}_P(\Lambda)$ and all X in $\underline{L}_P(\Lambda^{\mathrm{op}})$.*

d) *The equivalence of categories $D\operatorname{Tr}_L\colon \underline{L}(\Lambda) \to \overline{L}(\Lambda^{\mathrm{op}})$ induces a representation equivalence $D\operatorname{Tr}_L\colon \operatorname{Ob}(\underline{L}_P(\Lambda)) \to \operatorname{Ob}(\overline{L_I}(\Lambda^{\mathrm{op}}))$ with the property $\operatorname{Tr}_L D(D\operatorname{Tr}_L(X)) \cong X$ in $L_P(\Lambda)$ for all X in $\underline{L}_P(\Lambda)$ and $D\operatorname{Tr}_L(\operatorname{Tr}_L D(Y)) \cong Y$ in $L_I(\Lambda)$ for all Y in $\overline{L_I}(\Lambda)$.*

The rest of this section is devoted to giving various analogues for lattices over an R-order Λ with R a complete local ring of results already given in Sections 5 and 6 for arbitrary noetherian R-algebras with R a complete local ring. In fact, the results concerning lattices are deduced from the earlier results using the following basic facts.

LEMMA 9.4. *Let* $d = \dim R$.

a) *The functor* $L(\Lambda) \to \operatorname{art}\Lambda$ *given by* $X \mapsto \operatorname{Hom}_R(D(X), I_d)$ *is fully faithful.*

b) *For* C *and* X *in* $L(\Lambda)$, *there are isomorphisms* $\operatorname{Ext}^1_\Lambda(C, X) \cong \operatorname{Ext}^1_\Lambda(\Omega^{-d}C, \operatorname{Hom}_R(D(X), I_d))$ *which are functorial in* C *and* X *in* $\underline{L}(\Lambda)$.

PROOF. a) The functor $L(\Lambda) \to \operatorname{art}\Lambda$ is the composition of dualities $D\colon L(\Lambda) \to L(\Lambda^{\mathrm{op}})$ given by $D(X) = \operatorname{Hom}_R(X, R)$ and $\operatorname{Hom}_R(\ , I_d)\colon \operatorname{noeth}\Lambda^{\mathrm{op}} \to \operatorname{art}\Lambda$ and is therefore fully faithful.

b) Proven in Section 8 using Lemma 8.4 (see argument preceding Proposition 8.7). □

As our first application we prove the following.

PROPOSITION 9.5. *Let* $0 \to A \xrightarrow{g} B \xrightarrow{f} C \to 0$ *be an exact sequence of lattices and let* $X = \operatorname{Tr}_L D(A)$. *Then*

a) f *is right* X*-determined in* $L(\Lambda)$.

b) *There is a summand* $0 \to A' \xrightarrow{g'} B' \xrightarrow{f'} C \to 0$ *over* C *of* $0 \to A \xrightarrow{g} B \xrightarrow{f} C \to 0$ *with* f' *a minimal right* X*-determined morphism in* $L(\Lambda)$ *such that* $\operatorname{Im}(\ , f')(X) = \operatorname{Im}(\ , f)(X)$.

c) f *is right minimal if and only if the zero module is the only summand* A' *of* A *such that* $g(A')$ *is a summand of* B.

d) *There are summands* A' *of* A *maximal with respect to* $g(A')$ *being a summand of* B. *Suppose* A_1 *and* A_2 *are two such summands of* A.

- i) *The induced exact sequences* $0 \to A/A_i \xrightarrow{g_i} B/g(A_i) \xrightarrow{f_i} C \to 0$ *in* $L(\Lambda)$ *have the property that* f_i *is minimal right* $\operatorname{Tr}_L DA$*-determined in* $L(\Lambda)$ *and* $\operatorname{Im}(\ , f_i)(X) = \operatorname{Im}(\ , f)(X)$ *for* $i = 1, 2$.
- ii) *The exact sequences* $0 \to A/A_i \xrightarrow{g_i} B/g(A_i) \xrightarrow{f_i} C \to 0$ *are isomorphic over* C.
- iii) A_1 *and* A_2 *are isomorphic summands of* A *and* A/A_1 *and* A/A_2 *are isomorphic summands of* A.

PROOF. a) Proven in Section 7.

b) Let y be the element of $\operatorname{Ext}^1_\Lambda(\Omega^{-d}C, \operatorname{Hom}_R(D(A), I_d))$ corresponding to $x\colon 0 \to A \xrightarrow{g} B \xrightarrow{f} C \to 0$ in $\operatorname{Ext}^1_\Lambda(C, A)$ under the functorial isomorphism $\operatorname{Ext}^1_\Lambda(C, A) \to \operatorname{Ext}^1_\Lambda(\Omega^{-d}C, \operatorname{Hom}_R(D(A), I_d))$. Then by Proposition 5.3 we know that $\operatorname{Hom}_R(D(A), I_d)$ can be written as $U_1 \amalg U_2$ such that the components y_i in $\operatorname{Ext}^1_\Lambda(\Omega^{-d}C, U_i)$ of y have the following properties a) $y_2 = 0$ and b) $y_1\colon 0 \to U_1 \xrightarrow{j} E \xrightarrow{h} C \to 0$ has the property h is right minimal or equivalently, if $t\colon U_1 \to U_1$ is such that $\operatorname{Ext}^1_\Lambda(\Omega^{-d}C, t)(y_1) = y_1$, then t is an isomorphism.

Since the functor $F\colon L(\Lambda) \to \operatorname{art}\Lambda$ given by $F(X) = \operatorname{Hom}_R(D(X), I_d)$ is fully faithful, there is a decomposition $A \cong A_1 \amalg A_2$ such that $F(A_i) \cong U_i$. Thus if we let x_i in $\operatorname{Ext}^1_\Lambda(C, A_i)$ be the components of x, then x_i corresponds to y_i under the isomorphism $\operatorname{Ext}^1_\Lambda(C, A_i) \cong \operatorname{Ext}^1_\Lambda(C, F(A_i))$. Hence $x_2 = 0$ and x_1 has the property that if $h\colon A_i \to A_i$ is such that $\operatorname{Ext}^1_\Lambda(C, u)(x_1) = x_1$, then u is an isomorphism. From this it follows that the exact sequence $x_1\colon A_1 \xrightarrow{g_1} B_1 \xrightarrow{f_1} C \to 0$ is right minimal. The fact that $x_2 = 0$ implies that f_1 is right X-determined with $\operatorname{Im}(\ , f_1)(X) = \operatorname{Im}(\ , f)(X)$. Thus the exact sequence $0 \to A_1 \xrightarrow{g_1} B_1 \xrightarrow{f_1} C \to 0$ is our desired exact sequence $0 \to A' \xrightarrow{g'} B' \xrightarrow{f'} C \to 0$.

c) and d). Proven in the same way as analogues in Proposition 5.4. □

For the convenience of the reader we state, without proof, the dual version of Proposition 9.5 which can easily be deduced from that proposition using the duality $D\colon L(\Lambda) \to L(\Lambda^{\mathrm{op}})$.

PROPOSITION 9.6. *Let $0 \to A \xrightarrow{g} B \xrightarrow{f} C \to 0$ be an exact sequence in $L(\Lambda)$ and let $X = D\,\mathrm{Tr}_L\, C$.*

a) *g is left X-determined in $L(\Lambda)$.*

b) *There is a summand $0 \to A \xrightarrow{g'} B' \xrightarrow{f'} C' \to 0$ over A of $0 \to A \xrightarrow{g} B \xrightarrow{f} C \to 0$ with g' a minimal left X-determined morphism in $L(\Lambda)$ such that $\mathrm{Im}(g',\ )(X) = \mathrm{Im}(g,\ )(X)$.*

c) *g is left minimal if and only if the zero module is the only summand C' of C such that the composition $B \to C \to C/C'$ is a splittable epimorphism.*

d) *There are summands C' of C minimal with respect to the composition $B \to C \to C/C'$ being a splittable epimorphism. Suppose C_1 and C_2 are two such submodules.*

i) *C_1 and C_2 are summands of C and there are summands B_1 and B_2 of B such that $f(B_i) = C_i$ and $f|_{B_i}\colon B_i \to C_i$ are isomorphisms for $i = 1, 2$. Moreover the exact sequences $0 \to A \xrightarrow{g_i} B/B_i \xrightarrow{f_i} C/C_i \to 0$ have the property that the g_i are minimal left X-determined and $\mathrm{Im}(g_i,\ )(X) = \mathrm{Im}(g,\ )(X)$ for $i = 1, 2$.*

ii) *The exact sequences $0 \to A \to B/B_i \to C/C_i \to 0$ are isomorphic over A.*

iii) *C/C_1 and C/C_2 are isomorphic summands of C and C_1 and C_2 are isomorphic summands of C.*

The following is the analogue for lattices of Proposition 5.6. It is stated without proof since the proof of Proposition 5.6 translates verbatim to the lattice situation.

PROPOSITION 9.7. *Let $0 \to A \xrightarrow{g} B \xrightarrow{f} C \to 0$ be exact in $L(\Lambda)$ such that f is right minimal. Suppose $A = \coprod_{i=1}^{k} V_i^{n_i}$ with the $n_i > 0$ and the $V_1, \dots, V_k$ nonisomorphic indecomposable Λ-lattices. Letting $V = \coprod_{i=1}^{k} V_i$, we have the following.*

a) *f is right $\mathrm{Tr}_L\, DV$-determined in $L(\Lambda)$.*

b) *If X is in $L(\Lambda)$ such that f is right X-determined but not right X'-determined for any proper summand X' of X, then $X \cong \mathrm{Tr}_L\, DV$.*

Again by duality we have the following dual of Proposition 9.7 which is the analogue of Proposition 6.7.

PROPOSITION 9.8. *Let $0 \to A \xrightarrow{g} B \xrightarrow{f} C \to 0$ be an exact sequence in $L(\Lambda)$ with g left minimal. Suppose $C = \coprod_{i=1}^{k} W_i^{n_i}$ with the $n_i > 0$ and the W_i nonisomorphic indecomposable Λ-lattices. Letting $W = \coprod_{i=1}^{k} W_i$ we have the following.*

a) *g is left $D\,\mathrm{Tr}_L\, W$-determined.*

b) *If X is in $L(\Lambda)$ such that g is left X-determined but not left X'-determined for any proper summand X' of X, then $X \cong D\,\mathrm{Tr}_L\, W$.*

As our final results of this section we have the following analogue of Theorem 5.7 and Theorem 6.6. Since the proofs are essentially the same as those for Theorems 5.7 and 6.6 we omit the proofs.

PROPOSITION 9.9. *Let A and C be in $L(\Lambda)$ and let $X = \mathrm{Tr}_L\, DA$. Then $(X, C)/P(X, C)$ is an $(\mathrm{End}\, X)^{\mathrm{op}}$-module of finite length. Let $S_1, \dots, S_d$ be a complete set of nonisomorphic simple $(\mathrm{End}\, X)^{\mathrm{op}}$-modules and for each $(\mathrm{End}\, X)^{\mathrm{op}}$-submodule H of (X, C) containing $P(X, C)$, let $n_1(H), \dots, n_d(H)$ be the uniquely determined nonnegative integers such that the $(\mathrm{End}\, X)^{\mathrm{op}}$-socle of $(X, C)/H$ is isomorphic to $\coprod_{i=1}^{d} S_i^{n_i(H)}$.*

Finally let $n = \max\{n_i(H)\}$ *as* i *runs through all* $(\operatorname{End} X)^{\mathrm{op}}$*-submodules of* (X, C) *containing* $P(X, C)$.

a) n *is finite.*

b) *If* $k > 0$ *and* $0 \to A^k \xrightarrow{g} B \xrightarrow{f} C \to 0$ *is exact, then there is a submodule* A' *of* A^k *isomorphic to* A^{k-n} *such that* $g(A')$ *is a summand of* B.

Finally, dualizing Proposition 9.9 we obtain the following.

PROPOSITION 9.10. *Let* A *and* C *be in* $L(\Lambda)$. *Then* $(A, X)/I(A, X)$ *is an* $\operatorname{End} X$*-module of finite length where* $X = D\operatorname{Tr}_L C$. *Let* $S_1, \dots, S_d$ *be a complete set of nonisomorphic simple* $\operatorname{End} X$*-modules and for each* $\operatorname{End} X$*-submodule* H *of* (A, X) *containing* $I(A, X)$, *let* $n_1(H), \dots, n_d(H)$ *be the uniquely determined nonnegative integers such that the* $(\operatorname{End} X)$*-socle of* (A, X) *is isomorphic to* $\coprod_{i=1}^{d} S_i^{n_i(H)}$. *Finally, let* $n = \max\{n_i(H)\}$ *as* i *runs through* $1, \dots, d$ *and* H *runs through all* $\operatorname{End} X$*-submodules* H *of* (X, C) *containing* $I(A, X)$.

a) n *is finite.*

b) *If* $k > n$ *and* $0 \to A \to B \to C^k \to 0$ *is an exact sequence of lattices, then there is an exact sequence* $0 \to A \to B \to C^{k-n} \to 0$ *which is a summand over* A *of* $0 \to A \to B \to C^k \to 0$.

10. A generalization

Combining the results in Sections 3 and 6, we have the following theorem.

Let Λ be a noetherian R-algebra with R a complete local ring. Suppose $0 \to A \xrightarrow{g} B \xrightarrow{f} C \to 0$ is an exact sequence of Λ-modules with A in $\operatorname{art} \Lambda$ and C in $\operatorname{noeth} \Lambda$. Then f is right $\operatorname{Tr} DA$-determined in $\operatorname{Mod} \Lambda$ and g is left $D\operatorname{Tr} C$-determined in $\operatorname{arno} \Lambda$.

Our main aim in this section is to strengthen this result by showing that g is left $D\operatorname{Tr} C$-determined in $\operatorname{Mod} \Lambda$, not just in $\operatorname{arno} \Lambda$. To accomplish this we need some preliminary results. We start by recalling some of the well known basic facts and definitions of pure exact sequences, pure injective, etc.

An exact sequence of Λ-modules $0 \to A \to B \to C \to 0$ is said to be *pure exact* if the sequence of functors $0 \to \otimes A \to \otimes B \to \otimes C \to 0$ is exact. A Λ-module A is said to be *pure injective* if every pure exact sequence $0 \to A \to B \to C \to 0$ splits. We now give some characterizations of when a Λ-module is pure injective. To this end it is convenient to make a few observations.

Let R be a commutative ring. An R-algebra consists of a ring morphism $R \to \Lambda$. Obviously every ring Λ can be considered as an R-algebra for some commutative ring R, for instance $R = Z$, the integers. Thus we loose no generality be considering R-algebras Λ rather than just rings Λ.

Suppose Λ is an R-algebra. Also suppose that I is an R-module which is an injective cogenerator for $\operatorname{Mod} R$. Clearly for each M in $\operatorname{Mod} \Lambda$, $\operatorname{Hom}_R(M, I)$ is a Λ^{op}-module by means of the operations of Λ on M. Thus we obtain a functor $\operatorname{Hom}_R(\ , I)\colon \operatorname{Mod} \Lambda \to \operatorname{Mod} \Lambda^{\mathrm{op}}$ which we denote by $D\colon \operatorname{Mod} \Lambda \to \operatorname{Mod} \Lambda^{\mathrm{op}}$. Since I is an injective cogenerator, D is a faithful exact functor. Moreover for the same reason, a sequence $A \to B \to C$ of Λ-modules is exact if and only if the sequence of R-modules $D(C) \to D(B) \to D(A)$ is exact. Returning to characterizing the Λ-modules which are pure injective, we have the following.

PROPOSITION 10.1. *The following statements are equivalent for a Λ-module A.*

a) *A is pure injective.*

b) *The Λ-morphism $\phi\colon A \to D^2(A)$ given by $\phi(a)(f) = f(a)$ for all a in A and f in $D(A)$, is a splittable monomorphism.*

c) *There is a Λ^{op}-Σ bimodule X for some ring Σ such that A is isomorphic to a summand of the Λ-module $\operatorname{Hom}_\Sigma(X, E)$ for some injective Σ-module E.*

d) *If $\{C_k\}_{k\in K}$ is a direct limit system of Λ-modules over a directed set K, then the natural morphism $\operatorname{Ext}^1_\Lambda(\varinjlim_{k\in K} C_k, A) \to \varprojlim_{k\in K} \operatorname{Ext}^1_\Lambda(C_k, A)$ is an isomorphism.*

e) *An exact sequence $0 \to A \to B \to C \to 0$ of Λ-modules splits if for each morphism $h\colon X \to C$ with X a finitely presented Λ-module the pull back exact sequence $0 \to A \to B\times_C X \to X \to 0$ splits.*

f) *An exact sequence of Λ-modules $0 \to A \to B \to C \to 0$ splits whenever $(X, B) \to (X, C) \to 0$ is exact for all X in* $\operatorname{mod}\Lambda$.

PROOF. a) implies b). Since I is an injective cogenerator, $A \to D^2(A)$ is a monomorphism. Let $0 \to A \to D^2(A) \to T \to 0$ be exact. Since we are assuming by a) that A is pure injective, to show that this exact sequence $0 \to A \to D^2(A) \to T \to 0$ splits, it suffices to show that it is pure exact.

From the exact sequence $0 \to A \to D^2(A) \to T \to 0$ we obtain the exact sequence of Λ^{op}-modules $0 \to D(T) \to D^3(A) \to D(A) \to 0$ (remember that D is exact). It is well know that the composition $D(A) \to D^3(A) \to D(A)$ is the identity where $D(A) \to D^3(A)$ is the natural morphism $D(A) \to D^2(D(A))$. Thus the exact sequence $0 \to D(T) \to D^3(A) \to D(A) \to 0$ splits. Hence for each X in $\operatorname{Mod}\Lambda^{\text{op}}$ we have that $0 \to (X, D(T)) \to (X, D^3(A)) \to (X, D(A))$ is exact. Applying a well known associative law we obtain that $0 \to D(X \underset{\Lambda}{\otimes} T) \to D(X \underset{\Lambda}{\otimes} D^2(A)) \to D(X \underset{\Lambda}{\otimes} D(A)) \to 0$ is exact. Therefore $0 \to X \underset{\Lambda}{\otimes} D(A) \to X \underset{\Lambda}{\otimes} D^2(A) \to X \underset{\Lambda}{\otimes} T \to 0$ is exact since D is fully faithful and exact on $\operatorname{Mod} R$. Since this is true for all X in $\operatorname{Mod}\Lambda^{\text{op}}$, the exact sequence $0 \to A \to D^2(A) \to T \to 0$ is pure exact. Therefore the fact that A is pure injective implies the sequence splits, showing that a) implies b).

b) implies c). Let $\Sigma = R$ and $E = I$.

c) implies d). We have the well known functorial isomorphism $\operatorname{Ext}^1_\Lambda(C, \operatorname{Hom}_\Sigma(X, E)) \to \operatorname{Hom}_\Sigma(\operatorname{Tor}^\Lambda_1(X, C), E)$ (see Proposition 6.2). Suppose $\{C_k\}_{k\in K}$ is a direct limit system over a directed set K. Then $\operatorname{Tor}^\Lambda_1(X, \varinjlim_{k\in K} C_k) = \varinjlim_{k\in K} \operatorname{Tor}^\Lambda_1(X, C_k)$ since Tor commutes with direct limits. Hence $\operatorname{Hom}_\Sigma(\operatorname{Tor}^\Lambda_1(X, \varinjlim_{k\in K} C_k), E) = \varprojlim \operatorname{Ext}^1_\Lambda(C_k, \operatorname{Hom}_\Sigma(X, E))$. This shows that the natural morphism $\operatorname{Ext}^1_\Lambda(\varinjlim_{k\in K} C_k, \operatorname{Hom}_\Sigma(X, E)) \to \varprojlim_{k\in K} \operatorname{Ext}^1_\Lambda(C_k, \operatorname{Hom}_\Sigma(X, E))$ is an isomorphism. From the fact that A is a summand of $\operatorname{Hom}_\Sigma(X, E)$, it now follows that $\operatorname{Ext}^1_\Lambda(\varinjlim_{k\in K} C_k, A) = \varprojlim \operatorname{Ext}^1_\Lambda(C_k, A)$.

d) implies e). Let x be the element $0 \to A \to B \to C \to 0$ in $\operatorname{Ext}^1_\Lambda(C, A)$. The hypothesis of e) is equivalent to given any morphism $f\colon X \to C$ with X finitely presented $\operatorname{Ext}^1_\Lambda(f, A)(x)$ in $\operatorname{Ext}^1_\Lambda(X, A)$ is zero. We know that $C = \varinjlim_{k\in K} C_k$ with the C_k in $\operatorname{mod}\Lambda$. Thus the hypothesis of e) tells us that x goes to zero under each of the morphisms $\operatorname{Ext}^1_\Lambda(C, A) \to \operatorname{Ext}^1_\Lambda(C_k, A)$. Since $\operatorname{Ext}^1_\Lambda(C, A) = \varprojlim \operatorname{Ext}^1_\Lambda(C_k, A)$, this means that $x = 0$, our desired result.

e) implies f). Suppose $x\colon 0 \to A \to B \to C \to 0$ is our exact sequence such that $(X, B) \to (X, C) \to 0$ is exact for all X in $\operatorname{mod}\Lambda$. Let $f\colon X \to C$. Then $\operatorname{Ext}^1(f, A)(x)$ in $\operatorname{Ext}^1_\Lambda(X, A)$ is the same as the image of f in $\operatorname{Ext}^1_\Lambda(X, A)$ given by the connecting morphism $(X, C) \to \operatorname{Ext}^1_\Lambda(X, A)$ induced by the exact sequence $0 \to A \to B \to C \to 0$. Since

$(X,B) \to (X,C) \to 0$ is exact, $(X,C) \to \operatorname{Ext}^1_\Lambda(X,A)$ is zero and so $\operatorname{Ext}^1_\Lambda(f,A)(x) = 0$. The fact that this is true for all morphisms $f\colon X \to C$ with X in $\operatorname{mod}\Lambda$ tells us by the hypothesis of e) that $0 \to A \to B \to C \to 0$ splits.

f) implies a). To show that A is pure injective we have to show that an exact sequence $0 \to A \to B \to C \to 0$ splits if it is pure exact. Suppose $0 \to A \to B \to C \to 0$ is pure exact and X is in $\operatorname{Mod}\Lambda^{\mathrm{op}}$. Then $0 \to X \underset{\Lambda}{\otimes} A \to X \underset{\Lambda}{\otimes} B \to X \underset{\Lambda}{\otimes} C \to 0$ is exact and so $0 \to D(X \underset{\Lambda}{\otimes} C) \to D(X \underset{\Lambda}{\otimes} B) \to D(X \underset{\Lambda}{\otimes} A) \to 0$ is exact. Applying a well known associative law we obtain that the sequence $0 \to (X, D(C)) \to (X, D(B)) \to (X, D(A)) \to 0$ is exact. Since this is true for all X in $\operatorname{Mod}\Lambda^{\mathrm{op}}$, we have that the exact sequence of Λ^{op}-modules $0 \to D(C) \to D(B) \to D(A) \to 0$ splits. Hence for X in $\operatorname{mod}\Lambda$ we have that $0 \to D(C) \underset{\Lambda}{\otimes} X \to D(B) \underset{\Lambda}{\otimes} X \to D(A) \underset{\Lambda}{\otimes} X \to 0$ is exact. Since X is a finitely presented Λ-module and I is R injective, it follows that there are isomorphisms $\operatorname{Hom}_R(Y,I) \otimes X \cong \operatorname{Hom}_R(\operatorname{Hom}(X,Y),I)$ functorial in Y (see [14, VI Proposition 5.2]. Thus the exact sequence $0 \to A \to B \to C \to 0$ has the property $0 \to (X,A) \to (X,B) \to (X,C) \to 0$ is exact for all X in $\operatorname{mod}\Lambda$ since $0 \to D((X,C)) \to D((X,B)) \to D((X,A)) \to 0$ is exact with D exact and faithful. Thus by the hypothesis of f) the sequence $0 \to A \to B \to C \to 0$ splits. Hence f) implies a), which finishes the proof of the proposition. □

Suppose $f\colon B \to C$ is a morphism in $\operatorname{Mod}\Lambda$ and X is in $\operatorname{Mod}\Lambda$. In order to check whether f is right X-determined it is necessary to show that if $h\colon L \to C$ is any morphism such that $(\operatorname{Im}(\ ,f))(X) \supset (\operatorname{Im}(\ ,h))(X)$, then there is a morphism $g\colon L \to B$ such that $h = fg$. We now show that under certain circumstances one only has to check if morphisms $h\colon L \to C$ with L finitely presented have this property.

Proposition 10.2. *Suppose* $0 \to A \xrightarrow{g} B \xrightarrow{f} C \to 0$ *is an exact sequence of* Λ*-modules with* A *pure injective and let* X *be in* $\operatorname{Mod}\Lambda$. *Then* $f\colon B \to C$ *is right* X*-determined if given any morphism* $h\colon L \to C$ *with* L *finitely presented such that* $(\operatorname{Im}(\ ,f))(X) \subset (\operatorname{Im}(\ ,f))(X)$, *there is a morphism* $g\colon L \to B$ *such that* $h = fg$.

Proof. Let $u\colon U \to C$ be an arbitrary morphism such that $(\operatorname{Im}(\ ,u))(X) \subset (\operatorname{Im}(\ ,f))(X)$. We want to show this implies that there is a $v\colon U \to B$ such that $u = fv$. Or equivalently, the pull back exact sequence $0 \to A \to U \times_C B \to U \to 0$ splits.

Since A is pure injective, to show that $0 \to A \to U \times_C B \to U \to 0$ splits it suffices to show that given any $g\colon L \to U$ with L finitely presented the pull back exact sequence $0 \to A \to (U \times_C B) \times_U L \to 0$ splits. From the exact commutative diagram

$$\begin{array}{ccccccccc}
0 & \longrightarrow & A & \longrightarrow & (U\times_C B)\times_U L & \longrightarrow & L & \longrightarrow & 0 \\
 & & \| & & \downarrow & & \downarrow{\scriptstyle g} & & \\
0 & \longrightarrow & A & \longrightarrow & U\times_C B & \longrightarrow & U & \longrightarrow & 0 \\
 & & \| & & \downarrow & & \downarrow{\scriptstyle u} & & \\
0 & \longrightarrow & A & \longrightarrow & B & \xrightarrow{f} & C & \longrightarrow & 0
\end{array}$$

we deduce the following:

a) $0 \to A \to (U \times_C B) \times_U L \to L \to 0$ is the pull back exact sequence for $ug\colon L \to C$.
b) $(\operatorname{Im}(\ ,ug))(X) \subset (\operatorname{Im}(\ ,u))(X) \subset (\operatorname{Im}(\ ,f))(X)$.

Since L is finitely presented, it follows from a) and b) that the exact sequence $0 \to A \to (U \times_C B) \times_U L \to L \to 0$ splits. Because this is true for all morphisms $L \to U$ with L finitely presented, it follows from Proposition 10.1 that the exact sequence $0 \to A \to$

$B \times_C U \to U \to 0$ splits. This finishes the proof that $0 \to A \to B \xrightarrow{f} C \to 0$ is right X-determined. □

With these preliminary results in mind, we turn our attention to establishing the results cited in the beginning of this section. So we assume throughout the rest of this section that Λ is a noetherian R-algebra with R a complete local ring. We begin with the following general observations.

PROPOSITION 10.3. *Let* $0 \to X \to Y \xrightarrow{h} C \to 0$ *be a nonsplit exact sequence in* $\operatorname{Mod}\Lambda$ *with* C *in* $\operatorname{noeth}\Lambda$.

a) *There is an exact commutative diagram*

$$\begin{array}{ccccccccc} 0 & \longrightarrow & X & \longrightarrow & Y & \longrightarrow & C & \longrightarrow & 0 \\ & & \downarrow & & \downarrow & & \| & & \\ 0 & \longrightarrow & D\operatorname{Tr}C & \longrightarrow & B & \longrightarrow & C & \longrightarrow & 0 \end{array}$$

where $0 \to D\operatorname{Tr}C \to B \to C \to 0$ *does not split.*

b) *There is an exact commutative diagram*

$$\begin{array}{ccccccccc} 0 & \longrightarrow & X & \longrightarrow & Y & \longrightarrow & C & \longrightarrow & 0 \\ & & \downarrow & & \downarrow & & \| & & \\ 0 & \longrightarrow & X'' & \longrightarrow & Y'' & \longrightarrow & C & \longrightarrow & 0 \\ & & \downarrow & & \downarrow & & & & \\ & & 0 & & 0 & & & & \end{array}$$

such that $0 \to X'' \to Y'' \to C \to 0$ *does not split with* X'' *an indecomposable module in* $\operatorname{art}_I \Lambda$.

PROOF. a) Since $h\colon Y \to C$ is an epimorphism we know that the right ideal $J = (\operatorname{Im}(\ ,h))(C)$ of $\operatorname{End}C$ contains $P(C,C)$. Further, J is a proper right ideal of $\operatorname{End}C$ since h is not a splittable epimorphism. Let H be a maximal right ideal of $\operatorname{End}C$ containing J. Since $(C,C)/H$ is a simple $(\operatorname{End}C)^{\mathrm{op}}$-module we have by Theorem 5.5 that there is an exact sequence $0 \to D\operatorname{Tr}C \to B \xrightarrow{f} C \to 0$ with f a right C-determined morphism and $\operatorname{Im}(\ ,f)(C) = H$. Since $J = \operatorname{Im}(\ ,h)(C) \subset$H we know there is a commutative exact diagram

$$\begin{array}{ccccccccc} 0 & \longrightarrow & X & \longrightarrow & Y & \xrightarrow{h} & C & \longrightarrow & 0 \\ & & \downarrow & & \downarrow & & \| & & \\ 0 & \longrightarrow & D\operatorname{Tr}C & \longrightarrow & B & \xrightarrow{f} & C & \longrightarrow & 0 \end{array}$$

with $0 \to D\operatorname{Tr}C \to B \xrightarrow{f} C \to 0$ not splittable.

b) By a) we know there is a commutative exact diagram

$$\begin{array}{ccccccccc} 0 & \longrightarrow & X & \longrightarrow & Y & \xrightarrow{h} & C & \longrightarrow & 0 \\ & & \downarrow{\scriptstyle t} & & \downarrow{\scriptstyle u} & & \| & & \\ 0 & \longrightarrow & D\operatorname{Tr}C & \longrightarrow & B & \xrightarrow{f} & C & \longrightarrow & 0 \end{array}$$

where $0 \to D\operatorname{Tr} C \to B \to C \to 0$ does not split. Hence $0 \to \operatorname{Im} t \to \operatorname{Im} u \to C \to 0$ is a nonsplit exact sequence. Since $D\operatorname{Tr} C$ is in $\operatorname{art}\Lambda$, we know that the submodule $\operatorname{Im} t$ of $D\operatorname{Tr} C$ is also in $\operatorname{art}\Lambda$. Hence $\operatorname{Im} t$ is isomorphic to a finite sum $\amalg A_i$ where each A_i is an indecomposable module in $\operatorname{art}\Lambda$. Therefore the fact that $0 \to \operatorname{Im} t \to \operatorname{Im} u \to C \to 0$ does not split means there is a projection $\operatorname{Im} t \to A_i$ for some i such that the exact push out diagram

$$\begin{array}{ccccccccc}
0 & \longrightarrow & \operatorname{Im} t & \longrightarrow & \operatorname{Im} u & \longrightarrow & C & \longrightarrow & 0 \\
 & & \downarrow & & \downarrow & & \| & & \\
0 & \longrightarrow & A_i & \longrightarrow & B & \longrightarrow & C & \longrightarrow & 0 \\
 & & \downarrow & & \downarrow & & & & \\
 & & 0 & & 0 & & & &
\end{array}$$

has the property $0 \to A_i \to B \to C \to 0$ does not split. Thus we have the exact commutative diagram

$$\begin{array}{ccccccccc}
0 & \longrightarrow & X & \longrightarrow & Y & \longrightarrow & C & \longrightarrow & 0 \\
 & & \downarrow & & \downarrow & & \| & & \\
0 & \longrightarrow & A_i & \longrightarrow & B & \longrightarrow & C & \longrightarrow & 0 \\
 & & \downarrow & & \downarrow & & & & \\
 & & 0 & & 0 & & & &
\end{array}$$

which finishes the proof of b). □

As a consequence of this result we have the following.

PROPOSITION 10.4. *Let* $0 \to X \xrightarrow{g} Y \xrightarrow{f} Z \to 0$ *be a nonsplit exact sequence in* $\operatorname{Mod}\Lambda$. *Then the following statements are equivalent.*

a) *There is a finitely generated submodule* Z' *of* Z *such that the exact sequence* $0 \to X \to f^{-1}(Z') \to Z' \to 0$ *does not split.*

b) *There is a submodule* X' *of* X *such that* X/X' *is in* $\operatorname{art}\Lambda$ *and the exact sequence* $0 \to X/X' \to Y/g(X') \to Z \to 0$ *does not split.*

PROOF. a) implies b). Let Z' be a finitely generated submodule of Z such that the element $x\colon 0 \to X \to f^{-1}(Z') \to Z' \to 0$ in $\operatorname{Ext}^1_\Lambda(Z', X)$ does not split. Then by Proposition 10.3 we know there is an epimorphism $t\colon X \to A$ with A in $\operatorname{art}\Lambda$ such that $\operatorname{Ext}^1_\Lambda(Z', t)(x)$ in $\operatorname{Ext}^1_\Lambda(Z', A)$ is not zero. Letting $0 \to X \to Y \to Z \to 0$ be the element y in $\operatorname{Ext}^1_\Lambda(Z, X)$ we have the commutative diagram

$$\begin{array}{ccc}
\operatorname{Ext}^1_\Lambda(Z, X) & \xrightarrow{\operatorname{Ext}^1_\Lambda(Z,t)} & \operatorname{Ext}^1_\Lambda(Z, A) \\
\downarrow{\scriptstyle \operatorname{Ext}^1_\Lambda(h,X)} & & \downarrow{\scriptstyle \operatorname{Ext}^1_\Lambda(h,A)} \\
\operatorname{Ext}^1_\Lambda(Z', X) & \xrightarrow[\operatorname{Ext}^1_\Lambda(Z',t)]{} & \operatorname{Ext}^1_\Lambda(Z', A)
\end{array}$$

where $h\colon Z' \to Z$ is the inclusion. Since $\operatorname{Ext}^1_\Lambda(Z', t)(\operatorname{Ext}^1_\Lambda(h, X))(y) = \operatorname{Ext}^1_\Lambda(Z', t)(x) \neq 0$, it follows that $\operatorname{Ext}^1_\Lambda(Z, t)(y)$ in $\operatorname{Ext}^1_\Lambda(Z, A)$ is not zero. This shows that a) implies b).

b) implies a). Suppose $0 \to X/X' \xrightarrow{h} Y/g(X') \to Z \to 0$ does not split and X/X' is in $\operatorname{art}\Lambda$. Then $X/X' \cong DD(X/X')$ and so by Proposition 10.1 X/X' is pure injective. Again by Proposition 10.1, there is a finitely generated submodule Z' of Z such that $0 \to X/X' \to j^{-1}(Z') \to Z' \to 0$ does not split. By an argument similar to that used in part a), it is not difficult to show that this implies that $0 \to X \to f^{-1}(Z') \to Z' \to 0$ does not split. This completes the proof of the proposition. □

This result enables us to establish the following.

PROPOSITION 10.5. *Let $0 \to A \xrightarrow{g} B \xrightarrow{f} C \to 0$ be an exact sequence of Λ-modules with C in* $\operatorname{noeth}\Lambda$. *Suppose X is a Λ-module. The morphism g is left X-determined if given any morphism $h\colon A \to Y$ with Y in* $\operatorname{art}\Lambda$ *such that* $\operatorname{Im}(h,\ )(X) \subset \operatorname{Im}(g,\ )(X)$, *there is a morphism $s\colon B \to Y$ such that $sg = h$.*

PROOF. Let $t\colon A \to L$ in $\operatorname{Mod}\Lambda$ have the property $(\operatorname{Im}(t,\ ))(X) \subset (\operatorname{Im}(g,\ ))(X)$. To prove the proposition we must show there is a morphism $v\colon B \to L$ such that $vg = t$.

Suppose no such $v\colon B \to L$ exists. This implies that if x in $\operatorname{Ext}^1_\Lambda(C, A)$ is the extension $0 \to A \to B \to C \to 0$, then $y = \operatorname{Ext}^1_\Lambda(C, t)(x)$ in $\operatorname{Ext}^1_\Lambda(C, L)$ is not zero. Since C is in $\operatorname{noeth}\Lambda$, we have by Proposition 10.3 that there is an epimorphism $u\colon L \to Y$ with Y in $\operatorname{art}\Lambda$ such that $\operatorname{Ext}^1_\Lambda(C, u)(y)$ in $\operatorname{Ext}^1_\Lambda(C, Y)$ is not zero. Hence the morphism $ut\colon A \to Y$ has the property that $\operatorname{Ext}^1_\Lambda(C, ut)(x)$ in $\operatorname{Ext}^1_\Lambda(C, Y)$ is not zero. Because Y is in $\operatorname{art}\Lambda$ we have by the hypothesis of the proposition that $\operatorname{Im}(ut,\ )(X)$ is not contained in $\operatorname{Im}(g,\ )(X)$. But $\operatorname{Im}(t,\ )(X) \supset \operatorname{Im}(ut,\ )(X)$. This contradiction shows that if $\operatorname{Im}(t,\ )(X) \subset \operatorname{Im}(g,\ )(X)$, then there is a $v\colon B \to L$ such that $t = vg$. This finishes the proof of the proposition. □

Combining these results we obtain the result announced in the beginning of this section.

THEOREM 10.6. *Let $0 \to A \xrightarrow{g} B \xrightarrow{f} C \to 0$ be an exact sequence in* $\operatorname{Mod}\Lambda$ *with A in* $\operatorname{art}_I\Lambda$ *and C in* $\operatorname{noeth}_P\Lambda$. *Then*

a) *f is right* $\operatorname{Tr} DA$*-determined in* $\operatorname{Mod}\Lambda$.

b) *g is left* $D\operatorname{Tr} C$*-determined in* $\operatorname{Mod}\Lambda$.

PROOF. a) See Section 3.

b) We have already seen in Section 6 that g is left $\operatorname{Tr} DC$-determined in $\operatorname{arno}\Lambda$. In particular, we know that if $j\colon A \to Y$ is a morphism in $\operatorname{art}\Lambda$ such that $\operatorname{Im}(j,\ )(\operatorname{Tr} D(C))$ is contained in $\operatorname{Im}(g,\ )(\operatorname{Tr} DC)$, then there is a morphism $s\colon B \to Y$ such that $sg = j$. Since C is in $\operatorname{noeth}\Lambda$, it follows from Proposition 10.5 that g is left $\operatorname{Tr} DC$-determined in $\operatorname{Mod}\Lambda$. □

Finally we point out, that as a consequence of this result, Theorem 6.6 remains valid if the phrases "determined in $\operatorname{arno}\Lambda$" are changed to "determined in $\operatorname{Mod}\Lambda$".

11. Appendix

Our purpose in this appendix is to give a proof of Proposition 3.8 which was used in establishing one of our basic existence theorems, Theorem 3.9.

Suppose X is a finitely presented Λ-module and $\Gamma = \operatorname{End} X$. Then X is a Γ-module in the obvious way. Since the operations of Λ and Γ commute, X is a Λ-Γ bimodule. Let I be an injective module. Then $\operatorname{Hom}_\Gamma(X, I)$ is a Λ^{op}-module by means of the operation of Λ on X. Now we have the morphism of functors

$$\psi\colon \operatorname{Hom}_\Gamma(X, I)\otimes_\Lambda \to \operatorname{Hom}_\Gamma(\operatorname{Hom}_\Lambda(\ , X), I)$$

which for each Y in $\operatorname{Mod}\Lambda$ is given by $\psi_Y(f \otimes y)(g) = f(g(y))$ for all f in $\operatorname{Hom}_\Gamma(X, I)$, y in Y, and g in $\operatorname{Hom}_\Lambda(Y, X)$. It is well known (see [14 VI, Proposition 5.2]) that since I is Γ-injective, ψ_Y is an isomorphism for all finitely presented Λ-modules Y. In particular, if $Y = X$, we have the isomorphism

$$\psi_X\colon \operatorname{Hom}_\Gamma(X, I) \otimes_\Lambda X \to \operatorname{Hom}_\Gamma(\operatorname{Hom}_\Lambda(X, X), I)$$

which is an isomorphism of Γ-modules where $\operatorname{Hom}_\Gamma(X, I) \otimes_\Lambda X$ is viewed as a Γ-module by means of the operation of Γ on X and the operation on $\operatorname{Hom}_\Gamma(\operatorname{Hom}_\Lambda(X, X), I)$ is given by the operation of Γ^{op} on $\operatorname{Hom}_\Lambda(X, X)$ defined by $yf(x) = f(yx)$ for all y in Γ^{op}, x in X and f in $\operatorname{Hom}_\Lambda(X, X)$. Since the morphism $\operatorname{Hom}_\Gamma(\operatorname{Hom}_\Lambda(X, X), I) \to I$ given by $f \mapsto f(1_x)$ is an isomorphism of Γ-modules, we obtain the Γ-module isomorphism

$$\alpha\colon \operatorname{Hom}_\Gamma(X, I) \otimes_\Lambda X \to I$$

given explicitly by $\alpha(f \otimes x) = f(x)$, which we will usually consider an identification.

The following result will be used constantly in the rest of this section.

PROPOSITION 11.1. *Let A be an arbitrary Λ^{op}-module. Then considering $A \otimes_\Lambda X$ and $\operatorname{Hom}_\Gamma(X, I) \otimes_\Lambda X$ as Γ-modules by means of the operation of Γ on X we have that the natural morphism*

$$\beta\colon \operatorname{Hom}_{\Lambda^{\mathrm{op}}}(A, \operatorname{Hom}_\Gamma(X, I)) \to \operatorname{Hom}_\Gamma(A \otimes_\Lambda X, \operatorname{Hom}_\Gamma(X, I) \otimes_\Lambda X)$$

given by $\beta(f) = f \otimes_\Lambda X$ is an isomorphism which is functorial in A.

PROOF. We have the standard isomorphism $\phi\colon \operatorname{Hom}_{\Lambda^{\mathrm{op}}}(A, \operatorname{Hom}_\Gamma(X, I)) \to \operatorname{Hom}_\Gamma(A \otimes_\Lambda X, I)$ which is functorial in A. We also have our identification $\alpha\colon \operatorname{Hom}_\Gamma(X, I) \otimes_\Lambda X \to I$ of Γ-modules. Hence we obtain a diagram

$$\begin{array}{ccc}
\operatorname{Hom}_{\Lambda^{\mathrm{op}}}(A, \operatorname{Hom}_\Gamma(X, I)) & \xrightarrow{\beta} & \operatorname{Hom}_\Gamma(A \otimes_\Lambda X, \operatorname{Hom}_\Gamma(X, I) \otimes_\Lambda X) \\
& \searrow^{\phi} & \downarrow{\scriptstyle \operatorname{Hom}_\Gamma(A \otimes_\Lambda X, X)} \\
& & \operatorname{Hom}_\Gamma(A \otimes_\Lambda X, I)
\end{array}$$

Since ϕ and $\operatorname{Hom}_\Gamma(A \otimes_\Lambda X, \alpha)$ are isomorphisms, we will have our desired result that β is an isomorphism if we show that the diagram commutes. But this is a routine calculation we leave to the reader. □

As an application of this result we obtain the following proof of Proposition 3.8.

PROPOSITION 11.2. *The restriction of the functor* $\operatorname{Hom}_\Gamma(X, \ )\colon \operatorname{Mod}\Gamma \to \operatorname{Mod}\Lambda^{\mathrm{op}}$ *to $\mathcal{I}$, the category of injective Γ-modules, is fully faithful. So letting $\mathcal{C}$ be the full subcategory of* $\operatorname{Mod}\Lambda^{\mathrm{op}}$ *consisting of all Λ^{op}-modules isomorphic to modules of the form* $\operatorname{Hom}_\Gamma(X, I)$ *for some I in $\mathcal{I}$, the functor $\mathcal{I} \to \mathcal{C}$ given by $I \mapsto \operatorname{Hom}_\Gamma(X, I)$ is an equivalence of categories.*

PROOF. Let I_1 and I_2 be two objects in $\mathcal{I}$. Then we have the following composition of morphisms $\operatorname{Hom}_\Gamma(I_1, I_2) \to \operatorname{Hom}_{\Lambda^{\mathrm{op}}}(\operatorname{Hom}_\Gamma(X, I_1), \operatorname{Hom}_\Gamma(X, I_2)) \xrightarrow{\sim} \operatorname{Hom}_\Gamma(\operatorname{Hom}_\Gamma(X, I_1) \underset{\Lambda}{\otimes} X, \operatorname{Hom}_\Gamma(X, I_2) \underset{\Lambda}{\otimes} X) \xrightarrow{\sim} \operatorname{Hom}_\Gamma(I_1, I_2)$ which it is easy to check is the identity. Hence $\operatorname{Hom}_\Gamma(I_1, I_2) \to \operatorname{Hom}_{\Lambda^{\mathrm{op}}}(\operatorname{Hom}_\Gamma(X, I_1), \operatorname{Hom}_\Gamma(X, I_2))$ is an isomorphism. □

As an immediate consequence of this result we have the following.

COROLLARY 11.3. a) *If* I_1 *and* I_2 *are two injective* Γ*-modules, then* $I_1 \cong I_2$ *if and only if the* Λ^{op}*-modules* $\operatorname{Hom}_\Gamma(X, I_1)$ *and* $\operatorname{Hom}_\Gamma(X, I_2)$ *are isomorphic* Λ^{op}*-modules.*

b) *If* I *is an injective* Γ*-module, then the natural morphism* $\operatorname{End} I \to \operatorname{End} \operatorname{Hom}_\Gamma(X, I)$ *is an isomorphism of rings.*

c) *Let* S *be a simple* Γ*-module and* I *an injective envelope of* S*. Then*

i) $f\colon I \to I$ *is not an isomorphism if and only if* $f(S) = 0$*, so*

ii) $\operatorname{End} I \cong \operatorname{End} \operatorname{Hom}_\Gamma(X, I)$ *are local rings.*

CHAPTER II: SIMPLE FUNCTORS

This chapter is devoted to describing the simple functors in $(\mathcal{C}^{\text{op}}, \text{Ab})$ and $(\mathcal{C}, \text{Ab})$ for various additive categories $\mathcal{C}$. Particular attention is given to when these simple functors are finitely presented in the case $\mathcal{C}$ is one of the categories of modules considered in Chapter I. Most of the results given in this section amount to little more than interpretations of the results on morphisms determined by modules developed in the previous chapter.

1. Simple functors

Let $\mathcal{C}$ be a preadditive category. We recall that a functor F in $(\mathcal{C}^{\text{op}}, \text{Ab})$ is said to be a *simple functor* if a) $F \neq 0$ and b) (0) and F are the only subfunctors of F. The simple functors in $(\mathcal{C}^{\text{op}}, \text{Ab})$ constitute the main concern of this chapter. This section is devoted to applying some of the results of Sections 1 and 2 of Chapter I to a preliminary study of the simple functors in $(\mathcal{C}^{\text{op}}, \text{Ab})$. A somewhat different approach to this subject can be found in [7].

As our first description of the simple functors in $(\mathcal{C}^{\text{op}}, \text{Ab})$ we have the following.

PROPOSITION 1.1. *A nonzero functor* S *in* $(\mathcal{C}^{\text{op}}, \text{Ab})$ *is simple if and only if for each* X *in* $\mathcal{C}$ *such that* $S(X) \neq 0$ *we have*

a) $S(X)$ *is a simple* $(\operatorname{End} X)^{\text{op}}$*-module.*

b) *The subfunctor* (0) *of* S *is determined by* X*.*

PROOF. Suppose S is a simple functor and suppose $S(X) \neq 0$. Let x be a nonzero element of $S(X)$. Then there is a morphism $\alpha\colon (\ , X) \to S$ such that $\alpha_X(1_X) = x$ (Yoneda's lemma). Since α is not zero, the fact that S is simple implies that α is an epimorphism. In particular, the $(\operatorname{End} X)^{\text{op}}$-morphism $\alpha_X\colon (X, X) \to S(X)$ is surjective and so x in $S(X)$ generates the $(\operatorname{End} X)^{\text{op}}$-module $S(X)$. Since this is true for each nonzero element x in $S(X)$, we have that $S(X)$ is a simple $(\operatorname{End} X)^{\text{op}}$-module.

Next let T be a subfunctor of S with the property that $T(X) = (0)$. Because $S(X) \neq (0)$, $T \neq S$ and so $T = (0)$ since S is simple. Thus if S is simple and $S(X) \neq 0$, then the zero subfunctor of S is determined by X.

Now suppose that S is a nonzero functor in $(\mathcal{C}^{\text{op}}, \text{Ab})$ such that if $S(X) \neq 0$, then a) and b) of the proposition hold. We want to show that S is simple. Suppose T is a proper subfunctor of S. Then there is an X in $\mathcal{C}$ such that $T(X) \neq S(X)$. Hence $T(X) = 0$, since $T(X)$ is a proper submodule of the simple $(\operatorname{End} X)^{\text{op}}$-module $S(X)$. Therefore $T = 0$, since by b) the zero subfunctor of S is determined by X. Thus S is simple which finishes the proof of the proposition. □

Recalling that a subfunctor F of G is said to be a *maximal subfunctor* of G if G/F is simple, we have the following easily verified consequence of Proposition 2.1 (Hint: Use I, Proposition 1.3).

COROLLARY 1.2. *Let F be a proper subfunctor of G in $(\mathcal{C}^{\mathrm{op}}, \mathrm{Ab})$. Then the following are equivalent where $S = G/F$.*

a) *F is a maximal subfunctor of G.*

b) *If $S(X) \neq 0$, then $F(X)$ is a maximal $(\mathrm{End}\, X)^{\mathrm{op}}$-submodule of $G(X)$ and F is determined by X, or equivalently, $F = G_{F(X)}$.*

We now apply these results to obtain a description of the maximal subfunctors of the representable functors $(\ , C)$ for each C in $\mathcal{C}$. To simplify terminology we observe that the $(\mathrm{End}\, C)^{\mathrm{op}}$-submodules of $(\ , C)(C)$ are the right ideals of $\mathrm{End}\, C$.

PROPOSITION 1.3. *Let C be in $\mathcal{C}$ and F a subfunctor of $(\ , C)$ in $(\mathcal{C}^{\mathrm{op}}, \mathrm{Ab})$.*

a) *$F = (\ , C)$ if and only if $1 \in F(C)$.*

b) *F is a maximal subfunctor of $(\ , C)$ if and only if*

i) *$F(C)$ is a maximal right ideal of $\mathrm{End}\, C$ and*

ii) *F is determined by C, or equivalently, $F = (\ , C)_{F(C)}$.*

PROOF. a) Obviously $1 \in F(C)$ if $F = (\ , C)$. Suppose now that 1 is in $F(C)$. Let $(\ , C) \to F$ be the morphism such that 1 in (C, C) is carried to 1 in $F(C)$. Then the composition $(\ , C) \to F \to (\ , C)$ is the identity where $F \to (\ , C)$ is the inclusion. Thus $F = (\ , C)$.

b) That F being a maximal subfunctor of $(\ , C)$ implies i) and ii), follows from a) and Corollary 1.2. Suppose F is a subfunctor of $G = (\ , C)$ satisfying i) and ii). Since $F(C)$ is a proper ideal of (C, C) by i) we know by part a) that F is a proper subfunctor of G. Suppose $F \subset F' \subset G$ and $F' \neq G$. Then $F(C) \subset F'(C)$ and so $F(C) = F'(C)$ since by a) the fact that $F' \neq G$ implies $F'(C) \neq G(C)$ and $F(C)$ is a maximal right ideal of $\mathrm{End}\, C$. Thus the assumption that F is a subfunctor of $(\ , C)$ determined by C implies $F' \subset F$, or equivalently $F' = F$. This shows that F is a maximal subfunctor of G which finishes the proof of b) and the proof of the proposition. □

As immediate consequences of this result we have the following.

COROLLARY 1.4. *Let C be in $\mathcal{C}$. The map which associates with each right ideal H of $\mathrm{End}\, C$ the subfunctor $(\ , C)_H$ of $(\ , C)$ induces a bijection between the maximal right ideals of $\mathrm{End}\, C$ and the maximal subfunctors of $(\ , C)$.*

COROLLARY 1.5. *Let F be a proper subfunctor of $(\ , C)$ for some C in $\mathcal{C}$. Then F is contained in a maximal subfunctor of $(\ , C)$.*

PROOF. We know by Proposition 1.3 that $F(C)$ is a proper right ideal of $\mathrm{End}\, C$ since $F \neq (\ , C)$. Let H be a maximal right ideal of $\mathrm{End}\, C$ containing $F(C)$. Then $(\ , C)_H$ is a maximal subfunctor of G such that $(\ , C)_H(C) = H$ (see Proposition 1.3). Since $(\ , C)_H$ is a subfunctor of $(\ , C)$ determined by C and $H \supset F(C)$, it follows that $(\ , C)_H \supset F$, giving us our desired result. □

We now apply these results to obtain more information about the simple functors in $(\mathcal{C}, \mathrm{Ab})$.

PROPOSITION 1.6. *Let C be in $\mathcal{C}$.*

a) *For each maximal right ideal H of $\mathrm{End}\, C$, the functor $S = (\ , C)/(\ , C)_H$ is a simple functor with the property $S(C) \neq (0)$.*

b) *If S is a simple functor such that $S(C) \neq 0$, then*

i) *$S(C)$ is a simple $(\mathrm{End}\, C)^{\mathrm{op}}$-module.*

ii) *If x is a nonzero element of $S(C)$ and $\operatorname{ann}(x)$ is the annihilator of x in $(\operatorname{End} C)^{\mathrm{op}}$, then $\operatorname{ann}(x)$ is a maximal right ideal of $\operatorname{End} C$ and $S \cong (\ ,C)/(\ ,C)_{\operatorname{ann}(x)}$.*

c) *If S_1 and S_2 are simple functors not vanishing on C, then $S_1 \cong S_2$ if and only if $S_1(C)$ and $S_2(C)$ are isomorphic simple $(\operatorname{End} C)^{\mathrm{op}}$-modules.*

d) *The map which associates with each simple functor S not vanishing on C the simple $(\operatorname{End} C)^{\mathrm{op}}$-module $S(C)$, induces a bijection between the isomorphism classes of simple functors not vanishing on C and the isomorphism classes of simple $(\operatorname{End} C)^{\mathrm{op}}$-modules.*

PROOF. a) Follows from Proposition 1.3.

b) Let x be a nonzero element of $S(C)$ and $\alpha\colon (\ ,C) \to S$ the morphism with the property $\alpha_C(1) = x$ in $S(C)$. Since S is a simple functor and α is not zero, we know that $\alpha\colon (\ ,C) \to S$ is an epimorphism. Moreover by Proposition 1.1, $S(C)$ is a simple $(\operatorname{End} C)^{\mathrm{op}}$-module and so $\operatorname{ann}(x)$ is a maximal right ideal in $\operatorname{End} C$. Since $\operatorname{ann}(x)$ is the kernel of $\alpha_C\colon (C,C) \to S(C)$ it follows that $(\operatorname{Ker}\alpha)(C) = \operatorname{ann}(x)$. Hence the maximal subfunctor $\operatorname{Ker}\alpha$ of $(\ ,C)$ is $(\ ,C)_{\operatorname{ann}(x)}$ by Proposition 1.3. Thus $S \cong (\ ,C)/(\ ,C)_{\operatorname{ann}(x)}$.

c) Suppose S_1 and S_2 are simple functors not vanishing on C and that $f\colon S_1(C) \to S_2(C)$ is an isomorphism of the simple $(\operatorname{End} C)^{\mathrm{op}}$-modules. Let x be a nonzero element in $S_1(C)$. Then $f(x)$ is a nonzero element of $S_2(C)$. Since $\operatorname{ann}(x) = \operatorname{ann} f(x)$, it follows from b) that $S_1 \cong (\ ,C)/(\ ,C)_{\operatorname{ann}(x)} \cong S_2$, and so $S_1 \cong S_2$.

d) Follows trivially from a) through c). □

As an immediate consequence of Proposition 1.6 we have the following which completes our description of the simple contravariant functors from $\mathcal{C}$ to Ab.

COROLLARY 1.7. *Let S be a simple functor in $(\mathcal{C}^{\mathrm{op}}, \mathrm{Ab})$. Then $S(C) \neq (0)$ for some C in $\mathcal{C}$ and so $S \cong (\ ,C)/(\ ,C)_H$ where H is a maximal right ideal of $\operatorname{End} C$.*

We end this discussion of simple functors in $(\mathcal{C}^{\mathrm{op}}, \mathrm{Ab})$ by recalling the definition of the radical of a functor [12]. In analogy with the situation for rings, the *radical of a functor* G in $(\mathcal{C}^{\mathrm{op}}, \mathrm{Ab})$, which we denote by $\underline{r}G$, is defined to be the intersection of the maximal subfunctors of G. Obviously $\underline{r}G = G$ if and only if G has no maximal subfunctors. The following general observations are useful in describing the radical of a functor.

Let $\{G_i\}_{i\in I}$ be a family of subfunctors of G in $(\mathcal{C}^{\mathrm{op}}, \mathrm{Ab})$. For each C in $\mathcal{C}$ define A_C to be the subgroup $\cap_{i\in I} G_i(C)$ of $G(C)$. It is easily checked that for each morphism $f\colon X \to Y$ in $\mathcal{C}$ we have $G(f)(A_Y) \subset A_X$. Thus by I, Lemma 1.1, there is a unique subfunctor F of G such that $F(C) = A_C$ for all C in $\mathcal{C}$. This subfunctor F of G is called the intersection of the G_i and is denoted by $\cap_{i\in I} G_i$.

PROPOSITION 1.8. *Suppose G is in $(\mathcal{C}^{\mathrm{op}}, \mathrm{Ab})$ and X is in $\mathcal{C}$. Let $\{H_i\}_{i\in I}$ be a family of $(\operatorname{End} X)^{\mathrm{op}}$-submodules of $G(X)$. Then $\cap_{i\in I} G_{H_i} = G_{\cap_{i\in I} H_i}$.*

PROOF. Since $(\cap_{i\in I} G_{H_i})(X) = \cap_{i\in I} H_i = G_{\cap H_i}(X)$, the fact that $G_{\cap H_i}$ is determined by X shows that $\cap_{i\in I} G_{H_i} \subset G_{\cap H_i}$. On the other hand, since for each i in I we have $G_{H_i}(X) = H_i \supset G_{\cap_{i\in I} H_i}(X)$, the fact that each G_{H_i} is determined by X implies that $G_{H_i} \supset G_{\cap_{i\in I} H_i}$ for each i in I. Thus $\cap_{i\in I} G_{H_i} \supset G_{\cap_{i\in I} H_i}$, which finishes the proof that $\cap_{i\in I} G_{H_i} = G_{\cap_{i\in I} H_i}$. □

Returning to our discussion of the radical of a functor $(\ ,C)$ we have as an easy consequence of Proposition 1.8 and Proposition 1.3 the following description of the radical of a representable functor $(\ ,C)$ in $(\mathcal{C}^{\mathrm{op}}, \mathrm{Ab})$.

PROPOSITION 1.9. *Let C be in $\mathcal{C}$. Then*

a) $\underline{r}(\ ,C) = (\ ,C)_{\operatorname{rad} \operatorname{End} C}$.

b) *The following are equivalent.*

i) $\operatorname{End} C$ *is local.*

ii) $(\ ,C)$ *has a unique maximal subfunctor.*

iii) $\underline{r}(\ ,C)$ *is a maximal subfunctor.*

PROOF. a) Let $\{\underline{m}_i\}_{i\in I}$ be the family of maximal right ideals of $\operatorname{End} C$. Then, by Proposition 1.3, $\{(\ ,C)_{\underline{m}_i}\}_{i\in I}$ is the family of maximal subfunctors of $(\ ,C)$. Hence $\underline{r}(\ ,C) = \cap_{i\in I}(\ ,C)_{\underline{m}_i}$. By Proposition 1.8 we know that $\cap_{i\in I}(\ ,C)_{\underline{m}_i} = (\ ,C)_{\cap \underline{m}_i}$. Since $\operatorname{rad} \operatorname{End} C = \cap \underline{m}_i$ we have that $\underline{r}(\ ,C) = (\ ,C)_{\operatorname{rad} \operatorname{End} C}$.

b) Follows easily from a). □

We end this preliminary discussion of simple functors by pointing out when they have projective covers.

PROPOSITION 1.10. *Let S be a simple functor in $(\mathcal{C}^{\mathrm{op}}, \mathrm{Ab})$ and $\alpha\colon (\ ,C) \to S$ a nonzero morphism. Then the following are equivalent.*

a) $\alpha\colon (\ ,C) \to S$ *is a projective cover.*

b) $\operatorname{End} C$ *is local.*

c) $\operatorname{Ker} \alpha = \underline{r}(\ ,C)$.

PROOF. a) implies b). To show that $\operatorname{End} C$ is local, it suffices to show that each f in $\operatorname{End} C$ which is not in the maximal right ideal $(\operatorname{Ker}\alpha)(C)$ has a right inverse (see [14, VIII, Proposition 2.1]). Suppose f is not in $(\operatorname{Ker}\alpha)(C)$. Then the composition $\alpha(\ ,f)\colon (\ ,C) \to S$ is not zero since $(\alpha(\ ,f))_C\colon (C,C) \to S(C)$ is not zero. From this it follows that there is a morphism $h\colon C \to C$ such that the diagram

$$\begin{array}{ccc} (\ ,C) & \xrightarrow{\alpha} & S \\ {\scriptstyle (\ ,h)}\downarrow & & \| \\ (\ ,C) & \xrightarrow{\alpha(\ ,f)} & S \end{array}$$

commutes. Therefore $\alpha = \alpha(\ ,f)(\ ,h) = \alpha(\ ,fh)$ and so $fh\colon C \to C$ is an isomorphism since $\alpha\colon (\ ,C) \to S$ is a projective cover. Hence h is a right inverse for f. This completes the proof that a) implies b).

b) implies a). Suppose $\operatorname{End} C$ is a local ring. Let $f\colon C \to C$ be an endomorphism such that $\alpha(\ ,f) = \alpha$. In particular $\alpha(\ ,f)_C\colon \operatorname{End} C \to S(C)$ is not zero so that f is not in the maximal ideal of the local ring $\operatorname{End} C$. Thus f is an isomorphism. This shows by I, Proposition 2.8 that $\alpha\colon (\ ,C) \to S$ is a projective cover.

b) equivalent to c). Special case of Proposition 1.8. □

As an immediate consequence of this result and the uniqueness properties of projective covers we have the following.

COROLLARY 1.11. *The map which associates with each C in $\mathcal{C}$ having the property $\operatorname{End} C$ is local the simple functor $(\ ,C)/\underline{r}(\ ,C)$, induces a bijection between the isomorphism classes of objects in $\mathcal{C}$ with local endomorphism rings and the isomorphism classes of simple functors in $(\mathcal{C}^{\mathrm{op}}, \mathrm{Ab})$ having projective covers.*

As usual, definitions and results analogous to those given for simple functors in $(\mathcal{C}^{\mathrm{op}}, \mathrm{Ab})$ can be given for simple functors in $(\mathcal{C}, \mathrm{Ab})$. Free use of these analogies will be made throughout the rest of the paper.

2. Presentations of simple functors

Throughout this section we assume that $\mathcal{C}$ is an additive category. We have already shown in the previous section that all simple functors in $(\mathcal{C}^{\mathrm{op}}, \mathrm{Ab})$ are finitely generated, i.e. if S is a simple functor, then there is an epimorphism $(\,, C) \to S \to 0$ for some C in $\mathcal{C}$. This section is devoted to studying what it means for a simple functor S in $(\mathcal{C}^{\mathrm{op}}, \mathrm{Ab})$ to be finitely presented, i.e. there is an exact sequence $(\,, B) \to (\,, C) \to S \to 0$ with B and C in $\mathcal{C}$.

Since a morphism $f\colon B \to C$ has the property $\mathrm{Coker}((\,, B) \xrightarrow{(\,, f)} (\,, C))$ is simple if and only if $\mathrm{Im}(\,, f)$ is a maximal subfunctor of $(\,, C)$, it is important to have criteria for when a morphism $f\colon B \to C$ has the property $\mathrm{Im}(\,, f)$ is a maximal subfunctor of $(\,, C)$. The following is a particularly useful result along these lines.

PROPOSITION 2.1. *Let $f\colon B \to C$ be a morphism in $\mathcal{C}$. Then the following statements are equivalent.*

a) $\mathrm{Im}(\,, f)$ *is a maximal subfunctor of* $(\,, C)$.

b) $f\colon B \to C$ *is right C-determined with* $(\mathrm{Im}(\,, f))(C)$ *a maximal right ideal of* $\mathrm{End}\, C$.

c) *One of the following holds for a morphism $g\colon L \to C$.*

i) *There is an $h\colon L \to B$ such that $fh = g$.*

ii) *The induced morphism $B \amalg L \to C$ is a splittable epimorphism.*

PROOF. a) implies b). We know from Proposition 1.3 that if $\mathrm{Im}(\,, f)$ is a maximal subfunctor of $(\,, C)$, then $\mathrm{Im}(\,, f)$ is determined by C and $(\mathrm{Im}(\,, f))(C)$ is a maximal right ideal of $\mathrm{End}\, C$. Thus a) implies b).

b) implies c). Let $g\colon L \to C$ be a morphism. If $(\mathrm{Im}(\,, g))(C) \subset (\mathrm{Im}(\,, f))(C)$, then there is a morphism $h\colon L \to B$ such that $fh = g$ since $f\colon B \to C$ is determined by C. If $(\mathrm{Im}(\,, g))(C)$ is not contained in $(\mathrm{Im}(\,, f))(C)$, then the right ideals $\mathrm{Im}(\,, g)(C)$ and $(\mathrm{Im}(\,, f))(C)$ generate all of $\mathrm{End}\, C$ since $(\mathrm{Im}(\,, f))(C)$ is a maximal right ideal of $\mathrm{End}\, C$. From this it follows that there are morphisms $u\colon C \to B$ and $v\colon C \to L$ such that $fu + gv = 1$ in $\mathrm{End}\, C$, or equivalently, the induced morphism $B \amalg L \to C$ is a splittable epimorphism.

c) implies a). It is easily verified that $\mathrm{Im}(\,, f)$ is a maximal subfunctor of $(\,, C)$ if given a morphism $g\colon L \to C$, then $\mathrm{Im}(\,, g)$ and $\mathrm{Im}(\,, f)$ together generate $(\,, C)$ whenever $\mathrm{Im}(\,, g)$ is not contained in $\mathrm{Im}(\,, f)$. The fact that c) implies a) is a trivial consequence of this observation. □

For the convenience of the reader we state the covariant version of the above proposition.

PROPOSITION 2.2. *Let $g\colon A \to B$ be a morphism in $\mathcal{C}$. Then the following are equivalent.*

a) $\mathrm{Im}(g, \,)$ *is a maximal subfunctor of* $(A, \,)$.

b) g *is left A-determined and* $\mathrm{Im}(g, \,)(A)$ *is a maximal left ideal of* $\mathrm{End}\, A$.

c) *One of the following holds for a morphism $h\colon A \to L$:*

i) *There is an $f\colon B \to L$ such that $fg = h$.*

ii) *The induced morphism $A \to B \amalg L$ is a splittable monomorphism.*

We now point out that right and left almost split morphisms are special cases of the types of morphisms described in the above propositions. For the convenience of the reader we begin by recalling the definitions of right and left almost split morphisms.

A morphism $f\colon B \to C$ is said to be *right almost split* if a) f is not a splittable epimorphism and b) given any morphism $h\colon L \to C$ which is not a splittable epimorphism then there is a morphism $g\colon L \to B$ such that $fg = h$. A morphism $g\colon A \to B$ is said to be *left almost split* if a) g is not a splittable monomorphism and b) given any morphism $h\colon A \to L$ which is not a splittable monomorphism, then there is an $f\colon B \to L$ such that $fg = h$.

PROPOSITION 2.3. *The following are equivalent for a morphism $f\colon B \to C$ in $\mathcal{C}$.*
a) $\operatorname{Im}(\ ,f)$ *is a maximal subfunctor of* $(\ ,C)$ *and* $\operatorname{End} C$ *is local.*
b) f *is right almost split.*

PROOF. a) implies b). Since there is a one-to-one correspondence between the maximal subfunctors of $(\ ,C)$ and the maximal right ideals of $\operatorname{End} C$ (see Corollary 1.4), the fact that $\operatorname{End} C$ is local (or equivalently, $\operatorname{End} C$ has a unique maximal right ideal) implies that $\operatorname{Im}(\ ,f)$ is the unique maximal subfunctor of $(\ ,C)$. In particular, f is not a splittable epimorphism.

Suppose now that $g\colon L \to C$ is not a splittable epimorphism, then $\operatorname{Im}(\ ,g) \neq (0)$. Hence $\operatorname{Im}(\ ,g) \subset \operatorname{Im}(\ ,f)$, the unique maximal subfunctor of $(\ ,C)$, since every proper subfunctor of a representable functor is contained in some maximal subfunctor (see Corollary 1.5). The fact that $(\ ,L)$ is a projective functor implies there exists some morphism $h\colon L \to B$ such that $(\ ,f)(\ ,h) = (\ ,g)$ or equivalently, $fh = g$. This shows that $f\colon B \to C$ is right almost split and finishes the proof of a) implies b).

b) implies a). We know by [7, Corollary 2.6] and [11, Lemma 2.3] that if $f\colon B \to C$ is right almost split, then $\operatorname{End} C$ is local and $\operatorname{Coker}(\ ,f)$ is simple, which shows that b) implies a). □

For the convenience of the reader we state the analogue of Proposition 2.3 for covariant functors.

PROPOSITION 2.4. *The following are equivalent for a morphism $g\colon A \to B$ in $\mathcal{C}$.*
a) $\operatorname{Im}(g,\ )$ *is a maximal subfunctor of* $(A,\ )$ *and* $\operatorname{End} A$ *is local.*
b) g *is left almost split.*

We now turn our attention to when simple functors are finitely presented. Recall that a functor F in $(\mathcal{C}^{\mathrm{op}}, \mathrm{Ab})$ is said to be *finitely presented* if there is an exact sequence $(\ ,B) \to (\ ,C) \to F \to 0$ with B and C in $\mathcal{C}$. Just as in the case of modules over a ring, it is not difficult to show the following.

LEMMA 2.5. *Let $(\ ,C) \to F \to 0$ be exact in $(\mathcal{C}^{\mathrm{op}}, \mathrm{Ab})$. Then F is finitely presented if and only if there is a morphism $B \to C$ such that the induced sequence $(\ ,B) \to (\ ,C) \to F \to 0$ is exact.*

Applying our remarks to simple functors in $(\mathcal{C}^{\mathrm{op}}, \mathrm{Ab})$ we obtain

PROPOSITION 2.6. *Let $\alpha\colon (\ ,C) \to S \to 0$ be exact with S a simple functor in $(\mathcal{C}^{\mathrm{op}}, \mathrm{Ab})$. Then S is finitely presented if and only if there is a right C-determined morphism $f\colon B \to C$ such that* $\operatorname{Im}(\ ,f)(C)$ *is the maximal right ideal* $(\operatorname{Ker}\alpha)(C)$ *of* $\operatorname{End} C$. *If such an $f\colon B \to C$ exists, then the exact sequence $(\ ,B) \xrightarrow{(\ ,f)} (\ ,C) \xrightarrow{\alpha} S \to 0$ is a projective presentation of S.*

Thus we see that the problem of when simple functors are finitely presented is intimately related to the existence of morphisms determined by objects in $\mathcal{C}$. We further emphasize this point in the proposition which follows the following definitions.

A morphism $f\colon B \to C$ in $\mathcal{C}$ is said to be *minimal right almost split* if it is right minimal and right almost split. The morphism $f\colon B \to C$ is said to be *minimal left almost split* if it is left minimal and left almost split.

PROPOSITION 2.7. *Let S be a simple functor in $(\mathcal{C}^{\text{op}}, \text{Ab})$ and suppose $f\colon B \to C$ is a morphism in $\mathcal{C}$ such that $(\ ,B) \xrightarrow{(\ ,f)} (\ ,C) \to S \to 0$ is exact. Then the following are equivalent.*

a) $(\ ,B) \xrightarrow{(\ ,f)} (\ ,C) \to S \to 0$ *is a minimal projective presentation of S.*

b) *f is a minimal right almost split morphism.*

As in previous sections analogues exist in $(\mathcal{C}, \text{Ab})$ for all results and definitions given here in $(\mathcal{C}^{\text{op}}, \text{Ab})$, even when not given explicitly in the text. These analogues will be used freely in the rest of the paper.

3. A duality theorem

Throughout this section we assume that we are given a fixed abelian category $\mathcal{A}$. Our first concern is to develop a duality theory between certain covariant and contravariant functors from $\mathcal{A}$ to abelian groups. This duality theory will then be used in the next section to obtain more information about morphisms which are determined by objects. These results in turn will be used in later sections in connection with establishing certain existence theorems. We begin with a review of some of the results in [2], where a more detailed discussion of this material is available.

It is well known that unless $\mathcal{A}$ is skeletally small (i.e. the isomorphism classes of objects in $\mathcal{A}$ form a set) the collections of functors $(\mathcal{A}^{\text{op}}, \text{Ab})$ and $(\mathcal{A}, \text{Ab})$ are not necessarily a category since the collections of morphisms between functors need not be a set. However $\text{f.p.}(\mathcal{A}^{\text{op}}, \text{Ab})$, the collection of finitely presented functors, do form a category since the morphisms from a finitely generated functor F to an arbitrary functor G do form a set. For if $(\ ,C) \to F \to 0$ is an epimorphism, then (F,G) is isomorphic to a subgroup of $((\ ,C),G) = G(C)$. Thus it makes sense to talk about the categories $\text{f.p.}(\mathcal{A}^{\text{op}}, \text{Ab})$ and $\text{f.p.}(\mathcal{A}, \text{Ab})$.

Suppose F is in $\text{f.p.}(\mathcal{A}^{\text{op}}, \text{Ab})$. Then there is an exact sequence $(\ ,B) \xrightarrow{(\ ,f)} (\ ,C) \to F \to 0$. If we let $A \xrightarrow{g} B$ be a kernel for f, then $0 \to (\ ,A) \to (\ ,B) \to (\ ,C) \to F \to 0$ is exact. From this fact one can deduce the following (see [2, Proposition 2.1]).

PROPOSITION 3.1. a) *Let $0 \to F_1 \to F_2 \to F_3 \to F_4 \to 0$ be an exact sequence in $(\mathcal{A}^{\text{op}}, \text{Ab})$. If F_2 and F_3 are in $\text{f.p.}(\mathcal{A}^{\text{op}}, \text{Ab})$, then F_1 and F_4 are in $\text{f.p.}(\mathcal{A}^{\text{op}}, \text{Ab})$.*

b) *If $0 \to F_1 \to F_2 \to F_3 \to 0$ is an exact sequence in $(\mathcal{A}^{\text{op}}, \text{Ab})$ and F_1 and F_3 are in $\text{f.p.}(\mathcal{A}^{\text{op}}, \text{Ab})$, then F_2 is in $\text{f.p.}(\mathcal{A}^{\text{op}}, \text{Ab})$.*

c) *Hence $\text{f.p.}(\mathcal{A}^{\text{op}}, \text{Ab})$ is an abelian category with enough projectives of global dimension at most 2.*

d) *By duality, $\text{f.p.}(\mathcal{A}\ \text{Ab})$ has the same properties a), b), c).*

Let $\text{f.p.}(\mathcal{A}^{\text{op}}, \text{Ab})_0$ be the full subcategory consisting of those F in $\text{f.p.}(\mathcal{A}^{\text{op}}, \text{Ab})$ having the property that there is an exact sequence $0 \to (\ ,A) \xrightarrow{(\ ,g)} (\ ,B) \xrightarrow{(\ ,f)} (\ ,C) \to F \to 0$ with $0 \to A \xrightarrow{g} B \xrightarrow{f} C \to 0$ exact. Similarly, let $\text{f.p.}(\mathcal{A}, \text{Ab})_0$ be the full subcategory of $\text{f.p.}(\mathcal{A}, \text{Ab})$ consisting of those G in $\text{f.p.}(\mathcal{A}^{\text{op}}, \text{Ab})$ having the property that there is an exact sequence $0 \to (C,\ ) \xrightarrow{(f,\ )} (B,\ ) \xrightarrow{(g,\ )} (A,\ ) \to G \to 0$ with $0 \to A \xrightarrow{g} B \xrightarrow{f} C \to 0$ exact. The following description of the objects in $\text{f.p.}(\mathcal{A}^{\text{op}}, \text{Ab})_0$ is quite useful (see [2, Proposition 3.2] for a proof).

PROPOSITION 3.2. *The following statements are equivalent for an F in $\text{f.p.}(\mathcal{A}^{\text{op}}, \text{Ab})$.*

a) *F is in $\text{f.p.}(\mathcal{A}^{\text{op}}, \text{Ab})_0$.*

b) *If* $0 \to A \to B \to C \to 0$ *is exact in* $\mathcal{A}$ *such that* $0 \to (\ ,A) \to (\ ,B) \to (\ ,C) \to F \to 0$ *is exact, then* $0 \to A \to B \to C \to 0$ *is exact.*

c) *Let* G *in* $(\mathcal{A}^{\mathrm{op}}, \mathrm{Ab})$ *be left exact, i.e. if* $X \to Y \to Z \to 0$ *is exact in* $\mathcal{A}$*, then* $0 \to G(Z) \to G(Y) \to G(X)$ *is exact. Then* $\mathrm{Ext}^i(F,G) = 0$ *for* $i = 0,1$.

d) $(F,G) = 0$ *for any* G *in* $(\mathcal{A}^{\mathrm{op}}, \mathrm{Ab})$ *which is left exact.*

e) $(F,(\ ,A)) = 0$ *for all* A *in* $\mathcal{A}$.

An analogous description of the objects in $\mathrm{f.p.}(\mathcal{A}, \mathrm{Ab})_0$ can be obtained by duality. As an easy consequence of Proposition 3.2 we have the following.

COROLLARY 3.3. *Let* $0 \to F_1 \to F_2 \to F_3 \to 0$ *be an exact sequence in* $\mathrm{f.p.}(\mathcal{A}^{\mathrm{op}}, \mathrm{Ab})$.

a) F_2 *is in* $\mathrm{f.p.}(\mathcal{A}^{\mathrm{op}}, \mathrm{Ab})_0$ *if and only if* F_1 *and* F_3 *are in* $\mathrm{f.p.}(\mathcal{A}^{\mathrm{op}}, \mathrm{Ab})_0$.

b) $\mathrm{f.p.}(\mathcal{A}^{\mathrm{op}}, \mathrm{Ab})_0$ *is an abelian category with the property that the inclusion functor* $\mathrm{f.p.}(\mathcal{A}^{\mathrm{op}}, \mathrm{Ab})_0 \to \mathrm{f.p.}(\mathcal{A}^{\mathrm{op}}, \mathrm{Ab})$ *is exact.*

Again a result analogous to this corollary holds for $\mathrm{f.p.}(\mathcal{A}, \mathrm{Ab})_0$. Our aim now is to describe a duality between $\mathrm{f.p.}(\mathcal{A}^{\mathrm{op}}, \mathrm{Ab})_0$ and $\mathrm{f.p.}(\mathcal{A}, \mathrm{Ab})_0$. To this end we introduce the contravariant functor $V\colon \mathrm{f.p.}(\mathcal{A}^{\mathrm{op}}, \mathrm{Ab}) \to \mathrm{f.p.}(\mathcal{A}, \mathrm{Ab})$ and $W\colon \mathrm{f.p.}(\mathcal{A}, \mathrm{Ab}) \to \mathrm{f.p.}(\mathcal{A}^{\mathrm{op}}, \mathrm{Ab})$ as follows.

For each F in $\mathrm{f.p.}(\mathcal{A}^{\mathrm{op}}, \mathrm{Ab})$ define $V(F)$ to be the covariant functor $\mathcal{A} \to \mathrm{Ab}$ given by $V(F)(X) = \mathrm{Ext}^2(F,(\ ,X))$ for all X in $\mathcal{A}$. The functor $V(F)$ can be described as follows. Let $0 \to A \xrightarrow{g} B \xrightarrow{f} C$ be an exact sequence in $\mathcal{A}$ such that $0 \to (\ ,A) \xrightarrow{(\ ,g)} (\ ,B) \xrightarrow{(\ ,f)} (\ ,C) \to F \to 0$ is exact. Since this exact sequence is a projective resolution of F, we have that $\mathrm{Ext}^2(F,(\ ,X)) = \mathrm{Coker}(((\ ,B),(\ ,X)) \to ((\ ,A),(\ ,X))) = \mathrm{Coker}((B,X) \to (A,X))$ for all X in $\mathcal{A}$. Thus $(B,\ ) \to (A,\ ) \to V(F) \to 0$ is exact. Hence $V(F)$ is in $\mathrm{f.p.}(\mathcal{A}, \mathrm{Ab})$. Moreover $V(F)$ is in $\mathrm{f.p.}(\mathcal{A}, \mathrm{Ab})_0$ if F is in $\mathrm{f.p.}(\mathcal{A}^{\mathrm{op}}, \mathrm{Ab})_0$. Therefore we obtain the contravariant functor $V\colon \mathrm{f.p.}(\mathcal{A}^{\mathrm{op}}, \mathrm{Ab}) \to \mathrm{f.p.}(\mathcal{A}, \mathrm{Ab})$ given by $F \mapsto V(F)$ for all F in $\mathrm{f.p.}(\mathcal{A}^{\mathrm{op}}, \mathrm{Ab})$. Since $V(F)$ is in $\mathrm{f.p.}(\mathcal{A}, \mathrm{Ab})_0$ whenever F is in $\mathrm{f.p.}(\mathcal{A}^{\mathrm{op}}, \mathrm{Ab})_0$, V induces a contravariant functor $V_0\colon \mathrm{f.p.}(\mathcal{A}^{\mathrm{op}}, \mathrm{Ab})_0 \to \mathrm{f.p.}(\mathcal{A}, \mathrm{Ab})_0$.

We define the contravariant functors $W\colon \mathrm{f.p.}(\mathcal{A}, \mathrm{Ab}) \to \mathrm{f.p.}(\mathcal{A}^{\mathrm{op}}, \mathrm{Ab})$ and $W_0\colon \mathrm{f.p.}(\mathcal{A}, \mathrm{Ab})_0 \to \mathrm{f.p.}(\mathcal{A}^{\mathrm{op}}, \mathrm{Ab})_0$ analogously. Namely, for G in $\mathrm{f.p.}(\mathcal{A}, \mathrm{Ab})$ define the contravariant functor $W(G)\colon \mathcal{A} \to \mathrm{Ab}$ by $W(G)(X) = \mathrm{Ext}^2(G,(X,\ ))$ for all X in $\mathcal{A}$. The functor $W(G)$ can be described as follows. Let $A \xrightarrow{g} B \xrightarrow{f} C \to 0$ be an exact sequence in $\mathcal{A}$ such that $0 \to (C,\ ) \xrightarrow{(f,\ )} (B,\ ) \xrightarrow{(g,\ )} (A,\ ) \to G \to 0$ is exact. Since this exact sequence is a projective resolution of G, we have that $\mathrm{Ext}^2(G,(X,\ )) = \mathrm{Coker}(((B,\ ),(X,\ )) \to ((C,\ ),(X,\ ))) = \mathrm{Coker}((X,B) \to (X,C))$ for all X in $\mathcal{A}$. Thus $(\ ,B) \to (\ ,C) \to W(G) \to 0$ is exact. Hence $W(G)$ is in $\mathrm{f.p.}(\mathcal{A}^{\mathrm{op}}, \mathrm{Ab})$. Moreover, $W(G)$ is in $\mathrm{f.p.}(\mathcal{A}^{\mathrm{op}}, \mathrm{Ab})_0$ if G is in $\mathrm{f.p.}(\mathcal{A}, \mathrm{Ab})_0$. Therefore we obtain the contravariant functor $W\colon \mathrm{f.p.}(\mathcal{A}, \mathrm{Ab}) \to \mathrm{f.p.}(\mathcal{A}^{\mathrm{op}}, \mathrm{Ab})$ given by $G \mapsto W(G)$ for all G in $\mathrm{f.p.}(\mathcal{A}, \mathrm{Ab})$. Since $W(G)$ is in $\mathrm{f.p.}(\mathcal{A}^{\mathrm{op}}, \mathrm{Ab})_0$ whenever G is in $\mathrm{f.p.}(\mathcal{A}, \mathrm{Ab})_0$, W induces a contravariant functor $W_0\colon \mathrm{f.p.}(\mathcal{A}, \mathrm{Ab})_0 \to \mathrm{f.p.}(\mathcal{A}^{\mathrm{op}}, \mathrm{Ab})_0$.

We can now state and prove our desired duality theorem.

THEOREM 3.4. *The contravariant functors* $V_0\colon \mathrm{f.p.}(\mathcal{A}^{\mathrm{op}}, \mathrm{Ab})_0 \to \mathrm{f.p.}(\mathcal{A}, \mathrm{Ab})_0$ *and* $W_0\colon \mathrm{f.p.}(\mathcal{A}, \mathrm{Ab})_0 \to \mathrm{f.p.}(\mathcal{A}^{\mathrm{op}}, \mathrm{Ab})_0$ *have the properties* $W_0V_0\colon \mathrm{f.p.}(\mathcal{A}^{\mathrm{op}}, \mathrm{Ab})_0 \to \mathrm{f.p.}(\mathcal{A}^{\mathrm{op}}, \mathrm{Ab})_0$ *and* $V_0W_0\colon \mathrm{f.p.}(\mathcal{A}, \mathrm{Ab})_0 \to \mathrm{f.p.}(\mathcal{A}, \mathrm{Ab})_0$ *are isomorphic to identity functors and so* V_0 *and* W_0 *are dualities which are inverses of each other.*

PROOF. We will just outline a proof that $V_0\colon \mathrm{f.p.}(\mathcal{A}^{\mathrm{op}}, \mathrm{Ab})_0 \to \mathrm{f.p.}(\mathcal{A}, \mathrm{Ab})_0$ is dense and fully faithful, in other words, that V_0 is a duality. The rest of the theorem is left to the reader.

We first show that V_0 is dense, i.e. given G in $\mathrm{f.p.}(\mathcal{A}, \mathrm{Ab})_0$ there is an F in $\mathrm{f.p.}(\mathcal{A}^{\mathrm{op}}, \mathrm{Ab})_0$ such that $V_0(F) \cong G$. Since G is in $\mathrm{f.p.}(\mathcal{A}, \mathrm{Ab})_0$ there is an exact sequence $0 \to A \to B \to C \to 0$ in $\mathcal{A}$ such that $0 \to (C, \ ) \to (B, \ ) \to (A, \ ) \to G \to 0$ is exact. Let F in $\mathrm{f.p.}(\mathcal{A}^{\mathrm{op}}, \mathrm{Ab})_0$ be such that $0 \to (\ , A) \to (\ , B) \to (\ , C) \to F \to 0$ is exact. Then from the description of $V(F)$ in terms of resolutions it is clear that $V(F) \cong G$.

We now want to show that V_0 is fully faithful. To do this it is useful to describe the action of V_0 on morphisms in terms of resolutions. Suppose F, F' are in $\mathrm{f.p.}(\mathcal{A}^{\mathrm{op}}, \mathrm{Ab})_0$ and $0 \to (\ , A) \xrightarrow{(\ ,g)} (\ , B) \xrightarrow{(\ ,f)} (\ , C) \to F \to 0$ and $0 \to (\ , A') \xrightarrow{(\ ,g')} (\ , B') \xrightarrow{(\ ,f')} (\ , C') \to F' \to 0$ are exact and hence projective resolutions. Let $\alpha\colon F \to F'$ be a morphism. Then there are morphisms $h_1\colon A \to A'$, $h_2\colon B \to B'$, and $h_3\colon C \to C'$ such that the exact diagram

$$\begin{array}{ccccccccccc}
0 & \longrightarrow & (\ , A) & \xrightarrow{(\ ,g)} & (\ , B) & \xrightarrow{(\ ,f)} & (\ , C) & \longrightarrow & F & \longrightarrow & 0 \\
 & & \downarrow{\scriptstyle (\ ,h_1)} & & \downarrow{\scriptstyle (\ ,h_2)} & & \downarrow{\scriptstyle (\ ,h_3)} & & \downarrow{\scriptstyle \alpha} & & \\
0 & \longrightarrow & (\ , A') & \xrightarrow{(\ ,g')} & (\ , B') & \xrightarrow{(\ ,f')} & (\ , C') & \longrightarrow & F' & \longrightarrow & 0
\end{array}$$

commutes. Hence

$$\begin{array}{ccccccccc}
0 & \longrightarrow & A & \xrightarrow{g} & B & \xrightarrow{f} & C & \longrightarrow & 0 \\
 & & \downarrow{\scriptstyle h_1} & & \downarrow{\scriptstyle h_2} & & \downarrow{\scriptstyle h_3} & & \\
0 & \longrightarrow & A' & \xrightarrow{g'} & B' & \xrightarrow{f'} & C' & \longrightarrow & 0
\end{array}$$

commutes, which gives rise to the exact commutative diagram

$$\begin{array}{ccccccccccc}
0 & \longrightarrow & (C', \ ) & \longrightarrow & (B', \ ) & \longrightarrow & (A', \ ) & \longrightarrow & V_0(F') & \longrightarrow & 0 \\
 & & \downarrow{\scriptstyle (h_3, \ )} & & \downarrow{\scriptstyle (h_2, \ )} & & \downarrow{\scriptstyle (h_1, \ )} & & & & \\
0 & \longrightarrow & (C, \ ) & \longrightarrow & (B, \ ) & \longrightarrow & (A, \ ) & \longrightarrow & V_0(F) & \longrightarrow & 0.
\end{array}$$

Hence there is a unique morphism $V_0(F') \to V_0(F)$ which makes this diagram commute. That this morphism is $V_0(\alpha)$ is not difficult to check. Using this description of the functor V_0, it follows from standard properties of projective resolutions that $V_0\colon (F, F') \to (V(F), V(F'))$ is an isomorphism for all F and F' in $\mathrm{f.p.}(\mathcal{A}^{\mathrm{op}}, \mathrm{Ab})_0$. This completes the outline of the proof that V_0 is a duality. □

We now apply this duality to some special situations of particular interest to us.

Throughout the rest of this section we assume that $\mathcal{C}$ is a full subcategory of $\mathcal{A}$ closed under extensions (i.e. if $0 \to A \to B \to C \to 0$ is exact with A and C in $\mathcal{C}$, then B is in $\mathcal{C}$) and summands. Let $\mathrm{f.p.}(\mathcal{C}^{\mathrm{op}}, \mathrm{Ab})$ be the category of finitely presented contravariant functors from $\mathcal{C}$ to Ab. We observe that every contravariant functor F in $\mathrm{f.p.}(\mathcal{C}^{\mathrm{op}}, \mathrm{Ab})$ has a canonical extension to $\mathcal{A}$ as follows. Let $(\ , B) \to (\ , C) \to F \to 0$ be an exact sequence in $\mathrm{f.p.}(\mathcal{C}^{\mathrm{op}}, \mathrm{Ab})$. Then define $\widetilde{F}\colon \mathcal{A} \to \mathrm{Ab}$ by $\widetilde{F}(X) = \mathrm{Coker}((X, B) \to (X, C))$ for all X in $\mathcal{A}$. Then $\widetilde{F}$ is in $\mathrm{f.p.}(\mathcal{A}^{\mathrm{op}}, \mathrm{Ab})$ and is independent (up to canonical isomorphism) of the projective presentation $(\ , B) \to (\ , C) \to F \to 0$ used. It is easily checked that we obtain a fully faithful functor $\mathrm{f.p.}(\mathcal{C}^{\mathrm{op}}, \mathrm{Ab}) \to \mathrm{f.p.}(\mathcal{A}^{\mathrm{op}}, \mathrm{Ab})$ by sending F to $\widetilde{F}$ for all F in $\mathrm{f.p.}(\mathcal{C}^{\mathrm{op}}, \mathrm{Ab})$. Moreover the fully faithful functor $\mathrm{f.p.}(\mathcal{C}^{\mathrm{op}}, \mathrm{Ab}) \to \mathrm{f.p.}(\mathcal{A}^{\mathrm{op}}, \mathrm{Ab})$ is right exact in the sense that if $0 \to F_1 \to F_2 \to F_3 \to 0$ is exact in $\mathrm{f.p.}(\mathcal{C}^{\mathrm{op}}, \mathrm{Ab})$ (i.e. $0 \to F_1(X) \to F_2(X) \to F_3(X) \to 0$ is exact for all X in $\mathcal{C}$), then $\widetilde{F}_1 \to \widetilde{F}_2 \to \widetilde{F}_3 \to 0$ is exact in $\text{f. p.}(\mathcal{A}^{\text{op}}, \text{Ab})$ (see [6, Proposition 3.1]). By duality there is an analogous fully faithful right exact functor $\text{f. p.}(\mathcal{C}, \text{Ab}) \to \text{f. p.}(\mathcal{A}, \text{Ab})$. Thus we can consider $\text{f. p.}(\mathcal{C}^{\text{op}}, \text{Ab})$ and $\text{f. p.}(\mathcal{C}, \text{Ab})$ full subcategories of $\text{f. p.}(\mathcal{A}^{\text{op}}, \text{Ab})$ and $\text{f. p.}(\mathcal{A}, \text{Ab})$ respectively.

Now define $\text{f. p.}(\mathcal{C}^{\text{op}}, \text{Ab})_0$ to be the full subcategory of $\text{f. p.}(\mathcal{A}, \text{Ab})_0$ consisting of all F in $\text{f. p.}(\mathcal{A}^{\text{op}}, \text{Ab})_0$ which have a projective resolution $0 \to (\ ,A) \to (\ ,B) \to (\ ,C) \to F \to 0$ with A, B, C in $\mathcal{C}$. Define the subcategory $\text{f. p.}(\mathcal{C}, \text{Ab})_0$ of $\text{f. p.}(\mathcal{A}, \text{Ab})_0$ similarly. It is then obvious from the definition of the dualities $V_0\colon \text{f. p.}(\mathcal{A}^{\text{op}}, \text{Ab})_0 \to \text{f. p.}(\mathcal{A}, \text{Ab})_0$ and $W_0\colon \text{f. p.}(\mathcal{A}, \text{Ab})_0 \to \text{f. p.}(\mathcal{A}^{\text{op}}, \text{Ab})_0$, that $V_0(F)$ is in $\text{f. p.}(\mathcal{C}^{\text{op}}, \text{Ab})_0$ whenever F is in $\text{f. p.}(\mathcal{C}^{\text{op}}, \text{Ab})_0$ and $W_0(G)$ is in $\text{f. p.}(\mathcal{C}^{\text{op}}, \text{Ab})_0$ whenever G is in $\text{f. p.}(\mathcal{C}, \text{Ab})_0$. As an easily verified consequence of these observations we have the following.

PROPOSITION 3.5. *The inverse dualities* $V_0\colon \text{f. p.}(\mathcal{A}^{\text{op}}, \text{Ab})_0 \to \text{f. p.}(\mathcal{A}, \text{Ab})_0$ *and* $W_0\colon \text{f. p.}(\mathcal{A}, \text{Ab})_0 \to \text{f. p.}(\mathcal{A}^{\text{op}}, \text{Ab})_0$ *induce inverse dualities* $V_0\colon \text{f. p.}(\mathcal{C}^{\text{op}}, \text{Ab})_0 \to \text{f. p.}(\mathcal{C}, \text{Ab})_0$ *and* $W_0\colon \text{f. p.}(\mathcal{C}, \text{Ab})_0 \to \text{f. p.}(\mathcal{C}^{\text{op}}, \text{Ab})_0$. *These dualities have the properties:*

a) *If* $0 \to (\ ,A) \xrightarrow{(\ ,g)} (\ ,B) \xrightarrow{(\ ,f)} (\ ,C) \to F \to 0$ *is exact in* $\text{f. p.}(\mathcal{C}^{\text{op}}, \text{Ab})_0$, *then* $0 \to (C,\ ) \xrightarrow{((f,\ )} (B,\ ) \xrightarrow{(g,\ )} (A,\ ) \to V_0(F) \to 0$ *is exact in* $\text{f. p.}(\mathcal{C}, \text{Ab})_0$.

b) *If* $0 \to (C,\ ) \xrightarrow{((f,\ )} (B,\ ) \xrightarrow{(g,\ )} (A,\ ) \to G \to 0$ *is exact in* $\text{f. p.}(\mathcal{C}, \text{Ab})_0$, *then* $0 \to (\ ,A) \xrightarrow{(\ ,g)} (\ ,B) \xrightarrow{(\ ,f)} (\ ,C) \to W_0(G) \to 0$ *is exact in* $\text{f. p.}(\mathcal{C}^{\text{op}}, \text{Ab})_0$.

Our applications of these dualities to morphisms determined by objects are based on the following.

PROPOSITION 3.6. a) *A functor* F *in* $\text{f. p.}(\mathcal{C}^{\text{op}}, \text{Ab})_0$ *is simple in* $\text{f. p.}(\mathcal{C}^{\text{op}}, \text{Ab})_0$ *if and only if* F *is simple in* $(\mathcal{C}^{\text{op}}, \text{Ab})$.

b) *A functor* G *in* $\text{f. p.}(\mathcal{C}, \text{Ab})_0$ *is simple in* $\text{f. p.}(\mathcal{C}, \text{Ab})_0$ *if and only if* G *is simple in* $(\mathcal{C}, \text{Ab})$.

PROOF. a) Suppose F is simple in $\text{f. p.}(\mathcal{C}^{\text{op}}, \text{Ab})_0$. To show that it is simple in $(\mathcal{C}^{\text{op}}, \text{Ab})$ it suffices to show that if X is in $\mathcal{C}$ and $\alpha\colon (\ ,X) \to F$ is not zero, then it is an epimorphism. Since F is in $\text{f. p.}(\mathcal{C}^{\text{op}}, \text{Ab})_0$, there is an exact sequence $0 \to A \xrightarrow{g} B \xrightarrow{f} C \to 0$ in $\mathcal{C}$ (i.e. $0 \to A \xrightarrow{g} B \xrightarrow{f} C \to 0$ is exact in $\mathcal{A}$ with A, B, C in $\mathcal{C}$) such that $0 \to (\ ,A) \xrightarrow{(\ ,g)} (\ ,B) \xrightarrow{(\ ,f)} (\ ,C) \xrightarrow{\beta} F \to 0$ is exact in $\text{f. p.}(\mathcal{C}^{\text{op}}, \text{Ab})_0$ and consequently, in $(\mathcal{C}^{\text{op}}, \text{Ab})$.

Let X be in $\mathcal{C}$ and $\alpha\colon (\ ,X) \to F$ a nonzero morphism. Then there is a morphism $h\colon X \to C$ such that $\alpha = \beta(\ ,h)$. Consider the pullback diagram in $\mathcal{A}$

$$\begin{array}{ccccccccc}
0 & \longrightarrow & A & \longrightarrow & B\times_C X & \longrightarrow & X & \longrightarrow & 0 \\
 & & \Vert & & \downarrow{\scriptstyle j} & & \downarrow{\scriptstyle h} & & \\
0 & \longrightarrow & A & \longrightarrow & B & \longrightarrow & C & \longrightarrow & 0.
\end{array}$$

Since X and A are in $\mathcal{C}$, we have that $B\times_C X$ is in $\mathcal{C}$. Hence we obtain the exact commutative diagram in $\text{f. p.}(\mathcal{C}^{\text{op}}, \text{Ab})_0$

$$\begin{array}{ccccccccccc}
0 & \longrightarrow & (\ ,A) & \longrightarrow & (\ ,B\times_C X) & \longrightarrow & (\ ,X) & \longrightarrow & G & \longrightarrow & 0 \\
 & & \Vert & & \downarrow{\scriptstyle (\ ,j)} & & \downarrow{\scriptstyle (\ ,h)} & & & & \\
0 & \longrightarrow & (\ ,A) & \longrightarrow & (\ ,B) & \longrightarrow & (\ ,C) & \xrightarrow{\beta} & F & \longrightarrow & 0.
\end{array}$$

Let $\gamma\colon G \to F$ be the unique morphism which makes the diagram commute. Consequently $\operatorname{Im}\gamma = \operatorname{Im}\beta(\ ,h) = \operatorname{Im}\alpha$. Thus the fact that $\alpha \neq 0$ implies $\gamma \neq 0$. Since F is simple in $\mathrm{f.p.}(\mathcal{C}^{\mathrm{op}}, \mathrm{Ab})_0$ and $\gamma\colon G \to F$ is not zero with G in $\mathrm{f.p.}(\mathcal{C}^{\mathrm{op}}, \mathrm{Ab})_0$, we have that $\operatorname{Im}\gamma = F$. Hence $\operatorname{Im}\alpha = \operatorname{Im}\gamma = F$. Since this holds for all nonzero morphisms $\alpha\colon (\ ,X) \to F$ with X in $\mathcal{C}$, it follows that F is simple in $(\mathcal{C}^{\mathrm{op}}, \mathrm{Ab})$.

That F in $\mathrm{f.p.}(\mathcal{C}^{\mathrm{op}}, \mathrm{Ab})_0$ being simple in $(\mathcal{C}^{\mathrm{op}}, \mathrm{Ab})$ implies F is simple in $\mathrm{f.p.}(\mathcal{C}^{\mathrm{op}}, \mathrm{Ab})_0$ is obvious.

b) Same as a). □

4. Almost split sequences

As in Section 3 we assume throughout this section that $\mathcal{C}$ is a full subcategory of an abelian category $\mathcal{A}$ and that $\mathcal{C}$ is closed under extensions and summands. We now apply the duality theory given in Section 3 to studying certain types of morphisms determined by objects in $\mathcal{C}$ including almost split morphisms.

We say that a sequence $C_1 \xrightarrow{f_1} C_2 \xrightarrow{f_2} C_3$ of morphisms in $\mathcal{C}$ is exact in $\mathcal{C}$ if it is exact in $\mathcal{A}$.

Proposition 4.1. *Let* $0 \to A \xrightarrow{g} B \xrightarrow{f} C \to 0$ *be an exact sequence in* $\mathcal{C}$. *Then* $\operatorname{Coker}(\ ,f)$ *is simple in* $(\mathcal{C}^{\mathrm{op}}, \mathrm{Ab})$ *if and only if* $\operatorname{Coker}(g,\ )$ *is simple in* $(\mathcal{C}, \mathrm{Ab})$.

Proof. Since $0 \to A \xrightarrow{g} B \xrightarrow{f} C \to 0$ is an exact sequence in $\mathcal{C}$, it follows that $\operatorname{Coker}(\ ,f)$ is in $\mathrm{f.p.}(\mathcal{C}^{\mathrm{op}}, \mathrm{Ab})_0$ and $\operatorname{Coker}(g,\ )$ is in $\mathrm{f.p.}(\mathcal{C}, \mathrm{Ab})_0$. Moreover $V_0(\operatorname{Coker}(\ ,f)) = \operatorname{Coker}(g,\ )$ where V_0 is the duality $V_0\colon \mathrm{f.p.}(\mathcal{C}^{\mathrm{op}}, \mathrm{Ab})_0 \to \mathrm{f.p.}(\mathcal{C}, \mathrm{Ab})_0$ described in Section 3. Therefore $\operatorname{Coker}(\ ,f)$ is simple in $\mathrm{f.p.}(\mathcal{C}^{\mathrm{op}}, \mathrm{Ab})_0$ if and only if $\operatorname{Coker}(g,\ )$ is simple in $\mathrm{f.p.}(\mathcal{C}, \mathrm{Ab})_0$. Since by Proposition 3.6 we know that $\operatorname{Coker}(\ ,f)$ is simple in $\mathrm{f.p.}(\mathcal{C}^{\mathrm{op}}, \mathrm{Ab})_0$ if and only if it is simple in $\mathrm{f.p.}(\mathcal{C}^{\mathrm{op}}, \mathrm{Ab})$ and $\operatorname{Coker}(g,\ )$ is simple in $\mathrm{f.p.}(\mathcal{C}, \mathrm{Ab})_0$ if and only if it is simple in $\mathrm{f.p.}(\mathcal{C}, \mathrm{Ab})$, it follows that $\operatorname{Coker}(\ ,f)$ is simple in $(\mathcal{C}^{\mathrm{op}}, \mathrm{Ab})$ if and only if $\operatorname{Coker}(g,\ )$ is simple in $(\mathcal{C}, \mathrm{Ab})$. □

The following application of Proposition 4.1 is of particular interest to us.

Proposition 4.2. *Let* $0 \to A \xrightarrow{g} B \xrightarrow{f} C \to 0$ *be exact in* $\mathcal{C}$.

a) *Suppose* $\operatorname{Coker}(\ ,f)$ *is simple in* $(\mathcal{C}^{\mathrm{op}}, \mathrm{Ab})$. *Then the following are equivalent.*

i) f *is right minimal.*

ii) g *is left almost split in* $\mathcal{C}$.

iii) $\operatorname{End} A$ *is local.*

b) *Suppose* $\operatorname{Coker}(g,\ )$ *is simple in* $(\mathcal{C}, \mathrm{Ab})$. *Then the following are equivalent.*

i) g *is left minimal.*

ii) f *is right almost split in* $\mathcal{C}$.

iii) $\operatorname{End} C$ *is local.*

Proof. a) Suppose f is right minimal. To show that $g\colon A \to B$ is left almost split in $\mathcal{C}$, we have to show that a morphism $h\colon A \to X$ is a splittable monomorphism provided there does not exist a morphism $j\colon B \to X$ such that $jg = h$.

Let the exact commutative diagram

$$\begin{array}{ccccccccc} 0 & \longrightarrow & A & \longrightarrow & B & \longrightarrow & C & \longrightarrow & 0 \\ & & \downarrow & & \downarrow & & \| & & \\ 0 & \longrightarrow & X & \longrightarrow & B\times^A X & \longrightarrow & C & \longrightarrow & 0 \end{array}$$

be a pushout diagram in $\mathcal{A}$. Since X and C are in $\mathcal{C}$, the bottom row is an exact sequence in $\mathcal{C}$. The hypothesis that there is no $j\colon B \to X$ such that $jf = h$ is equivalent to the bottom row not splitting. From this exact commutative diagram we deduce the exact commutative diagram of functors in $(\mathcal{C}^{\mathrm{op}}, \mathrm{Ab})$.

$$\begin{array}{ccccccccc}
0 & \longrightarrow & (\ ,A) & \longrightarrow & (\ ,B) & \longrightarrow & (\ ,C) & \longrightarrow & F & \longrightarrow & 0 \\
 & & \downarrow{\scriptstyle (\ ,h)} & & \downarrow{\scriptstyle (\ ,v)} & & \| & & \downarrow{\scriptstyle \beta} & & \\
0 & \longrightarrow & (\ ,X) & \longrightarrow & (\ ,B\times^A X) & \longrightarrow & (\ ,C) & \longrightarrow & G & \longrightarrow & 0.
\end{array}$$

Since $0 \to X \to B\times^A X \to C \to 0$ does not split, $G \neq 0$. Since β is an epimorphism and F is simple by hypothesis, $\beta\colon F \to G$ is an isomorphism. Therefore the subfunctor $H = \mathrm{Im}((\ ,B) \to (\ ,C))$ of $(\ ,C)$ is the same as $\mathrm{Im}((\ ,B\times^A X) \to (\ ,C))$. Since the exact sequences $0 \to (\ ,A) \to (\ ,B) \to H \to 0$ and $0 \to (\ ,X) \to (\ ,B\times^A X) \to H \to 0$ are projective resolutions, we have an exact commutative diagram

$$\begin{array}{ccccccccc}
0 & \longrightarrow & (\ ,A) & \xrightarrow{(\ ,g)} & (\ ,B) & \longrightarrow & H & \longrightarrow & 0 \\
 & & \downarrow{\scriptstyle (\ ,h)} & & \downarrow{\scriptstyle (\ ,v)} & & \| & & \\
0 & \longrightarrow & (\ ,X) & \longrightarrow & (\ ,B\times^A X) & \longrightarrow & H & \longrightarrow & 0 \\
 & & \downarrow{\scriptstyle (\ ,w)} & & \downarrow{\scriptstyle (\ ,u)} & & \| & & \\
0 & \longrightarrow & (\ ,A) & \longrightarrow & (\ ,B) & \longrightarrow & H & \longrightarrow & 0.
\end{array}$$

Since $(\ ,B)$ is a projective cover of H, we know that $(\ ,u)(\ ,v) = (\ ,uv)$ is an isomorphism and hence that $(\ ,w)(\ ,h) = (\ ,wh)$ is an isomorphism. Therefore the composition $A \xrightarrow{h} X \xrightarrow{w} A$ is an isomorphism, which gives us our desired result that h is a splittable monomorphism. This finishes the proof that i) implies ii).

ii) implies iii). The fact that $g\colon A \to B$ being left almost split implies that $\mathrm{End}\, A$ is local was shown in Proposition 2.4.

iii) implies i). This is a trivial consequence of the following more general result. □

LEMMA 4.3. *Let $0 \to A \xrightarrow{g} B \xrightarrow{f} C \to 0$ be a nonsplit exact sequence in $\mathcal{C}$.*
a) *If $\mathrm{End}\, A$ is local, then $f\colon B \to C$ is right minimal in $\mathcal{C}$.*
b) *If $\mathrm{End}\, C$ is local, then $g\colon A \to B$ is left minimal in $\mathcal{C}$.*

PROOF. a) Suppose we are given $h\colon B \to B$ such that $fh = f$. Then we have an exact commutative diagram in $\mathcal{C}$

$$\begin{array}{ccccccccc}
0 & \longrightarrow & A & \xrightarrow{g} & B & \xrightarrow{f} & C & \longrightarrow & 0 \\
 & & \downarrow{\scriptstyle j} & & \downarrow{\scriptstyle h} & & \| & & \\
0 & \longrightarrow & A & \longrightarrow & B & \xrightarrow{f} & C & \longrightarrow & 0.
\end{array}$$

Since $0 \to A \to B \to C \to 0$ does not split, we know that it represents a nonzero element x of the abelian group $\mathrm{Ext}^1(C, A)$ viewing $\mathrm{Ext}^1(C, A)$ as a module over $\mathrm{End}\, A$ by means of the operation $\mathrm{End}\, A$ on A. The fact that the above diagram commutes means that the endomorphism j has the property that $jx = x$. Hence j is not in $\mathrm{rad}\,\mathrm{End}\, A$. Thus j is an isomorphism since $\mathrm{End}\, A$ is local. From the fact that j is an isomorphism, it follows that $h\colon B \to B$ is an isomorphism, which shows that $f\colon B \to C$ is right minimal.

b) Dual to a). □

Note: We have tacitly assumed in the above proof that the objects in $\mathcal{C}$ have the property that $\operatorname{Ext}^1(C, A)$ is a set for all A and C in $\mathcal{C}$. This will always be true in the situations of interest to us (for instance when there are enough projectives in $\mathcal{A}$).

Returning to the proof of Proposition 4.2, it only remains to remark that b) is simply the dual of a) to finish the proof.

We now recall the definition of an almost split sequence (see [11, §2]).

We say that a morphism $f: B \to C$ is *right almost split in* $\mathcal{C}$ if for each morphism $h: X \to C$ in $\mathcal{C}$ which is not a splittable epimorphism there is a morphism $s: X \to B$ such that $fs = h$. A morphism $g: A \to B$ in $\mathcal{C}$ is said to be *left almost split in* $\mathcal{C}$ if for each morphism $h: A \to Y$ in $\mathcal{C}$ which is not a splittable monomorphism there is a morphism $t: B \to Y$ such that $tg = h$. Finally, we say that an exact sequence $0 \to A \xrightarrow{g} B \xrightarrow{f} C \to 0$ in $\mathcal{C}$ is an almost split sequence in $\mathcal{C}$ if f is right almost split in $\mathcal{C}$ and g is left almost split in $\mathcal{C}$.

We now describe some of the basic properties of almost split sequences.

PROPOSITION 4.4. *Let $0 \to A \xrightarrow{g} B \xrightarrow{f} C \to 0$ be an exact sequence in $\mathcal{C}$. Then the following are equivalent.*

a) *$0 \to A \xrightarrow{g} B \xrightarrow{f} C \to 0$ is an almost split sequence in $\mathcal{C}$.*
b) *f is minimal right almost split in $\mathcal{C}$.*
c) *f is right almost split in $\mathcal{C}$ and* $\operatorname{End} A$ *is local.*
d) *g is minimal left almost split in $\mathcal{C}$.*
e) *g is left almost split in $\mathcal{C}$ and* $\operatorname{End} C$ *is local.*
f) i) $\operatorname{Coker}(\ ,f)$ *is simple in* $(\mathcal{C}^{\mathrm{op}}, \mathrm{Ab})$ *and the exact sequence* $0 \to (\ ,A) \to (\ ,B) \to (\ ,C) \to \operatorname{Coker}(\ ,f) \to 0$ *is a minimal projective resolution of* $\operatorname{Coker}(\ ,f)$ *in* $(\mathcal{C}^{\mathrm{op}}, \mathrm{Ab})$ *and*
ii) $\operatorname{Coker}(g,\ )$ *is simple in* $(\mathcal{C}, \mathrm{Ab})$ *and* $0 \to (C,\ ) \to (B,\ ) \to (A,\ ) \to \operatorname{Coker}(g,\ ) \to 0$ *is a minimal projective resolution of* $\operatorname{Coker}(g,\ )$ *in* $(\mathcal{C}, \mathrm{Ab})$.

PROOF. The equivalence of a) through e) is an immediate consequence of Proposition 4.2 and Lemma 4.3. The equivalence of a) and f) follows easily from Proposition 2.7 and 3.6 and the equivalence of a) through e). □

COROLLARY 4.5. a) *A simple functor S in* $\mathrm{f.p.}(\mathcal{C}^{\mathrm{op}}, \mathrm{Ab})_0$ *has a minimal projective resolution in* $(\mathcal{C}^{\mathrm{op}}, \mathrm{Ab})$ *if and only if there is an exact sequence* $0 \to (\ ,A) \xrightarrow{(\ ,g)} (\ ,B) \xrightarrow{(\ ,f)} (\ ,C) \to S \to 0$ *with $0 \to A \xrightarrow{g} B \xrightarrow{f} C \to 0$ an almost split sequence in $\mathcal{C}$.*

b) *A simple functor T in* $\mathrm{f.p.}(\mathcal{C}, \mathrm{Ab})_0$ *has a minimal projective resolution in* $\mathrm{f.p.}(\mathcal{C}, \mathrm{Ab})$ *if and only if there is an exact sequence* $0 \to (C,\ ) \xrightarrow{(f,\ )} (B,\ ) \xrightarrow{(g,\ )} (A,\ ) \to T \to 0$ *with $0 \to A \xrightarrow{g} B \xrightarrow{f} C \to 0$ an almost split sequence.*

PROOF. a) If there is an almost split sequence $0 \to A \xrightarrow{g} B \xrightarrow{f} C \to 0$ in $\mathcal{C}$ such that $0 \to (\ ,A) \to (\ ,B) \to (\ ,C) \to S \to 0$ is exact, then we know by Proposition 4.4 that this exact sequence of functors is a minimal projective resolution.

On the other hand, suppose S is in $\mathrm{f.p.}(\mathcal{C}^{\mathrm{op}}, \mathrm{Ab})_0$ and has a minimal projective resolution in $(\mathcal{C}^{\mathrm{op}}, \mathrm{Ab})$, $0 \to (\ ,A) \xrightarrow{(\ ,g)} (\ ,B) \xrightarrow{(\ ,f)} (\ ,C) \to S \to 0$. Then $0 \to A \xrightarrow{g} B \xrightarrow{f} C \to 0$ is exact in $\mathcal{C}$ by the definition of $\mathrm{f.p.}(\mathcal{C}^{\mathrm{op}}, \mathrm{Ab})_0$ and is not split since $S \neq 0$.

Since $(\ ,B) \xrightarrow{(\ ,f)} (\ ,C) \to S \to 0$ is a minimal projective presentation in $(\mathcal{C}^{\mathrm{op}}, \mathrm{Ab})$ we know that $f\colon B \to C$ is minimal right almost split in $\mathcal{C}$ (see Proposition 2.7). By Proposition 4.2 we know that this implies that $\operatorname{End} A$ is local. Thus $0 \to A \xrightarrow{g} B \xrightarrow{f} C \to 0$ is almost split in $\mathcal{C}$ by Proposition 4.4. □

As our final comment on almost split sequences in this section, we point out the following uniqueness theorem.

THEOREM 4.6. *Let $0 \to A \xrightarrow{g} B \xrightarrow{f} C \to 0$ and $0 \to A' \xrightarrow{g'} B' \xrightarrow{f'} C' \to 0$ be two almost split sequences in $\mathcal{C}$. Then the following are equivalent:*

a) *The sequences are isomorphic.*
b) $\operatorname{Coker}(\ ,f) \cong \operatorname{Coker}(\ ,f')$.
c) $C \cong C'$.
d) $\operatorname{Coker}(g, \ ) \cong \operatorname{Coker}(g', \ )$.
e) $A \cong A'$.

PROOF. a) equivalent to b). Since by Proposition 4.4 the exact sequences $0 \to (\ ,A) \xrightarrow{(\ ,g)} (\ ,B) \xrightarrow{(\ ,f)} (\ ,C) \to \operatorname{Coker}(\ ,f) \to 0$ and $0 \to (\ ,A') \xrightarrow{(\ ,g')} (\ ,B') \xrightarrow{(\ ,f')} (\ ,C') \to \operatorname{Coker}(\ ,f') \to 0$ are minimal projective resolutions, the equivalence of a) and b) follows from the uniqueness of minimal projective resolutions.

a) equivalent to c). Since a) obviously implies c) we only have to show that c) implies a). Suppose C and C' are isomorphic. Since $\operatorname{Coker}(\ ,f)$ and $\operatorname{Coker}(\ ,f')$ are simple and $\operatorname{End} C$ and $\operatorname{End} C'$ are local, we know that $\operatorname{Coker}(\ ,f) \cong (\ ,C)/\underline{r}(\ ,C)$ and $\operatorname{Coker}(\ ,f') \cong (\ ,C')/\underline{r}(\ ,C')$. The fact that $C \cong C'$ now implies $\operatorname{Coker}(\ ,f) \cong \operatorname{Coker}(\ ,f')$ which implies a). The equivalence of a), c) and d) is the dual of the above argument. □

As our final result of this section we give the following generalization of the uniqueness of almost split sequences.

PROPOSITION 4.7. *Let $0 \to A \xrightarrow{g} B \xrightarrow{f} C \to 0$ and $0 \to A' \xrightarrow{g'} B' \xrightarrow{f'} C' \to 0$ be two exact sequences in $\mathcal{C}$.*

a) *Suppose f and f' are right minimal with the property that* $\operatorname{Coker}(\ ,f)$ *and* $\operatorname{Coker}(\ ,f')$ *are simple. Then the following are equivalent:*

i) $\operatorname{Coker}(\ ,f) \cong \operatorname{Coker}(\ ,f')$.
ii) $\operatorname{Coker}(g, \ ) \cong \operatorname{Coker}(g', \ )$.
iii) $A \cong A'$.

b) *Suppose g and g' are left minimal with the property that* $\operatorname{Coker}(g, \ )$ *and* $\operatorname{Coker}(g', \ )$ *are simple. Then the following are equivalent:*

i) $\operatorname{Coker}(g, \ ) \cong \operatorname{Coker}(g', \ )$.
ii) $\operatorname{Coker}(\ ,f) \cong \operatorname{Coker}(\ ,f')$.
iii) $C \cong C'$.

PROOF. a) i) equivalent to ii). Trivial consequence of the duality $V_0\colon \mathrm{f.p.}(\mathcal{C}^{\mathrm{op}}, \mathrm{Ab})_0 \to \mathrm{f.p.}(\mathcal{C}, \mathrm{Ab})_0$ since $V_0(\operatorname{Coker}(\ ,f)) = \operatorname{Coker}(g, \ )$ and $V_0(\operatorname{Coker}(\ ,f')) = \operatorname{Coker}(g', \ )$.

ii) equivalent to iii). Since f and f' are right minimal and $\operatorname{Coker}(\ ,f)$ and $\operatorname{Coker}(\ ,f')$ are simple, we know by Proposition 4.2 that g and g' are left almost split. Hence by Proposition 2.7 we know that $(A, \ ) \to \operatorname{Coker}(g, \ )$ and $(A', \ ) \to \operatorname{Coker}(g', \ )$ are projective covers with kernels $\underline{r}(A, \ )$ and $\underline{r}(A', \ )$ respectively. The equivalence ii) and iii) now follows trivially.

b) Dual to a). □

5. Simple functors not vanishing on $\operatorname{mod}\Lambda$

Let Λ be an arbitrary ring. This section is devoted to applying the results of the previous sections to studying the simple functors in $((\operatorname{Mod}\Lambda)^{\mathrm{op}}, \mathrm{Ab})$ which do not vanish on some finitely presented Λ-module. Analogous results for noetherian algebras over complete local rings and orders will be given later on. In the following proposition we summarize some of our previous results in a form particularly well suited to our present purposes.

PROPOSITION 5.1. *Let C be a nonprojective Λ-module in $\operatorname{mod}\Lambda$ and H a maximal right ideal in $\operatorname{End} C$ containing $P(C,C)$. Let $\Gamma = \operatorname{End}\operatorname{Tr} C$ and I an injective Γ-module which is an injective envelope for the simple Γ-module $\operatorname{End} C/H$ (remember that we have a ring isomorphism $(\underline{\operatorname{End}}\operatorname{Tr} C)^{\mathrm{op}} \cong \underline{\operatorname{End}} C$).*

Then there is a unique (up to isomorphism) exact sequence $0 \to \operatorname{Hom}_\Gamma(\operatorname{Tr} C, I) \xrightarrow{g} B \xrightarrow{f} C \to 0$ of Λ-modules having the property f is a minimal right C-determined morphism such that $(\operatorname{Im}(\ ,f))(C) = H$. Moreover this uniquely determined exact sequence has the following properties.

a) *$\operatorname{Coker}(\ ,f)$ is a simple functor in $((\operatorname{Mod}\Lambda)^{\mathrm{op}}, \mathrm{Ab})$ which vanishes on all projective Λ-modules.*

b) *g is left almost split and so*

i) *$\operatorname{Coker}(g,\ )$ is a simple functor in $((\operatorname{Mod}\Lambda)^{\mathrm{op}}, \mathrm{Ab})$ which vanishes on all injective modules, and*

ii) *$\operatorname{End}_\Lambda(\operatorname{Hom}_\Gamma(\operatorname{Tr} C, I)) \cong \operatorname{End}_\Gamma(I)$ is local.*

iii) *$(\operatorname{Hom}_\Gamma(\operatorname{Tr} C, I),\ ) \to \operatorname{Coker}(g,\ ) \to 0$ is a projective cover.*

c) *$0 \to \operatorname{Hom}_\Gamma(\operatorname{Tr} C, I) \xrightarrow{g} B \xrightarrow{f} C \to 0$ is an almost split sequence if and only if $\operatorname{End} C$ is local.*

PROOF. The existence and uniqueness of the exact sequence $0 \to \operatorname{Hom}_\Gamma(\operatorname{Tr} C, I) \xrightarrow{g} B \xrightarrow{f} C \to 0$ follows from I, Theorem 3.9 and I, Proposition 3.12.

a) Since H is a maximal right ideal in $\operatorname{End} C$ and $f\colon B \to C$ is right C-determined with $\operatorname{Im}(\ ,f)(C) = H$, it follows that $\operatorname{Coker}(\ ,f)$ is a simple functor in $((\operatorname{Mod}\Lambda)^{\mathrm{op}}, \mathrm{Ab})$. The fact that f is an epimorphism clearly implies that $\operatorname{Coker}(\ ,f)(P) = 0$ for all projective modules P.

b) Since I is the Γ-injective envelope of the simple Γ-module $\operatorname{End} C/H$ we know that f is a minimal right C-determined morphism (see I, Theorem 3.9). This implies that g is left almost split as well as $\operatorname{End}_\Lambda(\operatorname{Hom}_\Gamma(\operatorname{Tr} C, I))$ is local (see Proposition 4.2). The fact that g is left almost split implies that $\operatorname{Coker}(g,\ )$ is a simple functor in $(\operatorname{Mod}\Lambda, \mathrm{Ab})$ (see Proposition 2.4). Also $\operatorname{Coker}(g,\ )$ vanishes on injectives since g is a monomorphism. Finally, because $\operatorname{End}_\Lambda(\operatorname{Hom}_\Gamma(\operatorname{Tr} C, I))$ is local $(\operatorname{Hom}_\Gamma(\operatorname{Tr} C, I),\ ) \to \operatorname{Coker}(g,\ ) \to 0$ is a projective cover.

c) Because g is left almost split, it follows from Proposition 4.4 that the exact sequence $0 \to \operatorname{Hom}_\Gamma(\operatorname{Tr} C, I) \xrightarrow{g} B \xrightarrow{f} C \to 0$ is almost split if and only if $\operatorname{End} C$ is local. □

As a consequence of this proposition we have the following results.

PROPOSITION 5.2. *Let S be a simple functor in $((\operatorname{Mod}\Lambda)^{\mathrm{op}}, \mathrm{Ab})$ vanishing on Λ.*

a) *Suppose C is in $\operatorname{mod}\Lambda$ such that $S(C) \neq 0$. Then*

i) *$S(C)$ is a simple* $(\operatorname{End} C)^{\mathrm{op}}$*-module isomorphic to* $\operatorname{End} C/H$ *for some maximal right ideal H containing $P(C,C)$.*

ii) *Let* $\Gamma = \operatorname{End} \operatorname{Tr} C$ *and I a Γ-injective envelope for the simple Γ-module $S(C)$ and* $0 \to \operatorname{Hom}_\Gamma(\operatorname{Tr} C, I) \xrightarrow{g} B \xrightarrow{f} C \to 0$ *an exact Λ-sequence such that f is minimal right C-determined with* $\operatorname{Im}(\ ,f)(C) = H$. *Then* $0 \to (\ ,\operatorname{Hom}_\Gamma(\operatorname{Tr} C, I)) \xrightarrow{(\ ,g)} (\ ,B) \xrightarrow{(\ ,f)} (\ ,C) \to S \to 0$ *is exact.*

iii) *The exact sequence* $0 \to (\ ,\operatorname{Hom}_\Gamma(\operatorname{Tr} C, I)) \to (\ ,B) \to (\ ,C) \to S \to 0$ *is a minimal projective resolution of S if and only if* $\operatorname{End} C$ *is local or equivalently* $0 \to \operatorname{Hom}_\Gamma(\operatorname{Tr} C, I)) \xrightarrow{g} B \xrightarrow{f} C \to 0$ *is almost split.*

b) *If* $0 \to A' \xrightarrow{g'} B' \xrightarrow{f'} C' \to 0$ *is an exact sequence with f' a minimal right C'-determined morphism such that* $0 \to (\ ,A') \xrightarrow{(\ ,g')} (\ ,B') \xrightarrow{(\ ,f')} (\ ,C') \to S$ *is exact then* $A' \cong \operatorname{Hom}_\Gamma(\operatorname{Tr} C, I)$

PROOF. a) i) Since S is a simple functor, the fact that $S(C) \neq 0$ implies that $S(C)$ is a simple $(\operatorname{End} C)^{\mathrm{op}}$-module. Since $S(\Lambda) = 0$, we know that $S(P) = 0$ for all finitely generated projective Λ-modules P. Because C is finitely generated, it follows from this that $P(C,C)S(C) = 0$. Thus if H is a maximal right ideal in $\operatorname{End} C$ such that $\operatorname{End} C/H \cong S(C)$ it follows that $H \supset P(C,C)$.

ii) and iii) Follows from Proposition 5.1.

b) Follows from Proposition 4.7. □

As an immediate consequence of Proposition 5.2 we have:

COROLLARY 5.3. *Let S be a simple functor in* $((\operatorname{Mod} \Lambda)^{\mathrm{op}}, \mathrm{Ab})$. *If $S(C) \neq 0$ for some C in* $\operatorname{mod} \Lambda$ *and $S(\Lambda) = 0$, then S is finitely presented. Further S has a minimal projective presentation in* $((\operatorname{Mod} \Lambda)^{\mathrm{op}}, \mathrm{Ab})$ *if and only if there is a C in* $\operatorname{mod} \Lambda$ *with* $\operatorname{End} C$ *local such that $S(C) \neq 0$.*

In view of these results, in order to finish our description of the simple functors S in $((\operatorname{Mod} \Lambda)^{\mathrm{op}}, \mathrm{Ab})$, not vanishing on some C in $\operatorname{mod} \Lambda$, it suffices to consider those S such that $S(P) \neq 0$ for some projective module P in $\operatorname{mod} \Lambda$. We begin with the following somewhat more general result.

PROPOSITION 5.4. *Let M be a submodule of the projective Λ-module P (P not necessarily finitely generated). Then the following are equivalent.*

a) *M is a maximal submodule of P.*

b) *The subfunctor $(\ ,M)$ of $(\ ,P)$ is a maximal subfunctor.*

c) *The inclusion $f\colon M \to P$ is a right P-determined morphism with* $\operatorname{Im}(\ ,f)(P)$ *a maximal right ideal of* $\operatorname{End} P$.

PROOF. a) equivalent to b). By Proposition 2.1 we know that $(\ ,M)$ is a maximal subfunctor of $(\ ,P)$ if and only if the inclusion $f\colon M \to P$ has the property that given a morphism $g\colon L \to P$ either there exists $h\colon L \to M$ such that $g = fh$ or the induced morphism $M \amalg L \to P$ is a splittable epimorphism. It is easily checked that since P is projective this latter property of $f\colon M \to P$ is equivalent to M being a maximal submodule of P.

b) equivalent to c). Special case of Proposition 2.1. □

We now apply this result to obtain the following description of the maximal subfunctors of $(\ ,P)$ with P a finitely generated projective Λ-module.

PROPOSITION 5.5. *Let P be a finitely generated projective Λ-module and F a subfunctor of $(\ ,P)$. Then F is a maximal subfunctor of $(\ ,P)$ if and only if there is a maximal submodule M of P such that $F = (\ ,M)$. Thus there is a bijection between the maximal submodules of P and the maximal subfunctors of $(\ ,P)$ given by $M \mapsto (\ ,M)$ for each maximal submodule M of P.*

PROOF. We have already seen in Proposition 5.3 that if M is a maximal submodule of P, then $(\ ,M)$ is a maximal subfunctor of $(\ ,P)$.

Suppose F is a maximal subfunctor of $(\ ,P)$. Then we know that $F(P)$ is a maximal right ideal H of $\operatorname{End} P$. Let N be the submodule of P generated by the submodules $f(P)$ as f runs through H. We claim that N is a proper submodule of P. For suppose $N = P$. Then because P is finitely generated there would be a finite number of morphisms $f_1, \dots, f_n$ in H such that the induced morphism $\coprod_{i=1}^n P \to P$ is an epimorphism. The fact that P is projective implies that there is a morphism $P \to \coprod_{i=1}^n P$ such that the composition $P \to \coprod_{i=1}^n P \to P$ is the identity on P. But this is impossible since it implies that 1 is in $\operatorname{Im}((P, \coprod_{i=1}^n P) \to (P,P)) \subset H$, which contradicts the fact that H is a right proper ideal of $\operatorname{End} P$. Thus we have that N is a proper submodule of P.

Next it is obvious from the definition of N and the fact that N is a proper submodule of P that $H \subset \operatorname{Hom}_\Lambda(P,N) \neq \operatorname{End} P$. Since H is a maximal right ideal of $\operatorname{End} P$, it follows that $H = (P,N)$. Because P is finitely generated, we know there is a maximal submodule M of P containing N. Therefore $(P,M) = (P,N)$ since (P,M) is a proper right ideal of $\operatorname{End} P$ containing the maximal right ideal $(P,N) = H$. Therefore by Proposition 5.1 we know that $(\ ,M)$ is a P-determined subfunctor of $(\ ,P)$ with $(P,M) = H = F(P)$. Therefore $F \subset (\ ,M)$ which implies that $F = (\ ,M)$ since F is a maximal subfunctor of $(\ ,P)$. This concludes the proof of the first part of Proposition 5.5.

The rest of the proof of the proposition now follows easily. □

As a consequence of this description of the maximal subfunctors of $(\ ,P)$ with P a finitely generated projective Λ-module, we have the following.

PROPOSITION 5.6. *Let S be a simple functor in $((\operatorname{Mod} \Lambda)^{\mathrm{op}}, \mathrm{Ab})$.*

a) *The following are equivalent:*

i) *$S(\Lambda) \neq 0$.*

ii) *$S(P) \neq 0$ for some finitely generated projective Λ-module P.*

iii) *There exists a maximal submodule M of P in $\operatorname{mod} \Lambda$ such that there is an exact sequence $0 \to (\ ,M) \to (\ ,P) \to S \to 0$.*

b) *Suppose $S(P) \neq 0$ for some projective Λ-module P in $\operatorname{mod} \Lambda$. Then*

i) *The projective dimension of S in $((\operatorname{Mod} \Lambda)^{\mathrm{op}}, \mathrm{Ab})$ is at most 1.*

ii) *S is projective if and only if there is some simple projective Λ-module Q such that $S \cong (\ ,Q)$.*

iii) *S has a projective cover if and only if there is a projective Λ-module Q in $\operatorname{mod} \Lambda$ with $\operatorname{End} Q$ local such that $S \cong (\ ,Q)/(\ ,M)$ where M is the unique maximal submodule of Q.*

PROOF. Only parts ii) and iii) of b) do not follow immediately from the results cited before the statement of the proposition.

b) ii) Suppose Q is a simple projective module, then $(\ ,Q)$ is simple since (0) is the maximal submodule of Q (see Proposition 5.4). Suppose S is a projective simple functor and $S(P) \neq 0$ for some P in $\operatorname{mod} \Lambda$. Then there is an epimorphism $(\ ,P) \to S \to 0$ which implies that S is isomorphic to a summand of $(\ ,P)$. Thus $S \cong (\ ,Q)$ for some summand Q of P. From the fact that S is simple it follows trivially that Q is simple.

b) iii) Suppose $(\ ,C) \to S \to 0$ is a projective cover for S. Since $(\ ,P) \to S \to 0$ is exact, it follows that $(\ ,C)$ is isomorphic to a summand of P and so C is projective. By Proposition 1.10 we know that $\operatorname{End} C$ is local since $(\ ,C) \to S \to 0$ is a projective cover of the simple functor S. It is not difficult to see that this combined with the fact that C is a projective module implies that C has a unique maximal submodule M. Thus by Proposition 5.5 we have that $S \cong (\ ,C)/(\ ,M)$. The rest of the proof is a consequence of Proposition 1.10. $\square$

An easily verified consequence of Proposition 5.6 is the following.

COROLLARY 5.7. *The map $S \to S(\Lambda)$ establishes a bijection between the isomorphism classes of simple functors S in $((\operatorname{Mod} \Lambda)^{\mathrm{op}}, \mathrm{Ab})$ not vanishing on Λ and the isomorphism classes of simple Λ-modules. This correspondence has the following properties.*

a) *S has a projective cover in $((\operatorname{Mod} \Lambda)^{\mathrm{op}}, \mathrm{Ab})$ if and only if the simple Λ-module $S(\Lambda)$ has a projective cover P. Furthermore, P is a projective cover for $S(\Lambda)$ if and only if $(\ ,P)$ is a projective cover for S.*

b) *S is projective if and only if $S(\Lambda)$ is a simple projective Λ-module. If $S(\Lambda)$ is simple projective, then $S \cong (\ ,S(\Lambda))$.*

We now apply these results concerning simple functors S in $((\operatorname{Mod} \Lambda)^{\mathrm{op}}, \mathrm{Ab})$ which do not vanish on Λ to finish our description of the simple factors of $(\ ,C)$ with C in $\operatorname{mod} \Lambda$. It has been shown that the simple factors of $(\ ,C)$ are of the form $(\ ,C)/(\ ,C)_H$ for some maximal right ideal H of $\operatorname{End} C$. A projective resolution for $(\ ,C)/(\ ,C)_H$ in case $H \supset P(C,C)$ was described in Proposition 5.1. Our aim now is to give a projective resolution for $(\ ,C)/(\ ,C)_H$ in case H is a maximal right ideal in $\operatorname{End} C$ not containing $P(C,C)$.

PROPOSITION 5.8. *Let C be in $\operatorname{mod} \Lambda$ and H a maximal right ideal in $\operatorname{End} C$, and S the simple functor $(\ ,C)/(\ ,C)_H$ in $((\operatorname{Mod} \Lambda)^{\mathrm{op}}, \mathrm{Ab})$.*

a) *$S(\Lambda) \neq 0$ if and only if H does not contain $P(C,C)$.*

b) *$S(\Lambda) \neq 0$ if and only if there is a submodule M of C such that $0 \to (\ ,M) \to (\ ,C) \to S \to 0$ is exact, i.e., $(\ ,M) = (\ ,C)_H$.*

c) *Let M be a submodule of C. Then the following statements are equivalent.*

i) *$(\ ,C)/(\ ,M)$ is simple.*

ii) *If $h\colon L \to C$ is a morphism with $\operatorname{Im} h \not\subset M$, then the induced morphism $M \amalg L \to C$ is a splittable epimorphism.*

PROOF. a) We have already shown in Proposition 5.1 that if $S(\Lambda) = 0$, then $H \supset P(C,C)$. Hence if H does not contain $P(C,C)$, then $S(\Lambda) \neq 0$.

Suppose now that $S(\Lambda) \neq 0$. Then there is an epimorphism $(\ ,\Lambda) \to S \to 0$. Since we also have the epimorphism $(\ ,C) \to S \to 0$, it follows that there are morphisms $g\colon C \to \Lambda$ and $h\colon \Lambda \to C$ such that the diagram

$$\begin{array}{ccccc}
(\ ,C) & \longrightarrow & S & \longrightarrow & 0 \\
\big\downarrow{\scriptstyle (\ ,g)} & & \| & & \\
(\ ,\Lambda) & \longrightarrow & S & \longrightarrow & 0 \\
\big\downarrow{\scriptstyle (\ ,h)} & & \| & & \\
(\ ,C) & \longrightarrow & S & \longrightarrow & 0
\end{array}$$

is exact and commutes. From this it follows that the endomorphism $hg\colon C \to C$ in $P(C,C)$ has the property $\operatorname{Im}(\ ,hg)$ is not contained in $(\ ,C)_H$. Hence $(\operatorname{Im}(\ ,hf))(C) \not\subset H$ since $(\ ,C)_H$ is determined by C. In particular $hg \notin H$. Since hg is in $P(C,C)$, it follows that H does not contain $P(C,C)$. This finishes the proof of a).

b) Suppose $S(\Lambda) \neq 0$. Then there is an exact sequence $0 \to (\ ,\underline{m}) \to (\ ,\Lambda) \to S \to 0$ with $\underline{m}$ a maximal left ideal in Λ (see Proposition 5.6). Then there is a morphism $f\colon C \to \Lambda$ which makes the exact diagram

$$\begin{array}{ccccc}
(\ ,C) & \longrightarrow & S & \longrightarrow & 0 \\
\downarrow & & \| & & \\
(\ ,\Lambda) & \longrightarrow & S & \longrightarrow & 0
\end{array}$$

commute. From the exact commutative diagram

$$\begin{array}{ccccc}
 & & 0 & & 0 \\
 & & \downarrow & & \downarrow \\
 & & \operatorname{Ker} f & = & \operatorname{Ker} f \\
 & & \downarrow & & \downarrow \\
0 & \longrightarrow & f^{-1}(\underline{m}) & \longrightarrow & C \\
 & & \downarrow{\scriptstyle h} & & \downarrow{\scriptstyle f} \\
0 & \longrightarrow & \underline{m} & \xrightarrow{\text{inc}} & \Lambda
\end{array}$$

we deduce the exact commutative diagram

$$\begin{array}{ccccccccc}
 & & 0 & & 0 & & & & \\
 & & \downarrow & & \downarrow & & & & \\
0 & \longrightarrow & (\ ,\operatorname{Ker} f) & = & (\ ,\operatorname{Ker} f) & & & & \\
 & & \downarrow & & \downarrow & & & & \\
0 & \longrightarrow & (\ ,f^{-1}(\underline{m})) & \longrightarrow & (\ ,C) & \longrightarrow & F & \longrightarrow & 0 \\
 & & \downarrow{\scriptstyle (\ ,h)} & & \downarrow{\scriptstyle (\ ,f)} & & \downarrow & & \\
0 & \longrightarrow & (\ ,\underline{m}) & \longrightarrow & (\ ,\Lambda) & \longrightarrow & S & \longrightarrow & 0 \\
 & & \downarrow{\scriptstyle \alpha} & & \downarrow{\scriptstyle \beta} & & & & \\
 & & G & \xrightarrow[\gamma]{} & H & & & & \\
 & & \downarrow & & \downarrow & & & & \\
 & & 0 & & 0 & & & &
\end{array} \qquad (*)$$

Suppose x is in $\operatorname{Ker}(G(L) \xrightarrow{\gamma_L} H(L))$ for some L in $\operatorname{Mod}\Lambda$. Let $u\colon L \to \underline{m}$ be such that $\alpha_L(u) = x$. The fact that $\gamma_L(x) = 0$ implies there is a $v\colon L \to C$ such that $fv = \operatorname{inc} u$. Then $\operatorname{Im} v \subset f^{-1}(\underline{m})$ and so there is $w\colon L \to f^{-1}(\underline{m})$ such that $v = \operatorname{inc} w$. Thus $hw = u$ and so $0 = \alpha_L(u) = x$. Thus $\gamma\colon G \to H$ is a monomorphism. It is well known, because of the commutativity and exactness of the diagram $(*)$ that this implies that $F \to S$

is a monomorphism. Since the composition $(\ ,C) \xrightarrow{(\ ,f)} (\ ,\Lambda) \to S$ is an epimorphism, it follows that $F \to S$ is an epimorphism. Thus $F \to S$ is an isomorphism. Therefore $0 \to (\ ,f^{-1}(\underline{m})) \to (\ ,C) \to S \to 0$ is exact and so $(\ ,f^{-1}(\underline{m})) = (\ ,C)_H$, which completes the proof of b), since the other implication is trivial.

c) This is a special case of Proposition 2.1. □

As an immediate consequence of this proposition we have the following.

COROLLARY 5.9. *Let C be in* $\operatorname{mod}\Lambda$. *The two sided ideal $P(C,C)$ of* $\operatorname{End} C$ *is not contained in the radical of* $\operatorname{End} C$ *if and only if there is a submodule M of C having the property that if $h\colon L \to C$ is a morphism with* $\operatorname{Im} h$ *not contained in M, then the induced morphism $M \amalg L \to C$ is a splittable epimorphism.*

6. Simple functors in the complete case

Throughout this section we assume that R is a complete noetherian local ring and Λ is a noetherian R-algebra. Our purpose is to give some more information about the simple functors in $((\operatorname{Mod}\Lambda)^{\mathrm{op}}, \mathrm{Ab})$ which do not vanish on some module in $\operatorname{noeth}\Lambda$.

Suppose S is a simple functor in $((\operatorname{Mod}\Lambda)^{\mathrm{op}}, \mathrm{Ab})$ whose restriction to $\operatorname{noeth}\Lambda$ is not zero. Since S is additive and every module is a finite sum of indecomposable modules, S does not vanish on some indecomposable modules. Suppose C is an indecomposable module in $\operatorname{noeth}\Lambda$ such that $S(C) \neq 0$. Then there is a nonzero morphism $\alpha\colon (\ ,C) \to S$ which must be an epimorphism since S is simple. The fact that Λ is a noetherian R-algebra with R a complete local ring implies that every indecomposable C in $\operatorname{noeth}\Lambda$ has a local endomorphism ring. Since $\alpha\colon (\ ,C) \to S$ is an epimorphism with $\operatorname{End} C$ local, we know that $\alpha\colon (\ ,C) \to S$ is a projective cover of S (see Proposition 1.10). Moreover $0 \to \underline{r}(\ ,C) \to (\ ,C) \to S \to 0$ is exact, i.e. $(\ ,C)/\underline{r}(\ ,C) \cong S$.

On the other hand, suppose C is indecomposable in $\operatorname{noeth}\Lambda$. Since $\operatorname{End} C$ is local, $\underline{r}(\ ,C)$ is the unique maximal subfunctor of $(\ ,C)$. Therefore $(\ ,C)/\underline{r}(\ ,C)$ is a simple functor and the natural epimorphism $(\ ,C) \to (\ ,C)/\underline{r}(\ ,C)$ is a projective cover. Finally suppose C and C' are indecomposable modules in $\operatorname{noeth}\Lambda$ such that the simple functors $(\ ,C)/\underline{r}(\ ,C)$ and $(\ ,C')/\underline{r}(\ ,C')$ are isomorphic. Then the functors $(\ ,C)$ and $(\ ,C')$ are isomorphic since they are projective covers of the isomorphic functors $(\ ,C)/\underline{r}(\ ,C')$ and $(\ ,C')/\underline{r}(\ ,C')$ respectively. Hence two indecomposable modules C and C' in $\operatorname{noeth}\Lambda$ are isomorphic if and only if the simple functors $(\ ,C)/\underline{r}(\ ,C)$ and $(\ ,C')/\underline{r}(\ ,C')$ are isomorphic.

We summarize this discussion as follows.

PROPOSITION 6.1. *The map* $\operatorname{noeth}\Lambda \to ((\operatorname{Mod}\Lambda)^{\mathrm{op}}, \mathrm{Ab})$ *given by* $C \mapsto (\ ,C)/\underline{r}(\ ,C)$ *for all C in* $\operatorname{noeth}\Lambda$ *induces a bijection between the isomorphism classes of indecomposable modules in* $\operatorname{noeth}\Lambda$ *and the isomorphism classes of simple functors in* $((\operatorname{Mod}\Lambda)^{\mathrm{op}}, \mathrm{Ab})$ *which do not vanish on* $\operatorname{noeth}\Lambda$. *Moreover a simple functor S is isomorphic to* $(\ ,C)/\underline{r}(\ ,C)$ *for some indecomposable C in* $\operatorname{noeth}\Lambda$ *if and only if $S(C) \neq 0$.*

Having determined the projective covers for simple functors in $((\operatorname{Mod}\Lambda)^{\mathrm{op}}, \mathrm{Ab})$ which do not vanish on $\operatorname{noeth}\Lambda$ we now show that these functors have minimal projective resolutions. We first consider those simple functors which do not vanish on some projective in $\operatorname{noeth}\Lambda$.

PROPOSITION 6.2. *Let P be an indecomposable projective Λ-module and $\underline{r}$ the radical of Λ.*

a) *$\underline{r}P$ is the unique maximal submodule of P.*

b) *The inclusion* $\underline{r}P \to P$ *is a minimal right almost split morphism.*

c) $0 \to (\,, \underline{r}P) \to (\,, P) \to (\,, P)/\underline{r}(\,, P) \to 0$ *is a minimal projective resolution of the simple functor* $(\,, P)/\underline{r}(\,, P)$.

Suppose S *is a simple functor in* $((\operatorname{Mod}\Lambda)^{\mathrm{op}}, \mathrm{Ab})$ *such that* $S(P) \neq 0$.

d) *Then* $S \cong (\,, P)/\underline{r}(\,, P)$ *and there is an exact sequence* $0 \to (\,, \underline{r}P) \to (\,, P) \to S \to 0$ *which is a minimal projective resolution of* S.

e) $\operatorname{pd} S \leq 1$ *and* $\operatorname{pd} S = 0$ *if and only if* $\underline{r}P = 0$, *i.e.* P *is a simple projective. If* $\operatorname{pd} S = 0$, *then* $S \cong (\,, P)$ *with* P *a simple projective.*

PROOF. Follows easily by either direct calculations or by results given in Section 5. □

Having described minimal projective resolutions for simple functors in $((\operatorname{Mod}\Lambda)^{\mathrm{op}}, \mathrm{Ab})$ not vanishing on an indecomposable projective in $\operatorname{noeth}\Lambda$, we now turn our attention to describing minimal projective resolutions for the simple functors which vanish on the projectives in $\operatorname{noeth}\Lambda$ but not on all of $\operatorname{noeth}\Lambda$.

PROPOSITION 6.3. *Let* C *be an indecomposable module in* $\operatorname{noeth}_P \Lambda$.

a) *There exists an almost split sequence* $0 \to D\operatorname{Tr} C \xrightarrow{g} S \xrightarrow{f} C \to 0$. *This sequence has the following properties.*

i) $0 \to (\,, D\operatorname{Tr} C) \xrightarrow{(\,,g)} (\,, B) \xrightarrow{(\,,f)} (\,, C) \to \operatorname{Coker}(\,, f) \to 0$ *is a minimal projective resolution of the simple functor* $\operatorname{Coker}(\,, f)$ *which does not vanish on* C.

ii) $0 \to (\,, C) \xrightarrow{(\,,f)} (\,, B) \xrightarrow{(\,,g)} (D\operatorname{Tr} C, \,) \to \operatorname{Coker}(g, \,) \to 0$ *is a minimal projective resolution of the simple functor* $\operatorname{Coker}(g, \,)$ *which does not vanish on* $D\operatorname{Tr} C$.

b) *Suppose* S *is a simple functor in* $((\operatorname{Mod}\Lambda)^{\mathrm{op}}, \mathrm{Ab})$ *such that* $S(C) \neq 0$. *Then* $S \cong \operatorname{Coker}(\,, f)$, *so there is an exact sequence* $0 \to (\,, D\operatorname{Tr} C) \xrightarrow{(\,,g)} (\,, B) \xrightarrow{(\,,f)} (\,, C) \to S \to 0$ *which is a minimal projective resolution of* S.

c) *Suppose* T *is a simple functor in* $(\operatorname{Mod}\Lambda, \mathrm{Ab})$ *such that* $T(D\operatorname{Tr} C) \neq 0$. *Then* $T \cong \operatorname{Coker}(g, \,)$, *so there is an exact sequence* $0 \to (C, \,) \to (B, \,) \to (D\operatorname{Tr} C, \,) \to T \to 0$ *which is a minimal projective resolution of* T.

PROOF. a) Since $\operatorname{End} C$ is local, we have by Proposition 5.1 that there is an almost split sequence $0 \to \operatorname{Hom}_\Gamma(\operatorname{Tr} C, I) \to B \to C \to 0$ where $\Gamma = \operatorname{End}\operatorname{Tr} C$ and I is an injective envelope of the unique simple Γ-module (remember Γ is a local ring). But $I = D(\Gamma)$ where $D(\Gamma) = \operatorname{Hom}_R(\Gamma, I(R/m))$. Thus $\operatorname{Hom}_\Gamma(\operatorname{Tr} C, I) \cong \operatorname{Hom}_R(\operatorname{Tr} C, I(R/m)) = D\operatorname{Tr} C$. Thus the existence of the almost split sequence is established.

The rest of the proposition is just a recapitulation of earlier results. □

Having described the minimal projective resolutions of simple functors in $((\operatorname{Mod}\Lambda)^{\mathrm{op}}, \mathrm{Ab})$ which do not vanish on $\operatorname{noeth}\Lambda$, we now turn our attention to describing the minimal projective resolutions for the simple functors in $(\operatorname{Mod}\Lambda, \mathrm{Ab})$ which do not vanish on $\operatorname{art}\Lambda$. For the most part, proofs will be omitted since these results are essentially restatements of earlier results as were the results given in this section so far. Those readers wishing to supply proofs should consult the obvious analogues already established in this section.

We begin with the following analogue of Proposition 6.1.

PROPOSITION 6.4. a) *For each indecomposable* A *in* $\operatorname{art}\Lambda$, *we have that* $(A, \,)/\underline{r}(A, \,)$ *is a simple functor not vanishing on* A *and that the natural morphism* $(A, \,) \to (A, \,)/\underline{r}(A, \,)$ *is a projective cover.*

b) *The map* $\operatorname{art}\Lambda \to (\operatorname{Mod}\Lambda, \operatorname{Ab})$ *given by* $A \mapsto (A, \ )/\underline{r}(A, \ )$ *for all* A *in* $\operatorname{art}\Lambda$ *induces a bijection between the isomorphism classes of indecomposable modules in* $\operatorname{art}\Lambda$ *and the isomorphism classes of simple functors in* $(\operatorname{Mod}\Lambda, \operatorname{Ab})$ *which do not vanish on* $\operatorname{art}\Lambda$. *Moreover, a simple functor* T *is isomorphic to* $(A, \ )/\underline{r}(A, \ )$ *for some indecomposable* A *in* $\operatorname{art}\Lambda$ *if and only if* $T(A) \neq 0$.

Having determined the projective covers for the simple functors in $(\operatorname{Mod}\Lambda, \operatorname{Ab})$ not vanishing on $\operatorname{art}\Lambda$, we now show that these functors have minimal projective resolutions. We first consider those simple functors which do not vanish on some injective module in $\operatorname{art}\Lambda$.

PROPOSITION 6.5. *Let* I *be an indecomposable injective module in* $\operatorname{art}\Lambda$ *and* $\operatorname{soc} I$, *the socle of* I, *which is a simple* Λ*-module.*

a) *The epimorphism* $I \to I/\operatorname{soc} I$ *is a minimal left almost split morphism.*

b) $0 \to (I/\operatorname{soc} I, \ ) \to (I, \ ) \to (I, \ )/\underline{r}(I, \ ) \to 0$ *is a minimal projective resolution of the simple functor* $(I, \ )/\underline{r}(I, \ )$.

Suppose T *is a simple functor in* $(\operatorname{Mod}\Lambda, \operatorname{Ab})$ *such that* $T(I) \neq 0$.

c) *Then* $T \cong (I, \ )/(I/\operatorname{soc} I, \ )$ *and so there is an exact sequence* $0 \to (I/\operatorname{soc} I, \ ) \to (I, \ ) \to T \to 0$ *which is a minimal projective resolution of* T.

d) $\operatorname{pd} T \leq 1$ *and* $\operatorname{pd} T = 0$ *if and only if* I *is simple. If* $\operatorname{pd} T = 0$, *then* $T \cong (I, \ )$ *with* I *a simple injective.*

PROOF. Essentially the dual of Proposition 6.2. □

Having described minimal projective resolutions for simple functors in $(\operatorname{Mod}\Lambda, \operatorname{Ab})$ which do not vanish on some indecomposable injective in $\operatorname{art}\Lambda$, we now describe the minimal projective resolutions of simple functors in $(\operatorname{Mod}\Lambda, \operatorname{Ab})$ which do not vanish on $\operatorname{art}_I \Lambda$.

PROPOSITION 6.6. *Let* A *be an indecomposable module in* $\operatorname{art}_I \Lambda$.

a) *There exists an almost split sequence* $0 \to A \xrightarrow{g} B \xrightarrow{f} \operatorname{Tr} DA \to 0$. *This sequence has the following properties.*

i) $0 \to (\operatorname{Tr} DA, \ ) \xrightarrow{(f, \ )} (B, \ ) \xrightarrow{(g, \ )} (A, \ ) \to \operatorname{Coker}(g, \ ) \to 0$ *is a minimal projective resolution of the simple functor* $\operatorname{Coker}(g, \ )$ *which does not vanish on* A.

ii) $0 \to (\ , A) \xrightarrow{(\ ,g)} (\ , B) \xrightarrow{(\ ,f)} (\ , \operatorname{Tr} DA) \to \operatorname{Coker}(\ , f) \to 0$ *is a minimal projective resolution for the simple functor* $\operatorname{Coker}(\ , f)$ *which does not vanish on* $\operatorname{Tr} DA$.

b) *Suppose* T *is a simple functor in* $(\operatorname{Mod}\Lambda, \operatorname{Ab})$ *such that* $T(A) \neq 0$. *Then* $T \cong \operatorname{Coker}(g, \ )$, *so there is an exact sequence* $0 \to (\operatorname{Tr} DA) \xrightarrow{(f, \ )} (B, \ ) \xrightarrow{(g, \ )} (A, \ ) \to T \to 0$ *which is a minimal projective resolution of* T.

c) *Suppose* S *is a simple functor in* $((\operatorname{Mod}\Lambda)^{\mathrm{op}}, \operatorname{Ab})$ *such that* $S(\operatorname{Tr} DA) \neq 0$. *Then* $S \cong \operatorname{Coker}(\ , f)$, *so there is an exact sequence* $0 \to (\ , A) \xrightarrow{(\ ,g)} (\ , B) \xrightarrow{(\ ,f)} (\ , \operatorname{Tr} DA) \to S \to 0$ *which is a minimal projective resolution of* S.

PROOF. Simply a restatement of Proposition 6.3 using the fact that $A \cong D\operatorname{Tr}(\operatorname{Tr} DA)$. □

7. Simple functors for orders

We assume throughout this section that Λ is an R-order with $d = \dim R$. Our aim is to study the simple functors in $(L(\Lambda)^{\mathrm{op}}, \operatorname{Ab})$ and $(L(\Lambda), \operatorname{Ab})$. We begin with the simple functors in $(L(\Lambda)^{\mathrm{op}}, \mathrm{Ab})$ and $(L(\Lambda), \mathrm{Ab})$ which vanish on the projective and injective lattices respectively.

The following is simply a restatement of earlier results which we recall for the convenience of the reader.

LEMMA 7.1. a) *Let S be a simple functor in $(L(\Lambda)^{\mathrm{op}}, \mathrm{Ab})$. Then the following are equivalent.*

i) $S(P) = 0$ *for all projectives P in $L(\Lambda)$.*
ii) $S(C) \neq 0$ *for some nonprojective C in $L(\Lambda)$.*
iii) $S(C) \neq 0$ *if and only if there is a maximal right ideal H of* $\mathrm{End}\, C$ *containing $P(C, C)$ such that $S \cong (\ , C)/(\ , C)_H$.*

b) *Let T be a simple functor in $(L(\Lambda), \mathrm{Ab})$. Then the following are equivalent.*

i) $T(I) = 0$ *for all injective lattices I in $L(\Lambda)$.*
ii) $T(A) \neq 0$ *for some noninjective lattice A in $L(\Lambda)$.*
iii) $T(A) \neq 0$ *if and only if there is a maximal left ideal H of* $\mathrm{End}\, A$ *containing $I(A, A)$ such that $T \cong (A,\ )/(A,\ )_H$.*

We now describe projective resolutions for the simple functors satisfying the above equivalent conditions.

PROPOSITION 7.2. *Let C be a nonprojective lattice and H a maximal right ideal of* $\mathrm{End}\, C$ *containing $P(C, C)$. There is an exact sequence $0 \to \mathrm{Hom}_R(\mathrm{Tr}_L\, C, R) \xrightarrow{g} B \xrightarrow{f} C \to 0$ in $L(\Lambda)$ having the following properties.*

a) *f is right C-determined in $L(\Lambda)$ with* $\mathrm{Im}(\ , f)(C) = H$.

b) *The exact sequence $0 \to (\ , \mathrm{Hom}_R(\mathrm{Tr}_L\, C, R)) \xrightarrow{(\ , g)} (\ , B) \xrightarrow{(\ , f)} (\ , C) \to \mathrm{Coker}(\ , f) \to 0$ is a projective resolution of the simple functor* $\mathrm{Coker}(\ , f)$.

c) *$0 \to (C,\ ) \xrightarrow{(f,\ )} (B,\ ) \xrightarrow{(g,\ )} \mathrm{Hom}_R(\mathrm{Tr}_L\, C, R) \to \mathrm{Coker}(g,\ ) \to 0$ is a projective resolution of the simple functor* $\mathrm{Coker}(g,\ )$.

d) *g is a left determined $\mathrm{Hom}_R(\mathrm{Tr}_L\, C, R)$-morphism in $L(\Lambda)$ with $\mathrm{Im}(g,\ )(\mathrm{Hom}_R(\mathrm{Tr}_L\, C, R))$ a maximal left ideal in $\mathrm{End}(\mathrm{Hom}_R(\mathrm{Tr}_L\, C, R))$.*

e) *If S is a simple functor in $(L(\Lambda)^{\mathrm{op}}, \mathrm{Ab})$ such that $S \cong (\ , C)/(\ , C)_H$, then there is an exact sequence $0 \to (\ , \mathrm{Hom}_R(\mathrm{Tr}_L\, C, R)) \xrightarrow{(\ , g)} (\ , B) \xrightarrow{(\ , f)} (\ , C) \to S \to 0$ which is a projective resolution of S. In particular, S is finitely presented in $(L(\Lambda)^{\mathrm{op}}, \mathrm{Ab})$.*

PROOF. a) Since $(C, C)/H$ is a simple $\mathrm{End}(C)$-module, it follows from the basic existence theorems that there is an exact sequence $0 \to \mathrm{Hom}_R(\mathrm{Tr}_L\, C, R) \xrightarrow{g} B \xrightarrow{f} C \to 0$ in $L(\Lambda)$ with the desired properties.

b) Since $f\colon B \to C$ is a right C-determined morphism with $\mathrm{Im}(\ , f)(C)$ a maximal right ideal of $\mathrm{End}\, C$, we know that $\mathrm{Coker}(\ , f)$ is simple (see Proposition 2.1). The rest of b) follows trivially from this remark.

c) Since $0 \to \mathrm{Hom}_R(\mathrm{Tr}_L\, C, R) \xrightarrow{g} B \xrightarrow{f} C \to 0$ is an exact sequence in $L(\Lambda)$, we know by Proposition 4.1 that $\mathrm{Coker}(\ , f)$ is simple in $(L(\Lambda)^{\mathrm{op}}, \mathrm{Ab})$ if and only if $\mathrm{Coker}(g,\ )$ is simple in $(L(\Lambda), \mathrm{Ab})$. Thus the fact that $\mathrm{Coker}(\ , f)$ is simple implies $\mathrm{Coker}(g,\ )$ is simple.

d) Trivial consequence of c).

e) Trivial. □

Combining these results we have the following.

PROPOSITION 7.3. *Each simple functor S in $(L(\Lambda)^{\text{op}}, \text{Ab})$ vanishing on projectives has a projective resolution of the form* $0 \to (\ ,\text{Hom}_R(\text{Tr}_L C, R)) \xrightarrow{(\ ,g)} (\ ,B) \xrightarrow{(\ ,f)} (\ ,C) \to S \to 0$ *where* $0 \to \text{Hom}_R(\text{Tr}_L C, R) \xrightarrow{g} B \xrightarrow{f} C \to 0$ *is exact in* $L(\Lambda)$.

As usual we have the dual results.

PROPOSITION 7.4. *Let A be a noninjective lattice and H a maximal left ideal of* End A *containing* $I(A, A)$. *There is an exact sequence* $0 \to A \xrightarrow{g} B \xrightarrow{f} \text{Tr}_L \text{Hom}_R(A, R), \to 0$ *in* $L(\Lambda)$ *having the following properties.*

a) *g is left A-determined with* $\text{Im}(g,\ )(A) = H$.

b) *The exact sequence* $0 \to (\text{Tr}_L \text{Hom}_R(A, R),\ ) \xrightarrow{(f,\ )} (B,\ ) \xrightarrow{(g,\ )} (A,\ ) \to \text{Coker}(g,\ ) \to 0$ *is a projective resolution of the simple functor* $\text{Coker}(g,\ )$.

c) $0 \to (\ ,A) \xrightarrow{(\ ,g)} (\ ,B) \xrightarrow{(\ ,f)} (\ ,\text{Tr}_L \text{Hom}_R(A, R)) \to \text{Coker}(\ ,f) \to 0$ *is a projective resolution of the simple functor* $\text{Coker}(\ ,f)$.

d) *f is right* $\text{Tr}_L(\text{Hom}_R(A, R))$*-determined with* $\text{Im}(f,\ )(\text{Tr}_L(\text{Hom}_R(A, R))$ *a maximal right ideal of* End $\text{Tr}_L(\text{Hom}_R(A, R))$.

e) *If T is a simple functor in $(L(\Lambda), \text{Ab})$ such that $T \cong (A,\ )/(A,\ )_H$, then there is an exact sequence* $0 \to (\text{Tr}_L \text{Hom}_R(A, R),\ ) \xrightarrow{(f,\ )} (B,\ ) \xrightarrow{(g,\ )} (A,\ ) \to T \to 0$ *which is a projective resolution of T.*

COROLLARY 7.5. *Let T be a simple functor in $(L(\Lambda), \text{Ab})$ which vanishes on injective lattices. Then there is a projective resolution* $0 \to (\text{Tr}_L \text{Hom}_R(A, R),\ ) \xrightarrow{(f,\ )} (B,\ ) \xrightarrow{(g,\ )} (A,\ ) \to T \to 0$ *with* $0 \to A \xrightarrow{g} B \xrightarrow{f} \text{Tr}_L \text{Hom}_R(A, R) \to 0$ *exact in $L(\Lambda)$. In particular T is finitely presented.*

We now consider simple functors in $(L(\Lambda)^{\text{op}}, \text{Ab})$ which do not vanish on projectives. If S is a simple functor such that $S(P) \neq 0$, then we have an epimorphism $(\ ,P) \to S \to 0$. So it would be helpful to be able to determine the maximal subfunctors in $(L(\Lambda)^{\text{op}}, \text{Ab})$ of $(\ ,P)$. Suppose we consider $(\ ,P)$ a functor from all of Mod Λ to Ab. Then in Proposition 5.5 we saw that the maximal subfunctors of $(\ ,P)$ are the inclusion morphisms $(\ ,M) \to (\ ,P)$ as M ranges over all maximal submodules of P. This result suggests the following considerations.

Suppose $\dim R = d = 0$. Then each submodule of P is a lattice since $L(\Lambda) = \text{noeth}\,\Lambda$ in this case. In particular, each maximal submodule M of P is a lattice and it is obvious that each of the inclusions $0 \to (\ ,M) \to (\ ,P)$ is a maximal subfunctor of $(\ ,P)$ in $(L(\Lambda)^{\text{op}}, \text{Ab})$. The proof of Proposition 5.5 can be used to show these are precisely all the maximal subfunctors of $(\ ,P)$ in $(L(\Lambda)^{\text{op}}, \text{Ab})$ in this case.

Suppose now that $\dim R = d = 1$. Let C' be a nonzero submodule of the lattice C. Then C' has no nonzero submodules of finite length since this is true for C. Thus C' is Cohen-Macaulay since $\dim R = 1$. Hence we know that C' is a lattice if $(C')_{\underline{p}}$ is $\Lambda_{\underline{p}}$-projective for all nonmaximal prime ideals $\underline{p}$ of R. Suppose C/C'' is a module of finite length. Then $(C/C'')_{\underline{p}} = 0$ for all nonmaximal prime ideals $\underline{p}$ in R. Hence $C''_{\underline{p}} = C_{\underline{p}}$ for all nonmaximal primes $\underline{p}$ of R and so $C''_{\underline{p}}$ is $\Lambda_{\underline{p}}$-projective for all nonmaximal prime ideals $\underline{p}$ of R. Hence if C' is a submodule of the lattice C such that C/C' has finite length then C' is a lattice. In particular, every maximal submodule of a lattice is a lattice. This observation enables us to show, as in Proposition 5.5, that if P is a projective lattice, then the maximal subfunctors of $(\ ,P)$ in $(L(\Lambda)^{\text{op}}, \text{Ab})$ are precisely the subfunctors of

$(\ , P)$ of the form $\operatorname{Im}((\ , M) \to (\ , P))$ given by the inclusion of the maximal sublattices M of P.

Summarizing this discussion we have the following.

PROPOSITION 7.6. *Let Λ be an R-order with $\dim R \leq 1$.*

a) *If C' is a submodule of the lattice C such that C/C' has finite length, then C' is a lattice.*

b) *If P is a projective lattice, then the maximal subfunctors of $(\ , P)$ in $(L(\Lambda)^{\text{op}}, \text{Ab})$ are precisely the subfunctors $\operatorname{Im}((\ , M) \to (\ , P))$ as $M \to P$ ranges over all inclusion morphisms of maximal sublattices M of P.*

c) *If S is a simple functor in $(L(\Lambda)^{\text{op}}, \text{Ab})$ such that $S(P) \neq 0$ for some projective lattice P, then there is a maximal sublattice M of P such that the sequence $0 \to (\ , M) \to (\ , P) \to S \to 0$ is exact and hence a projective resolution of S in $(L(\Lambda)^{\text{op}}, \text{Ab})$. In particular, S is finitely presented in $(L(\Lambda)^{\text{op}}, \text{Ab})$ and $\operatorname{pd} S \leq 1$.*

Continuing with our assumption that $\dim R = d \leq 1$, it is possible, using the duality $(\ , R)\colon L(\Lambda) \to L(\Lambda^{\text{op}})$, to deduce from Proposition 7.7 analogous results concerning the simple functors in $(L(\Lambda), \text{Ab})$ not vanishing on injective lattices.

Suppose A in $L(\Lambda)$ is an injective lattice. Then $\operatorname{Hom}_R(A, R)$ in $L(\Lambda^{\text{op}})$ is projective. Suppose $0 \to M \xrightarrow{g} \operatorname{Hom}_R(A, R)$ is a maximal sublattice of $\operatorname{Hom}_R(A, R)$. This induces a morphism of lattices $A \xrightarrow{\operatorname{Hom}_R(g,R)} \operatorname{Hom}_R(M, R)$ which is an epimorphism with simple kernel if $\dim R = 0$ and is a monomorphism with a simple Λ-module as cokernel if $\dim R = 1$. But in either case $\operatorname{Im}((\operatorname{Hom}_R(M, R), \ ) \to (A, \ ))$ is a maximal subfunctor of $(A, \ )$ in $(L(\Lambda), \text{Ab})$, as we now show.

Since $\operatorname{Hom}_R(\ , R)\colon L(\Lambda) \to L(\Lambda^{\text{op}})$ is a duality, we know by I, Lemma 2.3 and Proposition 2.4 that $A \xrightarrow{\operatorname{Hom}_R(g,R)} \operatorname{Hom}_R(M, R)$ is left A-determined with $\operatorname{Im}(\operatorname{Hom}_R(g, R), \ )(A)$ a maximal left ideal of $\operatorname{End} A$ because $M \xrightarrow{g} \operatorname{Hom}_R(A, R)$ is right $\operatorname{Hom}_R(A, R)$-determined with $\operatorname{Im}(\ , g)(\operatorname{Hom}_R(A, R))$ a maximal right ideal of $\operatorname{End}\operatorname{Hom}_R(A, R)$. Therefore $\operatorname{Im}(\operatorname{Hom}_R(g, R), \ )$ is a maximal subfunctor of $(A, \ )$ by Proposition 2.1. We now show that we obtain all the maximal subfunctors of $(A, \ )$ in this way.

Suppose $F \subset (A, \ )$ is a maximal subfunctor of $(A, \ )$. Then we know by Corollary 1.2 that $F(A)$ is a maximal left ideal of $\operatorname{End} A$ and F is determined by A. Under the ring isomorphism $\operatorname{End} A \to (\operatorname{End}\operatorname{Hom}_R(A, R))^{\text{op}}$ we know that $\operatorname{Hom}_R(F(A), R)$ is a maximal right ideal of $(\operatorname{End}(\operatorname{Hom}_R(A, R))^{\text{op}}$. Thus $(\ , \operatorname{Hom}_R(A, R))_{\operatorname{Hom}_R(F(A),R)}$ is a maximal subfunctor of $(\ , \operatorname{Hom}_R(A, R))$. Hence by Proposition 7.6, there is a maximal sublattice $0 \to M \xrightarrow{g} \operatorname{Hom}_R(A, R)$ of $\operatorname{Hom}_R(A, R)$ which is right $\operatorname{Hom}_R(A, R)$-determined with $\operatorname{Im}(g, \ )(\operatorname{Hom}_R(A, R)) = DF(A)$. Hence by I, Lemma 2.3 and Proposition 2.4, $A \xrightarrow{\operatorname{Hom}_R(g,R)} \operatorname{Hom}_R(M, R)$ is left A-determined with $\operatorname{Im}(\operatorname{Hom}_R(g, R), \ )(A) = F(A)$. Therefore $\operatorname{Im}(\operatorname{Hom}_R(g, R), \ ) = F$.

Summarizing this discussion we have the following.

PROPOSITION 7.7. *Let Λ be an R-order with $\dim R \leq 1$.*

a) *Let A be an injective lattice. A subfunctor F of $(A, \ )$ is maximal if and only if there is a morphism of lattices $A \xrightarrow{g} B$ satisfying the following:*

i) *$0 \to \operatorname{Hom}_R(B, R) \xrightarrow{\operatorname{Hom}_R(g,R)} \operatorname{Hom}_R(A, R)$ is exact with $\operatorname{Im}(\operatorname{Hom}_R(g, R))$ a maximal sublattice of $\operatorname{Hom}_R(A, R)$, and $\operatorname{Im}(g, \ ) = F$.*

b) *$0 \to (B, \ ) \xrightarrow{(g, \ )} (A, \ )$ is exact and $\operatorname{Im}(g, \ ) = F$.*

c) *If $\dim R = 0$, then g is an epimorphism with $\operatorname{Ker} g$ simple.*

If $\dim R = 1$, *then* g *is a monomorphism with* $\operatorname{Coker} g$ *simple. If* T *is a simple functor in* $(L(\Lambda), \mathrm{Ab})$ *such that* $T(A) \neq 0$, *then there is a morphism* $A \xrightarrow{g} B$ *in* $L(\Lambda)$ *such that* $0 \to (B, \ ) \to (A, \ ) \to T \to 0$ *is a projective resolution of* T. *Thus* T *is a finitely presented functor in* $(L(\Lambda), \mathrm{Ab})$ *and* $\operatorname{pd} T \leq 1$.

PROOF. The only parts of this proposition that need any explanation are parts b) and c).

By a) we have that $(*)$ $0 \to \operatorname{Hom}_R(B, R) \to \operatorname{Hom}_R(A, R) \to V \to 0$ is exact with V a simple Λ-module. If $\dim R = 0$, then R is selfinjective. Applying $\operatorname{Hom}_R(\ , R)$ to $(*)$ we have the exact sequence $0 \to \operatorname{Hom}_R(V, R) \to A \xrightarrow{g} B \to 0$ in $\operatorname{noeth} \Lambda$. Since $\operatorname{Hom}_R(\ , R)\colon \operatorname{noeth} \Lambda \to \operatorname{noeth} \Lambda^{\mathrm{op}}$ we know that $\operatorname{Hom}_R(V, R)$ is a simple Λ-module since V is a simple Λ-module. Also $0 \to (B, \ ) \to (A, \ )$ is obviously exact in $(L(\Lambda), \mathrm{Ab})$.

Suppose $\dim R = 1$. Then applying $\operatorname{Hom}_R(\ , R)$ to $(*)$ we obtain the exact sequence $0 \to A \xrightarrow{g} B \to \operatorname{Ext}^1_R(V, R) \to 0$ since $\operatorname{Hom}_R(V, R) = 0$ because V is of finite length and $\operatorname{depth} R = 1$ and $\operatorname{Ext}^1_R(X, R) = 0$ for all lattices X. Because R is a Gorenstein ring of dimension 1, the functor $\operatorname{Ext}^1_R(\ , R)\colon \operatorname{noeth} \Lambda^{\mathrm{op}} \to \operatorname{noeth} \Lambda$ is a duality on modules of finite length. Hence the fact that V is a simple Λ^{op}-module implies $\operatorname{Ext}^1_R(V, R)$ is a simple Λ-module. Hence $0 \to (B, X) \to (A, X)$ is exact for all X in $L(\Lambda)$, since $\operatorname{Hom}_\Lambda(\operatorname{Ext}^1_\Lambda(V, R), X) = 0$ for all lattices X because $\operatorname{Ext}^1_\Lambda(V, R)$ has finite length and no lattice has a nonzero submodule of finite length because $\dim R = 1$. Thus $0 \to (B, \ ) \xrightarrow{(g, \ )} (A, \ )$ is exact in $(L(\Lambda), \mathrm{Ab})$. □

Summarizing this discussion, we have the following.

COROLLARY 7.8. *Let* Λ *be an* R*-order with* $\dim R \leq 1$. *Then the simple functors in* $(L(\Lambda)^{\mathrm{op}}, \mathrm{Ab})$ *not vanishing on projectives and the simple functors in* $(L(\Lambda), \mathrm{Ab})$ *not vanishing on injectives are finitely presented and have projective dimension at most* 1.

We next point out the following analogue of Proposition 5.8, which we state without proof since it can either be easily deduced from Proposition 5.8 or proven using essentially the same proof as that given for Proposition 5.8.

PROPOSITION 7.9. *Let* Λ *be an* R*-order with* $\dim R \leq 1$. *Suppose* C *is in* $L(\Lambda)$ *and* H *a maximal right ideal in* $\operatorname{End} C$ *and* S *the simple functor* $(\ , C)/(\ , C)_H$ *in* $(L(\Lambda)^{\mathrm{op}}, \mathrm{Ab})$.

a) $S(\Lambda) \neq 0$ *if and only if* H *does not contain* $P(C, C)$.

b) $S(\Lambda) \neq 0$ *if and only if there is a sublattice* M *of* C *such that* $0 \to (\ , M) \to (\ , C) \to S \to 0$ *is exact in* $(L(\Lambda)^{\mathrm{op}}, \mathrm{Ab})$, *i.e.* $(\ , M) = (\ , C)_H$.

c) *Let* M *be a sublattice of* C. *Then the following statements are equivalent.*

i) $(\ , C)/(\ , M)$ *is simple.*

ii) *If* $h\colon L \to C$ *is a morphism in* $L(\Lambda)$ *with* $\operatorname{Im} h \not\subset M$, *then the induced morphism* $M \amalg L \to C$ *is a splittable epimorphism.*

The following is an immediate consequence of this proposition.

COROLLARY 7.10. *Suppose* Λ *is an* R*-order with* $\dim R \leq 1$. *Let* C *be in* $L(\Lambda)$. *The two sided ideal* $P(C, C)$ *of* $\operatorname{End} C$ *is not contained in the radical of* $\operatorname{End} C$ *if and only if there is a sublattice* M *of* C *having the property that if* $h\colon L \to C$ *is a morphism in* $L(\Lambda)$ *with* $\operatorname{Im} h \not\subset M$, *then the induced morphism* $M \amalg L \to C$ *is a splittable epimorphism.*

There are obvious analogues of Proposition 7.9 and Corollary 7.10 dealing with the two sided ideal $I(A, A)$ of $\operatorname{End} A$ for A in $L(\Lambda)$ where Λ is an R-order such that $\dim R \leq 1$. The statements and proofs of these analogues are left to the reader to give.

8. Simple functors for complete orders

Throughout this section we assume that Λ is an R-order with R a complete local ring. Our main purpose is to point out the influence the special properties of lattices have when R is a complete local ring on the properties of simple functors in $(L(\Lambda)^{\text{op}}, \text{Ab})$ and $(L(\Lambda), \text{Ab})$. Most of these special features of the theory in this case are obvious analogues of results given in Section 6 for the case Λ is a noetherian R-algebra with R a complete local ring. We omit those proofs which are essentially the same as those given in Section 6 if one keeps in mind the results of Section 7.

Let S be a simple functor in $(L(\Lambda)^{\text{op}}, \text{Ab})$. Since S is additive and every lattice is a finite sum of indecomposable lattices, S does not vanish on some indecomposable lattice C. Hence there is a nonzero morphism $\alpha\colon (\ ,C) \to S$ which must be an epimorphism since S is simple. Because R is complete, we know that $\operatorname{End} C$ is local. Therefore the epimorphism $\alpha\colon (\ ,C) \to S$ is a projective cover of S (see Proposition 1.10). Moreover $0 \to \underline{r}(\ ,C) \to (\ ,C) \to S \to 0$ is exact, i.e. $(\ ,C)/\underline{r}(\ ,C) \cong S$.

On the other hand, suppose C is indecomposable in $L(\Lambda)$. Since $\operatorname{End} C$ is local, $\underline{r}(\ ,C)$ is the unique maximal subfunctor of $(\ ,C)$ in $(L(\Lambda)^{\text{op}}, \text{Ab})$. Therefore $(\ ,C)/\underline{r}(\ ,C)$ is a simple functor and the natural epimorphism $(\ ,C) \to (\ ,C)/\underline{r}(\ ,C)$ is a projective cover. Finally, suppose C and C' are indecomposable lattices such that the simple functors $(\ ,C)/\underline{r}(\ ,C)$ and $(\ ,C')/\underline{r}(\ ,C')$ in $(L(\Lambda)^{\text{op}}, \text{Ab})$ are isomorphic. Then the functors $(\ ,C)$ and $(\ ,C')$ are isomorphic since they are projective covers of the isomorphic functors $(\ ,C)/\underline{r}(\ ,C)$ and $(\ ,C')/\underline{r}(\ ,C')$ respectively. Hence two indecomposable lattices C and C' are isomorphic if and only if the simple functors $(\ ,C)/\underline{r}(\ ,C)$ and $(\ ,C')/\underline{r}(\ ,C')$ are isomorphic.

We summarize this discussion as follows.

PROPOSITION 8.1. *The map $L(\Lambda) \to (L(\Lambda)^{\text{op}}, \text{Ab})$ given by $C \mapsto (\ ,C)/\underline{r}(\ ,C)$ for all C in $L(\Lambda)$ induces a bijection between isomorphism classes of indecomposable lattices and the isomorphism classes of simple functors in $(L(\Lambda)^{\text{op}}, \text{Ab})$. Moreover a simple functor S in $(L(\Lambda)^{\text{op}}, \text{Ab})$ is isomorphic to $(\ ,C)/\underline{r}(\ ,C)$ for some indecomposable C in $L(\Lambda)$ if and only if $S(C) \neq 0$.*

Having determined the projective covers for simple functors in $(L(\Lambda)^{\text{op}}, \text{Ab})$, we now show that these functors have minimal projective resolutions, at least in certain cases. We first consider those simple functors which vanish on projectives.

PROPOSITION 8.2. *Let C be indecomposable in $L_P(\Lambda)$.*

a) *There exists an almost split sequence $0 \to D\operatorname{Tr}_L C \xrightarrow{g} B \xrightarrow{f} C \to 0$ (where $D(X) = \operatorname{Hom}_R(X,R)$) in $L(\Lambda)$. This sequence has the following properties:*

i) *$0 \to (\ ,D\operatorname{Tr}_L C) \xrightarrow{(\ ,g)} (\ ,B) \xrightarrow{(\ ,f)} (\ ,C) \to \operatorname{Coker}(\ ,f) \to 0$ is a minimal projective resolution in $(L(\Lambda)^{\text{op}}, \text{Ab})$ of the simple functor $\operatorname{Coker}(\ ,f)$ which does not vanish on C.*

ii) *$0 \to (C,\ ) \xrightarrow{(f,\ )} (B,\ ) \xrightarrow{(g,\ )} (D\operatorname{Tr}_L C,\ ) \to \operatorname{Coker}(g,\ ) \to 0$ is a minimal projective resolution in $(L(\Lambda), \text{Ab})$ of the simple functor $\operatorname{Coker}(g,\ )$ which does not vanish on $D\operatorname{Tr}_L C$.*

b) *Suppose S is a simple functor in $(L(\Lambda)^{\text{op}}, \text{Ab})$ such that $S(C) \neq 0$. Then $S \cong \operatorname{Coker}(\ ,f)$, so there is an exact sequence $0 \to (\ ,D\operatorname{Tr}_L C) \xrightarrow{(\ ,g)} (\ ,B) \xrightarrow{(\ ,f)} S \to 0$ which is a minimal projective resolution of S in $(L(\Lambda^{\text{op}}), \text{Ab})$.*

c) *Suppose* T *is a simple functor in* $(L(\Lambda), \mathrm{Ab})$ *such that* $T(D\,\mathrm{Tr}_L\,C) \neq 0$. *Then* $T \cong \mathrm{Coker}(g,\ )$, *so there is an exact sequence* $0 \to (C,\ ) \to (B,\ ) \to (D\,\mathrm{Tr}_L\,C,\ ) \to T \to 0$ *which is a minimal projective resolution of* T *in* $(L(\Lambda), \mathrm{Ab})$.

PROOF. Analogous to the proof of Proposition 6.3 . □

We now turn our attention to the simple functors in $(L(\Lambda)^{\mathrm{op}}, \mathrm{Ab})$ which do not vanish on projectives. We restrict our discussion to the classical situations of $\dim R \leq 1$.

PROPOSITION 8.3. *Suppose* $\dim R \leq 1$ *and* $\underline{r}$ *is the radical of* Λ. *Let* P *be an indecomposable projective* Λ*-lattice.*

a) $\underline{r}P$ *is the unique maximal sublattice of* P.

b) *The inclusion* $\underline{r}P \to P$ *is a minimal right almost split morphism in* $L(\Lambda)$.

c) $0 \to (\ ,\underline{r}P) \to (\ ,P) \to (\ ,P)/\underline{r}(\ ,P) \to 0$ *is a minimal projective resolution in* $(L(\Lambda)^{\mathrm{op}}, \mathrm{Ab})$ *of the simple functor* $(\ ,P)/\underline{r}(\ ,P)$.

Suppose S *is a simple functor in* $(L(\Lambda)^{\mathrm{op}}, \mathrm{Ab})$ *such that* $S(P) \neq 0$.

d) *Then* $S \cong (\ ,P)/\underline{r}(\ ,P)$ *and there is an exact sequence* $0 \to (\ ,\underline{r}P) \to (\ ,P) \to S \to 0$ *which is a minimal projective resolution of* S *in* $(L(\Lambda)^{\mathrm{op}}, \mathrm{Ab})$.

e) $\mathrm{pd}\, S \leq 1$.

PROOF. The proof is the same as that for Proposition 6.2, if one keeps in mind the fact established in Proposition 7.6 that a submodule C' of a lattice C is a sublattice provided C/C' is a module of finite length. □

Having described the minimal projective resolutions of simple functors in $(L(\Lambda)^{\mathrm{op}}, \mathrm{Ab})$ (at least in case $\dim R \leq 1$), we now turn our attention to describing the minimal projective resolutions for the simple functors in $(L(\Lambda), \mathrm{Ab})$. We begin with the following analogue of Proposition 8.1.

PROPOSITION 8.4. a) *For each indecomposable* A *in* $L(\Lambda)$, *we have that* $(A,\ )/\underline{r}(A,\ )$ *is a simple functor not vanishing on* A *and that the natural morphism* $(A,\ ) \to (A,\ )/\underline{r}(A,\ )$ *is a projective cover.*

b) *The map* $L(\Lambda) \to (L(\Lambda), \mathrm{Ab})$ *given by* $A \mapsto (A,\ )/\underline{r}(A,\ )$ *for all* A *in* $L(\Lambda)$ *induces a bijection between the isomorphism classes of indecomposable lattices and the isomorphism classes of simple functors in* $(L(\Lambda), \mathrm{Ab})$. *Moreover a simple functor* T *is isomorphic to* $(A,\ )/\underline{r}(A,\ )$ *for an indecomposable* A *if and only if* $T(A) \neq 0$.

Having determined the projective cover for simple functors in $(L(\Lambda), \mathrm{Ab})$ we now show that those which vanish on injective lattices have minimal projective resolutions in $(L(\Lambda), \mathrm{Ab})$. We do this in the following analogue of Proposition 6.6.

PROPOSITION 8.5. *Let* A *be an indecomposable lattice in* $L_I(\Lambda)$.

a) *There exists an almost split sequence* $0 \to A \xrightarrow{g} B \xrightarrow{f} \mathrm{Tr}_L\,DA \to 0$ *in* $L(\Lambda)$. *This sequence has the following properties.*

i) $0 \to (\mathrm{Tr}_L\,DA,\ ) \xrightarrow{(f,\ )} (B,\ ) \xrightarrow{(g,\ )} (A,\ ) \to \mathrm{Coker}(g,\ ) \to 0$ *is a minimal projective resolution of the simple functor* $\mathrm{Coker}(g,\ )$ *which does not vanish on* A.

ii) $0 \to (\ ,A) \xrightarrow{(\ ,g)} (\ ,B) \xrightarrow{(\ ,f)} (\ ,\mathrm{Tr}_L\,DA) \to \mathrm{Coker}(\ ,f) \to 0$ *is a minimal projective resolution for the simple functor* $\mathrm{Coker}(\ ,f)$ *which does not vanish on* $\mathrm{Tr}_L\,DA$.

b) *Suppose* T *is a simple functor in* $(L(\Lambda), \mathrm{Ab})$ *such that* $T(A) \neq 0$. *Then* $T \cong \mathrm{Coker}(g,\ )$, *so there is an exact sequence* $0 \to (\mathrm{Tr}_L\,DA,\ ) \xrightarrow{(f,\ )} (B,\ ) \xrightarrow{(g,\ )} (A,\ ) \to T \to 0$ *which is a projective resolution of* T.

c) *Suppose S is a simple functor in $(L(\Lambda)^{\text{op}}, \text{Ab})$ such that $S(\operatorname{Tr}_L DA) \neq 0$. Then $S \cong \operatorname{Coker}(\ ,f)$, so there is an exact sequence $0 \to (\ ,A) \xrightarrow{(\ ,g)} (\ ,B) \xrightarrow{(\ ,f)} (\ ,\operatorname{Tr}_L DA) \to S \to 0$ which is a minimal projective resolution of S.*

Having described minimal projective resolutions for simple functors T in $(L(\Lambda), \text{Ab})$ vanishing on injective lattices, we now describe minimal projective resolutions for those that do not vanish on injective lattices in the case $\dim R \leq 1$. This is accomplished in the following analogue of Proposition 6.5.

PROPOSITION 8.6. *Suppose $\dim R \leq 1$. Let I be an indecomposable injective lattice. Then $D(I)$ is an indecomposable projective Λ^{op}-lattice with unique maximal sublattice $\underline{r}D(I)$. Let $I_0 = D(\underline{r}D(I))$ and let $I \to I_0$ be the dual of the inclusion $\underline{r}D(I) \to D(I)$. Then the morphism $I \to I_0$ has the following properties.*

a) *$I \to I_0$ is a minimal left almost split morphism.*

b) *$0 \to (I_0,\ ) \to (I,\ ) \to (I,\ )/\underline{r}(I,\ ) \to 0$ is a minimal projective resolution of the simple functor $(I,\ )/\underline{r}(I,\ )$ in $(L(\Lambda), \text{Ab})$.*

Suppose T is a simple functor in $(L(\Lambda), \text{Ab})$ such that $T(I) \neq 0$.

c) *Then $T \cong (I,\ )/I_0,\ )$ and so there is an exact sequence $0 \to (I_0,\ ) \to (I,\ ) \to T \to 0$ which is a minimal projective resolution of T.*

PROOF. Easily deduced from Proposition 7.7. □

CHAPTER III: SOME SPECIAL ORDERS

This chapter is devoted mainly to pointing out some special features of our general theory of lattices for particular types of orders including Gorenstein orders and R-orders where R is a complete local ring of dimension 1 or 2. In the last section we return to some general considerations which put in a more general setting are our basic existence theorem.

1. Gorenstein orders

Let R be an equidimensional Gorenstein ring. A noetherian R-algebra Λ is said to be a *Gorenstein R-order* if a) Λ is a Cohen-Macaulay R-module and b) $\Lambda \cong \operatorname{Hom}_R(\Lambda, R)$ as two-sided Λ-modules. It is obvious that a Gorenstein R-order is an R-order. The following are some important examples of Gorenstein R-orders.

a) R itself is a Gorenstein R-order.

b) If G is a finite group, then $R[G]$ is a Gorenstein R-order.

c) If Λ is a symmetric finite dimensional algebra over a field k, then Λ is a Gorenstein k-order.

This section is devoted primarily to pointing out some of the special features the category $L(\Lambda)$ has when Λ is a Gorenstein R-order. These properties will be used in this and the next section to prove some special properties of lattices over Gorenstein R-orders. We assume throughout this section that Λ is a Gorenstein R-order and that $\alpha\colon \Lambda \to \operatorname{Hom}_R(\Lambda, R)$ is a fixed two-sided isomorphism. We begin our discussion with the following preliminaries.

PROPOSITION 1.1. a) *The R-morphism* $\alpha^{\mathrm{op}}\colon \Lambda^{\mathrm{op}} \to \mathrm{Hom}_R(\Lambda^{\mathrm{op}}, R)$ *given by* $\alpha^{\mathrm{op}}(\lambda) = \alpha(\lambda)$ *for all* $\lambda \in \Lambda$, *is a two-sided* Λ^{op}*-isomorphism which we also denote by* α. *Thus* Λ^{op} *is a Gorenstein R-order.*

b) *The functors* $\mathrm{Hom}_\Lambda(\ ,\Lambda)$, $\mathrm{Hom}_\Lambda(\ ,R)\colon \mathrm{noeth}\,\Lambda \to \mathrm{noeth}\,\Lambda^{\mathrm{op}}$ *are isomorphic.*

c) *If M is in* $L(\Lambda)$, *then* $\mathrm{Hom}_\Lambda(M,\Lambda) \cong \mathrm{Hom}_R(M,R)$ *is in* $L(\Lambda^{\mathrm{op}})$.

d) *If M is in* $L(\Lambda)$, *then the natural morphism* $M \to \mathrm{Hom}_{\Lambda^{\mathrm{op}}}(\mathrm{Hom}_\Lambda(M,\Lambda),\Lambda^{\mathrm{op}})$ *is an isomorphism of Λ-lattices.*

e) *The functors* $\mathrm{Hom}_\Lambda(\ ,\Lambda)\colon L(\Lambda) \to L(\Lambda^{\mathrm{op}})$ *and* $\mathrm{Hom}_{\Lambda^{\mathrm{op}}}(\ ,\Lambda^{\mathrm{op}})\colon L(\Lambda^{\mathrm{op}}) \to L(\Lambda)$ *are inverse dualities.*

f) *If M is in* $L(\Lambda)$, *then* $\mathrm{Ext}^i_\Lambda(M,\Lambda) = 0 = \mathrm{Ext}^i_{\Lambda^{\mathrm{op}}}(\mathrm{Hom}_\Lambda(M,\Lambda),\Lambda^{\mathrm{op}})$ *for all* $i > 0$.

g) *For each* $i > 0$, *the R-algebras* $\underline{\mathrm{End}}(M)$ *and* $\underline{\mathrm{End}}(\Omega^i M)$ *are isomorphic R-algebras.*

h) *Let* $0 \to \Omega^2(M) \to P_1 \to P_0 \to M \to 0$ *be exact with the* P_i *finitely generated projective Λ-modules and M a lattice.*

i) $\Omega^2(M)$ *is in* $L(\Lambda)$.

ii) $\mathrm{Tr}(M) \cong \mathrm{Hom}_\Lambda(\Omega^2(M),\Lambda)$ *and is therefore in* $L(\Lambda)$.

iii) $\mathrm{Hom}_R(\mathrm{Tr}\,M, R) \cong \mathrm{Hom}_\Lambda(\mathrm{Tr}(M),\Lambda) \cong \Omega^2(M)$.

iv) Λ *is an injective as well as projective lattice. Thus* $P(\Lambda,B) = I(\Lambda,B)$ *for all lattices and so* $\underline{L}(\Lambda) = \overline{L}(\Lambda)$.

PROOF. a) Easily checked.

b) Let $\alpha\colon \Lambda \to \mathrm{Hom}_R(\Lambda,R)$ be our given two-sided isomorphism. Then for each M in $\mathrm{noeth}\,\Lambda$, the isomorphism α induces an isomorphism of Λ^{op}-modules $\mathrm{Hom}_\Lambda(M,\Lambda) \to \mathrm{Hom}_\Lambda(M,\mathrm{Hom}_R(\Lambda,R))$ which is functorial in M. Standard associative laws give isomorphisms $\mathrm{Hom}_\Lambda(M,\mathrm{Hom}_R(\Lambda,R)) \cong \mathrm{Hom}_R(M,R)$ functorial in M. Hence we obtain isomorphisms $\mathrm{Hom}_\Lambda(M,\Lambda) \cong \mathrm{Hom}_R(M,R)$ functorial in M.

c) Let M be in $L(\Lambda)$. Then $\mathrm{Hom}_R(M,R)$ is a Λ^{op}-lattice. Since by b) we have that $\mathrm{Hom}_R(M,R) \cong \mathrm{Hom}_\Lambda(M,\Lambda)$, it follows that $\mathrm{Hom}_\Lambda(M,\Lambda)$ is a lattice.

d) If M is in $L(\Lambda)$, then we know that the natural morphism $M \to \mathrm{Hom}_R(\mathrm{Hom}_R(M,R),R)$ is an isomorphism. From this it follows, using our fixed two-sided isomorphisms $\Lambda \to \mathrm{Hom}_R(\Lambda,R)$ and $\Lambda^{\mathrm{op}} \to \mathrm{Hom}_R(\Lambda^{\mathrm{op}},R)$, that $M \to \mathrm{Hom}_{\Lambda^{\mathrm{op}}}(\mathrm{Hom}_\Lambda(M,\Lambda),\Lambda^{\mathrm{op}})$ is an isomorphism.

e) Easy consequence of d).

f) Let $0 \to R \to I_0 \to I_1 \to \cdots \to I_d \to 0$ be minimal injective resolution of R. Since $\mathrm{Hom}_R(\Lambda,R) \cong \Lambda$ is an R-lattice, we know that $\mathrm{Ext}^i_R(\mathrm{Hom}_R(\Lambda,R),R) = 0$ for all $i > 0$. Thus $(*)$ $0 \to \mathrm{Hom}_R(\mathrm{Hom}_R(\Lambda,R),R) \to \mathrm{Hom}_R(\mathrm{Hom}_R(\Lambda,R),I_0) \cdots \to \mathrm{Hom}_R(\mathrm{Hom}_R(\Lambda,R),I_d) \to 0$ is exact. Since $\mathrm{Hom}_R(\Lambda,R) \cong \Lambda$ is a two-sided Λ-module, it follows that $\mathrm{Hom}_R(\mathrm{Hom}_R(\Lambda,R),I_2)$ is an injective Λ-module for all $i > 0$. Thus $(*)$ is an injective Λ-resolution of Λ. Since $\mathrm{Hom}_\Lambda(M,\mathrm{Hom}_R(\mathrm{Hom}_R(\Lambda,R),I_i)) \cong \mathrm{Hom}_R(\mathrm{Hom}_R(\Lambda,R)\otimes_\Lambda M, I_i)$ for all i and $\mathrm{Hom}_R(\Lambda,R) \cong \Lambda$ as a Λ^{op}-module, we have that $\mathrm{Hom}_\Lambda(M,\mathrm{Hom}_R(\mathrm{Hom}_R(\Lambda,R),I_i)) \cong \mathrm{Hom}_R(M,I_i)$ for all i. Therefore $\mathrm{Ext}^i_\Lambda(M,\Lambda) \cong \mathrm{Ext}^i_R(M,R)$ for all $i > 0$. The fact that M is an R-lattice, implies that $\mathrm{Ext}^i_R(M,R) = 0$ for $i > 0$. Hence $\mathrm{Ext}^i_\Lambda(M,\Lambda) = 0$ for all $i > 0$. The rest of f) follows trivially from what has been proven and the fact that $\mathrm{Hom}_\Lambda(M,\Lambda)$ is in $L(\Lambda^{\mathrm{op}})$ if M is in $L(\Lambda)$.

g) Follows from the fact that $\mathrm{Ext}^i_\Lambda(M,\Lambda) = 0$ for all $i > 0$ (see I, Proposition 7.4).

h) i) Since $0 \to \Omega^2(M) \to P_1 \to P_0 \to M \to 0$ is exact with the P_i and M in $L(\Lambda)$, it follows that $\Omega^2(M)$ is in $L(\Lambda)$.

ii) Since $\mathrm{Ext}^i_\Lambda(M,\Lambda) = 0$ for $i > 0$, it follows that $0 \to \mathrm{Hom}_\Lambda(M,\Lambda) \to \mathrm{Hom}_\Lambda(P_0,\Lambda) \to \mathrm{Hom}_\Lambda(P_1,\Lambda) \to \mathrm{Hom}_\Lambda(\Omega^2(M),\Lambda) \to 0$ is an exact sequence of Λ^{op}-modules. Thus $\mathrm{Hom}_\Lambda(\Omega^2(M),\Lambda) \cong \mathrm{Tr}\,M$.

iii) Since $\operatorname{Hom}_R(\operatorname{Tr} M, R) \cong \operatorname{Hom}_\Lambda(\operatorname{Tr} M, \Lambda)$, it follows that $\operatorname{Hom}_R(\operatorname{Tr} M, R) \cong \operatorname{Hom}_{\Lambda^{\mathrm{op}}}(\operatorname{Hom}_\Lambda(\Omega^2(M), \Lambda), \Lambda^{\mathrm{op}}) \cong \Omega^2(M)$.

iv) Obvious from the fact that $\Lambda \cong \operatorname{Hom}_R(\Lambda, R)$ as a two-sided Λ-module. □

In order to state the main point of this discussion of lattices over Gorenstein R-orders, it is convenient to make the following definitions.

Suppose we have chosen for each M in $\operatorname{noeth}\Lambda$ an epimorphism $P_M \to M \to 0$ with P_M a finitely generated projective Λ-module. Suppose we have done a similar thing for Λ^{op}-modules in $\operatorname{noeth}\Lambda^{\mathrm{op}}$. This enables us to construct particular projective resolutions for noetherian Λ and Λ^{op}-modules. For each M in $\operatorname{noeth}\Lambda$ and $\operatorname{noeth}\Lambda^{\mathrm{op}}$, we define $\Omega^i(M)$ using these projective resolutions. In particular, if M is in $L(\Lambda)$, then we know that $\Omega^i(M)$ is in $L(\Lambda)$ for all $i \geq 0$. Thus given M in $L(\Lambda)$, we have the Λ^{op}-lattices $\Omega^i(\operatorname{Hom}_\Lambda(M, \Lambda))$ associated with M for all $i > 0$. We define $\Omega^{-j}(M)$ to be the Λ-lattice $\operatorname{Hom}_{\Lambda^{\mathrm{op}}}(\Omega^j(\operatorname{Hom}_\Lambda(M, \Lambda), \Lambda^{\mathrm{op}}))$ for all $j \geq 0$. The following alternative description of $\Omega^{-1}(M)$ will perhaps make a little clearer what is going on.

Let $\cdots \xrightarrow{g_{i+1}} Q_i \to \cdots \xrightarrow{g_1} Q_0 \to \operatorname{Hom}_\Lambda(M, \Lambda) \to 0$ be our chosen projective resolution of the Λ^{op}-lattice $\operatorname{Hom}_\Lambda(M, \Lambda)$. Since $\operatorname{Ext}^i_{\Lambda^{\mathrm{op}}}(\operatorname{Hom}_\Lambda(M, \Lambda), \Lambda^{\mathrm{op}}) = 0$ for all $i > 0$, the sequence $0 \to \operatorname{Hom}_\Lambda(\operatorname{Hom}_\Lambda(M, \Lambda), \Lambda^{\mathrm{op}}) \to \operatorname{Hom}_{\Lambda^{\mathrm{op}}}(Q_0, \Lambda^{\mathrm{op}}) \to \cdots \xrightarrow{\operatorname{Hom}_{\Lambda^{\mathrm{op}}}(g_i, \Lambda^{\mathrm{op}})} \operatorname{Hom}_{\Lambda^{\mathrm{op}}}(Q_i, \Lambda^{\mathrm{op}}) \to \cdots$ is an exact sequence with the $\operatorname{Hom}_{\Lambda^{\mathrm{op}}}(Q_i, \Lambda^{\mathrm{op}})$ projective Λ-modules in $L(\Lambda)$. Then $\operatorname{Ker}(g_i, \Lambda^{\mathrm{op}}) = \Omega^{1-i}(M)$ for all $i > 0$. In this way we get associated with the lattice M an exact sequence of finitely generated projective Λ-modules

$$(*) \quad \cdots \to P_i \to \cdots \xrightarrow{f_2} P_1 \xrightarrow{f_1} P_0 \to \operatorname{Hom}_{\Lambda^{\mathrm{op}}}(Q_0, \Lambda^{\mathrm{op}}) \xrightarrow{\operatorname{Hom}_{\Lambda^{\mathrm{op}}}(g_1, \Lambda^{\mathrm{op}})} \operatorname{Hom}_{\Lambda^{\mathrm{op}}}(Q_1, \Lambda^{\mathrm{op}}) \to \cdots$$

with the following properties:

a) $\operatorname{Im} f_i = \Omega^i(M)$ for all $i \geq 0$.

b) $\operatorname{Im}(g_i, \Lambda^{\mathrm{op}}) = \Omega^{-i}(M)$ for all $i > 0$.

c) The sequence $(*)$ remains exact if we apply the functor $\operatorname{Hom}_\Lambda(\ , \Lambda)$ to it.

Using this description of the $\Omega^i(M)$ for all i in $\mathbb{Z}$, the group of all integers, it is not difficult to show the following.

LEMMA 1.2. *For each M in $L(\Lambda)$ we have*

a) $\Omega^j(\Omega^k(M)) \cong \Omega^{j+k}(M)$ *in $\underline{L}(\Lambda)$ for all j and k in $\mathbb{Z}$, the group of integers.*

b) $\Omega^j(\operatorname{Hom}_\Lambda(A, \Lambda)) \cong \operatorname{Hom}_\Lambda(\Omega^{-j}A, \Lambda)$ *in $\underline{L}(\Lambda)$ for all integers j.*

We now come to the main point of this section.

PROPOSITION 1.3. *Suppose* $\dim R = d$ *and A is in $L(\Lambda)$.*

a) $\operatorname{Hom}_R(\operatorname{Tr}_L A, R) \cong \Omega^{2-d}A$ *in $\underline{L}(\Lambda)$.*

b) $\operatorname{Tr}_L(\operatorname{Hom}_R(A, R)) \cong \Omega^{d-2}A$ *in $\underline{L}(\Lambda)$.*

PROOF. a) We recall (see I, Section 7) that $\operatorname{Tr}_L \colon \underline{L}(\Lambda) \to \underline{L}(\Lambda^{\mathrm{op}})$ is defined to be the composition $\underline{L}(\Lambda) \xrightarrow{\operatorname{Tr}} \underline{J}(\Lambda^{\mathrm{op}}) \xrightarrow{\Omega^d} \underline{L}(\Lambda^{\mathrm{op}})$. Let $0 \to \Omega^2 A \to P_1 \to P_0 \to A \to 0$ be exact with the P_i finitely generated projective Λ-modules. Then $0 \to \operatorname{Hom}_\Lambda(A, \Lambda) \to \operatorname{Hom}_\Lambda(P_0, \Lambda) \to \operatorname{Hom}_\Lambda(P_1, \Lambda) \to \operatorname{Hom}_\Lambda(\Omega^2 A, \Lambda) \to 0$ is exact since $\operatorname{Ext}^i_\Lambda(A, \Lambda) = 0$ for all $i > 0$. Thus $\operatorname{Tr}(A) = \Omega^{-2}\operatorname{Hom}_\Lambda(A, \Lambda)$. Hence $\operatorname{Tr}_L(A) = \Omega^d(\Omega^{-2}\operatorname{Hom}_\Lambda(A, \Lambda)) = \Omega^{d-2}(\operatorname{Hom}_\Lambda(A, \Lambda))$. Thus $\operatorname{Hom}_R(\operatorname{Tr}_L(A), R) \cong \operatorname{Hom}_\Lambda(\Omega^{d-2}(\operatorname{Hom}_\Lambda(A, \Lambda), \Lambda)$ (see Proposition 1.1). But by Lemma 1.2, we know that $\operatorname{Hom}_\Lambda(\Omega^{d-2}(\operatorname{Hom}_\Lambda(A, \Lambda), \Lambda) \cong \Omega^{2-d}(\operatorname{Hom}_{\Lambda^{\mathrm{op}}}(\operatorname{Hom}_\Lambda(A, \Lambda), \Lambda^{\mathrm{op}}))$. Since A is reflexive it follows that $\operatorname{Hom}_R(\operatorname{Tr}_L(A), R) \cong \Omega^{2-d}(A)$ in $\underline{L}(\Lambda)$.

b) By a) $\operatorname{Hom}_R(\operatorname{Tr}_L(\operatorname{Hom}_R(A,R)),R) \cong \Omega^{2-d}(\operatorname{Hom}_R(A,R))$. But $\operatorname{Hom}_R(\Omega^{d-2}(A),R) \cong \Omega^{2-d}(\operatorname{Hom}_R(A,R))$ by Lemma 1.2. Thus $\operatorname{Tr}_L(\operatorname{Hom}_R(A,R)) \cong \Omega^{d-2}(A)$ in $\underline{L}(\Lambda)$. □

We now point out a few consequences of Proposition 1.3.

PROPOSITION 1.4. *Suppose* $\dim R = d \geq 2$ *and* A *is a nonprojective* Λ*-lattice. Then* $\operatorname{Ext}_\Lambda^{d-1}(A,A) \cong \operatorname{Hom}_R(\underline{\operatorname{End}}(A), I_d) \neq 0$ *as an* $\underline{\operatorname{End}}(A)$*-*$\underline{\operatorname{End}}(A)^{\mathrm{op}}$*-bimodule.*

PROOF. First we observe that $\operatorname{Ext}_\Lambda^{d-1}(A,A) = \operatorname{Ext}_\Lambda^1(\Omega^{d-2}A,A)$. By I, Proposition 8.7 we have that $\operatorname{Ext}_\Lambda^1(\Omega^{d-2}A,A) \cong \operatorname{Hom}_R(\underline{\operatorname{Hom}}_\Lambda(\operatorname{Tr}_L \operatorname{Hom}_R(A,R), \Omega^{d-2}A), I_d)$. By Proposition 1.3 we have $\operatorname{Tr}_L \operatorname{Hom}_R(A,R) \cong \Omega^{d-2}A$. Thus $\operatorname{Ext}_\Lambda^1(\Omega^{d-2}A,A) \cong \operatorname{Hom}_R(\underline{\operatorname{Hom}}_\Lambda(\Omega^{d-2}A, \Omega^{d-2}A), I_d)$. But $\underline{\operatorname{End}}(\Omega^{d-2}A) \cong \underline{\operatorname{End}}(A)$ by Proposition 1.1. Since all these isomorphisms are $\operatorname{End}(A)$-$(\operatorname{End} A)^{\mathrm{op}}$-bimodule isomorphisms, we are done. □

As a consequence of this proposition we have the following.

THEOREM 1.5. *Suppose* $\dim R = d \geq 2$. *A Cohen-Macaulay* Λ*-module* A *is* Λ*-projective if and only if it satisfies both of the following conditions.*

a) $A_{\underline{p}}$ *is* $\Lambda_{\underline{p}}$*-projective for all prime ideals* $\underline{p}$ *of* R *such that* $\dim R_{\underline{p}} \geq 1$.

b) $\operatorname{Ext}_\Lambda^i(A,A) = 0$ *for* $i = 1, \dots, d-1$.

PROOF. It is easy to see that if Λ is a Gorenstein R-order, then $\Lambda_{\underline{p}}$ is a Gorenstein $R_{\underline{p}}$-order for all prime ideals $\underline{p}$ of R. Furthermore a Λ-module A is Cohen-Macaulay and satisfies a) and b) if and only if $A_{\underline{p}}$ is a Cohen-Macaulay $\Lambda_{\underline{p}}$-module satisfying a) and b) for all prime ideals $\underline{p}$ of R. Finally, these observations together with the fact that a module A in $\operatorname{noeth}\Lambda$ is projective if and only if $A_{\underline{p}}$ is projective for all prime ideals $\underline{p}$ of R, show that we may assume without loss of generality that R is a local ring with $\dim R \geq 2$. We now proceed by induction on $d = \dim R$.

Assume $d = 2$. Since A is Cohen-Macaulay and $A_{\underline{p}}$ is $\Lambda_{\underline{p}}$-free for all nonmaximal ideals $\underline{p}$ of R (since $\dim R_{\underline{p}} \leq 1$ if $\underline{p}$ is not maximal) it follows that A is a Λ-lattice. The fact that $\operatorname{Ext}_\Lambda^1(A,A) = 0$ then implies by Proposition 1.4 that A is Λ-projective. The rest of the induction proof proceeds in the same way. □

Before giving our final application of Proposition 1.3, it is convenient to make a few observations. Suppose $\dim R = 1$ and M is an R-module of finite length. Then $\operatorname{ann}(M)$, the annihilator of M, contains some ideal $\underline{a}$ such that $R/\underline{a}$ is a Gorenstein artin ring. For instance, since $R/\operatorname{ann}(M)$ is an artin ring (remember M is a module of finite length) $\operatorname{ann}(M)$ must contain some R-regular element x and R/xR is a Gorenstein artin ring. In particular if C is a Λ-lattice, then $\underline{\operatorname{End}}C$ is an R-module of finite length and so there is an ideal $\underline{a}$ of R such that $\underline{a}\,\underline{\operatorname{End}}C = 0$ and $R/\underline{a}$ is a Gorenstein artin ring.

THEOREM 1.6. *Suppose* $\dim R = 1$, Λ *is a Gorenstein* R*-order, and* C *is nonprojective lattice over* Λ. *Let* $\underline{a}$ *be an ideal in* R *such that* $\underline{a}\,\underline{\operatorname{End}}C = 0$ *and* $R/\underline{a}$ *is an artin Gorenstein ring. Then* $\underline{\operatorname{End}}C$ *is a Gorenstein* $R/\underline{a}$*-algebra.*

PROOF. By our remarks preceding the statement of the proposition we know that such an ideal $\underline{a}$ exists. Since $R/\underline{a}$ is selfinjective, it follows that $\operatorname{Hom}_R(R/\underline{a}, I_1) \cong R/\underline{a}$ as R-modules. Hence $\operatorname{Hom}_R(\underline{\operatorname{End}}C, I_1) \cong \operatorname{Hom}_{R/\underline{a}}(\underline{\operatorname{End}}C, R/\underline{a})$ as a two-sided $\underline{\operatorname{End}}C$-module. Therefore to prove the proposition it suffices to show that $\underline{\operatorname{End}}C$ is isomorphic to $\operatorname{Hom}_R(\underline{\operatorname{End}}C, I_1)$ as a two-sided $\underline{\operatorname{End}}C$-module.

Since $\operatorname{Hom}_R(\ , I_1)$ is a duality on the category of R-modules of finite length, we know that $\operatorname{Hom}_R(\underline{\operatorname{End}}C, I_1)$ and $\underline{\operatorname{End}}C$ have the same length over R. Now because Λ is a Gorenstein R-order we know by Proposition 1.3 that $\operatorname{Hom}_R(\operatorname{Tr}_L C, R) = \Omega C$ and so by I, Proposition 8.8 we have an isomorphism $\operatorname{Ext}^1_\Lambda(C, \Omega(C)) \cong \operatorname{Hom}_\Lambda(\underline{\operatorname{Hom}}_\Lambda(C, C), I_1)$ functorial in both variables. Thus this isomorphism is a two-sided $\underline{\operatorname{End}}C$-isomorphism under the following operations.

The operation of $\underline{\operatorname{End}}C^{\mathrm{op}}$ is that induced by $\operatorname{End} C$ on C. The operation of $\underline{\operatorname{End}}C$ on $\operatorname{Ext}^1_\Lambda(C, \Omega C)$ is given by the isomorphism $\Omega\colon \underline{\operatorname{End}}C \to \underline{\operatorname{End}}(\Omega C)$ and the operation of $\underline{\operatorname{End}}\Omega C$ induced by the natural operation of $\operatorname{End} \Omega C$ on ΩC (remember that since injective and projective lattices over a Gorenstein order are the same, $P(C, C)\cdot(\operatorname{Ext}^1_\Lambda(C, \Omega C)) = 0$). Thus to show that $\underline{\operatorname{End}}C$ and $\operatorname{Hom}_R(\underline{\operatorname{End}}(C), I_1)$ are isomorphic bimodules over $\underline{\operatorname{End}}C^{\mathrm{op}}$-$\underline{\operatorname{End}}C$, it suffices to show that $\underline{\operatorname{End}}C$ and $\operatorname{Ext}^1_\Lambda(C, \Omega(C))$ are isomorphic $(\underline{\operatorname{End}}C)^{\mathrm{op}}$-$\underline{\operatorname{End}}C$-bimodules.

Let $x\colon 0 \to \Omega C \to P \to C \to 0$ be the extension in $\operatorname{Ext}^1_\Lambda(C, \Omega C)$ with P projective. Since P is projective, given any extension $0 \to \Omega C \to E \to C \to 0$, there exists a commutative diagram

$$\begin{array}{ccccccccc}
0 & \longrightarrow & \Omega E & \longrightarrow & P & \longrightarrow & C & \longrightarrow & 0 \\
 & & \downarrow{\scriptstyle f} & & \downarrow & & \| & & \\
0 & \longrightarrow & \Omega E & \longrightarrow & E & \longrightarrow & C & \longrightarrow & 0
\end{array}$$

so that x generates $\operatorname{Ext}^1_\Lambda(C, \Omega C)$ over $\underline{\operatorname{End}}\Omega C$ and hence over $\underline{\operatorname{End}}C$. Since P is also an injective lattice, a similar argument shows that x generates $\operatorname{Ext}^1_\Lambda(C, \Omega C)$ over $(\underline{\operatorname{End}}C)^{\mathrm{op}}$. Hence the $\operatorname{End} C$ morphism $g\colon \underline{\operatorname{End}}C \to \operatorname{Ext}^1_\Lambda(C, \Omega C)$ given by $g(\underline{f})(x) = \operatorname{Ext}^1_\Lambda(C, \Omega \underline{f})(x)$ is an epimorphism and thus an isomorphism since $\underline{\operatorname{End}}C$ and $\operatorname{Ext}^1_\Lambda(C, \Omega C)$ have the same length over R. A similar argument shows that the $(\operatorname{End} C)^{\mathrm{op}}$-morphism $h\colon (\underline{\operatorname{End}}C)^{\mathrm{op}} \to \operatorname{Ext}^1(C, \Omega(C))$ given by $h(f) = \operatorname{Ext}^1_\Lambda(f, \Omega C)(x)$ is also an isomorphism. Therefore if we show for each f in $\operatorname{End} C$ that $\operatorname{Ext}^1_\Lambda(f, \Omega C)(x) = \operatorname{Ext}^1_\Lambda(C, \Omega \underline{f})(x)$, then the maps g and h would be the same and so $h\colon \underline{\operatorname{End}}C \to \operatorname{Ext}^1_\Lambda(C, \Omega C)$ would be a $(\underline{\operatorname{End}}C)^{\mathrm{op}}$-$\underline{\operatorname{End}}C$-bimodule isomorphism, which is our desired result.

We first show that if s and t in $\operatorname{End} C$ are such that $\operatorname{Ext}^1_\Lambda(s, \Omega(C))(x)$ and $\operatorname{Ext}^1_\Lambda(C, \Omega(\underline{t}))(x)$ have the same value y, then $\underline{s} = \underline{t}$. Let $0 \to \Omega C \to E \to C \to 0$ represent y. Then our hypothesis means that there is a commutative exact diagram

$$\begin{array}{ccccccccc}
0 & \longrightarrow & \Omega C & \longrightarrow & P & \longrightarrow & C & \longrightarrow & 0 \\
 & & \downarrow{\scriptstyle \Omega(t)} & & \downarrow & & \| & & \\
0 & \longrightarrow & \Omega C & \longrightarrow & E & \longrightarrow & C & \longrightarrow & 0 \\
 & & \| & & \downarrow & & \downarrow{\scriptstyle s} & & \\
0 & \longrightarrow & \Omega C & \longrightarrow & P & \longrightarrow & C & \longrightarrow & 0
\end{array}$$

from which we deduce the commutative exact diagram

$$\begin{array}{ccccccccc}
0 & \longrightarrow & \Omega C & \longrightarrow & P & \longrightarrow & C & \longrightarrow & 0 \\
 & & \downarrow{\scriptstyle \Omega(t)} & & \downarrow & & \downarrow{\scriptstyle s} & & \\
0 & \longrightarrow & \Omega C & \longrightarrow & P & \longrightarrow & C & \longrightarrow & 0
\end{array}$$

which shows that $\Omega(\underline{s}) = \Omega(\underline{t})$. Since $\Omega\colon \underline{\operatorname{End}}C \to \underline{\operatorname{End}}(\Omega C)$ is an isomorphism we know that $\underline{s} = \underline{t}$. Hence if s and t are in $\underline{\operatorname{End}}C$ such that $\operatorname{Ext}^1_\Lambda(s, \Omega(C))(x) = \operatorname{Ext}^1_\Lambda(C, \Omega(t))(x)$, then $\underline{s} = \underline{t}$.

To see that $\operatorname{Ext}^1_\Lambda(s, \Omega(C))(x) = \operatorname{Ext}^1_\Lambda(C, \Omega(\underline{s}))(x)$ for all s in $\operatorname{End} C$ is now trivial. For we know that $\operatorname{Ext}^1_\Lambda(s, \Omega(C))(x) = \operatorname{Ext}^1_\Lambda(C, \Omega(\underline{t}))(x)$ for some t. Then by our previous result we know that $\underline{t} = \underline{s}$. So $\operatorname{Ext}^1_\Lambda(s, \Omega(C))(x) = \operatorname{Ext}^1(C, \Omega(\underline{s}))(x)$ for all s in $\operatorname{End} C$. This finishes the proof that $\underline{\operatorname{End}}(C)$, $\operatorname{Ext}^1_\Lambda(C, \Omega(C))$ and $\operatorname{Hom}_{R/\underline{a}}(\underline{\operatorname{End}}(C), R/\underline{a})$ are all isomorphic $(\underline{\operatorname{End}}C)^{\text{op}}$-$\underline{\operatorname{End}}C$-bimodules and thus the fact that $\underline{\operatorname{End}}C$ is a Gorenstein $R/\underline{a}$-algebra is complete. □

We end this section with a few remarks about lattices over Gorenstein R-orders in case R is a complete local ring.

Suppose Λ is a Gorenstein R-order with R a complete local ring of dimension d. As usual, we assume that we use only minimal projective resolutions in making our definitions and computations. With these conventions in mind we have the following result.

PROPOSITION 1.7. *Suppose Λ is a Gorenstein R-order with R complete local ring of dimension d. Let M be in $L_P(\Lambda)$ and $\cdots \to P_i \xrightarrow{f_i} P_{i-1} \to \cdots \to P_0 \xrightarrow{f_0} M \to 0$ a minimal projective resolution of M.*

a) *$\Omega^i(M) = \operatorname{Im} f_i$ is in $L_P(\Lambda)$ for all $i \geq 0$.*

b) *The following statements are equivalent:*

i) *M is indecomposable.*

ii) *$\Omega^j(M)$ is indecomposable for some $j \geq 0$.*

iii) *$\Omega^j(M)$ is indecomposable for all $j \geq 0$.*

c) *$\operatorname{Hom}_\Lambda(M, \Lambda)$ is in $L_P(\Lambda^{\text{op}})$.*

d) *Let $\cdots \xrightarrow{g_i} Q_i \xrightarrow{g_{i-1}} \cdots \to Q_0 \xrightarrow{g_0} \operatorname{Hom}_\Lambda(M, \Lambda) \to 0$ be a minimal projective Λ^{op}-resolution of $\operatorname{Hom}_\Lambda(M, \Lambda)$. Then the exact sequence of Λ-modules $0 \to \operatorname{Hom}_\Lambda(\operatorname{Hom}_\Lambda(M, \Lambda), \Lambda) \xrightarrow{\operatorname{Hom}_\Lambda(g_0, \Lambda)} \operatorname{Hom}_\Lambda(Q_0, \Lambda) \to \cdots \xrightarrow{\operatorname{Hom}_\Lambda(g_i, \Lambda)} \operatorname{Hom}_\Lambda(Q_i, \Lambda) \to \cdots$ has the following properties:*

i) *The epimorphisms $\operatorname{Hom}_\Lambda(Q_i, \Lambda) \to \operatorname{Im} \operatorname{Hom}_\Lambda(g_{i+1}, \Lambda)$ are projective covers for all $i \geq 0$.*

ii) *$\operatorname{Im} \operatorname{Hom}_\Lambda(g_i, \Lambda)$ is in $L_P(\Lambda)$ for all $i \geq 0$.*

PROOF. All these statements follow from the fact that $\operatorname{Ext}^i_\Lambda(M, \Lambda) = 0$ for all $i > 0$ using arguments similar to those given in I, Section 9. See in particular I, Lemma 9.1. □

As a readily verified consequence of this result we have the following strengthening of Lemma 1.2 and Proposition 1.3.

PROPOSITION 1.8. *Let Λ be an R-order with R a complete local ring of dimension d. For each Λ-lattice M in $L_P(\Lambda)$ we have*

a) *$\Omega^j(\Omega^k(M)) \cong \Omega^{j+k}(M)$ in $L(\Lambda)$ for all integers j and k.*

b) *$\Omega^j(\operatorname{Hom}_\Lambda(M, \Lambda)) \cong \operatorname{Hom}_\Lambda(\Omega^{-j}(M), \Lambda)$ in $L(\Lambda)$ for all integers j.*

c) *$\operatorname{Hom}_R(\operatorname{Tr}_L M, R) \cong \Omega^{2-d} M$ in $L(\Lambda)$.*

d) *$\operatorname{Tr}_L(\operatorname{Hom}_R(M, R)) \cong \Omega^{d-2} M$ in $L(\Lambda)$.*

2. Orders over complete one-dimensional local rings

Throughout this section we assume that R is a complete one dimensional Gorenstein local ring and Λ is an R-order. Our purpose in this section is to develop some results concerning $L(\Lambda)$ which are direct analogues of results already established for finitely generated modules over artin algebras (see [7] and [11]). For other analogies between these theories the reader should consult some of the recent work of K. Roggenkamp and J. Schmidt. Our first aim is to give a criterion for when $L(\Lambda)$ is of finite representation type (i.e. when $L(\Lambda)$ has only a finite number of nonisomorphic indecomposable objects) in terms of the category of functors in $(L(\Lambda^{\mathrm{op}}), \mathrm{Ab})$ vanishing on projectives. We begin with the following preliminary result.

PROPOSITION 2.1. *Let Λ be an R-order.*

a) *$\underline{r}$, the radical of Λ, is a lattice.*

b) *Let A be in $L(\Lambda)$. Then the restriction of the functor $\mathrm{Ext}^1_\Lambda(\ ,A)\colon \mathrm{Mod}\,\Lambda \to \mathrm{Ab}$ to $L(\Lambda)$ is a finitely presented functor in $(L(\Lambda)^{\mathrm{op}}, \mathrm{Ab})$ which vanishes on the projective Λ-lattices.*

c) *A lattice C is projective if and only if $\mathrm{Ext}^1_\Lambda(C, \underline{r}) = 0$.*

d) *If F is a finitely generated functor in $(L(\Lambda)^{\mathrm{op}}, \mathrm{Ab})$ which vanishes on projectives, then $F(A)$ is an $\underline{\mathrm{End}}(A)^{\mathrm{op}}$-module of finite length for all A in $L(\Lambda)$.*

PROOF. a) It was shown in II, Proposition 7.6 that if $\dim R \leq 1$ and C' is a submodule of a lattice C such that C/C' has finite length, then C' is a lattice. Since Λ is a lattice and $\Lambda/\underline{r}$ has finite length, it follows that $\underline{r}$ is a lattice.

b) This is an easy consequence of I, Proposition 8.2.

c) It is well known and easily shown that a module C in $\mathrm{noeth}\,\Lambda$ is projective if $\mathrm{Ext}^1_\Lambda(C, \underline{r}) = 0$. In particular a lattice C is projective if $\mathrm{Ext}^1_\Lambda(C, \underline{r}) = 0$.

d) Since F in $(L(\Lambda^{\mathrm{op}}), \mathrm{Ab})$ is finitely generated there is an epimorphism $\phi\colon (\ ,C) \to F \to 0$. The fact that $F(P) = 0$ for all projective P in $L(\Lambda)$ implies that if $f\colon P \to C \to 0$ is exact with P projective, then the composition $(\ ,P) \to (\ ,C) \to F \to 0$ is zero. Hence the epimorphism $(\ ,C) \to F \to 0$ induces an epimorphism $\underline{\mathrm{Hom}}_\Lambda(\ ,C) \to F \to 0$. Let X be in $L(\Lambda)$. We know that $\underline{\mathrm{Hom}}_\Lambda(X, C)$ is an R-module of finite length and consequently an $(\underline{\mathrm{End}}X)^{\mathrm{op}}$-module of finite length. Therefore $F(X)$ is an $(\underline{\mathrm{End}}X)^{\mathrm{op}}$-module of finite length for all X in $L(\Lambda)$. □

We now give our first application in this section. We know that an artin algebra Λ is of finite representation type if and only if every finitely presented functor in $((\mathrm{mod}\,\Lambda)^{\mathrm{op}}, \mathrm{Ab})$ has finite length (see [7]). Based on this result, it is not difficult to show that Λ is of finite representation type if and only if every finitely presented functor in $((\mathrm{mod}\,\Lambda)^{\mathrm{op}}, \mathrm{Ab})$ vanishing on projectives has finite length. We now prove an analogue to this latter result for when $L(\Lambda)$ is of finite representation type. To this end, it is convenient to have the following result.

LEMMA 2.2. *A finitely generated functor F in $(L(\Lambda)^{\mathrm{op}}, \mathrm{Ab})$ which vanishes on projectives is of finite length if and only if there are only a finite number of nonisomorphic indecomposable A in $L(\Lambda)$ such that $F(A) \neq 0$.*

PROOF. By [7, Theorem 2.12] we know that F is of finite length if and only if $F(A)$ is of finite length over $(\mathrm{End}\,A)^{\mathrm{op}}$ for each indecomposable A in $L(\Lambda)$ and there are only a finite number of nonisomorphic indecomposable A in $L(\Lambda)$ such that $F(A) \neq 0$. Since $F(A)$ is of finite length over $(\mathrm{End}\,A)^{\mathrm{op}}$ for all A in $L(\Lambda)$ by Proposition 1.1, the lemma is proven. □

THEOREM 2.3. *The following statements are equivalent for an R-order Λ.*

a) *$L(\Lambda)$ is of finite representation type.*

b) *$\underline{\mathrm{Hom}}(\ ,X)$ is a functor in $(L(\Lambda)^{\mathrm{op}}, \mathrm{Ab})$ of finite length for each X in $L(\Lambda)$.*

c) *Each finitely generated functor F in $(L(\Lambda)^{\mathrm{op}}, \mathrm{Ab})$ which vanishes on projectives is of finite length.*

d) *$\mathrm{Ext}^1_\Lambda(\ ,\underline{r})$ is of finite length.*

PROOF. a) implies b). Since for each X in $L(\Lambda)$, $\underline{\mathrm{Hom}}(\ ,X)$ is a finitely generated functor vanishing on projectives, the fact that $L(\Lambda)$ is of finite representation type implies that there are only a finite number of nonisomorphic indecomposable A in $L(\Lambda)$ such that $\underline{\mathrm{Hom}}(A,X) \neq 0$. Hence by Lemma 1.2, $\underline{\mathrm{Hom}}(\ ,X)$ is of finite length.

b) implies c). Since F is finitely generated, there is an epimorphism $(\ ,X) \to F \to 0$. Since F vanishes on projectives, $(\ ,X) \to F \to 0$ induces an epimorphism $\underline{\mathrm{Hom}}_\Lambda(\ ,X) \to F \to 0$. The fact that $\underline{\mathrm{Hom}}_\Lambda(\ ,X)$ is of finite length obviously implies that F is of finite length.

c) implies d). By Proposition 1.1 we know that $\mathrm{Ext}^1_\Lambda(\ ,\underline{r})$ is a finitely generated functor in $(L(\Lambda)^{\mathrm{op}},\mathrm{Ab})$ vanishing on projectives. Hence $\mathrm{Ext}^1_\Lambda(\ ,\underline{r})$ is of finite length.

d) implies a). Assume that $\mathrm{Ext}^1_\Lambda(\ ,\underline{r})$ has finite length. Thus there are only a finite number of nonisomorphic nonprojective indecomposable A in $L(\Lambda)$ such that $\mathrm{Ext}^1_\Lambda(A,\underline{r}) \neq 0$. Since A in $L(\Lambda)$ has $\mathrm{Ext}^1_\Lambda(A,\underline{r}) = 0$ only if A is projective (see Proposition 1.1), there are only a finite number of nonisomorphic nonprojective indecomposable A in $L(\Lambda)$. But there are only a finite number of nonisomorphic indecomposable projective modules in $L(\Lambda)$. Therefore there are only a finite number of nonisomorphic indecomposable A in $L(\Lambda)$ if $\mathrm{Ext}^1_\Lambda(\ ,\underline{r})$ in $(L(\Lambda)^{\mathrm{op}},\mathrm{Ab})$ is a functor of finite length. □

Before giving our final result of this section we recall the notion of an irreducible morphism. A morphism $f\colon A \to B$ in $L(\Lambda)$ is said to be *irreducible in* $L(\Lambda)$ if a) f is not a splittable monomorphism or epimorphism and b) if f is the composition $A \xrightarrow{g} X \xrightarrow{h} B$ in $L(\Lambda)$, then either h is a splittable epimorphism or g is a splittable monomorphism. The reader is referred to [11] for a more complete discussion of irreducible morphisms. The only facts concerning irreducible morphisms we will need we give in the following result.

PROPOSITION 2.4. *Let Λ be an R-order.*

a) *Let $g\colon A \to B$ be a minimal left almost split morphism in $L(\Lambda)$. A morphism $h\colon A \to X$ with $X \neq 0$ is irreducible if and only if h can be written as a composition $A \xrightarrow{g} B \xrightarrow{t} X$ with t a splittable epimorphism.*

b) *Let $f\colon B \to C$ be a minimal right almost split morphism in $L(\Lambda)$. A morphism $h\colon X \to C$ with $X \neq 0$ is irreducible if and only if h can be written as a composition $X \xrightarrow{t} B \xrightarrow{f} C$ with t a splittable monomorphism.*

d) *Suppose $0 \to A \xrightarrow{g} B \xrightarrow{f} C \to 0$ is an almost split sequence in $L(\Lambda)$. Then B has an indecomposable projective summand if and only if A is isomorphic to a noninjective summand of $\underline{r}P$ for some indecomposable projective P.*

PROOF. a) and b). See [11, Theorem 2.4].

c) We include a proof of c) for the convenience of the reader even though it is the same as the proof of [11, Theorem 4.2].

Suppose $0 \to A \xrightarrow{g} B \xrightarrow{f} C \to 0$ is an almost split sequence in $L(\Lambda)$. Since the sequence does not split, A is not an injective lattice. Let P be an indecomposable projective summand of B and suppose $t\colon B \to P$ is an epimorphism. Then by a), the composition $tg\colon A \to P$ is irreducible. Since by II, Proposition 8.3 the inclusion $i\colon \underline{r}P \to P$ is a minimal right almost split morphism, there is by b) a splittable monomorphism $u\colon A \to \underline{r}P$ such that tg is iu. Thus A is isomorphic to a noninjective summand of $\underline{r}P$.

On the other hand, suppose A is isomorphic to a noninjective summand of $\underline{r}P$ for some indecomposable projective P. Then there is an irreducible morphism $h\colon A \to P$. Hence by a) the morphism h can be written as a composition $A \xrightarrow{g} B \xrightarrow{t} P$ such that t

is a splittable epimorphism. Thus B has the indecomposable projective module P as a summand. □

With these preliminary results in mind, we now prove the following.

THEOREM 2.5. *Let Λ be an R-order. Suppose C is an indecomposable nonprojective lattice. Then the following are equivalent.*

a) *If $0 \to A \xrightarrow{g} B \xrightarrow{f} C \to 0$ is an almost split sequence in $L(\Lambda)$, then B has an indecomposable projective summand.*

b) *If $X \to C$ is an epimorphism with X in $L(\Lambda)$ which is not a splittable epimorphism, then X has an indecomposable projective summand.*

c) *A is isomorphic to a noninjective summand of $\underline{r}P$ for some indecomposable projective P.*

PROOF. a) implies b). Suppose P is an indecomposable projective summand of B. Then $h = f|_P$ is an irreducible morphism $P \to C$. Suppose $u\colon X \to C$ is an epimorphism with X in $L(\Lambda)$ which is not a splittable epimorphism. Because u is an epimorphism and P is projective, there is a morphism $s\colon P \to X$ such that $us = h$. Since u is not a splittable epimorphism and h is irreducible, $s\colon P \to X$ is a splittable monomorphism. Hence X has an indecomposable projective summand.

b) implies a). Trivial.

a) equivalent to c). Contained in Proposition 3.11. □

THEOREM 2.6. *Let Λ be an R-order.*

a) *There is a one-to-one correspondence between the isomorphism classes of indecomposable nonprojective C in $L(\Lambda)$ having the property if X is in $L(\Lambda)$ and $X \to C$ is an epimorphism which is not splittable, then X has an indecomposable projective summand and the isomorphism classes of nonprojective indecomposable summands of $\underline{r}$ given by $C \mapsto (\operatorname{Tr}_L C)^* = \operatorname{Hom}_R(\operatorname{Tr}_L C, R)$.*

b) *Λ is not an hereditary ring if and only if there is an indecomposable nonprojective C in $L(\Lambda)$ such that an X in $L(\Lambda)$ must have an indecomposable projective summand if there is an epimorphism $X \to C$ which is not splittable.*

PROOF. a) Since C is indecomposable and not projective, we know that there is an almost split sequence in $L(\Lambda)$ $0 \to (\operatorname{Tr}_L C)^* \to B \to C \to 0$. By Theorem 2.5 we know C satisfies the condition of a) if and only if $(\operatorname{Tr}_L C)^*$ is isomorphic to a noninjective summand of $\underline{r}$. Since, up to isomorphism, C is uniquely determined by $(\operatorname{Tr}_L C)^*$, the map $C \mapsto (\operatorname{Tr}_L C)^*$ gives an injection between the isomorphism classes of C satisfying the condition of a) and noninjective indecomposable summands of $\underline{r}$. On the other hand, if A is a noninjective indecomposable summand of $\underline{r}$, then the almost split sequence $0 \to A \to B \to C \to 0$ has the property that B has an indecomposable projective summand (see Theorem 2.5). Since $(\operatorname{Tr}_L C)^* \cong A$, the map $C \mapsto (\operatorname{Tr}_L C)^*$ gives a bijection between the isomorphism classes of C in $L(\Lambda)$ satisfying the condition of a) and the isomorphism classes of noninjective indecomposable summands of $\underline{r}$.

b) Since Λ is a semi-perfect noetherian ring, it is well known that $\operatorname{gl.dim.}\Lambda = \operatorname{pd}\Lambda/\underline{r}$. Therefore Λ is hereditary if and only if $\underline{r}$ is a projective Λ-lattice. Because all Λ-lattices are submodules of projective Λ-modules, it follows that Λ is hereditary if and only if every Λ-lattice is projective. But we know that a Λ-lattice C is projective if and only if $\operatorname{Ext}^1_\Lambda(C, \underline{r}) = 0$ (see Proposition 1.1). Therefore Λ is hereditary if and only if $\operatorname{Ext}^1_\Lambda(C, \underline{r}) = 0$ for all Λ-lattices C. But $\operatorname{Ext}^1_\Lambda(C, \underline{r}) = 0$ for all Λ-lattices C if and only if $\underline{r}$ is an injective lattice. Thus by a), Λ is not hereditary if and only if there is an indecomposable nonprojective Λ-lattice C such that an X in $L(\Lambda)$ has an indecomposable projective summand whenever there is an epimorphism $X \to C$ which is not splittable. □

As an immediate consequence of this result we have the following result about complete commutative one-dimensional local integral domains.

COROLLARY 2.7. *Let S be a complete commutative one-dimensional local domain which is not a discrete valuation ring. Then there is a unique, up to isomorphism, indecomposable nonprojective torsion-free module C such that if $X \to C$ is an epimorphism which is not a splittable epimorphism with X a torsion-free module, then X has S as a summand.*

PROOF. It is well known that there is a complete discrete valuation ring R which is a subring of S such that S is a finitely generated free R-module (see [17, Corollary 31.6]). Hence S is an R-order. S is not hereditary since it is not a discrete valuation ring. The result is now a trivial consequence of Theorem 2.6. □

Added in proof: Suppose Λ is an R-order with R an arbitrary 1-dimensional Gorenstein ring (not necessarily local or complete). Let C be a Λ-lattice. In general, while a Λ-submodule B of C is Cohen-Macaulay since $\dim R = 1$, it need not be a Λ-lattice since $B_{\underline{p}}$ need not be $\Lambda_{\underline{p}}$-projective for all nonmaximal prime ideals $\underline{p}$ of R. However if $\Lambda_{\underline{p}}$ is semi-simple for all nonmaximal prime ideals $\underline{p}$ of R, then this difficulty disappears and we have that every Λ-submodule B of a Λ-lattice C is again a Λ-lattice. For the rest of this discussion we assume that the R-order Λ has the property $\Lambda_{\underline{p}}$ is semi-simple for all nonmaximal prime ideals $\underline{p}$ of R. Clearly the R-order Λ^{op} also has the property $(\Lambda^{\mathrm{op}})_{\underline{p}} = (\Lambda_{\underline{p}})^{\mathrm{op}}$ is semi-simple for all nonmaximal prime ideals $\underline{p}$ of R. Hence $L(\Lambda)$ and $L(\Lambda^{\mathrm{op}})$ are closed under taking submodules. This fact enables us to prove the following analogue of I, Theorem 3.19.

THEOREM 2.8. *Let X and C be in $L(\Lambda)$. Suppose H is an $(\operatorname{End} X)^{\mathrm{op}}$-submodule of (X, C). Then there is an exact sequence $0 \to A \xrightarrow{g} B \xrightarrow{f} C$ of Λ-lattices with f a right X-determined morphism in $L(\Lambda)$ such that $\operatorname{Im}(\ ,f)(X) = H$. Moreover, if we let $C' = \operatorname{Im} f$ and let $0 \to A \xrightarrow{g} B \xrightarrow{f'} C' \to 0$ be the induced exact sequence, then*

a) *H is an $(\operatorname{End} X)^{\mathrm{op}}$-submodule of (X, C') containing $P(X, C')$.*

b) *f' is right X-determined in $L(\Lambda)$ with $\operatorname{Im}(\ ,f')(X) = H$.*

PROOF. The proof is entirely analogous to that given for I, Theorem 3.19 and is left to the reader. □

Using the usual duality between $L(\Lambda)$ and $L(\Lambda^{\mathrm{op}})$ we obtain the following dual of Theorem 2.8.

THEOREM 2.9. *Let Y and A be in $L(\Lambda)$ and H an $\operatorname{End} Y$-submodule of (A, Y). Then there is an exact sequence $A \xrightarrow{g} B \xrightarrow{f} C \to 0$ in $L(\Lambda)$ with g left Y-determined in $L(\Lambda)$ such that $\operatorname{Im}(g,\ )(Y) = H$. Moreover, if we let $\operatorname{Im} g = A'$, and let $0 \to A' \xrightarrow{g'} B \to C \to 0$ be the induced exact sequence in $L(\Lambda)$, then*

a) *H is an $\operatorname{End} Y$-submodule of (A', Y) containing $I(A', Y)$.*

b) *g' is left Y-determined in $L(\Lambda)$ with $\operatorname{Im}(g',\ )(Y) = H$.*

It should be noted that if R is a complete local ring, then the morphism $f \colon B \to C$ in Theorem 2.8 can be chosen to be right minimal while the morphism $g \colon A \to B$ in Theorem 2.9 can be chosen to be left minimal.

3. Orders over two dimensional rings

Suppose Λ is an R-order with $\dim R = d$. In II, Section 7, we showed that if S is a simple functor in $(L(\Lambda)^{\text{op}}, \text{Ab})$ vanishing on projectives, then there is an exact sequence $0 \to A \to B \to C \to 0$ in $L(\Lambda)$ such that $0 \to (\ ,A) \to (\ ,B) \to (\ ,C) \to S \to 0$ is exact. Similarly, if T is a simple functor in $(L(\Lambda), \text{Ab})$ vanishing on injective lattices, we showed that there is an exact sequence $0 \to A \to B \to C \to 0$ in $L(\Lambda)$ such that $0 \to (C,\ ) \to (B,\ ) \to (A,\ ) \to T \to 0$ is exact. This leaves open the question of describing projective resolutions for simple functors in $(L(\Lambda)^{\text{op}}, \text{Ab})$ and $(L(\Lambda), \text{Ab})$ which do not vanish on projectives or injective lattices respectively. This question was solved in II, Section 7 in the case $\dim R \leq 1$. In this section we give a solution of the problem for $\dim R = 2$. As far as we know, the question is open for $\dim R \geq 3$.

Let S be a simple functor in $(L(\Lambda)^{\text{op}}, \text{Ab})$ not vanishing on a projective lattice P. Then there is an epimorphism $\alpha\colon (\ ,P) \to S$. This induces an epimorphism of Λ-modules $\alpha_\Lambda\colon (\Lambda, P) \to S(\Lambda) \to 0$. Since S is a simple functor and $S(\Lambda) \neq 0$, we know that $S(\Lambda)$ is a simple Λ-module. Hence $\operatorname{Ker} \alpha_\Lambda$ is a maximal submodule M of $(\Lambda, P) = P$. We have seen in II, Section 2, that if $\dim R \leq 1$, then M is a lattice. However, M fails to be a lattice for $\dim R \geq 2$ because, although it has the property $M_{\underline{p}}$ is $\Lambda_{\underline{p}}$-projective for all nonmaximal prime ideals $\underline{p}$ of R, it is not a Cohen-Macaulay module. Nonetheless, the restriction of the representable functor $(\ ,M)$ in $((\operatorname{Mod} \Lambda)^{\text{op}}, \text{Ab})$ to $L(\Lambda)$, which we also denote by $(\ ,M)$, plays an important role in describing a projective presentation for S since, as we now show, the sequence $0 \to (\ ,M) \xrightarrow{\beta} (\ ,P) \xrightarrow{\alpha} S \to 0$ is exact in $(L(\Lambda)^{\text{op}}, \text{Ab})$.

Obviously, β is a monomorphism. Suppose $\alpha\beta\colon (\ ,M) \to S$ is not zero. Then it is an epimorphism since S is simple. In particular $\alpha\beta_\Lambda\colon (\Lambda, M) \to S(\Lambda)$ is an epimorphism, i.e. the composition $(\Lambda, M) \xrightarrow{\beta_\Lambda} (\Lambda, P) \xrightarrow{\alpha_\Lambda} S(\Lambda)$ is an epimorphism. But this contradicts the fact that $0 \to (\Lambda, M) \to (\Lambda, P) \xrightarrow{\alpha_\Lambda} S(\Lambda) \to 0$ is exact since $M = \operatorname{Ker} \alpha_\Lambda$. Therefore $\alpha\beta = 0$ or equivalently $\operatorname{Im} \beta \subset \operatorname{Ker} \alpha$. Thus we are done if we show that $\operatorname{Im} \beta$ is a maximal subfunctor of $(\ ,P)$. To see this, we only have to show that if $X \to P$ is a morphism in $L(\Lambda)$ such that $\operatorname{Im}((\ ,X) \to (\ ,P))$ is not contained in $\operatorname{Im} \beta$, then the induced morphism $(\ ,X) \amalg (\ ,M) \to (\ ,P)$ is an epimorphism. But such an $f\colon X \to B$ has to have the property that $\operatorname{Im} f \not\subset M$. Since M is a maximal submodule of P, this implies that the induced morphism of modules $X \amalg M \to P$ is an epimorphism and thus a splittable epimorphism because P is projective. Therefore $(\ ,X) \amalg (\ ,M) \to (\ ,P)$ is an epimorphism, which is our desired result. As a consequence of this discussion we have the following.

PROPOSITION 3.1. a) *Suppose M is a maximal submodule of the finitely generated projective module P and $(\ ,M)$ denotes the restriction of the representable functor $(\ ,M)$ in $((\operatorname{Mod} \Lambda)^{\text{op}}, \text{Ab})$ to $L(\Lambda)$. Then* $\operatorname{Coker}((\ ,M) \to (\ ,P))$ *is a simple functor in* $(L(\Lambda)^{\text{op}}, \text{Ab})$.

b) *Suppose S is a simple functor in $(L(\Lambda)^{\text{op}}, \text{Ab})$ and P a projective lattice such that $S(P) \neq 0$. Then there is a maximal submodule M of P such that there is an exact sequence $0 \to (\ ,M) \to (\ ,P) \to S \to 0$ in $(L(\Lambda)^{\text{op}}, \text{Ab})$, where $(\ ,M) \to (\ ,P)$ is given by the inclusion $M \to P$.*

As an easily verified consequence of this result we have the following.

COROLLARY 3.2. *Every simple functor S in $(L(\Lambda)^{\text{op}}, \text{Ab})$ not vanishing on projectives is finitely presented if and only if given any maximal submodule M of a projective lattice*

P, there is a lattice E and an epimorphism $E \to M$ *of modules such that* $\mathrm{Hom}_\Lambda(X,E) \to \mathrm{Hom}_\Lambda(X,M) \to 0$ *is exact for all lattices X. Moreover any morphism* $E \to M$ *such that* $\mathrm{Hom}_\Lambda(X,E) \to \mathrm{Hom}_\Lambda(X,M) \to 0$ *is exact for all lattices X is an epimorphism of modules.*

Our aim now is to show that if $\dim R = 2$ and M is a maximal submodule of a projective lattice P, then there is an epimorphism $E \to M \to 0$ with E a lattice such that $(X,E) \to (X,M) \to 0$ is exact for all lattices X.

Suppose such an epimorphism $E \xrightarrow{f} M \to 0$ exists. This gives rise to an exact sequence $0 \to A \xrightarrow{g} E \xrightarrow{f} P \to P/M \to 0$. Since $\dim R = 2$ and P and E are Cohen-Macaulay modules, it follows that A is Cohen-Macaulay. Also since P/M is a simple Λ-module and P and E are lattices, it follows that $A_{\underline{p}}$ is $\Lambda_{\underline{p}}$-projective for all nonmaximal prime ideals $\underline{p}$ of R. Thus A is a lattice with the property that $0 \to (X,A) \to (X,E) \to (X,M) \to 0$ is exact for all lattices X, or equivalently, the monomorphism $0 \to A \xrightarrow{g} E$ of lattices has the property that $0 \to \mathrm{Ext}^1_\Lambda(X,A) \to \mathrm{Ext}^1_\Lambda(X,E)$ is exact for all lattices X. Since this latter condition is satisfied if $\mathrm{Ext}^1_\Lambda(X,A) = 0$ for all lattices X, i.e. if A is an injective lattice, this analysis suggests trying to find an exact sequence $0 \to A \xrightarrow{g} E \xrightarrow{f} P \to P/M \to 0$ of Λ-modules with A and E lattices with A an injective lattice. On the basis of our preceding discussion, it is not difficult to see that such an exact sequence has the property that the induced epimorphism $E \to M$ satisfies the condition that $(X,E) \to (X,M) \to 0$ is exact for all lattices X. We now proceed to show that such an exact sequence exists.

First we show that if $\dim R = 2$ and M is a maximal submodule of a projective lattice P, then there is an exact sequence $0 \to A \xrightarrow{g} E \to P \to P/M \to 0$ of Λ-modules with A an injective lattice such that g is not a splittable monomorphism. We will then show that such an exact sequence has our desired properties by showing that E is a lattice. The existence of such an exact sequence will follow from the following more general result.

PROPOSITION 3.3. *Suppose* $\dim R = d \geq 1$. *If X is a lattice and A is a* Λ-*module of finite length, then* $\mathrm{Ext}^i_\Lambda(A,X) = 0$ *for* $i = 0,\dots,d-1$ *and* $\mathrm{Ext}^d_\Lambda(A,X) \cong \mathrm{Hom}_R(\mathrm{Hom}_R(X,R)\otimes_\Lambda A, I_d)$. *Thus* $\mathrm{Ext}^d_\Lambda(A,X) = 0$ *if and only if* $\mathrm{Hom}_R(X,R)\otimes_\Lambda A = 0$.

PROOF. Let $0 \to R \to I_0 \to \cdots \to I_d \to 0$ be a minimal injective resolution of R over R. Let $Y = \mathrm{Hom}_R(X,R)$. Then $0 \to \mathrm{Hom}_R(Y,R) \to \mathrm{Hom}_R(Y,I_0) \to \cdots \to \mathrm{Hom}_R(Y,I_d) \to 0$ is exact with the $\mathrm{Hom}_R(Y,I_j)$ injective Λ-modules for $j = 0,\dots,d-1$ since Y is a Λ^{op}-lattice (see I, Lemma 8.4). Now $\mathrm{Hom}_\Lambda(A,\mathrm{Hom}_R(Y,I_j)) \cong \mathrm{Hom}_R(Y\otimes_\Lambda A, I_j)$ for all $j = 0,\dots,d-1$. Since A is of finite length, $Y \otimes_\Lambda A$ is an R-module of finite length and so $\mathrm{Hom}_R(Y\otimes_\Lambda A, I_j) = 0$ for all $j = 0,\dots,d-1$ (see I, section 7). From this it follows $\mathrm{Ext}^i_\Lambda(A,\mathrm{Hom}_R(Y,R)) = 0$ for $i = 0,\dots,d-1$ and $\mathrm{Ext}^d_\Lambda(A,\mathrm{Hom}_R(Y,R)) \cong \mathrm{Hom}_R(Y\otimes_\Lambda A, I_d)$. The first part of the proposition now follows since $\mathrm{Hom}_R(Y,R) \cong X$ and $Y = \mathrm{Hom}_R(X,R)$.

Since I_d is an injective cogenerator for R, we know that $\mathrm{Hom}_R(X,R)\otimes_\Lambda A = 0$ if and only if $\mathrm{Hom}_R(\mathrm{Hom}_R(X,R)\otimes_\Lambda A, I_d) = 0$. This finishes the proof of the proposition. □

As an easy consequence of this proposition, we have the following result.

COROLLARY 3.4. *Suppose* $\dim R = 2$, *S is a simple* Λ-*module, and P is a projective lattice such that there is an epimorphism* $\epsilon\colon P \to S \to 0$. *Then there is an exact sequence* $0 \to A \xrightarrow{g} E \xrightarrow{f} P \xrightarrow{\epsilon} S \to 0$ *such that g is not a splittable monomorphism and A is the injective lattice* $\mathrm{Hom}_R(\mathrm{Hom}_\Lambda(P,\Lambda),R)$.

PROOF. Since P is a projective Λ-module, the elements of $\operatorname{Ext}^2_\Lambda(S,A)$ correspond to exact sequences $0 \to A \xrightarrow{g} E \xrightarrow{f} P \xrightarrow{\epsilon} S \to 0$ with $0 \to A \xrightarrow{g} E \xrightarrow{f} P \xrightarrow{\epsilon} S \to 0$ corresponding to the zero element of $\operatorname{Ext}^2_\Lambda(S,A)$ if and only if g is a splittable monomorphism. Now $\operatorname{Hom}_R(A,R) \cong \operatorname{Hom}_\Lambda(P,\Lambda)$, while by Proposition 3.3 we know that $\operatorname{Ext}^2_\Lambda(S,A) \neq 0$ if $\operatorname{Hom}_R(\Lambda,R) \otimes_\Lambda S \neq 0$. But $\operatorname{Hom}_\Lambda(P,\Lambda) \otimes_\Lambda S = \operatorname{Hom}_\Lambda(P,S)$ since P is a finitely generated projective Λ-module. Therefore $\operatorname{Ext}^2_\Lambda(S,A) \neq 0$ since the epimorphism $\epsilon \colon P \to S$ is in $\operatorname{Hom}_\Lambda(P,S)$. Consequently there is an exact sequence $0 \to A \xrightarrow{g} E \xrightarrow{f} P \xrightarrow{\epsilon} S \to 0$ with g not a splittable monomorphism. □

We now proceed to the next step of the proof of our main result.

PROPOSITION 3.5. *Suppose* $\dim R = 2$ *and* $0 \to A \xrightarrow{g} B \xrightarrow{f} P \xrightarrow{\epsilon} S \to 0$ *is an exact sequence in* noeth Λ *with g not a splittable monomorphism. Then B is a lattice if A and P are lattices with P projective and S a simple Λ-module.*

PROOF. Since S is a simple Λ-module, then S is an R-module of finite length so $S_{\underline{p}} = 0$ for all nonmaximal prime ideals $\underline{p}$ of R. Thus $0 \to A_{\underline{p}} \to B_{\underline{p}} \xrightarrow{p} P_{\underline{p}} \to 0$ is an exact sequence of $\Lambda_{\underline{p}}$-modules for all nonmaximal prime ideals $\underline{p}$ of R. Since $P_{\underline{p}}$ and $A_{\underline{p}}$ are $\Lambda_{\underline{p}}$-projective for all nonmaximal prime ideals $\underline{p}$ of R (remember A and P are lattices), $B_{\underline{p}}$ is $\Lambda_{\underline{p}}$-projective for all nonmaximal prime ideals of R. Therefore to show that B is a lattice, we must show that B is a Cohen-Macaulay module over R. Because $\dim R = 2$, this amounts to showing that B is a reflexive R-module.

We first split the exact sequence of Λ-modules $0 \to A \xrightarrow{g} B \xrightarrow{f} P \xrightarrow{\epsilon} S \to 0$ into the two exact sequences $0 \to A \xrightarrow{g} B \to M \to 0$ and $0 \to M \to P \to S \to 0$ where $M = \operatorname{Im} f = \operatorname{Ker} \epsilon$, which is a maximal submodule of P since S is simple. The fact that S is a simple Λ-module implies that S is an R-module of finite length. Therefore $\operatorname{Ext}^i_R(S,R) = 0$ for $i = 0, 1$ since R is a Gorenstein equidimensional ring of dimension 2. Consequently the inclusion $0 \to M \to P$ induces an isomorphism $\operatorname{Hom}_R(P,R) \cong \operatorname{Hom}_R(M,R)$. Also because S is an R-module of finite length we have that $M_{\underline{p}} \cong P_{\underline{p}}$ and is therefore $\Lambda_{\underline{p}}$-projective for all nonmaximal prime ideals $\underline{p}$ of R.

On the other hand the exact sequence $0 \to A \to B \to M \to 0$ induces the exact sequence $0 \to \operatorname{Hom}_R(M,R) \to \operatorname{Hom}_R(B,R) \to \operatorname{Hom}_R(A,R) \to \operatorname{Ext}^1_R(M,R)$. Since $M_{\underline{p}}$ is $\Lambda_{\underline{p}}$-projective and therefore a Cohen-Macaulay $R_{\underline{p}}$-module for all nonmaximal prime ideals $\underline{p}$ of R, we have that $\operatorname{Ext}^1_R(M,R)_{\underline{p}} \cong \operatorname{Ext}^1_{R_{\underline{p}}}(M_{\underline{p}}, R_{\underline{p}}) = 0$ for all nonmaximal prime ideals $\underline{p}$ of R. Therefore $\operatorname{Ext}^1_R(M,R)$ is an R-module of finite length. As in the argument above, this implies that $0 \to \operatorname{Hom}_R(\operatorname{Hom}_R(A,R),R)) \to \operatorname{Hom}_R(\operatorname{Hom}_R(B,R),R) \to \operatorname{Hom}_R(\operatorname{Hom}_R(M,R),R)$ is exact. Using the fact that the monomorphism $M \to P$ induces an isomorphism $\operatorname{Hom}_R(P,R) \to \operatorname{Hom}_R(M,R)$ and therefore an isomorphism $\operatorname{Hom}_R(\operatorname{Hom}_R(M,R),R) \to \operatorname{Hom}_R(\operatorname{Hom}_R(P,R),R)$ and the fact that A and P are R-reflexive we have the following exact commutative diagram of Λ-modules

$$\begin{array}{ccccccccc}
 & & & & 0 & & 0 & & \\
 & & & & \downarrow & & \downarrow & & \\
0 & \longrightarrow & A & \xrightarrow{\;g\;} & B & \xrightarrow{\;u\;} & M & \longrightarrow & 0 \\
 & & \| & & \downarrow & & \downarrow & & \\
0 & \longrightarrow & A & \longrightarrow & \operatorname{Hom}_R(\operatorname{Hom}_R(B,R),R) & \xrightarrow{\;t\;} & P. & &
\end{array}$$

Now if t were onto, then the bottom row would split since P is Λ-projective. But this would imply that g is a splittable monomorphism, which contradicts the hypothesis of the proposition. Thus t is not onto. Since $M \subset \operatorname{Im} t \subset P$ and M is a maximal submodule of P, this means that $M = \operatorname{Im} t$. It then follows from the commutative diagram that $B \to \operatorname{Hom}_R(\operatorname{Hom}_R(B, R), R)$ is an isomorphism or in other words B is R-reflexive. This establishes the proposition. $\square$

Combining our previous remarks we obtain our desired result.

THEOREM 3.6. *Suppose* $\dim R = 2$.

a) *Let M be a maximal submodule of a projective lattice P. Then there is an exact sequence of Λ-modules* $0 \to A \xrightarrow{g} E \xrightarrow{f} P \to P/M \to 0$ *having the following properties.*

i) *A, E, P are in $L(\Lambda)$.*

ii) *A is isomorphic to the injective lattice* $\operatorname{Hom}_R(\operatorname{Hom}_\Lambda(P, \Lambda), R)$.

iii) *g is not a splittable monomorphism.*

iv) *The exact sequence* $0 \to (\ , A) \xrightarrow{(\ ,g)} (\ , E) \xrightarrow{(\ ,f)} (\ , P) \to \operatorname{Coker}(\ , f) \to 0$ *in* $(L(\Lambda)^{\mathrm{op}}, \mathrm{Ab})$ *has the property that* $\operatorname{Coker}(\ , f)$ *is a simple functor not vanishing on P.*

b) *Let S be a simple functor in* $(L(\Lambda)^{\mathrm{op}}, \mathrm{Ab})$ *such that $S(P) \neq 0$ for some projective P in $L(\Lambda)$. Then there is an exact sequence* $0 \to A \xrightarrow{g} E \xrightarrow{f} P$ *in $L(\Lambda)$ with g not a splittable monomorphism with* $A \cong \operatorname{Hom}_R(\operatorname{Hom}_\Lambda(P, \Lambda), R)$ *such that there is an exact sequence* $0 \to (\ , A) \xrightarrow{(\ ,g)} (\ , E) \xrightarrow{(\ ,f)} (\ , P) \to S \to 0$ *in* $(L(\Lambda)^{\mathrm{op}}, \mathrm{Ab})$.

We now study the projective resolutions of simple functors in $(L(\Lambda)^{\mathrm{op}}, \mathrm{Ab})$ which do not vanish on some injectives.

PROPOSITION 3.7. *Suppose* $\dim R = 2$. *Let* $0 \to A \xrightarrow{g} E \xrightarrow{f} P \xrightarrow{\epsilon} S \to 0$ *be an exact sequence in* noeth Λ *having the following properties:*

i) *P is projective and S is simple.*

ii) $A \cong \operatorname{Hom}_R(\operatorname{Hom}_\Lambda(P, \Lambda), R)$.

Then the sequence $0 \to (P,\ ) \xrightarrow{(f,\ )} (E,\ ) \xrightarrow{(g,\ )} (A,\ ) \to \operatorname{Coker}(g,\ ) \to 0$ *of functors in* $(L(\Lambda), \mathrm{Ab})$ *is exact with* $\operatorname{Coker}(g,\ )$ *a simple functor which does not vanish on the injective lattice A.*

PROOF. We first show that the sequence $0 \to (P,\ ) \xrightarrow{(f,\ )} (E,\ ) \xrightarrow{(g,\ )} (A,\ )$ is exact. Let X be a Λ-lattice. Since S is a simple Λ-module, we have by Proposition 3.3 that $\operatorname{Ext}^i_\Lambda(S, X) = 0$ for $i = 0, 1$. From this it follows that the inclusion $\operatorname{Im} f \to P$ induces an isomorphism $(P, X) \to (\operatorname{Im} f, X)$. Since the exact sequence $0 \to A \xrightarrow{g} E \to \operatorname{Im} f \to 0$ induces the exact sequence $0 \to (\operatorname{Im} f, X) \to (E, X) \to (A, X)$, we have that $0 \to (P, X) \xrightarrow{(f,X)} (E, X) \to (A, X)$ is exact. The fact that this is true for all X in $L(\Lambda)$ shows that $0 \to (P,\ ) \xrightarrow{(f,\ )} (E,\ ) \xrightarrow{(g,\ )} (A,\ )$ is exact in $(L(\Lambda), \mathrm{Ab})$. We now show that $\operatorname{Coker}(g,\ )$ is a simple functor.

We know by II, Proposition 2.2, that a morphism $g\colon A \to E$ which is not a splittable monomorphism has the property $\operatorname{Coker}(g,\ )$ is simple if given any morphism $h\colon A \to X$ in $L(\Lambda)$ such that there is no morphism $t\colon E \to X$ such that $tg = h$, then the induced morphism $A \to E \amalg X$ is a splittable monomorphism. Suppose $h\colon A \to X$ is a morphism in $L(\Lambda)$ such that there is no morphism $t\colon E \to X$ such that $tg = h$. Consider the exact pushout diagram

$$\begin{array}{ccccccccc} 0 & \longrightarrow & A & \xrightarrow{g} & E & \longrightarrow & \operatorname{Im} f & \longrightarrow & 0 \\ & & \downarrow{\scriptstyle h} & & \downarrow{\scriptstyle v} & & \| & & \\ 0 & \longrightarrow & X & \xrightarrow{w} & E\times^{A}X & \longrightarrow & \operatorname{Im} f & \longrightarrow & 0. \end{array}$$

By the hypothesis on h, the bottom row does not split. Since $0 \to X \xrightarrow{w} E\times^{A}X \to P \xrightarrow{\epsilon} S \to 0$ is exact with w a nonsplittable monomorphism, X a lattice, P a projective lattice and S simple, it follows from Proposition 3.5 that $E\times^{A}X$ is a lattice. Therefore there exists an exact commutative diagram

$$\begin{array}{ccccccccc} 0 & \longrightarrow & X & \xrightarrow{w} & E\times^{A}X & \longrightarrow & \operatorname{Im} f & \longrightarrow & 0 \\ & & \downarrow{\scriptstyle j} & & \downarrow{\scriptstyle k} & & \| & & \\ 0 & \longrightarrow & A & \xrightarrow{g} & E & \longrightarrow & \operatorname{Im} f & \longrightarrow & 0 \end{array}$$

which gives rise to the commutative diagram

$$\begin{array}{ccccccccc} 0 & \longrightarrow & A & \xrightarrow{g} & E & \longrightarrow & \operatorname{Im} f & \longrightarrow & 0 \\ & & {\scriptstyle jh}\downarrow & & {\scriptstyle kv}\downarrow & & \| & & \\ 0 & \longrightarrow & A & \longrightarrow & E & \longrightarrow & \operatorname{Im} f & \longrightarrow & 0. \end{array}$$

From this it follows that there is a morphism $s\colon E \to A$ such that $1_A - jh = sg$. The fact that $1_A = jh + sg$ shows that the composition $A \xrightarrow{(g,h)} E\coprod X \xrightarrow{(s,j)} A$ is the identity, or in other words, $A \to E\coprod X$ is a splittable monomorphism. This shows that $\operatorname{Coker}(g,\ )$ is simple which finishes the proof of the proposition. □

Proposition 3.7 describe projective resolutions for certain simple functors in $(L(\Lambda), \mathrm{Ab})$ which do not vanish on injective lattices. We now show that in fact one actually gets projective resolutions for all simple functors in $(L(\Lambda), \mathrm{Ab})$ not vanishing on injective lattices in this way.

PROPOSITION 3.8. *Let T be a simple functor in $(L(\Lambda), \mathrm{Ab})$ which does not vanish on an injective lattice A. Then there is an exact sequence $0 \to A \xrightarrow{g} E \xrightarrow{f} P \xrightarrow{\epsilon} S \to 0$ with E a lattice, P the projective Λ-module $\operatorname{Hom}_{\Lambda^{\mathrm{op}}}(\operatorname{Hom}_R(A, R), \Lambda^{\mathrm{op}})$ and S a simple Λ-module such that there is an exact sequence $0 \to (P,\ ) \to (E,\ ) \to (A,\ ) \to T \to 0$ in $(L(\Lambda), \mathrm{Ab})$.*

PROOF. Since $T(A) \neq 0$, we know that there is a maximal left ideal H of $\operatorname{End} A$ such that $T \cong (A,\ )/(A,\ )_H$. Let $D\colon L(\Lambda) \to L(\Lambda^{\mathrm{op}})$ be the duality given by $\operatorname{Hom}_R(\ , R)$. Then $D(A)$ is a projective Λ^{op}-module and $D(H)$ is a maximal right ideal in $\operatorname{End} D(A)$. Let S be the simple functor $(\ , D(A))/(\ , D(A))_H$ in $(L(\Lambda)^{\mathrm{op}}, \mathrm{Ab})$. Therefore by Proposition 3.6 there is an exact sequence

$$(*) \quad 0 \to U \xrightarrow{g} V \xrightarrow{f} D(A) \xrightarrow{\epsilon} W \to 0$$

with $U \cong \operatorname{Hom}_R(\operatorname{Hom}_{\Lambda^{\mathrm{op}}}(D(A), \Lambda^{\mathrm{op}}), R)$, V a lattice and W a simple Λ^{op}-module such that $0 \to (\ , U) \to (\ , V) \to (\ , D(A)) \to S \to 0$ is exact. Thus by the usual duality arguments it follows that $A \xrightarrow{D(f)} D(V)$ has the property that $(D(V),\ ) \xrightarrow{(Df,\ )} (A,\ ) \to T \to 0$ is exact. Since W is an R-module of finite length and $D(A)$ is a Λ^{op}-lattice, we have that $\operatorname{Ext}^i_R(W,R)=0$ for $i=0,1$ and $\operatorname{Ext}^i_R(D(A),R)=0$ for $i>0$. Therefore the exact sequence $0\to\operatorname{Im}f\to D(A)\to W\to 0$ has the property $D^2(A)\to D(\operatorname{Im}f)$ is an isomorphism and $\operatorname{Ext}^1_R(\operatorname{Im}f,R)\cong\operatorname{Ext}^2_R(W,R)$. Therefore applying D to the exact sequence $(*)$ we obtain the exact sequence

$$(**)\quad 0\to A\to D(V)\to D(U)\to\operatorname{Ext}^2_R(W,R)\to 0.$$

Since $U=\operatorname{Hom}_R(\operatorname{Hom}_{\Lambda^{\mathrm{op}}}(D(A),\Lambda^{\mathrm{op}}),R)$, we have that $DU=\operatorname{Hom}_{\Lambda^{\mathrm{op}}}(D(A),\Lambda^{\mathrm{op}})=P$. Finally, since the functor $\operatorname{Ext}^2_R(\ ,R)\colon \operatorname{noeth}\Lambda^{\mathrm{op}}\to\operatorname{noeth}\Lambda$ induces a duality between the Λ^{op}- and Λ-modules of finite length because R is a Gorenstein ring of $\dim 2$, we have that W being a simple Λ^{op}-module implies that $\operatorname{Ext}^2_R(W,R)$ is a simple Λ-module. This finishes the proof of the proposition since we have already seen that the properties of $(**)$ imply that $0\to(D(U),\ )\to(D(V),\ )\to(A,\ )\to T\to 0$ is exact in $(L(\Lambda),\mathrm{Ab})$ (see Proposition 3.7). □

Throughout the rest of this section we assume that Λ is an R-order with $\dim R=2$. Our aim is to show that in the case R is a complete local ring then $L(\Lambda)$ is of finite representation type if and only if the finitely presented functors in $(L(\Lambda)^{\mathrm{op}},\mathrm{Ab})$ which vanish on Λ have finite length. This result is based on the following observation.

Let $\{\underline{m}_i\}_{i\in I}$ be a family of maximal left ideals such that $\{\Lambda/\underline{m}_i\}_{i\in I}$ is a complete set of nonisomorphic simple Λ-modules. Then by Theorem 3.6 we know that for each i in I there is an epimorphism $t_i\colon E_i\to\underline{m}_i\to 0$ of Λ-modules with E_i a lattice such that $(X,E_i)\to(X,\underline{m}_i)\to 0$ is exact for all lattices X. We now point out some other properties of these lattices $\{E_i\}_{i\in I}$ and the epimorphisms $t_i\colon E_i\to\underline{m}_i$.

PROPOSITION 3.9. *Let $\{E_i\}_{i\in I}$ be the family of lattices described above. Let X be a nonzero lattice. Then*

a) *$(X,E_i)\neq 0$ for each i in I.*

b) *X is projective if and only if $\underline{\operatorname{Hom}}_\Lambda(X,E_i)=0$ for all i in I.*

PROOF. a) Since R is a Gorenstein ring, we know that the prime ideals associated to (0) are all minimal prime ideals. Combining this with the fact that R is equidimensional of dimension 2, we have that each maximal ideal contains a regular element of R. Because each $\underline{m}_i$ is a maximal left ideal of Λ, we know that $\underline{m}_i\cap R$ is a maximal ideal of R and therefore contains an R-regular element r_i. Since Λ is a lattice and therefore a reflexive R-module, we know that each r_i is regular in Λ, i.e. $0\to\Lambda\xrightarrow{r_i}\Lambda$ is exact.

Suppose now that X is a nonzero lattice. Since $\dim R=2$, we know that X is a reflexive Λ-module. From this it follows that $\operatorname{Hom}_\Lambda(X,\Lambda)\neq 0$. Let $f\colon X\to\Lambda$ be a nonzero morphism. Then for each i in I, the Λ-morphism $r_if\colon X\to\Lambda$ is not zero with $\operatorname{Im}(r_if)\subset\Lambda r_i\subset\underline{m}_i$. Thus we have shown that for each i in I, there is a nonzero morphism $f_i\colon X\to\underline{m}_i$. Since $(X,E_i)\xrightarrow{(X,t_i)}(X,\underline{m}_i)\to 0$ is exact for each i in I, we have that $(X,E_i)\neq 0$ for each i in I.

b) Suppose X is a lattice such that $\underline{\operatorname{Hom}}_\Lambda(X,E_i)=0$. Since the epimorphism $t_i\colon E_i\to\underline{m}_i$ has the property that $(X,E_i)\xrightarrow{(X,t_i)}(X,\underline{m}_i)\to 0$ is exact, the fact that $\underline{\operatorname{Hom}}_\Lambda(X,E_i)=0$ implies that $\underline{\operatorname{Hom}}_\Lambda(X,\underline{m}_i)=0$. But $\underline{\operatorname{Hom}}_\Lambda(X,\underline{m}_i)\cong\operatorname{Tor}^\Lambda_1(\operatorname{Tr}X,\underline{m}_i)$. Since $\operatorname{Tor}^\Lambda_1(\operatorname{Tr}X,\underline{m}_i)\cong\operatorname{Tor}^\Lambda_2(\operatorname{Tr}X,\Lambda/\underline{m}_i)$, we have that $\operatorname{Tor}^\Lambda_2(\operatorname{Tr}X,\Lambda/\underline{m}_i)=0$.

Let $P_1\to P_0\to X\to 0$ be a projective presentation of X used to construct $\operatorname{Tr}X$. Then we have the exact sequence

$$0\to(X,\Lambda)\to(P_0,\Lambda)\to(P_1,\Lambda)\to\operatorname{Tr}X\to 0$$

Hence if we assume that $\underline{\mathrm{Hom}}_\Lambda(X, E_i) = 0$ for all i, then $\mathrm{Tor}_1^\Lambda(V, \Lambda/\underline{m}_i) = 0$ for all i in I where $V = \mathrm{Ker}((P_1, \Lambda) \to \mathrm{Tr}\, X)$. Thus V is a projective Λ^{op}-module since $\mathrm{Tor}_1^\Lambda(V, S)$ being zero for all simple Λ-modules S implies that V is projective. Since $0 \to (X, \Lambda) \to (P_0, \Lambda) \to V \to 0$ is exact, it follows that (X, Λ) is a projective Λ^{op}-module. This implies that $X \cong ((X, \Lambda), \Lambda)$ is projective. Hence if $\underline{\mathrm{Hom}}_\Lambda(X, E_i) = 0$ for all i in I, then X is projective. Because it is obvious that $\underline{\mathrm{Hom}}_\Lambda(X, E_i) = 0$ for X projective, the proof of b) is complete. □

We now apply this result to prove our previously cited result.

THEOREM 3.10. *For an R-order Λ with R a complete local ring of dimension* 2, *the following statements are equivalent.*

a) *Λ is of finite lattice type.*

b) *For each X in $L(\Lambda)$, the functor $\underline{\mathrm{Hom}}_\Lambda(\ , X)$ in $(L(\Lambda)^{\mathrm{op}}, \mathrm{Ab})$ is of finite length.*

c) *Each finitely generated F in $(L(\Lambda)^{\mathrm{op}}, \mathrm{Ab})$ such that $F(\Lambda) = 0$ is of finite length.*

d) *$\underline{\mathrm{Hom}}_\Lambda(\ , E_i)$ is of finite length, where $\{E_i\}_{i \in I}$ is the family of lattices described prior to the statement of Proposition 3.9.*

PROOF. a) implies b). Since for Y in $L(\Lambda)$, we have that $\underline{\mathrm{Hom}}_\Lambda(Y, X)$ is an $(\mathrm{End}\, Y)^{\mathrm{op}}$-submodule of finite length, it follows that $\underline{\mathrm{Hom}}_\Lambda(\ , X)$ is of finite length if and only if there are only a finite number of nonisomorphic indecomposable Y in $L(\Lambda)$ such that $\underline{\mathrm{Hom}}_\Lambda(Y, X) \neq 0$ ([7, Proposition 2.11]). Thus if $L(\Lambda)$ is of finite type, then $\underline{\mathrm{Hom}}_\Lambda(\ , X)$ is of finite length for all X in $L(\Lambda)$.

b) implies c). Let F be a finitely generated functor such that $F(\Lambda) = 0$. Since F is finitely generated there is an epimorphism $(\ , X) \to F \to 0$ for some X in $L(\Lambda)$. Since $F(\Lambda) = 0$, it is not hard to see that $(\ , X) \to F \to 0$ factors through $\underline{\mathrm{Hom}}_\Lambda(\ , X)$ and hence there is an epimorphism $\underline{\mathrm{Hom}}_\Lambda(\ , X) \to F \to 0$. Therefore F is of finite length since $\underline{\mathrm{Hom}}_\Lambda(\ , X)$ is of finite length.

c) implies d). Obvious.

d) implies a). Since R is complete, there are only a finite number of simple Λ-modules. Thus the set I is finite. Since each $\underline{\mathrm{Hom}}_\Lambda(\ , E_i)$ is of finite length, there are only a finite number of nonisomorphic indecomposable lattices X such that $\underline{\mathrm{Hom}}_\Lambda(X, E_i) \neq 0$ for each i in I. Therefore there are only a finite number of nonisomorphic indecomposable lattices X such that $\underline{\mathrm{Hom}}_\Lambda(X, E_i) \neq 0$ for some i in I. This means there are only a finite number of nonisomorphic nonprojective lattices since by Proposition 3.9 we know that a lattice X is projective if and only if $\underline{\mathrm{Hom}}_\Lambda(X, E_i) = 0$ for all i in I. Combining this with the fact that R being a complete local ring implies there are only a finite number of nonisomorphic indecomposable projective Λ-modules, we have that Λ is of finite lattice type, which finishes the proof that d) implies a). □

4. A generalization

Let Λ be an arbitrary ring, X a module in $\mathrm{mod}\, \Lambda$, the category of finitely presented Λ-modules. In addition, suppose the finitely presented Λ^{op}-module $\mathrm{Tr}\, X$ is a Λ^{op}-Σ-bimodule and I is an injective Σ-module. Considering $\mathrm{Hom}_\Sigma(\mathrm{Tr}\, X, I)$ a Λ-module in the obvious way, we saw in I, Section 3, that for each Λ-module C the duality $\mathrm{Tr}\colon \underline{\mathrm{mod}}\Lambda \to \underline{\mathrm{mod}}\Lambda^{\mathrm{op}}$ gave rise to a Σ-module structure on $\underline{\mathrm{Hom}}_\Lambda(X, C)$ and that we have isomorphisms $\mathrm{Ext}_\Lambda^1(C, \mathrm{Hom}_\Sigma(\mathrm{Tr}\, X, I)) \to \mathrm{Hom}_\Sigma(\underline{\mathrm{Hom}}_\Lambda(X, C), I)$ which are functorial in C and I (see I, Proposition 3.4). These functorial isomorphisms played a fundamental role in establishing our basic existence theorems for morphisms determined by modules. It is our aim in this section to give some applications in a very different direction.

In II, Section 3, we established a duality $D\colon \mathrm{f.p.}((\mathrm{Mod}\,\Lambda)^{\mathrm{op}}, \mathrm{Ab})_0 \to \mathrm{f.p.}(\mathrm{Mod}\,\Lambda, \mathrm{Ab})_0$. Recall that F is in $\mathrm{f.p.}((\mathrm{Mod}\,\Lambda)^{\mathrm{op}}, \mathrm{Ab})_0$ if and only if F is finitely presented and vanishes on all projective Λ-modules. Suppose, as above, X is in $\mathrm{mod}\,\Lambda$ and $\mathrm{Tr}\,X$ is a Λ^{op}-Σ-bimodule. Since $F(P) = 0$ for all projective Λ-modules P, we have that the $(\mathrm{End}_\Lambda\, X)^{\mathrm{op}}$-module $F(X)$ is annihilated by $P(X, X)$ and so is an $(\mathrm{End}_\Lambda\, X)^{\mathrm{op}}$-module. Also the duality $\mathrm{Tr}\colon \underline{\mathrm{mod}}\Lambda \to \underline{\mathrm{mod}}\Lambda^{\mathrm{op}}$ gives rise to a ring isomorphism $\mathrm{Tr}\colon (\underline{\mathrm{End}}_\Lambda X)^{\mathrm{op}} \to \underline{\mathrm{End}}_\Lambda\, \mathrm{Tr}\,X$. Therefore the ring morphism $\Sigma \to \mathrm{End}_\Lambda\, \mathrm{Tr}\,X$ given by the bimodule structure Λ^{op}-Σ of $\mathrm{Tr}\,X$ induces a ring morphism $\Sigma \to (\underline{\mathrm{End}}_\Lambda X)^{\mathrm{op}}$. Hence $F(X)$ is a Σ-module in a natural way. As already noted $\mathrm{Hom}_\Sigma(\mathrm{Tr}\,X, I)$ is a Λ-module in a natural way. With these definitions and notations in mind, we establish the following relationship between F and DF for all F in $\mathrm{f.p.}((\mathrm{Mod}\,\Lambda)^{\mathrm{op}}, \mathrm{Ab})$.

THEOREM 4.1. *Let X be in* $\mathrm{mod}\,\Lambda$ *and suppose that the finitely presented Λ^{op}-module* $\mathrm{Tr}\,X$ *is a Λ^{op}-Σ-bimodule. Then for each F in* $\mathrm{f.p.}((\mathrm{Mod}\,\Lambda)^{\mathrm{op}}, \mathrm{Ab})$ *and each injective Σ-module I we have isomorphisms* $\mathrm{Hom}_\Sigma(F(X), I) \cong DF(\mathrm{Hom}_\Sigma(\mathrm{Tr}\,X, I))$ *which are functorial in F and I.*

PROOF. Since F is in $\mathrm{f.p.}((\mathrm{Mod}\,\Lambda)^{\mathrm{op}}, \mathrm{Ab})$, there is an exact sequence $0 \to A \to B \to C \to 0$ of Λ-modules such that $0 \to (\ , A) \to (\ , B) \to (\ , C) \to F \to 0$ is exact. Now there is an exact commutative diagram

$$\begin{array}{ccccccccc}
0 & \longrightarrow & P_0 & \longrightarrow & P_1 & \longrightarrow & P_2 & \longrightarrow & 0 \\
 & & \downarrow & & \downarrow & & \downarrow & & \\
0 & \longrightarrow & A & \longrightarrow & B & \longrightarrow & C & \longrightarrow & 0 \\
 & & \downarrow & & \downarrow & & \downarrow & & \\
 & & 0 & & 0 & & 0 & &
\end{array}$$

with the P_i projective Λ-modules. This gives rise to the following exact commutative diagram of functors

$$\begin{array}{ccccccccc}
0 & \longrightarrow & (\ , P_0) & \longrightarrow & (\ , P_1) & \longrightarrow & (\ , P_2) & \longrightarrow & 0 \\
 & & \downarrow & & \downarrow & & \downarrow & & \\
0 & \longrightarrow & (\ , A) & \longrightarrow & (\ , B) & \longrightarrow & (\ , C) & \longrightarrow & 0 \\
 & & \downarrow & & \downarrow & & \downarrow & & \\
 & & \underline{\mathrm{Hom}}_\Lambda(\ , A) & \longrightarrow & \underline{\mathrm{Hom}}_\Lambda(\ , B) & \longrightarrow & \underline{\mathrm{Hom}}_\Lambda(\ , C) & & \\
 & & \downarrow & & \downarrow & & \downarrow & & \\
 & & 0 & & 0 & & 0 & &
\end{array}$$

which gives rise to the exact sequence $\underline{\mathrm{Hom}}_\Lambda(\ , B) \to \underline{\mathrm{Hom}}_\Lambda(\ , C) \to F \to 0$. Thus $\underline{\mathrm{Hom}}_\Lambda(X, B) \to \underline{\mathrm{Hom}}_\Lambda(X, C) \to F(X) \to 0$ is an exact sequence of Σ-modules by means of the ring morphism $\Sigma \to (\underline{\mathrm{End}} X)^{\mathrm{op}}$ described earlier. Thus we obtain the exact sequence $0 \to \mathrm{Hom}_\Sigma(F(X), I) \to \mathrm{Hom}_\Sigma(\underline{\mathrm{Hom}}_\Lambda(X, C), I) \to \mathrm{Hom}_\Sigma(\underline{\mathrm{Hom}}_\Lambda(X, R), I)$. Using the isomorphism of functors $\mathrm{Ext}^1_\Lambda(\ , \mathrm{Hom}_\Sigma(\mathrm{Tr}\,X, I)) \to \mathrm{Hom}_\Sigma(\underline{\mathrm{Hom}}_\Lambda(X,\ ), I)$ we obtain the exact sequence $(*)$ $0 \to \mathrm{Hom}_\Sigma(F(X), I) \to \mathrm{Ext}^1_\Lambda(C, \mathrm{Hom}_\Sigma(\mathrm{Tr}\,X, I)) \to \mathrm{Ext}^1_\Lambda(B, \mathrm{Hom}_\Sigma(\mathrm{Tr}\,X, I))$. Since $0 \to A \to B \to C \to 0$ is exact, we have that $(B,\ ) \to (A,\ ) \to \mathrm{Ext}^1_\Lambda(C,\ ) \to \mathrm{Ext}^1_\Lambda(B,\ )$ is exact, which implies that $0 \to DF \to$

$\operatorname{Ext}^1_\Lambda(C,\ ) \to \operatorname{Ext}^1_\Lambda(B,\ )$ is exact, since $(B,\ ) \to (A,\ ) \to DF \to 0$ is exact (see II, Section 3). Therefore

$$(**)\ 0 \to DF(\operatorname{Hom}_\Sigma(\operatorname{Tr} X, I)) \to \operatorname{Ext}^1_\Lambda(C, \operatorname{Hom}_\Sigma(\operatorname{Tr} X, I)) \to \operatorname{Ext}^1_\Lambda(B, \operatorname{Hom}_\Sigma(\operatorname{Tr} X, I))$$

is exact. Combining $(*)$ and $(**)$ we obtain an isomorphism $\operatorname{Hom}_\Sigma(F(X), I) \cong DF(\operatorname{Hom}_\Sigma(\operatorname{Tr} X, I))$. The fact that this isomorphism is functorial in F and I can be verified in a straightforward way by checking through the proof and seeing that it is functorial in F and I at each stage. □

As an application of this result we have the following.

COROLLARY 4.2. *Let $0 \to A \xrightarrow{g} B \xrightarrow{f} C \to 0$ be an exact sequence of Λ-modules. Suppose X is in $\operatorname{mod}\Lambda$ and $\operatorname{Tr} X$ in $\operatorname{mod}\Lambda^{\mathrm{op}}$ is a Λ^{op}-Σ-bimodule. Further suppose I is an injective cogenerator for Σ. Then the following statements are equivalent.*

a) *For each $h\colon X \to C$ there is a $s\colon X \to B$ such that $fs = h$.*

b) *For each $j\colon A \to \operatorname{Hom}_\Sigma(\operatorname{Tr} X, I)$ there is a $t\colon B \to \operatorname{Hom}_\Sigma(\operatorname{Tr} X, I)$ such that $j = tg$.*

PROOF. From the exact sequences $(\ ,B) \to (\ ,C) \to F \to 0$ and $(B,\ ) \to (A,\ ) \to DF \to 0$ we deduce that a) is equivalent to $F(X) = 0$ and b) is equivalent to $DF(\operatorname{Hom}_\Sigma(\operatorname{Tr} X, I)) = 0$. Since $\operatorname{Hom}_\Sigma(F(X), I) \cong DF(\operatorname{Hom}_\Sigma(\operatorname{Tr} X, I))$ by Theorem 4.1 and I is injective cogenerator, it follows that $F(X) = 0$ if and only if $DF(\operatorname{Hom}_\Sigma(\operatorname{Tr} X, I)) = 0$. This finishes the proof of the corollary. □

As another application of Theorem 4.1, we have the following.

COROLLARY 4.3. *Let X be in $\operatorname{mod}\Lambda$ and suppose $\operatorname{Tr} X$ in $\operatorname{mod}\Lambda^{\mathrm{op}}$ is a Λ^{op}-Σ-bimodule. Then for A in $\operatorname{Mod}\Lambda$ and injective Σ-module I we have isomorphisms $\operatorname{Hom}_\Sigma(\operatorname{Ext}^1_\Lambda(X, A), I) \cong \overline{\operatorname{Hom}}_\Lambda(A, \operatorname{Hom}_\Sigma(\operatorname{Tr} X, I))$ which are functorial in A and I.*

PROOF. Let $0 \to A \to B \to C \to 0$ be an exact sequence of Λ-modules with B an injective Λ-module. Then $(\ ,B) \to (\ ,C) \to \operatorname{Ext}^1_\Lambda(\ ,A) \to 0$ and $(B,\ ) \to (A,\ ) \to \overline{\operatorname{Hom}}_\Lambda(A,\ ) \to 0$ are exact since B is injective. Therefore $D(\operatorname{Ext}^1_\Lambda(\ ,A)) = \overline{\operatorname{Hom}}_\Lambda(A,\ )$ and so applying Theorem 4.1 we have the isomorphism $\operatorname{Hom}_\Sigma(\operatorname{Ext}^1_\Lambda(X, A), I) \cong \overline{\operatorname{Hom}}_\Lambda(A, \operatorname{Hom}_\Sigma(\operatorname{Tr} X, I))$ which is functorial in A and I. □

The rest of this section is devoted to pointing out the analogues of these results for lattices over an R-order.

Suppose Λ is an R-order. Let $\mathrm{f.p.}(L(\Lambda)^{\mathrm{op}}, \mathrm{Ab})$ denote the category of finitely presented functors from $L(\Lambda^{\mathrm{op}})$ to Ab. Further let $\mathrm{f.p.}(L(\Lambda)^{\mathrm{op}}, \mathrm{Ab})_0$ denote the full subcategory of $\mathrm{f.p.}(L(\Lambda)^{\mathrm{op}}, \mathrm{Ab})$ consisting of all F in $\mathrm{f.p.}(L(\Lambda)^{\mathrm{op}}, \mathrm{Ab})$ such that $F(P) = 0$ for projective lattices P. It is easily checked that a functor F is in $\mathrm{f.p.}(L(\Lambda)^{\mathrm{op}}, \mathrm{Ab})$ if and only if there is an exact sequence of lattices $0 \to A \to B \to C \to 0$ such that $0 \to (\ ,A) \to (\ ,B) \to (\ ,C) \to F \to 0$ is exact.

Similarly we denote by $\mathrm{f.p.}(L(\Lambda), \mathrm{Ab})$ the category of finitely presented functors from $L(\Lambda)$ to Ab and denote by $\mathrm{f.p.}(L(\Lambda), \mathrm{Ab})_0$ the full subcategory of $\mathrm{f.p.}(L(\Lambda), \mathrm{Ab})$ consisting of all G such that $G(I) = 0$ for all injective lattices I. It is easily checked that a functor G is in $\mathrm{f.p.}(L(\Lambda), \mathrm{Ab})_0$ if and only if there is an exact sequence of lattices $0 \to A \to B \to C \to 0$ such that $0 \to (C,\ ) \to (B,\ ) \to (A,\ ) \to G \to 0$ is exact.

Now we define a duality $D\colon \mathrm{f.p.}(L(\Lambda)^{\mathrm{op}}, \mathrm{Ab})_0 \to \mathrm{f.p.}(L(\Lambda), \mathrm{Ab})_0$ as follows. For each F in $\mathrm{f.p.}(L(\Lambda)^{\mathrm{op}}, \mathrm{Ab})_0$ choose an exact sequence $0 \to A \to B \to C \to 0$ in $L(\Lambda)$ such that $0 \to (\ ,A) \to (\ ,B) \to (\ ,C) \to F \to 0$ is exact. Then define DF by the exact sequence $0 \to (C, \ ) \to (B, \ ) \to (A, \ ) \to DF \to 0$. It is not difficult to see that D is a duality which, up to a unique isomorphism, is independent of the particular exact sequence $0 \to A \to B \to C \to 0$ of lattices chosen for the F in $\mathrm{f.p.}(L(\Lambda)^{\mathrm{op}}, \mathrm{Ab})$. With these preliminaries in mind, we have the following analogue of Theorem 4.1.

THEOREM 4.4. *Let Λ be an R-order. For each F in $\mathrm{f.p.}(L(\Lambda)^{\mathrm{op}}, \mathrm{Ab}$ and X in $L(\Lambda)$ we have isomorphisms $\mathrm{Hom}_R(F(X), I_d^n) \cong DF(\mathrm{Hom}_R(\mathrm{Tr}_L X, R^n))$ which are functorial in F and X.*

PROOF. Recall that for lattices C and X we have isomorphisms $\mathrm{Ext}^1_\Lambda(C, \mathrm{Hom}_R(\mathrm{Tr}_L X, R^n)) \to \mathrm{Hom}_R(\underline{\mathrm{Hom}}_\Lambda(X, C), I_d^n)$ which are functorial in C and X (see I, Proposition 8.9). Using these functorial isomorphisms, the proof of the theorem proceeds almost verbatim the same way as the proof of Theorem 4.1. □

As an analogue of Corollary 4.2 we have the following.

COROLLARY 4.5. *Let Λ be an R-order and let $0 \to A \xrightarrow{g} B \xrightarrow{f} C \to 0$ be an exact sequence of Λ-lattices. Then for a Λ-lattice X, the following statements are equivalent.*

a) *For each $h: X \to C$ there is an $s: X \to B$ such that $h = fs$.*

b) *For each $j: A \to \mathrm{Hom}_R(\mathrm{Tr}_L X, R)$, there is a $t: B \to \mathrm{Hom}_R(\mathrm{Tr}_L X, R)$ such that $j = tg$.*

PROOF. Analogous to proof of Corollary 4.2 once one observes that for all F in $\mathrm{f.p.}(L(\Lambda)^{\mathrm{op}}, \mathrm{Ab})$ we have that $F(X)$ is an R-module of finite length and so $F(X) = 0$ if and only if $\mathrm{Hom}_R(F(X), I_d) = 0$. □

As our final result, we have the following analogue of Corollary 4.3

COROLLARY 4.6. *Let Λ be an R-order. For all lattices X and A we have isomorphisms $\mathrm{Hom}_R(\mathrm{Ext}^1_\Lambda(X, A), I_d) \cong \overline{\mathrm{Hom}}_\Lambda(A, \mathrm{Hom}_R(\mathrm{Tr}_L X, R))$ functorial in X and A.*

References

[1] Auslander, M., *Applications of almost split sequences*, these Proceedings.

[2] Auslander, M., *Coherent functors*, Proceedings of the Conference on Categorical Algebra, Springer-Verlag, Berlin, Heidelberg, New York, 1966.

[3] Auslander, M., *Existence theorems for almost split sequences*, Ring Theory II: Proceedings of the Second Oklahoma Conference, Marcel Dekker, New York and Basel, 1977.

[4] Auslander, M., *Large modules over Artin algebras*, Algebra, Topology and Categories, Academic Press, New York, 1976.

[5] Auslander, M., *Representation dimension of Artin algebras*, Queen Mary College Mathematics Notes, London.

[6] Auslander, M. *Representation theory of Artin algebras,* I, Comm. in Algebra, (1974), 177–268.

[7] Auslander, M., *Representation theory of Artin algebras,* II, Comm. in Algebra, (1974), 269–310.

[8] Auslander, M. and Bridger, M., *Stable module theory*, Memoirs of the American Mathematical Society, No. 94, 1969.

[9] Auslander, M. and Platzeck, M. I., *Representation theory of hereditary Artin algebras*, these Proceedings.

[10] Auslander, M. and Reiten, I., *Representation theory of Artin algebras,* III, Comm. in Algebra, (1975), 239–294.

[11] Auslander, M. and Reiten. I., *Representation theory of Artin algebras,* IV, Comm. in Algebra, (1977), 443–518.

[12] Auslander, M. and Reiten, I., *Stable equivalence of dualizing R-varieties*, Advances in Math., (1974), 300–366.

[13] Bass, H. *On the ubiquity of Gorenstein rings*, Math. Zeitschr., (1963), 8–28.

[14] Eilenberg, S. and Cartan, E., *Homological Algebra*, Princeton University Press, Princeton, New Jersey, 1956.

[15] Eckman, B. and Hilton, P. J., *Homotopy groups of maps and exact sequences*, Comment. Math. Helv., (1960), 271–304.
[16] Herzog, J. and Kung. E., *Der Kanonische Modul eines Cohen-Macauley Rings*, Springer-Verlag, New York, 1971.
[17] Nagata, M., *Local Rings*, Interscience Publishers, New York, 1962.

BRANDEIS UNIVERSITY, WALTHAM, MASSACHUSETTS

Applications of morphisms determined by modules[1]

Maurice Auslander

In [2] a general theory of morphisms determined by modules was developed. This paper begins with a summary of this general theory specialized to artin algebras. Basic definitions and the important results are given, but no effort is made to duplicate proofs already given in [2].

After these preliminaries, various applications are given of morphisms determined by modules to the study of finitely generated modules. In particular, much of the paper is devoted to showing the connections between almost split morphisms, a notion introduced in [4], and morphisms determined by modules. The paper ends with a discussion of some variations of the Brauer-Thrall Conjecture II.

1. The transpose

We assume throughout this section that R is a commutative artin ring. By an *artin R-algebra* we mean an R-algebra Λ which is a finitely generated R-module. Obviously an artin R-algebra is a two-sided artin ring. Our main purpose in this section is to recall some basic facts concerning the category $\operatorname{mod}\Lambda$, the category of finitely generated left Λ-modules. Of particular concern is the equivalence of categories $D\operatorname{Tr}\colon \underline{\operatorname{mod}}\Lambda \to \overline{\operatorname{mod}}\Lambda$ where $\underline{\operatorname{mod}}\Lambda$ denotes the category $\operatorname{mod}\Lambda$ modulo projectives and $\overline{\operatorname{mod}}\Lambda$ denotes the category $\operatorname{mod}\Lambda$ modulo injectives. In addition to recalling the basic properties of $D\operatorname{Tr}$ we give various elementary applications. The basic facts concerning $D\operatorname{Tr}$ are given without proof. The reader is referred to [5] and [6] for more details.

Since R is a commutative artin ring, $I = I(R/\operatorname{rad}R)$, the injective envelope of $R/\operatorname{rad}R$, is a finitely generated R-module which is an injective cogenerator for $\operatorname{Mod}R$, the category of all R-modules.

Suppose Λ is an artin algebra. Then Λ^{op}, the opposite ring of Λ, is also an artin R-algebra in an obvious way. Then the functor $D\colon \operatorname{mod}\Lambda \to \operatorname{mod}\Lambda^{\mathrm{op}}$ given by $D(M) = \operatorname{Hom}_R(M, I)$ is well known to be a duality with dual inverse $D\colon \operatorname{mod}\Lambda^{\mathrm{op}} \to \operatorname{mod}\Lambda$. Since the objects in $\operatorname{mod}\Lambda^{\mathrm{op}}$ have projective covers, it follows from the duality $D\colon \operatorname{mod}\Lambda^{\mathrm{op}} \to \operatorname{mod}\Lambda$ that every object in $\operatorname{mod}\Lambda$ has an injective envelope in $\operatorname{mod}\Lambda$.

Associated with the category $\operatorname{mod}\Lambda$ are the categories $\underline{\operatorname{mod}}\Lambda$ and $\overline{\operatorname{mod}}\Lambda$, called respectively the *category* $\operatorname{mod}\Lambda$ *modulo projectives* and $\operatorname{mod}\Lambda$ *modulo injectives*. Since these categories $\underline{\operatorname{mod}}\Lambda$ and $\overline{\operatorname{mod}}\Lambda$ play such an important role in the entire paper, we briefly recall their definitions (see [5] and [6]).

For each pair of objects A and B in $\operatorname{mod}\Lambda$, denote by $P(A,B)$ the R-submodule of $(A,B) = \operatorname{Hom}_\Lambda(A,B)$ consisting of all morphisms $f\colon A \to B$ which can be factored through a projective in $\operatorname{mod}\Lambda$, i.e. $f\colon A \to B$ is in $P(A,B)$ if and only if f is a composition

[1] Lecture Notes in Pure Appl. Math., Vol. 37, 245–327, Dekker, New York, 1978

1

$A \xrightarrow{g} P \xrightarrow{h} B$ with P a projective module in $\operatorname{mod}\Lambda$. It is easily checked that given morphisms $f\colon A \to B$ and $g\colon B \to C$ in $\operatorname{mod}\Lambda$, then gf is in $P(A,B)$ if either f is in $P(A,B)$ or g is in $P(B,C)$. For each pair of objects A and B in $\operatorname{mod}\Lambda$ we denote the R-module $(A,B)/P(A,B)$ by $\underline{\operatorname{Hom}}_\Lambda(A,B)$ and for each f in (A,B) we denote its image in $\underline{\operatorname{Hom}}_\Lambda(A,B)$ under the canonical epimorphism $(A,B) \to (A,B)/P(A,B)$ by $\underline{f}$. Then by our previous remarks it is clear that given A, B, C in $\operatorname{mod}\Lambda$, the R-bilinear map $(A,B) \times (B,C) \to (A,C)$ given by $(f,g) \mapsto gf$ induces an R-bilinear map $\underline{\operatorname{Hom}}_\Lambda(A,B) \times \underline{\operatorname{Hom}}_\Lambda(B,C) \to \underline{\operatorname{Hom}}_\Lambda(A,C)$ given by $(\underline{f},\underline{g}) \mapsto \underline{fg}$. Then $\underline{\operatorname{mod}}\Lambda$ is defined to be the category given by the following data. The objects of $\underline{\operatorname{mod}}\Lambda$ are the same as those of $\operatorname{mod}\Lambda$. For each pair of objects A, B, in $\underline{\operatorname{mod}}\Lambda$, the morphisms from A to B in $\underline{\operatorname{mod}}\Lambda$ are $\underline{\operatorname{Hom}}_\Lambda(A,B)$. Finally, the composition in $\underline{\operatorname{mod}}\Lambda$ is given by the R-bilinear maps $\underline{\operatorname{Hom}}_\Lambda(A,B) \times \underline{\operatorname{Hom}}_\Lambda(B,C) \to \underline{\operatorname{Hom}}_\Lambda(A,C)$ defined above by $(\underline{f},\underline{g}) \to \underline{gf}$. It is not difficult to check that $\underline{\operatorname{mod}}\Lambda$ has the following properties.

PROPOSITION 1.1. a) *$\underline{\operatorname{mod}}\Lambda$ is an additive category.*

b) *The functor $F\colon \operatorname{mod}\Lambda \to \underline{\operatorname{mod}}\Lambda$ given by $F(A) = A$ for all A in $\operatorname{mod}\Lambda$ and $F\colon (A,B) \to \underline{\operatorname{Hom}}_\Lambda(F(A),F(B))$ is the canonical morphism $(A,B) \to (A,B)/P(A,B)$ for all A, B in $\operatorname{mod}\Lambda$ has the following properties:*
   i) *F is a full additive functor.*
   ii) *$F(A) \approx 0$ in $\underline{\operatorname{mod}}\Lambda$ if and only if A is projective.*
   iii) *$F(A) \approx F(B)$ if and only if there are projective modules P and Q in $\operatorname{mod}\Lambda$ such that $A \amalg P \approx B \amalg Q$.*
   iv) *If $\mathcal{C}$ is an additive category and $G\colon \operatorname{mod}\Lambda \to \mathcal{C}$ is an additive functor such that $G(P) = 0$ for all projective modules P in $\operatorname{mod}\Lambda$, then there is a unique (up to isomorphism) additive functor $H\colon \underline{\operatorname{mod}}\Lambda \to \mathcal{C}$ such that $G = HF$.*

Next we recall that $\operatorname{mod}\Lambda$ is a Krull-Schmidt category, i.e. each indecomposable object in $\operatorname{mod}\Lambda$ has a local endomorphism ring and each object A in $\operatorname{mod}\Lambda$ can be written as a finite sum $A_1 \amalg \cdots \amalg A_n$ of indecomposable modules. From this it follows that if $A_1 \amalg \cdots \amalg A_m \approx B_1 \amalg \cdots \amalg B_n$ with the A_i and B_i indecomposable, then $m = n$ and after suitable relabeling $A_i \approx B_i$ for all $i = 1, \ldots, n$. Thus if we let $\operatorname{mod}_P \Lambda$ be the full subcategory of $\operatorname{mod}\Lambda$ consisting of all A in $\operatorname{mod}\Lambda$ with no projective summands, then $\operatorname{mod}_P \Lambda$ is closed under summands and finite sums and every A in $\operatorname{mod}\Lambda$ can be written uniquely (up to isomorphism) as $A_1 \amalg P$ with A_1 in $\operatorname{mod}_P \Lambda$ and P a projective in $\operatorname{mod}\Lambda$. From this it follows that if we denote by $\underline{\operatorname{mod}}_P\Lambda$ the full subcategory of $\underline{\operatorname{mod}}\Lambda$ whose objects are the objects in $\operatorname{mod}_P \Lambda$, then the inclusion $\underline{\operatorname{mod}}_P\Lambda \to \underline{\operatorname{mod}}\Lambda$ is an equivalence of categories which we will often consider an identification.

Clearly the functor $F\colon \operatorname{mod}\Lambda \to \underline{\operatorname{mod}}\Lambda$ induces a full functor $F\colon \operatorname{mod}_P \Lambda \to \underline{\operatorname{mod}}_P\Lambda$. It is not difficult to show that for each A in $\operatorname{mod}_P \Lambda$, the kernel of the surjective R-algebra morphism $F\colon \operatorname{End} A \to \underline{\operatorname{End}} A$ is contained in the radical of $\operatorname{End} A$ (see [1, Proposition 2.5]). Thus f in $\operatorname{End} A$ is an isomorphism if and only if $\underline{f}$ in $\underline{\operatorname{End}} A$ is an isomorphism. Further, combining this observation with the fact that $\operatorname{End} A$ is an artin ring since it is an artin R-algebra in a natural way, it is not difficult to show the following.

PROPOSITION 1.2. *Let $F\colon \operatorname{mod}_P \Lambda \to \underline{\operatorname{mod}}_P\Lambda$ be the canonical functor.*

a) *If A and B are in $\operatorname{mod}_P \Lambda$, then $A \approx B$ if and only if $F(A) \approx F(B)$.*

b) *If e in $\underline{\operatorname{End}} F(A)$ is an idempotent, then there is an idempotent f in $\operatorname{End} A$ such that $\underline{f} = e$.*

c) *For A in $\operatorname{mod}_P \Lambda$, the following are equivalent:*
   i) *A is indecomposable.*

ii) $\mathrm{End}\, A$ *is local.*

iii) $\underline{\mathrm{End}}A$ *is local.*

iv) $F(A)$ *is indecomposable in* $\underline{\mathrm{mod}}_P\Lambda$.

d) *For each* A *in* $\mathrm{mod}_P\,\Lambda$, *we have that* $A \approx A_1 \amalg \cdots \amalg A_n$ *in* $\mathrm{mod}_P\,\Lambda$ *if and only if* $F(A) \approx F(A_1) \amalg \cdots \amalg F(A_n)$ *in* $\underline{\mathrm{mod}}_P\Lambda$ *and the* A_i *are indecomposable in* $\mathrm{mod}\,\Lambda$ *if and only if the* $F(A_i)$ *are indecomposable in* $\underline{\mathrm{mod}}\Lambda$.

We next recall there is a duality $\mathrm{Tr}\colon \underline{\mathrm{mod}}\Lambda \to \underline{\mathrm{mod}}\Lambda^{\mathrm{op}}$ called the *transpose* which has the following property. Let M be in $\mathrm{mod}\,\Lambda$ and suppose that $P_1 \xrightarrow{f} P_0 \to M \to 0$ is a minimal projective presentation of M. Then denoting by X^* the Λ^{op}-module $\mathrm{Hom}_\Lambda(X,\Lambda)$ for all X in $\mathrm{mod}\,\Lambda$, we have that $\mathrm{Tr}\, M \approx \mathrm{Coker}(P_0^* \xrightarrow{f^*} P_1^*)$ in $\mathrm{mod}\,\Lambda^{\mathrm{op}}$. Combining this property of $\mathrm{Tr}\colon \underline{\mathrm{mod}}\Lambda \to \underline{\mathrm{mod}}\Lambda^{\mathrm{op}}$ with the properties of the functor $F\colon \mathrm{mod}\,\Lambda \to \underline{\mathrm{mod}}\Lambda$ cited above, it is not difficult to show that $\mathrm{Tr}\colon \underline{\mathrm{mod}}\Lambda \to \underline{\mathrm{mod}}\Lambda^{\mathrm{op}}$ has the following properties.

PROPOSITION 1.3. a) *If* A *is in* $\mathrm{mod}\,\Lambda$, *then* $\mathrm{Tr}\, A = 0$ *in* $\mathrm{mod}\,\Lambda$ *if and only if* A *is projective.*

b) *Given a finite family* $\{A_i\}_{i\in I}$ *of modules in* $\mathrm{mod}\,\Lambda$, *then* $\mathrm{Tr}(\coprod_{i\in I} A_i) \approx \coprod_{i\in I} \mathrm{Tr}\, A_i$.

c) $\mathrm{Tr}\, A$ *is in* $\mathrm{mod}_P\,\Lambda^{\mathrm{op}}$ *for each* A *in* $\mathrm{mod}\,\Lambda$.

d) *For each* A *in* $\mathrm{mod}\,\Lambda$, *we have that* $A \approx \mathrm{Tr}\,\mathrm{Tr}\, A \amalg P$ *where* $\mathrm{Tr}\,\mathrm{Tr}\, A$ *is in* $\mathrm{mod}_P\,\Lambda$ *and* P *is projective.*

e) *The induced duality* $\mathrm{Tr}\colon \underline{\mathrm{mod}}_P\Lambda \to \underline{\mathrm{mod}}_P\Lambda^{\mathrm{op}}$ *has the following properties:*

i) $\mathrm{Tr}\,\mathrm{Tr}\, A \approx A$ *in* $\mathrm{mod}_P\,\Lambda$ *for all* A *in* $\mathrm{mod}_P\,\Lambda$.

ii) $A \approx B$ *in* $\mathrm{mod}_P\,\Lambda$ *if and only if* $\mathrm{Tr}\, A \approx \mathrm{Tr}\, B$ *in* $\mathrm{mod}_P\,\Lambda^{\mathrm{op}}$.

iii) *If* $\{A_i\}_{i\in I}$ *is a finite family of objects in* $\mathrm{mod}_P\,\Lambda$, *then* $\mathrm{Tr}(\coprod_{i\in I} A_i) \approx \coprod_{i\in I} \mathrm{Tr}\, A_i$ *in* $\mathrm{mod}_P\,\Lambda^{\mathrm{op}}$.

iv) A *in* $\mathrm{mod}_P\,\Lambda$ *is indecomposable if and only if* $\mathrm{Tr}\, A$ *is indecomposable in* $\mathrm{mod}_P\,\Lambda^{\mathrm{op}}$.

With this preliminary discussion of the category $\underline{\mathrm{mod}}\Lambda$ in mind, we turn our attention to the category $\overline{\mathrm{mod}}\Lambda$, the category of $\mathrm{mod}\,\Lambda$ modulo injectives. The objects of $\overline{\mathrm{mod}}\Lambda$ are the same as those of $\mathrm{mod}\,\Lambda$. For each pair of objects A, B in $\mathrm{mod}\,\Lambda$, the R-module of morphisms from A to B in $\overline{\mathrm{mod}}\Lambda$ which we denote by $\overline{\mathrm{Hom}}_\Lambda(A,B)$ is defined to be $\mathrm{Hom}_\Lambda(A,B)/I(A,B)$, where $I(A,B)$ is the R-submodule of $\mathrm{Hom}_\Lambda(A,B)$ consisting of those morphisms $f\colon A \to B$ which can be factored through an injective Λ-module. The composition of morphisms in $\overline{\mathrm{mod}}\Lambda$ is given by the formula $\overline{g}\overline{f} = \overline{gf}$ where by $\overline{f}$ we mean the image of f in (A,B) in $\overline{\mathrm{Hom}}_\Lambda(A,B)$ under the canonical epimorphism $(A,B) \to (A,B)/I(A,B)$.

In analogy with the category $\mathrm{mod}_P\,\Lambda$, we define $\mathrm{mod}_I\,\Lambda$ to be the full subcategory of $\mathrm{mod}\,\Lambda$ consisting of all A in $\mathrm{mod}\,\Lambda$ having no nonzero injective summands. Then denoting by $\overline{\mathrm{mod}}_I\Lambda$ the full subcategory of $\overline{\mathrm{mod}}\Lambda$ whose objects are those of $\mathrm{mod}_I\,\Lambda$, we have that the inclusion $\overline{\mathrm{mod}}_I\Lambda \to \overline{\mathrm{mod}}\Lambda$ is an equivalence of categories which we sometimes consider an identification.

Next we observe that the duality $D\colon \mathrm{mod}\,\Lambda \to \mathrm{mod}\,\Lambda^{\mathrm{op}}$ induces dualities $D\colon \mathrm{mod}_P\,\Lambda \to \mathrm{mod}_I\,\Lambda$ and $D\colon \underline{\mathrm{mod}}\Lambda \to \overline{\mathrm{mod}}\Lambda^{\mathrm{op}}$. These dualities enable us to establish properties of $\mathrm{mod}_I\,\Lambda$ and $\overline{\mathrm{mod}}_I\Lambda$ analogous to the results cited above for $\mathrm{mod}_P\,\Lambda$ and $\underline{\mathrm{mod}}_P\Lambda$. For this reason we do not state these analogues explicitly even though they will be used freely throughout the paper.

Since the functors $\operatorname{Tr}\colon \underline{\operatorname{mod}}_P\Lambda \to \underline{\operatorname{mod}}_P\Lambda^{\mathrm{op}}$ and $D\colon \underline{\operatorname{mod}}_P\Lambda^{\mathrm{op}} \to \overline{\operatorname{mod}}_I\Lambda$ are dualities, it is obvious that their composition $D\operatorname{Tr}\colon \underline{\operatorname{mod}}_P\Lambda \to \overline{\operatorname{mod}}_I\Lambda$ is an equivalence of categories. Similarly the functor $\operatorname{Tr}D\colon \overline{\operatorname{mod}}_I\Lambda \to \underline{\operatorname{mod}}_P\Lambda$ which is the composition $\overline{\operatorname{mod}}_I\Lambda \xrightarrow{D} \underline{\operatorname{mod}}_P\Lambda^{\mathrm{op}} \xrightarrow{\operatorname{Tr}} \underline{\operatorname{mod}}_P\Lambda$ is an equivalence of categories. Furthermore, $\operatorname{Tr}D(D\operatorname{Tr})\colon \underline{\operatorname{mod}}_P\Lambda \to \underline{\operatorname{mod}}_P\Lambda$ is isomorphic to the identity on $\underline{\operatorname{mod}}_P\Lambda$ while $D\operatorname{Tr}(\operatorname{Tr}D)\colon \overline{\operatorname{mod}}_I\Lambda \to \overline{\operatorname{mod}}_I\Lambda$ is isomorphic to the identity on $\overline{\operatorname{mod}}_I\Lambda$. These various dualities and equivalences play a critical role throughout the paper. In fact, the whole paper is essentially devoted to the study of these functors and their connections with the module theory of artin algebras. The rest of this section is devoted to pointing out some of their more elementary properties.

PROPOSITION 1.4. *Let Λ be an artin algebra with $\underline{r}$ the radical of Λ.*

a) *Let M be in $\operatorname{mod}_P\Lambda$ and let $0 \to \Omega M \to P_0 \to M \to 0$ be exact with $P_0 \to M \to 0$ a projective cover.*
   i) *$D(\Omega M/\underline{r}\Omega M) \approx \operatorname{Tr} M/\underline{r}\operatorname{Tr} M$ and M is an indecomposable Λ-module if and only if $\operatorname{Tr} M$ is an indecomposable Λ^{op}-module.*
   ii) *$\operatorname{Soc}(D\operatorname{Tr} M) \approx \Omega M/\underline{r}\Omega M$ and M is an indecomposable Λ-module if and only if $D\operatorname{Tr} M$ is an indecomposable Λ-module.*

b) *Let M be in $\operatorname{mod}_I\Lambda$ and let $0 \to M \to I_0 \to \Omega^{-1}M \to 0$ be exact with $0 \to M \to I_0$ an injective envelope of M. Then $\operatorname{Soc}(\Omega^{-1}M) \approx \operatorname{Tr}DM/\underline{r}\operatorname{Tr}M$. Furthermore, M is indecomposable if and only if $\operatorname{Tr}DM$ is indecomposable.*

PROOF. a) i) Let $P_1 \to \Omega M \to 0$ be a projective cover for ΩM. Then $P_1/\underline{r}P_1 \approx \Omega M/\underline{r}\Omega M$ and $P_1 \to P_0 \to M \to 0$ is a minimal projective presentation for M. From the fact that $P_1 \to P_0 \to M \to 0$ is a minimal projective presentation for M, it follows easily that $P_0^* \to P_1^* \to \operatorname{Tr} M \to 0$ is a minimal projective presentation for $\operatorname{Tr} M$. Hence $\operatorname{Tr} M/\underline{r}\operatorname{Tr} M \approx P_1^*/\underline{r}P_1^*$. But it is well known that $P^*/\underline{r}P^* \approx D(P/\underline{r}P)$ for all projective Λ-modules P. Hence $\operatorname{Tr} M/\underline{r}\operatorname{Tr} M \approx D(\Omega M/\underline{r}\Omega M)$. The fact that M is an indecomposable Λ-module if and only if $\operatorname{Tr} M$ is an indecomposable Λ^{op}-module follows from Proposition 1.3.

a) ii) Follows from a) i) using the duality $D\colon \operatorname{mod}\Lambda^{\mathrm{op}} \to \operatorname{mod}\Lambda$.

b) Follows from a) using the duality $D\colon \operatorname{mod}\Lambda \to \operatorname{mod}\Lambda^{\mathrm{op}}$. □

It is worthwhile observing that Proposition 1.4 can be used to show that certain types of indecomposable modules exist. We illustrate this point in the following easily verified consequence of Proposition 1.4.

COROLLARY 1.5. a) *Let $\underline{a}$ be a proper nonzero submodule of an indecomposable projective module P. Then there is an indecomposable Λ-module M such that $\operatorname{Soc} M \approx \underline{a}/\underline{ra}$, for instance $M = D\operatorname{Tr}(P/\underline{a})$.*

b) *Let M be a proper nonzero submodule of an indecomposable injective Λ-module I. Then there is an indecomposable Λ-module N such that $N/\underline{r}N \approx \operatorname{Soc}(I/M)$, for instance $N \approx \operatorname{Tr}D(M)$.*

We now give some estimates concerning the relationship between $L(M)$, the length of M, and $L(\operatorname{Tr} M)$ which has some interesting applications.

PROPOSITION 1.6. *Let Λ be an artin R-algebra and suppose m is the maximum of $L(\Lambda)$ and $L(\Lambda^{\mathrm{op}})$.*

a) *If M is a Λ-module and $n = L(M/\underline{r}M)$, then $L(\operatorname{Tr} M) \le nm^2$.*

b) *Suppose $\{M_i\}_{i\in I}$ is a family of Λ-modules in $\operatorname{mod}_P\Lambda$, then the following statements are equivalent:*

i) $\{L(M_i)\}_{i\in I}$ *is bounded.*
ii) $\{L(\operatorname{Tr} M_i)\}_{i\in I}$ *is bounded.*
ii) $\{L(D\operatorname{Tr} M_i)\}_{i\in I}$ *is bounded.*
iv) $\{L(\operatorname{Tr} DM_i)\}_{i\in I}$ *is bounded.*

PROOF. a) Let $0 \to \Omega M \to P \to M \to 0$ be exact with $P \to M \to 0$ a projective cover. Then $L(P) \leq L(M/\underline{r}M)L(\Lambda) \leq nm$. Since $L(\Omega M/\underline{r}\Omega M) \leq L(\Omega M) \leq L(P) \leq nm$, the fact that $D(\Omega M/\underline{r}\Omega M) \approx \operatorname{Tr} M/\underline{r}\operatorname{Tr} M$ implies that $L(\operatorname{Tr} M/\underline{r}\operatorname{Tr} M) \leq nm$. Hence $L(\operatorname{Tr} M) \leq nm^2$, our desired result.

b) i) implies ii). Follows from a).

ii) implies i). Since each M_i is in $\operatorname{mod}_P \Lambda$, we know that $\operatorname{Tr}(\operatorname{Tr} M_i) \approx M_i$ for all i in I. Hence ii) implies i) also follows from a).

ii) equivalent to iii). Easy consequence of previous results.

ii) equivalent to iv). Easy consequence of previous results. □

As a straightforward application of this result we have the following.

COROLLARY 1.7. a) *Let $\{M_i\}_{i\in I}$ be a family of Λ-modules in $\operatorname{mod}_P \Lambda$ and let $0 \to \Omega M \to P_i \to M_i \to 0$ be exact sequences with the $P_i \to M_i \to 0$ projective covers. Then $\{L(M_i)\}_{i\in I}$ is bounded if and only if $\{L(\Omega M_i)\}_{i\in I}$ is bounded.*

b) *Let $\{M_i\}_{i\in I}$ be a family of Λ-modules in $\operatorname{mod}_I \Lambda$ and let $0 \to M_i \to I_i \to \Omega^{-1}M_i \to 0$ be exact with the $0 \to M_i \to I_i$ injective envelopes. Then $\{L(M_i)\}_{i\in I}$ is bounded if and only if $\{L(\Omega^{-1}M_i)\}_{i\in I}$ is bounded.*

PROOF. a) It is obvious that if $\{L(M_i)\}_{i\in I}$ is bounded, then $\{L(\Omega M_i)\}_{i\in I}$ is bounded. Assume now that $\{L(\Omega M_i)\}_{i\in I}$ is bounded. Then $\{L(\Omega M_i/\underline{r}\Omega M_i)\}_{i\in I}$ is bounded and so $\{L(\operatorname{Soc}(D\operatorname{Tr} M_i))\}_{i\in I}$ is bounded since $\Omega M_i/\underline{r}\Omega M_i \approx \operatorname{Soc}(D\operatorname{Tr} M_i)$ for all i. Hence $\{L(D\operatorname{Tr} M_i)\}_{i\in I}$ is bounded which implies by Proposition 1.6 that $\{L(M_i)\}_{i\in I}$ is bounded since each of the M_i are in $\operatorname{mod}_P \Lambda$. Thus if $\{L(\Omega M_i)\}_{i\in I}$ is bounded, then $\{L(M_i)\}_{i\in I}$ is bounded which finishes the proof of a).

b) Just the dual of a). □

We conclude this section by pointing out that in the case of symmetric artin algebras the functors $D\operatorname{Tr}$ and $\operatorname{Tr} D$ are familiar functors. In order to do this it is convenient to recall the definition of the functors $\Omega\colon \underline{\operatorname{mod}}\Lambda \to \underline{\operatorname{mod}}\Lambda$ and $\Omega^{-1}\colon \overline{\operatorname{mod}}\Lambda \to \overline{\operatorname{mod}}\Lambda$.

The functor $\Omega\colon \underline{\operatorname{mod}}\Lambda \to \underline{\operatorname{mod}}\Lambda$ is defined on objects as follows. For each M in $\operatorname{mod}\Lambda$ choose an exact sequence $0 \to K_M \to P_M \to M \to 0$ with $P_M \to M \to 0$ a projective cover. Then $\Omega M = K_M$ by definition. We now define Ω on morphisms. Given a morphism $f\colon M \to N$ in $\operatorname{mod}\Lambda$ there is a commutative diagram

$$\begin{array}{ccccccccc} 0 & \longrightarrow & \Omega M & \longrightarrow & P_M & \longrightarrow & M & \longrightarrow & 0 \\ & & \downarrow{\scriptstyle f_1} & & \downarrow{\scriptstyle f_0} & & \downarrow{\scriptstyle f} & & \\ 0 & \longrightarrow & \Omega N & \longrightarrow & P_N & \longrightarrow & N & \longrightarrow & 0. \end{array}$$

While the morphism $f_1\colon \Omega M \to \Omega N$ is uniquely determined by $f_0\colon P_M \to P_N$, the morphism $f_0\colon P_M \to P_N$ is not uniquely determined by $f\colon M \to N$. However the different morphisms $\Omega M \to \Omega N$ which result from different choices of $P_M \to P_N$, which give a commutative diagram, differ from f_1 by morphisms in $P(\Omega M, \Omega N)$. Thus if given $f\colon M \to N$ we define $\Omega(f)$ in $\underline{\operatorname{Hom}}_\Lambda(\Omega M, \Omega N)$ to be $\underline{f}$, then $\Omega(f)$ is well defined and we obtain a morphism of groups $\Omega\colon (M,N) \to \underline{\operatorname{Hom}}_\Lambda(\Omega M, \Omega N)$ given by $f \mapsto \Omega(f)$ for all f in (M,N). Since it is easily seen that $\operatorname{Ker}\Omega \supset P(M,N)$, the morphism $\Omega\colon (M,N) \to$

$\underline{\mathrm{Hom}}_\Lambda(\Omega M, \Omega N)$ induces a morphism $\Omega\colon \underline{\mathrm{Hom}}_\Lambda(M,N) \to \underline{\mathrm{Hom}}_\Lambda(\Omega M, \Omega N)$. The fact that the morphisms $\Omega\colon \underline{\mathrm{Hom}}_\Lambda(M,N) \to \underline{\mathrm{Hom}}_\Lambda(\Omega M, \Omega N)$ satisfy the requirements for Ω to be a functor is readily verified. Having defined the functor $\Omega\colon \underline{\mathrm{mod}}\Lambda \to \underline{\mathrm{mod}}\Lambda$, we define $\Omega^i\colon \underline{\mathrm{mod}}\Lambda \to \underline{\mathrm{mod}}\Lambda$ for all nonnegative integers i by induction as follows: $\Omega^0\colon \underline{\mathrm{mod}}\Lambda \to \underline{\mathrm{mod}}\Lambda$ is the identity and Ω^{i+1} is defined to be the composition $\underline{\mathrm{mod}}\Lambda \xrightarrow{\Omega^i} \underline{\mathrm{mod}}\Lambda \xrightarrow{\Omega} \underline{\mathrm{mod}}\Lambda$.

By analogy we define $\Omega^{-1}\colon \overline{\mathrm{mod}}\Lambda \to \overline{\mathrm{mod}}\Lambda$ as follows. For each M we choose an exact sequence $0 \to M \to I_M \to L_M \to 0$ with $0 \to M \to I_M$ an injective envelope. We define $\Omega^{-1}M$ to be L_M for each M in $\mathrm{mod}\,\Lambda$ so Ω^{-1} is defined on objects. The morphisms $\Omega^{-1}\colon \overline{\mathrm{Hom}}_\Lambda(M,N) \to \overline{\mathrm{Hom}}_\Lambda(\Omega^{-1}M, \Omega^{-1}N)$ are defined in a manner dual to the definition of Ω. Similarly we define $\Omega^{-i}\colon \overline{\mathrm{mod}}\Lambda \to \overline{\mathrm{mod}}\Lambda$ for all nonnegative integers i by induction on i as follows: $\Omega^{-0}\colon \overline{\mathrm{mod}}\Lambda \to \overline{\mathrm{mod}}\Lambda$ is the identity and define $\Omega^{-(i+1)}$ to be the composition $\Omega^{-1}\Omega^{-i}$.

Suppose now that Λ is selfinjective. Then the projective and injective Λ-modules are the same. Hence $P(M,N) = I(M,N)$ for all N and N in $\mathrm{mod}\,\Lambda$, so $\underline{\mathrm{mod}}\Lambda = \overline{\mathrm{mod}}\Lambda$. Also it is well known that under these circumstances $\Omega\colon \underline{\mathrm{mod}}\Lambda \to \underline{\mathrm{mod}}\Lambda$ and $\Omega^{-1}\colon \underline{\mathrm{mod}}\Lambda \to \underline{\mathrm{mod}}\Lambda$ are inverse equivalences of categories. Further it is well known that since Λ is selfinjective, the functor $(\ ,\Lambda)\colon \mathrm{mod}\,\Lambda \to \mathrm{mod}\,\Lambda^{\mathrm{op}}$ is a duality with inverse duality $(\ ,\Lambda^{\mathrm{op}})\colon \mathrm{mod}\,\Lambda^{\mathrm{op}} \to \mathrm{mod}\,\Lambda$. In general the duality $(\ ,\Lambda)\colon \mathrm{mod}\,\Lambda \to \mathrm{mod}\,\Lambda^{\mathrm{op}}$ need not be isomorphic to the duality $D\colon \mathrm{mod}\,\Lambda \to \mathrm{mod}\,\Lambda^{\mathrm{op}}$. In fact, it is not difficult to see that the dualities $(\ ,\Lambda)$ and $D\colon \mathrm{mod}\,\Lambda \to \mathrm{mod}\,\Lambda^{\mathrm{op}}$ are isomorphic functors if and only if Λ is a symmetric R-algebra, i.e. there is a two-sided Λ-isomorphism $\Lambda \to \mathrm{Hom}_R(\Lambda, I)$. Straightforward calculations now suffice to show the following.

PROPOSITION 1.8. *Let Λ be a symmetric R-algebra. Then the functors $D\,\mathrm{Tr}$, $\Omega^2\colon \underline{\mathrm{mod}}\Lambda \to \underline{\mathrm{mod}}\Lambda$ are isomorphic and the functors $\mathrm{Tr}\,D$, $\Omega^{-2}\colon \underline{\mathrm{mod}}\Lambda \to \underline{\mathrm{mod}}\Lambda$ are isomorphic.*

2. Morphisms determined by modules

We assume as usual that R is a commutative artin ring and Λ is an artin R-algebra. Also unless stated to the contrary, we assume that all Λ-modules are finitely generated. Our aim in this section is to study the question: Given morphisms $f\colon B \to C$ and $h\colon L \to C$, when does there exist a morphism $g\colon L \to B$ such that $fg = h$? The answer to this question that we give is in terms of the notion of a morphism being determined by a module, a notion we introduced in [2]. In fact, this section consists of a summary of some of the results in [2] specialized to artin algebras, with a somewhat different organization. While few proofs are given in detail, some outlines of proofs are discussed. The material in this section will be used freely in the rest of the paper.

Suppose we are given a pair of morphisms $f\colon B \to C$ and $h\colon L \to C$. A morphism $g\colon L \to B$ is called a *lifting of h to B* if $fg = h$. Suppose $g\colon L \to B$ is a lifting of h to B. Then given a morphism $t\colon X \to L$ the composition $ht\colon X \to C$ can be lifted to B since $ht = fgt$. On the other hand, if $h\colon L \to C$ has the property that for each $t\colon X \to L$, the composition ht can be lifted to B, then $h\colon L \to B$ can be lifted to B. Simply let $X = L$ and $t\colon L \to L$ be the identity. Then the morphism $u\colon L \to B$ such that $ht = fu$ is our desired lifting of h to B.

These observations suggest the following question. For a morphism $f\colon B \to C$ does there exist a fixed module X such that a morphism $h\colon L \to C$ can be lifted to B if for each $t\colon X \to L$ the composition ht can be lifted to B? In connection with this question, it is convenient to make the following definitions.

A morphism $f\colon B \to C$ in $\operatorname{mod}\Lambda$ is said to be *right determined by a module* X in $\operatorname{mod}\Lambda$ if a morphism $h\colon L \to C$ in $\operatorname{mod}\Lambda$ can be lifted to B whenever for each $t\colon X \to L$ the composition $ht\colon X \to B$ can be lifted to B.

We restate this definition in more compact form using the map $(X,d)\colon (X,E) \to (X,F)$ induced for each X in $\operatorname{mod}\Lambda$ by a morphism $d\colon E \to F$ in $\operatorname{mod}\Lambda$. A morphism $f\colon B \to C$ is right determined by a module X if and only if a morphism $h\colon L \to C$ can be lifted to B whenever $\operatorname{Im}(X,f) \supset \operatorname{Im}(X,h)$. We say that a morphism $f\colon B \to C$ is *right X-determined* to mean that f is right determined by the module X. Thus our original question becomes in this language: Is every morphism $f\colon B \to C$ in $\operatorname{mod}\Lambda$ right X-determined for some X in $\operatorname{mod}\Lambda$?

In order to discuss the notion dual to a morphism being right determined by a module, it is convenient to introduce the following definitions.

Suppose $f\colon A \to B$ and $h\colon A \to L$ are morphisms in $\operatorname{mod}\Lambda$. A morphism $g\colon B \to L$ is called an *extension of h to B* if $gf = h$. A morphism $f\colon A \to B$ in $\operatorname{mod}\Lambda$ is said to be *left determined by a module* Y in $\operatorname{mod}\Lambda$ if a morphism $h\colon A \to L$ can be extended to B whenever given $t\colon L \to Y$ the composition th can be extended to B.

We restate this definition in a more compact form using the map $(d,Y)\colon (F,Y) \to (E,Y)$ induced for each Y in $\operatorname{mod}\Lambda$ by a morphism $d\colon E \to F$ in $\operatorname{mod}\Lambda$. A morphism $f\colon A \to B$ is left determined by a module Y in $\operatorname{mod}\Lambda$ if and only if $h\colon A \to L$ can be extended to B whenever $\operatorname{Im}(f,Y) \supset \operatorname{Im}(h,Y)$.

We say that a morphism $f\colon A \to B$ is *left Y-determined* to mean that $f\colon A \to B$ is left determined by the module Y in $\operatorname{mod}\Lambda$. In view of our previous discussion it is natural to ask: Is every $f\colon A \to B$ left Y-determined for some Y in $\operatorname{mod}\Lambda$?

Surprisingly, every morphism $f\colon B \to C$ in $\operatorname{mod}\Lambda$ is right X-determined for some X in $\operatorname{mod}\Lambda$ as well as left Y-determined for some Y in $\operatorname{mod}\Lambda$. That this is the case is shown in Chapter I of [2]. We briefly outline the main elements of the proof for the convenience of the reader.

We first point out the following easily verified result.

LEMMA 2.1. *Let $f\colon B \to C$ be a morphism in* $\operatorname{mod}\Lambda$. *Then*

a) *f is right X-determined if and only if the Λ^{op}-morphism $D(f)\colon D(C) \to D(B)$ is left $D(X)$-determined.*
b) *If f is a monomorphism, then f is right P-determined where P is a projective cover for* $\operatorname{Coker} f$.
c) *If f is an epimorphism, then f is left I-determined where I is an injective envelope of* $\operatorname{Ker} f$.
d) *If the induced epimorphism $B \to$* $\operatorname{Im} f$ *is right X-determined, then $f\colon B \to C$ is right $X \amalg P$-determined where P is projective cover for* $\operatorname{Coker} f$.
e) *If the induced monomorphism* $\operatorname{Im} f \to C$ *is left Y-determined, then $f\colon B \to C$ is left $Y \amalg I$-determined where I is an injective envelope for* $\operatorname{Ker} f$.

As a consequence of parts d) and e) of this lemma, we have that in order to show that every morphism in $\operatorname{mod}\Lambda$ is both right and left determined by modules in $\operatorname{mod}\Lambda$, it suffices to show that if $0 \to A \xrightarrow{g} B \xrightarrow{f} C \to 0$ is exact in $\operatorname{mod}\Lambda$, then g is left Y-determined and f is right X-determined for some Λ-modules Y and X. This follows from the following somewhat more explicit result.

PROPOSITION 2.2. *Let* $0 \to A \xrightarrow{g} B \xrightarrow{f} C \to 0$ *be an exact sequence in* $\operatorname{mod}\Lambda$. *Then*

a) g *is left determined by* $D\operatorname{Tr} C$.

b) f *is right determined by* $\operatorname{Tr} DA$.

The following are some useful consequences of this result.

PROPOSITION 2.3. *Let* $0 \to A \xrightarrow{g} B \xrightarrow{f} C \to 0$ *be an exact sequence in* $\operatorname{mod}\Lambda$.

a) *The following statements are equivalent:*

i) $0 \to A \xrightarrow{g} B \xrightarrow{f} C \to 0$ *splits.*

ii) $(g, D\operatorname{Tr} C)\colon (B, D\operatorname{Tr} C) \to (A, D\operatorname{Tr} C)$ *is an epimorphism.*

iii) $(\operatorname{Tr} DA, f)\colon (\operatorname{Tr} DA, B) \to (\operatorname{Tr} DA, C)$ *is an epimorphism.*

b) *Let* X *be in* $\operatorname{mod}\Lambda$. *The following are equivalent:*

i) $(X, f)\colon (X, B) \to (X, C)$ *is an epimorphism.*

ii) $(g, D\operatorname{Tr} X)\colon (B, D\operatorname{Tr} X) \to ((A, D\operatorname{Tr} X))$ *is an epimorphism.*

PROOF. a) That i) implies ii) and iii) is obvious.

ii) implies i). By Proposition 2.2, we know that $A \xrightarrow{g} B$ is left $D\operatorname{Tr} C$-determined. Therefore, the identity morphism $\mathrm{id}_A\colon A \to A$ has the property that there is an $s\colon B \to A$ such that $sg = \mathrm{id}_A$ if $\operatorname{Im}(g, D\operatorname{Tr} C) \supset (A, D\operatorname{Tr} C)$, i.e. if the morphism $(g, D\operatorname{Tr} C)\colon (B, D\operatorname{Tr} C) \to (A, D\operatorname{Tr} C)$ is an epimorphism. Since we are assuming by ii) that $(g, D\operatorname{Tr} C)$ is an epimorphism, we have that there is an $s\colon B \to A$ such that $sg = \mathrm{id}_A$, or equivalently, the exact sequence $0 \to A \xrightarrow{g} B \xrightarrow{f} C \to 0$ splits. Thus we have shown that ii) implies i).

ii) implies i). The proof proceeds in an obviously analogous way to the proof that ii) implies i).

b) i) implies ii). Suppose $(X, f)\colon (X, B) \to (X, C) \to 0$ is exact and let h be in $(A, D\operatorname{Tr} X)$. Then we have the following exact pushout diagram

$$\begin{array}{ccccccccc}
0 & \longrightarrow & A & \xrightarrow{g} & B & \xrightarrow{f} & C & \longrightarrow & 0 \\
 & & \big\downarrow{\scriptstyle h} & & \big\downarrow{\scriptstyle t} & & \| & & \\
0 & \longrightarrow & D\operatorname{Tr} X & \xrightarrow{g'} & B' & \xrightarrow{f'} & C & \longrightarrow & 0.
\end{array}$$

From this it follows that (X, f') is an epimorphism since (X, f) is an epimorphism. Because $\operatorname{Tr} D(D\operatorname{Tr} X) \amalg P \approx X$ for some projective P (possibly zero), the fact that (X, f') is an epimorphism implies that $(\operatorname{Tr} D(D\operatorname{Tr} X), f')$ is an epimorphism. Hence by part a), we have that $0 \to D\operatorname{Tr} X \xrightarrow{g'} B' \to C \to 0$ splits. Thus there is an $s\colon B \to D\operatorname{Tr} X$ such that $sg = h$. Since h was an arbitrary element of $(A, D\operatorname{Tr} X)$, this argument shows that (X, f) being an epimorphism implies that $(g, D\operatorname{Tr} X)$ is an epimorphism. Thus we have shown that i) implies ii).

ii) implies i). The proof is an obvious analogue of the proof given for i) implies ii). □

In connection with this result, we point out the following generalization of Proposition 2.3, which was established in Chapter III, Section 4 of [2].

PROPOSITION 2.4. *Let* $0 \to A \xrightarrow{g} B \xrightarrow{f} C \to 0$ *be an exact sequence in* $\operatorname{mod}\Lambda$. *For each* X *in* $\operatorname{mod}\Lambda$, *there is an isomorphism*

$$\operatorname{Hom}_R(\operatorname{Coker}(X, f), I) \approx \operatorname{Coker}(g, D\operatorname{Tr} X)$$

which is functorial in X, *where* $I = I(R/\operatorname{rad} R)$.

Before proceeding further, it is convenient to introduce the following notation. Suppose M is in $\operatorname{mod}\Lambda$ and $M = \coprod_{i=1}^{k} M_i^{n_i}$ with the M_i indecomposable and the n_i positive integers (where M^n denotes the sum of n copies of M). Then we denote the sum $\coprod_{i=1}^{k} M_i$ by $[M]$. The following easily verified results make clear the reason for introducing this notation.

LEMMA 2.5. *Let $f\colon B \to C$ be a morphism in* $\operatorname{mod}\Lambda$.

a) *If f is right X-determined, then f is right $X \amalg V$-determined for any V in* $\operatorname{mod}\Lambda$.
b) *f is right X-determined if and only if f is right $[X]$-determined.*
c) *If f is left Y-determined, then f is left $Y \amalg V$-determined for any V in* $\operatorname{mod}\Lambda$.
d) *f is left Y-determined if and only if f is left $[Y]$-determined.*

Summarizing and extending slightly some of our results so far, we have the following.

THEOREM 2.6. a) *Let $0 \to A \xrightarrow{g} B \xrightarrow{f} C$ be exact in* $\operatorname{mod}\Lambda$. *Then f is right $(\operatorname{Tr} DA \amalg P)$-determined, where P is a projective cover for* $\operatorname{Coker} f$.

b) *Let $A \xrightarrow{g} B \xrightarrow{f} C \to 0$ be exact in* $\operatorname{mod}\Lambda$. *Then g is left $(D\operatorname{Tr} C \amalg I)$-determined, where I is an injective envelope for* $\operatorname{Ker} g$.

While most of the results given so far follow fairly easily from the basic definitions, the proof of Proposition 2.2 is not so straightforward. For this reason we now outline its proof.

Let X be in $\operatorname{mod}\Lambda$. For each C in $\operatorname{mod}\Lambda$ we view (X, C) as an $(\operatorname{End} X)^{\mathrm{op}}$-module by means of the action of $\operatorname{End} X$ on X. Since it is easily verified that $P(X, C)$ is an $(\operatorname{End} X)^{\mathrm{op}}$-submodule of (X, C) (remember that $P(X, C)$ denotes those morphisms from X to C which factor through projectives), we have that $\underline{\operatorname{Hom}}_\Lambda(X, C) = (X, C)/P(X, C)$ is an $(\operatorname{End} X)^{\mathrm{op}}$-module. But the two sided ideal $P(X, X)$ of $(\operatorname{End} X)^{\mathrm{op}}$ is contained in the annihilator of $\underline{\operatorname{Hom}}_\Lambda(X, C)$, as is easily verified. Hence $\underline{\operatorname{Hom}}_\Lambda(X, C)$ is an $(\underline{\operatorname{End}} X)^{\mathrm{op}}$-module. Since the duality $\operatorname{Tr}\colon \underline{\operatorname{mod}}\Lambda \to \underline{\operatorname{mod}}\Lambda^{\mathrm{op}}$ gives a ring isomorphism $(\underline{\operatorname{End}} X)^{\mathrm{op}} \to \underline{\operatorname{End}} \operatorname{Tr} X$ we have that $\underline{\operatorname{Hom}}_\Lambda(X, C)$ is an $\underline{\operatorname{End}} \operatorname{Tr} X$-module. Therefore the ring surjection $\operatorname{End} \operatorname{Tr} X \to \underline{\operatorname{End}} \operatorname{Tr} X$ makes $\underline{\operatorname{Hom}}_\Lambda(X, C)$ an $\operatorname{End} \operatorname{Tr} X$-module for each C in $\operatorname{mod}\Lambda$. With these observations in mind we have the following fundamental result (see [2, Chapter I, Proposition 3.4]).

PROPOSITION 2.7. *Let C and X be in* $\operatorname{mod}\Lambda$ *and let E be an injective* $\Gamma = \operatorname{End}\operatorname{Tr} X$-*module. Then there are isomorphisms*

$$v\colon \operatorname{Ext}^1_\Lambda(C, \operatorname{Hom}_\Gamma(\operatorname{Tr} X, E)) \to \operatorname{Hom}_\Gamma(\underline{\operatorname{Hom}}_\Lambda(X, C), E)$$

which are functorial in C and E, where the operation of Λ on $\operatorname{Hom}_\Gamma(\operatorname{Tr} X, E)$ *is the one induced by the Λ^{op}-module structure on* $\operatorname{Tr} X$. *Moreover if x in* $\operatorname{Ext}^1_\Lambda(C, \operatorname{Hom}_\Gamma(\operatorname{Tr} X, E))$ *is represented by the exact sequence* $0 \to \operatorname{Hom}_\Gamma(\operatorname{Tr} X, E) \xrightarrow{g} B \xrightarrow{f} C \to 0$ *then*

a) *f is right X-determined.*
b) $\operatorname{Im}(X, f) \supset P(X, C)$ *and* $\operatorname{Im}(X, f)/P(X, C) = \operatorname{Ker} v(x)$.

As one application of this result, we obtain a proof of Proposition 2.2. For suppose $0 \to A \xrightarrow{g} B \xrightarrow{f} C \to 0$ is an exact sequence in $\operatorname{mod}\Lambda$. Then $A = A_0 \amalg I$, where A_0 is in

$\operatorname{mod}_I \Lambda$ and I is an injective Λ-module. Hence $0 \to A \xrightarrow{g} B \xrightarrow{f} C \to 0$ is isomorphic to

$$\begin{array}{ccccccccc} & & I & = & I & & & & \\ & & \amalg & & \amalg & & & & \\ 0 & \longrightarrow & A_0 & \xrightarrow{g_0} & B_0 & \xrightarrow{f_0} & C & \longrightarrow & 0. \end{array}$$

From this it follows that $\operatorname{Im}(X, f) = \operatorname{Im}(X, f_0)$ for all X in $\operatorname{mod} \Lambda$ and that X in $\operatorname{mod} \Lambda$ has the property that f is right X-determined if and only if f_0 is right X-determined. Since $\operatorname{Tr} DA = \operatorname{Tr} DA_0$, if we show that f_0 is right $\operatorname{Tr} DA_0$-determined, we will also have shown that f is right $\operatorname{Tr} DA$-determined. This will establish part b) or Proposition 2.2.

Hence we can assume without loss of generality that A is in $\operatorname{mod}_I \Lambda$. Then we know $A = D\operatorname{Tr}(\operatorname{Tr} A)$. Hence letting $X = \operatorname{Tr} DA$, we have that $A = D\operatorname{Tr} X$. Thus letting $\Gamma = \operatorname{End} \operatorname{Tr} X$ we have that $A = \operatorname{Hom}_R(\Gamma \otimes_\Gamma \operatorname{Tr} X, I) \approx \operatorname{Hom}_\Gamma(\operatorname{Tr} X, \operatorname{Hom}_R(\Gamma, I))$, where $I = I(R/\operatorname{rad} R)$. Since $E = \operatorname{Hom}_R(\Gamma, I)$ is an injective Γ-module, $A = \operatorname{Hom}_\Gamma(\operatorname{Tr} X, E)$ with E an injective Γ-module. Thus the exact sequence $0 \to A \xrightarrow{g} B \xrightarrow{f} C \to 0$ can be written in the form $0 \to \operatorname{Hom}_\Gamma(\operatorname{Tr} X, E) \xrightarrow{g} B \xrightarrow{f} C \to 0$. It then follows from Proposition 2.7 that f is right X-determined. Substituting $X = \operatorname{Tr} DA$, we have that f is right $\operatorname{Tr} DA$-determined which establishes part b) of Proposition 2.2. Since part a) of Proposition 2.2 follows from part b) by duality, the proof of Proposition 2.2 is complete.

Before giving our next application of Proposition 2.2, it is convenient to make the following convention. Suppose Γ is an artin R-algebra. Since $D\colon \operatorname{mod}\Gamma \to \operatorname{mod}\Gamma^{\mathrm{op}}$ is a duality, the morphisms $D_{A,B}\colon \operatorname{Hom}_\Gamma(A, B) \to \operatorname{Hom}_{\Gamma^{\mathrm{op}}}(D(B), D(A))$ given by $D_{A,B}(f) = D(f)$ are isomorphisms functorial in A and B in $\operatorname{mod}\Gamma$. We will often consider these isomorphisms identifications. With these remarks in mind, we have the following important consequence of Proposition 2.7.

PROPOSITION 2.8. *Suppose X and C are in* $\operatorname{mod}\Lambda$ *and H is an* $(\operatorname{End} X)^{\mathrm{op}}$*-submodule of (X, C) containing $P(X, C)$. Let E be a finitely generated injective $\Gamma = \operatorname{End}\operatorname{Tr} X$-module such that there is a Γ-morphism $t\colon \underline{\operatorname{Hom}}_\Lambda(X, C) \to E$ with* $\operatorname{Ker} t = H/P(X, C)$ *(since Γ is an artin R-algebra, such an E exists). Then $D(E)$ is a finitely generated projective Γ^{op}-module and the exact sequence*

$$x\colon 0 \to \operatorname{Hom}_{\Gamma^{\mathrm{op}}}(D(E), D\operatorname{Tr} X) \xrightarrow{g} B \xrightarrow{f} C \to 0$$

in $\operatorname{Ext}^1_\Lambda(C, \operatorname{Hom}_{\Gamma^{\mathrm{op}}}(D(E), D\operatorname{Tr} X)$ *corresponding to t under the isomorphism*

$$v\colon \operatorname{Ext}^1_\Lambda(C, \operatorname{Hom}_{\Gamma^{\mathrm{op}}}(D(E), D\operatorname{Tr} X) \to \operatorname{Hom}_\Gamma(\underline{\operatorname{Hom}}_\Lambda(X, C), E)$$

described in Proposition 2.5 *has the following properties:*

a) *f is right X-determined.*
b) $\operatorname{Im}(X, f) = H$.
c) *x splits if and only if $H = (X, C)$.*
d) *If E is an injective envelope for* $\operatorname{Im} t$*, then in any commutative diagram*

$$\begin{array}{ccccccccc} 0 & \longrightarrow & \operatorname{Hom}_{\Gamma^{\mathrm{op}}}(D(E), D\operatorname{Tr} X) & \xrightarrow{g} & B & \xrightarrow{f} & C & \longrightarrow & 0 \\ & & \downarrow & & \downarrow & & \| & & \\ 0 & \longrightarrow & \operatorname{Hom}_{\Gamma^{\mathrm{op}}}(D(E), D\operatorname{Tr} X) & \xrightarrow{g} & B & \xrightarrow{f} & C & \longrightarrow & 0 \end{array}$$

the vertical morphisms are isomorphisms.

In connection with this result it should be noted that under the hypothesis of the previous proposition $D(E)$ is a finitely generated projective Γ^{op}-module with the property there is an epimorphism $D(E) \to \text{Hom}_R((X,C)/H, I)$ and $D(E)$ is a projective cover for $\text{Hom}_R((X,C)/H, I)$ if and only if E is an injective envelope for $\text{Im}\, t$.

So far we have been looking primarily at epimorphisms right determined by a module. We now develop the analogous results for monomorphisms left determined by a module. While much of this information could be obtained by duality, we indicate the exact analogues.

Let A be in module. For each X in $\text{mod}\,\Lambda$ we view (A, X) as an $\text{End}\,X$-module by means of the operation of $\text{End}\,X$ on X. Since it is easily checked that $I(A, X)$ is an $\text{End}\,X$-submodule of (A, X) we have that $\overline{\text{Hom}}_\Lambda(A, X)$ is an $\text{End}\,X$-module. It is also easily checked that the two sided ideal $I(X, X)$ of $\text{End}\,X$ is contained in the annihilator of $\overline{\text{Hom}}_\Lambda(A, X)$. Thus $\overline{\text{Hom}}_\Lambda(A, X)$ is an $\overline{\text{End}}X$-module. The equivalence $\text{Tr}\,D\colon \overline{\text{mod}}\Lambda \to \underline{\text{mod}}\Lambda$ gives an isomorphism $\overline{\text{End}}X \to \underline{\text{End}}\,\text{Tr}\,DX$. Hence $\overline{\text{Hom}}_\Lambda(A, X)$ is an $(\underline{\text{End}}\,\text{Tr}\,DX)$-module and thus an $(\text{End}\,\text{Tr}\,DX)$-module. Letting $\Sigma = \text{End}\,\text{Tr}\,DX$ we have that Σ is an artin R-algebra in a natural way. Hence the duality $D\colon \text{mod}\,\Sigma \to \text{mod}\,\Sigma^{\text{op}}$ induces a duality between the category of finitely generated injective Σ-modules and the category of finitely projective Σ^{op}-modules. With these observations and notations we have the following analogue of Proposition 2.7.

PROPOSITION 2.9. *Let A and X be in* $\text{mod}\,\Lambda$. *Let E be a finitely generated injective $\Sigma = (\text{End}\,\text{Tr}\,DX)$-module. Then there are isomorphisms*

$$w\colon \text{Ext}^1_\Lambda(D(E) \otimes_\Sigma \text{Tr}\,DX, A) \to \text{Hom}_\Sigma(\overline{\text{Hom}}(A, X), E)$$

which are functorial in A and X, where the operation of Λ on $D(E) \otimes_\Sigma \text{Tr}\,DX$ is the one induced by the Λ-module structure on $\text{Tr}\,DX$. Moreover if x in $\text{Ext}^1_\Lambda(D(E) \otimes \text{Tr}\,DX, A)$ is represented by the exact sequence $0 \to A \xrightarrow{g} B \to D(E) \otimes_\Sigma \text{Tr}\,DX \to 0$ then

a) *g is left X-determined.*
b) $\text{Im}(g, X) \supset I(A, X)$ *and* $\text{Im}(g, X)/I(A, X) = \text{Ker}\,w(x)$.

PROOF. We only sketch the proof of the existence of the isomorphisms $w\colon \text{Ext}^1_\Lambda((D(E) \otimes_\Sigma \text{Tr}\,DX, A) \to \text{Hom}_\Sigma(\overline{\text{Hom}}_\Lambda(A, X), E)$. The rest of the proposition can be proved using arguments similar to those used in Chapter I, Section 3 of [2] to prove Proposition 2.7 of this paper.

Since A is finitely generated we know that $A \approx DD(A)$. Hence by [11, Chapter VI, Proposition 5.1] we have the isomorphisms

$$\text{Ext}^1_\Lambda(D(E) \otimes_\Sigma \text{Tr}\,DX, A) \to \text{Hom}_R(\text{Tor}^\Lambda_1(DA, D(E) \otimes_\Sigma \text{Tr}\,DX, I).$$

Since $D(E)$ is a projective Σ^{op}-module, we have that $\text{Tor}^\Lambda_1(DA, D(E) \otimes_\Sigma \text{Tr}\,DX) \approx D(E) \otimes_\Sigma \text{Tor}^\Lambda_1(DA, \text{Tr}\,DX)$. We also know that $\text{Tor}^\Lambda_1(DA, \text{Tr}\,DX) \approx \underline{\text{Hom}}_\Lambda(\text{Tr}\,DA, \text{Tr}\,DX)$ (see [6, Proposition 2.2]). But the equivalence $\text{Tr}\,D\colon \overline{\text{mod}}\Lambda \to \underline{\text{mod}}\Lambda$, gives a Σ-isomorphism $\underline{\text{Hom}}_\Lambda(\text{Tr}\,DA, \text{Tr}\,DX) \to \overline{\text{Hom}}_\Lambda(A, X)$. Hence we have the isomorphism $\text{Hom}_R(\text{Tor}^\Lambda_1(DA, D(E) \otimes_\Sigma \text{Tr}\,DX), I) \to \text{Hom}_R(D(E) \otimes_\Sigma \overline{\text{Hom}}_\Lambda(A, X), I)$. But $\text{Hom}_R(D(E) \otimes_\Sigma \overline{\text{Hom}}_\Lambda(A, X), I) \approx \text{Hom}_\Sigma(\overline{\text{Hom}}_\Lambda(A, X), DD(E))$. Since $E \approx DD(E)$ we have our desired isomorphism $w\colon \text{Ext}^1_\Lambda(D(E) \otimes_\Sigma \text{Tr}\,DX, A) \to \text{Hom}_\Sigma(\overline{\text{Hom}}_\Lambda(A, X), E)$. □

As an application of Proposition 2.9 we obtain the following analogue of Proposition 2.8.

PROPOSITION 2.10. *Suppose X and A are in* $\operatorname{mod}\Lambda$ *and H is an* $\operatorname{End}X$*-submodule of (A,X) containing $I(A,X)$. Let E be a finitely generated injective* $\Sigma = \operatorname{End}\operatorname{Tr}DX$*-module such that there is a Σ-morphism $t\colon \overline{\operatorname{Hom}_\Lambda}(A,X) \to E$ with* $\operatorname{Ker}t = H/I(A,X)$ *(since Σ is an artin R-algebra such an E exists). Then the exact sequence*

$$x\colon 0 \to A \xrightarrow{g} B \xrightarrow{f} D(E)\otimes_\Sigma \operatorname{Tr}DX \to 0$$

in $\operatorname{Ext}^1_\Lambda(D(E)\otimes_\Sigma \operatorname{Tr}DX, A)$ *corresponding to t under the isomorphism*

$$w\colon \operatorname{Ext}^1_\Lambda(D(E)\otimes_\Sigma \operatorname{Tr}DX, A) \to \operatorname{Hom}_\Sigma(\overline{\operatorname{Hom}_\Lambda}(A,X), I)$$

described in Proposition 2.9 *has the following properties:*

a) *g is left X-determined.*
b) $\operatorname{Im}(g,X) = H$.
c) *x splits if and only if $H = (A,X)$.*
d) *If E is an injective envelope for* $\operatorname{Im}t$, *then in any commutative diagram*

$$\begin{array}{ccccccccc} 0 & \longrightarrow & A & \xrightarrow{g} & B & \longrightarrow & D(E)\otimes_\Sigma \operatorname{Tr}DX & \longrightarrow & 0 \\ & & \| & & \downarrow & & \downarrow & & \\ 0 & \longrightarrow & A & \longrightarrow & B & \longrightarrow & D(E)\otimes_\Sigma \operatorname{Tr}DX & \longrightarrow & 0 \end{array}$$

the vertical morphisms are isomorphisms.

PROOF. Follows by arguments similar to those used in Chapter I, Section 3 of [2] to derive Proposition 2.8 from 2.7. □

In connection with this result it should be noted that under the hypothesis of the previous proposition $D(E)$ is a finitely generated projective Σ^{op}-module with the property there is an epimorphism $D(E) \to \operatorname{Hom}_R((X,C)/H, I)$ and $D(E)$ is a projective cover for $\operatorname{Hom}_R((X,C)/H, I)$ if and only if E is an injective envelope for $\operatorname{Im}t$.

Suppose we are given a morphism $f\colon B \to C$. We know that f is right X-determined for some X in $\operatorname{mod}\Lambda$. In view of our previous discussion, the $(\operatorname{End}X)^{\mathrm{op}}$-submodule $\operatorname{Im}(X,f)$ of (X,C) is an interesting invariant of f. It is therefore natural to wonder to what extent a morphism $f\colon B \to C$ is determined by knowing that $f\colon B \to C$ is right X-determined and knowing the $(\operatorname{End}X)^{\mathrm{op}}$-submodule $\operatorname{Im}(X,f)$ of (X,C). Obviously a morphism $f\colon B \to C$ is not determined by this data, for if B' is any nonzero module and $f'\colon B \amalg B' \to C$ is the morphism such that $f'|_B = f$ and $f'|_{B'} = 0$, then f' is also right X-determined and $\operatorname{Im}(X,f') = \operatorname{Im}(X,C)$ even though f and f' are certainly not the same. Nonetheless, knowing that a morphism $f\colon B \to C$ is right X-determined together with knowing the $(\operatorname{End}X)^{\mathrm{op}}$-submodule $\operatorname{Im}(X,f)$ of (X,C) does tell us quite a bit about the morphism f as we shall soon see. As a preliminary step in this direction, we have the following easily verified result.

LEMMA 2.11. *Suppose $f\colon B \to C$ and $f'\colon B' \to C$ are two morphisms in* $\operatorname{mod}\Lambda$ *with f right X-determined.*

a) $\operatorname{Im}(X,f') \subset \operatorname{Im}(X,f)$ *if and only if there is a morphism $g\colon B' \to B$ such that $f' = fg$.*
b) *If f' is also right X-determined, then* $\operatorname{Im}(X,f') = \operatorname{Im}(X,f)$ *if and only if there are morphisms $g\colon B' \to B$ and $h\colon B \to B'$ such that $f' = gf$ and $f = f'h$.*

Similarly, we have the following analogue for left determined morphisms.

LEMMA 2.12. *Suppose $g\colon A \to B$ and $g'\colon A \to B'$ are two morphisms in* $\operatorname{mod}\Lambda$ *with g left X-determined.*

a) *$\operatorname{Im}(g', X) \subset \operatorname{Im}(g, X)$ if and only if there is a morphism $h\colon B \to B'$ such that $hg = g'$.*
b) *If g' is also left X-determined, then $\operatorname{Im}(g, X) = \operatorname{Im}(g', X)$ if and only if there are morphisms $h\colon B \to B'$ and $h'\colon B' \to B$ such that $hg = g'$ and $h'g' = g$.*

Suppose now that $0 \to A \xrightarrow{g} B \xrightarrow{f} C \to 0$ is exact with f right X-determined. Since f is an epimorphism it is easily seen that the $(\operatorname{End} X)^{\mathrm{op}}$-submodule $\operatorname{Im}(X, f)$ contains $P(X, C)$. Because (X, C) is a finitely generated $(\operatorname{End} X)^{\mathrm{op}}$-module, there is a finitely generated injective $(\operatorname{End} X)^{\mathrm{op}}$-module E such that there is a morphism $t\colon \underline{\operatorname{Hom}}_\Lambda(X, C) \to E$ with the property that $\operatorname{Ker} t = \operatorname{Im}(X, f)/P(X, C)$ and the induced monomorphism $(X, C)/\operatorname{Im}(X, f) \to E$ is an injective envelope. Then by Proposition 2.8 there is an exact sequence in $\operatorname{mod}\Lambda$

$$0 \to \operatorname{Hom}_{\Gamma^{\mathrm{op}}}(D(E), D\operatorname{Tr} X) \xrightarrow{g'} B \xrightarrow{f'} C \to 0$$

with the following properties:

a) f' is right X-determined;
b) $\operatorname{Im}(X, f') = \operatorname{Im}(X, f)$;
c) In any commutative diagram

$$\begin{array}{ccccccccc}
0 & \longrightarrow & \operatorname{Hom}_{\Gamma^{\mathrm{op}}}(D(E), D\operatorname{Tr} X) & \xrightarrow{g'} & B' & \xrightarrow{f'} & C & \longrightarrow & 0 \\
 & & \downarrow & & \downarrow & & \| & & \\
0 & \longrightarrow & \operatorname{Hom}_{\Gamma^{\mathrm{op}}}(D(E), D\operatorname{Tr} X) & \xrightarrow{g'} & B' & \xrightarrow{f'} & C & \longrightarrow & 0
\end{array}$$

the vertical morphisms are isomorphisms.

Since f and f' are both right X-determined and $\operatorname{Im}(X, f') = \operatorname{Im}(X, f)$ we have by Lemma 2.11 that there is a commutative exact diagram

$$\begin{array}{ccccccccc}
0 & \longrightarrow & \operatorname{Hom}_{\Gamma^{\mathrm{op}}}(D(E), D\operatorname{Tr} X) & \xrightarrow{g'} & B' & \xrightarrow{f'} & C & \longrightarrow & 0 \\
 & & \downarrow{\scriptstyle s} & & \downarrow{\scriptstyle t} & & \| & & \\
0 & \longrightarrow & A & \xrightarrow{g} & B & \xrightarrow{f} & C & \longrightarrow & 0 \\
 & & \downarrow{\scriptstyle u} & & \downarrow{\scriptstyle v} & & \| & & \\
0 & \longrightarrow & \operatorname{Hom}_{\Gamma^{\mathrm{op}}}(D(E), D\operatorname{Tr} X) & \xrightarrow{g'} & B' & \xrightarrow{f'} & C & \longrightarrow & 0
\end{array}$$

with us and vt isomorphisms. From this it follows that A and B have decompositions $A_1 \amalg A_2$ and $B_1 \amalg B_2$ such that $g(A_1) \subset B_1$ and $g(A_2) = B_2$. This implies that $0 \to A_1 \xrightarrow{g|A_1} B_1 \xrightarrow{f|B_1} C \to 0$ is exact and $h\colon A_2 \to B_2$, given by $h(a_2) = g(a_2)$ for all a_2 in A_2, is an isomorphism. So the exact sequence $0 \to A \xrightarrow{g} B \xrightarrow{f} C \to 0$ can be written as

$$\begin{array}{ccccccccc}
 & & A_2 & \xrightarrow{h} & B_2 & & & & \\
 & & \amalg & & \amalg & & & & \\
0 & \longrightarrow & A_1 & \xrightarrow{g|A_1} & B_1 & \xrightarrow{f|B_1} & C & \longrightarrow & 0.
\end{array}$$

Moreover there is a commutative exact diagram

$$\begin{array}{ccccccccc}
0 & \longrightarrow & A_1 & \xrightarrow{g|_{A_1}} & B_1 & \xrightarrow{f|_{B_1}} & C & \longrightarrow & 0 \\
 & & \downarrow{\scriptstyle w} & & \downarrow{\scriptstyle z} & & \| & & \\
0 & \longrightarrow & \operatorname{Hom}_\Gamma(\operatorname{Tr} X, E) & \xrightarrow{g'} & B' & \xrightarrow{f'} & C & &
\end{array}$$

with the w and the z isomorphisms. Hence the exact sequence $0 \to A_1 \xrightarrow{g|_{A_1}} B_1 \xrightarrow{f|_{B_1}} C \to 0$ has the following properties:

a) $f|_{B_1}$ is right X-determined.
b) $\operatorname{Im}(X, f|_{B_1}) = \operatorname{Im}(X, f)$.
c) In any commutative diagram

$$\begin{array}{ccccccccc}
0 & \longrightarrow & A_1 & \xrightarrow{g|_{A_1}} & B_1 & \xrightarrow{f|_{B_1}} & C & \longrightarrow & 0 \\
 & & \downarrow & & \downarrow & & \| & & \\
0 & \longrightarrow & A_1 & \xrightarrow{g|_{A_1}} & B_1 & \xrightarrow{f|_{B_1}} & C & \longrightarrow & 0
\end{array}$$

the vertical morphisms are isomorphisms.

It should be observed that c) is equivalent to the following statement. If $h\colon B_1 \to B_1$ has the property $(f|_{B_1})h = f|_{B_1}$, then h is an isomorphism. This discussion suggests introducing the following definitions (see [7, Section 2]).

A morphism $f\colon B \to C$ is said to be *right minimal* if each endomorphism $h\colon B \to B$ satisfying $fh = f$ is an isomorphism. A morphism $f\colon B \to C$ which is both right minimal and right X-determined is said to be *minimal right X-determined.*

A morphism $g\colon A \to B$ is said to be *left minimal* if each endomorphism $h\colon B \to B$ satisfying $hg = g$ is an isomorphism. A morphism $g\colon A \to B$ which is both left minimal and left X-determined is said to be *minimal left X-determined.*

Our previous discussion yields the following result in this new terminology.

PROPOSITION 2.13. *Let $0 \to A \xrightarrow{g} B \xrightarrow{f} C \to 0$ be an exact sequence in* $\operatorname{mod} \Lambda$.

a) *Suppose f is right X-determined. Then there is an exact sequence $0 \to A' \xrightarrow{g'} B' \xrightarrow{f'} C \to 0$ in $\operatorname{mod} \Lambda$ with f' a minimal right X-determined morphism such that $\operatorname{Im}(X, f) = \operatorname{Im}(X, f')$. Further there is a commutative diagram*

$$\begin{array}{ccccccccc}
0 & \longrightarrow & A' & \xrightarrow{g'} & B' & \xrightarrow{f'} & C & \longrightarrow & 0 \\
 & & \downarrow{\scriptstyle u} & & \downarrow{\scriptstyle v} & & \| & & \\
0 & \longrightarrow & A & \longrightarrow & B & \xrightarrow{f} & C & \longrightarrow & 0 \\
 & & \downarrow{\scriptstyle w} & & \downarrow{\scriptstyle z} & & \| & & \\
0 & \longrightarrow & A' & \longrightarrow & B' & \longrightarrow & C & \longrightarrow & 0
\end{array}$$

with wu and zv isomorphisms.

b) *Suppose g is left X-determined. Then there is an exact sequence $0 \to A \xrightarrow{g''} B'' \xrightarrow{f''} C'' \to 0$ in $\operatorname{mod} \Lambda$ with g'' a minimal left Y-determined morphism such* *that* $\mathrm{Im}(g'', Y) = \mathrm{Im}(g, Y)$. *Further there is a commutative diagram*

$$\begin{array}{ccccccccc}
0 & \longrightarrow & A & \xrightarrow{g''} & B'' & \xrightarrow{f''} & C'' & \longrightarrow & 0 \\
 & & \| & & \downarrow{\scriptstyle s} & & \downarrow{\scriptstyle t} & & \\
0 & \longrightarrow & A & \xrightarrow{g} & B & \xrightarrow{f} & C & \longrightarrow & 0 \\
 & & \| & & \downarrow{\scriptstyle q} & & \downarrow{\scriptstyle r} & & \\
0 & \longrightarrow & A & \xrightarrow{g''} & B'' & \xrightarrow{f''} & C'' & \longrightarrow & 0
\end{array}$$

such that qs *and* rt *are isomorphisms.*

In view of our discussion so far, the following uniqueness result is of interest.

PROPOSITION 2.14. a) *Let* $f\colon B \to C$ *and* $f'\colon B' \to C$ *be two minimal right* X*-determined morphisms. Then the following are equivalent:*

i) $\mathrm{Im}(X, f) = \mathrm{Im}(X, f')$.

ii) *There is an* $h\colon B \to B'$ *having the property that* $f = f'h$ *and any such* h *is an isomorphism.*

b) *Let* $g\colon A \to B$ *and* $g'\colon A \to B'$ *be two minimal left* Y*-determined morphisms. Then the following are equivalent:*

i) $\mathrm{Im}(g, Y) = \mathrm{Im}(g', Y)$.

ii) *There are* $h\colon B \to B'$ *such that* $hg = g'$, *and any such* h *is an isomorphism.*

In view of these results, it is clear that right minimal epimorphisms and left minimal monomorphisms are the basic type of epimorphisms and monomorphisms. For this reason characterizations and properties of such morphisms are of importance. We now give some which have proven to be useful.

PROPOSITION 2.15. *Let* $0 \to A \xrightarrow{g} B \xrightarrow{f} C \to 0$ *be exact in* $\mathrm{mod}\,\Lambda$.

a) *The following are equivalent:*

i) f *is right minimal.*

ii) *If* B' *is a summand of* B *such that* $f|_{B'} = 0$, *then* $B' = 0$.

iii) *If* $A' \neq 0$ *and* $t\colon A \to A'$ *is a splittable epimorphism, i.e. there is an* $s\colon A' \to A$ *such that* $ts = \mathrm{id}_{A'}$, *then the exact pushout sequence* $0 \to A' \to B \times^A A' \to C \to 0$ *does not split.*

b) *If* f *is right minimal, then* f *is right* X*-determined if and only if* X *has a summand isomorphic to* $[\mathrm{Tr}\, DA]$.

c) *The following are equivalent:*

i) g *is left minimal.*

ii) *If* $B' \subset B$ *is a summand of* B *such that* $\mathrm{Im}\, g \subset B'$, *then* $B' = B$.

iii) *If* $C' \subset C$ *is a summand of* C *such that the pullback exact sequence* $0 \to A \to B \times_C C' \to C' \to 0$ *splits, then* $C' = 0$.

d) *If* g *is left minimal, then* g *is left* Y*-determined if and only if* Y *has a summand isomorphic to* $[D\,\mathrm{Tr}\, C]$.

e) *If* $0 \to A \xrightarrow{g} B \xrightarrow{f} C \to 0$ *does not split, then*

i) f *is right minimal if* A *is indecomposable.*

ii) g *is left minimal if* C *is indecomposable.*

These results can be readily applied to arbitrary morphisms and not just epimorphisms and monomorphisms once one observes the following.

LEMMA 2.16. a) *A morphism $f\colon B \to C$ is right minimal if and only if the induced epimorphism $B \to \operatorname{Im} f$ is right minimal.*

b) *If a morphism $f\colon B \to C$ is right X-determined, then the induced epimorphism $B \to \operatorname{Im} f$ is right X-determined.*

c) *A morphism $g\colon A \to B$ is left minimal if and only if the induced monomorphism $\operatorname{Im} g \to B$ is left minimal.*

d) *If $g\colon A \to B$ is left Y-determined, then the induced monomorphism $\operatorname{Im} g \to B$ is left Y-determined.*

As a consequence of this lemma and the previous results, we have the following.

PROPOSITION 2.17. *Let $0 \to A \xrightarrow{g} B \xrightarrow{f} C$ be an exact sequence in $\operatorname{mod} \Lambda$.*

a) *There exists an exact sequence $0 \to A' \xrightarrow{g'} B' \xrightarrow{f'} C$ with f' right minimal such that there is a commutative diagram*

$$\begin{CD}
0 @>>> A' @>g'>> B' @>f'>> C\\
@. @VuVV @VvVV @|\\
0 @>>> A @>g>> B @>f>> C\\
@. @VwVV @VzVV @|\\
0 @>>> A' @>g'>> B' @>f'>> C
\end{CD}$$

such that wu and zv are isomorphisms. Consequently f is right X-determined if and only if f' is right X-determined and moreover $\operatorname{Im}(X,f) = \operatorname{Im}(X,f')$.

b) *If $0 \to A'' \xrightarrow{g'} B'' \xrightarrow{f''} C$ is exact with f'' right minimal such that there exist morphisms $s\colon B'' \to B$ and $t\colon B \to B''$ such that $fs = f''$ and $f''t = f$, then there exists a commutative diagram*

$$\begin{CD}
0 @>>> A'' @>g''>> B'' @>f''>> C\\
@. @VVV @VVV @|\\
0 @>>> A' @>g'>> B' @>f'>> C
\end{CD}$$

with the vertical morphisms isomorphisms.

Similarly, we have the following.

PROPOSITION 2.18. *Let $A \xrightarrow{g} B \xrightarrow{f} C \to 0$ be exact in $\operatorname{mod} \Lambda$.*

a) *There exists an exact sequence $A \xrightarrow{g'} B' \xrightarrow{f'} C' \to 0$ with g' left minimal such that there is a commutative diagram*

$$\begin{CD}
A @>g'>> B' @>f'>> C @>>> 0\\
@| @VsVV @VtVV @.\\
A @>g>> B @>f>> C @>>> 0\\
@| @VqVV @VrVV @.\\
A @>g'>> B' @>f'>> C' @>>> 0
\end{CD}$$

with qs and rt isomorphisms. Consequently g is left Y-determined if and only if g' is left Y-determined and moreover $\operatorname{Im}(g, Y) = \operatorname{Im}(g, Y')$.

b) *If $A \xrightarrow{g''} B'' \xrightarrow{f''} C \to 0$ is an exact sequence with g'' left minimal such that there exist morphisms $u\colon B'' \to B$ and $v\colon B \to B''$ with $ug'' = g$ and $vg = g''$, then there exists a commutative diagram*

$$\begin{array}{ccccccc} A & \xrightarrow{g''} & B'' & \xrightarrow{f''} & C'' & \longrightarrow & 0 \\ \| & & \downarrow & & \downarrow & & \\ A & \xrightarrow{g'} & B' & \xrightarrow{f'} & C' & \longrightarrow & 0 \end{array}$$

with the vertical morphisms isomorphisms.

So far we have been primarily concerned with $(\operatorname{End} X)^{\mathrm{op}}$-submodules of (X, C) containing $P(X, C)$. We can generalize some of our previous results in the following way.

PROPOSITION 2.19. *Let X and C be in* $\operatorname{mod}\Lambda$. *For each* $(\operatorname{End} X)^{\mathrm{op}}$*-submodule H of (X, C) there is an exact sequence $0 \to A \xrightarrow{g} B \xrightarrow{f} C$ in* $\operatorname{mod}\Lambda$ *satisfying the following.*

a) *f is a minimal right X-determined morphism with* $\operatorname{Im}(X, f) = H$.
b) *A is in* $\operatorname{add}(D \operatorname{Tr} X)$, *the full additive subcategory consisting of summands of finite sums of copies of $D \operatorname{Tr} X$.*
c) *f is an epimorphism if and only if $H \supset P(X, C)$.*

We also have the following analogue of Proposition 2.19.

PROPOSITION 2.20. *Let A and X be modules in* $\operatorname{mod}\Lambda$. *For each* $\operatorname{End} X$*-submodule H of (A, X) there is an exact sequence $A \xrightarrow{g} B \xrightarrow{f} C \to 0$ in* $\operatorname{mod}\Lambda$ *satisfying the following.*

a) *g is a minimal left X-determined morphism with* $\operatorname{Im}(g, X) = H$.
b) *C is in* $\operatorname{add}(\operatorname{Tr} DX)$.
c) *g is a monomorphism if and only if $H \supset I(A, X)$.*

In order to state the final results of this section it is convenient to have the following definitions.

Two morphisms $f\colon B \to C$ and $f'\colon B' \to C$ are said to be *isomorphic over C* if there is an isomorphism $h\colon B \to B'$ such that $f'h = f$. Two morphisms $g\colon A \to B$ and $g'\colon A \to B'$ are said to be *isomorphic over A*, if there is an isomorphism $h\colon B \to B'$ such that $hg = g'$.

PROPOSITION 2.21. *Let X and C be in* $\operatorname{mod}\Lambda$ *and let $0 \to A_i \xrightarrow{g_i} B_i \xrightarrow{f_i} C$ be the family of exact sequences in* $\operatorname{mod}\Lambda$ *such that each f_i is right minimal and each A_i is in* $\operatorname{add}(D \operatorname{Tr} X)$. *Then the map which sends each f_i to (X, f_i) induces a bijection between the over C isomorphism classes of the f_i and the* $(\operatorname{End} X)^{\mathrm{op}}$*-submodules of (X, C).*

Similarly, let X and A be in $\operatorname{mod}\Lambda$ *and let $A \xrightarrow{g_i} B_i \xrightarrow{f_i} C_i \to 0$ be the family of exact sequences in* $\operatorname{mod}\Lambda$ *such that each g_i is left minimal and each C_i is in* $\operatorname{add}(\operatorname{Tr} DX)$. *Then the map which sends each g_i to (g_i, X) induces a bijection between the over A isomorphism classes of the g_i and the* $\operatorname{End} X$*-submodules of (A, X).*

3. Some applications

As usual, we assume that R is a commutative artin ring, Λ is an artin R-algebra, and all Λ-modules are finitely generated. In this section we begin exploring some of the implications of the general theory outlined in Section 2.

An important question in representation theory is the following: When is an indecomposable module X isomorphic to a summand of a module M? The notion of morphisms determined by a module gives the following novel formulation of this problem.

PROPOSITION 3.1. *Let X be an indecomposable nonprojective Λ-module and let $A = D\operatorname{Tr} X$. Then the following statements are equivalent for a Λ-module M.*

a) *X is isomorphic to a summand of M.*

b) *If $0 \to A \xrightarrow{g} B \xrightarrow{f} C \to 0$ is a nonsplit exact sequence, then f is right M-determined.*

c) *There is a nonsplit exact sequence $0 \to A \xrightarrow{g} B \xrightarrow{f} C \to 0$ such that f is right M-determined.*

PROOF. a) implies b). Since X is a nonprojective indecomposable module, we know that $\operatorname{Tr} X$ is a nonprojective indecomposable Λ^{op}-module (see Proposition 1.3). Hence $A = D\operatorname{Tr} X$ is a noninjective indecomposable Λ-module. Let $0 \to A \xrightarrow{g} B \xrightarrow{f} C \to 0$ be a nonsplit exact sequence. Then f is right minimal since A is indecomposable (see Proposition 2.15). Therefore f is right M-determined if and only if $X = \operatorname{Tr} DA$ is isomorphic to a summand of M (see Proposition 2.15). Hence the assumption that X is isomorphic to a summand of M implies that f is right M-determined.

b) implies c). Since X not being projective implies that $A = D\operatorname{Tr} X$ is not injective, we know that there is a nonsplit extension $0 \to A \xrightarrow{g} B \xrightarrow{f} C \to 0$. The hypothesis of b) implies that for this extension f is right M-determined.

c) implies a). Suppose $0 \to A \xrightarrow{g} B \xrightarrow{f} C \to 0$ is a nonsplit exact sequence with f right M-determined. We have already seen in the proof of a) implies b) that under these circumstances $X = \operatorname{Tr} DA$ is isomorphic to a summand of M. □

As an immediate consequence of this result we obtain the following characterization of the $\operatorname{Tr} DA$ for all noninjective indecomposable Λ-modules A.

COROLLARY 3.2. *Let A be a noninjective indecomposable Λ-module. Then the following statements are equivalent for an indecomposable Λ-module X.*

a) *$X \approx \operatorname{Tr} DA$.*

b) *If $0 \to A \xrightarrow{g} B \xrightarrow{f} C \to 0$ is a nonsplit exact sequence, then f is right X-determined.*

c) *There is a nonsplit exact sequence $0 \to A \xrightarrow{g} B \xrightarrow{f} C \to 0$ with f right X-determined.*

Applying the duality $D\colon \operatorname{mod}\Lambda \to \operatorname{mod}\Lambda^{\mathrm{op}}$ to the above results we obtain the following results.

PROPOSITION 3.3. *Let Y be a noninjective indecomposable Λ-module and let $C = \operatorname{Tr} DY$. Then the following statements are equivalent for a module N.*

a) *Y is isomorphic to a summand of N.*

b) *If $0 \to A \xrightarrow{g} B \xrightarrow{f} C \to 0$ is a nonsplit exact sequence, then g is left N-determined.*

c) *There is a nonsplit exact sequence $0 \to A \xrightarrow{g} B \xrightarrow{f} C \to 0$ with g left N-determined.*

COROLLARY 3.4. *Let C be a nonprojective indecomposable Λ-module. Then the following statements are equivalent for an indecomposable module Y.*

a) *$Y \approx D\operatorname{Tr} C$.*

b) *If* $0 \to A \xrightarrow{g} B \xrightarrow{f} C \to 0$ *is a nonsplit exact sequence, then* g *is left* Y*-determined.*

c) *There is a nonsplit exact sequence* $0 \to A \xrightarrow{g} B \xrightarrow{f} C \to 0$ *with* g *left* Y*-determined.*

The following is another general question which is intimately related to our general theory of morphisms determined by modules. Given two exact sequences

$$0 \to A' \xrightarrow{g'} B' \xrightarrow{f'} C \to 0$$

$$0 \to A \xrightarrow{g} B \xrightarrow{f} C \to 0$$

in $\operatorname{mod}\Lambda$, when does there exist a commutative diagram

$$\begin{array}{ccccccccc} 0 & \longrightarrow & A' & \xrightarrow{g'} & B' & \xrightarrow{f'} & C & \longrightarrow & 0 \\ & & \downarrow & & \downarrow & & \| & & \\ 0 & \longrightarrow & A & \xrightarrow{g} & B & \xrightarrow{f} & C & \longrightarrow & 0 \end{array} \quad ?$$

This question is answered in the following result.

PROPOSITION 3.5. *Let* $0 \to A \xrightarrow{g} B \xrightarrow{f} C \to 0$ *and* $0 \to A' \xrightarrow{g'} B' \xrightarrow{f'} C \to 0$ *be exact sequences in* $\operatorname{mod}\Lambda$. *There exists a commutative diagram*

$$(*) \quad \begin{array}{ccccccccc} 0 & \longrightarrow & A' & \xrightarrow{g'} & B' & \xrightarrow{f'} & C & \longrightarrow & 0 \\ & & \downarrow & & \downarrow & & \| & & \\ 0 & \longrightarrow & A & \xrightarrow{g} & B & \xrightarrow{f} & C & \longrightarrow & 0 \end{array}$$

if and only if $\operatorname{Im}(\operatorname{Tr} DA, f) \supset \operatorname{Im}(\operatorname{Tr} DA, f')$.

PROOF. Clearly $\operatorname{Im}(\operatorname{Tr} DA, f) \supset \operatorname{Im}(\operatorname{Tr} DA, f')$ if such a commutative diagram $(*)$ exists.

Suppose now that $\operatorname{Im}(\operatorname{Tr} DA, f) \supset \operatorname{Im}(\operatorname{Tr} DA, f')$. The fact that f is right $\operatorname{Tr} DA$-determined (see Proposition 2.2) means that there is a morphism $g\colon B' \to B$ such that $fg = f'$. This implies trivially that a commutative diagram of the type $(*)$ exists. □

The dual of this result is the following.

PROPOSITION 3.6. *Let* $0 \to A \xrightarrow{g} B \xrightarrow{f} C \to 0$ *and* $0 \to A \xrightarrow{g'} B' \xrightarrow{f'} C' \to 0$ *be exact sequences in* $\operatorname{mod}\Lambda$. *There exists a commutative diagram*

$$\begin{array}{ccccccccc} 0 & \longrightarrow & A & \xrightarrow{g} & B & \xrightarrow{f} & C & \longrightarrow & 0 \\ & & \| & & \downarrow & & \downarrow & & \\ 0 & \longrightarrow & A & \xrightarrow{g'} & B' & \xrightarrow{f'} & C' & \longrightarrow & 0 \end{array}$$

if and only if $\operatorname{Im}(D \operatorname{Tr} C, g) \supset \operatorname{Im}(D \operatorname{Tr} C, g')$.

A question that is often of interest is the following. Let $0 \to A' \xrightarrow{f} A \xrightarrow{g} A'' \to 0$ be exact in $\operatorname{mod}\Lambda$ and x an element of $\operatorname{Ext}^1_\Lambda(C, A)$. When is there an element x' in $\operatorname{Ext}^1_\Lambda(C, A')$ such that $\operatorname{Ext}^1_\Lambda(C, f)(x') = x$? While we cannot answer this question in this degree of generality, our theory gives an answer in case $A' = \operatorname{Hom}_\Lambda(\Lambda/\underline{a}, A)$ for some two sided ideal $\underline{a}$ of Λ as we now demonstrate. To do this we need the following lemma.

LEMMA 3.7. *Let $\underline{a}$ be a two sided ideal in Λ and A a $\Lambda/\underline{a}$-module. Let C be an arbitrary Λ-module and $0 \to \Omega C \to P \to C \to 0$ an exact sequence with $P \to C$ a projective cover. If $0 \to A \xrightarrow{g} B \xrightarrow{f} C \to 0$ is exact in* $\operatorname{mod}\Lambda$, *then there is a commutative exact diagram*

$$\begin{array}{ccccccccc} 0 & \longrightarrow & \Omega C/\underline{a}\Omega C & \longrightarrow & P/\underline{a}\Omega C & \longrightarrow & C & \longrightarrow & 0 \\ & & \downarrow & & \downarrow & & \| & & \\ 0 & \longrightarrow & A & \longrightarrow & B & \longrightarrow & C & \longrightarrow & 0 \end{array}$$

PROOF. Since P is projective we know there is an exact commutative diagram

$$\begin{array}{ccccccccc} 0 & \longrightarrow & \Omega C & \longrightarrow & P & \longrightarrow & C & \longrightarrow & 0 \\ & & \downarrow{\scriptstyle u} & & \downarrow{\scriptstyle v} & & \| & & \\ 0 & \longrightarrow & A & \longrightarrow & B & \longrightarrow & C & \longrightarrow & 0. \end{array}$$

Since $\underline{a}A = 0$, we have that $\underline{a}\Omega C \subset \operatorname{Ker} u$ and hence that $\underline{a}\Omega C$ is also contained in $\operatorname{Ker} v$. The result now follows trivially. □

PROPOSITION 3.8. *Suppose $\underline{a}$ is a two sided ideal in Λ. Let $0 \to \Omega C \to P \to C \to 0$ be an exact sequence in* $\operatorname{mod}\Lambda$ *with $P \to C$ a projective cover and let $0 \to \Omega C/\underline{a}\Omega C \to P/\underline{a}\Omega C \xrightarrow{t} C \to 0$ be the induced exact sequence. Then the following statements are equivalent for an exact sequence $x\colon 0 \to A \xrightarrow{g} B \xrightarrow{f} C \to 0$ in* $\operatorname{mod}\Lambda$.

a) $\operatorname{Im}(\operatorname{Tr} DA, t) \subset \operatorname{Im}(\operatorname{Tr} DA, f)$.

b) *There is a commutative exact diagram*

$$\begin{array}{ccccccccc} 0 & \longrightarrow & A' & \longrightarrow & B' & \longrightarrow & C & \longrightarrow & 0 \\ & & \downarrow & & \downarrow & & \downarrow & & \\ 0 & \longrightarrow & A & \xrightarrow{g} & B & \xrightarrow{f} & C & \longrightarrow & 0 \end{array}$$

with A' in $\operatorname{mod}\Lambda/\underline{a}$.

c) *If we let $0 \to A' \xrightarrow{h} A$ be the inclusion of the submodule $A' = \operatorname{Hom}_\Lambda(\Lambda/\underline{a}, A)$ of A, then there is an x' in $\operatorname{Ext}^1_\Lambda(C, A')$ such that $\operatorname{Ext}^1_\Lambda(C, h)(x') = x$.*

PROOF. a) implies b). Trivial consequence of the fact that f is right $\operatorname{Tr} DA$-determined.

b) implies c). Trivial.

c) implies a). Follows easily from Lemma 3.7. □

For the convenience of the reader we state the dual of Proposition 3.8.

PROPOSITION 3.9. *Suppose $\underline{a}$ is a two sided ideal in Λ. Let $0 \to A \xrightarrow{u} I \xrightarrow{v} \Omega^{-1}(A) \to 0$ be exact in* $\operatorname{mod}\Lambda$ *with u an injective envelope and let $0 \to A \xrightarrow{t} u^{-1}(L) \to L \to 0$ be the induced exact sequence where $L = \operatorname{Hom}_\Lambda(\Lambda/\underline{a}, \Omega^{-1}A)$. Then the following statements are equivalent for an exact sequence $x\colon 0 \to A \xrightarrow{g} B \xrightarrow{f} C \to 0$ in* $\operatorname{mod}\Lambda$.

a) $\operatorname{Im}(t, D\operatorname{Tr} C) \subset \operatorname{Im}(f, D\operatorname{Tr} C)$.

b) *There is a commutative exact diagram*

$$\begin{CD} 0 @>>> A @>f>> B @>g>> C @>>> 0 \\ @. @| @VVV @VVV @. \\ 0 @>>> A @>>> B'' @>>> C'' @>>> 0 \end{CD}$$

with C'' *in* $\operatorname{mod}\Lambda/\underline{a}$.

c) *If* $h\colon C \to C/\underline{a}C$ *is the canonical epimorphism, then there is an* x' *in* $\operatorname{Ext}^1_\Lambda(C/\underline{a}C, A)$ *such that* $\operatorname{Ext}^1_\Lambda(h, A)(x') = x$.

Using similar arguments to those used to prove Proposition 3.8, we establish the following.

PROPOSITION 3.10. *Let* $\underline{a}$ *be a two sided ideal in* Λ. *Let* C *be a* $\Lambda/\underline{a}$*-module and* $P \xrightarrow{t} C \to 0$ *a projective cover of* C *in* $\operatorname{mod}\Lambda/\underline{a}$. *Then the following statements are equivalent for an exact sequence* $0 \to A \xrightarrow{g} B \xrightarrow{f} C \to 0$ *in* $\operatorname{mod}\Lambda$.

a) $\operatorname{Im}(\operatorname{Tr} DA, f) \supset \operatorname{Im}(\operatorname{Tr} DA, t)$.

b) *There is a commutative exact diagram*

$$\begin{CD} 0 @>>> A' @>>> B' @>>> C @>>> 0 \\ @. @VVV @VVV @| @. \\ 0 @>>> A @>>> B @>>> C @>>> 0 \end{CD}$$

with A', B', C *in* $\operatorname{mod}\Lambda/\underline{a}$.

c) *The sequence of* $\Lambda/\underline{a}$*-modules*

$$0 \to (\Lambda/\underline{a}, A) \to (\Lambda/\underline{a}, B) \to (\Lambda/\underline{a}, C) \to 0$$

is exact.

We now state the dual of Proposition 3.10.

PROPOSITION 3.11. *Let* $\underline{a}$ *be a two sided ideal in* Λ. *Let* A *be a* $\Lambda/\underline{a}$*-module and let* $0 \to A \xrightarrow{t} I$ *be an injective envelope of* A *in* $\operatorname{mod}\Lambda/\underline{a}$. *Then the following statements are equivalent for an exact sequence* $0 \to A \xrightarrow{g} B \xrightarrow{f} C \to 0$ *in* $\operatorname{mod}\Lambda$.

a) $\operatorname{Im}(f, D\operatorname{Tr} C) \supset \operatorname{Im}(t, D\operatorname{Tr} C)$.

b) *There is an exact commutative diagram*

$$\begin{CD} 0 @>>> A @>>> B @>>> C @>>> 0 \\ @. @| @VVV @VVV @. \\ 0 @>>> A @>>> B' @>>> C' @>>> 0 \end{CD}$$

with the A, B', C' *in* $\operatorname{mod}\Lambda/\underline{a}$.

c) *The sequence of* $\Lambda/\underline{a}$*-modules*

$$0 \to \Lambda/\underline{a} \otimes_\Lambda A \to \Lambda/\underline{a} \otimes_\Lambda B \to \Lambda/\underline{a} \otimes_\Lambda C \to 0$$

is exact.

We now turn our attention to the following problem which is obviously related but somewhat different from the problem discussed above. Suppose X is in $\operatorname{mod}\Lambda$ and we are given an element x in $\operatorname{Ext}^1_\Lambda(C, A)$ with C and A in $\operatorname{mod}\Lambda$. When does there exist a morphism $h\colon X' \to A$ with X' in $\operatorname{add}(X)$ such that x is contained in the image of $\operatorname{Ext}^1_\Lambda(C,h)\colon \operatorname{Ext}^1_\Lambda(C,X') \to \operatorname{Ext}^1_\Lambda(C,A)$? The answer to this question is based on the following.

PROPOSITION 3.12. *The following are equivalent for an exact sequence* $0 \to A \xrightarrow{g} B \xrightarrow{f} C \to 0$.

a) $\operatorname{Im}(\operatorname{Tr} DA, f) = P(\operatorname{Tr} DA, C)$.

b) *If* $0 \to A' \to B' \to C \to 0$ *is exact with* A' *in* $\operatorname{add}(A)$, *then there is an exact commutative diagram*

$$\begin{array}{ccccccccc} 0 & \longrightarrow & A & \longrightarrow & B & \longrightarrow & C & \longrightarrow & 0 \\ & & \downarrow & & \downarrow & & \| & & \\ 0 & \longrightarrow & A' & \longrightarrow & B' & \longrightarrow & C & \longrightarrow & 0. \end{array}$$

c) *Let* $0 \to X \to E \to C \to 0$ *be exact. Then there is an exact commutative diagram*

$$\begin{array}{ccccccccc} 0 & \longrightarrow & A' & \longrightarrow & B' & \longrightarrow & C & \longrightarrow & 0 \\ & & \downarrow & & \downarrow & & \| & & \\ 0 & \longrightarrow & X & \longrightarrow & E & \longrightarrow & C & \longrightarrow & 0 \end{array}$$

with A' *in* $\operatorname{add} A$ *if and only if there is an exact commutative diagram*

$$\begin{array}{ccccccccc} 0 & \longrightarrow & A & \longrightarrow & B & \longrightarrow & C & \longrightarrow & 0 \\ & & \downarrow & & \downarrow & & \downarrow & & \\ 0 & \longrightarrow & X & \longrightarrow & E & \longrightarrow & C & \longrightarrow & 0. \end{array}$$

PROOF. a) implies b). Let $0 \to A' \xrightarrow{g'} B' \xrightarrow{f'} C \to 0$ be exact with A' in $\operatorname{add} A$. Then f' is right $\operatorname{Tr} DA'$-determined. Since $[\operatorname{Tr} DA']$ is a summand of $[\operatorname{Tr} DA]$, we have that f' is right $\operatorname{Tr} DA$-determined. Since f' is an epimorphism we know that $\operatorname{Im}(\operatorname{Tr} DA, f') \supset P(\operatorname{Tr} DA, C) = \operatorname{Im}(\operatorname{Tr} DA, f)$. Therefore the fact that f' is right $\operatorname{Tr} DA$-determined implies that there is an exact commutative diagram

$$\begin{array}{ccccccccc} 0 & \longrightarrow & A & \xrightarrow{g} & B & \xrightarrow{f} & C & \longrightarrow & 0 \\ & & \downarrow & & \downarrow & & \| & & \\ 0 & \longrightarrow & A' & \xrightarrow{g'} & B' & \xrightarrow{f'} & C & \longrightarrow & 0 \end{array}$$

b) implies a). We know by our general existence theorem that there is an exact sequence $0 \to A' \xrightarrow{g'} B' \xrightarrow{f'} C \to 0$ with A' in $\operatorname{add} A$ such that $\operatorname{Im}(\operatorname{Tr} DA, f') = P(\operatorname{Tr} DA, C)$. Since by the hypothesis of b) there is an exact commutative diagram

$$\begin{array}{ccccccccc} 0 & \longrightarrow & A & \xrightarrow{g} & B & \xrightarrow{f} & C & \longrightarrow & 0 \\ & & \downarrow & & \downarrow & & \| & & \\ 0 & \longrightarrow & A' & \xrightarrow{g'} & B' & \xrightarrow{f'} & C & \longrightarrow & 0 \end{array}$$

we have that $\operatorname{Im}(\operatorname{Tr} DA, f) \subset \operatorname{Im}(\operatorname{Tr} DA, f') = P(\operatorname{Tr} DA, C)$. Hence $\operatorname{Im}(\operatorname{Tr} DA, f) = P(\operatorname{Tr} DA, C)$, since the fact that f is an epimorphism implies that $\operatorname{Im}(\operatorname{Tr} DA, f) \supset P(\operatorname{Tr} DA, C)$. Therefore b) implies a).

a) equivalent to c). This is a straightforward consequence of the fact that a) is equivalent to b). □

As an easy consequence of this proposition, we have the following result which answers the question posed before Proposition 3.12.

PROPOSITION 3.13. *Let* $0 \to A \xrightarrow{g} B \xrightarrow{f} C \to 0$ *be exact with* f *a minimal right* $\operatorname{Tr} DA$*-determined morphism such that* $\operatorname{Im}(\operatorname{Tr} DA, f) = P(\operatorname{Tr} DA, C)$. *Then the following are equivalent for an exact sequence* $x\colon 0 \to X \xrightarrow{j} E \xrightarrow{h} C \to 0$ *in* $\operatorname{Ext}^1_\Lambda(C, X)$.

a) $\operatorname{Im}(\operatorname{Tr} DX, f) \subset \operatorname{Im}(\operatorname{Tr} DX, h)$.

b) *There is a morphism* $t\colon A' \to X$ *with* A' *in* $\operatorname{add} A$ *such that* x *is in* $\operatorname{Im}\operatorname{Ext}^1_\Lambda(C, t)$.

We now state the duals of these results. We begin with the dual of Proposition 3.12.

PROPOSITION 3.14. *The following statements are equivalent for an exact sequence* $0 \to A \xrightarrow{g} B \xrightarrow{f} C \to 0$.

a) $\operatorname{Im}(g, D \operatorname{Tr} C) = I(A, D \operatorname{Tr} C)$.

b) *If* $0 \to A \xrightarrow{g'} B' \xrightarrow{f'} C' \to 0$ *is exact with* C' *in* $\operatorname{add} C$, *then there is an exact commutative diagram*

$$\begin{array}{ccccccccc} 0 & \longrightarrow & A & \xrightarrow{g'} & B' & \xrightarrow{f'} & C' & \longrightarrow & 0 \\ & & \| & & \downarrow & & \downarrow & & \\ 0 & \longrightarrow & A & \longrightarrow & B & \longrightarrow & C & \longrightarrow & 0 \end{array}$$

c) *Let* $0 \to A \to E \to Y \to 0$ *be exact. Then there is an exact commutative diagram*

$$\begin{array}{ccccccccc} 0 & \longrightarrow & A & \longrightarrow & E & \longrightarrow & Y & \longrightarrow & 0 \\ & & \| & & \downarrow & & \downarrow & & \\ 0 & \longrightarrow & A & \longrightarrow & B' & \longrightarrow & C' & \longrightarrow & 0 \end{array}$$

with C' *in* $\operatorname{add} C$ *if and only if there is an exact commutative diagram*

$$\begin{array}{ccccccccc} 0 & \longrightarrow & A & \longrightarrow & E & \longrightarrow & Y & \longrightarrow & 0 \\ & & \| & & \downarrow & & \downarrow & & \\ 0 & \longrightarrow & A & \xrightarrow{g} & B & \xrightarrow{f} & C & \longrightarrow & 0. \end{array}$$

The following is the dual of Proposition 3.13.

PROPOSITION 3.15. *Let* $0 \to A \xrightarrow{g} B \to C \to 0$ *be exact with* g *a minimal left* $D \operatorname{Tr} C$*-determined morphism such that* $\operatorname{Im}(g, D \operatorname{Tr} C) = I(A, D \operatorname{Tr} C)$. *Then the following statements are equivalent for an exact sequence* $x\colon 0 \to A \xrightarrow{h} E \xrightarrow{j} Y \to 0$ *in* $\operatorname{Ext}^1_\Lambda(Y, A)$.

a) $\operatorname{Im}(h, D \operatorname{Tr} Y) \supset \operatorname{Im}(g, D \operatorname{Tr} Y)$.

b) *There is a morphism* $t\colon Y \to C'$ *with* C' *in* $\operatorname{add} C$ *such that* y *is in* $\operatorname{Im}(\operatorname{Ext}^1(t, A)$.

As an application of Proposition 3.12 and 3.14 we have the following result.

PROPOSITION 3.16. *Suppose* A *and* C *are in* $\operatorname{mod} \Lambda$.

a) *If* $P(\operatorname{Tr} DA, C) = 0$, *then for each submodule* C' *of* C, *the induced morphism* $\operatorname{Ext}^1_\Lambda(C, A) \to \operatorname{Ext}^1_\Lambda(C', A)$ *is an epimorphism.*

b) *If* $I(A, D\operatorname{Tr} C) = 0$, *then for each epimorphism* $A \to A''$, *the induced morphism* $\operatorname{Ext}^1_\Lambda(C, A) \to \operatorname{Ext}^1_\Lambda(C, A'')$ *is an epimorphism.*

PROOF. a) Since $P(\operatorname{Tr} DA, C) = 0$, we know by our general existence theorem that there is an exact sequence $x\colon 0 \to A' \xrightarrow{g} B \xrightarrow{f} C \to 0$ with A' in $\operatorname{add}(A)$ such that f is right $\operatorname{Tr} DA$-determined and $\operatorname{Im}(\operatorname{Tr} DA, f) = 0$. Suppose C' is a submodule of C. Then it is easily checked that the exact sequence $y\colon 0 \to A' \to f^{-1}(C) \xrightarrow{h} C' \to 0$ has the property that $\operatorname{Im}(\operatorname{Tr} DA, h) = 0$. Since A' is in $\operatorname{add} A$, it follows that h is right $\operatorname{Tr} DA$-determined. Moreover if $u\colon C' \to C$ is the inclusion we have that $\operatorname{Ext}^1_\Lambda(u, A')(x) = y$.

Suppose now that z is an element of $\operatorname{Ext}^1_\Lambda(C', A')$. Then by Proposition 3.13 we know that there is a morphism $t\colon A' \to A$ such that the morphism $\operatorname{Ext}^1_\Lambda(C', t)\colon \operatorname{Ext}^1_\Lambda(C', A') \to \operatorname{Ext}^1_\Lambda(C', A)$ has the property that $\operatorname{Ext}^1_\Lambda(C', t)(y) = z$. From the commutative diagram

$$\begin{array}{ccc}
\operatorname{Ext}^1_\Lambda(C, A') & \xrightarrow{\operatorname{Ext}^1_\Lambda(C,t)} & \operatorname{Ext}^1_\Lambda(C, A) \\
{\scriptstyle \operatorname{Ext}^1_\Lambda(u,A')}\downarrow & & \downarrow{\scriptstyle \operatorname{Ext}^1_\Lambda(u,A)} \\
\operatorname{Ext}^1_\Lambda(C', A') & \xrightarrow{\operatorname{Ext}^1_\Lambda(C',t)} & \operatorname{Ext}^1_\Lambda(C', A)
\end{array}$$

and the fact that $\operatorname{Ext}^1_\Lambda(u, A')(x) = y$, it follows that $\operatorname{Ext}^1_\Lambda(u, A)(\operatorname{Ext}^1_\Lambda(C, t)(x)) = z$. Thus z is in $\operatorname{Im} \operatorname{Ext}^1_\Lambda(u, A)$. Since this is true for all z in $\operatorname{Ext}^1_\Lambda(C', A)$, it follows that the morphism $\operatorname{Ext}^1_\Lambda(u, A)\colon \operatorname{Ext}^1_\Lambda(C, A) \to \operatorname{Ext}^1_\Lambda(C', A)$ is an epimorphism, which finishes the proof of a).

b) Proven similar to a). □

Before going on to the last topic in this section, we point out the following easily verified fact. Let A be an indecomposable Λ-module and suppose $\operatorname{Ext}^1_\Lambda(C, A) \neq 0$. Suppose $x_1, \ldots, x_t$ in $\operatorname{Ext}^1_\Lambda(C, A)$ are a minimal set of generators for $\operatorname{Ext}^1_\Lambda(C, A)$ over the local ring $\operatorname{End} A$. Let $x\colon 0 \to A^t \xrightarrow{g} B \xrightarrow{f} C \to 0$ represent the element $(x_1, \ldots, x_t)$ in $\operatorname{Ext}^1_\Lambda(C, A)^t = \operatorname{Ext}^1_\Lambda(C, A^t)$. Then f is a right minimal $\operatorname{Tr} DA$-determined morphism with $\operatorname{Im}(\operatorname{Tr} DA, f) = P(\operatorname{Tr} DA, C)$. The obvious dual of this also holds.

The rest of this section is devoted to generalizing Proposition 3.16 by giving a description of the kernel and cokernel of a morphism $\operatorname{Ext}^1_\Lambda(C, A) \to \operatorname{Ext}^1_\Lambda(C', A)$ induced by a morphism $f\colon C' \to C$. This description is based on the isomorphism

$$(*) \qquad \operatorname{Ext}^1_\Lambda(C, A) \approx \operatorname{Hom}_R(\underline{\operatorname{Hom}}_\Lambda(\operatorname{Tr} DA, C), I)$$

which is functorial in C ([2, Chapter I, Proposition 8.7]).

Suppose $f\colon C' \to C$ is a morphism. Then f induces a morphism of $(\operatorname{End} \operatorname{Tr} DA)^{\mathrm{op}}$-modules $\underline{\operatorname{Hom}}(\operatorname{Tr} DA, f)\colon \underline{\operatorname{Hom}}(\operatorname{Tr} DA, C') \to \underline{\operatorname{Hom}}(\operatorname{Tr} DA, C)$. Letting $K = \operatorname{Ker} \underline{\operatorname{Hom}}(\operatorname{Tr} DA, f)$ and $L = \operatorname{Coker} \underline{\operatorname{Hom}}(\operatorname{Tr} DA, f)$ we obtain from isomorphism $(*)$ the exact sequence

$$0 \to \operatorname{Hom}_R(L, I) \to \operatorname{Ext}^1_\Lambda(C, A) \to \operatorname{Ext}^1_\Lambda(C', A) \to \operatorname{Hom}_R(K, I) \to 0.$$

We now give an explicit description of the $(\operatorname{End} \operatorname{Tr} DA)^{\mathrm{op}}$-modules K and L. The morphism $f\colon C' \to C$ induces the morphism $(\operatorname{Tr} DA, f)\colon (\operatorname{Tr} DA, C') \to (\operatorname{Tr} DA, C)$. It then follows that $K = (\operatorname{Tr} DA, f)^{-1}(P(\operatorname{Tr} DA, C))/P(\operatorname{Tr} DA, C')$ while $L = (\operatorname{Tr} DA, C)/(\operatorname{Im}(\operatorname{Tr} DA, f), P(\operatorname{Tr} DA, C))$. Summarizing this discussion we have:

PROPOSITION 3.17. *Let* $f\colon C' \to C$ *be a morphism in* $\operatorname{mod} \Lambda$ *and let* A *be in* $\operatorname{mod} \Lambda$. *Then* f *induces the morphism* $h = (\operatorname{Tr} DA, f)\colon (\operatorname{Tr} DA, C') \to (\operatorname{Tr} DA, C)$. *Set* $K = h^{-1}(P(\operatorname{Tr} DA, C'))/P(\operatorname{Tr} DA, C')$ *and* $L = (\operatorname{Tr} DA, C)/(\operatorname{Im}(h, P(\operatorname{Tr} DA, C))$. *Then*

a) *there is an exact sequence* $0 \to \operatorname{Hom}_R(L, I) \to \operatorname{Ext}^1_\Lambda(C, A) \xrightarrow{\operatorname{Ext}^1_\Lambda(f,A)} \operatorname{Ext}^1_\Lambda(C', A) \to \operatorname{Hom}_R(K, I) \to 0$.
b) $\operatorname{Ext}^1_\Lambda(f, A)\colon \operatorname{Ext}^1_\Lambda(C, A) \to \operatorname{Ext}^1_\Lambda(C', A)$ *is an epimorphism if and only if a morphism* $u\colon \operatorname{Tr} DA \to C'$ *is in* $P(\operatorname{Tr} DA, C')$ *whenever* $fu\colon \operatorname{Tr} DA \to C$ *is in* $P(\operatorname{Tr} DA, C)$.
c) $\operatorname{Ext}^1_\Lambda(f, A)\colon \operatorname{Ext}^1_\Lambda(C, A) \to \operatorname{Ext}^1_\Lambda(C', A)$ *is a monomorphism if and only if given any* $v\colon \operatorname{Tr} DA \to C$ *there is a morphism* $u\colon \operatorname{Tr} DA \to C'$ *such that* $v = fu + s$ *with* s *in* $P(\operatorname{Tr} DA, C)$.

As an immediate consequence of this result we have the following.

COROLLARY 3.18. *Suppose* A *and* C *are in* $\operatorname{mod} \Lambda$ *with* $P(\operatorname{Tr} DA, C) = 0$. *Let* $f\colon C' \to C$ *be a monomorphism. Then the morphism* $\operatorname{Ext}^1_\Lambda(f, A)\colon \operatorname{Ext}^1_\Lambda(C, A) \to \operatorname{Ext}^1_\Lambda(C', A)$ *has the following properties.*

a) $\operatorname{Ext}^1_\Lambda(f, A)$ *is an epimorphism.*
b) $\operatorname{Ext}^1_\Lambda(f, A)$ *is an isomorphism if and only if every morphism* $\operatorname{Tr} DA \to C$ *has its image in* C'.

Similarly (or by duality) one can obtain a description of the kernel and cokernel of a morphism $\operatorname{Ext}^1_\Lambda(C, A) \to \operatorname{Ext}^1_\Lambda(C, A')$ induced by a morphism $f\colon A \to A'$. This description is based on the isomorphism

$$\operatorname{Ext}^1_\Lambda(C, A) \to \operatorname{Hom}_R(\overline{\operatorname{Hom}}_\Lambda(A, D \operatorname{Tr} C), I)$$

which is functorial in A ([2, Chapter I, Proposition 8.17]) and proceeds exactly as above, an exercise left to the interested reader to carry out.

4. Almost split sequences

This section is devoted to giving some applications of the theory of morphisms determined by objects to the minimal left and right almost split morphisms and almost split sequences. We start by pointing out some of the elementary connections between these various notions.

Suppose that C is an indecomposable module. Then $\operatorname{End} C$ is a local ring and has a unique maximal ideal $\underline{m}$. We know that there is a right minimal C-determined morphism $f\colon B \to C$ such that $\operatorname{Im}(C, f) = \underline{m}$. From this it follows that f has the following properties:

a) f is not a splittable epimorphism since $\operatorname{Im}(C, f) \neq (C, C)$.
b) If $h\colon B \to B$ is such that $fh = f$, then h is an isomorphism.
c) If $t\colon L \to C$ is not a splittable epimorphism, then there is an $h\colon L \to B$ such that $fh = t$. For $g\colon L \to C$ is not a splittable epimorphism precisely when $\operatorname{Im}(C, t) \subset \underline{m}$ and if $\operatorname{Im}(C, t) \subset \underline{m} = \operatorname{Im}(C, f)$, then there is an $h\colon L \to B$ such that $fh = t$ since f is right C-determined.

We recall (see [7, Section 2]) that a morphism $f\colon B \to C$ is called a minimal right almost split morphism if it satisfies conditions a), b), and c). On the other hand it is not difficult to show that if a morphism $f\colon B \to C$ is minimal right almost split, i.e. satisfies a), b), c), then C is indecomposable and f is a minimal right C-determined morphism with $\operatorname{Im}(C, f)$ the maximal ideal of $\operatorname{End} C$ (see [7, Section 2]).

Similarly, suppose A is an indecomposable module and $\underline{m}$ the maximal ideal of the local ring $\operatorname{End} A$. Then there is a morphism $g\colon A \to B$ which is a minimal left A-determined morphism with $\operatorname{Im}(g, A) = \underline{m}$. From this it follows that g has the following properties.

a') g is not a splittable monomorphism.
b') If $h\colon B \to B$ is a morphism such that $g = hg$, then h is an isomorphism.
c') If $h\colon A \to L$ is not a splittable monomorphism, then there is a morphism $s\colon B \to L$ such that $sg = h$.

We recall (see [7, Section 2]) that a morphism $g\colon A \to B$ is said to be minimal left almost split if it satisfies a'), b'), and c'). On the other hand it is not difficult to see that if $g\colon A \to B$ is minimal left almost split, then A is indecomposable and g is minimal left A-determined with $\operatorname{Im}(g, A)$ the maximal ideal of $\operatorname{End} A$ (see [7, Section 2]).

Suppose C is indecomposable and $f\colon B \to C$ is a minimal right C-determined morphism with $\operatorname{Im}(C, f)$ the maximal ideal of $\operatorname{End} C$. If C is projective, then $P(C, C) = \operatorname{End} C$ and so $\operatorname{Im}(C, f)$ does not contain $P(C, C)$. Hence $f\colon B \to C$ is not an epimorphism. In fact it is easily seen that f is isomorphic over C to the inclusion morphism $\underline{r}C \to C$ where $\underline{r}$ is the radical of Λ. If C is not projective, then $P(C, C) \subset \operatorname{Im}(C, f)$. Hence letting $\Gamma = \operatorname{End} \operatorname{Tr} C$ and E be an injective envelope for the Γ-module $(C, C)/\operatorname{Im}(C, f)$ we have that there is an exact sequence $0 \to \operatorname{Hom}_\Gamma(\operatorname{Tr} C, E) \xrightarrow{g} B' \xrightarrow{f'} C \to 0$ (see Proposition 2.9) where f' is a minimal right C-determined morphism with $\operatorname{Im}(C, f') = \operatorname{Im}(C, f)$. Since $(C, C)/\operatorname{Im}(C, f)$ is a simple $(\operatorname{End} C)^{\mathrm{op}}$-module, it is a simple Γ-module and so $E = \operatorname{Hom}_R(\Gamma, I)$. Thus $\operatorname{Hom}_\Gamma(\operatorname{Tr} C, E) \approx \operatorname{Hom}_\Gamma(\operatorname{Tr} C, \operatorname{Hom}_R(\Gamma, I)) \approx \operatorname{Hom}_R(\operatorname{Tr} C, I) = D \operatorname{Tr} C$. Hence $\operatorname{Ker} f' \approx D \operatorname{Tr} C$. But $\operatorname{Ker} f \approx \operatorname{Ker} f'$ since $f\colon B \to C$ and $f'\colon B' \to C$ are isomorphic over C because they are both minimal right C-determined morphisms with $\operatorname{Im}(C, f') = \operatorname{Im}(C, f)$.

So we have the exact sequence $0 \to D \operatorname{Tr} C \xrightarrow{g} B \xrightarrow{f} C \to 0$ with f minimal right C-determined and $\operatorname{Im}(C, f)$ the maximal ideal of $\operatorname{End} C$. Since C is a nonprojective indecomposable module, we know that $D \operatorname{Tr} C$ is indecomposable. Because f is not a splittable epimorphism, we know that $g\colon D \operatorname{Tr} C \to B$ is not a splittable monomorphism. Thus $\operatorname{Im}(g, D \operatorname{Tr} C)$ is contained in the maximal ideal $\underline{m}$ of $\operatorname{End} D \operatorname{Tr} C$. Suppose t is in $\underline{m}$, i.e. t is an endomorphism of $D \operatorname{Tr} C$ which is not an isomorphism. Then we have the commutative diagram

$$\begin{array}{ccccccccc}
0 & \longrightarrow & D \operatorname{Tr} C & \xrightarrow{g} & B & \xrightarrow{f} & C & \longrightarrow & 0 \\
 & & \downarrow{\scriptstyle t} & & \downarrow{\scriptstyle u} & & \| & & \\
0 & \longrightarrow & D \operatorname{Tr} C & \xrightarrow{g'} & B' & \xrightarrow{f'} & C & \longrightarrow & 0.
\end{array}$$

Then f' is right determined by $C = \operatorname{Tr} D(D \operatorname{Tr} C)$. Also $\operatorname{Im}(C, f) \subset \operatorname{Im}(C, f')$. If $\operatorname{Im}(C, f) = \operatorname{Im}(C, f')$, then f' is not a splittable epimorphism and f' is right minimal since $D \operatorname{Tr} C$ is indecomposable (see Proposition 2.16). But this implies that u and hence t is an isomorphism. This contradicts the fact that t is in $\underline{m}$. Therefore $\operatorname{Im}(C, f')$ contains $\operatorname{Im}(C, f)$ properly and thus $\operatorname{Im}(C, f') = \operatorname{End}(C)$ since $\operatorname{Im}(C, f)$ is the maximal ideal of $\operatorname{End} C$. Hence f' is a splittable epimorphism which implies there is an $h\colon B \to D \operatorname{Tr} C$ such that $t = hg$. Therefore $\operatorname{Im}(g, D \operatorname{Tr} C)$ is the maximal ideal of $\operatorname{End}(D \operatorname{Tr} C)$.

Thus the exact sequence $0 \to D \operatorname{Tr} C \xrightarrow{g} B \xrightarrow{f} C \to 0$ has the following properties.

a) $\operatorname{Im}(g, D \operatorname{Tr} C)$ is the maximal ideal of $\operatorname{End} D \operatorname{Tr} C$.
b) g is left minimal since C is indecomposable and $0 \to D \operatorname{Tr} C \xrightarrow{g} B \xrightarrow{f} C \to 0$ does not split.
c) g is left $D \operatorname{Tr} C$-determined.

In other words the exact sequence $0 \to D\operatorname{Tr} C \xrightarrow{g} B \xrightarrow{f} C \to 0$ has the property that g is minimal left almost split and f is minimal right almost split. We recall that a sequence $0 \to A \xrightarrow{g} B \xrightarrow{f} C \to 0$ is said to be *almost split* if g is minimal left almost split and f is minimal right almost split (see [7, Section 2]). Thus the sequence $0 \to D\operatorname{Tr} C \xrightarrow{g} B \xrightarrow{f} C \to 0$ is almost split.

Having indicated some of the connections between minimal left and right almost split morphisms, we summarize the main features of these notions. The reader is referred to [6] and [7] for proofs not already given.

PROPOSITION 4.1. a) *If P is an indecomposable projective Λ-module, the inclusion $\underline{r}P \to P$ is a minimal right almost split morphism.*

b) *Let C be an indecomposable nonprojective module.*
   - i) *There is an almost split sequence $0 \to A \xrightarrow{g} B \xrightarrow{f} C \to 0$.*
   - ii) *If $0 \to A \xrightarrow{g} B \xrightarrow{f} C \to 0$ is almost split, then $A \approx D\operatorname{Tr} C$.*
   - iii) *If $0 \to A' \xrightarrow{g'} B' \xrightarrow{f'} C \to 0$ and $0 \to A \xrightarrow{g} B \xrightarrow{f} C \to 0$ are almost split, they are isomorphic over C.*

c) *Let A be an indecomposable noninjective Λ-module.*
   - i) *There exists an almost split sequence $0 \to A \xrightarrow{g} B \xrightarrow{f} C \to 0$.*
   - ii) *If $0 \to A \xrightarrow{g} B \xrightarrow{f} C \to 0$ is almost split, then $C \approx \operatorname{Tr} DA$.*
   - iii) *If $0 \to A \xrightarrow{g} B \xrightarrow{f} C \to 0$ and $0 \to A \xrightarrow{g'} B' \xrightarrow{f'} C \to 0$ are almost split, then they are isomorphic over A.*

d) *If I is an indecomposable injective Λ-module, then the epimorphism $I \to I/\operatorname{Soc} I$ is minimal left almost split.*

e) *If $0 \to A \xrightarrow{g} B \xrightarrow{f} C \to 0$ and $0 \to A' \xrightarrow{g'} B' \xrightarrow{f'} C' \to 0$ are almost split, then the following are equivalent.*
   - i) *$A \approx A'$.*
   - ii) *$C \approx C'$.*
   - iii) *There exists a commutative diagram*

$$\begin{array}{ccccccccc}
0 & \longrightarrow & A & \xrightarrow{g} & B & \xrightarrow{f} & C & \longrightarrow & 0 \\
 & & \downarrow & & \downarrow & & \downarrow & & \\
0 & \longrightarrow & A' & \xrightarrow{g'} & B' & \xrightarrow{f'} & C' & \longrightarrow & 0
\end{array}$$

with the vertical morphisms isomorphisms.

The reader should observe that part e) of the above proposition states that an almost split sequence $0 \to A \to B \to C \to 0$ is an invariant of the indecomposable module A and the indecomposable module C. Consequently any invariant of the almost split sequence is an invariant of A and C. For example if $B \approx \coprod_{i=1}^{n} B_i$ with the B_i indecomposable, then the following are all numerical invariants of C: the number n, the number of isomorphic B_i, the number of B_i with $L(B_i) > L(C)$, the number of B_i with $L(B_i) < L(C)$. Very little is known at this time about what the module theoretic properties of an almost split sequence $0 \to A \to B \to C \to 0$ tells one about A and C. The reader should see [6] and [7] for more details concerning this question.

Before going on with our applications, it is convenient to recall the notion of an irreducible morphism (see [7, Section 2]).

A morphism $f\colon B \to C$ in $\operatorname{mod}\Lambda$ is said to be *irreducible* if f is not a splittable epimorphism or a splittable monomorphism and if f is the composition $B \xrightarrow{g} X \xrightarrow{h} C$, then either h is a splittable epimorphism or g is a splittable monomorphism. The following are some of the basic facts concerning irreducible morphisms (see [7, Section 2]).

PROPOSITION 4.2. *Let $f\colon B \to C$ be a morphism in* $\operatorname{mod}\Lambda$.

a) *If f is irreducible, then f is either a proper epimorphism (i.e. is an epimorphism but not an isomorphism) or a proper monomorphism.*
b) *If C is indecomposable and B is not zero, then f is irreducible if and only if there is a morphism $f'\colon B' \to C$ such that the induced morphism $(f, f')\colon B \amalg B' \to C$ is right minimal almost split.*
c) *If B is indecomposable, then f is irreducible if and only if there is a morphism $f'\colon B \to C'$ such that the induced morphism $B \to C \amalg C'$ is minimal left almost split.*

We now point out some easily verified consequences of this result.

PROPOSITION 4.3. a) *Suppose C is an indecomposable module. There are only a finite number of nonisomorphic modules B which have the property that there is an irreducible morphism $B \to C$ or an irreducible morphism $C \to B$.*
b) *Let $0 \to A \to B \to C \to 0$ be an almost split sequence and let B' be isomorphic to a summand of B. Then $L(B')$ is different from $L(A)$ and $L(C)$ where $L(X)$ denotes the length of X. Hence the indecomposable summands of B are not isomorphic to C or A.*

We now point out another useful characterization of irreducible morphisms (see [7, Section 2]).

PROPOSITION 4.4. *Let $0 \to A \xrightarrow{g} B \xrightarrow{f} C \to 0$ be a nonsplit exact sequence in* $\operatorname{mod}\Lambda$.

a) *f is irreducible if and only if g has the property that given any $h\colon A \to X$ either there is a morphism $s\colon B \to X$ such that $sg = h$ or there is a $t\colon X \to B$ such that $th = g$.*
b) *g is irreducible if and only if f has the property that given any $h\colon X \to C$ either there is a morphism $s\colon X \to B$ such that $gs = h$ or there is a $t\colon B \to X$ such that $f = ht$.*

In connection with this result it is important to make the following observation. Suppose the nonsplit exact sequence $0 \to A \xrightarrow{g} B \xrightarrow{f} C \to 0$ representing the element x in $\operatorname{Ext}^1_\Lambda(C, A)$ has the property that f is irreducible. Let $h\colon A \to A'$ be a proper epimorphism. The fact that g is a monomorphism and h has a nontrivial kernel means that there is no morphism $s\colon A' \to B$ such that $sh = g$. Hence by Proposition 4.4 we know there is a morphism $t\colon B \to X$ such that $tg = h$, or equivalently, $\operatorname{Ext}^1_\Lambda(C, h)(x)$ in $\operatorname{Ext}^1_\Lambda(C, X)$ is zero. In other words, the nonzero element x in $\operatorname{Ext}^1_\Lambda(C, A)$ has the property that a morphism $h\colon A \to X$ is a monomorphism if $\operatorname{Ext}^1_\Lambda(C, h)(x)$ in $\operatorname{Ext}^1_\Lambda(C, X)$ is not zero. It is easily checked that this property of x implies that A is indecomposable. This observation leads to the following notions (see [4, Section 2]).

An element x in $\operatorname{Ext}^1_\Lambda(C, \ )(A)$ is said to be *minimal* if x is not zero and any morphism $h\colon A \to X$ is a monomorphism if $\operatorname{Ext}^1_\Lambda(C, h)(x)$ in $\operatorname{Ext}^1_\Lambda(C, X)$ is not zero. An element y in $\operatorname{Ext}^1_\Lambda(\ , A)(C)$ is said to be *minimal* if y is not zero and any morphism $h\colon Y \to C$ is an epimorphism if $\operatorname{Ext}^1_\Lambda(h, A)(y)$ in $\operatorname{Ext}^1_\Lambda(Y, A)$ is not zero. We now summarize some of the basic properties of these special types of extensions.

PROPOSITION 4.5. a) *If x in $\operatorname{Ext}^1_\Lambda(C,\ )(A)$ is minimal, then A is indecomposable. If y in $\operatorname{Ext}^1_\Lambda(\ ,A)(C)$ is minimal, then C is indecomposable.*

b) *Let $x\colon 0 \to A \xrightarrow{g} B \xrightarrow{f} C \to 0$ be exact.*

i) *If f is irreducible, then x viewed as an element in $\operatorname{Ext}^1_\Lambda(C,\ )(A)$ is minimal and so A is indecomposable.*

ii) *If g is irreducible, then x viewed as an element of $\operatorname{Ext}^1_\Lambda(\ ,A)(C)$ is minimal and so C is indecomposable.*

c) *Let x be a nonzero element in $\operatorname{Ext}^1_\Lambda(C,A)$.*

i) *There is an epimorphism $h\colon A \to A''$ such that the element $\operatorname{Ext}^1_\Lambda(C,h)(x)$ in $\operatorname{Ext}^1_\Lambda(C,\ )(A'')$ is minimal.*

ii) *There is a monomorphism $h\colon C' \to C$ such that the element $\operatorname{Ext}^1_\Lambda(h,A)(x)$ in $\operatorname{Ext}^1_\Lambda(\ ,A)(C')$ is minimal.*

It should be observed that the last part of the above proposition shows that there are lots of exact sequences $x\colon 0 \to A \to B \to C \to 0$ with x a minimal element of $\operatorname{Ext}^1_\Lambda(C,\ )(A)$, or in $\operatorname{Ext}^1_\Lambda(\ ,A)(C)$, or in both $\operatorname{Ext}^1_\Lambda(C,\ )(A)$ and $\operatorname{Ext}^1_\Lambda(\ ,A)(C)$.

The applications in the rest of this section are based on the following formulation of Proposition 2.3 which is particularly appropriate for our present purposes.

PROPOSITION 4.6. *Suppose $0 \to A \xrightarrow{g} B \xrightarrow{f} C \to 0$ is an exact sequence in $\operatorname{mod}\Lambda$ representing the element x in $\operatorname{Ext}^1_\Lambda(C,A)$. Further, suppose X is in $\operatorname{mod}\Lambda$. Then the following statements are equivalent.*

a) *$(X,f)\colon (X,B) \to (X,C) \to 0$ is exact.*

b) *For each $h\colon X \to C$ we have that $\operatorname{Ext}^1_\Lambda(h,A)(x)$ in $\operatorname{Ext}^1_\Lambda(X,A)$ is zero.*

c) *$(g, D\operatorname{Tr}X)\colon (B, D\operatorname{Tr}X) \to (A, D\operatorname{Tr}X) \to 0$ is exact.*

d) *For each $h\colon A \to D\operatorname{Tr}X$ we have that $\operatorname{Ext}^1_\Lambda(C,h)(x)$ in $\operatorname{Ext}^1_\Lambda(C, D\operatorname{Tr}X)$ is zero.*

As our first application of Proposition 4.6 we have the following.

PROPOSITION 4.7. *Suppose the exact sequence $x\colon 0 \to A \xrightarrow{g} B \xrightarrow{f} C \to 0$ is a minimal element in $\operatorname{Ext}^1_\Lambda(C,\ )(A)$. Let X be in $\operatorname{mod}\Lambda$. Then*

a) *$(X,f)\colon (X,B) \to (X,C) \to 0$ is exact if and only if $\operatorname{Im}((B, D\operatorname{Tr}X) \xrightarrow{(g, D\operatorname{Tr}X)} (A, D\operatorname{Tr}X))$ contains all monomorphisms $A \to D\operatorname{Tr}X$.*

b) *Suppose $\operatorname{Im}(X,f) \neq (X,C)$.*

i) *Then there is a monomorphism $h\colon A \to D\operatorname{Tr}X$ such that $\operatorname{Ext}^1_\Lambda(C,h)(x)$ in $\operatorname{Ext}^1_\Lambda(C, D\operatorname{Tr}X)$ is not zero.*

ii) *$\operatorname{Soc}A$ is isomorphic to a summand of $\Omega(X)/r\Omega(X)$.*

iii) *If $f\colon B \to C$ is irreducible, then the monomorphism $0 \to A \xrightarrow{g} B$ can be written as a composition $A \xrightarrow{h} D\operatorname{Tr}X \to B$. In particular, $(D\operatorname{Tr}X, B) \neq 0$.*

PROOF. a) Suppose $h\colon A \to D\operatorname{Tr}X$ is not a monomorphism. Then because x in $\operatorname{Ext}^1_\Lambda(C,\ )(A)$ is minimal we know that $\operatorname{Ext}^1_\Lambda(C,h)(x)$ in $\operatorname{Ext}^1_\Lambda(C, D\operatorname{Tr}X)$ is zero, but $\operatorname{Ext}^1_\Lambda(C,h)(x)$ being zero is equivalent to there being a morphism $t\colon B \to D\operatorname{Tr}X$ such that $tg = h$. Thus $\operatorname{Im}((B, D\operatorname{Tr}X) \xrightarrow{(g, D\operatorname{Tr}X)} (A, D\operatorname{Tr}X))$ contains all nonmonomorphisms $A \to D\operatorname{Tr}X$. Thus if $\operatorname{Im}(g, D\operatorname{Tr}X)$ also contains all monomorphisms, then $\operatorname{Im}(g, D\operatorname{Tr}X) = (A, D\operatorname{Tr}X)$, which finishes the proof of a).

b) i) Since $\operatorname{Im}(X,f) \neq (X,C)$, there is a monomorphism $h\colon A \to D\operatorname{Tr}X$ such that h is not in $\operatorname{Im}((B, D\operatorname{Tr}X) \to (A, D\operatorname{Tr}X))$. Then h has the property that $\operatorname{Ext}^1_\Lambda(C,h)(x)$ in $\operatorname{Ext}^1_\Lambda(C, D\operatorname{Tr}X)$ is not zero.

ii) The monomorphism $h\colon A \to D\operatorname{Tr} X$ given in part i) induces a monomorphism $\operatorname{Soc} A \to \operatorname{Soc} D\operatorname{Tr} X$. Since $\operatorname{Soc} D\operatorname{Tr} X$ is semisimple, $\operatorname{Soc} A$ is isomorphic to a summand of $\operatorname{Soc} D\operatorname{Tr} X$. But we know by Proposition 1.4 that $\operatorname{Soc} D\operatorname{Tr} X \approx \Omega(X)/\underline{r}\Omega(X)$. Hence $\operatorname{Soc} A$ is isomorphic to a summand of $\Omega(X)/\underline{r}\Omega(X)$.

iii) Let

$$\begin{array}{ccccccccc} 0 & \longrightarrow & A & \xrightarrow{\quad g \quad} & B & \xrightarrow{\quad f \quad} & C & \longrightarrow & 0 \\ & & \downarrow{\scriptstyle h} & & \downarrow & & \| & & \\ 0 & \longrightarrow & D\operatorname{Tr} X & \longrightarrow & B \times^A D\operatorname{Tr} X & \longrightarrow & C & \longrightarrow & 0 \end{array}$$

be an exact commutative pushout diagram. Then by i) the exact pushout sequence $0 \to D\operatorname{Tr} X \to B \times^A D\operatorname{Tr} X \to C \to 0$ does not split. Hence there is no morphism $t\colon B \to D\operatorname{Tr} X$ such that $tg = h$. Therefore there must be a morphism $t\colon D\operatorname{Tr} X \to B$ such that $g = th$ since we are assuming that f is irreducible (see Proposition 4.4).

This finishes the proof of the proposition. □

As an immediate consequence of this result, we have the following.

Corollary 4.8. *Let $x\colon 0 \to A \xrightarrow{g} B \xrightarrow{f} C \to 0$ be exact with x a minimal element of $\operatorname{Ext}^1_\Lambda(C,\ )(A)$. Suppose $t\colon X \to C$ is not in $\operatorname{Im}(X, f)$ and the composition $X \xrightarrow{u} Y \xrightarrow{v} C$ is t, for instance $Y = C$ and $v = 1_C$.*

a) *v is not in $\operatorname{Im}(Y, f)$.*

b) i) *There is a monomorphism $h\colon A \to D\operatorname{Tr} Y$ such that $\operatorname{Ext}^1_\Lambda(C, h)(x)$ in $\operatorname{Ext}^1_\Lambda(C, D\operatorname{Tr} Y)$ is not zero.*

ii) *$\operatorname{Soc} A$ is isomorphic to a summand of $\Omega Y/\underline{r}\Omega Y$.*

iii) *If f is irreducible, the monomorphism $A \xrightarrow{g} B$ can be written as a composition $A \xrightarrow{h} D\operatorname{Tr} Y \to B$. In particular $(D\operatorname{Tr} Y, B) \neq 0$.*

Proof. a) Trivial consequence of the fact that $t\colon X \to C$ is not in $\operatorname{Im}(X, f)$.

b) Follows trivially from a) and Proposition 4.7. □

We now explore some consequences of Corollary 4.8.

Proposition 4.9. *Suppose $x\colon 0 \to A \xrightarrow{g} B \xrightarrow{f} C \to 0$ is minimal in $\operatorname{Ext}^1_\Lambda(C,\ )(A)$ and $f\colon B \to C$ does not generate (B, C) over $(\operatorname{End} B)^{\mathrm{op}}$.*

a) *There are monomorphisms $0 \to A \to D\operatorname{Tr} B$ and $0 \to A \to D\operatorname{Tr} C$.*

b) *$\operatorname{Soc} A$ is a summand of $\Omega(C)/\underline{r}\Omega(C)$ and $\Omega(B)/\underline{r}\Omega(B)$.*

c) *If f is irreducible, then there is a factorization $A \to D\operatorname{Tr} B \to B$ of the monomorphism $g\colon A \to B$. In particular $(D\operatorname{Tr} B, B) \neq 0$.*

Proof. This is an immediate consequence of Corollary 4.9, once one observe that if $f\colon B \to C$ does not generate (B, C) over $(\operatorname{End} B)^{\mathrm{op}}$, then $\operatorname{Im}(B, f) \neq (B, C)$. □

As our final example of how these ideas can be applied, we have the following.

Proposition 4.10. *Let $0 \to A \xrightarrow{g} B \xrightarrow{f} C \to 0$ be exact with f irreducible and $g \notin P(A, B)$. Further, suppose X is in $\operatorname{mod}\Lambda$ such that $\operatorname{Im}(X, f) \neq (X, C)$. Then*

a) *$\operatorname{Ext}^1_\Lambda(D\operatorname{Tr} X, D\operatorname{Tr} A) \neq 0$;*

b) *$\operatorname{Ext}^1_\Lambda(B, (D\operatorname{Tr})^2 X) \neq 0$;*

c) *$D\operatorname{Tr} X$ is not projective.*

PROOF. a) Since f is irreducible and $\operatorname{Im}(X, f) \neq (X, C)$, we know that $g\colon A \to B$ can be written as a composition $A \xrightarrow{h} D\operatorname{Tr} X \xrightarrow{j} B$. Since $g \notin P(A, B)$, it follows that $h\colon A \to D\operatorname{Tr} X$ is not in $P(A, D\operatorname{Tr} X)$ and $j\colon D\operatorname{Tr} X \to B$ is not in $P(D\operatorname{Tr} X, B)$. Hence $\underline{\operatorname{Hom}}(A, D\operatorname{Tr} X) \neq 0$ and $\underline{\operatorname{Hom}}(D\operatorname{Tr} X, B) \neq 0$. Since $\operatorname{Ext}^1_\Lambda(D\operatorname{Tr} X, D\operatorname{Tr} A) \approx \operatorname{Hom}_R(\underline{\operatorname{Hom}}(A, D\operatorname{Tr} X), I)$ and $\operatorname{Ext}^1_\Lambda(B, (D\operatorname{Tr})^2(X)) \approx \operatorname{Hom}_R(\underline{\operatorname{Hom}}(D\operatorname{Tr} X, B), I)$, a) and b) are established.

Part c) is an obvious consequence of a). □

Some examples of when the hypothesis of Proposition 4.10 is satisfied are the following.

a) Λ is hereditary and A is not projective.

b) Λ is selfinjective and B has no nonzero injective summands.

Propositions 4.7 through 4.10 have been concerned with exact sequences $x\colon 0 \to A \xrightarrow{g} B \xrightarrow{f} C \to 0$ with x minimal in $\operatorname{Ext}^1_\Lambda(C, \ )(A)$. Dual results hold for exact sequences with x minimal in $\operatorname{Ext}^1_\Lambda(\ , A)(C)$. The statements and proofs of these results are left to the reader to consider.

5. Brauer-Thrall $1\frac{1}{2}$

As usual Λ is an artin R-algebra. We denote by $\operatorname{ind}\Lambda$ the full subcategory of $\operatorname{mod}\Lambda$ whose objects are the indecomposable objects in $\operatorname{mod}\Lambda$. Further, for each integer $i \geq 0$ we denote by $\operatorname{ind}_i \Lambda$ the full subcategory of $\operatorname{ind}\Lambda$ consisting of those indecomposable Λ-modules A with $L(A) = i$, where $L(A)$ denotes the length of A over Λ. Also for each integer $n \geq 0$ we denote the full subcategory $\cup_{i \geq n} \operatorname{ind}_i \Lambda$ of $\operatorname{ind}\Lambda$ by $\mathcal{C}_n$. Finally, for each full subcategory $\mathcal{C}$ of $\operatorname{ind}\Lambda$ we denote by $|\mathcal{C}|$ the cardinality of the set of isomorphism classes of objects in $\mathcal{C}$. With this notation in mind, we can state the main result we prove in this section.

BRAUER-THRALL $1\frac{1}{2}$. *Λ is not of finite representation type (i.e. $|\operatorname{ind}\Lambda| \geq \aleph_0$) if and only if $|\mathcal{C}_n| = |\operatorname{ind}\Lambda|$ for all $n \geq 0$.*

Clearly this result is stronger than Brauer-Thrall 1 which states that Λ is not of finite representation type if and only if $\operatorname{ind}_i \Lambda$ is not empty for an infinite number of i. On the other hand, it is weaker than Brauer-Thrall 2 which states that if $\operatorname{card}\Lambda \geq \aleph_0$, then there are an infinite number of integers i such that $|\operatorname{ind}_i \Lambda| = \operatorname{card}\Lambda$. Hence the name Brauer-Thrall $1\frac{1}{2}$. While Brauer-Thrall 1 was first proven for artin algebras by Roiter [14] and for arbitrary artin rings by Auslander [4], the situation for Brauer-Thrall 2 is not quite so complete.

For finite dimensional algebras over a field k Brauer-Thrall 2 was first proven by Nazarova-Roiter in the case k is perfect [12] and extended by Ringel (unpublished) to k an arbitrary field. To the best of my knowledge there is no proof of Brauer-Thrall 2 for arbitrary artin algebras. Unlike the situation for Brauer-Thrall 1, Brauer-Thrall 2 does not hold for arbitrary artin rings, as was shown by Ringel [13].

Finally, before beginning our proof of Brauer-Thrall $1\frac{1}{2}$, we would like to say that Smalø has recently given a short elegant proof (unpublished) of the following result. If Λ is an artin algebra, then for each $i \geq 0$ there is a $j > i$ such that $|\operatorname{ind}_i \Lambda| \leq |\operatorname{ind}_j \Lambda|$, which is a generalization to artin algebras of earlier results of Nazarova-Roiter and Ringel. This clearly implies Brauer-Thrall $1\frac{1}{2}$. In view of all these developments, our proof of Brauer-Thrall $1\frac{1}{2}$ is being given mainly to illustrate some ideas and techniques which are of interest in their own right and have proven useful in other contexts. It would also be interesting to know if Brauer-Thrall $1\frac{1}{2}$ holds for arbitrary artin rings.

For the convenience of the reader, we start by recalling some basic facts concerning the radical and socle of finitely presented functors from $\operatorname{mod}\Lambda$ to abelian groups. The reader is referred to [10] for details. We recall that a functor $F\colon \operatorname{mod}\Lambda \to \mathrm{Ab}$ is said to be finitely presented if there is an exact sequence $(A,\ ) \to (B,\ ) \to F \to 0$ with A and B in $\operatorname{mod}\Lambda$. We denote by $\mathrm{f.p.}(\operatorname{mod}\Lambda, \mathrm{Ab})$ the full subcategory of $(\operatorname{mod}\Lambda, \mathrm{Ab})$, the category of all additive functors from $\operatorname{mod}\Lambda$ to abelian groups, whose objects are the finitely presented functors. Finally the *radical* of an arbitrary functor $F\colon \operatorname{mod}\Lambda \to \mathrm{Ab}$, which we denote by $\underline{r}F$, is the intersection of the maximal subfunctors of F.

PROPOSITION 5.1. *Let F be in* $\mathrm{f.p.}(\operatorname{mod}\Lambda, \mathrm{Ab})$. *Then $\underline{r}F$ has the following properties:*

a) *$\underline{r}F$ is a finitely presented functor.*
b) *$\underline{r}F = F$ if and only if $F = 0$.*
c) *$F/\underline{r}F$ is a semisimple functor of finite length.*
d) *If F' is a subfunctor of F, then F/F' is semisimple if and only if $F' \supset \underline{r}F$.*
e) *If $f\colon F \to G$ is a morphism of finitely presented functors, then $f(\underline{r}F) \subset \underline{r}G$ and $f(\underline{r}F) = \underline{r}G$ if f is an epimorphism.*

PROOF. See [10, Section 3]. □

We define $\underline{r}^iF$ for all nonnegative integers i by induction on i. Define $\underline{r}^0F = F$ and $\underline{r}^{i+1}F = \underline{r}(\underline{r}^iF)$. Also we define $\underline{r}^\infty F$, the *infinite radical of* F, by $\underline{r}^\infty F = \cap_{i=0}^\infty \underline{r}^iF$. We now list some of the basic properties of the descending chain of subfunctors

$$F \supset \underline{r}F \supset \underline{r}^2F \supset \cdots \supset \underline{r}^iF \supset \cdots$$

in the case F is finitely presented. In this connection we recall that the *support* of a functor F, $\operatorname{Supp} F$, is the full subcategory of $\operatorname{ind}\Lambda$ consisting of all A in $\operatorname{ind}\Lambda$ such that $F(A) \neq 0$. Also if F is finitely generated, then $|\operatorname{Supp} F| < \aleph_0$ if and only if $L(F) < \infty$, where $L(F)$ is the length of F (see [3] and [4]).

PROPOSITION 5.2. *Suppose F is in* $\mathrm{f.p.}(\operatorname{mod}\Lambda, \mathrm{Ab})$.

a) *For each nonnegative integer i, we have that $\underline{r}^iF$ is in* $\mathrm{f.p.}(\operatorname{mod}\Lambda, \mathrm{Ab})$ *and $\underline{r}^iF/\underline{r}^{i+1}F$ is semisimple of finite length.*
b) *If $F' \subset F$, then $L(F/F') < \infty$ if and only if $F' \subset \underline{r}^iF$ for some $i < \infty$.*
c) *The following statements are equivalent for F.*
    i) *$L(F) < \infty$.*
    ii) *$\underline{r}^iF = 0$ for some $i < \infty$.*
    iii) *$L(F/\underline{r}^\infty F) < \infty$.*
d) *For A in $\operatorname{ind}\Lambda$, there is an integer i such that $((A,\ ), \underline{r}^iF) = ((A,\ ), \underline{r}^\infty F)$.*
e) *$|\operatorname{Supp}(F/\underline{r}^\infty F)| \leq \aleph_0$ and $|\operatorname{Supp}(F/\underline{r}^\infty F)| < \infty$ if and only if $L(F) < \infty$.*
f) *If $f\colon F \to G$ is a morphism of finitely presented functors, then $f(\underline{r}^\infty F) \subset \underline{r}^\infty G$ and so f induces a morphism $F/(\underline{r}^\infty F) \to G/(\underline{r}^\infty G)$.*

PROOF. a), b), and c) follows easily from Proposition 5.1.

d) Since F is finitely presented we know that $F(A)$ is a finitely generated $\operatorname{End} A$-module and is therefore an artin $\operatorname{End} A$-module. Hence the descending chain

$$F(A) \supset \underline{r}F(A) \supset \cdots \supset \underline{r}^tF(A) \supset \cdots$$

has the property that there is an i such that $\underline{r}^iF(A) = \underline{r}^jF(A)$ for all $j \geq i$. Thus $\underline{r}^iF(A) = \underline{r}^\infty F(A)$ or equivalently $((A,\ ), \underline{r}^iF) = ((A,\ ), \underline{r}^\infty F)$.

e) From d) it follows that $\operatorname{Supp}(F/\underline{r}^{\infty}F) = \cup_{i=0}\operatorname{Supp}(\underline{r}^iF/\underline{r}^{i+1}F)$. Since each $\underline{r}^iF/\underline{r}^{i+1}F$ is of finite length we have that $|\operatorname{Supp}(\underline{r}^iF/\underline{r}^{i+1}F)| < \aleph_0$ and so $|\operatorname{Supp}(F/\underline{r}^{\infty}F)| \leq \aleph_0$.

Since $F/\underline{r}^{\infty}F$ is a finitely generated functor, we know that $L(F/\underline{r}^{\infty}F) < \infty$ if and only if $|\operatorname{Supp}(F/\underline{r}^{\infty}F)| < \infty$.

f) Follows from Proposition 5.1 □

For a functor $F\colon \operatorname{mod}\Lambda \to \operatorname{Ab}$ define $\operatorname{Soc} F$, the *socle* of F, to be the subfunctor of F generated by the simple subfunctors of F. We now recall some basic facts concerning the socle of finitely presented functors, which are essentially the "duals" of those given for the radical of finitely presented functors (see [10, Section 3]).

PROPOSITION 5.3. *Let F be in* f. p.(mod Λ, Ab).

a) $\operatorname{Soc} F$ *is semisimple of finite length so that* $F/\operatorname{Soc} F$ *is in* f. p.(mod Λ, Ab).
b) F *is an essential extension of* $\operatorname{Soc} F$ *so that* $F = 0$ *if and only if* $\operatorname{Soc} F = 0$.
c) *A subfunctor F' of F is semisimple if and only if* $F' \subset \operatorname{Soc} F$.

Let F be a finitely presented functor. Define $\operatorname{Soc}^i F$ for all nonnegative integers i by induction as follows: $\operatorname{Soc}^0 F = F$ and $\operatorname{Soc}^{i+1} F$ is the kernel of the natural epimorphism $F \to F/\operatorname{Soc}^i F/\operatorname{Soc}(F/\operatorname{Soc}^i F)$. Also define $\operatorname{Soc}^{\infty} F$, the *infinite socle* of F, by $\operatorname{Soc}^{\infty} F = \cup_{i=0}^{\infty}\operatorname{Soc}^i F$. We now list some of the basic properties of the ascending chain of subfunctors

$$\operatorname{Soc} F \subset \operatorname{Soc}^2 F \subset \cdots \subset \operatorname{Soc}^i F \subset \cdots \qquad (i < \infty).$$

Since they are essentially dual to those given for the radical series in Proposition 5.2, we omit proofs.

PROPOSITION 5.4. *Let F be in* f. p.(mod Λ, Ab).

a) *For all nonnegative i we have that* $\operatorname{Soc}^i F$ *is finitely presented and* $\operatorname{Soc}^{i+1} F/\operatorname{Soc}^i F$ *is semisimple of finite length.*
b) *If $F' \subset F$, then $L(F') < \infty$ if and only if $F' \subset \operatorname{Soc}^i F$ for some $i < \infty$.*
c) *The following are equivalent:*
    i) $L(F) < \infty$.
    ii) $\operatorname{Soc}^i F = F$ *for some* i.
    iii) $\operatorname{Soc}^{\infty} F = F$.
d) *If A is in* ind Λ, *then there is an integer i such that* $((A,\ ), \operatorname{Soc}^i F) = ((A,\ ), \operatorname{Soc}^{\infty} F)$.
e) $|\operatorname{Supp}\operatorname{Soc}^{\infty} F| \leq \aleph_0$ *and* $|\operatorname{Supp}\operatorname{Soc}^{\infty} F| < \aleph_0$ *if and only if* $L(F) < \infty$.
f) *If $f\colon F \to G$ is a morphism of finitely presented functors, then* $f(\operatorname{Soc}^{\infty} F) \subset \operatorname{Soc}^{\infty} G$ *and so f induces a morphism* $F/\operatorname{Soc}^{\infty} F \to G/\operatorname{Soc}^{\infty} G$.

As a consequence of the above we have:

PROPOSITION 5.5. *Let $\mathcal{C}$ be the full subcategory of* ind Λ *consisting of those A in* ind Λ *such that $L(A,\ ) < \infty$. Then* $\mathcal{C} \subset \operatorname{Supp}\operatorname{Soc}^{\infty}(\Lambda,\ )$ *and so* $|\mathcal{C}| \leq \aleph_0$.

PROOF. Let $\Lambda^n \to A \to 0$ be surjective. Then $0 \to (A,\ ) \to (\Lambda^n,\ ) = \coprod(\Lambda,\ )$. If $L((A,\ )) < \infty$, then $(A,\ ) \subset \operatorname{Soc}^{\infty}(\coprod_n(\Lambda,\ )) = \coprod_n(\operatorname{Soc}^{\infty}(\Lambda,\ ))$. So $\operatorname{Soc}^{\infty}(\Lambda,\ )(A) \neq 0$. Therefore it follows that $\mathcal{C} \subset \operatorname{Supp}(\operatorname{Soc}^{\infty}(\Lambda,\ ))$ and so $|\mathcal{C}| \leq \aleph_0$. □

Before proceeding to proving the main results of this section, we need the following very useful fact.

LEMMA 5.6. *Suppose A and B are in* ind Λ *and $f\colon A \to B$ is not an isomorphism. Then $f = \sum_{i=1}^n f_i + h$ where*

a) *each* f_i *is a finite nonzero composition of irreducible morphisms* $A = A_0^{(i)} \to A_2^{(i)} \to \cdots \to A_{n_i}^{(i)} = B$ *with the* $A_j^{(i)}$ *indecomposable modules.*

b) h *is in* $\underline{r}^\infty(A,\ )(B)$.

As a consequence of this lemma we have the following.

PROPOSITION 5.7. *Let* A *and* B *be in* $\operatorname{ind}\Lambda$.

a) *Suppose we are given a morphism* $t\colon (B,\ )/\underline{r}^\infty(B,\ ) \to (A,\ )/\underline{r}^\infty(A,\ )$ *which is not an isomorphism. Then there is an* $f\colon A \to B$ *such that* $f = \Sigma f_i$ *where each* f_i *is a finite composition of irreducible morphisms between indecomposable modules such that the diagram*

$$\begin{array}{ccc} (B,\ ) & \xrightarrow{(f,\ )} & (A,\ ) \\ \downarrow & & \downarrow \\ (B,\ )/\underline{r}^\infty(B,\ ) & \xrightarrow{t} & (A,\ )/\underline{r}^\infty(A,\ ) \end{array}$$

commutes.

b) *If* $((B,\ )/\underline{r}^\infty(B,\ ),(A,\ )/\underline{r}^\infty(A,\ )) \neq 0$ *and* $A \not\approx B$ *then there must be a chain of irreducible morphisms* $A = A_0 \xrightarrow{f_0} A_1 \to \cdots \xrightarrow{f_{n-1}} A_n = B$ *with the* A_i *indecomposable such that* $f_{n-1}\cdots f_0 \neq 0$.

c) *If* $\mathcal{U}(B)$ *is the full subcategory of* $\operatorname{ind}\Lambda$ *consisting of those* A *in* $\operatorname{ind}\Lambda$ *such that* $((B,\ )/\underline{r}^\infty(B,\ ),(A,\ )/\underline{r}^\infty(A,\ )) \neq 0$ *then* $|\mathcal{U}(B)| \leq \aleph_0$.

PROOF. a) Since $(B,\ )$ is projective, there is a morphism $f'\colon A \to B$ such that the diagram

$$\begin{array}{ccc} (B,\ ) & \xrightarrow{(f',\ )} & (A,\ ) \\ \downarrow & & \downarrow \\ (B,\ )/\underline{r}^\infty(B,\ ) & \xrightarrow{t} & (A,\ )/\underline{r}^\infty(A,\ ) \end{array}$$

commutes. Furthermore $f'\colon A \to B$ is not an isomorphism since t is not. Thus by Lemma 5.6 we know that $f' = f + h$ where h is in $\underline{r}^\infty(A,\ )(B)$ and $f = \sum_{i=1}^n f_i$ where each f_i is a finite composition of irreducible morphisms between indecomposable modules. Since h is in $\underline{r}^\infty(A,\ )(B)$, the morphism $h\colon A \to B$ has the property that $(h,\ )((B,\ )) \subset \underline{r}^\infty(A,\ )$. Therefore

$$\begin{array}{ccc} (B,\ ) & \xrightarrow{(f,\ )} & (A,\ ) \\ \downarrow & & \downarrow \\ (B,\ )/\operatorname{rad}^\infty(B,\ ) & \xrightarrow{t} & (A,\ )/\operatorname{rad}^\infty(A,\ ) \end{array}$$

commutes, which completes the proof of a).

b) Easy consequence of a).

c) Given any indecomposable module X we know (see Proposition 4.3) that there is only a finite number of nonisomorphic X_i in $\operatorname{ind}\Lambda$ such that there is an irreducible morphism $X_i \to X$. Thus for each integer there are only a finite number of nonisomorphic X_i in $\operatorname{ind}\Lambda$ such that there is a chain of irreducible morphisms $X_i = Y_0 \xrightarrow{f_0} Y_1 \to \cdots \xrightarrow{f_{n-1}} Y_n = B$ with the Y_i in $\operatorname{ind}\Lambda$. Hence there are at most a denumerable number of nonisomorphic A in $\operatorname{ind}\Lambda$ such that there is a finite chain of irreducible morphisms $A = A_0 \xrightarrow{f_0} A_1 \to \cdots \xrightarrow{f_{n-1}} A_n = B$ with each A_i in $\operatorname{ind}\Lambda$. Since by part b), for each A in $\mathcal{U}(B)$ with $A \neq B$, there is a finite chain of irreducible morphisms in $\operatorname{ind}\Lambda$ $A = A_0 \xrightarrow{f_0} A_1 \xrightarrow{f_1} \cdots \xrightarrow{f_{n-1}} A_n = B$ with the A_i in $\operatorname{ind}\Lambda$, we have that $|\mathcal{U}(B)| \leq \aleph_0$. $\square$

For each integer $i \geq 0$ let $\operatorname{ind}_i \Lambda$ be the full subcategory of $\operatorname{ind}\Lambda$ consisting of those A in $\operatorname{ind}\Lambda$ such that $L(A) = i$. For each n let $\mathcal{C}_n = \cup_{i \geq n} \operatorname{ind}_i \Lambda$. Also for each integer n let $\mathcal{A}_n$ be the full subcategory of $\operatorname{ind}\Lambda$ consisting of all A in $\operatorname{ind}\Lambda$ such that $((B,\ )/\underline{r}^\infty(B,\ ), (A,\ )/\underline{r}^\infty(A,\ )) = 0$ for all B in $\mathcal{C}_n$.

PROPOSITION 5.8. *Let A be in $\operatorname{ind}\Lambda$, then A is in $\mathcal{A}_n$ for some n if and only if $L(A,\ ) < \infty$.*

PROOF. If $L(A,\ ) < \infty$, then $|\operatorname{Supp}(A,\ )| < \aleph_0$. Hence there is some integer n such that $(A, B) = 0$ for all B in $\mathcal{C}_n$. Thus $((B,\ )/\underline{r}^\infty(B,\ ), (A,\ )) = 0$ for all B in $\mathcal{C}_n$. But $(A,\ ) = (A,\ )/\underline{r}^\infty(A,\ )$ since $L(A,\ ) < \infty$. Hence we have that A is in $\mathcal{A}_n$.

Suppose A is in $\mathcal{A}_n$. We want to show that $L(A,\ ) < \infty$. We do this by showing that $L((A,\ )/\underline{r}^\infty(A,\ )) < \infty$ and so $L(A,\ ) < \infty$.

We first observe that the functor $F = (A,\ )/\underline{r}^\infty(A,\ )$ has the following properties.

a) $F(X) \neq 0$ for some X in $\mathcal{A}_n$ (namely $X = A$).

b) If $X \xrightarrow{f} Y$ is a morphism such that $f \in \underline{r}^\infty(Y,\ )(X)$, then $F(f) = 0$.

a) is obvious.

b) Suppose $f\colon X \to Y$ is a morphism in $\underline{r}^\infty(X,\ )(Y)$. Then the composition $(Y,\ ) \xrightarrow{(f,\ )} (X,\ ) \to (X,\ )/\underline{r}^\infty(X,\ )$ is zero. Thus the composition

$$\begin{array}{ccc} ((X,\ )/\underline{r}^\infty(X,\ ), (A,\ )/\underline{r}^\infty(A,\ )) & \xrightarrow{\sim} & ((X,\ ), (A,\ )/\underline{r}^\infty(A,\ )) \\ & & \downarrow \\ & & ((Y,\ ), (A,\ )/\underline{r}^\infty(A,\ )) \end{array}$$

is zero. Hence the vertical morphism is the zero morphism. But we have the commutative diagram

$$\begin{array}{ccc} ((X,\ ), (A,\ )/\underline{r}^\infty(A,\ )) & \xrightarrow{(f,\ )} & ((Y,\ ), (A,\ )/\underline{r}^\infty(A,\ )) \\ \downarrow \wr & & \downarrow \wr \\ (A,\ )/\underline{r}^\infty(A,\ )(X) & \xrightarrow{(A,\ )/\underline{r}^\infty(A,\ )(f)} & (A,\ )/\underline{r}^\infty(A,\ )(Y) \end{array}$$

so $(A,\ )/\underline{r}^\infty(A,\ )(f) = 0$.

We now show that if $F\colon \operatorname{mod}\Lambda \to \mathrm{Ab}$ satisfies a) and b), then F must have a simple subfunctor. Suppose F has no simple subfunctors. Let X be in $\mathcal{A}_n$ such that $F(X) \neq 0$. Then we know that there is a chain of morphisms

$$X \xrightarrow{f} X_0 \xrightarrow{f_0} X_1 \to \cdots \xrightarrow{f_{n-1}} X_n \to \cdots$$

between indecomposable modules with the f_i proper monomorphisms and f an epimorphism such that $F(f_j \cdots f_0) \neq 0$ for each j (see [3, Theorem 1.5]). Since the $L(X_i)$ get arbitrarily large, $X_i \in \mathcal{C}_n$ for some i. From the fact that $F(f_{i-1} \cdots f_0) \neq 0$ we know that $X \xrightarrow{f_{i-1}\cdots f_0 f} X_i$ is not in $\underline{r}^\infty(X,\ )(X_i)$. Hence $((X_i,\ )/\underline{r}^\infty(X_i,\ ), (X,\ )/\underline{r}^\infty(X,\ ))$ is not zero. This contradicts the fact that $X_i \in \mathcal{C}_n$. Thus F must have a simple subfunctor.

Returning to $(A,\ )/\underline{r}^{\infty}(A,\ )$ we know that every proper factor of $(A,\ )/\underline{r}^{\infty}(A,\ )$ has a simple subfunctor since it satisfies a) and b) above. Thus $(A,\ )/\underline{r}^{\infty}(A,\ )$ is locally finite. Since it is finitely generated, it must be of finite length. Therefore $L(A,\ ) < \infty$ which is our desired result. □

As an easy consequence of Proposition 5.7 we have the following.

COROLLARY 5.9. *Suppose A is in* $\operatorname{ind}\Lambda$ *and* $L(A,\ ) = \infty$. *Then given any integer* $n \geq 0$ *there is a nonzero composition of irreducible morphisms* $A \xrightarrow{f_0} A_1 \to \cdots \xrightarrow{f_{t-1}} A_t$ *with the* $A_1, \dots, A_t$ *indecomposable modules such that* $L(A_t) \geq n$.

PROOF. Suppose we are given $n \geq 0$ and A in $\operatorname{ind}\Lambda$ such that $L(A,\ ) = \infty$. Since $L(A) = \infty$ we know by Proposition 5.8 that A is not in $\mathcal{A}_k$ where $k > \max(n, L(A))$. Hence we know there is a B in $\mathcal{C}_k$ such that $((B,\ )/\underline{r}^{\infty}(B,\ ), (A,\ )/\underline{r}^{\infty}(A,\ )) \neq 0$. Since $B \not\approx A$, it follows from Proposition 5.7 that there is a nonzero composition of irreducible morphisms $A \to A_1 \to \cdots \to A_t = B$ with the $A_1, \dots, A_t$ indecomposable modules. □

As another application of Proposition 5.8 we prove Brauer-Thrall $1\frac{1}{2}$.

THEOREM 5.10. *Suppose Λ is not of finite representation type. Then* $|\mathcal{C}_n| = |\operatorname{ind}\Lambda|$ *for all n.*

PROOF. From Brauer-Thrall 1 we know that $\aleph_0 \leq |\mathcal{C}_n|$. So we have $\aleph_0 \leq |\mathcal{C}_n| \leq |\operatorname{ind}\Lambda| \leq \sum_{B\in\mathcal{C}_n} |\mathcal{U}(B)| + |\mathcal{A}_n| = \sum_{B\in\mathcal{C}_n} |\mathcal{U}(B)| + \aleph_0 \leq \aleph_0 \cdot |\mathcal{C}_n| + \aleph_0$ since $|\mathcal{U}(B)| \leq \aleph_0$. From the fact that $|\mathcal{C}_n| \geq \aleph_0$ we have that $\aleph_0 \cdot |\mathcal{C}_n| + \aleph_0 = |\mathcal{C}_n|$ and so $|\operatorname{ind}\Lambda| = |\mathcal{C}_n|$. □

As an easy consequence of Theorem 5.10 we have

COROLLARY 5.11. *Suppose* $|\operatorname{ind}\Lambda| \geq \aleph_0$. *Then there are an infinite number of n such that* $|\operatorname{ind}_n \Lambda| > \aleph_0$.

6. A partition of ind Λ

As usual we assume that Λ is an artin R-algebra. Our purpose in this section is to introduce a certain equivalence relation in the category $\operatorname{ind}\Lambda$ and point out some of its properties.

In Section 5 we dealt only with covariant additive functors from $\operatorname{mod}\Lambda$ to abelian groups. Since the duality $D\colon \operatorname{mod}\Lambda \to \operatorname{mod}\Lambda^{\mathrm{op}}$ gives an equivalence $(\operatorname{mod}\Lambda)^{\mathrm{op}} \to \operatorname{mod}\Lambda^{\mathrm{op}}$, the category of contravariant additive functors from $\operatorname{mod}\Lambda$ to abelian groups $((\operatorname{mod}\Lambda)^{\mathrm{op}}, \mathrm{Ab})$ is equivalent to $(\operatorname{mod}\Lambda^{\mathrm{op}}, \mathrm{Ab})$ which shows that the results in Section 5 for covariant functors have exact analogues for contravariant functors. We will use these analogues freely, without proof.

Our immediate concern is to show that the relations R_1, R_2, and R_3 which we are about to define on $\operatorname{ind}\Lambda$ generate the same equivalence on $\operatorname{ind}\Lambda$.

Definition of R_1. Given A and B in $\operatorname{ind}\Lambda$, we say that AR_1B if one of the following holds: i) $A \approx B$; ii) there is an irreducible morphism $A \to B$; iii) there is an irreducible morphism $B \to A$.

Definition of R_2. Given A and B in $\operatorname{ind}_P \Lambda$ we say that AR_2B if and only if either

$$((\ ,A)/\underline{r}^{\infty}(\ ,A), (\ ,B)/\underline{r}^{\infty}(\ ,B)) \neq 0 \text{ or } ((\ ,B)/\underline{r}^{\infty}(\ ,B)), (\ ,A)/\underline{r}^{\infty}(\ ,A)) \neq 0.$$

Definition of R_3. Given A and B in $\operatorname{ind}\Lambda$ we say that AR_3B if either

$$((A,\ )/\underline{r}^{\infty}(A,\ ), (B,\ )/\underline{r}^{\infty}(B,\ )) \neq 0 \text{ or } ((B,\ )/\underline{r}^{\infty}(B,\ ), (A,\ )/\underline{r}^{\infty}(A,\ )) \neq 0.$$

Since all three of these relations obviously have the properties AR_iA, and AR_iB implies BR_iA for all A and B in $\operatorname{ind}\Lambda$, they each generate an equivalence on $\operatorname{ind}\Lambda$.

PROPOSITION 6.1. *The relations R_1, R_2, and R_3 generate the same equivalence relation on* $\operatorname{ind}\Lambda$.

PROOF. We show that R_1 and R_2 generate the same equivalence. A similar proof shows that R_1 and R_3 generate the same equivalence relation.

First we observe that if AR_1B, then AR_2B. Obviously, if $A \approx B$, then AR_2B. Suppose there is an irreducible morphism $f\colon A \to B$. Then $(\ ,f)\colon (\ ,A) \to (\ ,B)$ has the property that the composition $(\ ,A) \xrightarrow{(\ ,f)} (\ ,B) \to (\ ,B)/\underline{r}^2(\ ,B)$ is not zero. Since $\underline{r}^\infty(\ ,B) \subset \underline{r}^2(\ ,B)$, we have that the composition $(\ ,A) \xrightarrow{(\ ,f)} (\ ,B) \to (\ ,B)/\underline{r}^\infty(\ ,B)$ is not zero. Since this composition has $\underline{r}^\infty(\ ,A)$ in its kernel, it induces a nonzero morphism $(\ ,A)/\underline{r}^\infty(\ ,A) \to (\ ,B)/\underline{r}^\infty(\ ,B)$. Thus AR_1B implies AR_2B.

Suppose AR_2B, i.e. $((\ ,A)/\underline{r}^\infty(\ ,A), (\ ,B)/\underline{r}^\infty(\ ,B)) \neq 0$. If $A \approx B$, then AR_1B and we are done. Suppose $A \not\approx B$. Then by Proposition 5.7, we know there is a nonzero morphism $f\colon A \to B$ which is a finite composition of irreducible morphisms $A = A_0 \xrightarrow{f_0} A_1 \to \cdots \xrightarrow{f_{n-1}} A_n = B$ with the A_i indecomposable modules. Thus $AR_1A_1, A_2R_1A_3, \dots, A_{n-1}R_1B$ and so A and B are in the same equivalence class of the equivalence relation generated by R_1, if AR_2B. Therefore R_1 and R_2 generate the same equivalence relation on $\operatorname{ind}\Lambda$. □

We denote the common equivalence relation on $\operatorname{ind}\Lambda$ generated by R_1, R_2, and R_3 by R. For each A in $\operatorname{ind}\Lambda$ we denote by $[A]$ the full subcategory of $\operatorname{ind}\Lambda$ whose objects are the equivalence classes of A under the equivalence relation R. We denote by $\{\mathcal{E}_i\}_{i\in I}$ the family of full subcategories of $\operatorname{ind}\Lambda$ where the objects of each $\mathcal{E}_i$ consists of one equivalence class under R.

PROPOSITION 6.2. a) $|\mathcal{E}_i| \leq \aleph_0$ *for all i in I.*

b) *If $A \in \mathcal{C}_i$ and $B \in \mathcal{C}_j$ with $i \neq j$, then*

$$((\ ,A)/\underline{r}^\infty(\ ,A), (\ ,B)/\underline{r}^\infty(\ ,B)) = 0 = ((\ ,B)/\underline{r}^\infty(\ ,B), (\ ,A)/\underline{r}^\infty(\ ,A))$$

and

$$((A,\ )/\underline{r}^\infty(A,\ ), (B,\ )/\underline{r}^\infty(B,\ )) = 0 = ((B,\ )/\underline{r}^\infty(B,\ ), (A,\ )/\underline{r}^\infty(A,\ )).$$

c) *For each $\mathcal{E}_i$ denote by $D(\mathcal{E}_i)$ the full subcategory of* $\operatorname{mod}\Lambda^{\mathrm{op}}$ *whose objects are of the form $D(X)$ with X in $\mathcal{E}_i$. Then the $D(\mathcal{E}_i)$ are the full subcategories of* $\operatorname{mod}\Lambda^{\mathrm{op}}$ *whose objects constitute an equivalence class for the relation R on* $\operatorname{mod}\Lambda^{\mathrm{op}}$.

PROOF. a) Follows immediately from the fact that for each A in $\operatorname{ind}\Lambda$, there are only a finite number of nonisomorphic indecomposable modules which have irreducible morphisms to A or from A.

b) Just a restatement of the definition of R.

c) This is an easy consequence of the fact that a morphism $f\colon A \to B$ in $\operatorname{ind}\Lambda$ is irreducible if and only if $D(f)\colon D(B) \to D(A)$ in $\operatorname{ind}\Lambda^{\mathrm{op}}$ is irreducible. □

As our first result concerning the partition $\{\mathcal{E}_i\}_{i\in I}$ of $\operatorname{mod}\Lambda$, we have the following.

PROPOSITION 6.3. *The following statements are equivalent for an $\mathcal{E}_i$.*

a) $|\mathcal{E}_i| < \aleph_0$.

b) *There is an integer n such that $L(A) < n$ for all A in $\mathcal{E}_i$.*

c) *$L(A,\ ) < \infty$ and $L(\ ,A) < \infty$ for all A in $\mathcal{E}_i$.*

PROOF. a) implies b). Obvious.

b) implies c). Suppose $L(A) \leq n$ for all A in $\mathcal{E}_i$ and let A be in $\mathcal{E}_i$. If $B \in \mathcal{C}_{n+1}$, i.e. $L(B) \geq n+1$, then B is not in $\mathcal{E}_i$. Hence by Proposition 6.2 we have that $((B,\ )/\underline{r}^\infty(B,\ ), (A,\ )/\underline{r}^\infty(A,\ )) = 0$. Since this is true for all B in $\mathcal{C}_{n+1}$, we have by Proposition 5.8 that $L(A,\ ) < \infty$. Applying the same argument to $D(\mathcal{E}_i)$, we deduce that $L(DA,\ ) < \infty$ or equivalently $L(\ , A) < \infty$. Thus b) implies c).

c) implies a). Suppose A is in $\mathcal{E}_i$ and let $f\colon A \to S$ be an epimorphism with S simple. Since $L(A,\ ) < \infty$, we have that $\underline{r}^\infty(A,\ ) = 0$. Hence by Lemma 5.6 we have that $f = \sum_{i=1}^{j} f_i$ with the f_i a finite composition of irreducible morphisms between indecomposable modules. Thus S is in $\mathcal{E}_i$.

Let $S_1, \dots, S_t$ be the simple modules in $\mathcal{E}_i$. Then we have that $L(\ , S_i) < \infty$ for each i. Thus $|\operatorname{Supp}(\ , \coprod_{i=1}^t S_i)| < \infty$. But our previous argument shows that $\mathcal{E}_i \subseteq \operatorname{Supp}(\ , \coprod_{i=1}^t S_i)$ and so $|\mathcal{E}_i| < \aleph_0$. □

As a consequence of this result, we have the following.

PROPOSITION 6.4. *Suppose $|\mathcal{E}_i| < \aleph_0$. Let $\widetilde{\mathcal{E}_i}$ be the additive category generated by $\mathcal{E}_i$ and let $\mathcal{J}$ be the additive category generated by $\bigcup_{j\neq i} \mathcal{E}_j$.*

a) *The category* $\operatorname{mod}\Lambda$ *is isomorphic to the product of categories $\widetilde{\mathcal{E}_i} \times \mathcal{J}$.*

b) *Λ is isomorphic to a product $\Lambda_1 \times \Lambda_2$ of artin R-algebras such that*

  i) *Λ_1 is an indecomposable R-algebra of finite type with* $\operatorname{mod}\Lambda_1$ *equivalent to $\widetilde{\mathcal{E}_i}$ and*

  ii) $\operatorname{mod}\Lambda_2$ *is equivalent to $\mathcal{J}$.*

PROOF. a) We first show that if A is in $\widetilde{\mathcal{E}_i}$ and B is in $\mathcal{J}$, then $(A, B) = 0 = (B, A)$. It clearly suffices to do this in the case A and B are indecomposable. Since $L(A,\ ) < \infty$ we have that $(A,\ )/\underline{r}^\infty(A,\ ) = (A,\ )$. Hence

$$(A, B) = ((B,\ ), (A,\ )) = ((B,\ ), (A,\ )/\underline{r}^\infty(A,\ )) = ((B,\ )/\underline{r}^\infty(B,\ ), (A,\ )/\underline{r}^\infty(A,\ )).$$

But $((B,\ )/\underline{r}^\infty(B,\ ), (A,\ )/\underline{r}^\infty(A,\ )) = 0$ since B is not in $\mathcal{E}_i$. Thus $(A, B) = 0$. The fact that $(B, A) = 0$ can be established similarly. Combining this with the fact that every object of $\operatorname{mod}\Lambda$ is isomorphic to a sum $A \amalg B$ with A in $\widetilde{\mathcal{E}_i}$ and B in $\mathcal{J}$ we obtain our desired result, that $\operatorname{mod}\Lambda$ is equivalent to $\widetilde{\mathcal{E}_i} \times \mathcal{J}$.

b) Corresponding to the decomposition $\widetilde{\mathcal{E}_i} \times \mathcal{J}$ of $\operatorname{mod}\Lambda$ is a decomposition of Λ into a product $\Lambda_1 \times \Lambda_2$ with $\operatorname{mod}\Lambda_1$ equivalent to $\widetilde{\mathcal{E}_i}$ and $\operatorname{mod}\Lambda_2$ is equivalent to $\operatorname{mod}\Lambda_2$. Since $|\mathcal{E}_i| < \aleph_0$, it follows that $\operatorname{mod}\Lambda_1$ is of finite type. Finally, it follows easily from the definition of $\mathcal{E}_i$ that $\mathcal{E}_i$, and hence $\widetilde{\mathcal{E}_i}$, is an indecomposable category, which means that Λ_1 is an indecomposable R-algebra. □

As an easily verified consequence of this result we have the following.

THEOREM 6.5. *Let $\{\mathcal{E}_i\}_{i\in I}$ be the decomposition of* $\operatorname{ind}\Lambda$ *given by the equivalence relation R on* $\operatorname{ind}\Lambda$.

a) *K, the set of all i such that $|\mathcal{E}_i| < \aleph_0$, is finite.*

b) *For each i in K let $\widetilde{\mathcal{E}_i}$ be the additive category generated by $\mathcal{E}_i$ and let $\mathcal{J}$ be the additive category generated by the $\mathcal{E}_j$ with j not in K. Then*

  i) $\operatorname{mod}\Lambda$ *is equivalent to the product $\prod_{i\in K} \widetilde{\mathcal{E}_i} \times J$.*

  ii) *Λ is isomorphic to a product of artin R-algebras $\prod_{j\in K} \Lambda_j \times \Sigma$ with the properties:*

1) Λ_j *is an indecomposable artin algebra of finite type with* $\operatorname{mod}\Lambda_j$ *equivalent to* $\widetilde{\mathcal{E}}_j$ *for each* j *in* K*;*
2) $\operatorname{mod}\Sigma$ *is equivalent to* $\mathcal{J}$*;*
3) *If* $\Sigma_1\times\cdots\times\Sigma_t$ *is the decomposition of* Σ *into indecomposable algebras, then no* Σ_i *is of finite type.*

c) Λ *is of finite type if and only if* $|\mathcal{E}_i|<\aleph_0$ *for all* i *in* I.

d) *If* Λ *is an indecomposable algebra, then* Λ *is of finite type if and only if* $|\mathcal{E}_i|<\aleph_0$ *for some* i.

In [3] it was shown that an artin algebra Λ is not of finite representation type if and only if there are denumerably, nonfinitely generated indecomposable Λ-modules. We now show how the results of this section can be used to obtain somewhat stronger existence theorems for denumerably, nonfinitely generated indecomposable Λ-modules. We begin with the following preliminary result.

PROPOSITION 6.6. *Let* Λ *be an indecomposable artin* R*-algebra of infinite representation type. Suppose* $\mathcal{E}_i$ *has the property that* $L(A,\ )<\infty$ *for all* A *in* $\mathcal{E}_i$.

a) *If* $f\colon A\to B$ *is a nonzero morphism in* $\operatorname{ind}\Lambda$ *and* A *is in* $\mathcal{E}_i$, *then* B *is in* $\mathcal{E}_i$.

b) *If* $0\to A\to I_0\to I_1\to\cdots\to I_n\to\cdots$ *is a minimal injective resolution for* A *and* A *is in* $\mathcal{E}_i$*, then* I_j *is in* $\widetilde{\mathcal{E}}_i$ *for all* $j=0,1,\ldots$.

c) *If* $0\to A\xrightarrow{g}B\xrightarrow{f}C\to 0$ *is exact with* A *and* C *in* $\widetilde{\mathcal{E}}_i$, *then* B *is in* $\widetilde{\mathcal{E}}_i$.

PROOF. Let $f\colon A\to B$ be a nonzero morphism in $\operatorname{ind}\Lambda$ with A in $\mathcal{E}_i$. Since $\underline{r}^\infty(A,\ )=0$, we know that $f=\sum f_i$ where the f_i are finite compositions of irreducible morphisms between indecomposable modules. Then $f_i\neq 0$ for some i since $f\neq 0$. This implies that B is in $\mathcal{E}_i$.

b) Let $0\to A\xrightarrow{g}I_0$ be an injective envelope of A with A in $\mathcal{E}_i$. Let $p\colon I_0\to I'_j$ be a split epimorphism with I'_j indecomposable. Then the composition $pg\colon A\to I'_j$ is not zero, since $g\colon A\to I$ is essential. So I_j is in $\mathcal{E}_i$ by part a). Since this is true for all projections $p\colon I_0\to I'_j$, we have that I_j is in $\widetilde{\mathcal{E}}_i$. Using this fact, together with part a), it is easy to establish b).

c) Suppose $0\to A\xrightarrow{g}B\xrightarrow{f}C\to 0$ is exact with A and C in $\widetilde{\mathcal{E}}_i$. Let $p\colon B\to B'$ be a splittable epimorphism with B' indecomposable. If the composition $pg=0$, then $\operatorname{Im}g\subset\operatorname{Ker}p$ and so B' is isomorphic to a summand of C and is therefore in $\mathcal{E}_i$. If $pg\neq 0$, then $pg(A')\neq 0$ for some indecomposable summand of A. Since A' is in $\mathcal{E}_i$, it follows that B' is in $\mathcal{E}_i$. Therefore every indecomposable summand of B is in $\mathcal{E}_i$ and so B is in $\widetilde{\mathcal{E}}_i$. □

Since there are only a finite number of nonisomorphic injective modules in $\operatorname{ind}\Lambda$, we have the following as an immediate consequence of Proposition 6.6.

COROLLARY 6.7. *Let* Λ *be an indecomposable artin algebra which is not of finite representation type. Then there are at most a finite number of* i *such that* $L(A,\ )<\infty$ *for all* A *in* $\mathcal{E}_i$.

PROPOSITION 6.8. *Let* Λ *be an indecomposable artin algebra of infinite representation type. Let* $\{\mathcal{E}_i\}_{i\in I}$ *be the usual decomposition of* $\operatorname{mod}\Lambda$. *Let* K *be the finite subset of* I *consisting of all* i *such that* $L(A,\ )<\infty$ *for all* A *in* $\mathcal{E}_i$.

a) $I-K$ *is not empty.*

b) *For each* j *in* $I-K$ *there exists a denumerably generated (not finitely generated) indecomposable* Λ*-module* X_j *with the following properties:*

i) *For each j, there is a proper ascending chain of submodules of X_j*

$$A_0 \subset A_1 \subset \cdots \subset A_n \subset \cdots$$

with the A_k in $\mathcal{E}_j$ such that $\cup_{k=0}^{\infty} A_k = X_j$.

ii) *X_j and $X_{j'}$ are not isomorphic if $j \neq j'$.*

PROOF. a) If $K = I$, then $L(A, \ ) < \infty$ for all A in $\operatorname{ind}\Lambda$. But this means that Λ is of finite representation type (see [4, Theorem 3.1]), which is a contradiction. Hence $K \neq I$.

b) Let j be in $I - K$. Then there is an A in $\mathcal{E}_j$ such that $L(A, \ )$ is not finite. Hence $L((A, \ )/\underline{r}^{\infty}(A, \ ))$ is also not finite (see Proposition 5.2). Therefore if we let $F = G/\operatorname{l.f.}G$, where $G = (A, \ )/\underline{r}^{\infty}(A, \ )$ and $\operatorname{l.f.}G$ is the maximal locally finite subfunctor of G, we have that $F \neq 0$ and F has no simple subfunctors (see [4, Section 1]). Therefore we know by [3, Theorem 1.5] that there is an infinite sequence of morphisms $A \xrightarrow{f} A_0 \xrightarrow{f_0} A_1 \to \cdots \xrightarrow{f_{n-1}} A_n \xrightarrow{f_n} \cdots$ having the following properties:

i) For each $i = 0, 1, \ldots$ there are nonzero x_i in $F(A_i)$ such that $F(f_i)(x_i) = x_{i+1}$.

ii) Each f_i is a proper monomorphism and each A_i is indecomposable.

iii) $\varinjlim A_i = X_j$ is indecomposable but not finitely generated and $F(X_j) \neq 0$.

Since $F(A_i) \neq 0$ it follows that $(A, \ )/\underline{r}^{\infty}(A, \ )(A_i) \neq 0$, or equivalently $((A_i, \ )/\underline{r}^{\infty}(A_i, \ ), (A, \ )/\underline{r}^{\infty}(A, \ )) \neq 0$. Thus each A_i is in $\mathcal{E}_j$. Consequently the first part of b) is established.

Suppose $j \neq j'$ are in $I - K$. Then by the above construction there is an A in $\mathcal{E}_j$ such that $(A, \ )/\underline{r}^{\infty}(A, \ )(X_j) \neq 0$. Since $\mathcal{E}_j \neq \mathcal{E}_{j'}$, we know that $(A, \ )/\underline{r}^{\infty}(A, \ )(Y) = 0$ for all Y in $\mathcal{E}_{j'}$. Since $X_{j'} = \varinjlim A_i'$ with the A_i' in $\mathcal{E}_j$, it follows that $(A, \ )/\underline{r}^{\infty}(A, \ )(X_j) = 0$ since $(A, \ )/\underline{r}^{\infty}(A, \ )(X_{j'}) = \varinjlim (A, \ )/\underline{r}^{\infty}(A, \ )(A_i) = 0$ (see [3, page 5, 6]). Because $(A, \ )/\underline{r}^{\infty}(A, \ )(X_j) \neq 0$ and $(A, \ )/\underline{r}^{\infty}(A, \ )(X_{j'}) = 0$, it follows that X_j is not isomorphic to $X_{j'}$. □

COROLLARY 6.9. *Let Λ be an indecomposable finite dimensional algebra over a field k of infinite representation type and suppose that* $\operatorname{card} k > \aleph_0$. *Then the cardinality of the isomorphism classes of denumerably generated but not finitely generated indecomposable Λ-modules is* $\operatorname{card} k$.

PROOF. By Brauer-Thrall 2, we know that $|\operatorname{ind}\Lambda| = \operatorname{card} k$. Hence if $\{\mathcal{E}_i\}_{i\in I}$ is the usual partition of $\operatorname{ind}\Lambda$, we have that $\operatorname{card} I = \operatorname{card} k$ since $|\mathcal{E}_i| = \aleph_0$ for all i and $\operatorname{card} k > \aleph_0$. Thus the cardinality of the set of all i with the property that $\mathcal{E}_i$ contains an A with $L(A, \ ) = \infty$ is $\operatorname{card} k$. The corollary now follows immediately from Proposition 6.8. □

References

[1] Auslander, M. *Comments on the functor* Ext, Topology, (1969), 151–166.

[2] Auslander, M. *Functors and morphisms determined by objects*, these proceedings.

[3] Auslander, M. *Large modules over Artin algebras*, Algebra, Topology and Categories, Academic Press, 1976.

[4] Auslander, M. *Representation theory of Artin algebras,* II, Comm. in Algebra, (1974), 269–310.

[5] Auslander, M. and Bridger, M. *Stable module theory,* Memoirs of the American Mathematical Society, No. 94, 1969.

[6] Auslander, M. and Reiten, I. *Representation theory of Artin algebras,* III, Comm. in Algebra, (1975), 239–294.

[7] Auslander, M. and Reiten. I. *Representation theory of Artin algebras,* IV, Comm. in Algebra, (1977), 443–518.

[8] Auslander, M. and Reiten. I. *Representation theory of Artin algebras,* V, Comm. in Algebra, (1977), 519–554.

[9] Auslander, M. and Reiten. I. *Representation theory of Artin algebras,* VI, Comm. in Algebra, (to appear).
[10] Auslander, M. and Reiten. I. *Stable equivalence of dualizing R-varieties*, Advances in Math., (1974), 300–366.
[11] Cartan, H, and Eilenberg, S., *Homological Algebra*, Princeton University Press, Princeton, New Jersey, 1956.
[12] Nazarova, L. A. and Roiter, A. V., *Categorical matrix problems and the Brauer-Thrall conjecture*, Preprint, Inst. Math. Acad. Sci., Kiev, 1973.
[13] Ringel, C. M., *Representations of K-species and bimodules*, J. of Algebra, (1976), 269–302.
[14] Roiter, A. V., *Unboundness of the dimension of the indecomposable representation of an algebra which has infinitely many indecomposable representations*, Izv. Akad. SR, Sev. Math. 32, (1968) 1275–1282.

BRANDEIS UNIVERSITY, WALTHAM, MASSACHUSETTS

A survey of existence theorems for almost split sequences[1]

Maurice Auslander

This paper is devoted to describing existence theorems for almost split sequences in various settings, ranging from the category of all modules over an arbitrary ring to locally free coherent sheaves on projective nonsingular curves. We follow closely the first of three lectures we gave at the Durham symposium on the representation theory of algebras. Since these lectures were purely expository, no proofs are given. References to past or future papers where proofs can be found will be given.

We recall that an exact sequence $0 \to A \xrightarrow{f} B \xrightarrow{g} C \to 0$ is called an almost split sequence if:

a) it is not split;
b) If $h\colon X \to C$ is not a splittable epimorphism, then h can be lifted to B;
c) If $t\colon A \to Y$ is not a splittable monomorphism, then t can be extended to B.

If $0 \to A \xrightarrow{f} B \xrightarrow{g} C \to 0$ is an almost split sequence, then it is easily seen that $\operatorname{End} C$ and $\operatorname{End} A$ are local rings, which places a serious limitation on the C and A that can be the end of an almost split sequence. Moreover, the following uniqueness theorem shows that the existence of an almost split sequence $0 \to A \to B \to C \to 0$ establishes a strong connection between A and C.

THEOREM 1. *Let $0 \to A_i \to B_i \to C_i \to 0$ be two almost split sequences for $i = 1, 2$. Then the following statements are equivalent:*

i) *The sequences are isomorphic;*
ii) $A_1 \cong A_2$*;*
iii) $C_1 \cong C_2$.

So the ends of an almost split sequence determine each other, up to isomorphism.

There are basically two different types of existence theorems for almost split sequences. One type proves their existence without describing how the ends determine each other. The other not only proves the existence of almost split sequences but also gives an independent description of one end in terms of the other. Since a priori knowledge of the ends of an almost split sequence is often helpful in actually finding the sequence, we deal in this talk only with the second type of existence theorem. Basic to describing the ends of an almost split sequence in terms of each other is the notion of the transpose of a finitely presented module. So we begin with this concept.

Let Λ be an arbitrary ring. We denote by $\operatorname{Mod} \Lambda$ the category of all Λ-modules and by $\operatorname{mod} \Lambda$ the category of finitely presented Λ-modules. Also we denote by $\underline{\operatorname{mod}}\Lambda$ the category of finitely presented Λ-modules modulo projectives, the category whose objects are the

---

[1] REPRESENTATIONS OF ALGEBRAS, LMS Lec. Notes Ser. 116 (1985) 81–89

same as those of $\operatorname{mod}\Lambda$ with $\underline{\operatorname{Hom}}_\Lambda(A,B) = \operatorname{Hom}_\Lambda(A,B)/P_\Lambda(A,B)$, where $P_\Lambda(A,B)$ is the subgroup of $\operatorname{Hom}_\Lambda(A,B)$ consisting of those morphisms factoring through finitely generated projective Λ-modules. The transpose is a duality $\operatorname{Tr}\colon \underline{\operatorname{mod}\Lambda} \to \underline{\operatorname{mod}\Lambda^{\mathrm{op}}}$ which on objects is defined as follows: For each C in $\operatorname{mod}\Lambda$ choose a fixed finite projective presentation $P_1 \xrightarrow{f} P_0 \to C \to 0$, with $P_0 \to C$ being the identity if C is projective. Then $\operatorname{Tr} C$ is defined by the exact sequence

$$\operatorname{Hom}_\Lambda(P_0,\Lambda) \xrightarrow{(f,\Lambda)} \operatorname{Hom}_\Lambda(P_1,\Lambda) \to \operatorname{Tr} C \to 0.$$

This notion was first introduced by us at the Stockholm International Congress of Mathematics in 1962 in the context of studying finitely generated modules over regular local rings. It was then used by several people as a means of comparing left and right finitely presented modules over noncommutative rings, as well as in comparing covariant and contravariant finitely presented functors. Thus its introduction to algebra long antidates its use in representation theory. We now describe a basic property of the transpose which underlies its connection with describing the ends of almost split sequences.

Suppose C is an arbitrary Λ-module, A a finitely presented Λ-module. Further, suppose $\operatorname{Tr} A$ is a Λ^{op}-Σ-bimodule. Then the composition of morphisms $\Sigma \to \underline{\operatorname{End}}(\operatorname{Tr} A) \to \underline{\operatorname{End}}(A)^{\mathrm{op}}$ makes $\underline{\operatorname{Hom}}(A,C)$ a Σ-module and we have the following basic result which may have a familiar ring to people working with almost split sequences [1].

PROPOSITION 2. *For each injective Σ-module I we have an isomorphism*

$$\phi\colon \operatorname{Ext}^1_\Lambda(C, \operatorname{Hom}_\Sigma(\operatorname{Tr} A, I)) \to \operatorname{Hom}_\Sigma(\underline{\operatorname{Hom}}(A,C), I)$$

which is functorial in C and I. Moreover this isomorphism has the following property: If $x = 0 \to \operatorname{Hom}_\Sigma(\operatorname{Tr} A, I) \xrightarrow{f} B \xrightarrow{g} C \to 0$, then $\phi(x)\colon \underline{\operatorname{Hom}}(A,C) \to I$ has the property that $\operatorname{Im}(\operatorname{Hom}(A,B) \xrightarrow{(A,g)} \operatorname{Hom}(A,C) \to \underline{\operatorname{Hom}}(A,C)) = \operatorname{Ker}(\underline{\operatorname{Hom}}(A,C) \xrightarrow{\phi(x)} I)$.

As an immediate consequence of this proposition we have the following, which can often be used to actually construct almost split sequences.

COROLLARY 3. *An exact sequence $0 \to \operatorname{Hom}_\Sigma(\operatorname{Tr} A, I) \to Y \to Z \to 0$ of Λ-modules splits if and only if every morphism $A \to Z$ can be lifted to Y.*

It is Proposition 2 which constitutes the starting point for the existence theorems for almost split sequences given here. We begin with the most general existence theorem for almost split sequences for modules over rings [1, 2, 4].

THEOREM 4. *Let Λ be an arbitrary ring, C a finitely presented nonprojective Λ-module with $\operatorname{End} C$ a local ring and let $\Gamma = \operatorname{End} \operatorname{Tr} C$. We consider the unique simple $\operatorname{End} C^{\mathrm{op}}$-module a Γ-module by means of the ring surjection which is the composition $\Gamma \to \underline{\operatorname{End}}(\operatorname{Tr} C) \xrightarrow{\operatorname{Tr}} \underline{\operatorname{End}} C^{\mathrm{op}}$ and let I be its Γ-injective envelope. Then we have an almost split sequence*

$$0 \to \operatorname{Hom}_\Gamma(\operatorname{Tr} C, I) \to B \to C \to 0.$$

REMARK. It should be noted that the uniqueness properties of almost split sequences show that $\operatorname{Hom}_\Gamma(\operatorname{Tr} C, I)$ depends only on C and not on the particular projective presentation of C used in constructing $\operatorname{Tr} C$.

We now apply this result to the situation that Λ is an algebra over a complete noetherian local ring R which is a finitely generated R-module. Then both $\operatorname{mod}\Lambda$ and $\operatorname{art}\Lambda$, the category of artinian modules, are Krull-Schmidt categories and the functor

$\operatorname{Hom}_R(\ ,I)\colon \operatorname{mod}\Lambda \to \operatorname{art}\Lambda$ gives a duality where I is the R-injective envelope of $R/\underline{m}$, and $\underline{m}$ is the maximal ideal of R. Since $\operatorname{mod}\Lambda$ is a Krull-Schmidt category we have that every C in $\operatorname{mod}\Lambda$ has a minimal projective presentation which we use in computing $\operatorname{Tr} C$. Finally C in $\operatorname{mod}\Lambda$ or $\operatorname{art}\Lambda$ is indecomposable if and only if $\operatorname{End} C$ is local. With these remarks and notations in mind we have the following easy consequence of Theorem 4 [1, 2].

THEOREM 5. *Let C be an indecomposable nonprojective module in* $\operatorname{mod}\Lambda$. *Then there is an almost split sequence* $0 \to \operatorname{Hom}_R(\operatorname{Tr} C, I) \to B \to C \to 0$ *where* $\operatorname{Hom}_R(\operatorname{Tr} C, I)$ *is in* $\operatorname{art}\Lambda$.

REMARKS. a) If $\dim R = 0$ (Krull dimension of R), then $\operatorname{mod}\Lambda = \operatorname{art}\Lambda$ and the almost split sequence is in $\operatorname{mod}\Lambda$, and so we obtain one of the original existence theorems of Auslander-Reiten [4] which has played an important role in much of the representation theory of artin algebras.

b) If $\dim R > 0$ and $\operatorname{depth} R \geq 1$, then $\operatorname{Hom}_R(\operatorname{Tr} C, I)$ is noetherian, or equivalently of finite length, if and only if $\operatorname{pd} C = 1$. From this it follows that if R is a local ring (complete or not) with $\operatorname{depth} R \geq 1$, then the category of modules of finite length has almost split sequences with respect to itself if and only if R is a 1-dimensional regular local ring. We shall have need of this remark when discussing the existence of almost split sequences for coherent sheaves.

Our next existence theorem is a generalization of the one just given. Again we assume that R is a complete local ring with maximal ideal $\underline{m}$ and I the injective envelope of $R/\underline{m}$. We now let $\mathcal{P}$ be an additive R-category in which idempotents split and whose isomorphism classes of objects constitute a set. That is, we assume that $\operatorname{Hom}(P_1, P_2)$ is a finitely generated R-module for all P_1, P_2 in $\mathcal{P}$ and that the composition of morphisms is R-bilinear. For example: R is a field and $\mathcal{P}$ is a quiver with relations over R such that $\operatorname{Hom}(P_1, P_2)$ is of finite dimension for all points P_1 and P_2. We denote by $\operatorname{Mod}\mathcal{P}$ the category of all additive contravariant functors from $\mathcal{P}$ to $\operatorname{Mod} R$ and we denote by $\operatorname{mod}\mathcal{P}$ the full subcategory of $\operatorname{Mod}\mathcal{P}$ consisting of all finitely presented functors where $C\colon \mathcal{P} \to \operatorname{Ab}$ is said to be a finitely presented functor if there is an exact sequence $\operatorname{Hom}_{\mathcal{P}}(\ , P_1) \to \operatorname{Hom}_{\mathcal{P}}(\ , P_0) \to C \to 0$ with the P_1 and P_0 in $\mathcal{P}$. As before $\operatorname{mod}\mathcal{P}$ is a Krull-Schmidt category so that each C in $\operatorname{mod} R$ has a (unique up to isomorphism) minimal projective presentation and C is indecomposable if and only if $\operatorname{End} C$ is a local R-algebra. Let $(\ , P_1) \to (\ , P_0) \to C \to 0$ be a minimal projective presentation. We define $\operatorname{Tr} C$ in $\operatorname{mod}\mathcal{P}^{\mathrm{op}}$ by the exact sequence $(P_0,\ ) \to (P_1,\ ) \to \operatorname{Tr} C \to 0$. Then C in $\operatorname{mod}\mathcal{P}$ is indecomposable and nonprojective if and only if $\operatorname{Tr} C$ is indecomposable and nonprojective. Finally for an F in $\operatorname{Mod}\mathcal{P}$ we denote by $\operatorname{Hom}_R(F, I)$ the object in $\operatorname{Mod}\mathcal{P}^{\mathrm{op}}$ which is the functor defined by $\operatorname{Hom}_R(F, I)(P) = \operatorname{Hom}_R(F(P), I)$ for all P in $\mathcal{P}$.

We now have the following generalization of Theorem 5.

THEOREM 6. *Let C in* $\operatorname{mod}\mathcal{P}$ *be nonprojective and indecomposable. Then we have the almost split sequence*

$$0 \to \operatorname{Hom}_R(\operatorname{Tr} C, I) \to B \to C \to 0.$$

We now give examples of such categories $\mathcal{P}$. In these examples we assume that $\dim R = 0$, for instance when R is a field.

a) If $\mathcal{P}$ has the property that $\operatorname{Hom}_R((P,\ ), I)$ and $\operatorname{Hom}_R((\ , P), I)$ are finitely presented for all P in $\mathcal{P}$ we obtain the notion of a dualizing R-variety [3]. In this case $\operatorname{Hom}_R(\operatorname{Tr} C, I)$ is in $\operatorname{mod}\mathcal{P}$ for all C in $\operatorname{mod}\mathcal{P}$ and so the almost split sequence $0 \to \operatorname{Hom}_R(\operatorname{Tr} C, I) \to B \to C \to 0$ lies in $\operatorname{mod}\mathcal{P}$.

b) If we assume that $(\ , P)$ and $(P, \ )$ have finite support for each indecomposable P in $\mathcal{P}$, we obtain R-varieties such that $\operatorname{mod}\mathcal{P}$ consists of the $\mathcal{P}$-modules of finite length. This includes for instance locally bounded k-categories for a field k [7].

c) Let k be a field and let $\Lambda = \Lambda_0 \amalg \Lambda_1 \amalg \ldots$ be a positively graded 2-sided noetherian k-algebra with the Λ_i finite dimensional k-vector spaces. Denote by $\operatorname{Gr}\Lambda$ the category of finitely generated $\mathbb{Z}$-graded Λ-modules with degree zero morphisms. Finally, let $\mathcal{P}$ be the full subcategory of $\operatorname{Gr}\Lambda$ consisting of the projective objects in $\operatorname{Gr}\Lambda$. Then $\mathcal{P}$ has our desired properties and $\operatorname{mod}\mathcal{P}$ is equivalent to $\operatorname{Gr}\Lambda$.

We will return to the situation covered by Theorem 6 later on when discussing graded orders.

So far our existence theorems were for almost split sequences which had the almost split properties relative to the whole category of modules or functors. We now give some existence theorems for almost split sequences for lattices over orders which are almost split relative to lattices only and not all modules. There are similar existence theorems for subcategories of $\operatorname{mod}\Lambda$ when Λ is an artin algebra which were used in proving existence of preprojective partitions [6], [7]. We begin by recalling the definition of an order and lattices over an order.

Let R be an equidimensional commutative Gorenstein ring of dimension d (Krull dimension). By an R-order we mean an R-algebra Λ which is a finitely generated maximal Cohen-Macaulay R-module having the property that $\operatorname{Hom}_R(\Lambda, R)_{\underline{p}}$ is $\Lambda^{\text{op}}_{\underline{p}}$-projective for all nonmaximal prime ideals $\underline{p}$ of R. A Λ-lattice is a finitely generated Λ-module C which is a maximal Cohen-Macaulay R-module such that $C_{\underline{p}}$ is $\Lambda_{\underline{p}}$-projective for all nonmaximal prime ideals $\underline{p}$ of R.

Suppose now that Λ is an R-order where R is a complete local Gorenstein ring of dimension d. Then $\operatorname{mod}\Lambda$ is a Krull-Schmidt category. Let C be an indecomposable nonprojective in $\mathcal{L}(\Lambda)$, the category of Λ-lattices. Then $\operatorname{Tr} C$ (computed using a minimal projective presentation for C) is indecomposable and nonprojective but not necessarily a lattice. However $\Omega^d \operatorname{Tr} C$ is an indecomposable lattice where $\Omega^d \operatorname{Tr} C$ is defined to be $\operatorname{Ker} f_{d-1}$ in a minimal projective resolution $\cdots \to P_d \to P_{d-1} \xrightarrow{f_{d-1}} \cdots \to P_0 \xrightarrow{f_0} C \to 0$ of C. The lattice $\Omega^d \operatorname{Tr} C$ plays the role of the transpose of C in the category of lattices and is denoted by $\operatorname{Tr}_{\mathcal{L}} C$. Also it is not difficult to see that $\operatorname{Hom}_R(\ , R)\colon \operatorname{mod}\Lambda \to \operatorname{mod}\Lambda^{\text{op}}$ gives a duality $\operatorname{Hom}_R(\ , R)\colon \mathcal{L}(\Lambda) \to \mathcal{L}(\Lambda^{\text{op}})$. Keeping these definitions and notations in mind, we have the following [1, 2].

THEOREM 7. *Let C in $\mathcal{L}(\Lambda)$ be indecomposable and nonprojective. Then there is an almost split sequence $0 \to \operatorname{Hom}_R(\operatorname{Tr}_{\mathcal{L}} C, R) \xrightarrow{f} B \xrightarrow{g} C \to 0$ in the category $\mathcal{L}(\Lambda)$, i.e. the terms of the exact sequence are in $\mathcal{L}(\Lambda)$, the sequence does not split and any $X \to C$ in $\mathcal{L}(\Lambda)$ which is not a splittable epimorphism can be lifted to B as well as any $\operatorname{Hom}_R(\operatorname{Tr}_{\mathcal{L}} C, R) \to Y$ in $\mathcal{L}(\Lambda)$ which is not a splittable monomorphism can be extended to B.*

As our final general existence theorem we consider graded orders and graded lattices. We begin with some general remarks and notations.

Let k be a field. By a graded k-algebra Λ we mean a k-algebra Λ together with a k-vector space decomposition $\Lambda = \amalg_{i \in \mathbb{Z}} \Lambda_i$ satisfying the following

a) $\dim_k \Lambda_i < \infty$ for all i and $\Lambda_i = 0$ for $i < 0$.

b) Λ_0 is semisimple.

c) $\Lambda_i = \Lambda_1^i$ and so $\Lambda_i \Lambda_j = \Lambda_{i+j}$.
d) Λ is a 2-sided noetherian ring.

By a graded Λ-module M we mean a Λ-module M together with a k-vector space decomposition $M = \amalg_{i \in \mathbb{Z}} M_i$ such that $\Lambda_i M_j \subset M_{i+j}$ for all i and j in $\mathbb{Z}$. Moreover with each graded Λ-module M and every n in $\mathbb{Z}$ we have the shifted graded module $M(n)$ given by $M(n)_d = M_{n+d}$ for all d in $\mathbb{Z}$.

By a graded order Λ over a polynomial ring $R = k[X_1, \dots, X_n]$ we mean a graded k-algebra Λ together with $X_1, \dots, X_n$ in Λ_1 algebraically independent elements over k which lie in the center of Λ, having the following properties:

a) Λ is a finitely generated free R-module i.e. $\Lambda = \amalg_{i<\infty} R(n_i)$ as an R-module.
b) $\operatorname{Hom}_R(\Lambda, R)_{\underline{p}}$ is $\Lambda_{\underline{p}}$-projective for each graded prime ideal $\underline{p} \neq (X_1, \dots, X_n)$.

Now the category of finitely generated graded Λ-modules with degree zero morphisms is Krull-Schmidt because $\operatorname{Hom}_{\operatorname{gr}\Lambda}(A, B)$ is finite dimensional over k for all A and B. If we consider $\mathcal{L}(\operatorname{gr}\Lambda)$, the subcategory of graded Λ-modules with degree zero morphisms consisting of all L such that L is R-free and $L_{\underline{p}}$ is $\Lambda_{\underline{p}}$-projective for all graded $\underline{p} \neq \underline{m}$, then we have the following [5].

THEOREM 8. *If C in $\mathcal{L}(\operatorname{gr}\Lambda)$ is an indecomposable nonprojective lattice, then there is an almost split sequence*

$$0 \to \operatorname{Hom}_R(\Omega^n \operatorname{Tr} C, R(-n)) \to B \to C \to 0$$

in the category $\mathcal{L}(\operatorname{gr}\Lambda)$ where Tr *and* Ω^n *are computed using minimal projective presentations and resolutions.*

As one consequence of this we have the following [5].

THEOREM 9. *Let X be a projective curve with only Gorenstein singularities (for instance, X is nonsingular), then the category of coherent sheaves in X has almost split sequences.*

REMARK. A. Schofield has independently given a sheaf theoretic proof of Theorem 9. Geigle and Lenzing have also obtained some special cases of this result.

References

[1] M. Auslander, *Existence theorems for almost split sequences*, Proc. Conf. on Ring Theory II, Oklahoma (Marcel Dekker, 1977) pp. 1–44.
[2] M. Auslander, *Functors and morphisms determined by objects*, Proc. Conf. on Representation Theory (Philadelphia 1976), (Marcel Dekker, 1978) pp. 1–244.
[3] M. Auslander and I. Reiten, *Stable equivalence of dualizing R-varieties I*, Adv. in Math. 12 (1974) 306–366.
[4] M. Auslander and I. Reiten, *Representation theory of artin algebras III. Almost split sequences*, Comm. Algebra 3 (1975) 239–294.
[5] M. Auslander and I. Reiten, *Almost split sequences for graded modules.*
[6] M. Auslander and S. O. Smalø, *Almost split sequences in subcategories*, J. Algebra 69 (1981) 426–454.
[7] K. Bongartz and P. Gabriel, *Coverings in representation theory*, Invent. Math. 65 (1982) 331–378.

BRANDEIS UNIVERSITY, WALTHAM, MA 02254, USA

CHAPTER VII

Tilting theory and homologically finite subcategories

[57] Coxeter functors without diagrams (with M. I. Platzeck, I. Reiten, 1979).
[62] Preprojective modules over artin algebras (with S. O. Smalø, 1980).
[63] Almost split sequences in subcategories (with S. O. Smalø, 1981).
[92] Applications of contravariantly finite subcategories (with I. Reiten, 1991).
[96] Homological theory of idempotent ideals (with M. I. Platzeck, G. Todorov, 1992).

In this chapter we cover two important developments in the representation theory of artin algebras where Auslander's work played an essential role: tilting theory and the theory of homologically finite subcategories. These theories, as well as the interplay between them, have had interesting applications also beyond artin algebras.

The reflection functors of Bernstein-Gelfand-Ponomarev were essential in their approach to the classification of quivers with only a finite number of indecomposable representations. The main importance of [57] was the interpretation of the reflection functors as functors of the form $\mathrm{Hom}_\Lambda(T,\ )$ between module categories, for a Λ-module T, together with the identification of the crucial properties of T. This work was generalized by Brenner-Butler and Happel-Ringel, leading to what is now called tilting theory. These Λ-modules T provided the first module theoretic examples of what came to be known as tilting modules. The theory of tilting modules and tilted algebras plays a central role in the representation theory of artin algebras, mainly because many questions about arbitrary algebras can be reduced to tilted algebras. Further generalizations have also been important; in particular the tilting (cotilting) modules of finite projective (injective) dimension and the connections with derived categories established by D. Happel. A further development was J. Richard's work on tilting complexes, which has proved to be useful in the representation theory of finite groups. In a different direction, the class of tilted algebras has been generalized to what is called quasitilted algebras. Tilting has also been studied beyond artin algebras, and others have introduced the related concept of star modules.

The concepts of contravariantly, covariantly and functorially finite subcategories were introduced in [62]. These properties can be viewed as weaker versions of the inclusion functors of a subcategory having left or right adjoints. There were originally two main applications of the theory [62, 63]. If a category has almost split sequences, then any functorially finite subcategory was shown to have right and left almost split morphisms and also relative almost split sequences under the additional assumption that the subcategory is closed under extensions. Existence theorems for almost split sequences in subcategories were of interest as a basis for extending some of the theory based on almost split sequences from the whole module category to subcategories. This yields for instance existence of almost split sequences for representations of partially ordered sets; a special case which had been treated earlier. Another influence was the general theory of lattices over orders, where the lattices form only subcategories of the whole module category.

Basically all existence theorems for almost split sequences in subcategories established later are, or can be based on, the main theorem from these papers.

Previous work of A. V. Roiter on finite representation type, together with the preprojective and preinjective modules over hereditary algebras, inspired the other main application of functorially finite subcategories. This was to prove existence of preprojective and preinjective partitions in general, and there is still ongoing research in this direction.

Even though the concepts were not yet formulated, some information on contravariantly and covariantly finite subcategories can be found in [27], for example sufficient conditions to ensure that the category of modules of projective dimension at most n for some nonnegative integer n, is covariantly finite in the category of finitely generated modules. Similar concepts appear in the literature under the name of covers, and the question of existence of flat covers has attracted attention. Contravariantly and covariantly finite subcategories along with the local versions of contravariant (or right) and covariant (or left) approximations continue to play an important role.

In addition to tilting and functorially (and contravariantly and covariantly) finite each being an important concept, there is a fruitful interplay between them. Such a connection was already noticed in [63], and was generalized to obtain a correspondence between certain contravariantly finite subcategories and cotilting modules (of finite injective dimension) [92]. An additional source of inspiration came from the commutative theory, to be discussed in Chapter IX [89]. Through a beautiful application by C. M. Ringel to quasihereditary algebras, this has led to new developments for Schur algebras and algebraic groups.

The paper [96] is also homological in nature, and was motivated by the theory of quasihereditary algebras.

TRANSACTIONS OF THE
AMERICAN MATHEMATICAL SOCIETY
Volume 250, June 1979

COXETER FUNCTORS WITHOUT DIAGRAMS

BY

MAURICE AUSLANDER[1], MARÍA INÉS PLATZECK[2] AND IDUN REITEN

ABSTRACT. The first part of this paper is devoted to a generalization of the notion of partial Coexeter functor from diagrams to certain types of artin rings and artin algebras. The rest of the paper is devoted to a discussion of the connection between the various Coxeter functors which exist for diagrams as well as for artin rings and artin algebras.

Introduction. For the study of the representation theory of hereditary artin algebras Λ in [9], [13], [17], in particular for the investigation of when Λ has only a finite number of nonisomorphic indecomposable finitely generated modules, certain types of functors between module categories over hereditary algebras called partial Coxeter functors and Coxeter functors played an important role.

In this paper we develop a theory of partial Coxeter functors for an artin algebra Λ with a simple projective noninjective module S. We shall not recall the definition of the partial Coxeter functors in the diagramatic setting here (see §5), but it is shown in [13] that such a functor S^+ from mod Λ to mod Λ', where Λ and Λ' are hereditary tensor algebras and mod Λ is the category of finitely generated (left) Λ-modules, is left exact and has the following properties. (i) There is a simple projective noninjective Λ-module S and a simple injective nonprojective Λ'-module T such that $S^+(S) = 0$ and S^+ induces an equivalence of categories between the full subcategory of mod Λ whose objects are the A in mod Λ such that S is not a summand of A, and the full subcategory of mod Λ' whose objects are the A in mod Λ' such that T is not a summand of A. (ii) If $0 \to A \to B \to C \to 0$ is an exact sequence in mod Λ and A has no summand isomorphic to S, then $0 \to S^+(A) \to S^+(B) \to S^+(C) \to 0$ is exact. For left artin rings Λ and Λ', we shall call a functor from mod Λ to mod Λ' with property (i) a left partial Coxeter functor, and show that it must be left exact and have property (ii).

Let Λ be an artin algebra. We will denote by D the ordinary duality for artin algebras, and by Tr M the transpose of the Λ-module M, that is the Λ^{op}-module defined by the exact sequence $\mathrm{Hom}_\Lambda(Q_0, \Lambda) \to \mathrm{Hom}_\Lambda(Q_1, \Lambda) \to$

Received by the editors September 14, 1977.
AMS (MOS) subject classifications (1970). Primary 16A46, 16A64; Secondary 16A62.
[1]Supported by N.S.F. Grant PCM 77-04951.
[2]Supported by N.S.F. Grant MCS 77-03478.

$\mathrm{Tr}\, M \to 0$, induced by the minimal projective presentation $Q_1 \to Q_0 \to M \to 0$. Let $\underline{\mathrm{mod}}\,\Lambda$ and $\overline{\mathrm{mod}}\,\Lambda$ denote, respectively, the category of finitely generated modules modulo projectives and the category of finitely generated modules modulo injectives. We recall that Tr induces a duality Tr: $\underline{\mathrm{mod}}\,\Lambda \to \underline{\mathrm{mod}}\,\Lambda^{\mathrm{op}}$, and $D\mathrm{Tr}$: $\underline{\mathrm{mod}}\,\Lambda \to \overline{\mathrm{mod}}\,\Lambda$ is an equivalence of categories (see [**6**, §2]). When Λ is hereditary then $\underline{\mathrm{mod}}\,\Lambda$ is just the full subcategory $\mathrm{mod}_P\,\Lambda$ of mod Λ, whose objects are the modules A with no nonzero projective summands and $\overline{\mathrm{mod}}\,\Lambda$ is the full subcategory $\mathrm{mod}_I\,\Lambda$ of mod Λ whose objects are the modules A with no nonzero injective summands. So in this case $D\mathrm{Tr}$ defines a functor from mod Λ to itself (see [**5**, §2]).

Now let Λ be a basic artin algebra and let S be a simple projective noninjective Λ-module. Let P be defined by $\Lambda = S \amalg P$, $X = P \amalg \mathrm{Tr}\, DS$ and $\Gamma = \mathrm{End}_\Lambda(X)^{\mathrm{op}}$. In §1 we study the functor $F = \mathrm{Hom}_\Lambda(X, \ )$: mod $\Lambda \to$ mod Γ. Under the assumption that $\mathrm{Hom}_\Lambda(\mathrm{Tr}\, DS, \Lambda)$ is zero we show that F is a left partial Coxeter functor. Since the partial Coxeter functors have turned out to be useful in the study of the representation theory of hereditary tensor algebras, it is hoped that these partial Coxeter functors, or more generally, the functor $F = \mathrm{Hom}_\Lambda(X, \ )$, even when $\mathrm{Hom}(\mathrm{Tr}\, DS, \Lambda)$ is not zero, should be useful. They have already been successfully applied by R. Bautista in [**8**], in his study of algebras where every map between indecomposable projectives is zero or a monomorphism. Clearly, if $F = \mathrm{Hom}(X, \ )$ is a partial Coxeter functor, then Λ has only a finite number of nonisomorphic indecomposable modules if and only if Γ does, whereas Bautista has shown that this is not true in general. However, it would be interesting to know when this is true, and also to know the class of algebras Λ such that after applying a finite number of functors of the above type we reach an hereditary algebra, since much is known about the representation theory of hereditary algebras.

The method we use is the theory of almost split sequences. We recall from [**6**, §4], that for an artin ring Λ, an exact sequence $0 \to A \xrightarrow{f} B \xrightarrow{g} C \to 0$ in mod Λ is said to be almost split if it is not split, A and C are indecomposable and in addition it has the following properties. Given any map h: $U \to C$ which is not a splittable epimorphism, there is a map t: $U \to B$ such that $gt = h$, and given any map h: $A \to V$ which is not a splittable monomorphism, there is a map s: $B \to V$ such that $sf = h$. It follows from the proof of Proposition 4.1 in [**6**] that a nonsplit exact sequence $0 \to A \xrightarrow{f} B \to C \to 0$ with A and C indecomposable is almost split if and only if given a map h: $A \to V$ which is not a split monomorphism, there is a map t: $B \to V$ such that $tf = h$. For an artin algebra Λ existence and uniqueness of an almost split sequence $0 \to A \to B \to C \to 0$ for every indecomposable nonprojective Λ-module C (every indecomposable noninjective Λ-module A) has been proven in [**6**, Proposition 4.3]. It has also been proven there for artin algebras that if the sequence $0 \to A \to B \to C \to 0$ is almost split then $A \cong D\,\mathrm{Tr}\, C$.

In §1 we only need the existence of an almost split sequence $0 \to A \to B \to C \to 0$ when the module A is simple projective. We also need that in this case B is a projective Λ-module. These properties are known for artin algebras (see [**6**, Theorem 5.5]). At the end of §1 we give an independent proof of the existence of the almost split sequence $0 \to A \to B \to C \to 0$ and of the fact that B is projective for the special case when A is simple projective noninjective, and for this we only need to assume that Λ is an artin ring that is not necessarily an artin algebra.

In §2 we study the notion of left partial Coxeter functors between left artin rings defined above, and also the notion of right partial Coxeter functors. We show that both the left and right partial Coxeter functors are unique up to "equivalence" of functors in the following sense. For left artin rings Λ_1, Λ_1', Λ_2, Λ_2' we say that two functors F_1: mod $\Lambda_1 \to$ mod Λ_1', F_2: mod $\Lambda_2 \to$ mod Λ_2' are equivalent if there are equivalences of categories u: mod $\Lambda_1 \to$ mod Λ_2 and v: mod $\Lambda_1' \to$ mod Λ_2' such that $vF_1 = F_2u$. We also show, up to equivalence, what the left partial Coxeter functors have to look like if the first ring is artin. In particular, the left and right partial Coxeter functors studied in §1 for artin algebras are the only ones up to equivalence.

In §4 we study certain functors between the module categories over generalized triangular matrix rings. Specializing our assumptions, our functors will be partial Coxeter functors. Using the results of §2 we can then describe all left partial Coxeter functors from an artin ring to a left artin ring. As we pointed out before, the partial Coxeter functors defined in [**9**], [**13**] using diagrams, and also the ones defined in [**14**] fit into our general theory of partial Coxeter functors. By our uniqueness result we get a strong relationship between these various cases.

In [**9**] and [**13**] a Coxeter functor C^+ from mod Λ to mod Λ', for an hereditary tensor ring Λ with duality conditions (see §5 for definition), is defined as a composite of partial Coxeter functors (see §5). In §3 we show for an hereditary artin algebra Λ, not necessarily a tensor algebra, how to construct ring isomorphisms ϕ: $\Lambda \to \Lambda_n$ and a composition of partial Coxeter functors $F_n \cdots F_1$: mod $\Lambda \to$ mod Λ_n such that $\bar{\phi}F_n \cdots F_1 = D\,\mathrm{Tr}$, where $\bar{\phi}$: mod $\Lambda_n \to$ mod Λ is the equivalence of categories induced by the isomorphism ϕ: $\Lambda \to \Lambda_n$. This does not mean, however, that a given Coxeter functor C^+ is isomorphic to the functor $D\,\mathrm{Tr}$ for hereditary tensor algebras. In fact, it is not possible to prove this, because we show that C^+ is not uniquely defined up to isomorphism, since it depends on the choice of certain bimodule isomorphisms. However, it has been proven in [**10**] that there is a special choice of these isomorphisms such that the corresponding functor C^+ is isomorphic to $D\,\mathrm{Tr}$.

The connections of our work with the diagramatic results of **[9]** and **[13]** are studied in §5. We also prove in this section that for an artin hereditary tensor ring with duality conditions any Coxeter functor C^+ is isomorphic to $D_1\mathrm{Tr}$, for some duality D_1, but there might be dualities D_1 such that $D_1\mathrm{Tr}$ is not isomorphic to any Coxeter functor C^+.

We shall only be dealing with basic left artin rings. This is no loss of generality since every left artin ring is Morita equivalent to a basic left artin ring. Further we have that if two basic left artin rings are Morita equivalent, they are isomorphic.

1. Functors and endomorphism rings. Throughout this paper we shall assume that all artin rings are basic, that is, the ring modulo its radical $\mathbf{r}$ is a product of division rings.

Let Λ be an artin algebra, that is, an artin ring which is a finitely generated module over its center, which is also an artin ring, and denote by mod Λ the category of finitely generated (left) Λ-modules. Assume that Λ has a simple projective module S which is not injective and denote by P the direct sum of one copy of each of the other indecomposable Λ-modules. Let further $X = \mathrm{Tr}\, DS \amalg P$, where D denotes the ordinary duality between mod Λ and mod Λ^{op}, and Tr denotes the transpose. Let $\Gamma = \mathrm{End}_\Lambda(X)^{\mathrm{op}}$, and denote by F the functor $\mathrm{Hom}_\Lambda(X, \ ) = (X, \ )$ from mod Λ to mod Γ. Denote by $\mathcal{C} = \mathcal{C}_S$ the full subcategory of mod Λ whose objects are the A in mod Λ such that $\mathrm{Hom}_\Lambda(A, S) = 0$, in other words the A in mod Λ having no summand isomorphic to S. We shall also let F denote the restriction functor from $\mathcal{C}$ to mod Γ. For a simple injective nonprojective Γ-module T, denote by $\mathfrak{D} = \mathfrak{D}_T$ the full subcategory of mod Γ whose objects are the B in mod Γ such that $\mathrm{Hom}_\Gamma(T, B)$ is zero, that is, the B in mod Γ having no summand isomorphic to T.

In this section we shall study the functor F: mod $\Lambda \to$ mod Γ, and relationships between the rings Λ and Γ. We also study a certain type of right exact functors $G = Y \otimes \ $: mod $\Gamma \to$ mod Λ, having properties "dual" to the properties of F.

More specifically, we show that F: $\mathcal{C} \to$ mod Γ is a fully faithful functor. Under the assumption that $(\mathrm{Tr}\, DS)^* = \mathrm{Hom}_\Lambda(\mathrm{Tr}\, DS, \Lambda)$ is zero, we show that $T = \mathrm{Ext}^1_\Lambda(X, S)$ is a simple injective Γ-module and that F induces an equivalence of categories between $\mathcal{C}_S$ and $\mathfrak{D}_T$. We go on to study some of the relationships between the algebras Λ and Γ. We show that if Λ is hereditary then Γ is. Actually, we show more generally that gl. dim. $\Lambda - 1 \leqslant$ gl. dim. $\Gamma \leqslant$ gl. dim. Λ and that gl. dim. $\Lambda =$ gl. dim. Γ if $\mathrm{id}_\Lambda S <$ gl. dim. Λ, where $\mathrm{id}_\Lambda S$ denotes the injective dimension of S, or if $(\mathrm{Tr}\, DS)^* = 0$. We also show that if $(\mathrm{Tr}\, DS)^* = 0$, then Λ is a factor of an hereditary algebra if and only if Γ is.

We end the section with a construction of an almost split sequence $0 \to S \to Q \to K \to 0$, when S is a simple projective noninjective module over an artin ring Λ, which is not necessarily an artin algebra.

We start out with some preliminary results.

LEMMA 1.1. *Let A be in* mod Λ. *Then* $\mathrm{Ext}^1_\Lambda(\mathrm{Tr}\, DS, A) = 0$ *if and only if A is in* $\mathcal{C}_S = \mathcal{C}$.

PROOF. This follows directly from [6]. However, for the convenience of the reader we outline a proof. Let A be indecomposable in mod Λ. If $A = S$, we have an almost split sequence $0 \to S \to E \to \mathrm{Tr}\, DS \to 0$, so that $\mathrm{Ext}^1(\mathrm{Tr}\, DS, S) \neq 0$. Assume that A is not isomorphic to S, and assume that there is some nonsplit exact sequence $0 \to A \to B \to \mathrm{Tr}\, DS \to 0$. Then we have a commutative diagram

$$\begin{array}{ccccccccc} 0 & \to & A & \to & B & \to & \mathrm{Tr}\, DS & \to & 0 \\ & & \downarrow i & & \downarrow j & & \| & & \\ 0 & \to & S & \to & E & \to & \mathrm{Tr}\, DS & \to & 0 \end{array}$$

where the bottom sequence is almost split. Since i must be zero because S is simple projective, it follows that the bottom sequence splits, which is a contradiction.

LEMMA 1.2. *Let* $0 \to A \to B \to C \to 0$ *be an exact sequence in* mod Λ.

(a) *If A is in $\mathcal{C}$, then* $0 \to (X, A) \to (X, B) \to (X, C) \to 0$ *is exact.*

(b) *If B is in $\mathcal{C}$ and* $0 \to (X, A) \to (X, B) \to (X, C) \to 0$ *is exact, then A is in* $\mathcal{C}$.

PROOF. Consider the long exact sequence $0 \to (X, A) \to (X, B) \to (X, C) \to \mathrm{Ext}^1_\Lambda(X, A) \to \mathrm{Ext}^1_\Lambda(X, B)$. By Lemma 1.1, $\mathrm{Ext}^1_\Lambda(X, A)$ is zero, and B is in $\mathcal{C}$ if and only if $\mathrm{Ext}^1_\Lambda(X, B)$ is zero. Hence both (a) and (b) follow.

LEMMA 1.3. *Let Λ be an artin algebra, Y some finitely generated Λ-module,* $\Gamma = \mathrm{End}_\Lambda(Y)^{\mathrm{op}}$, *and* $F = (Y, \ )$: mod $\Lambda \to$ mod Γ.

(a) *F induces an equivalence of categories between* add(Y), *whose objects are the summands of finite sums of copies of Y, and the category of finitely generated projective Γ-modules.*

(b) *The restriction of F to the full subcategory of* mod Λ *whose objects are the M such that there is an exact sequence* $Y_1 \to Y_0 \to M \to 0$, *with* Y_0, Y_1 *in* add(Y) *with the property that* $(Y, Y_0) \to (Y, Y_1) \to (Y, M) \to 0$ *is exact, is fully faithful.*

See [4, §1] and [2, Proposition 3.2] for the proof.

We can now prove the following.

PROPOSITION 1.4. *Let the notation be as given before the previous lemmas. Then* F: $\mathcal{C} \to$ mod Γ *is a fully faithful functor.*

PROOF. By Lemma 1.3 it is enough to show that if M is in $\mathcal{C}$, then there is an exact sequence $X_2 \to X_1 \to M \to 0$ with X_2, X_1 in add(X), such that $F(X_2) \to F(X_1) \to F(M) \to 0$ is exact.

Since M is in $\mathcal{C}$, it is clear that the projective cover P of M has no summand isomorphic to S. Hence we have some epimorphism f: $X' \to M$ with X' in add(X). It is easy to see, by using that $\mathrm{Hom}_\Lambda(X, M)$ is a finitely generated module over the center of Λ, that we have some epimorphism f: $X_1 \to M$ with X_1 in add(X) such that $F(X_1) \to F(M) \to 0$ is exact. By Lemma 1.2(b) we have that, since X_1 is in $\mathcal{C}$, $K = \mathrm{Ker}(f)$ is in $\mathcal{C}$. As above we choose an epimorphism g: $X_2 \to K$ with X_2 in add(X), such that $F(X_2) \to F(K) \to 0$ is exact. Then $X_2 \to X_1 \to M \to 0$ has the property that $F(X_2) \to F(X_1) \to F(M) \to 0$ is exact, by Lemma 1.3(b). This finishes the proof of the proposition.

To prove a dual result we introduce the following notation. Assume that a basic artin algebra Γ has a simple injective module T which is not projective. Let I denote the direct sum of one copy of each of the other indecomposable injective Γ-modules. Let Y denote the Γ^{op}-module $D(I)\amalg\mathrm{Tr}(T)$, and let $\Lambda = \mathrm{End}(Y)$. Denote by G the functor $Y \otimes \quad$ from mod Γ to mod Λ, and also its restriction to $\mathcal{D}_T = \mathcal{D}$. We then have the following.

PROPOSITION 1.5. *Let the notation be as above. Then* G: $\mathcal{D}_T \to \mathrm{mod}\ \Lambda$ *is a fully faithful functor.*

PROOF. Let F: mod $\Gamma^{\mathrm{op}} \to$ mod Λ^{op} be the functor $\mathrm{Hom}_{\Gamma^{\mathrm{op}}}(\mathrm{Tr}\ T\amalg D(I),\ )$. Here Tr T = Tr $D(DT)$, where $D(T)$ is a simple projective Γ^{op}-module and $D(I)$ is the direct sum of one copy of each of the other indecomposable projective Γ^{op}-modules. Let M be in mod Γ. By [**11**, p. 119], we have a natural isomorphism

$$\begin{aligned} DG(M) &= D((\mathrm{Tr}\ T\amalg D(I)) \otimes_\Gamma M) \\ &\cong \mathrm{Hom}_\Gamma(\mathrm{Tr}\ T\amalg D(I), D(M)) = FD(M). \end{aligned}$$

If we consider this isomorphism an identification, we then have a commutative diagram:

$$\begin{array}{ccc} \mathrm{mod}\ \Gamma & \xrightarrow{D} & \mathrm{mod}\ \Gamma^{\mathrm{op}} \\ G\downarrow & & F\downarrow \\ \mathrm{mod}\ \Lambda & \xrightarrow{D} & \mathrm{mod}\ \Lambda^{\mathrm{op}} \end{array}$$

Since we know from Proposition 1.4 that F: $\mathcal{C}_{D(T)} \to$ mod Λ^{op} is fully faithful, it follows that G: $\mathcal{D}_T \to$ mod Λ is fully faithful.

Before we give the main result of this section we need some more preliminary results.

LEMMA 1.6. *Let* $0 \to S \xrightarrow{g} Q \xrightarrow{f} \operatorname{Tr} DS \to 0$ *be the almost split sequence in* mod Λ. *Then*:

(a) f *is a projective cover.*

(b) *The induced sequence of* Γ-*modules* $0 \to (X, Q) \to (X, \operatorname{Tr} DS) \to \operatorname{Ext}^1(X, S) \to 0$ *is exact and is a minimal projective presentation for* $\operatorname{Ext}^1_\Lambda(X, S)$.

(c) $T = \operatorname{Ext}^1_\Lambda(X, S)$ *is a simple* Γ-*module.*

PROOF. Since S is simple projective we know from [6, Theorem 5.5], that we have an almost split sequence $0 \to S \to Q \to \operatorname{Tr} DS \to 0$, where $Q \to \operatorname{Tr} DS$ is a projective cover. We also know from [7, Theorem 2.4], that in an almost split sequence no end term is isomorphic to a summand of the middle term. Hence S is not isomorphic to a summand of Q. By Lemma 1.1 we then have $\operatorname{Ext}^1_\Lambda(\operatorname{Tr} DS, Q) = 0$, and consequently $\operatorname{Ext}^1_\Lambda(X, Q) = 0$. Since also $(X, S) = 0$, we have the exact sequence $0 \to (X, Q) \xrightarrow{s} (X, \operatorname{Tr} DS) \xrightarrow{t} \operatorname{Ext}^1_\Lambda(X, S) \to 0$. Since $0 \to S \to Q \to \operatorname{Tr} DS \to 0$ is almost split, we have from [6, Proposition 3.1], that $\operatorname{Ext}^1_\Lambda(X, S)$ is a simple Γ-module. The given projective resolution of T is minimal because $(X, \operatorname{Tr} DS)$ is indecomposable. Therefore (a), (b) and (c) have been proven.

LEMMA 1.7. *If* $(\operatorname{Tr} DS)^* = 0$, *then* $\operatorname{End}(\operatorname{Tr} DS)^{\mathrm{op}}$ *is a division algebra.*

PROOF. Let $0 \to S \xrightarrow{f} Q \xrightarrow{g} \operatorname{Tr} DS \to 0$ be an almost split sequence. Assume that there is a nonzero morphism h: $\operatorname{Tr} DS \to \operatorname{Tr} DS$ which is not an isomorphism. Then there would be a nonzero morphism t: $\operatorname{Tr} DS \to Q$ such that $gt = h$. Since $(\operatorname{Tr} DS)^* = 0$ and Q is projective, we have a contradiction. Hence every nonzero morphism h: $\operatorname{Tr} DS \to \operatorname{Tr} DS$ is an isomorphism, so that $\operatorname{End}(\operatorname{Tr} DS)^{\mathrm{op}}$ is a division algebra.

LEMMA 1.8. *Assume that* $(\operatorname{Tr} DS)^* = 0$. *If* α: $(X, \operatorname{Tr} DS) \to (X, X_1)$, *with* X_1 *in* add(X), *is a nonzero morphism, then* α *is a split monomorphism.*

PROOF. Since X_1 is in add(X), we have $X_1 = Q \amalg (\operatorname{Tr} DS)^r$, where Q is projective, $r \geqslant 0$ and $(\operatorname{Tr} DS)^0 = 0$. Given α: $(X, \operatorname{Tr} DS) \to (X, X_1)$, there is by Lemma 1.3 a morphism α_1: $\operatorname{Tr} DS \to X_1 = Q \amalg (\operatorname{Tr} DS)^r$ in mod Λ such that $(X, \alpha_1) = \alpha$. Since $(\operatorname{Tr} DS)^* = 0$, we have $\operatorname{Im} \alpha_1 \subseteq (\operatorname{Tr} DS)^r$. There is then a morphism $(\operatorname{Tr} DS)^r \to \operatorname{Tr} DS$ such that the composite morphism $\operatorname{Tr} DS \to (\operatorname{Tr} DS)^r \to \operatorname{Tr} DS$ is not zero. Since, by Lemma 1.7, $\operatorname{End}(\operatorname{Tr} DS)^{\mathrm{op}}$ is a division algebra, α_1: $\operatorname{Tr} DS \to (\operatorname{Tr} DS)^r$ is a split monomorphism, so that $\alpha = (X, \alpha_1)$: $(X, \operatorname{Tr} DS) \to (X, X_1)$ is a split monomorphism.

LEMMA 1.9. *If* $(\operatorname{Tr} DS)^* = 0$, *then the simple* Γ-*module* $T = \operatorname{Ext}^1(X, S)$ *is injective.*

PROOF. To prove the lemma it is enough to show that every monomorphism $\alpha\colon T \to N$ splits, where N is in mod Γ. So let $\alpha\colon T \to N$ be a monomorphism. Let $(X, X_1) \to N$ be a projective cover, where X_1 is in add(X). Then we have a commutative diagram:

$$\begin{array}{ccccc} (X, \operatorname{Tr} DS) & \to & T & \to & 0 \\ \downarrow\beta & & \downarrow\alpha & & \\ (X, X_1) & \to & N & \to & 0 \end{array}$$

Since β is not zero, it is a split monomorphism by Lemma 1.8. Hence the composite morphism $(X, \operatorname{Tr} DS) \to (X, X_1) \to N \to N/\mathbf{r}N$ is not zero, where $\mathbf{r}N$ denotes the radical of the module N. This means that $T \to N \to N/\mathbf{r}N$ is not zero. It follows that α is a split monomorphism, and this shows that T is injective.

LEMMA 1.10. *Let N be indecomposable in* mod Γ. *Then there is some M in* mod Λ *and a nonnegative integer r such that there is an exact sequence* $0 \to T^r \to N \to (X, M) \to 0$.

PROOF. Let $(X, X_2) \xrightarrow{s} (X, X_1) \to N \to 0$ be a minimal projective presentation for N in mod Γ, where X_1 and X_2 are in add(X). Consider the exact sequence $X_2 \xrightarrow{h} X_1 \xrightarrow{g} M \to 0$ in mod Λ, where $h\colon X_2 \to X_1$ is such that $F(h) = s$. Let $L = \operatorname{Ker}(g)$ and let $0 \to K \to X_2 \to L \to 0$ be exact. Then $0 \to (X, K) \to (X, X_2) \xrightarrow{t} (X, L) \to \operatorname{Ext}^1_\Lambda(X, K)$ is exact. But by Lemma 1.1, $\operatorname{Ext}^1_\Lambda(X, K) \simeq T^j$, where j is the multiplicity of S as a summand of K. Hence, $\operatorname{Coker}(t) \simeq T^r$, for some $r \leq j$. Since L is clearly in $\mathcal{C}$, we have by Lemma 1.2 an exact sequence $0 \to (X, L) \to (X, X_1) \to (X, M) \to 0$. We then have a commutative diagram:

$$\begin{array}{ccccccccc} & & 0 & & & & & & \\ & & \downarrow & & & & & & \\ 0 & \to & (X, X_2)/(X, K) & \to & (X, X_1) & \to & N & \to & 0 \\ & & \downarrow & & \| & & \downarrow u & & \\ 0 & \to & (X, L) & \to & (X, X_1) & \to & (X, M) & \to & 0 \\ & & \downarrow & & & & \downarrow & & \\ & & T^r & & & & 0 & & \\ & & \downarrow & & & & & & \\ & & 0 & & & & & & \end{array}$$

Hence $\operatorname{Ker}(u) = T^r$, that is, we have the exact sequence $0 \to T^r \to N \xrightarrow{u} (X, M) \to 0$, as desired.

We are now ready to prove the main results of this section.

THEOREM 1.11. *Let* Λ *be a basic algebra with a simple projective noninjective* Λ*-module* S, *such that* $(\mathrm{Tr}\, DS)^* = 0$. *Let* $X = P \amalg \mathrm{Tr}\, DS$, *where* P *is given by* $\Lambda = P \amalg S$. *Let* $\Gamma = \mathrm{End}_\Lambda(X)^{\mathrm{op}}$ *and consider the functor* $F = \mathrm{Hom}_\Lambda(X,\ )$: $\mathrm{mod}\, \Lambda \to \mathrm{mod}\, \Gamma$. *Then* $T = \mathrm{Ext}^1_\Lambda(X, S)$ *is a simple injective* Γ*-module, and* F *induces an equivalence of categories between the full subcategory* $\mathcal{C} = \mathcal{C}_S$ *of* $\mathrm{mod}\, \Lambda$ *whose objects are the* A *in* $\mathrm{mod}\, \Lambda$ *not having* S *as a summand, and the full subcategory* $\mathcal{D} = \mathcal{D}_T$ *of* $\mathrm{mod}\, \Gamma$ *whose objects are the* B *in* $\mathrm{mod}\, \Gamma$ *not having* T *as a summand.*

PROOF. By Proposition 1.4 we know that the restriction of F to $\mathcal{C}$ is fully faithful, and by Lemma 1.9 we know that $T = \mathrm{Ext}^1_\Lambda(X, S)$ is a simple injective Γ-module. Let now N be an object in $\mathcal{D}$. By Lemma 1.10 there is then some exact sequence $0 \to T' \to N \to (X, M) \to 0$. Since T is injective and cannot be a summand of N, we have an isomorphism $N \to (X, M)$. This finishes the proof of the theorem.

THEOREM 1.12. *Let* Γ *be a basic artin algebra with a simple injective nonprojective module* T *such that* $(\mathrm{Tr}\, T)^* = 0$. *Let* I *denote the sum of one copy of each of the other indecomposable injective* Γ*-modules. Let* $Y = D(I) \amalg \mathrm{Tr}\, T$, *and* $\Lambda = \mathrm{End}_\Gamma(Y)$. *Then* $S = \mathrm{Tor}^\Gamma_1(Y, T)$ *is a simple projective* Λ*-module and the functor* $G = Y \otimes\ \ $: $\mathrm{mod}\, \Gamma \to \mathrm{mod}\, \Lambda$ *induces an equivalence of categories between* $\mathcal{D} = \mathcal{D}_T$ *and* $\mathcal{C} = \mathcal{C}_S$.

PROOF. This follows directly from the proof of Proposition 1.5 and from Theorem 1.11. We only need to point out that $\mathrm{Ext}^1_\Gamma(Y, DT)$ is a simple injective Λ^{op}-module and that $D(\mathrm{Ext}^1_\Lambda(Y, DT)) = \mathrm{Tor}^\Gamma_1(Y, T) = S$ by [**11**, p. 120].

We point out that if Λ is an hereditary basic artin algebra which is not semisimple, there is always some simple projective noninjective Λ-module S and, since submodules of projective modules are projective, the condition $(\mathrm{Tr}\, DS)^* = 0$ automatically holds. Since this is an important special case of Theorem 1.11 we formulate it as a special result.

PROPOSITION 1.13. *Let* Λ *be a basic hereditary artin algebra and* S *a simple projective noninjective* Λ*-module. With the previous notations,* $T = \mathrm{Ext}^1_\Lambda(X, S)$ *is a simple injective* Λ*-module, and* $F = \mathrm{Hom}_\Lambda(X,\ )$: $\mathrm{mod}\, \Lambda \to \mathrm{mod}\, \Gamma$ *induces an equivalence of categories between* $\mathcal{C}_S$ *and* $\mathcal{D}_T$. *Further,* Γ *is again hereditary.*

PROOF. In view of our above results, we only need to prove that Γ is hereditary. This will be a consequence of the first part of the next lemma, since we know that $\mathrm{pd}_\Gamma T = 1$.

We shall now investigate the connection between the algebras Λ and Γ with respect to global dimension and with respect to the property of being a factor of an hereditary artin algebra.

LEMMA 1.14. *Let, as before,* Λ *be a basic artin algebra with a simple projective noninjective module* S, $X = P \amalg \operatorname{Tr} DS$, $\Gamma = \operatorname{End}_\Lambda(X)^{\mathrm{op}}$. *Let* M *be an indecomposable* Λ*-module not isomorphic to* S.

(a) $\operatorname{pd}_\Lambda M - 1 \leqslant \operatorname{pd}_\Gamma(X, M) \leqslant \operatorname{pd}_\Lambda M$.

(b) *If* $\operatorname{id}_\Lambda S < \operatorname{pd}_\Lambda M$ *or if* $(\operatorname{Tr} DS)^* = 0$ *and* M *is not isomorphic to* $\operatorname{Tr} DS$, *then* $\operatorname{pd}_\Gamma(X, M) = \operatorname{pd}_\Lambda M$.

PROOF. (a) For $i \geqslant 1$, let $\Omega^i M$ be defined by the exact sequence $0 \to \Omega^i M \to P_{i-1} \to \cdots \to P_0 \to M \to 0$, where $P_{i-1} \to \cdots \to P_0 \to M \to 0$ is part of a minimal projective resolution for M. Write $\Omega^0 M = M$. We first prove that if B is a Λ-module such that $(X, \Omega^1 B) \neq 0$, then $\operatorname{pd}(X, B) = \operatorname{pd}(X, \Omega^1 B) + 1$. To see this, consider the exact sequence $0 \to \Omega^1 B \to P_0 \to B \to 0$. We can clearly assume that B has no nonzero projective summand, and hence P_0 can have no summand isomorphic to S. We then have $\operatorname{Ext}^1_\Lambda(X, P_0) = 0$ by Lemma 1.1, and consequently an exact sequence

$$0 \to (X, \Omega^1 B) \to (X, P_0) \to (X, B) \to \operatorname{Ext}^1_\Lambda(X, \Omega^1 B) \to 0.$$

This gives rise to the exact sequences

$$0 \to (X, \Omega^1 B) \to (X, P_0) \to N \to 0$$

and

$$0 \to N \to (X, B) \to \operatorname{Ext}^1_\Lambda(X, \Omega^1 B) \to 0.$$

Since by our assumption $(X, \Omega^1 B) \neq 0$, it is easy to see, using that $(X, \ ): \mathcal{C} \to \operatorname{mod} \Gamma$ is fully faithful, that $0 \to (X, \Omega^1 B) \to (X, P_0)$ is not a split monomorphism, so that N is not projective, and hence $\operatorname{pd} N = \operatorname{pd}(X, \Omega^1 B) + 1$. By Lemmas 1.1 and 1.6 we have that if $\operatorname{Ext}^1_\Lambda(X, \Omega^1 B) \neq 0$, then $\operatorname{pd} \operatorname{Ext}^1_\Lambda(X, \Omega^1 B) = 1$. Since $\operatorname{pd} N \geqslant 1$, we get that $\operatorname{pd} N = \operatorname{pd}(X, B)$, so that $\operatorname{pd}(X, B) = \operatorname{pd}(X, \Omega^1 B) + 1$. If $(X, B) \neq 0$ and $(X, \Omega^1 B) = 0$, we would get $\operatorname{pd}(X, B) \leqslant 1$.

Assume now that M is an indecomposable Λ-module not isomorphic to S. If M is projective we have $\operatorname{pd}(X, M) = 0 = \operatorname{pd} M$. Assume now that $\operatorname{pd}_\Lambda M = n \geqslant 1$. For $i < n$ $\Omega^i M$ is then not projective, so that $(X, \Omega^i M) \neq 0$. By repeated application of the above we then get that $\operatorname{pd}(X, M) = \operatorname{pd}(X, \Omega^{n-1} M) + n - 1$, and if also $(X, \Omega^n M) \neq 0$, we get $\operatorname{pd}(X, M) = \operatorname{pd}(X, \Omega^n M) + n = 0 + n = n$. If $(X, \Omega^n M) = 0$, it follows from the above that $\operatorname{pd}(X, \Omega^{n-1} M) \leqslant 1$, so that $n - 1 \leqslant \operatorname{pd}(X, M) \leqslant n$. If $\operatorname{pd} M = \infty$, we clearly have $(X, \Omega^i M) \neq 0$ for all i, so that $\operatorname{pd}(X, M) = \operatorname{pd}(X, \Omega^i M) + i \geqslant i$ for all i, and consequently $\operatorname{pd}(X, M) = \infty$. This finishes the proof of (a).

(b) Let now M be an indecomposable Λ-module not isomorphic to S, so that (X, M) is not zero. Assume that $\operatorname{pd} M = n$ and $\operatorname{pd}(X, M) \neq n$. By the above, we must then have $n \geqslant 1$ and $(X, \Omega^n M) = 0$. Since $\Omega^n M \neq 0$, we must then have $S | \Omega^n M$, and hence $\operatorname{Ext}^n_\Lambda(M, S) \neq 0$. This shows that $\operatorname{id}_\Lambda S \geqslant n$.

Further, if $\operatorname{pd}(X, M) \neq n$, we also see by the above that we must have $\operatorname{pd}(X, \Omega^{n-1}M) = 0$. Since $\Omega^{n-1}M$ is not projective, this means that $\operatorname{Tr} DS | \Omega^{n-1}M$. If $n > 1$, $\operatorname{Tr} DS$ is a submodule of P_{n-2}, and hence $(\operatorname{Tr} DS)^* \neq 0$. If $n = 1$, $\operatorname{Tr} DS \simeq M = \Omega^0 M$. This finishes the proof of (b).

We use the above lemma to get the following relationship between the global dimensions of Λ and Γ.

PROPOSITION 1.15. *Let Λ be a basic artin algebra with a simple projective noninjective module S, and let X, Γ, T and F be as before.*

(a) gl. dim. $\Lambda - 1 \leqslant$ gl. dim. $\Gamma \leqslant$ gl. dim. Λ.

(b) *If* $\operatorname{id}_\Lambda S <$ gl. dim. Λ *or if* $(\operatorname{Tr} DS)^* = 0$, *then* gl. dim. Λ = gl. dim. Γ.

PROOF. If N is an indecomposable Γ-module, there is by Lemma 1.10 an exact sequence $0 \to T' \to N \to (X, M) \to 0$ with M in mod Λ. Since $\operatorname{pd}_\Gamma T = 1$, we have $\operatorname{pd}_\Gamma N \leqslant \max(1, \operatorname{pd}_\Gamma(X, M))$, hence $\operatorname{pd}_\Gamma N \leqslant \operatorname{pd}_\Gamma(X, M)$, for if (X, M) is projective, then $N = (X, M)$. The desired result is then a direct consequence of Lemma 1.14.

An example of an artin algebra Λ such that gl. dim. Γ = gl. dim. $\Lambda - 1$ is

$$\Lambda = \begin{pmatrix} K & 0 & 0 \\ K & K & 0 \\ 0 & K & K \end{pmatrix},$$

where K is a field. Here gl. dim. $\Lambda = 2$, S is the unique simple projective Λ-module and

$$\Gamma = \begin{pmatrix} K & 0 & 0 \\ K & K & 0 \\ K & K & K \end{pmatrix}$$

which has global dimension 1.

We next study the relationship between Λ and Γ with respect to being factors of hereditary artin algebras. For this we shall use that Λ is a factor of an hereditary algebra if and only if the following condition holds. Given an infinite chain $\cdots \to P_n \xrightarrow{f_n} P_{n-1} \to \cdots \to P_1 \xrightarrow{f_1} P_0$ of nonzero morphisms between indecomposable projective modules, there is some i such that f_n is an isomorphism for $n > i$ (see [**19**]).

PROPOSITION 1.16. *Let Λ be a basic artin algebra with a simple projective noninjective module S, and let X, Γ, T and F be as before.*

(a) *If Γ is a factor of an hereditary algebra, then Λ is.*

(b) *If* $(\operatorname{Tr} DS)^* = 0$ *and Λ is a factor of an hereditary algebra, then Γ is.*

PROOF. (a) Assume that Γ is a factor of an hereditary artin algebra, and consider a chain $\cdots \to P_n \xrightarrow{f_n} P_{n-1} \to \cdots \to P_1 \xrightarrow{f_1} P_0$ of nonzero maps between indecomposable projective Λ-modules. We can assume that no P_i is isomorphic to S, since otherwise f_n would be an isomorphism for $n > i$. We then have a chain

$$\cdots \to (X, P_n) \xrightarrow{(X, f_n)} (X, P_{n-1}) \to \cdots \to (X, P_1) \xrightarrow{(X, f_1)} (X, P_0)$$

of nonzero maps between indecomposable projective Γ-modules. Since Γ is a factor of an hereditary artin algebra, there is some i such that (X, f_n) is an isomorphism for $n > i$. Hence f_n is an isomorphism for $n > i$. This finishes the proof of (a).

(b) Assume that $(\mathrm{Tr}\, DS)^* = 0$ and that Λ is a factor of an hereditary algebra. The proof of this part is similar, using now that there is no nonzero map from $(X, \mathrm{Tr}\, DS)$ to (X, Q) when Q is an indecomposable projective Λ-module.

Let the notation be as before Lemma 1.1. It would be interesting to know the connection between the representation theory of Λ and Γ. In particular, when does the fact that one of these algebras is of finite representation type imply that the other one also is. From Proposition 1.4 it follows that if Γ is of finite representation type, then Λ is, and it follows from Theorem 1.11 that the converse holds if $(\mathrm{Tr}\, DS)^* = 0$. On the other hand, Bautista has given an example showing that Γ may be of infinite representation type and Λ of finite type.

Since a complete classification of when an hereditary artin algebra is of finite representation type is known **[13]**, [3], it would also be interesting to know for which artin algebras Λ we reach an hereditary algebra after applying our construction a finite number of times. By the above we know that it is necessary that Λ is a factor of an hereditary artin algebra, and if Λ is not hereditary it is necessary that there is a simple projective Λ-module S whose injective dimension is equal to gl. dim. Λ.

We end this section with some comments on the possibility of generalizing the material in this section to arbitrary artin rings (i.e. left and right artin rings). One of the essential tools used is the existence of an almost split sequence $0 \to S \to Q \to K \to 0$, with Q projective, when S is a simple projective noninjective Λ-module. While in general almost split sequences may not exist in mod Λ, we can prove that an almost split sequence of the above type exists when S is simple projective and not injective. Actually we show how to construct this almost split sequence, and this at the same time gives a direct approach in this special case also for artin algebras. For S simple projective we denote $S^*/\mathbf{r}S^*$ by $D'S$. Note that if Λ is an artin algebra, then $D'S = DS$.

PROPOSITION 1.17. *Let S be a simple projective noninjective module over an artin ring Λ, and let $Q^* \xrightarrow{t^*} S^* \to D'S \to 0$ be a minimal projective presentation* *in* mod Λ^{op}, *where* Q *is projective in* mod Λ. *Then we get an almost split sequence* $0 \to S \xrightarrow{t} Q \to \operatorname{Tr} D'S \to 0$.

PROOF. The sequence $0 \to S \xrightarrow{t} Q \to \operatorname{Tr} D'S \to 0$ is clearly exact since t is not zero and hence a monomorphism. Also it is easy to see that it does not split. Let now $j\colon S \to U$ be a map which is not a split monomorphism. Assume first that U is projective. Then $j^*\colon U^* \to S^*$ is not a split epimorphism. Hence $\operatorname{Im}(j^*) \subsetneq \mathbf{r}S^*$, so there is a map $q^*\colon U^* \to Q^*$ such that $t^*q^* = j^*$, and consequently $qt = j$. If U is not projective we have, since S is projective, a commutative diagram

$$\begin{array}{ccc} S & \xrightarrow{\ i\ } & P_U \\ \big\downarrow{\scriptstyle j} & \swarrow{\scriptstyle p} & \\ U & & \end{array}$$

where $p\colon P_U \to U$ is a projective cover. Since j is not a split monomorphism, i is not, because S is simple projective. By the above we then have a map q: $Q \to P_U$ such that $qt = i$, so that $pq\colon Q \to U$ verifies that $(pq)t = j$. So t: $S \to Q$ is left almost split. Since $\operatorname{Tr} D'S$ is indecomposable, it follows as in the proof of Proposition 4.1 in [6], that $0 \to S \xrightarrow{t} Q \to \operatorname{Tr} D'S \to 0$ is an almost split sequence.

On the basis of this proposition the Lemmas 1.1, 1.2 and 1.7 still hold if Λ is an artin ring which is not necessarily an artin algebra. We do not know exactly how many of the other results carry over. We would also like to know if Γ is necessarily left artin and if $(X,\ )$ takes mod Λ to mod Γ. If $(\operatorname{Tr} D'S)^* = 0$, we shall see as a consequence of our results in §4, where we consider related functors, that this turns out to be the case. Hence also Theorem 1.11 carries over to arbitrary artin rings.

2. General study of partial Coxeter functors. Let Λ be a basic left artin ring with a simple projective noninjective module S, and Γ a basic left artin ring with a simple injective nonprojective module T. Denote as before by $\mathcal{C} = \mathcal{C}_S$ the full subcategory of mod Λ whose objects are the modules in mod Λ having no summand isomorphic to S, and by $\mathfrak{D} = \mathfrak{D}_T$ the full subcategory of mod Γ whose objects are the objects in mod Γ with no summand isomorphic to T. We define a functor $F = F_S$: mod $\Lambda \to$ mod Γ to be a *left partial Coxeter functor* if $F(S) = 0$ and F induces an equivalence between $\mathcal{C}_S$ and $\mathfrak{D}_T$, for a simple projective noninjective Λ-module S and a simple injective nonprojective Γ-module T. We also define a functor $G = G_T$: mod $\Gamma \to$ mod Λ to be a right partial Coxeter functor if $G(T) = 0$ and G induces an equivalence of categories between $\mathfrak{D}_T$ and $\mathcal{C}_S$ for a simple injective nonprojective Γ-module T and a simple projective noninjective Λ-module S.

In §5 we shall study the connection with representations of diagrams, where these concepts were originally defined. In the notation of §1, we have seen there for an artin algebra that if $(\mathrm{Tr}\, DS)^* = 0$, then $(X, \ )$ is a left partial Coxeter functor and that if $(\mathrm{Tr}\, T)^* = 0$, then $Y \otimes$ is a right partial Coxeter functor. So a study of left and right partial Coxeter functors in general will give in particular more information about these functors, and this will be used in §3.

In this section we show that a left and a right partial Coxeter functor between artin rings is uniquely determined up to "equivalence" of functors in the following sense. For left artin rings Λ, Γ, Λ', Γ' we say that two functors F: mod $\Lambda \to$ mod Γ and F': mod $\Lambda' \to$ mod Γ' are *equivalent* if there are equivalences of categories u: mod $\Lambda \to$ mod Λ' and v: mod $\Gamma \to$ mod Γ' such that $vF = F'u$. To prove this uniqueness we shall need some preliminary results which will be useful also in §3. We further show that up to equivalence a left partial Coxeter functor F is of the following type if Λ is artin. $F = (X, \ )$: mod $\Lambda \to$ mod Γ, where $X = P \amalg \mathrm{Tr}\, D'S$, P is given by $S \amalg P = \Lambda$, $\Gamma = \mathrm{End}(X)^{\mathrm{op}}$ and $D'S$ is, as in §1, the simple Λ^{op}-module with the same annihilator as S. We also show that the left partial Coxeter functors studied in §1 are essentially the only ones for artin algebras, and the same will be the case for right partial Coxeter functors. In §4 we shall show that for an artin ring $(X, \ )$ as defined above is a left partial Coxeter functor when $(\mathrm{Tr}\, D'S)^* = 0$.

At the end of the section we get some more information about partial Coxeter functors for artin algebras, which we shall need in §3. The reason for the restriction to artin algebras is that we use the existence of almost split sequences and duality.

We start out with some preliminary results.

LEMMA 2.1. *Let Λ be a left artin ring, $\mathcal{C}$ a full additive subcategory of* mod Λ. *Let $f: A \to B$ and $g: B \to C$ be morphisms in $\mathcal{C}$.*

(a) *If $\mathcal{C}$ is closed under kernels, then $0 \to A \xrightarrow{f} B \xrightarrow{g} C$ is exact in* mod Λ *if and only if $0 \to (\ , A)|\mathcal{C} \to (\ , B)|\mathcal{C} \to (\ , C)|\mathcal{C}$ is exact, where $(\ , V)|\mathcal{C}$ denotes the restriction of the functor $(\ , V)$ to $\mathcal{C}$.*

(b) *If $\mathcal{C}$ is closed under cokernels, then $A \xrightarrow{f} B \xrightarrow{g} C \to 0$ is exact if and only if $0 \to (C, \ )|\mathcal{C} \to (B, \ )|\mathcal{C} \to (A, \ )|\mathcal{C}$ is exact.*

(c) *If $\mathcal{C}$ is closed under submodules or under factor modules, then $0 \to A \xrightarrow{f} B \xrightarrow{g} C \to 0$ is exact in* mod Λ *if and only if $0 \to (\ , A)|\mathcal{C} \to (\ , B)|\mathcal{C} \to (\ , C)|\mathcal{C}$ and $0 \to (C, \ )|\mathcal{C} \to (B, \ )|\mathcal{C} \to (A, \ )|\mathcal{C}$ are exact.*

PROOF. (a) Assume that $0 \to (\ , A)|\mathcal{C} \to (\ , B)|\mathcal{C} \to (\ , C)|\mathcal{C}$ is exact. Consider the exact sequence of Λ-modules $0 \to \mathrm{Ker}(g) \to B \to C$, where by assumption Ker(g) is in $\mathcal{C}$. We then have the exact sequence $0 \to (\ , \mathrm{Ker}(g))|\mathcal{C} \to (\ , B)|\mathcal{C} \to (\ , C)|\mathcal{C}$. Since also $0 \to (\ , A)|\mathcal{C} \to (\ , B)|\mathcal{C} \to (\ , C)|\mathcal{C}$ is exact, we get a commutative diagram

$$\begin{array}{ccccccc} 0 & \to & A & \xrightarrow{f} & B & \xrightarrow{g} & C \\ & & \downarrow\wr & & \| & & \| \\ 0 & \to & \mathrm{Ker}\, g & \to & B & \xrightarrow{g} & C \end{array}$$

which shows that $0 \to A \xrightarrow{f} B \xrightarrow{g} C$ is exact. The other implication is trivial.

(b) The proof is similar to the proof of (a).

(c) Assume that $0 \to (\ , A)|\mathcal{C} \to (\ , B)|\mathcal{C} \to (\ , C)|\mathcal{C}$ and $0 \to (C, \ )|\mathcal{C} \to (B, \ )|\mathcal{C} \to (A, \ )|\mathcal{C}$ are exact. Assume that $\mathcal{C}$ is closed under submodules. Then it follows from (a) that $0 \to A \xrightarrow{f} B \xrightarrow{g} C$ is exact. Consider the exact sequence $0 \to A \xrightarrow{f} B \xrightarrow{g} \mathrm{Im}(g) \to 0$. Then $0 \to (\mathrm{Im}(g), \ )|\mathcal{C} \to (B, \ )|\mathcal{C} \to (A, \ )|\mathcal{C}$ is exact. Since Im(g) is contained in C, Im(g) is in $\mathcal{C}$. Using that $0 \to (C, \ )|\mathcal{C} \to (B, \ )|\mathcal{C} \to (A, \ )|\mathcal{C}$ is exact we get our desired conclusion. If $\mathcal{C}$ is closed under factor modules, the proof is similar. The proof of the converse is trivial.

LEMMA 2.2. *Let Λ and Λ' be left artin rings, $\mathcal{C}$ a full additive subcategory of* mod Λ, *$\mathcal{C}'$ a full additive subcategory of* mod Λ'. *Assume that $\mathcal{C}$ and $\mathcal{C}'$ are closed under submodules or factor modules, and that there is an equivalence of categories α: $\mathcal{C} \to \mathcal{C}'$.*

(a) *Let f: $A \to B$ and g: $B \to C$ be morphisms in $\mathcal{C}$. Then $0 \to A \xrightarrow{f} B \xrightarrow{g} C \to 0$ is exact in* mod Λ *if and only if $0 \to \alpha A \xrightarrow{\alpha f} \alpha B \xrightarrow{\alpha g} \alpha C \to 0$ is exact in* mod Λ'.

(b) *Let C in $\mathcal{C}$ be indecomposable.*

(i) *If $\mathcal{C}'$ is closed under subobjects and projective covers then $\alpha(C)$ is projective if C is projective.*

(ii) *If $\mathcal{C}$ is closed under subobjects and projective covers, then C is projective if αC is projective.*

(c) *If $\mathcal{C}$ is closed under extensions and all the indecomposable nonprojective objects are in $\mathcal{C}$, then C is injective if αC is injective.*

(d) *If $\mathcal{C}$ and $\mathcal{C}'$ are closed under subobjects and projective covers, then α can be extended to an equivalence α_1: $\mathcal{C}_1 \to \mathcal{C}_1'$. Here the objects of $\mathcal{C}_1$ are the M in* mod Λ *such that there is an exact sequence $P_1 \to P_0 \to M \to 0$ with P_1, P_0 projective and in $\mathcal{C}$, and the objects in $\mathcal{C}_1'$ are defined analogously.*

PROOF. (a) is an immediate consequence of Lemma 2.1.

(b)(i) Assume that C is indecomposable projective and in $\mathcal{C}$. Since $\mathcal{C}'$ is closed under subobjects and projective covers, we have an exact sequence $0 \to \alpha A \to \alpha B \to \alpha C \to 0$, where αB is projective in mod Λ'. By (a) we then have an exact sequence $0 \to A \to B \to C \to 0$, which must split since C is projective. Hence $\alpha B \to \alpha C$ is a split epimorphism, which shows that αC is projective.

(ii) The proof is analogous to the proof of (i).

(c) Let C be in $\mathcal{A}$ such that αC is injective in mod Λ'. Suppose K is indecomposable in $\mathcal{A}$ and consider an exact sequence $0 \to C \to E \to K \to 0$. By assumption E is in $\mathcal{A}$, so that by (a) $0 \to \alpha C \to \alpha E \to \alpha K \to 0$ is exact. This splits if C is injective, hence $0 \to C \to E \to K \to 0$ splits. This shows that $\operatorname{Ext}^1_\Lambda(K, C) = 0$. Since every indecomposable object in mod Λ not in $\mathcal{A}$ is projective, we conclude that $\operatorname{Ext}^1_\Lambda(\ , C)$ is zero, which implies that C is injective.

(d) Assume that $\mathcal{A}$ and $\mathcal{A}'$ are closed under subobjects and projective covers. By (b) there is induced an equivalence between the full subcategory of projective objects in $\mathcal{A}$ and in $\mathcal{A}'$, which induces an equivalence α_1: $\mathcal{A}_1 \to \mathcal{A}'_1$. Here α_1 is defined in the following way. If M is indecomposable and in $\mathcal{A}_1$ we have by assumption an exact sequence $P_1 \xrightarrow{f} P_0 \to M \to 0$ with P_1, P_0 projective and in $\mathcal{A}$. We then define $\alpha_1(M) = \operatorname{Coker} \alpha(f)$. We now want to show that α_1 extends α. Let C be indecomposable in $\mathcal{A}$, and $P_1 \xrightarrow{f} P_0 \xrightarrow{g} C \to 0$ a minimal projective presentation. By the assumption on $\mathcal{A}$ we know that P_1 and P_0 are in $\mathcal{A}$. It is then not hard to see that $\alpha P_1 \to \alpha P_0 \to \alpha C \to 0$ is exact, and this finishes the proof.

We are now ready to deduce some properties about partial Coxeter functors, in particular, to get uniqueness and a description of what they look like up to equivalence.

We first recall that if t: $\Lambda_1 \to \Lambda_2$ is a ring isomorphism, then t induces naturally an equivalence of categories t^*: mod $\Lambda_2 \to$ mod Λ_1 in the following way. For a Λ_2-module M, $t^*(M)$ is the abelian group M with the Λ_1-module structure induced by t. That is, $\lambda_1 m = t(\lambda_1)m$, for Λ_1 in Λ_1, m in M.

THEOREM 2.3. *Let Λ and Γ be left artin rings, S a simple projective noninjective Λ-module, $F = F_S$:* mod $\Lambda \to$ mod Γ *a left partial Coxeter functor, and let T denote the simple injective nonprojective Γ-module associated to F.*

(a) *F is left exact and if $0 \to A \to B \to C \to 0$ is exact with A in $\mathcal{C} = \mathcal{C}_S$, then $0 \to F(A) \to F(B) \to F(C) \to 0$ is exact.*

(b) *If $F' = F'_S$:* mod $\Lambda \to$ mod Γ' *is a left partial Coxeter functor, then there is a ring isomorphism t: $\Gamma' \to \Gamma$ such that we have a commutative diagram*

$$\begin{array}{ccc} \text{mod } \Lambda & \xrightarrow{F} & \text{mod } \Gamma \\ & \searrow^{F'} & \downarrow t^* \\ & & \text{mod } \Gamma' \end{array}$$

where t^* is an equivalence of categories induced by t: $\Gamma' \to \Gamma$.

(c) *F has a left adjoint G which is a right partial Coxeter functor.*

(d) *If* Λ *is artin, then* F *is equivalent to* $(X,\ )$: $\text{mod}\ \Lambda \to \text{mod}\ \text{End}(X)^{\text{op}}$, *where* P *is given by* $\Lambda = S \amalg P$, $X = P \amalg \text{Tr}\ D'S$, *and* $(\text{Tr}\ D'S)^* = 0$.

(e) *Assume that the indecomposable injective modules over* Λ *and* Γ *are finitely generated. Let* I *denote the direct sum of one copy of each of the indecomposable injective* Λ*-modules and* E *the direct sum of one copy of each of the indecomposable injective* Γ*-modules. Let* I_1 *be an injective envelope of* S, $I = I_1 \amalg I_2$, $E = T \amalg E_1$. *Then* $F(I) = F(I_1) \amalg E_1$ *and* $E = T \amalg F(I_2)$.

PROOF. (a) Let $0 \to A \xrightarrow{f} B \xrightarrow{g} C \to 0$ be an exact sequence in mod Λ. We want to show that $0 \to F(A) \to F(B) \to F(C)$ is exact and that if A is in $\mathcal{C}$, then $0 \to F(A) \to F(B) \to F(C) \to 0$ is exact. We can clearly assume that the sequence has no split exact summands. Then it is not hard to see that B and C must be in $\mathcal{C}$. Since $\mathcal{C}$ is clearly closed under factor modules and $\mathcal{D} = \mathcal{D}_T$ is closed under submodules, we conclude by Lemma 2.2 that if A is in $\mathcal{C}$, then $0 \to F(A) \to F(B) \to F(C) \to 0$ is exact. If A is not in $\mathcal{C}$ we write $A = A_1 \amalg S^t$ for some $t > 0$. Since S is simple projective, $(\ , S)|\mathcal{C}$ is zero, so we get the exact sequence $0 \to (\ , A_1)|\mathcal{C} \to (\ , B)|\mathcal{C} \to (\ , C)|\mathcal{C}$. Hence we get an exact sequence $0 \to (\ , F(A_1))|\mathcal{D} \to (\ , F(B))|\mathcal{D} \to (\ , F(C))|\mathcal{D}$. Since $\mathcal{D}_T$ is closed under submodules, it follows from Lemma 2.1 that $0 \to F(A_1) \to F(B) \to F(C)$ is exact. Since $F(S) = 0$ we then have that $0 \to F(A) \to F(B) \to F(C)$ is exact.

(b) Let $F' = F'_S$: $\text{mod}\ \Lambda \to \text{mod}\ \Gamma'$ be a left partial Coxeter functor. We want to show that there is a close connection between F' and the left partial Coxeter functor F: $\text{mod}\ \Lambda \to \text{mod}\ \Gamma$.

Denote by T' the simple injective nonprojective Γ'-module such that the left partial Coxeter F' induces an equivalence between $\mathcal{C}_S$ and $\mathcal{D}_{T'}$. We have a commutative diagram

$$\begin{array}{ccc} \text{mod}\ \Lambda & \xrightarrow{F} & \mathcal{D}_T \subset \text{mod}\ \Gamma \\ & \underset{F'}{\searrow} & \downarrow \alpha \\ & & \mathcal{D}_{T'} \subset \text{mod}\ \Gamma' \end{array}$$

where α is an equivalence of categories. Since $\mathcal{D}_T$ and $\mathcal{D}_{T'}$ are closed under submodules and contain all finitely generated projective modules, it follows from Lemma 2.2 that α can be extended to an equivalence t^*: $\text{mod}\ \Gamma \to \text{mod}\ \Gamma'$. Since the rings are basic, t^* is induced by a ring isomorphism t: $\Gamma' \to \Gamma$.

(c) We want to show that the left partial Coxeter functor F has a left adjoint G which is a right partial Coxeter functor. Let β: $\mathcal{D}_T \to \mathcal{C}_S$ be an inverse equivalence, and define G as an extension of β, by defining $G(T) = 0$. It is then easy to see that F and G are adjoint functors.

(d) We want to show that the left partial Coxeter functor F is equivalent to $(X,\ )$: $\operatorname{mod}\Lambda \to \operatorname{mod}\operatorname{End}(X)^{\mathrm{op}}$. F is isomorphic to $(G(\Gamma),\ )$: $\operatorname{mod}\Lambda \to \operatorname{mod}\Gamma$, where $G(\Gamma)$ is a λ-Γ-bimodule in a natural way. Since $\Gamma \to \operatorname{End}(\Gamma)^{\mathrm{op}} \to \operatorname{End} G(\Gamma)^{\mathrm{op}}$ is a ring isomorphism, we have a commutative diagram

$$\begin{array}{ccc} \operatorname{mod}\Lambda & \xrightarrow{(G(\Gamma),\ )} & \operatorname{mod}\Gamma \\ & \underset{(G(\Gamma),\ )}{\searrow} & \downarrow \beta \\ & & \operatorname{mod}\operatorname{End}(G(\Gamma))^{\mathrm{op}} \end{array}$$

where β is an equivalence of categories. Hence F is equivalent to $(G(\Gamma),\ )$: $\operatorname{mod}\Lambda \to \operatorname{mod}\operatorname{End}(G(\Gamma))^{\mathrm{op}}$.

To get our desired result we want to show that $G(\Gamma)$ is isomorphic to X. To compute $G(\Gamma)$ we want to find the indecomposable C in $\operatorname{mod}\Lambda$ such that $F(C)$ is projective. Let Q be indecomposable projective not isomorphic to S. Then it follows from Lemma 2.2(b) that $F(Q)$ is projective.

Write $\Lambda = P \amalg S$. Then we can write $G(\Gamma) = P \amalg Y$. If $Y = 0$, $G(\Gamma)$ would be projective, and it is then not hard to see that every object in $\operatorname{mod}\operatorname{End} G(\Gamma)^{\mathrm{op}}$ would be isomorphic to an object of the form $(G(\Gamma), U)$, for U in $\operatorname{mod}\Lambda$. We want to show that $Y = \operatorname{Tr} D'S$. Let Y_1 be an indecomposable summand of Y. Since (Y_1, S) is zero, Y_1 is not isomorphic to S, and consequently Y_1 is not projective. We claim that $\operatorname{Ext}^1(Y_1, A) = 0$ for each indecomposable A in $\mathcal{C}$. For otherwise we would have a nonsplit exact sequence $0 \to A \to E \to Y_1 \to 0$. But since, by (a), $0 \to (Y_1, A) \to (Y_1, E) \to (Y_1, Y_1) \to 0$ would be exact, we would get a contradiction. Since Y_1 is not projective, we must then have $\operatorname{Ext}^1(Y_1, S) \neq 0$. Consider now the almost split sequence $0 \to S \to Q \to \operatorname{Tr} D'S \to 0$ (see Proposition 1.17). This gives rise to the exact sequence $0 \to (Y_1, S) \to (Y_1, Q) \to (Y_1, \operatorname{Tr} D'S) \to \operatorname{Ext}^1(Y_1, S) \to 0$. There is then a map h: $Y_1 \to \operatorname{Tr} D'S$ which does not factor through Q. Since $0 \to S \to Q \to \operatorname{Tr} D'S \to 0$ is almost split and Y_1 is indecomposable, h: $Y_1 \to \operatorname{Tr} D'S$ must be an isomorphism. This shows that $G(\Gamma) = P \amalg \operatorname{Tr} D'S = X$. To finish the proof of (d) it only remains to show that $(\operatorname{Tr} D'S)^*$ is zero.

To see this, consider again the almost split sequence $0 \to S \to Q \to \operatorname{Tr} D'S \to 0$, and the minimal projective resolution $0 \to (X, Q) \to (X, \operatorname{Tr} D'S) \to T_1 \to 0$ in $\operatorname{mod}\Gamma$. We want to show that T_1 is isomorphic to the simple injective Γ-module T. We know that $T_1 = \operatorname{Ext}^1_\Lambda(\operatorname{Tr} D'S, S)$. To show that T_1 is simple, we first show that any nonzero element in $\operatorname{Ext}^1_\Lambda(\operatorname{Tr} D'S, S)$ is an almost split sequence. To see this, let $0 \to S \to E \to \operatorname{Tr} D'S \to 0$ be a nonsplit exact sequence, and consider the commutative diagram

$$\begin{array}{ccccccccc} 0 & \to & S & \to & E & \to & \operatorname{Tr} D'S & \to & 0 \\ & & \downarrow t & & \downarrow s & & \downarrow \| & & \\ 0 & \to & S & \to & Q & \to & \operatorname{Tr} D'S & \to & 0 \end{array}$$

where t must clearly be an isomorphism, and hence also s. Using this it is not hard to see that $T_1 = \operatorname{Ext}^1_\Lambda(\operatorname{Tr} D'S, S)$ is a simple Γ-module. To show that T_1 is isomorphic to T it is then sufficient to show that $(X, \operatorname{Tr} D'S)$ is a projective cover for T. Assume to the contrary that $(X, P_1) \to T$ is a projective cover, where P_1 is an indecomposable projective Λ-module. Since every submodule of (X, P_1) is in $\mathfrak{D}_T$, we have an exact sequence $0 \to (X, K) \xrightarrow{u} (X, P_1) \to T \to 0$, with K in mod Λ. Since $F|\mathcal{C}$ is fully faithful, $u = (X, u_1)$, for some u_1: $K \to P_1$. Consider the exact sequence $K \to P_1 \to C \to 0$. We then get an exact sequence $0 \to ((X, P_1),)|\mathfrak{D} \to ((X, K),)|\mathfrak{D}$, hence an exact sequence $0 \to (P_1,)|\mathcal{C} \to (K,)|\mathcal{C}$. Since we also have the exact sequence $0 \to (C,)|\mathcal{C} \to (P,)|\mathcal{C} \to (K,)|\mathcal{C}$ and C is obviously in $\mathcal{C}$, we conclude that C is zero, so that $K \to P$ is an epimorphism. Hence we have our desired contradiction, so that $T = T'$. Since T is simple injective, T cannot occur in the composition series of any indecomposable projective (X, P_1), where P_1 is an indecomposable projective Λ-module. This implies that $\operatorname{Hom}((X, \operatorname{Tr} D'S), (X, P))$ is zero, and consequently $(\operatorname{Tr} D'S)^*$ is zero. This finishes the proof of (d).

(e) Assume that the indecomposable injective modules over Λ and Γ are finitely generated. We first want to show that if I_1 is an injective envelope for S, then $F(I_1)$ is not injective. Consider the exact sequence $0 \to S \to I_1 \to I_1/S \to 0$. Since F is left exact, we get the exact sequence $0 \to F(I_1) \to F(I_1/S)$. This does not split since $I_1 \to I_1/S$ is not a split monomorphism. Hence $F(I_1)$ is not injective. If C is indecomposable and $F(C)$ is injective, we know by Lemma 2.2(c) that C is injective. Let now I_2 denote the direct sum of the indecomposable injective Λ-modules not isomorphic to I_1. By (c) we know that Λ and Γ have the same number of indecomposable projective modules. Hence a counting argument implies that $F(I_2)$ is injective. The rest of (e) is now easy.

We get the following analogue for right partial Coxeter functors.

THEOREM 2.4. *Let Γ and Λ be left artin rings, T a simple injective nonprojective Γ-module, $G = G_T$:* mod $\Gamma \to$ mod Λ *a right partial Coxeter functor, and S the associated simple projective noninjective Λ-module.*

(a) *G is right exact and if $0 \to A \to B \to C \to 0$ is exact in* mod Γ *with C in $\mathfrak{D} = \mathfrak{D}_T$, then $0 \to G(A) \to G(B) \to G(C) \to 0$ is exact.*

(b) *If $G' = G'_T$:* mod $\Gamma \to$ mod Λ' *is a right partial Coxeter functor and Λ and Λ' are artin, then there is a ring isomorphism t: $\Lambda' \to \Lambda$ and a commutative diagram*

$$\begin{array}{ccc} \operatorname{mod}\Gamma & \xrightarrow{G} & \operatorname{mod}\Lambda \\ & \searrow^{G'} & \downarrow t^* \\ & & \operatorname{mod}\Lambda' \end{array}$$

where t^* is an equivalence of categories induced by $t\colon \Lambda' \to \Lambda$.

(c) *G has as an adjoint a left partial Coxeter functor F.*

PROOF. (a) and (c) follow as before.

(b) We have a commutative diagram

$$\begin{array}{ccc} \mathcal{D}_T & \xrightarrow{G} & \mathcal{C}_S \subset \operatorname{mod}\Lambda \\ & \searrow^{G'} & \downarrow \alpha \\ & & \mathcal{C}_{S'} \subset \operatorname{mod}\Lambda' \end{array}$$

where α is an equivalence of categories. Let $0 \to S \to Q \xrightarrow{g} \operatorname{Tr} D'S \to 0$ be an almost split sequence. Then we have seen that $0 \to F(Q) \to F(\operatorname{Tr} D'S) \to T \to 0$ is a minimal projective presentation in $\operatorname{mod}\Gamma$, where F denotes an inverse equivalence for G. Using the converse result, we then get an almost split sequence $0 \to S' \to \alpha(Q) \xrightarrow{\alpha(g)} \alpha(\operatorname{Tr} D'S) \to 0$ in $\operatorname{mod}\Lambda'$. Since $0 \to S \to Q \to \operatorname{Tr} D'S \to 0$ is almost split, we have for all K in $\mathcal{C}_S$ an exact sequence $0 \to (\operatorname{Tr} D'S, K) \to (Q, K) \to (S, K) \to 0$. Hence $0 \to (\alpha(\operatorname{Tr} D'S), \alpha(K)) \to (\alpha(Q), \alpha(K)) \to (S', \alpha(K)) \to 0$ is also exact. This gives a natural isomorphism $(S, K) \to (S', \alpha(K))$. To define a map $(S, S) \to (S', S')$ consider for $u\colon S \to S$ the diagram

$$\begin{array}{ccccccccc} 0 & \to & S & \to & Q & \xrightarrow{g} & \operatorname{Tr} D'S & \to & 0 \\ & & \downarrow u & & \downarrow v & & \downarrow w & & \\ 0 & \to & S & \to & Q & \xrightarrow{g} & \operatorname{Tr} D'S & \to & 0 \end{array}$$

and the induced diagram:

$$\begin{array}{ccccccccc} 0 & \to & S' & \to & \alpha(Q) & \xrightarrow{\alpha(g)} & \alpha(\operatorname{Tr} D'S) & \to & 0 \\ & & \downarrow & & \downarrow \alpha(v) & & \downarrow \alpha(w) & & \\ 0 & \to & S' & \to & \alpha(Q) & \xrightarrow{\alpha(g)} & \alpha(\operatorname{Tr} D'S) & \to & 0 \end{array}$$

We can now extend α to an equivalence $t^*\colon \operatorname{mod}\Lambda \to \operatorname{mod}\Lambda'$, by defining $t^*(S) = S'$ and using the natural isomorphisms $(S, K) \to (S', \alpha(K))$ and $(S, S) \to (S', S')$ described above. Since the rings are basic, t^* is induced by a ring isomorphism $t\colon \Lambda' \to \Lambda$.

We shall now get some additional information about partial Coxeter functors in the case of artin algebras, which will be useful in §3. Actually, for parts (a) and (b) it will be sufficient to assume that our artin rings have a duality and that almost split sequences exist in mod Λ and mod Γ. D Tr would then be replaced by a map γ from the indecomposable nonprojective objects to the indecomposable noninjective objects in mod Λ, given by the almost split sequence $0 \to \gamma(C) \to B \to C \to 0$.

PROPOSITION 2.5. *Let* F: mod $\Lambda \to$ mod Γ *be a left partial Coxeter functor, where* Λ *and* Γ *are artin algebras. Let* S *be the associated simple projective* Λ*-module and* T *the associated simple injective* Γ*-module. Then we have the following.*

(a) *If* A *is an indecomposable noninjective* Λ*-module not isomorphic to* S *and* $0 \to A \to B \to C \to 0$ *is an almost split sequence, then* $0 \to F(A) \to F(B) \to F(C) \to 0$ *is also an almost split sequence,* $F(\operatorname{Tr} DA) \cong \operatorname{Tr} DF(A)$, *and* $F(D \operatorname{Tr} C) \cong D \operatorname{Tr} F(C)$.

(b) $F(I_1) \cong D \operatorname{Tr} T$, *where* I_1 *is an injective envelope for* S.

(c) $\Lambda \cong \operatorname{End}_\Gamma(Y)$, *where* $Y = \operatorname{Tr} T \amalg D(I)$, *and* I *is the sum of one copy of each of the indecomposable injective* Γ*-modules not isomorphic to* T.

PROOF. Let A be an indecomposable Λ-module which is not injective and not isomorphic to S, and let $0 \to A \xrightarrow{f} B \xrightarrow{g} C \to 0$ be an almost split sequence. Since the sequence has no split exact summands, we know that B and C are in $\mathcal{C}$, so that $0 \to F(A) \xrightarrow{F(f)} F(B) \xrightarrow{F(g)} F(C) \to 0$ is exact by Theorem 2.3(a). It does not split since $0 \to A \xrightarrow{f} B \xrightarrow{g} C \to 0$ does not split.

Let h: $F(A) \to M$ be a map which is not a splittable monomorphism. If M is isomorphic to the simple injective Γ-module T then there is a map g: $F(B) \to M$ such that $gf = h$. If M is indecomposable and not isomorphic to T then $M = F(N)$ for some Λ-module N. Since $F|\mathcal{C}$: $\mathcal{C} \to \mathcal{D}$ is an equivalence of categories, there is a map h_1: $A \to N$ such that $F(h_1) = h$. Then h_1 is not a splittable monomorphism, because h is not a splittable monomorphism. The sequence $0 \to A \xrightarrow{f} B \xrightarrow{g} C \to 0$ is almost split, so there is some v: $B \to N$ such that $vf = h_1$. Then $F(v)F(f) = F(h_1) = h$, and therefore the sequence $0 \to F(A) \xrightarrow{F(f)} F(B) \xrightarrow{F(g)} F(C) \to 0$ is almost split. The rest of (a) follows from the above and the fact that since $0 \to F(A) \to F(B) \to F(C) \to 0$ and $0 \to A \to B \to C \to 0$ are almost split, then $A \cong D \operatorname{Tr} C$ and $F(A) \cong D \operatorname{Tr} F(C)$.

(b) $D \operatorname{Tr} T$ is not injective and is, therefore, in $\mathcal{D}$. So there is some E in $\mathcal{C}$ such that $D \operatorname{Tr} T \cong F(E)$. If E is not injective then we know by (a) that $F(\operatorname{Tr} DE) \cong \operatorname{Tr} DF(E)$. Hence we get $T \cong F(\operatorname{Tr} DE)$, and this is a contradiction. So E is injective and, since $D \operatorname{Tr} T \cong F(E)$ is not injective, it follows from Theorem 2.3(b) that $E = I_1$ is the injective envelope of S.

(c) Let $\{I_1, \ldots, I_n\}$ be a complete set of nonisomorphic injective Λ-modules. We know by Theorem 2.3(e) that the sum of one copy of each of the indecomposable injective Γ-modules not isomorphic to T is isomorphic to

$F(I_2)\amalg \cdots \amalg F(I_n)$. And by (b) we know that $F(I_1) \cong D\operatorname{Tr} T$. Therefore for $Y = \operatorname{Tr} T \amalg D(I)$, where I is the sum of one copy of each of the indecomposable injective Γ-modules not isomorphic to T, $D(Y) = D\operatorname{Tr} T \amalg I = F(I_1)\amalg F(I_2)\amalg \cdots \amalg F(I_n) = F(D(\Lambda))$. Then $\operatorname{End}_\Gamma(Y) \cong \operatorname{End}_\Gamma(DY)^{\mathrm{op}} \cong \operatorname{End}_\Gamma(F(D\Lambda))^{\mathrm{op}} \cong \operatorname{End}_\Lambda(D\Lambda)^{\mathrm{op}} \cong (\operatorname{End}_\Lambda(\Lambda_\Lambda)) \cong \Lambda$.

For right partial Coxeter functors we have the following result which is analogous to Proposition 2.5.

PROPOSITION 2.6. *Let Λ and Γ be artin algebras. Let G: mod $\Gamma \to$ mod Λ be a right partial Coxeter functor. Then we have the following.*

(a) *If C is an indecomposable nonprojective Γ-module not isomorphic to T and the sequence $0 \to A \to B \to C \to 0$ is almost split, then $0 \to G(A) \to G(B) \to G(C) \to 0$ is also an almost split sequence. Therefore $D\operatorname{Tr} G(C) \cong G(D\operatorname{Tr} C)$ and $\operatorname{Tr} DG(A) \cong G(\operatorname{Tr} DA)$.*

(b) *$G(D\operatorname{Tr} T)$ is an injective Λ-module. If $\{T, I_2, \ldots, I_n\}$ is a complete set of nonisomorphic indecomposable injective Γ-modules, then $\{G(D\operatorname{Tr} T), G(I_2), \ldots, G(I_n)\}$ is a complete set of nonisomorphic indecomposable injective Λ-modules.*

(c) *$\Lambda \cong \operatorname{End}_\Gamma(Y)$, where $Y = \operatorname{Tr} T\amalg DI$, and I is the sum of one copy of each of the indecomposable injective Γ-modules not isomorphic to T.*

We now apply some of the above results to obtain more information about the partial Coxeter functors F and G studied in §1.

PROPOSITION 2.7. *Let Λ be an artin algebra with a simple projective noninjective module S and assume that $(\operatorname{Tr} DS)^* = 0$. Let $\Gamma = \operatorname{End}_\Lambda(X)^{\mathrm{op}}$, where $X = P\amalg \operatorname{Tr} DS$ and P is determined by $\Lambda = P\amalg S$. Let I be the sum of one copy of each of the indecomposable injective Γ-modules not isomorphic to $T = \operatorname{Ext}^1_\Lambda(X, S)$, and let $Y = \operatorname{Tr} T\amalg DI$. Then we have the following.*

(a) *${}_\Lambda X_\Gamma$ and ${}_\Lambda Y_\Gamma$ are isomorphic Λ-Γ-bimodules, where X and Y are considered as modules over their endomorphism rings in the usual way.*

(b) *The functors $F = \operatorname{Hom}_\Lambda(X, \ )$: mod $\Lambda \to$ mod Γ and $G = Y \otimes$: mod $\Gamma \to$ mod Λ are adjoint functors.*

(c) *Let $\mathcal{C}$ and $\mathcal{D}$ be the categories considered at the beginning of the section and let F and G be as in (b). Then $F|\mathcal{C}: \mathcal{C} \to \mathcal{D}$ and $G|\mathcal{D}: \mathcal{D} \to \mathcal{C}$ are inverse equivalences of categories.*

PROOF. (a) We have seen that F is a left partial Coxeter functor, so applying Proposition 2.5 we have that $F(D\Lambda) \cong D\operatorname{Tr} T\amalg I$. Then $Y = \operatorname{Tr} T\amalg DI = DF(D\Lambda) = D\operatorname{Hom}_\Lambda(X, D\Lambda)$. Since there is an isomorphism $X \to D\operatorname{Hom}_\Lambda(X, D\Lambda)$ functorial in X (see [**11**, p. 119]), X and Y are isomorphic Λ-Γ-bimodules.

(b) is an immediate consequence of (a).

(c) follows from the fact that an inverse equivalence of $F\colon \mathcal{C} \to \mathcal{D}$ extends to a right partial Coxeter functor which is adjoint to F.

3. Hereditary artin algebras. In this section we study hereditary basic artin algebras. If S is a simple projective noninjective module over an hereditary artin algebra Λ, then we have $(\operatorname{Tr} DS)^* = 0$, so that any functor of the type $(X, \ )$: $\operatorname{mod} \Lambda \to \operatorname{mod} \Gamma$, where $X = P \amalg \operatorname{Tr} DS$ and as usual $\Lambda = P \amalg S$, is a partial Coxeter functor. In the introduction we pointed out that for an hereditary artin algebra Λ, $D \operatorname{Tr}$ is a functor from mod Λ to mod Λ, taking projectives to zero and inducing an equivalence of categories between $\operatorname{mod}_P \Lambda$ and $\operatorname{mod}_I \Lambda$. Our main result in this section is that $D \operatorname{Tr}$ can be written as a composition of partial Coxeter functors in a natural way.

We recall that all the artin algebras that we consider are basic. Throughout this section Λ will denote an indecomposable hereditary basic artin algebra which is not simple.

Since Λ is hereditary we can define a partial order in the set of indecomposable projective Λ-modules by writing $P \leqslant Q$ if and only if $\operatorname{Hom}_\Lambda(P, Q) \neq 0$. In analogy with admissible sequences of simple modules (see [3, §3]) we say that $P_1, \ldots, P_n$ is an *admissible sequence* of projective Λ-modules if $P_1, \ldots, P_n$ is a complete set of nonisomorphic indecomposable projective Λ-modules such that $i < j$ implies $P_j \nleqslant P_i$, that is, $\operatorname{Hom}_\Lambda(P_j, P_i) = 0$. When Λ is hereditary admissible sequences always exist. For example if $P_1, \ldots, P_n$ are the nonisomorphic indecomposable projective Λ-modules and length $P_i \leqslant$ length P_j for $i < j$, then the sequence $P_1, \ldots, P_n$ is admissible.

We shall denote a pair consisting of an hereditary artin algebra Λ and an admissible sequence $P_1, \ldots, P_n$ of projective Λ-modules by $(\Lambda; P_1, \ldots, P_n)$. We now describe a particular way of associating with a pair $(\Lambda; P_1, \ldots, P_n)$ another pair $(\Lambda_1; P_1', \ldots, P_n')$, where Λ_1 is an hereditary artin algebra and $P_1', \ldots, P_n'$ is an admissible sequence of projective Λ_1-modules.

The projective Λ-module P_1 is simple, because $(P_i, P_1) = 0$ for $i > 1$ and is, therefore, not injective since we have assumed that Λ is indecomposable. Let $P = P_2 \amalg \cdots \amalg P_n$, let $X = P \amalg \operatorname{Tr} DP_1$ and let $\Lambda_1' = \operatorname{End}_\Lambda(X)^{\mathrm{op}}$. Then, by Proposition 1.13, Λ_1 is an hereditary artin algebra. Let F denote the partial Coxeter functor $(X, \ )$: $\operatorname{mod} \Lambda \to \operatorname{mod} \Lambda_1$. Then, because F restricted to the full subcategory of mod Λ whose objects do not contain a summand isomorphic to $P_1 = S$ is fully faithful (Proposition 1.4) and $(\operatorname{Tr} DS)^* = 0$, we know that the sequence $P_1', \ldots, P_n'$ of projective Λ_1-modules, where $P_1' = F(P_2)$, $P_2' = F(P_3), \ldots, P_{n-1}' = F(P_n)$, $P_n' = F(\operatorname{Tr} DP_1)$, is an admissible sequence of nonisomorphic projective modules. Thus associated with the pair $(\Lambda; P_1, \ldots, P_n)$ there is the pair $(\Lambda_1, P_1', \ldots, P_n')$.

Iterating this process we obtain from the pair $(\Lambda; P_1, \ldots, P_n)$ a sequence of pairs $(\Lambda_0; P_1^{(0)}, \ldots, P_n^{(0)})$, $(\Lambda_1, P_1^{(1)}, \ldots, P_n^{(1)}), \ldots, (\Lambda_n; P_1^{(n)}, \ldots, P_n^{(n)})$, where $(\Lambda_0; P_1^{(0)}, \ldots, P_n^{(0)}) = (\Lambda; P_1, \ldots, P_n)$. We denote by F_i the partial Coxeter functor

$$(X_i, \ ): \operatorname{mod} \Lambda_i \to \operatorname{mod} \Lambda_{i+1},$$

where $X_i = P_2^{(i)} \amalg \cdots \amalg P_n^{(i)} \amalg \operatorname{Tr} DP_1^{(i)}$.

Before we state the main result of this section we need some preliminary results.

LEMMA 3.1. *Given* $(\Lambda; P_1, \ldots, P_n)$ *and the pair* $(\Lambda_i; P_1^{(i)}, \ldots, P_n^{(i)})$ *with* $i \geqslant 1$ *we have that* $P_1^{(i)} = F_i \ldots F_1(P_{i+1}), \ldots, P_{n-i}^{(i)} = F_i \cdots F_1(P_n), P_{n-i+1}^{(i)} = F_i \cdots F_1(\operatorname{Tr} DP_1), \ldots, P_{n-i+j}^{(i)} = F_i \cdots F_j \operatorname{Tr} DF_{j-1} \cdots F_1(P_j), \ldots, P_n^{(i)} = F_i \operatorname{Tr} DF_{i-1} \cdots F_1(P_i)$.

PROOF. Follows from the definition.

LEMMA 3.2. *Given the pair* $(\Lambda; P_1, \ldots, P_n)$ *the composition* $F_n \cdots F_1$ *of the functors* F_i: $\operatorname{mod} \Lambda_{i-1} \to \operatorname{mod} \Lambda_i$ *restricted to* $\operatorname{mod}_P \Lambda$ *induces an equivalence of categories between* $\operatorname{mod}_P \Lambda$ *and* $\operatorname{mod}_I \Lambda_n$.

PROOF. We know that $F_i|\mathcal{C}_{P_1^{(i-1)}}: \mathcal{C}_{P_1^{(i-1)}} \to \mathfrak{D}_{T_i}$ is an equivalence of categories, where $\mathcal{C}_{P_1^{(i-1)}}$ and $\mathfrak{D}_{T_i}$ are, respectively, the full subcategories of $\operatorname{mod} \Lambda_{i-1}$ and $\operatorname{mod} \Lambda_i$ whose objects do not contain $P_1^{(i-1)}$ and T_i as a summand, respectively. Since $F_i(P_1^{(i-1)}) = 0$, the module $F_n \cdots F_1(P_i) = F_n \cdots F_1(F_{i-1} \cdots F_1(P_i))$ which by Lemma 3.1 is equal to $F_n \cdots F_i(P_1^{(i-1)})$, is zero. So $F_n \cdots F_1$ vanishes on projectives and the composition $F_n \cdots F_1|\operatorname{mod}_P \Lambda$: $\operatorname{mod}_P \Lambda \to \operatorname{mod}_I \Lambda$ is fully faithful. We know by Theorem 2.3 that the modules $F_n \cdots F_2(T_1)$, $F_n \cdots F_3(T_2), \ldots, T_n$ are injective. Since there are n indecomposable injective Λ_n-modules, and these are exactly the indecomposable Λ_n-modules that are not of the form $F_n \cdots F_1(X)$, for some X in $\operatorname{mod} \Lambda$, we have that $F_n \cdots F_1|\operatorname{mod}_P \Lambda$: $\operatorname{mod}_P \Lambda \to \operatorname{mod}_I \Lambda$ is an equivalence of categories.

PROPOSITION 3.3. *Let* Λ *be an indecomposable hereditary artin ring,* α: $\operatorname{mod} \Lambda \to \operatorname{mod} \Lambda$ *an equivalence of categories. If the restriction of* α *to* $\operatorname{mod}_I \Lambda$ *is isomorphic to the identity, then* α *is isomorphic to the identity.*

PROOF. It has been proven in [5, Chapter III, Propositions 1.1 and 1.3], that if Λ is an indecomposable hereditary artin algebra either Λ has no projective injectives or is isomorphic to the lower triangular matrix ring $T_n(K)$ over a division algebra K. It is not hard to see that the same is true for an hereditary ring.

This remark enables us to divide the proof into two cases. The proof in both cases is based on the fact that α is isomorphic to the identity if it carries projectives to projectives and restricted to the full subcategory of projective modules is isomorphic to the identity. Since $\operatorname{mod}_I \Lambda$ is closed under submodules and projective covers, we are done by Lemma 2.2 when Λ has no nonzero projective injectives.

Assume now that $\Lambda \cong T_n(K)$, for a division ring K. Then Λ has only one indecomposable projective injective module P_n, and the other indecomposable projective modules are $P_i = \mathbf{r}^{n-i}P_n$, $i = 1, \ldots, n-1$. Then P_i is in $\operatorname{mod}_I \Lambda$, for $i < n$. As already remarked, it is enough to show that the restriction of α to the full subcategory of projective objects is isomorphic to the identity. Let μ: $\mathrm{id}|\operatorname{mod}_I \Lambda \to \alpha|\operatorname{mod}_I \Lambda$ be an isomorphism. Consider the inclusion i: $P_{n-1} \to P_n$, and the diagram:

$$\begin{array}{ccc} \mathrm{id}(P_{n-1}) & \xrightarrow{\mu_{P_{n-1}}} & \alpha(P_{n-1}) \\ \downarrow \mathrm{id}(i) & & \downarrow \alpha(i) \\ \mathrm{id}(P_n) & & \alpha(P_n) \end{array}$$

Choose μ_{P_n}: $\mathrm{id}(P_n) \to \alpha(P_n)$ such that $\mu_{P_n}\mathrm{id}(i) = \alpha(i)\mu_{P_{n-1}}$. This is possible, because $\alpha(P_n)$ is injective. Now any map f: $P_j \to P_n$ factors as $P_j \xrightarrow{f} P_{n-1} \xrightarrow{j} P_n$, so $\alpha(f)\mu_{P_j} = \mu_{P_n}\mathrm{id}(f)$. Let now f: $P_n \to P_n$ be a map. One can prove that the diagram

$$\begin{array}{ccc} \mathrm{id}(P_n) & \xrightarrow{\mu_{P_n}} & \alpha(P_n) \\ \mathrm{id}(f)\downarrow & & \downarrow \alpha(f) \\ \mathrm{id}(P_n) & \xrightarrow{\mu_{P_n}} & \alpha(P_n) \end{array}$$

commutes, using the following facts: $\mu_{P_n}\mathrm{id}(i) = \alpha(i)\mu_{P_{n-1}}$, μ: $\mathrm{id}|\operatorname{mod}_I \Lambda \to \alpha|\operatorname{mod}_I \Lambda$ is a morphism of functors, and $\operatorname{End}_\Lambda(P_n)$ is a division ring.

We have the following consequence of the above proposition.

COROLLARY 3.4. *Let K be a division ring. Then all the automorphisms of $T_n(K)$ are inner.*

PROOF. Let Φ: $T_n(K) \to T_n(K)$ be a ring isomorphism. Then the equivalence of categories $\overline{\Phi}$: $\operatorname{mod} T_n(K) \to \operatorname{mod} T_n(K)$ induced by Φ is isomorphic to the identity if and only if Φ is inner (see [**15**, pp. 454–456]). So it is enough to prove that every equivalence σ: $\operatorname{mod} T_n(K) \to \operatorname{mod} T_n(K)$ is isomorphic to the identity. To prove this we proceed by induction on n. The result is true if $n = 1$. If $n > 1$, the restriction of Φ to $\operatorname{mod}_I \Lambda$ induces an equivalence $\Phi|\operatorname{mod}_I \Lambda$: $\operatorname{mod}_I \Lambda \to \operatorname{mod}_I \Lambda$. Since $\operatorname{mod}_I \Lambda$ is equivalent to $\operatorname{mod} T_{n-1}(K)$ we have a commutative diagram

$$\begin{array}{ccc} \operatorname{mod}_I \Lambda & \xrightarrow{\Phi|\operatorname{mod}_I \Lambda} & \operatorname{mod}_I \Lambda \\ \alpha\downarrow & & \downarrow\alpha \\ \operatorname{mod} T_{n-1}(K) & \xrightarrow{\Phi'} & \operatorname{mod} T_{n-1}(K) \end{array}$$

where α and Φ' are equivalences. By the induction hypothesis Φ' is isomorphic to the identity, so $\Phi|\text{mod}_I\ \Lambda$ is also isomorphic to the identity. Then it follows, by Proposition 3.3, that σ is isomorphic to the identity. This ends the proof of the corollary.

LEMMA 3.5. *Let* Λ_1, Λ_2 *be indecomposable hereditary artin rings which are not simple and let* τ: $\text{mod}_I\ \Lambda_1 \to \text{mod}_I\ \Lambda_2$ *be an equivalence of categories. Then there is an equivalence* α: $\text{mod}\ \Lambda_1 \to \text{mod}\ \Lambda_2$ *whose restriction to* $\text{mod}_I\ \Lambda_1$ *is* τ.

PROOF. Since $\text{mod}_I\ \Lambda$ is closed under submodules we know by Lemma 2.2 that a Λ_1-module P is projective noninjective if and only if $\alpha(P)$ is projective. Since τ has an inverse, it follows that the number of indecomposable projective noninjective Λ_1-modules is the same as the number of indecomposable projective noninjective Λ_2-modules, and the restriction of τ to the full subcategory $\mathcal{P}_1$ of projective noninjective Λ_1-modules induces an equivalence of categories between $\mathcal{P}_1$ and the full subcategory $\mathcal{P}_2$ of projective noninjective Λ_2-modules.

To prove the lemma it is enough, as before, to consider the case when Λ_1 has no nonzero projective injectives and the case when $\Lambda_1 = T_n(K)$.

Case 1. There are no nonzero projective injective Λ_1-modules. If Λ_2 has a nonzero projective injective module, then $\Lambda_2 \cong T_m(K)$, for a division ring K. Then $\mathcal{P}_2$ is equivalent to the full subcategory of mod $T_{m-1}(K)$ of projective modules, and α: $\mathcal{P}_1 \to \mathcal{P}_2$ induces an equivalence of categories between mod Λ and mod $T_{m-1}(K)$. Therefore Λ is isomorphic to $T_{m-1}(K)$, because both rings are basic, and this contradicts the assumption that there are no nonzero projective injective Λ_1-modules. This proves that there are no nonzero projective injective Λ_2-modules, so $\tau|\mathcal{P}_1$: $\mathcal{P}_1 \to \mathcal{P}_2$ induces an equivalence α: $\text{mod}\ \Lambda_1 \to \text{mod}\ \Lambda_2$. If M is in $\text{mod}_I\ \Lambda$ and $0 \to P_1 \to P_0 \to M \to 0$ is a projective resolution of M, we have that $\tau(P_1) \to \tau(P_0) \to \tau(M) \to 0$ is exact by Lemma 2.2. This shows that α extends τ and finishes the proof of the lemma in this case.

Case 2. Assume now that $\Lambda_1 = T_n(K)$, for a division ring K. If Λ_2 has no nonzero projective injective modules then we know by the first case that Λ_1 and Λ_2 are isomorphic, and this is a contradiction. So $\Lambda_2 \cong T_m(K')$, where K' is a division ring. We will see now that $K \cong K'$. We recall that K is the endomorphism ring of an arbitrary indecomposable Λ_1-module, K' is the endomorphism ring of an arbitrary indecomposable Λ_2-module. So let A be a nonzero indecomposable noninjective Λ_1-module. Since τ: $\text{mod}_I\ \Lambda_1 \to \text{mod}_I\ \Lambda_2$ is an equivalence of categories, then $\text{End}_\Lambda(A) \cong \text{End}_\Lambda(\tau A)$, so $K \cong K'$. Since $\alpha|\mathcal{P}_1$: $\mathcal{P}_1 \to \mathcal{P}_2$ is an equivalence and every ring has only one indecomposable projective injective module, then $n - 1 = m - 1$, so $n = m$. The map φ: $T_n(K) \to T_{n-1}(K)$ defined by

$$\varphi\begin{bmatrix} a_{11} & 0 & & \cdots & 0 \\ a_{21} & a_{22} & 0 & \cdots & 0 \\ \vdots & & & & \vdots \\ a_{n1} & & & \cdots & a_{nn} \end{bmatrix} = \begin{bmatrix} a_{22} & 0 & \cdots & 0 \\ \vdots & & & \vdots \\ a_{n2} & & \cdots & a_{nn} \end{bmatrix}$$

induces a fully faithful functor $\bar{\varphi}$: mod $T_{n-1}(K) \to$ mod $T_n(K)$ whose image is $\text{mod}_I(T_n(K))$. So it is enough to prove that given any equivalence β: mod $T_{n-1}(K) \to$ mod $T_{n-1}(K)$ there is an equivalence $\bar{\beta}$: mod $T_n(K) \to$ mod $T_n(K)$ such that $\bar{\beta}\bar{\varphi} = \bar{\varphi}\beta$. So let β: mod $T_{n-1}(K) \to$ mod $T_{n-1}(K)$ be an equivalence, and let ρ: $T_{n-1}(K) \to T_{n-1}(K)$ be the ring isomorphism defining β, that is, such that $\bar{\rho} = \beta$. We know by Corollary 3.4 that ρ is inner. So $\rho(x) = AxA^{-1}$, for some $A \in T_{n-1}(K)$, and for all $x \in T_{n-1}(K)$. Let ρ': $T_n(K) \to T_n(K)$ be the automorphism defined by conjugation by $\left(\begin{smallmatrix} 1 & 0 \\ 0 & A \end{smallmatrix}\right)$. Then the equivalence $\bar{\beta} = \bar{\rho}'$: mod $T_n(K) \to$ mod $T_n(K)$ verifies the required condition.

We can now prove the following result.

PROPOSITION 3.6. *Let Λ, Λ_1, Λ_2, be hereditary artin rings with no semisimple ring summands, let G: $\text{mod}_P\, \Lambda \to \text{mod}_I\, \Lambda_1$, H: $\text{mod}_P\, \Lambda \to \text{mod}_I\, \Lambda_2$ be equivalences of categories. Then:*

(a) *There is a ring isomorphism ϕ: $\Lambda_2 \to \Lambda_1$ such that $\bar{\phi}G = H$.*

(b) *ϕ': $\Lambda_2 \to \Lambda_1$ is a ring isomorphism such that $\bar{\phi}'G$ is isomorphic to H if and only if $\phi' = \alpha\phi$, where α is an inner automorphism of Λ_2.*

PROOF. (a) Follows from Lemma 3.5.

(b) Follows from Proposition 3.3 and from the fact that if α: $\Lambda \to \Lambda$ is a ring isomorphism, then the equivalence $\bar{\alpha}$: mod $\Lambda \to$ mod Λ defined by α is isomorphic to the identity if and only if α is inner.

As an easy consequence of Proposition 3.6 we obtain the following important result.

THEOREM 3.7. *We use the notations of the beginning of this section. Given a pair $(\Lambda; P_1, \ldots, P_n)$ where $n > 1$ there is a ring isomorphism ϕ: $\Lambda_n \to \Lambda$ such that the functors $\bar{\phi}F_n \ldots F_1$ and D Tr are isomorphic. Moreover, ϕ': $\Lambda_n \to \Lambda$ is another ring isomorphism such that $\bar{\phi}'F_n \ldots F_1$ and D Tr are isomorphic if and only if $\phi' = \alpha\phi$, where α is an inner automorphism of Λ.*

PROOF. By Lemma 3.2 we know that $F_n \cdots F_1$: $\text{mod}_P\, \Lambda \to \text{mod}_I\, \Lambda$ is an equivalence of categories. We also know that D Tr: $\text{mod}_P\, \Lambda \to \text{mod}_I\, \Lambda$ is an equivalence of categories. Then Theorem 3.7 is obtained by applying Proposition 3.6 to these two equivalences.

In view of this result, in order to describe the isomorphisms ϕ: $\Lambda \to \Lambda_n$ which have the property that $\bar{\phi}F_n \cdots F_1 \cong D$ Tr it is enough to give one isomorphism from Λ to Λ_n with this property.

To do this it is enough, as we remarked above, to consider the case when Λ has no indecomposable projective injective modules and the case when $\Lambda \cong T_n(K)$.

Case 1. Λ has no indecomposable projective injective modules.

If U is an indecomposable noninjective Λ_{i-1}-module not isomorphic to the simple projective Λ_{i-1}-module $F_{i-1} \cdots F_1(P_i)$, then we know by Proposition 2.1 that $F_i(U)$ is not injective and $F_i(\operatorname{Tr} DU) \cong \operatorname{Tr} DF_i(U)$. Although these isomorphisms are not natural, once they have been chosen the rest of the construction of the isomorphism $\psi\colon \Lambda \to \Lambda_n$ is natural. Iterating this argument we have, since $P_1, \ldots, P_n$ are not injective, that $\operatorname{Tr} DF_j \cdots F_1(P_{j+1}) \cong F_j \cdots F_1(\operatorname{Tr} DP_{j+1})$, $j = 1, \ldots, n-1$. From this and Lemma 3.1 we have that the nonisomorphic projective indecomposable Λ_n-modules are $F_n \cdots F_1(\operatorname{Tr} DP_i)$, $i = 1, \ldots, n$. Since Λ_n is basic, Λ_n is isomorphic, as a Λ_n-module, to $F_n \cdots F_1(\operatorname{Tr} D\Lambda)$. Let ψ be the composition of the natural ring isomorphisms

$$\begin{aligned}\Lambda_n &\cong \operatorname{End}_{\Lambda_n}(\Lambda_n)^{\mathrm{op}} \cong \operatorname{End}_{\Lambda_n}(F_n \cdots F_1(\operatorname{Tr} D\Lambda))\\ &\cong \operatorname{End}_\Lambda(\operatorname{Tr} D\Lambda) \overset{D\operatorname{Tr}}{\cong} \operatorname{End}_\Lambda(\Lambda),\end{aligned}$$

where the last map is an isomorphism because Λ is in $\operatorname{mod}_I \Lambda$ and $D\operatorname{Tr}\colon \operatorname{mod}_P\Lambda \to \operatorname{mod}_I \Lambda$ is an equivalence of categories.

Both $\bar{\psi}F_n \cdots F_1$ and $D\operatorname{Tr}$ are functors from $\operatorname{mod}\Lambda$ to itself that are zero on projectives, so to prove that they are isomorphic we consider M in $\operatorname{mod}_P \Lambda$. There are functorial group isomorphisms

$$\begin{aligned}F_n \cdots F_1(M) &\cong \operatorname{Hom}_{\Lambda_n}(\Lambda_n, F_n \cdots F_1(M))\\ &\cong \operatorname{Hom}_{\Lambda_n}(F_n \cdots F_1(\operatorname{Tr} D\Lambda), F_n \cdots F_1(M)).\end{aligned}$$

Since $\operatorname{Tr} D\Lambda$, and M are in $\operatorname{mod}_P \Lambda$ and $F_n \cdots F_1|\operatorname{mod}_P \Lambda$ is fully faithful we have that

$$\begin{aligned}&\operatorname{Hom}_{\Lambda_n}(F_n \cdots F_1(\operatorname{Tr} D\Lambda), F_n \cdots F_1(M))\\ &\qquad\cong \operatorname{Hom}_\Lambda(\operatorname{Tr} D\Lambda, M) \overset{D\operatorname{Tr}}{\cong} \operatorname{Hom}_\Lambda(\Lambda, D\operatorname{Tr} M) \cong D\operatorname{Tr} M.\end{aligned}$$

With a straightforward verification using the definition of $\bar{\psi}$ one can now check that the composition $\delta(M)\colon \bar{\psi}F_n \cdots F_1(M) \to D\operatorname{Tr} M$ of the isomorphisms described above is a Λ-module homomorphism. Since δ is functorial in M, we have that ψ verifies the desired conditions.

Case 2. $\Lambda \cong T_n(K)$. Then P_n is the only indecomposable projective injective module, and is the injective envelope of the simple projective module P_1. So $F_1(P_n) \cong \operatorname{Tr} DT$, where T is the simple injective $\operatorname{End}_\Lambda(\amalg_2^n P_i \amalg \operatorname{Tr} DP_1)^{\mathrm{op}}$-module that is not in the image of F_1 (see Proposition 2.1). Therefore $F_1(P_n)$ is not injective. Since the modules $P_1, \ldots, P_{n-1}$ are not injective either, we know that $\operatorname{Tr} DF_j \cdots F_1(P_{j+1}) \cong F_j \cdots F_1(\operatorname{Tr} DP_{j+1})$, $j = 1, \ldots, n-2$, and $\operatorname{Tr} DF_j \cdots F_1(P_n) \cong F_j \cdots F_2(\operatorname{Tr} DF_1(P_n))$, for $j < n$. To prove this we use, as in Case 1, Proposition 2.1. Λ_n is isomorphic, as a Λ_n-module, to the sum of one copy of each of the indecomposable projective Λ_n-modules. That is,

$$\Lambda_n \cong \coprod_{i=1}^{n-1} F_n \cdots F_1(\operatorname{Tr} DP_i) \amalg F_n \cdots F_2(\operatorname{Tr} DF_1(P_n)).$$

We write $Q = \amalg_{i=1}^{n-1} P_i$. Then $\Lambda \cong P_n \amalg Q$,

$$\Lambda_n \cong F_n \cdots F_2 \operatorname{Tr} DF_1(P_n) \amalg F_n \cdots F_1(\operatorname{Tr} DQ),$$

and there is a natural ring isomorphism

$$\varphi\colon \operatorname{End}_\Lambda(Q) \to \operatorname{End}_{\Lambda_n}(F_n \cdots F_1 \operatorname{Tr} DQ),$$

defined by $\varphi(f) = F_n \cdots F_1 \operatorname{Tr} D(f)$. By hypothesis, the ring Λ is isomorphic to $T_n(K)$, so Λ_n is also isomorphic to $T_n(K)$, by Theorem 3.6. There are then ring isomorphisms $\pi\colon \Lambda \to T_n(K)$, $\pi'\colon \Lambda_n \to T_n(K)$ such that

$$\pi\left(\operatorname{End}_\Lambda(Q)^{\text{op}}\right) = \pi'\left(\operatorname{End}_{\Lambda_n}(Q')^{\text{op}}\right) = \left\{\begin{pmatrix} A & 0 \\ 0 & 0 \end{pmatrix}, A \in T_{n-1}(K)\right\}.$$

Then $\pi'\varphi\pi^{-1}$ can be considered as an automorphism of $T_{n-1}(K)$ and is, therefore, inner, by Corollary 3.4, and is given by conjugation by some element $A_1 \in T_{n-1}(K)$. Let α be the inner automorphism of $T_n(K)$ defined by conjugation by the matrix $\left(\begin{smallmatrix} A_1 & 0 \\ 0 & 1 \end{smallmatrix}\right)$. We will now prove that $\bar{\alpha}F_n \cdots F_1 \cong D \operatorname{Tr}$. As we recalled above, $\operatorname{Tr} DF_1(P_n) \cong T$ is simple injective. Therefore, $\operatorname{Hom}_{\Lambda_1}(\operatorname{Tr} DF_1(P_n), F_1(M)) = 0$, because $F_1(M) \not\cong T$. On the other hand, $\operatorname{Hom}_\Lambda(P_n, D \operatorname{Tr} M) = 0$, because P_n is injective and therefore all its factors are injective. So $F_n \cdots F_1(M)$ and $D \operatorname{Tr} M$ are, respectively, modules over $\operatorname{End}_{\Lambda_n}(F_n \cdots F_1(Q))^{\text{op}}$ and $\operatorname{End}_\Lambda(Q)^{\text{op}}$, and we can proceed as in Case 1. $\delta(M)$ is the composition of the group isomorphisms

$$\begin{aligned} \operatorname{Hom}_{\Lambda_1}(F_1(\operatorname{Tr} DQ), F_1(M)) &\cong \operatorname{Hom}_\Lambda(\operatorname{Tr} DQ, M) \\ &\cong \operatorname{Hom}_\Lambda(Q, \operatorname{Tr} DM) \cong \operatorname{Hom}_\Lambda(\Lambda, D \operatorname{Tr} M) \cong D \operatorname{Tr} M. \end{aligned}$$

One can again verify that $\delta(M)\colon \bar{\psi}F_n \cdots F_1(M) \to D \operatorname{Tr} M$ is a Λ-homomorphism. Since δ is functorial in M, this ends our discussion.

4. Functors between modules over triangular matrix rings. In this section we study certain functors F', G' between modules over triangular matrix rings, which induce equivalences between subcategories of module categories. Specializing our assumptions, our functors F' and G' will be partial Coxeter functors between left artin rings. We show that if Λ is a basic artin ring with a simple projective noninjective Λ-module S_i, $\Lambda = P \amalg S_1$, $X = P \amalg \operatorname{Tr} D'S_1$, where $D'S_1 = S_1^*/\mathbf{r}S_1^*$ and $(\operatorname{Tr} D'S_1)^* = 0$, then our functor F' will be equivalent to $(X, \ )$: mod $\Lambda \to$ mod $\operatorname{End}_\Lambda(X)^{\mathrm{op}}$, so that in particular $(X, \ )$ is a partial Coxeter functor.

We study functors between modules over triangular matrix rings of the form $\begin{pmatrix} R & 0 \\ M & S \end{pmatrix}$, where R and S are rings and ${}_SM_R$ is an S-R-bimodule. We consider the (left) $\begin{pmatrix} R & 0 \\ M & S \end{pmatrix}$-modules as triples (A, B, f), where A is an R-module, B is an S-module and f: $M \otimes A \to B$ is an S-homomorphism **[16]**. A morphism between two objects (A, B, f) and (A', B', f') is a pair (α, β), where α: $A \to A'$ is an R-homomorphism and β: $B \to B'$ is an S-homomorphism such that the diagram

$$\begin{array}{ccc} M \otimes A & \xrightarrow{1_M \otimes \alpha} & M \otimes A' \\ \downarrow f & & \downarrow f' \\ B & \xrightarrow{\beta} & B' \end{array}$$

commutes. When it is more convenient we consider the $\begin{pmatrix} R & 0 \\ M & S \end{pmatrix}$-modules as triples (g, U, V), where U is an R-module, V an S-module and g: $U \to \operatorname{Hom}_S(M, V)$ is an R-homomorphism (see **[16]**). A morphism between two objects (g, U, V) and (g', U', V') is then a pair (α, β), where α: $U \to U'$ is an R-homomorphism, β: $V \to V'$ is an S-homomorphism such that the diagram

$$\begin{array}{ccc} U & \xrightarrow{\alpha} & U' \\ \downarrow g & & \downarrow g' \\ \operatorname{Hom}_S(M, V) & \xrightarrow{\operatorname{Hom}_S(1_M, \beta)} & \operatorname{Hom}_S(M, V') \end{array}$$

commutes. If ${}_SM$ is finitely generated projective, then there is an isomorphism φ between the functors $\operatorname{Hom}_S(M, S) \otimes$ and $\operatorname{Hom}_S(M, \ )$ from Mod S to Mod R, where φ_X: $\operatorname{Hom}_S(M, S) \otimes X \to \operatorname{Hom}_S(M, X)$ is given by $[\varphi_X(h \otimes x)](m) = h(m)x$, for $m \in M$, $h \in \operatorname{Hom}_S(M, S)$ and $x \in X$. Hence the modules can be considered as triples (g, U, V), where g: $U \to M^{*S} \otimes_S V$ is an R-map. Here M^{*S} denotes the R-S-bimodule $\operatorname{Hom}_S(M, S)$, and for a ring Δ, Mod Δ denotes the category of all (left) Δ-modules. A morphism between two objects (g, U, V) and (g', U', V') is then a pair of morphisms (γ, δ), where γ: $U \to U'$ is an R-homomorphism and δ: $V \to V'$ is an S-homomorphism such that the diagram

$$\begin{array}{ccc} U & \xrightarrow{\gamma} & U' \\ \downarrow g_1 & & \downarrow g' \\ M^{*S} \otimes_S V & \xrightarrow{1_{M^{*S}} \otimes \delta} & M^{*S} \otimes_S V' \end{array}$$

commutes. It will be clear from the context which description of the modules we use in each case.

Let now R and S be rings and ${}_SM_R$ a bimodule such that M_R is finitely generated and projective. Then ${}_RN_S = M^{*R}$ is a finitely generated projective (left) R-module and N^{*R} is isomorphic to ${}_SM_R$.

Let $\Lambda = \begin{pmatrix} R & 0 \\ M & S \end{pmatrix}$ and $\Gamma' = \begin{pmatrix} S & 0 \\ N & R \end{pmatrix}$. We shall now define a functor F': Mod $\Lambda \to$ Mod Γ', which for artin rings will be closely related to the functor F we studied in §1. We shall also define a functor G': Mod $\Gamma' \to$ Mod Λ which will be related to the functor G defined in §1.

For an $\begin{pmatrix} R & 0 \\ M & S \end{pmatrix}$-module (A, B, f), define $F'(A, B, f)$ to be $(j, \operatorname{Ker}(f), A)$, where j: $\operatorname{Ker}(f) \to M \otimes A$ is the natural morphism in the exact sequence $0 \to \operatorname{Ker}(f) \xrightarrow{j} M \otimes A \xrightarrow{f} B$. For a morphism (α, β): $(A, B, f) \to (A', B', f')$, we define $F'(\alpha, \beta) = (\gamma, \alpha)$, where γ: $\operatorname{Ker}(f) \to \operatorname{Ker}(f')$ is induced by $1_M \otimes \alpha$: $M \otimes A \to M \otimes A'$.

We define the functor G': $\operatorname{Mod}\begin{pmatrix} S & 0 \\ N & R \end{pmatrix} \to \operatorname{Mod}\begin{pmatrix} R & 0 \\ M & S \end{pmatrix}$ as follows, $G'(g, U, V) = (V, \operatorname{Coker}(g), p)$, where p: $M \otimes V \to \operatorname{Coker}(g)$ is the natural morphism such that $U \xrightarrow{g} M \otimes V \xrightarrow{p} \operatorname{Coker}(g) \to 0$ is exact. For a morphism (γ, δ): $(g, U, V) \to (g', U', V')$, we define $G(\gamma, \delta) = (\delta, \beta)$: $(V, \operatorname{Coker}(g), p) \to (V', \operatorname{Coker}(g'), p')$, where β: $\operatorname{Coker}(g) \to \operatorname{Coker}(g')$ is induced by $1_M \otimes \delta$: $M \otimes V \to M \otimes V'$.

We denote by $\mathcal{C}$ the full subcategory of $\operatorname{Mod}\begin{pmatrix} R & 0 \\ M & S \end{pmatrix}$ whose objects are the (A, B, f), where f: $M \otimes A \to B$ is an epimorphism, and by $\mathcal{D}$ the full subcategory of $\operatorname{Mod}\begin{pmatrix} S & 0 \\ N & R \end{pmatrix}$ whose objects are the (g, U, V) where g: $U \to M \otimes V$ is a monomorphism.

We have the following relationship between the functors F' and G'.

PROPOSITION 4.1. *Let the notation and assumptions be as above.*

(a) *F' is a right adjoint functor to G', hence F' is left exact and G' is right exact.*

(b) *The natural transformation $G'F' \to I$ given by the adjointness of F' and G' is an isomorphism on an object (A, B, f) if and only if f: $M \otimes A \to B$ is an epimorphism, and the natural transformation $I \to F'G'$ is an isomorphism on an object (g, U, V) if and only if g: $U \to M \otimes V$ is a monomorphism.*

(c) *F': $\mathcal{C} \to \mathcal{D}$ and G': $\mathcal{D} \to \mathcal{C}$ are inverse equivalences of categories.*

(d) *If $0 \to X \to Y \to Z \to 0$ is an exact sequence in* $\operatorname{Mod}\begin{pmatrix} R & 0 \\ M & S \end{pmatrix}$, *where X lies in $\mathcal{C}$, then $0 \to F'(X) \to F'(Y) \to F'(Z) \to 0$ is exact, and if $0 \to X \to Y \to Z \to 0$ is an exact sequence in* $\operatorname{Mod}\begin{pmatrix} S & 0 \\ N & R \end{pmatrix}$ *where Z lies in $\mathcal{D}$, then $0 \to G'(X) \to G'(Y) \to G'(Z) \to 0$ is exact.*

PROOF. (a) Let (A, B, f) be in Mod Λ and (g, U, V) in Mod Γ'. We want to define a natural isomorphism between $\operatorname{Hom}(G'(g, U, V), (A, B, f))$ and

$\mathrm{Hom}((g, U, V), F'(A, B, f))$. A morphism from $G'(g, U, V) = (V, \mathrm{Coker}(g), f')$ to (A, B, f) is a pair (α, β), where $\alpha\colon V \to A$, $\beta\colon \mathrm{Coker}(g) \to B$ are such that the diagram

$$\begin{array}{ccc}
 & & 0 \\
 & & \downarrow \\
U & & \mathrm{Ker}(f) \\
\downarrow g & & \downarrow \\
M \otimes V & \xrightarrow{1_M \otimes \alpha} & M \otimes A \\
\downarrow & & \downarrow f \\
\mathrm{Coker}(g) & \xrightarrow{\beta} & B \\
\downarrow & & \\
0 & &
\end{array}$$

commutes. There is induced a morphism $\gamma\colon U \to \mathrm{Ker}(f)$ such that (γ, α) is a morphism from (g, U, V) to $(g', \mathrm{Ker}(f), A) = F'(A, B, f)$. It is not hard to see that in this way we get a natural isomorphism, so that F' is a right adjoint functor to G'. Hence F' is left exact and G' is right exact.

(b) It is easy to see that $G'F'(A, B, f) = (A, \mathrm{Im}(f), f)$, and the natural morphism $(A, \mathrm{Im}(f), f) \to (A, B, f)$ is $(1_A, j)$, where $j\colon \mathrm{Im}(f) \to B$ is the inclusion morphism. From this it follows that the natural morphism $G'F'(A, B, f) \to (A, B, f)$ is an isomorphism if and only if $f\colon M \otimes A \to B$ is an epimorphism. Similarly $F'G'(g, U, V) = (g', \mathrm{Im}(g), V)$, where $g'\colon \mathrm{Im}(g) \to M \otimes V$ is the natural morphism induced by $g\colon U \to M \otimes V$, such that the diagram

$$\begin{array}{ccc}
U & \xrightarrow{g} & M \otimes V \\
 & {\scriptstyle p}\searrow \quad \nearrow{\scriptstyle g'} & \\
 & \mathrm{Im}(g) &
\end{array}$$

commutes. We then have the natural morphism $(p, 1_V)$ from (g, U, V) to $(g', \mathrm{Im}(g), V)$, which is clearly an isomorphism if and only if $g\colon U \to M \otimes V$ is a monomorphism.

(c) Follows directly from (b), observing in addition that $F'(A, B, f)$ is always in $\mathcal{D}$ and $G'(g, U, V)$ is always in $\mathcal{C}$.

(d) Let $0 \to (A', B', f') \to (A, B, f) \to (A'', B'', f'') \to 0$ be an exact sequence in $\mathrm{Mod}\,\Lambda$, where $f'\colon M \otimes A' \to B'$ is an epimorphism. That the above sequence is exact means that $0 \to A' \to A \to A'' \to 0$ and $0 \to B' \to B \to B'' \to 0$ are exact sequences. Using that M_R is projective and using the snake lemma, we get the following commutative diagram:

$$\begin{array}{ccccccccc} & & 0 & & 0 & & 0 & & \\ & & \downarrow & & \downarrow & & \downarrow & & \\ 0 & \to & \operatorname{Ker}(f') & \to & \operatorname{Ker}(f) & \to & \operatorname{Ker}(f'') & \to & 0 \\ & & \downarrow & & \downarrow & & \downarrow & & \\ 0 & \to & M \otimes A' & \to & M \otimes A & \to & M \otimes A'' & \to & 0 \\ & & \downarrow & & \downarrow & & \downarrow & & \\ 0 & \to & B' & \to & B & \to & B'' & \to & 0 \\ & & \downarrow & & & & & & \\ & & 0 & & & & & & \end{array}$$

This shows that we get an exact sequence

$$0 \to F'(A', B', f') \to F'(A, B, f) \to F'(A'', B'', f'') \to 0.$$

The second part of (d) follows similarly.

Assume now that R and S are basic artin rings, not necessarily artin algebras, and that ${}_S M_R$, in addition to being a finitely generated projective R-module, is a finitely generated S-module. Then $\Lambda = \left(\begin{smallmatrix} R & 0 \\ M & S \end{smallmatrix}\right)$ is an artin ring, and $\Gamma' = \left(\begin{smallmatrix} S & 0 \\ M^{*R} & R \end{smallmatrix}\right)$ is a left artin ring. In this case it is easy to see that if (A, B, f) is finitely generated, then $F'(A, B, f)$ is finitely generated, so that F' induces a functor from mod Λ to mod Γ'. If we assume that S is a division ring and M is not zero, F will be a (left) partial Coxeter functor in the sense of §2, as follows from Proposition 4.1. For it is not hard to see that in this case the only indecomposable Λ-module not in $\mathcal{C}$ is $S_1 = (0, S, 0)$, which is simple projective noninjective, and the only indecomposable Γ'-module not in $\mathcal{D}$ is $(S, 0, 0)$, which is a simple injective nonprojective Γ'-module.

If we start with a basic artin ring Λ with a simple projective module S_1 and we write $\Lambda = S_1 \amalg P$, then Λ is isomorphic to the triangular matrix ring $\left(\begin{smallmatrix} R & 0 \\ M & S \end{smallmatrix}\right)$ where $R = \operatorname{End}_\Lambda(P)^{\mathrm{op}}$, $S = \operatorname{End}_\Lambda(S_1)^{\mathrm{op}}$ and $M = \operatorname{Hom}_\Lambda(S_1, P)$. Since we get a partial Coxeter functor when M_R is finitely generated projective, we shall investigate when this condition is satisfied.

PROPOSITION 4.2. *Let Λ be a basic artin ring with a simple projective noninjective module S_1 and let P be given by $\Lambda = S_1 \amalg P$. Then $M = (S_1, P)$ is a right projective* $\operatorname{End}(P)^{\mathrm{op}}$*-module if and only if* $(\operatorname{Tr} D'S_1)^* = 0$, *where as in* §1, $D'S_1 = S_1^*/\mathbf{r}S_1^*$. *In particular, M is a projective* $\operatorname{End}(P)^{\mathrm{op}}$*-module if Λ is hereditary.*

PROOF. Let $0 \to S_1 \to Q \to \operatorname{Tr} D'S_1 \to 0$ be an almost split sequence, as described at the end of §1. Consider the exact sequence

$$0 \to (\operatorname{Tr} D'S_1, P) \xrightarrow{t} (Q, P) \to (S_1, P) \to \operatorname{Ext}^1_\Lambda(\operatorname{Tr} D'S_1, P) \to 0.$$

Since S_1 is simple and not isomorphic to a summand of P, it follows, as for algebras in Lemma 1.1, that $\operatorname{Ext}^1_\Lambda(\operatorname{Tr} D'S, P)$ is zero. Assume now that

$(\mathrm{Tr}\, D'S_1)^* = 0$. Then we have an isomorphism $(Q, P) \to (S_1, P)$. Since every indecomposable summand of Q is a summand of P, we conclude that (Q, P), and hence (S_1, P), is a projective $\mathrm{End}(P)$-module.

Assume conversely that (S_1, P) is a projective $\mathrm{End}(P)$-module. Then there is a map $v\colon (S_1, P) \to (Q, P)$ such that $(Q, P) = \mathrm{Im}(t) \amalg \mathrm{Im}(v)$. Since Q is in $\mathrm{add}(P)$ and $(\ , P)$ induces an equivalence of categories between $\mathrm{add}(P)$ and the finitely generated projective $\mathrm{End}(P)$-modules, there is a decomposition $Q = Q_1 \amalg Q_2$ inducing the decomposition $(Q, P) = (Q_1, P) \amalg (Q_2, P)$, where $(Q_1, P) = \mathrm{Im}(t)$ and $(Q_2, P) = \mathrm{Im}(v)$. But then the induced morphism $(Q_1, P) \to (S_1, P)$ is zero. Since Q_1 is in $\mathrm{add}(P)$ this implies that the map $S_1 \to Q_1$ is zero. This is a contradiction since the map $S_1 \to Q_1$ is irreducible, because Q_1 is a summand of Q, and $0 \to S_1 \to Q \to \mathrm{Tr}\, D'S_1 \to 0$ is an almost split sequence [7, Theorem 2.4]. Hence $(\mathrm{Tr}\, D'S_1, P)$ is zero, and since obviously $(\mathrm{Tr}\, D'S_1, S_1)$ is zero, we conclude that $(\mathrm{Tr}\, D'S_1)^* = 0$.

We can now combine our results so far in this section to get the following proposition, which in particular gives us examples of Coxeter functors between left artin rings which are not necessarily artin algebras.

PROPOSITION 4.3. *Let Λ be a basic artin ring with a simple projective noninjective module S_1, let $\Lambda = S_1 \amalg P$, and assume that $(\mathrm{Tr}\, D'S_1)^*$ is zero. Then $\Lambda = \left(\begin{smallmatrix} R & 0 \\ M & S \end{smallmatrix}\right)$, where $R = \mathrm{End}(P)^{\mathrm{op}}$, $S = \mathrm{End}(S_1)^{\mathrm{op}}$ and $M_R = (S_1, P)_R$ is finitely generated projective. Let $\Gamma' = \left(\begin{smallmatrix} S & 0 \\ M^{*R} & R \end{smallmatrix}\right)$. Then the functors $F'\colon \mathrm{mod}\, \Lambda \to \mathrm{mod}\, \Gamma'$ and $G'\colon \mathrm{mod}\, \Gamma' \to \mathrm{mod}\, \Lambda$ are adjoint partial Coxeter functors which induce inverse equivalences of categories between $\mathcal{C}_{S_1}$ and $\mathfrak{D}_T$, where $S_1 = (0, S, 0)$ and $T = (S, 0, 0)$.*

COROLLARY 4.4. *Let Λ be a basic artin ring and S_1 a simple projective noninjective Λ-module such that $(\mathrm{Tr}\, D'S_1)^*$ is zero, $\Lambda = P \amalg S_1$, $X = P \amalg \mathrm{Tr}\, D'S_1$, $\Gamma = \mathrm{End}(X)^{\mathrm{op}}$. Then $(X,)\colon \mathrm{mod}\, \Lambda \to \mathrm{mod}\, \Gamma$ is a left partial Coxeter functor.*

PROOF. By Proposition 4.3 there is some left partial Coxeter functor $F = F_S\colon \mathrm{mod}\, \Lambda \to \mathrm{mod}\, \Lambda'$, which is by Theorem 2.3 equivalent to $(X,)$, which is then a left partial Coxeter functor.

5. Connections with diagrams. For a certain class of hereditary artin tensor rings, namely hereditary rings with duality conditions, partial Coxeter functors and Coxeter functors have been studied in [9] and [13], as a tool for studying the representation theory of these rings. Actually, these functors are defined between categories of representations of diagrams, but can be interpreted to give functors between module categories. They are left or right partial Coxeter functors in the sense of §2, and are by our uniqueness results from §2, closely related to the functors studied in §§1 and 4. We shall see here that the connection with the triangular matrix case studied in §4 is particularly natural. A Coxeter functor C^+ in the diagram case is defined to be a composite of partial Coxeter functors defined for diagrams, and is, by our results in §2, closely related to $D\,\mathrm{Tr}$ in the case of artin algebras. However, we can not conclude in general that C^+ is isomorphic to $D\,\mathrm{Tr}$. In particular, C^+ is not uniquely defined up to isomorphism of functors, and we include a discussion of this problem. However, it has been shown in **[10]** that for an artin tensor algebra Λ there is some Coxeter functor C^+ which is isomorphic to $D\,\mathrm{Tr}$. Here we show that for an arbitrary tensor ring with duality conditions, a given Coxeter functor C_1^+ is isomorphic to $D_1\,\mathrm{Tr}$ for some duality D_1.

For the convenience of the reader, we start out by recalling some background material from **[9]**, **[13]**.

A *valued graph* $(\Gamma, \mathbf{d})$ is a finite set Γ of vertices together with pairs of nonnegative integers (d_{ij}, d_{ji}) with $d_{ii} = 0$ for all $i \in \Gamma$, and such that there exist natural numbers f_i satisfying $d_{ij}f_j = d_{ji}f_i$ for all i, j in Γ. We shall assume that $(\Gamma, \mathbf{d})$ is *connected*, in the sense that for every k, t in Γ there is a sequence $k, \ldots, i, j, \ldots, t$ of vertices of Γ such that $d_{ij} \neq 0$ for each pair of subsequent vertices i, j. Pairs $\{i, j\}$ where $d_{ij} \neq 0$ are the *edges* and i and j are said to be neighbors.

An *orientation* Ω of a valued graph $(\Gamma, \mathbf{d})$ is given by prescribing for each edge i, j of $(\Gamma, \mathbf{d})$ an order, indicated by an arrow $i \to j$. Given an orientation Ω and a vertex k in Γ, we get a new orientation $s_k\Omega$ of $(\Gamma, \mathbf{d})$ by reversing the direction of arrows along all edges containing k. A vertex k in Γ is said to be a *sink* with respect to Ω if $k \leftarrow i$ for all neighbors $i \in \Gamma$ of k, and a *source* if $k \to i$ for all neighbors $i \in \Gamma$ of k. An orientation Ω of $(\Gamma, \mathbf{d})$ is said to be *admissible* if there is an ordering $k_1, \ldots, k_n$ of the vertices of Γ such that each vertex k_r is a sink with respect to the orientation $s_{k_{r-1}} \cdots s_{k_2}s_{k_1}\Omega$ for all r, $1 \leqslant r \leqslant n$. Such an ordering of the k_i is called an admissible ordering for Ω.

A *modulation* $\mathbf{M}$ of a valued graph $(\Gamma, \mathbf{d})$ is a set of division rings F_i, $i \in \Gamma$, together with finite dimensional F_i-F_j-bimodules ${}_iM_j$ and ${}_jM_i$ for all edges $\{i, j\}$ of $(\Gamma, \mathbf{d})$ such that

(1) there are F_j-F_i-bimodule isomorphisms ${}_jM_i \cong \mathrm{Hom}_{F_i}({}_iM_j, F_i) \cong \mathrm{Hom}_{F_j}({}_iM_j, F_j)$,

(2) $\dim({}_iM_j)_{F_i} = d_{ij}$.

A *realization* $(\mathbf{M}, \Omega)$ of a valued graph $(\Gamma, \mathbf{d})$ is a modulation $\mathbf{M}$ of $(\Gamma, \mathbf{d})$ together with an admissible orientation Ω.

A *representation* $\mathbf{X} = ({}_iX, \varphi_{ij})$ of a realization $(\mathbf{M}, \Omega)$ of $(\Gamma, \mathbf{d})$ is a set of finite dimensional left F_i-spaces ${}_iX$, $i \in \Gamma$, together with F_j-linear maps ${}_j\varphi_i$: ${}_jM_i \otimes_{F_i} {}_iX \to {}_jX$ for all oriented edges $i \to j$. A morphism from a representation $\mathbf{X} = ({}_iX, {}_j\varphi_i)$ to a representation $\mathbf{X}' = ({}_iX', {}_j\varphi_i')$ is a set $\boldsymbol{\alpha} = (\alpha_i)$ of

F_i-linear maps α_i: ${}_iX \to {}_iX'$ for $i \in \Gamma$, such that ${}_j\varphi_i'(1_{{}_jM_i} \otimes \alpha_i) = \alpha_j\ {}_j\varphi_i$. The representations of $(\mathbf{M}, \Omega)$ form an abelian category which we denote by $\mathfrak{L}(\mathbf{M}, \Omega)$.

For a sink k of the realization $(\mathbf{M}, \Omega)$ there is a partial Coxeter functor S_k^+: $\mathfrak{L}(\mathbf{M}, \Omega) \to \mathfrak{L}(\mathbf{M}, s_k\Omega)$ defined as follows. Let $\mathbf{X} = ({}_iX, {}_j\varphi_i)$ be an object in $\mathfrak{L}(\mathbf{M}, \Omega)$. Then $S_k^+(\mathbf{X})$ is defined to be $\mathbf{Y} = ({}_iY, {}_j\psi_i)$, where ${}_iY = {}_iX$ for all $i \neq k$ and ${}_kY$ is the kernel in the sequence

$$0 \to {}_kY \xrightarrow{({}_j\pi_k)_j} \coprod_{j \in \Gamma} {}_kM_j \otimes {}_jX \xrightarrow{({}_k\varphi_j)_j} {}_kX.$$

Further ${}_j\psi_i = {}_j\varphi_i$ for $i \neq k$, and ${}_j\psi_k = {}_j\bar{\pi}_k$: ${}_jM_k \otimes {}_kY \to {}_jY$ is determined by ${}_j\pi_k$: ${}_kY \to {}_kM_j \otimes {}_jY$, using the natural isomorphism

$$\operatorname{Hom}_{F_j}({}_jM_k \otimes {}_kY, {}_jY) \to \operatorname{Hom}_{F_k}\big({}_kY, \operatorname{Hom}_{F_j}({}_jM_k, {}_jY)\big)$$

and the inverse of the natural isomorphism

$$\operatorname{Hom}_{F_k}({}_jM_k, {}_jY) \to \operatorname{Hom}_{F_k}({}_jM_k, F_k) \otimes {}_jY,$$

together with the given isomorphism $\operatorname{Hom}_{F_k}({}_jM_k, F_k) \to {}_kM_j$. S_k^+ is defined on morphisms in a natural way. Now a Coxeter functor C^+: $\mathfrak{L}(\mathbf{M}, \Omega) \to \mathfrak{L}(\mathbf{M}, \Omega)$ is defined to be the composite functor $S_{k_n}^+ \cdots S_{k_1}^+$, where $k_1, \ldots, k_n$ is an admissible ordering with respect to the orientation Ω. The definition of C^+ depends on the choice of bimodule isomorphisms. We shall see later in this section that different choices may give rise to nonisomorphic Coxeter functors.

For the convenience of the reader we also recall the connection between the categories $\mathfrak{L}(\mathbf{M}, \Omega)$ and the categories of finitely generated modules over certain hereditary artin rings. (See **[12]**, **[18]**.) Assume that $(\Gamma, \mathbf{d})$ is a graph with no oriented cycles, and let $(\mathbf{M}, \Omega)$ be a realization of $(\Gamma, \mathbf{d})$. Then there is a natural equivalence of categories between $\mathfrak{L}(\mathbf{M}, \Omega)$ and mod Λ, where Λ is the following hereditary basic artin ring. Let $i = 1, \ldots, m$ be the vertices of Γ and F_i the corresponding division rings. Let $\Sigma = \coprod_{i=1}^m F_i$, and let $\mathbf{r} = \coprod {}_jM_i$, where ${}_jM_i$ runs through all modules where there is an arrow $i \to j$. Since $(\Gamma, \mathbf{d})$ has no oriented cycles, the n-fold tensor product of $\mathbf{r}$ is zero for some smallest n. Λ is then the ring $\Sigma + \mathbf{r} + \mathbf{r} \otimes \mathbf{r} + \cdots + \mathbf{r} \otimes \cdots \otimes \mathbf{r}$, where the last term is an $(n-1)$-fold tensor product. Conversely, let Λ be a basic hereditary artin ring of this form, where $\Sigma = \coprod_{i=1}^m F_i$, and where we have a two-sided decomposition of $\mathbf{r}$ over Σ as $\coprod {}_jM_i$, where we mean that for ${}_jM_i$ only F_j has a nonzero action on the left and only F_i has a nonzero action on the right. Assuming that $\operatorname{Hom}_{F_i}({}_jM_i, F_i)$ is isomorphic to $\operatorname{Hom}_{F_j}({}_jM_i, F_j)$, we get a realization of a valued graph $(\Gamma, \mathbf{d})$, where $1, 2, \ldots, m$ are the vertices of Γ, and there is an arrow $i \to j$ if and only if there is some module ${}_jM_i \neq 0$ in the two-sided decomposition of $\mathbf{r}$. When an hereditary artin ring is of this type, we say that it is a tensor ring with duality conditions.

If $\mathbf{X} = ({}_iX, {}_i\varphi_j)$ is in $\mathfrak{L}(\mathbf{M}, \Gamma)$, then the corresponding Λ-module is $\amalg\, {}_iX$ as Σ-module, and the action of $\mathbf{r} = \amalg\, {}_jM_i$ on $\amalg\, {}_iX$ is given via the morphisms ${}_i\varphi_j$.

If Λ' denotes the hereditary artin ring associated with $(\mathbf{M}, s_k\Omega)$, where k is a sink for Ω, then the partial Coxeter functor S_k^+ defines a functor from mod Λ to mod Λ'. And for fixed bimodule isomorphisms the corresponding Coxeter functor C^+ defines a functor from mod Λ to mod Λ. By the properties of S_k^+ listed in [**13**, §2], we know that S_k^+ is a left partial Coxeter functor in the sense defined in §2. Hence we know by Theorem 2.3 that S_k^+ is equivalent to the other left partial Coxeter functors we have studied in this paper. We shall see here that we have a particularly natural connection between S_k^+ and the left partial Coxeter functor defined for triangular matrix rings in §4.

Let now $(\mathbf{M}, \Omega)$ be a realization of a valued graph $(\Gamma, \mathbf{d})$ with no oriented cycles, let k be a sink and let $1, 2, \ldots, k-1$ be the other vertices. Let R be the hereditary ring determined by the realization (M', Ω') of the valued graph $(\Gamma', \mathbf{d}')$ which we get by removing the vertex k and the bimodules of the form ${}_kM_i$ and ${}_iM_k$. Let $K = \amalg_i\, {}_kM_i$, $M = K \otimes_\Sigma R$ and $S = F_k$. Let $\mathbf{X} = ({}_iX, \varphi_{ij})$ be a representation of $(\mathbf{M}, \Omega)$. Then $\mathbf{X}$ determines in a natural way a module (A, B, f) over the triangular matrix ring $\left(\begin{smallmatrix} R & 0 \\ {}_SM_R & S\end{smallmatrix}\right)$, where the R-module A is $\amalg_{i<k}\, {}_iX$ as a Σ-module and the R-action is given by the φ_{ij}. $B = {}_kX$, and the morphisms ${}_k\varphi_i$: ${}_kM_i \otimes {}_iX \to {}_kX$ determine a unique S-morphism $K \otimes_\Sigma A \to B$, hence a morphism f: $M \otimes_R A \to B$, since we have a natural isomorphism $K \otimes_\Sigma R \otimes_R A \to K \otimes_\Sigma A$. It is not hard to see that we get a natural equivalence of categories H_1: $\mathfrak{L}(\mathbf{M}, \Omega) \to \operatorname{mod}\left(\begin{smallmatrix} R & 0 \\ {}_SM_R & S\end{smallmatrix}\right)$.

Let now $Y = ({}_iY, \psi_{ij})$ be a representation of $(\mathbf{M}, s_k\Omega)$. Let $C = {}_kY$, D the R-module which is $\amalg_{i<k}\, {}_iY$ as a Σ-module and the R-action is given by the ψ_{ij}. Then (g, C, D) is a module over $\left(\begin{smallmatrix} S & 0 \\ M^{*R} & R\end{smallmatrix}\right)$, where g: $C \to M \otimes D$ is given as follows. The morphisms ${}_j\psi_k$: ${}_jM_k \otimes {}_kY \to {}_jY$ determine morphisms ${}_j\pi_k$: ${}_kY \to {}_kM_j \otimes {}_jY$, and hence determine morphisms g: $C \to \amalg_j\, {}_kM_j \otimes {}_jY = K \otimes_\Sigma D = M \otimes_R D$. It is not hard to see that we get a natural equivalence of categories

$$H_2\colon \mathfrak{L}(\mathbf{M}, s_k\Omega) \to \operatorname{mod}\begin{pmatrix} S & 0 \\ M^{*R} & R\end{pmatrix}.$$

Considering the definition of the functors S_k^+: $\mathfrak{L}(\mathbf{M}, \Omega) \to \mathfrak{L}(\mathbf{M}, s_k\Omega)$ and

$$F'\colon \operatorname{mod}\begin{pmatrix} R & 0 \\ M & S\end{pmatrix} \to \operatorname{mod}\begin{pmatrix} S & 0 \\ M^{*R} & R\end{pmatrix},$$

it is not hard to see that we have a commutative diagram:

$$\begin{array}{ccc} \mathfrak{L}(\mathbf{M}, \Omega) & \overset{H_1}{\rightarrow} & \operatorname{mod}\begin{pmatrix} R & 0 \\ M & S \end{pmatrix} \\ \downarrow S_k^+ & & \downarrow F' \\ \mathfrak{L}(M, s_k\Omega) & \overset{H_2}{\rightarrow} & \operatorname{mod}\begin{pmatrix} S & 0 \\ M^{*R} & R \end{pmatrix} \end{array}$$

We collect some of the above in the following proposition.

PROPOSITION 5.1. *Let Λ be a tensor ring with duality conditions, and S_1 a simple projective noninjective Λ-module. Then the corresponding partial Coxeter functors S_1^+ and F' from §4 are equivalent functors in a natural way.*

Let Λ be an hereditary tensor algebra. Then we know that a given Coxeter functor C^+: mod $\Lambda \rightarrow$ mod Λ, which is by definition a composite of partial Coxeter functors, is equivalent to D Tr by Theorem 3.6. For we know from [**13**, §2] that C^+ takes projectives to zero and induces an equivalence of categories $\operatorname{mod}_P \Lambda \rightarrow \operatorname{mod}_I \Lambda$. However, this does not necessarily mean that D Tr and C^+ are isomorphic functors. In fact, the definition of C^+ was made using fixed bimodule isomorphisms $\operatorname{Hom}_{F_i}({}_iM_j, F_i) \rightarrow \operatorname{Hom}_{F_j}({}_iM_j, F_j)$. As remarked in [**13**], the theory in [**13**] is independent of this choice. We shall see that different choices may give rise to nonisomorphic Coxeter functors. We shall, in fact, devote some time to a discussion of what this depends on.

Choosing bimodule isomorphisms $\operatorname{Hom}_{F_i}({}_iM_j, F_i) \rightarrow \operatorname{Hom}_{F_j}({}_iM_j, F_j)$ is equivalent to choosing bimodule isomorphisms $\operatorname{Hom}_{F_j}(({}_iM_j, F_i), F_j) \rightarrow {}_iM_j$. Two different choices of these isomorphisms give rise to bimodule isomorphisms ${}_iM_j \rightarrow {}_iM_j$, which induce a ring isomorphism t: $\Lambda \rightarrow \Lambda$ by using the identity on the F_i. It is then not hard to see that if C_1^+ and C_2^+ are two Coxeter functors, then $C_1^+ = t^* C_2^+$, where t^*: mod $\Lambda \rightarrow$ mod Λ is an equivalence of categories induced by a ring isomorphism which is given by bimodule isomorphisms ${}_iM_j \rightarrow {}_iM_j$. Since C_1^+ and C_2^+ both induce equivalences of categories between $\operatorname{mod}_P \Lambda$ and $\operatorname{mod}_I \Lambda$, it is easy to see that C_1^+ and C_2^+ are isomorphic if and only if the restriction of t^* to $\operatorname{mod}_I \Lambda$ is isomorphic to the identity.

We can then get the following result.

PROPOSITION 5.2. *Let Λ be an artin hereditary tensor ring with duality conditions. Then the isomorphism classes of Coxeter functors are in one to one correspondence with the elements of the quotient group G/H, where G is the group of automorphisms induced by bimodule isomorphisms ${}_iM_j \rightarrow {}_iM_j$, and H the subgroup consisting of those of this type which are inner.*

PROOF. Since an isomorphism t: $\Lambda \rightarrow \Lambda$ induces a functor t^*: mod $\Lambda \rightarrow$ mod Λ which is isomorphic to the identity if and only if t is an inner automorphism [**15**, pp. 454–456], this is a direct consequence of Proposition 3.3 and the above remarks.

As an immediate consequence we have the following.

COROLLARY 5.3. *Let Λ be an artin hereditary tensor ring with duality conditions. Then all Coxeter functors are isomorphic if and only if every ring isomorphism $\Lambda \to \Lambda$ induced by bimodule isomorphisms ${}_iM_j \to {}_iM_j$ is an inner automorphism.*

To give a concrete example for when not all Coxeter functors are isomorphic consider $\Lambda = \left(\begin{smallmatrix} K & 0 \\ K\amalg K & K \end{smallmatrix}\right)$, for a field K, where the K-K-bimodule structure of $K\amalg K$ is given by $a(b, c) = (ab, ac)$, $(b, c)d = (bd, cd)$. Then the isomorphism t defined by

$$t\begin{pmatrix} a & 0 \\ (b, c) & d \end{pmatrix} = \begin{pmatrix} a & 0 \\ (c, b) & d \end{pmatrix}$$

is induced by a bimodule isomorphism $K\amalg K \to K\amalg K$, which is easily checked not to be inner. Hence by the above there are two Coxeter functors C_1^+ and C_2^+ which are not isomorphic.

Let

$$P = \begin{pmatrix} K & 0 \\ K\amalg K & 0 \end{pmatrix}, \qquad N_1 = \begin{pmatrix} 0 & 0 \\ K\amalg 0 & 0 \end{pmatrix}, \qquad N_2 = \begin{pmatrix} 0 & 0 \\ 0\amalg K & 0 \end{pmatrix}.$$

Since $t^*(P/N_1) = P/N_2$ and P/N_1, P/N_2 are obviously not isomorphic, C_1^+ and C_2^+ are also not isomorphic on objects.

We shall now give a direct interpretation of what a Coxeter functor C^+ for an artin hereditary tensor ring Λ looks like.

We have already seen that for artin algebras there is a close relationship between a given C^+ and D Tr.

We shall now see that if C_1^+ is a Coxeter functor for an arbitrary hereditary tensor ring Λ with duality conditions, then there is some duality D_1: mod Λ^{op} $\to$ mod Λ such that C_1^+ is isomorphic to D_1Tr. In particular, this shows that the hereditary tensor rings Λ with duality conditions have a duality. Our result will be a consequence of the following more general result.

THEOREM 5.4. *Let Λ and Λ' be indecomposable hereditary artin rings and α: $\text{mod}_P\,\Lambda \to \text{mod}_I\,\Lambda'$ an equivalence of categories. Then $\alpha = D_1\text{Tr}|\text{mod}_P\,\Lambda$ for some duality D_1*: mod $\Lambda^{\text{op}} \to$ mod Λ'.

PROOF. Since Tr: mod $\Lambda^{\text{op}} \to$ mod Λ induces a duality Tr: $\text{mod}_P\,\Lambda^{\text{op}} \to \text{mod}_P\,\Lambda$ we get a duality β: $\text{mod}_P\,\Lambda^{\text{op}} \xrightarrow{\text{Tr}} \text{mod}_P\,\Lambda \xrightarrow{\alpha} \text{mod}_I\,\Lambda'$. Since $\text{mod}_P\,\Lambda^{\text{op}}$ is closed under factors and $\text{mod}_I\,\Lambda'$ is closed under subobjects, we can use Lemma 2.1 to conclude that a sequence $0 \to A \xrightarrow{f} B \xrightarrow{g} C \to 0$ with A, B and C in $\text{mod}_P\,\Lambda^{\text{op}}$ is exact if and only if $0 \to \beta C \xrightarrow{\beta g} \beta B \xrightarrow{\beta f} \beta A \to 0$ is exact.

Let A be indecomposable and not projective in mod Λ^{op}. If A is injective, we can show as in Lemma 2.2 that $\beta(A)$ is projective since $\mathrm{mod}_I\ \Lambda'$ is closed under subobjects and projective covers, using that Λ' is hereditary. Assume next that $\beta(A)$ is projective. We will prove that then A is injective. Let i: $A \to B$ be a monomorphism where B is in $\mathrm{mod}_P\ \Lambda^{\mathrm{op}}$, and consider the exact sequence $0 \to A \xrightarrow{i} B \to C \to 0$. Since A, B and C are in $\mathrm{mod}_P\ \Lambda^{\mathrm{op}}$, we know that $\beta(i)$: $\beta(B) \to \beta(A)$ is an epimorphism, which splits since $\beta(A)$ is projective. Hence i: $A \to B$ is a split monomorphism. Using now the fact that all monomorphisms i: $A \to B$ in $\mathrm{mod}_P\ \Lambda^{\mathrm{op}}$ split we will prove that A is injective. Let j: $K \to L$ be an arbitrary monomorphism in mod Λ^{op}, and t: $K \to A$ a morphism. Consider the pushout diagram:

$$\begin{array}{ccccccccc} 0 & \to & K & \xrightarrow{j} & L & \to & L/K & \to & 0 \\ & & t\downarrow & & \downarrow & & \| & & \\ 0 & \to & A & \xrightarrow{s} & E & \to & L/K & \to & 0 \end{array}$$

Since A is not projective and Λ^{op} is hereditary, there is a decomposition $E = E_1 \amalg E_2$ such that $\mathrm{Im}(s) \subset E_1$ and E_1 has no nonzero projective summands. Since E and hence E_1 is finitely generated, we know that $A \to E_1$, and hence s: $A \to E$ is a split monomorphism. This shows that there is a map u: $L \to A$ such that $uj = t$, and it follows that A is injective.

We point out that the above also shows that all indecomposable injectives in mod Λ^{op} are finitely generated. For let A be indecomposable and not projective, and consider $0 \to \beta(K) \to \beta(B) \to \beta(A) \to 0$, where $\beta(B) \to \beta(A)$ is a projective cover. Then $0 \to A \to B$ is exact, and we have just seen that B is then injective, so the injective envelope of A is finitely generated. Let now S be simple. If the injective envelope $I(S)$ of S is not finitely generated then there is a finitely generated nonprojective module E such that $S \subset E \subset I(S)$. By the above we know that the injective envelope of E, which is $I(S)$, is finitely generated, and this is a contradiction. So for every simple module S the injective envelope of S is finitely generated.

We can now conclude from the above that β induces a duality between the finitely generated injective nonprojective Λ^{op}-modules and the finitely generated projective noninjective Λ-modules. If Λ and Λ' have no projective injectives, this can be extended to a duality D_1: mod $\Lambda^{\mathrm{op}} \to$ mod Λ'. D_1 extends β since β preserves exact sequences $0 \to A \to B \to C \to 0$ with A, B and C in $\mathrm{mod}_P\ \Lambda^{\mathrm{op}}$. This shows that $\alpha = D_1\mathrm{Tr}$ in this case.

If Λ has a nonzero projective injective, we know that $\Lambda^{\mathrm{op}} = T_n(K)$ for a division ring K. Then $\mathrm{mod}_P\ \Lambda^{\mathrm{op}}$ is equivalent to mod $T_{n-1}(K)$, and we would get a duality between the finitely generated injective $T_{n-1}(K)$-modules and the finitely generated projective noninjective Λ'-modules. If Λ' had no projective injectives we would get a duality between mod $T_{n-1}(K)$ and mod Λ', and get a contradiction. Hence we get that $\Lambda' = T_m(K')$, where clearly $n = m$ and $K' = K^{\mathrm{op}}$. Let Q be the projective injective indecomposable Λ'-module and let P be the projective injective indecomposable Λ^{op}-module. We define $\beta(P) = Q$. Using an argument similar to that used in Proposition 3.3 one can extend the restriction of β to the full subcategory of finitely generated injective nonprojective Λ^{op}-modules to a duality β from the full subcategory of finitely generated projective Λ^{op}-modules to the full subcategory of finitely generated projective Λ'-modules. Then, as in the first case, one can extend β to a duality D_1; mod $\Lambda^{\mathrm{op}} \to$ mod Λ'.

COROLLARY 5.5. *Let Λ be an hereditary artin tensor ring with duality conditions, and C_1^+ a Coxeter functor. Then $C_1^+ = D_1\mathrm{Tr}$ for some duality D_1.*

PROOF. This follows directly from Theorem 5.4, using that C_1^+ and $D_1\mathrm{Tr}$ take projectives to zero.

Clearly we can not expect more specific information about the duality D_1 unless we have more information about the Coxeter functor C_1^+. It would be interesting to know in general how to describe the duality D_1 such that $C_1^+ = D_1\mathrm{Tr}$ directly. Actually, the following gives a natural candidate for this duality D_1.

Let $(\mathbf{M}, \Omega)$ be a realization corresponding to an hereditary tensor ring Λ with duality conditions. As before, we denote by F_i the associated division rings. Then the graph corresponding to Λ^{op} has the same number of edges and arrows, all arrows have opposite direction, and the bimodules are the same, if we consider a representation as a collection of right F_i-spaces. Let $\mathbf{X} = ({}_iX, \varphi_{ij})$ be a representation of $(\mathbf{M}, \Omega)$. We define $\mathbf{Y} = D_1\mathbf{X}$ to be the representation $(Y_i, {}_i\psi_j)$, where $Y_i = \mathrm{Hom}_{F_i}({}_iX, F_i)$, and ${}_i\psi_j$: $\mathrm{Hom}_{F_i}({}_iX, F_i) \otimes {}_iM_j \to \mathrm{Hom}_{F_j}({}_jX, F_j)$ is the map that we will describe now. The given map ${}_i\varphi_j$: ${}_iM_j \otimes {}_jX \to {}_iX$ gives by adjointness a map ${}_jX \to \mathrm{Hom}_{F_i}({}_iM_j, {}_iX)$. Let then $f \in \mathrm{Hom}_{F_i}({}_iX, F_i)$, $m \in {}_iM_j$. Then we get the following map from ${}_jX$ to F_j:

$${}_jX \to \mathrm{Hom}_{F_i}({}_iM_j, {}_iX) \overset{(\mathrm{id}, f)}{\to} \mathrm{Hom}_{F_i}({}_iM_j, F_i) \overset{m^*}{\to} F_j,$$

where m^* is the image of m under the bimodule isomorphism ${}_iM_j \to \mathrm{Hom}_{F_j}(\mathrm{Hom}_{F_i}({}_iM_j, F_i), F_j)$. There is also a natural definition on morphisms, in such a way that we get a duality D_1.

We conjecture that $C_1^+ = D_1\mathrm{Tr}$, where D_1 is defined as above. In fact, if Λ is an artin algebra, this turns out to be the case. To see this we use that Brenner and Butler have shown that for a particular choice of bimodule isomorphisms the corresponding Coxeter functor C^+ is isomorphic to $D\,\mathrm{Tr}$. We recall which C^+ they consider in **[10]**. With the previous notation we assume that the hereditary artin tensor ring Λ is an algebra over the field K, and we fix F_i-bimodule isomorphisms t_i: $\mathrm{Hom}_K(F_i, K) \to F_i$. Consider the natural induced isomorphism

$$\mathrm{Hom}_K({}_iM_j, K) \to \mathrm{Hom}_{F_i}({}_iM_j, \mathrm{Hom}_K(F_i, K)) \to \mathrm{Hom}_{F_i}({}_iM_j, F_i)$$

and $\mathrm{Hom}_K({}_iM_j, K) \to \mathrm{Hom}_{F_j}({}_iM_j, F_j)$. Then the Coxeter functor defined using the induced isomorphisms $\mathrm{Hom}_{F_i}({}_iM_j, F_i) \to \mathrm{Hom}_{F_j}({}_iM_j, F_j)$ is isomorphic to D Tr **[10]**.

It is not hard to see that the duality determined by the above bimodule isomorphisms is in fact isomorphic to the ordinary duality D for artin algebras. Since changing a bimodule isomorphism amounts to composing D Tr and C^+ with the same equivalence of categories, we have the claimed result for artin algebras.

We point out that for hereditary tensor rings with duality conditions, the collection of isomorphism classes of functors of the form $D_1\mathrm{Tr}$ for some duality D_1 can be larger than the collection of isomorphism classes of functors C_1^+. This will be the case for Λ if there is an isomorphism t: $\Lambda \to \Lambda$ such that the induced equivalence t^*: mod $\Lambda \to$ mod Λ is not the identity on simples, because if C_1^+ and C_2^+ are Coxeter functors then $C_1^+(A)$ and $C_2^+(A)$ clearly have the same composition factors. For example, consider a field K and the ring

$$\Lambda = \begin{pmatrix} K & 0 & 0 \\ K & K & 0 \\ K & 0 & K \end{pmatrix},$$

and let t: $\Lambda \to \Lambda$ be given by

$$t\begin{pmatrix} a & 0 & 0 \\ b & c & 0 \\ d & 0 & e \end{pmatrix} = \begin{pmatrix} a & 0 & 0 \\ d & e & 0 \\ b & 0 & c \end{pmatrix}.$$

It is not hard to see that t is a ring isomorphism with the required property.

We also point out that for any duality D_1 there is an isomorphism s: $\Lambda_n \to \Lambda$ (in the notation of §3) such that $D_1\mathrm{Tr}$ is isomorphic to a composition of partial Coxeter functors $F_n \cdots F_1$ with the equivalence of categories mod $\Lambda_n \to$ mod Λ induced by s. This follows from Theorem 3.6. Further we point out that in the more general case of an hereditary ring with a duality D_1, we still have the functors $D_1\mathrm{Tr}$: mod $\Lambda \to$ mod Λ, even though the Coxeter functors C^+ may not be defined, since their definition is dependent on a diagramatic description of the rings. In this case, $D_1\mathrm{Tr}$ still has the property that it takes projectives to zero and induces an equivalence of categories $\mathrm{mod}_P\ \Lambda \to \mathrm{mod}_I\ \Lambda$.

More generally, for an artin ring Λ, not necessarily hereditary, with some duality D_1, $D_1\mathrm{Tr}$ is still defined on objects, but is not in general an equivalence between $\mathrm{mod}_P\ \Lambda$ and $\mathrm{mod}_I\ \Lambda$, but between $\underline{\mathrm{mod}}\ \Lambda$ and $\overline{\mathrm{mod}}\ \Lambda$.

For an artin algebra Λ, the ordinary duality D plays a particularly important role in the theory of almost split sequences, since all almost split sequences are of the form $0 \to D\,\mathrm{Tr}\,C \to E \to C \to 0$. If Λ is an artin ring, not much is known about almost split sequences unless Λ is of finite type. It would be interesting to know when almost split sequences exist for hereditary tensor rings with duality conditions, and if when they do there is a particular duality D_1 such that the almost split sequences are of the form $0 \to D_1\mathrm{Tr}\,C \to E \to C \to 0$. We point out that Ringel has shown that there are hereditary tensor rings with duality conditions Λ having an indecomposable module C in mod Λ such that there is no almost split sequence $0 \to A \to B \to C \to 0$ in mod Λ (see **[20]**).

To get some more information about what the hereditary tensor rings with duality conditions look like, we prove the following preliminary result.

LEMMA 5.6. *Let Λ be a basic artin ring with a duality D_1 and let ${}_\Lambda I_\Lambda$ be a two-sided injective Λ-module such that D_1 is isomorphic to* $\mathrm{Hom}_\Lambda(\ , I)$ *(see* **[1,** *Chapter* 6]). *Let $\Lambda/\mathbf{r} = F_1 \times \cdots \times F_n$, and let $\mathbf{r}/\mathbf{r}^2 = \amalg\, {}_iM_j$ be a decomposition into indecomposable bimodules, with notation as before. If* $\mathrm{soc}(I)$ *and* $\Lambda/\mathbf{r}$ *are isomorphic bimodules, then* $\mathrm{Hom}_{F_i}({}_iM_j, F_i)$ *and* $\mathrm{Hom}_{F_j}({}_iM_j, F_j)$ *are isomorphic bimodules.*

PROOF. Since $D_1 = \mathrm{Hom}_\Lambda(\ , I)$ is a duality, it takes simple modules to simple modules, and consequently modules of Loewy length 2 to modules of Loewy length 2. Hence we get that $\mathrm{soc}({}_\Lambda I) = \mathrm{Hom}_\Lambda(\Lambda/\mathbf{r}, I)$ is a semisimple Λ^{op}-module, and consequently $\mathrm{soc}({}_\Lambda I) = \mathrm{soc}(I_\Lambda)$. Similarly, $\mathrm{Hom}_\Lambda({}_\Lambda\Lambda/\mathbf{r}^2, {}_\Lambda I)$ is isomorphic as a bimodule to the submodule L of I consisting of those elements y in I such that $\mathbf{r}^2 y = 0$. Since $\Lambda/\mathbf{r}^2$ has Loewy length 2, then $D_1(\Lambda/\mathbf{r}^2) = \mathrm{Hom}_\Lambda(\Lambda/\mathbf{r}^2, {}_\Lambda I)$ is a Λ^{op}-module of Loewy length 2. Thus L is contained in the submodule L' of I consisting of those elements y' such that $y'\mathbf{r}^2 = 0$. With a dual argument we prove that $L' \subset L$, so $L = L'$ and we have then a natural isomorphism $\mathrm{Hom}_\Lambda({}_\Lambda\Lambda/\mathbf{r}^2, {}_\Lambda I) \to \mathrm{Hom}_\Lambda(\Lambda/\mathbf{r}^2_\Lambda, I_\Lambda)$. Consider now the exact sequence of bimodules $0 \to \mathbf{r}/\mathbf{r}^2 \to \Lambda/\mathbf{r}^2 \to \Lambda/\mathbf{r} \to 0$. This gives rise to the exact sequences

$$0 \to \mathrm{Hom}({}_\Lambda\Lambda/\mathbf{r}, {}_\Lambda I) \to \mathrm{Hom}({}_\Lambda\Lambda/\mathbf{r}^2, {}_\Lambda I) \to \mathrm{Hom}({}_\Lambda\mathbf{r}/\mathbf{r}^2, {}_\Lambda I) \to 0$$

and

$$0 \to \mathrm{Hom}(\Lambda/\mathbf{r}_\Lambda, I_\Lambda) \to \mathrm{Hom}(\Lambda/\mathbf{r}^2_\Lambda, I_\Lambda) \to \mathrm{Hom}(\mathbf{r}/\mathbf{r}^2_\Lambda, I_\Lambda) \to 0.$$

From the above remarks we get induced an isomorphism of bimodules between $\mathrm{Hom}_{F_i}({}_iM_j, F_i)$ and $\mathrm{Hom}_{F_j}({}_iM_j, F_j)$.

For hereditary tensor rings we now get the following description of when the duality condition holds.

PROPOSITION 5.7. *Let Λ be an artin hereditary tensor ring. Then Λ satisfies the duality conditions if and only if there is a duality D_1*: mod $\Lambda \to$ mod Λ^{op} *such that $D_1({}_\Lambda\Lambda/\mathbf{r})$ and $\Lambda/\mathbf{r}$ are isomorphic bimodules.*

PROOF. Assume that Λ satisfies the duality conditions. For fixed bimodule isomorphisms between the ${}_iM_j$ let D_1 be the duality after Corollary 5.5. Then $D_1(F_i) = \operatorname{Hom}_{F_i}(F_i, F_i)$ and F_i are isomorphic bimodules, so $D_1({}_\Lambda\Lambda/\mathbf{r})$ and $\Lambda/\mathbf{r}$ are isomorphic bimodules.

Assume conversely that $D_1({}_\Lambda\Lambda/\mathbf{r})$ and $\Lambda/\mathbf{r}$ are isomorphic bimodules. Write $D_1 = \operatorname{Hom}_\Lambda(\ , I)$, for a two-sided injective Λ-module I. $D_1({}_\Lambda\Lambda/\mathbf{r}) = \operatorname{soc}_\Lambda(I)$, so $\operatorname{soc}_\Lambda(I)$ and $\Lambda/\mathbf{r}$ are isomorphic bimodules. Our desired result then follows from Lemma 5.6.

It is proven in [9], [13] that the composition factors for the partial Coxeter functors applied to modulus are given by the symmetries of the Weyl group. We end this section with a direct proof of this fact, proving that this is the case for partial Coxeter functors defined on an hereditary algebra Λ, even when Λ is not a tensor algebra.

Let Λ be an hereditary artin algebra and let $S_1, \ldots, S_n$ be an admissible indexing of the simples of Λ. Let Gr Λ denote the Grothendieck group of Λ and let $[M]$ denote the class of the module M in Gr Λ. There is a bilinear form B: Gr $\Lambda \times$ Gr $\Lambda \to Q$, where Q denotes the field of rational numbers, defined as follows. We assume that Λ is indecomposable, so the center of Λ is a field K. Then

$$B([M], [N]) = \tfrac{1}{2}(\dim_K(M, N) + \dim_K(N, M) - \dim_K\operatorname{Ext}^1_\Lambda(M, N) - \dim_K\operatorname{Ext}^1_\Lambda(N, M)).$$

(See [20, §2], [3, §3].) Gr Λ is a free abelian group, and we fix the basis $[S_1]$, $[S_2], \ldots, [S_n]$. We will then identify Gr Λ with Z^n, where Z denotes the ring of integers. One can then uniquely extend B to a map from $Q^n \times Q^n$ to Q. Let σ_i: $Q^n \to Q^n$ be the symmetry defined by

$$\sigma_i(x) = x - 2\frac{B(x, [S_i])}{B([S_i], [S_i])}[S_i].$$

Let P_i denote the projective cover for S_i, and let as in §1 $X = \operatorname{Tr} DS \amalg P_2 \amalg \cdots \amalg P_n$, $\Gamma = \operatorname{End}_\Lambda(X)^{\text{op}}$ and $F = (X, \ )$: mod $\Lambda \to$ mod Γ. We now prove the following result.

PROPOSITION 5.8. *We keep the above notations. We identify* Gr Λ *and* Gr Γ *by means of the map transforming the basis $[S_1], \ldots, [S_n]$ of* Gr Λ *onto the basis $[F(\operatorname{Tr} DS_1)/\mathbf{r}F(\operatorname{Tr} DS_1)]$, $[F(P_2)/\mathbf{r}F(P_2)], \ldots, [F(P_n)/\mathbf{r}F(P_n)]$. Then $[F(M)] = \sigma_1([M])$.*

PROOF. We assume that Λ is an indecomposable algebra over the field K. It is easy to see that we have the following formula:

$$[M] = \sum_{i=1}^{n} \frac{\dim_K(P_i, M)}{\dim_K(S_i, S_i)}[S_i].$$

We know from Proposition 2.1 that the indecomposable projective Λ-modules are $F(\operatorname{Tr} DS_1)$, $F(P_2), \ldots, F(P_n)$, and that $T = F(\operatorname{Tr} DS_1)/\mathbf{r}F(\operatorname{Tr} DS_1)$ is simple injective.

Since $\operatorname{Tr} D$: $\operatorname{mod}_I \Lambda \to \operatorname{mod}_P \Lambda$ is an equivalence of categories and S_1 is not injective we have that $\dim_K(S_1, S_1) = \dim_K(\operatorname{Tr} DS_1, \operatorname{Tr} DS_1) = \dim_K(F(\operatorname{Tr} DS_1), F_1(\operatorname{Tr} DS_1)) = \dim_K(T, T)$.

On the other hand, $\dim_K(F(P_i), F(P_i)) = \dim_K(P_i, P_i)$ for $i = 2, \ldots, n$. Hence we have

$$\begin{aligned}[F(M)] &= \sum_{i=2}^{n} \frac{\dim_K(F(P_i), F(M))}{\dim_K(F(S_i)/\mathbf{r}F(S_i), F(S_i)/\mathbf{r}F(S_i))}[S_i] \\ &\quad + \frac{\dim_K(F(\operatorname{Tr} DS_1), F(M))}{\dim_K(T, T)}[S_1] \\ &= \sum_{i=1}^{n} \frac{\dim_K(P_i, M)}{\dim_K(S_i, S_i)}[S_i] + \frac{\dim_K(\operatorname{Tr} DS_1, M)}{\dim_K(S_1, S_1)}[S_1].\end{aligned}$$

Hence

$$[F(M)] - [M] = \frac{\dim_K(\operatorname{Tr} DS_1, M) - \dim_K(S_1, M)}{\dim_K(S_1, S_1)}[S_1].$$

Using that $D(\operatorname{Tr} DS_1, M) = \operatorname{Ext}^1_\Lambda(M, S_1)$ (see [6, Theorem 3.3], [3,§1]) we have that

$$[F(M)] = [M] - \frac{B([S_1], [M])}{B([S_1], [S_1])}[S_1] = \sigma_1(M).$$

This finishes the proof of the proposition.

Note. In the above proposition we dealt with the functor $F = (X, \;)$. Using the results of the end of §2 about the uniqueness of the partial Coxeter functor up to equivalence, one can prove that Proposition 5.8 holds for an arbitrary partial Coxeter functor F. But we observe that this does not mean that if F, F': $\operatorname{mod} \Lambda \to \operatorname{mod} \Gamma$ are two partial Coxeter functors, then $F(M)$ and $F'(M)$ have the same composition factors. It means that there is a permutation β of the simples of Λ such that if $[F(M)] = \sum_{i=1}^{n} n_i[S_i]$, then $[F'(M)] = \sum_{i=1}^{n} n_i[\beta(S_i)]$.

References

1. F. W. Anderson and K. R. Fuller, *Rings and categories of modules*, Graduate Texts in Mathematics, no. 13, Springer-Verlag, Berlin and New York, 1974.

2. M. Auslander, *Representation theory of artin algebras*. I, Comm. Algebra **1** (1974), 177–268.

3. M. Auslander and M. I. Platzeck, *Representation theory of hereditary artin algebras*, Lecture Notes in Pure and Appl. Math., vol. 37 Dekker, New York, 1978.

4. M. Auslander and I. Reiten, *Notes on representation theory of artin algebras*, Brandeis Univ., 1972.

5. ______, *Stable equivalence of artin algebras*, Lecture Notes in Math., vol. 353, Springer-Verlag, Berlin and New York, 1973.

6. ______, *Representation theory of artin algebras*. III, *Almost split sequences*, Comm. Algebra **3** (1975), 239–294.

7. ______, *Representation theory of artin algebras*. IV, Comm. Algebra **5** (1977), 443–518.

8. R. Bautista, *Algebras close to hereditary algebras* (to appear).

9. I. N. Bernstein, I. M. Gelfand and V. A. Ponomarev, *Coxeter functors and a theorem of Gabriel*, Uspehi Mat. Nauk **28** (1973), 19–33 = Russian Math. Surveys **28** (1973), 17–32.

10. S. Brenner and M. C. R. Butler, *The equivalence of certain functors occurring in the representation theory of artin algebras and species*, J. London Math. Soc. **14** (1976), 183–187.

11. E. Cartan and S. Eilenberg, *Homological algebra*, Princeton Univ. Press, Princeton, N. J., 1956.

12. V. Dlab and C. M. Ringel, *Representations of graphs and algebras*, J. Algebra **33** (1975), 306–394.

13. ______, *Representations of graphs and algebras*, Mem. Amer. Math. Soc., No. 173 (1976).

14. ______, *Representation theory of algebras*, Lecture Notes in Pure and Appl. Math., vol. 37, Dekker, New York, 1978.

15. C. Faith, *Algebra: Rings, modules and categories*. I, Die Grundlehren der math. Wissenschaften, Band 190, Springer-Verlag, Berlin and New York, 1973.

16. R. Fossum, P. Griffith and I. Reiten, *Trivial extensions of abelian categories*, Lecture Notes in Math., vol. 456, Springer-Verlag, Berlin and New York, 1975.

17. P. Gabriel, *Réprésentations indécomposables*, Séminaire Bourbaki, 26e année, (1973/74), no. 444.

18. E. Green, *The representation theory of tensor algebras*, J. Algebra **34** (1975), 136–171.

19. M. Harada, *Hereditary semiprimary rings and triangular matrix rings*, Nagoya Math. J. **27** (1966), 463–484.

20. C. M. Ringel, *Representations of K-species and bimodules*, J. Algebra **41** (1976), 269–302.

Department of Mathematics, Brandeis University, Waltham, Massachusetts 02154 (Current address of M. Auslander)

Department of Mathematics, University of Illinois, Urbana, Illinois 61801

Department of Mathematics, University of Trondheim, NLHT, 7000 Trondheim, Norway (Current address of I. Reiten)

Current address (M. I. Platzeck): Corrientes 211.9°B, 8000 Bahia Blanca, Argentina

Reprinted from JOURNAL OF ALGEBRA
All Rights Reserved by Academic Press, New York and London

Vol. 66, No. 1, September 1980
Printed in Belgium

Preprojective Modules over Artin Algebras

M. AUSLANDER* AND SVERRE O. SMALØ†

Brandeis University, Waltham, Massachusetts 02154; and University of Trondheim, NLHT, Trondheim, Norway

Communicated by W. Feit

Received July 6, 1979

In their study of the representation theory of finite-dimensional tensor algebras Dlab and Ringel [11] described certain modules which they called preprojective and preinjective modules. Platzeck in her work on the representation theory of artin algebras stably equivalent to hereditary artin algebras [14], which includes among other things the tensor, hereditary, and square radical zero artin algebras, found and studied modules which are clearly analogs of the preprojective and preinjective modules introduced by Dlab and Ringel. Our purpose in this paper is to develop a general theory of preprojective and preinjective modules over arbitrary artin algebras. As would be expected, the preprojective and preinjective modules described by the general theory coincide with the modules considered earlier by Dlab and Ringel and Platzeck in their respective situations.

In view of these remarks it is perhaps surprising that the original impetus for our work did not come from the theory of hereditary artin algebras or those stably equivalent to hereditary artin algebras. Rather it came from an effort to explain a much older result of Gabriel and Roiter [12, 15] concerning artin algebras of finite representation type in terms of the technics and ideas developed by Auslander and Reiten in connection with almost split sequences and irreducible morphisms [6, 7]. Our generalization of the Gabriel–Roiter result is as follows.

Let Λ be an artin algebra (for instance, a finite-dimensional algebra over a field) and ind Λ the full subcategory of the category of finitely generated Λ-modules consisting of the indecomposable Λ-modules. Then there is a unique collection of full subcategories $\{\mathbf{P}_i\}_{i\in N\cup\infty}$ of ind Λ where N is the non-negative integers having the following properties:

(a) If $A \in \mathbf{P}_i$ and $B \approx A$, then $B \in \mathbf{P}_i$ for all i.

(b) $\mathbf{P}_i \cap \mathbf{P}_j = \varnothing$ for $i \neq j$ and $\bigcup_{i\in N\cup\infty} \mathbf{P}_i = \text{ind}\,\Lambda$.

* Written while a Guggenheim Fellow with the partial support of NSF MCS 77 04 951.
† Supported by the Norwegian Research Council.

0021-8693/80/090061-62$02.00/0

(c) $\mathbf{P}_i$ has only a finite number of nonisomorphic objects for each i in N.

(d) For each i in N, an indecomposable module A is in $\mathbf{P}_i$ if and only if every surjective morphism $f: B \to A$ is a splittable surjection whenever every indecomposable summand of B is in $\bigcup_{k \geqslant i} \mathbf{P}_k$.

This uniquely determined collection $\{\mathbf{P}_i\}_{i \in N \cup \infty}$ of subcategories of ind Λ is called the preprojective partition of ind Λ. In this terminology the Gabriel–Roiter theorem says that if Λ is of finite representation type, i.e., ind Λ has only a finite number of nonisomorphic modules, then ind Λ has a preprojective partition.

It is in terms of preprojective partitions that preprojective modules are defined; namely, a finitely generated Λ-module M is preprojective if each indecomposable summand of M is in $\mathbf{P}_i$ for some $i < \infty$. While preprojective modules are defined in terms of the preprojective partition, they can also be given descriptions which do not depend on the preprojective partition. For example, an indecomposable module M is preprojective if and only if there is a maximal submodule M' of M such that there are only a finite number of nonisomorphic indecomposable modules $B_1, \ldots, B_n$ having a morphism $f: B_i \to M$ with $\operatorname{Im} f$ not contained in M'. The definition and existence of preinjective partitions and preinjective modules is given by duality.

The proofs of the main existence theorem for preprojective and preinjective partitions are based on the following result concerning full subcategories $\mathbf{C}$ of ind Λ. Suppose $\mathbf{C}$ has the property that there are a finite number of indecomposable modules $B_1, \ldots, B_n$ such that if C is in ind Λ and not isomorphic to any B_i, then C is in $\mathbf{C}$. Then given an X in ind Λ we have:

(a) There is a morphism $f: C \to X$ where every indecomposable summand of C is in $\mathbf{C}$ such that for each morphism $g: L \to X$ with L in $\mathbf{C}$ there is an $h: L \to C$ satisfying $g = fh$.

(b) There is a morphism $f: X \to C'$ where every indecomposable summand of C' is in $\mathbf{C}$ such that for each morphism $g: X \to L$ with L in $\mathbf{C}$, there is an $h: C' \to L$ satisfying $hf = g$.

For technical as well as conceptual reasons, the initial discussion in this paper of preprojective and preinjective partitions as well as preprojective and preinjective modules is concerned with full subcategories of ind Λ rather than ind Λ itself. While this necessitates slightly more abstract and longer proofs, we feel this approach is justified on two counts.

First, some of the seemingly abstract notions have intrinsic value, especially the notion of mod Λ being contravariantly or covariantly finite over a full subcategory $\mathbf{C}$. As can be seen from the discussion here, the question of whether or not a subcategory $\mathbf{C}$ has left or right almost split morphisms is intimately connected with mod Λ being contravariantly or covariantly finite over $\mathbf{C}$. In another paper [8] we will show how these notions are connected with **C** having almost split sequences.

Second, the greater generality shows that many subcategories of ind Λ and not just ind Λ itself have preprojective and/or preinjective partitions: hence the greater applicability of the theory. This is particularly important in view of the work of Bautista and Martinez [9] showing that various classification problems, for instance, for representations of partially ordered sets, can be viewed as classification problems for specific subcategories in ind Λ for suitably chosen artin algebras Λ.

Roughly speaking the paper is divided into three parts. The first four sections deal with basic concepts and existence theorems. In Sections 5 through 7, the preprojective and preinjective partitions of subcategories of ind Λ are studied. The rest of the paper is devoted to applying these results to ind Λ as well as giving more detailed information concerning the structure of preprojective and preinjective modules.

A preliminary announcement of this work was given at the Antwerp Ring Theory Conference in the summer of 1978 [3].

After suitable reformulation, many of the notions and results of this paper are also applicable to lattices over orders over complete discrete valuation rings. In particular one obtains preprojective and preinjective partitions for lattices as well as preprojective and preinjective lattices. These matters will be discussed in another publication.

1. Minimal Morphisms

Throughout this section we assume that Λ is a left artin ring and all modules and morphisms are in mod Λ, the category of finitely generated left Λ-modules.

Suppose $f: B \to C$ and $g: X \to C$ are morphisms. We say that a morphism $h: X \to B$ is a *lifting of g to f* if $fh = g$. The morphism $g: X \to C$ is said to be *liftable to f* if there is a lifting $h: X \to B$ of g to f. Finally, we say that two morphisms $f_1: B_1 \to C$ and $f_2: B_2 \to C$ are *lifting equivalent* if a morphism $g: X \to C$ can be lifted to f_1 if and only if it can be lifted to f_2.

Clearly for a fixed C in mod Λ, the relation on the collection of all morphisms $X \to C$ given by $f_1: B_1 \to C$ is related to $f_2: B_2 \to C$ if f_1 and f_2 are lifting equivalent is an equivalence relation. Our purpose in this section is to describe how the morphisms in one equivalence class, called a lifting equivalence class, are related. Even though this material has been discussed previously (see [2, pp. 28–31]), we give here a self-contained account for the sake of completeness and ease of reference. We assume through out this discussion that C is a fixed module in mod Λ.

We begin with the following easily verified observations.

LEMMA 1.1. *Let $f_1\colon B_1 \to C$ and $f_2\colon B_2 \to C$ be morphisms in* mod Λ:

(a) *f_1 can be lifted to f_2 if and only if each $g\colon X \to C$ which can be lifted to f_1 can also be lifted to f_2.*

(b) *f_1 and f_2 are lifting equivalent if and only if f_1 can be lifted to f_2 and f_2 can be lifted to f_1.*

(c) *Suppose f_1 and f_2 are lifting equivalent and the diagram*

$$\begin{array}{ccc} B_1 & \xrightarrow{f_1} & C \\ \downarrow & & \| \\ B & \xrightarrow{f} & C \\ \downarrow & & \| \\ B_2 & \xrightarrow{f_2} & C \end{array}$$

commutes. Then f, f_1, and f_2 are lifting equivalent.

Proof. Left as an exercise to the reader.

Before giving the main result of this discussion, it is convenient to give the following definition.

A morphism $B \to^f C$ is said to be *right minimal* if every lifting of f to f is an isomorphism, i.e., $h\colon B \to B$ is an isomorphism if $fh = f$.

PROPOSITION 1.2. *Let $\mathscr{E}$ be a lifting equivalence class in the class of all morphisms $X \to C$ in* mod Λ.

(a) *If $f\colon B \to C$ in $\mathscr{E}$ has the property that $l(B) \leqslant l(X)$ (where $l(Y)$ means the length of Y) for all X such that there is $g\colon X \to C$ in $\mathscr{E}$, then $f\colon B \to C$ is right minimal. Thus there is an $f\colon B \to C$ in $\mathscr{E}$ which is right minimal.*

(b) *If $f_1\colon B_1 \to C$ and $f_2\colon B_2 \to C$ are two right minimal morphisms in $\mathscr{E}$, then every lifting of f_1 to f_2 is an isomorphism. Hence, there is an isomorphism $h\colon B_1 \to B_2$ such that $f_1 = f_2 h$.*

(c) *Let $f\colon B \to C$ be a morphism in $\mathscr{E}$. If $B = B_1 \coprod B_2$ (direct sum) such that $f(B_2) = 0$, then $f \mid B_1 : B_1 \to C$ in $\mathscr{E}$.*

(d) *Let $f\colon B \to C$ be in $\mathscr{E}$. Then there is a decomposition $B = B_1 \coprod B_2$ such that $f(B_2) = 0$ and $f \mid B_1\colon B_1 \to C$ is a right minimal morphism in $\mathscr{E}$.*

Proof (a) Let $h\colon B \to B$ be such that $fh = f$. Then we have the commutative diagram

$$\begin{array}{ccc} B & \xrightarrow{f} & C \\ \downarrow & & \| \\ h(B) & \xrightarrow{f \mid h(B)} & C \\ \downarrow & & \| \\ B & \xrightarrow{f} & C \end{array}$$

By Lemma 1.1(c), we know that $f \mid h(B)\colon h(B) \to C$ is in $\mathscr{E}$. Since $l(B) \leqslant l(X)$ for all $g\colon X \to C$ in $\mathscr{E}$, we have that $l(h(B)) \leqslant l(B) \leqslant l(h(B))$. Hence $h(B) = B$ and so $h\colon B \to B$ is surjective and thus an isomorphism since B has finite length. The rest of (a) is obvious.

(b) Suppose $f_1\colon B_1 \to C$ and $f_2\colon B_2 \to C$ are two right minimal morphisms in $\mathscr{E}$. Since they are both in $\mathscr{E}$ we know by Lemma 1.1(b) that there are morphisms $h_1\colon B_1 \to B_2$ and $h_2\colon B_2 \to B_1$ such that $f_2 h_1 = f_1$ and $f_2 = f_1 h_2$. Hence $f_1 h_2 h_1 = f_1$ and so $h_1 h_2\colon B_1 \to B_2$ is an isomorphism since f_1 is right minimal. Also $f_2 = f_2 h_1 h_2$ and so $h_1 h_2\colon B_2 \to B_1$ is an isomorphism since f_2 is right minimal. Hence $h_1\colon B_1 \to B_2$ is an isomorphism such that $f_2 h_1 = f_1$.

(c) Suppose $f\colon B \to C$ is in $\mathscr{E}$ and $B = B_1 \coprod B_2$ a decomposition such that $f(B_2) = 0$. Then we have the commutative diagram

$$\begin{array}{ccc} B_1 \coprod B_2 & \xrightarrow{f} & C \\ \downarrow{\scriptstyle p} & & \| \\ B_1 & \xrightarrow{f \mid B_1} & C \\ \downarrow{\scriptstyle i} & & \| \\ B_1 \coprod B_2 & \xrightarrow{f} & C \end{array}$$

where $p\colon B_1 \coprod B_2 \to B_1$ is the projection morphism and $i\colon B_1 \to B_1 \coprod B_2$ is the injection morphism. Therefore by Lemma 1.1(c) we have that $f \mid B_1\colon B_1 \to C$ is in $\mathscr{E}$.

(d) Let $f\colon B \to C$ be in $\mathscr{E}$. Then by part (a) we know there is a right minimal $g\colon B' \to C$ in $\mathscr{E}$. Thus there is a commutative diagram

$$\begin{array}{ccc} B' & \xrightarrow{g} & C \\ \downarrow{\scriptstyle h_1} & & \| \\ B & \xrightarrow{f} & C \\ \downarrow{\scriptstyle h_2} & & \| \\ B' & \xrightarrow{g} & C \end{array}$$

Since $g: B' \to C$ is right minimal, we know that $h_2h_1: B' \to B'$ is an isomorphism. Therefore $B = \text{Ker}\, h_2 \coprod \text{Im}\, h$, and it is easy to see that $f(\text{Ker}\, h_2) = 0$ and $f \mid \text{Im}\, h_1$ is a right minimal morphism in $\mathscr{E}$ since $h_1: B' \to B$ induces an isomorphism $j: B' \to \text{Im}\, h_1$ with the property $g = (f \mid \text{Im}\, h_1)j$. Then $B = B_1 \coprod B_2$, where $B_1 = \text{Im}\, h_1$ and $B_2 = \text{Ker}\, h_2$ is our desired decomposition of B.

In order to simplify our terminology we give the following definition:

Let C be in $\text{mod}\, \Lambda$. Two morphisms $f_1: X_1 \to C$ and $f_2: X_2 \to C$ are said to be *isomorphic* if there is an isomorphism $h: X_1 \to X_2$ which is a lifting of f_1 to f_2. Obviously if a lifting $h: X_1 \to X_2$ of f_1 to f_2 is an isomorphism, then $h^{-1}: X_2 \to X_1$ is a lifting of f_2 to f_1.

Thus, in this terminology, Proposition 1.2(b) has the following form. Let $\mathscr{E}$ be a lifting equivalence class in the class of all morphisms $X \to C$ in $\text{mod}\, \Lambda$. Then any two right minimal morphisms in $\mathscr{E}$ are isomorphic.

In the rest of this paper, expecially Sections 2 and 3, we will often be interested in finding a solution to the following type of problem. Let C be in $\text{mod}\, \Lambda$ and $\mathscr{S}$ a family of morphisms $Y \to C$. Is there a morphism $f: B \to C$ in $\mathscr{S}$ with the property that each morphism $g: X \to C$ in $\mathscr{S}$ is liftable to f? While our discussion so far gives no information concerning the existence of solutions to this problem, it does give considerable information concerning the structure of the solutions when they do exist, at least in the case $\mathscr{S}$ satisfies some mild additional conditions.

PROPOSITION 1.3. *Let C be in* $\text{mod}\, \Lambda$ *and $\mathscr{S}$ a family of morphisms $X \to C$ satisfying the condition that if $f: X \to C$ is in $\mathscr{S}$ and $g: Y \to X$ is a splittable injection (i.e., there is a $t: X \to Y$ such that $tg = id_Y$), then the composition fg: $Y \to C$ is in $\mathscr{S}$. Suppose there is an $f: B \to C$ in $\mathscr{S}$ such that every g in $\mathscr{S}$ is liftable to f. Then*

(a) *There is a right minimal $f': B' \to C$ in $\mathscr{S}$ such that every g in $\mathscr{S}$ can be lifted to f.*

(b) *Any two right minimal morphisms $f': B' \to C$ in $\mathscr{S}$ such that every g in $\mathscr{S}$ can be lifted to f' are isomorphic.*

(c) *Suppose $f: B \to C$ in $\mathscr{S}$ has the property that every g in $\mathscr{S}$ can be lifted to f. Then there is a decomposition $B = B_1 \coprod B_2$ such that* (i) *$f \mid B_1: B_1 \to C$ is right minimal with the property every g in $\mathscr{S}$ can be lifted to $f \mid B_1$ and* (ii) $f(B_2) = 0$.

Proof. (a) and (c). Let $f: B \to C$ in $\mathscr{S}$ be such that every g in $\mathscr{S}$ can be lifted to f. We know by Proposition 1.2(c) that there is a decomposition $B = B_1 \coprod B_2$ with $f \mid B_1: B_1 \to C$ right minimal and $f(B_2) = 0$. Since the inclusion $i_1: B_1 \to B$ is a splittable injection, $f \mid B_1 = fi_1$ is in $\mathscr{S}$. Because $f(B_2) = 0$, we have that the projection $p_1: B \to B_1$ has the property $f = (f \mid B_1)p_1$. Thus every g in $\mathscr{S}$ can be lifted to $f \mid B_1$. This proves (a) and (c).

(b) Suppose $f_1: B_1 \to C$ and $f_2: B_2 \to C$ are two right minimal morphisms in $\mathscr{S}$ such that every g in $\mathscr{S}$ can be lifted to f_1 and f_2. Then there are h_1: $B_1 \to B_2$ and $h_2: B_2 \to B_1$ such that $f_1 = f_2h_1$ and $f_2 = f_1h_2$. From this it follows that h_1 and h_2 are isomorphisms since f_1 and f_2 are right minimal. This completes the proof of (b) and the entire proposition.

In order to fix definitions and for ease of reference, we now state the duals of the above concepts and results. No proofs are given since they follow easily from the above using dual arguments.

Suppose $f: B \to C$ and $g: B \to Y$ are morphisms. We say that a morphism $h: C \to Y$ is an *extension of g to f* if $hf = g$. The morphism $g: B \to Y$ is said to be *extendable to* $f: B \to C$ if there is an extension $h: C \to Y$ from f to g. Finally, we say that two morphisms $f_1: B_1 \to C_1$ and $f_2: B \to C_2$ are *extension equivalent* if a morphism $B \to Y$ can be extended to f_1 if and only if it can be extended to f_2.

Clearly for a fixed B in mod Λ, the relation on the collection of morphisms $B \to Y$ given by $f_1: B \to C_1$ is related to $f_2: B \to C_2$ if f_1 and f_2 are extension equivalent is an equivalence relation. The equivalence classes under this relation are called the *extension classes*.

While we leave the dual of Lemma 1.1 to the reader to state and prove, we will state the dual of Proposition 1.2 as soon as we define left minimal morphisms, which are the dual of right minimal morphisms.

A morphism $f: B \to C$ is *left minimal* if every extension of f to f is an isomorphism.

PROPOSITION 1.4. *Let B be in* mod Λ *and* $\mathscr{F}$ *an extension class of the class of all morphisms* $B \to Y$ *(B fixed) in* mod Λ.

(a) *If* $f: B \to C$ *has the property that* $l(C) \leqslant l(Y)$ *with* $B \to Y$ *in* $\mathscr{F}$, *then* $f: B \to C$ *is left minimal. Thus there is a left minimal* $f: B \to C$ *in* $\mathscr{F}$.

(b) *If* $f_1: B \to C_1$ *and* $f_2: B \to C_2$ *in* $\mathscr{F}$ *are left minimal, then every extension from* f_1 *to* f_2 *is an isomorphism. Hence there is an isomorphism* $h: C_1 \to C_2$ *such that* $hf_1 = f_2$.

(c) *Let* $f: B \to C$ *be in* $\mathscr{F}$. *If* $C = C_1 \amalg C_2$ *and the projections* $p_i: C \to C_i$ *have the property that* $p_2f: B \to C_2$ *is zero, then* $p_1f: B \to C_1$ *is in* $\mathscr{F}$.

(d) *Let* $f: B \to C$ *be in* $\mathscr{F}$. *Then there is a decomposition* $C = C_1 \amalg C_2$ *such that* $p_2f: B \to C_2$ *is zero and* $p_1f: B \to C_1$ *is a left minimal morphism in* $\mathscr{F}$.

Let B be in mod Λ. Two morphisms $f_1: B \to X_1$ and $f_2: B \to X_2$ are said to be *isomorphic* if there is an isomorphism $h: X_1 \to X_2$ which is an extension of f_2 to f_1.

With this definition in mind we give the dual of Proposition 1.3.

PROPOSITION 1.5. *Let B be in* mod Λ *and* $\mathscr{T}$ *a family of morphisms* $B \to X$

satisfying the condition that if $f: B \to X$ *is in* $\mathcal{T}$ *and* $g: X \to Y$ *is a splittable surjection (i.e., there is an* $s: Y \to X$ *such that* $gs = id_Y$*), then the composition* $gf: B \to Y$ *is in* $\mathcal{T}$. *Suppose there is an* $f: B \to C$ *in* $\mathcal{T}$ *such that each g in* $\mathcal{T}$ *can be extended to f. Then*

(a) *There is a left minimal* $f': B \to C'$ *in* $\mathcal{T}$ *such that each g in* $\mathcal{T}$ *can be extended to* f'.

(b) *Any two right minimal morphisms* $f': B \to C'$ *in* $\mathcal{T}$ *such that each g in* $\mathcal{T}$ *can be extended to* f' *are isomorphic.*

(c) *Suppose* $f: B \to C$ *in* $\mathcal{T}$ *has the property that each g in* $\mathcal{T}$ *can be extended to f. Then there is a decomposition* $C = C_1 \coprod C_2$ *such that the projections* $p_i: C \to C_i$ *have the property that* $p_1 f$ *is a left minimal morphism in* $\mathcal{T}$ *such that each g in* $\mathcal{T}$ *can be extended to* $p_1 f$ *and* $p_2 f = 0$.

2. Covers and Splitting Projectives

Throughout the rest of this paper we assume that R is a commutative artin ring and Λ an R-algebra which is a finitely generated R-module. It is obvious that the R-algebra structure on Λ induces an R-algebra structure on Λ^{op} in such a way that Λ^{op} is also a finitely generated R-module where Λ^{op} is the opposite ring of Λ. We denote the category of finitely generated left Λ-modules by mod Λ. Clearly mod Λ^{op} is the category of finitely generated right Λ-modules. Further, letting I be the injective envelope of $R/\mathrm{rad}\, R$, the contravariant functor $\mathrm{Hom}_R(\ , I)$: mod $\Lambda \to$ mod Λ^{op} given by $M \to \mathrm{Hom}_R(M, I)$ is a duality with inverse $\mathrm{Hom}_R(\ , I)$: mod $\Lambda^{\mathrm{op}} \to$ mod Λ. We denote both of these dualities by D: mod $\Lambda \to$ mod Λ^{op} and D: mod $\Lambda^{\mathrm{op}} \to$ mod Λ.

By a subcategory **C** of mod Λ we always mean a full subcategory having the property that if M in mod Λ is isomorphic to a summand (direct) of C in **C**, then M is in **C**. Associated with each subcategory **C** of mod Λ is the subcategory D(**C**) of mod Λ^{op} consisting of all the Λ^{op}-modules isomorphic to $D(C)$ for some C in **C**. The duality D: mod $\Lambda \to$ mod Λ^{op} clearly induces a duality D: **C** $\to D$(**C**) which gives an equivalence of $\mathbf{C}^{\mathrm{op}}$, the opposite category of **C**, with D(**C**). We will usually identify the categories $\mathbf{C}^{\mathrm{op}}$ and D(**C**) by means of this equivalence.

In studying a subcategory **C** of mod Λ, it is often convenient to also consider the categories add **C** and ind **C** which we now describe.

The category add **C** is the subcategory of mod Λ consisting of all Λ-modules isomorphic to summands of finite sums of modules in **C**. Thus add **C** is an additive subcategory of mod Λ in which idempotents split.

The category ind **C** is the subcategory of **C** consisting of all the indecomposable modules in **C**. Clearly add ind **C** = add **C** and ind add **C** = ind **C**. Also D(add **C**) = add D(**C**) and D(ind **C**) = ind D(**C**).

Of primary concern to us throughout this paper is the kinds of covers and cocovers a subcategory of mod Λ has; notions we now introduce.

A *cover* for a subcategory $\mathbf{C}$ of mod Λ is a subcategory $\mathbf{A}$ of ind $\mathbf{C}$ such that for each C in $\mathbf{C}$ there is a surjective morphism $f: A \to C$ with A in add $\mathbf{A}$. A cover $\mathbf{A}$ of $\mathbf{C}$ is said to be a *minimal cover* for $\mathbf{C}$ if no proper subcategory of $\mathbf{A}$ is a cover for $\mathbf{C}$. Clearly a subcategory $\mathbf{A}$ of $\mathbf{C}$ is a (minimal) cover for $\mathbf{C}$ if and only if $\mathbf{A}$ is a (minimal) cover for ind $\mathbf{C}$.

Dualizing, we obtain the dual notions of a cocover and minimal cocover for a subcategory of mod Λ.

A *cocover* for a subcategory $\mathbf{C}$ of mod Λ is a subcategory $\mathbf{B}$ of ind $\mathbf{C}$ such that for each C in $\mathbf{C}$ there is an injective morphism $f: C \to B$ with B in add $\mathbf{B}$. A cocover $\mathbf{B}$ of $\mathbf{C}$ is said to be a *minimal cocover* for $\mathbf{C}$ if no proper subcategory of $\mathbf{B}$ is a cocover for $\mathbf{C}$. Clearly a subcategory $\mathbf{B}$ of $\mathbf{C}$ is a (minimal) cocover for $\mathbf{C}$ if and only if $\mathbf{B}$ is a (minimal) cocover for ind $\mathbf{C}$. Finally, we observe that a subcategory $\mathbf{A}$ of $\mathbf{C}$ is a (minimal) cover for $\mathbf{C}$ if and only if $D(\mathbf{A})$ is a (minimal) cocover for $D(\mathbf{C})$. Thus the theory of covers of subcategories of mod Λ is the dual of the theory of cocovers of subcategories of mod Λ^{op}.

In order to motivate our discussion of covers and cocovers for subcategories $\mathbf{C}$ of mod Λ we consider the special case $\mathbf{C} =$ mod Λ. Let $\mathbf{P}_0$ be the subcategory of ind mod Λ consisting of all the indecomposable projective modules in mod Λ. Then it is easily seen that $\mathbf{P}_0$ has the following properties.

(a) $\mathbf{P}_0$ is a cover for mod Λ.

(b) If $f: X \to P$ is surjection with P in $\mathbf{P}_0$, then f is a splittable surjection.

(c) If $\mathbf{A}$ is a cover for mod Λ, then $\mathbf{A}$ contains $\mathbf{P}_0$.

(d) $\mathbf{A}$ is a minimal cover for mod Λ if and only if $\mathbf{A} = \mathbf{P}_0$.

Since (a) and (b) are well known, it only remains to establish (c) and (d). But (c) follows readily from (b) (see Proposition 2.1) and (d) follows from (a) and (c). Thus $\mathbf{P}_0$ is the unique minimal cover for mod Λ which gives a complete characterization of the subcategory of mod Λ consisting of the indecomposable projective modules in terms of the notion of covers.

Obviously, a similar discussion can be carried out for cocovers of mod Λ, giving results dual to those given for covers. In particular, mod Λ has a unique minimal cocover which is the subcategory $\mathbf{I}_0$ of mod Λ consisting of the indecomposable injective modules in mod Λ.

This discussion suggests that for an arbitrary subcategory $\mathbf{C}$ of mod Λ there might be some connection between the covers (cocovers) of $\mathbf{C}$ and some sort of suitably defined notion of "projective" ("injective") objects in $\mathbf{C}$. The right notion of "projective" in $\mathbf{C}$ seems to be the one given in property (b) for the projectives.

Let $\mathbf{C}$ be a subcategory of mod Λ. We say that a C in $\mathbf{C}$ is a *splitting projective* in $\mathbf{C}$ if each surjective morphism $X \to C$ with X in add $\mathbf{C}$ is a splittable surjection.

Clearly C in $\mathbf{C}$ is a splitting projective in $\mathbf{C}$ if and only if each indecomposable summand of C is a splitting projective in $\mathbf{C}$. We denote by $\mathbf{P}_0(\mathbf{C})$ the subcategory of ind $\mathbf{C}$ consisting of the indecomposable splitting projectives in $\mathbf{C}$. It is clear that $\mathbf{P}_0(\mathbf{C}) = \mathbf{P}_0(\text{ind } \mathbf{C}) = \mathbf{P}_0(\text{add } \mathbf{C})$.

Dually, we say that C in $\mathbf{C}$ is a *splitting injective* in $\mathbf{C}$ if each injective morphism $C \to Y$ with Y in add $\mathbf{C}$ is a splittable injection. Obviously C in $\mathbf{C}$ is a splitting injective in $\mathbf{C}$ if and only if each indecomposable summand of C is a splitting injective in $\mathbf{C}$. We denote by $\mathbf{I}_0(\mathbf{C})$ the subcategory of ind $\mathbf{C}$ consisting of all the indecomposable splitting injectives in $\mathbf{C}$. It is clear that $\mathbf{I}_0(\mathbf{C}) = \mathbf{I}_0$ (ind $\mathbf{C}$) $= \mathbf{I}_0(\text{add } \mathbf{C})$. Moreover it is easy to see that C in $\mathbf{C}$ is a splitting projective in $\mathbf{C}$ if and only if $D(C)$ is a splitting injective in $D(\mathbf{C})$. Hence $D(\mathbf{P}_0(\mathbf{C})) = \mathbf{I}_0(D(\mathbf{C}))$ and $D(\mathbf{I}_0(\mathbf{C})) = \mathbf{P}_0(D(\mathbf{C}))$.

While the notions of splitting projectives and splitting injectives are obviously intimately related to the usual notions of projectives and injectives there is the following important difference. Let $f: X \to Y$ be a surjective morphism in $\mathbf{C}$ and P a splitting projective in $\mathbf{C}$. Then it is *not* necessarily true that each g: $P \to Y$ can be lifted to f. A similar remark holds for splitting injectives in $\mathbf{C}$. On the other hand, if $\mathbf{C} = \text{mod } \Lambda$, then it is obvious that $\mathbf{P}_0(\mathbf{C})$ is the subcategory of indecomposable projective Λ-modules, while $\mathbf{I}_0(\mathbf{C})$ is the subcategory of indecomposable injective Λ-modules. It should also be noted, as we will show later, that there are subcategories $\mathbf{C}$ of mod Λ for which $\mathbf{P}_0(\mathbf{C})$ or $\mathbf{I}_0(\mathbf{C})$ is empty.

Let $\mathbf{A}$ be a cover for the subcategory $\mathbf{C}$ of mod Λ. In the case $\mathbf{C} = \text{mod } \Lambda$ we saw that $\mathbf{A}$ is a minimal cover for $\mathbf{C}$ if and only if $\mathbf{A} = \mathbf{P}_0(\mathbf{C})$. Our aim is to show that this result holds for arbitrary subcategories $\mathbf{C}$ of mod Λ. We begin with the following easily established preliminary result.

PROPOSITION 2.1. *Let* $\mathbf{C}$ *be a subcategory of* mod Λ.

(a) *If* $\mathbf{A}$ *is a cover for* $\mathbf{C}$, *then* $\mathbf{A}$ *contains* $\mathbf{P}_0(\mathbf{C})$.

(b) *If* $\mathbf{B}$ *is a cocover for* $\mathbf{C}$, *then* $\mathbf{B}$ *contains* $\mathbf{I}_0(\mathbf{C})$.

Proof. (a) Let P be in $\mathbf{P}_0(\mathbf{C})$. Since $\mathbf{A}$ is a cover for $\mathbf{C}$, there is a surjective morphism $f: A \to P$ with A in add $\mathbf{A}$. Because P is a splitting projective in $\mathbf{C}$, the surjection $f: A \to P$ is a splitting surjection. Hence P is isomorphic to an indecomposable summand of A and thus must be in $\mathbf{A}$. Therefore $\mathbf{A} \supset \mathbf{P}_0(\mathbf{C})$.

(b) Dual of (a).

The rest of the proof that a cover $\mathbf{A}$ of $\mathbf{C}$ is minimal if and only if $\mathbf{A} = \mathbf{P}_0(\mathbf{C})$ is based on the following general definitions and results which are also of interest in their own right.

Let A and B be in mod Λ. The *trace of B in A* is the submodule $\tau_B(A)$ of A generated by all the homomorphic images of B in A. Since A is noetherian, it follows that there is a morphism $f: nB \to A$, where nB is a sum of n copies of B, such that $\operatorname{Im} f = \tau_B(A)$. Also, viewing A as an End A-module, where End A

is the endomorphism ring of A, we have that $\tau_B(A)$ is an End A-submodule of A.

We now use these observations to prove the following general result.

PROPOSITION 2.2. *Let A and B be in* mod Λ *with A indecomposable.*

(a) *Suppose $f_i\colon A \to A$, $i = 1,\ldots, n$ is a finite family of morphisms and f_0: $B \to A$ is a morphism such that the induced morphism $f\colon B \coprod nA \to A$ is a surjection which is not a splittable surjection. Then $\tau_B(A) = A$ and so there is a surjection $mB \to A$ of m copies of B(m finite) to A.*

(b) *Suppose $f_i\colon A \to A$, $i = 1,\ldots, n$ is a finite family of morphisms and f_0: $A \to B$ a morphism such that the induced morphism $f\colon A \to nA \coprod B$ is an injection which is not a splittable injection. Then there is an injection $A \to mB$, where mB is a finite sum of copies of B.*

Proof. (a) Since the surjection $f\colon B \coprod nA \to A$ is not a splittable surjection, none of the $f_i\colon A \to A$ is an isomorphism. The fact that A is indecomposable implies that End A is local. Hence each of the $f_i\colon A \to A$ is in rad End A. Now $f\colon B \coprod nA \to A$ being surjective implies that the submodule $\sum_{i=1}^{n} f_i(A) + f_0(B)$ of A generated by the $f_i(A)$, $i = 1,\ldots, n$ and $f_0(B)$ is all of A. Since each $f_i(A) \subset$ (rad End A) $\cdot$ A and $f_0(B) \subset \tau_B(A)$, we have that $A =$ (rad End A)$\cdot$ $A + \tau_B(A)$. Thus viewing A as an End A-module, we have by Nakayama's lemma that $\tau_B(A) = A$ since $\tau_B(A)$ is an End A-submodule of A and A is a finitely generated End A-module.

(b) Dual of (a).

We now prove our main result connecting minimal covers and cocovers of a subcategory $\mathbf{C}$ of mod Λ with $\mathbf{P}_0(\mathbf{C})$ and $\mathbf{I}_0(\mathbf{C})$.

THEOREM 2.3. *Let* $\mathbf{C}$ *be a subcategory of* mod Λ.

(a) *A cover* $\mathbf{A}$ *of* $\mathbf{C}$ *is a minimal cover of* $\mathbf{C}$ *if and only if* $\mathbf{A} = \mathbf{P}_0(\mathbf{C})$.

(b) *A cocover* $\mathbf{B}$ *of* $\mathbf{C}$ *is a minimal cocover of* $\mathbf{C}$ *is and only if* $\mathbf{B} = \mathbf{I}_0(\mathbf{C})$.

Proof. (a) We have already seen, Proposition 2.1, that every cover of $\mathbf{C}$ contains $\mathbf{P}_0(\mathbf{C})$. Hence if $\mathbf{A}$ is a cover of $\mathbf{C}$ and $\mathbf{A} = \mathbf{P}_0(\mathbf{C})$, then $\mathbf{A}$ is a minimal cover for $\mathbf{C}$ We now finish the proof of (a) showing that if $\mathbf{A}$ is a cover of $\mathbf{C}$ and $\mathbf{A} \neq \mathbf{P}_0(\mathbf{C})$, then $\mathbf{A}$ is not minimal.

Suppose A in $\mathbf{A}$ is not in $\mathbf{P}_0(\mathbf{C})$. Then there is a surjective morphism $f\colon C \to A$ in $\mathbf{C}$ which is not a splittable surjection. Since $\mathbf{A}$ is a cover for $\mathbf{C}$, we know there is a surjection $g\colon X \to C$ with X in $\mathbf{A}$. Then the composition $h = gf\colon X \to A$ is a surjection which is not a splittable surjection since f is not a splittable surjection. We can write X as a sum $B \coprod nA$ where B has no summands isomorphic to A. Then by Proposition 2.2, we know there is a surjection $mB \to A$. Since B is in add($\mathbf{A} - \{A\}$), where $\{A\}$ is the subcategory of $\mathbf{A}$ of all modules isomorphic to A, it follows that $\mathbf{A} - \{A\}$ is a cover for $\mathbf{A}$ and hence for $\mathbf{C}$. Hence, if there is an A in $\mathbf{A}$ which is not in $\mathbf{P}_0(\mathbf{C})$, then $\mathbf{A}$ is not a minimal cover for $\mathbf{C}$. Therefore $\mathbf{A}$ being a minimal cover for $\mathbf{C}$ implies $\mathbf{P}_0(\mathbf{C}) = \mathbf{A}$, which finishes the proof of (a).

(b) Dual of (a).

As an immediate consequence of Theorem 2.3 we have the following.

COROLLARY 2.4. *Let* $\mathbf{C}$ *be a subcategory of* mod Λ.

(a) If $\mathbf{A}_1$, $\mathbf{A}_2$ *are two minimal covers of* $\mathbf{C}$, *then* $\mathbf{A}_1 = \mathbf{A}_2 = \mathbf{P}_0(\mathbf{C})$.

(b) *If* $\mathbf{A}$ *is a finite cover for* $\mathbf{C}$ (*i.e.*, $\mathbf{A}$ *has only a finite number of nonisomorphic objects*), *then* $\mathbf{P}_0(\mathbf{C})$ *is a finite minimal cover for* $\mathbf{C}$.

(c) *If* $\mathbf{B}_1$ *and* $\mathbf{B}_2$ *are minimal cocovers for* $\mathbf{C}$, *then* $\mathbf{B}_1 = \mathbf{B}_2 = \mathbf{I}_0(\mathbf{C})$.

(d) *If* $\mathbf{C}$ *has a finite cocover, then* $\mathbf{I}_0(\mathbf{C})$ *is a finite minimal cocover for* $\mathbf{C}$.

Proof. (a) By Theorem 2.3 we know that $\mathbf{A}_1 = \mathbf{P}_0(\mathbf{C}) = \mathbf{A}_2$ and so $\mathbf{A}_1 = \mathbf{A}_2$.

(b) If $\mathbf{A}$ is a finite cover for $\mathbf{C}$, then there is a subcategory $\mathbf{A}'$ of $\mathbf{A}$ with the smallest number of nonisomorphic objects such that $\mathbf{A}'$ is a cover for $\mathbf{C}$. Then by Proposition 2.3, $\mathbf{A}' = \mathbf{P}_0(\mathbf{C})$ which shows that $\mathbf{P}_0(\mathbf{C})$ is a finite cover for $\mathbf{C}$.

(c) and (d). Duals of (a) and (b), respectively.

Corollary 2.4 plays a critical role in the rest of this paper. We end this section with our first application of this result.

We recall that a module M of finite length over a principal ideal domain R has two standard representations: (a) $M \approx \coprod_{i=1}^{s} R/p_i^{n_i} R$, where each p_i is a prime, nonzero element in R and (b) $M \approx \coprod_{i=1}^{t} R/a_i R$, where $a_1 \mid a_2 \mid \cdots \mid a_t$. Moreover each of these representations is unique in the sense that the $p_i^{n_i}$ and $a_1, \ldots, a_t$ are unique for any such representations. Since the R-modules of the form $R/p^n R$ with a p prime element in R are precisely the indecomposable modules of finite length, representation (a) has a ready analog for modules in mod Λ, namely, the usual representation as a sum of indecomposable modules. We now show that each module M in mod Λ has two other decompositions each of which is the analog of the decomposition (b). Results similar to these were given by Roiter [15].

We say that a module L in mod Λ is *multiplicity free* if its representation $\coprod_{i=1}^{n} M_i$ as a sum of indecomposable modules M_i has the property $M_i \approx M_j$ implies $i = j$ for all $1 \leqslant i \leqslant n$ and $1 \leqslant j \leqslant n$. A module L is said to be *covering indecomposable* if $\tau_K(L) \neq L$ for each proper summand K of L. Clearly if L is covering indecomposable, then L is multiplicity free.

By a *covering decomposition* of a module M we mean a representation of $M \approx \coprod_{i=1}^{t} L_i$ having the following properties: (a) each L_i is covering indecomposable and (b) $\tau_{L_i}(L_{i+1}) = L_{i+1}$ for all $i = 1, \ldots, t-1$.

Dually, a module L is said to be *cocovering indecomposable* if there is no injective morphism $L \to nK$ with K a proper summand of L. Clearly a cocovering indecomposable module is multiplicity free. By a *cocovering decomposition* of a module M we mean a representation of $M \approx \coprod_{i=1}^{s} L_i$ having the following properties: (a) each L_i is cocovering indecomposable and (b) for each $i = 1,\ldots, s-1$ there is an injection $L_{i+1} \to n_i L_i$ for some n_i.

Before showing that every module in mod Λ has both a covering and cocovering decomposition, we show that these representations are essentially unique. To this end it is convenient to introduce the following definition.

Let M be a module in mod Λ. We denote the subcategory of ind mod Λ consisting of those indecomposable modules which are isomorphic to summands of M by ind M. Clearly ind M is a finite category.

PROPOSITION 2.5. *Let M be a nonzero module in* mod Λ.

(a) *Let $M \approx \coprod_{i=1}^{n} L_i$ be a covering decomposition. Then for each $i = 1,\ldots, n$ we have that* $\operatorname{ind} L_i = \mathbf{P}_0(\operatorname{ind}(M/\coprod_{j<i} L_i))$. *Thus if $M \approx \coprod_{j=1}^{m} L_j'$ is another covering decomposition, then $n = m$ and $L_i \approx L_i'$ for all $i = 1,\ldots, n$.*

(b) *Let $M \approx \coprod_{j=1}^{s} J_j$ be a cocovering decomposition. Then for each $j = 1,\ldots, s$ we have that* ind $J_j = \mathbf{I}_0(\operatorname{ind}(M/\coprod_{i<j} J_i))$. *Thus if $M \approx \coprod_{j=1}^{t} J_j'$ is another cocovering decomposition, then $s = t$ and $J_j \approx J_j'$ for all $j = 1,\ldots, s$.*

Proof. (a) Suppose $M \approx \coprod_{i=1}^{n} L_i$ is a covering decomposition. Since $M/\coprod_{j<i} L_i = \coprod_{k=i}^{n} L_k$, we have to show that $\mathbf{P}_0(\operatorname{ind}(\coprod_{k=i}^{n} L_k)) = \operatorname{ind} L_i$. Because $\operatorname{ind}(\coprod_{k=i}^{n} L_k)$ is finite, we have by Corollary 2.4 that it suffices to show that $\operatorname{ind} L_i$ is a minimal cover for $\operatorname{ind}(\coprod_{k=i}^{n} L_k)$. Since $\tau_{L_k}(L_{k+1}) = L_{k+1}$ for all $k = 1,\ldots, n-1$, it follows that $\operatorname{ind}(L_i)$ is a cover for $\operatorname{ind}(\coprod_{i=k}^{n} L_i)$. But $\operatorname{ind}(L_i)$ is a minimal cover for $\operatorname{ind}(\coprod_{k=i}^{n} L_k)$. Suppose L_1' is the sum of a complete set of nonisomorphic representatives of the objects of a proper subcategory of $\operatorname{ind}(L_i)$. Because L_i is multiplicity free, it follows that L_i' is a proper summand of L_i. Therefore $\tau_{L_i'}(L_i) \neq L_i$ and so $\operatorname{ind}(L_i')$ is not a cover for $\operatorname{ind}(\coprod_{k=i}^{n} L_k)$. Hence $\operatorname{ind}(L_i)$ is a minimal cover for $\operatorname{ind}(\coprod_{k=i}^{n} L_k)$ and so $\operatorname{ind}(L_i) = \mathbf{P}_0(M/\coprod_{j<i} L_j)$ for all $i = 1,\ldots, n$. This proves the first part of (a).

Since each L_i is multiplicity free, the fact that $\operatorname{ind}(L_i) = \mathbf{P}_0 \operatorname{ind}(M/\coprod_{j<i} L_j)$ implies that L_i is isomorphic to the sum of any complete set of nonisomorphic representations of objects in $\mathbf{P}_0 \operatorname{ind}(M/\coprod_{j<i} L_j)$. Hence L_1 is isomorphic to the sum of any complete set of nonisomorphic representatives of objects in $\mathbf{P}_0$ $(\operatorname{ind}(M))$. Thus if $M \approx \coprod_{j=1}^{m} L_j'$ is another covering decomposition for M, then $L_1 \approx L_1'$. Then $M/L_1 \approx M/L_1'$, and so $L_2 \amalg \cdots \amalg L_n$ and $L_2' \amalg \cdots \amalg L_m'$ are covering decompositions of the same module M/L_1. Hence by induction on n, we have that $n = m$ and $L_i \approx L_i'$ for $i = 2,\ldots, n$. Combining this with the fact that $L_1 \approx L_1$, we have finished the proof of (a).

(b) Dual of (a).

PROPOSITION 2.6. *Let M be a nonzero module in* mod Λ.

(a) *M has a covering decomposition $\coprod_{i=1}^{n} L_i$ which is unique in the sense that the L_i are uniquely determined up to isomorphism by M. Moreover* $\operatorname{ann}(L_i) \subset \operatorname{ann}(L_2) \subset \cdots \subset \operatorname{ann}(L_n)$.

(b) *M has a cocovering decomposition $\coprod_{j=1}^{m} J_j$ which is unique in the sense that the J_j are uniquely determined up to isomorphism by M. Moreover* $\operatorname{ann}(J_1) \subset \operatorname{ann}(J_2) \subset \cdots \subset \operatorname{ann}(J_m)$.

Proof. (a) Suppose $M = \coprod_{i=1}^{t} n_i M_i$ with each M_i indecomposable and $n_i \geqslant 1$. Proceed by induction on $\sum_{i=1}^{t} n_i$. If $\sum_{i=1}^{t} n_i = 1$, then M is indecomposable and $L_1 = M$ is our desired decomposition. Suppose (a) holds for all $M = \coprod_{i=1}^{t} n_i M_i$ with $1 \leqslant \sum_{i=1}^{t} n_i < k$ and suppose $\sum_{i=1}^{t} n_i = k$. Define L_1 to be the multiplicity-free module $\coprod N_j$ where N_j ranges over a complete set of nonisomorphic representatives of the objects in the finite subcategory $\mathbf{P}_0(\operatorname{ind} M)$. Since $\operatorname{ind}(M)$ is finite, we know by Corollary 2.4 that $\mathbf{P}_0(\operatorname{ind} M)$ is a finite minimal cover for ind M. In particular $\mathbf{P}_0(\operatorname{ind} M)$ is not empty so L_1 is not zero. Hence by the inductive hypothesis $M_2 = M/L_1$ has a unique covering decomposition $L_2 \coprod \cdots \coprod L_n$. We now claim that $L_1 \coprod L_2 \coprod \cdots \coprod L_n$ is a covering decomposition for M.

Since $\mathbf{P}_0(\operatorname{ind} M)$ is a minimal cover for ind M, it follows that L_1 has the following properties: (a) for each X in ind M there is a surjective morphism $nL_1 \twoheadrightarrow X$ for some n in N and (b) if L_1' is a proper summand of L_1, then there is an indecomposable X in ind M such that there is no surjection $nL_1' \twoheadrightarrow X$ for any n in N. Therefore there is certainly a surjective morphism $nL_1 \twoheadrightarrow L_2$. Moreover if there were a surjective morphism $nL_1' \twoheadrightarrow L_1$ with L_1' a proper summand of L_1, then $\operatorname{ind}(L_1')$ would be a cover for ind M since, by Proposition 2.5, $\operatorname{ind}(L_2)$ is a cover for $\operatorname{ind}(L_2 \coprod \cdots \coprod L_n)$. But the fact that L_1' is a proper summand of L_1 which is multiplicity free means that $\operatorname{ind} L_1'$ is a proper subcategory of $\operatorname{ind} L_1$. Therefore we would have that $\operatorname{ind} L_1$ is not a minimal cover for ind M. But by Corollary 2.4 we know that $\mathbf{P}_0(\operatorname{ind} M) = \operatorname{ind} L_1$ is a minimal cover for ind M. This contradiction shows that $\tau_{L_1'}(L_1) \neq L_1$ and so we have that $L_1 \coprod L_2 \coprod \cdots \coprod L_n$ is a covering decomposition of M.

The uniqueness of the L_i was shown in Proposition 2.5.

Since for each $i = 1,..., n-1$ we have that there is a surjection $n_i L_i \twoheadrightarrow L_{i+1}$, it follows that $\operatorname{ann}(L_i) \subset \operatorname{ann}(L_{i+1})$ for all $i = 1,..., n-1$. This completes the proof of (a).

(b) Dual of (a).

It is worth noting that in case Λ is a proper factor of a principal ideal domain and M is in mod Λ, then the covering and cocovering decompositions of M are the same, namely, the usual invariant factor decomposition of M. This is because modules over such a Λ are self-dual. However, for arbitrary Λ the covering and cocovering decomposition of M are usually not the same.

3. Preprojective Partitions

Let **C** be a subcategory of mod Λ. In this section we are primarily concerned with giving our basic existence theorem for when **C** has a preprojective partition or a preinjective partition; notions we introduce after the following notation.

Suppose **C** is a subcategory of mod Λ and **X** a subcategory of **C**. We denote by $\mathbf{C}_\mathbf{X}$ the subcategory of **C** consisting of those objects in **C** with no summands in **X**. Clearly $\mathbf{C}_\mathbf{X} = \mathbf{C}_{\mathrm{ind}\mathbf{X}}$ and if **C** is a subcategory of ind Λ, then $\mathbf{C}_\mathbf{X} = \mathbf{C} - \mathbf{X}$.

Let **C** be a subcategory of mod Λ. We have already defined $\mathbf{P}_0(\mathbf{C})$ to be the subcategory of ind **C** consisting of the splitting projectives in ind **C**. We define $\mathbf{P}_1(\mathbf{C}) = \mathbf{P}_0(\mathbf{C}_{\mathbf{P}_0(\mathbf{C})})$ and, by induction, $\mathbf{P}_k(\mathbf{C}) = \mathbf{P}_0(\mathbf{C}_{\mathbf{P}_0(\mathbf{C})\cup\cdots\cup\mathbf{P}_{k-1}(\mathbf{C})})$. Finally we denote the subcategory $\bigcup_{i<\infty} \mathbf{P}_i(\mathbf{C})$ of ind **C** by $\mathbf{P}(\mathbf{C})$ and $\mathrm{ind}(\mathbf{C}_{\mathbf{P}(\mathbf{C})})$ by $\mathbf{P}_\infty(\mathbf{C})$. We now list some easily verified properties of these subcategories of **C**.

(a) $\mathbf{P}_i(\mathbf{C}) \cap \mathbf{P}_j(\mathbf{C}) = \varnothing$ for all $i, j \leqslant \infty$ if $i \neq j$.

(b) If $\mathbf{P}_i(\mathbf{C}) = \varnothing$, then $\mathbf{P}_j(\mathbf{C}) = \varnothing$ for $i \leqslant j < \infty$.

We say that the collection $\{\mathbf{P}_i(\mathbf{C})\}_{i=0,\ldots,\infty}$ is a *preprojective partition of* **C** if $\mathbf{P}_i(\mathbf{C})$ is a finite cover for $\mathbf{C}_{\mathbf{P}_0(\mathbf{C})\cup\cdots\cup\mathbf{P}_{i-1}(\mathbf{C})}$ for each $i < \infty$. We say that **C** has a *preprojective partition* if $\{\mathbf{P}_i(\mathbf{C})\}_{i=0,\ldots,\infty}$ is a preprojective partition of **C**. If **C** has a preprojective partition, then we say that C in **C** is *preprojective* if every indecomposable summand of C is in $\mathbf{P}(\mathbf{C}) = \bigcup_{i<\infty} \mathbf{P}_i(C)$.

Dually, we have already defined $\mathbf{I}_0(\mathbf{C})$ to be the subcategory of ind **C** consisting of the splitting injectives in ind **C**. By induction we define $\mathbf{I}_k(\mathbf{C}) = \mathbf{I}_0(\mathbf{C}_{\mathbf{I}_0(\mathbf{C})\cup\cdots\cup\mathbf{I}_{k-1}(\mathbf{C})})$. Finally we denote the subcategory $\bigcup_{i<\infty} \mathbf{I}_i(\mathbf{C})$ of ind **C** by $\mathbf{I}(\mathbf{C})$ and $\mathrm{ind}(\mathbf{C})_{\mathbf{I}(\mathbf{C})}$ by $\mathbf{I}_\infty(\mathbf{C})$. The following properties of these subcategories of **C** are easily verified.

(a) $\mathbf{I}_i(\mathbf{C}) \cap \mathbf{I}_j(\mathbf{C}) = \varnothing$ for $i, j \leqslant \infty$ with $i \neq j$.

(b) If $\mathbf{I}_i(\mathbf{C}) = \varnothing$, then $\mathbf{I}_j(\mathbf{C}) = \varnothing$ for $i \leqslant j < \infty$.

We say that the collection $\{\mathbf{I}_i(\mathbf{C})\}_{i=0,\ldots,\infty}$ is a *preinjective partition* of **C** if $\mathbf{I}_i(\mathbf{C})$ is a finite cocover for $\mathbf{C}_{\mathbf{I}_0(\mathbf{C})\cup\cdots\cup\mathbf{I}_{i-1}(\mathbf{C})}$ for each $i < \infty$. We say that **C** *has a preinjective partition* if $\{\mathbf{I}_i(\mathbf{C})\}_{i=0,\ldots,\infty}$ is a preinjective partition of **C**. If **C** has a preinjective partition, then we say that a C in **C** is preinjective if every indecomposable summand of C is in $\mathbf{I}(\mathbf{C}) = \bigcup_{i<\infty} \mathbf{I}_i(\mathbf{C})$.

We have already seen that $D(\mathbf{P}_0(\mathbf{C})) = \mathbf{I}_0(D(\mathbf{C}))$ and $D(\mathbf{I}_0(\mathbf{C})) = \mathbf{P}_0(D(\mathbf{C}))$. Consequently it is easily seen that $D(\mathbf{P}_i(\mathbf{C})) = \mathbf{I}_i(D(\mathbf{C}))$ and $D(\mathbf{I}_i(\mathbf{C})) = \mathbf{P}_i(D(\mathbf{C}))$ for all $i \leqslant \infty$ and $D(\mathbf{P}(\mathbf{C})) = \mathbf{I}(D(\mathbf{C}))$ and $D(\mathbf{I}(\mathbf{C})) = \mathbf{P}(D(\mathbf{C}))$. From these remarks it follows that $\{\mathbf{P}_i(\mathbf{C})\}_{i=0,\ldots,\infty}$ is a preprojective partition of **C** if and only if $\{\mathbf{I}_i(D(\mathbf{C}))\}_{i=0,\ldots,\infty}$ is a preinjective partition of $D(\mathbf{C})$.

Before stating our general criterion for when a subcategory **C** of mod Λ has a preprojective or a preinjective partition, we give the following result due to

Gabriel and Roiter [12, 15] which was also the starting point of our own investigations.

THEOREM 3.1. *Let* **C** *be a subcategory of* mod Λ *of finite type* (*i.e.*, ind **C** *is finite*). *Then* **C** *has both preprojective and preinjective partitions.*

Proof. Since ind **C** is finite, ind **X** is finite for all subcategories **X** of **C**. Hence by Corollary 2.4, we know that $\mathbf{P}_0(\mathbf{X})$ is a finite cover for **X** for each subcategory **X** of **C**. Since by definition, $\mathbf{P}_i(\mathbf{C}) = \mathbf{P}_0(\mathbf{C}_{\mathbf{P}_0(\mathbf{C})\cup\cdots\cup\mathbf{P}_{i-1}(\mathbf{C})})$, it follows that each $\mathbf{P}_i(\mathbf{C})$ is a finite cover for $\mathbf{C}_{\mathbf{P}_0(\mathbf{C})\cup\cdots\cup\mathbf{P}_{i-1}(\mathbf{C})}$ for all $i < \infty$. Hence $\{\mathbf{P}_i(\mathbf{C})\}_{i=0,\ldots,\infty}$ is a preprojective partition of **C** which shows that **C** has a preprojective partition. That **C** also has a preinjective partition follows by duality.

In connection with Theorem 3.1 we point out the following relationship between the preprojective and preinjective partitions of subcategories of mod Λ of finite type and the covering and cocovering decompositions of modules introduced in Section 2.

PROPOSITION 3.2. *Let* **C** *be a subcategory of* mod Λ *of finite type. Let* $C_1, \ldots, C_t$ *be a complete set of nonisomorphic representatives of objects in* ind **C** *and let* $C = \coprod_{i=1}^{t} C_i$.

(a) *If* $C \approx \coprod_{i=1}^{r} L_j$ *is a covering decomposition for* C, *then* $\mathbf{P}_i(\mathbf{C}) =$ ind L_i *for* $i = 1, \ldots, r$ *and* $\mathbf{P}_j(\mathbf{C}) = \varnothing$ *for* $j > r$.

(b) *If* $C \approx \coprod_{i=1}^{s} J_j$ *is a cocovering decomposition of* C, *then* ind $J_j = \mathbf{I}_j(\mathbf{C})$ *for* $j = 1, \ldots, s$ *and* $\mathbf{I}_i(\mathbf{C}) = \varnothing$ *for* $i > s$.

Proof. Left as an exercise.

We now state our main criterion for when a subcategory **C** of mod Λ has a preprojective partition or a preinjective partition.

THEOREM 3.3. *Let* **C** *be a subcategory of* mod Λ.

(a) *Suppose for each* M *in* mod Λ *there is a morphism* $g: M \to C$ *with* C *in* add **C** *such that* $(g, X): (C, X) \to (M, X)$ *is surjective for all* X *in* **C**, *where* $(C, X) = \mathrm{Hom}_\Lambda(C, X)$. *Then* **C** *has a preprojective partition.*

(b) *Suppose for each* M *in* mod Λ *there is a morphism* $f: C \to M$ *with* C *in* add **C** *such that* $(X, f): (X, C) \to (X, M)$ *is surjective for all* X *in* **C**, *where* $(X, C) = \mathrm{Hom}_\Lambda(X, C)$. *Then* **C** *has a preinjective partition.*

The proof of Theorem 3.3 will occupy the rest of this section. We begin with the following lemmas which, when suitably specialized, give a criterion for when a subcategory **C** of mod Λ has a finite cocover. This criterion is especially well suited to proving Theorem 3.3.

LEMMA 3.4. *Let* $\mathbf{D} \supset \mathbf{C}$ *be subcategories of* mod Λ *and assume that* $\mathbf{I}_0(\mathbf{D})$ *is a finite cocover for* **D**. *Then a finite subcategory* **A** *of* ind **C** *is a cocover for* **C** *if for each* I *in* $\mathbf{I}_0(\mathbf{D})$ *there is a morphism* $f_I: A_I \to I$ *with* A_I *in* add **A** *such that* (X, f_I): $(X, A_I) \to (X, I)$ *is surjective for all* X *in* **C**. *Moreover if* $\mathbf{D} = \text{mod}\, \Lambda$ *and* **A** *in* ind Λ *is a finite cocover for* **C**, *then there exists for each* I *in* $\mathbf{I}_0(\text{mod}\, \Lambda)$ *a morphism* $f_I: A_I \to I$ *with* A_I *in* add A *such that* $(X, f_I): (X, A_I) \to (X, I)$ *is surjective for all* X *in* **C**.

Proof. Suppose that for each I in $\mathbf{I}_0(\mathbf{D})$ there exists a morphism $f_I: A_I \to I$ with A_I in add **A** such that $(X, f_I): (X, A_I) \to (X, I)$ is surjective for all X in **C**. Let X be in **C** and let $g: X \to I$ be an injective morphism with I in add $\mathbf{I}_0(\mathbf{D})$. Then there is an $h: X \to A_I$ such that $g = f_I h$. Therefore $h: X \to A_I$ is injective with A_I in add **A**. Since X is an arbitrary object in **C**, we have shown that **A** is a cocover for **C**. Now let $\mathbf{D} = \text{mod}\, \Lambda$ and assume that **A** is a finite cocover for **C**. Let $A_1, \ldots, A_n$ be a complete set of nonisomorphic objects in **A**. Let I be in $\mathbf{I}_0(\text{mod}\, \Lambda)$. Since (A_i, I) is a finitely generated End A^{op}-module, there is for each $i = 1, \ldots, n$ a morphism $f_i: n_i A_i \to I$ such that $(A_i, f): (A_i, n_i A_i) \to (A_i, I)$ is surjective. Therefore the induced morphism $f: \coprod n_i A_i \to I$ has the property that $(A, f): (A, \coprod n_i A_i) \to (A, I)$ is surjective for each A in **A**. Thus we have shown that for each indecomposable module I in $\mathbf{I}_0(\mathbf{D})$ there is a morphism $f_I: A_I \to I$ with A_I in add **A** such that $(A, f_I): (A, A_I) \to (A, I)$ is exact for each A in **A** and hence for each A in add **A**.

Let C be in **C**. Since we are assuming that **A** is a cocover for **C**, we know that there is a monomorphism: $g: C \to A$ with A in add **A**. Suppose $h: C \to I$ is a morphism with I in $\mathbf{I}_0$ (mod Λ). Then there is a $j: A \to I$ such that $h = jg$. Since A is in add **A**, we know by our previous remarks that there is $t: A \to A_I$ such that $j = f_I t$. Therefore $h = f_I t g$. Because tg is in (C, A_I) we have that h is in the image of $(C, f_I): (C, A_I) \to (C, I)$. Therefore $(C, f_I): (C, A_I) \to (C, I)$ is surjective for all C in **C**. This finishes the proof of the lemma.

Assume that **C** is a subcategory of mod Λ which has a finite cocover. As a consequence of Lemma 3.4 we have the following description of the minimal cocover $\mathbf{I}_0(\mathbf{C})$ of the subcategory **C**.

LEMMA 3.5. *Suppose* **C** *is a subcategory of* mod Λ *which has a finite cocover. Then the following hold for a finite cocover* **A** *of* **C**.

(a) *For each indecomposable injective* I *there is a unique (up to isomorphism) right minimal morphism* $f_I: A_I \to I$ *with* A_I *in* add **A** *such that* $(X, f_I): (X, A_I) \to (X, I)$ *is surjective for all* X *in* **C**.

(b) $\mathbf{I}_0(\mathbf{C}) = \bigcup_I \text{ind}\, A_I$ *as* I *runs through a complete set of nonisomorphic objects in* $\mathbf{I}_0$ (mod Λ).

Proof. (a) Since **A** is a finite cocover for **C**, we know by Lemma 3.4 that for each I in $\mathbf{I}_0$ (mod Λ) there is a morphism $f': A_I \to I$ with A_I in add **A** such that $(X, f'): (X, A_I) \to (X, I)$ is surjective for all X in **C**. Then by Proposition

1.3 we know there are unique (up to isomorphism) right minimal morphisms f_I: $A_I \to I$ with A_I isomorphic to a summand of A_I such that (X, f_1): $(X, A_I) \to (X, I)$ is surjective for all X in add **A**. Since the A'_I are summands of the A_I, it follows that the A_I are in add **A** which finishes the proof of (a).

(b) Since the f_I: $A_I \to I$ have the property that (X, f_I): $(X, A_I) \to (X, I)$ is surjective for all X in **C**, it follows from Lemma 3.4, that $\bigcup_I \operatorname{ind}(A_I)$, as I runs through a complete set of nonisomorphic objects in $\mathbf{I}_0$ (mod Λ), is a finite cocover for **C**. Hence $\mathbf{I}_0(\mathbf{C}) \subset \bigcup_I \operatorname{ind}(A_I)$ by Proposition 2.1. We now show that $\mathbf{I}_0(\mathbf{C}) \supset \bigcup_I \operatorname{ind}(A_I)$ by showing that $\mathbf{I}_0(\mathbf{C}) \supset \operatorname{ind}(A_I)$ for each indecomposable injective I.

Since **C** has a finite cocover, we know by Corollary 2.4 that $\mathbf{I}_0(\mathbf{C})$ is the minimal cocover of **C**. Therefore by Lemma 3.4 we know that given an I in $\mathbf{I}_0$ (mod Λ) there is a morphism g: $C_I \to I$ with C_I in $\mathbf{I}_0(\mathbf{C})$ such that (X, g) is surjective for all X in **C**. In particular, there is an h: $A_I \to C_I$ such that $f_I = gh$. On the other hand there is an h': $C_I \to A_I$ such that $g = f_I h'$. Hence $f_I h' h = gh = f_I$ so that $h'h$: $A_I \to A_I$ is an isomorphism since f_I is right minimal. Thus A_I is isomorphic to a summand of C_I and so ind $A_I \subset \mathbf{I}_0(\mathbf{C})$. Therefore we have our desired result $\mathbf{I}_0(\mathbf{C}) = \bigcup_I \operatorname{ind} A_I$ as I runs through a complete set of nonisomorphic objects in $\mathbf{I}_0$(mod Λ).

Summarizing Lemmas 3.4 and 3.5 we have the following result.

PROPOSITION 3.6. *Let* **C** *be a subcategory of* mod Λ. *Then the following statements are equivalent.*

(a) **C** *has a finite cocover.*

(b) *For each indecomposable injective module I there is a morphism f_I: $A_I \to I$ with A_I in* add **C** *such that (C, f_I) is surjective for all C in* **C**.

(c) *For each indecomposable injective module I there is a unique (up to isomorphism) right minimal f_I: $A_I \to I$ with A_I in* add **C** *such that (C, f_I) is surjective for all C in* **C**.

Moreover, if **C** *has a finite cocover, then $\mathbf{I}_0(\mathbf{C})$, its minimal cocover, can be described as follows. Let $I_1, \ldots, I_t$ be a complete set of nonisomorphic indecomposable injective modules and let f_j: $A_j \to I_j$ be the unique right minimal morphisms with A_j in* add **C** *such that (C, f_j) is surjective for all C in* **C**. *Then* $\mathbf{I}_0(\mathbf{C}) = \bigcup_{j=1}^t \operatorname{ind} A_j$.

Dualizing Proposition 3.6 we obtain.

PROPOSITION 3.7. *Let* **C** *be a subcategory of* mod Λ. *Then the following are equivalent.*

(a) **C** *has a finite cover.*

(b) *For each indecomposable projective module P there is a morphism g_P: $P \to A_P$ with A_P in* add **C** *such that (g_P, C) is surjective for all C in* **C**.

(c) *For each indecomposable projective module P there is a unique (up to isomorphism) left minimal morphism $g_P: P \to A_P$ with A_P in* add **C** *such that (g_P, C) is surjective for all C in* **C**.

Moreover, if **C** *has a finite cover, then* $\mathbf{P}_0(\mathbf{C})$, *its minimal cover, can be described as follows. Let $P_1, \ldots, P_t$ be a complete set of nonisomorphic indecomposable projective modules and let $g_j: P_j \to A_j$ be the unique left minimal morphisms with A_j in* add **C** *such that (g_j, C) is surjective for all C in* **C**. *Then* $\mathbf{P}_0(\mathbf{C}) = \bigcup_{j=1}^{s} \operatorname{ind} A_j$.

Theorem 3.3, the result we want to prove, states: Suppose **C** is a subcategory of mod Λ with the property that for each M in mod Λ there is a C in **C** and a morphism $f: C \to M$ such that (X, f) is surjective for all X in **C**. Then **C** has a preinjective partition. It obviously follows from Proposition 3.6 that $\mathbf{I}_0(\mathbf{C})$ is a finite cocover for any subcategory **C** of mod Λ satisfying the hypothesis of the theorem. Suppose we can show that **C** satisfying the hypothesis of Theorem 3.3 implies $\mathbf{C}_{\mathbf{I}_0(\mathbf{C})}$ also does. Then, proceeding by induction on n, it would follow trivially that $\mathbf{I}_n(\mathbf{C})$ is a finite cocover for $\mathbf{C}_{\mathbf{I}_0(\mathbf{C}) \cup \cdots \cup \mathbf{I}_{n-1}(\mathbf{C})}$ for all n, which is simply a restatement of Theorem 3.3. Hence to finish the proof of Theorem 3.3 it remains to show that if **C** is a subcategory of mod Λ satisfying the hypothesis of Theorem 3.3, then $\mathbf{C}_{\mathbf{I}_0(\mathbf{C})}$ also does. This will be a consequence of some of the following considerations.

We begin by recalling some of the basic facts concerning the radical of a category that we shall need in this section. Suppose **C** is a subcategory of mod Λ. For each A and B in **C** the radical of (A, B), denoted by $\mathrm{r}(A, B)$ consists of all $f: A \to B$ such that for each Z in ind C, every composition $Z \to A \to B \to Z$ is in rad End Z or, equivalently, is not an isomorphism. The radical of **C**, denoted by $\mathrm{r}(\mathbf{C})$, consists of all the morphisms in $\mathrm{r}(A, B)$ for some A and B in **C**. Further, for each integer $n \geq 0$, we define $\mathrm{r}^n(A, B)$ to consist of all morphisms $f: A \to B$ which can be written as a sum $\sum f_i$ where each f_i is a composition of at least n morphisms in $\mathrm{r}(\mathbf{C})$. We denote by $\mathrm{r}^n(\mathbf{C})$ the collection of all morphisms in $\mathrm{r}^n(A, B)$ for some A and B in **C**. We now list some easily verified properties of $\mathrm{r}^n(A, B)$ for all integers $n \geq 0$ and A and B in **C**.

(a) $\mathrm{r}^n(A, B)$ is a subgroup of (A, B).

(b) $\mathrm{r}^0(A, B) = (A, B)$, $\mathrm{r}^1(A, B) = \mathrm{r}(A, B)$, and $\mathrm{r}^n(A, B) \supset \mathrm{r}^{n+1}(A, B)$.

(c) The composition $U \to A \xrightarrow{f} B \to V$ is in $\mathrm{r}^n(U, V)$ if f is in $\mathrm{r}^n(A, B)$.

(d) $\mathrm{r}^n(A, B)$ is functorial in A and B so that $\mathrm{r}^n(\coprod_{i \in I} A_i, B) = \coprod_{i \in I} \mathrm{r}^n(A_i, B)$ and $\mathrm{r}^n(A, \coprod_{j \in J} B_j) = \coprod_{j \in J} \mathrm{r}^n(A, B_j)$ for all finite families of modules $\{A_i\}_{i \in I}$ and $\{B_j\}_{j \in J}$ in **C**.

(e) $\mathrm{r}^n(A, B)$ is the same whether we view A and B as objects in **C** or add **C**.

(f) $\mathrm{r}^n(A, A) \supset (\mathrm{rad}\ \mathrm{End}\ A)^n$.

(g) If f is in $\mathrm{r}^i(A, B)$ and g is in $\mathrm{r}^j(B, C)$, then gf is in $\mathrm{r}^{i+j}(A, C)$.

(h) $\mathrm{r}(A, B) = (A, B)$ if and only if no indecomposable summand of A is isomorphic to an indecomposable summand of B.

In connection with this last property we point out the following easily deduced facts.

LEMMA 3.8. *Suppose* $\mathbf{C}$ *is a subcategory of* $\operatorname{mod} \Lambda$ *and* B *is in* $\operatorname{ind} \mathbf{C}$.

(a) *For a morphism* $f: A \to B$ *in* $\mathbf{C}$ *the following are equivalent*:

(i) f *is not a splittable surjection*,

(ii) f *is in* $\mathrm{r}(A, B)$,

(iii) *for each* X *in* $\mathbf{C}$, *the induced morhism* $(X, f): (X, A) \to (X, B)$ *has* $\operatorname{Im}(X, f) \subset \mathrm{r}(X, B)$.

(b) *For a morphism* $g: B \to C$ *in* $\mathbf{C}$, *the following are equivalent*:

(i) g *is not a splittable injection*,

(ii) g *is in* $\mathrm{r}(B, C)$,

(iii) *for each* Y *in* $\mathbf{C}$, *the induced morphism* $(g, Y): (C, Y) \to (B, Y)$ *has* $\operatorname{Im}(g, Y) \subset \mathrm{r}(B, Y)$.

Lemma 3.8 naturally raises the question: given a C in $\operatorname{ind} \mathbf{C}$, is there a morphism $f: B \to C$ in $\operatorname{add} \mathbf{C}$ such that $\operatorname{Im}((X, f): (X, B) \to (X, C)) = \mathrm{r}(X, C)$ for all X in $\operatorname{add} \mathbf{C}$ or, equivalently, for all X in $\operatorname{ind} \mathbf{C}$? If $\mathbf{C} = \operatorname{mod} \Lambda$, it was shown in [6] that such an $f: B \to C$ always exists and it was called a right almost split morphism.

Similarly, one can ask: given an indecomposable B in $\operatorname{ind} \mathbf{C}$, is there a g: $B \to C$ in $\operatorname{add} \mathbf{C}$ such that $\operatorname{Im}((g, Y): (C, Y) \to (B, Y)) = \mathrm{r}(B, Y)$ for all Y in $\operatorname{add} \mathbf{C}$ or, equivalently, for all Y in $\operatorname{ind} \mathbf{C}$? Again, if $\mathbf{C} = \operatorname{mod} \Lambda$, it was shown in [6] that such a $g: B \to C$ always exists and it was called a left almost split morphism.

These remarks suggest the following terminology.

Let $\mathbf{C}$ be a subcategory of $\operatorname{mod} \Lambda$ and let C be in $\operatorname{ind} \mathbf{C}$. A morphism $B \to C$ in $\operatorname{add} \mathbf{C}$ is said to be *right almost split* in $\mathbf{C}$ if for each X in $\operatorname{add} \mathbf{C}$ we have that $\operatorname{Im}((X, f): (X, B) \to (X, C)) = \mathrm{r}(X, C)$. Clearly if $f: B \to C$ is right almost split, then f is in $\mathrm{r}(B, C)$. We say that $\mathbf{C}$ *has right almost split morphisms* if for each indecomposable C in $\operatorname{ind} \mathbf{C}$ there is a right almost split morphism $f: B \to C$. It is easy to see that $\mathbf{C}$ has right almost split morphisms if and only if given any C in $\mathbf{C}$ (not necessarily indecomposable) there is a morphism $f: B \to C$ in $\operatorname{add} \mathbf{C}$ such that $\operatorname{Im}(X, f) = \mathrm{r}(X, C)$ for all X in $\operatorname{add} \mathbf{C}$. Clearly such an f: $B \to C$ is in $\mathrm{r}(B, C)$.

Similarly, if B is in $\operatorname{ind} \mathbf{C}$ a morphism $g: B \to C$ in $\operatorname{add} \mathbf{C}$ is said to be *left almost split* in $\mathbf{C}$ if for each Y in $\operatorname{add} \mathbf{C}$ we have that $\operatorname{Im}((g, Y): (C, Y) \to (B, Y)) = \mathrm{r}(B, Y)$. We say that $\mathbf{C}$ *has left almost split morphisms* if for each indecomposable B in ind $\mathbf{C}$ there is a left almost split morphism $f: B \to C$. It is easy to see that $\mathbf{C}$ has left almost split morphisms if and only if given any B in $\mathbf{C}$ (not necessarily indecomposable) there is a $g: B \to C$ in add $\mathbf{C}$ such that $\mathrm{Im}(g, Y) = \mathrm{r}(B, Y)$ for all Y in add $\mathbf{C}$. Clearly such a $g: B \to C$ is in $\mathrm{r}(B, C)$.

The fact that certain subcategories $\mathbf{C}$ of mod Λ have right and/or left almost split morphisms plays an important role in our proof of Theorem 3.3. While we do not have a complete description of such subcategories, we do have a sufficient condition which is based on the following notion.

Let $\mathbf{C} \supset \mathbf{D}$ be subcategories of mod Λ. We say that a module C in $\mathbf{C}$ is *contravariantly finite over* $\mathbf{D}$ if there is a morphism $f: D \to C$ with D in add $\mathbf{D}$ such that $(X, f): (X, D) \to (X, C)$ is surjective for all X in add $\mathbf{D}$. We say that C is *covariantly finite over* $\mathbf{D}$ if there is a morphism $g: C \to D$ with D in add $\mathbf{D}$ such that $(g, X): (D, X) \to (C, X)$ is surjective for all X in add $\mathbf{D}$. It is not hard to see that a C in $\mathbf{C}$ is contravariantly (covariantly) finite over $\mathbf{D}$ if and only if every indecomposable summand of C is contravariantly (covariantly) finite over $\mathbf{D}$.

The reason for this terminology is the following. Let C be in $\mathbf{C}$ and let $(\ , C) \mid \mathbf{D}$ denote the restriction of the representable functor $(\ , C)$ to $\mathbf{D}$. By definition, the functor $(\ , C) \mid \mathbf{D}$ is finitely generated if and only if there is an epimorphism $\alpha: (\ , D) \to (\ , C) \mid \mathbf{D}$ with D in add $\mathbf{D}$. But every morphism $\alpha: (\ , D) \to (\ , C) \mid \mathbf{D}$ is of the form $(\ , f): (\ , D) \to (\ , C) \mid \mathbf{D}$ for some morphism $f: D \to C$ since D is in add $\mathbf{D}$. Thus the condition that there exists a morphism $f: D \to C$ such that $(X, f): (X, D) \to (X, C)$ is surjective for all X in add $\mathbf{D}$ is equivalent to $(\ , C) \mid \mathbf{D}$ being finitely generated over $\mathbf{D}$. Hence the terminology C is contravariantly finite over $\mathbf{D}$. Similar remarks hold for C being covariantly finite over $\mathbf{D}$. Moreover it should be observed that C is contravariantly finite over $\mathbf{D}$ if and only if $D(C)$ in $D(\mathbf{C})$ is covariantly finite over $D(\mathbf{D})$.

Familiar arguments, based on Section 1, give the following result (see Propositions 1.2 and 1.3.)

PROPOSITION 3.9. *Let* $\mathbf{C} \supset \mathbf{D}$ *be subcategories of* mod Λ.

(a) *C in $\mathbf{C}$ is contravariantly finite over $\mathbf{D}$ if and only if there is a right minimal morphism $f: D \to C$ with D in* add $\mathbf{D}$ *having the property that (X, f): $(X, D) \to (X, C)$ is surjective for all X in* add D. *Moreover such an $f: D \to C$ is unique (up to isomorphism) and we denote by $f_C: C_{\mathbf{D}} \to C$ one particular such morphism chosen once and for all.*

(b) *C in $\mathbf{C}$ is covariantly finite over $\mathbf{D}$ if and only if there is a left minimal morphism $g: C \to D$ with D in* add $\mathbf{D}$ *which has the property that (g, X): $(D, X) \to (C, X)$ is surjective for all X in* add $\mathbf{D}$. *Moreover such a $g: C \to D$ is unique (up to isomorphism) and we denote by $g^C: C \to C^{\mathbf{D}}$ one particular such morphism chosen once and for all.*

Globalizing these remarks we have the following.

Let $\mathbf{C} \supset \mathbf{D}$ be subcategories of mod Λ. We say that $\mathbf{C}$ is *contravariantly* (covariantly) *finite over* $\mathbf{D}$ if every C in $\mathbf{C}$ is contravariantly (covariantly) finite over $\mathbf{D}$. Finally we say that $\mathbf{C}$ is functorially finite over $\mathbf{D}$ if it is both contravariantly and covariantly finite over $\mathbf{D}$. The reader should observe that Theorem 3.3 can be restated as follows using this new terminology. Let $\mathbf{C}$ be a subcategory of mod Λ. Then (a) $\mathbf{C}$ has a preprojective partition if mod Λ is covariantly finite over $\mathbf{C}$ and b) $\mathbf{C}$ has a preinjective partition if mod Λ is contravariantly finite over $\mathbf{C}$.

Returning to the question of when a subcategory $\mathbf{C}$ of mod Λ has right and/or left almost split morphisms, we have the following result.

PROPOSITION 3.10. *Suppose* $\mathbf{C} \supset \mathbf{D}$ *are subcategories of* mod Λ.

(a) *If* $\mathbf{C}$ *has right almost split morphisms and is contravariantly finite over* $\mathbf{D}$, *then* $\mathbf{D}$ *has right almost split morphisms.*

(b) *If* $\mathbf{C}$ *has left almost split morphisms and is covariantly finite over* $\mathbf{D}$, *then* $\mathbf{D}$ *has left almost split morphisms.*

Proof. (a) Let D be an indecomposable object in $\mathbf{D}$ and $f\colon C \to D$ a right almost split morphism in $\mathbf{C}$, i.e., $(X, f)\colon (X, C) \to (X, D)$ has image $\mathrm{r}(X, D)$ for all X in $\mathbf{C}$. Since C is in add $\mathbf{C}$ and $\mathbf{C}$ is contravariantly finite over $\mathbf{D}$, there is a morphism $f_C\colon \mathbf{C}_{\mathbf{D}} \to C$ with $C_{\mathbf{D}}$ in add $\mathbf{D}$ such that $\mathrm{Im}(X, f_C) = (X, C)$ for all X in $\mathbf{D}$. Consequently the composition $C_{\mathbf{D}} \to C \to D$ which we denote by $g\colon C_{\mathbf{D}} \to D$ has the property $\mathrm{Im}(X, g) = \mathrm{r}(X, D)$ for all X in add $\mathbf{D}$. Therefore $g\colon C_{\mathbf{D}} \to D$ is a right almost aplit morphism in $\mathbf{D}$. Since D was an arbitrary indecomposable in $\mathbf{D}$, this shows that $\mathbf{D}$ has right almost split morphisms.

(b) Dual of (a).

Letting $\mathbf{C} =$ mod Λ in Proposition 3.10 and recalling Propositions 3.6 and 3.7 we obtain the following step toward proving Theorem 3.3.

LEMMA 3.11. *Let* $\mathbf{C}$ *be a subcategory of* mod Λ.

(a) *Suppose* mod Λ *is contravariantly finite over* $\mathbf{C}$. *Then* $\mathbf{C}$ *has a finite cocover and right almost split morphisms.*

(b) *Suppose* mod Λ *is covariantly finite over* $\mathbf{C}$. *Then* $\mathbf{C}$ *has a finite cover and left almost split morphisms.*

In order to give our next result which, combined with Lemma 3.11, will finish the proof of Theorem 3.3, it is convenient to make the following definitions.

Let $\mathbf{C}$ be a subcategory of mod Λ. A C in $\mathbf{C}$ is said to be *contravariantly nilpotent* if there is an integer n such that $\mathrm{r}^n(\ , C) = 0$, i.e., $\mathrm{r}^n(X, C) = 0$ for all X in $\mathbf{C}$. $\mathbf{C}$ is said to be *locally contravariantly nilpotent* if each C in $\mathbf{C}$ is contravariantly nilpotent. $\mathbf{C}$ is said to be *contravariantly nilpotent* if there is an n such that $r^n(\ , C) = 0$ for all C in **C**. Obviously if **C** is contravariantly nilpotent it is locally contravariantly nilpotent, but the converse is not necessarily true.

Dually, a C in **C** is said to be *covariantly nilpotent* if there is an n such that $r^n(C, \ ) = 0$. **C** is said to be *locally covariantly nilpotent* if each C in **C** is covariantly nilpotent. **C** is said to be *covariantly nilpotent* if there is an integer n such that $r^n(C, \ ) = 0$ for all C in **C**. Obviously if **C** is covariantly nilpotent, it is locally covariantly nilpotent, but the converse is not necessarily true.

Finally, we say that **C** is nilpotent if it is both covariantly and contravariantly nilpotent. Obviously **C** is nilpotent if and only if there is an integer n such that $r^n(C_1, C_2) = 0$ for all C_1, C_2 in **C**. If **C** is of finite type, then the fact that the radical of the endomorphism ring of the sum of a complete set of nonisomorphic indecomposable modules in ind Λ is nilpotent implies that **C** is nilpotent. More generally, we have the following result of Harada and Sai [13].

LEMMA 3.12. *A subcategory* **C** *of* mod Λ *is nilpotent if it is bounded, i.e., there is an integer n such that the length of each indecomposable module in* **C** *is at most n.*

Returning to the proof of Theorem 3.3 we have the following result.

PROPOSITION 3.13. *Let* $\mathbf{C} \supset \mathbf{D}$ *be subcategories of* mod Λ.

(a) *Suppose* **C** *has right almost split morphisms and* $\mathbf{C_D}$ *is locally contravariantly nilpotent. Then* **C** *is contravariantly finite over* **D**.

(b) *Suppose* **C** *has left almost split morphisms and* $\mathbf{C_D}$ *is locally covariantly nilpotent. Then* **C** *is covariantly finite over* **D**.

(c) *If* **C** *has right and left almost split morphisms and* $\mathbf{C_D}$ *is nilpotent, then* **C** *is functorially finite over* **D**.

Proof. Since (c) is a trivial consequence of (a) and (b), and (b) is the dual of (a), it suffices to prove (a).

(a) Suppose we can show that **C** having right almost split morphisms implies that for each C in $\mathbf{C_D}$ and each integer $n \geqslant 0$ there is a morphism $f: D \coprod Y \to C$ with Y in add $\mathbf{C_D}$, D in add **D**, and $f \mid Y$ in $r^n(Y, C)$ such that $(X, f): (X, D \coprod Y) \to (X, C)$ is surjective for all X in add **D**. Since we are assuming that C is contravariantly nilpotent, there is an n such that $r^n(Y, C) = 0$ for all Y in $\mathbf{C_D}$. Therefore for this value of n we have that there is an $f: D \to C$ with D in add **D** such that $(X, f): (X, D) \to (X, C)$ is surjective for all X in add **D**. Hence we would have that each C in $\mathbf{C_D}$ is covariantly finite over **D**. Since each D in **D** is obviously covariantly finite over **D**, we would have shown that **C** is covariantly finite over **D**, under the hypothesis of the proposition. This would prove the proposition. Hence Proposition 3.13 follows from the following.

LEMMA 3.14. *Suppose* $\mathbf{C} \supset \mathbf{D}$ *are subcategories of* mod Λ.

(a) *Suppose* **C** *has right almost split morphisms. For each* C *in* $\mathbf{C_D}$ *and integer* $n \geqslant 0$, *there is a morphism* $f\colon D \coprod Y \to C$ *with* D *in* add **D** *and* Y *in* add $C_{\mathbf{D}}$ *such that* $f \mid Y$ *is in* $\mathrm{r}^n(Y, C)$ *and* $\mathrm{Im}(X, f) = (X, C)$ *for all* X *in* **D**.

(b) *Suppose* **C** *has left almost split morphisms. For each* C *in* $\mathbf{C_D}$ *and integer* $n \geqslant 0$, *there is a morphism* $g\colon C \to D \coprod Z$ *with* D *in* add **D** *and* Z *in* add $\mathbf{C_D}$ *such that* pg *is in* $\mathrm{r}^n(C, Z)$ *and* $\mathrm{Im}(g, X) = (C, X)$ *for all* X *in* **D**, *where* $p\colon D \coprod Z \to Z$ *is the projection morphism.*

Proof. Proceed by induction on n. For $n = 0, f = id_C$ works. Suppose $k > 0$ and the lemma is true for $0 \leqslant n < k$. Let C be in $\mathbf{C_D}$. By the induction hypothesis we know there is a morphism $f'\colon D' \coprod Y' \to C$ with D' in add **D**, Y' in add $\mathbf{C_D}$ such that $f' \mid Y'$ is in $\mathrm{r}^{k-1}(Y', C)$ and $\mathrm{Im}(X, f') = (X, C)$ for all X in **D**. Since **C** has right almost split morphisms, we know that there is a morphism $g\colon Z \to Y'$ such that $\mathrm{Im}(X, g) = \mathrm{r}(X, Y')$ for all X in **C**. Since Y' is in add $\mathbf{C_D}$, no indecomposable summand of Y' is in **D**. Hence $\mathrm{r}(X, Y') = (X, Y')$ for all X in add **D** and so $\mathrm{Im}(X, g) = (X, Y')$ for all X in **D**. Thus the composition $D' \coprod Z \xrightarrow{(idD', g)} D' \coprod Y' \xrightarrow{f'} C$ which we denote by h has the property $\mathrm{Im}(X, h) = (X, C)$ for all X in add **D**.

Write Z as a sum $D'' \coprod Y''$ where D'' is in add **D** and Y'' is in add $\mathbf{C_D}$. Then $h \mid Y''$ is in $\mathrm{r}^k(Y'', C)$ since $h \mid Y''$ is the composition of $g \mid Y''$ which is in $\mathrm{r}(Y'', Y')$ and $f \mid Y'$ which is in $\mathrm{r}^{k-1}(Y', C)$. Therefore the morphism $h\colon D' \coprod D'' \coprod Y'' \to C$ is our desired morphism since $D' \coprod D''$ is in add **D**, Y'' is in add $\mathbf{C_D}$, $hh \mid Y''$ is in $\mathrm{r}^k(Y'', C)$, and $\mathrm{Im}(X, h) = (X, C)$ for all X in add **D**.

(b) Dual of (a).

We are now ready to prove Theorem 3.3. To do this we must show that if mod Λ is contravariantly finite over **C**, then **C** has a preinjective partition. Since mod Λ is contravariantly finite over **C**, we know by Lemma 3.11 that **C** has a finite cocover and right almost split morphisms. Therefore $\mathbf{I}_0(\mathbf{C})$ is the finite minimal cocover for **C**. Since **C** also has right almost split morphisms and $\mathbf{I}_0(\mathbf{C})$ is finite and hence nilpotent, it follows from Proposition 3.13 that **C** is contravariantly finite over $\mathbf{C}_{\mathbf{I}_0(\mathbf{C})}$. Therefore by Proposition 3.6, $\mathbf{C}_{\mathbf{I}_0(\mathbf{C})}$ has a finite cocover and so $\mathbf{I}_1(\mathbf{C}) = \mathbf{I}_0(\mathbf{C}_{\mathbf{I}_0(\mathbf{C})})$ is a finite cocover for $\mathbf{C}_{\mathbf{I}_0(\mathbf{C})}$. The fact that $\mathbf{I}_j(\mathbf{C})$ is a finite cocover for $\mathbf{C}_{\mathbf{I}_0(\mathbf{C}) \cup \cdots \cup \mathbf{I}_{j-1}(\mathbf{C})}$ for all $j \geqslant 0$ now follows easily by induction on j, proving part (b) of Theorem 3.3. The rest of Theorem 3.3 follows by duality.

4. Subcategories over Which mod Λ Is Functorially Finite

This section is devoted to giving various criteria for mod Λ to be covariantly or contravariantly finite over **C**. In view of Theorem 3.3 such subcategories automatically have preprojective or preinjective partitions depending on whether mod Λ is covariantly or contravariantly finite over **C**. Our main result is that if **C** has images (i.e., if $f: C_1 \to C_2$ is a morphism in **C**, then $\operatorname{Im} f$ is in **C**), then mod Λ is covariantly (contravariantly) finite over **C** if and only if **C** has a finite cover (cocover). Before taking up the proof of this result we point out the following results which are essentially recapitulations of results in Section 3.

We begin with the following, which is essentially Proposition 3.13 with $\mathbf{C} =$ mod Λ.

THEOREM 4.1. *Let* **C** *be a subcategory of* ind Λ *such that* ind $\Lambda_{\mathbf{C}}$ *is of bounded length. Then* mod Λ *is functorially finite over* **C** *and so* **C** *has the following properties.*

(a) **C** *has preprojective and preinjective partitions.*

(b) **C** *has right and left almost split morphisms.*

In particular, mod Λ *has preprojective and preinjective partitions.*

In some of our arguments in Section 3 we implicitely used the following.

PROPOSITION 4.2. *Let* **C** *be a subcategory of* mod Λ *of finite type. Then* mod Λ *is functorially finite over* **C**.

Proof. Let $C_1, \ldots, C_n$ be a complete set of nonisomorphic indecomposable objects in **C** and let $C = \coprod_{i=1}^{n} C_i$. Then (C, M) is a finitely generated End C^{op}-module for each M in mod Λ. Let $f_1, \ldots, f_t$ be a set of generators for (C, M) over End C^{op} and let $f: tC \to M$ be the morphism induced by the $f_i: C \to M$. Then $(C, f): (C, tC) \to (C, M)$ is surjective. Since each C_i is isomorphic to a summand of C, it follows that $(C_i, f): (C_i, tC) \to (C_i, M)$ is surjective for all $i = 1, \ldots, n$ and hence $(X, f): (X, tC) \to (X, M)$ is surjective for all X in **C**. Hence mod Λ is contravariantly finite over **C**.

The fact that mod Λ is also covariantly finite over **C** follows by duality. Therefore we have that mod Λ is functorially finite over **C** if **C** is of finite type.

Our next criterion for mod Λ being contravariantly or covariantly finite over a subcategory **C** concerns subcategories **C** which have the property that if $f: C_1 \to C_2$ is a morphism in **C**, then $\operatorname{Im} f$ is in **C**. Such subcategories **C** will be said to *have images*. It is easily seen that a subcategory **C** of mod Λ has images if and only $D(\mathbf{C})$ in mod Λ^{op} has images. This follows from the fact that if $f: C_1 \to C_2$ is a morphism in **C**, then $\operatorname{Im}(D(f): D(C_2) \to D(C_1)) = D(\operatorname{Im} f)$.

We now show that associated with each subcategory **C** of mod Λ is a smallest additive subcategory of mod Λ having images and containing **C**. For each subcategory **C** of mod Λ let $\operatorname{Im} \mathbf{C}$ denote the collection of all M in mod Λ having the property that $M \approx \operatorname{Im}(f: C_1 \to C_2)$ for some C_1, C_2 in add **C**. Then we have the following.

PROPOSITION 4.3. *Let* **C** *be a subcategory of* mod Λ. *Then* Im **C** *has the following properties.*

(a) Im **C** *is an additive subcategory of* mod Λ *having images.*

(b) Add **C** *has images if and only if* add **C** = Im **C**.

(c) Im **C** = Im ind **C**.

(d) Im $D(\mathbf{C}) = D(\operatorname{Im} \mathbf{C})$.

(e) *A subcategory* **A** *of* ind **C** *is a cover (cocover) for* **C** *if and only if it is a cover (cocover) for* Im **C**. *Hence* **C** *has a finite cover (cocover) if and only if* Im **C** *has a finite cover (cocover).*

Proof. (a) Suppose $f\colon C_1 \to C_2$ is a morphism in add **C** and $M = \operatorname{Im} f$. Suppose M' is a summand of M. Then there is a morphism $g\colon M \to M'$ such that the composition $M' \to^{\mathrm{inc}} M \to^{g} M'$ is the identity where $M' \to^{\mathrm{inc}} M$ is the inclusion morphism. Then the composition $C_1 \to^{f} M \to^{g} M' \to^{\mathrm{inc}} C_2$ has M' as image. Thus M' is in Im **C**. So Im **C** is a subcategory of mod Λ. Since Im **C** is obviously closed under finite sums, Im **C** is an additive subcategory of mod Λ. So it only remains to show that Im **C** has images.

Suppose $g\colon M_1 \to M_2$ is a morphism where $M_i \approx \operatorname{Im}(f_i\colon C_i \to C_i')$ for $i = 1, 2$ with the C_i and C_i' in add **C**. Since the image of the composition $C_1 \to M_1 \to \operatorname{Im} g \to^{\mathrm{inc}} C_2'$ is clearly $\operatorname{Im} g$ and C_1 and C_2' are in add **C**, it follows that $\operatorname{Im} g$ is in Im **C**. Thus Im **C** has images. This completes the proof of (1).

(b)–(e) Trivial.

As our final preliminary remark concerning additive subcategories **C** of mod Λ with images we point out the following.

LEMMA 4.4. *Suppose* **C** *is an additive subcategory of* mod Λ *with images. Suppose* $\{f_i\colon X_i \to C\}_{i\in I}$ *is a family of morphisms in* **C** *and* $f\colon \coprod_{i\in I} X_i \to C$ *the induced morphism. Then* $\operatorname{Im} f$ *is in* **C**.

Proof. Since C is a finitely generated module over an artin ring, C is noetherian, and so $\operatorname{Im} f$ is finitely generated. Hence there is a finite subset J of I such that $f(\coprod_{j\in J} X_j) = \operatorname{Im} f$. But $\coprod_{j\in J} X_j$ is in **C** since **C** is additive. Hence $\operatorname{Im} f = \operatorname{Im}(f \mid \coprod_{j\in J} X_j)$ is in **C** since **C** has images.

We now describe precisely when mod Λ is contravariantly or covariantly finite over an additive subcategory **C** with images.

THEOREM 4.5. *Suppose* **C** *is an additve subcategory of* mod Λ *with images.*

(a) mod Λ *is contravariantly finite over* **C** *if and only if* **C** *has a finite cocover.*

(b) mod Λ *is a covariantly finite over* **C** *if and only if* **C** *has a finite cover.*

(c) mod Λ *is functorially finite over* **C** *if and only if* **C** *has a finite cover and a finite cocover.*

Proof. (a) By Proposition 3.6 we know that **C** has a finite cocover if and only if I is contravariantly finite over **C** for all injective modules. Hence if mod Λ is contravariantly finite over **C**, then **C** has a finite cocover.

Suppose now that **C** has a finite cocover **A**. Then again by Proposition 3.6, we know that for each injective module I there is a morphism $f_I\colon A_I \to I$ with A_I in add **A** such that $(X, f)\colon (X, A_I) \to (X, I)$ is surjective for all X in **C**. Let M be a Λ-module and inc: $M \to I(M)$ its injective envelope. Since $\operatorname{Im}(X, f_{I(M)}) = (X, I(M))$ for all X in **C**, it follows that the induced morphism $g\colon f^{-1}_{I(M)}(M) \to M$ has the property that $\operatorname{Im}(X, g) = (X, M)$ for all X in **C**. Letting $\{h_j\colon X_j \to A_{I(M)}\}_{j\in J}$ be the family of all morphisms in **C** with $\operatorname{Im} h_j \subset f^{-1}_{I(M)}(M)$, we know by Lemma 4.4 that the induced morphism $h\colon \coprod X_j \to A_{I(M)}$ has the property that $\operatorname{Im} h \subset f^{-1}_{I(M)}(M)$ is contained in **C** since **C** is an additive category with images. It is now easily seen that the induced morphism $f \mid \operatorname{Im} h\colon \operatorname{Im} h \to M$ has the property $\operatorname{Im}(X, f \mid \operatorname{Im} h) = (X, M)$ for all X in **C**. Therefore M is contravariantly finite over **C**. Because this is true for arbitrary Λ-modules M, it follows that mod Λ is contravariantly finite over **C** if **C** has a finite cocover.

(b) Dual of (a).

(c) Trivial consequence of (a) and (b).

We now give various examples of additive subcategories **C** closed under images with finite covers or cocovers.

We say that a subcategory **C** *has factor modules* if whenever there is a surjection $C \to C'$ with C in **C**, then C' is in **C**. Clearly if **C** has factor modules, then **C** has images. If **C** is an arbitrary subcategory of mod Λ, then the collection of modules M such that there is a surjective $C \to M$ with C in add **C** is easily seen to be an additive subcategory of mod Λ with factor modules which we denote by Fac **C**. Obviously **C** $\subset$ Fac **C** and **C** $=$ Fac **C** if and only if **C** is an additive subcategory of mod Λ with factor modules. In particular, if M is a module we will denote Fac add M more simply by Fac M.

PROPOSITION 4.6. *Suppose* **C** *is an additive subcategory of* mod Λ *with factor modules.*

(a) **C** *has a finite cocover so* mod Λ *is contravariantly finite over* **C**. *Hence* **C** *has a preinjective partition and right almost split morphisms.*

(b) **C** *has a finite cover if and only if* **C** $=$ Fac C *for some module* C *in* **C**.

(c) mod Λ *is functorially finite over* **C** *if and only if* **C** $=$ Fac C *for some* C *in* **C**.

Proof. (a) Let I be an indecomposable injective Λ-module. Let $\{f_i\colon C_i \to I\}_{i\in J}$ be the family of all morphisms with the C_i a complete set of nonisomorphic objects in **C**. Then by Lemma 4.4, there is a finite subset J' of J such that induced morphisms $f'\colon \coprod_{j\in J'} C_j \to I$ and $f\colon \coprod_{i\in J} C_i \to I$ have the same images which denote by $\tau_{\mathbf{C}}(I)$. Then $\tau_{\mathbf{C}}(I)$ is in $\mathbf{C}$ since $\coprod_{j\in J} C_j$ is in $\mathbf{C}$ and $\mathbf{C}$ is closed under factor objects. Clearly $\tau_{\mathbf{C}}(I)$ has the property that if $g: C \to I$ is a morphism with C in $\mathbf{C}$, then $\operatorname{Im} g \subset \tau_{\mathbf{C}}(I)$.

Let $I_1, \ldots, I_n$ be a complete set of nonisomorphic indecomposable injective Λ-modules. We claim that the family $C_1 = \tau_{\mathbf{C}}(I), \ldots, C_n = \tau_{\mathbf{C}}(I_n)$ of objects in $\mathbf{C}$ has the property that if C is in $\mathbf{C}$, then there is an injective morphism $C \to \coprod_{i=1}^{n} n_i C_i$. For there is an injection $f: C \to \coprod_{i=1}^{n} n_i I_i$ and for each projection $p_j: \coprod_{i=1}^{n} n_i I_i \to I_j$ we have that $\operatorname{Im} p_j f \subset \tau_{\mathbf{C}}(I_j) = C_j$. Therefore $\operatorname{Im} f \subset \coprod_{i=1}^{n} n_i C_i$. This shows that the subcategory of ind $\mathbf{C}$ consisting of the modules isomorphic to the C_j is a finite cocover for $\mathbf{C}$. Since $\mathbf{C}$ has images and a finite cocover we have by Theorem 4.5 that mod Λ is contravariantly finite over $\mathbf{C}$. Hence the proof of (a) is complete.

(b) and (c) Trivial.

The dual to a subcategory $\mathbf{C}$ having factor modules is that $\mathbf{C}$ have submodules. We say that $\mathbf{C}$ *has submodules* if whenever there is an injection $C' \to C$ with C in $\mathbf{C}$, then C' is in $\mathbf{C}$. Clearly if $\mathbf{C}$ has submodules, then $\mathbf{C}$ has images. Also it is obvious that $\mathbf{C}$ has submodules if and only if $D(\mathbf{C})$ has factor modules.

If $\mathbf{C}$ is an arbitrary subcategory of mod Λ, then the collection of modules M such that there is an injection $M \to C$ with C in add $\mathbf{C}$ is easily seen to be an additive subcategory of mod Λ with submodules which we denote by Sub $\mathbf{C}$. Obviously $\mathbf{C} \subset$ Sub $\mathbf{C}$ and $\mathbf{C} =$ Sub $\mathbf{C}$ if and only if $\mathbf{C}$ is an additive subcategory of mod Λ with submodules. In particular, if M is a module, we will denote Sub add M more simply by Sub M.

We now state without proof the dual of Proposition 4.6.

PROPOSITION 4.7. *Suppose* $\mathbf{C}$ *is an additive subcategory of* mod Λ *with submodules.*

(a) $\mathbf{C}$ *has a finite cover so* mod Λ *is covariantly finite over* $\mathbf{C}$. *Hence* $\mathbf{C}$ *has a preprojective partition and left almost split morphisms.*

(b) $\mathbf{C}$ *has a finite cocover if and only if* $\mathbf{C} =$ sub C *for some* C *in* $\mathbf{C}$.

(c) mod Λ *is functorially finite over* $\mathbf{C}$ *if and only if* $\mathbf{C} =$ Sub C *for some* C *in* $\mathbf{C}$.

In the course of proving Propositions 4.6 *and* 4.7 *we implicitly proved the following which we now state without proof.*

PROPOSITION 4.8. *Let* $I_1, \ldots, I_n$ *and* $P_1, \ldots, P_n$ *be complete sets of nonisomorphic indecomposable injective and projective modules respectively and let* $\mathbf{C}$ *be an additive subcategory of* mod Λ.

(a) *Suppose* $\mathbf{C}$ *has factor modules.*

(i) *Each* Λ*-module* M *has a unique maximal submodule* $\tau_{\mathbf{C}}(M)$ *in* $\mathbf{C}$.

(ii) *For each Λ-module M. the inclusion $\tau_{\mathbf{C}}(M) \to M$ induces an isomorphism $(C, \tau_{\mathbf{C}}(M)) \to (C, M)$ for all C in $\mathbf{C}$.*

(iii) *$\mathbf{I}_0(\mathbf{C})$ consists of the nonzero modules isomorphic to $\tau_{\mathbf{C}}(I_j)$ for $j = 1, \ldots, n$.*

(iv) *If $C_1 \to C_2$ is an injection in $\mathbf{C}$, then $(C_2; V) \to (C_1, V)$ is surjective for all V in* add $\mathbf{I}_0(\mathbf{C})$.

(b) *Suppose $\mathbf{C}$ has submodules.*

(i) *Each Λ-module M has a unique submodule $M_{\mathbf{C}}$ minimal with respect to $M/M_{\mathbf{C}}$ being in $\mathbf{C}$.*

(ii) *The canonical morphism $M \to M/M_{\mathbf{C}}$ has the property that $(M/M_{\mathbf{C}}, C) \to (M, C)$ is an isomorphism for all C in $\mathbf{C}$.*

(iii) *$\mathbf{P}_0(\mathbf{C})$ consists of the nonzero modules isomorphic to $P_i/(P_i)_{\mathbf{C}}$ for some $i = 1, \ldots, n$.*

(iv) *If $C_1 \to C_2$ is a surjection in $\mathbf{C}$, then $(U, C_1) \to (U, C_2)$ is surjective for all U in* add $\mathbf{P}_0(\mathbf{C})$.

As a corollary to this proposition we want to point out the following.

COROLLARY 4.9. *Let $\mathbf{C}$ be an additive subcategory of* mod Λ.

(1) *Assume $\mathbf{C}$ has factor modules. Then the functor $M \to \tau_{\mathbf{C}}(M)$ is the right adjoint of the inclusion from $\mathbf{C}$ to* mod Λ.

(2) *Assume $\mathbf{C}$ has submodules. Then the functor $M \to M/M_{\mathbf{C}}$ is the left adjoint of the inclusion from $\mathbf{C}$ to* mod Λ.

Finally we point out the following connection between arbitrary categories closed under images and those closed under submodules and factor modules.

PROPOSITION 4.10. *Let $\mathbf{C}$ be an additive subcategory of* mod Λ. *Then*

(i) *$\mathbf{C}$ has images if and only if* $\mathbf{C} = \operatorname{Sub} \mathbf{C} \cap \operatorname{Fac} \mathbf{C}$.

(ii) *$\mathbf{C}$ has images and a finite cover if and only if* $\mathbf{C} = \operatorname{Sub} \mathbf{C} \cap \operatorname{Fac} M$ *for some M in* mod Λ.

(iii) *$\mathbf{C}$ has images and a finite cocover if and only if* $\mathbf{C} = \operatorname{Fac} \mathbf{C} \cap \operatorname{Sub} M$ *for some M in* mod Λ.

(iv) *$\mathbf{C}$ has images and a finite cover and cocover if and only if* $\mathbf{C} = \operatorname{Sub} M \cap \operatorname{Fac} N$ *for some M and N in* mod Λ.

Proof. The proof of this is straightforward and left to the reader.

5. PREPROJECTIVE AND PREINJECTIVE MODULES

Let, as usual, mod Λ denote the category of finitely generated modules over the artin algebra Λ. Let $\mathbf{C}$ be a subcategory of mod Λ. We recall from Section 3 that $\mathbf{C}$ is said to have a preprojective partition if there exists a partition $\mathbf{P}_i$, $i = 0, 1, \ldots, \infty$, of ind $\mathbf{C}$, the category of indecomposable modules in $\mathbf{C}$, satisfying the following properties.

(i) $\bigcup_{i=0}^{\infty} \mathbf{P}_i = \operatorname{ind} \mathbf{C}$.

(ii) $\mathbf{P}_i \cap \mathbf{P}_j = \varnothing$ when $i \neq j$.

(iii) $\mathbf{P}_i$ is finite when $i < \infty$ and is a minimal cover for $\bigcup_{j=i}^{\infty} \mathbf{P}_j$.

An indecomposable module A in $\mathbf{C}$ was defined to be preprojective in $\mathbf{C}$ if $A \in \bigcup_{i<\infty} \mathbf{P}_i$ and a module M in $\mathbf{C}$ was defined to be preprojective in $\mathbf{C}$ if $M \in \operatorname{add} \bigcup_{i<\infty} \mathbf{P}_i$. We denote by $\mathbf{P}$ the subcategory of ind $\mathbf{C}$ consisting of the indecomposable preprojective modules.

Dually, $\mathbf{C}$ is said to have a preinjective partition if there exists a partition $\mathbf{I}_i$, $i = 0, 1, 2, \ldots, \infty$, of ind $\mathbf{C}$ satisfying the following properties.

(i) $\bigcup_{i=0}^{\infty} \mathbf{I}_i = \operatorname{ind} \mathbf{C}$.

(ii) $\mathbf{I}_i \cap \mathbf{I}_j = \varnothing$ *when* $i \neq j$.

(iii) $\mathbf{I}_i$ is finite when $i < \infty$ and is a minimal cocover for $\bigcup_{j=i}^{\infty} \mathbf{I}_j$.

A module M in $\mathbf{C}$ is defined to be preinjective if M is in add $\bigcup_{i<\infty} \mathbf{I}_i$. We denote by $\mathbf{I}$ the subcategory of ind $\mathbf{C}$ consisting of the indecomposable preinjective modules.

Our main purpose in this section is to give various characterizations of the preprojective and preinjective modules as well as some structure theorems for these types of modules.

Before going on we recall the definition of the trace of a subcategory $\mathbf{C}$ in mod Λ in a module M. Let A and B be two finitely generated Λ-modules, then $\tau_B(A)$ is defined to be the submodule of A generated by $\{\operatorname{Im} f \mid f \in \operatorname{Hom}(B, A)\}$. Similarly, if A is a finitely generated Λ-module and $\mathbf{C}$ is a subcategory of mod Λ, then $\tau_{\mathbf{C}}(A)$ is defined to be the submodule of A generated by $\{\operatorname{Im} f \mid f \in \operatorname{Hom}(B, A), B \in \mathbf{C}\}$. Observe also that $\tau_{\mathbf{C}}(\ )$ is a subfunctor of the identity functor.

We now start out with the following characterization of the preprojective modules.

THEOREM 5.1. *Let* $\mathbf{C}$ *be a subcategory of* mod Λ *having a preprojective partition denoted by* $\mathbf{P}_i$, $i = 0, 1, 2, \ldots, \infty$. *The following are equivalent for an indecomposable module* A *in* $\mathbf{C}$.

(i) *A is preprojective, i.e.,* $A \in \mathbf{P} = \bigcup_{i<\infty} \mathbf{P}_i$.

(ii) *There exists an* $n < \infty$ *such that* $A \in \mathbf{P}_n$.

(iii) *There exists an* $n < \infty$ *such that if there is a nonsplittable surjective morphism* $B \to A$ *in* $\mathbf{C}$, *then* B *has a summand from* $\bigcup_{i=0}^{n-1} \mathbf{P}_i$.

(vi) *There exists a finite subcategory* $\mathbf{A}$ *of* ind $\mathbf{C}$ *such that if there is a nonsplittable surjective morphism* $B \to A$ *in* $\mathbf{C}$, *then* B *contains a summand isomorphic to a module in* $\mathbf{A}$.

(v) *There exists a simple module S and a nonzero morphism $f\colon A \to S$ such that $\tau_M(A) \subset \operatorname{Ker} f$ for all but a finite number of modules M in* ind **C**.

(vi) $\bigcup_{i<\infty} \tau_{\mathbf{P}_i}(A) \neq A$ [$\tau_{\varnothing}(A) = 0$ *by definition*].

(vii) $\tau_{\mathbf{P}_\infty}(A) \neq A$.

Proof. We want to prove the theorem by proving (i) $\Rightarrow$ (ii), (ii) $\Rightarrow$ (iii), (iii) $\Rightarrow$ (iv), (iv) $\Rightarrow$ (v), (v) $\Rightarrow$ (vi), (vi) $\Rightarrow$ (vii), and (vii) $\Rightarrow$ (i).

(i) $\Rightarrow$ (ii). This is just the definition that $\mathbf{P} = \bigcup_{i<\infty} \mathbf{P}_i$.

(ii) $\Rightarrow$ (iii). Let $B \to A$ be a nonsplittable surjective morphism in **C**. By (ii) there exists an n such that $A \in \mathbf{P}_n$. But then A is an indecomposable splitting projective module in $\bigcup_{i=n}^{\infty} \mathbf{P}_i$. Therefore, since $B \to A$ is a nonsplittable surjective morphism B is not in add $\bigcup_{i=n}^{\infty} \mathbf{P}_i$ and hence has to contain a summand isomorphic to a module in $\bigcup_{i=0}^{n-1} \mathbf{P}_i$.

(iii) $\Rightarrow$ (iv). Let $B \to A$ be a nonsplittable surjective morphism in **C**. Then by (iii), B contains a summand of $\bigcup_{i=0}^{n-1} \mathbf{P}_i$ for some $n < \infty$ depending only on A. Each $\mathbf{P}_i$, $i < \infty$, is finite so $\bigcup_{i=0}^{n-1} \mathbf{P}_i$ is finite and hence (iv) follows.

(iv) $\Rightarrow$ (v). To prove this, assume **A** is a finite subcategory of ind **C** such that for each nonsplittable surjective morphism $B \to A$ in **C**, B contains a summand isomorphic to a module in **A**. Let $A' = \tau_{\mathbf{C}'}(A)$ where $\mathbf{C}' = \operatorname{ind} \mathbf{C} - (\{A\} \cup \mathbf{A})$. We now claim that $A' \neq A$. Assume for a moment that this was proven. Let A'' be a maximal submodule of A containing A'. Then $\tau_M(A) \subseteq A' \subseteq A''$ for all M in ind **C** which are not in $\mathbf{A} \cup \{A\}$ and $\mathbf{A} \cup \{A\}$ is a finite subcategory of ind **C**. Hence $\tau_M(A) \subset \operatorname{Ker} f$ for all but a finite number of modules M in ind **C** where $f\colon A \to A/A'' = S$ is the natural surjective morphism.

This shows that if we are able to prove $A' \neq A$ we have proven (vi) $\Rightarrow$ (v). Assume to the contrary that $A' = A$. Then there exists a B in add $\mathbf{C}'$ and a surjective morphism $B \to A$ which is not a splitting surjective morphism since A is not in $\mathbf{C}'$. Hence B has to contain a summand isomorphic to module in **A**. This is a contradiction since $\mathbf{A} \cap \mathbf{C}' = \varnothing$.

(v) $\Rightarrow$ (vi) Assume there exists a simple module S and a nonzero morphism $f\colon A \to S$ such that $\tau_M(A) \subset \operatorname{Ker} f$ for all but a finite number of modules M in ind **C**. Under this hypothesis we want to prove $\bigcup_{i<\infty} \tau_{\mathbf{P}_i}(A) \neq A$.

Assume to the contrary that $\bigcap_{i<\infty} \tau_{\mathbf{P}_i}(A) = A$. Then $\tau_{\mathbf{P}_i}(A) \not\subset \operatorname{Ker} f$ for any $i < \infty$. Hence, for each i, there exists an indecomposable module B_i in $\mathbf{P}_i$ such that $\tau_{B_i}(A) \not\subset \operatorname{Ker} f$. Since the $\mathbf{P}_i$ are disjoint, the B_i are all nonisomorphic. This is contradiction and proves (v) $\Rightarrow$ (vi).

(vi) $\Rightarrow$ (vii). By assumption $\bigcap_{i<\infty} \tau_{\mathbf{P}_i}(A) \neq A$ but then $\tau_{\mathbf{P}_\infty}(A) \neq A$ since $\tau_{\mathbf{P}_\infty}(A) \subset \tau_{\mathbf{P}_i}(A)$ for all $i < \infty$.

(vii) $\Rightarrow$ (i). This is trivial.

Before we state the dual result we want to introduce the dual of the trace, called the cotrace, which is a quotient functor of the identity functor.

Let A and B be in mod Λ. Then the cotrace $\mathrm{Co}\,\tau_B(A)$ of B in A is A/A' where $A' = \bigcap\{\mathrm{Ker}\, f \mid f \in (A, B)\}$. Now this quotient is determined by A' and $A' = \bigcap\{\mathrm{Ker}\, f \mid f \in (A, B)\}$ is denoted by $\mathrm{Rej}_B(A)$, the reject of B in A. Similarly if A is in mod Λ and $\mathbf{B}$ is a subcategory of mod Λ, $\mathrm{Rej}_{\mathbf{B}}(A)$ is defined to be $\bigcap\{\mathrm{Ker}\, f \mid f \in (A, B), B \in \mathbf{B}\}$ and $\mathrm{Co}\,\tau_{\mathbf{B}}(A) = A/\mathrm{Rej}_{\mathbf{B}}(A)$. The $\mathrm{Rej}_{\mathbf{B}}(\ )$ is a subfunctor of the identity functor on mod Λ. The reject is not the dual of the trace, but rather the kernel of the map from the identity onto the cotrace and hence all statements about the cotrace may be transformed into statements about the reject. This justifies that we are using the trace and reject in dual statements even though the reject is not the formal dual of the trace. By definition, $\mathrm{Rej}_{\varnothing}(A) = A$.

We now state the dual of Theorem 5.1.

THEOREM 5.2. *Let* $\mathbf{C}$ *be a subcategory of* mod Λ *having a preinjective partition denoted by* $\mathbf{I}_i$, $i = 0, 1, 2,..., \infty$. *The following are equivalent for an indecomposable module* A *in* $\mathbf{C}$.

(i) *A is preinjective, i.e.,* $A \in \mathbf{I} = \bigcup_{i<\infty} \mathbf{I}_i$.

(ii) *There exists an* $n < \infty$ *such that* $A \in \mathbf{I}_n$.

(iii) *There exists an* $n < \infty$ *such that if there is a nonsplittable injective morphism* $A \to B$ *in* $\mathbf{C}$, *then* B *contains a summand from* $\bigcup_{i=0}^{n-1} \mathbf{I}_i$.

(iv) *There exists a finite subcategory of indecomposable modules* $\mathbf{A}$ *of* $\mathbf{C}$ *such that if there is a nonsplittable injective morphism* $A \to B$ *in* $\mathbf{C}$, *then* B *contains a summand isomorphic to a module in* $\mathbf{A}$.

(v) *There exists a simple module* S *and a nonzero morphism* $f: S \to A$ *such that* $f(S) \subset \mathrm{Rej}_B(A)$ *for all but a finite number of indecomposable modules* B *in* $\mathbf{C}$.

(vi) *The submodule* A' *of* A *generated by* $\mathrm{Rej}_{\mathbf{I}_i}(A)$, $i < \infty$ *is different from zero.*

(vii) $\mathrm{Rej}_{\mathbf{I}_\infty}(A) \neq 0$.

Proof. The proof of this theorem is dual to the proof of Theorem 5.1 and is left to the reader.

Let $\mathbf{C}$ be a subcategory of mod Λ having a preprojective partition. We have seen in Theorem 5.1 that the submodules $A_0 = \bigcap_{i<\infty} \tau_{\mathbf{P}_i}(A)$ and $A_0' = \tau_{\mathbf{P}_\infty}(A)$ determine whether an indecomposable module A in $\mathbf{C}$ is preprojective or not. Since $\tau_{\mathbf{A}}(\ )$ is a subfunctor of the identity functor on mod Λ for any subcategory $\mathbf{A}$ we know that $\tau_{\mathbf{A}}(M \coprod M') = \tau_{\mathbf{A}}(M) \coprod \tau_{\mathbf{A}}(M')$. From this observation with a similar remark for the reject the following consequence of Theorems 5.1 and 5.2 telling when an arbitrary module M has a preprojective or preinjective summand respectively.

COROLLARY 5.3. *Let* $\mathbf{C}$ *be a subcategory of* mod Λ.

(a) *Assume* $\mathbf{C}$ *has a preprojective partition and let* M *be a module in* $\mathbf{C}$. *Then the following are equivalent.*

(i) M *has a preprojective summand.*

(ii) $M_0 = \bigcap_{i<0} \tau_{\mathbf{P}_i}(M) \neq M$.

(iii) $M_0' = \tau_{\mathbf{P}_\infty}(M) \neq M$.

(b) *Assume* $\mathbf{C}$ *has a preinjective partition and let* M *be a module in* $\mathbf{C}$. *Then the following are equivalent.*

(i) M *has a preinjective summand.*

(ii) *The submodule* M' *of* M *generated by* $\mathrm{Rej}_{\mathbf{I}_i}(M)$ $i < \infty$ *is different from zero.*

(iii) $\mathrm{Rej}_{\mathbf{I}_\infty}(M) \neq 0$.

Proof. This is a direct consequence of Theorem 5.1 and Theorem 5.2, respectively.

We are now going to study the submodules $A_0 = \bigcap_{i<\infty} \tau_{\mathbf{P}_i}(A)$ and $A_0' = \tau_{\mathbf{P}_\infty}(A)$ of an indecomposable preprojective module A in a subcategory $\mathbf{C}$ of mod Λ having a preprojective partition and give different characterizations of them.

LEMMA 5.4. *Let* $\mathbf{C}$ *be a subcategory of* mod Λ *having a preprojective partition and let* A *be an indecomposable preprojective module in* $\mathbf{C}$. *Then* $A_0' = \tau_{\mathbf{P}_\infty}(A) \subset A_0 = \bigcap_{i<\infty} \tau_{\mathbf{P}_i}(A)$.

Proof. Observe that for each module M in $\mathbf{P}_\infty$ and $i < \infty$ there exists a module M_i in add $\mathbf{P}_i$ and a surjective morphism $f_i\colon M_i \to M$. Therefore $\tau_{\mathbf{P}_\infty}(A) \subset \tau_{\mathbf{P}_i}(A)$ for all $i < \infty$ which completes the proof of the lemma.

PROPOSITION 5.5. *Let* $\mathbf{C}$ *be subcategory of* mod Λ *having a preprojective partition and let* M *be a module in* $\mathbf{C}$. *Then* $M_0 = \bigcap_{i<\infty} \tau_{\mathbf{P}_i}(M)$ *is characterized by each of the following properties.*

(i) M_0 *is the unique minimal submodule of* M *containing* $\tau_B(M)$ *for all but a finite number of indecomposable modules* B *in* $\mathbf{C}$.

(ii) M_0 *is the unique minimal submodule of* M *such that for all but a finite number of indecomposable modules* B *in* $\mathbf{C}$ *there exists no nonzero morphism to* M/M_0 *factoring through the natural morphism* $M \to M/M_0$.

Proof. (i) First we observe that there is a unique minimal submodule M' such $\tau_B(M) \subset M'$ for all but a finite number of indecomposable modules in $\mathbf{C}$ since the set of submodules of M with this property is closed under intersections.

Now let $N_i = \tau_{\mathbf{P}_i}(M)$. Then $N_0 \supset N_1 \supset N_2 \supset \cdots$ is a descending chain of submodules of M, hence stops say at k $< \infty$, since M is artinian. Then $\tau_B(M) \subset N_k = M_0$ for all B in $\bigcup_{i=k}^{\infty} \mathbf{P}_i$. Hence $M_0 \supset M'$. We now want to prove that this is an equality. Assume to the contrary that this inclusion is proper. Then $\tau_{\mathbf{P}_i}(M) \not\subset M'$ for any $i < \infty$, hence there exist infinitely many B in ind $\mathbf{C}$ with $\tau_B(M) \not\subset M'$ since $\mathbf{P}_i \cap \mathbf{P}_j = \varnothing$ when $i \neq j$. This is a contradiction and hence $M' = M_0$.

(ii) This is just a reformulation from submodules to kernels of morphisms.

If we assume that the subcategory $\mathbf{C}$ of mod Λ with a preprojective partition is also closed under images, there are some other characterizations of the submodule $M_0 = \bigcap_{i<\infty} \tau_{\mathbf{P}_i}(M)$ for a module M in $\mathbf{C}$.

Proposition 5.6. *Let* $\mathbf{C}$ *be a subcategory of* mod Λ *closed under images and having a preprojective partition. Let M be a module in C. Then the following are true for* $M_0 = \bigcap_{i<\infty} \tau_{\mathbf{P}_i}(M)$.

(i) M_0 *is in* $\mathbf{C}$ *and is the unique minimal submodule of M such that* $\tau_B(M) \subset$ *for all but a finite number of modules in* ind $\mathbf{C}$.

(ii) $M_0 = \tau_{\mathbf{P}_\infty}(M)$.

(iii) M_0 *is the unique submodule of M maximal with respect to the properties that it is in* $\mathbf{C}$ *and does not contain any preprojective summands.*

Proof. (i) Since $\mathbf{C}$ is closed under images we have that M_0 is in $\mathbf{C}$. The rest is just Proposition 5.5.(i)

(ii) Since $\bigcap_{i<\infty} \tau_{\mathbf{P}_i}(M_0) = \bigcap_{i<\infty} \tau_{\mathbf{P}_i}(M) = M_0$ and $M_0 \in \mathbf{C}$, we have by Proposition 5.3 that M_0 does not contain any preprojective summands. Hence $M_0 \supset \tau_{\mathbf{P}_\infty}(M)$. The other inclusion is clear from Lemma 5.4.

(iii) Since add $\mathbf{P}_\infty$ is closed under images it is clear that there exists a unique submodule M' of M maximal with respect to the property that M' is in $\mathbf{C}$ and without preprojective summands. Now the claim that this submodule is M_0 follows trivially from (ii).

Corollary 5.7. *Let* $\mathbf{C}$ *be a subcategory of* mod Λ *having a preprojective partition, let A be a preprojective module in* $\mathbf{C}$, *and let* $A_0 = \bigcap_{i<\infty} \tau_{\mathbf{P}_i}(A)$.

(i) *Let M be a module in* $\mathbf{C}$. *Now, if there exists an* $f\colon M \to A$ *such that* $\operatorname{Im} f \not\subset A_0$, *then M contains a preprojective summand.*

(ii) $A_0 = 0$ *if and only if there is only a finite number of nonisomorphic indecomposable modules B in* $\mathbf{C}$ *such that* $\operatorname{Hom}(B, A) \neq 0$.

(iii) *If* $\mathbf{C}$ *is closed under submodules, then any submodule of A not contained in* A_0 *contains a preprojective summand. In particular, if* $x \notin A_0$, Λx, *the submodule generated by x, has a preprojective summand.*

Proof The proof is left to the reader.

We now want to collect the dual statements of Proposition 5.5, Proposition 5.6, and Corollary 5.7 in one proposition for a subcategory **C** of mod Λ having a preinjective partition.

PROPOSITION 5.8. *Let* **C** *be a subcategory of* mod Λ *having a preinjective partition and let* M *be a module in* **C**. *Let* $M^0 = \sum_{i<\infty} \mathrm{Rej}_{\mathbf{I}_i}(M)$ *be the submodule of* M *generated by all* $\mathrm{Rej}_{\mathbf{I}_i}(M)$, $i < \infty$.

(a) *Then each of the following properties characterizes* M^0.

(i) M^0 *is the unique maximal submodule of* M *contained in* $\mathrm{Rej}_B(M)$ *for all but a finite number of nonisomorphic indecomposable modules* B *in* **C**.

(ii) M^0 *is the unique maximal submodule of* M *such that for all but a finite number of nonisomorphic indecomposable modules* B *in* **C** *there exists no nonzero morphism* M^0 *to* B *factoring through the inclusion* $M^0 \to M$.

(b) *If* **C** *in addition to having a preinjective partition is closed under images we have that*

(i) $M^0 = \mathrm{Rej}_{\mathbf{I}_\infty}(M)$ *and* M/M^0 *is in* **C**.

(ii) M^0 *is the submodule of* M *minimal with respect to the property* M/M^0 *is in* **C** *and does not contain any preinjective summands*.

(c) *Let* A *be a preinjective module in* **C**.

(i) *If* B *is a module in* **C** *and there exists a morphism* $f: A \to B$ *with* $A^0 \not\subset \mathrm{Ker} f$, *then* B *contains a preinjective summand*.

(ii) $A^0 = A$ *if and only if there are only a finite number of nonisomorphic indecomposable modules* B *in* **C** *such that* $\mathrm{Hom}(A, B) \neq 0$.

(d) *If* **C** *is closed under factors and* A *is a preinjective module in* **C** *and* x *an element of* A *such that* $A^0 \not\subset \Lambda x$, *the submodule of* A *generated by* x, *then* $A/\Lambda x$ *has a preinjective summand*.

Assume now that **C** is a subcategory of mod Λ which is closed under images and having a preprojective partition. We then know by Theorem 4.5(b) that mod Λ is covariantly finite over **C** and hence **C** has left almost split morphisms. We are now able to characterize the indecomposable modules in **C** which are summands of $A_0 = \bigcap_{i<\infty} \tau_{\mathbf{P}_i}(A)$ for some preprojective module A in **C** in terms of these minimal left almost split morphisms.

PROPOSITION 5.9. *Let* **C** *be a subcategory of* mod Λ *closed under images and having a preprojective partition. Let* B *be a module in* $\mathbf{P}_\infty$ *and let* $B \to M$ *be a minimal left almost split morphism. Then* B *is a summand of* $A_0 = \bigcap_{i<\infty} \tau_{\mathbf{P}_i}(A)$ *for some preprojective module* A *if and only if* M *contains a preprojective summand*.

Proof. Observe that we may assume that A is an indecomposable preprojective module. Suppose that B in $\mathbf{P}_\infty$ is a summand of A_0 for an indecomposable preprojective module A. Then the composed morphism $B \to A_0 \to A$ is not a splittable injective morphism and can therefore be factored through the minimal left almost split morphism $B \to M$. We then get the diagram

$$\begin{array}{ccc} B & \longrightarrow & M \\ \downarrow & & \\ A_0 & \swarrow_{f} & \\ \downarrow & & \\ A & & \end{array}$$

If $\operatorname{Im} f \subset A_0$, $B \to M$ will be a splittable injective morphism since $B \to A_0$ is a splittable injective morphism. This is a contradiction, hence $\operatorname{Im} f \not\subset A_0$ and therefore by Corollary 5.7(i), M contains a preprojective summand.

For the other implication, assume B is a module in $\mathbf{P}_\infty$ and M contains a preprojective summand where $B \to M$ is a minimal left almost split morphism. Decompose M as $M' \coprod A$ where A is an indecomposable preprojective summand of M. Then the image of the composed morphism $B \to M \to A$ is in A_0. Hence, the minimal left almost split morphism $B \to M$ can be factored to give the commutative diagram

$$\begin{array}{ccc} B & \longrightarrow & M \\ \downarrow & \nearrow & \\ M' \coprod A_0 & & \end{array}$$

Since the morphism $B \to M$ is not a splittable injective morphism, the induced morphism $B \to M \to M'$ is not a splittable injective morphism either.

Now look at the induced morphism $B \to M' \coprod A_0$. If this morphism is not a splittable injective morphism we get a commutative diagram

$$\begin{array}{ccc} B & \longrightarrow & M \\ \downarrow & \swarrow & \\ M' \coprod A_0 & & \\ \downarrow & & \\ M & & \end{array}$$

But $B \to M$ a left minimal morphism; hence, this is a contradiction since $A_0 \neq A$. This shows that $B \to M' \coprod A_0$ is a splittable monomorphism. Since B is indecomposable, the induced morphism $B \to M' \coprod A_0 \to A_0$ has to be a splittable monomorphism, since the induced morphism $B \to M' \coprod A_0 \to M'$ is not. This completes the proof of the proposition.

The dual statement is as follows.

PROPOSITION 5.10. *Let* **C** *be a subcategory of* mod Λ *closed under images and having a preinjective partition. Let B be a module in* $\mathbf{I}_\infty$ *and let* $M \to B$ *be a minimal right almost split morphism in* **C**. *Then B is a summand of* A/A^0 *for a preinjective module A if and only if M contains a preinjective summand.*

As the final general result in this section we have the following.

PROPOSITION 5.11. (i) *Assume* **C** *is a subcategory of* mod Λ *having a preprojective partition. If* ind **C** *is infinite, then there is no bound on the length of the preprojective modules in* ind **C**.

(ii) *Assume C is a subcategory of* mod Λ *having a preinjective partition. If* ind **C** *is infinite, then there is no bound on the length of the preinjective modules n* ind **C**.

Proof. We will prove (i). (ii) follows by duality.

(i) Let $\mathbf{P}_i$, $i = 1, 2, \ldots, \infty$, denote the preprojective partition of **C**. Since for all n, $\mathbf{P}_n$ is a cover for $\bigcup_{i=n}^{\infty} \mathbf{P}_i$, we can get for all $n < \infty$ a chain of epimorphisms $P_0 \to P_1 \to \cdots \to P_n$ with P_i in add $\mathbf{P}_i$. This gives that for each n there exists chains of maps $P_0' \to P_1' \to P_2' \to \cdots \to P_n'$ with each $P_i' \in \mathbf{P}_i$ $i = 0, 1, \ldots, n$ and the composition of the maps nonzero. Since $\mathbf{P}_i \cap \mathbf{P}_j = \varnothing$ if $i \neq j$, the maps $P_i' \to P_{i+1}'$ are not isomorphisms. Then using the following Lemma of Harada and Sai [13] we get our desired result.

LEMMA 5.12. *Let* Λ *be any ring and* $\{M_i\}_{i=1,2,\ldots}$ *a collection of indecomposable modules of length less than or equal to m and let* $f_i \colon M_i \to M_{i+1}$ *be nonisomorphisms between these modules. Then there exists an m' such that* $f_{m'} \circ \cdots \circ f_2 \circ f_1 = 0$.

6. SUBCATEGORIES OF FINITE TYPE

Throughout this section **C** will denote a subcategory of mod Λ closed under images and having both a preprojective and a preinjective partition denoted by $\mathbf{P}_i$, $i = 0, 1, \ldots, \infty$, and $\mathbf{I}_i$, $i = 0, 1, \ldots, \infty$, respectively.

The main purpose of this section is to use the results of Section 5 to give characterizations of when **C** is of finite type in terms of the preprojective and preinjective partitions of **C**. The main result is the following.

THEOREM 6.1. *The following are equivalent for* **C**.

(i) **C** *is of finite type.*

(ii) $\mathbf{I}_\infty = \varnothing$.

(ii′) $\mathbf{P}_\infty = \varnothing$.

(iii) *All modules in* **C** *are preinjective in* **C**.

(iii′) *All modules in* **C** *are preprojective in* **C**.

(iv) *All preprojective modules in* **C** *are preinjective in* **C**.

(iv′) *All preinjective modules in* **C** *are preprojective in* **C**.

(v) *All preprojective modules in* **C** *which are quotients of some module in* $\mathbf{P}_0$ *are preinjective in* **C**.

(v′) *All preinjective modules in* **C** *which are submodules of some module in* $\mathbf{I}_0$ *are preprojective in* **C**.

Proof. The proof of the theorem will go as follows: (i) ⇒ (ii) ⇒ (iii) ⇒ (iv) ⇒ (v) ⇒ (i). The proof of the rest is then the dual cycle and is left to the reader.

(i) ⇒ (ii) is trivial since **C** is of finite type.

(ii) ⇒ (iii). This is just the definition of $\mathbf{I}_\infty$.

(iii) ⇒ (iv). This follows directly from the fact that if every indecomposable module in **C** is preinjective, then every preprojective module in **C** is preinjective.

(iv) ⇒ (v) is also a triviality.

This leaves only (v) ⇒ (i) which will be proven after some intermediate results occupying the rest of this section.

Associated with **C** we have for each $n = 0, 1, 2,..., n < \infty$ the subcategories $\mathbf{P}^n = \bigcup_{i<n} \mathbf{P}_i$ and $\mathbf{I}^n = \bigcup_{i<n} \mathbf{I}_i$, and for $n = \infty$ let $\mathbf{P}^\infty = \mathbf{P} = \bigcup_{i<\infty} \mathbf{P}_i$ and $\mathbf{I}^\infty = \mathbf{I} = \bigcup_{i<\infty} \mathbf{I}_i$. Now $\mathbf{C}_{\mathbf{P}^n}$ and $\mathbf{C}_{\mathbf{I}^n}$ denote as usual the subcategories of **C** consisting of modules without summands from $\mathbf{P}^n = \bigcup_{i<n} \mathbf{P}_i$ and $\mathbf{I}^n = \bigcup_{i<n} \mathbf{I}_i$, respectively. Since $\mathbf{P}^n = \bigcup_{i<n} \mathbf{P}_i$ and $\mathbf{I}^n = \bigcup_{i<n} \mathbf{I}_i$ are finite for all $n < \infty$, $\mathbf{C}_{\mathbf{P}^n}$ and $\mathbf{C}_{\mathbf{I}^n}$ have their own preprojective and preinjective partitions which we will denote by $\mathbf{P}_i(\mathbf{C}_{\mathbf{P}^n})$, $\mathbf{I}_i(\mathbf{C}_{\mathbf{P}^n})$, $\mathbf{P}_i(\mathbf{C}_{\mathbf{I}^n})$, and $\mathbf{I}_i(\mathbf{C}_{\mathbf{I}^n})$, $i = 0, 1, 2,..., \infty$, respectively. In order not to make the statements too complicated we will only deal with the subcategories $\mathbf{C}_{\mathbf{I}^n}$, $n < \infty$ since the results for $\mathbf{C}_{\mathbf{P}^n}$ will follow by duality.

PROPOSITION 6.2. *Let* **C** *be as before.*

(i) *Let A be any module in* **C**, *let B be a module in $\mathbf{C}_{\mathbf{I}^n}$ for some $n = 0, 1,..., \infty$, and let $f: A \to B$ be a morphism. Then* $\operatorname{Im} f \in \mathbf{C}_{\mathbf{I}^n}$.

(ii) $\mathbf{C}_{\mathbf{I}^n}$ *has a preprojective partition for $n = 0, 1,..., \infty$ and also a preinjective partition if $n < \infty$.*

Remark. In the next section we will show that $\mathbf{C}_{\mathbf{I}^\infty}$ does not have any preinjective partition if $\mathbf{C}_{\mathbf{I}^\infty}$ is nonempty.

Proof. (i) Let $f: A \to B$ be a morphism with A in $\mathbf{C}$ and B in $\mathbf{C}_{\mathbf{I}^n}$. Then $\operatorname{Im} f$ is in $\mathbf{C}$ since $\mathbf{C}$ is closed with respect to images. Assume for a moment that $\operatorname{Im} f$ contains a summand M' from $\mathbf{I}^n$. The composed morphism $M' \to \operatorname{Im} f \to B$ is a nonsplittable injective morphism since B does not contain any summand from $\mathbf{I}^n$, but then we know by Theorem 5.2(iii) that B has to contain a summand from $\mathbf{I}^n$, which gives us the desired contradiction.

(ii) If $n < \infty$, then $\mathbf{I}^n$ is a finite category and hence by Lemma 3.12, $\mathbf{I}^n$ is nilpotent. Proposition 3.13(iii) then states that $\mathbf{C}$ is functorially finite over $\mathbf{C}_{\mathbf{I}^n}$ and hence $\mathbf{C}_{\mathbf{I}^n}$ has both a preprojective and a preinjective partition.

To prove the rest of (ii) we will use Theorem 4.5(b) which says that if a subcategory of $\operatorname{mod} \Lambda$ is closed under images and has a finite cover, then $\operatorname{mod} \Lambda$ is covariantly finite over $\mathbf{C}$ and hence $\mathbf{C}$ has a preprojective partition. From part (i) of the proposition we have that $\mathbf{C}_{\mathbf{I}}$ is not only closed under images but also contains $\operatorname{Im} f$ if $f: A \to B$ is a morphism in $\operatorname{mod} \Lambda$ with $A \in \mathbf{C}$ and B in $\mathbf{C}_{\mathbf{I}}$. Now let P_0 be the sum of a complete set of nonisomorphic modules in $\mathbf{P}_0$. It is not hard to see that $\operatorname{add}(P_0/\operatorname{Rej}_{\mathbf{I}_\infty} P_0)$ is a finite cover for $\mathbf{C}_{\mathbf{I}} = \operatorname{add} \mathbf{I}_\infty$. This then together with Theorem 4.5(b) completes the proof of the proposition.

We now study the minimal covers for $\mathbf{C}_{\mathbf{I}^n}$ as n varies and find some relations between them. Before we start we make the following definition.

DEFINITION. Let $\mathbf{A}$ be a finite subcategory of $\operatorname{ind} \Lambda$. Define $l(\mathbf{A})$ to be $\sum l(A)$ where the sum is taken over a complete set of nonisomorphic objects A in $\mathbf{A}$ and $l(A)$ denotes the length of the module A.

PROPOSITION 6.3. *Let $\mathbf{C}$ be as before. Then $\mathbf{P}_0$ is a cover for $\mathbf{C}_{\mathbf{I}_0}$ if and only if $\mathbf{P}_0 \cap \mathbf{I}_0 = \varnothing$. Moreover, if $\mathbf{P}_0 \cap \mathbf{I}_0 \neq \varnothing$, then $l\mathbf{P}_0 > l(\mathbf{P}_0(\mathbf{C}_{\mathbf{I}_0}))$ and the modules in $\mathbf{P}_0(\mathbf{C}_{\mathbf{I}_0})$ are quotients of modules in $\mathbf{P}_0$.*

Proof. The proof of the first assertion in the proposition is trivial.

To prove the second one, assume $\mathbf{P}_0 \cap \mathbf{I}_0 \neq \varnothing$. Then we know that P_0, the sum of a complete set of nonisomorphic modules in $\mathbf{P}_0$, contains a summand from $\mathbf{I}_0$. Hence $A = \operatorname{Rej}_{\bigcup_{i=1}^\infty \mathbf{I}_i} P_0 \neq 0$. Further, by Proposition 6.2(i) we know that $P_0/A \in \mathbf{C}_{\mathbf{I}^0}$ since P_0/A is the image of morphism $f: P_0 \to M$ with M in $\mathbf{C}_{\mathbf{I}^0}$. Since all modules in $\mathbf{C}_{\mathbf{I}^0}$ are quotients of a direct sum of copies of P_0 and $A \subset \operatorname{Ker} g$ for all morphism $g: P_0 \to C$ with C in $\mathbf{C}_{\mathbf{I}^0}$, we have that all modules in $\mathbf{C}_{\mathbf{I}^0}$ are quotients of sums of copies of P_0/A. Therefore, $\operatorname{add}(P_0/A)$ is a finite cover for $\mathbf{C}_{\mathbf{I}^0}$ and the first claim follows from the inequalities $l(\mathbf{P}_0) = l(P_0) > l(P_0/A) \geqslant l(\mathbf{P}_0(\mathbf{C}_{\mathbf{I}^0}))$.

Since the $\operatorname{Rej}_{\mathbf{C}_{\mathbf{I}^0}}$ is a subfunctor of the identity functor, we know that $A = \operatorname{Rej}_{\mathbf{C}_{\mathbf{I}^0}}(P_0)$ is mapped to itself by all morphisms $f: P_0 \to P_0$. Therefore every indecomposable summand of P_0/A is a quotient of a module in $\mathbf{P}_0$. This finishes the proof of the proposition.

As a corollary to this proposition we have the following.

COROLLARY 6.4. *Let* $\mathbf{C}$ *be as before. Then*

(i) $l(\mathbf{P}_0(\mathbf{C}_{\mathbf{I}^n})) \geqslant l(\mathbf{P}_0(\mathbf{C}_{\mathbf{I}^{n+1}}))$ *with equality if and only if* $\mathbf{P}_0(\mathbf{C}_{\mathbf{I}^n}) = P_0(\mathbf{C}_{\mathbf{I}^{n+1}})$.

(ii) *For* $n < \infty, l(\mathbf{P}_0(\mathbf{C}_{\mathbf{I}^n})) > l(\mathbf{P}_0(\mathbf{C}_{\mathbf{I}^{n+1}}))$ *if and only if* $\mathbf{P}_0(\mathbf{C}_{\mathbf{I}^n}) \cap \mathbf{I}_n \neq \varnothing$.

(iii) *For* $n < \infty, \mathbf{P}_0(\mathbf{C}_{\mathbf{I}^n})$ *is a minimal cover for* $\mathbf{C}_{\mathbf{I}^k}$ *for all* $k \geqslant n$ *if and only if* $P_0(\mathbf{C}_{\mathbf{I}^n}) \cap I_k = \varnothing$ *for all* $n \leqslant k < \infty$.

Proof. (i), (ii), and (iii) are all trivial consequences of the proposition.

We are now in position to complete the proof of Theorem 6.1 by proving (v) implies (i), the last unproven implication. To do this, we assume all preprojective modules in $\mathbf{C}$ which are quotients of some module in $\mathbf{P}_0$ are preinjective and show that $\mathbf{C}$ is of finite type.

By Proposition 6.3 we have that the modules in $\mathbf{P}_0(\mathbf{C}_{\mathbf{I}^k})$ are all quotients of modules in $\mathbf{P}_0$ for all $k < \infty$. Further all these modules are preprojective in $\mathbf{C}$ by Theorem 5.1(iii) since $\mathbf{I}^k$ is finite when $k < \infty$ and therefore by the assumption also preinjective. This is then the same as saying that $\mathbf{P}_0(\mathbf{C}_{\mathbf{I}^n}) \cap (\bigcup_{n<j<\infty} \mathbf{I}_i) \neq \varnothing$ for all $n < \infty$, which by Corollary 6.4(iii) is the same as saying that there exists no $n < \infty$ such that $\mathbf{P}_0(\mathbf{C}_{\mathbf{I}^n})$ is a minimal cover for $\mathbf{C}_{\mathbf{I}^k}$ for all $k \geqslant n$.

This shows by Corollary 6.4(i) that there is no lower bound greater than zero for $l(\mathbf{P}_0(\mathbf{C}_{\mathbf{I}^n}))$. This then implies that $l(\mathbf{P}_0(\mathbf{C}_{\mathbf{I}^n})) = 0$ for some $n < \infty$ and hence ind $\mathbf{C} = \mathbf{I}^n$ for some $n < \infty$ which proves that $\mathbf{C}$ is of finite type. This completes the proof of the theorem.

7. THE CATEGORIES $\mathbf{P}_\infty$ AND I_∞

In Section 6 we were studying subcategories of mod Λ closed under images with preprojective and a preinjective partition and characterized those of finite type. We will also in this section assume that $\mathbf{C}$ is a subcategory of mod Λ closed under images and with a preprojective and a preinjective partition denoted by $\mathbf{P}_i$, $i = 0, 1, \ldots, \infty$, and $\mathbf{I}_i$, $i = 0, 1, \ldots, \infty$, respectively. Contrary to Section 6 we will now be mainly interested in the case where $\mathbf{C}$ is of infinite type. We will be especially interested in properties of the subcategories $\mathbf{P}_\infty$ and $\mathbf{I}_\infty$ of $\mathbf{C}$.

In Proposition 6.2(ii) we showed that $\mathbf{C}_\mathbf{I} = \operatorname{add} I_\infty$ has a preprojective partition and, by duality, that $\mathbf{C}_\mathbf{P} = \operatorname{add} \mathbf{P}_\infty$ has a preinjective partition. We will now show that $\mathbf{C}_\mathbf{I}$ does not have a preinjective partition and $\mathbf{C}_\mathbf{P}$ does not have a preprojective partition.

PROPOSITION 7.1. *Let* $\mathbf{C}$ *be as before and assume* $\mathbf{C}$ *is of infinite type. Then*

(i) $\mathbf{C_P}$ *does not contain any splitting projective modules and hence no monimal cever.*

(ii) $\mathbf{C_I}$ *does not contain any splitting injective modules and hence no minimal cocever.*

Proof. We will prove (i). (ii) follows then by duality. Since $\mathbf{C}$ is closed under images and has a preprojective and preinjective partition, mod Λ is functorially finite over $\mathbf{C}$ by Theorem 4.5 and hence $\mathbf{C}$ has both left and right almost split morphism. $\mathbf{P}_\infty$ is nonempty by Theorem 6.1 so let A be a module in $\mathbf{P}_\infty$ and look at the right almost split morphism $M \to A$ in $\mathbf{C}$. Since A is in $\mathbf{P}_\infty$ we know that for each $i = 0, 1, 2,...$ there exists a nonsplittable surjective morphism $f_i\colon P_i \to A$ with P_i in add($\mathbf{P}_i$). These morphisms factor through the right almost split morphism $M \to A$. Let $M_0 = \bigcap_{i<\infty} \tau_{\mathbf{P}_i}(M)$ which is in $\mathbf{C_P}$ by Proposition 5.6(i) since $\mathbf{C}$ is closed under images. This implies that there exists a $k < \infty$ such that $f_j\colon P_j \to A$ factors through the composed morphism $M_0 \to M \to A$ for all $j \geqslant k$. From this we get that $M_0 \to A$ is a surjective morphism in $\mathbf{C_P}$. Since the morphism $M \to A$ is not a splittable surjective morphism, it follows that $M_0 \to A$ is also not a splittable surjective morphism. This shows that A is not a splitting projective in $\mathbf{C_P}$. Since A was an arbitrary module in $\mathbf{P}_\infty =$ ind $\mathbf{C_P}$ we have shown that $\mathbf{C_P}$ does not contain any splitting projective modules. This also shows that $\mathbf{C_P}$ does not have a minimal cover since all modules in a minimal cover are splitting projectives by Theorem 3.4(a). The first part of the proposition is now proven. The second part follows by duality.

As a consequence of Proposition 7.1 we have:

COROLLARY 7.2. *Let* $\mathbf{C}$ *be as before. Then the following are equivalent.*

(i) $\mathbf{C}$ *is of infinite type.*

(ii) $\mathbf{P}_\infty \neq \varnothing$.

(ii′) $\mathbf{I}_\infty \neq \varnothing$.

(iii) $\mathbf{P}_\infty$ *is of infinite type.*

(iii′) $\mathbf{I}_\infty$ *is of infinite type.*

Let $\mathbf{C}$ be as before and of infinite type. From Proposition 6.2 we know that $\mathbf{C_P}$ has a preinjective partition and by duality $\mathbf{C_I}$ has a preprojective partition. Denote these by $\mathbf{I}_i(\mathbf{P}_\infty)$, $i = 0, 1,..., \infty$, and $\mathbf{P}_i(\mathbf{I}_\infty)$, $i = 0, 1,..., \infty$, respectively. From Proposition 7.1 we know that $\mathbf{P}_\infty$ does not contain any splitting projective modules and hence has no preprojective partition and $\mathbf{I}_\infty$ does not contain any splitting injective modules and hence has no preinjective partition. Some natural questions in this situation are

(i) If $\mathbf{I}_\infty(\mathbf{P}_\infty)$ is different from zero, does it contain any splitting projective or splitting injective modules?

(i′) If $\mathbf{P}_\infty(\mathbf{I}_\infty)$ is different from zero, does it contain any splitting injective or splitting projective modules?

(ii) What is the relation between $\mathbf{I}_\infty(\mathbf{P}_\infty)$, $\mathbf{P}_\infty(\mathbf{I}_\infty)$ and $\mathbf{P}_\infty \cap \mathbf{I}_\infty$?

We know that all preinjective modules M in $\mathbf{C}$ which are in $\mathbf{P}_\infty$ are preinjective in $\mathbf{P}_\infty$. The problem is, can there be any other preinjective modules in $\mathbf{P}_\infty$ except those which are already preinjective in $\mathbf{C}$? The next proposition deals with this problem and gives a necessary condition for this to happen.

PROPOSITION 7.3. *Let $\mathbf{C}$ be as before and of infinite type.*

(i) *If there exists a module M in $\mathbf{P}_\infty$ which is preinjective in $\mathbf{P}_\infty$ but is not preinjective in $\mathbf{C}$, then there exist infinitely many indecomposable nonisomorphic modules A in $\mathbf{C}$ which are both preprojective and preinjective in $\mathbf{C}$.*

(ii) *If there exists a module N in $\mathbf{I}_\infty$ which is preprojective in $\mathbf{I}_\infty$ but is not preprojective in $\mathbf{C}$, then there exist infinitely many indecomposable nonisomorphic modules B in $\mathbf{C}$ which are both preprojective and preinjective in $\mathbf{C}$.*

Proof. (i) Let M be a preinjective module in $\mathbf{P}_\infty$ which is not preinjective in $\mathbf{C}$. Then for each $i = 0, 1, 2,\ldots$ there exists a module B_i in add $\mathbf{I}_i$ and an injective morphism $f: M \to B_i$. B_i and B_j have no common summands if $i \neq j$, therefore only a finite number of the B_i are in $\mathbf{C}_\mathbf{P} = \text{add}\,\mathbf{P}_\infty$, i.e., all but a finite number of the B_i contains a preprojective direct summand which then is both a preprojective and a preinjective module in $\mathbf{C}$.

(ii) Dual to (i).

THEOREM 7.4. *Let $\mathbf{C}$ be as before of infinite type satisfying one of the following conditions.*

(i) *If A is a preprojective module in* ind $\mathbf{C}$ *and $A \to B$ is a minimal left almost split morphism in $\mathbf{C}$, then all summands of B are preprojective.*

(ii) *If C is a preinjective module in* ind $\mathbf{C}$ *and $B \to C$ is a minimal right almost split morphism, then all summands of B are preinjective.*

Under these hypothesis $\mathbf{P}_\infty\mathbf{I}_\infty = \mathbf{I}_\infty\mathbf{P}_\infty = \mathbf{P}_\infty \cap \mathbf{I}_\infty$.

Proof. Assume $\mathbf{C}$ satisfies (ii).

If now $\mathbf{P}_\infty\mathbf{I}_\infty \neq \mathbf{P}_\infty \cap \mathbf{I}_\infty$ we know by Proposition 7.3 that there are infinitely many preprojective modules which are preinjective. Therefore by the assumption each minimal cover of mod $\Lambda_{\mathbf{I}^n}$ contain a preinjective module. But then by Corollary 6.4 we know that $l(\mathbf{P}_0(\mathbf{C}_{\mathbf{I}^n})) = 0$ for some $n < \infty$ and hence $\mathbf{C}$ is of finite type. This is a contradiction and therefore $\mathbf{P}_\infty\mathbf{I}_\infty = \mathbf{P}_\infty \cap \mathbf{I}_\infty$ which finish the first part of the proposition.

If now $\mathbf{I}_\infty\mathbf{P}_\infty \neq \mathbf{P}_\infty \cap \mathbf{I}_\infty$ we know again by Proposition 7.3 that there are infinitely many preprojective modules which are preinjective, hence by the same argument as above $\mathbf{C}$ has to be of finite type. This is again a contradiction and hence $\mathbf{I}_\infty \mathbf{P}_\infty = \mathbf{P}_\infty \cap \mathbf{I}_\infty$.

We have now proven half of the theorem. The other half follows by duality.

The rest of this section is devoted to applying some of our results to the situation that $\Lambda = R$, a commutative local artin ring and $\mathbf{C} = \operatorname{mod} \Lambda$. In particular, we develop several criteria for R to be Gorenstein, i.e., self-injective.

To do this, we need a result from [10] about the indecomposability of quotient modules of a sum of copies of the same module M. For the convenience of the reader, we state this result here.

THEOREM 7.5. *Let R be any ring with identity, A and B be two unitary modules and Φ and ψ be two homomorphisms from A to B satisfying the following*:

(i) $\operatorname{Ext}^1(B, \phi)$: $\operatorname{Ext}^1(B, A) \to \operatorname{Ext}^1(B, B)$ *is a monomorphism.*

(ii) *If h is an idempotent endomorphism of A with $\Phi h = t\Phi$ for some endomorphism t of B, then $h = 0$ or $h = 1$.*

(iii) *If t is an endomorphism of B with $t\phi = 0$, then t is in the radical of* $\operatorname{End}(B)$.

(iv) *For any $f, g \in \operatorname{End}(B)$ and $h \in \operatorname{End}(A)$ if $f\psi = g\phi + \phi h$, then $f\psi = 0$.*

(v) *If $f \in \operatorname{End}(B)$ and $f\psi = 0$, then $f\phi = 0$.*

Let g_n: $A \coprod A \coprod \cdots \coprod A \to B \coprod B \coprod B \coprod \cdots \coprod B$ (n copies of each) be given by the matrix

$$\begin{pmatrix} \phi & 0 & 0 & \cdots & 0 & 0 \\ \psi & \phi & 0 & \cdots & 0 & 0 \\ 0 & \psi & \phi & \cdots & 0 & 0 \\ & & & & & \\ 0 & 0 & 0 & \cdots & \phi & 0 \\ 0 & 0 & 0 & \cdots & \psi & \phi \end{pmatrix}$$

Then coker g_n *is indecomposable for all $n \in N$.*

Now if R is a commutative local artinian ring and $l(\operatorname{Soc} R) \geqslant 2$, we know that there are two different simple ideals in R. Let ϕ and ψ be embeddings of $R/\mathrm{r} = S$ (r is the radical of R) into R such that $\phi(S) \cap \psi(S) = 0$. Then it is clear that ϕ and ψ satisfy the conditions in Theorem 7.5 and we can use this to construct infinitely many indecomposable modules.

THEOREM 7.6. *Let R be a commutative local ring. Then the following statements are equivalent.*

(i) *R is self-injective.*

(ii) *R is preinjective as an R-module.*

(iii) *I, the indecomposable injective R-module, is preprojective.*

Proof. If R is a commutative artinian ring, the duality D is a duality from mod R to mod R. This induces a duality between the preprojective and the preinjective R-modules. Hence (ii) is equivalent to (iii).

(i) $\Rightarrow$ (ii) is trivial.

(ii) $\Rightarrow$ (i). Assume R is preinjective in mod R. To show that R is self-injective it suffices to show that R has simple socle. Assume to the contrary that R does not have a simple socle. Then we can find morphisms ϕ and ψ from $R/\mathrm{r} = S$ to R satisfying the conditions of Theorem 7.5. So let g_n: $S \amalg S \amalg \cdots \amalg S \to \Lambda \amalg \Lambda \amalg \cdots \amalg \Lambda$ (n copies of each) be given by the matrix

$$\begin{pmatrix} \phi & 0 & \cdots & 0 & 0 \\ \psi & \phi & \cdots & 0 & 0 \\ & & & & \\ 0 & 0 & \cdots & \phi & 0 \\ 0 & 0 & \cdots & \psi & \phi \end{pmatrix}$$

Then by Theorem 7.5 coker g_n is indecomposable for all $n \in N$. Note also that the composed morphism

$$R \xrightarrow{(1,0,\ldots,0)} R \amalg R \amalg \cdots \amalg R \to \operatorname{coker} g_n$$

is a monomorphism. Hence R can be embedded in infinitely many nonisomorphic indecomposable modules which gives the desired contradiction since R was assumed to be preinjective. Therefore the socle of R is simple. But then $R \subset I$ (I the indecomposable injective R-module) and the length of them are the same, so $R = I$.

As a consequence of Theorem 7.6 we have:

COROLLARY 7.7. *Suppose R is a commutative local self-injective artinian ring. Then*

(i) r, *the radical of R, is preprojective if and only if R is Nakayama.*

(ii) *$R/\mathrm{Soc}\, R$ is preinjective if and only if R is Nakayama.*

Proof. (i) Suppose r, the radical of R, is preprojective. Then it is also a preprojective $R/\mathrm{Soc}\, R$-module since $\mathrm{Soc}\, R \cdot \mathrm{r} = 0$. But since r is an injective $R/\mathrm{Soc}\, R$-module we have by Theorem 7.6 that $R/\mathrm{Soc}\, R$ is self-injective. It is well-known result that this is equivalent to the ring being Nakayama.

The other implication is trivial since all modules in ind R are preprojective if R is of finite representation type and R is Nakayama if and only if it is of finite representation type.

(ii) This is just the dual of (i) and left to the reader.

THEOREM 7.8. *Suppose R is a commutative local artin ring and let* $\mathfrak{a}_i = \operatorname{Rej}_{\mathbf{I}_i} R$. *Then*

(i) $\mathfrak{a}_n = 0$ *for all n if R is not self-injective.*

(ii) $\mathfrak{a}_n = \operatorname{Soc} R$ *for all $n \geqslant 1$ if R is a self-injective ring and not a Nakayama ring.*

(iii) $\mathfrak{a}_n = \operatorname{Soc}^n R$ *if R is a Nakayama ring where* $\operatorname{Soc}^n R = p_{n-1}^{-1} \operatorname{Soc}(R/\operatorname{Soc}^{n-1} R)$ *with p_{n-1} the natural morphism $R \to R/\operatorname{Soc}^{n-1} R$.*

Proof. (i) Assume R is not self-injective. Then by Theorem 7.6, R as an R-module is not preinjective, and hence $\mathfrak{a}_n = \operatorname{Rej}_{\mathbf{I}_i}(R) = 0$ for all n.

(ii) Suppose R is a self-injective ring which is not Nakayama. Then by Corollary 7.7(ii), $R/\operatorname{Soc} R$ is not preinjective and therefore $\mathfrak{a}_n = \operatorname{Soc} R$ for all n.

(iii) This follows since $R/\operatorname{Soc}^n R$ is injective as $R/\operatorname{Soc}^n R$-module when R is Nakayama.

It would be nice to know if a similar result holds for arbitrary indecomposable self-injective rings.

8. THE PREPROJECTIVE AND PREINJECTIVE IRREDUCIBLE MAPPING PROPERTY

In the rest of this paper we will deal with the preprojective and preinjective partitions of mod Λ which we know exist by Theorem 3.3. We will denote the preprojective partition by $\mathbf{P}_i$, $i = 0, 1, \ldots, \infty$, and the preinjective partition by $\mathbf{I}_i$, $i = 0, 1, \ldots, \infty$, throughout the rest of the paper. Some of the earlier results will also be restated for mod Λ. Before we start we recall some constructions and notion from [6, 7].

If Λ is an artin algebra over the commutative artin ring R, we will let D denote the functor $\operatorname{Hom}(\ , I_0(R/\operatorname{rad} R))$ from mod Λ to mod Λ^{op}, where Λ^{op} is the opposite ring of Λ, I_0 stands for the injective envelope, and rad R is the radical of R. D is a duality. Another way to obtain a module in mod Λ^{op} from a module in mod Λ is by means of the following construction. Let M be a module in mod Λ and let $P_1 \xrightarrow{f} P_0 \to M \to 0$ be a minimal projective presentation of M. We now apply the functor $\operatorname{Hom}_\Lambda(\ , \Lambda)$ and we denote by $\operatorname{Tr} M$ the Cokernel of $\operatorname{Hom}_\Lambda(f, \Lambda)$. This gives a duality Tr: $\underline{\operatorname{mod}}\,\Lambda \to \underline{\operatorname{mod}}\,\Lambda^{\mathrm{op}}$, the transpose, where $\underline{\operatorname{mod}}\,\Lambda$ is the category mod Λ modulo projectives.

DEFINITION. An exact sequence $0 \to A \to B \to C \to 0$ in mod Λ is called an almost split sequence if

(i) A and C are indecomposable,

(ii) the morphism $B \to C$ is a minimal right almost split morphism.

(iii) The morphism $A \to B$ is a minimal left almost split morphism.

From [6] we have the following existence and uniqueness theorem for almost split sequences.

THEOREM 8.1. (a) *Let C be a finitely generated nonprojective indecomposable Λ-module. Then there exists a unique up to isomorphism, almost split sequence* $0 \to A \to B \to C \to 0$. *Moreover* $A \simeq D\,\mathrm{Tr}\,C$.

(b) *Let A' be a finitely generated noninjective indecomposable Λ-module. Then there exists a unique up to isomorphism almost split sequence* $0 \to A' \to B' \to C' \to 0$. *Moreover* $C' \simeq \mathrm{Tr}\,DA'$.

Another notion taken from [7] which is closely connected to minimal left and right almost split morphism is that of irreducible morphism.

DEFINITION. Let A and B be in mod Λ. A morphism $f\colon A \to B$ is called an irreducible morphism if f is neither a splittable epimorphism nor a splittable monomorphism but whenever the diagram

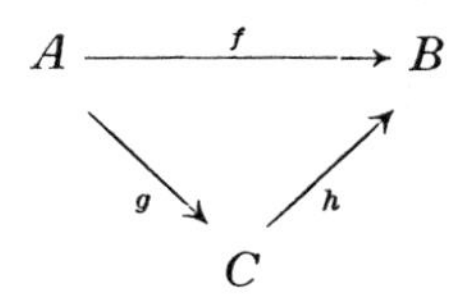

commutes either g is a splittable monomorphism or h is a splittable epimorphism.

The connection between irreducible morphisms and minimal left and right almost split morphisms is given in the next proposition taken from [7].

PROPOSITION 8.2. *Let A be an indecomposable module in* mod Λ. *A morphism $f\colon A \to C$ is irreducible if and only if there exists a morphism $f'\colon A \to C'$ such that the induced morphism $A \to C \coprod C'$ is a minimal left almost split morphism. A morphism $g\colon B \to A$ is irreducible if and only if there exists a morphisms $g'\colon B' \to A$ such that the induced morphism $B \coprod B' \to A$ is a minimal right almost split morphism.*

By using the notion of irreducible morphism we are now able to give a necessary condition on an indecomposable module A in mod Λ to be preprojective.

PROPOSITION 8.3. *Let A be an indecomposable preprojective module in* mod Λ. *Then there exist indecomposable modules $A = M_1, M_2, \ldots, M_k$ and irreducible morphisms $M_{i+1} \to M_i$, $i = 1, 2, \ldots, k-1$, with each M_i, $i = 2, \ldots, k-1$, preprojective and M_k projective.*

Note. This proposition can be strengthened to the result that the composition of the irreducible morphisms is nonzero, but we leave this out here since the proof is based on some other technics.

Proof. The proof goes by induction on i where $A \in \mathbf{P}_i$. If $A \in \mathbf{P}_0$, A is projective and the result holds with $k = 1$ and $M_1 = A$. Assume now that if $B \in \mathbf{P}_i$, $i < n$, then there exists an integer k and indecomposable modules $B = M_1, \ldots, M_k$ with irreducible morphisms $M_{j+1} \to M_j$, $j = 1, \ldots, k-1$, M_j, $j = 2, \ldots, k-1$, preprojective and M_k projective. Let A be a module in $\mathbf{P}_{n+1}$. A is not projective so the right almost split morphism $C \to A$ in $\operatorname{mod} \Lambda$ is surjective and it is not a splittable surjection, hence C contains a summand from $\mathbf{P}^{n-1} = \bigcup_{i=0}^{n-1} \mathbf{P}_i$ since A is a splitting projective module in $(\operatorname{mod} \Lambda)_{\mathbf{P}^{n-1}}$. Let B be such a preprojective module, then there exists an irreducible morphism $B \to A$ and by induction hypothesis there exist indecomposable modules $M_1', \ldots, M_k'$ and irreducible morphisms $M_{j+1}' \to M_j'$, $j = 1, \ldots, k-1$, with $M_1' = B$, M_j' preprojective for $j = 2, \ldots, k-1$, and M_k' projective. Then $M_1, \ldots, M_{k+1}$, where $M_1 = A$ and $M_j = M_{j-1}'$ for $j = k+1, \ldots, 2$, have the right properties. This completes the proof of the proposition.

We know that in many cases the converse of Proposition 8.3 also holds. However, this is not true in general. Alperin [1] has shown that this is not the case for the group algebra $F[A_5]$ where F is an algebraically closed field of characteristic 2. It would be nice to know a classification of the artin algebras where Proposition 8.1 completely describes the preprojective modules; i.e., all artin algebras with the property that if A is an indecomposable preprojective module, B is an indecomposable module, and there exists an irreducible morphism $A \to B$, then B is preprojective. We will say that an artin algebra with this property satisfies the *irreducible mapping property* for *preprojective modules*.

In their work on the representation theory for hereditary tensor algebras, Dlab and Ringel [11] defined an indecomposable module A to be preprojective if $A = \operatorname{Tr} D^n P$ for some $n \in N$ (the natural numbers) and some indecomposable projective module P. In [5] Auslander and Platzeck proved that these modules could also be described as indecomposable modules A such that $\operatorname{Hom}(B, A) \neq 0$ for only a finite number of indecomposable modules B. They also showed that this was equivalent to the existence of indecomposable modules M_i, $i = 1, \ldots, k$, with M_k projective, $M_1 = A$, and irreducible morphisms $f_i\colon M_{i+1} \to M_i$.

By now combining the result in Theorem 6.1(iv) specialized to $\operatorname{mod} \Lambda$ and the result of Auslander and Platzeck, we obtain that the modules we have defined to be preprojective in $\operatorname{mod} \Lambda$ coincide with Dlab and Ringel's definition of preprojective modules for hereditary artin algebras. Moreover, the second description of these modules by Auslander and Platzeck shows that the hereditary artin algebras satisfy the irreducible mapping property for preprojective modules.

We now state the dual result of Proposition 8.3, giving a necessary condition, on an indecomposable module A to be preinjective in $\operatorname{mod} \Lambda$.

PROPOSITION 8.4. *Let A be an indecomposable preinjective module in* $\operatorname{mod} \Lambda$. *Then there exists indecomposable modules $A = M_1, M_2, \ldots, M_k$ and irreducible* *morphisms* $M_i \to M_{i+1}$, $i = 1, 2, \ldots, k-1$, *with each* M_i, $i = 1, 2, \ldots, k-1$, *preinjective and* M_k *injective.*

Proof. The proof is dual to the proof of Proposition 8.3 and is left to the reader.

Again, the converse of Proposition 8.4 is not valid in general but holds in many cases. We will say that an artin algebra Λ satisfies the *irreducible mapping property* for *preinjective modules* if whenever $f: A \to B$ is an irreducible morphism between indecomposable modules A and B and B is preinjective, A is also preinjective. By the duality this is the same as Λ^{op} (the opposite ring of Λ) satisfies the irreducible mapping property for preprojective modules.

PROPOSITION 8.5. *Let A be an indecomposable preprojective module in* mod Λ *and let B be an indecomposable module with an irreducible morphism $A \to B$. Then B or D* Tr *B is preprojective. Moreover, if $A \in \mathbf{P}_n$ and B is not preprojective, then D* Tr *$B \in \mathbf{P}_j$ for some $j \leq n-1$.*

Proof. If B is preprojective, there is nothing to prove, so we may assume that $\mathbf{B} \in \mathbf{P}_\infty$, hence B, is not projective. Let A be in $\mathbf{P}_n$ and look at the almost split sequence

$$0 \to D \operatorname{Tr} B \to C \to B \to 0.$$

Since we have an irreducible morphism $A \to B$, A is a summand of C. Since mod Λ is closed under images and has both a preprojective and a preinjective partition, we know by Proposition 7.1 that add $\mathbf{P}_\infty = \mathbf{C_P}$ does not contain any splitting projective modules. Therefore there exists a nonsplittable surjective morphism $F \to B$ in $\mathbf{C_P}$. Since $C \to B$ is minimal right almost split, the surjective morphism $F \to B$ can be lifted to C. Hence we have a surjective morphism $D \operatorname{Tr} B \coprod F \to C$. Composing this morphism with the splittable surjective morphism $C \to A$ which represents A as a summand of C, we obtain a nonsplittable surjective morphism $D \operatorname{Tr} B \coprod F \to A$. Since A is a splitting projective module in mod $\Lambda_{\mathbf{P}^n}$, $D \operatorname{Tr} B \coprod F$ has to contain a summand from $\mathbf{P}^{n-1}$ But all summands in F are in $\mathbf{P}_\infty$; hence $D \operatorname{Tr} B$ is in $\mathbf{P}^{n-1}$ which completes the proof of the proposition.

As a corollary to this proposition we have the following characterization of the case when Λ satisfies the irreducible mapping property for preprojective modules.

COROLLARY 8.6. *For an artin algebra Λ the following are equivalent.*

(i) *Λ satisfies the irreducible mapping property for preprojective modules.*

(ii) Tr *DB is preprojective for all noninjective indecomposable preprojective modules B.*

Proof. (i) $\Rightarrow$ (ii). Assume B is an indecomposable preprojective noninjective module in mod Λ and that Λ satisfies the irreducible mapping property for preprojective modules. If $0 \to B \to C \to \operatorname{Tr} DB \to 0$ is an almost split sequence, we get by choosing any indecomposable summand C' of C a chain of irreducible morphisms $B \to C' \to \operatorname{Tr} DB$ with B', C', and $\operatorname{Tr} DB$ indecomposable. Hence C', and therefore also $\operatorname{Tr} DB$, is preprojective.

(ii) $\Rightarrow$ (i) Now assume $\operatorname{Tr} DB$ is preprojective for all indecomposable noninjective preprojective modules B and let $f: A \to C$ be an irreducible morphism with A preprojective and C indecomposable. If C is projective, there is nothing to prove, so we may assume C is not projective. Then by Proposition 8.5, C or $D \operatorname{Tr} C$ is preprojective. In particular by the hypothesis, $C = \operatorname{Tr} D(D \operatorname{Tr} C)$ is preprojective since $D \operatorname{Tr} C$ is not injective.

We now state the dual results for preinjective modules in mod Λ without proofs.

PROPOSITION 8.7. *Let A be a preinjective module and assume there is an irreducible morphism $B \to A$ with B indecomposable. Then B or* $\operatorname{Tr} DB$ *is preinjective. Moreover if $A \in \mathbf{I}_n$ and B is not preinjective, then* $\operatorname{Tr} DB \in \mathbf{I}_j$ *for some $j \leq n - 1$.*

COROLLARY 8.8. *For an artin algebra Λ the following are equivalent:*

(i) *Λ satisfies the irreducible mapping property for preinjective modules.*

(ii) *$D \operatorname{Tr} B$ is preinjective for all indecomposable nonprojective preinjective modules B.*

9. THE SUBMODULE A_0 OF A

In this section we continue our study of the preprojective and preinjective partitions of mod Λ.

In Theorem 5.1 we gave several characterizations of the preprojective modules for an arbitrary subcategory of mod Λ which, of course, holds for the preprojective modules in mod Λ itself. Keeping the same notation as in Section 5 we know that the submodule $A_0 = \bigcap_{i<\infty} \tau_{\mathbf{P}_i} A$ for an indecomposable module A plays an important role in determinating whether A is preprojective or not, and this submodule together with its dual will be studied in this section. At the end of this section we give some conditions on the preprojective modules which are equivalent to A_0 being 0.

We start out by recalling the different characterizations of the submodule A_0 of a preprojective module A in mod Λ.

THEOREM 9.1. *Let A be a preprojective module in* $\operatorname{mod}\Lambda$ *and let* $A_0 = \bigcap_{i<\infty} \tau_{\mathbf{P}_i}(A)$. *Then*

(i) $A_0 = \tau_{\mathbf{P}_\infty}(A)$ ($\tau_\varnothing(A) = 0$ *by definition*).

(ii) A_0 *is the unique submodule of A maximal among submodules A' of A without preprojective summands.*

(iii) A_0 *is the unique submodule of A minimal among submodule A' of A such that $\tau_M(A) \subset A'$ for all but a finite number of indecomposable modules M in* $\operatorname{mod}\Lambda$.

(iv) A_0 *is the unique submodule of A minimal among submodules A' of A such that for all but a finite number of modules B in* $\operatorname{ind}\Lambda$ *there is no nonzero morphism from B to A/A' factoring through the natural morphism $A \to A/A'$.*

From [4] we have the following way of calculating the length of a functor from $\operatorname{mod}\Lambda$ to Ab, the category of abelian groups.

If F is a functor from $\operatorname{mod}\Lambda$ to abelian groups, then $l(F) = \sum lF(A)$, where the sum is taken over all nonisomorphic indecomposable modules A in $\operatorname{mod}\Lambda$ with $F(A) \neq 0$ and $lF(A)$ is the length of $F(A)$ as $\operatorname{End}(A)$-module. In particular, if F is a factor of a representable functor, then $l(F) < \infty$ if and only if there are only a finite number of nonisomorphic modules A in $\operatorname{ind}\Lambda$ such that $F(A) \neq 0$. These observations enable us to reformulate parts of Proposition 9.1 as follows.

PROPOSITION 9.2. *Let A be a preprojective module in* $\operatorname{mod}\Lambda$ *and let A_0 be as usual. Then the canonical surjection $p: A \to A/A_0$ has the following properties*:

(i) $l(\operatorname{Im}(\ ,p)) < \infty$.

(ii) *If $f: A \to B$ has $l(\operatorname{Im}(\ ,f)) < \infty$, then there is a unique morphism $g: A/A_0 \to B$ such that $f = gp$.*

Proof. The proof of this is straightforward and left to the reader.

We can now write down some consequences of these propositions.

COROLLARY 9.3. *Let A be a preprojective module in* $\operatorname{mod}\Lambda$ *and let* $A_0 = \bigcap_{i<\infty} \tau_{\mathbf{P}_i}(A)$.

(i) *Let B be a submodule of A such that $B \not\subset A_0$. Then B contains a preprojective summand.*

(ii) *If $x \in A$ and $x \notin A_0$, then the submodule of A generated by x has a preprojective summand.*

(iii) $A_0 = 0$ *if and only if* $\operatorname{Hom}(B, A) \neq 0$ *for only a finite number of modules B in* $\operatorname{ind}\Lambda$.

(iv) $A_0 = 0$ *if and only if all submodules of A are preprojective.*

(v) $A_0 = 0$ *if and only if* $l((\ ,A)) < \infty$.

Proof. All these statements follows easily from the proposition.

Using Proposition 8.7 we give an analog of Proposition 5.9, characterizing the summands of the submodules A_0 of the preprojective modules A in mod Λ.

PROPOSITION 9.4. *Let B be in $\mathbf{P}_\infty$. Then B is a summand of A_0 for some preprojective module A if and only if* $\operatorname{Tr} DB$ *is preprojective.*

Proof. Let B be a noninjective module in $\mathbf{P}_\infty$ and let $0 \to B \to M \to \operatorname{Tr} DB \to 0$ be an almost split sequence. Then we know by Proposition 5.9 that B is summand of A_0 for a preprojective module A if and only if M contains a preprojective summand. But this happens if and only if there exist a preprojective module C and an irreducible morphism $C \to \operatorname{Tr} DB$. By Proposition 8.7 this happens if and only if $\operatorname{Tr} DB$ is preprojective.

We have been studying the submodule A_0 of a preprojective module A. By dualizing these results we get information about $B^0 = \sum_{i<\infty} \operatorname{Rej}_{\mathbf{I}_i}(B)$, the submodule of B generated by $\operatorname{Rej}_{\mathbf{I}_i}(B)$, $i < \infty$, for preinjective modules B.

PROPOSITION 9.5. *Let B be a preinjective module in* mod Λ *and let* $B^0 = \sum_{i<\infty} \operatorname{Rej}_{\mathbf{I}_i}(B)$. *Then*

(i) $B^0 = \operatorname{Rej}_{\mathbf{I}_\infty}(B)$ ($\operatorname{Rej}_\varnothing(B) = B$ *by definition*).

(ii) *B^0 is the unique submodule of B minimal along submodules B' of B such that B/B' does not contain any preinjective summands.*

(iii) *B^0 is the unique submodule of B maximal among submodules B' of B such that $B' \subset \operatorname{Rej}_M(B)$ for all but a finite number of indecomposable modules M in* mod Λ.

(iv) *B^0 is the unique submodule of B maximal among submodules B' of B such that for all but a finite number of modules M in* ind Λ *there are no nonzero morphisms from B' to M factoring through the natural inclusion $B' \to B$.*

PROPOSITION 9.6. *Let B be a preinjective module in* mod Λ *and let B^0 be as usual. Then the canonical inclusion $i: B^0 \to B$ has the following properties.*

(i) $l(\operatorname{Im}(i, \ )) < \infty$.

(ii) *If $f: C \to B$ is such that $l(\operatorname{Im}(f, \ )) < \infty$, then there exists a morphism $g: C \to B^0$ such that $f = ig$.*

COROLLARY 9.7. *Let B be a preinjective module in* mod Λ *and let* $B^0 = \sum_{i<\infty} \operatorname{Rej}_{\mathbf{I}_i}(B)$.

(i) *If B' is a submodule of B such that $B^0 \not\subset B'$, then B/B' contains a preinjective summand.*

(ii) *If $x \in B$ such that $B^0 \not\subset (x)$, then $B/(x)$ contains a preinjective summand where (x) denotes the submodule of B generated by x.*

(iii) $B^0 = B$ *if and only if* $\operatorname{Hom}(B, M) \neq 0$ *for only a finite number of modules* M *in* $\operatorname{ind} \Lambda$.

(iv) $B^0 = B$ *if and only if* $l((B, \ )) < \infty$.

The dual of Proposition 9.4 states:

PROPOSITION 9.8. *Let* A *be in* $\mathbf{I}_\infty$. *Then* A *is a summand of* B/B^0 *where* $B^0 = \sum_{i<\infty} \operatorname{Rej}_{\mathbf{I}_i} B$ *for a preinjective module* B *if and only if* $D \operatorname{Tr} A$ *is preinjective.*

We have been giving descriptions of the summands of A_0 for the preprojective modules A and the summands of B/B^0 for the preinjective modules B. We will now take a look at the summands of A/A_0 for the preprojective modules A and the summands of B^0 for the preinjective modules B.

Let A be a preprojective module and let $P \xrightarrow{f} A/A_0$ be a projective cover for A/A_0. We have then that all nonzero morphisms from any module M in $\operatorname{ind} \Lambda$ which factors through f factors through the natural epimorphism $A \to A/A_0$. If we denote by $P(M, A/A_0)$ the subgroup of $\operatorname{Hom}_\Lambda(M, A/A_0)$ consisting of morphisms factoring through a projective module, we get that all $f \in P(M, A/A_0)$ factor through the natural epimorphism $A \to A/A_0$. But from Proposition 9.1(iv) we have that there are only a finite number of indecomposable modules M with a nonzero morphism $M \to A/A_0$ which factor through the natural morphism $A \to A/A_0$. From this discussion we have the following proposition.

PROPOSITION 9.9. *Let* A *be a preprojective module,* A_0 *as usual, and* $P(M, N)$ *be the subgroup of* $\operatorname{Hom}_\Lambda(M, N)$ *consisting of morphisms factoring through a projective module. Then* $P(M, A/A_0)$ *is different from zero for only a finite number of modules* M *in* $\operatorname{ind} \Lambda$. *Moreover if* $P(M, A/A_0) \neq 0$, *then* M *is preprojective.*

We do not know a complete characterization of the indecomposable modules occurring as summands of A/A_0 for the preprojective modules A in $\operatorname{mod} \Lambda$, but we have a complete description of all indecomposable modules B such that $P(M, B)$ is different from zero for only a finite number of indecomposable modules M in $\operatorname{mod} \Lambda$.

PROPOSITION 9.10. *Let* M *be a module in* $\operatorname{mod} \Lambda$. *Then* $P(N, M)$ *is different from zero for only a finite number of modules* N *in* $\operatorname{ind} \Lambda$ *if and only if* $\Lambda_0 \cdot M = 0$, *where* $\Lambda_0 = \bigcap_{i<\infty} \tau_{\mathbf{P}}(\Lambda)$, i.e., $l(P(\ , M)) < \infty$ *if and only if* $\Lambda_0 \cdot M = 0$.

Proof. Assume first that $\Lambda_0 \cdot M = 0$ and let $P \xrightarrow{f} M$ be a projective cover for M. Then we know that $\Lambda_0 \cdot P = \bigcap_{i<\infty} \tau_{\mathbf{P}_i}(P) \subset \operatorname{Ker} f$. Hence, there is only a finite number of indecomposable modules N such that $\tau_N(P)$ is not in $\operatorname{Ker} f$; i.e., $P(N, M)$ is different from zero for only a finite number of modules N in $\operatorname{ind} \Lambda$. This finishes one of the implications.

For the other implication, assume $P(N, M)$ is different from zero for only a finite number of modules N in ind Λ and assume to the contrary that $\Lambda_0 \cdot M \neq 0$. This is the same as saying that there exists a morphism $f: \Lambda \to M$ such that $\Lambda_0 \not\subset \operatorname{Ker} f$. But then for each $i = 0, 1, \ldots$, there exists a module N in $\mathbf{P}_i$ and a nonzero morphism $g: N \to M$ factoring through f, i.e., $P(N, M)$ is different from zero for infinitely many modules N in ind Λ. This is a contradiction; hence $\Lambda_0 \cdot M = 0$ which completes the proof of the proposition.

COROLLARY 9.11. *$\Lambda_0 = 0$ if and only if $l(P(\ , M)) < \infty$ for all modules M in* mod Λ.

For an hereditary artin algebra Λ, $A_0 = \bigcap_{i<\infty} \tau_{\mathbf{P}_i}(A) = 0$ for all preprojective modules A in mod Λ. This shows that ind(mod(Λ/Λ_0) is by no means a complete characterization of the indecomposable modules occurring as summands of A/A_0 for preprojective modules A in mod Λ. It would be nice to know if the summands of A/A_0 for the preprojective modules A can be completely described. For instance, is it possible that these modules are the preprojective modules over Λ/Λ_0?

We now discuss when $\Lambda_0 = 0$.

Let $\underline{\text{mod}}\ \Lambda$ be the category given by the following. The objects are the same as in mod Λ and if $A, B \in \underline{\text{mod}}\ \Lambda$, then $\underline{\text{Hom}}(A, B) = \text{Hom}(A, B)/P(A, B)$.

Similarly, let $\overline{\text{mod}}\ \Lambda$ be the category given by the following: The objects are the same as in mod Λ and if $A, B \in \overline{\text{mod}}\ \Lambda$, then $\overline{\text{Hom}}(A, B) = \text{Hom}(A, B)/I(A, B)$, where $I(A, B)$ is the subgroup of $\text{Hom}(A, B)$ consisting of morphism factoring through an injective module.

From [6] we have the following.

THEOREM 9.12. *$D\,\mathrm{Tr}$: $\underline{\text{mod}}\ \Lambda \to \overline{\text{mod}}\ \Lambda$ is an equivalence with inverse* $\mathrm{Tr}\, D$: $\overline{\text{mod}}\ \Lambda \to \underline{\text{mod}}\ \Lambda$.

In what is coming up we will need the following lemma.

LEMMA 9.13. *Let $0 \to D\,\mathrm{Tr}\, C \to B \to C \to 0$ be an exact sequence and assume $l(P(\ , C)) < \infty$ and $l((\ , D\,\mathrm{Tr}\, C)) < \infty$. Then $l((\ , C))$ and $l((\ , B))$ are also finite.*

Proof. By Theorem 9.12, $D\,\mathrm{Tr}$: $\underline{\text{mod}}\ \Lambda \to \overline{\text{mod}}\ \Lambda$ is an equivalence and we have therefore that $D\,\mathrm{Tr}$: $\underline{\text{Hom}}(X, C) \simeq \overline{\text{Hom}}(D\,\mathrm{Tr}\, X, D\,\mathrm{Tr}\, C)$ for all X in ind Λ. Since $l((\ , D\,\mathrm{Tr}\, C)) < \infty$, we have that $l(\overline{\text{Hom}}(\ , D\,\mathrm{Tr}\, C)) < \infty$ and therefore also $l(\underline{\text{Hom}}(\ , C)) < \infty$. But, $0 \to P(\ , C) \to (\ , C) \to \underline{\text{Hom}}(\ , C) \to 0$ is exact so $l((\ , C)) < \infty$ since $l(P(\ , C))$ and $l(\underline{\text{Hom}}(\ , C))$ are finite. Finally, the exact sequence $0 \to (\ , D\,\mathrm{Tr}\, C) \to (\ , B) \to (\ , C)$ gives that $l((\ , B)) < \infty$.

PROPOSITION 9.14. *Let Λ be an artin algebra such that $\Lambda_0 = 0$.*

(a) *The following are equivalent for a module A in* ind Λ.

(i) *A is preprojective.*

(ii) $l((\ ,A)) < \infty$ *and if $f: B \to A$ is an irreducible morphism then* $l((\ ,B)) < \infty$.

(iii) *There exist an n and modules M_i, $i = 1,..., n$, with M_1 projective, $M_n \approx A$, and irreducible morphisms f_i: $M_i \to M_{i+1}$, $i = 1,..., n-1$.*

(b) *Λ satisfies the irreducible mapping property for preprojective modules.*

Proof. (a) That (ii) implies (i) follows directly from Theorem 5.1(iv) and that (i) implies (iii) is just Proposition 8.3. The proof of (iii) implies (ii) will be by induction on n where n is the number of modules M_i in ind Λ needed to connect $M_n \approx A$ to a projective module $P \approx M_1$ by irreducible morphisms $M_i \to M_{i+1}$. If $n = 1$, then A is a projective module; hence by assumption $l((\ ,A)) < \infty$. Further if $f: B \to A$ is irreducible, B is a summand of rA; hence $(\ ,B) \subset (\ ,A)$ and therefore $l((\ ,B)) \leqslant l((\ ,A)) < \infty$. Assume now the claim is true for $n = k$. We want to prove it for $n = k + 1$. By assumption there exists a module M_k in ind Λ and an irreducible morphism $M_k \xrightarrow{f} A$ such that $l((\ ,M_k)) < \infty$ and whenever there is an irreducible morphism $g: B \to M_k$, then $l((\ ,B)) < \infty$. We may assume that A is not projective, so look at the almost split sequence $0 \to D\operatorname{Tr} A \to C \to A \to 0$. Since there is an irreducible morphism $f: M_k \to A$, we have that M_k is a summand in C and there exists an irreducible morphism h: $D\operatorname{Tr} A \to M_k$. Therefore by induction hypothesis $l((\ ,D\operatorname{Tr} A)) < \infty$. By combining Corollary 9.11 and Lemma 9.13 we get that $l((\ ,A)) < \infty$ and $l((\ ,C)) < \infty$. If now h': $B' \to A$ is an irreducible morphism, then B' is a summand of C and hence $l((\ ,B')) < \infty$ which completes the proof that (iii) implies (ii).

(b) is a trivial consequence of (a).

THEOREM 9.15. *For an artin algebra Λ the following are equivalent*:

(i) $\Lambda_0 = 0$.

(ii) $l((\ ,\Lambda)) < \infty$.

(ii) $l(P(\ ,M)) < \infty$ *for all modules M in* mod Λ.

(iv) *A module A in* ind Λ *is preprojective if and only if* $l((\ ,A)) < \infty$.

(v) *A module A in* ind Λ *is preprojective if and only if* $A_0 = 0$.

Proof. The equivalence of (i), (ii), and (iii) is trivial by Corollary 9.11. Similarly, the equivalence of (iv) and (v) is just Corollary 9.3(v). Since in particular Λ is preprojective, (iv) implies (ii) and finally, by Proposition 9.14, we have that (ii) $\Rightarrow$ (iv).

From [2] we have the following relation on ind Λ. Let A and B be in ind Λ, then $A \sim B$ if $A \approx B$ or there exists an irreducible morphism $f: A \to B$ or an irreducible morphism $g: B \to A$. This relation generates an equivalence relation on ind Λ. Let $[A]$ denote the equivalence class of A with respect to this relation. With this notation in mind we have the following result.

PROPOSITION 9.16. *Let Λ be an artin algebra. Then the following are equivalent.*

(i) $\Lambda_0 = 0$.

(ii) *A module A in* ind Λ *is preprojective if and only if $[A] = [P]$ for some projective module P in* ind Λ.

Proof. Observe first that the "only if" part of (ii) is always satisfied by Proposition 8.3. Therefore we only have to prove the equivalence of (i) $\Lambda_0 = 0$ and (ii′) A module A in ind Λ is preprojective if $[A] = [P]$ for some projective module P in ind Λ.

We want to prove that (i) implies (ii′). So assume $\Lambda_0 = 0$ and let A be a module in ind Λ such that $[A] = [P]$ for a projective module P in ind Λ. If there are modules M_i, $i = 1,\ldots, n$, and irreducible morphisms $M_{i+1} \to M_i$ with M_n projective and $M_1 \approx A$, Proposition 9.14 gives that A is preprojective. Also, if B is preprojective and $C \to B$ is an irreducible morphism, it follows that C is preprojective since $l((\ , B)) < \infty$ and all modules X in ind Λ with $\operatorname{Hom}(X, B) \neq 0$ is preprojective. Now by an easy induction argument we get that A is preprojective which proves that (i) implies (ii).

For the other implication, assume $\Lambda_0 \neq 0$. Then there exists a projective module P in ind Λ with $P_0 \neq 0$. Let X be an indecomposable summand of P_0. Then by Proposition 9.4 we know that $\operatorname{Tr} DX$ is preprojective, but then $[X] = [\operatorname{Tr} DX] = [P']$ for some projective module P' in ind Λ. By (ii), this implies that X is preprojective which is a contradiction since all indecomposable summands of P_0 are in $\mathbf{P}_\infty$ by Theorem 9.1. This finishes the proof of the proposition.

For the sake of completeness we now state the dual of the results from Proposition 9.9 to Proposition 9.16.

PROPOSITION 9.17. *Let B be a preinjective module, $B^0 = \sum_{i<\infty} \operatorname{Rej}_{\mathbf{I}_i} B$ and $I(M, N)$ be the subgroup of* $\operatorname{Hom}_\Lambda(M, N)$ *consisting of morphisms factoring through an injective module. Then $I(B^0, M)$ is different from zero for only a finite number of modules M in* ind Λ *which, in addition, are all preinjective.*

PROPOSITION 9.18. *Let M be a module in* ind Λ. *Then $I(M, X)$ is different from zero for only a finite number of modules X in* ind Λ *if and only if $\tau_M(I) \subset I^0$ for all injective modules I in* mod Λ, *i.e., $l(I(M,\ )) < \infty$ if and only if $\tau_M(I) \subset I^0$ for all injective modules in* ind Λ.

COROLLARY 9.19. *$I^0 = I$ for all injective module I in* ind Λ *if and only if $l(I(M, -)) < \infty$ for all modules M in* mod Λ.

LEMMA 9.20. *Let* $0 \to A \to B \to \operatorname{Tr} DA \to 0$ *be an exact sequence and assume* $l(I(A, \ )) < \infty$ *and* $l(\operatorname{Tr} DA, \ )) < \infty$. *Then* $l((A, \ )) < \infty$ *and* $l((B, \ )) < \infty$.

THEOREM 9.21. *For an artin algebra* Λ, *the following are equivalent:*

(i) $I^0 = I$ *for all injective* Λ*-modules* I.

(ii) $l((I, \ )) < \infty$ *for all injective* Λ*-modules* I.

(iii) $l(I(M, \ )) < \infty$ *for all modules* M *in* mod Λ.

(iv) *A module* B *is preinjective in* ind Λ *if and only if* $l((B, \ )) < \infty$.

(v) *A module* B *is preinjective in* ind Λ *if and only if* $B^0 = B$.

PROPOSITION 9.22. *Let* Λ *be an artin algebra such that* $I^0 = I$ *for all injective modules in* mod Λ. *Then*

(i) Λ *satisfies the irreducible mapping property for preinjective modules.*

(ii) B *in* ind Λ *is preinjective if and only if* $[B] = [I]$ *for some injective module* I *in* ind Λ.

The obvious examples where $l((\ , \Lambda)) < \infty$ are the rings of finite type and the hereditary artin algebras. However, these rings are not the only artin algebras Λ satisfying the property that $l((\ , \Lambda)) < \infty$. Green has, in private correspondence, given a method of constructing many such algebras. Hereditary and artin algebras of finite type also have $l((I, \ )) < \infty$ for all injective I. Dualizing Green's construction it is possible to obtain artin algebras satisfying this condition. It would be interesting to have a description of the artin algebras Λ such that $l((\ , \Lambda)) < \infty$ and also a description of the artin algebras satisfying both $l((\ , \Lambda)) < \infty$ and $l((I, \ )) < \infty$ for I injective.

10. STRONGLY PREPROJECTIVE MODULES

We now return to the study of the submodule $A_0 = \bigcap_{i<\infty} \tau_{\mathbf{P}_i}(A)$ for a preprojective module A in ind Λ. For hereditary artin algebras and more generally for all artin algebras with $l((\ , \Lambda)) < \infty$ we know that $A_0 = 0$ for all preprojective modules A in ind Λ, but in general we only know that $A_0 \neq A$. The result we are going to prove is that if $A_0 \subset \mathrm{r}A$ for all preprojective modules A in ind Λ where r is the radical of Λ, then Λ satisfies the irreducible mapping property for preprojective modules. The dual result holds for preinjective modules and is as follows: Λ satisfies the irreducible mapping property for preinjective modules if $\operatorname{Soc} B \subset B^0 = \sum_{i<\infty} \operatorname{Rej}_{\mathbf{I}_i} B$ for all preinjective modules B in ind Λ.

DEFINITION. A module A is called strongly preprojective if $A_0 \subset \mathrm{r}A$ where r is the radical of Λ.

Note. All strongly preprojective modules are preprojective.

DEFINITION. A module B is called strongly preinjective if $B^0 \supset \operatorname{Soc} B$ is the socle of B.

Note. All strongly preinjective modules are preinjective.

We may reformulate these definitions as follows.

PROPOSITION 10.1. (a) *Let A be in* ind Λ. *Then the following are equivalent:*

(i) *A is strongly preprojective.*

(ii) *For any simple module S in* ind Λ *there exist only a finite number of modules M in* ind Λ *with a nonzero morphism $f: M \to S$ factoring through A.*

(iii) *For all simple modules S in* ind Λ *and morphisms $g: A \to S$, $l(\operatorname{Im}(\ ,g)) < \infty$*

(b) *Let B be in* ind Λ. *Then the following are equivalent.*

(i) *B is strongly preinjective.*

(ii) *For any simple module S in* ind Λ *there exist only a finite number of modules M in* ind Λ *with a nonzero morphism $S \to M$ factoring through B.*

(iii) *For all simple modules S in* ind Λ *and morphisms $g: S \to B$, $l(\operatorname{Im}(g,\ )) < \infty$.*

Proof. The proof of this is straightforward and left to the reader.

Let $\underline{\operatorname{Hom}}(M, N)$ be the quotient group $\operatorname{Hom}(M, N)/P(M, N)$ where $P(M, N)$ is the subgroup of $\operatorname{Hom}(M, N)$ consisting of morphisms factoring through a projective module. From Corollary 9.3(iii) we have that $A_0 = 0$ for a preprojective module A in mod Λ if and only if $\operatorname{Hom}(N, A) \neq 0$ for only a finite number of modules N in ind Λ. By using the quotient group $\underline{\operatorname{Hom}}(M, A)$ of $\operatorname{Hom}(M, A)$ for a preprojective module A, we give a condition for A_0 to be in $\mathrm{r}A$.

PROPOSITION 10.2. *Let A be a module in* ind Λ. *If $\underline{\operatorname{Hom}}(M, A) \neq 0$ for only a finite number of modules M in* ind Λ, *then $A_0 \subset \mathrm{r}A$ and hence A is strongly preprojective.*

Proof. Assume $\underline{\operatorname{Hom}}(M, A) \neq 0$ for only a finite number of modules M in ind Λ. Then there is an n, $0 < n < \infty$, such that every $f: B \to A$ with $B \in \mathbf{P}_m$, $n \leqslant m < \infty$, factors through a projective module, i.e., every $f: B \to A$ is equal to some hg where $g: B \to P$ and $h: P \to A$ with P projective. Therefore $\operatorname{Im} f = \operatorname{Im} hg$, $\operatorname{Im} f = \operatorname{Im} hg \subset h(\mathrm{r}P) \subset \mathrm{r}A$ since B is not projective. This shows that $\tau_{\mathbf{P}_m}(A) \subset \mathrm{r}A$ for $n \leqslant m < \infty$ and hence $A_0 \subset \mathrm{r}A$.

Another condition on a module A in ind Λ which implies that A is strongly preprojective is given in the next proposition.

PROPOSITION 10.3. *Let A be in* ind Λ. *Then $\underline{\operatorname{Hom}}(M, A) \neq 0$ for only a* *finite number of modules* M *in* $\operatorname{ind} \Lambda$ *if and only if* $\operatorname{Ext}_\Lambda^1(A, N) \neq 0$ *for only a finite number of modules* N *in* $\operatorname{ind} \Lambda$. *Hence,* A *is strongly projective if* $\operatorname{Ext}_\Lambda^1(A, N) \neq 0$ *for only a finite number of modules* N *in* $\operatorname{ind} \Lambda$.

Proof. The proof is easily obtained by using the isomorphism

$$\operatorname{Hom}_R(\underline{\operatorname{Hom}}_\Lambda(M, A), I_0(R/\operatorname{rad} R)) \approx \operatorname{Ext}_\Lambda^1(A, D\operatorname{Tr} M),$$

where R is the center of Λ and $I_0(R/\operatorname{rad} R)$ stands for the injective envelope of R modulo the radical of R. See [2, p. 19.]

From Platzeck's work on the representation theory of an artin algebra Λ stably equivalent to a hereditary artin algebra [14] we know that the following are equivalent for a module A in $\operatorname{ind} \Lambda$:

(i) $\underline{\operatorname{Hom}}(M, A) \neq 0$ for only a finite number of modules M in $\operatorname{ind} \Lambda$.

(ii) There exist modules M_i, $i = 0, 1,..., n$, in $\operatorname{mod} \Lambda$ with M_0 projective, $M_n \approx A$, and irreducible morphisms $M_i \to M_{i+1}$.

This result together with Proposition 8.3 and Proposition 10.2 then gives that $\underline{\operatorname{Hom}}(M, A) \neq 0$ for only a finite number of modules M in $\operatorname{ind} \Lambda$ is a complete description of the preprojective modules in $\operatorname{mod} \Lambda$ when Λ is stably equivalent to a hereditary artin algebra.

Let $\overline{\operatorname{Hom}}(M, N) = \operatorname{Hom}(M, N)/I(M, N)$ where $I(M\, N)$ is the subgroup of $\operatorname{Hom}(M, N)$ consisting of morphisms $f: M \to N$ that factors through an injective module. The duals of Propositions 10.2 and 10.3 now state:

PROPOSITION 10.4. *Let* B *be in* $\operatorname{ind} \Lambda$.

(i) *If* $\overline{\operatorname{Hom}}(B, N) \neq 0$ *for only a finite number of modules* N *in* $\operatorname{ind} \Lambda$, $\operatorname{Soc} B \subset B^0 = \sum_{i<\infty} \operatorname{Rej}_{\mathbf{I}_i} B$ *and hence* B *is strongly preinjective.*

(ii) $\overline{\operatorname{Hom}}(B, N) \neq 0$ *for only a finite number of modules* N *in* $\operatorname{ind} \Lambda$ *if and only if* $\operatorname{Ext}^1(M, B) \neq 0$ *for only a finite number of modules* M *in* $\operatorname{ind} \Lambda$.

For an artin algebra Λ *stably equivalent to a hereditary artin algebra we get by dualizing the result of Platzeck that the preinjective modules* B *in* $\operatorname{ind} \Lambda$ *are completely described by the property that* $\overline{\operatorname{Hom}}(B, M) \neq 0$ *for only a finite number of indecomposable modules* M *in* $\operatorname{ind} \Lambda$. *Therefore, all preinjective modules are strongly preinjective in* $\operatorname{ind} \Lambda$ *when* Λ *is stably equivalent to an hereditary artin algebra.*

Before proving that Λ satisfies the irreducible mapping property for preprojective modules if all preprojective modules are strongly preprojective, we give a couple of general results.

PROPOSITION 10.5. *Let* A *be in* $\operatorname{ind} \Lambda$. *Then*

(i) $A \in \mathbf{P}_0$ *if and only if* A *is projective in* $\operatorname{mod} \Lambda$.

(ii) *If A is not projective, then $A \in \mathbf{P}_1$ if and only if there exists an irreducible morphism $C \to A$ with C projective.*

(iii) $A \in \mathbf{P}_1$ *if and only if* $A = \operatorname{Tr} DB$ *where B is a summand of* r *as a left module.*

Proof. (i) This is just the definition of $\mathbf{P}_0$.

(ii) By Proposition 4.1 in [7] we know for a nonprojective module A in ind Λ that the following are equivalent. (a) $D \operatorname{Tr} A$ is a summand of r as a left module. (b) There exists an irreducible morphism $C \to A$ with C projective.

This shows that the statements in (ii) and (iii) are equivalent, so we need only prove (iii).

Assume first that $A \in \mathbf{P}_1$. We then know that there is an irreducible morphism $C \to A$ with C in $\mathbf{P}_0$. Since A is not projective we know that $A = \operatorname{Tr} D(D \operatorname{Tr} A)$ and there exists an irreducible morphism $D \operatorname{Tr} A \to C$ with C projective. But then we know that $D \operatorname{Tr} A$ is isomorphic to a summand in rC; i.e., $D \operatorname{Tr} A$ is isomorphic to a summand in r.

To prove the other implication assume $A = \operatorname{Tr} DB$ where B is an indecomposable summand of r. Then A is not projective and there exists a projective module P in ind Λ and an irreducible morphism $P \to A$. Now let $A' \to A$ be an epimorphism with A' in mod $\Lambda_{\mathbf{P}_0}$. Since P is projective, we know that the irreducible morphism $P \to A$ factors through the epimorphism $A' \to A$; i.e., we have obtained a commuting diagram

$$\begin{array}{ccc} P & \longrightarrow & A \\ \downarrow & \nearrow & \\ A' & & \end{array}$$

Since A' is in mod $\Lambda_{\mathbf{P}_0}$, $P \to A'$ is not a splittable injective morphism; hence $A' \to A$ is a splittable surjective morphism. This shows that A is a splitting projective module in mod $\Lambda_{\mathbf{P}_0}$, hence A is in $\mathbf{P}_1$. This completes the proof of the proposition.

In general, very little is known about which modules appear in $\mathbf{P}_n$, $n \geqslant 2$. The only case where one has a good description is in the hereditary case where Todorov [16] has given the following description of the modules in $\mathbf{P}_n$: $A \in \mathbf{P}_n$ if and only if the following two conditions hold. (a) For every irreducible morphism $f: M \to A$ with M in ind Λ, M is in $\mathbf{P}_{n-1} \cup \mathbf{P}_n$. (b) There is an M_0 in $\mathbf{P}_{n-1}$ and an irreducible morphism $f: M_0 \to A$.

Dualizing Proposition 10.5 we obtain the following.

PROPOSITION 10.6. *Let B be in* ind Λ.

(i) $B \in \mathbf{I}_0$ *if and only if B is injective.*

(ii) *$B \in \mathbf{I}_1$ if and only if B is not injective but there exists an irreducible morphism $f: B \to N$ with N injective.*

(iii) *$B \in \mathbf{I}_1$ if and only if B is not injective but* $\operatorname{Tr} DB$ *is a summand of* $N/\operatorname{Soc} N$ *for some injective module N where as usual* $\operatorname{Soc} N$, *the socle of N, is the maximal semisimple submodule of N.*

The main result in this section is a consequence of these observations and the following proposition.

PROPOSITION 10.7. *Let A be a nonprojective module in* $\operatorname{ind} \Lambda$. *If the middle term B of the almost split sequence*

$$0 \to D \operatorname{Tr} A \to B \to A \to 0$$

is strongly preprojective, then A is strongly preprojective and $D \operatorname{Tr} A$ is also strongly preprojective if $D \operatorname{Tr} A$ is not simple.

Proof. Let A be a nonprojective module in $\operatorname{ind} \Lambda$ and assume that B, the middle term of the almost split sequence $0 \to D \operatorname{Tr} A \to B \xrightarrow{f} A \to 0$, is a strongly projective module. Then there exists an n such that $A \notin \bigcup_{n<i<\infty} \mathbf{P}_i$. Therefore $\tau_{\mathbf{P}_i}(A) \subset f(\tau_{\mathbf{P}_i}(B))$ for all i, $n \leqslant i < \infty$. Hence $A_0 \subset f(\bigcap_{i<\infty} \tau_{\mathbf{P}_i}(B)) \subset f(\mathrm{r}\, B) \subset \mathrm{r}\, A$, i.e., A is a strongly preprojective module in $\operatorname{mod} \Lambda$.

For the other part of the proposition assume that $D \operatorname{Tr} A$ is not simple, but assume that $(D \operatorname{Tr} A)_0 \not\subset \mathrm{r}(D \operatorname{Tr} A)$. Then there exists a C in $\mathbf{P}_\infty$ and a simple module S in $\operatorname{mod} \Lambda$ with a nonzero morphism $C \to S$ factoring through $D \operatorname{Tr} A$.

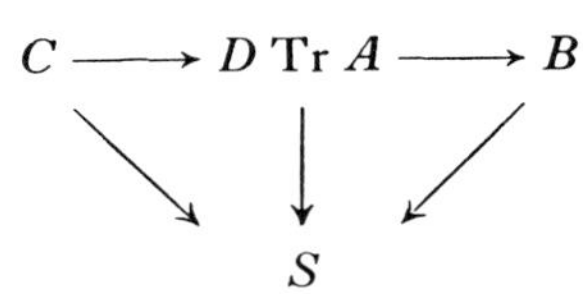

Since the morphism $D \operatorname{Tr} A \to S$ is not a splittable injective morphism, there exists a morphism $B \to S$ making the diagram commute. But then $\tau_C(B) \not\subset \mathrm{r}\, B$ which is a contradiction since $C \in \mathbf{P}_\infty$ and $\tau_{\mathbf{P}_\infty}(B) \subset \mathrm{r}\, B$. This shows that $D \operatorname{Tr} A$ is strongly preprojective if it is not simple.

We now prove the main result of this section.

THEOREM 10.8. *Let Λ be an artin algebra. If all preprojective modules in* $\operatorname{mod} \Lambda$ *are strongly preprojective, then Λ satisfies the irreducible mapping property for preprojective modules.*

Proof. We are going to prove the theorem by proving that if we have a preprojective module A with an irreducible morphism $A \to B$ with B in $\mathbf{P}_\infty$, there is a preprojective module which is not strongly preprojective. So assume there exists an $n < \infty$ with $A \in \mathbf{P}_n$ and an irreducible morphism $A \to B$ with B in $\mathbf{P}_\infty$. Choose n minimal with this property. By using Proposition 8.5 we then know that $D\,\mathrm{Tr}\,B$ is in $\mathbf{P}_j$ for some $j \leqslant n-1$, and hence the middle term A' of the almost split sequence $0 \to D\,\mathrm{Tr}\,B \to A' \to B \to 0$ is preprojective. Now if A' was strongly preprojective, B would also be strongly preprojective by Proposition 10.6. This is a contradiction, so A' is preprojective but not strongly preprojective.

We do not know if the converse of the last theorem is true or false. In any event it would be interesting to know which artin algebras have the property that all preprojective modules are strongly preprojective.

We now state the dual of the two last results.

PROPOSITION 10.9. *Let A be a noninjective module in* ind Λ. *If the middle term of the almost split sequence*

$$0 \to A \to B \to \mathrm{Tr}\,DA \to 0$$

is strongly preinjective in mod Λ, *then A is strongly preinjective and* Tr DA *is also strongly preinjective if* Tr DA *is not simple.*

THEOREM 10.10. *Let Λ be an artin algebra. If all preinjective modules in* mod Λ *are strongly preinjective, then Λ satisfies the irreducible mapping property for preinjective modules.*

REFERENCES

1. J. L. ALPERIN, A preprojective module which is not strongly preprojective, *Comm. Algebra*, in press.
2. M. AUSLANDER Applications of morphisms determinated by modules, *in* "Representation Theory of Algebras," Lecture Notes in Pure and Applied Mathematics, Vol. 37, Dekker, New York, 1978.
3. M. AUSLANDER, Preprojective modules over artin algebras, *in* "Ring Theory: Proceedings of the 1978 Antwerp Conference," Dekker, New York, 1979.
4. M. AUSLANDER, Representation theory of artin algebras, II, *Comm. Algebras* **2** (1974), 269–310.
5. M. AUSLANDER AND M. I. PLATZECK, Representation theory of hereditary algebras, *in* "Representation theory of algebras," Lecture. Notes in Pure and Applied Mathematics, Vol. 37, Dekker, New York, 1978.
6. M. AUSLANDER AND I. REITEN, Represntation theory of artin algebras III. Almost split sequences, *Comm. Algebra* **3** (1975), 239–294.
7. M. AUSLANDER AND I. REITEN, Representation theory of artin algebras. IV. Invariants given by almost split sequences, *Comm. Algebra* **5** (1977), 441–518.
8. M. AUSLANDER AND S. O. SMALØ, Almost split sequences in subcategories, *J. Algebra*, in press.

9. R. BAUTISTA AND R. MARTINEZ, Representations of partially ordered sets and 1-Gorenstein algebras, "Ring Theory: Proceedings of the 1978 Antwerp Conference," Dekker, New York, 1979.
10. S. E. DICKSON AND G. M. KELLY, Interlacing methods and large indecomposables, *Bull. Austral. Math. Soc.* **3** (1970), 257–300.
11. V. DLAB AND C. M. RINGEL, The representation of tame hereditary artin algebra, *in* "Representation Theory of Algebras," Lecture Notes in Pure and Applied Mathematics, Vol. 37, Dekker, New York, 1978.
12. P. GABRIELL, "Indecomposable Representation II," Symposia Mathematica, Vol. XI, pp. 81–104, Academic Press, London/New York, 1973.
13. H. HARADA AND Y. SAI, On Categories of indecomposable Modules, I, *Osaka J. Math.* **7** (1970), 323–344.
14. M. I. PLATZECK, Representation theory of algebras stably equivalent to an hereditary artin algebra, *Trans. Amer. Math. Soc.* **238** (1978), 89–128.
15. A. V. ROITER, Unboundness of the dimension of the indecomposable representations of an algebra which has infinitely many indecomposable representations, *Izv. Akad. Nauk SSSR. Ser. Mat.* **32** (1968), 1275–1282.
16. G. TODOROV, Almost split sequences in the representation theory of certain classes of artin algebras, Thesis, Brandeis University, 1978.

Printed by the St. Catherine Press Ltd., Tempelhof 37, Bruges, Belgium

Reprinted from JOURNAL OF ALGEBRA
All Rights Reserved by Academic Press, New York and London
Vol. 69, No. 2, April 1981
Printed in Belgium

Almost Split Sequences in Subcategories

M. AUSLANDER* AND SVERRE O. SMALØ†

Brandeis University, Waltham, Massachusetts, 02154 and University of Trondheim, NLHT, Trondheim, Norway

Communicated by Walter Feit

Received June 9, 1980

INTRODUCTION

Let R be a commutative artin ring and let Λ be an R-algebra which is a finitely generated R-module. In [5] Auslander and Reiten introduced the notion of an almost split sequence in mod Λ, the category of finitely generated modules over Λ, and the existence and uniqueness of such short exact sequences were established there. The theory of almost split sequences developed further, and similar results for certain subcategories of mod Λ were proved by Bautista and Martinez [8] and Roggenkamp [9].

The main purpose of this paper is to develop a more general theory for subcategories of mod Λ having almost split sequences, which has the previous examples as special cases.

Notions closely related to that of almost split sequences are those of minimal left and right almost split morphisms. These notions were introduced in [6] for mod Λ and further developed in [7] for subcategories of mod Λ. This paper is based upon [7] and we therefore use the same notations and conventions as used there.

We now give some definitions and then proceed to describe the content of the paper section by section. Let $\mathbf{C}$ be a subcategory of mod Λ. Then a morphism $g: B \to C$ in $\mathbf{C}$ is said to be a right almost split morphism in $\mathbf{C}$ if (i) g is not a splittable epimorphism and (ii) whenever there is a nonsplittable epimorphism $h: C' \to C$ in $\mathbf{C}$, there exists an $h': C' \to B$ such that $gh' = h$. Dually, a morphism $f: A \to B$ in $\mathbf{C}$ is said to be left almost split if (i) f is not a splittable monomorphism and (ii) whenever there is a nonsplittable monomorphism $h: A \to A'$ there exists an $h': B \to A'$ such that $h'f = h$. We say that $\mathbf{C}$ has right almost split morphisms if for all indecomposable objects

* Written while a Guggenheim Fellow with the partial support of NSF MCS 77 04 951.
† Supported by the Norwegian Research Council.

0021-8693/81/040426–29$02.00/0

C in **C** there is a B in **C** and an $f: B \to C$ which is right almost split in **C**. **C** having left almost split morphisms is defined dually. Finally we say that **C** has almost split morphisms if **C** has both left and right almost split morphisms. Subcategories **C** of mod Λ having right almost split morphisms, left almost split morphisms and almost split morphisms were studied in [7] and some sufficient conditions for this to happen were given there.

Now assume **C** is closed with respect to extensions; i.e., if $0 \to A \to B \to C \to 0$ is exact in mod Λ with A and C in **C**, then B is also in **C**. A module C in **C** is said to be Ext-projective in **C** if $\operatorname{Ext}^1_\Lambda(C, X) = 0$ for all X in **C** and a module B in **C** is said to be Ext-injective in **C** if $\operatorname{Ext}^1_\Lambda(X, B) = 0$ for all X in **C**. We say that **C** has almost split sequences if

(i) **C** has almost split morphisms.

(ii) For each indecomposable non-Ext-projective C in **C** there exists an exact sequence $0 \to A \to^f B \to^g C \to 0$ in **C** with f a left almost split morphism in **C** and g a right almost split morphism in **C**.

(iii) For each indecomposable non-Ext-injective module A in **C** there exists an exact sequence $0 \to A \to^f B \to^g C \to 0$ in **C** with f a left almost split morphism in **C** and g a right almost split morphism in **C**.

In Section 1 the basic existence and uniqueness theorem for almost split sequences in subcategories **C** of mod Λ which are closed under extensions is given. The criterion developed in Section 1 for the existence of almost split sequences is not easy to verify, so in Section 2, by restricting the class of subcategories considered, we get an equivalent condition, which is easier to verify. Section 3 is devoted to subcategories **C** of mod Λ which are closed either under submodules or under factormodules. In the case **C** is closed under submodules, the Ext-projective modules are determined and dually, if **C** is closed under factor modules the Ext-injective modules are determined. The study of the Ext-projective and Ext-injective modules continues in Section 4. Before we state the main result of that section we need one definition. A subcategory **C** of mod Λ is said to have a finite cocover if there exists a module C in **C** such that all modules X in **C** can be embedded in a direct sum of copies of C. We are now able to state the theorem. If **C** is closed under extensions and submodules and has a finite cocover, then there are only a finite number of indecomposable nonisomorphic Ext-projective modules and Ext-injective modules in **C**. We also describe these modules explicitly. The dual results are also stated.

So far we have assumed that **C** is closed under extensions. Let M be a module in mod Λ. Denote by Sub M the subcategory of mod Λ consisting of all objects which are submodules of finite direct sums of copies of M. The subcategory Fac M of mod Λ is defined dually. In Section 5 amongst other things we have the following result giving sufficient conditions for Sub M

and Fac M for an M in mod Λ to be closed under extensions. Let M and N be in mod Λ. Then (i) Sub M is closed under extensions if $\mathrm{Hom}(\mathrm{Tr}\, DM, M) = 0$. (ii) Fac N is closed under extensions if $\mathrm{Hom}(N, D\,\mathrm{Tr}\, N) = 0$. (iii) Sub $M \cap$ Fac N is closed under extensions if $\mathrm{Hom}(\mathrm{Tr}\, DM, M) = 0 = \mathrm{Hom}(N, D\,\mathrm{Tr}\, N)$.

In Section 6 we apply the theory developed in Sections 1 through 5 of subcategories of mod Λ having almost split sequences. For example, we show that if S is a simple Λ-module such that $\mathrm{Ext}^1(S, S) = 0$, then Sub(D Tr S) and Fac(Tr DS) has almost split sequences. Further, if I is an injective in mod Λ and P is a projective in mod Λ then it is proven that Sub I and Fac P have almost split sequences. In these cases also the Ext-injective and Ext-projective modules are determined in fully.

In Section 7 we show that the full subcategory of mod Λ consisting of the objects C such that $\mathrm{Ext}^1(C, \Lambda) = 0$ has almost split sequences.

1. Basic Existence Theorem

We assume throughout this paper that Λ is an artin R-algebra with R a commutative artin ring. Following the definitions and notation of [7], a subcategory $\mathbf{C}$ of mod Λ, the category of finitely generated Λ-modules, is always a full subcategory of mod Λ closed under isomorphisms and nonzero summands. By an exact sequence in $\mathbf{C}$ we mean an exact sequence of Λ-modules $\cdots \to C_{i-1} \to C_i \to C_{i+1}$ with the nonzero C_i in $\mathbf{C}$. An object C in $\mathbf{C}$ is said to be Ext-projective if each exact sequence $0 \to A \to B \to C \to 0$ in $\mathbf{C}$ splits. Dually an object A in $\mathbf{C}$ is said to be Ext-injective if each exact sequence $0 \to A \to B \to C \to 0$ in $\mathbf{C}$ splits.

Next we recall some of the definitions and results given in [7] concerning right and left almost split morphisms in $\mathbf{C}$ as well as almost split sequences in $\mathbf{C}$.

A morphism $f\colon B \to C$ is right almost split in $\mathbf{C}$ if f is not a splittable epimorphism and any morphism $g\colon X \to C$ in $\mathbf{C}$ which is not a splittable epimorphism can be lifted to B; i.e., there is an $h\colon X \to B$ such that $hg = f$. We recall that if f if right almost split then C is indecomposable. Dually, a morphism $g\colon A \to B$ in $\mathbf{C}$ is said to be left almost split in $\mathbf{C}$ if g is not a splittable monomorphism and any morphism $h\colon A \to Y$ in $\mathbf{C}$ which is not a splittable monomorphism can be extended to B; i.e., there is a $j\colon B \to Y$ such that $h = jg$. Finally, an almost split sequence in $\mathbf{C}$ is an exact sequence $0 \to A \to^g B \to^f C \to 0$ in $\mathbf{C}$ such that g is a left almost split morphism and f is a right almost split morphism. We recall the uniqueness properties of almost split sequences in $\mathbf{C}$. The following statements are equivalent for two almost split sequences $0 \to A \to B \to C \to 0$ and $0 \to A' \to B' \to C' \to 0$ in $\mathbf{C}$: (a) The sequences are isomorphic, (b) $A \approx A'$ and (c) $C \approx C'$.

We say that a subcategory $\mathbf{C}$ of $\operatorname{mod} \Lambda$ has almost split sequences if it satisfies the following conditions:

(a) If C in $\mathbf{C}$ is indecomposable, then there is a right almost split morphism $f: B \to C$ in $\mathbf{C}$ and there is a left almost split morphism $g: C \to B$ in $\mathbf{C}$.

(b) If A is a indecomposable non-Ext-injective in $\mathbf{C}$, then there is an almost split sequence $0 \to A \to B \to C \to 0$ in $\mathbf{C}$.

(c) If C is an indecomposable non-Ext-projective in $\mathbf{C}$, then there is an almost split sequence $0 \to A \to B \to C \to 0$ in $\mathbf{C}$.

Our main purpose in this section is to give a sufficient condition for a subcategory $\mathbf{C}$ of $\operatorname{mod} \Lambda$ to have almost split sequences. This result serves as our basic existence theorem in the rest of the paper. However, before stating and proving this result, we recall some facts about dualizing R-varieties, a notion introduced in [4].

Let $I = I(R/\mathbf{r})$, the injective envelope over R of $R/\mathbf{r}$ where $\mathbf{r}$ is the radical of R. Suppose F: $\operatorname{mod} \Lambda \to Ab$ is an additive functor. Then for each C in $\operatorname{mod} \Lambda$ the abelian group $F(C)$ has a natural R-module structure. Define DF: $(\operatorname{mod} \Lambda)^{\mathrm{op}} \to Ab$ by $(DF)(C) = \operatorname{Hom}_R(F(C), I)$. We then obtain the contravariant functor D: $(\operatorname{mod} \Lambda, Ab) \to ((\operatorname{mod} \Lambda)^{\mathrm{op}}, Ab)$ where $(\operatorname{mod} \Lambda, Ab)$ and $((\operatorname{mod} \Lambda)^{\mathrm{op}}, Ab)$ are the categories of covariant and contravariant additive functors from $\operatorname{mod} \Lambda$ to Ab, the category of abelian group. Similarly we have the contravariant functor D: $((\operatorname{mod} \Lambda)^{\mathrm{op}}, Ab) \to (\operatorname{mod} \Lambda, Ab)$.

Next we recall that a functor F: $\operatorname{mod} \Lambda \to Ab$ is said to be finitely presented if there is an exact sequence of functors $(C_1, \ ) \to (C_2, \ ) \to F \to 0$ where $(C_i, \ ) = \operatorname{Hom}(C_i, \ )$ in $\operatorname{mod} \Lambda$ for $i = 1, 2$. Similarly a functor G: $(\operatorname{mod} \Lambda)^{\mathrm{op}} \to Ab$ is said to be finitely presented if there is an exact sequence of functors $(\ , C_1) \to (\ , C_2) \to G \to 0$. The full subcategories f.p.$(\operatorname{mod} \Lambda, Ab)$ and f.p.$((\operatorname{mod} \Lambda)^{\mathrm{op}}, Ab)$ of $(\operatorname{mod} \Lambda, Ab)$ and $((\operatorname{mod} \Lambda)^{\mathrm{op}}, Ab)$ respectively consisting of the finitely presented functors have the following basic properties:

(a) If $0 \to F_1 \to F_2 \to F_3 \to F_4 \to 0$ is an exact sequence with F_2 and F_3 finitely presented, then F_1 and F_4 are also finitely presented.

(b) If $0 \to F_1 \to F_2 \to F_3 \to 0$ is an exact sequence of functors with F_1 and F_3 finitely presented, then F_2 is finitely presented.

(c) A functor F is finitely presented if and only if DF is finitely presented.

(d) The induced contravariant functors D: f.p.$(\operatorname{mod} \Lambda, Ab) \to$ f.p.$((\operatorname{mod} \Lambda)^{\mathrm{op}}, Ab)$ and D: f.p.$((\operatorname{mod} \Lambda)^{\mathrm{op}}, Ab) \to$ f.p.$(\operatorname{mod} \Lambda, Ab)$ are dualities which are dual inverses.

Suppose now that $\mathbf{C}$ is an additive subcategory of $\operatorname{mod} \Lambda$; i.e., $\mathbf{C} = \operatorname{add} \mathbf{C}$, the subcategory of $\operatorname{mod} \Lambda$ consisting of all finite sums of objects in $\mathbf{C}$. Then the contravariant functors D: $(\operatorname{mod} \Lambda, Ab) \to ((\operatorname{mod} \Lambda)^{\text{op}}, Ab)$ and D: $((\operatorname{mod} \Lambda)^{\text{op}}, Ab) \to (\operatorname{mod} \Lambda, Ab)$ induce contravariant functors D: $(\mathbf{C}, Ab) \to (\mathbf{C}^{\text{op}}, Ab)$ and D: $(\mathbf{C}^{\text{op}}, Ab) \to (\mathbf{C}, Ab)$ in an obvious way. We say that $\mathbf{C}$ is a dualizing R-subvariety of $\operatorname{mod} \Lambda$ if F: $\mathbf{C} \to Ab$ is finitely presented in $(\mathbf{C}, Ab)$ if and only if DF: $\mathbf{C}^{\text{op}} \to Ab$ is finitely presented in $(\mathbf{C}^{\text{op}}, Ab)$ and G: $\mathbf{C}^{\text{op}} \to Ab$ is finitely presented in $(\mathbf{C}^{\text{op}}, Ab)$ if and only in DG: $\mathbf{C} \to Ab$ is finitely presented. If $\mathbf{C}$ is a dualizing R-subvariety of $\operatorname{mod} \Lambda$, then D: $\text{f.p.}(\mathbf{C}, Ab) \to \text{f.p.}(\mathbf{C}^{\text{op}}, Ab)$ and D: $\text{f.p.}(\mathbf{C}^{\text{op}}, Ab) \to \text{f.p.}(\mathbf{C}, Ab)$ are dualities which are dual inverses. In other words, $\mathbf{C}$ is a dualizing R-subvariety of $\operatorname{mod} \Lambda$ if and only if $\mathbf{C}$ is a dualizing R-variety in the sense of [4]. The reader is referred to [4] for details concerning the basic properties of dualizing R-varieties we use in this paper.

We now state and prove our main existence theorem.

THEOREM 1.1. *If* $\mathbf{C}$ *is a dualizing R-subvariety of* $\operatorname{mod} \Lambda$ *closed under extensions, then* $\mathbf{C}$ *has almost split sequences.*

Proof. Since $\mathbf{C}$ is a dualizing R-variety, then we know by [4, Proposition 3.2], that all simple functors in $(\mathbf{C}, Ab)$ and $(\mathbf{C}^{\text{op}}, Ab)$ are finitely presented. In other words, if C is an indecomposable object in $\mathbf{C}$, then there is a right almost split morphism $B \to C$ in $\mathbf{C}$ and a left almost split morphism $C \to B'$ in $\mathbf{C}$.

Suppose A in $\mathbf{C}$ is an indecomposable non-Ext-injective object. Then there is a nonsplit exact sequence $0 \to A \to^{u} B' \to^{v} C' \to 0$ in $\mathbf{C}$ which induces the exact sequence of functors $0 \to (\ , A) \to^{(\ ,u)} (\ , B') \to^{(\ ,v)} (\ , C') \to F \to 0$ when $F = \operatorname{Coker}(\ , v)$. Since the exact sequence $0 \to A \to^{u} B' \to^{v} C' \to 0$ does not split F is not zero. Therefore we know by [4, p. 324] that $\mathbf{C}$ being a dualizing R-variety implies that F contains a simple subfunctor S.

Let C be the uniquely determined indecomposable in $\mathbf{C}$ such that $S(C) \neq 0$. Therefore there is a nonzero morphism $(\ , C) \to^{t} S$ which is an epimorphism since S is simple. Because $(\ , C)$ is projective in $(\mathbf{C}^{\text{op}}, Ab)$ we know there is a morphism h: $C \to C'$ such that the exact diagram

$$\begin{array}{ccccc} & & 0 & & \\ & & \downarrow & & \\ (\ , C) & \xrightarrow{t} & S & & \\ \downarrow{\scriptstyle (\ ,h)} & & \downarrow & & \\ (\ , C') & \longrightarrow & F & \longrightarrow & 0 \end{array}$$

commutes where $S \to F$ is the inclusion of S into F. Let

$$\begin{array}{ccccccccc} 0 & \longrightarrow & A & \xrightarrow{g} & B & \xrightarrow{f} & C & \longrightarrow & 0 \\ & & \| & & \downarrow & & \downarrow{\scriptstyle h} & & \\ 0 & \longrightarrow & A & \xrightarrow{u} & B' & \xrightarrow{v} & C' & \longrightarrow & 0 \end{array} \qquad (*)$$

be the pullback diagram in $\bmod \Lambda$. Since $\mathbf{C}$ is closed under extensions, B is in $\mathbf{C}$. Furthermore it is straightforward to check that $(*)$ induces the exact commutative diagram

$$\begin{array}{ccccccccccc} & & & & & & & & 0 & & \\ & & & & & & & & \downarrow & & \\ 0 & \longrightarrow & (\ ,A) & \longrightarrow & (\ ,B) & \longrightarrow & (\ ,C) & \xrightarrow{t} & S & \longrightarrow & 0 \\ & & \| & & \downarrow & & \downarrow{\scriptstyle (\ ,h)} & & \downarrow{\scriptstyle \text{inc}} & & \\ 0 & \longrightarrow & (\ ,A) & \longrightarrow & (\ ,B') & \longrightarrow & (\ ,C') & \longrightarrow & F & \longrightarrow & 0 \end{array}$$

Since S is simple and $\operatorname{End} A$ is a local ring, the fact that $\mathbf{C}$ is closed under extensions implies that the exact sequence $0 \to A \to B \to C \to 0$ in $\mathbf{C}$ is almost split in $\mathbf{C}$ [2, Chap II, Proposition 4.4].

Consequently we have shown that there is an almost split sequence $0 \to A \to B \to C \to 0$ in $\mathbf{C}$ if A is an indecomposable non-Ext-injective module in $\mathbf{C}$. A similar argument shows that there is an almost split sequence $0 \to A \to B \to C \to 0$ in $\mathbf{C}$ if C is an indecomposable non-Ext-projective module in $\mathbf{C}$. Therefore we have established our desired result that $\mathbf{C}$ has almost split sequence if $\mathbf{C}$ is a dualizing R-subvariety of $\bmod \Lambda$ closed under extensions.

2. Main Existence Theorem

In view of our main existence theorem for a subcategory $\mathbf{C}$ of $\bmod \Lambda$ to have almost split sequences, it is importante to know when $\mathbf{C}$ is a dualizing R-subvariety of $\bmod \Lambda$. Unfortunately it does not seem to be easy in practise to check whether or not a subcategory $\mathbf{C}$ of $\bmod \Lambda$ satisfies the general description of dualizing R-varieties given in [4]. For this reason this section is devoted to exploring the connection between $\mathbf{C}$ being a dualizing R-subvariety of $\bmod \Lambda$ and the more easily verified condition of $\bmod \Lambda$ being functorially finite over $\mathbf{C}$.

The notion of $\bmod \Lambda$ being functorially finite over $\mathbf{C}$, or $\mathbf{C}$ being functorially finite in $\bmod \Lambda$, was first introduced in [7], which serves as the foundation for much of this paper. Recall that $\mathbf{C}$ is said to be contravariantly finite in mod Λ if for each X in mod Λ the restriction $(\ , X) \mid \mathbf{C}$ to $\mathbf{C}$ of the representable functor $(\ , X)$ is a finitely generated functor on $\mathbf{C}$; i.e., there is an epimorphism $\mathrm{Hom}_{\mathbf{C}}(\ , C) \to (\ , X) \mid \mathbf{C}$ for some C in $\mathbf{C}$. Equivalently, $\mathbf{C}$ is contravariantly finite in mod Λ if for each X in mod Λ, there is a morphism $C \to X$ with C in add $\mathbf{C}$ such that $(C', C) \to (C', X) \to 0$ is exact for all C' in $\mathbf{C}$. Dually $\mathbf{C}$ is covariantly finite in mod Λ if for each X in mod Λ the restriction $(X, \ ) \mid \mathbf{C}$ of the representable functor $(X, \ )$ is a finitely generated functor on $\mathbf{C}$. Equivalently $\mathbf{C}$ is covariantly finite in mod Λ if for each X in mod Λ there is a morphism $X \to C$ with C in add $\mathbf{C}$ such that $(C, C') \to (X, C') \to 0$ is exact for all C' in $\mathbf{C}$. Finally $\mathbf{C}$ is said to be functorially finite in mod Λ if it is both covariantly and contravariantly finite in mod Λ.

Next we recall that a subcategory $\mathbf{C}'$ of mod Λ consisting of indecomposable objects in $\mathbf{C}$ is said to be a finite cover for $\mathbf{C}$ if (a) there is only a finite number of nonisomorphic objects in $\mathbf{C}'$ and (b) for each C in $\mathbf{C}$ there is a surjection $C' \to C$ with C' in add $\mathbf{C}'$. It was shown in [7, Proposition 3.7] that a subcategory $\mathbf{C}'$ of mod Λ consisting of only a finite number of nonisomorphic objects in $\mathbf{C}$ is a finite cover for $\mathbf{C}$ if and only if there is a morphism $\Lambda \to C'$ with C' in add $\mathbf{C}'$ such that $(C', C) \to (\Lambda, C) \to 0$ is exact for all C in $\mathbf{C}$. Thus if $\mathbf{C}$ is covariantly finite in mod Λ, then $\mathbf{C}$ has a finite cover.

Dually a subcategory $\mathbf{C}''$ of mod Λ consisting of indecomposable objects in $\mathbf{C}$ is said to be a finite cocover for $\mathbf{C}$ if (a) $\mathbf{C}''$ has only a finite number of nonisomorphic objects and (b) for each C in $\mathbf{C}$ there is an injection $C \to C''$ with C'' in add $\mathbf{C}''$. It was shown in [7, Proposition 3.6] that a subcategory $\mathbf{C}''$ of mod Λ consisting of only a finite number of nonisomorphic indecomposable objects in $\mathbf{C}$ is a cocover for $\mathbf{C}$ if and only if there is a morphism $C'' \to D(\Lambda) = \mathrm{Hom}_R(\Lambda, I)$ with C'' in add $\mathbf{C}''$ such that $(C, C'') \to (C, D(\Lambda)) \to 0$ is exact for all C in $\mathbf{C}$. Thus if $\mathbf{C}$ is contravariantly finite in mod Λ, then $\mathbf{C}$ has a finite cocover.

Before starting to prove the main result of this section we recall one more bit of terminology.

Let $\mathbf{C}$ be an additive subcategory of mod Λ; i.e., $\mathbf{C} = \mathrm{add}\,\mathbf{C}$. And let $C_1 \xrightarrow{f_1} C_2 \xrightarrow{f_2} C_3$ be morphisms in $\mathbf{C}$. Then f_1 is said to be a pseudokernel of f_2 in $\mathbf{C}$ if $(C, C_1) \to (C, C_2) \to (C, C_3)$ is exact for all C in $\mathbf{C}$ and f_2 is said to be a pseydocokernel of f_1 if $(C_3, C) \to (C_2, C) \to (C_1, C)$ is exact for all C in $\mathbf{C}$. We say that $\mathbf{C}$ has pseudokernels if each morphism in $\mathbf{C}$ has a pseudokernel. Similarly, we say that $\mathbf{C}$ has pseudocokernels if each morphism in $\mathbf{C}$ has a pseudocokernel.

In connection with the above definitions we point out the following which will be useful in establishing the main result of this section.

PROPOSITION 2.1. *Let* $\mathbf{C}$ *be an additive subcategory of* mod Λ *which is contravariantly finite in* mod Λ. *Then*

(a) **C** *has pseudokernels.*

(b) *If* F: $(\text{mod}\,\Lambda)^{\text{op}} \to Ab$ *is a finitely psesented functor, then* $F \mid \mathbf{C}$ *is finitely presented over* **C**.

Proof. (a) Let $0 \to K \to C_1 \to C_2$ be exact with C_1 and C_2 in **C**. Since **C** is an additive subcategory of mod Λ which is contravariantly finite in mod Λ, there is a morphism $C_0 \to K$ with C_0 in **C** such that $(C, C_0) \to (C, K) \to 0$ is exact for all C in **C**. Therefore the induced morphism $C_0 \to C_1$ is a pseudokernel for $C_1 \to C_2$. Hence **C** has pseudokernels.

(b) We first show that $(\ ,X) \mid \mathbf{C}$ is finitely presented for all X in mod Λ. Since **C** is contravariantly finite in mod Λ, there is a morphism $C_0 \to X$ with C_0 in **C** such that $(C, C_0) \to (C, X) \to 0$ is exact for all C in **C**. Applying the same argument to $\text{Ker}(C_0 \to X)$ we obtain that there is a sequence $C_1 \to C_0 \to X$ such that $(C, C_1) \to (C, C_0) \to (C, X) \to 0$ is exact for all C in **C**. Thus $(\ ,X) \mid \mathbf{C}$ is finitely presented.

Suppose F: $(\text{mod}\,\Lambda)^{\text{op}} \to Ab$ is finitely presented and $(\ ,X_1) \to (\ ,X_0) \to F \to 0$ is exact. Then $(\ ,X_1) \mid \mathbf{C} \to (\ ,X_0) \mid \mathbf{C} \to F \mid \mathbf{C} \to 0$ is exact. By our previous remarks above we know that the $(\ ,X_i) \mid \mathbf{C}$ are finitely presented. Since **C** has pseudokernels, f.p. $(\mathbf{C}^{\text{op}}, Ab)$ is closed under cokernels [4, p. 315], so $F \mid \mathbf{C}$ is finitely presented.

For the sake of completeness we state the dual of this result.

PROPOSITION 2.2. *Let* **C** *be an additive subcategory of* mod Λ *which is covariantly finite in* mod Λ. *Then*

(a) **C** *has pseudocokernels.*

(b) *If* F: mod $\Lambda \to Ab$ *is finitely presented, then* $F \mid \mathbf{C}$ *is finitely presented.*

We now turn our attention to the main result of this section.

THEOREM 2.3. *The following are equivalent for an additive subcategory* **C** *of* mod Λ.

(a) **C** *is functorially finite in* mod Λ.

(b) **C** *is a dualizing R-subvariety of* mod Λ *with a finite cover and a finite cocover.*

(c) *The functors* $(\ , D(\Lambda)) \mid \mathbf{C}$ *and* $(\Lambda\ ,\ ) \mid \mathbf{C}$ *are finitely presented and* **C** *has pseudocokernels and pseudokernels.*

Proof. (a) implies (b). We first show that **C** is a dualizing R-subvariety of mod Λ. Let F: $\mathbf{C}^{\text{op}} \to Ab$ be a finitely presented functor. Then there is a morphism f: $C_1 \to C_0$ in **C** such that $\text{Hom}_{\mathbf{C}}(\ , C_1) \xrightarrow{(\ ,f_1)} \text{Hom}_{\mathbf{C}}(\ , C_0) \to F \to 0$ is exact. The morphism f: $C_1 \to C_0$ also gives us an exact sequence of functors in $((\operatorname{mod} \Lambda)^{\mathrm{op}}, Ab)$ $\operatorname{Hom}_\Lambda(\ , C_1) \xrightarrow{(\ , f_1)} \operatorname{Hom}_\Lambda(\ , C_0) \to G \to 0$ which has the property that the exact sequence

$$\operatorname{Hom}_\Lambda(\ , C_1) \mid \mathbf{C} \to \operatorname{Hom}_\Lambda(\ , C_0) \mid \mathbf{C} \to G \mid \mathbf{C} \to 0$$

is isomorphic to $\operatorname{Hom}_{\mathbf{C}}(\ , C_1) \xrightarrow{(\ , f_1)} \operatorname{Hom}_{\mathbf{C}}(\ , C_0) \to F \to 0$. Since $\operatorname{mod} \Lambda$ is a dualizing R-variety we know that DG is a finitely presented functor. Hence by Proposition 2.1 we have that $(DG) \mid \mathbf{C}$ is a finitely presented functor. But $(DG) \mid \mathbf{C} = D(G \mid \mathbf{C}) = D(F)$. Therefore if F: $\mathbf{C}^{\mathrm{op}} \to Ab$ is finitely presented, then DF is finitely presented.

The fact that if F: $\mathbf{C} \to Ab$ is finitely presented, then DF is finitely presented follows by a similar, actually dual, argument. Hence we have shown that $\mathbf{C}$ is a dualizing R-subvariety of $\operatorname{mod} \Lambda$. We have already remarked that $\mathbf{C}$ being functorially finite in $\operatorname{mod} \Lambda$ implies that $\mathbf{C}$ has a finite cocover and a finite cover.

(b) implies (c). As was remarked earlier, the fact that $\mathbf{C}$ has finite cover and cocover means that the functors $(\Lambda, \ ) \mid \mathbf{C}$ and $(\ , D(\Lambda)) \mid \mathbf{C}$ are finitely generated. But $(\Lambda, X) = \Lambda \otimes_\Lambda X$ and so $D(\Lambda, X) = \operatorname{Hom}_R(\Lambda \otimes_\Lambda X, I) \approx \operatorname{Hom}_\Lambda(X, D(\Lambda))$. Hence $D((\Lambda, \ ) \mid \mathbf{C}) = (\ , D(\Lambda)) \mid \mathbf{C}$. Since $\mathbf{C}$ is a dualizing R-variety, we know that a functor F: $\mathbf{C} \to Ab$ is finitely presented if and only if F and DF are finitely generated [4, Proposition 3.1]. Therefore $(\Lambda, \ ) \mid \mathbf{C}$ and $(\ , D(\Lambda)) \mid \mathbf{C}$ are finitely presented functors. The fact that $\mathbf{C}$ has pseudokernels and pseudocokernels is also a consequence of $\mathbf{C}$ being a dualizing R-variety [4, Theorem 2.4].

(c) implies (a). We first show that $\mathbf{C}$ is covariantly finite in $\operatorname{mod} \Lambda$ by showing that if X is in $\operatorname{mod} \Lambda$, then $(X, \ ) \mid \mathbf{C}$ is finitely generated. Let $P_1 \to P_0 \to X \to 0$ be in $\operatorname{mod} \Lambda$ with the P_i projective. Then the exact sequence $0 \to (X, \ ) \to (P_0, \ ) \to (P_1, \ )$ gives rise to the exact sequence $0 \to (X, \ ) \mid \mathbf{C} \to (P_0, \ ) \mid \mathbf{C} \to (P_1, \ ) \mid \mathbf{C}$. Since $(\Lambda, \ ) \mid \mathbf{C}$ is finitely presented, the $(P_i, \ ) \mid \mathbf{C}$ are finitely presented. But $\mathbf{C}$ having pseudocokernels implies that the kernel of a morphism between finitely presented functors is also finitely presented. Therefore $(X, \ ) \mid \mathbf{C}$ is finitely generated since it is finitely presented.

The rest of the implication (c) implies (a) follows by a similar, actually dual, argument. Thus the theorem is established.

As an immediate consequence of Theorems 1.1 and 2.3 we have the following.

THEOREM 2.4. *Let* $\mathbf{C}$ *be a functorially finite subcategory of* $\operatorname{mod} \Lambda$ *which is closed under extensions. Then*

(a) $\mathbf{C}$ *has a finite cover and a finite cocover.*

(b) $\mathbf{C}$ *has almost split sequences.*

Thus we see that subcategories **C** of mod Λ which are closed under extensions and are functorially finite in mod Λ are similar to categories of finitely generated modules over artin algebras in quite a few respects. These similarities suggests the following problems about such subcategories **C** of mod Λ.

(a) Describe the Ext-projective and Ext-injective objects in **C**. In particular, need there be only a finite number of isomorphism classes of indecomposable objects which are Ext-projective or Ext-injective? Also, need these numbers be the same when finite?

(b) Suppose $0 \to A \to B \to C \to 0$ is an almost split sequence in **C**. Does there exist a method of constructing A from C and vice versa similar to the construction given by the functors $D\,\mathrm{Tr}$ and $\mathrm{Tr}\,D$?

Although we cannot answer these problems in general, we do have information along these lines in some special situations as we now proceed to show.

3. Subcategories Closed under Submodules

Throughout this section we assume that **C** is a subcategory of mod Λ closed under extensions. Clearly an indecomposable C in **C** is Ext-projective (Ext-injective) if and only if $\mathrm{Ext}^1_\Lambda(C, X) = 0$ ($\mathrm{Ext}^1_\Lambda(X, C) = 0$) for all X in **C**. Suppose **C** has a minimal finite cover $\mathbf{P}_0(\mathbf{C})$; i.e., $\mathbf{P}_0(\mathbf{C})$ is a finite cover of **C** which is contained in all other covers for **C**. Then C is in $\mathbf{P}_0(\mathbf{C})$ if and only if C is a splitting projective in **C**; i.e., if $f\colon Y \to C$ is a surjection in **C**, then f is a splittable surjection (see [7, pp. 14–17]). Hence the objects in $\mathbf{P}_0(\mathbf{C})$ are Ext-projective in **C**. However, in general, there may be Ext-projective objects in **C** which are not in $\mathbf{P}_0(\mathbf{C})$. Dually, suppose **C** has a minimal finite cocover $\mathbf{I}_0(\mathbf{C})$; i.e., $\mathbf{I}_0(\mathbf{C})$ is a finite cocover of **C** which is contained in all cocovers of **C**. Then C is in $\mathbf{I}_0(\mathbf{C})$ if and only if C is a splitting injective in **C**; i.e., if $g\colon C \to Y$ is an injection in **C**, then g is a splittable injection (see [7, pp. 14–17]). Hence the objects of $\mathbf{I}_0(\mathbf{C})$ are Ext-injective in **C**. However, in general, there may be Ext-injective objects in **C** which are not in $\mathbf{I}_0(\mathbf{C})$.

Suppose now that **C** is closed under submodules; i.e., if C' is a submodule of C in **C**, then C' is in **C**. Then it is easily seen (see [7, Proposition 4.8.]) that for each X in mod Λ there is a unique submodule $t_{\mathbf{C}}X$ minimal with respect to the property $X/t_{\mathbf{C}}X$ is in **C**. Moreover the surjection $X \to X/t_{\mathbf{C}}X$ induces an isomorphism $(X/t_{\mathbf{C}}X, C) \approx (X, C)$ for all C in **C**. From this it follows that if $P_1,\ldots, P_n$ is a complete set of nonisomorphic indecomposable projective Λ-modules then **C** has a minimal finite cover $\mathbf{P}_0(\mathbf{C})$ which consists of the modules isomorphic to the nonzero $P_i/t_{\mathbf{C}}P_i$ for $i = 1,\ldots, n$. Finally, if an indecomposable C in $\mathbf{C}$ is Ext-projective, then C is in $\mathbf{P}_0(\mathbf{C})$. For we have an exact sequence $0 \to A \to B \to C \to 0$ with B in add $\mathbf{P}_0(\mathbf{C})$. Since $\mathbf{C}$ is closed under submodules, A is in $\mathbf{C}$. Therefore if C is Ext-projective, the sequence $0 \to A \to B \to C \to 0$ splits, so C is in $\mathbf{P}_0(\mathbf{C})$. Thus we have shown the following.

PROPOSITION 3.1. *Suppose* $\mathbf{C}$ *is closed under submodules. Then*

(a) $\mathbf{C}$ *has a minimal finite cover* $\mathbf{P}_0(\mathbf{C})$.

(b) C *in* $\mathbf{C}$ *is in* $\mathbf{P}_0(\mathbf{C})$ *if and only if* $C \approx P/t_{\mathbf{C}}P$ *for some indecomposable projective* Λ*-module* P *such that* $P/t_{\mathbf{C}}P \neq 0$.

(c) *An indecomposable* C *in* $\mathbf{C}$ *is* Ext*-projective if and only if* C *is in* $\mathbf{P}_0(\mathbf{C})$.

Dually, suppose $\mathbf{C}$ is closed under factor modules; i.e., if C is in $\mathbf{C}$, C/C' is in $\mathbf{C}$ for all submodules C' of C. Then it is easily seen (see [7, Proposition 4.8]) that for each X in mod Λ, there is a unique submodule $\tau_{\mathbf{C}}(X)$ of X maximal with respect to being in $\mathbf{C}$. Moreover the injection $\tau_{\mathbf{C}}(X) \to X$ induces an isomorphism $(C, \tau_{\mathbf{C}}(X)) \to (C, X)$ for all C in $\mathbf{C}$. We also have the following dual version of Proposition 3.1.

PROPOSITION 3.2. *Suppose* $\mathbf{C}$ *is closed under factor modules. Then*

(a) $\mathbf{C}$ *has a minimal finite cocover* $\mathbf{I}_0(\mathbf{C})$.

(b) C *is in* $\mathbf{I}_0(\mathbf{C})$ *if and only if* $C \approx \tau_{\mathbf{C}}(I)$ *for some indecomposable injective* Λ*-module* I *such that* $\tau_{\mathbf{C}}(I) \neq 0$.

(c) *An indecomposable* C *in* $\mathbf{C}$ *is* Ext*-injective if and only if* C *is in* $\mathbf{I}_0(\mathbf{C})$.

Suppose once again that $\mathbf{C}$ is closed under submodules. Having described the indecomposable Ext-projectives in $\mathbf{C}$, we now turn our attention to characterizing the indecomposable Ext-injective modules in $\mathbf{C}$.

PROPOSITION 3.3. *Suppose* $\mathbf{C}$ *is closed under submodules. Then the following statements are equivalent for an indecomposable object* A *in* $\mathbf{C}$.

(a) A *is not* Ext*-injective in* $\mathbf{C}$.

(b) $t_{\mathbf{C}}(\operatorname{Tr} DA) \neq \operatorname{Tr} DA$.

(c) *The almost split sequence* $0 \to A \to B \to \operatorname{Tr} DA \to 0$ *in* mod Λ *has the property that the induced sequence in* $\mathbf{C}$, $0 \to A \to B/t_{\mathbf{C}}B \to \operatorname{Tr} DA/t_{\mathbf{C}}(\operatorname{Tr} DA) \to 0$, *is exact and therefore nonsplit.*

Proof. (a) implies (b). Since A is not Ext-injective in $\mathbf{C}$, there is a nonzero element x in $\mathrm{Ext}^1_\Lambda(X, A)$ for some X in $\mathbf{C}$. Then there is a morphism $f\colon \mathrm{Tr}\, DA \to X$ such that $\mathrm{Ext}^1(f, A)(x)$ in $\mathrm{Ext}^1(\mathrm{Tr}\, DA, A)$ is the almost split sequence. Hence $f\colon \mathrm{Tr}\, DA \to X$ is not zero, so $t_\mathbf{C}\, \mathrm{Tr}\, DA \neq \mathrm{Tr}\, DA$.

(b) implies (c). Since $0 \to A \to^f B \to^g \mathrm{Tr}\, DA \to 0$ is an almost split sequence and $t_\mathbf{C}(\mathrm{Tr}\, DA) \neq \mathrm{Tr}\, DA$, there is a morphism $h\colon t_\mathbf{C}(\mathrm{Tr}\, DA) \to B$ such that $(g \mid \mathrm{Im}\, h)h = \mathrm{id}_{t_\mathbf{C}(\mathrm{Tr}\, DA)}$. Letting $B' = \mathrm{Im}\, h$, we have the exact sequence $0 \to A \to B/B' \to \mathrm{Tr}\, DA/t_\mathbf{C}(\mathrm{Tr}\, DA) \to 0$. Since $\mathbf{C}$ is closed under extensions B/B' is in $\mathbf{C}$. Therefore $B' \supset t_\mathbf{C} B$ and so $t_\mathbf{C}(\mathrm{Tr}\, DA) = g(B') \supset g(t_\mathbf{C} B)$. So we have the exact commutative diagram

$$\begin{array}{ccccccccc}
 & & & & 0 & & 0 & & \\
 & & & & \downarrow & & \downarrow & & \\
 & & & & B'/t_\mathbf{C}B & \xrightarrow{\sim} & B'/t_\mathbf{C}B & & \\
 & & & & \downarrow & & \downarrow & & \\
0 & \longrightarrow & A & \longrightarrow & B/t_\mathbf{C}B & \longrightarrow & \mathrm{Tr}\, DA/g(t_\mathbf{C}B) & \longrightarrow & 0 \\
 & & \| & & \downarrow & & \downarrow & & \\
0 & \longrightarrow & A & \longrightarrow & B/B' & \longrightarrow & \mathrm{Tr}\, DA/t_\mathbf{C}(\mathrm{Tr}\, DA) & \longrightarrow & 0 \\
 & & & & \downarrow & & \downarrow & & \\
 & & & & 0 & & 0 & &
\end{array}$$

Since $B/t_\mathbf{C} B$ is in $\mathbf{C}$, we have that $B'/t_\mathbf{C} B$ is in $\mathbf{C}$ and hence that $\mathrm{Tr}\, DA/g(t_\mathbf{C} B)$ is in $\mathbf{C}$ (remember that $\mathbf{C}$ is closed under submodules and extensions).

Hence $g(t_\mathbf{C} B) \supset t_\mathbf{C}(\mathrm{Tr}\, DA)$ and so $g(t_\mathbf{C} B) = t_\mathbf{C}(\mathrm{Tr}\, DA)$. Therefore $B'/t_\mathbf{C} B = 0$, which shows that $0 \to A \to B/t_\mathbf{C} B \to \mathrm{Tr}\, DA/t_\mathbf{C}\, \mathrm{Tr}\, DA \to 0$ is exact. The fact that this sequences does not split follows from the fact that $0 \to A \to B \to \mathrm{Tr}\, DA \to 0$ does not split.

(c) implies (a). Trivial.

As an immediate consequence of Proposition 3.3, we have the following characterization of the indecomposable Ext-injective objects in $\mathbf{C}$.

COROLLARY 3.4. *Suppose* $\mathbf{C}$ *is closed under submodules. Then an indecomposable* A *in* $\mathbf{C}$ *is* Ext-*injective in* $\mathbf{C}$ *if and only if* $t_\mathbf{C}(\mathrm{Tr}\, DA) = \mathrm{Tr}\, DA$.

As another consequence of Proposition 3.3 we have the following result concerning the existence and structure of almost split sequence in $\mathbf{C}$.

COROLLARY 3.5. *Suppose* $\mathbf{C}$ *is closed under submodules. Let* A *be an indecomposable object in* $\mathbf{C}$ *which is not* Ext-*injective in* $\mathbf{C}$ *and let*

$0 \to A \to B \to \operatorname{Tr} DA \to 0$ be an almost split sequence in $\operatorname{mod} \Lambda$. *Then there is a commutative exact diagram*

$$\begin{array}{ccccccccc}
 & & & & 0 & & 0 & & \\
 & & & & \downarrow & & \downarrow & & \\
 & & & & C & = & C & & \\
 & & & & \downarrow & & \downarrow & & \\
0 & \longrightarrow & A & \xrightarrow{g} & B/t_{\mathbf{C}}B & \xrightarrow{f} & \operatorname{Tr} DA/t_{\mathbf{C}} \operatorname{Tr} DA & \longrightarrow & 0 \\
 & & \| & & \downarrow & & \downarrow & & \\
0 & \longrightarrow & A & \longrightarrow & B' & \longrightarrow & A' & \longrightarrow & 0 \\
 & & & & \downarrow & & \downarrow & & \\
 & & & & 0 & & 0 & &
\end{array}$$

satisfying:

(a) *$0 \to A \to B' \to A' \to 0$ is an almost split sequence in* $\mathbf{C}$;

(b) *$C \to B/t_{\mathbf{C}}B$ and $C \to \operatorname{Tr} DA/t_{\mathbf{C}} \operatorname{Tr} DA$ are splittable monomorphisms.*

Proof. The fact that $0 \to A \to B/t_{\mathbf{C}}B \to \operatorname{Tr} DA/t_{\mathbf{C}} \operatorname{Tr} DA \to 0$ is exact was established in Proposition 3.3. The fact that $0 \to A \to B$ is left almost split in $\operatorname{mod} \Lambda$ implies that $0 \to A \to B/t_{\mathbf{C}}B$ is left almost split in $\mathbf{C}$. Now the exact sequence $0 \to A \to B/t_{\mathbf{C}}B \to \operatorname{Tr} DA/t_{\mathbf{C}} \operatorname{Tr} DA \to 0$ can be written as a sum of exact sequences

$$\begin{array}{ccccccccc}
0 & \longrightarrow & A & \xrightarrow{f'} & B' & \xrightarrow{g'} & A' & \longrightarrow & 0 \\
 & & & & \amalg & & \amalg & & \\
 & & & & C & = & C & &
\end{array}$$

where f' is left minimal [7, Proposition 1.2]). Hence the exact sequence $0 \to A \to^{f'} B' \to^{g'} A' \to 0$ in $\mathbf{C}$ has the property that f' is minimal left almost split in $\mathbf{C}$, which by [2, Chap. II, Proposition 4.4] implies that it is almost split in $\mathbf{C}$.

For the sake of completeness we state without proof the duals of these last results if $\mathbf{C}$ has factor modules instead of submodules.

PROPOSITION 3.6. *Suppose $\mathbf{C}$ is closed under factor modules. Then the following statements are equivalent for an indecomposable object C in $\mathbf{C}$.*

(a) *C is not* Ext-*projective.*

(b) $\tau_{\mathbf{C}}(D \operatorname{Tr} C) \neq 0$.

(c) *The almost split sequence $0 \to D \operatorname{Tr} C \to B \to C \to 0$ in* $\operatorname{mod} \Lambda$ *has the property that the induced sequence in $\mathbf{C}$ $0 \to \tau_{\mathbf{C}}(D \operatorname{Tr} C) \to \tau_{\mathbf{C}}(B) \to C \to 0$ is exact and therefore not split.*

COROLLARY 3.7. *Suppose* **C** *is closed under factor modules. Then an indecomposable* C *in* **C** *is* Ext-*projective if and only if* $\tau_{\mathbf{C}}(D\operatorname{Tr} C) = 0$.

COROLLARY 3.8. *Suppose* **C** *is closed under factor modules. Let* C *be an indecomposable in* **C** *which is not* Ext-*projective and let* $0 \to D\operatorname{Tr} C \to B \to C \to 0$ *an almost split sequence in* $\operatorname{mod} \Lambda$. *Then there is a commutative exact diagram*

$$\begin{array}{ccccccccc} & & 0 & & 0 & & & & \\ & & \downarrow & & \downarrow & & & & \\ 0 & \longrightarrow & C' & \longrightarrow & B' & \longrightarrow & C & \longrightarrow & 0 \\ & & \downarrow & & \downarrow & & \| & & \\ 0 & \longrightarrow & \tau_{\mathbf{C}}(D\operatorname{Tr} C) & \longrightarrow & \tau_{\mathbf{C}}(B) & \longrightarrow & C & \longrightarrow & 0 \\ & & \downarrow & & \downarrow & & & & \\ & & A & = & A & & & & \\ & & \downarrow & & & & & & \\ & & 0 & & & & & & \end{array}$$

satisfying:

(a) $0 \to C' \to B' \to C \to 0$ *is an almost split sequence in* **C**;

(b) $0 \to C' \to \tau_{\mathbf{C}}(D\operatorname{Tr} C)$ *and* $0 \to B' \to \tau_{\mathbf{C}}(B)$ *are splittable monomorphisms.*

4. THE Ext-PROJECTIVE AND Ext-INJECTIVE MODULES IN Sub M

We assume throughout this section that **C** is a subcategory of $\operatorname{mod} \Lambda$ which is closed under extensions. In the previous section we examined the Ext-injective and Ext-projective objects in **C** under the additional hypothesis that **C** was closed under either submodules or factor modules. In this section we obtain somewhat sharper results by adding the hypothesis that if **C** is closed under submodules, then it also has a finite cocover and that if **C** is closed under factor modules, then it also has a finite cover. Amongst other things we show that under either of these additional hypothesis **C** has only a finite number of nonisomorphic, indecomposable Ext-projective and Ext-injective modules.

Suppose **C** is a subcategory of $\operatorname{mod} \Lambda$ closed under submodules. Then **C** has a finite cocover if and only if there is a C in **C** such that $\mathbf{C} = \operatorname{Sub} C$ where $\operatorname{Sub} C$ is the subcategory of $\operatorname{mod} \Lambda$ consisting of all modules isomorphic to submodules of finite sums of copies of C. We now assume that $\mathbf{C} = \operatorname{Sub} C$ for some C in **C**. Then by [7, Proposition 4.7] we know that **C** is functorially finite in $\operatorname{mod} \Lambda$. Therefore **C** has almost split sequences by

Theorem 2.4 since we have the blanket assumption that **C** is closed under extensions. Before describing the indecomposable Ext-injective modules in **C** we introduce some notation.

Since **C** is functorially finite in $\operatorname{mod} \Lambda$ we know by [7, Proposition 3.9] that for each X in $\operatorname{mod} \Lambda$ there is a unique (up to isomorphism) right minimal morphism $f: C \to X$ with C in **C** such that $(C', C) \to (C', X) \to 0$ is exact for all C' in **C**. For each X we denote by $f_X: X_{\mathbf{C}} \to X$ one fixed such morphism. The fact that **C** is closed under submodules implies $\operatorname{Ker} f_X$ is contained in **C** and we denote $\operatorname{Ker} f_X$ by $A^X_{\mathbf{C}}$. Finally, we recall that if **D** is a subcategory of $\operatorname{mod} \Lambda$, we denote by $\operatorname{Ind} \mathbf{D}$ the subcategory of $\operatorname{mod} \Lambda$ consisting of the indecomposable modules in **D** and if X is a module in $\operatorname{mod} \Lambda$ we denote by $\operatorname{Ind} X$ the subcategory of $\operatorname{mod} \Lambda$ consisting of the indecomposable modules isomorphic to summands of X.

THEOREM 4.1. *Suppose* $\mathbf{C} = \operatorname{Sub} C$ *for some* C *in* **C**. *Denoting* $\mathbf{I}_0(\operatorname{mod} \Lambda)$ *by* $\mathbf{I}_0$ *we have*

(a) $\mathbf{I}_0(\mathbf{C}) = \bigcup_{I \in \mathbf{I}_0} \operatorname{Ind} I_{\mathbf{C}}$,

(b) $\bigcup_{I \in \mathbf{I}_0} \operatorname{Ind} A^I_{\mathbf{C}} \cap \mathbf{I}_0(\mathbf{C}) = \varnothing$,

(c) $\mathbf{I}_0(\mathbf{C}) \cup \bigcup_{I \in \mathbf{I}_0} \operatorname{Ind} A^I_{\mathbf{C}}$ *is the subcategory of* $\operatorname{Ind}(\operatorname{mod} \Lambda)$ *consisting of the* Ext-*injective modules in* $\operatorname{Ind} \mathbf{C}$.

Proof. (a) See [7, Lemma 3.5].

(b) Suppose X is in $\mathbf{I}_0(\mathbf{C}) \cap \bigcup_{I \in \mathbf{I}_0} \operatorname{Ind} A^I_{\mathbf{C}}$. Then there is a I in $\mathbf{I}_0$ such that X is a summand of $A^I_{\mathbf{C}}$. Thus the composition of monomorphisms $X \to A^I_{\mathbf{C}} \to I_{\mathbf{C}}$ is a splittable monomorphism. But then $f_I: I_{\mathbf{C}} \to I$ is not right minimal, which is a contradiction. Hence $\mathbf{I}_0(\mathbf{C}) \cap \bigcup_{I \in \mathbf{I}_0} \operatorname{Ind} A^I_{\mathbf{C}} = \varnothing$.

(c) We first show that if X is in $\mathbf{I}_0(\mathbf{C}) \cup \bigcup_{I \in \mathbf{I}_0} \operatorname{Ind} A^I_{\mathbf{C}}$, then X is Ext-injective in **C**, or what is the same thing, $\operatorname{Ext}^1_\Lambda(C, X) = 0$ for all C in **C**. We have already seen at the beginning of Section 3 that $\operatorname{Ext}^1_\Lambda(C, X) = 0$ for all C in **C** if X is in $\mathbf{I}_0(\mathbf{C})$. Let I be in $\operatorname{Ind} \mathbf{I}_0$ and let $T = \operatorname{Im}(I_{\mathbf{C}} \to^{f_I} I)$. Since (C, f_I): $(C, I_{\mathbf{C}}) \to (C, I) \to 0$ is exact for all C in **C**, we have that $(C, I_{\mathbf{C}}) \to (C, T) \to 0$ is exact for all C in **C**. Hence the exact sequence $0 \to A^I_{\mathbf{C}} \to I_{\mathbf{C}} \to T \to 0$ gives the exact sequence $0 \to \operatorname{Ext}^1_\Lambda(C, A^I_{\mathbf{C}}) \to \operatorname{Ext}^1_\Lambda(C, I_{\mathbf{C}})$. Therefore $\operatorname{Ext}^1_\Lambda(C, A^I_{\mathbf{C}}) = 0$ for all C in **C** since $\operatorname{Ext}^1_\Lambda(C, I_{\mathbf{C}}) = 0$ for all C in **C**. Therefore we have shown that $\mathbf{I}_0(\mathbf{C}) \cup \bigcup_{I \in \mathbf{I}_0} \operatorname{Ind} A^I_{\mathbf{C}}$ consists of Ext-injective objects in **C**.

Suppose now that X in **C** is an indecomposable Ext-injective module which is not in $\mathbf{I}_0(C)$. We want to show that X is in $\bigcup_{I \in \mathbf{I}_0} \operatorname{Ind} A^I_{\mathbf{C}}$. Since $\mathbf{I}_0(\mathbf{C})$ is a cocover for **C**, we know there is an exact sequence $0 \to X \to^h Y \to U \to 0$ with Y in $\operatorname{add} \mathbf{I}_0(\mathbf{C})$ such that h is left minimal. Since $\operatorname{Ext}^1_\Lambda(C, X) = 0$ for all C in **C**, it follows that $(C, Y) \to (C, U) \to 0$ is exact for all C in **C**. Hence $0 \to (\ , X) \mid \mathbf{C} \to (\ , Y) \mid \mathbf{C} \to (\ , U) \mid \mathbf{C} \to 0$ is a minimal projective presentation of $(\ , U) \mid \mathbf{C}$ in $(\mathbf{C}^{\mathrm{op}}, Ab)$.

Let $U \to I$ be an injective envelope of U in mod Λ. Then we have the exact commutative diagram

$$\begin{array}{ccccccccc} & & & & 0 & & 0 & & \\ & & & & \downarrow & & \downarrow & & \\ 0 & \longrightarrow & A_{\mathbf{C}}^{I} & \longrightarrow & f^{-1}(U) & \longrightarrow & U & \longrightarrow & 0 \\ & & \| & & \downarrow & & \downarrow & & \\ 0 & \longrightarrow & A_{\mathbf{C}}^{I} & \xrightarrow{g} & I_{\mathbf{C}} & \xrightarrow{f} & I & & \end{array} \qquad (*)$$

which implies $f^{-1}(U) \to U$ has the property $(\ ,f^{-1}(U))\,|\,\mathbf{C} \to (\ ,U)\,|\,\mathbf{C} \to 0$ is exact. Since $\mathbf{C}$ is closed under submodules, $f^{-1}(U)$ is in $\mathbf{C}$, so we have that $(\ ,f^{-1}(U))\,|\,\mathbf{C}$ is projective. Hence the fact that $(\ ,Y)\,|\,\mathbf{C} \to (\ ,U)\,|\,\mathbf{C} \to 0$ is a projective cover means that there is an exact commutative diagram

$$\begin{array}{ccccccccc} & & 0 & & 0 & & & & \\ & & \downarrow & & \downarrow & & & & \\ 0 & \longrightarrow & X & \longrightarrow & Y & \longrightarrow & U & \longrightarrow & 0 \\ & & \downarrow & & \downarrow & & \| & & \\ 0 & \longrightarrow & A_{\mathbf{C}}^{I} & \longrightarrow & f^{-1}(U) & \longrightarrow & U & \longrightarrow & 0 \end{array} \qquad (**)$$

with $Y \to f^{-1}(U)$ a splittable monomorphism.

Combining $(*)$ and $(**)$ we obtain the exact commutative diagram

$$\begin{array}{ccccccccc} & & 0 & & 0 & & 0 & & \\ & & \downarrow & & \downarrow & & \downarrow & & \\ 0 & \longrightarrow & X & \longrightarrow & Y & \longrightarrow & U & \longrightarrow & 0 \\ & & \downarrow & & \downarrow & & \downarrow & & \\ 0 & \longrightarrow & A_{\mathbf{C}}^{I} & \longrightarrow & I_{\mathbf{C}} & \longrightarrow & I & & \end{array}$$

Since Y is in $\mathbf{I}_0(C)$, we know that $Y \to I_{\mathbf{C}}$ is a splittable monomorphism. So we obtain the exact sequence of functors on $\mathbf{C}^{\mathrm{op}}$

$$\begin{array}{ccccccccc} & & 0 & & 0 & & 0 & & \\ & & \downarrow & & \downarrow & & \downarrow & & \\ 0 & \longrightarrow & (\ ,X) & \longrightarrow & (\ ,Y) & \longrightarrow & (\ ,U)\,|\,\mathbf{C} & \longrightarrow & 0 \\ & & \downarrow & & \downarrow & & \downarrow & & \\ 0 & \longrightarrow & (\ ,A_{\mathbf{C}}^{I}) & \longrightarrow & (\ ,I_{\mathbf{C}}) & \longrightarrow & (\ ,I)\,|\,\mathbf{C} & \longrightarrow & 0 \\ & & \downarrow & & \downarrow & & \downarrow & & \\ 0 & \longrightarrow & G & \longrightarrow & (\ ,I_{\mathbf{C}}/Y)\,|\,\mathbf{C} & \longrightarrow & F & \longrightarrow & 0 \\ & & \downarrow & & \downarrow & & \downarrow & & \\ & & 0 & & 0 & & 0 & & \end{array}$$

where $G = \mathrm{Coker}((\ ,X) \to (\ ,A_{\mathbf{C}}^{I}))$ and $F = \mathrm{Coker}((\ ,U)\,|\,\mathbf{C} \to (\ ,I)\,|\,\mathbf{C})$. Since $Y \to I_{\mathbf{C}}$ is a splittable monomorphism, $I_{\mathbf{C}}/Y$ is in $\mathbf{C}$ so $(\ ,I_{\mathbf{C}}/Y)\,|\,\mathbf{C} = (\ ,I_{\mathbf{C}}/Y)$ is projective in $(\mathbf{C}^{\mathrm{op}}, Ab)$.

Next we observe that F is a subfunctor of $(\ ,I/U)\,|\,\mathbf{C}$. Therefore $\operatorname{pd} F \leqslant 1$ where $\operatorname{pd} F$ means projective dimension of F in $(\mathbf{C}^{\mathrm{op}}, Ab)$. To see this we first observe that if M is in $\operatorname{mod} \Lambda$, then the exact sequence $0 \to A^M_{\mathbf{C}} \to M_{\mathbf{C}} \to M$ induces the exact sequence of functors $0 \to (\ ,A^M_{\mathbf{C}}) \to (\ ,M_{\mathbf{C}}) \to (\ ,M)\,|\,\mathbf{C} \to 0$ with the $(\ ,A^M_{\mathbf{C}})$ and $(\ ,M_{\mathbf{C}})$ projective in $(\mathbf{C}^{\mathrm{op}}, Ab)$. Thus $\operatorname{pd}(\ ,M)\,|\,\mathbf{C} \leqslant 1$ for all M in $\operatorname{mod} \Lambda$. In particular, $\operatorname{pd}(\ ,I/U)\,|\,\mathbf{C} \leqslant 1$. But the fact that $\mathbf{C}$ is closed under submodules implies $\operatorname{gl\,dim} \operatorname{f.p}(\mathbf{C}^{\mathrm{op}}, Ab) \leqslant 2$. Hence the fact that $F \subset (\ ,I/U)\,|\,\mathbf{C}$ and $\operatorname{pd}(\ ,I/U)\,|\,\mathbf{C} \leqslant 1$, implies $\operatorname{pd} F \leqslant 1$.

Consequently, it follows from the exact sequence $0 \to G \to (\ , I_{\mathbf{C}}/Y) \to F \to 0$ that G is projective since $(\ ,I_{\mathbf{C}}/Y))$ is projective. Hence the exact sequence $0 \to (\ ,X) \to (\ ,A^I_{\mathbf{C}}) \to G \to 0$ splits, which means that $X \to A^I_{\mathbf{C}}$ is a splittable monomorphism. Therefore X is in $\operatorname{Ind} A^I_{\mathbf{C}}$, our desired result. This completes the proof of the theorem.

As an immediate consequence of this theorem we have the following.

COROLLARY 4.2. *Suppose* $\mathbf{C} = \operatorname{Sub} C$ *for some* C *in* $\mathbf{C}$. *Then there are a finite number of nonisomorphic indecomposable objects in* $\mathbf{C}$ *which are* Ext-*injective as well as only a finite number which are* Ext-*projective.*

We end this section with the statements of the duals of Theorem 4.1 and Corollary 4.2.

Suppose $\mathbf{C}$ is a subcategory of $\operatorname{mod} \Lambda$ which is closed under factor modules. Then $\mathbf{C}$ has a finite cover if and only if $\mathbf{C} = \operatorname{Fac} C$ for some C in $\mathbf{C}$, where $\operatorname{Fac} C$ is the subcategory of $\operatorname{mod} \Lambda$ consisting of all modules isomorphic to factor modules of finite sums of C. Assume $\mathbf{C} = \operatorname{Fac} C$. Then by [7, Proposition 4.6] we know that $\mathbf{C}$ is functorially finite in $\operatorname{mod} \Lambda$. Hence for each X in $\operatorname{mod} \Lambda$, there is a unique, up to isomorphism, left minimal morphism $g\colon X \to C$ with C in $\mathbf{C}$ such that $(C, C') \to (X, C') \to 0$ is exact for all C' in $\mathbf{C}$. For each X in $\operatorname{mod} \Lambda$ we denote by $g^X\colon X \to X^{\mathbf{C}}$ one fixed such morphism. The fact that $\mathbf{C}$ is closed under factor modules implies $\operatorname{Coker} g^X$ is in $\mathbf{C}$ and we denote $\operatorname{Coker} g^X$ by $A^{\mathbf{C}}_X$.

We now give the dual of Theorem 4.1.

THEOREM 4.3. *Suppose* $\mathbf{C} = \operatorname{Fac} C$ *for some* C *in* $\mathbf{C}$. *Denoting* $\mathbf{P}_0(\operatorname{mod} \Lambda)$ *by* $\mathbf{P}_0$ *we have*

(a) $\mathbf{P}_0(\mathbf{C}) = \bigcup_{P \in \mathbf{P}_0} \operatorname{Ind} P^{\mathbf{C}}$.

(b) $\bigcup_{P \in \mathbf{P}_0} \operatorname{Ind} A^{\mathbf{C}}_P \cap \mathbf{P}_0(\mathbf{C}) = \varnothing$.

(c) $\mathbf{P}_0(\mathbf{C}) \cup \bigcup_{P \in \mathbf{P}_0} \operatorname{Ind} A^{\mathbf{C}}_P$ *is the subcategory of* $\operatorname{Ind}(\operatorname{mod} \Lambda)$ *consisting of the* Ext-*projective modules in* $\operatorname{Ind} \mathbf{C}$.

Finally, we have the following dual of Corollary 4.2.

COROLLARY 4.4. *Suppose* $\mathbf{C} = \operatorname{Fac} C$ *for some* C *in* $\mathbf{C}$. *Then* $\mathbf{C}$ *has only a finite number of nonisomorphic indecomposable modules which are* Ext-*projective as well as only a finite number which are* Ext-*injective.*

5. When Is Sub M Closed under Extensions?

In view of Theorems 4.1 and 4.3 just established it is of interest to know when for a module M in mod Λ the subcategory Sub M of mod Λ is closed under extensions as well as when the subcategory Fac M is closed under extensions. We begin our discussion of these problems with some observations concerning the subcategories of mod Λ of the forms Sub M and Fac M for some M in mod Λ.

Let M be in mod Λ. Then add M has a unique minimal finite cocover $\mathbf{I}_0(\text{add } M)$. Let $\bar{M}$ be a sum of a complete set of nonisomorphic modules in $\mathbf{I}_0(\text{add } M)$. It is now easy to see that $\bar{M}$ has the following properties:

(i) $\bar{M}$ is a sum of nonisomorphic indecomposable modules;

(ii) ind $\bar{M} = \mathbf{I}_0(\text{add } \bar{M})$;

(iii) Sub $\bar{M}$ = sub M;

(iv) The above properties of $\bar{M}$ uniquely determine $\bar{M}$ up to isomorphism.

In view of these remarks, we see that in discussing the subcategories of mod Λ of the form Sub M for some M we might as well assume that $M = \bar{M}$. Unless stated to the contrary, we make the convention that when we write Sub M we are automatically assuming that $M = \bar{M}$.

Similarly, add M has a unique finite minimal cover $\mathbf{P}_0(\text{add } M)$. Let $\underline{M}$ be a sum of a complete set of nonisomorphic modules in $\mathbf{P}_0(\text{add } M)$. Then we have

(i) $\underline{M}$ is a sum of nonisomorphic indecomposable modules;

(ii) Ind $\underline{M} = \mathbf{P}_0(\text{add } \underline{M})$

(iii) Fac M = Fac $\underline{M}$

(iv) The above properties of $\underline{M}$ uniquely determine $\underline{M}$ up to isomorphism.

Unless stated to the contrary, we will assume $M = \underline{M}$ when we write Fac M.

One final notation. For a subcategory $\mathbf{C}$ of mod Λ we denote by Ext($\mathbf{C}$) the subcategory of mod Λ consisting of summands of modules M which have a sequence of submodules $0 = M_0 \subset M \subset \cdots \subset M_n = M$ such that M_{i+1}/M_i is in $\mathbf{C}$ for $i = 0, \ldots, n-1$.

With these conventions and notations in mind, we have the following result which is basic to our entire discussion in this section.

PROPOSITION 5.1. *Suppose* $\mathbf{C}$ *is a subcategory of* Sub M *for some* M *in* mod Λ.

(a) *If* $\operatorname{Ext}^1_\Lambda(C, M) = 0$ *for all* C *in* $\mathbf{C}$, *then* $\operatorname{Ext}(\mathbf{C})$ *is contained in* Sub M.

(b) *Moreover if* Ind M *is contained in* $\mathbf{C}$, *then* $\operatorname{Ext}(\mathbf{C})$ *being contained in* Sub M *implies that* $\operatorname{Ext}^1_\Lambda(C, M) = 0$ *for all* C *in* $\mathbf{C}$.

Proof. (a) Suppose $\operatorname{Ext}^1_\Lambda(C, M) = 0$ for all C in $\mathbf{C}$. Let X be in $\operatorname{Ext}(\mathbf{C})$ and suppose $0 = X_0 \subset X_1 \subset \cdots \subset X_n = X$ is a chain of submodules of X such that X_{i+1}/X_i is in $\mathbf{C}$. The proof goes by induction on the minimal length of these chains. If $n = 1$, then X is in $\mathbf{C}$ and hence in Sub M. Assume now that the claim is proved for all modules in $\operatorname{Ext}(\mathbf{C})$ with chains of length $n < k$ and suppose X in $\operatorname{Ext}(\mathbf{C})$ has a chain of length $n = k$. Then there is an exact sequence $0 \to X_{n-1} \to X \to X/X_{n-1} \to 0$ with X_{n-1} and X/X_{n-1} in Sub M. Then we have that there is an exact commutative diagram

$$\begin{array}{ccccccccc}
 & & 0 & & 0 & & & & \\
 & & \downarrow & & \downarrow & & & & \\
0 & \longrightarrow & X_{n-1} & \longrightarrow & X & \longrightarrow & X/X_{n-1} & \longrightarrow & 0 \\
 & & \downarrow & & \downarrow & & \| & & \\
0 & \longrightarrow & M' & \longrightarrow & E & \longrightarrow & X/X_{n-1} & \longrightarrow & 0
\end{array}$$

with M' in add M since Ind M is a cocover for Sub M. Since X/X_{n-1} is in $\mathbf{C}$, we have that $\operatorname{Ext}^1(X/X_{n-1}, M') = 0$, so $E = M' \amalg X/X_{n-1}$. But $X/X_{n-1} \subset M''$ in add M so $E \subset M' \amalg M''$. Hence X is contained in $M' \amalg M''$ and is therefore in Sub M. So part (a) is proven.

(b) This is a trivial consequence of the following.

LEMMA 5.2. *Let* M *be a module such that* Ind $M = \mathbf{I}_0(\text{add } M)$. *If* $0 \to M \to E \to A \to 0$ *is a nonsplit exact sequence in* mod Λ, *then* E *is not in* Sub M.

Proof. Suppose E is in Sub M. Then there is an injection $E \to M'$ with M' in add M. Thus the composition $M \to E \to M'$ is an injection which is a splittable monomorphism since Ind $M = \mathbf{I}_0(\text{add } M)$. Therefore $0 \to M \to E$ is a splittable monomorphism. This contradiction shows that E is not in Sub M.

As a consequence of Proposition 5.1 we have the following criterion for Sub M closed under extensions.

PROPOSITION 5.3. *The following are equivalent for* $\mathbf{C} =$ Sub M.

(a) **C** *is closed under extensions.*

(b) $\mathrm{Ext}^1_\Lambda(C, M) = 0$ *for all* C *in* **C**.

(c) $\mathrm{Ext}^1_\Lambda(M', M)$ *for all submodules* M' *of* M.

Proof. (a) equivalent to (b). Trivial consequence of Proposition 5.1.

(b) implies (c). Trivial.

(c) implies (b). Suppose $C \subset nM$. If $n = 1$, there is nothing to prove. Assume true for $n < k$. Then the natural projection $kM \to (k-1)M$ induces an exact sequence $0 \to C' \to C \to C'' \to 0$ with $C' \subset M$ and $C'' \subset (k-1)M$. The exact sequence

$$\mathrm{Ext}^1_\Lambda(C'', M) \to \mathrm{Ext}^1_\Lambda(C, M) \to \mathrm{Ext}^1_\Lambda(C', M)$$

gives that $\mathrm{Ext}^1_\Lambda(C, M) = 0$ since $\mathrm{Ext}^1_\Lambda(C'', M) = 0 = \mathrm{Ext}^1_\Lambda(C', M)$.

For the sake of completeness we state the duals of Propositions 5.1 and 5.3.

PROPOSITION 5.4. *Suppose* **C** *is a subcategory of* Fac M *for some* M *in* mod Λ.

(a) *If* $\mathrm{Ext}^1_\Lambda(M, C) = 0$ *for all* C *in* **C**, *then* Ext(**C**) *is contained in* Fac M.

(b) *Moreover if* Ind M *is contained in* **C**, *then* Ext(**C**) *being contained in* Fac M *implies that* $\mathrm{Ext}^1_\Lambda(M, C) = 0$ *for all* C *in* **C**.

The dual of Proposition 5.3 is the following.

PROPOSITION 5.5. *The following are equivalent for* **C** = Fac M.

(a) **C** *is closed under extensions.*

(b) $\mathrm{Ext}^1_\Lambda(M, C) = 0$ *for all* C *in* **C**.

(c) $\mathrm{Ext}^1_\Lambda(M, M'') = 0$ *for all factor modules* M'' *of* M.

In view of these results it is of interest to know when a module M has the property $\mathrm{Ext}^1_\Lambda(M', M) = 0$ for all submodules M' of M or the property $\mathrm{Ext}^1_\Lambda(M, M'') = 0$ for all factor modules M'' of M. In this connection we have the following results.

PROPOSITION 5.6. *The following statements are equivalent for a pair of modules* M, N *in* mod Λ.

(a) $\mathrm{Ext}^1_\Lambda(N', M) = 0$ *for all submodules* N' *of* N.

(b) $\underline{\mathrm{Hom}}_\Lambda(\mathrm{Tr}\, DM, N') = 0$ *for all submodules* N' *of* N.

(c) $\mathrm{Hom}_\Lambda(\mathrm{Tr}\, DM, N) = 0$.

Proof. (a) equivalent to (b). We know by [2, Chap. I, Proposition 5.4] that $\mathrm{Ext}^1_\Lambda(Y, X) \approx D(\underline{\mathrm{Hom}}(\mathrm{Tr}\, DX, Y))$ for all X and Y in mod Λ. From this the equivalence of (a) and (b) follows trivially.

(b) implies (c). Suppose f: $\mathrm{Tr}\, DM \to N$ is a nonzero morphism with $N' = \mathrm{Im}\, f$. Then the induced epimorphism f': $\mathrm{Tr}\, DM \to N'$ does not factor through a projective. For let $P \to N'$ be a projective cover of N' and h: $\mathrm{Tr}\, DM \to P$ any morphism. Then since $\mathrm{Tr}\, DM$ has no nontrivial projective summands, $h(\mathrm{Tr}\, DM) \subset \mathbf{r}P$ and so the composition $\mathrm{Tr}\, DM \to^h P \to N'$ is not an epimorphism. Therefore the image of f' in $\underline{\mathrm{Hom}}_\Lambda(\mathrm{Tr}\, DM, N')$ is not zero. Hence if $\underline{\mathrm{Hom}}_\Lambda(\mathrm{Tr}\, DM, N') = 0$ for all submodules N' of N, then $\mathrm{Hom}_\Lambda(\mathrm{Tr}\, DM, N) = 0$.

(c) implies (b). Trivial.

As an immediate application of this proposition we have the following.

COROLLARY 5.7. *The following statements are equivalent for a module M in* mod Λ.

(a) $\mathrm{Ext}^1_\Lambda(M', M) = 0$ *for all submodules M' of M.*

(b) $\underline{\mathrm{Hom}}_\Lambda(\mathrm{Tr}\, DM, M') = 0$ *for all submodules M' of M.*

(c) $\mathrm{Hom}_\Lambda(\mathrm{Tr}\, DM, M) = 0$.

For the sake of completeness we give the duals of Proposition 5.6 and Corollary 5.7.

PROPOSITION 5.8. *The following statements are equivalent for a pair of modules M, N in* mod Λ.

(a) $\mathrm{Ext}^1_\Lambda(M, N'') = 0$ *for all factor modules N'' of N.*

(b) $\overline{\mathrm{Hom}}_\Lambda(N'', D\,\mathrm{Tr}\, M) = 0$ *for all factor modules N'' of N.*

(c) $\mathrm{Hom}_\Lambda(N, D\,\mathrm{Tr}\, M) = 0$.

COROLLARY 5.9. *The following statements are equivalent for a module M in* mod Λ.

(a) $\mathrm{Ext}^1_\Lambda(M, M'') = 0$ *for all factor modules M'' of M.*

(b) $\overline{\mathrm{Hom}}_\Lambda(M'', D\,\mathrm{Tr}\, M) = 0$ *for all factors modules M'' of M.*

(c) $\mathrm{Hom}_\Lambda(M, D\,\mathrm{Tr}\, M) = 0$.

Summarizing some of these results we obtain the following existence theorem for subcategories $\mathbf{C}$ of mod Λ which are closed under extensions and are functorially finite in mod Λ.

THEOREM 5.10. *Let M and N be in* mod Λ.

(a) *If* $\operatorname{Hom}_\Lambda(\operatorname{Tr} DM, M) = 0$, *then* Sub M *is closed under extensions and is functorially finite in* mod Λ.

(b) *If* $\operatorname{Hom}_\Lambda(N, D \operatorname{Tr} N) = 0$, *then* Fac N *is closed under extensions and is functorially finite in* mod Λ.

(c) *If* $\operatorname{Hom}_\Lambda(\operatorname{Tr} DM, M) = 0 = \operatorname{Hom}_\Lambda(N, D \operatorname{Tr} N)$, *then* $\mathbf{C} = \operatorname{Sub} M \cap \operatorname{Fac} N$ *is closed under extensions and is functorially finite in* mod Λ.

Proof. (a) and (b) Already established.

(c) The fact that $\mathbf{C}$ is closed under extensions follows from (a) and (b). From the fact that $\mathbf{C} = \operatorname{Sub} M \cap \operatorname{Fac} N$, it follows easily that if f: $C_1 \to C_2$ is a morphism in $\mathbf{C}$ then $\operatorname{Im} f$ is in $\mathbf{C}$. Thus $\mathbf{C}$ is closed under images. Therefore we know by [7, Propositions 4.5, 4.10] that $\mathbf{C}$ is functorially finite in mod Λ if $\mathbf{C}$ has a finite cocover and a finite cover. But it is not difficult to see that M has a unique submodule M' maximal with respect to being in $\mathbf{C}$ and that Ind M' is a finite cocover for $\mathbf{C}$. Similarly, it is not difficult to see that there is a unique submodule N' of N minimal with respect to N/N' being in $\mathbf{C}$ and that $\operatorname{Ind}(N/N')$ is a finite cover for $\mathbf{C}$.

6. EXAMPLES

This section is devoted to applying our previous results to obtain examples of subcategories $\mathbf{C}$ of mod Λ which have almost split sequences because they are closed under extensions and are functorially finite in mod Λ.

We begin with subcategories of mod Λ which are hereditary torsion or cotorsion theories. Our first result was first obtained by Bautista and Martinez [8] in the case of the torsionless modules over 1-Gorenstein artin algebras and by Roggenkamp [9] for certain other cases.

Let Λ be an arbitrary artin algebra. For each simple Λ-module S we choose fixed projective covers $P(S)$ and injective envelopes $I(S)$. We also assume that we also have a fixed complete set $S_1, \ldots, S_n$ of nonisomorphic simple modules and $P_i = P(S_i)$ and $I_i = I(S_i)$.

Let J be a subset of $\{1, \ldots, n\}$ and let P_J denote $\coprod_{i \in J} P_i$ and let I_J denote $\coprod_{i \in J} I_i$. Sub I_J consists of all modules M with the property that each simple submodule of M is isomorphic to S_j for some j in J and Fac P_J consists of all modules M with the property that each simple submodule of $M/\mathbf{r}M$ is isomorphic to S_j for some j in J. Hence the subcategories of mod Λ which are of the form Sub I_J for some subset J of $\{1, \ldots, n\}$ are precisely the hereditary torsion theories of mod Λ and the subcategories Fac P_J are precisely the hereditary cotorsion theories of mod Λ.

PROPOSITION 6.1. *For each subset J of $\{1,\dots,n\}$ the subcategories* Sub I_J *and* Fac P_J *have almost split sequences since they are functorially finite in* mod Λ *and are closed under extensions.*

Proof. Since $\operatorname{Tr} DI_J = 0$, we have $(\operatorname{Tr} DI_J, I_J) = 0$, so by Theorem 5.10 Sub I_J is closed under extensions and is functorially finite in mod Λ. Similarly, since $D \operatorname{Tr} P_J = 0$, we have $(P_J, D \operatorname{Tr} P_J) = 0$, so again by Theorem 5.10 Fac P_J is closed under extensions and is functorially finite in mod Λ.

We next want to determine the Ext-projective and Ext-injective objects in the subcategories Sub I_J and Fac P_J. This result it based on the following.

LEMMA 6.2. *Let M be a Λ-module.*

(a) *Suppose* Sub M *is closed under extensions. Then X in* Sub M *is* Ext*-injective in* Sub M *if and only if* $(\operatorname{Tr} DX, M) = 0$.

(b) *Suppose* Fac M *is closed under extensions. Then Y in* Fac M *is* Ext*-projective if and only if* $(M, D \operatorname{Tr} Y) = 0$.

Proof. (a) By Proposition 5.6, we know that $\operatorname{Ext}^1_\Lambda(M', X) = 0$ for all submodules M' of M if and only if $(\operatorname{Tr} DX, M) = 0$. Hence if X is Ext-injective in Sub M, then $(\operatorname{Tr} DX, M) = 0$. On the other hand it is not hard to show that $\operatorname{Ext}^1_\Lambda(C, X) = 0$ for all C in Sub M if $\operatorname{Ext}^1_\Lambda(M', X) = 0$ for all submodules M' of M using the same argument as that given in proving Proposition 5.3. Hence if $(\operatorname{Tr} DX, M) = 0$, then X is Ext-injective in Sub M.

(b) Dual of (a).

PROPOSITION 6.3. *Let J be a subset of $\{1,\dots,n\}$, let K be the subset of $\{1,\dots,n\}$ consisting of all i in $\{1,\dots,n\}$ such that $(P_i, I_J) \neq 0$ and let $\mathbf{C} =$ Sub I_J. Then*

(a) $K \supset J$.

(b) *The set of all $P_k/t_{\mathbf{C}}P_k$ with k in K is a complete set of nonisomorphic indecomposable* Ext*-projectives in* Sub $I_J = \mathbf{C}$.

(c) *The set* $\{I_j\}_{j\in J} \cup \{D \operatorname{Tr}(P_i/\tau_{P_J}(P_i))\}_{i\in K-J}$ *is a complete set of nonisomorphic indecomposable* Ext*-injective modules in* Sub I_J.

(d) *The number of isomorphism classes of indecomposable* Ext*-projective is the same as the number of isomorphism classes of indecomposable* Ext*-injective modules.*

Proof. (a) Trivial.

(b) Follows from the general description of the Ext-projective objects in subcategories of the form Sub M.

(c) Clearly each I_j with $j \in J$ is Ext-injective in Sub I_J. Suppose X is an indecomposable noninjective module. By Lemma 6.2 we know that an indecomposable X is Ext-injective in Sub I_J if and only if $(\operatorname{Tr} DX, I_J) = 0$. Hence X is Ext-injective if and only if Tr DX has no simple composition factors S_i with i in J. Therefore we want to determine the indecomposable nonprojective modules Y in mod Λ which satisfy: (i) Y has no composition factors in $\{S_j\}_{j \in J}$ and (ii) $D \operatorname{Tr} Y$ is in Sub I_J or equivalently, such that the composition factors of Soc $D \operatorname{Tr} Y$ are in $\{S_j\}_{j \in J}$.

Suppose $P_1 \to^g P_0 \to^f Y \to 0$ is a minimal projective presentation of Y. Now it is easily seen that Soc $D \operatorname{Tr} Y \approx \coprod_{j \in J} n_j S_j$ if and only if $P_1 \approx \coprod_{j \in J} n_j P_j$. Also, Y has no composition factors in $\{S_j\}_{j \in J}$ if and only if Im $g \supset \tau_{P_J}(P_0)$. Therefore Y satisfies (i) and (ii) above if and only if Im $g = \tau_{P_J}(P_0)$.

Suppose Y satisfies Im $g = \tau_{P_J}(P_0)$. Since τ_{P_J} is an additive functor, Y is indecomposable if and only if P_0 is indecomposable, i.e., if and only if $Y \approx P_i/\tau_{P_J}(P_i)$ with i not in J. But $\tau_{P_J}(P_i) \neq 0$ if and only if $(P_i, I_J) \neq 0$. Therefore Y is an indecomposable nonprojective module satisfying (i) and (ii) if and only if $Y \approx P_i/\tau_{P_J}(P_i)$ with i in $K - J$. Hence a noninjective indecomposable X in **C** is Ext-injective in **C** if and only if $X \approx D \operatorname{Tr}((P_i/\tau_{P_J}(P_i))$ for some i in $K - J$. This finishes the proof of (c).

(d) Follows trivially from (c).

For the sake of completeness we state without proof the following description of the indecomposable Ext-projectives and Ext-injectives in Fac P_J. This can be obtained by duality from Proposition 6.3.

PROPOSITION 6.4. *Let J be a subset of $\{1, \ldots, n\}$, let K be the subset of $\{1, \ldots, n\}$ consisting of all i in $\{1, \ldots, n\}$ such that $(P_J, I_i) \neq 0$, let* $\mathbf{C} =$ Fac P_J *and let* $\mathbf{A} =$ Sub I_J.

(a) $K \supset J$.

(b) *The set of all $\tau_{\mathbf{C}}(I_k)$ with k in K is a complete set of nonisomorphic indecomposable* Ext-*injectives in* Fac $P_J = \mathbf{C}$.

(c) *The set* $\{P_j\}_{j \in J} \cup \{\operatorname{Tr} D(t_{\mathbf{A}} I_i)\}_{i \in K-J}$ *is a complete set of nonisomorphic* Ext-*projective modules in* Fac P_J.

(d) *The number of isomorphism classes of indecomposable* Ext-*projective is the same as the number of isomorphism classes of indecomposable* Ext-*injective modules.*

As a consequence of Propositions 6.3 and 6.4, we have the following result which was also obtained independently by R. Bautista.

PROPOSITION 6.5. *Let $S_1, \ldots, S_n$ be a complete set of nonisomorphic simple Λ-modules and J and K subsets of $\{1, \ldots, n\}$. Then the subcategory* **C** *of* mod Λ *consisting of all* M *such that every simple submodule of* M *is in* J *and every simple submodule of* $M/\mathbf{r}M$ *is in* K *has almost split sequences.*

Proof. Clearly $\mathbf{C} = \operatorname{Sub} I_J \cap \operatorname{Fac} P_K$. Since $\operatorname{Tr} DI_J = 0$, we have that $\operatorname{Hom}_\Lambda(\operatorname{Tr} DI_J, I_J) = 0$ and since $D \operatorname{Tr} P_K = 0$, we have that $\operatorname{Hom}(P_K, D \operatorname{Tr} P_K) = 0$. Hence by Theorem 5.10, $\mathbf{C}$ is closed under extensions and is functorially finite in mod Λ. Hence $\mathbf{C}$ has almost split sequences.

We now give some examples involving preprojective and preinjective modules over hereditary artin algebras Λ.

PROPOSITION 6.6. *Suppose* Λ *is a hereditary artin algebra. Let* $M = \coprod_{i=1}^{t} M_i$ *and* $N = \coprod_{j=1}^{t} N_j$ *where the* M_i *and* N_j *are indecomposable modules such that there are integers* $m \geqslant 0$ *and* $n \geqslant 0$ *with the property that* $D \operatorname{Tr}^m M_i$ *is projective and* $D \operatorname{Tr}^n N_i$ *is projective. Then the subcategories* $\operatorname{Sub} M$, $\operatorname{Fac} M$, $\operatorname{Sub} N \cap \operatorname{Fac} M$ *and* $\operatorname{Sub} M \cap \operatorname{Fac} N$ *of* mod Λ *have almost split sequences.*

Proof. The hypothesis on M implies that $(M, D \operatorname{Tr} M) = 0 = (\operatorname{Tr} DM, M) =$ and $(N, D \operatorname{Tr} M) = 0 = (\operatorname{Tr} DN, N)$ (see [3, Sect. 1]). The desired result now follows from Theorem 5.10.

For reasons similar to those used to establish Proposition 6.6 we have the following.

PROPOSITION 6.7. *Suppose* Λ *is a hereditary artin algebra. Let* $A = \coprod_{i=1}^{t} A_i$ *and* $B = \coprod_{i=1}^{s} B_j$ *where* A_i *and* B_j *are indecomposable modules such that there are integers* $m \geqslant 0$ *and* $n \geqslant 0$ *such that* $\operatorname{Tr} D^m M_i$ *and* $\operatorname{Tr} D^n N_j$ *are injective. Then the subcategories* $\operatorname{Sub} M$, $\operatorname{Fac} M$, $\operatorname{Sub} N \cap \operatorname{Fac} M$ *and* $\operatorname{Sub} M \cap \operatorname{Fac} N$ *of* mod Λ *have almost split sequences.*

C. Riedtmann has informed us of another source of examples. Namely, she knows infinite families of selfinjective algebras of finite type Λ such that $(\operatorname{Tr} DM, M) = 0 = (M, D \operatorname{Tr} M)$ for all indecomposable Λ-modules M.

As another example of modules M satisfying $(\operatorname{Tr} DM, M) = 0$ or $(M, D \operatorname{Tr} M) = 0$, we have the following.

PROPOSITION 6.8. *Let* S *be a simple* Λ*-module such that* $\operatorname{Ext}^1_\Lambda(S, S) = 0$. *Then the subcategories* $\operatorname{Sub}(D \operatorname{Tr} S)$, $\operatorname{Fac}(\operatorname{Tr} DS)$ *and* $\operatorname{Sub}(D \operatorname{Tr} S) \cap \operatorname{Fac}(\operatorname{Tr} DS)$ *of* mod Λ *have almost split sequences.*

Proof. Let $0 \to \mathbf{r}P \to P \to S \to 0$ be exact with $P \to S \to 0$ a projective cover. Then $\operatorname{Ext}^1(S, S) = 0$ if and only if S is not a submodule of $\mathbf{r}P/\mathbf{r}^2P$. Since $\mathbf{r}P/\mathbf{r}^2P \approx \operatorname{Soc} D \operatorname{Tr} S$ (see [5, Proposition 5.3]), the assumption $\operatorname{Ext}^1(S, S) = 0$ implies $(S, D \operatorname{Tr} S) = 0$. Because $\operatorname{Tr} D(D \operatorname{Tr} S) = S$, we have that $(\operatorname{Tr} D(D \operatorname{Tr} S), D \operatorname{Tr} S)=0$. Hence $\operatorname{Sub}(D \operatorname{Tr} S)$ has almost split sequences. The dual argument shows that $\operatorname{Ext}_\Lambda^1(S, S)=0$ implies $(\operatorname{Tr} DS, S)=0$. So $(\operatorname{Tr} DS, D \operatorname{Tr}(\operatorname{Tr} DS))=0$ and hence $\operatorname{Fac}(\operatorname{Tr} DS)$ has almost split sequences as does $\operatorname{Sub}(D \operatorname{Tr} S) \cap \operatorname{Fac}(\operatorname{Tr} DS)$.

There are many examples of artin algebras Λ having simple modules S satisfying $\operatorname{Ext}^1(S, S)=0$. We just cite one family of examples. Suppose Λ is an artin algebra of finite type with $M_1, \ldots, M_n$ a complete set of nonisomorphic indecomposable Λ-modules. Then $\Gamma=\operatorname{End}_\Lambda(\coprod_{i=1}^n M_i)$ has the property that $\operatorname{Ext}_\Gamma^1(S, S)=0$ for all simple Γ-modules S.

We now point out another source of examples of subcategories of $\bmod \Lambda$ having almost split sequences.

Let Λ be an arbitrary artin algebra. The subcategory $\operatorname{Sub} \Lambda$ consists of the submodules of projective Λ-modules. Then $I_0(\operatorname{Sub} \Lambda)$ consists of those indecomposable projective modules P with the property that if $0 \rightarrow P \rightarrow T$ is exact with T in $\operatorname{Sub} \Lambda$, then $0 \rightarrow P \rightarrow T$ is a splittable monomorphism. We say that a projective module is maximal if it is in $I_0(\operatorname{Sub} \Lambda)$. We know by Proposition 5.3 that $\operatorname{Sub} \Lambda$ is closed under extensions if and only if for each maximal projective P we have that $\operatorname{Ext}_\Lambda^1(C, P)=0$ for all C in $\operatorname{Sub} \Lambda$. But $\operatorname{Ext}_\Lambda^1(C, P)=0$ for all C in $\operatorname{Sub} \Lambda$ if and only if $\operatorname{Ext}_\Lambda^2(X, P)=0$ for all X in $\bmod \Lambda$, or equivalently, $\operatorname{inj} \dim P \leqslant 1$. Therefore we have shown that $\operatorname{Sub} \Lambda$ is closed under extensions if and only if $\operatorname{inj} \dim P \leqslant 1$ for all maximal projective modules P. Hence we have proven

PROPOSITION 6.9. *Let Λ be an arbitrary artin algebra. Then* $\operatorname{Sub} \Lambda$ *has almost split sequences if and only if* $\operatorname{inj} \dim P \leqslant 1$ *for all maximal projective modules P.*

As an easy consequence of this result we have the following due to Bautista and Martinez [8].

COROLLARY 6.10. *Let Λ be a 1-Gorenstein artin algebra*; *i.e.*, $\operatorname{pd} I_0(\Lambda)=0$ *where $I_0(\Lambda)$ is the injective envelope of Λ. Then* $\operatorname{Sub} \Lambda$ *has almost split sequences.*

Proof. Since $I_0(\Lambda)$ is projective, $I_0(\operatorname{Sub} \Lambda)=\operatorname{Ind} I_0(\Lambda)$. Therefore $\operatorname{inj} \dim P=0$ for all maximal projective modules P. Our desired result now follows trivially from Proposition 6.9.

We now end this section by pointing out how, given a subcategory $\operatorname{Sub} M$ of $\bmod \Lambda$ having almost split sequences, we can obtain other such subcategories. Suppose $\mathbf{C}=\operatorname{Sub} M$ has almost split sequences or, what is the same thing, is closed under extensions. Then for each indecomposable injective module I in $\bmod \Lambda$ we have the exact sequence $0 \rightarrow A_{\mathbf{C}}^I \rightarrow I_{\mathbf{C}} \rightarrow I$ described in Section 4, where $\bigcup_{I \in \mathbf{I}_0} \operatorname{Ind} A_{\mathbf{C}}^I$ consists of the indecomposable Ext-injective modules in $\mathbf{C}$ which are not part of a minimal cocover for $\mathbf{C}$.

Let M_1 be the sum of a complete set of nonisomorphic modules in $\bigcup_{I \in \mathbf{I}_0} \operatorname{Ind} A^I_{\mathbf{C}}$. Then $\operatorname{Ext}^1_\Lambda(C, M_1) = 0$ for all C in $\mathbf{C}$ since certainly $\operatorname{Ext}^1_\Lambda(C, M) = 0$ for all C in $\operatorname{Sub} M_1 \subset \operatorname{Sub} M$. Hence $\operatorname{Sub} M_1$ has almost split sequences. Proceeding by induction we obtain a sequence of subcategories with almost split sequences $\operatorname{Sub} M \supset \operatorname{Sub} M_1 \supset \cdots \supset \operatorname{Sub} M_i \supset \cdots$ naturally associated with the subcategory $\operatorname{Sub} M$.

Finally it should be observed that $M_1 = 0$ or equivalently $A^I_{\mathbf{C}} = 0$ for all I in $\mathbf{I}_0$ if and only if $\tau_{\mathbf{C}}(I)$ is in $\mathbf{C}$ for all I in $\mathbf{I}_0$ if and only if the indecomposable Ext-injective modules in $\mathbf{C}$ are precisely the modules in $\mathbf{I}_0(\mathbf{C})$.

A similar discussion can be carried out for the subcategories $\operatorname{Fac} M$ with almost split sequences.

7. Further Examples

The examples of subcategories of $\operatorname{mod} \Lambda$ having almost split sequences we gave in Section 6 were all of the form $\operatorname{Sub} M$ or $\operatorname{Fac} M$. In this section we give a different type of example, namely:

PROPOSITION 7.1. *Let Λ be an arbitrary artin algebra and $\mathbf{C}$ the subcategory of* $\operatorname{mod} \Lambda$ *consisting of all modules M such that* $\operatorname{Ext}^1_\Lambda(M, \Lambda) = 0$. *Then $\mathbf{C}$ is functorially finite in* $\operatorname{mod} \Lambda$ *and has almost split sequences.*

Proof. Since $\mathbf{C}$ is clearly closed under extensions, if we show that $\mathbf{C}$ is functorially finite in $\operatorname{mod} \Lambda$ we will also have that $\mathbf{C}$ has almost split sequences. Now for each Λ^{op}-module N we have a natural exact sequence $0 \to \operatorname{Ext}^1_\Lambda(\operatorname{Tr} N, \Lambda) \to N \to N^{**}$ where $N \to N^{**}$ is the usual morphism $N \to \operatorname{Hom}_\Lambda(\operatorname{Hom}_\Lambda(N, \Lambda), \Lambda)$ (see [1, Proposition 6.3]). Hence $N \to N^{**}$ is a monomorphism if and only if $\operatorname{Ext}^1_\Lambda(\operatorname{Tr} N, \Lambda) = 0$. Therefore M is in $\mathbf{C}$ if and only if $\operatorname{Tr} M$ is in $\operatorname{Sub} \Lambda^{\mathrm{op}}$. Or equivalently $\mathbf{C}$ is $\operatorname{add}(\operatorname{Tr}(\operatorname{Sub} \Lambda^{\mathrm{op}}) \cup \mathbf{P}_0(\Lambda))$. We now want to use this description combined with the fact that $\operatorname{Sub} \Lambda^{\mathrm{op}}$ is functorially finite in $\operatorname{mod} \Lambda^{\mathrm{op}}$, to show that $\mathbf{C}$ is functorially finite in $\operatorname{mod} \Lambda$. This is a trivial consequence of the following general considerations which are of interest in their own right.

We recall that a subcategory $\mathbf{C}$ of $\operatorname{mod} \Lambda$ is contravariantly (covariantly) finite in $\operatorname{mod} \Lambda$ if and only if $\operatorname{Ind} \mathbf{C}$ is contravariantly (covariantly) finite in $\operatorname{Ind}(\operatorname{mod} \Lambda)$ when $\operatorname{Ind} \mathbf{C}$ being contravariantly (coveriantly) finite in $\operatorname{Ind}(\operatorname{mod} \Lambda)$ means that for each X in $\operatorname{Ind}(\operatorname{mod} \Lambda)$ there is a morphism $C' \to X$ $(X \to C')$ with C' in $\operatorname{add} \operatorname{Ind} \mathbf{C}$ such that $(C, C') \to (C, X) \to 0$ $((C', C) \to (X, C) \to 0)$ is exact for all C in $\operatorname{Ind} \mathbf{C}$ (see [7, Sect. 3]). Hence to prove Proposition 7.1 it suffices to show $\operatorname{Ind} \mathbf{C}$ is functorially finite in $\operatorname{Ind}(\operatorname{mod} \Lambda)$.

PROPOSITION 7.2. *Let* Λ *be an arbitrary artin algebra and* $\mathbf{B}$ *a subcategory of* Ind(mod Λ).

(a) *If* $\mathbf{B}$ *is contravariantly finite in* Ind(mod Λ), *then* $\operatorname{Tr} \mathbf{B} \cup \mathbf{P}_0(\Lambda^{op})$ *is covariantly finite in* Ind (mod Λ^{op}).

(b) *If* $\mathbf{B}$ *is covariantly finite in* Ind(mod Λ), *then* $\operatorname{Tr} \mathbf{B} \cup \mathbf{P}_0(\Lambda^{op})$ *is contravariantly finite in* Ind(mod Λ^{op}).

Proof. (a) Let X be in Ind(mod Λ^{op}). If X is in $\mathbf{P}_0(\Lambda^{op})$ there is nothing to prove. Suppose X is not projective and let $Y = \operatorname{Tr} X$. Since $\mathbf{B}$ is contravariantly finite in Ind(mod Λ), there is a morphism $f\colon B \to Y$ with B in add $\mathbf{B}$ such that $(B', B) \to (B', Y) \to 0$ is exact for all B' in $\mathbf{B}$. Because Tr: $\underline{\operatorname{mod}}\, \Lambda \to \underline{\operatorname{mod}}\, \Lambda^{op}$ is a duality, we have that $\operatorname{Tr}(f)$: $\operatorname{Tr} Y \to \operatorname{Tr} B$ has the property that the induced morphism $\underline{\operatorname{Hom}}(\operatorname{Tr} B, V) \to \underline{\operatorname{Hom}}(\operatorname{Tr} Y, V) \to 0$ is exact for all V in $\operatorname{Tr} \mathbf{B}$. Now we know that there is a morphism $\operatorname{Tr} Y \to Q$ with Q projective such that $(Q, M) \to (\operatorname{Tr} Y, M) \to \underline{\operatorname{Hom}}(\operatorname{Tr} Y, M) \to 0$ is exact for all M in mod Λ^{op}. Namely, if $P \to (\operatorname{Tr} Y)^* \to 0$ is a projective cover, then the composition $\operatorname{Tr} Y \to (\operatorname{Tr} Y)^{**} \to P^*$ is our desired morphism $\operatorname{Tr} Y \to Q$. Hence the induced morphism $\operatorname{Tr} Y \to \operatorname{Tr} B \sqcup Q$ has the property that $(\operatorname{Tr} B \sqcup Q, V) \to (\operatorname{Tr} Y, V) \to 0$ is exact for all V in mod Λ^{op}. Since $X = \operatorname{Tr} Y$ was an arbitrary nonprojective module in Ind(mod Λ^{op}), we have shown that $\operatorname{Tr} \mathbf{C} \cup P_0(\Lambda^{op})$ is covariantly finite in Ind(mod Λ).

(b) Proven similarly.

COROLLARY 7.3. *Let* $\mathbf{B}$ *be a subcategory of* Ind(mod Λ). *Then the following are equivalent*:

(a) $\mathbf{B}$ *is contravariantly (covariantly) finite in* Ind(mod Λ).

(b) $\operatorname{Tr} \mathbf{B} \cup \mathbf{P}_0(\Lambda^{op})$ *is covariantly (contravariantly) finite in* Ind(mod Λ).

(c) $\operatorname{Tr} \mathbf{B}$ *is covariantly (contravariantly) finite in* Ind(mod Λ^{op}).

(d) $D \operatorname{Tr} \mathbf{B}$ *is contravariantly (covariantly) finite in* Ind(mod Λ).

Proof. (a) equivalent to (b). Shown in Proposition 7.2.

(b) equivalent to (c). We know by [7, Proposition 3.13] that since $\mathbf{P}_0(\Lambda^{op})$ has only a finite number of nonisomorphic modules, $\operatorname{Tr} \mathbf{C} \cup \mathbf{P}_0(\Lambda^{op})$ is covariantly (contravariantly) finite in Ind(mod Λ^{op}) if and only if $\operatorname{Tr} \mathbf{C}$ is covariantly (contravariantly) finite in Ind(mod Λ^{op}).

(c) equivalent to (d). Follows by duality.

Clearly Proposition 7.1 follows from Corollary 7.3.

References

1. M. Auslander, Coherent functors, *in* "Proceedings Conference on Categorical Algebra, La Jolla, 1965," pp. 189–231, Springer-Verlag, New York/Berlin.
2. M. Auslander, Functors and morphisms determined by objects, *in* "Proceedings, Conference at Temple University, 1976," Lecture Notes in Pure and Applied Mathematics, Vol. 37, Dekker, New York, 1978.
3. M. Auslander and M. I. Platzeck, Representation theory of hereditary artin algebras, *in* "Proceedings, Conference at Temple University, 1976," Lecture Notes in Pure and Applied Mathematics, Vol. 37, Dekker, New York, 1978.
4. M. Auslander and I. Reiten, Stable equivalence of dualizing R-varieties, *Advances in Math.* **12**, No. 3 (1974).
5. M. Auslander and I. Reiten, Representation theory of artin algebras. III. Almost split sequences, *Comm. Algebra* **3**, No. 3 (1975), 239–294.
6. M. Auslander and I. Reiten, Representation theory of artin algebras. IV. Invariants given by almost split sequences, *Comm. Algebra* **5** (1977), 443–518.
7. M. Auslander and S. O. Smalø, Preprojective modules over artin algebras, *J. Algebra* **66** (1980), 61–122.
8. R. Bautista and R. Martinez, Representations of partially ordered sets and 1-Gorenstein Artin Algebras, *in* "Proceedings, Conference on Ring Theory, Antwerp, 1978," Dekker, New York, 1979.
9. K. W. Roggenkamp, Auslander–Reiten sequences for "nice" torsion theories of Artin algebras, *Canad. Math. Bull.* **23**, No. 1 (1980), 61–65.

Printed by the St. Catherine Press Ltd., Tempelhof 41, Bruges, Belgium

Reprinted from ADVANCES IN MATHEMATICS
All Rights Reserved by Academic Press, New York and London

Vol. 86, No. 1, March 1991
Printed in Belgium

Applications of Contravariantly Finite Subcategories

MAURICE AUSLANDER*,† AND IDUN REITEN‡

**Department of Mathematics, Brandeis University, Waltham, Massachusetts 02254; and*
Department of Mathematics and Statistics, The University of Trondheim, Dragvoll, Norway

INTRODUCTION

Throughout this paper we assume that all modules are finitely generated over an artin algebra Λ. We denote by mod Λ the category of all finitely generated Λ-modules.

The notions of contravariantly and covariantly finite subcategories of mod Λ (see Sect. 1 for definitions) were introduced in [8, 9] by Auslander and Smalø in connection with studying the problem of which subcategories of mod Λ have almost split sequences. While subcategories of mod Λ which are either covariantly or contravariantly finite in mod Λ are also the main objects of study in this paper, the point of view is quite different from [8] or [9]. Here we show that these notions are intimately related to questions about (a) the subcategory $\mathscr{P}^\infty(\Lambda)$ of mod Λ consisting of the modules of finite projective dimension and (b) tilting and cotilting theories.

As far as $\mathscr{P}^\infty(\Lambda)$ is concerned, two problems are discussed: when is $\mathscr{P}^\infty(\Lambda)$ contravariantly finite in mod Λ and what does its being contravariantly finite imply about $\mathscr{P}^\infty(\Lambda)$? If Λ is of finite representation type, then $\mathscr{P}^\infty(\Lambda)$ is contravariantly finite since every subcategory of mod Λ is contravariantly finite. In fact, Λ is of finite representation type if and only if every subcategory of mod Λ is contravariantly (covariantly) finite in mod Λ (see Prop. 1.2). We also show that $\mathscr{P}^\infty(\Lambda)$ is contravariantly finite when Λ is stably equivalent to a hereditary algebra. However, an example due to Igusa *et al.* [13] shows that $\mathscr{P}^\infty(\Lambda)$ is not always contravariantly finite in mod Λ. It is not hard to see that $\mathscr{P}^\infty(\Lambda)$ is contravariantly finite in mod Λ if and only if $\mathscr{P}^\infty(T_2(\Lambda))$ is contravariantly finite in mod $T_2(\Lambda)$, where $T_2(\Lambda) = \left(\begin{smallmatrix}\Lambda & 0\\ \Lambda & \Lambda\end{smallmatrix}\right)$. A more general result is proved by Smalø in [18]. Combining these observations we get a very wide variety of algebras Λ

* Partially supported by NAVF (Norwegian Research Council for Science and Humanities) and NSF Grant DMS 8603487.
† Partially supported by NAVF.

0001-8708/91 $7.50

with $\mathscr{P}^\infty(\Lambda)$ contravariantly finite in mod Λ and of algebras with $\mathscr{P}^\infty(\Lambda)$ not contravariantly finite in mod Λ. An interesting problem is to characterize those Λ with $\mathscr{P}^\infty(\Lambda)$ contravariantly finite in mod Λ.

Suppose now that $\mathscr{P}^\infty(\Lambda)$ is contravariantly finite. Then we show two things about $\mathscr{P}^\infty(\Lambda)$. One is that the projective dimensions of the modules in $\mathscr{P}^\infty(\Lambda)$ are bounded, i.e., the finiteness of the finitistic dimension of Λ holds. The other result is that there are a finite number of modules $A_1, ..., A_n$ in $\mathscr{P}^\infty(\Lambda)$ such that the modules in $\mathscr{P}^\infty(\Lambda)$ are precisely the direct summands of modules M having filtrations $M = M_0 \supset M_1 \supset \cdots \supset M_n = 0$ with each $M_i/M_{i+1} \cong A_j$ for some $j = 1, ..., n$. This sort of says that, in the sense just described, $\mathscr{P}^\infty(\Lambda)$ is finitely generated.

An obvious, but important, property of $\mathscr{P}^\infty(\Lambda)$ is that it is a resolving subcategory, i.e., is closed under extensions, kernels of surjective morphisms, and contains the projective modules. These types of subcategories arise in cotilting theory in the following way. Suppose with each module T having the property $\operatorname{Ext}^i_\Lambda(T, T) = 0$ for all $i > 0$ we associate the subcategory ${}^\perp T$ consisting of all X in mod Λ such that $\operatorname{Ext}^i_\Lambda(X, T) = 0$ for all $i > 0$. It is obvious that ${}^\perp T$ is a resolving subcategory of mod Λ. We show that T is a cotilting module of finite injective dimension (see Section 5) if and only if the resolving subcategory ${}^\perp T$ is contravariantly finite in mod Λ and every Λ-module has a finite resolution in ${}^\perp T$; i.e., for each C in mod Λ there is an exact sequence $0 \to X_n \to \cdots \to X_0 \to C \to 0$ with the X_i in ${}^\perp T$. In fact sending T to ${}^\perp T$ gives a one–one correspondence between isomorphism classes of basic cotilting modules and contravariantly finite resolving subcategories of mod Λ such that every module has a finite resolution in the subcategory. Here we say that a module C is basic if in a direct sum decomposition into indecomposable modules, no indecomposable module appears more than once. When the global dimension of Λ is finite, this gives a one–one correspondence between the isomorphism classes of basic cotilting modules and all contravariantly finite resolving subcategories of mod Λ. As a consequence of this, we show that when Λ is hereditary all the contravariantly finite resolving subcategories of mod Λ are of the form Sub T for some cotilting module T, where Sub T consists of all submodules of finite sums (direct) of copies of T.

In the final section of the paper, some special types of cotilting modules are introduced in terms of which the notions of Cohen–Macaulay and Gorenstein artin algebras are defined. As the names suggest, these are analogues for artin algebras of the familiar notions of Cohen–Macaulay and Gorenstein rings which are commutative complete local noetherian rings. The essential point here is that the dualizing modules for such commutative rings are cotilting modules if one gives a slightly different, but equivalent, definition for cotilting modules over artinian algebras.

Some of the results in this paper were announced at the May, 1988

Conference on Representations of Algebras at the Banach Center, and some results have been obtained independently by Wakamatsu [20].

1. Basic Results

This section consists of variations on the following basic result. Let $\mathscr{T}$ be a subcategory of mod Λ which is closed under extensions and is covariantly finite. Then the subcategory of mod Λ of all X such that $\operatorname{Ext}^1_\Lambda(X, T)=0$ for all T in $\mathscr{T}$ is contravariantly finite in mod Λ. These variations consist of showing how this result can be used to construct contravariantly finite or covariantly finite subcategories from other covariantly or contravariantly finite subcategories. The rest of the paper is based in an essential way on the results in this section.

Before recalling some basic facts about contravariantly and covariantly finite subcategories of mod Λ, we discuss the notions of right and left minimal morphisms introduced in [8].

A morphism $f\colon B \to C$ in mod Λ is said to be *right minimal* if an endomorphism $g\colon B \to B$ is an automorphism whenever $f = fg$. The importance of right minimal morphisms lies in the following (see [8]).

Proposition 1.1. *Let $f\colon B \to C$ be a morphism.*

(a) *There is a decomposition $B = B_0 \amalg B_1$ (direct sum) such that $f \mid B_0$ is right minimal and $f \mid B_1 = 0$.*

(b) *If $B = B_0^{(i)} \amalg B_1^{(i)}$ for $i = 0, 1$ are such that $f \mid B_0^{(i)}$ is right minimal and $f \mid B_1^{(i)} = 0$ for $i = 0, 1$, then there is an automorphism $g\colon B \to B$ such that $f = fg$ and $g(B_j^{(0)}) = B_j^{(1)}$ for $j = 0, 1$ and $i = 0, 1$.*

Given a morphism $f\colon B \to C$, we will often denote by $f_0\colon B_0 \to C$ a right minimal version of f, i.e., $B = B_0 \amalg B_1$ and $f_0 = f \mid B_0$ is right minimal and $f \mid B_1 = 0$. It is not difficult to see that two right minimal versions f_0 and f_0' of f are isomorphic in the sense that there is an isomorphism $g\colon B_0 \to B_0'$ such that $f_0 = f_0' g$.

The notion of a left minimal morphism is dual to that of a right minimal morphism. We will often use, without stating them, results for left minimal morphisms which are dual to those already given for right minimal morphisms.

We now begin our discussion of contravariantly and covariantly finite subcategories with the notion of right and left approximations of modules. By a subcategory $\mathscr{X}$ of mod Λ we always mean a full subcategory closed under isomorphisms and summands. A morphism $f\colon X \to C$ is said to be a right $\mathscr{X}$-approximation of C if $(\mathscr{X}, X) \xrightarrow{(\mathscr{X}, f)} (\mathscr{X}, C) \to 0$ is exact, i.e.,

$\operatorname{Hom}_\Lambda(X', X) \to \operatorname{Hom}_\Lambda(X', C) \to 0$ is exact for all X' in $\mathscr{X}$. It is easily seen that if $f\colon X \to C$ is a right $\mathscr{X}$-approximation of C, then any right minimal version $f_0\colon X_0 \to C$ of f is also right $\mathscr{X}$-approximation of C. A right $\mathscr{X}$-approximation $h\colon X \to C$ is said to be a minimal right $\mathscr{X}$-approximation of C if h is a right minimal morphism. Hence a module C has a minimal right $\mathscr{X}$-approximation if it has a right $\mathscr{X}$-approximation. It follows directly from the definition that two minimal right $\mathscr{X}$-approximations $h_i\colon X_i \to C$, $i = 1, 2$, are isomorphic in the sense that there is an isomorphism $g\colon X_1 \to X_2$ such that $h_1 = h_2 g$. We will often use the notation $f_C\colon X_C \to C$ to denote a minimal right $\mathscr{X}$-approximation of C. The subcategory $\mathscr{X}$ is said to be *contravariantly finite* in mod Λ if every C in mod Λ has a right $\mathscr{X}$-approximation. Stated in terms of functors, a morphism $h\colon X \to C$ with X in $\mathscr{X}$ is a right $\mathscr{X}$-approximation if and only if $(\;, X)|\mathscr{X} \xrightarrow{(\;,f)|\mathscr{X}} (\;, C)|\mathscr{X} \to 0$ is exact, where $(\;, C)|\mathscr{X}$ means the functor $\operatorname{Hom}_\Lambda(\;, C) = (\;, C)$ restricted to $\mathscr{X}$. Therefore C has a right $\mathscr{X}$-approximation if and only if $(\;, C)|\mathscr{X}$ is a finitely generated contravariant functor from $\mathscr{X}$ to abelian groups. This explains the terminology $\mathscr{X}$ is contravariantly finite in mod Λ if every C in mod Λ has a right $\mathscr{X}$-approximation.

Again we will use freely the notions of left $\mathscr{X}$-approximations, minimal left $\mathscr{X}$-approximations, and $\mathscr{X}$ being covariantly finite in mod Λ, which are the obvious duals of the notions given above, without explicitly giving their definitions.

While we will see later on various examples of subcategories of mod Λ which are not contravariantly finite and subcategories which are not covariantly finite, we have the following result which should help to put in perspective the general question of subcategories being contravariantly (covariantly) finite in mod Λ.

PROPOSITION 1.2. *The following are equivalent for* mod Λ.

(a) *Λ is of finite representation type;*

(b) *every subcategory of* mod Λ *is covariantly finite;*

(c) *every subcategory of* mod Λ *is contravariantly finite.*

Proof. (a) $\Rightarrow$ (b), (c). It is easily seen [8, Prop. 4.2] that a subcategory $\mathscr{X}$ of mod Λ with only a finite number of nonisomorphic indecomposable objects is both covariantly and contravariantly finite.

(b) $\Rightarrow$ (a). Suppose Λ is of infinite representation type. Then it was shown in [8, Prop. 7.1] that add $\mathscr{P}_\infty$, where $\mathscr{P}_\infty$ is the nonpreprojective indecomposable modules, has no finite cover. We recall that a category $\mathscr{X}$ is a *cover* for $\mathscr{Y}$ if every object in $\mathscr{Y}$ is a factor of an object in $\mathscr{X}$. This implies that add $\mathscr{P}_\infty$ is not covariantly finite.

(c) $\Rightarrow$ (a). Dual of (b) implies (a).

While we cannot say very much about the properties of $\mathscr{X}$-approximations for arbitrary subcategories $\mathscr{X}$, we obtain some interesting results when we assume that $\mathscr{X}$ is closed under extensions; i.e., if $0 \to X \to Y \to Z \to 0$ is an exact sequence in mod Λ with X and Z in $\mathscr{X}$, then Y is in $\mathscr{X}$. We now give some results along these lines which we will be using throughout the rest of this paper.

The first of these results is due to Wakamatsu [20] and will be referred to as Wakamatsu's Lemma. For the convenience of the reader, we state it here without proof.

LEMMA 1.3. *Let $\mathscr{X}$ be a subcategory of* mod Λ *which is closed under extensions. Let C be an arbitrary Λ-module.*

(a) *If $0 \to Y \to X \xrightarrow{f} C$ is exact with f a right minimal $\mathscr{X}$-approximation of C, then* $\mathrm{Ext}^1_\Lambda(\mathscr{X}, Y) = 0$, *i.e.,* $\mathrm{Ext}^1_\Lambda(X', C) = 0$ *for all X' in $\mathscr{X}$.*

(b) *If $C \xrightarrow{g} X \to Z \to 0$ is exact with g a left $\mathscr{X}$-approximation, then* $\mathrm{Ext}^1_\Lambda(Z, \mathscr{X}) = 0$.

Our next result along these lines is the following.

PROPOSITION 1.4. *Suppose $\mathscr{Y}$ is a subcategory of* mod Λ *closed under extensions and let $\mathscr{X}$ be the subcategory of* mod Λ *consisting of all X such that* $\mathrm{Ext}^1(X, \mathscr{Y}) = 0$. *The following are equivalent for a C in* mod Λ.

(a) $\mathrm{Ext}^1_\Lambda(C, \ )|\mathscr{Y}$ *is a finitely generated functor from $\mathscr{Y}$ to abelian groups.*

(b) *There is an exact sequence $0 \to Y \to X \xrightarrow{f} C \to 0$ with f a minimal right $\mathscr{X}$-approximation of C and Y in $\mathscr{Y}$.*

Proof. (b) $\Rightarrow$ (a). From the exact sequence $(X, \mathscr{Y}) \to (Y, \mathscr{Y}) \to \mathrm{Ext}^1_\Lambda(C, \mathscr{Y}) \to \mathrm{Ext}^1_\Lambda(X, \mathscr{Y})$ and the fact that $\mathrm{Ext}^1_\Lambda(X, \mathscr{Y}) = 0$, it follows that $(Y, \mathscr{Y}) \to \mathrm{Ext}^1_\Lambda(C, \mathscr{Y}) \to 0$ is exact. Hence $(Y, \ )|\mathscr{Y} \to \mathrm{Ext}^1_\Lambda(C, \ )|\mathscr{Y} \to 0$ is exact with Y in $\mathscr{Y}$, which means that $\mathrm{Ext}^1_\Lambda(C, \ )|\mathscr{Y}$ is finitely generated.

(a) $\Rightarrow$ (b). Since $\mathrm{Ext}^1_\Lambda(C, \ )|\mathscr{Y}$ is finitely generated and $\mathscr{Y}$ is closed under summands, it follows that $\mathrm{Ext}^1_\Lambda(C, \ )|\mathscr{Y}$ has a projective cover $(Y, \ )\ \mathscr{Y} \xrightarrow{\varphi} \mathrm{Ext}^1_\Lambda(C, \ )|\mathscr{Y}$. Let $0 \to Y \to X \xrightarrow{f} C \to 0$ be $\varphi(1_X)$. That $f: X \to C$ is right minimal is an immediate consequence of the fact that $\varphi: (Y, \ )|\mathscr{Y} \to \mathrm{Ext}^1_\Lambda(C, \ )|\mathscr{Y}$ is a projective cover. Suppose now we show that X is in $\mathscr{X}$. Then we would have the exact sequence $(\mathscr{X}, X) \to (\mathscr{X}, C) \to \mathrm{Ext}^1_\Lambda(\mathscr{X}, Y)$ with $\mathrm{Ext}^1_\Lambda(\mathscr{X}, Y) = 0$. Therefore the right minimal morphism $f: X \to C$ would also be a right $\mathscr{X}$-approximation of C. Hence the proof of the proposition would be complete since Y is in $\mathscr{Y}$ by definition. So we now turn our attention to showing that the X in the exact sequence $0 \to Y \to X \to C \to 0$ is in $\mathscr{X}$.

Since we want to show that X is in $\mathscr{X}$, we want to show that any exact sequence $0 \to Y' \to U \xrightarrow{g} X \to 0$ with Y' in $\mathscr{Y}$ splits.

Given such an exact sequence, we obtain the exact commutative pullback diagram

$$\begin{array}{ccccccccc}
 & & & & & & 0 & & \\
 & & & & & & \downarrow & & \\
0 & \longrightarrow & Y' & \longrightarrow & g^{-1}(Y) & \longrightarrow & Y & \longrightarrow & 0 \\
 & & \| & & \downarrow & & \downarrow & & \\
0 & \longrightarrow & Y' & \longrightarrow & U & \xrightarrow{g} & X & \longrightarrow & 0 \\
 & & & & \downarrow & & \downarrow & & \\
 & & & & C & = & C & & \\
 & & & & \downarrow & & \downarrow & & \\
 & & & & 0 & & 0 & &
\end{array} \qquad (*)$$

Since $\mathscr{Y}$ is closed under extensions, it follows that $g^{-1}(Y)$ is in $\mathscr{Y}$. Also the fact that $(Y, \ )|\mathscr{Y} \to \operatorname{Ext}^1_\Lambda(C, \ )|\mathscr{Y} \to 0$ is exact implies that there is a commutative diagram

$$\begin{array}{ccccccccc}
0 & \longrightarrow & Y & \longrightarrow & X & \xrightarrow{f} & C & \longrightarrow & 0 \\
 & & \downarrow & & \downarrow & & \| & & \\
 & & g^{-1}(Y) & \longrightarrow & U & \longrightarrow & C & \longrightarrow & 0
\end{array} \qquad (**)$$

We therefore obtain from $(*)$ and $(**)$ the commutative diagram

$$\begin{array}{ccccccccc}
0 & \longrightarrow & Y & \longrightarrow & X & \xrightarrow{f} & C & \longrightarrow & 0 \\
 & & \downarrow & & \downarrow & & \| & & \\
0 & \longrightarrow & g^{-1}(Y) & \longrightarrow & U & \longrightarrow & C & \longrightarrow & 0 \\
 & & \downarrow & & \downarrow{\scriptstyle g} & & \| & & \\
0 & \longrightarrow & Y & \longrightarrow & X & \xrightarrow{f} & C & \longrightarrow & 0
\end{array}$$

with exact rows. From the fact that f is right minimal, it follows that the composition $Y \to g^{-1}(Y) \to Y$ is an isomorphism; hence the composition

$X \to U \xrightarrow{g} X$ is an isomorphism. Therefore $0 \to Y' \to U \to X \to 0$ splits, giving our desired result that $\mathrm{Ext}^1_\Lambda(X, \mathscr{Y}) = 0$, i.e., X is in $\mathscr{X}$.

As an immediate consequence of this result we have the following.

COROLLARY 1.5. *Let $\mathscr{Y}$ be a subcategory of* mod Λ *closed under extensions and let* $\mathscr{X} = \{X$ in mod $\Lambda \mid \mathrm{Ext}^1_\Lambda(X, \mathscr{Y}) = 0\}$. *Then the following are equivalent*:

(a) $\mathrm{Ext}^1_\Lambda(C, \ )|\mathscr{Y}$ *is finitely generated for all C in* mod Λ.

(b) *$\mathscr{X}$ is contravariantly finite in* mod Λ *and if* $0 \to Y \to X \xrightarrow{f} C \to 0$ *is exact with f a minimal right $\mathscr{X}$-approximation of C, then Y is in $\mathscr{Y}$.*

It should be noted that Λ is in $\mathscr{X}$ and so all right $\mathscr{X}$-approximations are epimorphisms. It should also be noted that $\mathscr{X}$ is closed under extensions.

We now state without proofs the duals of Proposition 1.4 and Corollary 1.5. The dual of Proposition 1.4 is as follows.

PROPOSITION 1.6. *Suppose $\mathscr{X}$ is a subcategory of* mod Λ *closed under extensions and let* $\mathscr{Y} = \{Y$ *in* mod $\Lambda \mid \mathrm{Ext}^1_\Lambda(\mathscr{X}, Y) = 0\}$. *The following are equivalent for a C in* mod Λ.

(a) $\mathrm{Ext}^1_\Lambda(\ , C)|\mathscr{X}$ *is a finitely generated functor from $\mathscr{X}$ to abelian groups.*

(b) *There is an exact sequence* $0 \to C \xrightarrow{g} Y \to X \to 0$ *with g a minimal left $\mathscr{Y}$-approximation of C and X in $\mathscr{X}$.*

We also have the following dual of Corollary 1.5.

COROLLARY 1.7. *Let $\mathscr{X}$ be a subcategory of* mod Λ *closed under extensions and let* $\mathscr{Y} = \{Y$ *in* mod $\Lambda \mid \mathrm{Ext}^1_\Lambda(\mathscr{X}, Y) = 0\}$. *Then the following are equivalent*:

(a) $\mathrm{Ext}^1_\Lambda(\ , C)|\mathscr{X}$ *is finitely generated for all C in* mod Λ.

(b) *$\mathscr{Y}$ is covariantly finite in* mod Λ *and if* $0 \to C \xrightarrow{g} Y \to X \to 0$ *is exact with g a minimal left $\mathscr{Y}$-approximation of C, then X is in $\mathscr{X}$.*

It should be noted that $\mathscr{Y}$ contains all the injective modules and so all left $\mathscr{Y}$-approximations are monomorphisms. It should also be noted that $\mathscr{Y}$ is closed under extensions.

The following is an easy consequence of these results.

PROPOSITION 1.8. *Suppose $\mathscr{T}$ is a subcategory of* mod Λ *which is closed under extensions and such that* $\mathrm{Ext}^1_\Lambda(C, \ )|\mathscr{T}$ *is finitely generated for all C in* mod Λ. *Then we have the following*:

.(a) *The subcategory* $\mathscr{X}=\{X$ *in* $\operatorname{mod} \Lambda \mid \operatorname{Ext}_\Lambda^1(X, \mathscr{T})=0\}$ *is contravariantly finite in* $\operatorname{mod} \Lambda$, *is closed under extensions, and contains all projective* Λ*-modules.*

(b) *The subcategory* $\mathscr{Y}=\{Y$ *in* $\operatorname{mod} \Lambda \mid \operatorname{Ext}^1(\mathscr{X}, Y)=0\}$ *is covariantly finite in* $\operatorname{mod} \Lambda$, *contains* $\mathscr{T}$ *as well as all injectives, and is closed under extensions.*

(c) *For each* C *in* $\operatorname{mod} \Lambda$, *there are exact sequences* $0 \rightarrow T_C \rightarrow X_C \xrightarrow{f} C \rightarrow 0$ and $0 \rightarrow C \xrightarrow{g} Y^C \rightarrow X^C \rightarrow 0$ *with* f *a minimal right* $\mathscr{X}$*-approximation of* C *and* T_C *in* $\mathscr{T}$ *and with* g *a minimal left* $\mathscr{Y}$*-approximation of* C *and* X^C *in* $\mathscr{X}$.

(d) $\mathscr{X}=\{C$ *in* $\operatorname{mod} \Lambda \mid \operatorname{Ext}_\Lambda^1(C, \mathscr{Y})=0\}$.

Remark. If $\mathscr{T}$ is covariantly finite in mod Λ, then it is well known (see [8]) that $\operatorname{Ext}_\Lambda^1(C, \ ) \mid \mathscr{T}$ is finitely generated for all C in mod Λ. Hence if $\mathscr{T}$ is also closed under extensions, then $\mathscr{T}$ satisfies the hypothesis of Proposition 1.8. Therefore $\mathscr{X}=\{X$ in $\operatorname{mod} \Lambda \mid \operatorname{Ext}_\Lambda^1(X, \mathscr{T})=0\}$ is contravariantly finite.

While we do not state the dual of Proposition 1.8, we will use it when necessary.

As an application of Proposition 1.8 and its dual, we prove the following.

PROPOSITION 1.9. *Let* $\mathscr{X}$ *and* $\mathscr{Y}$ *be subcategories of* mod Λ *satisfying* X *is in* $\mathscr{X}$ *if and only if* $\operatorname{Ext}^1(X, \mathscr{Y})=0$ *and* Y *is in* $\mathscr{Y}$ *if and only if* $\operatorname{Ext}^1(\mathscr{X}, Y)=0$ *and let* $\omega=\mathscr{X} \cap \mathscr{Y}$. *Then we have the following.*

(a) $\mathscr{X}$ *is contravariantly finite in* mod Λ *if and only if* $\mathscr{Y}$ *is covariantly finite in* mod Λ.

(b) If $\mathscr{X}$ *is contravariantly finite, then for each* C *in* mod Λ *there are exact sequences* $0 \rightarrow Y_C \rightarrow X_C \xrightarrow{f} C \rightarrow 0$ *and* $0 \rightarrow C \xrightarrow{g} Y^C \rightarrow X^C \rightarrow 0$ *where* f *is a minimal right* $\mathscr{X}$*-approximation of* C *and* Y_C *is in* $\mathscr{Y}$ *and where* g *is a minimal left* $\mathscr{Y}$*-approximation of* C *and* X^C *is in* $\mathscr{X}$.

(c) $\operatorname{Ext}_\Lambda^1(\omega, \omega)=0$.

(d) *Suppose* $\mathscr{X}$ *is contravariantly finite in* mod Λ. *Then*

(i) *For each* X *in* $\mathscr{X}$, *there is an exact sequence* $0 \rightarrow X \rightarrow w \rightarrow X' \rightarrow 0$ *in* $\mathscr{X}$ *with* $w \in \omega$.

(ii) *For each* Y *in* $\mathscr{Y}$ *there is an exact sequence* $0 \rightarrow Y' \rightarrow w \rightarrow Y \rightarrow 0$ *in* $\mathscr{Y}$ *with* w *in* ω.

Proof. (a) To use Proposition 1.8 to show that $\mathscr{Y}$ being covariantly finite in mod Λ implies that $\mathscr{X}$ is contravariantly finite, it is only necessary to observe that $\mathscr{Y}$ being covariantly finite in mod Λ implies that

$\mathrm{Ext}^1_\Lambda(C, \ )|\mathscr{Y}$ is finitely generated for all C in mod Λ. The rest of (a) follows by duality.

(b) Trivial consequence of Proposition 1.8.

(c) Trivial consequence of definition of ω.

(d)(i) Let X be in $\mathscr{X}$. Since $\mathscr{X}$ is contravariantly finite we know by (c) that there is an exact sequence $0 \to X \to Y \to X' \to 0$ with Y in $\mathscr{Y}$ and X' in $\mathscr{X}$. Since $\mathscr{X}$ is closed under extensions, it follows that Y is in $\mathscr{X}$. Hence Y is in ω.

(ii) Analogous proof to (i).

As an example of when the hypothesis of Proposition 1.9 is satisfied, we point out the following.

PROPOSITION 1.10. *Suppose $\mathscr{X}$ is a contravariantly finite subcategory of* mod Λ *which is closed under extensions and contains Λ. Let $\mathscr{Y} = \{Y$ in* mod $\Lambda \mid \mathrm{Ext}^1_\Lambda(\mathscr{X}, Y) = 0\}$. *Then $\mathscr{X} = \{C$ in* mod $\Lambda \mid \mathrm{Ext}^1_\Lambda(C, \mathscr{Y}) = 0\}$.

Proof. Clearly $\mathscr{X}$ is contained in $\{C$ in mod $\Lambda \mid \mathrm{Ext}^1_\Lambda(C, \mathscr{Y}) = 0\}$ by definition. Suppose $\mathrm{Ext}^1_\Lambda(C, \mathscr{Y}) = 0$. Since $\mathscr{X}$ is contravariantly finite in mod Λ, there is a right minimal $\mathscr{X}$-approximation $f_C\colon X_C \to C$. Since Λ is in $\mathscr{X}$, we have that f_C is in epimorphism. Since $\mathscr{X}$ is also closed under extensions we know by Wakamatsu's Lemma that $Y_C = \mathrm{Ker}\, f_C$ is in $\mathscr{Y}$. Therefore the exact sequence $0 \to Y_C \to X_C \xrightarrow{f_C} C \to 0$ splits since $\mathrm{Ext}^1_\Lambda(C, \mathscr{Y}) = 0$. Hence C is in $\mathscr{X}$ and we are done.

While we do not give the statement of the dual of Proposition 1.10, we will feel free to use it.

2. MODULES T WITH $\mathrm{Ext}^1_\Lambda(T, T) = 0$

In this section we illustrate and apply the results of Section 1 in the course of studying modules T with $\mathrm{Ext}^1_\Lambda(T, T) = 0$. We start out by giving various characterizations of when $\mathrm{id}_\Lambda \Lambda \leqslant 1$ (where $\mathrm{id}_\Lambda M$ is the injective dimension of M) which lead to the following result. If $\mathrm{id}_\Lambda \Lambda \geqslant 2$, then there are modules M which are not submodules of projective modules such that $\mathrm{Ext}^1_\Lambda(M \amalg \Lambda, M \amalg \Lambda) = 0$. The following known result [7, Prop. 3.5] is part of the reason for interest in such modules M.

Suppose M is a Λ-module with the property that $\mathrm{Ext}^1_\Lambda(M \amalg \Lambda, M \amalg \Lambda) = 0$. Let $\underline{\mathrm{End}}_\Lambda(M)$ be the endomorphism ring of M modulo the ideal consisting of endomorphisms which factor through projective modules. Then for each indecomposable $\underline{\mathrm{End}}(M)^{\mathrm{op}}$-module A, there is an indecomposable Λ-module B such that the $\underline{\mathrm{End}}(M)^{\mathrm{op}}$-module $\mathrm{Ext}^1_\Lambda(M, B)$

is isomorphic to Λ. In particular if Λ is of finite representation type, then $\underline{\mathrm{End}}(M)$ is of finite representation type. Until now it was not very clear that there were many nonprojective Λ-modules M satisfying $\mathrm{Ext}^1_\Lambda(M \amalg \Lambda, M \amalg \Lambda)=0$. The result stated above gives a sort of existence theorem for such modules. For this reason it would be interesting to explore further the connection between $\underline{\mathrm{End}}(M)^{\mathrm{op}}$-modules and Λ-modules just described when $\mathrm{Ext}^1_\Lambda(M \amalg \Lambda, M \amalg \Lambda)=0$.

Another reason for interest in modules M with the property $\mathrm{Ext}^1_\Lambda(M \amalg \Lambda, M \amalg \Lambda)=0$ is the generalized Nakayama Conjecture which is equivalent to saying that a module M is projective if $\mathrm{Ext}^i_\Lambda(M \amalg \Lambda, M \amalg \Lambda)=0$ for all $i>0$ [6]. Perhaps a better understanding of modules M satisfying $\mathrm{Ext}^1_\Lambda(M \amalg \Lambda, M \amalg \Lambda)=0$ will lead to a better understanding of the generalized Nakayama Conjecture.

Suppose T is a module in mod Λ and let add T be the subcategory of mod Λ consisting of summands of finite sums of T. Since add T is of finite type, i.e., has only a finite number of nonisomorphic indecomposable summands, we know that add T is both covariantly and contravariantly finite in mod Λ [8, Prop. 4.2]. Assume now that T satisfies $\mathrm{Ext}^1_\Lambda(T, T)=0$. Then add T is closed under extensions and therefore satisfies the hypothesis of Proposition 1.8. We denote by $\mathscr{X}(T)$ the contravariantly finite subcategory whose objects are the X in mod Λ such that $\mathrm{Ext}^1_\Lambda(X, T)=0$. We denote by $\mathscr{Y}(T)$ the covariantly finite subcategory of mod Λ consisting of all Y such that $\mathrm{Ext}^1_\Lambda \mathscr{X}(T)=0$. Finally we denote $\mathscr{X}(T) \cap \mathscr{Y}(T)$ by $\omega(T)$.

The result about modules M such that $\mathrm{Ext}^1_\Lambda(M \amalg \Lambda, M \amalg \Lambda)=0$ given in the Introduction is an immediate consequence of these definitions and the following result, as we shall see later on.

THEOREM 2.1. *The following are equivalent*:

(a) $\mathrm{id}_\Lambda \Lambda \leqslant 1$.

(b) $\omega(\Lambda)$ *consists only of torsionless modules, i.e., submodules of free Λ-modules.*

The proof of this theorem will take several steps and is based on the following characterization of when $\mathrm{id}_\Lambda \Lambda \leqslant 1$.

PROPOSITION 2.2. *The following are equivalent for Λ.*

(a) $\mathrm{id}_{\Lambda^{\mathrm{op}}} \Lambda^{\mathrm{op}} \leqslant 1$.

(b) *A Λ-module X is torsionless if* $\mathrm{Ext}^1_\Lambda(X, \Lambda)=0$.

Proof. (a) $\Rightarrow$ (b). Suppose $\mathrm{Ext}^1_\Lambda(X, \Lambda)=0$. Then $\mathrm{Tr}\, X$ is a torsionless Λ^{op}-module since we have the exact sequence $0 \to \mathrm{Ext}^1_\Lambda(X, \Lambda) \to \mathrm{Tr}\, X \to (\mathrm{Tr}\, X)^{**}$, where $\mathrm{Tr}\, X$ is the transpose of X and $M^* = \mathrm{Hom}_\Lambda(M, \Lambda)$ [1,

Prop. 6.3]. Therefore $\mathrm{Ext}^1_{\Lambda^{\mathrm{op}}}(\mathrm{Tr}\,X, \Lambda^{\mathrm{op}})=0$ since $\mathrm{Tr}\,X$ is a submodule of a free module and $\mathrm{id}_{\Lambda^{\mathrm{op}}}\Lambda^{\mathrm{op}}\leqslant 1$. This implies that $\mathrm{Tr}(\mathrm{Tr}\,X)$ is torsionless, which implies that X is torsionless.

(b) ⇒ (a). Let A be a torsionless Λ^{op}-module. Then $\mathrm{Ext}^1_{\Lambda}(\mathrm{Tr}\,A, \Lambda)=0$. Therefore by (b) we have that $\mathrm{Tr}\,A$ is torsionless, which implies $\mathrm{Ext}^1_{\Lambda^{\mathrm{op}}}(\mathrm{Tr}\,\mathrm{Tr}\,A, \Lambda^{\mathrm{op}})=0$. Therefore $\mathrm{Ext}^1_{\Lambda^{\mathrm{op}}}(A, \Lambda^{\mathrm{op}})=0$ for all torsionless Λ^{op}-modules A, which is equivalent to $\mathrm{id}_{\Lambda^{\mathrm{op}}}\Lambda^{\mathrm{op}}\leqslant 1$.

Combining this proposition with Proposition 1.9 we obtain the following.

PROPOSITION 2.3. *The following are equivalent for Λ.*

(a) $\mathrm{id}_{\Lambda^{\mathrm{op}}}\Lambda^{\mathrm{op}}\leqslant 1$.

(b) $\omega(\Lambda)$ *consists only of torsionless modules.*

Proof. (a) ⇒ (b). Let M be in $\omega(\Lambda)$. Since Λ is in $\omega(\Lambda)$ and clearly $\mathrm{Ext}^1_{\Lambda}(\omega(\Lambda), \omega(\Lambda))=0$, it follows that $\mathrm{Ext}^1_{\Lambda}(M, \Lambda)=0$. Therefore M is torsionless by Proposition 2.2 since $\mathrm{id}_{\Lambda^{\mathrm{op}}}\Lambda^{\mathrm{op}}\leqslant 1$.

(b) ⇒ (a). By Proposition 2.2 we have that $\mathrm{id}_{\Lambda^{\mathrm{op}}}\Lambda^{\mathrm{op}}\leqslant 1$, if $\mathrm{Ext}^1_{\Lambda}(M, \Lambda)=0$ implies M is torsionless. Suppose $\mathrm{Ext}^1_{\Lambda}(M, \Lambda)=0$. Then M is in $\mathscr{X}(\Lambda)$. Therefore by Proposition 1.9, we know that there is a monomorphism $0\to X\to w$ with w in $\omega(\Lambda)$. Therefore X is torsionless since all modules in $\omega(\Lambda)$ are torsionless.

Theorem 2.1 is now a trivial consequence of Proposition 2.3 and the well-known result that $\mathrm{id}_{\Lambda}\Lambda\leqslant 1$ if and only if $\mathrm{id}_{\Lambda^{\mathrm{op}}}\Lambda^{\mathrm{op}}\leqslant 1$. This can be seen by observing that if $\mathrm{id}_{\Lambda}\Lambda\leqslant 1$, then since $\mathrm{Ext}^1_{\Lambda}(\Lambda, \Lambda)=0$ and Λ has the same number of nonisomorphic indecomposable summands as the number of simple modules, then Λ is a classical cotilting module by [10]. Hence there is an exact sequence $0\to P_1\to P_0\to D(\Lambda^{\mathrm{op}})\to 0$ with P_0 and P_1 projective Λ-modules, so that $\mathrm{id}_{\Lambda^{\mathrm{op}}}\Lambda^{\mathrm{op}}\leqslant 1$. Here D denotes the ordinary duality with respect to the commutative artin ground ring.

As an easy consequence of Theorem 2.1 we have the following.

COROLLARY 2.4. *Suppose* $\mathrm{id}_{\Lambda}\Lambda\geqslant 2$. *Then there are Λ-modules M which are not torsionless such that* $\mathrm{Ext}^1_{\Lambda}(M\amalg\Lambda, M\amalg\Lambda)=0$. *In particular, there are nonprojective Λ-modules M such that* $\mathrm{Ext}^1_{\Lambda}(M\amalg\Lambda, M\amalg\Lambda)=0$.

Proof. Since Λ is in $\omega(\Lambda)$ and $\mathrm{Ext}^1_{\Lambda}(\omega(\Lambda), \omega(\Lambda))=0$ it follows that $\mathrm{Ext}^1_{\Lambda}(M\amalg\Lambda, M\amalg\Lambda)=0$ for all M in $\omega(\Lambda)$. Since $\mathrm{id}_{\Lambda}\Lambda\geqslant 2$, it follows from Theorem 2.1 that there are M in $\omega(\Lambda)$ which are not torsionless. Hence we obtain our desired result.

We end this discussion of modules M such that $\operatorname{Ext}^1_\Lambda(M \amalg \Lambda, M \amalg \Lambda) = 0$ by pointing out another way of constructing such modules. This will also give a way of characterizing when a module T with $\operatorname{Ext}^1_\Lambda(T, T) = 0$ has $\operatorname{id}_\Lambda T \leqslant 1$, i.e., is a summand of a cotilting module of injective dimension at most one [12].

Suppose T is a module such that $\operatorname{Ext}^1_\Lambda(T, T) = 0$. We define $\mathscr{Z}(T) = \{Z \text{ in } \operatorname{mod} \Lambda \mid \operatorname{Ext}^1_\Lambda(Z, \mathscr{X}(T)) = 0\}$. Since T is in $\mathscr{X}(T)$ we have that $\mathscr{Z}(T) \subset \mathscr{X}(T)$ and so $\operatorname{Ext}^1_\Lambda(\mathscr{Z}(T), \mathscr{Z}(T)) = 0$. It then follows that $\operatorname{Ext}^1_\Lambda(M \amalg \Lambda, M \amalg \Lambda) = 0$ for all M in $\mathscr{Z}(T)$ since Λ is in $\mathscr{Z}(T)$. Therefore there is some interest in knowing when $\mathscr{Z}(T)$, which contains the projective Λ-modules, consists only of projective Λ-modules. We give an answer to this question under the additional hypothesis that $\mathscr{X}(T)$ is covariantly finite in addition to being contravariantly finite.

Proposition 2.5. *The following are equivalent for a module T satisfying* $\operatorname{Ext}^1_\Lambda(T, T) = 0$.

(a) $\operatorname{id}_\Lambda T \leqslant 1$.

(b) *$\mathscr{X}(T)$ is covariantly finite in* $\operatorname{mod} \Lambda$ *and $\mathscr{Z}(T)$ consists only of projective Λ-modules.*

Proof. (a) $\Rightarrow$ (b). We shall see in Section 5 that if $\operatorname{Ext}^1_\Lambda(T, T) = 0$ and $\operatorname{id}_\Lambda T \leqslant 1$, then $\mathscr{X}(T)$ is covariantly finite in $\operatorname{mod} \Lambda$. We show now that $\mathscr{Z}(T)$ consists solely of projective modules. Since $\operatorname{id}_\Lambda T \leqslant 1$, we have that $\mathscr{X}(T)$ contains ΩC, the first syzygy of C, for all C in $\operatorname{mod} \Lambda$. Suppose Z in $\mathscr{Z}(T)$ is indecomposable. Let $f: Z \to S$ be an epimorphism with S a simple Λ-module. Let $0 \to \Omega S \to P \to S \to 0$ be exact with $P \to S$ a projective cover. Since ΩS is in $\mathscr{X}(T)$, we have that $\operatorname{Ext}^1_\Lambda(Z, \Omega S) = 0$. Therefore $f: Z \to S$ can be lifted to $P \to S$, which implies that Z is projective since it is indecomposable.

(b) $\Rightarrow$ (a). Since we are assuming that $\mathscr{X}(T)$ is covariantly finite and closed under extensions, we know by Proposition 1.8 that for each C in $\operatorname{mod} \Lambda$ there is an exact sequence $0 \to X_C \to Z_C \to C \to 0$ with X_C in $\mathscr{X}(T)$ and Z_C in $\mathscr{Z}(T)$. But we are also assuming that Z_C is projective and therefore $\operatorname{Ext}^2_\Lambda(C, T) \cong \operatorname{Ext}^1_\Lambda(X_C, T) = 0$ for all C in $\operatorname{mod} \Lambda$. Hence $\operatorname{id}_\Lambda T \leqslant 1$.

As our final remark along these lines, we point out that it is known [8, Prop. 7.1] that $\mathscr{X}(\Lambda)$ is always covariantly finite. Therefore we have the following.

Corollary 2.6. *The following are equivalent for Λ.*

(a) $\operatorname{id} \Lambda \leqslant 1$.

(b) *$\mathscr{Z}(\Lambda)$ consists of the projective Λ-modules.*

3. Resolving Subcategories

In Section 1 we were mainly concerned with subcategories of mod Λ closed under extensions. This section is devoted to pointing out the form some of these results take if we assume the subcategories involved are not only closed under extensions but are resolving or coresolving subcategories. These results will be used throughout the rest of the paper. We begin by giving the definitions of resolving and coresolving subcategories, notions introduced in [2].

A subcategory $\mathscr{X}$ of mod Λ is said to be a *resolving* subcategory if it satisfies the following three conditions: (a) closed under extensions, (b) closed under kernels of surjections, and (c) contains the projective Λ-modules. If $\mathscr{C}$ is a subcategory of mod Λ we define ${}^{\perp}\mathscr{C}$ to be the subcategory of mod Λ consisting of those X such that $\operatorname{Ext}^i_\Lambda(X, \mathscr{C}) = 0$ for all $i > 0$. It is easily checked that ${}^{\perp}\mathscr{C}$ is a resolving subcategory for all subcategories $\mathscr{C}$ of mod Λ. It is these types of resolving subcategories which will be our main concern.

There is also the dual notion of a coresolving subcategory. A subcategory $\mathscr{Y}$ is said to be *coresolving* if it satisfies the following three conditions: (a) closed under extensions, (b) closed under cokernels of injections, and (c) contains all injective Λ-modules. If $\mathscr{C}$ is a subcategory of mod Λ we denote by $\mathscr{C}^{\perp}$ the subcategory consisting of all Y such that $\operatorname{Ext}^i_\Lambda(\mathscr{C}, Y) = 0$ for all $i > 0$. It is easily checked that $\mathscr{C}^{\perp}$ is a coresolving subcategory of mod Λ for all subcategories $\mathscr{C}$ of mod Λ.

We now point out an important but easily proven property of resolving and coresolving subcategories.

Lemma 3.1. *Suppose $\mathscr{Y}$ is a coresolving subcategory of* mod Λ. *Let* $\mathscr{X} = \{X \text{ in mod } \Lambda \mid \operatorname{Ext}^1_\Lambda(X, \mathscr{Y}) = 0\}$. *Then*

(a) $\mathscr{X} = {}^{\perp}\mathscr{Y}$.

(b) *$\mathscr{X}$ is a resolving subcategory of* mod Λ.

Proof. (a) Since ${}^{\perp}\mathscr{Y} \subset \mathscr{X}$, we only have to show that if $\operatorname{Ext}^1_\Lambda(X, \mathscr{Y}) = 0$, then $\operatorname{Ext}^i_\Lambda(X, \mathscr{Y}) = 0$ for all $i > 0$. Suppose Y is in $\mathscr{Y}$ and let $0 \to Y \xrightarrow{\delta_0} I_0 \xrightarrow{\delta_1} I_1 \xrightarrow{\delta_2} \cdots$ be a minimal injective resolution of Y. Since $\mathscr{Y}$ is coresolving, we know that $\operatorname{Im} \delta_i = \Omega^{-i} Y$ is in $\mathscr{Y}$ for all $i \geqslant 0$, and our claim follows easily from this.

(b) Easy consequence of (a).

We now state the dual of Lemma 3.1.

LEMMA 3.2. *Suppose* $\mathscr{X}$ *is a resolving subcategory of* mod Λ *and let* $\mathscr{Y} = \{Y$ *in* mod $\Lambda \mid \operatorname{Ext}^1_\Lambda(\mathscr{X}, Y) = 0\}$. *Then*

(a) $\mathscr{Y} = \mathscr{X}^\perp$.

(b) $\mathscr{Y}$ *is a coresolving subcategory of* mod Λ.

Combining Lemma 3.1 and 3.2 with the results of Section 1, we obtain the following.

PROPOSITION 3.3. *Let* $\mathscr{X}$ *be a resolving, contravariantly finite subcategory of* mod Λ.

(a) $\mathscr{Y} = \mathscr{X}^\perp$ *is a coresolving, covariantly finite subcategory of* mod Λ.

(b) $\mathscr{X} = {}^\perp\mathscr{Y} = {}^\perp(\mathscr{X}^\perp)$.

(c) *For each* C *in* mod Λ, *there is a unique up to isomorphism exact sequence* $0 \to Y_C \to X_C \xrightarrow{f} C \to 0$ *satisfying*

(i) f *is a minimal right* $\mathscr{X}$*-approximation of* C.

(ii) Y_C *is in* $\mathscr{Y}$.

(iii) f *induces isomorphisms* $\operatorname{Ext}^i_\Lambda(X, X_C) \simeq \operatorname{Ext}^i_\Lambda(X, C)$ *for all* X *in* $\mathscr{X}$ *and* $i > 0$.

(d) *For all* C *in* mod Λ, *there is a unique up to isomorphism exact sequence* $0 \to C \xrightarrow{g} Y^C \to X^C \to 0$ *satisfying*

(i) g *is a left minimal* $\mathscr{Y}$*-approximation of* C.

(ii) X^C *is in* X.

(iii) g *induces isomorphisms* $\operatorname{Ext}^i_\Lambda(Y^C, Y) \simeq \operatorname{Ext}^i_\Lambda(C, Y)$ *for all* $i > 0$ *and all* Y *in* $\mathscr{Y}$.

Proof. (a) Since $\mathscr{X}$ is closed under extensions and is contravariantly finite, we know by Proposition 1.8 that $\mathscr{Y} = \{Y$ in mod $\Lambda \mid \operatorname{Ext}^1_\Lambda(\mathscr{X}, Y) = 0\}$ is covariantly finite. By Lemma 3.2 we have that $\mathscr{Y} = \mathscr{X}^\perp$ and is coresolving since $\mathscr{X}$ is resolving.

(b) Since $\mathscr{X}$ is contravariantly finite, closed under extensions, and contains Λ, we know by Proposition 1.10 that $\mathscr{X} = \{X$ in mod $\Lambda \mid \operatorname{Ext}^1_\Lambda(X, \mathscr{Y}) = 0\}$. Since $\mathscr{Y}$ is coresolving, we know by Lemma 3.1 that $X = {}^\perp\mathscr{Y}$.

(c) The fact that there is an exact sequence $0 \to Y_C \to X_C \to C \to 0$ satisfying (i) and (ii) was shown in Proposition 1.8.

That it also satisfies (iii) is a trivial consequence of the fact that $\mathscr{Y} = \mathscr{X}^\perp$.

(d) dual of (c).

Given a resolving contravariantly finite subcategory $\mathscr{X}$, we will usually denote $\mathscr{X}^{\perp}$ by $\mathscr{Y}$ and denote $\mathscr{X} \cap \mathscr{Y}$ by ω. With this notation in mind, we have the following result. Here we say that a category or a module C is *selforthogonal* if $\operatorname{Ext}^i_\Lambda(C, C) = 0$ for all $i > 0$.

PROPOSITION 3.4. *Let $\mathscr{X}$ be a resolving contravariantly finite subcategory of* mod Λ. *Then $\omega = \mathscr{X} \cap \mathscr{Y}$ has the following properties.*

(a) *ω is selforthogonal.*

(b) *For each X in $\mathscr{X}$ there is an exact sequence $0 \to X \to w \to X' \to 0$ with w in ω and X' in $\mathscr{X}$.*

(c) *For each Y in $\mathscr{Y}$ there is an exact sequence $0 \to Y' \to w \to Y \to 0$ with Y' in $\mathscr{Y}$ and w in ω.*

Proof. (a) is a trivial consequence of the definition of ω. (b) and (c) are proven in Proposition 1.9.

Remark. Instead of starting in Proposition 3.3 with a resolving, contravariantly finite subcategory $\mathscr{X}$, we could have started with a coresolving, covariantly finite subcategory $\mathscr{Y}$ and obtained the dual result to Proposition 3.3. This shows that there is a bijection between resolving, contravariantly finite subcategories $\mathscr{X}$ and coresolving, covariantly finite subcategories $\mathscr{Y}$ of mod Λ given by $\mathscr{X} \mapsto \mathscr{X}^{\perp}$ with inverse $\mathscr{Y} \mapsto {}^{\perp}\mathscr{Y}$.

Our aim now is to show that if $\mathscr{X}$ is a resolving subcategory of mod Λ, then the subcategory consisting of all Λ-modules having right $\mathscr{X}$-approximations is closed under extensions. This has the important consequence that in order to check that a resolving subcategory $\mathscr{X}$ is contravariantly finite, one has only to check that the simple modules have right $\mathscr{X}$-approximations instead of having to check this for all modules.

We begin with a preliminary observation.

LEMMA 3.5. *Suppose $\mathscr{X}$ is a resolving subcategory of* mod Λ *and $0 \to Y_C \to X_C \xrightarrow{f} C \to 0$ is an exact sequence with $f\colon X_C \to C$ a minimal right $\mathscr{X}$-approximation of C. Then Y_C is in $\mathscr{X}^{\perp}$.*

Proof. By Wakamatsu's lemma (Lemma 1.3) we know that $\operatorname{Ext}^1_\Lambda(\mathscr{X}, Y_C) = 0$ since $\mathscr{X}$ is closed under extensions. Since $\mathscr{X}$ is also resolving it follows by Lemma 3.2 that Y_C is in $\mathscr{X}^{\perp}$.

Our desired result will be a trivial consequence of the following result whose proof we include for completeness even though it is essentially the same as the proof of Lemma 3.1 in [3].

PROPOSITION 3.6. *Suppose $\mathscr{X}$ is a resolving subcategory of* mod Λ. *Let*

$$\begin{array}{ccccccccc}
 & & 0 & & & & 0 & & \\
 & & \downarrow & & & & \downarrow & & \\
 & & Y_1 & & & & Y_3 & & \\
 & & \downarrow & & & & \downarrow & & \\
 & & X_1 & & & & X_3 & & \\
 & & \downarrow{\scriptstyle f_1} & & & & \downarrow{\scriptstyle f_3} & & \\
0 & \longrightarrow & C_1 & \longrightarrow & C_2 & \longrightarrow & C_3 & \longrightarrow & 0 \\
 & & \downarrow & & & & \downarrow & & \\
 & & 0 & & & & 0 & &
\end{array}$$

be an exact commutative diagram of Λ-modules with the X_i in $\mathscr{X}$ and Y_i in $\mathscr{Y} = \mathscr{X}^{\perp}$. Then there is an exact commutative diagram

$$\begin{array}{ccccccccc}
 & & 0 & & 0 & & 0 & & \\
 & & \downarrow & & \downarrow & & \downarrow & & \\
0 & \longrightarrow & Y_1 & \longrightarrow & Y_2 & \longrightarrow & Y_3 & \longrightarrow & 0 \\
 & & \downarrow & & \downarrow & & \downarrow & & \\
0 & \longrightarrow & X_1 & \longrightarrow & X_2 & \longrightarrow & X_3 & \longrightarrow & 0 \\
 & & \downarrow{\scriptstyle f_1} & & \downarrow{\scriptstyle f_2} & & \downarrow{\scriptstyle f_3} & & \\
0 & \longrightarrow & C_1 & \longrightarrow & C_2 & \longrightarrow & C_3 & \longrightarrow & 0 \\
 & & \downarrow & & \downarrow & & \downarrow & & \\
 & & 0 & & 0 & & 0 & &
\end{array}$$

with X_2 in $\mathscr{X}$ and Y_2 in $\mathscr{Y}$.

Proof. Let

$$\begin{array}{ccccccccc}
0 & \longrightarrow & C_1 & \longrightarrow & X_3 \times_{C_3} C_2 & \longrightarrow & X_3 & \longrightarrow & 0 \\
 & & \| & & \downarrow & & \downarrow & & \\
0 & \longrightarrow & C_1 & \longrightarrow & C_2 & \longrightarrow & C_3 & \longrightarrow & 0
\end{array}$$

be a pullback diagram. Because Y_1 is in $\mathscr{X}^{\perp}$ and X_1 is in $\mathscr{X}$, we know that the epimorphism $f_1 : X_1 \to C_1$ induces an isomorphism $\operatorname{Ext}^1_{\Lambda}(X_3, X_1) \simeq \operatorname{Ext}^1_{\Lambda}(X_3, C_1)$. Therefore we have a commutative exact diagram

$$
\begin{array}{ccccccccc}
 & & 0 & & & & & & \\
 & & \downarrow & & & & & & \\
 & & Y_1 & & & & & & \\
 & & \downarrow & & & & & & \\
0 & \longrightarrow & X_1 & \longrightarrow & X_2 & \longrightarrow & X_3 & \longrightarrow & 0 \\
 & & \downarrow & & \downarrow & & \| & & \\
0 & \longrightarrow & C_1 & \longrightarrow & X_3 \times C_3 & \longrightarrow & X_3 & \longrightarrow & 0 \\
 & & \downarrow & & \downarrow & & & & \\
 & & 0 & & 0 & & & &
\end{array}
$$

This gives rise to the exact commutative diagram

$$
\begin{array}{ccccccccc}
 & & 0 & & 0 & & 0 & & \\
 & & \downarrow & & \downarrow & & \downarrow & & \\
0 & \longrightarrow & Y_1 & \longrightarrow & Y_2 & \longrightarrow & Y_3 & \longrightarrow & 0 \\
 & & \downarrow & & \downarrow & & \downarrow & & \\
0 & \longrightarrow & X_1 & \longrightarrow & X_2 & \longrightarrow & X_3 & \longrightarrow & 0 \\
 & & \downarrow{\scriptstyle f_1} & & \downarrow{\scriptstyle f_2} & & \downarrow{\scriptstyle f_3} & & \\
0 & \longrightarrow & C_1 & \longrightarrow & C_2 & \longrightarrow & C_3 & \longrightarrow & 0 \\
 & & \downarrow & & \downarrow & & \downarrow & & \\
 & & 0 & & 0 & & 0 & &
\end{array}
$$

Since $\mathscr{X}$ and $\mathscr{Y} = \mathscr{X}^{\perp}$ are closed under extensions, it follows that X_2 is in $\mathscr{X}$ and Y_2 is in $\mathscr{Y}$. This finishes the proof.

Combining Lemma 3.5 and Proposition 3.6, we have the following, which includes our promised result.

PROPOSITION 3.7. *Suppose $\mathscr{X}$ is a resolving subcategory of* mod Λ. *Then we have the following.*

(a) *The subcategory of all modules having right $\mathscr{X}$-approximations is closed under extensions.*

(b) *$\mathscr{X}$ is contravariantly finite in* mod Λ *if and only if the simple Λ-modules have right $\mathscr{X}$-approximations.*

Proof. Let $0 \to C_1 \to C_2 \to C_3 \to 0$ be exact and suppose $X_i \xrightarrow{p_i} C_i \to 0$ are minimal $\mathscr{X}$-approximations for C_i for $i = 1$ and 3. Then we have the commutative exact diagram

$$\begin{array}{ccccccccc}
 & & 0 & & & & 0 & & \\
 & & \downarrow & & & & \downarrow & & \\
 & & Y_1 & & & & Y_3 & & \\
 & & \downarrow & & & & \downarrow & & \\
 & & X_1 & & & & X_3 & & \\
 & & \downarrow & & & & \downarrow & & \\
0 & \longrightarrow & C_1 & \longrightarrow & C_2 & \longrightarrow & C_3 & \longrightarrow & 0 \\
 & & \downarrow & & & & \downarrow & & \\
 & & 0 & & & & 0 & &
\end{array}$$

which by Lemma 3.5 has the property that the Y_i are in $\mathscr{Y} = \mathscr{X}^{\perp}$ and the X_i are in $\mathscr{X}$ for $i = 1, 3$. Therefore by Proposition 3.6, we know there is an exact sequence $0 \to Y_2 \to X_2 \xrightarrow{f_2} C_2 \to 0$ with Y_2 in $\mathscr{Y}$ and X_2 in $\mathscr{X}$. Hence $f_2 \colon X_2 \to C_2$ is a right $\mathscr{X}$-approximation of C_2.

(b) Trivial consequence of (a).

As another interesting consequence of Proposition 3.6, we have the following.

PROPOSITION 3.8. *Suppose $\mathscr{X}$ is a resolving, contravariantly finite subcategory of* mod Λ. *Let $S_1, ..., S_t$ be a complete set of nonisomorphic simple Λ-modules and let $f_i \colon X_i \to S_i$ be minimal right $\mathscr{X}$-approximations of the S_i for $i = 1, ..., t$. Then the modules in $\mathscr{X}$ consist of the summands of modules M with the property that there is a filtration $M = M_0 \supset M_1 \supset \cdots \supset M_n = 0$ such that for each $i = 0, ..., n-1$ we have $M_i/M_{i+1} \cong X_j$ for some $j = 1, ..., t$.*

Proof. Proceeding by induction on the length of a module, it follows from Proposition 3.6 that for each Λ-module C, there is an exact sequence $0 \to Y \to X \to C \to 0$ with Y in $\mathscr{Y} = \mathscr{X}^{\perp}$ and X in $\mathscr{X}$ such that X has a filtration $X = X_0 \supset X_1 \supset \cdots \supset X_n = 0$ with each $X_i/X_{i+1} \cong X_j$ for some $j = 1, ..., t$.

Suppose now that C is in $\mathscr{X}$. Then the exact sequence $0 \to Y \to X \to C \to 0$ splits since Y is in $\mathscr{Y}$. This gives our desired result.

The following is an easy but important consequence of Proposition 3.8. Here $\operatorname{pd}_\Lambda X$ denotes the projective dimension of X.

COROLLARY 3.9. *Let $\mathscr{X}$ be resolving contravariantly finite subcategory of* mod Λ. *Let $f_i: X_i \to S_i$, $i = 1, \ldots, t$, be minimal right $\mathscr{X}$-approximations for a complete set of nonisomorphic simple modules $S_1, \ldots, S_t$. Then the following are equivalent.*

(a) $k = \max\{\operatorname{pd}_\Lambda X_i \mid i = 1, \ldots, t\} < \infty$,

(b) $\max\{\operatorname{pd}_\Lambda X \mid X \textit{ in } \mathscr{X}\} = k < \infty$,

(c) $\operatorname{pd}_\Lambda X < \infty$ *for each X in $\mathscr{X}$.*

Proof. (a) $\Rightarrow$ (b). Let C be in $\mathscr{X}$. Then we know by Proposition 3.8 that C is a summand of a module M which has a filtration $M = M_0 \supset M_1 \supset \cdots \supset M_n = 0$ such that each $M_i/M_{i+1} \cong X_j$ for some $j = 1, \ldots, t$. Therefore $\operatorname{pd} M \leqslant k$ which implies $\operatorname{pd} C \leqslant k$ since C is a summand of M.

(b) $\Rightarrow$ (c). Trivial.

(c) $\Rightarrow$ (a). Trivial.

As a trivial consequence of Corollary 3.9, we have the following.

COROLLARY 3.10. *Let $\mathscr{X}$ be the subcategory of* mod Λ *consisting of the modules with* $\operatorname{pd}_\Lambda X < \infty$. *If $\mathscr{X}$ is contravariantly finite in* mod Λ, *then* $\max\{\operatorname{pd}_\Lambda X \mid X \textit{ in } \mathscr{X}\} < \infty$.

Proof. Since it is well known that $\mathscr{X}$ is a resolving subcategory of mod Λ, our desired result is an immediate consequence of Corollary 3.9.

If the subcategory $\mathscr{X}$ consisting of the modules of finite projective dimension were always contravariantly finite, it would then follow from Corollary 3.10 that the finiteness of the finitistic dimension of artin algebras is valid. Unfortunately, there are examples of artin algebras where the subcategory of modules of finite projective dimension is not contravariantly finite, as well as many examples where it is. We give some of these examples in the next section. Little is known about the general question of when the subcategory of modules of finite projective dimension is contravariantly finite.

The dual statements of the results in this section are left to the reader to give.

4. Subcategories of Modules of Finite Projective Dimension

Until now we have not placed any condition on the modules in the subcategories we have been dealing with. In this section we will be dealing primarily with subcategories all of whose objects have either finite projective dimension or finite injective dimension. As usual, we will be studying these subcategories from the point of view of being contravariantly or covariantly finite in mod Λ. We first introduce some notations.

We denote by $\operatorname{pd}_\Lambda C$ or $\operatorname{pd} C$ the projective dimension of the Λ-module C. For each nonnegative integer k we denote the subcategory consisting of all modules M with $\operatorname{pd}_\Lambda M \leqslant k$ by $\mathscr{P}^k(\Lambda)$ and the subcategory consisting of all modules N with $\operatorname{id}_\Lambda M \leqslant k$ by $\mathscr{I}^k(\Lambda)$. We denote by $\mathscr{P}^\infty(\Lambda)$ the subcategory of all modules of finite projective dimension, and by $\mathscr{I}^\infty(\Lambda)$ the subcategory of modules of finite injective dimension. It is easily checked that $\mathscr{P}^k(\Lambda)$ is a resolving and $\mathscr{I}^k(\Lambda)$ is a coresolving subcategory of mod Λ for all $k = 1, ..., \infty$. In view of our previous results, it is interesting to know for which k the subcategory $\mathscr{P}^k(\Lambda)$ is contravariantly finite, and for which k the subcategory $\mathscr{I}^k(\Lambda)$ is covariantly finite.

Clearly $\mathscr{P}^0(\Lambda)$ is contravariantly finite and $\mathscr{I}^0(\Lambda)$ is covariantly finite since they have only a finite number of nonisomorphic indecomposable modules. For a similar reason, when Λ is of finite representation type, all the $\mathscr{P}^k(\Lambda)$ are contravariantly finite and all the $\mathscr{I}^k(\Lambda)$ are covariantly finite.

We now turn our attention to the question of whether $\mathscr{P}^1(\Lambda)$ is contravariantly finite. The fact that it is not in general the case has been shown by Igusa *et al.* [13]. They show that if Λ is given by the quiver

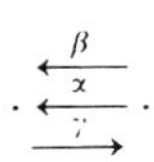

with relations $\alpha\gamma = \gamma\alpha = \gamma\beta = 0$, then $\mathscr{P}^1(\Lambda) = \mathscr{P}^\infty(\Lambda)$ and $\mathscr{P}^1(\Lambda)$ is not contravariantly finite. On the other hand, they show that if Λ has the property that $\operatorname{pd}_\Lambda I_0(\Lambda) \leqslant 1$ where $I_0(\Lambda)$ is the injective envelope of Λ, then $\mathscr{P}^1(\Lambda)$ is contravariantly finite in mod Λ. In view of these examples it is of interest to have a better understanding of when $\mathscr{P}^1(\Lambda)$ is contravariantly finite.

We now give some characterizations of when $\mathscr{P}^1(\Lambda)$ is contravariantly finite which are of interest in themselves as well as being useful for showing later on that $\mathscr{P}^\infty(\Lambda)$ is contravariantly finite for all Λ which are stably equivalent to hereditary algebras.

To this end, it is convenient to have the following notation. For each subcategory $\mathscr{C}$ of mod Λ, we denote the category $\mathscr{C}$ modulo projectives by $\underline{\mathscr{C}}$. We say that a module M is torsion if $M^* = \operatorname{Hom}_\Lambda(M, \Lambda) = 0$.

Finally, we denote by Tr: $\underline{\text{mod}}\,\Lambda \to \underline{\text{mod}}\,\Lambda^{\text{op}}$ the duality given by the transpose. Keeping this notation in mind, we begin with the following preliminary observation.

LEMMA 4.1. *The duality* Tr: $\underline{\text{mod}}\,\Lambda \to \underline{\text{mod}}\,\Lambda^{\text{op}}$ *induces a duality* Tr: $\underline{\mathscr{P}^1(\Lambda)} \to \underline{\mathscr{T}}$, *where* $\mathscr{T}$ *is the category of torsion* Λ^{op}*-modules.*

Proof. Suppose $\text{pd}_\Lambda M \leqslant 1$ and let $0 \to P_1 \to P_0 \to M \to 0$ be a minimal projective resolution. Then $P_0^* \to P_1^* \to \text{Ext}^1_\Lambda(M, \Lambda) \to 0$ is exact. Therefore $0 \to \text{Ext}^1_\Lambda(M, \Lambda)^* \to P_1^{**} \to P_0^{**}$ is exact. This means that $\text{Ext}^1_\Lambda(M, \Lambda)^* = 0$ since the P_i are reflexive Λ-modules. Hence $\text{Tr}\, M = \text{Ext}^1_\Lambda(M, \Lambda)$ is a torsion module.

Suppose now that U is in $\mathscr{T}$. From the minimal projective presentation $Q_1 \to Q_0 \to U \to 0$, we deduce the exact sequence $0 \to U^* \to Q_0^* \to Q_1^* \to \text{Tr}\, U \to 0$. Then $\text{pd}_\Lambda \text{Tr}\, U \leqslant 1$ since $U^* = 0$.

As a consequence of this lemma, we have the following.

PROPOSITION 4.2. *The subcategory* $\mathscr{P}^1(\Lambda)$ *of* mod Λ *has the following properties.*

(a) $\mathscr{P}^1(\Lambda)$ *is covariantly finite in* mod Λ.

(b) $\mathscr{P}^1(\Lambda)$ *is contravariantly finite in* mod Λ *if and only if* $\mathscr{T}$ *is covariantly finite in* mod Λ.

Proof. This proof is based on the following fact.

Suppose $\mathscr{C}$ and $\mathscr{D}$ are subcategories of mod Λ and mod Λ^{op}, respectively, such that the duality Tr: $\underline{\text{mod}}\,\Lambda \to \underline{\text{mod}}\,\Lambda^{\text{op}}$ induces a duality between $\underline{\mathscr{C}}$ and $\underline{\mathscr{D}}$. Then $\mathscr{C}$ is contravariantly finite in mod Λ if and only if $\mathscr{D}$ is covariantly finite in mod Λ^{op} [8, Cor. 7.3].

(a) Since $\mathscr{T}$ is closed under factor modules, it is easily seen that $\mathscr{T}$ is contravariantly finite in mod Λ^{op} [8]. Hence by our above remark $\mathscr{P}^1(\Lambda)$ is covariantly finite since the transpose induces an equivalence between $\underline{\mathscr{P}^1(\Lambda)}$ and $\underline{\mathscr{T}}$ by Lemma 4.1.

(b) This is a trivial consequence of Lemma 4.1 and the above remark.

As an easy consequence of Proposition 4.2, we have the following.

COROLLARY 4.3. *The subcategory* $\{X$ *in* mod $\Lambda \mid \text{Ext}^1_\Lambda(X, \mathscr{P}^1(\Lambda)) = 0\}$ *is a contravariantly finite subcategory of* mod Λ.

Proof. Since $\mathscr{P}^1(\Lambda)$ is closed under extensions and is covariantly finite by Proposition 4.2, we have by the remark following Proposition 1.8 that the subcategory of all X such that $\operatorname{Ext}^1_\Lambda(X, \mathscr{P}^1(\Lambda))=0$ is contravariantly finite.

It would be interesting to have a description of the modules X such that $\operatorname{Ext}^1(X, \mathscr{P}^1(\Lambda))=0$.

We now want to give another description of when $\mathscr{P}^1(\Lambda)$ is contravariantly finite. This is based on the following.

PROPOSITION 4.4. *$\mathscr{P}^1(\Lambda)^\perp$ consists of all Λ-modules M which have a filtration $M=M_0 \supset M_1 \supset \cdots \supset M_n=0$ such that M_i/M_{i+1} is a factor of an injective module for all $i=0, ..., n-1$.*

Proof. It is clear that $\mathscr{P}^1(\Lambda)^\perp$ consists of all M such that $\operatorname{Ext}^1_\Lambda(\mathscr{P}^1(\Lambda), M)=0$. It is also clear that if $\operatorname{pd} X \leqslant 1$, then $\operatorname{Ext}^1_\Lambda(X, M)=0$ if M is a factor of an injective module. Hence if M has a filtration $M=M_0 \supset M_1 \supset \cdots \supset M_n=0$ such that M_i/M_{i+1} is a factor of an injective module, then $\operatorname{Ext}^1_\Lambda(\mathscr{P}^1(\Lambda), M)=0$.

Suppose $\operatorname{pd} X \leqslant 1$. Then for each Λ-module M, we have $\operatorname{Ext}^1_\Lambda(X, M) \cong D\,\underline{\operatorname{Hom}}(\operatorname{Tr} DM, X) \cong D(\underline{\operatorname{Hom}}_\Lambda(\operatorname{Tr} X, D(M)))$ [7, Sect. 3]. But $\underline{\operatorname{Hom}}_\Lambda(\operatorname{Tr} DM, X) \cong \operatorname{Hom}_\Lambda(\operatorname{Tr} X, D(M))$ since by Lemma 4.1 we have that $\operatorname{Tr} X$ is torsion. Again by Lemma 4.1, as X runs through $\mathscr{P}^1(\Lambda)$, we have that $\operatorname{Tr} X$ runs through all torsion Λ^{op}-modules. Hence $\operatorname{Ext}^1_\Lambda(\mathscr{P}^1(\Lambda), M)=0$ if and only if $\operatorname{Hom}_{\Lambda^{\mathrm{op}}}(T, D(M))=0$ for all torsion Λ^{op}-modules T. Using duality, our result is an immediate consequence of the following.

LEMMA 4.5. *A Λ^{op}-module N has the property that $\operatorname{Hom}_{\Lambda^{\mathrm{op}}}(\mathscr{T}, N)=0$ if and only if N has a filtration $N=N_0 \supset N_1 \supset \cdots \supset N_n=0$ such that N_i/N_{i+1} is a torsionless Λ^{op}-module for $i=0, ..., n-1$.*

Proof. Suppose N is a torsionless module. Then $\operatorname{Hom}_{\Lambda^{\mathrm{op}}}(\mathscr{T}, N)=0$ since N is a submodule of a free Λ-module. Hence $\operatorname{Hom}_{\Lambda^{\mathrm{op}}}(T, N)=0$ if N has a filtration $N=N_0 \supset N_1 \supset \cdots \supset N_n=0$ such that N_i/N_{i+1} is torsionless for $i=0, ..., n-1$.

Suppose now that $\operatorname{Hom}_{\Lambda^{\mathrm{op}}}(\mathscr{T}, N)=0$ and $N \neq 0$. Since N is not in $\mathscr{T}$, we have that $f\colon N \to N^{**}$ is not zero. Letting $N_1=\ker f$, we have that N/N_1 is torsionless. If $N_1=0$, we are done. If $N_1 \neq 0$, we may proceed as before since $\operatorname{Hom}_{\Lambda^{\mathrm{op}}}(\mathscr{T}, N_1)=0$. Therefore we obtain a descending chain of submodules $N=N_0 \supset N_1 \supset \cdots \supset N_n=0$ with N_i/N_{i+1} torsionless.

As an easy consequence of Proposition 4.4, we obtain the following, where we recall that factor modules of injective modules are called cotorsionless modules.

COROLLARY 4.6. *The following are equivalent.*

(a) *$\mathscr{P}^1(\Lambda)$ is contravariantly finite.*

(b) *The subcategory of* mod Λ *consisting of all Λ-modules M having filtrations $M = M_0 \supset M_1 \supset \cdots \supset M_n = 0$ such that M_i/M_{i+1} is a cotorsionless module for all $i = 0, ..., n-1$ is covarantly finite.*

Proof. It is easily seen that $\mathrm{pd}_\Lambda X \leqslant 1$ if and only if $\mathrm{Ext}^1_\Lambda(X, M) = 0$ for all cotorsionless modules M. Therefore $\mathscr{P}^1(\Lambda) = {}^\perp((\mathscr{P}^1(\Lambda))^\perp)$. Combining this with the fact that $\mathscr{P}^1(\Lambda)$ is a resolving subcategory, we know by Proposition 1.9 that $\mathscr{P}^1(\Lambda)$ is contravariantly finite if and only if $\mathscr{P}^1(\Lambda)^\perp$ is covariantly finite. By Proposition 4.4, we know that $\mathscr{P}^1(\Lambda)^\perp$ is precisely the subcategory of mod Λ described in (b). Hence the equivalence of (a) and (b) is proven.

The rest of this section is devoted to proving that if Λ is stably equivalent to a hereditary algebra, then $\mathscr{P}^\infty(\Lambda)$ is contravariantly finite in mod Λ. We begin with the following more general result.

PROPOSITION 4.7. *Let $\underline{a}$ be an ideal in Λ with* $\mathrm{pd}_\Lambda \Lambda/\underline{a} < \infty$ *such that if M is a Λ-module with* $\mathrm{pd}_\Lambda M < \infty$, *then* $\mathrm{pd}_{\Lambda/\underline{a}} M/\underline{a}M < \infty$. *Let C be a $\Lambda/\underline{a}$-module. Then we have the following.*

(a) *A map $B \to C$ in* mod $\Lambda/\underline{a}$ *is a right $\mathscr{P}^\infty(\Lambda/\underline{a})$-approximation of C if and only if it is a right $\mathscr{P}^\infty(\Lambda)$-approximation of C.*

(b) *If $A \to C$ is a right $\mathscr{P}^\infty(\Lambda)$-approximation of C, then $A/\underline{a}A \to C$ is a right $\mathscr{P}^\infty(\Lambda/\underline{a})$-approximation of C.*

Proof. Suppose $B \xrightarrow{f} C$ is a right $\mathscr{P}^\infty(\Lambda/\underline{a})$-approximation of C. Since $\mathrm{pd}_\Lambda \Lambda/\underline{a} < \infty$ and $\mathrm{pd}_{\Lambda/\underline{a}} B < \infty$, we know that $\mathrm{pd}_\Lambda B < \infty$. Suppose now that we have a morphism $X \xrightarrow{g} C$ with X in $\mathscr{P}^\infty(\Lambda)$. Then g is the composition $X \xrightarrow{h} X/\underline{a}X \xrightarrow{j} C$. Since $\mathrm{pd}_{\Lambda/\underline{a}} X < \infty$, j can be lifted to B. Therefore g can be lifted to B.

The fact that if a map $B \to C$ in mod $\Lambda/\underline{a}$ is a right $\mathscr{P}^\infty(\Lambda)$-approximation, then it is a right $\mathscr{P}^\infty(\Lambda/\underline{a})$-approximation, follows trivially from the fact that $\mathscr{P}^\infty(\Lambda/\underline{a}) \subset \mathscr{P}^\infty(\Lambda)$ since $\mathrm{pd}_\Lambda \Lambda/\underline{a} < \infty$.

(b) We have that $A \to C$ is the composition $A \to A/\underline{a}A \to C$. Let $X \to C$ be a morphism with X in $\mathscr{P}^\infty(\Lambda/\underline{a})$. Then X is in $P^\infty(\Lambda)$ and so $X \to C$ factors through $A \to C$ and therefore through $A/\underline{a}A \to C$. Since $\mathrm{pd}_{\Lambda/\underline{a}} A/\underline{a}A < \infty$, we have that $A/\underline{a}A \to C$ is a right $\mathscr{P}^\infty(\Lambda/\underline{a})$-approximation.

As an easy consequence of this proposition we have the following.

COROLLARY 4.8. *Let $\underline{a}$ be an ideal in Λ satisfying the hypothesis of Proposition* 4.7. *Then we have the following.*

(a) *If* $\mathscr{P}^\infty(\Lambda)$ *is contravariantly finite in* mod Λ, *then* $\mathscr{P}^\infty(\Lambda/\underline{a})$ *is contravariantly finite in* mod Λ.

(b) *If* $\underline{a}S=0$ *for each simple* Λ*-module* S *with* $\mathrm{pd}_\Lambda S=\infty$, *then* $\mathscr{P}^\infty(\Lambda/\underline{a})$ *contravariantly finite in* mod $\Lambda/\underline{a}$ *implies* $\mathscr{P}^\infty(\Lambda)$ *contravariantly finite in* mod Λ.

Proof. (a) This is an immediate consequence of part (b) of the above proposition.

(b) Since the category of Λ-modules which have right $\mathscr{P}^\infty(\Lambda)$-approximations is closed under extensions, it follows that $\mathscr{P}^\infty(\Lambda)$ is contravariantly finite in mod Λ if each simple Λ-module has a right $\mathscr{P}^\infty(\Lambda)$-approximation (see Proposition 3.7). Suppose S is a simple Λ-module. If $\mathrm{pd}_\Lambda S<\infty$, we are done. Suppose $\mathrm{pd}_\Lambda S=\infty$. Then by assumption S is a $\Lambda/\underline{a}$-module and therefore has a right $\mathscr{P}^\infty(\Lambda/\underline{a})$-approximation $A\to S$, since $\mathscr{P}^\infty(\Lambda/\underline{a})$ is contravariantly finite in mod Λ by assumption. But $A\to S$ is also a right $\mathscr{P}^\infty(\Lambda)$-approximation by part (a) of the above proposition. This finishes the proof of the corollary.

We give an application of this corollary after introducing the following notation. We denote by Ω (mod Λ) the subcategory consisting of the syzygy modules $\Omega(C)$ for all C in mod Λ. It should be noted that Ω(mod Λ) is nothing more than the submodules of $\underline{r}P$ for all projective modules P, where $\underline{r}$ is the radical of Λ. Further we denote by $\tau_{\mathscr{C}}(A)$ the trace of a category $\mathscr{C}$ in the module A, that is, the submodule of A generated by the images of all maps $C\to A$ with C in $\mathscr{C}$.

PROPOSITION 4.9. *Suppose* $\mathrm{pd}\ \underline{a}<\infty$ *where* $\underline{a}$ *is the ideal* $\tau_{\Omega(\mathrm{mod}\,\Lambda)\cap\mathscr{P}^\infty(\Lambda)}(\underline{r})$. *Then*

(a) $\underline{a}\subset\underline{r}$.

(b) *If* M *is a* Λ*-module with* $\mathrm{pd}_\Lambda M<\infty$, then $\mathrm{pd}_{\Lambda/\underline{a}} M/\underline{a}M=0$.

(c) $\mathscr{P}^\infty(\Lambda/\underline{a})=\mathscr{P}^0(\Lambda/\underline{a})$ *and is therefore contravariantly finite in* mod $\Lambda/\underline{a}$.

(d) $\mathscr{P}^\infty(\Lambda)$ *is covariantly finite in* mod Λ.

Proof. (a) Obvious.

(b) We first observe that if P is a projective module, then $\underline{a}P=\tau_{\Omega(\mathrm{mod}\,\Lambda)\cap\mathscr{P}^\infty(\Lambda)}\underline{r}P$. Suppose now that $\mathrm{pd}_\Lambda M<\infty$ and that $0\to\Omega M\to P\xrightarrow{f} M\to 0$ is exact with f a projective cover of M. Since $\mathrm{pd}_\Lambda\Omega M<\infty$, we have that ΩM is in $\Omega(\mathrm{mod}\,\Lambda)\cap\mathscr{P}^\infty(\Lambda)$. Therefore $\Omega M\subset\tau_{\Omega(\mathrm{mod}\,\Lambda)\cap\mathscr{P}^\infty(\Lambda)}\underline{r}P=\underline{a}P$, and we have $P/\underline{a}P\simeq M/\underline{a}M$.

(c) Suppose M is in $\mathscr{P}^\infty(\Lambda/\underline{a})$. Then $\mathrm{pd}_\Lambda M<\infty$ and so $M=M/\underline{a}M$ is $\Lambda/\underline{a}$-projective by (b).

(d) Since the ideal $\underline{a}$ satisfies the hypothesis of Proposition 4.7 and $\underline{a} \subset \underline{r}$, it follows from Corollary 4.8 that $\mathscr{P}^{\infty}(\Lambda)$ is contravariantly finite since $\mathscr{P}^{\infty}(\Lambda/\underline{a})$ is contravariantly finite.

As an easy consequence of this proposition we have the following.

COROLLARY 4.10. *Suppose* pd $\underline{a} < \infty$ *when* $\underline{a}$ *is the ideal* $\tau_{\Omega(\text{mod}\,\Lambda) \cap \mathscr{P}^{\infty}(\Lambda)}(\underline{r})$. *Then the* Λ*-modules of finite projective dimension are the summands of modules* M *which have filtrations* $M = M_0 \supset M_1 \supset \cdots \supset M_n = 0$ *where each* M_i/M_{i+1} *is an indecomposable projective* $\Lambda/\underline{a}$*-module.*

Proof. Since $\mathscr{P}^0(\Lambda/\underline{a}) = \mathscr{P}^{\infty}(\Lambda/\underline{a})$, we know that the right $\mathscr{P}^{\infty}(\Lambda/\underline{a})$-approximation is its projective cover. It then follows from the fact that $\underline{a}$ satisfies the hypothesis of Proposition 4.7 (see Proposition 4.9) that the $\Lambda/\underline{a}$-projective covers of the simple Λ-modules are the right $\mathscr{P}^{\infty}(\Lambda)$-approximations of the simple Λ-modules. The corollary follows from this observation by Proposition 3.7.

We now give some criteria for when $\text{pd}_{\Lambda}\underline{a} < \infty$ for the ideal $\underline{a} = \tau_{\Omega(\text{mod}\,\Lambda) \cap \mathscr{P}^{\infty}(\Lambda)}(\underline{r})$.

PROPOSITION 4.11. *Suppose* Λ *satisfies the following*:

(i) *Each indecomposable torsionless module is simple or has finite projective dimension.*

(ii) *If* S *is a simple composition factor of* $\underline{r}P/\text{soc}\,P$ *for* P *indecomposable projective, then* S *is either a torsion module or has finite projective dimension.*

Then $\text{pd}_{\Lambda}\underline{a} < \infty$ *where* $\underline{a} = \tau_{\Omega(\text{mod}\,\Lambda) \cap \mathscr{P}^{\infty}(\Lambda)}(\underline{r})$.

Proof. Suppose pd $\underline{a} = \infty$. Then by (i) we know that $\underline{a} = A \amalg S_1 \amalg S_2 \amalg \cdots \amalg S_n$ where pd $A < \infty$ and the S_i are simple modules with pd $S_i = \infty$ for all $i = 1, \dots, n$. *Let* $S = S_i$ for some i. Then there is an epimorphism $L \xrightarrow{f} S$ with L indecomposable in $\Omega(\text{mod}\,\Lambda) \cap \mathscr{P}^{\infty}$. It follows that $f(\text{soc}\,L) = 0$. For otherwise f would be a splittable epimorphism and so $\text{pd}_{\Lambda} S < \infty$, since $\text{pd}_{\Lambda} L < \infty$; but this is impossible. Hence S is a composition factor of $L/\text{soc}\,L$. Since L is also in $\Omega(\text{mod}\,\Lambda)$, we have that $L \subset \underline{r}P$ for some projective module P. Therefore $L/\text{soc}\,L \subset \underline{r}P/\text{soc}\,P$ which implies by (ii) that S is torsion since pd $S = \infty$. But this contradicts the fact that $S \subset \underline{a}$. Therefore $\underline{a} = A$, that is, pd $\underline{a} < \infty$.

We now show that algebras stably equivalent to hereditary algebras satisfy the hypothesis of Proposition 4.11.

LEMMA 4.12. *The following are equivalent for an artin algebra* Λ.

(a) *Λ satisfies the following.*

(i) *A simple module S is a torsion module if it is a composition factor of $\underline{r}P/\mathrm{soc}\, P$ for some indecomposable projective module P.*

(ii) *Every indecomposable torsionless module is simple or projective.*

(b) *Λ is stably equivalent to a hereditary algebra.*

Proof. We first recall that Λ is stably equivalent to a hereditary algebra if and only if it satisfies (i) every simple torsionless nonprojective module is cotorsionless and (ii) every indecomposable torsionless module is simple or projective [5].

(a)$\Rightarrow$(b). We only have to show that every simple torsionless nonprojective module is a cotorsionless module. We do this by showing that part (i) of (a) implies what we want.

By Proposition 4.4 we know that a simple module S is cotorsionless if $\mathrm{Ext}^1_\Lambda(\mathscr{P}^1(\Lambda), S)=0$. Let S be a simple torsionless nonprojective module and let $\mathrm{pd}_\Lambda M \leqslant 1$. Let $0 \to P_1 \xrightarrow{h} P_0 \to M \to 0$ be a minimal projective resolution of M. Then $\mathrm{Im}\, h \subset \underline{r}P_0$. Let P_1' be an indecomposable summand of P_1. Then $P_1'/\mathrm{soc}\, P_1' \hookrightarrow \underline{r}P_0/\mathrm{soc}\, P_0$ and so the composition factors of $P_1'/\mathrm{soc}\, P_1'$ are torsion modules. So $\mathrm{Hom}_\Lambda(P_1'/\mathrm{soc}\, P_1', S)=0$ since S is torsionless. Also if $f\colon P_1' \to S$ is not zero, then $f(\mathrm{soc}\, P_1')=0$. Otherwise, f would be a splittable epimorphism which implies S is projective. Therefore $\mathrm{Hom}_\Lambda(P_1, S)=0$, which implies that $\mathrm{Ext}^1_\Lambda(M, S)=0$. Therefore S is cotorsionless.

(b)$\Rightarrow$(a). We only have to show that (b) implies part (i) of (a). Let S be a simple composition factor of $\underline{r}P_0/\mathrm{soc}\, P_0$ for some indecomposable projective module P_0. Then S is not projective since all the simple projective composition factors of a module are in the socle of the module. Let P_1 be the projective cover of S. Then there is a morphism $g\colon P_1 \to P_0$ with $\mathrm{Im}\, g$ not simple. Therefore g is a nonisomorphism since (b) implies that $\mathrm{Im}\, g$ is projective, and P_1 is indecomposable. Consider the exact sequence $0 \to P_1 \xrightarrow{g} P_0 \to M \to 0$. This gives rise to the exact sequence $\mathrm{Hom}_\Lambda(P_0, S) \xrightarrow{(g,S)} \mathrm{Hom}_\Lambda(P_1, S) \to \mathrm{Ext}^1_\Lambda(M, S) \to 0$. Since $\mathrm{Im}\, g \subset \underline{r}P_0$, we have that $(g, S)=0$. Therefore $\mathrm{Ext}^1_\Lambda(M, S) \neq 0$ since $\mathrm{Hom}_\Lambda(P_1, S) \neq 0$. Since $\mathrm{pd}\, M=1$, we know that S is not cotorsionless, which by (b) means that S is a torsion module because we have already seen that it is not projective.

As an immediate consequence of Lemma 4.12 we have the following.

PROPOSITION 4.13. *Suppose Λ is stably equivalent to a hereditary algebra. Then $\underline{a}=\tau_{\Omega(\mathrm{mod}\,\Lambda)\cap\mathscr{P}^\infty(\Lambda)}(\underline{r})$ has finite projective dimension and so $\mathscr{P}^\infty(\Lambda)$ is contravariantly finite in* mod Λ.

We point out that it is easy to find examples of algebras not stably equivalent to hereditary algebras where pd $\underline{a} < \infty$ for $\underline{a} = \tau_{\Omega(\operatorname{mod} \Lambda) \cap \mathscr{P}^\infty(\Lambda)}(\underline{r})$. For example, consider the path algebra of the quiver ${}^{\alpha}\circlearrowright \cdot \xrightarrow{\beta} \cdot \xrightarrow{\gamma} \cdot \xrightarrow{\delta} \cdot$ modulo the relations $\alpha^2 = 0$, $\delta\gamma\beta = 0$.

We also note that if Λ satisfies the hypothesis of Proposition 4.11, then Λ is ultimately closed; that is for each C in mod Λ there are only a finite number of indecomposable modules which are summands of syzygy modules for C. But our results cannot be generalized to ultimately closed algebras, since the example in [13] where $\mathscr{P}^\infty(\Lambda)$ is not contravariantly finite is ultimately closed.

Since $\mathscr{P}^\infty(\Lambda)$ is contravariantly finite if and only if $\mathscr{P}^\infty\begin{pmatrix} \Lambda & 0 \\ \Lambda & \Lambda \end{pmatrix}$ is contravariantly finite (see [18]), one can create classes of examples where $\mathscr{P}^\infty(\Lambda)$ is contravariantly finite or not.

5. Tilting and Cotilting Modules

We have seen in Section 3 that $\mathscr{X} \mapsto \mathscr{X}^\perp$ gives a one–one correspondence between contravariantly finite resolving and covariantly finite coresolving subcategories of mod Λ. In this section we study the connection between cotilting modules and contravariantly finite resolving subcategories, and show amongst other things that $T \mapsto {}^\perp T$ gives a one–one correspondence between basic cotilting modules of finite injective dimension and contravariantly finite resolving subcategories, when Λ has finite global dimension. Special cases of such connections were used in [9]. We say that a module is basic if in a direct sum decomposition no indecomposable module appears more than once.

Note that ${}^\perp T$ is the largest subcategory $\mathscr{X}$ of mod Λ where T is strong Extinjective, in the sense that $\operatorname{Ext}^i_\Lambda(\mathscr{X}, T) = 0$ for all $i > 0$. We start by studying a slightly different category associated with a selforthogonal Λ-module T, namely the subcategory $\mathscr{X}_T$ of ${}^\perp T$ whose objects are the C such that there is an exact sequence $0 \to C \to T_0 \to T_1 \to \cdots \to T_n \xrightarrow{f_n} T_{n+1} \cdots$ with T_i in add T and $\operatorname{Im} f_n$ in ${}^\perp T$ for all $n \geqslant 0$. For a selforthogonal additive category ω $\mathscr{X}_\omega$ is defined similarly.

We shall see in the first proposition that $\mathscr{X}_T$ is closed under direct summands, so that this standard assumption is satisfied. $\mathscr{X}_T$ is the largest subcategory of mod Λ where T is strong Extinjective and a cogenerator in the sense that each object in the category has an embedding into an object in add T with the cokernel again in the category. We shall investigate when these categories are contravariantly finite and resolving. We first prove some properties of $\mathscr{X}_\omega$, which are needed to apply the theory in [3] (see also [20]). Then we go on to give connections with tilting theory.

PROPOSITION 5.1. *For a selforthogonal category ω the category $\mathscr{X}_\omega$ is closed under extensions, kernels of epimorphisms, and direct summands.*

Proof. We first prove that $\mathscr{X}_\omega$ is closed under extensions. So let $0 \to A \to B \to C \to 0$ be an exact sequence with A and C in $\mathscr{X}_\omega$. Then we have exact sequences $0 \to A \to T_0 \to L \to 0$ and $0 \to C \to T_0' \to K \to 0$, with T_0 and T_0' in ω and L and K in $\mathscr{X}_\omega$.

Consider the pushout diagram

$$
\begin{array}{ccccccccc}
 & & 0 & & 0 & & & & \\
 & & \downarrow & & \downarrow & & & & \\
0 & \longrightarrow & A & \longrightarrow & B & \longrightarrow & C & \longrightarrow & 0 \\
 & & \downarrow & & \downarrow & & \| & & \\
0 & \longrightarrow & T_0 & \longrightarrow & U & \longrightarrow & C & \longrightarrow & 0 \\
 & & \downarrow & & \downarrow & & & & \\
 & & L & = & L & & & & \\
 & & \downarrow & & \downarrow & & & & \\
 & & 0 & & 0 & & & &
\end{array}
$$

Since C is in $\mathscr{X}_\omega \subset {}^\perp T$, we have $\operatorname{Ext}_\Lambda^1(C, T_0) = 0$ so that $U \simeq T_0 \amalg C$. We then have the commutative exact diagram

$$
\begin{array}{ccccccccc}
 & & & & 0 & & 0 & & \\
 & & & & \downarrow & & \downarrow & & \\
0 & \longrightarrow & B & \longrightarrow & T_0 \amalg C & \longrightarrow & L & \longrightarrow & 0 \\
 & & \| & & \downarrow & & \downarrow & & \\
0 & \longrightarrow & B & \longrightarrow & T_0 \amalg T_0' & \longrightarrow & V & \longrightarrow & 0 \\
 & & & & \downarrow & & \downarrow & & \\
 & & & & K & = & K & & \\
 & & & & \downarrow & & \downarrow & & \\
 & & & & 0 & & 0 & &
\end{array}
$$

It follows that V is in ${}^\perp\omega$ since ${}^\perp\omega$ is closed under extensions. Using that L and K are in $\mathscr{X}_\omega$ and repeating the process, we see that B is in $\mathscr{X}_\omega$. Hence $\mathscr{X}_\omega$ is closed under extensions.

Now let $0 \to A \to B \to C \to 0$ be an exact sequence with B in $\mathscr{X}_\omega$. We then have an exact commutative diagram

$$
\begin{array}{ccccccccccc}
 & & & & 0 & & 0 & & & & \\
 & & & & \downarrow & & \downarrow & & & & \\
 & & 0 & \longrightarrow & A & \longrightarrow & B & \longrightarrow & C & \longrightarrow & 0 \\
 & & & & \downarrow & & \downarrow & & & & \\
 & & 0 & \longrightarrow & T_0 & = & T_0 & & & & \\
 & & & & \downarrow & & \downarrow & & & & \\
0 & \longrightarrow & C & \longrightarrow & K & \longrightarrow & E & \longrightarrow & 0 & & \\
 & & & & \downarrow & & \downarrow & & & & \\
 & & & & 0 & & 0 & & & &
\end{array}
\qquad (*)
$$

with T_0 in ω and E in $\mathscr{X}_\omega$.

If C is also in $\mathscr{X}_\omega$, we have that K is in $\mathscr{X}_\omega$ since $\mathscr{X}_\omega$ is closed under extensions. Hence A is in $\mathscr{X}_\omega$, so that $\mathscr{X}_\omega$ is closed under kernels of epimorphisms.

If $B \simeq A \amalg C$, we see from the exact sequence $0 \to A \amalg C \to A \amalg K \to E \to 0$ coming from $(*)$ that K is a summand of an object in $\mathscr{X}_\omega$, since $\mathscr{X}_\omega$ is closed under extensions. Since summands of objects in $\mathscr{X}_\omega$ are clearly in ${}^\perp\omega$, we see by repeating the process that A is in $\mathscr{X}_\omega$. This shows that $\mathscr{X}_\omega$ is closed under summands.

For a subcategory $\mathscr{X}$ of $\operatorname{mod} \Lambda$ we denote by $\hat{\mathscr{X}}$ the subcategory of $\operatorname{mod} \Lambda$ whose objects are the C for which there is an exact sequence $0 \to X_n \to \cdots \to X_0 \to C \to 0$ with X_i in $\mathscr{X}$. For a further investigation of $\mathscr{X}_\omega$ the following result is useful [3, Th. 2.3, Prop. 3.6].

PROPOSITION 5.2. *Let $\mathscr{X}$ be a resolving subcategory of* $\operatorname{mod} \Lambda$ *with Extinjective cogenerator ω. If $\hat{\mathscr{X}} = \operatorname{mod} \Lambda$, we have the following.*

(a) *$\mathscr{X}$ is contravariantly finite in* $\operatorname{mod} \Lambda$.

(b) $\mathscr{Y} = \mathscr{X}^\perp = \hat{\omega}$.

Since when ω is selforthogonal, $\mathscr{X}_\omega$ is a resolving subcategory of $\operatorname{mod} \Lambda$ with cogenerator ω, we would like to investigate when $\hat{\mathscr{X}}_\omega = \operatorname{mod} \Lambda$. This leads to a close connection with tilting theory.

We recall that a selforthogonal module T is a tilting module if $\operatorname{pd} T < \infty$ and Λ is in $\widecheck{\operatorname{add} T}$, where for a subcategory $\mathscr{X}$ of $\operatorname{mod} \Lambda$ we denote by $\check{\mathscr{X}}$ the subcategory whose objects are the C for which there is an exact sequence $0 \to C \to X_0 \to \cdots \to X_n \to 0$ with all X_i in $\mathscr{X}$. A selforthogonal module T is a cotilting module if id $T < \infty$ and all injective modules are in $\widehat{\operatorname{add} T}$. In establishing our desired connection between cotilting modules and contravariantly finite resolving subcategories and between cotilting modules and covariantly finite coresolving subcategories, the following result is useful.

PROPOSITION 5.3. *Let $\mathscr{X}$ be a contravariantly finite resolving subcategory of* mod Λ, *and $\mathscr{Y} = \mathscr{X}^{\perp}$ the associated covariantly finite coresolving subcategory of* mod Λ.

Then $\hat{\mathscr{X}} = \operatorname{mod} \Lambda$ if and only if $\mathscr{Y} \subset \mathscr{I}^{\infty}(\Lambda)$.

Proof. Assume first that $\hat{\mathscr{X}} = \operatorname{mod} \Lambda$, and let for each simple Λ-module S, n_s denote the maximal number such that there is an exact sequence $0 \to X_{n_s} \to \cdots X_1 \to X_0 \to S \to 0$ with the X_i in $\mathscr{X}$. Let n denote the maximum of the numbers n_s. Since for Y in $\mathscr{Y}$, $\operatorname{Ext}^i_\Lambda(\mathscr{X}, Y) = 0$ for $i > 0$, we get that $\operatorname{Ext}^i_\Lambda(S, Y) = 0$ for all $i > n$. This shows that id $Y \leqslant n$, so that $\mathscr{Y} \subset \mathscr{I}^{\infty}(\Lambda)$.

Assume conversely that $\mathscr{Y} \subset \mathscr{I}^{\infty}(\Lambda)$. Since $\mathscr{Y}$ is a covariantly finite coresolving subcategory of mod Λ, every object in $\mathscr{Y}$ is a direct summand of a module which has a filtration with factors amongst the Y^S, where $S \to Y^S$ is a left $\mathscr{Y}$-approximation for the simple Λ-module S, by Proposition 3.8. Hence $\mathscr{Y} \subset \mathscr{I}^d(\Lambda)$ for some integer d, where d is the maximum of the injective dimensions of the modules Y^S. Then for C in mod Λ, $\operatorname{Ext}^{i+d}_\Lambda(C, \mathscr{Y}) = 0$ for $i > 0$, so that $\Omega^d C$ is in $\mathscr{X}$. Since $\mathscr{X}$ contains the projectives, this shows that C is in $\hat{\mathscr{X}}$, and consequently $\hat{\mathscr{X}} = \operatorname{mod} \Lambda$.

We now get the following characterization of a category ω to be add T for a cotilting module T, in terms of the associated category $\mathscr{X}_\omega$.

THEOREM 5.4. *Let ω be a selforthogonal additive subcategory of* mod Λ.

(a) *$\omega = \operatorname{add} T$ for a cotilting module T if and only if $\hat{\mathscr{X}}_\omega = \operatorname{mod} \Lambda$.*

(b) *If T is a cotilting module, then $\mathscr{X}_T = {}^{\perp}T$.*

Proof. Assume that $\omega = \operatorname{add} T$ for a cotilting module T. We then have id $T = n < \infty$ and we want to show that $\mathscr{X}_T = {}^{\perp}T$. Assume C is in ${}^{\perp}T$, and let $C \to I$ be an injective envelope. Since I is in $\widehat{\operatorname{add} T}$, we have the following exact diagram

$$\begin{array}{ccccccccccccc}
 & & & & 0 & \longrightarrow & L & \longrightarrow & f^{-1}(C) & \longrightarrow & C & \longrightarrow & 0 \\
 & & & & & & & & \downarrow & & \downarrow & & \\
0 & \longrightarrow & T_n & \longrightarrow & \cdots & \longrightarrow & T_1 & \xrightarrow{g} & T_0 & \xrightarrow{f} & I & \longrightarrow & 0
\end{array}$$

with the T_i in ω and where $L = \operatorname{Coker} g$.

Since $L \in \widehat{\operatorname{add} T}$, we have $\operatorname{Ext}^i_\Lambda({}^\perp T, L) = 0$ for $i > 0$, so that $0 \to L \to f^{-1}(C) \to C \to 0$ splits. Hence we get a monomorphism $0 \to C \to T_0$. Then choose a map $C \to T_0'$ such that any map $C \to T$ factors through $C \xrightarrow{h} T_0 \amalg T_0'$. Hence we have $\operatorname{Ext}^i_\Lambda(\operatorname{Coker} h, T) = 0$ for $i > 0$, so that $\operatorname{Coker} h$ is in ${}^\perp T$. Continuing in this way, we see that C is in $\mathscr{X}_T$, so that ${}^\perp T = \mathscr{X}_T = X_\omega$. Then we have $\operatorname{Ext}^i_\Lambda(C, T) = 0$ for all $i > n$ and C in mod Λ. This shows that $\Omega^n C$ is in ${}^\perp T = \mathscr{X}_T$, so that $\hat{\mathscr{X}}_T = \operatorname{mod} \Lambda$.

Assume conversely that $\hat{\mathscr{X}}_\omega = \operatorname{mod} \Lambda$. Since $\mathscr{X}_\omega$ is resolving and has ω as a cogenerator, we get that $\mathscr{X}_\omega$ is contravariantly finite and $\mathscr{X}_\omega^\perp = \hat{\omega}$ by Proposition 5.2. Further id $T < \infty$ for T in ω by Proposition 5.3. Since $D(\Lambda)$ is in $\mathscr{X}_\omega^\perp$, we have an exact sequence $0 \to \omega_n \to \cdots \to \omega_1 \to \omega_0 \to D(\Lambda) \to 0$ with the ω_i in ω. Then $T = \omega_0 \amalg \cdots \amalg \omega_n$ is a cotilting module. Since any module in ω having T as a direct summand must be a cotilting module, and the number of nonisomorphic indecomposable summands of a cotilting module is the number of nonisomorphic simple Λ-modules [11, 15], we conclude that $\omega = \operatorname{add} T$.

We now return to the correspondences $T \mapsto {}^\perp T$ and $T \mapsto \widehat{\operatorname{add} T}$ for a selforthogonal module T. They give connections between cotilting modules and contravariantly finite resolving subcategories, and consequently also between cotilting modules and covariantly finite coresolving subcategories.

THEOREM 5.5. *Let T be a selforthogonal Λ-module.*

(a) *$T \mapsto {}^\perp T$ gives a one–one correspondence between isomorphism classes of basic cotilting modules and contravariantly finite resolving subcategories $\mathscr{X}$ with $\hat{\mathscr{X}} = \operatorname{mod} \Lambda$.*

(b) *$T \mapsto \widehat{\operatorname{add} T}$ gives a one–one correspondence between isomorphism classes of basic cotilting modules and covariantly finite coresolving subcategories of $\mathscr{I}^\infty(\Lambda)$.*

Proof. We have seen that if T is cotilting, then ${}^\perp T = \mathscr{X}_T$ is contravariantly finite resolving with $\widehat{{}^\perp T} = \operatorname{mod} \Lambda$ and that $\widehat{\operatorname{add} T}$ is covariantly finite coresolving.

If $\mathscr{X}$ is contravariantly finite resolving, we know that $\mathscr{X} \subset \mathscr{X}_\omega$ by Proposition 1.9, where $\omega = \mathscr{X} \cap \mathscr{X}^\perp$. If $\hat{\mathscr{X}} = \operatorname{mod} \Lambda$, we have then $\hat{\mathscr{X}}_\omega = \operatorname{mod} \Lambda$, so that $\omega = \operatorname{add} T$ for a cotilting module T by Theorem 5.4, and $\mathscr{X}_\omega$ is also contravariantly finite resolving, by Propositions 5.1 and 5.2. Then $\mathscr{X}_\omega^\perp = \mathscr{X}^\perp = \hat{\omega}$ by Proposition 5.2, so that $\mathscr{X}_\omega = \mathscr{X}$ by Proposition 1.10. This finishes the proof of (a). Part (b) follows from (a) and Proposition 5.3.

For algebras of finite global dimension we have $\hat{\mathscr{X}} = \operatorname{mod} \Lambda$ for any category $\mathscr{X}$ containing all projective modules. This gives the following.

COROLLARY 5.6. *Assume that* gl. dim $\Lambda < \infty$. *Then* $T \mapsto {}^{\perp}T$ *gives a one–one correspondence between isomorphism classes of basic cotilting modules and contravariantly finite resolving subcategories, and* $T \mapsto \widehat{\operatorname{add} T}$ *gives a one–one correspondence between isomorphism classes of basic cotilting modules and covariantly finite coresolving subcategories.*

If T is a classical cotilting module, that is, a cotilting module with id $T \leqslant 1$, it is well known that ${}^{\perp}T = \operatorname{Sub} T$ [12]. We get the following, since all cotilting modules over hereditary algebras are classical.

COROLLARY 5.7. *Let* Λ *be an hereditary algebra. Then* $T \mapsto \operatorname{Sub} T$ *gives a one–one correspondence between isomorphism classes of (classical) basic cotilting modules and contravariantly finite resolving subcategories of* mod Λ.

Since it is often of interest to decide when $\mathscr{X} = {}^{\perp}\omega$, we give the following criterion.

PROPOSITION 5.8. *Let* $\mathscr{X}$ *be a contravariantly finite resolving subcategory of* mod Λ, *let* $\mathscr{Y} = \mathscr{X}^{\perp}$, *and* $\omega = \mathscr{X} \cap \mathscr{Y}$. *Then* $\mathscr{X} = {}^{\perp}\omega$ *if and only if* $\mathscr{Y} \cap {}^{\perp}\omega = \omega$.

Proof. If $\mathscr{X} = {}^{\perp}\omega$, it is clear that $\mathscr{Y} \cap {}^{\perp}\omega = \omega$. Assume then that $\mathscr{Y} \cap {}^{\perp}\omega = \omega$ and let C be in ${}^{\perp}\omega$. Since $\mathscr{X}$ is contravariantly finite resolving, we have an exact sequence $0 \to Y \to X \to C \to 0$ with X in $\mathscr{X}$ and Y in $\mathscr{Y}$ by Lemma 1.3. Since C and X are in ${}^{\perp}\omega$, it follows that Y is in ${}^{\perp}\omega$. Hence the sequence splits, so that C is in $\mathscr{X}$. This shows that $\mathscr{X} = {}^{\perp}\omega$.

Dually, we associate with a selforthogonal Λ-module T the categories $T^{\perp} = \{C \mid \operatorname{Ext}^i(T, C) = 0, \text{ for } i > 0\}$ and $\mathscr{Y}_T = D\mathscr{X}_{DT}$. We then have dual versions of all the above results. In particular, $\mathscr{Y}_T$ is covariantly finite and coresolving when T is a tilting module, and add T is contravariantly finite resolving.

We shall now show that if T is cotilting, then $\mathscr{X}_T$ is in addition to being contravariantly finite also covariantly finite, and if T is tilting, then $\mathscr{Y}_T$ is also contravariantly finite. Since $\mathscr{X}_T$ and $\mathscr{Y}_T$ are then by definition functorially finite and are in addition closed under extensions, they have almost split sequences [9, Th. 2.4]. For this we shall need the following general result.

PROPOSITION 5.9. *Let* (F, G) *be an adjoint pair of functors with* $F\colon \operatorname{mod} \Lambda \to \operatorname{mod} \Gamma$ *and* $G\colon \operatorname{mod} \Gamma \to \operatorname{mod} \Lambda$. *Let* $\mathscr{Y} \subset \operatorname{mod} \Lambda$ *be such that the natural morphism* $GF \to I$, *where* I *is the identity, is an isomorphism on the objects of* $\mathscr{Y}$ *and* $\mathscr{Z} \subset \operatorname{mod} \Gamma$ *such that the natural morphism* $I \to FG$ *is an isomorphism on the objects of* $\mathscr{X}$, *and assume that* $F(\mathscr{Y}) \subset \mathscr{X}$ *and* $G(\mathscr{X}) \subset \mathscr{Y}$.

(a) *If* $\mathscr{X}$ *is contravariantly finite in* $\operatorname{mod} \Gamma$, *then* $\mathscr{Y}$ *is contravariantly finite in* $\operatorname{mod} \Lambda$.

(b) *If* $\mathscr{Y}$ *is covariantly finite in* $\operatorname{mod} \Lambda$, *then* $\mathscr{X}$ *is covariantly finite in* $\operatorname{mod} \Gamma$.

Proof. (a) Let C be in $\operatorname{mod} \Lambda$, and let $f: Z \to FC$ be a right $\mathscr{X}$-approximation of FC. Then we have the induced maps $GZ \to GFC \to C$ in $\operatorname{mod} \Lambda$, and we claim that this composition is a right $\mathscr{Y}$-approximation for C. Let $g: Y \to C$ be a map with $Y \in \mathscr{Y}$. We then get the commutative diagram

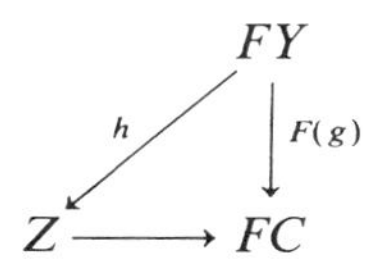

This gives rise to the commutative diagram

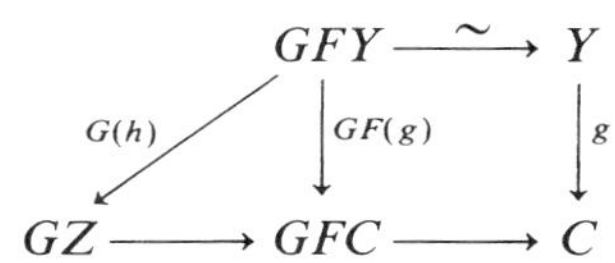

and we are done.

(b) is dual to (a).

We now have the following consequence.

COROLLARY 5.10. (a) *If* T *is a cotilting module, then* $\mathscr{X}_T$ *is functorially finite and resolving.*

(b) *If* T *is a tilting module, then* $\mathscr{Y}_T$ *is functorially finite and coresolving.*

Proof. If ${}_\Lambda T$ is a (co)tilting module and $\Gamma = \operatorname{End}_\Lambda(T)^{\mathrm{op}}$, then T_Γ is also a (co)tilting module and ${}_\Gamma(DT)$ a cotilting (tilting) module [15]. Further, $(T, \ )$: $\operatorname{mod} \Lambda \to \operatorname{mod} \Gamma$ induces an equivalence between $\mathscr{Y}_T$ and $\mathscr{X}_{DT}$ when T is a tilting module, and $D(DT, \ )D = DT \otimes_\Lambda$: $\operatorname{mod} \Lambda \to \operatorname{mod} \Gamma$ induces an equivalence between $\mathscr{X}_T$ and $\mathscr{Y}_{DT}$ when T is a cotilting module. Further $F = T_\Gamma \otimes -$ and $G = (T, \ )$ is an adjoint pair of functors in the first case, and $DT \otimes_\Lambda$ and $(DT, \ )$ in the second case, satisfying the hypothesis of Proposition 5.8 (see [15]).

Our results in this section suggest generalizing the concept of cotilting beyond finite injective dimension, by defining a general cotilting module to be a selforthogonal module T such that ${}^{\perp}T$ is contravariantly finite and add $T = {}^{\perp}T \cap ({}^{\perp}T)^{\perp}$. Similarly we define a general tilting module to be a selforthogonal module T such that $T^{\perp}$ is covariantly finite and add $T = T^{\perp} \cap {}^{\perp}(T^{\perp})$. (Note that Wakamatsu has defined a generalized tilting module to be, in our terminology, a selforthogonal module T such that $\mathscr{X}_T$ is resolving [19], and this is a strictly more general concept [20].) We now prove that our definition of a general cotilting module coincides with the usual one in the case of finite dimension.

PROPOSITION 5.11. *Assume* id $T < \infty$. *Then T is a cotilting module if and only if ${}^{\perp}T$ is contravariantly finite and* add $T = {}^{\perp}T \cap ({}^{\perp}T)^{\perp}$.

Proof. Assume that T is such that ${}^{\perp}T$ is contravariantly finite, add $T = {}^{\perp}T \cap ({}^{\perp}T)^{\perp}$, and id $T = d < \infty$. Then for C in mod Λ, $\operatorname{Ext}^i_\Lambda(\Omega^d C, T) \simeq \operatorname{Ext}^{i+d}_\Lambda(C, T) = 0$ for $i > 0$, so that $\Omega^d C$ is in ${}^{\perp}T$. Then ${}^{\perp}\hat{T} = \operatorname{mod} \Lambda$, and since ${}^{\perp}T$ is clearly resolving, we conclude that T is a cotilting module by Theorem 5.4.

If conversely T is a cotilting module, then ${}^{\perp}T = \mathscr{X}_T$ is contravariantly finite by Theorem 5.4. We also have that add T is the category of Extinjectives in $\mathscr{X}_T = {}^{\perp}T$.

To construct tilting or cotilting modules, it is of interest to have criteria for a module to be a summand of a tilting or cotilting module, also in the case of general tilting and cotilting modules. In the classical case it is well known that if $\operatorname{Ext}^1_\Lambda(T, T) = 0$ and id $T \leqslant 1$, then there is some T' such that $T \amalg T'$ is a cotilting module [10]. We shall see that in general this problem is closely related to the question of ${}^{\perp}T$ being contravariantly finite. We note that in general T being selforthogonal and id $T < \infty$ is not sufficient for T to be extended to a cotilting module [16].

We have the following sufficient condition, which is also given in [16] with a different proof.

PROPOSITION 5.12. *If T is selforthogonal,* id $T = d < \infty$ *and ${}^{\perp}T$ is contravariantly finite, then there is some T' such that $T'' = T \amalg T'$ is a cotilting module and* id $T'' = d$.

Proof. Let ω be the additive category of Extinjective modules in ${}^{\perp}T$. It is easy to see that ${}^{\perp}T = {}^{\perp}\omega$, and since id $T < \infty$, we have ${}^{\perp}\hat{\omega} = {}^{\perp}\hat{T} = \operatorname{mod} \Lambda$. Since ${}^{\perp}\hat{\omega}$ is contravariantly finite, it follows from Theorem 5.3 that $\omega = \operatorname{add} T''$ for a cotilting module T''.

For C in $\operatorname{mod} \Lambda$, $\operatorname{Ext}^i(\Omega^d C, T) = 0$ implies $\operatorname{Ext}^i(\Omega^d C, T'') \simeq \operatorname{Ext}^{i+d}(C, T'') = 0$ so that $\operatorname{id} T'' \leqslant d$.

Note that the fact that T can be extended to a cotilting module when $\operatorname{id} T \leqslant 1$ is a special case of Proposition 5.12, since ${}^{\perp}T = \{C \mid \operatorname{Ext}^1(C, T) = 0\}$. This category is contravariantly finite by the remark after Proposition 1.8, since add T is covariantly finite.

It would be interesting to know if conversely for any summand T of a (general) cotilting module, ${}^{\perp}T$ is contravariantly finite. Also we do not know if when T is selforthogonal and ${}^{\perp}T$ is contravariantly finite, there is only a finite number of indecomposable Extinjective modules in ${}^{\perp}T$.

If T is selforthogonal, we do not know if the number of nonisomorphic indecomposable summands of T is at most the number of simple modules, even if $\operatorname{id} T < \infty$. This is the case when ${}^{\perp}T$ is contravariantly finite and $\operatorname{id} T < \infty$ by Proposition 5.12 and [11, 16]. In particular, it is not known if $\operatorname{Ext}^i(\Lambda \amalg T, \Lambda \amalg T) = 0$ for $i > 0$ implies that T is projective. This problem is equivalent to the generalized Nakayama conjecture (see [6]). It holds for finite representation type, as proved in [6]. Here we give another proof in the spirit of contravariantly finite subcategories. Note that Proposition 5.11 cannot be applied since we do not know if the number of nonisomorphic indecomposable summands of a general cotilting module is the same as the number of nonisomorphic simple modules. The hypothesis in the next result is satisfied for algebras of finite representation type.

PROPOSITION 5.13. *Let T in* mod Λ *be such that $T \amalg \Lambda$ is selforthogonal. If $T^{\perp}$ is contravariantly finite, then T is projective.*

Proof. Clearly $T^{\perp}$ contains T, the projectives and the injectives. $\mathscr{Y} = T^{\perp\perp}$ also contains all the injectives. Assume Y is in $\mathscr{Y}$ and consider $0 \to Y \to I \to Z \to 0$ where I is injective and hence in $T^{\perp}$. We have $\operatorname{Ext}^i(T, Y) = 0$ for $i > 0$, since T is in $T^{\perp}$. Hence we get $\operatorname{Ext}^i(T, Z) = 0$ for $i > 0$, so that Z is in $T^{\perp}$. But then $\operatorname{Ext}^1(Z, Y) = 0$, so that the sequence splits, showing that Y is injective. When $T^{\perp}$ is contravariantly finite, we have ${}^{\perp}\mathscr{Y} = T^{\perp}$ by Proposition 1.10. Since $\mathscr{Y}$ is the category of injectives, we then conclude $T^{\perp} = \operatorname{mod} \Lambda$. In particular, $\operatorname{Ext}^1(T, C) = 0$ for all C in $\operatorname{mod} \Lambda$, so that T is projective.

It is worth noting that if $T \amalg \Lambda$ selforthogonal implies T projective for a Λ-module T when Λ is a commutative local selfinjective artin ring, we have the following consequence: If R is a commutative local Gorenstein ring and M is a Cohen–Macaulay module with $\operatorname{Ext}^i(M, M) = 0$ for $i > 0$, then M is projective. For if M is not projective, consider $M/(x_1, \ldots, x_n)M$ over the artin ring $R/(x_1, \ldots, x_n)$, where $x_1, \ldots, x_n$ is an R-sequence of maximal length.

6. Strong Cotilting Modules and Cohen–Macaulay Rings

Throughout this section let R denote a commutative complete local noetherian Cohen–Macaulay ring. We show that the dualizing R-module w is closely connected with cotilting modules over artin algebras. Special properties of the dualizing module motivate the definition of a special type of cotilting module which we call strong cotilting module. This leads us to an analogue of Cohen–Macaulay and Gorenstein rings for artin algebras, and we give some properties and examples of these.

To get a common setting for the dualizing R-module w and a cotilting module over an artin algebra Λ, we give the following definition.

Throughout this section Γ denotes a noetherian algebra, i.e., Γ is an algebra over a commutative noetherian ring S which is a finitely generated S-module. From this it follows that add M is covariantly finite in mod Γ for all M in mod Γ. We say that T in mod Γ is a cotilting module if T is selforthogonal, id $T < \infty$, and add T is a cogenerator for mod Γ.

It is easy to see that this definition coincides with the usual definition for artin algebras, and it follows from [3, Ex. 4, Th. 11] that the dualizing R-module w is a cotilting module in this sense. The category $\mathscr{X}_w$ is then the Cohen–Macaulay modules. Our previous proofs give the following.

Proposition 6.1. *If T is a cotilting module over the noetherian algebra Γ, then $\mathscr{X}_T = \operatorname{mod} \Gamma$.*

The notion of tilting module generalizes verbatim to the rings Γ, and Γ is clearly itself a tilting module. We now show that for the rings R there are no tilting or cotilting modules other than the ones we have mentioned.

Proposition 6.2. *Let R be a commutative complete local noetherian Cohen–Macaulay ring with dualizing module w, and let T be a basic R-module.*

(a) *T is a cotilting module $\Leftrightarrow T \simeq w$.*

(b) *T is a tilting module $\Leftrightarrow T \simeq R$.*

Proof. (a) Let T be a basic selforthogonal R-module with id $T < \infty$. We first show that T is maximal Cohen–Macaulay. If $d = \dim R$, we know id $T = d$ since id $T < \infty$. If depth $C > 0$, it is easy to see that $\operatorname{Ext}^d(C, T) = 0$. Let B be such that $\operatorname{Ext}^d(B, T) \neq 0$. Then for the maximal submodule A of finite length, we have $\operatorname{Ext}^d(A, T) \neq 0$. Hence $\operatorname{Ext}^d(M, T) \neq 0$ for any module M of finite length and consequently for any M with depth $M = 0$. Let $x_1, \ldots, x_s$ be a maximal regular T-sequence. Since $\operatorname{Ext}^i(T, T) = 0$ for $i > 0$, it follows that $\operatorname{Ext}^i(T/(x_1, \ldots, x_s)T, T) = 0$ for $i > s$. Since depth $T/(x_1, ..., x_s)T = 0$, we get $d = \text{id}\, T = s$, showing that T is (maximal) Cohen–Macaulay.

Since $\text{id}\, T < \infty$, we have an exact sequence $0 \to w_n \to \cdots \to w_0 \to T \to 0$ with all w_i in add w [17, 3]. Since T is Cohen–Macaulay, we have $\text{Ext}^i(T, w) = 0$ for $i > 0$. This shows that T is a summand of w_0, and hence $T \simeq w$ since T is basic.

(b) Let T be a basic orthogonal R-module with $\text{pd}\, T = n < \infty$. Then $\text{Ext}^n(T, R) \neq 0$, and since $\text{Ext}^n(T, \ )$ is right exact, we have $\text{Ext}^n(T, T) \simeq \text{Ext}^n(T, R) \otimes_R T$. Since for a local ring the tensor product of two nonzero modules is not zero, we must have $n = 0$. This shows that $T \simeq R$.

For the dualizing R-module w, we know from [17; 3, Cor. 6.4] that $\widehat{\text{add}\, w} = \mathscr{I}^\infty(R)$. Motivated by this we call a cotilting Λ-module T with $\widehat{\text{add}\, T} = \mathscr{I}^\infty(\Lambda)$ a *strong cotilting module*. Dually we say that a tilting module T is strong if add $T = \mathscr{P}^\infty(\Lambda)$. By our results from Section 5 we have the following.

PROPOSITION 6.3. (a) *An artin algebra Λ has a strong cotilting module if and only if $\mathscr{I}^\infty(\Lambda)$ is contravariantly finite in* mod Λ.

(b) *If Λ has a basic strong cotilting module, it is a direct sum of one copy of each indecomposable Extprojective module in $\mathscr{I}^\infty(\Lambda)$, and is hence unique.*

We note that we get the following description of a strong cotilting module from [3, Prop. 4.6].

PROPOSITION 6.4. *A cotilting module T is strong if and only if T is a minimal cover for* add T.

Note that this means that if for a cotilting module T we consider the preprojective partition of add T, then all summands of T are in the first layer $\mathscr{P}_0(\text{add}\, T)$ if and only if T is strong.

If ${}_\Lambda T$ is a (co)tilting module and $\Gamma = \text{End}_\Lambda(T)^{\text{op}}$. then T_Γ is also a (co)tilting module. But ${}_\Lambda T$ may be strong without T_Γ being strong, as the following example shows.

EXAMPLE. Let Λ be an artin algebra whose finitistic dimension is 0, and where Λ^{op} has nonzero finitistic dimension. Such examples exist by Small (see [14]). Then $\mathscr{P}^\infty(\Lambda) = \mathscr{P}^0(\Lambda)$, so that ${}_\Lambda \Lambda$ is a strong tilting module. But if $\text{pd}_{\Lambda^{\text{op}}} C = n > 0$, then $\text{Ext}^n_{\Lambda^{\text{op}}}(C, \Lambda^{\text{op}}) \neq 0$, so that Λ_Λ is not Extinjective in $\mathscr{P}^\infty(\Lambda^{\text{op}})$.

There is, however, the following connection between ${}_\Lambda T$ and T_Γ [4].

PROPOSITION 6.5. *${}_{\Lambda}T$ is a strong cotilting module if and only if every simple Γ^{op}-module is contained in T_{Γ}.*

Like for the rings R we get the following when ${}_{\Lambda}T$ and T_{Γ} are both strong cotilting modules.

PROPOSITION 6.6. *Assume that ${}_{\Lambda}T$ and T_{Γ} are both strong cotilting modules, where $\Gamma = \operatorname{End}_{\Lambda}(T)^{\text{op}}$. Then the functor $F = (T, \ ): \operatorname{mod} \Lambda \to \operatorname{mod} \Gamma$ induces an equivalence of categories $F: \mathscr{I}^{\infty}(\Lambda) \to \mathscr{P}^{\infty}(\Gamma)$.*

Proof. Assume $\operatorname{pd}_{\Gamma} B < \infty$. Since ${}_{\Gamma}(DT)$ is a strong tilting module, we have an exact sequence $0 \to B \to DT_0 \to \cdots \to DT_n \to 0$ with the DT_i in $\operatorname{add} DT$. Since $\operatorname{Ext}^i_{\Gamma}(T, T) = 0$ for $i > 0$, we have $D(\operatorname{Ext}^i_{\Gamma}(T, T)) = \operatorname{Tor}^{\Gamma}_i(T, DT) = 0$, and hence by dimension shift we have $\operatorname{Tor}^{\Gamma}_i(T, B) = 0$. Let then $0 \to P_n \to \cdots \to P_0 \to B \to 0$ be exact in $\operatorname{mod} \Gamma$, with the P_i projective. Then $0 \to T \otimes_{\Gamma} P_n \to \cdots \to T \otimes_{\Gamma} P_0 \to T \otimes_{\Gamma} B \to 0$ is exact, showing that $T \otimes_{\Gamma} B \in \mathscr{I}^{\infty}(\Lambda)$. Consider then

$$\begin{array}{ccccccccccc}
0 & \longrightarrow & (T, T\otimes_{\Gamma} P_n) & \longrightarrow \cdots \longrightarrow & (T, T\otimes_{\Gamma} P_0) & \longrightarrow & (T, T\otimes_{\Gamma} B) & \longrightarrow & 0 \\
 & & \downarrow{\scriptstyle \alpha_n} & & \downarrow{\scriptstyle \alpha_0} & & \downarrow{\scriptstyle \alpha} & & \\
0 & \longrightarrow & P_n & \longrightarrow \cdots \longrightarrow & P_0 & \longrightarrow & B & \longrightarrow & 0
\end{array}$$

Here we write (A, B) for $\operatorname{Hom}_{\Lambda}(A, B)$.

The first sequence is exact since $\operatorname{Ext}^i_{\Lambda}(T, T \otimes_{\Gamma} B) = 0$ for $i > 0$, using that T is Extprojective in $\mathscr{I}^{\infty}(\Lambda)$. Since $\operatorname{Hom}_{\Gamma}(T, T \otimes \Lambda) \to \Lambda$ is an isomorphism, we get that $\alpha: (T, T \otimes B) \to B$ is an isomorphism.

Assume now that $\operatorname{id}_{\Lambda} C < \infty$. Since ${}_{\Lambda}T$ is a strong cotilting module, we have an exact sequence $0 \to T_n \to \cdots \to T_0 \to C \to 0$ with the T_i in $\operatorname{add} T$. Since $\operatorname{Ext}^i_{\Lambda}(T, C) = 0$ for $i > 0$, we get an exact sequence $0 \to (T, T_n) \to \cdots \to (T, T_0) \to (T, C) \to 0$ in $\operatorname{mod} \Gamma$, so that $\operatorname{pd}_{\Gamma}(T, C) < \infty$. Since then $\operatorname{Tor}^{\Gamma}_i(T, (T, C)) = 0$ for $i > 0$, we get an exact sequence

$$0 \to T \otimes_{\Gamma} (T, T_n) \to \cdots \to T \otimes_{\Gamma} (T, T_0) \to T \otimes_{\Gamma} (T, C) \to 0.$$

Since $T \otimes_{\Gamma} (T, T) \to T$ is an isomorphism, we get that $T \otimes_{\Gamma} (T, C) \to C$ is an isomorphism. This finishes the proof.

Motivated by complete local noetherian Cohen–Macaulay rings we define a Λ-bimodule ${}_{\Lambda}T_{\Lambda}$ to be a *dualizing module* if ${}_{\Lambda}T$ and T_{Λ} are both strong cotilting modules and the natural map $\Lambda \to \operatorname{End}({}_{\Lambda}T)$ is an isomorphism. We say that Λ is a *Cohen–Macaulay algebra* if Λ has a dualizing module. We define the Cohen–Macaulay modules to be the category $CM(\Lambda) = \mathscr{X}_T$. As an immediate consequence of Proposition 6.6 we then have the following, where the second part follows directly from tilting theory [15].

PROPOSITION 6.7. *Let Λ be a Cohen–Macaulay algebra with dualizing module ${}_\Lambda T_\Lambda$.*

(a) $F=(T,\ )\colon \operatorname{mod}\Lambda \to \operatorname{mod}\Lambda$ *induces an equivalence of categories between $\mathscr{I}^\infty(\Lambda)$ and $\mathscr{P}^\infty(\Lambda)$.*

(b) $G=(\ ,T)\colon \operatorname{mod}\Lambda \to \operatorname{mod}\Lambda^{\mathrm{op}}$ *induces a duality between $CM(\Lambda)$ and $CM(\Lambda^{\mathrm{op}})$.*

We point out that there are strong cotilting modules ${}_\Lambda T$ such that T_Γ is also strong, but $\Gamma \not\simeq \Lambda$. It is easy to see that this happens when Λ is the algebra with $\underline{r}^2=0$ given by the following quiver: $\circlearrowright\!\cdot \to \circlearrowright\!\cdot \to \cdot$.

We now give some examples of Cohen–Macaulay algebras.

Let Λ be a local artin algebra. Then it is easy to see that the finitistic dimension of Λ is 0, so that $\mathscr{I}^\infty(\Lambda)$ consists of the injective modules, which are then the Extprojective modules in $\mathscr{I}^\infty(\Lambda)$. Hence ${}_\Lambda(D\Lambda)_\Lambda$ is a dualizing module, so that Λ is Cohen–Macaulay.

The following result gives a procedure for constructing new examples from old ones.

PROPOSITION 6.8. *If ${}_\Lambda T_\Lambda$ is a dualizing module, then $\begin{pmatrix} T & 0 \\ T & T\end{pmatrix}$ is a dualizing module for $T_2(\Lambda)=\begin{pmatrix} \Lambda & 0 \\ \Lambda & \Lambda\end{pmatrix}$.*

Proof. If ${}_\Lambda T$ is a strong cotilting module, it is easy to see that we have a minimal covariant approximation $\Lambda \to T'$ in $\mathscr{I}^\infty(\Lambda)$, with $\operatorname{add} T' = \operatorname{add} T$. Considering the $T_2(\Lambda)$-modules as maps between Λ-modules, we have that

$$\begin{pmatrix} 0 \\ \downarrow \\ \Lambda \end{pmatrix} \to \begin{pmatrix} 0 \\ \downarrow \\ T' \end{pmatrix} \quad \text{and} \quad \begin{pmatrix} \Lambda \\ \| \\ \Lambda \end{pmatrix} \to \begin{pmatrix} T' \\ \| \\ T' \end{pmatrix}$$

are covariant $\mathscr{I}^\infty(T_2(\Lambda))$-approximations. It is easy to see that this means that $\begin{pmatrix} T & 0 \\ T & T\end{pmatrix}$ is a strong cotilting module, as a left $T_2(\Lambda)$-module, and similarly as a right $T_2(\Lambda)$-module. Further the isomorphism $\Lambda \simeq \operatorname{End}(T)^{\mathrm{op}}$ induces an isomorphism

$$\begin{pmatrix} \Lambda & 0 \\ \Lambda & \Lambda \end{pmatrix} \xrightarrow{\sim} \operatorname{End}\begin{pmatrix} T & & 0 \\ \| & \amalg & \downarrow \\ T & & T \end{pmatrix}^{\mathrm{op}}$$

Another interesting class of Cohen–Macaulay algebras are the algebras Λ with $\mathrm{id}_\Lambda \Lambda < \infty$ and $\mathrm{id}\, \Lambda_\Lambda < \infty$, which we call *Gorenstein algebras*. We first make the following well-known easy observation.

LEMMA 6.9. *Let Λ be a Gorenstein ring.*

(a) $\mathrm{id}_\Lambda \Lambda = \mathrm{id}\, \Lambda_\Lambda$.

(b) $\mathscr{I}^\infty(\Lambda) = \mathscr{P}^\infty(\Lambda)$.

Proof. (a) If $\mathrm{id}_\Lambda \Lambda = n$, then $\mathrm{Ext}^i_\Lambda(D(\Lambda^{\mathrm{op}}), \Lambda) = 0$ for $i > n$. Since $\mathrm{pd}_\Lambda D(\Lambda^{\mathrm{op}}) = \mathrm{id}_{\Lambda^{\mathrm{op}}} \Lambda^{\mathrm{op}} < \infty$, this shows that $\mathrm{id}_{\Lambda^{\mathrm{op}}} \Lambda^{\mathrm{op}} \leqslant n = \mathrm{id}_\Lambda \Lambda$. Dually we get $\mathrm{id}_\Lambda \Lambda \leqslant \mathrm{id}_{\Lambda^{\mathrm{op}}} \Lambda^{\mathrm{op}}$.

(b) This follows directly using $\mathrm{id}_\Lambda \Lambda < \infty$ and $\mathrm{pd}_\Lambda D(\Lambda^{\mathrm{op}}) < \infty$.

To see that a Gorenstein algebra Λ is Cohen–Macaulay we observe that Λ is a cotilting module, and hence add $\Lambda = \mathscr{P}^\infty(\Lambda) = \mathscr{I}^\infty(\Lambda)$ is covariantly finite. By duality it follows that $\mathscr{P}^\infty(\Lambda) = \mathscr{I}^\infty(\Lambda)$ is also contravariantly finite.

It would be interesting to know if $\mathrm{id}_\Lambda \Lambda < \infty$ implies that Λ is Gorenstein, and we devote the rest of the section to a discussion of this. This problem is connected with two other unsolved problems:

If T is a selforthogonal Λ-module with $\mathrm{id}\, T < \infty$ and the number of nonisomorphic summands of T is equal to the number of nonisomorphic simple Λ-modules, is then T a cotilting module? Λ is clearly selforthogonal and the number of nonisomorphic indecomposable summands is the same as the number of simple Λ-modules. If $\mathrm{id}_\Lambda \Lambda < \infty$, then Λ is clearly a cotilting module if and only if $\mathrm{id}\, \Lambda_\Lambda < \infty$.

The next result gives a connection with the finitistic dimension conjecture.

PROPOSITION 6.10. *If $\mathrm{id}\, \Lambda_\Lambda < \infty$, then $\mathrm{id}_\Lambda \Lambda < \infty$ if and only if the finitistic dimension of Λ is finite.*

Proof. Assume that $\mathrm{id}\, \Lambda_\Lambda = n < \infty$, and let $0 \to \Lambda \to I_0 \xrightarrow{f_1} I_1 \to I_2 \to \cdots \to I_{i-1} \xrightarrow{f_i} I_i \to \cdots$ be a minimal injective resolution of Λ. Then we have $\mathrm{Ext}^i(\mathrm{Im}\, f_i, \Lambda) \neq 0$ if $\mathrm{Im}\, f_i \neq 0$. Since $\mathrm{pd}(\mathrm{Im}\, f_i) < \infty$, both implications follow from this.

As a consequence of our results in Sections 5 we have the following additional information on the problem.

PROPOSITION 6.11. *If $\mathrm{id}_{\Lambda^{\mathrm{op}}} \Lambda^{\mathrm{op}} < \infty$, the following are equivalent.*

(a) $\mathrm{id}_\Lambda \Lambda < \infty$.

(b) $\hat{\mathscr{X}}_\Lambda = \text{mod}\ \Lambda$.

(c) $\mathscr{X}_\Lambda$ *is contravariantly finite in* mod Λ.

Proof. Since we have $\text{id}_{\Lambda^{\text{op}}}\Lambda^{\text{op}} < \infty$, we know that Λ is a cotilting module if and only if $\text{id}_\Lambda \Lambda < \infty$, and hence (a) is equivalent to (b).

We have by our previous results that (b) implies (c).

Assume now that $\mathscr{X}_\Lambda$ is contravariantly finite. Then $\mathscr{Y} = \text{add}\ \Lambda \subset \mathscr{P}^\infty(\Lambda)$ is covariantly finite. For each simple Λ-module S consider the exact sequence $0 \to S \to Y^S \to X^S \to 0$ with Y^S in $\mathscr{Y}$ and X^S in $\mathscr{X}_\Lambda$. Since $\text{Ext}^i(X^S, \Lambda) = 0$ for $i > 0$, we have $\text{Ext}^i(Y^s, \Lambda) \xrightarrow{\sim} \text{Ext}^i(S, \Lambda) = 0$ for $i > n$ for some n, using pd $Y^S < \infty$. This shows $\text{id}_\Lambda \Lambda < \infty$.

PROPOSITION 6.12. *If* $\text{id}_\Lambda \Lambda = n < \infty$, *the following are equivalent.*

(a) $\text{id}_{\Lambda^{\text{op}}}\Lambda^{\text{op}} < \infty$.

(b) ${}^\perp\Lambda$ *is contravariantly finite.*

Proof. If (a) holds, we have seen that Λ is a cotilting module, so that ${}^\perp\Lambda$ is contravariantly finite.

If ${}^\perp\Lambda$ is contravariantly finite, we have by Theorem 5.2 that ${}^\perp\Lambda = \text{mod}\ \Lambda$, so that Λ is a cotilting module by Theorem 5.4. Hence we have that $\text{id}_{\Lambda^{\text{op}}}\Lambda^{\text{op}}$ is finite.

REFERENCES

1. M. AUSLANDER, Coherent functors, *in* "Proceedings of the Conference on Categorical Algebra, La Jolla, 1965," pp. 189–231, Springer-Verlag, New York.
2. M. AUSLANDER AND M. BRIDGER, Stable module theory, *Mem. Amer. Math. Soc.* **94** (1969).
3. M. AUSLANDER AND R. O. BUCHWEITZ, Maximal Cohen–Macaulay approximations, *Soc. Math. France* **38** (1989), 5–37.
4. M. AUSLANDER AND E. L. GREEN, Modules over endomorphism rings, preprint.
5. M. AUSLANDER AND I. REITEN, Stable equivalence of artin algebras, *in* "Proc. Conf. on Orders Group Rings and Related Topics," pp. 8–71, Lecture Notes in Mathematics, Vol. 353, Springer-Verlag, New York/Berlin, 1972.
6. M. AUSLANDER AND I. REITEN, On a generalized version of the Nakayama conjecture, *Proc. Amer. Math. Soc.* **52** (1972), 69–74.
7. M. AUSLANDER AND I. REITEN, Representation theory of artin algebras III, Almost split sequences, *Comm. Algebra* **3** (1975), 239–294.
8. M. AUSLANDER AND S. O. SMALØ, Preprojective modules over artin algebras, *J. Algebra* **66** (1980), 61–122.
9. M. AUSLANDER AND S. O. SMALØ, Almost split sequences in subcategories, *J. Algebra* **69** (1981), 426–454; Addendum *J. Algebra* **71** (1981), 592–594.
10. K. BONGARTZ, Tilted algebras, *in* "Proc. ICRAIII, Puebla, 1980," pp. 26–38, Lecture Notes in Mathematics, Vol. 903, Springer-Verlag, New York/Berlin, 1981.
11. D. HAPPEL, On the derived category of a finite dimensional algebra, *Comm. Math. Helv.* **62** No. 3 (1987), 339–389.

12. D. Happel and C. M. Ringel, Tilted algebras, *Trans. Amer. Math. Soc.* **274** (1982), 399–443.
13. K. Igusa, S. O. Smalø, and G. Todorov, Finite projectivity and contravariant finiteness, *Proc. Amer. Math. Soc.*
14. C. U. Jensen and H. Lenzing, Homological dimension and representation type of algebras under base field extension, *Manuscripta Math.* **39** (1982), 1–13.
15. T. Miayshita, Tilting modules of finite projective dimension, *Math. Z.* **193** (1986), 113–146.
16. J. Rickart and A. Scofield, Cocovers and tilting modules, preprint.
17. R. Y. Sharp, Finitely generated modules of finite injective dimension over certain Cohen–Macaulay rings, *Proc. London Math. Soc. (3)* **25** (1972), 303–328.
18. S. O. Smalø, Functorial finite subcategories over triangular matrix rings, *Proc. Amer. Math. Soc.*
19. T. Wakamatsu, On modules with trivial selfextensions, *J. Algebra* **114** (1988), 106–114.
20. T. Wakamatsu, Stable equivalence of selfinjective algebras and a generalization of tilting modules, preprint.

Printed by Catherine Press, Ltd., Tempelhof 41, B-8000 Brugge, Belgium

TRANSACTIONS OF THE
AMERICAN MATHEMATICAL SOCIETY
Volume 332, Number 2, August 1992

HOMOLOGICAL THEORY OF IDEMPOTENT IDEALS

M. AUSLANDER, M. I. PLATZECK AND G. TODOROV

ABSTRACT. Let Λ be an artin algebra $\mathfrak{A}$ and a two-sided ideal of Λ. Then $\mathfrak{A}$ is the trace of a projective Λ-module P in Λ. We study how the homological properties of the categories of finitely generated modules over the three rings $\Lambda/\mathfrak{A}$, Λ and the endomorphism ring of P are related. We give some applications of the ideas developed in the paper to the study of quasi-hereditary algebras.

INTRODUCTION

Throughout this paper we assume that Λ is an artin algebra and all Λ-modules are assumed to lie in $\operatorname{mod}\Lambda$, the category of finitely generated left Λ-modules.

In connection with their definition of quasi-hereditary algebras, Cline, Parshall and Scott [CPS] introduced the notion of an heredity ideal which is defined as follows. An heredity ideal of Λ is a two-sided ideal $\mathfrak{A}$ satisfying the following conditions: (i) $\mathfrak{A}$ is idempotent (i.e. $\mathfrak{A}^2 = \mathfrak{A}$), (ii) $\mathfrak{A}$ is projective as a left Λ-module and (iii) $\operatorname{End}_\Lambda(\mathfrak{A})$ is semisimple. The results concerning heredity ideals proven in [CPS, DR1,2, BF] suggested studying the homological properties of idempotent ideals in a broader context. This paper is a preliminary step in this direction.

Our starting point is the following easily checked well-known observation. Let P be a projective Λ-module. Then $\mathfrak{A} = \tau_P(\Lambda)$, the trace of P in Λ which is the ideal generated by the homomorphic images of P in Λ, is an idempotent ideal in Λ and one obtains all the idempotent ideals of Λ this way. Moreover if P and Q are projective Λ-modules, then $\tau_P(\Lambda) = \tau_Q(\Lambda)$ if and only if $\operatorname{add} P = \operatorname{add} Q$, where $\operatorname{add} M$, for an arbitrary Λ-module M, is the full subcategory of $\operatorname{mod}\Lambda$ consisting of all modules isomorphic to summands of finite sums (direct) of M. Therefore associated with a projective Λ-module P is the idempotent ideal $\mathfrak{A}_P = \tau_P(\Lambda)$ and the artin algebras $\Lambda/\mathfrak{A}_P$ and $\Gamma_P = \operatorname{End}(P)^{\mathrm{op}}$. Our basic aim is to study how the homological properties of the three categories $\operatorname{mod}\Lambda/\mathfrak{A}_P$, $\operatorname{mod}\Lambda$ and $\operatorname{mod}\Gamma_P$ are related. It is worth noting, that

Received by the editors August 1, 1989 and, in revised form, May 11, 1990.
1980 *Mathematics Subject Classification* (1985 *Revision*). Primary 16A64; Secondary 16A35, 16A62.
Key words and phrases. idempotent ideal, projective resolution, quasihereditary algebra.
The first author's work supported by NSF Grant No. DMS-8603487.
The second author's work supported by a fellowship of the Consejo Nacional de Investigaciones Científicas y Técnicas de la República Argentina.

since $\Lambda/\mathfrak{A}_P$ and Γ_P have fewer simple modules than Λ, a good understanding of the relationship between the categories $\operatorname{mod}\Lambda/\mathfrak{A}_P$, $\operatorname{mod}\Lambda$ and $\operatorname{mod}\Gamma_P$ should lead to a way of studying $\operatorname{mod}\Lambda$ in terms of algebras with fewer simple modules.

One relationship of interest is the inclusion functor $\operatorname{mod}\Lambda/\mathfrak{A}_P \to \operatorname{mod}\Lambda$. Saying that $\mathfrak{A}_P$ is an idempotent ideal is equivalent to saying that $\operatorname{mod}\Lambda/\mathfrak{A}_P$ is a Serre subcategory of $\operatorname{mod}\Lambda$. Moreover the exact functor $(P,\ ): \operatorname{mod}\Lambda \to \operatorname{mod}\Gamma_P$ given by $M \mapsto \operatorname{Hom}_\Lambda(P, M)$ gives an equivalence of categories between $\operatorname{mod}\Gamma_P$ and the quotient of $\operatorname{mod}\Lambda$ by $\operatorname{mod}\Lambda/\mathfrak{A}_P$. In other words, we have an exact sequence of categories

$$\operatorname{mod}\Lambda/\mathfrak{A}_P \to \operatorname{mod}\Lambda \to \operatorname{mod}\Gamma_P.$$

It is this exact sequence which is our main object of study.

We first concentrate on the inclusion $\operatorname{mod}\Lambda/\mathfrak{A}_P \to \operatorname{mod}\Lambda$. For any two-sided ideal $\mathfrak{A}$ in Λ, we have a morphism of connected sequences of functors $\operatorname{Ext}^i_{\Lambda/\mathfrak{A}}(X, Y) \to \operatorname{Ext}^i_\Lambda(X, Y)$ for all X and Y in $\operatorname{mod}\Lambda/\mathfrak{A}$ induced by the canonical isomorphism of functors $\operatorname{Hom}_{\Lambda/\mathfrak{A}}(X, Y) \xrightarrow{\sim} \operatorname{Hom}_\Lambda(X, Y)$ for all X and Y in $\operatorname{mod}\Lambda/\mathfrak{A}$. It is well known that $\operatorname{Ext}^1_{\Lambda/\mathfrak{A}}(X, Y) \to \operatorname{Ext}^1_\Lambda(X, Y)$ is an isomorphism if and only if $\mathfrak{A}$ is an idempotent ideal. This suggests the following definitions. A two-sided ideal $\mathfrak{A}$ is a k-idempotent for $k \geq 1$ if $\operatorname{Ext}^i_{\Lambda/\mathfrak{A}}(X, Y) \to \operatorname{Ext}^i_\Lambda(X, Y)$ is an isomorphism for all $0 \leq i \leq k$ and all X and Y in $\operatorname{mod}\Lambda/\mathfrak{A}$. We say that $\mathfrak{A}$ is a strong idempotent if $\operatorname{Ext}^i_\Lambda(X, Y) \to \operatorname{Ext}^i_{\Lambda/\mathfrak{A}}(X, Y)$ is an isomorphism for all $i \geq 0$ and all X and Y in $\operatorname{mod}\Lambda/\mathfrak{A}$. It is worth noting that if $\mathfrak{A}$ is an heredity ideal, then $\mathfrak{A}$ is a strong idempotent [DR1, Part 2]. We give characterizations of k-idempotents in terms of projective and injective Λ-resolutions of $\Lambda/\mathfrak{A}$-modules.

Next we turn our attention to the functor $(P,\ ): \operatorname{mod}\Lambda \to \operatorname{mod}\Gamma_p$. Since $(P,\ )$ is exact we obtain a morphism of connected sequences of functors $\operatorname{Ext}^i_\Lambda(X, Y) \to \operatorname{Ext}^i_{\Gamma_p}((P, X), (P, Y))$ for all X and Y in $\operatorname{mod}\Lambda$. In connection with studying when this is an isomorphism we introduce the following subcategories of $\operatorname{mod}\Lambda$. For $k \geq 0$, we define $\mathbf{I}_{k+1}$ to be the full subcategory of $\operatorname{mod}\Lambda$ consisting of all Y in $\operatorname{mod}\Lambda$ such that $\operatorname{Ext}^i_\Lambda(X, Y) \to \operatorname{Ext}^i_{\Gamma_P}((P, X), (P, Y))$ is an isomorphism for all $0 \leq i \leq k$ and all X in $\operatorname{mod}\Lambda$. And we define $\mathbf{I}_\infty$ to be the full subcategory of $\operatorname{mod}\Lambda$ consisting of all Y in $\operatorname{mod}\Lambda$ such that $\operatorname{Ext}^i_\Lambda(X, Y) \to \operatorname{Ext}^i_{\Gamma_P}((P, X), (P, Y))$ is an isomorphism for all $i \geq 0$ and all X in $\operatorname{mod}\Lambda$. We characterize the subcategories $\mathbf{I}_k$ and $\mathbf{I}_\infty$ in terms of injective coresolutions of Γ-modules. In particular we consider the problem of when $\mathbf{I}_k = \mathbf{I}_\infty$. In this connection we show that an idempotent ideal $\mathfrak{A}$ is a projective Λ-module if and only if $\mathfrak{A}$ is a strong idempotent and $\mathbf{I}_1 = \mathbf{I}_\infty$.

As an illustration of how these ideas and results can be used to deduce information about Λ from the algebras $\Lambda/\mathfrak{A}$ and Γ_P we prove the following generalization of a similar theorem about quasi-hereditary algebras proved in [CPS and DR1].

Let $\mathfrak{A}$ be an idempotent ideal. Suppose that $\operatorname{pd}\mathfrak{A} = r$, $\operatorname{gl.dim}\Lambda/\mathfrak{A} = s$ and $\operatorname{gl.dim}\Gamma = t$. Then $\operatorname{gl.dim}\Lambda \leq r + s + t + 2$.

The paper ends with applications of these ideas to the study of quasihereditary algebras. In addition to giving proofs of some known results, we give a new description of quasihereditary algebras in terms of l-hereditary algebras.

1. STRONG AND k-IDEMPOTENT IDEALS

In this section we study the inclusion functor $\operatorname{mod}\Lambda/\mathfrak{A} \to \operatorname{mod}\Lambda$, where $\mathfrak{A}$ is an idempotent ideal in Λ. By an idempotent ideal we mean a two-sided ideal which is idempotent.

Let $\mathfrak{A}$ be a two-sided ideal of Λ. There is a map between connected sequences of functors $\varphi = (\varphi^i)_{i\geq 0}$ with $\varphi^i_{X,Y}\colon \operatorname{Ext}^i_{\Lambda/\mathfrak{A}}(X, Y) \to \operatorname{Ext}^i_\Lambda(X, Y)$ for $i \geq 0$ and X, Y in $\operatorname{mod}\Lambda/\mathfrak{A}$, induced by the canonical isomorphism $\varphi^0_{X,Y}\colon \operatorname{Hom}_{\Lambda/\mathfrak{A}}(X, Y) \to \operatorname{Hom}_\Lambda(X, Y)$. Notice that the ideal $\mathfrak{A}$ is idempotent precisely when the map $\varphi^1_{X,Y}\colon \operatorname{Ext}^1_{\Lambda/\mathfrak{A}}(X, Y) \to \operatorname{Ext}^1_\Lambda(X, Y)$ is an isomorphism for all X, Y in $\operatorname{mod}\Lambda/\mathfrak{A}$. We say that a two-sided ideal $\mathfrak{A}$ is k-idempotent if the maps $\varphi^i_{X,Y}$ are isomorphisms for all $i = 0, 1, \dots, k$, and all X, Y in $\operatorname{mod}\Lambda/\mathfrak{A}$. We say that $\mathfrak{A}$ is a strong idempotent ideal if the maps $\varphi^i_{X,Y}$ are isomorphisms for all $i \geq 0$ and all X, Y in $\operatorname{mod}\Lambda/\mathfrak{A}$. In this section we give several characterizations of strong and k-idempotent ideals which are central to the rest of the paper.

We observe that being a k-idempotent is a symmetric condition. That is ${}_\Lambda\mathfrak{A}$ is a k-idempotent ideal if and only if $\mathfrak{A}_\Lambda$ is a k-idempotent ideal. This follows by duality, since $\operatorname{Ext}^i_{\Lambda^{\mathrm{op}}}(DX, DY) \simeq \operatorname{Ext}^i_\Lambda(Y, X)$ for X, Y in $\operatorname{mod}\Lambda$, where $D\colon \operatorname{mod}\Lambda \to \operatorname{mod}\Lambda^{\mathrm{op}}$ denotes the usual duality for artin algebras.

As another consequence of the definition we have that $\operatorname{gl.dim}\Lambda/\mathfrak{A} \leq \operatorname{gl.dim}\Lambda$ when $\mathfrak{A}$ is a strong idempotent.

Now we study the behavior of injective coresolutions in $\operatorname{mod}\Lambda$, when we apply the right adjoint $\operatorname{Hom}_\Lambda(\Lambda/\mathfrak{A}, \ )\colon \operatorname{mod}\Lambda \to \operatorname{mod}\Lambda/\mathfrak{A}$ of the inclusion $\operatorname{mod}\Lambda/\mathfrak{A} \to \operatorname{mod}\Lambda$.

We start by recalling some facts about traces of modules. Let M, N be Λ-modules. We denote by $\tau_M(N)$ the trace of M in N. That is, $\tau_M(N)$ is the submodule of N generated by the images of morphisms from M to N. Let $\mathfrak{A}$ be a two-sided ideal of Λ. Then there is a natural isomorphism of $\Lambda/\mathfrak{A}$-modules $\Theta\colon \operatorname{Hom}_\Lambda(\Lambda/\mathfrak{A}, X) \xrightarrow{\sim} \tau_{\Lambda/\mathfrak{A}}(X)$ given by $\Theta(f) = f(\bar{1})$. If I is an injective Λ-modfule, then $\tau_{\Lambda/\mathfrak{A}}(I) \simeq \operatorname{Hom}(\Lambda/\mathfrak{A}, I)$ is an injective $\Lambda/\mathfrak{A}$-module. Moreover, if I is an injective envelope of a Λ-module X, then $\tau_{\Lambda/\mathfrak{A}}(I)$ is an injective envelope of the $\Lambda/\mathfrak{A}$-module $\tau_{\Lambda/\mathfrak{A}}(X)$.

We describe when the above property about injective envelopes can be extended to a statement about the first k terms, or about all of the terms, of a minimal injective coresolution of $\tau_{\Lambda/\mathfrak{A}}(X)$ in $\operatorname{mod}\Lambda/\mathfrak{A}$.

In order to avoid writing separate statements we will sometimes write $j \leq i \leq k$ where k is either a positive integer or ∞. By $j \leq i \leq \infty$ we mean all $i \geq j$.

Proposition 1.1. *Let X be a Λ-module, let $0 \to X \to I_0 \to I_1 \to \cdots$ be a minimal injective coresolution of X and $1 \leq k \leq \infty$. Then the following are equivalent*

(i) $0 \to \tau_{\Lambda/\mathfrak{A}}X \to \tau_{\Lambda/\mathfrak{A}}I_0 \to \cdots \to \tau_{\Lambda/\mathfrak{A}}I_k$ *is the beginning of a minimal injective coresolution of $\tau_{\Lambda/\mathfrak{A}}(X)$ in* $\operatorname{mod}\Lambda/\mathfrak{A}$.

(ii) $\operatorname{Ext}^i_\Lambda(\Lambda/\mathfrak{A}, X) = 0$ *for all i such that* $1 \leq i \leq k$.

(iii) *Let Y be in $\operatorname{mod}\Lambda/\mathfrak{A}$ and let* $\varphi^i_{X,Y}\colon \operatorname{Ext}^i_{\Lambda/\mathfrak{A}}(Y, \tau_{\Lambda/\mathfrak{A}}X) \to \operatorname{Ext}^i_\Lambda(Y, X)$ *be the map of connected sequences of functors induced by the canonical isomor-*

phism $\operatorname{Hom}_{\Lambda/\mathfrak{A}}(Y, \tau_{\Lambda/\mathfrak{A}}X) \simeq \operatorname{Hom}_\Lambda(Y, X)$. *Then* $\varphi^i_{\Lambda/\mathfrak{A}}$ *is an isomorphism for* $1 \le i \le k$.

Proof. We have that $\operatorname{Ext}^i_\Lambda(\Lambda/\mathfrak{A}, X) = 0$ for all $1 \le i \le k$ if and only if the complex obtained after applying $\operatorname{Hom}_\Lambda(\Lambda/\mathfrak{A},\ )$ to $0 \to X \to I_0 \to I_1 \to \cdots \to I_k$ is exact. The equivalence of (i) and (ii) follows, since $\operatorname{Hom}_\Lambda(\Lambda/\mathfrak{A},\ )$ and $\tau_{\Lambda/\mathfrak{A}}(\ )$ are isomorphic functors.

For a $\Lambda/\mathfrak{A}$-module Y the complex obtained by applying the functor $\operatorname{Hom}_\Lambda(Y,\ )$ to the complex $0 \to X \to I_0 \to I_1 \to \cdots$ is isomorphic to the complex obtained by applying $\operatorname{Hom}_{\Lambda/\mathfrak{A}}(Y,\ )$ to $0 \to \tau_{\Lambda/\mathfrak{A}}X \to \tau_{\Lambda/\mathfrak{A}}I_0 \to \tau_{\Lambda/\mathfrak{A}}I_1 \cdots$. So the homologies of the resulting complexes are isomorphic, proving that (i) $\Rightarrow$ (iii). Clearly (iii) $\Rightarrow$ (ii), so the proof is complete. □

Dually, we can consider projective resolutions of Λ-modules and apply the left adjoint functor $\Lambda/\mathfrak{A}\otimes_\Lambda\colon \operatorname{mod}\Lambda \to \operatorname{mod}\Lambda/\mathfrak{A}$ to the inclusion $\operatorname{mod}\Lambda/\mathfrak{A} \to \operatorname{mod}\Lambda$. We observe that $\Lambda/\mathfrak{A} \otimes_\Lambda X \simeq X/\mathfrak{A}X$. So we have the following proposition.

Proposition 1.2. *Let* X *be a* Λ*-module, let* $\cdots \to P_1 \to P_0 \to X \to 0$ *be a minimal projective resolution of* X *and* $1 \le k \le \infty$. *Then the following are equivalent:*

(i) $P_k/\mathfrak{A}P_k \to \cdots \to P_0/\mathfrak{A}P_0 \to X/\mathfrak{A}X \to 0$ *is the beginning of a minimal projective resolution of* $X/\mathfrak{A}X$ *in* $\operatorname{mod}(\Lambda/\mathfrak{A})$.

(ii) $\operatorname{Tor}^\Lambda_i(\Lambda/\mathfrak{A}, X) = 0$ *for all* i *such that* $1 \le i \le k$.

(iii) *Let* Y *in* $\operatorname{mod}(\Lambda/\mathfrak{A})^{\mathrm{op}}$ *and let* $\psi^{X,Y}_i\colon \operatorname{Tor}^\Lambda_i(Y, X) \to \operatorname{Tor}^{\Lambda/\mathfrak{A}}_i(Y, X/\mathfrak{A}X)$ *be the map of connected sequences of functors induced by the isomorphism* $Y\otimes_\Lambda X \simeq Y \otimes_{\Lambda/\mathfrak{A}} X/\mathfrak{A}X$. *Then* $\psi^{X,Y}_i$ *is an isomorphism for* $i = 1, \ldots, k$.

From the above results we have the following characterization of k-idempotent ideals in terms of Ext and Tor functors.

Proposition 1.3. *Let* $\mathfrak{A}$ *be a two-sided ideal. Let* $1 \le k \le \infty$. *Then the following conditions are equivalent:*

(i) *The ideal* $\mathfrak{A}$ *is* k*-idempotent.*

(ii) $\varphi^i_{X/Y}\colon \operatorname{Ext}^i_{\Lambda/\mathfrak{A}}(X, Y) \to \operatorname{Ext}^i_\Lambda(X, Y)$ *is an isomorphism for all* X, Y *in* $\operatorname{mod}\Lambda/\mathfrak{A}$ *and all* $0 \le i \le k$.

(iii) $\operatorname{Ext}^i_\Lambda(\Lambda/\mathfrak{A}, Y) = 0$ *for all* $\Lambda/\mathfrak{A}$*-modules* Y *and all* $1 \le i \le k$.

(iv) $\operatorname{Ext}^i_\Lambda(\Lambda/\mathfrak{A}, I) = 0$ *for all injective* I *in* $\operatorname{mod}\Lambda/\mathfrak{A}$ *and* $1 \le i \le k$.

(ii′) $\varphi^i_{X,Y}\colon \operatorname{Tor}^{\Lambda/\mathfrak{A}}_i(X, Y) \to \operatorname{Tor}^\Lambda_i(X, Y)$ *is an isomorphism for all* X *in* $\operatorname{mod}(\Lambda/\mathfrak{A})^{\mathrm{op}}$ *and* Y *in* $\operatorname{mod}\Lambda/\mathfrak{A}$ *and all* $0 \le i \le k$.

(iii′) $\operatorname{Tor}^\Lambda_i(\Lambda/\mathfrak{A}, Y) = 0$ *for all* Y *in* $\operatorname{mod}\Lambda/\mathfrak{A}$ *and all* $1 \le i \le k$.

(iv′) $\operatorname{Tor}^\Lambda_i(\Lambda/\mathfrak{A}, \Lambda/\mathfrak{A}) = 0$ *for all* $1 \le i \le k$.

Proof. (i) $\Leftrightarrow$ (ii) is just the definition of k-idempotent ideals. Obviously (ii) $\Rightarrow$ (iii) $\Rightarrow$ (iv). By Proposition 1.1 it follows that (iii) $\Rightarrow$ (ii). (iv) $\Rightarrow$ (iii) Let Y be a $\Lambda/\mathfrak{A}$-module, I its injective envelope in $\operatorname{mod}\Lambda/\mathfrak{A}$ and $0 \to Y \to I \to I/Y \to 0$ exact. Applying $\operatorname{Hom}_\Lambda(\Lambda/\mathfrak{A},\ )$ to this sequence we get a commutative diagram:

$$\begin{array}{ccccccccc}
0 & \to & \operatorname{Hom}_\Lambda(\Lambda/\mathfrak{A}, Y) & \to & \operatorname{Hom}_\Lambda(\Lambda/\mathfrak{A}, I) & \to & \operatorname{Hom}_\Lambda(\Lambda/\mathfrak{A}, I/Y) & \to & \operatorname{Ext}^1_\Lambda(\Lambda/\mathfrak{A}, Y) \\
 & & \uparrow & & \uparrow & & \uparrow & & \\
0 & \to & Y & \to & I & \to & I/Y & \to & 0
\end{array}$$

Since the vertical maps are isomorphisms we have a long exact sequence:

$$0 \longrightarrow \operatorname{Ext}^1_\Lambda(\Lambda/\mathfrak{A}, Y) \longrightarrow \operatorname{Ext}^1_\Lambda(\Lambda/\mathfrak{A}, I) \longrightarrow \operatorname{Ext}^1_\Lambda(\Lambda/\mathfrak{A}, I/Y) \longrightarrow \cdots$$

$$\cdots \longrightarrow \operatorname{Ext}^{k-1}_\Lambda(\Lambda/\mathfrak{A}, I/Y) \longrightarrow \operatorname{Ext}^k_\Lambda(\Lambda/\mathfrak{A}, Y) \longrightarrow \operatorname{Ext}^k_\Lambda(\Lambda/\mathfrak{A}, I) \longrightarrow \cdots$$

Then (iv) $\Rightarrow$ (iii) follows by induction on k.

For the other equivalences observe that $D(\operatorname{Ext}^i_\Lambda(X, Y)) \simeq \operatorname{Tor}^\Lambda_i(DY, X)$, where $D\colon \operatorname{mod}\Lambda \to \operatorname{mod}\Lambda^{\mathrm{op}}$ is the usual duality for artin algebras [CE, Chapter VI, §5]. □

We give a description of k-idempotent ideals where $k = 1$ and 2 in terms of the multiplication map $\mathfrak{A} \otimes_\Lambda \mathfrak{A} \to \mathfrak{A}$. We do not have a similar description for higher k.

Lemma 1.4. *Let $\mathfrak{A}$ be a two-sided ideal and let $m\colon \mathfrak{A} \otimes_\Lambda \mathfrak{A} \to \mathfrak{A}$ be the multiplication map. Then*

(a) *$\mathfrak{A}$ is idempotent if and only if m is an epimorphism,*

(b) *$\mathfrak{A}$ is 2-idempotent if and only if m is an isomorphism.*

Proof. (a) is clear, and (b) follows from the following result from [CE, Exercise 19, p. 126], which we state for ease of reference.

Lemma 1.5. *Let $\mathfrak{A}$ be an idempotent ideal. Then there is an exact sequence $0 \to \operatorname{Tor}^\Lambda_2(\Lambda/\mathfrak{A}, \Lambda/\mathfrak{A}) \to \mathfrak{A} \otimes_\Lambda \mathfrak{A} \xrightarrow{m} A \to 0$, where m is the multiplication map.*

We get the following characterization of k-idempotent ideals in terms of the projective and injective resolutions of $\Lambda/\mathfrak{A}$-modules.

Theorem 1.6. *The following conditions are equivalent for a two-sided ideal $\mathfrak{A}$:*

(i) *The ideal $\mathfrak{A}$ is k-idempotent.*

(ii) *Let X be a $\Lambda/\mathfrak{A}$-module and $0 \to X \to I_0 \to \cdots$ a minimal injective coresolution of X in $\operatorname{mod}\Lambda$. Then $0 \to X \to \tau_{\Lambda/\mathfrak{A}} I_0 \to \cdots \to \tau_{\Lambda/\mathfrak{A}} I_k$ is the beginning of a minimal injective coresolution of X in $\operatorname{mod}\Lambda/\mathfrak{A}$.*

(iii) *Let Y be a $\Lambda/\mathfrak{A}$-module and $\cdots \to P_0 \to Y \to 0$ a minimal projective resolution of Y in $\operatorname{mod}\Lambda$. Then $P_k/\mathfrak{A}P_k \to \cdots \to P_0/\mathfrak{A}P_0 \to Y \to 0$ is the beginning of a minimal projective resolution of Y in $\operatorname{mod}\Lambda/\mathfrak{A}$.*

Proof. It was proven in Proposition 1.3 that $\mathfrak{A}$ is k-idempotent is equivalent to $\operatorname{Ext}^i_\Lambda(\Lambda/\mathfrak{A}, Y) = 0$ for all $\Lambda/\mathfrak{A}$-modules Y and $1 \le i \le k$. The condition (ii) in the theorem is part of the statement of Proposition 1.1 for Y in $\operatorname{mod}\Lambda/\mathfrak{A}$. Similarly $\mathfrak{A}$ being k-idempotent is equivalent to $\operatorname{Tor}^\Lambda_i(\Lambda/\mathfrak{A}, Y) = 0$ for all Y in $\operatorname{mod}\Lambda/\mathfrak{A}$ and $1 \le i \le k$ which is equivalent to (iii) by Proposition 1.2. □

Next we give several examples of strong and k-idempotent ideals.

Example 1. Let $\mathfrak{A}$ be an idempotent ideal which is a projective left Λ-module. Then $\mathfrak{A}$ is a strong idempotent ideal. To see this, let Y be a $\Lambda/\mathfrak{A}$-module. Then $\operatorname{Ext}^i_\Lambda(\Lambda/\mathfrak{A}, Y) = 0$ for all $i \ge 1$. This is true for $i = 1$ because $\mathfrak{A}$ is an idempotent ideal, and for $i > 1$ because the projective dimension of $\Lambda/\mathfrak{A}$ is at most 1.

Therefore, if Λ is a hereditary algebra any idempotent ideal is strong idempotent. It would be nice to know what other algebras Λ contain nontrivial idempotent ideals which are projective left Λ-modules. Nonsimple quasi-hereditary algebras do, since heredity ideals are projective left modules, by definition. (See [CPS, DR1] or §6 for definitions.)

Example 2. The ideals which occur in a heredity chain in a quasihereditary algebra are all strong idempotent. More generally, let Λ be an artin algebra having a chain of idempotent ideals $0 = \mathfrak{A}_0 \subset \mathfrak{A}_1 \subset \cdots \subset \mathfrak{A}_n \subset \Lambda$ such that $\mathfrak{A}_i/\mathfrak{A}_{i-1}$ is projective $\Lambda/\mathfrak{A}_{i-1}$-module for all $i = 1, \ldots, n$. Then $\mathfrak{A}_i$ is strong idempotent for $i = 1, \ldots, n$.

This can be proven in a straightforward way by induction on n. Since it will also follow from results in §6 we omit the proof here.

In the above examples the projective dimension of the strong idempotent ideals was always finite. We give now an example of a strong idempotent ideal of infinite projective dimension.

Example 3. Let Λ be the path algebra (over a field) of the quiver

$$1 \bullet \underset{\beta}{\overset{\alpha}{\rightleftarrows}} \bullet\ 2,$$

modulo the ideal generated by the relation $\beta\alpha\beta = 0$. Let P be the projective corresponding to the vertex 1. Then $\mathfrak{A} = \tau_P(\Lambda)$ is a strong idempotent ideal and has infinite projective dimension.

The following example shows that for every natural number $k \geq 1$, there are k-idempotent ideals which are not $(k+1)$-idempotent.

Example 4. Let Λ be the path algebra (over a field) of the quiver

$$\underset{0}{\bullet} \longleftarrow \underset{1}{\bullet} \longleftarrow \cdots \longleftarrow \underset{k}{\bullet} \longleftarrow \underset{k+1}{\bullet}$$

modulo the ideal generated by the paths of length 2. Let P_i be the projective corresponding to the vertex i. Let $\mathfrak{A} = \tau_{P_2 \amalg \cdots \amalg P_k}(\Lambda)$. Then $\mathfrak{A}$ is k-idempotent and is not $(k+1)$-idempotent.

To see this we denote the simple Λ-modules by $S_j = P_j/\underline{r}P_j$, where $\underline{r}$ is the radical of Λ. Then $\Lambda/\mathfrak{A} \simeq S_0 \amalg S_{k+1}$ is semisimple. So the only indecomposable $\Lambda/\mathfrak{A}$-module which is not an injective Λ-module is S_0. We have that $\operatorname{Ext}^i_\Lambda(\Lambda/\mathfrak{A}, S_0) = \operatorname{Ext}^i_\Lambda(S_{k+1}, S_0)$ for $i \geq 1$. But $\operatorname{Ext}^i_\Lambda(S_{k+1}, S_0) = 0$ for $i \leq k$ and $\operatorname{Ext}^{k+1}_\Lambda(S_{k+1}, S_0) \neq 0$. It follows that $\mathfrak{A}$ is k-idempotent but is not $(k+1)$-idempotent.

Notice that the above algebra Λ can be given as the $(k+2) \times (k+2)$ lower triangular matrix ring modulo the square of the radical.

2. Projective resolutions of k-idempotent ideals

In the previous section we characterized k-idempotent ideals in terms of the projective resolutions of all $\Lambda/\mathfrak{A}$-modules. We show here that knowing the projective resolution of only one module, namely ${}_\Lambda\mathfrak{A}$, is enough to determine for which k the ideal $\mathfrak{A}$ is k-idempotent. In this section we prove the following theorem.

Theorem 2.1. *Let $\mathfrak{A} = \tau_P(\Lambda)$ for some projective Λ-module P and $1 \leq k \leq \infty$. Let $\cdots \to P_i \to \cdots \to P_0 \to \mathfrak{A} \to 0$ be a minimal projective resolution* *of* $\mathfrak{A}$. *Then* $\mathfrak{A}$ *is* $(k+1)$*-idempotent if and only if the* P_i *are in* $\operatorname{add} P$ *for* $i = 0, \ldots, k$.

In the proof of the theorem and throughout the rest of the paper the following facts and notation will be used. A two-sided ideal $\mathfrak{A}$ is idempotent if and only if $\mathfrak{A} = \tau_P(\Lambda)$ for some projective Λ-module P. Furthermore, $\tau_P(\Lambda) = \tau_{P'}(\Lambda)$ if and only if $\operatorname{add} P = \operatorname{add} P'$. From now on, $\mathfrak{A}$ will always denote an idempotent ideal, and P a fixed projective Λ-module such that $\mathfrak{A} = \tau_P(\Lambda)$. We now recall some properties of traces of projective modules.

Remark 2.2. Let $\mathfrak{A} = \tau_P(\Lambda)$. Then $\tau_P(X) = \tau_P(\Lambda)X = \mathfrak{A}X$ for any Λ-module X. The projective cover $P_0(X)$ of X is in $\operatorname{add} P$ if and only if $X = \tau_P(X) = \mathfrak{A}X$. We also have $\operatorname{add} P = \operatorname{add}(P_0(\mathfrak{A}))$.

In order to prove the theorem it will be convenient to introduce the following notation:

Definition 2.3. Let P be a projective Λ-module. For each $0 \le k \le \infty$ we define $\mathbf{P}_k$ to be the full subcategory of $\operatorname{mod}\Lambda$ consisting of the Λ-modules X having a projective resolution $\cdots \to P_1 \to P_0 \to X \to 0$ with P_i in $\operatorname{add} P$ for $0 \le i \le k$. Since $\mathbf{P}_k$ depends only on $\operatorname{add} P$, it is well defined for the idempotent ideal $\mathfrak{A} = \tau_P(\Lambda)$. It would be more precise to write $\mathbf{P}_{k,P}$ or $\mathbf{P}_{k,\mathfrak{A}}$, but whenever it is clear which projective or which idempotent ideal we are using, we will use the above notation $\mathbf{P}_k$.

Theorem 2.1 can now be restated.

Theorem 2.1′. *An idempotent ideal* $\mathfrak{A} = \tau_P(\Lambda)$ *is* $(k+1)$*-idempotent if and only if* $\mathfrak{A}$ *is in* $\mathbf{P}_k$, *for* $1 \le k \le \infty$.

In the following proposition we give a characterization of the modules in $\mathbf{P}_k$ needed in the proof of the above theorem as well as in the rest of the paper.

Proposition 2.4. *Let* $\mathfrak{A} = \tau_P(\Lambda)$ *and* $1 \le k \le \infty$. *Then the following conditions are equivalent for a* Λ*-module* X.

(i) X *is in* $\mathbf{P}_k$.

(ii) $\operatorname{Ext}^i_\Lambda(X, Y) = 0$ *for all* $\Lambda/\mathfrak{A}$*-modules* Y *and* $i = 0, \ldots, k$.

(iii) $\operatorname{Ext}^i_\Lambda(X, E) = 0$ *for all injective* $\Lambda/\mathfrak{A}$*-modules* E *and* $i = 0, \ldots, k$.

Proof. By induction on k. □

Proof of Theorem 2.1′ (and Theorem 2.1). Let $X = \mathfrak{A}$ in the above proposition. We only need to observe that $\operatorname{Ext}^{i+1}_\Lambda(\Lambda/\mathfrak{A}, Y) \simeq \operatorname{Ext}^i_\Lambda(\mathfrak{A}, Y)$ for all $i \ge 0$ and use that $\mathfrak{A}$ is $(k+1)$-idempotent if and only if $\operatorname{Ext}^{i+1}_\Lambda(\Lambda/\mathfrak{A}, Y) = 0$ for all Y in $\operatorname{mod}\Lambda/\mathfrak{A}$ and $i = 0, \ldots, k$. □

Dually, for $0 \le k \le \infty$ we define $\mathbf{I}_k$ to be the full subcategory of $\operatorname{mod}\Lambda$ consisting of the Λ-modules Y having an injective coresolution $0 \to Y \to I_0 \to I_1 \to \cdots$ with I_i in $\operatorname{add} I$ for $0 \le i \le k$. Here I denotes the injective envelope of $P/\underline{r}P$.

Using Theorem 2.1′ and the duality D, we see again that there is a module such that its injective resolution determines for which k the ideal $\mathfrak{A}$ is k-idempotent.

Theorem 2.5. *Let* $M = E/E'$, *where* E *is the sum of all nonisomorphic indecomposable injective* Λ*-modules and* E' *is the sum of all nonisomorphic* *indecomposable injective* $\Lambda/\mathfrak{A}$*-modules. Let* $1 \leq k \leq \infty$*. Then* M *is in* $\mathbf{I}_k$ *if and only if* $\mathfrak{A}$ *is* $(k+1)$*-idempotent.*

Notice that the module M is the dual of $\mathfrak{A}$ considered as a right Λ-module.

We now write the dual statement of Proposition 2.4, since it will be used in the rest of the paper.

Proposition 2.6. *Let* $\mathfrak{A} = \tau_P(\Lambda)$ *and* $1 \leq k \leq \infty$*. Then the following conditions are equivalent for a* Λ*-module* Y:

(i) Y *is in* $\mathbf{I}_k$.

(ii) $\operatorname{Ext}^i_\Lambda(X, Y) = 0$ *for all* X *in* $\operatorname{mod}\Lambda/\mathfrak{A}$ *and all* $i = 0, \dots, k$.

(iii) $\operatorname{Ext}^i_\Lambda(\Lambda/\mathfrak{A}, Y) = 0$ *for all* $i = 0, \dots, k$.

3. Extensions over the endomorphism ring of P

Let $\mathfrak{A} = \tau_P(\Lambda)$ be an idempotent ideal and $\Gamma = \operatorname{End}_\Lambda(P)$. It would be more precise to write Γ_P, but we have assumed in §2 that P is a fixed projective module throughout the rest of the paper, so we just write Γ.

In §1 we considered the inclusion $\operatorname{mod}\Lambda/\mathfrak{A} \to \operatorname{mod}\Lambda$, and studied conditions on $\mathfrak{A}$ so that $\operatorname{Ext}^i_\Lambda(X, Y) \simeq \operatorname{Ext}^i_{\Lambda/\mathfrak{A}}(X, Y)$ for all $i = 0, 1, \dots, k$. Now we turn our attention to the functor $(P, \ ): \operatorname{mod}\Lambda \to \operatorname{mod}\Gamma$. This functor induces natural morphisms $\rho^i_{X,Y}: \operatorname{Ext}^i_\Lambda(X, Y) \to \operatorname{Ext}^i_\Gamma((P, X), (P, Y))$, for all $i \geq 0$ and all X, Y in $\operatorname{mod}\Lambda$. We will study in this section conditions under which these morphisms are isomorphims for all $i = 0, \dots, k$. And, also in analogy with what we did for $\Lambda/\mathfrak{A}$-modules and Λ-modules, we will explore the relationship between injective coresolutions in $\operatorname{mod}\Gamma$ and in $\operatorname{mod}\Lambda$.

We start by recalling some well-known properties of the functor $(P, \ ): \operatorname{mod}\Lambda \to \operatorname{mod}\Gamma$.

Let $\mathbf{I}_k$ and $\mathbf{P}_k$ be as in §2, for $0 \leq k \leq \infty$. Then $\mathbf{P}_1$ consists of the Λ-modules with a projective presentation in $\operatorname{add}P$, and $\mathbf{I}_1$ of those with an injective copresentation in $\operatorname{add}I$. We will use the following results [A, §§5 and 6].

Lemma 3.1. (1) *The functor* $(P, \ )$ *induces equivalences of categories*

$$(P, \ )|_{\mathbf{P}_1}: \mathbf{P}_1 \to \operatorname{mod}\Gamma, \qquad (P, \ )|_{\mathbf{I}_1}: \mathbf{I}_1 \to \operatorname{mod}\Gamma.$$

(2) *Let* $\rho_{X,Y}: (X, Y) \to ((P, X), (P, Y))$ *be the map induced by* $(P, \ )$*. Then*

(i) $\rho_{X,Y}$ *is a monomorphism if either* X *is in* $\mathbf{P}_0$ *or* Y *is in* $\mathbf{I}_0$.

(ii) $\rho_{X,Y}$ *is an isomorphism if* X *is in* $\mathbf{P}_0$ *and* Y *is* $\mathbf{I}_0$.

(iii) $\rho_{X,Y}$ *is an isomorphism if either* X *is in* $\mathbf{P}_1$ *or* Y *is in* $\mathbf{I}_1$.

(3) $(P, \ )$ *induces equivalences of categories between* $\operatorname{add}P$ *and the category of projective* Γ *modules and between* $\operatorname{add}I$ *and the category of injective* Γ*-modules.*

Since $(P, \ ): \operatorname{mod}\Lambda \to \operatorname{mod}\Gamma$ is an exact functor, it induces morphisms $\rho^i_{X,Y}: \operatorname{Ext}^i_\Lambda(X, Y) \to \operatorname{Ext}^i_\Lambda((P, X), (P, Y))$, for X, Y in $\operatorname{mod}\Lambda$, $i \geq 0$, functorial in X and Y. These morphisms are defined as follows. Let $0 \to Y \to I_0 \to I_1 \to \cdots$ be an injective copresentation of Y in $\operatorname{mod}\Lambda$. Then the exact complex $0 \to (P, Y) \to (P, I_0) \to (P, I_1) \to \cdots$ maps into an exact complex $0 \to (P, Y) \to E_0 \to E_1 \to \cdots$, with E_i injective in $\operatorname{mod}\Gamma$ for all $i \geq 0$. For X in $\operatorname{mod}\Lambda$ we get maps of complexes

$$\begin{array}{ccccccccc} 0 & \longrightarrow & (X, Y) & \longrightarrow & (X, I_0) & \longrightarrow & (X, I_1) & \longrightarrow & \cdots \\ & & \downarrow & & \downarrow & & \downarrow & & \\ 0 & \longrightarrow & ((P, X), (P, Y)) & \longrightarrow & ((P, X), (P, I_0)) & \longrightarrow & ((P, X), (P, I_1)) & \longrightarrow & \cdots \end{array}$$

and

$$\begin{array}{ccccccccc} 0 & \longrightarrow & ((P, X), (P, Y)) & \longrightarrow & ((P, X), (P, I_0)) & \longrightarrow & ((P, X), (P, I_1) & \longrightarrow & \cdots \\ & & \downarrow & & \downarrow & & \downarrow & & \\ 0 & \longrightarrow & ((P, X), (P, Y)) & \longrightarrow & ((P, X), E_0) & \longrightarrow & ((P, X), E_1) & \longrightarrow & \cdots \end{array}$$

The composition of these maps induces a map $\rho_{X,Y}$ between the homologies of the involved complexes. This is, a family of maps $\rho^i_{X,Y}\colon \operatorname{Ext}^i_\Lambda(X, Y) \to \operatorname{Ext}^i_\Gamma((P, X)(P, Y))$, $i \geq 0$, with $\rho^0_{X,Y} = \rho_{X,Y}$.

The following result relates extensions in $\operatorname{mod}\Gamma$ and in $\operatorname{mod}\Lambda$.

Theorem 3.2. *Let $n \geq 0$. Then the map*

$$\rho^n_{X,Y}\colon \operatorname{Ext}^n_\Lambda(X, Y) \to \operatorname{Ext}^n_\Gamma((P, X), (P, Y))$$

above defined is an isomorphism provided one of the three following conditions holds:

(a) $\mathbf{X}$ *is in* $\mathbf{P}_i$, Y *is in* $\mathbf{I}_j$ *and* $n \leq i + j$.
(b) X *is in* $\operatorname{mod}\Lambda$, Y *in* $\mathbf{I}_{n+1}$.
(c) X *is in* $\mathbf{P}_{n+1}$, Y *in* $\operatorname{mod}\Lambda$.

Proof. In $n = 0$ the theorem states the same as Lemma 3.1,2, (ii) and (iii). So we assume $n \geq 1$ and prove the theorem by induction on n. Let X in $\mathbf{P}_0$ and consider an exact sequence $0 \to K \to P_0 \to X \to 0$ with P_0 is $\operatorname{add} P$. Then $0 \to (P, K) \to (P, P_0) \to (P, X) \to 0$ is exact and (P, P_0) is projective in $\operatorname{mod}\Gamma$. We get a diagram

$$\begin{array}{cccccc} 0 & \to & (X, Y) & \to & (P_0, Y) & \to \\ & & \downarrow \rho_{X,Y} & & \downarrow \rho_{P_0,Y} & \\ 0 & \to & ((P, X), (P, Y)) & \to & ((P, P_0), (P, Y)) & \to \end{array}$$

$$\begin{array}{ccccc} (K, Y) & \to & \operatorname{Ext}^1_\Lambda(X, Y) & \to & 0 \\ \downarrow \rho_{K,Y} & & \downarrow \rho^1_{X,Y} & & \\ ((P, K), (P, Y)) & \to & \operatorname{Ext}^1_\Lambda((P, X), (P, Y)) & \to & 0. \end{array}$$

If X is in $\mathbf{P}_2$ then K is in $\mathbf{P}_1$. So it follows from Lemma 3.1, (2) (iii) that the three left vertical maps are isomorphisms if either X is in $\mathbf{P}_2$ or Y is in $\mathbf{I}_1$. Thus $\rho^1_{X,Y}$ is an isomorphism for X in $\mathbf{P}_2$, Y in $\mathbf{I}_1$. Considering an exact sequence $0 \to Y \to I_0 \to I \to 0$ with I_0 in $\operatorname{add} I$ one proves that $\rho^1_{X,Y}$ is an isomorphism if either $X \in \mathbf{P}_1$ or $Y \in \mathbf{I}_2$. Thus the theorem holds for $n = 1$.

Assume now that $n > 1$. First we consider the case when X is in $\mathbf{P}_i$ and $i \geq 0$. Let $0 \to K \to P_0 \to X \to 0$ be an exact sequence with P_0 in $\operatorname{add} P$. Then the sequence $0 \to (P, K) \to (P, P_0) \to (P, X) \to 0$ is exact, and since X is in $\mathbf{P}_0$ we have that (P, P_0) is projective in $\operatorname{mod}\Gamma$. We get a commutative diagram

$$\begin{array}{ccc} \operatorname{Ext}^{n-1}_\Lambda(K, Y) & \longrightarrow & \operatorname{Ext}^n_\Lambda(X, Y) \\ \Big\downarrow \rho^{n-1}_{L,Y} & & \Big\downarrow \rho^n_{X,Y} \\ \operatorname{Ext}^{n-1}_\Gamma((P, K), (P, Y)) & \longrightarrow & \operatorname{Ext}^n_\Gamma((P, X), (P, Y)) \end{array}$$

where the horizontal maps are isomorphisms. Since X is in $\mathbf{P}_i$ and $i \geq 0$, then K is in $\mathbf{P}_{i-1}$. Let now Y be in $\mathbf{I}_j$ with $i + j = n$. We can apply the induction hypothesis to K and Y and conclude that $\rho^{n-1}_{K,Y}$ is an isomorphism. Thus $\rho^n_{X,Y}$ is also an isomorphism. If Y is any Λ-module and $i = n + 1$ the same argument holds.

The other cases follow in a similar way, considering an exact sequence $0 \to Y \to I_0 \to K \to 0$ with I_0 in $\operatorname{add} I$. $\square$

We observe that when X is in $\mathbf{P}_\infty$ the theorem states that $\rho^n_{X,Y}$: $\operatorname{Ext}^n_\Lambda(X, Y) \to \operatorname{Ext}^n_\Gamma((P, X), (P, Y))$ is an isomorphism for all $n \geq 0$ and for all Y in $\operatorname{mod}\Lambda$. Similarly, if Y is in $\mathbf{I}_\infty$ then $\rho^n_{X,Y}$ is an isomorphism for all $n \geq 0$ and all X in $\operatorname{mod}\Lambda$. Since $(P, \ ): \operatorname{mod}\Lambda \to \operatorname{mod}\Gamma$ is a dense functor we obtain the following corollaries.

Corollary 3.3. (a) *If X is in $\mathbf{P}_\infty$ then* $p\dim_\Lambda X = p\dim_\Gamma(P, X)$.
(b) *If X is in $\mathbf{I}_\infty$ then* $\operatorname{inj}\dim_\Lambda X = \operatorname{inj}\dim_\Gamma(P, X)$.

Corollary 3.4. *If $\mathbf{P}_1 = \mathbf{P}_\infty$ or $\mathbf{I}_1 = \mathbf{I}_\infty$ then* $\operatorname{gl.dim}\Gamma \leq \operatorname{gl.dim}\Lambda$.

We observe that in general it is not true that $\operatorname{gl.dim}\Gamma \leq \operatorname{gl.dim}\Lambda$, as the following example shows.

Example 3.5. Let k be any field and Λ the k-algebra associated to the quiver

$$\underset{\bullet}{1} \ \underset{\underset{\beta}{\longleftarrow}}{\overset{\alpha}{\longrightarrow}} \ \underset{\bullet}{2}$$

modulo the relation $\alpha\beta = 0$. Let P_1 be the projective associated to the vertex 1. Then $\operatorname{gl.dim}(\operatorname{End}_\Lambda(P_1))^{\mathrm{op}} = \infty$ and $\operatorname{gl.dim}\Lambda = 2$.

Assume now that Y is in $\mathbf{I}_k$ for some $0 \leq k \leq \infty$ and let $0 \to Y \to I_0 \to \cdots \to I_k \to \cdots$ be a minimal injective coresolution of Y. Then I_j is in $\operatorname{add} I$ for $j \leq k$, so that $0 \to (P, Y) \to (P, I_0) \to \cdots \to (P, I_k)$ is the beginning of an injective coresolution of (P, Y). In particular, when $\mathbf{I}_1 = \mathbf{I}_\infty$ the injective coresolutions of the Γ-modules are all obtained by applying the functor $(P, \ )$ to injective coresolutions of Λ-modules. So we want to know what modules are in $\mathbf{I}_k$, and give necessary and sufficient conditions for $\mathbf{I}_1$ to be equal to $\mathbf{I}_\infty$. We are interested also in dual results about $\mathbf{P}_k$ to study projective resolutions.

Using the preceding theorem we can give the following characterization of modules in $\mathbf{I}_k$.

Theorem 3.6. *Let* $1 \leq k \leq \infty$. *Then*

(a) *A* Λ*-module* Y *is in* $\mathbf{I}_k$ *if and only if the above defined map* $\rho^i_{X,Y}$: $\operatorname{Ext}^i_\Lambda(X, Y) \to \operatorname{Ext}^i_\Gamma((P, X), (P, Y))$ *is an isomorphism, for all* $0 \leq i \leq k-1$ *and all* X *in* $\operatorname{mod}\Lambda$.

(b) *A* Λ*-module* X *is in* $\mathbf{P}_k$ *if and only if*

$$\rho^i_{X,Y}\colon \operatorname{Ext}^i_\Lambda(X, Y) \to \operatorname{Ext}^i_\Gamma((P, X), (P, Y))$$

is an isomorphism, for all $0 \leq i \leq k-1$ *and all* Y *in* $\operatorname{mod}\Lambda$.

Proof. (a) The fact that Y in $\mathbf{I}_k$ implies that

$$\operatorname{Ext}^i_\Lambda(X, Y) \simeq \operatorname{Ext}^i_\Gamma((P, X), (P, Y))$$

for all X in $\operatorname{mod}\Lambda$ and all $0 \leq i \leq k-1$ is the statement (b) of Theorem 3.2. To prove the converse we will use the characterization of modules in $\mathbf{I}_k$ given in Proposition 2.6, stating that a Λ-module M is in $\mathbf{I}_k$ if and only if $\operatorname{Ext}^i_\Lambda(\Lambda/\mathfrak{A}, M) = 0$ for $i = 0, \dots, k$.

Let $i \geq 0$. We consider the commutative diagram

$$\begin{array}{ccc} \operatorname{Ext}^i_\Lambda(\Lambda, Y) & \longrightarrow & \operatorname{Ext}^i_\Lambda(\mathfrak{A}, Y) \\ \downarrow \rho^i_{\Lambda,Y} & & \downarrow \rho^i_{\mathfrak{A},Y} \\ \operatorname{Ext}^i_\Gamma((P, \Lambda), (P, Y)) & \longrightarrow & \operatorname{Ext}^i_\Gamma((P, \mathfrak{A}), (P, Y)) \end{array}$$

induced by the inclusion $\mathfrak{A} \to \Lambda$. The lower horizontal map is an isomorphism because $(P, \mathfrak{A}) \to (P, \Lambda)$ is an isomorphism.

We assume now that $\rho^i_{X,Y}\colon \operatorname{Ext}^i_\Lambda(X, Y) \to \operatorname{Ext}^i_\Gamma((P, X), (P, Y))$ is an isomorphism for all $0 \leq i \leq k-1$ and all X in $\operatorname{mod}\Lambda$. Then the vertical maps in the diagram are isomorphisms and therefore $\operatorname{Ext}^i_\Lambda(\Lambda, Y) \simeq \operatorname{Ext}^i_\Lambda(\mathfrak{A}, Y)$, for $0 \leq i \leq k-1$. Hence $\operatorname{Ext}^i_\Lambda(\mathfrak{A}, Y) = 0$ for $1 \leq i \leq k-1$. On the other hand, from the exact sequence $0 \to \mathfrak{A} \to \Lambda \to \Lambda/\mathfrak{A} \to 0$ we get an exact sequence $0 \to (\Lambda/\mathfrak{A}, Y) \to (\Lambda, Y) \to (\mathfrak{A}, Y) \to \operatorname{Ext}^1_\Lambda(\Lambda/\mathfrak{A}, Y) \to 0$. We just proved that $(\Lambda, Y) \to (\mathfrak{A}, Y)$ is an isomorphism, so $(\Lambda/\mathfrak{A}, Y) = 0$ and $\operatorname{Ext}^1_\Lambda(\Lambda/\mathfrak{A}, Y) = 0$. Thus $\operatorname{Ext}^i_\Lambda(\Lambda/\mathfrak{A}, Y) = 0$ for all $0 \leq i \leq k$, this is, Y is in $\mathbf{I}_k$.

(b) follows from (a) by duality. □

The following propositions gives another characterization of modules in $\mathbf{I}_k$.

Proposition 3.7. (a) *Let* X *be in* $\mathbf{I}_1$ *and* $k \geq 1$. *Then* X *is in* $\mathbf{I}_k$ *if and only if* $\operatorname{Ext}^i_\Gamma(P^*, (P, X)) = 0$ *for all* $1 \leq i \leq k-1$, *where* P^* *denotes the* Γ*-module* $\operatorname{Hom}_\Lambda(P, \Lambda)$.

(b) *Let* X *be in* $\mathbf{P}_1$ *and* $k \geq 1$. *Then* X *is in* $\mathbf{P}_k$ *if and only if*

$$\operatorname{Tor}^\Gamma_i(P, (P, X)) = 0$$

for all $1 \leq i \leq k-1$, *where* P *is considered as a module over* $\Gamma^{\mathrm{op}} = \operatorname{End}_\Lambda(P)$ *in the natural way.*

Proof. (a) Let X be in $\mathbf{I}_k$. Then we know by Proposition 3.6 (a) that

$$\operatorname{Ext}^i_\Gamma(P^*, (P, X)) = \operatorname{Ext}^i_\Gamma((P, \Lambda), (P, X)) \simeq \operatorname{Ext}^i_\Lambda(\Lambda, X),$$

for $0 \leq i \leq k-1$. This proves that $\operatorname{Ext}^i_\Gamma(P^*, (P, X)) = 0$ for $1 \leq i \leq k-1$.

To prove the converse we assume that $\operatorname{Ext}^i_\Gamma(P^*, (P, X)) = 0$ for $1 \le i \le k-1$. We will prove that X is in $\mathbf{I}_k$ by induction on k. The result is true for $k = 1$. So let $k > 1$. Then X is in $\mathbf{I}_1$, and since $\mathfrak{A}$ is in $\mathbf{P}_0$ we can apply Theorem 3.2 and obtain that $\operatorname{Ext}^1_\Gamma((P, \mathfrak{A}), (P, X)) \simeq \operatorname{Ext}^1_\Lambda(\mathfrak{A}, X)$. Since $(P, \mathfrak{A}) \simeq P^*$ and we are assuming that $\operatorname{Ext}^i_\Gamma(P^*, (P, X)) = 0$ for $i \le k-1$ we obtain that $\operatorname{Ext}^2_\Lambda(\Lambda/\mathfrak{A}, X) \simeq \operatorname{Ext}^1_\Lambda(\mathfrak{A}, X) = 0$. So $X \in \mathbf{I}_2$.

Consider now an exact sequence $0 \to X \to I_0 \to L \to 0$ with I_0 in $\operatorname{add} I$. Then $0 \to (P, X) \to (P, I_0) \to (P, L) \to 0$ is exact and (P, I_0) is an injective Γ-module. So $\operatorname{Ext}^i_\Gamma(P^*, (P, L)) = 0$ for $1 \le i \le k-2$. Since X is in I_2 it follows that L is in $\mathbf{I}_1$, so the induction hypothesis applies to L. Therefore L is in $\mathbf{I}_{k-1}$, so X is in $\mathbf{I}_k$, as we wanted to prove.

(b) follows from (a) by duality, observing that $\mathfrak{A} = \tau_{P^*}(\Lambda^{\mathrm{op}})$ as is also $\tau_P(\Lambda)$. □

We characterize now when $\mathbf{I}_1 = \mathbf{I}_\infty$, and when $\mathbf{P}_1 = \mathbf{P}_\infty$.

Corollary 3.8. (a) $\mathbf{I}_1 = \mathbf{I}_\infty$ *if and only if the* Γ*-module* P^* *is projective.*

(b) $\mathbf{P}_1 = \mathbf{P}_\infty$ *if and only if the* Γ^{op}*-module* P *is projective.*

Proof. (a) follows directly from the Proposition 3.7. To prove (b) we write $\mathfrak{A} = \tau_{P^*}(\Lambda^{\mathrm{op}})$. Since $\mathbf{P}_1 = \mathbf{P}_\infty$ we get that the subcategories $\mathbf{I}_{1, P^*}$ and $\mathbf{I}_{\infty, P^*}$ of $\operatorname{mod}\Lambda^{\mathrm{op}}$ are the same. By (a) we obtain that the Γ^{op}-module $(P^*)^* = P$ is projective. □

We recall now some properties of the functor $\operatorname{Hom}_\Gamma(P^*, \ ) = (P^*, \ ): \operatorname{mod}\Gamma \to \operatorname{mod}\Lambda$, which is the right adjoint of $(P, \ ): \operatorname{mod}\Lambda \to \operatorname{mod}\Gamma$ [A, §5]. The functor $(P^*, \ )$ is left exact, fully faithful and its image is the subcategory $\mathbf{I}_1$ of $\operatorname{mod}\Lambda$. The composition $\operatorname{mod}\Gamma \xrightarrow{(P^*, \ )} \operatorname{mod}\Lambda \xrightarrow{(P, \ )} \operatorname{mod}\Gamma$ is the identity of $\operatorname{mod}\Gamma$, and $(P^*, \ )$ preserves injective envelopes. However, $(P^*, \ )$ does not preserve injective coresolutions, not even injective copresentations. In fact, given an injective coresolution $0 \to M \to I_0 \to I_1 \to \cdots$ of a Γ-module M, then $0 \to (P^*, M) \to (P^*, I_0) \to \cdots \to (P^*, I_k)$ is the beginning of an injective coresolution of (P^*, M) if and only if $\operatorname{Ext}^i_\Gamma(P^*, M) = 0$ for $i = 1, \dots, k$. Therefore Proposition 3.7(a) can now be written in the following way.

Proposition 3.9. *Let* X *be a* Λ*-module and* $M = (P, X)$. *Then the following conditions are equivalent for* $1 \le k \le \infty$.

(i) X *is in* $\mathbf{I}_k$.

(ii) *If* $0 \to M \to I_0 \to I_1 \to \cdots$ *is an injective resolution of* M *in* $\operatorname{mod}\Gamma$ *then* $0 \to X \to (P^*, I_0) \to \cdots \to (P^*, I_{k-1})$ *is the beginning of an injective coresolution of* $X = (P^*, M)$ *in* $\operatorname{mod}\Lambda$.

It follows that $\mathbf{I}_1 = \mathbf{I}_\infty$ if and only if the functor $(P^*, \ ): \operatorname{mod}\Gamma \to \operatorname{mod}\Lambda$ carries injective coresolution of Λ-modules into injective coresolutions of Λ-modules.

On the other hand, $(P, \ ): \operatorname{mod}\Lambda \to \operatorname{mod}\Gamma$ has also a left adjoint, the functor $P\otimes_\Gamma: \operatorname{mod}\Gamma \to \operatorname{mod}\Lambda$ [A, §5]. This functor is right exact, fully faithful and its image is $\mathbf{P}_1$. The composition

$$\operatorname{mod}\Gamma \xrightarrow{P\otimes_\Gamma} \operatorname{mod}\Lambda \xrightarrow{(P\,,\ )} \operatorname{mod}\Gamma$$

is the identity of $\operatorname{mod}\Gamma$, and $P\otimes_\Gamma$ preserves projective covers. Proposition 3.7(b) can be written as follows.

Proposition 3.10. *Let X be a Λ-module and $M = (P, X)$. Then the following conditions are equivalent for $1 \le k \le \infty$.*

(i) *X is in $\mathbf{P}_k$.*

(ii) *If $\cdots \to P_k \to \cdots \to P_0 \to M \to 0$ is a projective resolution of M in $\operatorname{mod}\Gamma$ then $P\otimes_\Gamma P_k \to \cdots \to P\otimes_\Gamma P_0 \to X \to 0$ is the beginning of a projective resolution of $X = P\otimes_\Gamma M$ in $\operatorname{mod}\Lambda$.*

4. Conditions for $\mathbf{I}_1 = \mathbf{I}_\infty$

In this section we give other necessary and sufficient conditions for $\mathbf{I}_1$ to be equal to $\mathbf{I}_\infty$, and for $\mathbf{P}_1$ to be equal to $\mathbf{P}_\infty$. We prove that $\mathbf{I}_1 = \mathbf{I}_\infty$ if and only if $\mathfrak{A}\otimes_\Lambda \mathfrak{A}$ is a projective Λ-module, and $\mathbf{P}_1 = \mathbf{P}_\infty$ if and only if $\mathfrak{A}\otimes_\Lambda\mathfrak{A}$ is a projective Λ^{op}-module.

We start by recalling some further properties of the functor $P\otimes_\Gamma\colon \operatorname{mod}\Gamma \to \operatorname{mod}\Lambda$. (See, for example, [A, §5].) Since $P\otimes_\Gamma\colon \operatorname{mod}\Gamma \to \operatorname{mod}\Lambda$ is a left adjoint of $(P,\ )\colon \operatorname{mod}\Lambda \to \operatorname{mod}\Gamma$, there is a unique isomorphism of functors f from the composition $(P\otimes_\Gamma\ )(P,\ )$ to the identity of $\operatorname{mod}\Lambda$ such that $(P,\ )\,f$ is an isomorphism. For X in $\operatorname{mod}\Lambda$, $f_X\colon P\otimes_\Gamma(P, X) \to X$ is given by $f_X(p\otimes h) = h(p)$, for $h \in (P, X)$, $p\in P$. The following is a well-known fact.

Lemma 4.1. *Let X be in $\operatorname{mod}\Lambda$ and $f_X\colon P\otimes_\Gamma(P, X)\to X$ be as above. Then*

(1) *$P\otimes_\Gamma(P, X)$ is in $\mathbf{P}_1$ and $\ker f_X$, $\operatorname{Coker} f_X$ are in $\operatorname{mod}\Lambda/\mathfrak{A}$.*

(2) *Given an exact sequence $0\to L_1 \to X_1 \xrightarrow{\varphi} X \to L_2 \to 0$ with X_1 in $\mathbf{P}_1$ and L_1, L_2 in $\operatorname{mod}\Lambda/\mathfrak{A}$ we have:*

(a) *There is a commutative diagram*

$$\begin{array}{ccccccccccc} 0 & \longrightarrow & \operatorname{Ker} f_X & \longrightarrow & P\otimes_\Gamma(P, X) & \xrightarrow{f_X} & X & \longrightarrow & \operatorname{Coker} f_X & \longrightarrow & 0 \\ & & \downarrow & & \downarrow & & \| & & \downarrow & & \\ 0 & \longrightarrow & L_1 & \longrightarrow & X_1 & \xrightarrow{\varphi} & X & \longrightarrow & L_2 & \longrightarrow & 0 \end{array}$$

where the vertical arrows are isomorphisms.

(b) *$(P, \varphi)\colon (P, X_1)\to(P, X)$ is an isomorphism.*

(c) *Let Y in $\mathbf{P}_k$ and $1\le k\le\infty$. The maps $\operatorname{Ext}^i_\Lambda(Y, \varphi)\colon \operatorname{Ext}^i_\Lambda(Y, X_1)\to \operatorname{Ext}^i_\Lambda(Y, X)$ are isomorphisms for all $0\le i\le k-1$.*

Now we state the dual results (see, for example, [A, §5]). Since $(P^*,\ )\colon \operatorname{mod}\Gamma \to \operatorname{mod}\Lambda$ is a right adjoint of $(P,\ )\colon\operatorname{mod}\Lambda\to\operatorname{mod}\Gamma$ there is a unique isomorphism of functors g from the identity of $\operatorname{mod}\Lambda$ to the composition $(P^*,\ )(P,\ ) \simeq (P\otimes_\Gamma P^*,\ )$. For X in $\operatorname{mod}\Lambda$, $g_X\colon X\to (P^*, (P, X))$ is given by $(g_X(x)(f))(p) = f(p)x$, for x in X, f in P^* and p in P.

Lemma 4.2. *Let X in $\operatorname{mod}\Lambda$ and $g_X\colon X\to(P^*, (P, X))$ be as above. Then*

(1) *$(P^*, (P, X))$ is in $\mathbf{I}_1$ and $\operatorname{Ker} g_X$, $\operatorname{Coker} g_X$ are in $\operatorname{mod}\Lambda/\mathfrak{A}$.*

(2) *Given an exact sequence $0\to L_1\to X \xrightarrow{\psi} X_1 \to L_2\to 0$ with X_1 in $\mathbf{I}_1$ and L_1, L_2 in $\operatorname{mod}\Lambda/\mathfrak{A}$, we have:*

(a) *There is a commutative diagram*

$$\begin{array}{ccccccccccc} 0 & \longrightarrow & L_1 & \longrightarrow & X & \xrightarrow{\psi} & X_1 & \longrightarrow & L_2 & \longrightarrow & 0 \\ & & \downarrow & & \| & & \downarrow & & \downarrow & & \\ 0 & \longrightarrow & \ker g_X & \longrightarrow & X & \xrightarrow{g_X} & (P^*, (P, X)) & \longrightarrow & \operatorname{Coker} g_X & \longrightarrow & 0 \end{array}$$

where the vertical arrows are isomorphisms.

(b) $(P, \psi): (P, X) \to (P, X_1)$ *is an isomorphism.*

(c) *Let* Y *in* $\mathbf{I}_k$ *and* $1 \leq k \leq \infty$. *The maps* $\operatorname{Ext}^i_\Lambda(\psi, Y): \operatorname{Ext}^i_\Lambda(X_1, Y) \to \operatorname{Ext}^i_\Lambda(X, Y)$ *are isomorphisms for all* $0 \leq i \leq k-1$.

To simplify notation we denote $P \otimes_\Gamma (P, X)$ *by* X_{P_1}, *and* $(P^*, (P, X))$ *by* X_{I_1}.

In the following proposition we describe Λ_{P_1}.

Proposition 4.3. *There are isomorphisms of* Λ*-modules* $\Lambda_{P_1} \simeq \mathfrak{A}_{P_1} \simeq \mathfrak{A} \otimes_\Lambda \mathfrak{A}$.

Proof. To prove the result we will use the following lemma.

Lemma 4.4. *Let* P_0 *be the projective cover of* $\mathfrak{A}$. *Then* P_0 *is also the projective cover of* $\mathfrak{A} \otimes_\Lambda \mathfrak{A}$, *and there is an exact sequence*

$$0 \to \operatorname{Tor}^\Lambda_3(\Lambda/\mathfrak{A}, \Lambda/\mathfrak{A}) \to C \to P_0 \to \mathfrak{A} \otimes_\Lambda \mathfrak{A} \to 0$$

with C *in* $\mathbf{P}_0$. *So* $\mathfrak{A} \otimes_\Lambda \mathfrak{A}$ *is in* $\mathbf{P}_1$.

Proof. Consider the exact sequence $0 \to K \to P_0 \to \mathfrak{A} \to 0$. Tensoring with $\mathfrak{A}$ we obtain the exact sequence

$$0 \to \operatorname{Tor}^\Lambda_1(\mathfrak{A}, \mathfrak{A}) \to \mathfrak{A} \otimes K \to \mathfrak{A} \otimes P_0 \to \mathfrak{A} \otimes_\Lambda \mathfrak{A} \to 0.$$

On the other hand, we have a commutative diagram

$$\begin{array}{ccc} \mathfrak{A} \otimes_\Lambda P_0 & \longrightarrow & \mathfrak{A} \otimes_\Lambda \mathfrak{A} \\ \downarrow{\scriptstyle m_{P_0}} & & \downarrow{\scriptstyle m_{\mathfrak{A}}} \\ P_0 & \longrightarrow & \mathfrak{A} \end{array}$$

where the vertical maps are the multiplication maps. We know that $\mathfrak{A}P_0 = P_0$ because P_0 is in add P. So, tensoring the exact sequence $0 \to \mathfrak{A} \to \Lambda \to \Lambda/\mathfrak{A} \to 0$ with the projective module ${}_\Lambda P_0$ we obtain that $\mathfrak{A} \otimes_\Lambda P_0 \to \Lambda \otimes_\Lambda P_0$ is an isomorphism. Thus m_{P_0} is an isomorphism. From the commutativity of the diagram it follows that P_0 is the projective cover of $\mathfrak{A} \otimes_\Lambda \mathfrak{A}$. This ends the proof of the lemma, since $\operatorname{Tor}^\Lambda_1(\mathfrak{A}, \mathfrak{A}) \simeq \operatorname{Tor}^\Lambda_3(\Lambda/\mathfrak{A}, \Lambda/\mathfrak{A})$, and $C = \mathfrak{A} \otimes_\Lambda K$ is in $\mathbf{P}_0$ because $\mathfrak{A}C = C$. □

Now we prove the proposition. We have that $(P, \Lambda) \simeq (P, \tau_P \Lambda) = (P, \mathfrak{A})$. Thus $(P, \Lambda_{P_1}) \simeq (P, \mathfrak{A}_{P_1})$, as follows from Lemma 4.1.2(b). Since both Λ_P, $\mathfrak{A}_{P_1}$ are in $\mathbf{P}_1$ and the restriction of $(P, \ )$ to $\mathbf{P}_1$ is an equivalence we get that $\Lambda_{P_1} \simeq \mathfrak{A}_{P_1}$.

Consider the exact sequence $0 \to \operatorname{Tor}^\Lambda_2(\Lambda/\mathfrak{A}, \Lambda/\mathfrak{A}) \to \mathfrak{A} \otimes_\Lambda \mathfrak{A} \to \mathfrak{A} \to 0$, as in Lemma 1.5. The module $\operatorname{Tor}^\Lambda_2(\Lambda/\mathfrak{A}, \Lambda/\mathfrak{A})$ is in $\operatorname{mod} \Lambda/\mathfrak{A}$, and $\mathfrak{A} \otimes_\Lambda \mathfrak{A}$ is in $\mathbf{P}_1$, by the lemma. The uniqueness of a sequence $0 \to L_1 \to X_1 \to \mathfrak{A} \to$

$L_2 \to 0$ with X_1 in $\mathbf{P}_1$ and L_1, L_2 in $\operatorname{mod} \Lambda/\mathfrak{A}$ (Lemma 4.1) implies that $\mathfrak{A}_{P_1} \simeq \mathfrak{A} \otimes_\Lambda \mathfrak{A}$. □

We can now describe when $\mathfrak{A} \otimes_\Lambda \mathfrak{A}$ is a projective Λ-module.

Proposition 4.5. *The following are equivalent for the idempotent ideal* $\mathfrak{A}$

(a) $\mathbf{I}_1 = \mathbf{I}_\infty$.

(b) $\mathfrak{A} \otimes_\Lambda \mathfrak{A}$ *is a projective* Λ*-module.*

(c) $\mathfrak{A} \otimes_\Lambda \mathfrak{A}$ *is a projective* Λ*-module and* $\operatorname{Tor}_3^\Lambda(\Lambda/\mathfrak{A}, \Lambda/\mathfrak{A}) = 0$.

Proof. We proved in Corollary 3.8 that $\mathbf{I}_1 = \mathbf{I}_\infty$ if and only if P^* is a projective Γ-module. This is the case if and only if $P^* = (P, Q)$ for some Q in add P, as follows from Lemma 3.1,3. On the other hand $(P, \ )|_{\mathbf{P}_1} : \mathbf{P}_1 \to \operatorname{mod} \Gamma$ is an equivalence of categories and $P^* = (P, \Lambda) = (P, \Lambda_{P_1})$. So P^* is a projective Γ-module if and only if Λ_{P_1} is a projective Λ-module. Therefore (a) $\Leftrightarrow$ (b), since we have proven in Proposition 4.3 that Λ_{P_1} and $\mathfrak{A} \otimes \mathfrak{A}$ are isomorphic Λ-modules.

Clearly (c) $\Rightarrow$ (b). We prove now (b) $\Rightarrow$ (c). We consider the exact sequence $0 \to \operatorname{Tor}_3^\Lambda(\Lambda/\mathfrak{A}, \Lambda/\mathfrak{A}) \to C \to P_0 \to \mathfrak{A} \otimes \mathfrak{A} \to 0$ with C in $\mathbf{P}_0$ and $P_0 = P_0(\mathfrak{A} \otimes \mathfrak{A})$, as in Lemma 4.4. If $\mathfrak{A} \otimes \mathfrak{A}$ is projective then $\operatorname{Tor}_3^\Lambda(\Lambda/\mathfrak{A}, \Lambda/\mathfrak{A}) \to C$ is an isomorphism. From this it follows that $C = 0$. Because on one hand $\mathfrak{A}C = C$ since C is in $\mathbf{P}_0$, and on the other hand $C \simeq \operatorname{Tor}_3^\Lambda(\Lambda/\mathfrak{A}, \Lambda/\mathfrak{A})$ is a $\Lambda/\mathfrak{A}$ module. □

By the definition of Λ_{P_1} we know that $\Lambda_{P_1} = P \otimes_\Gamma (P, \Lambda) \simeq P \otimes_\Gamma P^*$. On the other hand Proposition 4.3 shows that $\Lambda_{P_1} \simeq \mathfrak{A} \otimes_\Lambda \mathfrak{A}$. Therefore $P \otimes_\Gamma P^*$ and $\mathfrak{A} \otimes_\Lambda \mathfrak{A}$ are isomorphic as left Λ-modules. We show next that they are, in fact, isomorphic Λ-Λ-bimodules by giving a bimodule isomorphism $t \colon P \otimes_\Gamma P^* \to \mathfrak{A} \otimes_\Lambda \mathfrak{A}$.

Proposition 4.6. *There is a* Λ*-*Λ*-bimodule isomorphism* $t \colon P \otimes_\Gamma P^* \to \mathfrak{A} \otimes_\Lambda \mathfrak{A}$ *such that the diagram*

$$\begin{array}{ccc} P \otimes_\Gamma P^* & \xrightarrow{t} & \mathfrak{A} \otimes_\Lambda \mathfrak{A} \\ & \searrow^{\theta} & \downarrow{\scriptstyle m_{\mathfrak{A}}} \\ & & \mathfrak{A} \end{array}$$

commutes where $m_{\mathfrak{A}}$ *is the multiplication map, and* $\theta(p \otimes f) = f(p)$, *for* $p \in P$, $f \in P^*$.

Proof. Let $\varepsilon \colon P^* \otimes_\Lambda P \to \Gamma$ be defined by $\varepsilon(f \otimes p)(p') = f(p')p$, for $f \in P^*$, $p, p' \in P$. Then ε is a Λ-Λ-bimodule isomorphism and induces an isomorphism $\alpha \colon P \otimes_\Gamma (P^* \otimes_\Lambda P) \otimes_\Gamma P^* \to P \otimes_\Gamma P^*$ of Λ-Λ-bimodules. We get a commutative diagram

$$\begin{array}{ccc} (P \otimes_\Gamma P^*) \otimes_\Lambda (P \otimes_\Gamma P^*) & \xrightarrow{\theta \otimes \theta} & \mathfrak{A} \otimes_\Lambda \mathfrak{A} \\ \downarrow{\scriptstyle \alpha} & & \downarrow{\scriptstyle m_{\mathfrak{A}}} \\ P \otimes_\Gamma P^* & \xrightarrow{\theta} & \mathfrak{A} \end{array}$$

We observed just before the proposition that $P \otimes_\Gamma P^*$ and $\mathfrak{A} \otimes_\Lambda \mathfrak{A}$ are isomorphic in $\operatorname{mod} \Lambda$. Thus the epimorphism $t = (\theta \otimes \theta)\alpha^{-1} \colon P \otimes_\Gamma P^* \to \mathfrak{A} \otimes_\Lambda \mathfrak{A}$ is an isomorphism and one easily checks that the required diagram commutes. □

Remark. We give an elementwise definition of t. Let $\mathrm{id}_P\colon P \to P$ be the identity map. Since $\varepsilon\colon P^* \otimes_\Lambda P \to \Gamma$ above defined is an isomorphism, there are $f_i \in P^*$, $p_i \in P$, $i = 1, \dots, n$, such that $\varepsilon(\sum_{i=1}^n f_i \otimes p_i) = \mathrm{id}_P$. Then $t(p \otimes f) = \sum_{i=1}^n f_i(p) \otimes f(p_i)$, for $p \in P$, $f \in P^*$.

As immediate consequences of the proposition we obtain the following two corollaries.

Corollary 4.7. *The idempotent ideal* $\mathfrak{A}$ *is 2-idempotent if and only if the map* $\theta\colon P \otimes_\Gamma P^* \to \mathfrak{A}$ *defined above is an isomorphism.*

Corollary 4.8. *Let* X *in* $\operatorname{mod}\Lambda$. *Then* $X_{I_1} = (\mathfrak{A} \otimes_\Lambda \mathfrak{A}, X)$ *and* $X_{P_1} = \mathfrak{A} \otimes_\Lambda \mathfrak{A} \otimes_\Lambda X$. *If* $\mathfrak{A}$ *is a 2-idempotent then* $X_{I_1} = (\mathfrak{A}, X)$ *and* $X_{P_1} = \mathfrak{A} \otimes_\Lambda X$.

We know that if $\mathfrak{A}$ is an idempotent ideal then $\mathfrak{A} \otimes_\Lambda \mathfrak{A}$ is not isomorphic to $\mathfrak{A}$, unless $\mathfrak{A}$ is 2-idempotent. However, we have the following result.

Corollary 4.9. *The map* $\mathrm{id}_{\mathfrak{A}} \otimes m_{\mathfrak{A}} : \mathfrak{A} \otimes_\Lambda \mathfrak{A} \otimes_\Lambda \mathfrak{A} \to \mathfrak{A} \otimes_\Lambda \mathfrak{A}$ *is an isomorphism, where* $\mathrm{id}_{\mathfrak{A}} : \mathfrak{A} \to \mathfrak{A}$ *is the identity map and* $m_{\mathfrak{A}} : \mathfrak{A} \otimes_\Lambda \mathfrak{A} \to \mathfrak{A}$ *is the multiplication map.*

Proof. We have a commutative diagram

$$\begin{array}{ccc}
\mathfrak{A} \otimes_\Lambda P \otimes_\Gamma P^* & \xrightarrow{\mathrm{id}_{\mathfrak{A}} \otimes t} & \mathfrak{A} \otimes_\Lambda \mathfrak{A} \otimes_\Lambda \mathfrak{A} \\
{\scriptstyle m_P \otimes \mathrm{id}_{P^*}} \downarrow & & \downarrow {\scriptstyle \mathrm{id}_{\mathfrak{A}} \otimes m_{\mathfrak{A}}} \\
P \otimes_\Gamma P^* & \xrightarrow{t} & \mathfrak{A} \otimes_\Lambda \mathfrak{A},
\end{array}$$

where $m_P\colon \mathfrak{A} \otimes_\Lambda P \to P$ is the multiplication map. Since both m_P and t are isomorphisms it follows that all the maps in the diagram are isomorphisms, proving the corollary. □

We give now a list of necessary and sufficient conditions for $\mathbf{I}_1$ to be equal to $\mathbf{I}_\infty$, summarizing many of the previous results.

Proposition 4.10. *The following are equivalent for an idempotent ideal* $\mathfrak{A}$ *of* Λ:

(1) $\mathbf{I}_1 = \mathbf{I}_\infty$.
(2) $\mathbf{I}_1 = \mathbf{I}_2$.
(3) P^* *is a projective* Γ*-module.*
(4) $\mathfrak{A} \otimes_\Lambda \mathfrak{A}$ *is a projective* Λ*-module.*
(5) $P \otimes_\Gamma P^*$ *is a projective* Λ*-module.*
(6) *The composition* $\operatorname{mod}\Lambda \xrightarrow{(P,\ )} \operatorname{mod}\Gamma \xrightarrow{(P^*),} \operatorname{mod}\Lambda$ *is exact.*

It would be nice to know what modules are in $\mathbf{I}_k$ and not in $\mathbf{I}_{k+1}$ for a given idempotent ideal $\mathfrak{A}$. We observe that $\mathbf{I}_k = \mathbf{I}_{k+1}$ implies $\mathbf{I}_k = \mathbf{I}_\infty$. It would also be nice to know which is the smallest k such that $\mathbf{I}_k = \mathbf{I}_\infty$, or, what is equivalent, such that $\mathbf{I}_k = \mathbf{I}_{k+1}$. We know that such k is 1 if and only if $\mathfrak{A} \otimes_\Lambda \mathfrak{A}$ is projective. Moreover, the following result is true

Proposition 4.11. *Let* $\mathfrak{A}$ *be an idempotent ideal and assume that*

$$\operatorname{p.dim}_\Lambda(\mathfrak{A} \otimes_\Lambda \mathfrak{A}) = r.$$

Then $\mathbf{I}_{r+1} = \mathbf{I}_{r+2} = \cdots = \mathbf{I}_\infty$.

Proof. Assume X is in $\mathbf{I}_{r+1}$. We know by Proposition 3.7 that this means that $\operatorname{Ext}^i_\Gamma(P^*, (P, X)) = 0$ for $i = 1, \dots, r$. To prove that X is in $\mathbf{I}_{r+2}$ we will prove that $\operatorname{Ext}_\Gamma^{r+1}(P^*, (P, X)) = 0$. We write $P^* = (P, \Lambda) = (P, \Lambda_{P_1})$. Since Λ_{P_1} is in $\mathbf{P}_1$ and X is in $\mathbf{I}_{r+1}$ it follows from Theorem 3.2 that

$$\operatorname{Ext}_\Gamma^{r+1}(P^*, (P, X)) \simeq \operatorname{Ext}_\Lambda^{r+1}(\Lambda_{P_1}, X).$$

And $\operatorname{Ext}_\Lambda^{r+1}(\Lambda_{P_1}, X) = 0$ because we are assuming that $\operatorname{p.dim}_\Lambda(\mathfrak{A} \otimes_\Lambda \mathfrak{A}) = r$, and $\Lambda_{P_1} \simeq \mathfrak{A} \otimes_\Lambda \mathfrak{A}$, by Proposition 4.3. So $\mathbf{I}_{r+1} = \mathbf{I}_{r+2}$. □

Remark. The converse of Proposition 4.11 is not true. Let $\Lambda, \mathfrak{A}, P$ be as in Example 3, §1. Then $\mathfrak{A}$ is a strong idempotent, so it is a 2-idempotent and thus $\mathfrak{A} \otimes \mathfrak{A} = \mathfrak{A}$. However, $\mathbf{I}_2 = \mathbf{I}_\infty$ and $\operatorname{p.dim} \mathfrak{A} = \infty$.

Using that $\mathfrak{A} = \tau_P(\Lambda)$ implies that $\mathfrak{A} = \tau_{P^*}(\Lambda^{\mathrm{op}})$ we obtain from Proposition 4.10 the following result.

Proposition 4.12. *The following conditions are equivalent for the idempoent ideal $\mathfrak{A}$ of Λ,*
(1) $\mathbf{P}_1 = \mathbf{P}_\infty$.
(2) $\mathbf{P}_1 = \mathbf{P}_2$.
(3) *P is a projective Γ^{op}-module.*
(4) *$\mathfrak{A} \otimes_\Lambda \mathfrak{A}$ is a projective Λ^{op}-module.*
(5) *$P \otimes_\Gamma P^*$ is a projective Λ^{op}-module.*
(6) *The composition $\operatorname{mod} \Lambda \xrightarrow{(P,\ )} \operatorname{mod} \Gamma \xrightarrow{P\otimes} \operatorname{mod} \Lambda$ is exact.*

The following result follows from Proposition 4.11.

Proposition 4.13. *Let $\mathfrak{A}$ be an idempotent ideal and assume that $\operatorname{p.dim}_{\Lambda^{\mathrm{op}}} \mathfrak{A} \otimes_\Lambda \mathfrak{A} = r$. Then $\mathbf{P}_{r+1} = \mathbf{P}_{r+2} = \cdots = \mathbf{P}_\infty$.*

5. Projectivity of $\mathfrak{A}$. Global dimension

In this section we characterize when an idempotent ideal $\mathfrak{A}$ is a projective Λ-module, and when it is a projective Λ^{op}-module. We also study relationships between the global dimensions of the algebras Λ, $\Lambda/\mathfrak{A}$ and Γ.

The following proposition combines results of the preceding sections.

Proposition 5.1. *Let $\mathfrak{A}$ be an idempotent ideal. Then the following are equivalent*
(a) *$\mathfrak{A}$ is a projective Λ-module.*
(b) *$\mathfrak{A}$ is a strong idempotent and $\mathbf{I}_1 = \mathbf{I}_\infty$.*
(c) *$\mathfrak{A}$ is a 2-idempotent and $\mathbf{I}_1 = \mathbf{I}_\infty$.*
(d) *The multiplication map $\mathfrak{A} \otimes_\Lambda \mathfrak{A} \to \mathfrak{A}$ is an isomorphism and $\mathfrak{A} \otimes_\Lambda \mathfrak{A}$ is a projective Λ-module.*

Proof. (a) $\Rightarrow$ (b) Assume that $\mathfrak{A}$ is a projective Λ-module. Then $\operatorname{p.dim} \Lambda/\mathfrak{A} \leq 1$, so it follows from the definitions that $\mathbf{I}_1 = \mathbf{I}_\infty$. And we observed in §1, Example 1, that $\mathfrak{A}$ is a strong idempotent. So (a) $\Rightarrow$ (b). Clearly (b) $\Rightarrow$ (c). We know by Proposition 1.4 that $\mathfrak{A}$ is a strong idempotent if and only if the multiplication map $\mathfrak{A} \otimes \mathfrak{A} \to \mathfrak{A}$ is an isomorphism. On the other hand, we proved in Proposition 4.5 that $\mathbf{I}_1 = \mathbf{I}_\infty$ if and only if $\mathfrak{A} \otimes_\Lambda \mathfrak{A}$ is projective in $\operatorname{mod} \Lambda$. Then (b) $\Leftrightarrow$ (d). □

We have the following characterization of when $\mathfrak{A}$ is a projective Λ^{op}-module.

Proposition 5.2. *Let $\mathfrak{A}$ be an idempotent ideal. Then the following are equivalent*:

(a) *$\mathfrak{A}$ is a projective Λ^{op}-module.*

(b) *$\mathfrak{A}$ is a strong idempotent and $\mathbf{P}_1 = \mathbf{P}_\infty$.*

(c) *$\mathfrak{A}$ is a 2-idempotent and $\mathbf{P}_1 = \mathbf{P}_\infty$.*

(d) *The multiplication map $\mathfrak{A} \otimes_\Lambda \mathfrak{A} \to \mathfrak{A}$ is an isomorphism and $\mathfrak{A} \otimes_\Lambda \mathfrak{A}$ is a projective Λ^{op}-module.*

When Γ is semisimple then $\mathbf{I}_1 = \mathbf{I}_\infty$ and $\mathbf{P}_1 = \mathbf{P}_\infty$, so we get the following result.

Corollary 5.3. *Assume Γ is semisimple. Then the following conditions are equivalent*:

(a) *$\mathfrak{A}$ is a projective Λ-module.*

(b) *$\mathfrak{A}$ is a 2-idempotent.*

(c) *$\mathfrak{A}$ is a projective Λ^{op}-module.*

The equivalence of (a) and (c) has been proven by Dlab and Ringel in [DR1, Part II, Statement 7]. We observe that the conditions $\mathfrak{A}$ is a strong idempotent, $\mathbf{I}_1 = \mathbf{I}_\infty$ and $\mathbf{P}_1 = \mathbf{P}_\infty$ are independent, as shown in the following examples.

Let Λ be as in Example 3, §1. That is, Λ is the path algebra (over a field) of the quiver

$$\overset{1}{\bullet} \underset{\beta}{\overset{\alpha}{\rightleftarrows}} \overset{2}{\bullet}$$

modulo the ideal generated by the relation $\beta\alpha\beta = 0$. Then the ideal $\mathfrak{A}_1 = \tau_{P_1}(\Lambda)$ is a strong idempotent which is not projective. Therefore, $\mathbf{I}_{1,\mathfrak{A}_1} \neq \mathbf{I}_{\infty,\mathfrak{A}_1}$. Let now $\mathfrak{A}_2 = \tau_{P_2}(\Lambda)$ be the trace of the projective module P_2 associated to the vertex 2. Then $\mathfrak{A}_2$ is a projective Λ-module, so $\mathfrak{A}_2$ is a strong idempotent and $\mathbf{I}_{1,\mathfrak{A}_2} = \mathbf{I}_{\infty,\mathfrak{A}_2}$. However $\mathbf{P}_{1,\mathfrak{A}_2} \neq \mathbf{P}_{\infty,\mathfrak{A}_2}$. This example shows that being projective is not a symmetric condition. One can check that directly or observe that $\mathbf{P}_{1,\mathfrak{A}_2} \neq \mathbf{P}_{\infty,\mathfrak{A}_2}$ implies that $\mathfrak{A}_2$ is not a projective Λ^{op}-module.

On the other hand, any nonprojective idempotent ideal $\mathfrak{A} = \tau_P\Lambda$ such that Γ is semisimple satisfies $\mathbf{I}_1 = \mathbf{I}_\infty$ and $\mathbf{P}_1 = \mathbf{P}_\infty$, but $\mathfrak{A}$ is not a 2-idempotent (Corollary 5.3). Such ideals are easy to find. One of them is the ideal $\mathfrak{A} = \tau_{P_2}(\Lambda)$ in $\Lambda = T_3(k)/r^2$, where $T_3(k)$ is the 3×3 lower triangular matrix ring with coefficients in a field k, $P_2 = \Lambda e_{22}$ and e_{22} is the matrix with 1 in the entry 22, 0 elsewhere.

We now study relationships between the global dimensions of the rings Λ, $\Lambda/\mathfrak{A}$ and Γ. We have seen that if the idempotent ideal $\mathfrak{A}$ is a strong idempotent then $\operatorname{gl.dim} \Lambda/\mathfrak{A} \leq \operatorname{gl.dim} \Lambda$. And if $\mathbf{I}_1 = \mathbf{I}_\infty$ then $\operatorname{gl.dim} \Gamma \leq \operatorname{gl.dim} \Lambda$.

We will prove now the following result.

Theorem 5.4. *Let $\mathfrak{A}$ be an idempotent ideal. Then* $\operatorname{gl.dim} \Lambda \leq \operatorname{p.dim}_\Lambda \Lambda/\mathfrak{A} + \operatorname{gl.dim} \Lambda/\mathfrak{A} + \operatorname{gl.dim} \Gamma + 1$.

We observe that when $\mathfrak{A}$ is projective and Γ is semisimple the proposition states that $\operatorname{gl.dim} \Lambda \leq \operatorname{gl.dim} \Lambda/\mathfrak{A} + 2$, an inequality proven by Dlab and Ringel in [DR1, Part 2, Statement 5].

To prove this proposition and the next we consider the full subcategory $\mathbf{T}$ of $\operatorname{mod} \Lambda$ consisting of the modules T such that $\operatorname{Ext}^i_\Lambda(\Lambda/\mathfrak{A}, T) = 0$ for all

$i \geq 1$. Then $\mathbf{I}_\infty \subseteq \mathbf{T}$, and a Λ-module T in $\mathbf{I}_\infty$ is in $\mathbf{T}$ if and only if $\tau_{\Lambda/\mathfrak{A}} T = \operatorname{Hom}_\Lambda(\Lambda/\mathfrak{A}, T) = 0$. Given a Λ-module X we denote by $\Omega^{-i}(X)$ the image of the map d_i in a minimal injective coresolution $0 \to X \to I_0 \xrightarrow{d_1} I_1 \xrightarrow{d_2} \cdots$ of X and we write $\Omega^0(X) = X$. We need the following lemma.

Lemma 5.5. *Let* $\mathbf{T}$ *be as above. Then*

(i) $\operatorname{p.dim}_\Lambda \Lambda/\mathfrak{A} = r$ *if and only if* Ω^{-r} $(\operatorname{mod} \Lambda) \subseteq \mathbf{T}$.

(ii) *If* $\operatorname{gl.dim} \Lambda/\mathfrak{A} = s < \infty$ *then* $\Omega^{-(s+1)}(\mathbf{T}) \subseteq \mathbf{I}_\infty$.

Proof. (i) Follows from the isomorphism

$$\operatorname{Ext}_\Lambda^{i+r}(\Lambda/\mathfrak{A}, X) \simeq \operatorname{Ext}_\Lambda^i(\Lambda/\mathfrak{A}, \Omega^{-r}X),$$

for all X in $\operatorname{mod} \Lambda$, $i \geq 1$.

(ii) Let T in $\mathbf{T}$. Clearly $\Omega^{-i}(T)$ is in $\mathbf{T}$ for all $i \geq 1$. So we only have to prove that $\tau_{\Lambda/\mathfrak{A}}(\Omega^{-(s+1)}(T)) = 0$. Let $0 \to T \to I_0 \to I_1 \to \cdots$ be a minimal injective coresolution of T. Since T is in $\mathbf{T}$ we have $0 \to \tau_{\Lambda/\mathfrak{A}} T \to \tau_{\Lambda/\mathfrak{A}} I_0 \to \tau_{\Lambda/\mathfrak{A}} I_1 \to \cdots$ is a minimal injective coresolution of $\tau_{\Lambda/\mathfrak{A}} T$ (Proposition 1.1). Since $\operatorname{gl.dim} \Lambda/\mathfrak{A} = s$ it follows that $\tau_{\Lambda/\mathfrak{A}} I_k = 0$ for $k > s$. So $\tau_{\Lambda/\mathfrak{A}}(\Omega^{-(s+1)}(T)) \subseteq \tau_{\Lambda/\mathfrak{A}}(I_{s+1}) = 0$, as we want to prove.

We prove now the theorem. Let $r = \operatorname{p.dim}_\Lambda \Lambda/\mathfrak{A}$, $s = \operatorname{gl.dim} \Lambda/\mathfrak{A}$, $t = \operatorname{gl.dim} \Gamma$. Let X be in $\operatorname{mod} \Lambda$. Using first (i) and then (ii) of Lemma 5.5, we conclude that $\Omega^{-(r+s+1)}(X)$ is in $\mathbf{I}_\infty$. Then

$$\operatorname{inj.dim}_\Lambda \Omega^{-(r+s+1)}(X) = \operatorname{inj.dim}_\Gamma(P, \Omega^{-(r+s+1)}(X)),$$

by Corollary 3.3. This number is at most $t = \operatorname{gl.dim} \Gamma$. So $\operatorname{inj.dim} X \leq r + s + t + 1$, and the proof of the theorem is complete. □

When $\mathfrak{A}$ is a projective Λ-module we obtain the following result

Corollary 5.6. *Assume the idempotent ideal* $\mathfrak{A}$ *is a projective* Λ*-module. Then the following are equivalent*

(a) $\operatorname{gl.dim} \Lambda < \infty$,

(b) $\operatorname{gl.dim} \Lambda/\mathfrak{A} < \infty$ *and* $\operatorname{gl.dim} \Gamma < \infty$.

Proof. (a) $\Rightarrow$ (b) follows from the fact that $\mathfrak{A}$ is projective if and only if $\mathfrak{A}$ is a strong idempotent and $\mathbf{I}_1 = \mathbf{I}_\infty$, proven in Proposition 5.1. (b) $\Rightarrow$ (a) is an immediate consequence of the proposition.

When the ideal $\mathfrak{A}$ is a strong idempotent we can prove a stronger inequality.

Proposition 5.7. *Let* $\mathfrak{A}$ *be a strong idempotent ideal. Then*

$$\operatorname{gl.dim} \Lambda \leq \operatorname{p.dim}_\Lambda \Lambda/\mathfrak{A} + \max\{\operatorname{p.dim}_{\Lambda^{\mathrm{op}}} \Lambda/\mathfrak{A} + \operatorname{gl.dim} \Lambda/\mathfrak{A}, \operatorname{gl.dim} \Gamma\}.$$

Proof. We start by stating the following lemma.

Lemma 5.8. *Let* $\mathfrak{A}$ *be a two-sided ideal and let* X *be a* $\Lambda/\mathfrak{A}$*-module. Then*

(i) $\operatorname{p.dim}_\Lambda X \leq \operatorname{p.dim}_{\Lambda/\mathfrak{A}} X + \operatorname{p.dim}_\Lambda \Lambda/\mathfrak{A}$,

(ii) $\operatorname{inj.dim}_\Lambda X \leq \operatorname{inj.dim}_{\Lambda/\mathfrak{A}} X + \operatorname{p.dim}_{\Lambda^{\mathrm{op}}} \Lambda/\mathfrak{A}$.

Proof. (i) follows by induction on $n = \operatorname{p.dim}_{\Lambda/\mathfrak{A}} X$, and (ii) follows from (i) by duality. □

Now we prove the proposition. We assume that $\mathfrak{A}$ is a strong idempotent ideal, and that X is a Λ-module. We may assume that $r = \operatorname{p.dim}_\Lambda \Lambda/\mathfrak{A} < \infty$. Then $T = \Omega^{-r}X$ is in $\mathbf{T}$. We consider the exact sequence $0 \to \tau_{\Lambda/\mathfrak{A}} T \to$

$T \to T/\tau_{\Lambda/\mathfrak{A}}T \to 0$. Since T is in $\mathbf{T}$, $\tau_{\Lambda/\mathfrak{A}}T$ is a $\Lambda/\mathfrak{A}$-module and $\mathfrak{A}$ is a strong idempotent, it follows that $T/\tau_{\Lambda/\mathfrak{A}}T$ is in $\mathbf{I}_\infty$. Thus $\text{inj.dim}\, T/\tau_{\Lambda/\mathfrak{A}}T \leq \text{gl.dim}\,\Gamma$. On the other hand we get by (ii) of Lemma 5.8 that $\text{inj.dim}_\Lambda(\tau_{\Lambda/\mathfrak{A}}T) \leq \text{gl.dim}\,\Lambda/\mathfrak{A} + \text{p.dim}_{\Lambda^{\text{op}}}\Lambda/\mathfrak{A}$. Combining these results we get

$$\text{inj.dim}_\Lambda T \leq \max\{\text{gl.dim}\,\Lambda/\mathfrak{A} + \text{p.dim}_{\Lambda^{\text{op}}}\Lambda/\mathfrak{A}\,,\ \text{gl.dim}\,\Gamma\}\,.$$

This inequality proves the proposition because $T = \Omega^{-r}X$, so $\text{inj.dim}\,X = r + \text{inj.dim}\,T$. □

We observe now that if $\mathfrak{A}$ is a strong idempotent then $\mathfrak{A}$ is in $\mathbf{P}_\infty$, so $\text{p.dim}_\Lambda\mathfrak{A} = \text{p.dim}_\Gamma(P\,,\mathfrak{A}) \leq \text{gl.dim}\,\Gamma$. Since being a strong idempotent is a symmetric condition we also have that $\text{p.dim}_{\Lambda^{\text{op}}}\mathfrak{A} \leq \text{gl.dim}\,\Gamma$. Combining the inequality in Proposition 5.7 with $\text{p.dim}_{\Lambda^{\text{op}}}\Lambda/\mathfrak{A} \leq \text{gl.dim}\,\Gamma + 1$, we obtain the inequality of Theorem 5.4.

As an immediate consequence of Proposition 5.7 we obtain the following corollary.

Corollary 5.9. *Assume that the idempotent ideal* $\mathfrak{A}$ *is projective in* $\text{mod}\,\Lambda$ *and in* $\text{mod}\,\Lambda^{\text{op}}$. *Then*

$$\text{gl.dim}\,\Lambda \leq \max\{\text{gl.dim}\,\Lambda/\mathfrak{A} + 2\,,\ \text{gl.dim}\,\Gamma + 1\}\,.$$

6. Quasihereditary algebras

In this section we apply some of our previous results to quasihereditary algebras. First we show that being quasihereditary is invariant under Morita equivalence. Second, we study the projective resolutions of the ideals in a heredity chain. We give another proof of the fact that hereditary artin algebras can be characterized as quasihereditary artin algebras for which each chain of idempotent ideals can be refined to a heredity chain, which was shown by Dlab and Ringel in [DR1]. Finally we give a characterization of quasihereditary algebras using l-hereditary algebras.

We recall now the definition of quasihereditary algebras [CPS]: An artin algebra Λ is quasihereditary if there is a chain of idempotent ideals of Λ, $0 = \mathfrak{A}_0 \subset \cdots \subset \mathfrak{A}_n = \Lambda$ satisfying the following conditions

(QH1) $\mathfrak{A}_i/\mathfrak{A}_{i-1}$ is projective in $\text{mod}(\Lambda/\mathfrak{A}_{i-1})$ for all $i = 1, \ldots, n$.

(QH2) $\Gamma_i = \text{End}_{\Lambda/\mathfrak{A}_{i-1}}(Q_i/\mathfrak{A}_{i-1}Q_i)^{\text{op}}$ is semisimple for all $i = 1, \ldots, n$, where Q_i is a projective Λ-module such that $\mathfrak{A}_i = \tau_{Q_i}(\Lambda)$.

Such chain is called an heredity chain.

Let $\mathfrak{A} = \tau_P\Lambda$ be an idempotent ideal of Λ. We proved in Corollary 5.3 that if $\Gamma = \text{End}(P)^{\text{op}}$ is a semisimple ring, then $\mathfrak{A}$ is projective in $\text{mod}\,\Lambda$ if and only if $\mathfrak{A}$ is 2-idempotent. So the conditions (QH1) in the definition of quasihereditary algebra can be replaced by either of the two following conditions

(QH1′) $\mathfrak{A}_i$ is 2-idempotent, for all $i = 1, \ldots, n$.

(QH1″) $\mathfrak{A}_i$ is strong idempotent, for all $i = 1, \ldots, n$.

We use this observation to prove that being quasihereditary is a Morita invariance, result which was proven in [CPS] using highest weight categories. As we noticed in the introduction giving an idempotent ideal $\mathfrak{A}$ of Λ is equivalent to giving a Serre subcategory of $\text{mod}\,\Lambda$. When $\mathfrak{A} = \tau_P(\Lambda)$ is an idempotent ideal of Λ then $S = \operatorname{mod}\Lambda/\mathfrak{A}$ is a Serre subcategory of $\operatorname{mod}\Lambda$ and $\operatorname{mod}\Lambda/\mathfrak{A} \to \operatorname{mod}\Lambda \to \operatorname{mod}\Gamma$ is an exact sequence of categories, where $\Gamma = \operatorname{End}(P)^{\mathrm{op}}$.

Since the properties of being strong idempotent and of being semisimple are Morita invariances it follows that being quasihereditary is also a Morita invariance.

In the following proposition we give a characterization of the chains $0 = \mathfrak{A}_0 \subset \mathfrak{A}_1 \subset \cdots \subset \mathfrak{A}_n$ of idempotent ideals of Λ such that $\mathfrak{A}_i$ is projective for $i = 1, \ldots, n$, in terms of the projective presentations of the ideals $\mathfrak{A}_i$ in the chain.

Proposition 6.1. *Let $0 = \mathfrak{A}_0 \subset \mathfrak{A}_1 \subset \cdots \subset \mathfrak{A}_n$ be a chain of idempotent ideals in Λ. Then there exist projective Λ-modules $P_1, \ldots, P_n$ such that $\mathfrak{A}_k = \tau_{P_1 \amalg \cdots \amalg P_k}(\Lambda)$ for all $k = 1, \ldots, n$. Moreover, the following conditions are equivalent:*

(a) *$\mathfrak{A}_k/\mathfrak{A}_{k-1}$ is projective for all $k = 1, \ldots, n$.*

(b) *$\mathfrak{A}_k$ has a projective resolution $\cdots \to P_{k,i} \to \cdots \to P_{k,0} \to \mathfrak{A}_k \to 0$, with $P_{k,i}$ in $\operatorname{add}(P_1 \amalg \cdots \amalg P_{k-i})$ for all $i < k$, $P_{k,i} = 0$ for $i \geq k$, for all $k = 1, \ldots, n$.*

(c) *$\mathfrak{A}_k$ has a projective presentation $P_{k,1} \to P_{k,0} \to \mathfrak{A}_k \to 0$, with $P_{k,1}$ in $\operatorname{add}(P_1 \amalg \cdots \amalg P_{k-1})$, for all $k = 1, \ldots, n$.*

Proof. The first statement follows from the following observation. If P, Q are projective Λ-modules and $\tau_P(\Lambda) \subset \tau_Q(\Lambda)$, then P is in $\operatorname{add} Q$.

To prove the equivalence of (a) and (b) we introduce the following notation. For $k \geq 1$, let $\mathbf{R}_k$ be the full subcategory of $\operatorname{mod}\Lambda$ consisting of the Λ-modules X having a projective resolution $\cdots \to Q_i \to \cdots \to Q_0 \to X \to 0$, with Q_i in $\operatorname{add}(P_1 \amalg \cdots \amalg P_{k-i})$ if $i \leq k-1$, $Q_i = 0$ otherwise. We observe that $\mathbf{R}_k$ is closed under extensions.

We prove that (a) $\Rightarrow$ (b) by induction on n It is clearly true for $n = 1$. So let $n > 1$ and consider the sequence $0 \to \mathfrak{A}_{n-1} \to \mathfrak{A}_n \to \mathfrak{A}_n/\mathfrak{A}_{n-1} \to 0$. To prove (b) we have to prove that $\mathfrak{A}_n$ is in $\mathbf{R}_n$. By the induction hypothesis we know that $\mathfrak{A}_{n-1}$ is in $\mathbf{R}_{n-1} \subset \mathbf{R}_n$. Since $\mathbf{R}_n$ is closed under extensions it is enough to see that $\mathfrak{A}_n/\mathfrak{A}_{n-1}$ is in $\mathbf{R}_n$. We know by (a) that $\mathfrak{A}_n/\mathfrak{A}_{n-1}$ is a projective $A/\mathfrak{A}_{n-1}$-module. So $\mathfrak{A}_n/\mathfrak{A}_{n-1} \simeq Q_n/\mathfrak{A}_{n-1}Q_n$, where Q_n is the projective cover of $\mathfrak{A}_n/\mathfrak{A}_{n-1}$ in $\operatorname{mod}\Lambda$. Since Q_n is in $\operatorname{add}\Lambda$, then $\mathfrak{A}_{n-1}Q_n$ is in $\operatorname{add}\mathfrak{A}_{n-1} \subseteq \mathbf{R}_n$. Thus from the exact sequence $0 \to \mathfrak{A}_{n-1}Q_n \to Q_n \to \mathfrak{A}_n/\mathfrak{A}_{n-1} \to 0$ with Q_n in $\operatorname{add} P_0(\mathfrak{A}_n) = \operatorname{add}(P_1 \amalg \cdots \amalg P_n)$ we get that $\mathfrak{A}_n/\mathfrak{A}_{n-1}$ is in $\mathbf{R}_n$. This finishes the proof of (a) $\Rightarrow$ (b).

(b) $\Rightarrow$ (c) is clear. So we prove now that (c) $\Rightarrow$ (a). Let $P_{k,1} \to P_{k,0} \to \mathfrak{A}_k \to 0$ be a presentation of $\mathfrak{A}_k$ with $P_{k,1}$ in $\operatorname{add}(P_1 \amalg \cdots \amalg P_{k-1})$, $1 \leq k \leq n$. We consider the exact sequence $0 \to K_k \to P_{k,0} \to \mathfrak{A}_k \to 0$. Tensoring with $\Lambda/\mathfrak{A}_{k-1}$ we get an exact sequence

$$K_k/\mathfrak{A}_{k-1}K_k \to P_{0,k}/\mathfrak{A}_{k-1}P_{k,0} \xrightarrow{\overline{\varphi}_k} \mathfrak{A}_k/\mathfrak{A}_{k-1} \to 0.$$

Then $P_0(K_k) \simeq P_{k,1}$ is in $\operatorname{add}(P_1, \ldots, P_{k-1})$, by (c). That is, $\mathfrak{A}_{k-1}K_k = K_k$. So $\overline{\varphi}_k$ is an isomorphism and $\mathfrak{A}_k/\mathfrak{A}_{k-1}$ is projective in $\operatorname{mod}(\Lambda/\mathfrak{A}_{k-1})$. □

The following corollary applies to heredity chains in a quasihereditary algebra.

Corollary 6.2. *Let* $0 = \mathfrak{A}_0 \subset \cdots \subset \mathfrak{A}_n$ *be a chain of idempotent ideals in* Λ *such that* $\mathfrak{A}_k/\mathfrak{A}_{k-1}$ *is a projective* $\Lambda/\mathfrak{A}_{k-1}$ *module for all* $k = 1, \ldots, n$. *Then* $\operatorname{p.dim}\mathfrak{A}_k \leq k-1$, $\mathfrak{A}_k$ *is strong idempotent, and* $\mathbf{I}_{k,\mathfrak{A}_k} = \mathbf{I}_{\infty,\mathfrak{A}_k}$, *for* $k = 1, \ldots, n$.

Proof. The equivalence of (a) and (b) in Proposition 6.1 implies that $\operatorname{p.dim}_\Lambda\mathfrak{A}_k \leq k-1$ and that $\mathfrak{A}_k$ has a projective resolution in $\operatorname{add}(P_0(\mathfrak{A}_k))$. By Theorem 2.1 we know that this implies that $\mathfrak{A}_k$ is a strong idempotent. Then $\mathfrak{A}_k$ is a 2-idempotent, so $\mathfrak{A}_k \otimes_\Lambda \mathfrak{A}_k \simeq \mathfrak{A}_k$, as proven in Proposition 1.4. Then $\operatorname{p.dim}\mathfrak{A}_k \otimes_\Lambda \mathfrak{A}_k \leq k$, so that $\mathbf{I}_{k,\mathfrak{A}_k} = \mathbf{I}_{\infty,\mathfrak{A}_k}$ (Proposition 4.11). □

As a consequence of the above proposition we prove the following result of Dlab and Ringel.

Proposition 6.3 [DR1, Part 1, Theorem 1]. *An artin algebra* Λ *is hereditary if and only if any chain of idempotent ideals in* Λ *can be refined to a heredity chain.*

Proof. Assume that any chain of idempotent ideals can be refined to a heredity chain. We will prove that Λ is hereditary.

Let Q be an indecomposable projective Λ-module. To prove that Λ is hereditary we prove that rQ is projective. Let P be the sum of all the indecomposable nonisomorphic projective modules not isomorphic to Q. Then $\mathfrak{A} = \tau_P(\Lambda)$ is a maximal proper idempotent ideal of Λ, so by hypothesis it is part of a heredity chain. By the maximality of $\mathfrak{A}$ it follows that it must be the last proper ideal in the chain, so that the ring $\Lambda/\mathfrak{A}$ is semisimple. Then $Q/\mathfrak{A}Q$ is a simple Λ-module, so $rQ = \mathfrak{A}Q$.

The module $\mathfrak{A}Q$ is in $\operatorname{add}\mathfrak{A}$ because Q is in $\operatorname{add}\Lambda$. Therefore to prove that $rQ = \mathfrak{A}Q$ is projective it is enough to prove that $\mathfrak{A}$ is projective. Let $P_1 \to P_0 \to \mathfrak{A} \to 0$ be a minimal projective presentation of $\mathfrak{A}$ in $\operatorname{mod}\Lambda$. We will prove that $P_1 = 0$. Let Q_1 be an indecomposable direct summand of P. We write $P = Q_1 \amalg P'$. The chain of idempotent ideals $\tau_{P'}(\Lambda) \subset \tau_P(\Lambda)$ can be refined to a heredity chain $0 = \mathfrak{A}_0 \subset \cdots \subset \mathfrak{A}_n = \Lambda$. Let k be such that $\mathfrak{A}_k = \tau_P(\Lambda) = \mathfrak{A}$. Since Q_1 is indecomposable we must have $\mathfrak{A}_{k-1} = \tau_{P'}(\Lambda)$. We can apply now Proposition 6.1 to the chain $\mathfrak{A}_0 \subset \cdots \subset \mathfrak{A}_k = \mathfrak{A}$ and the given projective presentation $P_1 \to P_0 \to \mathfrak{A} \to 0$ of $\mathfrak{A}$. We conclude that P_1 is in $\operatorname{add}P' \subseteq \operatorname{add}P$ and therefore has no summand isomorphic to Q_1. Therefore we have proven that no indecomposable summand Q_1 of P is in $\operatorname{add}P_1 \subseteq \operatorname{add}P$. So $P_1 = 0$. This finishes the proof that Λ is hereditary.

So we have proven that if any chain of idempotent ideals can be refined to a heredity chain then Λ is hereditary. The converse follows from the fact that any chain of idempotent ideals can be refined to a chain $0 \subset \tau_{P_1} \subset \cdots \subset \tau_{P_1 \amalg \cdots \amalg P_n} \subset \Lambda$, where all the projective Λ-modules P_i are indecomposable. □

We know by [BF, Proposition 1.6], that l-hereditary algebras are quasi-hereditary. We give now a characterization of quasihereditary algebras using l-hereditary algebras.

We recall that an artin algebra Λ is l-hereditary if the maps between indecomposable projective modules are either zero or monomorphisms. If Λ is an l-hereditary algebra and P is a projective Λ-module, then the algebra $\Gamma = \operatorname{End}_\Lambda(P)^{\mathrm{op}}$ is also l-hereditary.

Proposition 6.4. *Let Λ be an artin algebra with a chain of idempotent ideals $0 = \mathfrak{A}_0 \subset \mathfrak{A}_1 \subset \cdots \subset \mathfrak{A}_n = \Lambda$ such that (1) $\mathfrak{A}_i/\mathfrak{A}_{i-1}$ is a projective $\Lambda/\mathfrak{A}_{i-1}$ module for $i = 1, \dots, n$ and (2) $\operatorname{End}_{\Lambda/\mathfrak{A}_{i-1}}(Q_i/\mathfrak{A}_{i-1}Q_i)$ is l-hereditary for $i = 1, \dots, n$, where Q_i is a projective Λ-module such that $\tau_{Q_i}(\Lambda) = \mathfrak{A}_i$. Then Λ is quasihereditary.*

Proof. The proposition follows by induction from the following lemma.

Lemma 6.5. *Let P be a projective Λ-module such that $\mathfrak{A} = \tau_P(\Lambda)$ is a projective Λ-module and $\Gamma = \operatorname{End}_\Lambda(P)^{\mathrm{op}}$ is an l-hereditary algebra. If Γ is not semisimple, then there exists a direct summand P_1 of P such that $\mathfrak{A}_1 = \tau_{P_1}(\Lambda)$ is a projective Λ-module, $\operatorname{End}_\Lambda(P_1)$ is semisimple and $\operatorname{End}_{\Lambda/\mathfrak{A}_1}(P/\mathfrak{A}_1P)$ is an l-hereditary algebra.*

Proof. Let P_1 be an indecomposable summand of P such that the Γ-module (P, P_1) has maximum length. Let $P_1, \dots, P_n$ be the nonisomorphic indecomposable summands of P. Since Γ is l-hereditary it follows that

$$((P, P_1), (P, P_i)) = 0 \quad \text{for } i = 2, \dots, r.$$

We will see that P_1 has the desired properties.

Since

$$\operatorname{End}_\Lambda(P_1) \simeq \operatorname{End}_\Gamma((P, P_1))$$

and Γ is l-hereditary it follows that $\operatorname{End}_\Lambda(P_1)$ is a division ring. On the other hand, $\operatorname{Hom}_\Lambda(P_1, P_i) = \operatorname{Hom}_\Gamma((P, P_1), (P, P_i)) = 0$ for $i = 2, \dots, n$, so $\tau_{P_1}(P_i) = 0$ if $i \geq 2$. Thus $\tau_{P_1}(X)$ is in $\operatorname{add} P_1$, for any X in $\operatorname{add} P$. Now, $\tau_{P_1}(\Lambda) = \tau_{P_1}(\tau_P(\Lambda))$, and $\tau_P(\Lambda)$ is in $\operatorname{add} P$ since it is projective by hypothesis. So $\tau_{P_1}(\Lambda)$ is in $\operatorname{add} P_1$ and is thus projective.

Since $\mathfrak{A}_1P = \tau_{P_1}(P)$ is in $\operatorname{add} P_1$ it follows that $P/\mathfrak{A}_1P$ is in

$$\operatorname{add}(P_1 \amalg \cdots \amalg P_n) \subseteq \operatorname{add} P.$$

Therefore $\operatorname{End}_\Lambda(P/\mathfrak{A}_1P) = \operatorname{End}_\Gamma((P, P/\mathfrak{A}_1P))$ and $(P, P/\mathfrak{A}_1P)$ is in $\operatorname{add}\Gamma$. Since Γ is l-hereditary it follows then that $\operatorname{End}_\Lambda(P/\mathfrak{A}_1P)$ is l-hereditary. This ends the proof of the lemma. □

Using this characterization of quasihereditary algebras we obtain a bound for the global dimension of a quasihereditary algebra, different from the one given in [DR1, Statement 9].

Corollary 6.6. *Let Λ be a quasihereditary algebra. Let $0 = \mathfrak{A}_0 \subset \cdots \subset \mathfrak{A}_n = \Lambda$ with $\mathfrak{A}_i = \tau_{P_i}(\Lambda)$ be a chain of idempotent ideals such that $\mathfrak{A}_i/\mathfrak{A}_{i-1}$ is projective in $\operatorname{mod}(\Lambda/\mathfrak{A}_{i-1})$ and $\operatorname{End}_{\Lambda/\mathfrak{A}_i}(P_i/\mathfrak{A}_{i-1}P_i)$ is hereditary for $i = 1, \dots, n$. Then $\operatorname{gl.dim}\Lambda \leq 3n - 2$.*

Proof. We prove the corollary by induction on n. If $n = 1$ then Λ is hereditary and the formula holds. Let $n > 1$. We apply Theorem 5.4 to the idempotent ideal $\mathfrak{A} = \mathfrak{A}_1 = \tau_{P_1}(\Lambda)$ and we obtain that

$$\operatorname{gl.dim}\Lambda \leq \operatorname{p.dim}_\Lambda \Lambda/\mathfrak{A}_1 + \operatorname{gl.dim}\Lambda/\mathfrak{A}_1 + \operatorname{gl.dim}\Gamma + 1,$$

where $\Gamma = \operatorname{End}_\Lambda(P_1)$ is hereditary by hypothesis. The induction hypothesis applied to the chain $0 \subset \mathfrak{A}_2/\mathfrak{A}_1 \subset \cdots \subset \mathfrak{A}_n/\mathfrak{A}_1 = \Lambda/\mathfrak{A}_1$ in $\operatorname{mod}(\Lambda/\mathfrak{A}_1)$ implies that $\operatorname{gl.dim}\Lambda/\mathfrak{A}_1 \leq 3(n-1)-2$. Thus $\operatorname{gl.dim}\Lambda \leq 3(n-1)-2+3 = 3n-2$. □

7. A CLASS OF QUASIHEREDITARY ALGEBRAS

We define in this section a class of artin algebras which contains several well-known classes of algebras and prove that the algebras in that class are quasihereditary.

Definition 7.1. We say that an artin algebra Λ satisfies the descending Loewy length condition on projective resolutions if for every Λ-module M a minimal projective resolution $\to P_i(M) \to \cdots \to P_0(M) \to M \to 0$ satisfies $LL(P_{i+1}(M)) < LL(P_i(M))$ for all $i \geq 1$. Here LL denotes the Loewy length.

The following are examples of classes of algebras satisfying the descending Loewy length condition on projective resolutions:

(1) l-hereditary algebras, which were defined in the previous section.

(2) Artin algebras of global dimension two.

(3) The endomorphism algebra $\Gamma = \mathrm{End}_\Lambda(\Lambda/r \amalg \cdots \amalg \Lambda/r^n)$, where Λ is an artin algebra and n is the index of nilpotency of the radical r of Λ [A, Proof of Proposition 10.2].

By proving that algebras satisfying the descending Loewy length condition on projective resolutions are quasihereditary, we give, in particular, a unified proof of the results, proven by V. Dlab and C. M. Ringel in [DR1, DR2], that the algebras of global dimension 2 and the endomorphism algebra Γ of example (3) are quasihereditary.

We start by proving the following lemma.

Lemma 7.2. *Let Λ be an artin algebra satisfying the descending Loewy length condition on projective resolutions. Let X be a submodule of a projective Λ-module. Then $LL(P_1(X)) < LL(P_0(X))$.*

Proof. Follows easily from the definitions.

Theorem 7.3. *Let Λ be an artin algebra satisfying the descending Loewy length condition on projective resolutions. Then Λ is quasihereditary.*

Proof. To prove this theorem we will use the characterization of quasihereditary algebras given in Proposition 6.4 which states: Λ is quasihereditary if and only if there is a chain of idempotent ideals $0 = \mathfrak{A}_0 \subset \mathfrak{A}_1 \subset \cdots \subset \mathfrak{A}_n = \Lambda$ such that (1) $\mathfrak{A}_i/\mathfrak{A}_{i-1}$ is a projective $\Lambda/\mathfrak{A}_{i-1}$-module for $i = 1, \ldots, n$ and (2) $\mathrm{End}_{\Lambda/\mathfrak{A}_{n-1}}(Q_i/\mathfrak{A}_{i-1}Q_i)$ is l-hereditary for $i = 1, \ldots, n$. Here $\mathfrak{A}_i = \tau_{Q_i}(\Lambda)$, and all Q_i are projective Λ-modules.

Let $i_1 < \cdots < i_t$ be the Loewy lengths of the indecomposable projective Λ-modules. Let Q_k be the sum of the nonisomorphic indecomposable projective Λ-modules of Loewy length at most i_k, and $\mathfrak{A}_k = \tau_{Q_k}(\Lambda)$ for $k = 1, \ldots, t$, $\mathfrak{A}_0 = 0$. Since $\mathfrak{A}_k$ is a submodule of Λ it follows from Lemma 7.2 that $LL(P_1(\mathfrak{A}_k)) < LL(P_0(\mathfrak{A}_k)) \leq i_k$. So $P_1(\mathfrak{A}_k)$ is in $\mathrm{add}\, Q_{k-1}$. This proves that $\mathfrak{A}_k/\mathfrak{A}_{k-1}$ is projective in $\mathrm{mod}(\Lambda/\mathfrak{A}_{k-1})$.

We will show in the following lemma that $\mathrm{End}_{\Lambda/\mathfrak{A}_{k-1}}(Q_k/\mathfrak{A}_{k-1}Q_k)$ is l-hereditary for all k. This will end the proof of the theorem. □

Lemma 7.4. *Let Λ be an artin algebra satisfying the descending Loewy length condition on projective resolutions. Let r be the Loewy length of an indecomposable projective Λ-module. Let Q be the sum of the indecomposable nonisomorphic projective Λ-modules of Loewy length r. Let P be the sum of the* *indecomposable nonisomorphic projective Λ-modules of Loewy length smaller than r, and $\mathfrak{A} = \tau_P(\Lambda)$. Then $\operatorname{End}_{\Lambda/\mathfrak{A}}(Q/\mathfrak{A}Q)$ is an l-hereditary algebra.*

Proof. Let $\overline{f}\colon Q_1/\mathfrak{A}Q_1 \to Q_2/\mathfrak{A}Q_2$ be a $\Lambda/\mathfrak{A}$-morphism. Then $\overline{f}$ is induced by a Λ-morphism $f\colon Q_1 \to Q_2$. Let $Q_1 \xrightarrow{p} \operatorname{Im} f \xrightarrow{j} Q_2$ be the factorization of f through the image of f. Then $\overline{f} = \overline{j} \cdot \overline{p}$ where $p\colon Q_1/\mathfrak{A}Q_1 \to \operatorname{Im} f/\mathfrak{A}\operatorname{Im} f$ and $\overline{j}\colon \operatorname{Im} f/\mathfrak{A}\operatorname{Im} f \to Q_2/\mathfrak{A}Q_2$ are the maps induced by p and j. We will show first that if Q_1 is in $\operatorname{add} Q$ then $\overline{p}$ is an isomorphism. And we will show that if Q_2 is in $\operatorname{add} Q$ then $\overline{j}$ is either zero or a monomorphism. Thus if we assume that both Q_1 and Q_2 are in $\operatorname{add} Q$ this shows that $\overline{f} = \overline{j} \cdot \overline{p}$ is either zero or a monomorphism, proving the lemma.

From the exact sequence $0 \to \operatorname{Ker} f \to Q_1 \xrightarrow{p} \operatorname{Im} f \to 0$ we get an exact sequence $\operatorname{Ker} f/\mathfrak{A}\operatorname{Ker} f \to Q_1/\mathfrak{A}Q_1 \xrightarrow{\overline{p}} \operatorname{Im} f/\mathfrak{A}\operatorname{Im} f \to 0$. Since $\operatorname{Im} f$ is a submodule of the projective module Q_2 it follows from the above lemma that $LL(P_0(\operatorname{Ker} f)) = LL(P_1(\operatorname{Im} f)) < LL(P_0(\operatorname{Im} f)) = LL(Q_1)$. If we assume that Q_1 is in $\operatorname{add} Q$ then $LL(Q_1) = r$. So $LL(P_0(\operatorname{Ker} f)) < r$. That is, $P_0(\operatorname{Ker} f)$ is in $\operatorname{add} P$. Hence $\mathfrak{A}\operatorname{Ker} f = \operatorname{Ker} f$. This proves that $\overline{p}$ is an isomorphism.

To prove that $\overline{j}$ is either zero or a monomorphism we consider the following exact sequence:

$$\begin{aligned} 0 &\to (\operatorname{Im} f \cap \mathfrak{A}Q_2)/\mathfrak{A}\operatorname{Im} f \to \operatorname{Im} f/\mathfrak{A}\operatorname{Im} f \\ &\to (\operatorname{Im} f + \mathfrak{A}Q_2)/\mathfrak{A}Q_2 \to 0. \end{aligned}$$

The last module is precisely the image of $\overline{j}\colon \operatorname{Im} f/\mathfrak{A}\operatorname{Im} f \to Q_2/\mathfrak{A}Q_2$. It follows that $\overline{j} = 0$ if $(\operatorname{Im} f + \mathfrak{A}Q_2) \subset \mathfrak{A}(\operatorname{Im} f + \mathfrak{A}Q_2)$, which is the same as $P_0(\operatorname{Im} f + \mathfrak{A}Q_2)$ is in $\operatorname{add} P$. Similarly, $\overline{j}$ is a monomorphism if $\operatorname{Im} f \cap \mathfrak{A}Q_2 = \mathfrak{A}(\operatorname{Im} f \cap \mathfrak{A}Q_2)$ which is the same as $P_0(\operatorname{Im} f \cap \mathfrak{A}Q_2)$ is in $\operatorname{add} P$.

We now show that either $P_0(\operatorname{Im} f + \mathfrak{A}Q_2)$ is in $\operatorname{add} P$ or $P_0(\operatorname{Im} f \cap \mathfrak{A}Q_2)$ is in $\operatorname{add} P$. This is a consequence of the following result applied to $X = \tau_P Q_2$, $Y = \operatorname{Im} f$.

Claim. Let Q_2 be in $\operatorname{add} Q$ and let X and Y be submodules of Q_2. Suppose $P_0(X)$ is in $\operatorname{add} P$ and $P_0(Y)$ is indecomposable and in $\operatorname{add} Q$. Then either $P_0(X + Y)$ is in $\operatorname{add} P$ or $P_0(X \cap Y)$ is in $\operatorname{add} P$.

To prove this we assume that $P_0(X + Y)$ is not in $\operatorname{add} P$, and consider the diagram:

$$\begin{array}{ccccccccc} 0 & \longrightarrow & Z & \longrightarrow & P_0(X \amalg Y) & \longrightarrow & X + Y & \longrightarrow & 0 \\ & & \downarrow & & \downarrow & & \| & & \\ 0 & \longrightarrow & X \cap Y & \longrightarrow & X \amalg Y & \longrightarrow & X + Y & \longrightarrow & 0 \\ & & \downarrow & & \downarrow & & & & \\ & & 0 & & 0 & & & & \end{array}$$

To prove that $P_0(X \cap Y)$ is in $\operatorname{add} P$ it is enough to show that $P_0(Z)$ is in $\operatorname{add} P$. Consider now the diagram

$$\begin{array}{ccccccccc}
 & & 0 & & 0 & & & & \\
 & & \downarrow & & \downarrow & & & & \\
 & & L & = & L & & & & \\
 & & \downarrow & & \downarrow & & & & \\
0 & \longrightarrow & Z & \longrightarrow & P_0(X \amalg Y) & \longrightarrow & X+Y & \longrightarrow & 0 \\
 & & \downarrow & & \downarrow & & \| & & \\
0 & \longrightarrow & \Omega_1(X+Y) & \longrightarrow & P_0(X+Y) & \longrightarrow & X+Y & \longrightarrow & 0 \\
 & & \downarrow & & \downarrow & & & & \\
 & & 0 & & 0 & & & &
\end{array}$$

The second vertical column splits. Since $P_0(X+Y)$ is not in $\operatorname{add} P$ and $P_0(X \amalg Y)$ has only one indecomposable direct summand not in $\operatorname{add} P$, it follows that L is in $\operatorname{add} P$. On the other hand, we know from the preceding lemma that

$$LL(P_0(\Omega_1(X+Y))) = LL(P_1(X+Y)) < LL(P_0(X+Y))$$

since $X+Y$ is a submodule of the projective module Q_2. But $LL(P_0(X+Y)) \leq LL(P_0(X \amalg Y)) \leq r$ since $P_0(X)$ and $P_0(Y)$ are in $\operatorname{add} Q$. So

$$LL(P_0(\Omega_1(X+Y))) < r.$$

That is, $P_0(\Omega_1(X+Y))$ is in $\operatorname{add} P$. So $P_0(Z)$ is in $\operatorname{add} P$. This ends the proof of the claim and of the lemma. □

REFERENCES

[A] M. Auslander, *Representation theory of artin algebras*. I, Comm. Algebra **1** (1974), 177–268.

[BF] W. D. Burguess and K. R. Fuller, *On quasihereditary rings*, 1988 (preprint).

[CE] H. Cartan and S. Eilenberg, *Homological algebra*, Princeton Univ. Press, Princeton, N.J., 1956.

[CPS] E. Cline, B. Parshall and L. Scott, *Finite dimensional algebras and highest weight categories*, J. Reine Angew. Math. **391** (1988), 277–291.

[DR1] V. Dlab and C. M. Ringel, *Quasihereditary algebras*, Illinois J. Math. (to appear).

[DR2] ——, *Every semiprimary ring is the endomorphism ring of a projective module over a quasihereditary ring*, 1987 (preprint).

[PS] B. Parshall and L. Scott (Eds.), *Derived categories, quasihereditary algebras and algebraic groups*, Proc. Ottawa-Mosonee Workshop in Algebra, Math. Lecture Notes Series No. 3, Center for Research in Algebra and Related Topics.

DEPARTMENT OF MATHEMATICS, BRANDEIS UNIVERSITY, WALTHAM, MASSACHUSETTS 02254–9110

DEPARTAMENTO DE MATEMÁTICA, UNIVERSIDAD NACIONAL DEL SUR, 8000 BAHIA BLANCA, ARGENTINA

DEPARTMENT OF MATHEMATICS, NORTHEASTERN UNIVERSITY, BOSTON, MASSACHUSETTS 02115

CHAPTER VIII

Almost split sequences and commutative rings

[71] Isolated singularities and existence of almost split sequences (1986).
[74] Rational singularities and almost split sequences (1986).
[78] Almost split sequences for rational double points (with I. Reiten, 1987).
[79] The Cohen-Macaulay type of Cohen-Macaulay rings (with I. Reiten, 1989).
[83] Almost split sequences for Cohen-Macaulay modules (with I. Reiten, 1987).
[84] The what, where and why of almost split sequences (1987).
[85] Cohen-Macaulay modules for graded Cohen-Macaulay rings and their completions (with I. Reiten, 1989).
[90] Graded modules and their completions (with I. Reiten, 1990).

While the basic foundations for applying almost split sequences to commutative noetherian ring theory appear already in the mid seventies [55], the main applications started in the early eighties. Attempts to obtain results analogous to developments for artin algebras, in particular for finite representation type, led to connections with singularity theory.

There are two main directions, both being intimately connected with almost split sequences. In [71] it is proved that existence of almost split sequences implies that we have an isolated singularity, and a local version is given in [83]. As a consequence finite representation type can only occur for an isolated singularity.

It had been proved by J. Herzog that two-dimensional invariant rings are of finite representation type. Through showing that the Auslander-Reiten quiver of these invariant rings is isomorphic to the corresponding McKay quiver, a connection was established with the resolution graph of the corresponding isolated hypersurface singularity. An interesting consequence was the connection with (extended) Dynkin diagrams for these Auslander-Reiten quivers; a fact which was established directly in [78] for arbitrary rational double points.

A special feature for two-dimensional rings established in [74] was the existence of what Auslander called fundamental sequences, and which have their roots in [55]. They are sequences of the form $0 \to A \to B \to C$ with C projective, which are analogues of almost split sequences and make it possible to apply combinatorial arguments to the whole Auslander-Reiten quiver. This principle has been crucial in later work, for example for classifying the two-dimensional orders of finite representation type. In the local commutative case the middle term of the fundamental sequence, now called the Auslander module, has turned out to be interesting in its own right, with connections to modules of differentials. Also there are various recent generalizations.

The fact that the isolated hypersurface singularities of finite representation type were shown to be exactly the simple hypersurface singularities, was a further indication that finite representation type was a fruitful concept also for commutative rings. Even though Auslander was not directly involved in this result, he was a source of inspiration for the work being done. This development motivated further investigations on finite representation type, and only two other examples of such commutative rings are known, in dimension at least three [79]. Similar questions for graded rings, also using almost split sequences, were considered in [80, 85, 90]. These rings were interesting because of the connection between module theory and vector bundles. For nonsingular projective curves the Serre duality formula corresponds to the formula on which the existence of almost split sequences for the corresponding graded ring is based (see [80]).

A classification of commutative graded rings of finite representation type was established later. The main work along these lines has now shifted to the tame situation, where significant progress has been made. A treatment of the work in this chapter can be found in a monograph by Y. Yoshino.

Isolated singularities and existence of almost split sequences[1]

Maurice Auslander
Notes by Luise Unger

During May and June of 1984, I gave a series of lectures at the University of Bielefeld on various aspects of the theory of commutative and noncommutative isolated singularities, a notion I now define.

Let R be a regular local ring and let Λ be an R-algebra which is a finitely generated free R-module. We say that Λ is nonsingular if $\operatorname{gl.dim.}\Lambda = \dim R$. We say that R is an isolated singularity if Λ is singular (i.e. not nonsingular) and Λ_p is nonsingular for all nonmaximal prime ideals p of R. One of the main objects of the Bielefeld lectures was to prove the following characterization of isolated singularities when R is complete.

Let R be a complete regular local ring, Λ an R-algebra which is a finitely generated free R-module and let $\mathcal{P}_R(\Lambda)$ denote the category of all finitely generated Λ-modules which are R-free. Then Λ is an isolated singularity if and only if Λ is singular and there is an almost split sequence $0 \to A \to B \to C \to 0$ in $\mathcal{P}_R(\Lambda)$ for each indecomposable C in $\mathcal{P}_R(\Lambda)$ such that $\operatorname{Ext}^1_\Lambda(C, X) \neq 0$ for some X in $\mathcal{P}_R(\Lambda)$ and for each A in $\mathcal{P}_R(\Lambda)$ such that $\operatorname{Ext}^1_\Lambda(Y, A) \neq 0$ for some Y in $\mathcal{P}_R(\Lambda)$.

The main body of this paper is devoted to giving a slightly modified version of the proof I gave in my lectures of this result. I am most grateful to Luise Unger for the careful and thoughtful way she wrote up this material. In addition, I have added an appendix which is devoted to pointing out some connections between the notions of orders as defined in [**A2**] and isolated singularities.

In conclusion, I would like to thank the University of Bielefeld and Prof. Claus Michael Ringel for their kind invitation to give these lectures, the participants for their patience and encouragement, and the German Research Council for their financial support.

1. Basic definitions and the main theorem

Throughout these notes, R will be a complete local (commutative) ring, and Λ an R-algebra which is a finitely generated free R-module. We denote by $\mathcal{P}_R(\Lambda)$ the category of finitely generated Λ-modules which are free when considered as R-modules, and $\mathcal{P}(\Lambda)$ will denote the category of finitely generated projective Λ-modules. Note that $\mathcal{P}(\Lambda) \subseteq \mathcal{P}_R(\Lambda)$, since we assume that Λ is a free R-module. Since any module X in $\mathcal{P}_R(\Lambda)$ is, as an R-module, free of finite rank, it follows that X can be written as a finite direct sum of indecomposable Λ-modules which again belongs to $\mathcal{P}_R(\Lambda)$. Also, since we assume that R is complete, any indecomposable module in $\mathcal{P}_R(\Lambda)$ has a local endomorphism ring. As a

[1] Lecture Notes in Math., 1178, 194–242, Springer, Berlin-New York, 1986

consequence, the Krull-Schmidt theorem is valid in $\mathcal{P}_R(\Lambda)$. Let us list the basic properties of $\mathcal{P}_R(\Lambda)$.

LEMMA 1.1. *Let R be a complete local ring, and Λ an R-algebra which is finitely generated free as an R-module.*

(a) *Let $0 \to A \to B \to C \to 0$ be exact with A, B, C in* $\operatorname{Mod}\Lambda$, *and assume $C \in \mathcal{P}_R(\Lambda)$. Then $A \in \mathcal{P}_R(\Lambda)$ if and only if $B \in \mathcal{P}_R(\Lambda)$.*
(b) *Let $C \in \mathcal{P}_R(\Lambda)$. Then $C \in \mathcal{P}(\Lambda)$ if and only if any exact sequence $0 \to A \to B \to C \to 0$ with $A, B \in \mathcal{P}_R(\Lambda)$, splits.*
(c) $\operatorname{Hom}_R(\ ,R)\colon \mathcal{P}_R(\Lambda) \to \mathcal{P}_R(\Lambda^{\mathrm{op}})$ *defines a duality.*
(d) *Let $A \in \mathcal{P}_R(\Lambda)$. Then $A = \operatorname{Hom}_R(Q,R)$ for some $Q \in \mathcal{P}(\Lambda^{\mathrm{op}})$, if and only if every exact sequence $0 \to A \to B \to C \to 0$ with $B, C \in \mathcal{P}_R(\Lambda)$, splits.*

PROOF. (a) Assume $0 \to A \to B \to C \to 0$ is exact, and $C \in \mathcal{P}_R(\Lambda)$. As a sequence of R-modules, this sequence splits, since C as an R-module is free. Note that ${}_RC$ is free of finite rank. Then ${}_RA$ is free of finite rank if and only if ${}_RB$ is free of finite rank.

(b) If $C \in \mathcal{P}(\Lambda)$, then any sequence $0 \to A \to B \to C \to 0$ in $\operatorname{Mod}\Lambda$ splits. On the other hand, given $C \in \mathcal{P}_R(\Lambda)$, let $P \to C$ be exact with $P \in P(\Lambda)$, and let A be the kernel. According to (a), we know that with P also A belongs to $\mathcal{P}_R(\Lambda)$. If now the sequence $0 \to A \to P \to C \to 0$ splits, then C as a direct summand of P belongs to $\mathcal{P}(\Lambda)$.

(c) Here we only use that any module X in $\mathcal{P}_R(\Lambda)$ is free of finite rank, when considered as R-module, thus $\operatorname{Hom}_R(\operatorname{Hom}_R(X,R),R) \simeq X$, canonically.

(d) This statement is obtained from (b) by applying the duality $\operatorname{Hom}_R(\ ,R)$. Of course, (b) is considered for Λ^{op} instead of Λ. $\square$

REMARKS. 1) Because of $\mathcal{P}(\Lambda) \subseteq \mathcal{P}_R(\Lambda)$ and property (a), we can take projective resolutions inside $\mathcal{P}_R(\Lambda)$: given $X \in \mathcal{P}_R(\Lambda)$, there is an exact sequence $\cdots \to P_1 \to P_0 \to X \to 0$ with all $P_i \in P(\Lambda)$.

2) The modules of the form $\operatorname{Hom}_R(Q,R)$ with $Q \in P(\Lambda^{\mathrm{op}})$ will be called *relative injective* in $\mathcal{P}_R(\Lambda)$. Given $X \in \mathcal{P}_R(\Lambda)$, there is an exact sequence $0 \to X \to Q_0 \to Q_1 \to \cdots$ with all Q_i relative injective. (Namely, let $X^* = \operatorname{Hom}(X,R)$, take a projective resolution $\cdots \to P_1 \to P_0 \to X \to 0$ in $\mathcal{P}_R(\Lambda^{\mathrm{op}})$, and dualize $0 \to X^{**} \to P_0^* \to P_1^* \to \cdots$. Then this new sequence is exact, since the given sequence splits over R, all P_i^* are relative injective, and $X = X^{**}$.)

3) In order to calculate $\operatorname{Ext}^1_\Lambda(A,B)$ for $A, B \in \mathcal{P}_R(\Lambda)$, we either may take a projective resolution of A (which we can assume to lie in $\mathcal{P}_R(\Lambda)$), apply $\operatorname{Hom}_\Lambda(\ ,B)$ and take homology, or else we can take a relative injective resolution of B, apply $\operatorname{Hom}_\Lambda(A,\ )$ and take homology.

The following two lemmata are devoted to state some results concerning the injective dimension of Λ. In this way, we prepare the introduction of the notion of nonsingularity.

LEMMA 1.2. *Let Λ be an R-algebra, finitely generated as an R-module. Then* $\operatorname{inj.dim.}_\Lambda \Lambda \geq \operatorname{inj.dim.}_R R$.

PROOF. Assume $\operatorname{inj.dim.}_\Lambda \Lambda = n < \infty$, and take an injective resolution of ${}_\Lambda\Lambda$, say

$$(*) \qquad 0 \to {}_\Lambda\Lambda \to I_0 \to I_1 \to \cdots \to I_n \to 0.$$

Since Λ_R is projective, the modules I_i considered as R-modules are injective too (namely ${}_RI_i = \operatorname{Hom}_\Lambda({}_\Lambda\Lambda_R, I_i)$). Thus, the sequence $(*)$ is an injective resolution of the finitely generated free R-module ${}_R\Lambda$, thus $\operatorname{inj.dim.}_R R = \operatorname{inj.dim.}_R \Lambda \leq n$. $\square$

LEMMA 1.3. *If* $\operatorname{gl.dim.}\Lambda < \infty$, *then* $\operatorname{inj.dim.}_\Lambda \Lambda = \operatorname{gl.dim.}\Lambda$ *(compare with* [R]*).*

PROOF. Let $\text{gl. dim.}\,\Lambda = n < \infty$. Then $\text{inj. dim.}\,M \leq n$ for all Λ-modules M, in particular, $\text{inj. dim.}_\Lambda \Lambda \leq n$. To prove the converse inequality we can assume $n \geq 1$. We have to show that there is a Λ-module M such that $\text{Ext}^n(M, {}_\Lambda\Lambda) \neq 0$. Since $\text{gl. dim.}\,\Lambda = n$, there is a Λ-module N with $\text{proj. dim.}\,N = n$. Let

$$0 \to P_n \xrightarrow{\delta_n} P_{n-1} \xrightarrow{\delta_{n-1}} \cdots \to P_0 \to N \to 0$$

be a projective resolution of N. Consider the image N_{n-1} of δ_{n-1}. There is the following exact sequence

$$(*) \qquad 0 \to P_n \xrightarrow{\delta_n} P_{n-1} \xrightarrow{\delta_{n-1}} N_{n-1} \to 0.$$

Applying $(\ ,\Lambda)$ we get the exact sequence

$$0 \to (N_{n-1}, \Lambda) \to (P_{n-1}, \Lambda) \xrightarrow{(\delta_n,\Lambda)} (P_n, \Lambda) \to \text{Ext}^1(N_{n-1}, \Lambda) \to 0.$$

Note that $\text{Ext}^n(N, \Lambda) \simeq \text{Ext}^1(N_{n-1}, \Lambda)$. We assume $\text{Ext}^n(N, \Lambda) = 0$. Then (δ_n, Λ) is surjective. But then (δ_n, P) is surjective for any P in $\mathcal{P}(\Lambda)$, in particular $(\delta_n, P_n)\colon (P_{n-1}, P_n) \to (P_n, P_n)$ is surjective. Thus there is $\rho \in (P_{n-1}, P_n)$ with $\rho\delta_n = 1_{P_n}$. This shows that $(*)$ splits, therefore N_{n-1} is projective, therefore $\text{proj. dim.}\,N \leq n-1$, a contradiction. This shows that $\text{Ext}^n(N, \Lambda) \neq 0$. □

Recall that a local ring R is said to be *regular* provided there exists a system of parameters generating the maximal ideal of M (see for example [**N**]). There is a homological characterization of regular local rings: a local ring R is regular if and only if the global dimension of R is finite, and then $\text{gl. dim.}\,R = \dim R$ where $\dim R$ is the Krull-dimension of R (see for example [**K**]); we therefore will call $\dim R$ just the dimension of R.

COROLLARY 1.4 *Let R be a regular local ring of dimension d. Let Λ be an R-algebra which is finitely generated free as an R-module. Then* $\text{gl. dim.}\,\Lambda \geq d$ *(for a more general statement see* [V]*).*

PROOF. $\text{gl. dim.}\,\Lambda \geq \text{inj. dim.}_\Lambda \Lambda$, trivially; $\text{inj. dim.}_\Lambda \Lambda \geq \text{inj. dim.}\,R$, according to (1.2), and $\text{inj. dim.}\,R = d$, according to (1.3). □

In case R is a regular local ring of dimension d, and Λ is an R-algebra, then Λ is said to be *nonsingular* provided Λ is a finitely generated free R-module, and $\text{gl. dim.}\,\Lambda = d$. Note that Λ is nonsingular if and only if Λ^{op} is nonsingular.

According to the previous corollary, the definition of a nonsingular R-algebra requires that its global dimension has the smallest possible value d. Note that for any $n > d$, it is easy to construct R-algebras Λ which are finitely generated free R-modules and have global dimension n (and so are singular). For example, we recall that given any ring S, and $m > 1$, then $\text{gl. dim.}\,T_m(S) = 1 + \text{gl. dim.}\,S$, where $T_m(S)$ is the ring of all $m \times m$ upper triangular $m \times m$-matrices over S (see, for example [**M**]). Of course, if S is an R-algebra which is finitely generated free as an R-module, the same is true for $T_m(S)$. Thus, we may start with $T_m(S)$, and iterate the process of forming triangular matrix rings.

We can characterize the nonsingular R-algebras as follows:

PROPOSITION 1.5. *Let R be a complete regular local ring, and Λ an R-algebra which is a finitely generated free R-module. Then the following assertions are equivalent (see* [R]*):*

(i) *Λ is nonsingular.*
(ii) $\mathcal{P}_R(\Lambda) = \mathcal{P}(\Lambda)$.

PROOF. Let $d = \dim R$.

(i) $\Rightarrow$ (ii): Let M be a non-zero module in $\mathcal{P}_R(\Lambda)$. Let $x_1, \dots, x_d$ be a system of parameters of R. Then $x_1, \dots, x_d$ is an R-sequence: this means, that x_i is not a zero-divisor on $R/(x_1, \dots, x_{i-1})R$, for all $1 \le i \le d$. Since M is non-zero, and is free as an R-module, also x_i is not a zero-divisor on $M/(x_1, \dots, x_{i-1})M$. Note that by induction, we have that $\operatorname{proj.dim.}_\Lambda M/(x_1, \dots, x_i)M = i + \operatorname{proj.dim.}_\Lambda M$, for $i = 1, \dots, d$ (use Theorems 3 and 9 in Part III of [**K**]). By assumption (i), $\operatorname{gl.dim.}\Lambda = d$, thus $d \ge \operatorname{proj.dim.}_\Lambda M/(x_1, \dots, x_d)M = d + \operatorname{proj.dim.}_\Lambda M$ shows that $\operatorname{proj.dim.}_\Lambda M = 0$, thus $M \in \mathcal{P}(\Lambda)$.

(ii) $\Rightarrow$ (i): We want to calculate the global dimension of Λ. Since Λ is noetherian, we only have to check the projective dimension of the modules $M \in \operatorname{mod}\Lambda$ (see for example [**K**], Part III, Theorem 21). Let $M \in \operatorname{mod}\Lambda$, and take an exact sequence

$$0 \to \Omega^d M \to P_{d-1} \to \cdots \to P_0 \to M \to 0,$$

with $P_i \in \mathcal{P}(\Lambda)$. Since the global dimension of R is d, it follows that $\Omega^d M$ considered as an R-module is projective, thus free. Of course, $\Omega^d M$ also is a finitely generated R-module, since R is noetherian. Thus $\Omega^d M \in \mathcal{P}_R(\Lambda)$. By assumption (ii), we even have $\Omega^d M \in \mathcal{P}(\Lambda)$, thus $\operatorname{proj.dim.}_\Lambda M \le d$. Therefore $\operatorname{gl.dim.}\Lambda \le d$. However, then also $\operatorname{gl.dim.}\Lambda = \operatorname{inj.dim.}_\Lambda \Lambda$ and $\operatorname{inj.dim.}_\Lambda \Lambda \ge \operatorname{inj.dim.}_R R = d$, according to (1.3) and (1.2), thus $\operatorname{gl.dim.}\Lambda = d$. This finishes the proof. $\square$

Let R be a regular local ring, and Λ an R-algebra. Then Λ is said to be an *isolated singularity* provided Λ_p is nonsingular for all nonmaximal prime ideals p of R, (here Λ_p denotes the localization with respect to p), and Λ is not nonsingular. Of course, Λ is an isolated singularity if and only if Λ^{op} is an isolated singularity.

For the formulation of the main theorem we still need some additional notation and definitions. Suppose M is an R-module, then $\ell_R(M)$ denotes its length. Given Λ-modules A, B, we usually will write (A, B) instead of $\operatorname{Hom}_\Lambda(A, B)$. Also we denote by $\mathcal{P}(A, B)$ the set of maps $A \to B$ which factor through a projective module (of course, if B is finitely generated, then $A \to B$ factors through a projective module if and only if it factors through a finitely generated projective module). Also let $\underline{\operatorname{Hom}}_\Lambda(A, B) = (A, B)/\mathcal{P}(A, B)$.

A non-split exact sequence $0 \to A \xrightarrow{f} B \xrightarrow{g} C \to 0$ with A, B, C in $\mathcal{P}_R(\Lambda)$ will be called an *almost split sequence* in $\mathcal{P}_R(\Lambda)$ provided both A and C are indecomposable, and given any map $y\colon Y \to C$ with $Y \in \mathcal{P}_R(\Lambda)$, and not a split epimorphism, there exists $y'\colon Y \to B$ with $y = y'g$. Note that the last property is equivalent to the property that given any map $x\colon A \to X$ with $X \in \mathcal{P}_R(\Lambda)$ and not a split monomorphism, there exists $x'\colon B \to X$ with $x = fx'$. We say that $\mathcal{P}_R(\Lambda)$ *has right almost split sequences* provided for any indecomposable $C \in \mathcal{P}_R(\Lambda)$ which is not projective, there exists an almost split sequence $0 \to A \to B \to C \to 0$, and $\mathcal{P}_R(\Lambda)$ *has left almost split sequences* provided for any indecomposable module $A \in \mathcal{P}_R(\Lambda)$ which is not relative injective in $\mathcal{P}_R(\Lambda)$, there exists an almost split sequence $0 \to A \to B \to C \to 0$. Finally, $\mathcal{P}_R(\Lambda)$ is said to *have almost split sequences*, provided it has both left almost split sequences and right almost split sequences.

We are now able to state the main theorem:

THEOREM. *Let R be a complete regular local ring, and Λ an R-algebra which is finitely generated free as an R-module. Then the following are equivalent:*

(a) *$\mathcal{P}_R(\Lambda)$ has almost split sequences.*
(b) *For all modules $X, Y \in \mathcal{P}_R(\Lambda)$, we have $\ell_R(\underline{\operatorname{Hom}}(X, Y)) < \infty$.*

(c) *For all modules* $C \in \mathcal{P}_R(\Lambda)$, *and any nonmaximal prime ideal* p *of* R, *the module* C_p *is projective over* Λ_p.

(d) Λ *is an isolated singularity, or nonsingular.*

Actually, in (a) we only have to assume that $\mathcal{P}_R(\Lambda)$ has left almost split sequences or that $\mathcal{P}_R(\Lambda)$ has right almost split sequences and obtain again equivalent conditions.

We shortly want to outline the proof of the main theorem, referring for some of the definitions to the following sections.

The proof of the theorem will cover sections 2–9 of this paper. The implication (a) $\Rightarrow$ (b) will be given in section 3. Note that here we do not use the assumption on R of being regular. The proof is based on a rather detailed investigation in section 2 of the categories $\operatorname{mod}(\mathcal{P}_R(\Lambda))$ and $\operatorname{mod}(\mathcal{P}_R(\Lambda)/\mathcal{P}(\Lambda))$ of all finitely presented contravariant functors on $\mathcal{P}_R(\Lambda)$, or $\mathcal{P}_R(\Lambda)/\mathcal{P}(\Lambda)$, respectively. In particular, we will show that the category $\mathcal{P}_R(\Lambda)/\mathcal{P}(\Lambda)$ is coherent and that $\operatorname{mod}(\mathcal{P}_R(\Lambda)/\mathcal{P}(\Lambda))$ has injective envelopes. Actually, the injective objects in $\mathcal{P}_R(\Lambda)/\mathcal{P}(\Lambda)$ turn out to be just the functors of the form $\operatorname{Ext}^1_\Lambda(\ ,A)$, with $A \in \mathcal{P}_R(\Lambda)$. In addition, we will use some commutative algebra in order to show that for any R-algebra Σ which is finitely generated as an R-module, finitely generated injective Σ-modules are of finite length.

The implication (b) $\Rightarrow$ (c) is rather straightforward, it will be given in section 4. We only have to use that a finitely generated R-module M is of finite length if and only if the localization of M with respect to any non-maximal prime ideal of R vanishes. Note that we may show, in the same way, also the reverse implication, directly.

The next section 5 comprises the proof of the implication (c) $\Rightarrow$ (d). Actually, we show that in case (c) is satisfied for some fixed prime ideal p of R, then Λ_p is nonsingular.

Sections 6, 7, and 8 are preparations for the proof of the implication (d) $\Rightarrow$ (a) which is given in section 9. The main point of this proof is to show that for any module $C \in \mathcal{P}_R(\Lambda)$, the functor $\operatorname{rad}(\ ,C)$ is finitely presented. This is a consequence of the general result presented in 8.8 that for any finitely presented functor on $\operatorname{mod}\Lambda$, and any finitely generated Λ-module M, the restriction of the functor $\operatorname{Hom}_R(F(\ ),M)$ to $\mathcal{P}_R(\Lambda)$ is finitely presented. Actually, our method of proof forces us to deal more generally with all the corresponding functors $\operatorname{Ext}^i_\Lambda(F(\ ),M)$, with $i \geq 0$, at the same time; the restrictions of all these functors to $\mathcal{P}_R(\Lambda)$ turn out to be finitely presented. On the way to this result, various other functors on $\mathcal{P}_R(\Lambda)$ are shown to be finitely presented, and constructions are outlined in order to obtain finitely presented functors from finitely presented functors. In particular, section 6 shows that under the assumption on R of being regular, also the category $\mathcal{P}_R(\Lambda)$ itself is coherent, so that kernels and cokernels of maps between finitely presented functors on $\mathcal{P}_R(\Lambda)$ are finitely presented again. Section 7 shows that for an arbitrary noetherian ring Λ, there is a duality functor $D\colon \operatorname{mod}((\operatorname{mod}\Lambda)^{\mathrm{op}}) \to \operatorname{mod}((\operatorname{mod}\Lambda^{\mathrm{op}})^{\mathrm{op}})$ such that the image of the representable functor $(A,\ )$, with A a finitely generated Λ-module is just the tensor functor $A \otimes -$ on $\operatorname{mod}\Lambda^{\mathrm{op}}$. As a consequence, any functor F on $\operatorname{mod}\Lambda$ has a length 2-coresolution by tensor functors.

In the final section 10, we derive a consequence of the main theorem. Assume again that R is a complete regular local ring, and Λ an R-algebra which is a finitely generated R-module. Then if $\mathcal{P}_R(\Lambda)$ has only finitely many isomorphism classes of indecomposable modules, Λ has to be an isolated singularity or nonsingular. Namely, the finiteness assumption of $\mathcal{P}_R(\Lambda)$ immediately implies that $\mathcal{P}_R(\Lambda)$ has almost split sequences.

2. $\operatorname{mod}(\mathcal{P}_R(\Lambda))$ and $\operatorname{mod}(\mathcal{P}_R(\Lambda)/\mathcal{P}(\Lambda))$

Many parts of the proof of the theorem will rely on the use of functors. The following remark should explain why the use of functors is rather natural.

Let Λ be a ring, and $\mathcal{P}(\Lambda)$ the category of finitely generated projective Λ-modules; this is an additive category in which idempotents split. Given an additive category $\mathcal{A}$, denote by $\operatorname{Mod}\mathcal{A}$ the category of additive functors $\mathcal{A}^{\mathrm{op}} \to \mathrm{Ab}$, where Ab is the category of abelian groups. Note that the evaluation functor $\operatorname{Mod}\mathcal{P}(\Lambda) \to \operatorname{Mod}\Lambda$ given by $F \to F(\Lambda)$ is an equivalence of categories. Thus, if we are interested in the Λ-modules, we may as well consider the functor category $\operatorname{Mod} P(\Lambda)$. However, note that in $\mathcal{P}(\Lambda)$, it is difficult to find Λ. The object Λ has the following property: it is an additive generator of $\mathcal{P}(\Lambda)$ (this means that every $P \in \mathcal{P}(\Lambda)$ is a summand of some $n\Lambda$), however, there will be other additive generators in $\mathcal{P}(\Lambda)$. On the other hand, given an additive generator Q of $\mathcal{P}(\Lambda)$ (thus, every $P \in \mathcal{P}(\Lambda)$ is a summand of some nQ), then again $\operatorname{Mod}\mathcal{P}(\Lambda) \approx \operatorname{Mod}((\operatorname{End} Q)^{\mathrm{op}})$. This is the theory of Morita equivalence. It means that $\mathcal{P}(\Lambda)$ may be considered as a kind of "uncoordinized ring".

Let $\mathcal{A}$ be an additive category in which idempotents split. The functors $(\ ,A)$ with $A \in \mathcal{A}$ are projective objects in $\operatorname{Mod}\mathcal{A}$, and will be called the finitely generated projective functors. The Yoneda-lemma asserts that $\mathcal{A}$ is equivalent to the full subcategory of $\operatorname{Mod}\mathcal{A}$ given by the finitely generated projective functors, under the functor $A \mapsto (\ ,A)$. In general, a functor $M \in \operatorname{Mod}\mathcal{A}$ is said to be *finitely generated* provided there is an exact sequence $(\ ,A) \to M \to 0$ for some $A \in \mathcal{A}$, and, M is said to be *finitely presented* provided there exists an exact sequence $(\ ,B) \to (\ ,A) \to M \to 0$ with $A, B \in \mathcal{A}$. We denote by $\operatorname{mod}\mathcal{A}$ the category of finitely presented functors in $\operatorname{Mod}\mathcal{A}$. In the same way, given the ring Λ, we denote by $\operatorname{mod}\Lambda$ the category of finitely presented Λ-modules; of course, $\operatorname{mod}\Lambda$ is just the category of finitely generated modules in case Λ is noetherian.

Assume in addition that for A indecomposable in $\mathcal{A}$, $\operatorname{End}(A)$ is a local ring, and that any object of $\mathcal{A}$ is the direct sum of a finite number of indecomposable objects. Then any finitely generated object $M \in \operatorname{Mod}\mathcal{A}$ has a minimal projective cover $(\ ,A) \xrightarrow{p} M \to 0$ (this means that this sequence is exact and that for a proper direct summand A' of A the restriction of p to $(\ ,A')$ is not surjective), and the minimal projective cover of M is unique up to isomorphism. Note that both assumptions on $\mathcal{A}$ are satisfied for $\mathcal{A} = \mathcal{P}_R(\Lambda)$.

The category $\mathcal{A}$ is called *coherent* provided given $f\colon A_1 \to A_2$ in $\mathcal{A}$, there exists $g\colon A_0 \to A_1$ in $\mathcal{A}$ such that the induced sequence

$$(\ ,A_0) \xrightarrow{(\ ,g)} (\ ,A_1) \xrightarrow{(\ ,f)} (\ ,A_2)$$

is exact. If $\mathcal{A}$ is coherent, and $0 \to F_0 \to F_1 \to F_2 \to F_3 \to 0$ is an exact sequence in $\operatorname{Mod}\mathcal{A}$, with F_1, F_2 finitely presented, then also F_0, F_3 are finitely presented (easy to prove, or see [**A1**]). In particular, for $\mathcal{A}$ coherent, the category $\operatorname{mod}\mathcal{A}$ is abelian.

It will be necessary to consider besides the functor category $\operatorname{mod}\mathcal{P}_R(\Lambda)$ also the functor category $\operatorname{mod}(\mathcal{P}_R(\Lambda)/\mathcal{P}(\Lambda))$. The main advantage is that $\mathcal{P}_R(\Lambda)/\mathcal{P}(\Lambda)$ is always a coherent category. Note the definition of $\mathcal{P}_R(\Lambda)/\mathcal{P}(\Lambda)$: it has the same objects as $\mathcal{P}_R(\Lambda)$, but we use as set of maps from X to Y in $\mathcal{P}_R(\Lambda)/\mathcal{P}(\Lambda)$ the set $\underline{\operatorname{Hom}}(X,Y)$, i.e. the set of Λ-homomorphism (X,Y) from X to Y modulo the set of maps $P(X,Y)$, which factor through projective modules. The indecomposable objects of $\mathcal{P}_R(\Lambda)/\mathcal{P}(\Lambda)$ are given by those indecomposable objects of $\mathcal{P}_R(\Lambda)$ which do not belong to $\mathcal{P}(\Lambda)$. Before we can prove that $\mathcal{P}_R(\Lambda)/\mathcal{P}(\Lambda)$ is coherent, we show the following elementary facts.

LEMMA 2.1. *Let* $0 \to A \xrightarrow{f} B \xrightarrow{g} C \to 0$ *be an exact sequence of* Λ*-modules.*

(a) *The induced sequence* $\underline{\mathrm{Hom}}(\ ,A) \to \underline{\mathrm{Hom}}(\ ,B) \to \underline{\mathrm{Hom}}(\ ,C)$ *is exact.*
(b) *The induced sequence* $\mathcal{P}(\ ,B) \to \mathcal{P}(\ ,C) \to 0$ *is exact.*

PROOF. (a) Given a map $h\colon X \to B$ with $hg \in P(X,C)$, we have to show that there is a map $h'\colon X \to A$ such that $h - h'f$ belongs to $P(X,B)$. Thus, let $hg = h_1h_2$ with $h_1\colon X \to P$, $h_2\colon P \to C$, where P is projective. Since g is surjective, there is $h_2'\colon P \to B$ such that $h_2 = h_2'g$, thus $(h - h_1h_2')g = hg - h_1h_2 = 0$, therefore there is $h'\colon X \to A$ such that $h - h_1h_2' = h'f$. This can be rewritten as $h - h'f = h_1h_2'$ and $h_1h_2' \in \mathcal{P}(X,B)$.

(b) Given a map $h \in P(X,C)$, we have to show that there exists $h' \in P(X,B)$ with $h'g = h$. Thus, let $h = h_1h_2$, with $h_1\colon X \to P$, $h_2\colon P \to C$, where P is projective. Since g is surjective, there is $h_2'\colon P \to B$ with $h_2'g = h_2$. Thus $h' = h_1h_2' \in P(X,B)$ and $h'g = h_1h_2'g = h_1h_2 = h$. □

LEMMA 2.2. *Let R be a complete local ring, and Λ an R-algebra which is finitely generated free as an R-module. Then $\mathcal{P}_R(\Lambda)/P(\Lambda)$ is coherent.*

PROOF. Given a map $B \to C$ in $\mathcal{P}_R(\Lambda)$, we can add a projective module P from $\mathcal{P}(\Lambda)$ to B in order to obtain an exact sequence $P \amalg B \to C \to 0$.

According to (1.1), the kernel K of this map lies in $\mathcal{P}_R(\Lambda)$, thus we have the exact sequence $0 \to K \to P \amalg B \to C \to 0$ in $\mathcal{P}_R(\Lambda)$, and we obtain the induced exact sequence $\underline{\mathrm{Hom}}(\ ,K) \to \underline{\mathrm{Hom}}(\ ,P \amalg B) \to \underline{\mathrm{Hom}}(\ ,C)$. Note that $\underline{\mathrm{Hom}}(\ ,P \amalg B) = \underline{\mathrm{Hom}}(\ ,B)$. □

LEMMA 2.3. *Let R be a complete local ring, and Λ an R-algebra which is finitely generated as an R-module. Then we can identify* $\mathrm{mod}(\mathcal{P}_R(\Lambda)/\mathcal{P}(\Lambda))$ *with the full subcategory of* $F \in \mathrm{mod}(\mathcal{P}_R(\Lambda))$ *satisfying* $F(\Lambda) = 0$.

PROOF. Let $F \in \mathrm{mod}\,\mathcal{P}_R(\Lambda)$, say with an exact sequence $(\ ,B) \xrightarrow{(\ ,g)} (\ ,C) \to F \to 0$, and assume $F(\Lambda) = 0$. Inserting Λ in the exact sequence we see that g is surjective. According to (2.1) (b), also $\mathcal{P}(\ ,g)$ is surjective. We therefore have the following commutative diagram with exact rows and columns

$$\begin{array}{ccccccccc}
0 & \longrightarrow & \mathcal{P}(\ ,B) & \longrightarrow & (\ ,B) & \longrightarrow & \underline{\mathrm{Hom}}(\ ,B) & \longrightarrow & 0 \\
 & & \downarrow{\scriptstyle P(\ ,g)} & & \downarrow{\scriptstyle (\ ,g)} & & \downarrow{\scriptstyle \underline{\mathrm{Hom}}(\ ,g)} & & \\
0 & \longrightarrow & \mathcal{P}(\ ,C) & \longrightarrow & (\ ,C) & \longrightarrow & \underline{\mathrm{Hom}}(\ ,C) & \longrightarrow & 0 \\
 & & \downarrow & & \downarrow & & \downarrow & & \\
 & & 0 & \longrightarrow & F & \xrightarrow{1} & F & \longrightarrow & 0 \\
 & & & & \downarrow & & \downarrow & & \\
 & & & & 0 & & 0 & &
\end{array} \qquad (*)$$

The last column shows that $F \in \mathrm{mod}(\mathcal{P}_R(\Lambda)/\mathcal{P}(\Lambda))$. Conversely, given a presentation $\underline{\mathrm{Hom}}(\ ,B') \xrightarrow{\underline{\mathrm{Hom}}(\ ,g')} \underline{\mathrm{Hom}}(\ ,C) \to F \to 0$ with $g'\colon B' \to C$, we may add a projective P from $P(\Lambda)$ to B' in order to obtain a surjective map $g\colon B = P \amalg B' \to C$. Then we obtain the same commutative diagram $(*)$ with exact rows and columns as above, showing that $F \in \mathrm{mod}(\mathcal{P}_R(\Lambda))$, according to the middle exact column. Also $F(\Lambda) = 0$, since g, and therefore (Λ, g), is surjective. □

LEMMA 2.4. *Let R be a complete local ring and Λ be an R-algebra which is finitely generated free as an R-module. Given $A \in \mathcal{P}_R(\Lambda)$, the functors* $\mathrm{Ext}^i(\ ,A) \in \mathrm{Mod}\,\mathcal{P}_R(\Lambda)$ *are finitely presented.*

PROOF. According to (1.1), there is an exact sequence $0 \to A \to Q \to L \to 0$ in $\mathcal{P}_R(\Lambda)$, with Q relative injective. The induced long exact sequence gives rise to the sequence

$$(*) \qquad 0 \to (\ ,A) \to (\ ,Q) \to (\ ,L) \to \mathrm{Ext}^1_\Lambda(\ ,A) \to 0$$

and

$$(**) \qquad 0 \to \mathrm{Ext}^i_\Lambda(\ ,L) \to \mathrm{Ext}^{i+1}_\Lambda(\ ,A) \to 0$$

for $i \geq 1$, due to the fact that $\mathrm{Ext}^i_\Lambda(\ ,Q) = 0$ for all $i \geq 0$. The sequence $(*)$ shows that $\mathrm{Ext}^1_\Lambda(\ ,A)$ is finitely presented (and in fact, gives a projective resolution of $\mathrm{Ext}^1(\ ,A)$ of length 2), the sequences $(**)$ show that any functor $\mathrm{Ext}^i_\Lambda(\ ,A)$ with $i > 1$ is isomorphic to a functor of the form $\mathrm{Ext}^1_\Lambda(\ ,A')$. □

REMARKS. 1) Of course, $\mathrm{Ext}^i_\Lambda({}_\Lambda\Lambda, A) = 0$, for $i \geq 1$, thus $\mathrm{Ext}^i_\Lambda(\ ,A)$, for $i \geq 1$ is in fact in $\mathrm{mod}(\mathcal{P}_R(\Lambda)/\mathcal{P}(\Lambda))$, according to (2.3).

2) Note that the subcategory of functors $F \in \mathrm{mod}(\mathcal{P}_R(\Lambda))$ satisfying $F(\Lambda) = 0$ is closed under extensions. Therefore, given $F_1, F_2 \in \mathrm{mod}(\mathcal{P}_R(\Lambda)/P(\Lambda))$, we have

$$\mathrm{Ext}^1_{\mathcal{P}_R(\Lambda)/\mathcal{P}(\Lambda))}(F_1, F_2) = \mathrm{Ext}^1_{\mathcal{P}_R(\Lambda)}(F_1, F_2).$$

PROPOSITION 2.5. *Let R be a complete local ring, and Λ an R-algebra which is finitely generated free as an R-module. A functor $F \in \mathrm{mod}(\mathcal{P}_R(\Lambda)/\mathcal{P}(\Lambda))$ is an injective object in $\mathrm{mod}(\mathcal{P}_R(\Lambda)/\mathcal{P}(\Lambda))$ if and only if $F = \mathrm{Ext}^1_\Lambda(\ ,A)$ for some $A \in \mathcal{P}_R(\Lambda)$. If A is indecomposable in $\mathcal{P}_R(\Lambda)$, and is not relatively injective, then $\mathrm{Ext}^1_\Lambda(\ ,A)$ is indecomposable in $\mathrm{mod}(\mathcal{P}_R(\Lambda)/\mathcal{P}(\Lambda))$. Finally given $F \in \mathrm{mod}(\mathcal{P}_R(\Lambda)/\mathcal{P}(\Lambda))$, there is an exact sequence $0 \to F \to \mathrm{Ext}^1_\Lambda(\ ,A)$ with $A \in \mathcal{P}(\Lambda)$, which is an injective envelope in $\mathrm{mod}(\mathcal{P}_R(\Lambda)/\mathcal{P}(\Lambda))$.*

PROOF. First, let us show that for $A \in \mathcal{P}_R(\Lambda)$, the functor $\mathrm{Ext}^1(\ ,A)$ is injective in $\mathrm{mod}(\mathcal{P}_R(\Lambda)/\mathcal{P}(\Lambda))$. Let $F \in \mathrm{mod}(\mathcal{P}_R(\Lambda)/\mathcal{P}(\Lambda))$. We want to show that $\mathrm{Ext}^1_{\mathcal{P}_R(\Lambda)/\mathcal{P}(\Lambda))}(F, \mathrm{Ext}^1(\ ,A)) = 0$. Consider F as a functor in $\mathrm{mod}\,\mathcal{P}_R(\Lambda)$, using (2.3), and take a presentation $(\ ,Y) \xrightarrow{(\ ,g)} (\ ,Z) \to F \to 0$. Since $F(\Lambda) = 0$, we see that g is surjective, let X be its kernel, thus we have an exact sequence $(*)$ $0 \to X \xrightarrow{f} Y \xrightarrow{g} Z \to 0$, inducing the exact sequence of functors $0 \to (\ ,X) \to (\ ,Y) \to (\ ,Z) \to F \to 0$. Since this is a projective resolution of F in $\mathrm{mod}\,\mathcal{P}_R(\Lambda)$, we can use it to calculate $\mathrm{Ext}^1_{\mathcal{P}_R(\Lambda)}(F, \mathrm{Ext}^1_\Lambda(\ ,A))$, namely it is the homology of

$$((\ ,Z), \mathrm{Ext}^1_\Lambda(\ ,A)) \to ((\ ,Y), \mathrm{Ext}^1_\Lambda(\ ,A)) \to ((\ ,X), \mathrm{Ext}^1_\Lambda(\ ,A)).$$

However, under the Yoneda-isomorphism, this sequence is isomorphic to $\mathrm{Ext}^1_\Lambda(Z,A) \to \mathrm{Ext}^1_\Lambda(Y,A) \to \mathrm{Ext}^1(X,A)$, and this sequence is induced from the exact sequence $(*)$, thus it is exact. Therefore, $\mathrm{Ext}^1_{\mathcal{P}_R(\Lambda)}(F, \mathrm{Ext}^1_\Lambda(\ ,A)) = 0$, thus also $\mathrm{Ext}^1_{\mathcal{P}_R(\Lambda)/\mathcal{P}(\Lambda)}(F, \mathrm{Ext}^1_\Lambda(\ ,A)) = 0$.

Next given $F \in \mathrm{mod}(\mathcal{P}_R(\Lambda)/P(\Lambda))$, let us embed F into some $\mathrm{Ext}^1(\ ,A)$. As above, there is an exact sequence $0 \to X \to Y \to Z \to 0$ such that it induces an exact sequence $0 \to (\ ,X) \to (\ ,Y) \to (\ ,Z) \to F \to 0$. However, $0 \to X \to Y \to Z \to 0$ also gives rise to the exact sequence $0 \to (\ ,X) \to (\ ,Y) \to (\ ,Z) \xrightarrow{\delta} \mathrm{Ext}^1(\ ,X)$, thus F is the image of δ, therefore F is embedded into $\mathrm{Ext}^1(\ ,X)$.

Now, assume $A \in \mathcal{P}_R(\Lambda)$, and consider a direct decomposition $\mathrm{Ext}^1_\Lambda(\ ,A) = F_1 \amalg F_2$ in $\mathrm{Mod}(\mathcal{P}_R(\Lambda))$. Of course, both F_i satisfy $F_i(\Lambda) = 0$. Since F_i is the kernel of an endomorphism of $\mathrm{Ext}^1_\Lambda(\ ,A)$, and $\mathcal{P}_R(\Lambda)/\mathcal{P}(\Lambda)$ is coherent, it follows that F_i belongs to $\mathrm{mod}(\mathcal{P}_R(\Lambda)/\mathcal{P}(\Lambda))$, thus to $\mathrm{mod}(\mathcal{P}_R(\Lambda))$. Take a minimal projective resolution of F_i, it is of the form $0 \to (\ ,A_i) \to (\ ,B_i) \to (\ ,C_i) \to F_i \to 0$, and the direct sum gives a minimal projective resolution of $F_1 \amalg F_2$,

$$0 \to (\ ,A_1 \amalg A_2) \to (\ ,B_1 \amalg B_2) \to (\ ,C_1 \amalg C_2) \to F_1 \amalg F_2 \to 0.$$

On the other hand, $F_1 \amalg F_2 = \operatorname{Ext}^1(\ ,A)$ also has a projective resolution $0 \to (\ ,A) \to (\ ,B) \to (\ ,C) \to \operatorname{Ext}^1_\Lambda(\ ,A) \to 0$. It follows that $A_1 \amalg A_2$ is a direct summand of A, say $A = A_1 \amalg A_2 \amalg A_3$. But $\operatorname{Ext}^1_\Lambda(\ ,A) = F_1 \amalg F_2 \subseteq \operatorname{Ext}^1_\Lambda(\ ,A_1 \amalg A_2) \amalg \operatorname{Ext}^1_\Lambda(\ ,A_3) = \operatorname{Ext}^1_\Lambda(\ ,A)$ gives $\operatorname{Ext}^1_\Lambda(\ ,A_3) = 0$, thus $A_3 \in \mathcal{P}(\Lambda)$ according to (1.1), and $F_i = \operatorname{Ext}^1_\Lambda(\ ,A_i)$. As a consequence, if A is indecomposable in $\mathcal{P}_R(\Lambda)$ and not injective, then $\operatorname{Ext}^1_\Lambda(\ ,A)$ is indecomposable. Also, if F is an injective object in $\operatorname{mod}(\mathcal{P}_R(\Lambda)/\mathcal{P}(\Lambda))$, we embed F into some $\operatorname{Ext}^1_\Lambda(\ ,A)$, and conclude $F = \operatorname{Ext}^1_\Lambda(\ ,A')$ with A' a direct summand of A.

Let A be indecomposable in $\mathcal{P}_R(\Lambda)$, and not injective. It follows from our considerations that $\operatorname{Ext}^1_\Lambda(\ ,A)$ is indecomposable (we use (1.1) in order to see that $\operatorname{Ext}^1_\Lambda(\ ,A) \neq 0$). Since its endomorphism ring E is a finitely generated R-algebra, and R is complete, E is a local ring. As a consequence, any non-zero finitely presented subfunctor G of $\operatorname{Ext}^1_\Lambda(\ ,A)$ is indecomposable. Namely, assume there exists a proper direct decomposition $G = G_1 \amalg G_2$, denote by ϵ the projection map from G to G_1. Since $\operatorname{Ext}^1_\Lambda(\ ,A)$ is injective, we can extend ϵ to an endomorphism ϵ' of $\operatorname{Ext}^1_\Lambda(\ ,A)$. However, ϵ' vanishes on G_2 and $1-\epsilon'$ vanishes on G_1, thus neither ϵ' nor $1-\epsilon'$ is an automorphism of $\operatorname{Ext}^1_\Lambda(\ ,A)$. This shows that any non-zero finitely presented subfunctor of $\operatorname{Ext}^1_\Lambda(\ ,A)$ is essential in $\operatorname{Ext}^1_\Lambda(\ ,A)$.

Finally, given $F \in \operatorname{mod}(\mathcal{P}_R(\Lambda)/\mathcal{P}(\Lambda))$, choose $A \in \mathcal{P}_R(\Lambda)$ of minimal R-rank, with an exact sequence $0 \to F \to \operatorname{Ext}^1(\ ,A)$. We claim that this is an injective envelope. Let $A = \amalg A_i$ be a direct decomposition of A with all A_i indecomposable. The minimality implies that the intersection F_i of $\operatorname{Ext}^1_\Lambda(\ ,A_i)$ with the image of F is non-zero, thus F_i is essential in $\operatorname{Ext}^1_\Lambda(\ ,A_i)$, and therefore the image of F is essential in $\operatorname{Ext}^1_\Lambda(\ ,A)$. This finishes the proof. □

3. The implication (a) ⇒ (b) of the main theorem

We are now going to prove that if $\mathcal{P}_R(\Lambda)$ has left almost split sequences, then for all modules X and Y in $\mathcal{P}_R(\Lambda)$, the R-module $\underline{\operatorname{Hom}}(X,Y)$ is of finite length.

PROPOSITION 3.1. *Let $0 \to A \to B \to C \to 0$ be an almost split sequence in $\mathcal{P}_R(\Lambda)$. Let $0 \to A \to Q \to L \to 0$ be exact in $\mathcal{P}_R(\Lambda)$, with Q relative-injective. Let $A' = C \amalg Q \amalg L$. Then, for any $X \in \mathcal{P}_R(\Lambda)$, the $(\underline{\operatorname{End}}(X \amalg A'))^{\mathrm{op}}$-module $\operatorname{Ext}^1(X \amalg A', A)$ is injective.*

PROOF. Let $X \in \mathcal{P}_R(\Lambda)$ be fixed. Define $Y = X \amalg A'$, and let $\Gamma = (\operatorname{End} Y)^{\mathrm{op}}$, and $\underline{\Gamma} = (\underline{\operatorname{End}} Y)^{\mathrm{op}}$.

Given a functor F on $\mathcal{P}_R(\Lambda)/\mathcal{P}$, note that $F(Y)$ is a Γ-module, and even a $\underline{\Gamma}$-module, since F vanishes on $\mathcal{P}$. Also, given a natural transformation $\eta = (\eta_M)_M \in (\operatorname{Ext}^1(\ ,A), F)$, then $\eta_Y \colon \operatorname{Ext}^1(Y,A) \to F(Y)$ is Γ-linear, or, equivalently, $\underline{\Gamma}$-linear. We claim that the restriction map

$$\rho_F \colon (\operatorname{Ext}^1(\ ,A), F) \to \operatorname{Hom}_{\underline{\Gamma}}(\operatorname{Ext}^1(Y,A), F(Y))$$

with $\rho_F(\eta) = \eta_Y$, is bijective. Namely, the exact sequence $0 \to A \to Q \to L \to 0$ with Q relative-injective in $\mathcal{P}_R(\Lambda)$ shows that there is a presentation

$$(\ ,Q) \to (\ ,L) \to \operatorname{Ext}^1(\ ,A) \to 0.$$

Since both Q, L are direct summands of Y, we similarly can use a presentation

$$(\ ,n_1 Y) \to (\ ,n_0 Y) \to \operatorname{Ext}^1(\ ,A) \to 0.$$

We have a sequence of canonical isomorphisms

$$((\ ,nY),F) \approx ((\ ,nY),F) \approx nF(Y) \approx (n({}_\Gamma\Gamma),F(Y)) \approx \mathrm{Hom}_\Gamma((Y,nY),F(Y)),$$

the second one being given by the Yoneda lemma. We use this isomorphism twice in the following commutative diagram

$$\begin{array}{ccccccc}
0 & \longrightarrow & (\mathrm{Ext}^1(\ ,A),F) & \longrightarrow & ((\ ,n_0Y),F) & \longrightarrow & ((\ ,n_1Y),F) \\
 & & \downarrow{\scriptstyle \rho_F} & & \downarrow{\scriptstyle \approx} & & \downarrow{\scriptstyle \approx} \\
0 & \longrightarrow & (\mathrm{Ext}^1(Y,A),F(Y)) & \longrightarrow & ((Y,n_0Y),F(Y)) & \longrightarrow & ((Y,n_1Y),F(Y))
\end{array}$$

the first vertical map being the restriction map ρ_F. As a consequence, ρ_F is an isomorphism.

In order to show that $\mathrm{Ext}^1(Y,A)$ as a $\underline{\Gamma}$-module is injective, consider a monomorphism

$$h\colon \mathrm{Ext}^1(Y,A) \to M$$

of $\underline{\Gamma}$-modules, where M is a finitely generated $\underline{\Gamma}$-module. We want to prove that h splits. (This then implies that $\mathrm{Ext}^1_{\underline{\Gamma}}(M',\mathrm{Ext}^1(Y,A)) = 0$ for any finitely generated $\underline{\Gamma}$-module M', since $\mathrm{Ext}^1(Y,A)$ is itself finitely generated; thus that $\mathrm{Ext}^1(Y,A)$ is injective as a $\underline{\Gamma}$-module). Since M is a finitely generated $\underline{\Gamma}$-module, and $\underline{\Gamma}$ is noetherian, there is an exact sequence of $\underline{\Gamma}$-modules

$$\underline{\mathrm{Hom}}(Y,m_1Y) \xrightarrow{\mathrm{Hom}(1,f)} \underline{\mathrm{Hom}}(Y,m_0Y) \to M \to 0,$$

with $f\colon m_1Y \to m_0Y$. We define the functor F on $\mathcal{P}_R(\Lambda)/\mathcal{P}$ by the following exact sequence

$$\underline{\mathrm{Hom}}(\ ,m_1Y) \xrightarrow{\mathrm{Hom}(\ ,f)} \underline{\mathrm{Hom}}(\ ,m_0Y) \to F \to 0,$$

in particular, $F(Y) = M$. We consider now the restriction map $\rho_F\colon (\mathrm{Ext}^1(\ ,A),F) \to \mathrm{Hom}_{\underline{\Gamma}}(\mathrm{Ext}^1(Y,A),F(Y))$. By the discussion above ρ_F is bijective. Thus, let $\xi\colon \mathrm{Ext}^1(\ ,A) \to F$ be a natural transformation with $\rho_F(\xi) = h$. We claim that ξ is mono. The almost split sequence

$$0 \to A \to B \to C \to 0$$

gives a presentation of the simple functor S_C corresponding to C

$$0 \to (\ ,A) \to (\ ,B) \to (\ ,C) \xrightarrow{\pi} S_C \to 0;$$

of course, S_C has to be just the image of the connecting transformation $(\ ,C) \xrightarrow{\delta} \mathrm{Ext}^1(\ ,A)$, since also

$$0 \to (\ ,A) \to (\ ,B) \to (\ ,C) \xrightarrow{\delta} \mathrm{Ext}^1(\ ,A)$$

is exact. Thus, let $\delta = \pi\iota$, where we can assume ι to be an inclusion map of the simple functor S_C as a subfunctor of $\mathrm{Ext}^1(\ ,A)$. Since $\mathrm{Ext}^1(\ ,A)$ is an indecomposable injective object of $\mathrm{mod}(\mathcal{P}_R(\Lambda)/\mathcal{P})$, it follows that S_C is the socle of $\mathrm{Ext}^1(\ ,A)$, and since $\mathrm{Ext}^1(\ ,A)$ is the injective envelope for S_C, it is essential in $\mathrm{Ext}^1(\ ,A)$. Therefore, in order to show that ξ is mono, we only have to check that the composition

$$S_C \xrightarrow{\iota} \mathrm{Ext}^1(\ ,A) \xrightarrow{\xi} F$$

is non-zero. However, inserting Y, we obtain $\iota_Y\xi_Y$, and both ι_Y and $\xi_Y = h$ are monomorphisms, whereas $S_C(Y) \neq 0$, since by construction, C is a direct summand of Y. Thus

$$S_C(Y) \xrightarrow{\iota_Y\xi_Y} F(Y)$$

is non-zero, therefore $\iota\xi$ is non-zero, and consequently, ξ is mono.

Since $\operatorname{Ext}^1(\ ,A)$ is an injective object in $\operatorname{mod}(\mathcal{P}_R(\Lambda)/\mathcal{P})$, it follows that ξ splits, say $\xi\eta = 1_{\operatorname{Ext}^1(\ ,A)}$ for some η, thus $\xi_Y n_Y = 1_{\operatorname{Ext}^1(Y,A)}$. This shows that $h = \xi_Y$ is split mono. □

LEMMA 3.2. *Let S be a commutative local noetherian ring, Σ an S-algebra, finitely generated as S-module. If I is a finitely generated injective Σ-module, then I is of finite length over S (and over Σ).*

PROOF. Let $\underline{a}$ be the unique maximal S-submodule of Σ of finite length (it exists, since Σ is a noetherian R-module). Note that the socle of $\Sigma/\underline{a}$ is zero, and that $\underline{a}$ is a twosided ideal of Σ. If $\underline{a} = \Sigma$, then I as an S-module is of finite length. Thus, assume $\underline{a} \neq \Sigma$. The exact sequence $0 \to \underline{a} \xrightarrow{\iota} \Sigma \to \Sigma/\underline{a} \to 0$ with ι the inclusion map, gives the exact sequence

$$(*) \qquad 0 \to (\Sigma/\underline{a}, I) \to (\Sigma, I) \to (\underline{a}, I) \to 0,$$

using that I is injective. We claim that $(\Sigma/\underline{a}, I) = 0$. Since the socle of $\Sigma/\underline{a}$ is zero, $\operatorname{rad} S$ does not belong to $\operatorname{Ass}(\Sigma/\underline{a})$, (see, for example, [**Bou**]), thus there exists $x \in \operatorname{rad} S$ which is not a zero-divisor on $\Sigma/\underline{a}$. We denote by x also the corresponding multiplication map on $\Sigma/\underline{a}$, say with coker Q, and apply $(-, I)$ to the exact sequence

$$0 \to \Sigma/\underline{a} \xrightarrow{x} \Sigma/\underline{a} \to Q \to 0.$$

Since I is injective, we obtain the exact sequence

$$0 \to (Q, I) \to (\Sigma/\underline{a}, I) \xrightarrow{(x,I)} (\Sigma/\underline{a}, I) \to 0.$$

Now, the endomorphism (x, I) of $(\Sigma/\underline{a}, I)$ is again given by multiplication by x, and since $x \in \operatorname{rad} S$, we see that

$$\operatorname{rad} S \circ (\Sigma/\underline{a}, I) = (\Sigma/\underline{a}, I)$$

But $(\Sigma/\underline{a}, I)$ is a finitely generated S-module, thus $(\Sigma/\underline{a}, I) = 0$ according to the Nakayama lemma. It follows from $(*)$ that we have isomorphisms of Σ-modules

$$I \approx (\Sigma, I) \approx (\underline{a}, I).$$

However, $\underline{a}$ being an S-module of finite length, is annihilated by $(\operatorname{rad} S)^n$ for some n, thus also I is annihilated by $(\operatorname{rad} S)^n$. Thus, ${}_SI$ as a finitely generated $S/(\operatorname{rad} S)^n$-module is of finite length over S. Therefore, I also as a Σ-module, is of finite length. □

LEMMA 3.3. *Let $0 \to Y' \to P \xrightarrow{\pi} Y \to 0$ be exact, with P projective. Let X be arbitrary. Then the kernel of the connecting map*

$$\delta \colon (X, Y) \to \operatorname{Ext}^1(X, Y')$$

is the set of all maps $X \to Y$ factoring through a projective module, thus the image of δ is just $\underline{\operatorname{Hom}}(X, Y)$.

PROOF. There is the exact sequence

$$(X, P) \xrightarrow{(X,\pi)} (X, Y) \xrightarrow{\delta} \operatorname{Ext}^1(X, Y'),$$

and the image of (X, π) is the set of maps of the form $f\pi$, with $f \in (X, P)$. Thus, any element of the kernel of δ factors through P. Conversely, given $g_1 \colon X \to P'$, $g_2 \colon P' \to Y$, with P' projective. Since π is surjective, and P' projective, there is g_2' with $g_2'\pi = g_2$, thus $g_1 g_2 = (g_1 g_2')\pi$ lies in the image of (X, π), thus in the kernel of δ. □

With the above results in mind we are able to prove the first implication of the main theorem.

PROPOSITION 3.4. *Assume $\mathcal{P}_R(\Lambda)$ has left almost split sequences. Then $\ell_R(\underline{\mathrm{Hom}}(X,Y)) < \infty$ for all $X, Y \in \mathcal{P}_R(\Lambda)$.*

PROOF. Let $0 \to A \to P \xrightarrow{\pi} Y \to 0$ be exact in $\mathcal{P}_R(\Lambda)$, with $P \in \mathcal{P}(\Lambda)$. Then $\underline{\mathrm{Hom}}(X,Y)$ is an R-submodule of $\mathrm{Ext}^1(X,A)$, according to Lemma 3.3. Thus, it is sufficient to show that $\ell_R(\mathrm{Ext}^1(X,A)) < \infty$. If A is relative injective, then $\mathrm{Ext}^1(X,A) = 0$. Thus, assume A is not relative injective. Also, we may assume that A is indecomposable. Since we assume that $\mathcal{P}_R(\Lambda)$ has left almost split sequences, there is an almost split sequence $0 \to A \to B \to C \to 0$. According to Proposition 3.1, there is $A' \in \mathcal{P}_R(\Lambda)$, such that $\mathrm{Ext}^1(X \amalg A', A)$ is an injective $\underline{\Gamma}$-module, with $\underline{\Gamma} = (\underline{\mathrm{End}}(X \amalg A'))^{\mathrm{op}}$. However, $\underline{\Gamma}$ is an R-algebra and finitely generated as an R-module, and $\mathrm{Ext}^1(X \amalg A', A)$ is a finitely generated injective $\underline{\Gamma}$-module. Thus, $\mathrm{Ext}^1(X \amalg A', A)$ is of finite length over R, according to Lemma 3.2. Since $\mathrm{Ext}^1(X,A)$ can be considered as an R-submodule of $\mathrm{Ext}^1(X \amalg A', A)$, we also have $\ell_R(\mathrm{Ext}^1(X,A)) < \infty$. □

4. The implication (b) ⇒ (c) of the main theorem

We are now going to prove directly that if $\ell_R(\underline{\mathrm{Hom}}(X,Y)) < \infty$ for all $X, Y \in \mathcal{P}_R(\Lambda)$ then the localization C_p of C with respect to any non-maximal prime ideal p of R is Λ_p-projective.

LEMMA 4.1. *Assume $\ell_R(\underline{\mathrm{Hom}}(X,Y)) < \infty$, for all $X, Y \in \mathcal{P}_R(\Lambda)$. Then C_p is Λ_p-projective, for any $C \in \mathcal{P}_R(\Lambda)$, and any non-maximal prime ideal p of R.*

PROOF. Let $C \in \mathcal{P}_R(\Lambda)$, and p a non-maximal prime ideal of R. We want to show that C_p is Λ_p-projective. We may assume C to be indecomposable. Choose an exact sequence $0 \to K \to P \xrightarrow{\pi} C \to 0$ in $\mathcal{P}_R(\Lambda)$ with $P \in P(\Lambda)$. Applying $(C, -)$, we obtain the following exact sequence

$$(C,P) \xrightarrow{(1,\pi)} (C,C) \to \underline{\mathrm{Hom}}(C,C) \to 0,$$

according to Lemma 3.3. Now, localize with respect to p. Since $\ell_R(\underline{\mathrm{Hom}}(C,C)) < \infty$, it follows that $\underline{\mathrm{Hom}}(C,C)_p = 0$. Since localization is exact, and since we can identify $(X,Y)_p = \mathrm{Hom}_{\Lambda_p}(X_p, Y_p)$, we see that the map

$$\mathrm{Hom}_{\Lambda_p}(C_p, P_p) \xrightarrow{(1,\pi)} \mathrm{Hom}_{\Lambda_p}(C_p, C_p)$$

is surjective. As a consequence, the identity map of C_p factors through π, thus C_p is a direct summand of P_p, and therefore a projective Λ_p-module. □

5. The implication (c) ⇒ (d) of the main theorem

In this section we are going to prove that given a prime ideal p of R such that for all modules $C \in \mathcal{P}_R(\Lambda)$ the Λ_p-module C_p is projective, then Λ_p is nonsingular. Of course, applying this to all non-maximal prime ideals of R we obtain just the implication (c) ⇒ (d).

LEMMA 5.1. *Let R be a local noetherian ring, and Λ an R-algebra which is finitely generated free as an R-module. Then the following two assertions are equivalent:*

(i) $\operatorname{Hom}_R(\Lambda, R)$ *is a projective* Λ^{op}*-module.*
(ii) $\operatorname{Hom}_R(\Lambda^{\text{op}}, R)$ *is a projective* Λ*-module.*

PROOF. First, let us note that we may assume, in addition, that R is complete. Namely, let m be the maximal ideal of R. The m-completion R_m of the local noetherian ring R is faithfully flat (see for example [**Bou**], § 3.3. Prop. 6 and § 3.5, Prop. 9). It follows that a finitely generated Λ-module M is projective if and only if the Λ_m-module M_m $(= M \underset{R}{\otimes} R_m)$ is projective.

(i) $\Rightarrow$ (ii): Let ${}_\Lambda\Lambda = \amalg_{i=1}^n (P_i^{n_i})$, where the P_i form a complete set of nonisomorphic indecomposable projective Λ-modules. Since we assume that $\operatorname{Hom}_R(\Lambda, R)$ is Λ^{op}-projective, it follows that the Λ^{op}-modules $\operatorname{Hom}_R(P_i, R)$, $1 \leq i \leq n$ are pairwise non-isomorphic indecomposable projective Λ^{op}-modules. Since a complete set of non-isomorphic indecomposable projective Λ^{op}-modules has precisely n elements (Λ being semi-perfect), we see that $\operatorname{Hom}_R(P_i, R)$, $1 \leq i \leq n$, is such a complete set. Hence $\Lambda^{\text{op}} = \amalg_{i=1}^n \operatorname{Hom}_R(P_i, R)^{m_i}$ for some $m_i \geq 1$. Dualizing again with respect to $\operatorname{Hom}_R(-, R)$, we obtain that $\operatorname{Hom}_R(\Lambda^{\text{op}}, R) = \amalg_{i=1}^n P_i^{n_i}$ is a projective Λ-module.

(ii) $\Rightarrow$ (i): This follows from the first implication by replacing Λ by Λ^{op}. □

REMARK. Under the same assumptions, also the following conditions are equivalent:

(i) $\operatorname{Hom}_R(\Lambda, R)$ is a projective Λ^{op}-module.
(ii) A Λ-module $X \in \mathcal{P}_R(\Lambda)$ is Λ-projective if and only if $\operatorname{Hom}_R(X, R)$ is Λ^{op}-projective.
(iii) The duality $\operatorname{Hom}_R(-, R)\colon \mathcal{P}_R(\Lambda) \to \mathcal{P}_R(\Lambda^{\text{op}})$ induces a duality $\mathcal{P}(\Lambda) \to \mathcal{P}(\Lambda^{\text{op}})$.

This is a direct consequence of the lemma.

5.2. *Let* Λ *be an* R*-algebra which is a finitely generated free* R*-module, and assume* $\operatorname{Hom}_R({}_\Lambda\Lambda, R)$ *is* Λ^{op}*-projective. Then* $\operatorname{inj.dim.}_\Lambda \Lambda = \operatorname{inj.dim.}_R R$.

PROOF. The inequality "$\geq$" has been shown in (1.2). For the converse inequality let $\operatorname{inj.dim.}_R R \leq m$, and let

$$0 \to R \to J_0 \to \cdots \to J_m \to 0$$

be an injective resolution of ${}_R R$. Apply $\operatorname{Hom}_R(\operatorname{Hom}_R(\Lambda, R),\ )$ to this sequence: the sequence

$$(*) \quad 0 \to \operatorname{Hom}_R(\operatorname{Hom}_R(\Lambda, R), R) \to \operatorname{Hom}_R(\operatorname{Hom}_R(\Lambda, R), J_0) \to \cdots \\ \to \operatorname{Hom}_R(\operatorname{Hom}_R(\Lambda, R), J_m) \to 0$$

is exact, since $\operatorname{Hom}_R(\Lambda, R)$ is a projective R-module. Since J_i is R-injective and $\operatorname{Hom}_R({}_\Lambda\Lambda, R)$ is Λ^{op}-projective, we see that $\operatorname{Hom}_R(\operatorname{Hom}_R({}_\Lambda\Lambda, R), J_i)$ is an injective Λ-module. Finally, $\operatorname{Hom}_R(\operatorname{Hom}_R({}_\Lambda\Lambda, R), R) \approx {}_\Lambda\Lambda$, since Λ as an R-module is finitely generated free. Thus, the sequence $(*)$ is an injective resolution of ${}_\Lambda\Lambda$, therefore $\operatorname{inj.dim.}_\Lambda \Lambda \leq m = \operatorname{inj.dim.}_R R$. □

Now we are able to prove the implication (c) $\Rightarrow$ (d) of the main theorem.

PROPOSITION 5.3. *Let* R *be a complete regular local ring, and* p *a prime ideal of* R*. Let* Λ *be an* R*-algebra which is finitely generated free as an* R*-module. Assume* C_p *is* Λ_p*-projective for all* $C \in \mathcal{P}_R(\Lambda)$*. Then* Λ_p *is nonsingular.*

Proof. Since $\operatorname{Hom}_R(\Lambda^{\mathrm{op}}, R)$ belongs to $\mathcal{P}_R(\Lambda)$, our assumption says, in particular, that $(\operatorname{Hom}_R(\Lambda^{\mathrm{op}}, R))_p = \operatorname{Hom}_{R_p}(\Lambda_p^{\mathrm{op}}, R_p)$ is Λ_p-projective. According to (5.1), this is equivalent to $\operatorname{Hom}_{R_p}(\Lambda_p, R_p)$ being Λ_p^{op}-projective. Therefore, we can apply (5.2), and see that we have $\operatorname{inj.dim.}_{\Lambda_p} \Lambda_p = \operatorname{inj.dim.} R_p = \dim R_p$. According to (1.3), it is sufficient to show that $\operatorname{gl.dim.} \Lambda_p$ is finite, in order to conclude that $\operatorname{gl.dim.} \Lambda_p = \operatorname{inj.dim.}_{\Lambda_p} \Lambda_p = \dim R_p$.

Let M be a finitely generated Λ_p-module. We claim that $\operatorname{proj.dim.} M \leq \dim R = d$. Let $m_1, \dots, m_n$ be a generating set for M as an Λ_p-module, and let A be the Λ-submodule of M generated by $m_1, \dots, m_n$. Since localization with respect to p is exact, we get from $A \subseteq M$ that $A_p \subseteq M_p = M$. Since A_p contains a generating set of the Λ_p-module M, we even have $A_p = M$. Consider a presentation of the Λ-module A

$$0 \to \Omega^d A \to P_{d-1} \to \cdots \to P_0 \to A \to 0$$

with all $P_i \in \mathcal{P}_R(\Lambda)$. Since Λ is a free R-module, the modules P_i, considered as R-modules, are also projective. But $d = \dim R = \operatorname{gl.dim.} R$, since R is regular, therefore $\Omega^d A$, as an R-module, is also projective. Hence $\Omega^d A \in \mathcal{P}_R(\Lambda)$. By assumption, $(\Omega^d A)_p$ is Λ_p-projective, therefore the exact sequence

$$0 \to (\Omega^d A)_p \to (P_{d-1})_p \to \cdots \to (P_0)_p \to M \to 0$$

obtained by localization, is a projective resolution of M, therefore $\operatorname{proj.dim.} M \leq d$. □

6. $\mathcal{P}_R(\Lambda)$ is coherent

For the proof of the implication (d) ⇒ (a) of the main theorem, saying that if Λ is an isolated singularity then $\mathcal{P}_R(\Lambda)$ has almost split sequences, we need some preliminary steps. The first is to show that $\mathcal{P}_R(\Lambda)$ is a coherent category. This will be done in this section.

Proposition 6.1. *Let R be a complete local ring, Λ an R-algebra which is a finitely generated free R-module. Let $A \in \operatorname{mod} \Lambda$, and assume there exists an exact sequence*

$$0 \to B_n \to B_{n-1} \to \cdots \to B_0 \to A \to 0,$$

with all $B_j \in \mathcal{P}_R(\Lambda)$. Then the functors $\operatorname{Ext}^i_\Lambda(\ ,A)|_{\mathcal{P}_R(\Lambda)}$ are finitely presented, for all $i \geq 0$.

Remark. We are mainly interested in the case $i = 0$, thus in the fact, that $(\ ,A)|_{\mathcal{P}_R(\Lambda)}$ is finitely presented. However, the general result is easier to prove.

Proof. Induction on n. If $n = 0$, then $A \in \mathcal{P}_R(\Lambda)$, hence the functors $\operatorname{Ext}^i_\Lambda(\ ,A)|_{\mathcal{P}_R(\Lambda)}$ are finitely presented, see (2.4).

Now, for $n \geq 1$, there is an exact sequence $0 \to K \to B_0 \to A \to 0$, with $B_0 \in \mathcal{P}_R(\Lambda)$, and such that, by induction, all functors $\operatorname{Ext}^i_\Lambda(\ ,K)|_{\mathcal{P}_R(\Lambda)}$ are finitely presented. The exact sequence $0 \to K \to B_0 \to A \to 0$ gives rise to the following long exact sequence of functors:

$$0 \longrightarrow (\ ,K)|_{\mathcal{P}_R(\Lambda)} \xrightarrow{f_0} (\ ,B_0)|_{\mathcal{P}_R(\Lambda)} \longrightarrow (\ ,A)|_{\mathcal{P}_R(\Lambda)} \longrightarrow$$
$$\operatorname{Ext}^1(\ ,K)|_{\mathcal{P}_R(\Lambda)} \xrightarrow{f_1} \operatorname{Ext}^1(\ ,B_0)|_{\mathcal{P}_R(\Lambda)} \longrightarrow \operatorname{Ext}^1(\ ,A)|_{\mathcal{P}_R(\Lambda)} \longrightarrow \cdots$$
$$\longrightarrow \operatorname{Ext}^i(\ ,K)|_{\mathcal{P}_R(\Lambda)} \xrightarrow{f_i} \operatorname{Ext}^i(\ ,B_0)|_{\mathcal{P}_R(\Lambda)} \longrightarrow \operatorname{Ext}^i(\ ,A)|_{\mathcal{P}_R(\Lambda)} \longrightarrow \cdots$$

We therefore can include the functors $\operatorname{Ext}^1(\ ,A)|_{\mathcal{P}_R(\Lambda)}$ in the following short exact sequence

$$(*)\qquad 0 \to \operatorname{Cok} f_i \to \operatorname{Ext}^i(\ ,A)|_{\mathcal{P}_R(\Lambda)} \to \operatorname{Ker} f_{i+1} \to 0$$

and we claim that all the functors $\operatorname{Cok} f_i$ and $\operatorname{Ker} f_{i+1}$, $i \geq 0$, are finitely presented. Since $(\ ,K)|_{\mathcal{P}_R(\Lambda)}$ is finitely presented, there is $K' \in \mathcal{P}_R(\Lambda)$, with an exact sequence

$$(\ ,K')|_{\mathcal{P}_R(\Lambda)} \xrightarrow{g} (\ ,K)|_{\mathcal{P}_R(\Lambda)} \to 0,$$

thus, there is the exact sequence

$$(\ ,K')|_{\mathcal{P}_R(\Lambda)} \xrightarrow{gf_0} (\ ,B_0)|_{\mathcal{P}_R(\Lambda)} \to \operatorname{Coker} f_0 \to 0$$

which shows that $\operatorname{Cok} f_0$ is finitely presented. On the other hand, for $i \geq 1$, the functors $\operatorname{Ext}^i(\ ,K)|_{\mathcal{P}_R(\Lambda)}$ and $\operatorname{Ext}^i(\ ,B_0)|_{\mathcal{P}_R(\Lambda)}$ vanish on $\mathcal{P}(\Lambda)$, and thus they belong to $\operatorname{mod}(\mathcal{P}_R(\Lambda)/\mathcal{P}(\Lambda))$. Since we know that $\mathcal{P}_R(\Lambda)/\mathcal{P}(\Lambda)$ is coherent (2.2), it follows that both $\operatorname{Ker} f_i$, $\operatorname{Cok} f_i$ (for $i \geq 1$) also belong to $\operatorname{mod}(\mathcal{P}_R(\Lambda)/\mathcal{P}(\Lambda))$, and therefore to $\operatorname{mod}\mathcal{P}_R(\Lambda)$. The sequence $(*)$ now implies that all $\operatorname{Ext}^i(\ ,A)|_{\mathcal{P}_R(\Lambda)}$ belong to $\operatorname{mod}\mathcal{P}_R(\Lambda)$. □

PROPOSITION 6.2. *Let R be a complete regular local ring, Λ an R-algebra which is finitely generated free as an R-module. Let $A \in \operatorname{mod}\Lambda$. Then all functors $\operatorname{Ext}^i_\Lambda(\ ,A)|_{\mathcal{P}_R(\Lambda)}$ are finitely presented.*

PROOF. Let $d = \dim R = \operatorname{gl.dim.} R$. Given $A \in \operatorname{mod}\Lambda$, consider a projective resolution

$$0 \to B_d \to P_{d-1} \to \cdots \to P_1 \to P_0 \to A \to 0,$$

with all $P_i \in \mathcal{P}(\Lambda)$, $0 \leq i \leq d-1$. As a sequence of R-modules, it is exact, and all P_i are projective R-modules, thus also B_d, as an R-module, is projective. Since B_d also is finitely generated (as an R-module, or as a Λ-module), $B_d \in \mathcal{P}_R(\Lambda)$. Of course, all $P_i \in P(\Lambda) \subseteq \mathcal{P}_R(\Lambda)$, thus we can apply the previous result. □

6.3. Recall that a full subcategory $\mathcal{C}$ of the additive category $\mathcal{A}$ is said to be *contravariantly finite*, provided given $A \in \mathcal{A}$, there is a map $f\colon A' \to A$ with $A' \in \mathcal{C}$ such that any map $g\colon C \to A$ with $C \in \mathcal{C}$ factors through f, say $g = g'f$ with $g'\colon C \to A'$. Note that $\mathcal{C}$ is contravariantly finite in $\mathcal{A}$, if and only if for any $A \in \mathcal{A}$, the functor $(\ ,A)|_{\mathcal{C}}$ is finitely generated. The dual notion is as follows: The full subcategory $\mathcal{C}$ of the additive category $\mathcal{A}$ is said to be *covariantly finite* provided for any $A \in \mathcal{A}$ there is a map $h\colon A \to A''$ with $A'' \in \mathcal{C}$ such that any map from A to an object in $\mathcal{C}$ factors through h; or, equivalently, provided the functor $(A,\ )|_{\mathcal{C}}$ is finitely generated. In case $\mathcal{C}$ is both contravariantly finite and covariantly finite in $\mathcal{A}$, we say that $\mathcal{C}$ is *functorially finite* in $\mathcal{A}$. Note that if $\mathcal{A}$ is an additive category with kernels and $\mathcal{C}$ is contravariantly finite in $\mathcal{A}$, then $\mathcal{C}$ is coherent. (For, given a map $c\colon C_1 \to C_2$ in $\mathcal{C}$, let $k\colon K \to C$ be a kernel of c in $\mathcal{A}$. Since $\mathcal{C}$ is contravariantly finite in $\mathcal{A}$, there is $f\colon K' \to K$ with K' in $\mathcal{C}$ such that any map from an object in $\mathcal{C}$ to K factors through f. Thus the sequence

$$(\ ,K) \xrightarrow{(\ ,fh)} (\ ,C_1) \xrightarrow{(\ ,c)} (\ ,C_2)$$

of functors $\mathcal{C} \to \mathrm{Ab}$ is exact.)

COROLLARY 6.4. *Let R be a complete regular local ring, and Λ an R-algebra which is finitely generated free as an R-module. Then $\mathcal{P}_R(\Lambda)$ is contravariantly finite in* $\operatorname{mod}\Lambda$.

COROLLARY 6.5. *Let R be a complete regular local ring, and Λ an R-algebra which is finitely generated as an R-module. Then $\mathcal{P}_R(\Lambda)$ is coherent.*

PROOF. Since Λ is noetherian, $\operatorname{mod}\Lambda$ is abelian, in particular, it has kernels. The assertion now follows from the previous corollary. □

Let us end this section by showing that under our assumptions, $\mathcal{P}_R(\Lambda)$ is also covariantly finite in $\operatorname{mod}\Lambda$. We will obtain this assertion as a rather immediate consequence of the fact that $\mathcal{P}_R(\Lambda^{\mathrm{op}})$ is contravariantly finite in $\operatorname{mod}\Lambda^{\mathrm{op}}$.

PROPOSITION 6.6. *Let R be a complete regular local ring, and Λ an R-algebra which is finitely generated free as an R-module. Then $\mathcal{P}_R(\Lambda)$ is functorially finite in* $\operatorname{mod}\Lambda$.

PROOF. Let $A \in \operatorname{mod}\Lambda$. We consider $\operatorname{Hom}_R(A, R)$, this is a Λ^{op}-module which is finitely generated as an R-module, thus also finitely generated as Λ^{op}-module. Since $\mathcal{P}_R(\Lambda^{\mathrm{op}})$ is contravariantly finite in $\operatorname{mod}\Lambda^{\mathrm{op}}$, according to (6.4) applied to the R-algebra Λ^{op}, there is a map $f\colon X \to \operatorname{Hom}_R(A, R)$ with $X \in \mathcal{P}_R(\Lambda^{\mathrm{op}})$, such that any other map $X' \to \operatorname{Hom}_R(A, R)$ with $X' \in \mathcal{P}_R(\Lambda^{\mathrm{op}})$ factors through it. Now, assume there is given a map $g\colon A \to Y$, with $A \in \mathcal{P}_R(\Lambda)$. Then $\operatorname{Hom}(g, R)\colon \operatorname{Hom}_R(Y, R) \to \operatorname{Hom}_R(A, R)$ is a map with $\operatorname{Hom}_R(Y, R)$ in $\mathcal{P}_R(\Lambda^{\mathrm{op}})$, thus there is $g'\colon \operatorname{Hom}_R(Y, R) \to X$ such that $\operatorname{Hom}(g, R) = g'f$. For any Λ-module M, denote by $u_M\colon M \to \operatorname{Hom}_R(\operatorname{Hom}_R(M, R), R)$ the canonical map; note that u_M is an isomorphism for M in $\mathcal{P}_R(\Lambda)$. We have

$$u_A \operatorname{Hom}(f, R) \operatorname{Hom}(g', R) = u_A \operatorname{Hom}(\operatorname{Hom}(g, R), R) = g u_Y$$

thus

$$(u_A \operatorname{Hom}(f, R))(\operatorname{Hom}(g', R) u_Y^{-1}) = g$$

shows that we can factor g through the map $u_A \operatorname{Hom}(f, R)\colon A \to \operatorname{Hom}_R(X, R)$, and $\operatorname{Hom}_R(X, R)$ belongs to $\mathcal{P}_R(\Lambda)$. This finishes the proof. □

7. Coresolutions by tensor functors

In this section we want to introduce a duality D for arbitrary noetherian rings Λ. This duality $D\colon \operatorname{mod}((\operatorname{mod}\Lambda)^{\mathrm{op}}) \to \operatorname{mod}((\operatorname{mod}\Lambda^{\mathrm{op}})^{\mathrm{op}})$ will be defined below by $DF(X) = (F, X\otimes\ )$ for $F \in \operatorname{mod}((\operatorname{mod}\Lambda)^{\mathrm{op}})$ and $X \in \operatorname{mod}\Lambda^{\mathrm{op}}$. This duality will be used in the next section to prove that several functors on $\mathcal{P}_R(\Lambda)$ are finitely presented by means of the following lemma:

LEMMA 7.1. *Let Λ be a noetherian ring. Let $F \in \operatorname{mod}((\operatorname{mod}\Lambda)^{\mathrm{op}})$. Then there exists an exact sequence $A \xrightarrow{f} B \xrightarrow{g} C \to 0$ in $\operatorname{mod}\Lambda^{\mathrm{op}}$ such that there is an exact sequence*

$$0 \to F \to A\otimes \xrightarrow{f\otimes} B\otimes \xrightarrow{g\otimes} C\otimes \to 0.$$

PROOF. Since Λ is noetherian, both $\operatorname{mod}\Lambda$ and $\operatorname{mod}\Lambda^{\mathrm{op}}$ are abelian. □

(1) Given $B \in \operatorname{mod}\Lambda^{\mathrm{op}}$, the functor $B\otimes\colon \operatorname{mod}\Lambda \to \mathrm{Ab}$ belongs to $\operatorname{mod}((\operatorname{mod}\Lambda)^{\mathrm{op}})$.

PROOF. Since B is a finitely presented Λ^{op}-module, there is a projective presentation

$$P_1 \to P_0 \to B \to 0$$

with $P_0, P_1 \in \mathcal{P}(\Lambda^{\mathrm{op}})$. Since the tensor product is right exact, we obtain an exact sequence of functors

$$P_1\otimes \to P_0\otimes \to B\otimes \to 0.$$

Also, $P_i\otimes \simeq ((P_1, \Lambda),\ )$, since P_i is a finitely generated projective Λ^{op}-module. Since (P_i, Λ) is a finitely generated projective Λ-module, it belongs to $\operatorname{mod}\Lambda$, thus the exact sequence

$$((P_1, \Lambda),\ ) \to ((P_0, \Lambda),\ ) \to B\otimes \to 0$$

shows that $B\otimes$ is finitely presented. □

(2) Given $A, B \in \operatorname{mod}\Lambda^{\mathrm{op}}$, we have $(A\otimes\ , B\otimes\ ) \simeq (A, B)$.

PROOF. Any map $f\colon A \to B$ gives rise to a transformation $f\otimes\colon A\otimes \to B\otimes$. Conversely, given $\eta = (\eta_M)_M\colon A\otimes \to B\otimes$, let $f = \eta_\Lambda\colon A = A\otimes\Lambda \to B\otimes\Lambda = B$. It is easily checked that $\eta = f\otimes$. □

(3) Given $F \in \operatorname{mod}((\operatorname{mod}\Lambda)^{\mathrm{op}})$, and $X \in \operatorname{mod}\Lambda^{\mathrm{op}}$, then $\operatorname{Ext}^1(F, X\otimes\ ) = 0$.

PROOF. Since F is finitely presented, and $\operatorname{mod}\Lambda$ is abelian, there is an exact sequence

$$A \xrightarrow{u} B \xrightarrow{v} C \to 0$$

such that

$$0 \to (C,\ ) \to (B,\ ) \to (A,\ ) \to F \to 0$$

is a resolution of F. Thus we can use it to calculate $\operatorname{Ext}^1(F, X\otimes\ )$ from the homology of the induced complex, after applying $(\ , X\otimes\ )$:

$$\begin{array}{ccccccc}
0 \longrightarrow (F, X\otimes\ ) \longrightarrow & ((X,\ ), X\otimes\ ) & \longrightarrow & ((B,\ ), X\otimes\ ) & \longrightarrow & ((C,\ ), X\otimes\ ) \\
& \downarrow\approx & & \downarrow\approx & & \downarrow\approx \\
& X\otimes A & \xrightarrow{X\otimes u} & X\otimes B & \xrightarrow{X\otimes v} & X\otimes C
\end{array}$$

The isomorphisms are those of the Yoneda-lemma. Since the lower sequence is exact, the homology at $((B,\ ), X\otimes)$ is zero, and this is just $\operatorname{Ext}^1(F, X\otimes\ )$. □

(3′) Given $X \in \operatorname{mod}\Lambda^{\mathrm{op}}$, the functor $X\otimes$ is an injective object of $\operatorname{mod}((\operatorname{mod}\Lambda)^{\mathrm{op}})$.

PROOF. According to (1), $X\otimes$ belongs to $\operatorname{mod}((\operatorname{mod}\Lambda)^{\mathrm{op}})$, and it is injective, according to (3). □

Let us introduce a functor $D\colon \operatorname{Mod}((\operatorname{mod}\Lambda)^{\mathrm{op}}) \to \operatorname{Mod}((\operatorname{mod}\Lambda^{\mathrm{op}})^{\mathrm{op}})$, by $D(F)(X) = (F, X\otimes\ )$ where $F \in \operatorname{Mod}((\operatorname{mod}\Lambda)^{\mathrm{op}})$, thus $D(F)$ is a covariant functor from $(\operatorname{mod}\Lambda)^{\mathrm{op}}$ to Ab, and $X \in \operatorname{mod}\Lambda^{\mathrm{op}}$. Actually, we only will be interested in the restriction of D to $\operatorname{mod}((\operatorname{mod}\Lambda)^{\mathrm{op}})$.

(4) Given $A \in \operatorname{mod}\Lambda^{\mathrm{op}}$, we have $D(A\otimes\ ) = (A,\ )$.

PROOF. For $X \in \operatorname{mod}\Lambda^{\mathrm{op}}$, $D(A\otimes\ )(X) = (A\otimes\ , X\otimes\ ) = (A, X)$, according to (2). □

(5) Given $C \in \operatorname{mod}\Lambda$, we have $D((C,\ )) = \ \otimes C$.

PROOF. For $X \in \operatorname{mod}\Lambda^{\mathrm{op}}$, $D((C,\ ))(X) = ((C,\ ), X\otimes\ ) = X\otimes C$, according to Yoneda-lemma. □

(6) The restriction of D to $\operatorname{mod}((\operatorname{mod}\Lambda)^{\mathrm{op}})$ is exact.

PROOF. Suppose, given $0 \to F_1 \to F_2 \to F_3 \to 0$, exact in $\operatorname{mod}((\operatorname{mod}\Lambda)^{\mathrm{op}})$. Applying $(\ , X\otimes\ )$, we obtain

$$0 \to (F_3, X\otimes\ ) \to (F_2, X\otimes\ ) \to (F_1, X\otimes\ ) \to \operatorname{Ext}^1(F_3, X\otimes\ ),$$

and the last term is zero, according to (3). However, the remaining sequence is nothing else but $0 \to DF_3 \to DF_2 \to DF_1 \to 0$, evaluated at X, and this sequence is exact. □

(7) The restriction of D to $\operatorname{mod}((\operatorname{mod}\Lambda)^{\mathrm{op}})$ maps $\operatorname{mod}((\operatorname{mod}\Lambda)^{\mathrm{op}})$ into $\operatorname{mod}((\operatorname{mod}\Lambda^{\mathrm{op}})^{\mathrm{op}})$, and satisfies $D^2F \simeq F$.

PROOF. Let $F \in \operatorname{mod}((\operatorname{mod}\Lambda)^{\mathrm{op}})$, and take a resolution of F, say

$$0 \to (C,\ ) \to (B,\ ) \to (A,\ ) \to F \to 0.$$

Applying D we get the exact sequence

$$0 \to DF \to\ \otimes A \to\ \otimes B \to\ \otimes C \to 0.$$

Since $\otimes A$, $\otimes B$ are finitely presented functors, according to (1) (but applied to Λ^{op}), it follows that also DF is finitely presented. Also, applying D again (now, with Λ replaced by Λ^{op}), we obtain the exact sequence

$$0 \to (C,\ ) \to (B,\ ) \to (A,\ ) \to D^2F \to 0,$$

thus $D^2F \simeq F$. □

(8) Any $F \in \operatorname{mod}((\operatorname{mod}\Lambda)^{\mathrm{op}})$ belongs to an exact sequence

$$0 \to F \to A\otimes\ \to B\otimes\ \to C\otimes\ \to 0,$$

with A, B, $C \in \operatorname{mod}\Lambda^{\mathrm{op}}$.

PROOF. According to (7), DF is finitely presented, thus there is an exact sequence

$$0 \to (C,\ ) \to (B,\ ) \to (A,\ ) \to DF \to 0$$

in $\operatorname{mod}((\operatorname{mod}\Lambda^{\mathrm{op}})^{\mathrm{op}})$, with $A, B, C \in \operatorname{mod}\Lambda^{\mathrm{op}}$. Applying D, we obtain the desired exact sequence. □

REMARK. Note that we have shown above that $D\colon \operatorname{mod}((\operatorname{mod}\Lambda)^{\mathrm{op}}) \to \operatorname{mod}((\operatorname{mod}\Lambda^{\mathrm{op}})^{\mathrm{op}})$ is a duality of categories.

PROOF. In order to see that D is fully faithful, we check this for representable functors (which are mapped to tensor functors) and use resolutions for general (finitely presented) functors. For showing that D is dense, just start with $G \in \operatorname{mod}((\operatorname{mod}\Lambda^{\mathrm{op}})^{\mathrm{op}})$ and note that $G \approx D^2G$. Thus D is a contravariant equivalence. □

8. Construction of finitely presented functors on $\mathcal{P}_R(\Lambda)$

Let Λ be an isolated singularity and $C \in \mathcal{P}_R(\Lambda)$ indecomposable and nonprojective. The main tool in the proof that there is an almost split sequence $0 \to A \to B \to C \to 0$ is to show that the simple functor S_C (for the definition see section 9.) is finitely presented. For the proof of this fact we first have to show that several functors on $\mathcal{P}_R(\Lambda)$ are finitely presented.

LEMMA 8.1. *Let Λ be an isolated singularity. Let $L \in \mathcal{P}_R(\Lambda)$, $X \in \operatorname{mod}\Lambda^{\mathrm{op}}$. Then* $\ell_R(\operatorname{Tor}_1(X, L)) < \infty$.

PROOF. Let p be a nonmaximal prime ideal of R. Then L_p is projective, according to (1.5). Thus $(\operatorname{Tor}_1^{\Lambda}(X, L))_p = \operatorname{Tor}^{\Lambda_p}(X_p, L_p) = 0$, where we use that Tor commutes with localization. Since $\operatorname{Tor}_1^{\Lambda}(X, L)$ is a finitely generated R-module and vanishes after localization with respect to any nonmaximal prime ideal, it follows that $\operatorname{Tor}_1^{\Lambda}(X, L)$ is of finite length as an R-module. □

COROLLARY 8.2. *Let Λ be an isolated singularity. Let $L \in \mathcal{P}_R(\Lambda^{\mathrm{op}})$, and I an injective R-module with zero socle. Then* $\operatorname{Hom}_R(L, I)$ *is an injective Λ-module.*

PROOF. With Λ also Λ^{op} is an isolated singularity. Thus, $\ell_R(\operatorname{Tor}_1^\Lambda(L,X)) < \infty$ for any $X \in \operatorname{mod}\Lambda$ according to the lemma, applied to Λ^{op}. Since I has zero socle, $\operatorname{Hom}_R^\Lambda(\operatorname{Tor}_1(L,X),I) = 0$. However, according to [**CE**], Prop. VI. 5.1, $\operatorname{Ext}_\Lambda^1(X,\operatorname{Hom}_R(L,I)) \approx \operatorname{Hom}_R(\operatorname{Tor}_1^\Lambda(L,X),I) = 0$. Since this holds true for any finitely generated Λ-module X, it follows that $\operatorname{Hom}_R(L,I)$ is an injective Λ-module. □

REMARK 8.3. The calculation of $\operatorname{Ext}^i(X,Y)$ is usually achieved by considering an injective resolution of Y. In fact, let

$$0 \to Y \to I_0 \to I_1 \to \cdots \to I_{d-1} \xrightarrow{\pi} I_d \to 0$$

be exact with all I_i, $0 \le i \le d-1$, injective, but I_d not necessarily injective. Then, for $0 \le i \le d$, the i-th homology of the complex

$$0 \to \operatorname{Hom}(X,I_0) \to \operatorname{Hom}(X,I_1) \to \cdots \to \operatorname{Hom}(X,I_{d-1}) \to \operatorname{Hom}(X,I_d) \to 0$$

is just $\operatorname{Ext}^i(X,Y)$.

PROOF. Let I'_d be an injective envelope of I_d, say with inclusion map $\iota\colon I_d \to I'_d$ and let I'_{d+1} be an injective envelope of $\operatorname{Cok}\iota$, thus we have the following injective resolution

$$0 \to Y \to I_0 \to I_1 \to \cdots \to I_d \xrightarrow{\delta_{d-1}} I'_d \xrightarrow{\delta_d} I'_{d+1},$$

where $\delta_{d-1} = \pi\iota$. The kernel of (X,δ_{d-1}) coincides with the kernel of (X,π). Also, the exact sequence $0 \to I_d \xrightarrow{\iota} I'_d \xrightarrow{\delta_d} I'_{d+1}$ gives $0 \to (X,I_d) \xrightarrow{(X,\iota)} (X,I'_d) \xrightarrow{(X,\delta_d)} (X,I'_{d+1})$, thus the kernel of (X,δ_d) is just (X,I_d), and the image of (X,δ_{d-1}) is the image of the image of (X,π) under (X,ι). □

LEMMA 8.4. *Let R be a complete regular local ring of dimension d. Let Λ be an R-algebra which is an isolated singularity. Let $L \in \mathcal{P}_R(\Lambda^{\text{op}})$. Then $\operatorname{Ext}_R^i(L\otimes\ ,R) \approx \operatorname{Ext}_\Lambda^i(\ ,\operatorname{Hom}_R(L,R))$ for $0 \le i \le d$.*

PROOF. Choose a minimal injective resolution

$$0 \to R \to I_0 \to \cdots \to I_d \to 0$$

of R. Since R is regular, the socle of I_i, $0 \le i \le d-1$, is zero [**Ba**]. As a consequence, $\operatorname{Hom}_R(L,I_i)$ is an injective Λ-module, for $0 \le i \le d-1$, according to (8.2). The complex

$$0 \to \operatorname{Hom}_R(L,R) \to \operatorname{Hom}_R(L,I_0) \to \cdots \to \operatorname{Hom}_R(L,I_d) \to 0$$

is exact, since L, considered as an R-module, is free. According to (8.3), $\operatorname{Ext}_\Lambda^i(\ ,\operatorname{Hom}_R(L,R))$ is given as the i-th homology of the complex

$$0 \to (\ ,\operatorname{Hom}_R(L,I_0)) \to \cdots \to (\ ,\operatorname{Hom}_R(L,I_d)) \to 0,$$

for $0 \le i \le d$. On the other hand, this complex is isomorphic to the complex

$$0 \to \operatorname{Hom}_R(L\otimes\ ,I_0) \to \cdots \to \operatorname{Hom}_R(L\otimes\ ,I_d) \to 0,$$

and its i-th homology is just $\operatorname{Ext}_R^i(L\otimes\ ,R)$, for any $i \ge 0$. □

REMARK. Note that the functors $\operatorname{Ext}_R^i(L\otimes\ ,R)$ and $\operatorname{Ext}_\Lambda^i(\ ,\operatorname{Hom}(L,R))$ only coincide for $i \le d$. For $i \ge d+1$, the first functor vanishes, since gl. dim. $R = d$, whereas the second functor does not need to vanish.

Our further aim is to investigate certain functors $\mathcal{P}_R(\Lambda) \to \text{Ab}$, and to show that they are finitely presented. Since usually these functors are defined on larger categories than $\mathcal{P}_R(\Lambda)$, we stress the fact that F is considered as an element of $\operatorname{Mod}\mathcal{P}_R(\Lambda)$ by writing $F|_{\mathcal{P}_R(\Lambda)}$.

COROLLARY 8.5. *Let R be a complete regular local ring of dimension d, let Λ be an R-algebra which is an isolated singularity. Let $L \in \mathcal{P}_R(\Lambda^{\mathrm{op}})$. Then the functors $\operatorname{Ext}^i_R(L\otimes\ ,R)|_{\mathcal{P}_R(\Lambda)}$, $0 \le i \le d$, are finitely presented.*

PROOF. Note that $\operatorname{Hom}_R(L,R)$ belongs to $\mathcal{P}_R(\Lambda)$, thus $\operatorname{Ext}^i_\Lambda(\ ,\operatorname{Hom}_R(L,R))|_{\mathcal{P}_R(\Lambda)}$ is finitely presented, according to (2.4). □

LEMMA 8.6. *Let R be a complete regular local ring of dimension d, let Λ be an R-algebra which is an isolated singularity. Let $M \in \operatorname{mod}(\Lambda^{\mathrm{op}})$. Then*

(a) $\operatorname{Ext}^i_R(M\otimes\ ,R)|_{\mathcal{P}_R(\Lambda)}$ *is finitely presented for all $i \ge 0$.*

(b) $\operatorname{Ext}^i_R(\operatorname{Tor}^\Lambda_1(M,\ ),R)|_{\mathcal{P}_R(\Lambda)}$ *is finitely presented for all $i \ge 0$.*

REMARK. Actually, we are only interested in a), with $i = 0$, however, our induction proof will need as intermediate steps also the remaining assertions.

PROOF. Take an exact sequence $0 \to \Omega M \to P \to M \to 0$, with $P \in \mathcal{P}(\Lambda)$, we obtain the following exact sequence of functors

$$0 \to \operatorname{Tor}^\Lambda_1(M,\ ) \to \Omega M\otimes\ \to P\otimes\ \to M\otimes\ \to 0,$$

and we denote by T the image of $M\otimes\ \to P\otimes$, thus we deal with the following two exact sequences

$$0 \to \operatorname{Tor}^\Lambda_1(M,\ ) \to \Omega M\otimes\ \to T \to 0, \qquad 0 \to T \to P\otimes\ \to M\otimes\ \to 0.$$

From the second short exact sequence, we obtain the following long exact sequence

$$(*)\quad \begin{array}{l} 0 \longrightarrow \operatorname{Hom}_R(M\otimes\ ,R) \longrightarrow \operatorname{Hom}_R(P\otimes\ ,R) \longrightarrow \operatorname{Hom}_R(T,R) \longrightarrow \\ \operatorname{Ext}^1_R(M\otimes\ ,R) \longrightarrow \operatorname{Ext}^1_R(P\otimes\ ,R) \longrightarrow \operatorname{Ext}^1_R(T,R) \longrightarrow \cdots \\ \longrightarrow \operatorname{Ext}^d_R(M\otimes\ ,R) \longrightarrow \operatorname{Ext}^d_R(P\otimes\ ,R) \longrightarrow \operatorname{Ext}^d_R(T,R) \longrightarrow 0. \end{array}$$

From the first short exact sequence, we obtain similarly

$$(**)\quad \begin{array}{l} 0 \longrightarrow \operatorname{Hom}_R(T,R) \longrightarrow \operatorname{Hom}_R(\Omega M\otimes\ ,R) \longrightarrow \operatorname{Hom}_R(\operatorname{Tor}^\Lambda_1(M,\ ),R) \longrightarrow \\ \operatorname{Ext}^1_R(T,R) \longrightarrow \operatorname{Ext}^1_R(\Omega M\otimes\ ,R) \longrightarrow \operatorname{Ext}^1_R(\operatorname{Tor}^\Lambda_1(M,\ ),R) \longrightarrow \cdots \\ \longrightarrow \operatorname{Ext}^d_R(T,R) \longrightarrow \operatorname{Ext}^d_R(\Omega M\otimes\ ,R) \longrightarrow \operatorname{Ext}^d_R(\operatorname{Tor}^\Lambda_1(M,\ ),R) \longrightarrow 0. \end{array}$$

All functors in $(*)$ and $(**)$ will be supposed to be defined on $\mathcal{P}_R(\Lambda)$ only.

The proof of the assertions a) and b) will be by induction on $\operatorname{proj.dim.}_R M$. If $\operatorname{proj.dim.}_R M = 0$, then M belongs to $\mathcal{P}_R(\Lambda)$, thus assertion a) is given in (8.5). In $(*)$, the functors occurring in the first two columns are finitely presented, using again (8.5). Since $\mathcal{P}_R(\Lambda)$ is coherent, it follows that also the functors occurring in the third column are finitely presented. Now, consider $(**)$. We have just seen that the functors in the first column are finitely presented. Since ΩM belongs to $\mathcal{P}_R(\Lambda)$, those of the second column are finitely presented by (8.5). Thus, using again that $\mathcal{P}_R(\Lambda)$ is coherent, we conclude that assertion b) is satisfied in our case.

Now assume, a), b) are true for all $M \in \operatorname{mod}(\Lambda^{\mathrm{op}})$ with $\operatorname{proj.dim.}_R M \le j$, and consider now a module $M \in \operatorname{mod}(\Lambda^{\mathrm{op}})$ with $\operatorname{proj.dim.}_R M = j+1$. Since P is projective also as an R-module, $\operatorname{proj.dim.}_R \Omega M = j$, thus by induction the functors occurring in the second column of $(**)$ are finitely presented. Note that for any $L \in \mathcal{P}_R(\Lambda)$, the R-module $\operatorname{Tor}^\Lambda_1(M,L)$ is of finite length, according to (8.1), therefore $\operatorname{Ext}^i_R(\operatorname{Tor}^\Lambda_1(M,L),R) = 0$ for $0 \le i \le d-1$. (Namely, we calculate these Ext-groups by taking the minimal injective resolution of R, say $0 \to R \to I_0 \to \cdots \to I_d \to 0$, then $\mathrm{Hom}_R(\mathrm{Tor}_1^\Lambda(M,L),I_i) = 0$ for $0 \le i \le d-1$. As a consequence, the maps

$$\mathrm{Ext}^i_R(T,R) \to \mathrm{Ext}^i_R(\Omega M\otimes\ ,R)$$

are isomorphisms, for $0 \le i \le d-1$. Also, there is for any $L \in \mathcal{P}_R(\Lambda)$, the exact sequence $0 \to T(L) \to P \underset{\Lambda}{\otimes} L \to M \underset{\Lambda}{\otimes} L \to 0$, and $P \underset{\Lambda}{\otimes} L$ is free as an R-module, thus $\mathrm{proj.dim.}_R\, T(L) \le d-1$, and therefore $\mathrm{Ext}^d_R(T(L),R) = 0$. As a consequence, in the last row of $(**)$, we have the isomorphism

$$\mathrm{Ext}^d_R(\Omega M\otimes\ ,R) \to \mathrm{Ext}^d_R(\mathrm{Tor}_1^\Lambda(M,\ ),R).$$

In this way, we have shown that the functors occurring in $(**)$ are either zero, or are isomorphic to a functor occurring in the middle column, thus finitely presented. Now, take the long exact sequence $(*)$. The functors in the last column also occur in $(**)$, thus they are finitely presented, those in the middle column are finitely presented by (8.5). Using again that $\mathcal{P}_R(\Lambda)$ is coherent, we conclude that the remaining functors of $(*)$ are also finitely presented. □

LEMMA 8.7. *Let R be a complete regular local ring of dimension d, let Λ be an R-algebra which is an isolated singularity. Let $F \in \mathrm{mod}((\mathrm{mod}\,\Lambda)^{\mathrm{op}})$. Then the functors $\mathrm{Ext}^i_R(F,R)|_{\mathcal{P}_R(\Lambda)}$ are finitely presented for all $i \ge 0$.*

PROOF. We have seen in (7.1) that there is an exact sequence

$$0 \to F \to A\otimes\ \to B\otimes\ \to C\otimes\ \to 0,$$

with $A,B,C \in \mathrm{mod}\,\Lambda^{\mathrm{op}}$. Denote by T the image of $A\otimes\ \to B\otimes$, thus there are two short exact sequences

$$0 \to F \to A\otimes\ \to T \to 0, \qquad 0 \to T \to B\otimes\ \to C\otimes\ \to 0.$$

The first short exact sequence gives rise to the long exact sequence

$$\begin{array}{lccccccc}
 & 0 \longrightarrow & \mathrm{Hom}_R(T,R) & \longrightarrow & \mathrm{Hom}_R(A\otimes\ ,R) & \longrightarrow & \mathrm{Hom}_R(F,R) & \longrightarrow \\
(*) & & \mathrm{Ext}^1_R(T,R) & \longrightarrow & \mathrm{Ext}^1_R(A\otimes\ ,R) & \longrightarrow & \mathrm{Ext}^1_R(F,R) & \longrightarrow \cdots \\
 & \longrightarrow & \mathrm{Ext}^d_R(T,R) & \longrightarrow & \mathrm{Ext}^d_R(A\otimes\ ,R) & \longrightarrow & \mathrm{Ext}^d_R(F,R) & \longrightarrow 0,
\end{array}$$

the second short exact sequence gives rise to the long exact sequence

$$\begin{array}{lccccccc}
 & 0 \longrightarrow & \mathrm{Hom}_R(C\otimes\ ,R) & \longrightarrow & \mathrm{Hom}_R(B\otimes\ ,R) & \longrightarrow & \mathrm{Hom}_R(T,R) & \longrightarrow \\
(**) & & \mathrm{Ext}^1_R(C\otimes\ ,R) & \longrightarrow & \mathrm{Ext}^1_R(B\otimes\ ,R) & \longrightarrow & \mathrm{Ext}^1_R(T,R) & \longrightarrow \cdots \\
 & \longrightarrow & \mathrm{Ext}^d_R(C\otimes\ ,R) & \longrightarrow & \mathrm{Ext}^d_R(B\otimes\ ,R) & \longrightarrow & \mathrm{Ext}^d_R(T,R) & \longrightarrow 0.
\end{array}$$

In $(**)$, the first two columns consist of functors (on $\mathcal{P}_R(\Lambda)$) which are finitely presented, by (8.6), thus the third column contains also only finitely presented functors. These functors occur again in the first column of $(*)$, those in the second column are finitely presented by (8.6), thus we conclude that the functors occurring in the last columns of $(*)$ are finitely presented. □

PROPOSITION 8.8. *Let R be a complete regular local ring of dimension d, let Λ be an R-algebra which is an isolated singularity. Let $F \in \mathrm{mod}((\mathrm{mod}\,\Lambda)^{\mathrm{op}})$, and $M \in \mathrm{mod}\,\Lambda$. Then the functors $\mathrm{Ext}^i_R(F,M)|_{\mathcal{P}_R(\Lambda)}$ are finitely presented for all $i \ge 0$.*

PROOF. By induction on proj. dim. M. If proj. dim. $M = 0$, thus M is free of finite rank, then the assertion follows from (8.7). If proj. dim. $M > 0$, take an exact sequence $0 \to \Omega M \to P \to M \to 0$ with P free of finite rank. We obtain the following long exact sequence

$$\begin{array}{ccccccccc} 0 & \longrightarrow & \mathrm{Hom}_R(F,\Omega M) & \longrightarrow & \mathrm{Hom}_R(F,P) & \longrightarrow & \mathrm{Hom}_R(F,M) & \longrightarrow & \\ & & \mathrm{Ext}^1_R(F,\Omega M) & \longrightarrow & \mathrm{Ext}^1_R(F,P) & \longrightarrow & \mathrm{Ext}^1_R(F,M) & \longrightarrow & \cdots \\ & \longrightarrow & \mathrm{Ext}^d_R(F,\Omega M) & \longrightarrow & \mathrm{Ext}^d_R(F,P) & \longrightarrow & \mathrm{Ext}^d_R(F,M) & \longrightarrow & 0. \end{array}$$

The functors in the first column are finitely presented on $\mathcal{P}_R(\Lambda)$, by induction, those of the second column, according to (8.7). Thus also the functors in the third column are finitely presented on $\mathcal{P}_R(\Lambda)$. □

9. The implication (d) ⇒ (a) of the main theorem

We are now able to prove that if Λ is an isolated singularity then $\mathcal{P}_R(\Lambda)$ has almost split sequences.

Given an indecomposable module C in $\mathcal{P}_R(\Lambda)$, we know that $\mathrm{End}(C)$ is a local ring, thus $(\ ,C)$ has just one maximal subfunctor, and we denote it by $\mathrm{rad}(\ ,C)$. Recall that $\mathrm{rad}(\ ,C)(X)$ is the set of all maps $g\colon X \to C$ such that fg belongs to $\mathrm{rad}\,\mathrm{End}(C)$, for every map $f\colon C \to X$. Since $\mathrm{rad}(\ ,C)$ is a maximal subfunctor of $(\ ,C)$, the functor $(\ ,C)/\mathrm{rad}(\ ,C)$ is a simple functor, and it will be denoted by S_C. The functor S_C has the following property: $S_C(C) = \mathrm{End}(C)/\mathrm{rad}\,\mathrm{End}(C)$, and $S_C(X) = 0$ for every indecomposable module X in $\mathcal{P}_R(\Lambda)$ with X non-isomorphic to C.

LEMMA 9.1. *Let R be a complete regular local ring, and Λ an R-algebra which is an isolated singularity. Let C be indecomposable in $\mathcal{P}_R(\Lambda)$. Then the functor S_C on $\mathcal{P}_R(\Lambda)$ is finitely presented.*

PROOF. Define a map $\alpha\colon (\ ,C) \to \mathrm{Hom}_R((C,\ ),\mathrm{End}(C)/\mathrm{rad}\,\mathrm{End}(C))$ by $\alpha_X(u)(f) = \overline{fu}$ (= the residue class of $fu \in \mathrm{End}(C)$ in $\mathrm{End}(C)/\mathrm{rad}\,\mathrm{End}(C)$), where X is in $\mathcal{P}_R(\Lambda)$, $u \in (X,C)$ and $f \in (C,X)$. It is easily checked that α is indeed a natural transformation. We claim that the kernel of α is just $\mathrm{rad}(\ ,C)$. But $u\colon X \to C$ belongs to $\mathrm{rad}(\ ,C)$ if and only if the composition $C \xrightarrow{f} X \xrightarrow{u} U$ belongs to $\mathrm{rad}\,\mathrm{End}(C)$ for all $f\colon C \to X$, thus if and only if $\alpha_X(u)(f) = 0$ for all $f\colon C \to X$, and this just means that $\alpha_X(u) = 0$. Thus, we have an exact sequence of functors

$$0 \to \mathrm{rad}(\ ,C) \to (\ ,C) \xrightarrow{\alpha} \mathrm{Hom}_R((C,\ ),\mathrm{End}(C)/\mathrm{rad}\,\mathrm{End}(C)).$$

The right hand functor is finitely presented, according to (8.8), thus since $\mathcal{P}_R(\Lambda)$ is coherent, also $\mathrm{rad}(\ ,C)$ is finitely presented, and therefore also S_C is finitely presented. □

LEMMA 9.2. *Let R be a complete local ring, and Λ an R-algebra which is finitely generated as an R-module. Let C be indecomposable in $\mathcal{P}_R(\Lambda)$, and not projective. Assume that the functor S_C on $\mathcal{P}_R(\Lambda)$ is finitely presented. Then there exists an almost split sequence $0 \to A \to B \to C \to 0$ in $\mathcal{P}_R(\Lambda)$.*

PROOF. Take a presentation, say

$$(\ ,B') \xrightarrow{(\ ,g')} (\ ,C) \to S_C \to 0.$$

Since C is not projective, $S_C(\Lambda) = 0$, thus, inserting Λ into the exact sequence, we see that $g'\colon B' \to C$ is surjective, thus we get an exact sequence

$$0 \to A' \to B' \xrightarrow{g} C \to 0$$

in $\mathcal{P}_R(\Lambda)$. Take an indecomposable direct summand A of A' such that the induced exact sequence

$$\begin{array}{ccccccccc} 0 & \longrightarrow & A' & \longrightarrow & B' & \longrightarrow & C & \longrightarrow & 0 \\ & & \downarrow & & \downarrow & & \| & & \\ 0 & \longrightarrow & A & \longrightarrow & B & \longrightarrow & C & \longrightarrow & 0 \end{array}$$

does not split. Note that also B belongs to $\mathcal{P}_R(\Lambda)$, since $\mathcal{P}_R(\Lambda)$ is closed under extensions. It follows that $0 \to A \to B \to C \to 0$ is an almost split sequence. □

LEMMA 9.3. *Let R be a complete local ring, and Λ an R-algebra which is finitely generated free as an R-module. Then the following are equivalent:*

(i) *$\mathcal{P}_R(\Lambda)$ has left almost split sequences.*
(ii) *$\mathcal{P}_R(\Lambda^{\text{op}})$ has right almost split sequences.*

PROOF. (i) $\Rightarrow$ (ii). Let C be indecomposable in $\mathcal{P}_R(\Lambda^{\text{op}})$, and not projective. Then $\text{Hom}_R(C, R)$ is indecomposable in $\mathcal{P}_R(\Lambda)$, and not relative injective. Thus, there exists an almost split sequence

$$0 \to \text{Hom}_R(C, R) \to V \to U \to 0.$$

Applying again $\text{Hom}_R(\ , R)$, we obtain an exact sequence

$$0 \to \text{Hom}_R(U, R) \to \text{Hom}_R(V, R) \to C \to 0,$$

which obviously is an almost split sequence in $\mathcal{P}_R(\Lambda^{\text{op}})$.

Similarly, one obtains the other implication. □

PROPOSITION 9.4. *If Λ is an isolated singularity, then $\mathcal{P}_R(\Lambda)$ has almost split sequences.*

PROOF. We have seen in (9.2) that $\mathcal{P}_R(\Lambda)$ has right almost split sequences. $\mathcal{P}_R(\Lambda^{\text{op}})$ also has right almost split sequences, and therefore according to (9.3), $\mathcal{P}_R(\Lambda)$ also has left almost split sequences. □

REMARK. We have shown in 3., that the existence of left almost split sequences implies that for all modules $X, Y \in \mathcal{P}_R(\Lambda)$ we have $\ell_R(\underline{\text{Hom}}(X, Y)) < \infty$, and therefore that Λ is an isolated singularity. Using (9.3), we also see that the existence of right almost split sequences similarly implies that Λ is an isolated singularity. Namely, if $\mathcal{P}_R(\Lambda)$ has right almost split sequences, then $\mathcal{P}_R(\Lambda^{\text{op}})$ has left almost split sequences. Hence Λ^{op} is an isolated singularity, and therefore Λ is an isolated singularity.

10. $\mathcal{P}_R(\Lambda)$ of finite type

Let us say that $\mathcal{P}_R(\Lambda)$ is of *finite type* provided it contains only finitely many isomorphism classes of indecomposable objects. As a consequence of the main theorem, we want to show that under our usual assumption, $\mathcal{P}_R(\Lambda)$ can only be of finite type in case Λ is an isolated singularity or even nonsingular.

THEOREM. *Let R be a complete regular local ring, and Λ an R-algebra which is finitely generated free as an R-module. Assume that $\mathcal{P}_R(\Lambda)$ is of finite type. Then Λ is an isolated singularity or nonsingular.*

PROOF. It suffices to show that $\mathcal{P}_R(\Lambda)$ has right almost split sequences. Let $C \in \mathcal{P}_R(\Lambda)$ be indecomposable. We want to construct an exact sequence

$$(\ ,B) \xrightarrow{(\ ,f)} (\ ,C) \to S_C \to 0.$$

Since $\mathcal{P}_R(\Lambda)$ is supposed to be of finite type, let $B_1, \dots, B_m$ be a complete list of all indecomposable modules in $\mathcal{P}_R(\Lambda)$. Now, $\operatorname{rad}(B_i, C)$ is a finitely generated R-module, thus there is some $f_i \colon m_i B_i \to C$ in $\operatorname{rad}(m_i B_i, C)$ such that any map in $\operatorname{rad}(B_i, C)$ can be factored through f_i. Let

$$f = (f_i) \colon \coprod_i m_i B_i \to C.$$

Then, obviously, given any $X \in \mathcal{P}_R(\Lambda)$, the image of (X, f) is just $\operatorname{rad}(X, C)$, and therefore the cokernel of $(\ , f)$ is S_C. If C is in addition not projective, then we obtain from (9.2) an almost split sequence $0 \to A \to B \to C \to 0$. This shows that $\mathcal{P}_R(\Lambda)$ has right almost split sequences. □

Appendix

This appendix is devoted to showing how the notion of an isolated singularity and the notion of an order as defined in [**A2**] are related. We begin by considering orders over a complete regular local ring R. Recall that an order over R is an R-algebra Λ which is a finitely generated free R-module such that $\operatorname{Hom}_R(\Lambda, R)_p$ is Λ_p-projective for all nonmaximal prime ideals p in R. Also if Λ is an R-order, a Λ-module L is said to be a Λ-lattice if L is a finitely generated free R-module such that L_p is Λ_p-projective for all nonmaximal prime ideals p in R. We denote by $\mathcal{L}(\Lambda)$ the full subcategory of $\operatorname{mod} \Lambda$ consisting of the Λ-lattices. From our previous results it is easy to see that if Λ is an R-algebra which is an isolated singularity, then Λ is an R-order and $\mathcal{L}(\Lambda) = \mathcal{P}_R(\Lambda)$. We now give various descriptions of when an R-order is an isolated singularity.

THEOREM A.1. *The following are equivalent for an R-order Λ where R is a complete regular local ring of dimension d.*

(a) *Λ is an isolated singularity.*
(b) *$\mathcal{L}(\Lambda)$ is functorially finite in* $\operatorname{mod} \Lambda$.
(c) *$\mathcal{L}(\Lambda)$ is contravariantly finite in* $\operatorname{mod} \Lambda$.

Moreover, if any of these conditions hold, then $\mathcal{L}(\Lambda)$ is coherent and $\operatorname{gl.dim.} \operatorname{mod} \mathcal{L}(\Lambda) = \max(2, \dim R) < \infty$.

PROOF. First we observe that if $\dim R = 0$, i.e. R is a field, then the notions of an R-order and an isolated singularity over R coincide and also $\mathcal{L}(\Lambda) = \mathcal{P}_R(\Lambda) = \operatorname{mod} \Lambda$. This implies the validity of the theorem. We now assume that $\dim R \geq 1$.

(a) implies (b). As remarked above the fact that Λ is an isolated singularity implies that $\mathcal{L}(\Lambda) = \mathcal{P}_R(\Lambda)$. Then $\mathcal{L}(\Lambda)$ is functorially finite in $\operatorname{mod} \Lambda$ since by Proposition 6.6 we have that $\mathcal{P}_R(\Lambda)$ is functorially finite in $\operatorname{mod} \Lambda$.

(b) implies (a). Trivial.

(c) implies (a). We first prove that c) implies that $\mathcal{L}(\Lambda)$ is coherent and $\operatorname{gl.dim.} \operatorname{mod} \mathcal{L}(\Lambda) = \max(2, \dim R)$. Since $\mathcal{L}$ is contravariantly finite in $\operatorname{mod} \Lambda$ we know that $\mathcal{L}$ is coherent and $\operatorname{mod} \mathcal{L}$ is an abelian category with enough projectives, namely the representable functors. The proof that $\text{gl. dim. } \mathcal{L}(\Lambda) = \max(2, \dim R)$ goes in several steps.

Step 1. Since Λ is an R-order, it follows from Lemma 5.1, that Λ^{op} is also an R-order. Let S be a simple object in $\text{mod}\,\mathcal{L}(\Lambda^{\text{op}})$. Suppose $S(\Lambda^{\text{op}}) = 0$. Then there is an indecomposable nonprojective C in $\mathcal{L}(\Lambda^{\text{op}})$ such that $S(C) \neq 0$. We know by [**A2**] that since Λ^{op} is an R-order, there is an almost split sequence $0 \to A \to B \to C \to 0$ in $\mathcal{L}(\Lambda^{\text{op}})$ from which it follows that $0 \to (\ ,A) \to (\ ,B) \to (\ ,C) \to S \to 0$ is a minimal projective resolution of S in $\text{Mod}\,\mathcal{L}(\Lambda^{\text{op}})$. Hence S is in $\text{mod}(\Lambda^{\text{op}})$ and $\text{proj. dim. } S = 2$.

Suppose $S(\Lambda^{\text{op}}) \neq 0$ and P is an indecomposable projective module such that $S(P) \neq 0$. Then we have that $0 \to (\ ,\text{rad}\,P)|_{\mathcal{L}(\Lambda^{\text{op}})} \to (\ ,P)|_{\mathcal{L}(\Lambda^{\text{op}})} \to S \to 0$ is exact in $\text{Mod}\,\Lambda^{\text{op}}$. Since $(\text{rad}\,R)_p$ is Λ_p^{op}-projective for all nonmaximal prime ideals p, we have by [**AS**, Theorem (3.9)] that there is an exact sequence $\cdots \to L_j \to \cdots \to L_2 \to L_1 \to \text{rad}\,P \to 0$ with L_j in $\mathcal{L}(\Lambda^{\text{op}})$ such that $\cdots \to (\ ,L_j)|_{\mathcal{L}(\Lambda)} \to \cdots \to (\ ,L_1)|_{\mathcal{L}(\Lambda^{\text{op}})} \to (\ ,\text{rad}\,P)|_{\mathcal{L}(\Lambda^{\text{op}})} \to 0$ is a minimal projective resolution. Therefore we have the minimal projective resolution $\cdots \to (\ ,L_j) \to \cdots \to (\ ,L_1) \to (\ L_0) \to S \to 0$ in $\text{mod}\,\mathcal{L}(\Lambda^{\text{op}})$. In particular $\cdots \to L_j \to \cdots \to L_0 \to L_0/\text{rad}\,L_0 \to 0$ is exact, from which it follows that if p is a nonmaximal prime ideal in R, then $(L_0/\text{rad}\,L_0)_p$ is zero, hence $(L_i)_p$ is Λ_p-projective. This implies that $\text{Ker}(L_{d-1} \to L_{d-2})$ is in $\mathcal{L}(\Lambda^{\text{op}})$ for $d \geq 2$ and $\text{Ker}(L_0 \to L_0/\text{rad}\,L_0)$ is in $\mathcal{L}(\Lambda^{\text{op}})$ for $d = 1$. In either case, we have that $\text{proj. dim. } S \leq d = \dim R$. From the fact that $L_0/\text{rad}\,L_0$ has finite length over R, we know that $\text{proj. dim.}_R\, L_0/\text{rad}\,L_0 = d$. Since $\text{proj. dim. } S \geq \text{proj. dim.}_R\, L_0/\text{rad}\,L_0$, we have that $\text{proj. dim. } S = d$. Thus we have shown that the simple objects in $\text{Mod}\,\mathcal{L}(\Lambda^{\text{op}})$ are in $\text{mod}\,\mathcal{L}(\Lambda^{\text{op}})$, and $\max \text{proj. dim. } S = \max(2, \dim R)$ where S ranges over all the simples in $\text{mod}\,\mathcal{L}$. Since $\mathcal{L}(\Lambda)$ is dual to $\mathcal{L}(\Lambda^{\text{op}})$, we have that $\max \text{proj. dim. } S = \max(2, \dim R)$ as S ranges over the simples in $\text{Mod}\,\mathcal{L}(\Lambda^{\text{op}}) = \text{Mod}\,\mathcal{L}(\Lambda)^{\text{op}}$.

Step 2. We now show that if S is simple in $\text{Mod}\,\mathcal{L}(\Lambda)$, then $\text{inj. dim. } S \leq \max(2, \dim R)$ in $\text{Mod}\,\mathcal{L}(\Lambda)$.

Suppose S is a simple object in $\text{Mod}\,\mathcal{L}(\Lambda)$. Then by our previous results we have a finite projective resolution $0 \to (L_t,\ ) \to \cdots \to (L_0,\ ) \to D(S) \to 0$ in $\text{Mod}\,\mathcal{L}(\Lambda^{\text{op}})$ where $D(S) = \text{Hom}_R(S, I(R/\text{rad}\,R))$ and t is either 2 or $\dim R$. Applying $\text{Hom}_R(\ , I(R/\text{rad}\,R))$ to this sequence we obtain the exact sequence $0 \to S \to \text{Hom}_R((L_0,\ ), I(R/\text{rad}\,R)) \to \cdots \to \text{Hom}_R((L_t,\ ), I(R/\text{rad}\,R)) \to 0$. From this we have our desired result that $\text{inj. dim. } S \leq \max(2, \dim R)$ provided we show that each of the $\text{Hom}_R((L_j,\ ), I(R/\text{rad}\,R))$ is injective, which we now do.

Let L be in $\mathcal{L}(\Lambda)$. Given F in $\text{Mod}\,\mathcal{L}(\Lambda)$, we have the first three terms of a projective resolution of F

$$\coprod_{k \in K} (\ ,A_k) \to \coprod_{j \in J} (\ ,B_j) \to \coprod_{i \in I} (\ ,C_i) \to F \to 0.$$

Hence $\coprod_{k \in K}(L, A_k) \to \coprod_{j \in J}(L, B_j) \to \coprod_{i \in I}(L, C_i)$ is exact and so

$$\prod_{i \in I} \text{Hom}_R((L, C_i), I(R/\text{rad}\,R)) \to \prod_{j \in J} \text{Hom}_R((L, B_j), I(R/\text{rad}\,R)) \to \prod_{k \in K} \text{Hom}_R((L, A_k), I(R/\text{rad}\,R))$$

is exact. Consequently

$$(\coprod_{i\in I}(C_i,\ ), \mathrm{Hom}_R((L,\ ), I(R/\operatorname{rad} R))) \to (\coprod_{j\in J}(B_j,\ ), \mathrm{Hom}_R((L,\ ), I(R/\operatorname{rad} R))) \to$$
$$(\coprod_{k\in K}(A_k,\ ), \mathrm{Hom}_R((L,\ ), I(R/\operatorname{rad} R)))$$

is exact, which implies $\mathrm{Ext}^1(F, \mathrm{Hom}_R((L,\ ), I(R/\operatorname{rad} R))) = 0$. Therefore, $\mathrm{Hom}_R((L,\ ), I(R/\operatorname{rad} R))$ is injective in $\operatorname{Mod} \mathcal{L}(\Lambda)$.

Step 3. We now apply this last result to show that c) implies $\operatorname{gl.dim.} \operatorname{mod} \mathcal{L}(\Lambda) = \max(2, \dim R)$.

Suppose F is in $\operatorname{mod} \mathcal{L}(\Lambda)$ and that $\operatorname{proj.dim.} F > t = \max(2, d)$. Let the exact sequence $0 \to K \to (L_t,\ ) \to \cdots \to (L_0,\ ) \to F \to 0$ be the beginning of a minimal projective resolution of F. Since $\mathrm{Ext}^{t+1}(F, S) = 0$ for all simple S by our previous results, we have that $((L_t,\ ), K/\operatorname{rad} K) \to (K, \operatorname{rad} K) \to 0$ is exact. Since $K \subset \operatorname{rad}(L_t,\ )$, this implies $K = 0$, which is a contradiction. Therefore, $\operatorname{gl.dim.} \operatorname{mod} \mathcal{L}(\Lambda) = \max(2, \dim R) < \infty$.

We apply this to prove that (c) implies (a). Since Λ is an order we know that $\mathrm{Hom}_R(\Lambda, R)_p$ is Λ_p^{op}-projective for all nonmaximal prime ideals of R. Hence Λ is an isolated singularity if $\operatorname{gl.dim.} \Lambda_p < \infty$ for all nonmaximal prime ideals p (Lemma 5.2). Let p be a nonmaximal prime ideal and A a finitely generated Λ_p-module. Then there is a finitely generated Λ-module M such that $M_p \simeq A$. Since $\mathcal{L}(\Lambda)$ is contravariantly finite in $\operatorname{mod} \Lambda$, we have that there is a finite projective resolution $0 \to (\ , L_t) \to \cdots \to (\ , L_0) \to (\ , M)|_{\mathcal{L}} \to 0$ with the L_i in $\mathcal{L}(\Lambda)$ and $t \leq \max(2, \dim R) < \infty$. This gives rise to the exact sequence of Λ_p-modules $0 \to (L_t)_p \to \cdots \to (L_0)_p \to M_p = A \to 0$. Since the $(L_j)_p$ are Λ_p-projective, we have that $\operatorname{proj.dim.}_{\Lambda_p} A \leq t < \infty$. Hence $\operatorname{gl.dim.} \Lambda_p < \infty$ for all nonmaximal prime ideals p of R, which finishes the proof of (c) implies (a). This completes the proof of the theorem. □

As an immediate consequence of Theorem A.1 we have the following:

COROLLARY A.2. *Let Λ be an R-order where R is a complete regular local ring. If $\mathcal{L}(\Lambda)$ is of finite representation type (i.e. $\mathcal{L}(\Lambda)$ has only a finite number of nonisomorphic indecomposable modules), then Λ is an isolated singularity.*

PROOF. By Theorem A.1., it suffices to show that $\mathcal{L}(\Lambda)$ is contravariantly finite in $\operatorname{mod} \Lambda$. Let L be the (direct) sum of a complete set of nonisomorphic indecomposable Λ-lattices. Since $\mathcal{L}(\Lambda)$ is of finite representation type, L is in $\mathcal{L}(\Lambda)$. Let M be in $\operatorname{mod} \Lambda$. Then $\mathrm{Hom}_\Lambda(L, M)$ is a finitely generated R-module and suppose $f_1, \ldots, f_n$ is a set of R-generators. Then it is obvious that the morphism $f\colon nL \to M$ induced by the f_i, $i = 1, \ldots, n$, has the property that $\mathrm{Hom}_\Lambda(L, nL) \xrightarrow{(\mathrm{Hom}_\Lambda(1,f))} (\mathrm{Hom}_\Lambda(L, M))$ is surjective and hence $(\ , L)|_{\mathcal{L}(\Lambda)} \to (\ , M)|_{\mathcal{L}(\Lambda)}$ is surjective. Therefore $\mathcal{L}(\Lambda)$ is contravariantly finite in $\operatorname{mod} \Lambda$. □

In [**A2**] a slightly more general definition than we have given here for an order was used in discussing the existence of almost split sequences for orders. Instead of assuming that R is a complete regular local ring, it was assumed that R is a complete local Gorenstein ring and an R-order was defined to be an R-algebra Λ which is a finitely generated Cohen-Macaulay R-module such that $\mathrm{Hom}_R(\Lambda, R)_p$ is Λ_p^{op}-projective for all nonmaximal prime ideals p. Our purpose now is to point out that the assumption that R is regular rather than Gorenstein, is not as great a restriction as it might seem at first.

Suppose R is a complete Gorenstein local ring. Now there are a great many cases of interest in algebraic geometry and elsewhere when one knows that R contains a complete regular local ring T such that R is a finitely generated T-module, for instance, if R contains a field or is a domain [**N**]. So suppose R contains a complete regular local ring T such that R is a finitely generated T-module. Then because R is Cohen-Macaulay, since it is Gorenstein, we know that R is a free T-module. Furthermore, an R-module is Cohen-Macaulay if and only if it is T-free. Thus the functor $\operatorname{Hom}_T(\ ,T)$ induces a duality on the Cohen-Macaulay R-modules. Also, since R is Gorenstein, the functor $\operatorname{Hom}_R(\ ,R)$ also induces a duality on the Cohen-Macaulay R-modules which is well known to be isomorphic to the duality $\operatorname{Hom}_T(\ ,T)$ [**HK**].

Now let Λ be an R-algebra which is a finitely generated R-module. We first show that Λ is an R-order if and only if it is a T-order. Clearly, Λ is a Cohen-Macaulay R-module, if and only if it is T-free. Suppose Λ is a Cohen-Macaulay R-module. Then because the two dualities $\operatorname{Hom}_T(\ ,T)$ and $\operatorname{Hom}_R(\ ,R)$ on Cohen-Macaulay modules are isomorphic, we have that the Λ^{op}-modules $\operatorname{Hom}_R(\Lambda, R)$ and $\operatorname{Hom}_T(\Lambda, T)$ are isomorphic. From this it follows that $\operatorname{Hom}_R(\Lambda, R)_p$ is Λ_p^{op}-projective for all nonmaximal prime ideals p in R if and only if $\operatorname{Hom}(\Lambda, T)_p$ is Λ_p^{op}-projective for all nonmaximal prime ideals p of T. Hence, Λ is an R-order if and only if Λ is a T-order. Clearly, if Λ is an R-order, then a Λ-module is an R-order lattice if and only if it is a T-order lattice. Hence, studying the lattices over the R-order Λ is the same thing as studying the lattices over the T-order Λ.

References

[A1] Auslander, M.: *Coherent functors*, Proc. Conf. Categorical Algebra, La Jolla 1965 (Springer-Verlag), 189–231.

[A2] Auslander, M.: *Functors and Morphisms Determined by Objects and Applications of Morphisms Determined by Modules*, Proc. Conf. Representation Theory, Philadelphia 1976, pp. 1–244 and pp. 245–327, Marcel Dekker, 1978.

[AR] Auslander, M.; Reiten, I.: *Stable Equivalence of Artin Algebras*, Proc. Conf. on Orders, Group Rings and Related Topics, Springer Lecture Notes 353, pp. 8–71, 1973.

[AS] Auslander, M.; Smalø, S.: *Lattices Over Orders: Finitely Presented Functors and Preprojective Partitions*, Trans. AMS, 273, pp. 433–446, 1982.

[Ba] Bass, H.: *On the Ubiquity of Gorenstein Rings*, MZ, 82, pp. 8–26, 1963.

[Bou] Bourbaki, N.: Commutative Algebra, Paris, 1972.

[CE] Cartan, H.; Eilenberg, S.: Homological Algebra, Princeton, New Jersey, 1956.

[HK] Herzog, J.; Kunz, E.: *Der kanonische Modul eines Cohen-Macaulay-Rings*, Springer Lecture Notes, 238, 1971.

[K] Kaplansky, I.: *Fields and Rings*, Chicago, Ill., 1970.

[M] Mitchell, B.: Theory of Categories, New York and London, 1965.

[N] Nagata, N.: Local Rings, New York, 1962.

Received: 19.11.1984

Added in proof:

[R] Ramras M.: *Maximal orders over regular rings of dimension* 2, Trans. Amer. Math. Soc. 142 (1969), 457–474.

[V] Vasconcelos W. V.: *On quasi-local regular algebras*, Symposia Mathematica XI (1973), 11–22.

BRANDEIS UNIVERSITY, WALTHAM, MASS., USA

UNIVERSITY OF BIELEFELD, BIELEFELD, W. GERMANY

TRANSACTIONS OF THE
AMERICAN MATHEMATICAL SOCIETY
Volume 293, Number 2, February 1986

RATIONAL SINGULARITIES AND ALMOST SPLIT SEQUENCES

BY

MAURICE AUSLANDER

ABSTRACT. The main aim of this paper is to relate almost split sequences to singularity theory by showing that the McKay quiver built from the finite-dimensional representations of a finite subgroup G of $\mathrm{GL}(2, \mathbf{C})$, where $\mathbf{C}$ is the complex numbers, is isomorphic to the AR quiver of the reflexive modules of the quotient singularity associated with G.

Over the past decade, almost split sequences have been playing an increasingly important role in the representation theories of finite-dimensional algebras and classical orders (see [**5** and **3**, **8**] for basic existence theorems in these contexts). While they have been known for some time to exist in higher-dimensional situations [**3**], it has not been at all clear how they related to singularity theory, if at all. The main aim of this paper is to relate almost split sequences to singularity theory by showing that the McKay quiver built from the finite-dimensional complex representations of a finite subgroup G of $\mathrm{GL}(2, \mathbf{C})$, where $\mathbf{C}$ is the complex numbers, is isomorphic to the AR quiver of the reflexive modules over the quotient singularity R associated with G. As in the case of finite-dimensional algebras, the AR quiver of reflexive R-modules is defined in terms of the almost split sequences of reflexive R-modules.

In the case $G \subset \mathrm{SL}(2, \mathbf{C})$, McKay observed that the underlying graph of the McKay quiver, with the trivial module removed, is isomorphic to the desingularization graph of the associated singularity. Various explanations of this phenomenon have been given by Knörrer, Gonzalez-Sprinberg-Verdier [**6**] and Artin-Verdier [**1**], which along the way have established, most explicitly in [**1**], a natural one-to-one correspondence between the indecomposable reflexive R-modules and the nodes of the desingularization graph. But why the almost split sequences describe the edge of the desingularization graph still remains to be explained.

An effort has been made to make this paper as self-contained as possible. In particular, no prior knowledge of almost split sequences is required.

Before describing the contents of the six sections of this paper we fix some notation. Throughout this paper G is a finite group, k an algebraically closed field of characteristic not dividing the order of G and V a two-dimensional k-representation of G. Setting $S = k[[X, Y]]$, the k-algebra of formal power series, the two-dimensional representation V gives a linear action of G on S as a group of k-algebra automorphisms. We denote by $S[G]$ the skew group ring given by this action.

Received by the editors January 2, 1985.
1980 *Mathematics Subject Classification*. Primary 14J17.
[1] Partially supported by NSF Grant #MCS 8303348.

§1. In this section the McKay quiver of G with respect to V is defined and shown to be given by the minimal $S[G]$-projective resolutions of the simple $S[G]$-modules.

§2. This section is devoted to showing that the minimal $S[G]$-projective resolutions of simple $S[G]$-modules are essentially the same thing as the almost split sequences in the category of reflexive modules of the quotient singularity $R = S^G$ in case V is a faithful G-module having no pseudo-reflections. This is used to show that the McKay quiver of G with respect to V is isomorphic to the AR quiver of the reflexive R-modules.

§3. We continue with the hypothesis that V is a faithful G-module having no pseudo-reflections and show how the almost split sequences for the reflexive R-modules can be constructed from a fixed exact sequence of reflexive R-modules called the fundamental exact sequence.

§4. In this section we assume that k is $\mathbf{C}$, the complex numbers. We give a different proof of the following result in Artin-Verdier [**1**]. The $\mathbf{C}$-algebras which are integrally closed, complete, local two-dimensional noetherian domains with residue class field $\mathbf{C}$ having only a finite number of nonisomorphic reflexive indecomposable modules are precisely the two-dimensional quotient singularities.

§5. In this section the notion of reflexive splitting trace modules over arbitrary integrally closed noetherian domains is introduced and some elementary properties established.

§6. In this the last section of the paper, we use the notion of reflexive splitting trace modules to generalize the results of §3 about the construction of almost split sequences from the fundamental exact sequences to arbitrary complete, integrally closed two-dimensional noetherian local domains.

It should be noted that the results of §§5 and 6 have almost exact analogues in the theory of modular group representations [**4**].

Finally, I would like to thank M. Artin, R. Buchweitz, H. Knörrer, I. Reiten and G. Schwarz, for many helpful conversations.

1. The McKay quiver and projective resolutions. We assume throughout this section that G is a finite group and k a field of characteristic not dividing the order of G. Then $k[G]$, the group ring of G over k, is semisimple and we denote by $V_0, V_1, \ldots, V_d$ a complete set of nonisomorphic simple $k[G]$-modules with $V_0 \simeq k$, the trivial representation. Let V be a fixed two-dimensional $k[G]$-module. We recall that the McKay quiver of V is the directed graph with vertices the $V_0, \ldots, V_d$ and having m arrows from V_i to V_j if and only if the multiplicity of V_i in $V \otimes_k V_j$ is m. Our aim in this section is to interpret the McKay quiver in terms of the projective resolutions of simple modules over certain skew group rings.

Let $S = k[[X, Y]]$ be the formal power series in two variables over k. Then the two-dimensional representation V induces a linear action of G on S as a group of k-automorphisms. We recall that the skew group ring $S[G]$ of G over S is additively the free S-module with basis the elements of G and with multiplication given by $(s_1\sigma_1)(s_2\sigma_2) = s_1\sigma_1(s_2)\sigma_1\sigma_2$, where the s_i are in S and the σ_i are in G for $i = 1, 2$. An $S[G]$-module is an S-module M together with a G-module structure satisfying

$\sigma(sm) = \sigma(s)\sigma(m)$ for all σ in G, s in S and m in M. Also, if M_1 and M_2 are $S[G]$-modules, then the morphisms from M_1 to M_2 consist of the S-module morphisms which are also G-module morphisms.

Next we observe that if M and N are $S[G]$-modules, then $\mathrm{Hom}_S(M, N)$ will be considered an $S[G]$-module by means of the usual action of G on $\mathrm{Hom}_S(M, N)$ given by $(\sigma f) = \sigma(f(\sigma^{-1}m))$ for all σ in G, f in $\mathrm{Hom}_S(M, N)$ and m in M. As usual, we have that $\mathrm{Hom}_{S[G]}(M, N) = \mathrm{Hom}_S(M, N)^G$, the fixed points of the action of G on $\mathrm{Hom}_S(M, N)$. An easy consequence of these remarks is the following useful criterion for when an $S[G]$-module is projective.

LEMMA 1.1. *An $S[G]$-module X is projective if and only if it is a free S-module.*

PROOF. Since $S[G]$ is a free S-module, any projective $S[G]$-module is a projective and therefore free S-module.

Suppose X is an $S[G]$-module which is a free S-module. Let $0 \to A \to B \to C \to 0$ be an exact sequence of $S[G]$-modules. Then $0 \to \mathrm{Hom}_S(X, A) \to \mathrm{Hom}_S(X, B) \to \mathrm{Hom}_S(X, C) \to 0$ is an exact sequence of $S[G]$-modules. Since $k[G]$ is semisimple, the fixed point functor is exact, so $\mathrm{Hom}_{S[G]}(X, B) \to \mathrm{Hom}_{S[G]}(X, C) \to 0$ is exact. This shows that X is $S[G]$-projective.

Our next aim is to use Lemma 1.1 to obtain an explicit description of the projective $S[G]$-modules which will be useful in describing the projective resolutions of the simple $S[G]$-modules. The first step in this direction is based on the following definitions.

Let M be an $S[G]$-module and W a $k[G]$-module. We consider $M \otimes_k W$ an $S[G]$-module by means of the operation $s\sigma(m \otimes w) = s\sigma(m) \otimes \sigma(w)$ for all s in S, σ in G, m in M and w in W. Also, we denote by S the projective $S[G]$-module which is the S-module S together with our given operation of G on S. As an easy consequence of these definitions and Lemma 1.1 we have the following

LEMMA 1.2. *If P is a projective $S[G]$-module, then $P \otimes_k W$ is a projective $S[G]$-module for all $k[G]$-modules W. In particular, $S \otimes_k W$ is a projective $S[G]$-module for all $k[G]$-modules W.*

PROOF. It follows from the definition of $P \otimes_k W$, that $P \otimes_k W$ is S-free and therefore $S[G]$-projective.

We now show that all projective $S[G]$-modules can be written uniquely up to isomorphism in the form $S \otimes_k W$ with W in mod $k[G]$.

Let $\mathfrak{m}$ be the maximal ideal (X, Y) of S. Since $\sigma(\mathfrak{m}) = \mathfrak{m}$ for all σ in G, it follows that $\mathfrak{m}S[G]$ is a two-sided ideal in $S[G]$. This implies that $\mathfrak{m}S[G]$ is contained in the radical of $S[G]$ since $\mathfrak{m}$ is the radical of S and $S[G]$ is a finitely generated S-module. From this it follows that $\mathfrak{m}S[G]$ is the radical of $S[G]$ since $S[G]/\mathfrak{m}S[G] \simeq k[G]$ which is semisimple. Thus, if P is a projective $S[G]$-module, then $P/\mathfrak{m}P$ is a $k[G]$-module. Moreover, since S is complete, the natural epimorphism $P \to P/\mathfrak{m}P$ is a projective cover. Consequently, two projective $S[G]$-modules P and Q are isomorphic if and only if $P/\mathfrak{m}P$ is isomorphic to $Q/\mathfrak{m}Q$. As a consequence of these remarks we have the following

PROPOSITION 1.3. (a) *If P is a projective $S[G]$-module, then $P \simeq S \otimes_k P/\mathfrak{m}P$.*

(b) *If the projective $S[G]$-modules $S \otimes_k W$ and $S \otimes_k W'$ are isomorphic, then the $k[G]$-modules W and W' are isomorphic.*

PROOF. Follows from previous observations and the fact that $S \otimes_k W$ is the projective cover of $S/\mathfrak{m} \otimes_k W \simeq W$ for all $k[G]$-modules W.

As an immediate consequence of Proposition 1.3, we have the following

COROLLARY 1.4. *Let $V_0, V_1, \ldots, V_d$ be a complete set of nonisomorphic simple $k[G]$-modules. Then $S \otimes_k V_0, \ldots, S \otimes_k V_d$ is a complete set of nonisomorphic indecomposable projective $S[G]$-modules.*

We are now ready to describe the projective resolutions of the simple $S[G]$-modules. We begin with the simple $S[G]$-module k with trivial group action. Since $S = k[[X, Y]]$, we know that the minimal projective S-resolution $0 \to S \to S \amalg S \to S \to S/\mathfrak{m} \to 0$ is given by the Koszul complex which can be written as $0 \to S \otimes_k \Lambda^2 V \to S \otimes_k V \to S \to k \to 0$ where V has been identified with $kX \amalg kY$. But it can easily be checked that the morphisms involved are also G-morphisms where G acts on the modules involved as defined above. Therefore this sequence is also a minimal projective $S[G]$-resolution of k. Now to obtain the minimal projective $S[G]$-resolution of an arbitrary simple $S[G]$-module V_j, we simply tensor the Koszul complex with V_j over k, getting

$$0 \to S \otimes_k (\Lambda^2 V \otimes_k V_j) \to S \otimes_k (V \otimes_k V_j) \to S \otimes_k V_j \to V_j \to 0.$$

This shows that the McKay quiver for V is isomorphic to the following quiver Γ associated with $S[G]$ in analogy with the situation for artin algebras.

Let P_i be a $S[G]$-projective cover of the simple $S[G]$-module V_i. Then $P_0, \ldots, P_d$ is a complete set of nonisomorphic indecomposable projective $S[G]$-modules which we consider the vertices of Γ. For each $j = 0, \ldots, d$, let

$$0 \to Q_2^{(j)} \to Q_1^{(j)} \to Q_0^{(j)} \to V_j \to 0$$

be a minimal projective resolution of V_j. Then the number of arrows from P_i to P_j in Γ is the multiplicity of P_i in a decomposition into indecomposables of $Q_1^{(j)}$. Denote this quiver by $\Gamma(V)$. Clearly, we have an isomorphism of $\Gamma(V)$ with the McKay quiver of V by sending P_i to V_i for all $i = 0, \ldots, d$.

2. McKay quivers and *AR* quivers. In addition to the hypothesis of §1, we assume k is an algebraically closed field and that the two-dimensional $k[G]$-module V has no pseudo-reflections, i.e. no $\sigma \neq \{1\}$ in G has 1 as an eigenvalue. Then the invariant algebra $R = S^G$ is a complete, integrally closed, two-dimensional domain and S is a finitely generated R-module. The hypothesis that G has no pseudo-reflections in V is equivalent to the height one primes in R being unramified in S (cf. [2]). Our aim in this section is to give a description of the quiver $\Gamma(V)$, and therefore of the McKay quiver, in terms of the almost split sequences in the category $\mathfrak{L}$ of reflexive R-modules.

Recall that an R-module M is said to be reflexive if the natural morphism $M \to \mathrm{Hom}_R(\mathrm{Hom}_R(M, R), R)$ is an isomorphism. Since R is integrally closed of dimension two, it follows that the reflexive R-modules are the Cohen-Macaulay R-modules. Hence S is a reflexive R-module. We denote by add S the full subcategory of $\mathfrak{L}$ consisting of the summands of finite sums of S. Clearly, add S is an additive category in which idempotents split. For the sake of completeness, we prove the following well known fact [7]:

PROPOSITION 2.1. *Up to isomorphism, the indecomposable reflexive R-modules are precisely the indecomposable R-summands of S. Hence $\mathfrak{L} =$ add S has only a finite number of nonisomorphic indecomposable reflexive R-modules.*

PROOF. Let L be an indecomposable reflexive R-module. The fact that the characteristic of k is relatively prime to the order of G implies that the inclusion of R-modules $0 \to R \to S$ splits. Therefore, we have that the monomorphism of R-modules $0 \to \mathrm{Hom}_R(M^*, R) \to \mathrm{Hom}_R(M^*, S)$ also splits, where $M^* = \mathrm{Hom}_R(M, R)$. Since $\mathrm{Hom}_R(M^*, S)$ is a Cohen-Macaulay S-module, it is a free S-modules, so $M = M^{**}$ is a summand of a finite sum nS as R-modules. Because R is complete, mod R is a Krull-Schmidt category. Hence, the fact that L is indecomposable and a summand of a finite sum of copies of S implies that L is a summand of S. The rest of the proposition is trivial.

Our aim now is to describe an equivalence of categories between $\mathfrak{P}$, the category of projective $S[G]$-modules and $\mathfrak{L} = \mathrm{add}_R S$, the category of reflexive R-modules.

Straightforward calculations show that the R-submodule $S[G]^G$ of $S[G]$ consists of all elements of the form $\sum_{\sigma \in G} \sigma(s)\sigma$ with s in S. Thus, we obtain the R-module isomorphism $S \to S[G]^G$ given by $s \mapsto \sum_{\sigma \in G} \sigma(s)\sigma$ for all s in S which we view as an identification. Therefore, P^G is in $\mathrm{add}_R S$ for all P in $\mathfrak{P}$, and so the functor mod $S[G] \to$ mod R given by $A \mapsto A^G$ induces a functor $\mathfrak{P} \to \mathrm{add}_R S$ which we now show is an equivalence of categories.

First of all it is easy to see that $\mathfrak{P} \to \mathrm{add}_R S$ is an equivalence if and only if the R-algebra morphism $\mathrm{End}_{S[G]}(S[G]) \to \mathrm{End}_R(S)$ given by $f \mapsto f|S$ for all f in $\mathrm{End}_{S[G]}(S[G])$ is an isomorphism. Next we have the R-algebra isomorphism β: $S[G]^{\mathrm{op}} \to \mathrm{End}_{S[G]}(S[G])$ given by $\beta(s\sigma)(x) = x \cdot s\sigma$ for all s in S, σ in G and x in $S[G]$. Also, we have the R-algebra isomorphism $S[G] \to S[G]^{\mathrm{op}}$ given by $s\sigma \mapsto \sigma^{-1}(s)\sigma^{-1}$ for all s in S and σ in G. Finally, it is straightforward to check that the composition of R-algebra morphisms

$$S[G] \to S[G]^{\mathrm{op}} \xrightarrow{\beta} \mathrm{End}_{S[G]}(S[G]) \to \mathrm{End}_R(S)$$

is the standard R-algebra morphism γ: $S[G] \to \mathrm{End}_R(S)$ given by $\gamma(s\sigma)(x) = s\sigma(x)$ for all s in S, σ in G and x in S. Hence we have that γ: $S[G] \to \mathrm{End}_R(S)$ is an isomorphism if and only if $\mathrm{End}_{S[G]}(S[G]) \to \mathrm{End}_R(S)$ is an isomorphism if and only if $\mathfrak{P} \to \mathrm{add}_R S$ is an equivalence. But the fact that the height one primes in R are unramified in S implies that γ: $S[G] \to \mathrm{End}_R(S)$ is an isomorphism [**2**].

Therefore, we have established the following

PROPOSITION 2.2. *The functor* $\mathfrak{P} \to \mathfrak{L} = \operatorname{add}_R S$ *given by* $P \mapsto P^G$ *for* P *in* $\mathfrak{P}$ *is an equivalence of categories.*

Our aim now is to apply this proposition to construct in terms of the almost split sequences in $\mathfrak{L}$ a quiver which is isomorphic to the quiver $\Gamma(V)$ defined in the previous section. To this end it is convenient to introduce some notation. As usual, let $V_0, \ldots, V_d$ be a complete set of nonisomorphic simple $S[G]$-modules with $V_0 = k$, the trivial simple, and let P_i be a projective cover for V_i for all $i = 0, \ldots, d$ with $P_0 = S$. Then $P_0, \ldots, P_d$ is a complete set of nonisomorphic indecomposable projective $S[G]$-modules. Furthermore, we know that $P_i \simeq S \otimes_k V_i$ for all $i = 0, \ldots, d$. Also we obtained in the previous section the minimal projective resolutions

$$0 \to S \otimes_k \left(\bigwedge^2 V \otimes_k V_i \right) \to S \otimes_k (V \otimes_k V_i) \to S \otimes_k V_i \to V_i \to 0.$$

Since $\bigwedge^2 V$ is of dimension 1 over k, we have $\bigwedge^2 V \otimes_k V_i$ is a simple $k[G]$-module which we denote by $\tau(V_i)$. Clearly, $\tau(V_i) \simeq \tau(V_j)$ if and only if $V_i \simeq V_j$. Therefore, $S \otimes_k \tau(V_i)$ is an indecomposable projective $S[G]$-module. We define $\tau(P_i)$ to be the unique P_j isomorphic to $S \otimes_k (V_i)$. Then we can write the minimal projective resolutions of the V_i as

$$0 \to \tau(P_i) \xrightarrow{u_i} Q_i \xrightarrow{v_i} P_i \to V_i \to 0$$

where $Q_i \simeq S \otimes_k (V \otimes_k V_i)$. Now denoting P_L^G, Q_L^G and $\tau(P_i)^G$ by L_i, E_i and $\tau(L_i)$, respectively, we have that the $L_0, \ldots, L_d$ are a complete set of nonisomorphic indecomposable reflexive R-modules and so are the $\tau(L_0), \ldots, \tau(L_d)$.

Applying the fixed point functor to the exact sequence

$$0 \to \tau(P_0) \to Q_0 \xrightarrow{u_0} P_0 \xrightarrow{v_0} k \to 0,$$

we obtain the exact sequence of R-modules

$$0 \to \tau(R) \xrightarrow{g_0} E_0 \xrightarrow{f_0} R \to k \to 0.$$

We first observe that f_0: $E_0 \to R$ has the property that if a morphism h: $L \to R$ is not a splittable epimorphism then it can be lifted to E_0, i.e. there is j: $L \to E_0$ such that $f_0 j = h$. To see this we use the equivalence $\mathfrak{P} \to \mathfrak{L}$. Let h_p: $P \to S$ be a morphism in $\mathfrak{P}$ such that h_p^G: $P^G \to S^G$ is the morphism h: $L \to R$. Since h is not a splittable epimorphism, h_p is not a splittable epimorphism. Therefore, $\operatorname{Im} h_p \subset \mathfrak{m}S = \operatorname{Im} v_0$, which means that h_p can be lifted to v_0. Hence h can be lifted to E_0.

We now show that if

$$0 \to A \xrightarrow{g} B \xrightarrow{f} R \to k \to 0$$

is an exact sequence of R-modules with A, B and R in $\mathfrak{L}$ with A indecomposable having the property that any morphism h: $L \to R$ in $\mathfrak{L}$ which is not a splittable epimorphism can be lifted to B, then $0 \to A \to B \to R \to k \to 0$ is isomorphic to

$0 \to \tau(R) \to E_0 \to R \to k$. For by definition we have a commutative diagram

$$\begin{array}{ccccccccc} 0 \to & A & \to B & \to R & \to k & \to 0 \\ & s\downarrow & \downarrow & \| & \| & \\ 0 \to & \tau(R) & \to E_0 & \to R & \to k & \to 0 \\ & \downarrow t & \downarrow & \| & \| & \\ 0 \to & A & \to B & \to R & \to k & \to 0 \end{array}$$

Now the endomorphism $\operatorname{Ext}^2(k, ts)$: $\operatorname{Ext}^2(k, A) \to \operatorname{Ext}^2(k, A)$ induced by the endomorphism ts: $A \to A$ has the property that $\operatorname{Ext}^2(k, ts)(x) = x$, where x is the nonzero element $0 \to A \to B \to R \to k \to 0$ in $\operatorname{Ext}^2(k, A)$. Therefore by Nakayama's lemma, ts is not in the radical of $\operatorname{End}_R(A)$ and therefore is an isomorphism since $\operatorname{End}_R(A)$ is a local ring. Therefore, s: $A \to \tau(R)$ is an isomorphism since $\tau(R)$ is indecomposable, showing that $0 \to A \to B \to R \to k \to 0$ is isomorphic to $0 \to \tau(R) \to E_0 \to R \to k \to 0$. This uniquely determined exact sequence $0 \to \tau(R) \overset{v_0}{\to} E_0 \overset{u_0}{\to} R \to k \to 0$ in mod R is called the fundamental exact sequence.

Suppose now that $V_i \neq k$. Then $V_i^G = 0$. Hence applying the fixed point functor to the exact sequence

$$0 \to \tau(P_i) \overset{u_i}{\to} Q_i \overset{v_i}{\to} P_i \to V_i \to 0,$$

we obtain the exact sequence of reflexive R-modules

$$0 \to \tau(L_i) \overset{g_i}{\to} E_i \overset{f_i}{\to} L_i \to 0.$$

Obviously f_i is not a splittable epimorphism. Using an argument similar to the argument used to characterize the fundamental exact sequence, it is not difficult to show that an exact sequence $0 \to A \to B \to L_i \to 0$ with A and B in $\mathfrak{L}$ is isomorphic to $0 \to \tau(L_i) \overset{g_i}{\to} E_i \overset{f_i}{\to} L_i \to 0$ if and only if it satisfies the following conditions:

(a) A is indecomposable;

(b) it does not split;

(c) every morphism h: $L \to L_i$ in $\mathfrak{L}$ which is not a splittable epimorphism can be lifted to B.

Thus for each $L_i \neq R$ the exact sequence $0 \to \tau(L_i) \to E_i \to L_i \to 0$ is uniquely determined by L_i and is called the almost split sequence determined by L_i. Following the procedure used in the theory of artin algebras we define for R the following quiver, usually called the AR quiver of $\mathfrak{L}$ which we will denote by $AR(\mathfrak{L})$.

The vertices of the AR quiver of $\mathfrak{L}$ are the indecomposable reflexive R-modules $L_0, L_1, \ldots, L_d$ and the number of arrows from L_j to L_i is the multiplicity of L_j in a decomposition of E_i into a sum of indecomposable modules. It is obvious that by construction there is an isomorphism $AR(\mathfrak{L}) \to \Gamma(V)$ sending L_i to P_i for all i. Combining this isomorphism with the isomorphism between $\Gamma(V)$ and the McKay quiver of V already described in §1, we obtain an isomorphism between the AR and McKay quivers.

3. Construction of almost split sequences. We continue with the same hypothesis and notation as in the previous section. Our main aim in this section is to show how all the almost split sequences in $\mathfrak{L}$ can be derived from the fundamental exact sequence in a rather simple way. It is for this reason that the exact sequence $0 \to \tau(R) \to E_0 \to R \to k \to 0$ is called the fundamental exact sequence. We begin with some preliminaries.

Since R is a complete local ring, we know that it contains a complete regular local ring T of dimension 2 such that R is a finitely generated T-module. In [3] it was shown that $\tau(R) = \operatorname{Hom}_T(R, T)$. Therefore, $\tau(R) = w$, the dualizing R-module. Therefore, $\operatorname{Ext}^2_R(k, w) \simeq k$ and the fundamental exact sequence is any one of the nonzero elements in $\operatorname{Ext}^2_R(k, w)$. We will denote $\tau(R)$ by w from now on.

Now given two $S[G]$-modules A and B we consider the S-module $A \otimes_S B$ given by $\sigma(a \otimes b) = \sigma(a) \otimes \sigma(b)$ for all σ in G, a in A and b in B. It is easily seen that for all A, B and C in mod $S[G]$ we have the $S[G]$-isomorphisms

$$\operatorname{Hom}_S(A \otimes_S B, C) \to \operatorname{Hom}_S(A, \operatorname{Hom}_S(B, C))$$

which are functorial in A, B and C. Applying the G-fixed point functor we have functorial isomorphism

$$\operatorname{Hom}_{S[G]}(A \otimes_S B, C) \to \operatorname{Hom}_{S[G]}(A, \operatorname{Hom}_S(B, C)).$$

Now if P and Q are $S[G]$-projective, then $P \otimes_S Q$ is S-free and hence $S[G]$-projective. Since

$$S/\mathfrak{m}(P \otimes_S Q) \simeq P/\mathfrak{m}P \otimes_k Q/\mathfrak{m}Q,$$

the functor $\mathfrak{P} \to \operatorname{mod} k[G]$ commutes with tensor products. Also we have that $(S \otimes_k W_1) \otimes_S (S \otimes_k W_2) \simeq S \otimes_k (W_1 \otimes_k W_2)$ for all W_1 and W_2 in mod $k[G]$. Therefore the projective $S[G]$-resolutions

$$0 \to S \otimes_k \left(\bigwedge^2 V \otimes_k V_j \right) \to S \otimes_k (V \otimes_k V_j) \to S \otimes_k V_j \to V_j \to 0$$

can be rewritten as

$$0 \to \left(S \otimes_k \bigwedge^2 V \right) \otimes_S (S \otimes_k V_j) \to (S \otimes_k V) \otimes_S (S \otimes_k V_j) \to S \otimes_k V_j \to V_j \to 0$$

and therefore as

$$0 \to \tau(S) \otimes_S P_j \to Q_0 \otimes_S P_j \to P_j \to V_j \to 0.$$

Therefore it is of importance to know which reflexive R-module $(P \otimes_S Q)^G$ is for P and Q in $\mathfrak{P}$. To this end we review some well-known facts about reflexive R-modules.

Let L be a reflexive R-module. Then it is easily seen that $\operatorname{Hom}_R(X, L)$ is R-reflexive for all R-modules X. Hence X^* and X^{**} are reflexive R-modules for all R-modules X, where $X^* = \operatorname{Hom}_R(X, R)$. Also the canonical morphism $X \to X^{**}$ induces isomorphisms $\operatorname{Hom}_R(X^{**}, L) \to \operatorname{Hom}_R(X, L)$ for all reflexive R-modules L. This obviously uniquely determines the reflexive R-module X^{**}.

Suppose now that L_1 and L_2 are in $\mathfrak{L}$. We want to define the "tensor product" $L_1 \cdot L_2$ of L_1 and L_2 in $\mathfrak{L}$. In general, $L_1 \otimes_R L_2$ is not even torsion free, let alone reflexive. So we define $L_1 \cdot L_2$ to be $(L_1 \otimes_R L_2)^{**}$. Now for each reflexive R-module L, we know that the natural morphism $L_1 \otimes_R L_2 \to L_1 \cdot L_2$ induces an R-isomorphism $\operatorname{Hom}_R(L_1 \cdot L_2, L) \tilde{\to} \operatorname{Hom}_R(L_1 \otimes_R L_2, L)$ which is functorial in L_1, L_2 and L. Also we have the usual R-isomorphisms

$$\operatorname{Hom}_R(L_1 \otimes_R L_2, L) \to \operatorname{Hom}_R(L_1, \operatorname{Hom}_R(L_2, L))$$

which are functorial in L_1, L_2 and L. Hence we obtain the R-isomorphisms of reflexive R-modules $\operatorname{Hom}_R(L_1 \cdot L_2, L) \to \operatorname{Hom}_R(L_1, \operatorname{Hom}_R(L_2, L))$ which are functorial in L_1, L_2 and L. It is for this reason that $L_1 \cdot L_2$ is called the "tensor product" of L_1 and L_2 in $\mathfrak{L}$. With these observations in mind, we prove the following

LEMMA 3.1. *Let P and Q be projective $S[G]$-modules. Then we have the following*:
(a) $\operatorname{Hom}_S(P, Q)^G = \operatorname{Hom}_R(P^G, Q^G)$;
(b) $(P \otimes_S Q)^G = P^G \cdot Q^G$.

PROOF. (a) $\operatorname{Hom}_S(P, Q)^G = \operatorname{Hom}_{S[G]}(P, Q) \tilde{\to} \operatorname{Hom}_R(P^G, Q^G)$ since the fixed point functor $\mathfrak{P} \to \mathfrak{L}$ is an equivalence of categories.

(b) Let L be in $\mathfrak{L}$ and let U be a projective $S[G]$-module such that $U^G = L$. Now

$$\begin{aligned}\operatorname{Hom}_R\big((P \otimes_S Q)^G, U\big) &= \operatorname{Hom}_R\big((P \otimes_S Q)^G, U^G\big)\\ &= \operatorname{Hom}_{S[G]}(P \otimes_S Q, U) = \operatorname{Hom}_{S[G]}(P, \operatorname{Hom}_S(Q, U))\\ &= \operatorname{Hom}_R\big(P^G, \operatorname{Hom}_{S[G]}(Q, U)\big) = \operatorname{Hom}_R\big(P^G, \operatorname{Hom}_R(Q^G, U^G)\big)\\ &= \operatorname{Hom}_R\big(P^G, \operatorname{Hom}_R(Q^G, L)\big) = \operatorname{Hom}_R\big(P^G \cdot Q^G, L\big).\end{aligned}$$

Hence we have an isomorphism of R-modules

$$\operatorname{Hom}_R\big((P \otimes_S Q)^G, L\big) \to \operatorname{Hom}_R\big(P^G \cdot Q^G, L\big)$$

which is functorial in L. Therefore $(P \otimes_S Q)^G = P^G \cdot Q^G$.

As an immediate consequence of this lemma we have the following

PROPOSITION 3.2. *For each $L_j \neq R$ we have the following*:
(a) $\tau(L_j) = w \cdot L_j$;
(b) *tensoring the exact sequence $0 \to w \to E_0 \to R$ with L_j, we obtain the exact sequence $0 \to w \cdot L_j \to E_0 \cdot L_j \to L_j \to 0$ which is the almost split sequence for L_j.*

PROOF. Since (a) follows from (b), it suffices to prove (b). We have already seen that the minimal projective $S[G]$-resolution of V_j can be obtained by tensoring the minimal projective $S[G]$-resolution of k with P_j over S obtaining

$$0 \to \tau(S) \otimes_S P_j \to Q_0 \otimes_S P_j \to S \otimes_S P_j \to k \otimes_S P_j = V_j \to 0.$$

Applying the fixed point functor, we obtain the almost split sequence $0 \to w \cdot L_j \to E_0 \cdot L_j \to L_j \to 0$, giving our desired result.

As an immediate consequence of this result we have the following slightly different description of the AR quiver of $\mathfrak{L}$.

COROLLARY 3.3. *In the AR quiver of $\mathfrak{L}$ we have that the number of arrows from L_i to L_j is the multiplicity of L_i in the decomposition of $E_0 \cdot L_j$ into indecomposable modules.*

4. Algebras with finitely many indecomposable reflexive modules. Suppose $\mathbf{C}$ is the complex numbers, G a finite subgroup of $\mathrm{GL}(2, \mathbf{C})$ such that the induced linear action on $S = \mathbf{C}[[X, Y]]$ has the property that every height one prime in $R = S^G$ is unramified in S, or equivalently, G has no pseudo-reflections. Then we know that $\mathfrak{L}$, the category of reflexive R-modules is of finite representation type, i.e. $\mathfrak{L}$ has only a finite number of nonisomorphic indecomposable modules. In this section we show how the ideas developed in the previous sections can be used to prove the converse, which is also proved in [**1**] using different techniques. We begin by pointing out some generalizations of results given in the previous sections.

Let k be a field and S an arbitrary local k-algebra with maximal ideal $\mathfrak{m}$ such that $S/\mathfrak{m} = k$. Suppose G is a finite group of k-automorphisms of S such that the order of G is not divisible by the characteristic of k. Then, as before, $\mathfrak{m}S[G]$ is the radical of the skew group ring $S[G]$ and $S[G]/\mathfrak{m}S[G] = k[G]$, the ordinary group ring of G over k, which is semisimple. As in the previous section, given two $S[G]$-modules A and B, we can form the $S[G]$-modules $\mathrm{Hom}_S(A, B)$ and $A \otimes_S B$. Likewise the argument given before shows that an $S[G]$-module A is projective if and only if A is S-free. Therefore $\mathrm{Hom}_S(A, B)$ and $A \otimes_S B$ are $S[G]$-projective if A and B are $S[G]$-projective.

PROPOSITION 4.1. *Let $\mathfrak{P}$ be the category of projective $S[G]$-modules. Then the functor $\mathfrak{P} \to \mathrm{mod}\, k[G]$ given by $P \mapsto P/\mathfrak{m}P$ for all P in $\mathfrak{P}$ has the following properties*:

(a) *the natural epimorphism $P \to P/\mathfrak{m}P$ is a projective cover for all P in $\mathfrak{P}$*;

(b) *$\mathfrak{P} \to \mathrm{mod}\, k[G]$ is full and dense*;

(c) *P and Q in $\mathfrak{P}$ are isomorphic if and only if the $k[G]$-modules $P/\mathfrak{m}P$ and $Q/\mathfrak{m}Q$ are isomorphic*;

(d) *suppose P and Q are projective $S[G]$-modules*; *then $P \otimes_S Q \mapsto P/\mathfrak{m} \otimes_k Q/\mathfrak{m}Q$ and* $\mathrm{Hom}_S(P, Q) \mapsto \mathrm{Hom}_k(P/\mathfrak{m}P, Q/\mathfrak{m}Q)$;

(e) $\mathrm{rank}_S P = \dim_k P/\mathfrak{m}P$ *for all P in $\mathfrak{P}$.*

PROOF. (a) This is a consequence of the fact that the idempotents in $k[G]$ can be lifted to $S[G]$ since the composition $k[G] \to S[G] \to k[G]$ is the identity, where the first morphism is the inclusion.

(b) Obvious.

(c) and (d) Same as previous proofs in case $S = k[[X, Y]]$.

(e) Obvious.

As an immediate consequence of this proposition we have the following which we will need in this section:

COROLLARY 4.2. *Let S and G be as above.*

(a) *Let $V_0, V_1, \ldots, V_d$ be a complete set of nonisomorphic simple $k[G]$-modules and $P_0, P_1, \ldots, P_d$ their projective covers respectively. Then $P_0, \ldots, P_d$ is a complete set of nonisomorphic indecomposable projective $S[G]$-modules.*

(b) *$\mathfrak{P}$ is a Krull-Schmidt category, i.e. each P in $\mathfrak{P}$ is isomorphic to a unique sum $\coprod_{i=0}^{d} n_i P_i$.*

We now generalize some of the other results in a different direction.

Let R be an integrally closed noetherian domain with field of quotients K. Let L be a finite galois extension of K with group G such that S, the integral closure of R in L, is a finitely generated R-module and every height one prime in R is unramified in S. Then, as before, the R-morphism $S \to S[G]^G$ given by $s \mapsto \sum_{\sigma \in G} \sigma(s)\sigma$ is an isomorphism which we view as an identification. Hence we have the functor $\mathfrak{P}S[G] \to \operatorname{add}_R S$ given by $P \mapsto P^G$ for all P in $\mathfrak{P}S[G]$, the category of projective $S[G]$-modules. Also because every height one prime ideal in R is unramified in S, we have that the natural R-algebra morphism $S[G] \to \operatorname{End}_R(S)$ given by $s\sigma(x) = s(\sigma(x))$ for all s in S, σ in G and x in S is an isomorphism [**2**]. As before, this implies that the natural R-algebra morphism $\operatorname{End}_{S[G]}(S[G]) \to \operatorname{End}_R(S)$ given by $f \mapsto f|S$ for all f in $\operatorname{End}_{S[G]}(S[G])$ is an isomorphism. This yields the following result:

PROPOSITION 4.3. *The functor* $\mathfrak{P}S[G] \to \operatorname{add}_R S$ *given by* $P \mapsto P^G$ *for all* P *in* $\mathfrak{P}S[G]$ *is an equivalence of categories.*

As before, S is a reflexive R-module. Suppose L_1 and L_2 are reflexive R-modules. Then we have that $\operatorname{Hom}_R(L_1, L_2)$ is a reflexive R-module. Also, as before, we define the tensor product $L_1 \cdot L_2$ of L_1 and L_2 in the category of reflexive R-modules by $L_1 \cdot L_2 = \operatorname{Hom}_R(\operatorname{Hom}_R(L_1 \otimes_R L_2, R), R)$. Again, given two $S[G]$-modules A and B, we consider $\operatorname{Hom}_S(A, B)$ and $A \otimes_S B$ as $S[G]$-modules in the usual way. Moreover, $\operatorname{Hom}_S(A, B)$ and $A \otimes_S B$ are projective $S[G]$-modules if A and B are projective $S[G]$-modules since an $S[G]$-module X is projective is and only if X is a projective S-module.

As before we have the following

PROPOSITION 4.4. *The functor* $\mathfrak{P}S[G] \to \operatorname{add}_R S$ *has the following properties for* P *and* Q *in* $\mathfrak{P}S[G]$:

(a) $\operatorname{Hom}_S(P, Q) \mapsto \operatorname{Hom}_R(P^G, Q^G)$;

(b) $P \otimes_S Q \mapsto P^G \cdot Q^G$.

As an immediate consequence of Proposition 4.4 we have the following

COROLLARY 4.5. *Let* L_1, L_2 *be in* $\operatorname{add}_R S$. *Then* $\operatorname{Hom}_R(L_1, L_2)$ *and* $L_1 \cdot L_2$ *are in* $\operatorname{add}_R S$.

For the rest of this section we assume we have a fixed algebraically closed field k. By a k-algebra T we will always mean a k-algebra which is a complete integrally closed noetherian local domain with residue field k. As a consequence of the above discussion we have the following

THEOREM 4.6. *Let* R *be a* k*-algebra with field of quotients* K. *Let* S *be the integral closure of* R *in a finite galois extension* L *of* K *with galois group* G *whose order is not divisible by the characteristic of* k. *Moreover, suppose every height one prime in* R *is unramified in* S. *Then the functors* $\mathfrak{P}S[G] \to \operatorname{add}_R S$ *given by* $P \mapsto P^G$ *for all* P *in* $\mathfrak{P}S[G]$ *and* $\mathfrak{P}S[G] \to \operatorname{mod} k[G]$ *given by* $P \mapsto P/\mathfrak{m}P$ *for all* P *in* $\mathfrak{P}S[G]$ *induce the functor* F: $\operatorname{add}_R S \to \operatorname{mod} k[G]$ *given by* $F(L) = P/\mathfrak{m}P$, *where* P *is the object in*

$\mathfrak{P}S[G]$ *such that* $P^G = L$. *The functor* F: $\text{add}_R S \to \text{mod } k[G]$ *has the following properties*:

(a) F *is full and dense*;

(b) $F(L_1) \approx F(L_2)$ *if and only if* $L_1 \simeq L_2$ *for all* L_1, L_2 *in* $\text{add}_R S$;

(c) L *in* $\text{add}_R S$ *is indecomposable if and only if* $F(L)$ *is simple in* $\text{mod } k[G]$;

(d) *for all* L_1, L_2 *in* S, *we have* $F(\text{Hom}_R(L_1, L_2)) = \text{Hom}_k(F(L_1), F(L_2))$ *and* $F(L_1 \cdot L_2) = F(L_1) \otimes_k F(L_2)$;

(e) $F(R) = k$, *the trivial representation*;

(f) $\text{rank}_R L = \dim_k F(L)$ *for all* L *in* $\text{add}_R S$.

COROLLARY 4.7. *Under the same hypothesis as Theorem* 4.6, *the functor* F: $\text{add}_R S \to \text{mod } k[G]$ *induces a bijection between the isomorphism classes of indecomposable objects in* $\text{add}_R S$ *and the isomorphism classes of simple objects in* $\text{mod } k[G]$.

We now apply Corollary 4.7 to prove the result promised in the beginning of this section.

Let $k = \mathbf{C}$, the complex numbers, R a k-algebra with field of quotients K. Let Ω be the algebraic field extension of K generated by all finite field extensions L of K such that the integral closure S of R in L has the property that the height one prime ideals of R in S are unramified. Then Ω is galois over K.

As an application of Corollary 4.7 we have the following

PROPOSITION 4.8. *Let* $K \subset L \subset L' \subset \Omega$ *be field extensions of* K. *Suppose* $K \subset L$ *and* $K \subset L'$ *are finite galois extensions with galois groups* G *and* G', *respectively. Suppose* $S' \supset S \supset R$ *are the integral closures of* R *in* L' *and* L, *respectively. Then* $\text{add}_R S' \supset \text{add}_R S$ *and they are equal if and only if* $S = S'$.

PROOF. Since $\mathbf{C} \subset R$, the usual trace argument shows that S is an S-summand of S'. Hence $\text{add}_R S' \supset \text{add}_R S$. Now suppose $\text{add}_R S' = \text{add}_R S$. Then $\text{add}_R S$ and $\text{add}_R S'$ have the same number of isomorphism classes of indecomposable R-modules. Therefore by Corollary 4.7 we have that $\mathbf{C}[G]$ and $\mathbf{C}[G']$ have the same number of isomorphism classes of simple modules. Hence G and G' have the same number of conjugacy classes. But this implies that $G = G'$, since $G = G'/H$ where H is the normal subgroup of G' consisting of the elements of G leaving L fixed. Therefore $S = S'$.

We now prove our desired result.

THEOREM 4.9. *Suppose* $\dim R = 2$ *and the number of isomorphism classes of indecomposable reflexive* R*-modules is finite. Then*

(a) Ω *is a finite galois extension of* K *with group* G;

(b) *the integral closure* S *of* R *in* Ω *is regular*, $S^G = R$ *and we can choose analytically independent elements* X *and* Y *is* S *such that* $S = \mathbf{C}[[X, Y]]$ *and the action of* G *on* $\mathbf{C}[[X, Y]]$ *is linear*.

PROOF. (a) Suppose Ω is not a finite extension of K. Then there is an infinite proper ascending chain

$$K \subsetneqq L_1 \subsetneqq L_2 \subsetneqq \cdots \subset \Omega$$

of galois extensions of K. Letting S_i be the integral extension of R in L_i, we obtain the infinite proper ascending chain $R \subsetneqq S_1 \subsetneqq S_2 \subsetneqq \cdots$. Therefore, by Proposition 4.8 we obtain the infinite proper ascending chain add $S_1 \subsetneqq$ add $S_2 \subsetneqq \cdots$. Since each of the S_i is a reflexive R-module, this contradicts the fact that there is only a finite number of isomorphism classes of indecomposable reflexive R-modules.

(b) Let S be the integral closure of R in Ω. Then S satisfies the purity of the branch locus, i.e. if $L \supset \Omega$ is a finite field extension of Ω such that the height one primes of $\mathfrak{p}$ in S are unramified in the integral closure S' of S, then S is unramified in S' or equivalently $S = S'$. Hence by Mumford we know that $S \simeq \mathbf{C}[[X, Y]]$.

Clearly $S^G = R$. The rest of the theorem follows from the fact that since G is finite the action of G on $\mathbf{C}[[X, Y]]$ is similar to a linear action.

5. Reflexive splitting trace modules. In this section we introduce the notion of a reflexive splitting trace module which plays an essential role in the next section where we show to what extent the method of constructing almost split sequences for complete two-dimensional quotient singularities described in §3 can be generalized to arbitrary two-dimensional complete local integrally closed domains. We begin with some generalities.

Throughout this section R denotes a noetherian, integrally closed domain with field of quotients K and all R-modules are finitely generated. Suppose M is a torsion free R-module. Then $\operatorname{End}_R(M) \subset \operatorname{End}_K(K \otimes_R M)$ and the usual trace map t: $\operatorname{End}_K(K \otimes_R M) \to K$ has the property that $t(\operatorname{End}_R(M)) \subset R$ since R is integrally closed. We now point out some important facts concerning the trace t: $\operatorname{End}_R(M) \to R$ which we will need in our discussion of almost split sequences.

PROPOSITION 5.1. *Let R be a local ring with maximal ideal $\mathfrak{m}$ and M a torsion free R-module. Then t: $\operatorname{End}_R(M) \to R$ has the property that $t(\operatorname{rad}\operatorname{End}_R M) \subset \mathfrak{m}$.*

The proof of this fact goes in several steps.

LEMMA 5.2. *Let R be a local domain (not necessarily integrally closed) with field of quotient K. Suppose M is a torsion free R-module and f: $M \to M$ an R-morphism such that $f^k(M) \subset \mathfrak{m}M$ for some $k \geq 0$. Further suppose that x in R is an eigenvalue for $f \otimes_R K$: $K \otimes_R M \to K \otimes_R M$. Then x is in $\mathfrak{m}$.*

PROOF. Suppose v is a nonzero element of $K \otimes_R M$ such that $(K \otimes f)(v) = xv$. Then $v = y/s$ with y in M and $s \in R - \{0\}$, from which it follows that $f(y) = xy$. Since $y \neq 0$, there is a j such that y is not in $\mathfrak{m}^j M$. But $x^{k+(j-1)}y = f^{k+(j-1)}(y)$ is in $\mathfrak{m}^j M$. Therefore x is not a unit, i.e. x is in $\mathfrak{m}$.

PROPOSITION 5.3. *Let R be an integrally closed local domain with maximal ideal $\mathfrak{m}$ and field of quotient K. Let M be a torsion free R-module and f: $M \to M$ an R-endomorphism with $f^k(M) \subset \mathfrak{m}M$ for some $k \geq 0$. Then the characteristic polynomial $g(X) = X^n + a_1 X^{n-1} + \cdots + a_n$ of $K \otimes_R f$: $K \otimes_R M \to K \otimes_R M$ has the property that a_i is in $\mathfrak{m}$ for all $i \geq 1$. In particular, $t(f) = a_1$ is in $\mathfrak{m}$.*

PROOF. Suppose $u_1, \ldots, u_n$ are the roots of $g(X)$ in some finite field extension of K. Since $g(X)$ is in $R[X]$, the u_i are integral over R, and hence $R[u_1, \ldots, u_n]$ is a finitely generated R-module because R is noetherian. Therefore $R[u_1, \ldots, u_n]$ is noetherian. Let $\mathfrak{m}'$ be a maximal ideal of $R[u_1, \ldots, u_n]$ and let $S = R[u_1, \ldots, u_n]_{\mathfrak{m}'}$. Then S is a local noetherian domain with maximal ideal $\mathfrak{m}'S$ having the property that $\mathfrak{m}'S \cap R = \mathfrak{m}$ and $\mathfrak{m}\, S$ is $\mathfrak{m}'S$-primary. Let L be the field of quotients of S and let SM be the S-submodule of $L \otimes_R M$ generated by M. Then $L \otimes f$: $L \otimes_R M \to L \otimes_R M$ has the property that $(L \otimes f)(SM) \subset SM$ and $(L \otimes f)^k \subset (\mathfrak{m}'S)SM$. Since $g(X)$ is also the characteristic polynomial for $L \otimes f$, it follows that the $u_1, \ldots, u_n$ are the eigenvalues for $L \otimes f$. Hence, by our previous lemma, the $u_1, \ldots, u_n$ are in $\mathfrak{m}'S$. Since the a_i are in R and are sums of products of the u_i, they are in $\mathfrak{m}'S \cap R = \mathfrak{m}$, giving our desired result.

We now return to the proof of Proposition 5.1. Since $\Lambda = \operatorname{End}_R(M)$ is a finitely generated R-module, we have that $\Lambda/\mathfrak{m}\Lambda$ is a finite dimensional $R/\mathfrak{m}$-algebra. Hence $(\operatorname{rad} \Lambda)^n \subset \mathfrak{m}\Lambda$ for some $n > 0$. Therefore if f: $M \to M$ is in rad End M, then we have that $f^k(M) \subset \mathfrak{m}M$ for some $k > 0$. Hence by Proposition 5.3 we have that $t(f) \in \mathfrak{m}$, which finishes the proof of Proposition 5.1.

Of particular interest to us in this section are the reflexive modules M over an integrally closed domain R having the property that R is isomorphic to a summand of $\operatorname{End}_R(M)$ viewed as an R-module. We now develop various criteria for M to have this property.

We begin with the following preliminary observation:

LEMMA 5.4. *Let M be a reflexive R-module.*

(a) *The map α: $\operatorname{End}_R(M) \to \operatorname{Hom}_R(\operatorname{End}_R(M), R)$ given by $\alpha(f)(g) = t(fg)$ for all f and g in $\operatorname{End}_R(M)$ is an isomorphism of $\operatorname{End}_R(M)$ modules.*

(b) *$\alpha(\mathrm{id}) = t$ and so t is a free generator for $\operatorname{Hom}_R(\operatorname{End}_R(M), R)$ over $\operatorname{End}_R(M)$.*

PROOF. (a) It is easily checked that α is a morphism of $\operatorname{End}_R(M)$-modules which is an isomorphism whenever M is free. Thus

$$\alpha_{\mathfrak{p}}\colon \operatorname{End}_{R_{\mathfrak{p}}}(M) \to \operatorname{Hom}_{R_{\mathfrak{p}}}\big(\operatorname{End}_{R_{\mathfrak{p}}}(M)_{\mathfrak{p}}\big), R_{\mathfrak{p}}$$

is an isomorphism for all height one prime ideals $\mathfrak{p} \subset R$. The fact that α is an isomorphism now follows from the fact that $\operatorname{End}_R(M)$ and $\operatorname{Hom}_R(\operatorname{End}(M), R)$ are reflexive R-modules and R is integrally closed.

(b) Trivial consequence of (a).

As an easy consequence of this result we have the following

PROPOSITION 5.5. *Let M be a reflexive R-module. Then R is isomorphic to a summand of $\operatorname{End}_R(M)$ (notation: $R|\operatorname{End}_R(M)$) if and only if the trace t: $\operatorname{End}_R(M) \to R$ is onto.*

PROOF. Clearly if t: $\operatorname{End}_R(M) \to R$ is onto, then $R|\operatorname{End}_R(M)$. Suppose $R|\operatorname{End}_R(M)$. Then there is an R-module morphism f: $\operatorname{End}_R(M) \to R$ which is onto. But $f = t\alpha^{-1}(f)$ by Lemma 4. Therefore t is onto.

For obvious reasons we will say that a reflexive module M over an integrally closed domain R is a splitting trace module if $R|\operatorname{End}_R(M)$ or equivalently $t(\operatorname{End}_R(M)) = R$.

As an easy consequence of Proposition 5 we have the following

COROLLARY 5.6. (a) *A reflexive R-module M is a splitting trace module if* rank M *is a unit in R.*

(b) *If R contains the rational numbers, then every reflexive R-module is a splitting trace module.*

PROOF. (a) This is a trivial consequence of the fact that $t(\mathrm{id}) = \operatorname{rank} M$.

(b) Trivial consequence of (a).

Specializing R a bit, we get the following description of when a reflexive R-module is a splitting trace module:

PROPOSITION 5.7. *Let R be a complete local ring with maximal ideal $\mathfrak{m}$ such that $R/\mathfrak{m}$ is algebraically closed. Then an indecomposable reflexive R-module M is a splitting trace module if and only if p, the characteristic of $R/\mathfrak{m}$, and* rank M *are relatively prime when $p \neq 0$.*

PROOF. Clearly $(p, \operatorname{rank} M) = 1$ whenever $p \neq 0$ if and only if rank M is invertible in R. Thus if $(p, \operatorname{rank} M) = 1$ when $p \neq 0$, then M is a splitting trace module by Corollary 5.6.

Suppose now that M is a splitting trace module. If $p = 0$, there is nothing to prove, so suppose $p \neq 0$ and rank M and p are not relatively prime. Then $\operatorname{rank} M \cdot 1$ is in $\mathfrak{m}$, the maximal ideal of R. Since R is complete and M is indecomposable, we know that $\operatorname{End}_R(M)$ is a local ring. Because $R/\mathfrak{m}$ is algebraically closed we have that $\operatorname{End}_R(M)/\operatorname{rad}\operatorname{End}_R M = R/\mathfrak{m}$. Therefore each f in $\operatorname{End}_R M$ can be written as $aI + g$ where I is the identity, a is in R and g is in $\operatorname{rad}\operatorname{End}_R(M)$. Then $t(f) = at(I) + t(g)$. Since by Proposition 5.1 we have that $t(g)$ is in $\mathfrak{m}$ and, by hypothesis, $t(I) = \operatorname{rank} M$ is $\mathfrak{m}$, it follows that $t(f)$ is in $\mathfrak{m}$. Hence $t(\operatorname{End}_R(M)) \subset \mathfrak{m}$, which shows that M is not a splitting trace module.

We now want to investigate what $R|\operatorname{Hom}_R(M, N)$, where M and N are reflexive modules over R, means. We begin with the following

LEMMA 5.8. *Let M be a reflexive module over R and let Y be a module over $\Gamma = \operatorname{End}_R(M)$.*

(a) *The trace map $t\colon \Gamma \to R$ induces an R-isomorphism $t_Y\colon \operatorname{Hom}_\Gamma(Y, \Gamma) \to \operatorname{Hom}_R(Y, R)$.*

(b) *Suppose now that R is a complete local ring and M is indecomposable.*

If R is an R-summand of Y, then Γ is a Γ-summand of Y.

PROOF. (a) The Γ-isomorphism $\alpha\colon \Gamma \to \operatorname{Hom}_R(\Gamma, R)$ given $\alpha(f)(g) = t(fg)$ for all f, g in Γ induces an isomorphism

$$\operatorname{Hom}_\Gamma(Y, \Gamma) \xrightarrow{(Y,\alpha)} \operatorname{Hom}(Y, \operatorname{Hom}_R(\Gamma, R)).$$

Also we have the standard isomorphism of R-modules v: $\operatorname{Hom}_\Gamma(Y, \operatorname{Hom}_R(\Gamma, R)) \to \operatorname{Hom}_R(Y, R)$ given by $v(f)(y) = f(y)(1)$ for all f in $\operatorname{Hom}_\Gamma(Y, \operatorname{Hom}_R(\Gamma, R))$, y in Y. Since $t_Y = v(Y, \alpha)$, as can easily be checked, we have our desired result that t_Y is an R-isomorphism.

(b) If Γ is not a Γ-summand of Y, then for each Γ-morphism f: $Y \to \Gamma$ we have that $\operatorname{Im} f \subset \operatorname{rad} \Gamma$ since Γ is a local ring. Since t: $\Gamma \to R$ has the property that $t(\operatorname{rad} \Gamma) \subset \operatorname{rad} R$, we have by (a) that all R-morphisms g: $Y \to R$ have $\operatorname{Im} g \subset \operatorname{rad} R$. Therefore R is not an R summand of Y.

We now apply this lemma to prove the following

LEMMA 5.9. *Let M, X, Y be reflexive R-modules and suppose* $\Gamma = \operatorname{End}_R M$.

(a) *The R-morphism u*: $\operatorname{Hom}_R(X, Y) \to \operatorname{Hom}_{\Gamma^{\mathrm{op}}}(\operatorname{Hom}_R(M, X), \operatorname{Hom}_R(M, Y))$, *given by* $f \mapsto \operatorname{Hom}_R(M, f)$ *for all f in* $\operatorname{Hom}_R(X, Y)$, *is an isomorphism functorial in X and Y, where* $\operatorname{Hom}_R(M, X)$ *is considered an Γ^{op}-module by means of the operation of Γ on M.*

(b) *The R-morphism u*: $\operatorname{Hom}_R(X, Y) \to \operatorname{Hom}_\Gamma(\operatorname{Hom}_R(Y, M), \operatorname{Hom}_R(X, M))$, *given by* $f \mapsto \operatorname{Hom}_R(f, M)$ *for all f in* $\operatorname{Hom}_R(X, Y)$, *is an isomorphism functorial in X and Y.*

PROOF. (a) and (b) Since the morphisms u and v are obviously R-morphisms between reflexive R-modules it suffices to prove that they are isomorphisms when localized at height one prime ideals. But if $\mathfrak{p}$ is a height one prime, then $M_\mathfrak{p}$, $X_\mathfrak{p}$, $Y_\mathfrak{p}$ are free $R_\mathfrak{p}$-modules, from which it follows that $u_\mathfrak{p}$ and $v_\mathfrak{p}$ are isomorphisms by induction on the rank of the free $R_\mathfrak{p}$-modules $X_\mathfrak{p}$ and $Y_\mathfrak{p}$.

As a consequence of Lemmas 5.8 and 5.9 we obtain the following result which was the main point of this discussion:

PROPOSITION 5.10. *Let M and N be two reflexive indecomposable R-modules when R is a complete local ring.*

(a) *If* $R|\operatorname{Hom}_R(N, M)$, *then $N \simeq M$ and M is a splitting trace module.*

(b) *If M is a splitting trace module, then $M \simeq N$ if and only if* $R|\operatorname{Hom}_R(N, M)$.

PROOF. (a) Suppose $R|\operatorname{Hom}_R(N, M)$. Then by Lemma 5.8 we know that there is a Γ-surjection j: $\operatorname{Hom}_R(N, M) \to \Gamma$ where $\Gamma = \operatorname{End}_R(M)$. Then by Lemma 5.9 we know there is an R-morphism g: $M \to N$ such that $j(f) = fg$ for all f in $\operatorname{Hom}_R(N, M)$. Hence there is an f in $\operatorname{Hom}_R(N, M)$ such that $fg = 1_M$. Therefore g: $M \to N$ is a splittable monomorphism and hence an isomorphism since N is indecomposable.

(b) Trivial consequence of (a).

6. Almost split sequences. Throughout this section we assume that R is a complete two-dimensional integrally closed local domain. Our aim is to demonstrate to what extent the method of computing almost split sequences for reflexive R-modules given in §3 can be generalized to arbitrary R. While it is known that in this context almost split sequences always exist for reflexive R-modules under our hypothesis on R [**3**], this fact will not be used in our discussion here in order to make the paper as self-contained as possible.

We begin by pointing out that what we called the fundamental exact sequence for complete two-dimensional quotient singularities also exists for any R. Recall that the dualizing module for R, which we denote by w, is $\mathrm{Hom}_T(R, T)$, where T is a two-dimensional regular local subring of R such that R is a finitely generated Γ-module. Then the reflexive R-modules are precisely the R-modules which are T-free. Therefore $X \mapsto \mathrm{Hom}_T(X, T)$ gives an exact duality on the category Ref R of reflexive R-modules. From this the well-known fact that $\mathrm{Ext}^1_R(X, w) = 0$ for all reflexive R-modules X follows easily.

Another well-known property of w is that $\mathrm{Ext}^2(R/\mathfrak{m}, w) \simeq R/\mathfrak{m}$ where $\mathfrak{m}$ is the maximal ideal of R. Therefore all the exact sequences $0 \to w \to E \to R \to R/\mathfrak{m} \to 0$ representing nonzero elements of $\mathrm{Ext}^2_R(R/\mathfrak{m}, w)$ are isomorphic. We call any such sequence a fundamental exact sequence. Using the fact that depth $R/\mathfrak{m} = 2$, it is not difficult to show that the E in a fundamental exact sequence $0 \to w \to E \xrightarrow{p} R \to R/\mathfrak{m} \to 0$ is reflexive, see [3]. Now the importance of the fundamental exact sequence lies in the fact that given any morphism $f\colon X \to R$ of reflexive R-modules which is not a splittable epimorphism, or equivalently not an epimorphism, then there is a $g\colon X \to E$ such that $f = pg$. The reason for this is that in the pullback exact commutative diagram

$$\begin{array}{ccccccccc} 0 & \to & w & \to & E \times_R X & \to & X & \to & 0 \\ & & \| & & \downarrow & & \downarrow f & & \\ 0 & \to & w & \to & E & \xrightarrow{p} & R & & \end{array}$$

the exact sequence $0 \to w \to E \times_R X \to X \to 0$ splits, giving our desired result.

PROPOSITION 6.1. *Let $0 \to w \to E \xrightarrow{p} R \to R/\mathfrak{m} \to 0$ be the fundamental exact sequence. Then for each reflexive R-module X we have that*

$$\mathrm{Im}(\mathrm{Hom}_R(X, E) \xrightarrow{\mathrm{Hom}(X,p)} \mathrm{Hom}_R(X, R))$$

consists of the nonepimorphisms from X to R.

As an immediate consequence of Proposition 6.1, we have the following

COROLLARY 6.2. *Let M be a nonprojective indecomposable reflexive R-module. Then applying the functor* $\mathrm{Hom}_R(M^*, \ )$ *to the fundamental exact sequence* $0 \to w \to E \xrightarrow{p} R \to R/\mathfrak{m} \to 0$, *we obtain the exact sequence*

(*) $\qquad 0 \to \mathrm{Hom}_R(M^*, w) \to \mathrm{Hom}_R(M^*, E) \to M^{**} = M \to 0.$

Our aim now is to show that the exact sequence (*) in Corollary 6.2 does not split if and only if M is a splitting trace module. To this end we need the following

LEMMA 6.3. *Let M be a reflexive R-module and* $\Gamma = \mathrm{End}_R(M)$. *Then for each reflexive R-module, X, the morphism*

$$u_X\colon \mathrm{Hom}_R(\Gamma, X) \to \mathrm{Hom}_R(M, \mathrm{Hom}_R(M^*, X))$$

given by $(u_X(f))(m)(g) = f(\delta(g \otimes m))$ *for all f in* $\mathrm{Hom}_R(\Gamma, X)$, *m in M and g in* M^*, *is an isomorphism of two-sided* Γ*-modules which is functorial in X and where* δ: $M^* \otimes_R M \to \Gamma$ *is given by* $\delta(g \otimes m)(x) = g(x)m$ *for all g in* M^* *and m and x in M.*

PROOF. Since X is reflexive, we have that $\delta\colon M^* \otimes_R M \to \Gamma$ induces an isomorphism of two-sided Γ-modules,

$$\operatorname{Hom}_R(\Gamma, X) \overset{\operatorname{Hom}_R(\delta, X)}{\to} \operatorname{Hom}_R(M^* \otimes_R M, X)$$

which is functorial in X. The standard isomorphism $\alpha_X\colon \operatorname{Hom}_R(M^* \otimes_R M, X) \to \operatorname{Hom}_R(M, \operatorname{Hom}(M^*, X))$ is also a two-sided Γ-isomorphism functorial in X. Our desired result now follows from the fact that $u_x = \alpha_x \operatorname{Hom}_R(\delta, X)$.

We now prove our promised result which we restate as follows:

PROPOSITION 6.4. *Let M be an indecomposable reflexive nonprojective R-module. Then the following are equivalent*:

(a) *M is a splitting trace module.*

(b) *The exact sequence*

$$0 \to \operatorname{Hom}_R(M^*, w) \to \operatorname{Hom}_R(M^*, E) \overset{\operatorname{Hom}_R(M^*, p)}{\to} M^{**} \to 0$$

induced from the fundamental exact sequence does not split.

PROOF. By the previous lemma we have the following commutative diagram:

$$\begin{array}{ccc} \operatorname{Hom}_R(\Gamma, E) & \to & \operatorname{Hom}_R(\Gamma, R) \\ \downarrow u_E & & \downarrow u_R \\ \operatorname{Hom}(M, \operatorname{Hom}_R(M^*, E)) & \to & \operatorname{Hom}_R(M, M^{**}). \end{array}$$

Since the u_E and u_R are isomorphisms, it follows that the top horizontal map is not surjective if and only if the bottom horizontal map is not surjective. Because Γ is R-reflexive, the basic property of $p\colon E \to R$ tells us the top morphism is not surjective if and only if $R|\Gamma$ or equivalently if and only if M is a splitting trace module. On the other hand, the bottom row is not surjective if and only if $\operatorname{Hom}_R(M^*, E) \to M^{**} = M$ is not a splittable epimorphism. Hence we have established the proposition.

Next we want to show that if M is an indecomposable nonprojective R-module, then the exact sequence $0 \to \operatorname{Hom}_R(M^*, w) \to \operatorname{Hom}_R(M^*, E) \to M^{**} \to 0$ is almost split if and only if it does not split. Clearly all one has to show is that it is almost split if it does not split. This requires the following preliminary result which has some interest in its own right:

PROPOSITION 6.5. *Let M be an indecomposable nonprojective reflexive R-module which is a splitting trace module and let $\Gamma = \operatorname{End}_R M$. Then*

(a) $\operatorname{Hom}_R(\Gamma, R) \simeq \Gamma$ *as a two-sided Γ-module.*

(b) *The two-sided Γ-morphism*

$$\operatorname{Hom}_R(\Gamma, E) \overset{\operatorname{Hom}_R(\Gamma, p)}{\to} \operatorname{Hom}_R(\Gamma, R)$$

is not surjective.

(c) *Let A be a Γ-module which is a reflexive R-module, then*

$$\operatorname{Im}\operatorname{Hom}_\Gamma(A, \operatorname{Hom}_R(\Gamma, E)) \to \operatorname{Hom}(A, \operatorname{Hom}_R(\Gamma, R))$$

consists of the nonepimorphisms from A to $\operatorname{Hom}_R(\Gamma, R)$.

PROOF. (a) We have seen that $\operatorname{Hom}_R(\Gamma, R)$ is free with generator t.

(b) Trivial consequence of M be a splitting trace module.

(c) Let A be a Γ-module which is a reflexive R-module and let $f\colon A \to \operatorname{Hom}_R(\Gamma, R)$ be a Γ-morphism. If f is in $\operatorname{Im}\operatorname{Hom}_R(A, \operatorname{Hom}_R(\Gamma, p))$, then f is not an epimorphism since $\operatorname{Hom}_R(\Gamma, p)$ is not an epimorphism by (b). Suppose now that $f\colon A \to \operatorname{Hom}_R(\Gamma, R)$ is not onto. Then for each a in A we have that $f(a)$ is in $\operatorname{rad}\Gamma \cdot \operatorname{Hom}_R(\Gamma, R)$ and therefore there is $g \in \operatorname{rad}\Gamma$ such that $f(a) = g \cdot t = t \circ g$. Hence $\operatorname{Im} f(a) \subset \mathfrak{m}$, the maximal ideal of R by Proposition 5.1 for each a in A.

Now we have the commutative diagram

$$\begin{array}{ccc}
\operatorname{Hom}_\Gamma(A, \operatorname{Hom}_R(\Gamma, E)) & \xrightarrow{\operatorname{Hom}_\Gamma(A, \operatorname{Hom}_R(\Gamma, p))} & \operatorname{Hom}_\Gamma(A, \operatorname{Hom}_R(\Gamma, R)) \\
\downarrow \phi & & \downarrow \phi' \\
\operatorname{Hom}_R(A, E) & \xrightarrow{\operatorname{Hom}_R(A, p)} & \operatorname{Hom}_R(A, R)
\end{array}$$

where the vertical maps are the standard isomorphisms. Now $\phi(f)(a) = f(a)(1)$ which is in $\mathfrak{m}$. Therefore $\phi'(f)$ is in $\operatorname{Im}\operatorname{Hom}_R(A, p)$ which means that f can be lifted to $\operatorname{Hom}_R(\Gamma, E)$, which is our desired result.

We are now in position to finish the proof of the following

THEOREM 6.6. *Let* $0 \to w \to E \to R \to R/\mathfrak{m} \to 0$ *be a fundamental sequence. Then the following are equivalent for an indecomposable reflexive nonprojective R-module M*:

(a) *M is a splitting trace module.*

(b) *The induced exact sequence*

$$0 \to \operatorname{Hom}_R(M^*, w) \to \operatorname{Hom}_R(M^*, E) \to M \to 0$$

does not split.

(c) *The induced exact sequence*

$$0 \to \operatorname{Hom}_R(M^*, w) \to \operatorname{Hom}_R(M^*, E) \to M \to 0$$

is almost split.

PROOF. The equivalence of (a) and (b) has already been shown. Since (c) trivially implies (a), it only remains to show that (a) implies (c).

Let $g\colon N \to M$ be a morphism of reflexive modules which is not a splittable epimorphism. Then the induced morphism of Γ^{op}-modules $\operatorname{Hom}_R(M, f)\colon \operatorname{Hom}_R(M, N) \to \operatorname{Hom}_R(M, M)$ is not surjective. Now we have the commutative diagram of two-sided Γ-modules

$$\begin{array}{ccc}
\operatorname{Hom}_R(M, \operatorname{Hom}_R(M^*, E)) & \to & \operatorname{Hom}_R(M, M) \\
\downarrow u_E^{-1} & & \downarrow u_R^{-1} \\
\operatorname{Hom}_R(\Gamma, E) & \to & \operatorname{Hom}_R(\Gamma, R)
\end{array}$$

with vertical isomorphisms. Since the composition of Γ^{op}-modules

$$\operatorname{Hom}(M, N) \to \operatorname{Hom}_R(M, M) \stackrel{u_R^{-1}}{\to} \operatorname{Hom}_R(\Gamma, R)$$

is not surjective, it follows from Proposition 6.5 that it can be lifted to $\operatorname{Hom}_R(\Gamma, E)$. It then follows from the above commutative diagram that there is a Γ^{op}-morphism g: $\operatorname{Hom}_R(M, N) \to \operatorname{Hom}_R(M, \operatorname{Hom}(M^*, E))$ such that the diagram of Γ^{op}-modules

$$\begin{array}{ccc} & & \operatorname{Hom}_R(M, N) \\ & g \swarrow & \downarrow \operatorname{Hom}_R(M, f) \\ \operatorname{Hom}_R(M, \operatorname{Hom}_R(M^*, E)) & \to & \operatorname{Hom}(M, M) \end{array}$$

commutes. But by Lemma 5.9, we know that the functor $\operatorname{Hom}(M, \ )$: Ref $R \to$ mod Γ given by $X \mapsto \operatorname{Hom}(M, X)$ is fully faithful. Therefore there is a unique

$$h: N \to \operatorname{Hom}_R(M^*, E)$$

such that $\operatorname{Hom}_R(M, h)$: $\operatorname{Hom}_R(M, N) \to \operatorname{Hom}_R(M, \operatorname{Hom}_R(M^*, E))$ is g. Consequently we have that f: $N \to M$ is the composition

$$N \stackrel{h}{\to} \operatorname{Hom}_R(M^*, E) \stackrel{\operatorname{Hom}_R(M^*, p)}{\to} M.$$

Therefore

$$\operatorname{Hom}_R(M^*, E) \stackrel{\operatorname{Hom}_R(M^*, p)}{\to} M$$

is right almost split. Since this holds for all morphisms of reflexive R-modules f: $N \to M$ which are not splittable epimorphisms and $\operatorname{Hom}_R(M^*, w)$ is indecomposable because M^* is indecomposable and $\operatorname{Hom}_R(M^*, w) \simeq \operatorname{Hom}_T(M^*, T)$, it follows that $0 \to \operatorname{Hom}_R(M^*, w) \to \operatorname{Hom}_R(M^*, E) \to M \to 0$ is almost split, finishing the proof of the theorem.

We now conclude with some final remarks. For M an indecomposable nonprojective reflexive R-module which is a splitting trace module, we have given an independent proof of the existence of an almost split sequence with M at the right-hand end, as well as a method for constructing it. Therefore, if R contains the rational numbers, we have obtained new proof of the existence of almost split sequences for reflexive R-modules, as well as a method for constructing them from the fundamental exact sequence. The situation is different when the characteristic of $R/\mathfrak{m}$ is $p \neq 0$. While, as noted before, we know there are almost split sequences, the method of constructing almost split sequences from the fundamental sequence given here only proves the existence for indecomposable, nonprojective splitting trace modules.

References

1. M. Artin and J. L. Verdier, *Reflexive modules over rational double points*, Math. Ann. **270** (1985), 79–82.

2. M. Auslander, *On the purity of the branch locus*, Amer. J. Math. **84** (1962), 116–125.

3. ______, *Functors and morphisisms determined by objects*, Proc. Conf. Representation Theory (Philadelphia 1976), Dekker, New York, 1978, pp. 1–244.

4. M. Auslander and J. Carlson, *Almost-split sequences and group rings*, J. Algebra (to appear).

5. M. Auslander and I. Reiten, *Representation theory of artin algebras*. III: *Almost split sequences*, Comm. Algebra **3** (1975), 239–294.

6. G. Gonzalez-Sprinberg and J. L. Verdier, *Construction géométrique de la correspondance de McKay*, Ann. Sci. Ecole Norm. Sup. **16** (1983), 409–449.

7. J. Herzog, *Ringe mit nur endlich vielen Isomorphie-klassen von maximalen unzerlegbaren Cohen-Macaulay-Moduln*, Math Ann. **233** (1978), 21–34.

8. K. W. Roggenkamp and J. Schmidt, *Almost split sequences for integral group rings and orders*, Comm. Algebra **4** (1976), 693–917.

DEPARTMENT OF MATHEMATICS, BRANDEIS UNIVERSITY, WALTHAM, MASSACHUSETTS 02254

TRANSACTIONS OF THE
AMERICAN MATHEMATICAL SOCIETY
Volume 302, Number 1, July 1987

ALMOST SPLIT SEQUENCES FOR RATIONAL DOUBLE POINTS

MAURICE AUSLANDER[1] AND IDUN REITEN

ABSTRACT. Let R be a complete local two-dimensional integrally closed noetherian nonregular Gorenstein domain with maximal ideal $\mathfrak{m}$ over an algebraically closed field k, such that $R/\mathfrak{m} \simeq k$. Then the category of finitely generated reflexive modules is known to have almost split sequences, and our main result is that the AR-quiver is an extended Dynkin quiver $\tilde{A}_n$, $\tilde{D}_n$, $\tilde{E}_6$, $\tilde{E}_7$ or $\tilde{E}_8$.

Let R be a complete local two-dimensional integrally closed noetherian nonregular Gorenstein domain with maximal ideal $\mathfrak{m}$ over an algebraically closed field k, such that $R/\mathfrak{m} \simeq k$. Denote by $\mathbf{L}$ the category of finitely generated reflexive modules, or equivalently, the (maximal) Cohen-Macaulay modules over R. Then $\mathbf{L}$ is known to have almost split sequences [**6**, **3**, **4**], and we prove that if R is of finite Cohen-Macaulay type, then the AR-quiver for $\mathbf{L}$ is an extended Dynkin quiver $\tilde{A}_n$ $\tilde{D}_n$, $\tilde{E}_6$, $\tilde{E}_7$, or $\tilde{E}_8$, in a sense explained in §1. If the vertex corresponding to R is removed, we get the corresponding Dynkin quiver.

The rational double points (over an algebraically closed field k) satisfy the above hypothesis, including finite Cohen-Macaulay type [**2**]. If the characteristic of k is 0, these are known to be the fixed point rings $k[[X,Y]]^G$, where $G \subset \mathrm{SL}(2,k)$ is a finite nontrivial group and the action is the natural one. In this case a proof for the above result was given by showing that the AR-quiver is isomorphic to the McKay quiver for $G \subset \mathrm{SL}(2,k)$ [**4**], which was already known to be an extended Dynkin quiver [**17**]. Here we give a direct proof valid for any characteristic. We use additive functions, as in [**15**] for McKay quivers, the fundamental "almost split" sequence from [**3**, **4**], preprojective partitions from [**10**], and the computation of Grothendieck groups using almost split sequences (see [**8**]).

The extension to arbitrary characteristic is of interest in connection with resolution of singularities, where for a rational double point the resolution graph is known to be Dynkin. In particular this work is used in [**13**] to prove that there is an isomorphism between the graph given by the AR-quiver after removing R, and the resolution graph, also in characteristic 2, 3, and 5.

We also give a generalization to the noncommutative two-dimensional case, for lattices over a T-order Λ when T is a complete regular local ring with maximal ideal $\mathfrak{m}$ and $T/\mathfrak{m}$ is algebraically closed. We here get as possibilities some additional AR-quivers with loops. In the last section we compute the AR-quivers for some noncommutative rings shown to be of finite lattice type in [**1**]. These computations show that the cases with loops actually do occur.

Received by the editors January 31, 1986.
1980 *Mathematics Subject Classification* (1985 *Revision*). Primary 13H10; Secondary 16A18.
[1]Written with partial support from NSF.

We would like to thank Øyvind Solberg for his help with calculations and careful reading of the manuscript.

1. The main theorem. In this section we shall deal with a complete local nonregular integrally closed noetherian two-dimensional Gorenstein domain R, with maximal ideal $\mathfrak{m}$, and study the category $\mathbf{L}$ of reflexive R-modules, which is known to have almost split sequences [**3, 4**]. For such a category denote for A and B indecomposable in $\mathbf{L}$ by r_{AB} the multiplicity of A in the minimal right almost split map for B, and by s_{AB} the multiplicity of B in the minimal left almost split map for A. Then we recall that the AR-quiver Γ for $\mathbf{L}$ has as vertices the isomorphism classes of indecomposable reflexives, and there are $r_{AB} = s_{AB}$ arrows from A to B. This holds when all $\operatorname{End}(A)/\operatorname{rad}\operatorname{End}(A)$ are isomorphic to the same algebraically closed field k (see [**11**]), which is the case under the assumptions of the next theorem. We denote by Γ_s the quiver obtained by removing the vertex corresponding to the projective module R. If in a quiver Γ the arrows between different vertices occur in pairs $\cdot \rightleftarrows \cdot$, we denote by $|\Gamma|$ the graph obtained by replacing $\cdot \rightleftarrows \cdot$ by $\cdot$ — $\cdot$ and ↻$\cdot$ by ○$\cdot$.

We have the following result on the AR-quiver.

THEOREM 1. *Let R be a complete nonregular integrally closed local two-dimensional noetherian Gorenstein domain with $R/\mathfrak{m} = k$ algebraically closed, assume that R has only a finite number of isomorphism classes of indecomposable reflexive modules, and let Γ be the corresponding AR-quiver.*

The arrows in Γ occur in pairs $\cdot \rightleftarrows \cdot$, and $|\Gamma|$ is an extended Dynkin diagram $\tilde{A}_n$, $\tilde{D}_n$, $\tilde{E}_6$, $\tilde{E}_7$, or $\tilde{E}_8$. $|\Gamma_s|$ is the corresponding Dynkin diagram A_n, D_n, E_6, E_7, or E_8.

PROOF. We first note that since $R/\mathfrak{m} = k$ is algebraically closed, $\operatorname{End}(A)/\operatorname{rad}\operatorname{End}(A) \simeq k$ for each indecomposable R-module A. Since R is Gorenstein of dimension 2, we know that $\operatorname{DTr} M \simeq M$ when M is an indecomposable nonprojective reflexive module [3]. This shows that the arrows in the AR-quiver occur in pairs $\cdot \rightleftarrows \cdot$. By [**10**] there is a preprojective partition for the category of finitely generated reflexive R-modules, and since we have finite representation type, every reflexive is preprojective. This implies that for every indecomposable M there is a chain of irreducible maps from a projective module, that is, in this case from R. It follows that Γ is connected.

A crucial point in our proof is that even though there is no almost split sequence with R on the right, there is an exact sequence $0 \to R \xrightarrow{f} E \xrightarrow{g} R \to R/m \to 0$ [**3, 4**], called the fundamental sequence in [**4**], such that $g\colon E \to R$ is minimal right almost split. It also follows directly or from [3] that $f\colon R \to E$ is minimal left almost split. This shows that $\operatorname{rk} E = 2 \operatorname{rk} R = 2$, where rk denotes rank, so that with respect to the additivity formula of ranks the fundamental sequence behaves like an almost split sequence. Hence by assigning to each vertex in the AR-quiver the rank of the corresponding module, we get an additive function for $|\Gamma|$, that is a function α on the vertices such that $2\alpha(i) = \Sigma\alpha(j)$, where the sum is taken over all edges $i \cdot$ — $\cdot\, j$.

Then it is known (see [**14**]) that $|\Gamma|$ must be extended Dynkin $\tilde{A}_n$, $\tilde{D}_n$, $\tilde{E}_6$, $\tilde{E}_7$, $\tilde{E}_8$, or

 or 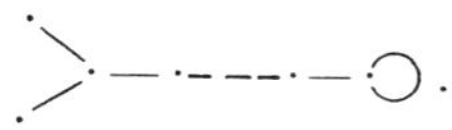.

We want to exclude the last two cases. The corresponding additive functions are known to be an integral multiple of the following.

 and 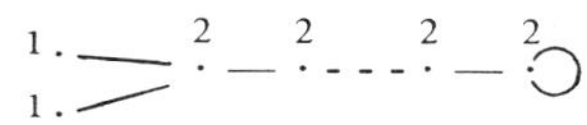 [**15**].

Since rk $R = 1$, the listed ones would be the additive functions. The class group $C(R)$ of R has as elements the isomorphism classes of indecomposable reflexive modules of rank 1. Since R is assumed not to be regular local, there are at least two nonisomorphic indecomposable reflexive R-modules. Hence if one of the above cases occurs, $C(R)$ must be nontrivial. On the other hand $C(R)$ can be computed in a different way, using that K_0 (mod R) $\simeq C(R) \amalg Z$, where K_0 (mod R) denotes the Grothendieck group of finitely generated R-modules modulo the relations generated by exact sequences (see [**8**]). In both cases we get one or two components of the form

by removing R. Consider the free group on the vertices $a_1, \ldots, a_n$, modulo the relations generated by almost split sequences, that is $2a_1 - a_2$, $2a_2 - a_1 - a_3, \ldots, 2a_{n-1} - a_{n-2} - a_n$, $2a_n - a_{n-1} - a_n = a_n - a_{n-1}$. From this it follows that $C(R) = (0)$, a contradiction.

Removing R means removing a vertex where the additive function given by ranks has value 1. In each of the cases this gives for the extended Dynkin diagrams $\tilde{\Delta}$ the Dynkin diagram Δ.

If for a category **L** having almost split sequences we drop the assumption $r_{AB} = s_{AB}$, we can consider the valued AR-quiver. Corresponding to an irreducible map from A to B we now write

$$A \cdot \xrightarrow{(r_{AB},\, s_{AB})} \cdot B.$$

If R is as before, but without the assumption that R/m is algebraically closed, we still have $\mathrm{DTr}\, A \simeq A$ for A indecomposable not isomorphic to R. The valued arrows between nonisomorphic modules then occur in pairs

$$A \cdot \underset{(s_{BA},\, r_{BA})}{\overset{(r_{AB},\, s_{AB})}{\rightleftarrows}} \cdot B,$$

where clearly $r_{AB} = s_{BA}$, $s_{AB} = r_{BA}$. Note that

 and 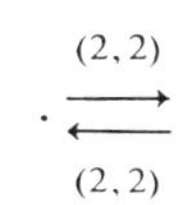

are the same AR-quivers. We then define $|\Gamma|$ to be the valued graph where each such pair is replaced by

$$A \cdot \xrightarrow{(r_{AB}, s_{AB})} \cdot B$$

and each

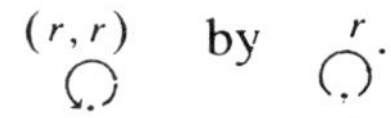

The arguments on the preprojective partition and the fundamental sequence are still valid. In this case the ranks give an additive function for the valued graph, defined in a natural way, and in addition to the previous possibilities we have the valued extended Dynkin diagrams or

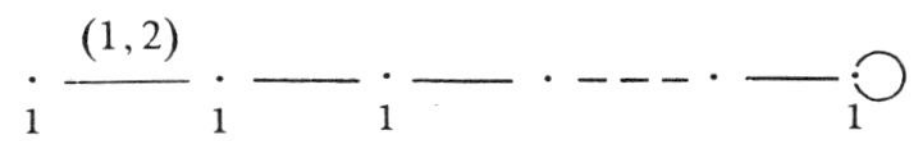

or

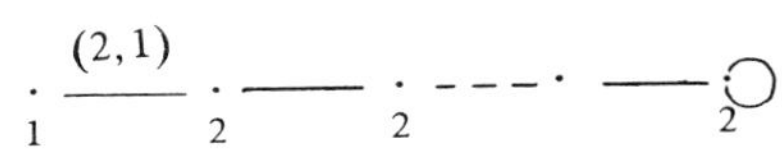

[**15**].

Here we have also listed the corresponding smallest additive function. The first of the last two cases can be excluded as before, but we do not know if the last one can occur.

Theorem 1 applies to rational double points over an algebraically closed field, since they are of finite representation type by [**2**], that is, have only a finite number of isomorphism classes of indecomposables in **L**. In this case the resolution graph for the corresponding isolated singularity is a Dynkin diagram. Based on a formula which we shall next explain and which follows from Theorem 1 (or alternatively, part of the proof), a direct proof of an isomorphism between $|\Gamma_s|$ and the resolution graph is given by Ésnault-Knörrer in [**13**].

When $|\Gamma|$ is extended Dynkin, we have $r_{AB} = s_{AB}$. Since DTr is the identity on nonprojective indecomposable reflexives, we have an additivity relation at each nonprojective vertex. Using $|\Gamma|$ being extended Dynkin, it is easy to see that this forces an additivity relation also at the vertex of R. In the proof of Theorem 1 these statements followed from the "symmetry" for almost split sequences when $R/\mathfrak{m}$ is algebraically closed, and from the existence and properties of the fundamental sequence. If now for each indecomposable reflexive A_i, $B_i \to A_i$ denotes a minimal right almost split map, the desired formula used in [**13**] is

$$\coprod_{i=1}^{n} (\operatorname{rk} A_i) B_i \simeq 2 \coprod_{i=1}^{n} (\operatorname{rk} A_i) A_i.$$

Here $A_1, \ldots, A_n$ are all the indecomposable reflexives. Clearly a given A_i occurs $2 \operatorname{rk} A_i$ times as a summand on the right. For each A_j, A_i occurs on the left with multiplicity $t_j \operatorname{rk} A_j$, where t_j is the multiplicity of A_i in B_j. By our symmetry t_j is then the multiplicity of A_j in C_i, when $A_i \to C_i$ is minimal left almost split. Summing up, we then get $2 \operatorname{rk} A_i$, by additivity.

If we drop the asumption that $R/m = k$ is algebraically closed, the formula holds if and only if the (smallest) additive function for $|\Gamma|$ is also an additive function for the valued graph obtained by replacing $A \cdot \xrightarrow{(r,s)} \cdot B$ by $A \cdot \xrightarrow{(s,r)} \cdot B$. By considering the different cases, listed in [**15**], we must then have $r = s$ for all edges, so that we are back in the situation when $|\Gamma|$ is extended Dynkin. Hence the formula does not hold in general when R/m is not algebraically closed. It holds, however, for the noncommutative two-dimensional rings treated in §3. In other dimensions we do not have fundamental sequences.

We now make some remarks on the construction of almost split sequences for rational double points. It is shown in [**4**] (in a generalized version) that if

$$0 \to R \to E \to R \to R/m \to 0$$

is the fundamental sequence, then for an indecomposable nonprojective reflexive R-module A, $0 \to A \to E \cdot A \to A \to 0$ is almost split if and only if $\operatorname{char} k \nmid \operatorname{rk} A$. If $\operatorname{char} k \mid \operatorname{rk} A$, the sequence splits. Here $E \cdot A = (E \otimes_R A)^{**}$, where $X^* = \operatorname{Hom}_R(X, R)$. Since by considering the smallest additive functions for extended Dynkin diagrams we know that $\operatorname{rk} A \leqslant 6$, all almost split sequences can be computed this way when $\operatorname{char} k > 5$. We note, however, that except for the case D_n for $\operatorname{char} k = 2$, this principle can still be used to compute the almost split sequences. For assume that for some indecomposable B, $\operatorname{char} k \mid \operatorname{rk} B$. Except for the case mentioned, for all neighbors A_i of B in $|\Gamma|$, $\operatorname{char} k \nmid \operatorname{rk} A_i$. Hence we have almost split sequences $0 \to A_i \to E \cdot A_i \to A_i \to 0$. $B \mid E \cdot A_i$, and the induced map $A_i \to B$ is irreducible, and we get a minimal right almost split map $\coprod A_i \to B$.

We end this section with observing that knowing that $|\Gamma|$ is extended Dynkin gives information on the relationship between rank and minimal number of generators.

PROPOSITION 2. *Let R be a rational double point over an algebraically closed field, and let A be an indecomposable reflexive module not isomorphic to R. Then the minimal number of generators of A is* $2 \operatorname{rk} A$.

PROOF. Ω induces an equivalence on the stable category, and as in the finite-dimensional case [**7**], a map between indecomposable nonprojectives is irreducible if and only if it is irreducible in the stable category. Hence Ω induces an automorphism on Γ_s. By considering the additive functions, we see that Ω must preserve rank. Our claim then follows by considering $0 \to \Omega M \to P \to M \to 0$.

2. Generalizations. In this section we first give a generalization of Theorem 1 to the two-dimensional noncommutative case, and for this we recall the definition of order and lattice [**3**]. §3 will shed some more light on this. We also consider the two-dimensional case when $D\operatorname{Tr}$ is not necessarily the identity, and also higher dimensions.

Let T be a complete regular local two-dimensional ring, and Λ an indecomposable T-order, that is, Λ is a finitely generated free T-module such that $\mathrm{Hom}_T(\Lambda, T)_{\mathfrak{p}}$ is $\Lambda_{\mathfrak{p}}$-projective for all nonmaximal prime ideals $\mathfrak{p}$ in T. A Λ-module L is said to be a lattice if L is a finitely generated free T-module such that $L_{\mathfrak{p}}$ is $\Lambda_{\mathfrak{p}}$-projective for all nonmaximal prime ideals $\mathfrak{p}$ [**3**]. Then we know that the category of lattices $\mathbf{L}$ has almost split sequences, and in this case $\tau = D\,\mathrm{Tr}$ is given by $\tau(X) = \mathrm{Hom}_T(\mathrm{Hom}_\Lambda(X, \Lambda), T)$ [**3**]. Considering τ as a functor τ: mod $\Lambda \to$ mod Λ, we see that τ is right exact, and hence $\tau \simeq \mathrm{Hom}_T(\Lambda, T) \otimes_\Lambda —$. It is important that if P is indecomposable projective, then we have an exact sequence

$$0 \to \tau P \xrightarrow{f} E \xrightarrow{g} P \to P/\mathrm{rad}\, P \to 0,$$

such that g is minimal right almost split, and f is minimal left almost split [**3**]. Using ranks (over T, or when possible, over Λ), this provides us with additivity relations as before.

As in the commutative case, the situation is especially nice when τ is the identity on indecomposable lattices. This is for example the case when Λ is a Gorenstein R-order, that is, we have a bimodule isomorphism $\Lambda \simeq \mathrm{Hom}_T(\Lambda, T)$. In §3 we will see interesting examples where Λ is not Gorenstein, but still $\tau = \mathrm{id}$ for indecomposable lattices.

If there is only one indecomposable projective lattice, it follows as before by using the preprojective partition that the AR-quiver Γ is connected. With the same arguments as before, where for simplicity we state the result for the case of nonvalued quivers, we get the following.

THEOREM 3. *Let T be a complete regular local two-dimensional ring with maximal ideal $\mathfrak{m}$, and Λ a T-order of finite lattice type. Assume that $\tau = \mathrm{id}$ on indecomposable lattices, and that $T/\mathfrak{m}$ is an algebraically closed field k. Then the following hold for the AR-quiver Γ and the stable AR-quiver Γ_s.*

(a) *A connected component of $|\Gamma|$ is extended Dynkin or*

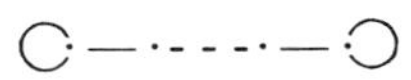 *or* 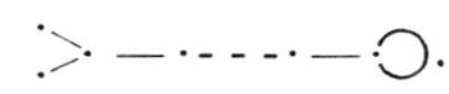

(b) *A connected component of $|\Gamma_s|$ is Dynkin or*

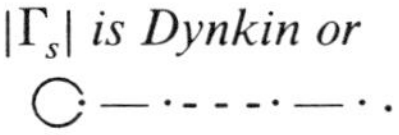

In §3 we shall see that contrary to the commutative two-dimensional case, the diagrams with loops actually do occur.

As before, there are restrictions on the valued graphs when we drop the assumption on $T/\mathfrak{m}$.

Even if τ is not the identity, considering ranks still gives additivity relations in the two-dimensional case. If there are no loops in the AR-quiver Γ and we have finite lattice type, combinatorics shows that Γ is of the form $Z\Delta/G$, where $Z\Delta$ is a translation quiver for an oriented tree Δ whose underlying graph is extended Dynkin or ${}_\infty A_\infty$, and G is a finite group (see [**18**, **14**]). An explicit computation for the commutative rings of the form $\Lambda = k[[X, Y]]^G$, where $G \subset \mathrm{GL}(2, k)$ is a finite group and k is an algebraically closed field with $\operatorname{char} k = 0$, is given in [**9**]. Other examples when τ is not the identity are given in §3.

For higher dimensions we do not have so satisfactory results since we do not have "almost split sequences" for the projectives. Then, if $\mathrm{D\,Tr}$ is the identity, the combinatorics only applies to $|\Gamma_s|$. For this recall that for a given graph a function α on the vertices is subadditive if for every vertex i, $2\alpha(i) \geqslant \Sigma\alpha(j)$, where the sum runs over all edges $i \cdot - \cdot j$. Let now R be a complete local integrally closed isolated hypersurface singularity with R/m algebraically closed. Then $\mathrm{D\,Tr}\, M \simeq \Omega^{2-2n} M$ by [**3**] and $\Omega^2 M \simeq M$ by [**12**], where $n = \dim R$. Hence $\mathrm{D\,Tr}$ is the identity for $M \not\simeq R$ in even dimensions, and also in some cases in odd dimensions. In these cases the rank function gives a subadditive function on $|\Gamma_s|$ which is not additive. Combinatorics then gives that if R is of finite lattice type, that is, has only a finite number of isomorphism classes of indecomposable Cohen-Macaulay modules [**5**], the connected components of $|\Gamma_s|$ are Dynkin or

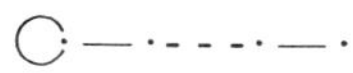 [**15**].

Now we point out that if $\operatorname{char} k \neq 2$ it is known from [**16**] that $|\Gamma_s|$ is actually Dynkin in the even-dimensional case, and the additional diagram occurs in any size in any odd dimension. In the odd-dimensional case $\mathrm{D\,Tr}\, M \simeq \Omega M$ for $M \not\simeq R$, and since $\Omega^2 M \simeq M$, Γ_s has a periodic vertex. Hence when there is no loop in Γ_s, $\Gamma_s \simeq Z\Delta/G$, where $Z\Delta$ is built from a Dynkin diagram Δ [**14**].

We point out that there is information along the same lines for infinite representation type, since there are still combinatorial results for additive and subadditive functions [**15**].

3. Computations for two-dimensional orders. Let T be the power series ring $T = k[[u, v]]$, where k is an algebraically closed field with $\operatorname{char} k = 0$. Let A be a maximal T-order with center T in a division ring D, and assume that A is not commutative. In [**1**] Artin has given a description of when A has only a finite number of isomorphism classes of indecomposable A-modules which are free T-modules. A is then a T-order in the sense of [**3**], and the A-modules which are free T-modules are the lattices. On the basis of Artin's classification, we compute the AR-quivers when A is of finite representation type. In some cases the equivalence

$$\tau = \operatorname{Hom}_T(\operatorname{Hom}_\Lambda(\ ,\Lambda), T)\colon \mathbf{L}(A) \to \mathbf{L}(A)$$

induces the identity on indecomposable objects, in other cases not. From the point of view of this paper the most interesting aspect of these computations is that both cases with loops which we could exclude in the commutative case actually do occur. At the same time we see that there is a close connection between the AR-quivers and the A-diagrams studied by Artin, thus giving an interpretation of Artin's diagrams in most cases.

We investigate the classes classified by Artin case by case (when there are at least two indecomposables). We use his description of indecomposables of A-rank 1, and the total distribution of ranks for all indecomposables. Note that the AR-quiver is connected since there is only one indecomposable projective lattice, by the argument using preprojective partitions as before.

Type (II_k). A is generated over T by x, y, satisfying

$$x^2 = u, \quad y^2 = u^k v, \quad xy + yx = 2v.$$

Here all indecomposable lattices have rank 1, and the nonprojective ones are $L_i = (u^i, y)$, $1 \leqslant i \leqslant k$.

We first observe that $\tau A \simeq A$. To see this, consider the T-basis $\{1, x, y, xy\}$ for A as left T-module, and let f_{xy}: $A \to T$ be the T-map given by

$$f_{xy}(r_1 + r_2 x + r_3 y + r_4 xy) = r_4.$$

Then $\tau(A) = \mathrm{Hom}_T(A, T) = Af_{xy} \simeq A$.

Let f: $A \to A$ be given by $f(a) = ax$, let g: $L_1 = (u, y) \to A$ be the inclusion, and consider the induced map (f, g): $A \amalg L_1 \to A$. It is not hard to get a good enough description of an arbitrary map h: $L_i \to A$ to see that it can be lifted to $A \amalg L_1$. Since $\mathrm{Im}(f, g) = (x, y)$, any nonisomorphism h: $A \to A$ can also be lifted. This shows that (f, g): $A \amalg L_1 \to A$ is right almost split, and a rank argument shows that it is minimal right almost split, so that we have the "almost split sequence" $0 \to A \to A \amalg L_1 \to A$. Applying τ, we see that $\tau L_1 \simeq L_1$. Hence we have an almost split sequence $0 \to L_1 \to A \amalg X \to L_1 \to 0$. Applying τ again shows that $\tau X \simeq X$. Continuing, and using that the AR-quiver Γ is connected, we get that τ is the identity on objects, and $|\Gamma|$ is

○—·—·····—○ ($k + 1$ vertices)

and $|\Gamma_s|$ is

○—·—·····—· (k vertices).

Type (III_k). A is generated over T by x, y, where

$$x^2 = v, \quad y^2 = uv, \quad xy + yx = 2u^{k+1}.$$

Here there is only one nonprojective lattice L of rank 1, and the k others are of rank 2.

$\{1, x, y, xy\}$ is T-basis for A as a left T-module, and it is easy to see that $\tau(A) = \mathrm{Hom}_T(A, T) = Af_{xy} \simeq A$, and $\mathrm{Hom}_T(A, T) \simeq f_{xy}A \simeq A$. Denoting also by τ the automorphism given by $\tau(a) = a'$, when $af_{xy} = f_{xy}a'$, we have $\tau(X) = X^\tau$. Here X^τ denotes the Λ-module with the same underlying group as X, and where $\lambda \cdot x = \tau(\lambda)x$. Since τ clearly preserves rank, we have $\tau L \simeq L$. Consider the "almost split sequence" $0 \to A \to K \to A$. Applying τ, we conclude $\tau K \simeq K$. Assume that $K \simeq K_1 \amalg K_2$, where K_1 and K_2 both have rank 1. The only possibilities for K_1 and K_2 are A and L, and in any case we would get a connected component of the AR-quiver with all indecomposables having rank 1. It follows that K is indecomposable. Consider the almost split sequence $0 \to K \to A \amalg X \to K \to 0$. X has rank 3, so X has one indecomposable summand of rank 1 and one of rank 2. Since $A/\mathfrak{m} = k$ is algebraically closed, where $\mathfrak{m}$ is the maximal ideal of A, A is not a summand of X, and hence we have almost split sequences

$$0 \to K \to A \amalg L \amalg K_1 \to K \to 0$$

and $0 \to L \to K \to L \to 0$. This shows that $\tau(K_1) \simeq K_1$. If $K_1 \simeq K$, we are done. If $K_1 \not\simeq K$, we have the almost split sequence $0 \to K_1 \to K \amalg K_2 \to K_1 \to 0$. K_2 can have no summand of rank 1, so K_2 is indecomposable of rank 2, and $\tau K_2 \simeq K_2$ follows. Continuing this way, we see that $\tau = \mathrm{id}$ on all indecomposable lattices, and $|\Gamma|$ is

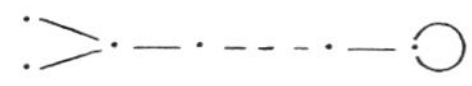 $(k + 2$ vertices$)$,

and $|\Gamma_s|$ is

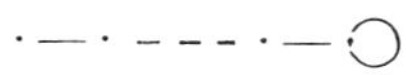 $(k + 1$ vertices$)$.

Type (IV). A is generated over T by x, y, where

$$x^3 = v, \quad y^3 = u^2 - v, \quad xy = \rho_3 yx,$$

ρ_3 a primitive third root of 1. The indecomposable nonprojective lattices of rank 1 are $L_1 = (u, x + y)$, $L_2 = (u, x + \rho_3 y)$ and $L_3 = (u, x + \rho_3^2 y)$. There is only one additional indecomposable lattice L, of rank 2.

$\{1, x, y, x^2, y^2, xy, xy^2, x^2y, x^2y^2\}$ is a T-basis for A as a left T-module. An easy computation gives $\mathrm{Hom}_T(A, T) = Af_{x^2y^2} = f_{x^2y^2}A$, and that the automorphism τ: $A \to A$ is determined by $\tau(x) = \rho_3^2 x$, $\tau(y) = \rho_3 y$. We further can compute $\tau(L_1) \simeq L_2$, $\tau(L_2) \simeq L_3$, $\tau(L_3) \simeq L_1$.

Using this, it is not hard to compute the AR-quiver. We have $\tau A \simeq A$, and an "almost split sequence" $0 \to A \to E \to A$, and hence $\tau E \simeq E$. Clearly $E \not\simeq A \amalg A$, and also no L_i can be a summand of E. Hence E must be the indecomposable of rank 2. Further we have the almost split sequence $0 \to E \to A \amalg F \to E \to 0$, where A is not a summand of F. Since F has rank 3, one of the L_i must then be a summand of F. Applying τ we get that we must have

$$0 \to E \to A \amalg L_1 \amalg L_2 \amalg L_3 \to E \to 0,$$

and hence we have almost split sequences $0 \to \tau(L_i) \to E \to L_i \to 0$. The AR-quiver is then

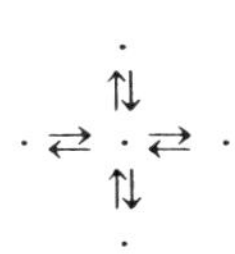

with τ permuting three end vertices cyclically. $|\Gamma|$ is then

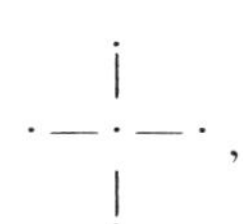

which is $\tilde{D}_4$.

Type (V). A is generated over T by x, y, where

$$x^2 = v, \quad y^2 = v^2 - u^3, \quad xy = -yx.$$

There are two nonprojective indecomposables of rank 1, $L_1 = (u, y + v)$ and $L_2 = (u, y - v)$. In addition there are three indecomposables of rank 2 and one of rank 3.

Here $\{1, x, y, xy\}$ is a T-basis for A as left T-module, and we have

$$\mathrm{Hom}_T(A, T) \simeq Af_{xy}$$

as left A-module, so that $\tau(A) \simeq A$. Also $\mathrm{Hom}_T(A, T) \simeq f_{xy}A$, and an easy computation shows that τ is induced by $\tau(x) = -x$, $\tau(y) = -y$. One can then further compute that $\tau(L_1) \simeq L_2$, $\tau(L_2) \simeq L_1$.

Consider the "almost split sequence" $0 \to A \to E \to A$. If E decomposes, we get a component with indecomposables of rank 1. Hence E is indecomposable and $\tau E \simeq E$. Consider the almost split sequence $0 \to E \to A \amalg F \to E \to 0$. A is not a summand of F, and since $\tau F \simeq F$, the indecomposable summands of F cannot all be of rank 1. Also there cannot be exactly one of rank 1. It follows that F is indecomposable of rank 3. Consider now the almost split sequence

$$0 \to F \to E \amalg G \to F \to 0.$$

A rank argument shows that G has no summand of rank 1, and consequently none of rank 3. Hence the indecomposable summands of G must have rank 2, and since there can be no repeats, we have $0 \to F \to E \amalg E' \amalg E'' \to F \to 0$, where E, E', E'' are the three indecomposables of rank 2. It now follows that the AR-quiver is

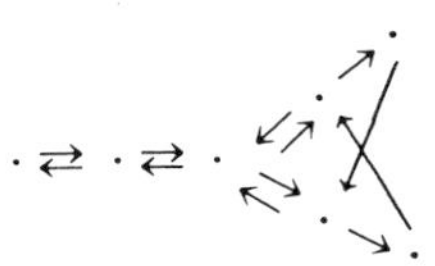

From the point of view of this paper, the computations for (II_k) and (III_k) are most interesting, since $\tau = \mathrm{id}$, and

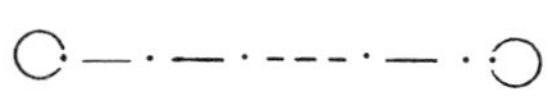 and 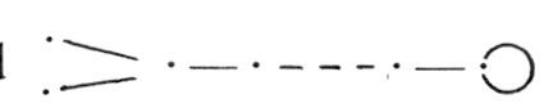

occur for $|\Gamma|$, for any size of these diagrams. (The rings are, however, not Gorenstein.) These diagrams also coincide with the A-diagrams of Artin in these cases, thus providing an interpretation of these A-diagrams. This is still the case for (IV). For (V) Artin's A-diagram is $\tilde{E}_6$, which is closely connected with the AR-quiver Γ, since Γ is of the form $Z\Delta/G$, where Δ is an oriented tree of type $\tilde{E}_6$.

References

1. M. Artin, *Maximal orders of global dimension and Krull dimension two*, Invent. Math. **84** (1986), 195–222.

2. M. Artin and M. L. Verdier, *Reflexive modules over rational double points*, Math. Ann. **270** (1985), 79–82.

3. M. Auslander, *Functors and morphisms determined by objects*, Proc. Conf. on Representation Theory (Philadelphia, 1976), Dekker, New York, 1978, pp. 1–244.

4. ______, *Rational singularities and almost split sequences*, Trans. Amer. Math. Soc. **293** (1986), 511–531.

5. ______, *Isolated singularities and existence of almost split sequences*, Proc. ICRA IV, Lecture Notes in Math., vol. 1178, Springer, 1986, pp. 194–241.

6. M. Auslander and I. Reiten, *Representation theory of artin algebras* III: *Almost split sequences*, Comm. Algebra **3** (1975), 239–294.

7. ______, *Stable equivalence of artin algebras*, Proc. Conf. on Orders, Group Rings and Related Topics, Lecture Notes in Math., vol. 353, Springer, 1972, pp. 8–71.

8. ______, *Grothendieck groups of algebras and orders*, J. Pure Appl. Algebra **39** (1986), 1–51.

9. ______, *McKay quivers and extended Dynkin diagrams*, Trans. Amer. Math. Soc. **293** (1986), 293–302.

10. M. Auslander and S. O. Smalø, *Lattices over orders*: *Finitely presented functors and preprojective partitions*, Trans. Amer. Math. Soc. **273** (1982), 433–446.

11. R. Bautista, *Irreducible maps and the radical of a category*, An. Inst. Mat. Univ. Nac. Autónoma México **22** (1982), 83–135.

12. D. Eisenbud, *Homological algebra on a complete intersection, with an application to group representations*, Trans. Amer. Math. Soc. **260** (1980), 35–64.

13. H. Esnault and H. Knörrer, *Reflexive modules over rational double points*, Math. Ann. **272** (1985), 545–548.

14. D. Happel, U. Preisel and C. M. Ringel, *Vinberg's characterization of Dynkin diagrams using subadditive functions with applications to* DTr-*periodic modules*, Proc. ICRA II, Ottawa 1979, Lecture Notes in Math., vol. 832, Springer, 1980, pp. 280–294.

15. ______, *Binary polyhedral groups and Euclidean diagrams*, Manuscripta Math. **31** (1980), 317–329.

16. H. Knörrer, *Cohen-Macaulay modules on hypersurface singularities* I (preprint)

17. J. McKay, *Graphs, singularities and finite groups*, Proc. Sympos. Pure Math., vol. 37, Amer. Math. Soc., Providence, R. I., 1980, pp. 183–186.

18. C. Riedtmann, *Algebren, Darstellungsköcher und Zurück*, Comment. Math. Helv. **55** (1980), 199–224.

DEPARTMENT OF MATHEMATICS, BRANDEIS UNIVERSITY, WALTHAM, MASSACHUSETTS 02254

INSTITUTE FOR MATHEMATICS AND STATISTICS, UNIVERSITY OF TRONDHEIM, AVH, 7055 DRAGVOLL, NORWAY (Current address of Idun Reiten)

Current address (Maurice Auslander): Department of Mathematics, Virginia Polytechnic Institute and State University, Blacksburg, Virginia 24061-4097

Reprinted from ADVANCES IN MATHEMATICS
All Rights Reserved by Academic Press, New York and London

Vol. 73, No. 1, January 1989
Printed in Belgium

The Cohen–Macaulay Type of Cohen–Macaulay Rings

MAURICE AUSLANDER

Department of Mathematics, Brandeis University, Waltham, Massachusetts 02254

AND

IDUN REITEN

Institutt for Matematikk og Statistikk, Universitetet i Trondheim, N-7055 Dragvoll, Norway

INTRODUCTION

Let R be an integrally closed complete local commutative noetherian Cohen–Macaulay domain with maximal ideal $\mathfrak{m}$ and $R/\mathfrak{m}=k$ an algebraically closed field. We abbreviate Cohen–Macaulay to CM and denote by $\mathrm{CM}(R)$ the category of finitely generated CM modules. In this paper we deal with the question of when R is of finite CM type, that is, has only a finite number of indecomposable CM modules. If $\dim R=2$, and R is a $\mathbb{C}$-algebra with $R/\mathfrak{m}=\mathbb{C}$, the complex numbers, then the R of finite CM type are exactly the fixed rings $\mathbb{C}[[X, Y]]^G$, where X, Y are indeterminates and $G\subset GL(2, \mathbb{C})$ is a finite group acting on $\mathbb{C}[[X, Y]]$ [17, 3, 15]. For hypersurfaces in characteristic zero the R of finite CM type are exactly the simple hypersurface singularities in the sense of Arnold [18, 13], so in this case there is a nice connection with algebraic geometry.

When $\dim R\geqslant 3$ and R is not a hypersurface, we know only two examples of finite CM type, and they both have dimension 3. These are $R_1=k[[X_0, X_1, X_2, Y_0, Y_1]]/(X_0X_2-X_1^2,\ X_0Y_1-X_1Y_0,\ X_1Y_1-X_2Y_0)$ and $R_2=k[[X, Y, Z]]^{Z_2}$, where the generator of Z_2 acts by sending each variable to its negative, and the characteristic of k is different from 2. Each of these rings belongs naturally to a larger class. R_1 is a scroll of type (2, 1). We define more generally a scroll of type $(m_1, ..., m_r)$ (see [14]) and show that when $\dim R\geqslant 3$, then all other scrolls (except type (1, 1), which is a hypersurface) are of infinite CM type. R_2 is among the rings $k[[X_1, ..., X_n]]^G$, where the order of $G\subset GL(n, k)$ is invertible in k and $n\geqslant 3$, and is the only one of finite CM type. These are the main results of this paper.

To prove that the scroll of type (2, 1) is of finite CM type we use the

0001-8708/89 $7.50

theory of almost split sequences, first developed for artin algebras [7, 8] and extended to this setting [2, 4]. We will assume that the reader is familiar with the basic theory of almost split sequences (see [19]). In particular, we generalize from artin algebras a criterion for having only a finite number of indecomposable modules, and this is a direct consequence of the theory of preprojective partitions from [12]. The same method can be used to show that our fixed ring is of finite CM type, but here we give in addition a different proof. Our study of the fixed rings is based on an investigation of the connection between the reflexive modules over R and the skew group ring $k[[X_1, ..., X_n]]\, G$.

We consider only faithful actions of G on $S=k[[X_1, ..., X_n]]$, and we also assume that the actions are linear.

The scrolls are investigated in Part I and the fixed rings in Part II. For the work on scrolls we are grateful to David Eisenbud and Finn Knudsen for several suggestions and helpful conversations.

Some of the results in this paper were anounced at Bielefeld (1984 and 1985), at the Seminaire Malliavin [20], and at the Durham Symposium on representations of algebras [5].

PART I: SCROLLS OF FINITE COHEN–MACAULAY TYPE

In this part we introduce scrolls (see [14]) and describe conditions under which they are of finite CM type.

1. A Criterion for Finite Representation Type

Let T be a complete regular local noetherian ring and let Λ be a T-order, that is, Λ is a finitely generated free T-module and $\mathrm{Hom}_{T_p}(\Lambda_p, T_p)$ is Λ_p^{op}-projective for all nonmaximal prime ideals p in T [2]. Denote by $\mathfrak{L}(\Lambda)$ the Λ-lattices, that is, the finitely generated Λ-modules M such that M is a free T-module and M_p is Λ_p-projective for all nonmaximal prime ideals p in T. The category $\mathfrak{L}(\Lambda)$ is known to have almost split sequences, that is, for each nonprojective indecomposable C in $\mathfrak{L}(\Lambda)$ (or each noninjective indecomposable A in $\mathfrak{L}(\Lambda)$), there is an almost split sequence $0 \to A \to B \to C \to 0$ in $\mathfrak{L}(\Lambda)$. And if C is indecomposable projective, there is a minimal right almost split map $B \to C$, and for A indecomposable injective there is a minimal left almost split map $A \to B$ [2, 12, 14].

For indecomposable finite dimensional algebras we know that if $\mathscr{C}$ is a finite set of indecomposable modules closed under irreducible maps, then $\mathscr{C}$ consists of all indecomposables [2]. There is a similar criterion for classical orders [21], and here we give a related criterion for T-orders Λ. We say that a set of indecomposable lattices $\mathscr{C}$ is closed under almost split sequences if for each indecomposable nonprojective object C in $\mathscr{C}$ (and each indecomposable noninjective object A in $\mathscr{C}$) all indecomposable summands of the terms in the almost split sequence $0 \to A \to B \to C \to 0$ are in $\mathscr{C}$. We denote A by τC and C by $\tau^{-1}A$.

THEOREM 1.1. *Let Λ be T-order and $\mathscr{C}$ a finite set of indecomposable lattices containing all indecomposable projectives.*

(a) *If $\mathscr{C}$ is closed under irreducible maps, then $\mathscr{C}$ consists of all indecomposable Λ-lattices.*

(b) *Assume that there are no projective injective Λ-lattices. If $\mathscr{C}$ is closed under almost split sequences, then $\mathscr{C}$ consists of all indecomposable Λ-lattices.*

Proof. Let C be an indecomposable nonprojective preprojective lattice in the sense of Auslander and Smalø [12]. Then there is a chain of irreducible maps between indecomposables $P \to C_1 \to \cdots \to C = C_n$, where P is projective and the C_i are not projective. We want to show that C is in $\mathscr{C}$. This is clear for (a). In case (b) we know that P is not injective, so we have an almost split sequence $0 \to P \to C_1 \amalg X \to \tau^{-1}P \to 0$, which shows that C_1 is in $\mathscr{C}$. If C_1 is not injective, the same argument shows that C_2 is in $\mathscr{C}$. If C_1 is injective, it is not projective, so we have an almost split sequence $0 \to \tau C_1 \to \tau C_2 \amalg Y \to C_1 \to 0$. Since C_2 is not projective, τC_2 is in $\mathscr{C}$. It then follows that C_2 is also in $\mathscr{C}$. Continuing this way, we get $C \in \mathscr{C}$. This shows that all indecomposable preprojectives are in $\mathscr{C}$, and hence there is only a finite number of them. By [12] we then know that all Λ-lattices are preprojective, and this finishes the proof.

Theorem 1.1 can be applied to the case of the CM modules $\mathrm{CM}(R)$ over a complete local integrally closed noetherian CM domain R. In this case R is the only indecomposable projective module and the dualizing module ω is the only indecomposable injective. Part (b) applies if $R \not\simeq \omega$, that is, if R is not Gorenstein. If R is Gorenstein, we can apply part (a). It is then not enough to show that a given set $\mathscr{C}$ is closed under almost split sequences, but we also need to find the minimal right almost split map $B \to R$ (and the minimal left almost split map $R \to C$, which can be obtained by duality). In practice $B \to R$ can be constructed using a construction of CM modules of Buchweitz, and the application of his construction to this problem appeared through discussions with him. The pertinent result is the following, where $\Omega^d X$ denotes the dth syzygy module for X.

PROPOSITION 1.2. *Let R be a complete local noetherian Gorenstein integrally closed domain of dimension $d+1 \geqslant 2$ and with maximal ideal $\mathfrak{m}$. If*

$B \to R$ is minimal right almost split, then the nonprojective part of B is isomorphic to $\Omega^{-d}(\Omega^d \mathfrak{m})$.

Proof. Consider the minimal projective resolution $P_{d-1} \to \cdots \to P_1 \to P_0 \to \mathfrak{m} \to 0$. Since $\dim R = d+1$, $\Omega^d \mathfrak{m} = \Omega^{d+1}(R/\mathfrak{m})$ is CM. Write $X^* = \operatorname{Hom}_R(X, R)$ and let $Q_d \to \cdots \to Q_1 \to Q_0 \to (\Omega^d \mathfrak{m})^* \to 0$ be a minimal projective resolution. Since $(\Omega^d \mathfrak{m})^*$ is CM, $\operatorname{Ext}^i_R((\Omega^d \mathfrak{m})^*, R) = 0$ for $i > 0$, so that we have an exact sequence $0 \to \Omega^d \mathfrak{m} \to Q_0^* \to Q_1^* \to \cdots \to Q_{d-1}^* \to \Omega^{-d}(\Omega^d M) = E \to 0$. Since $\operatorname{Ext}^i(\mathfrak{m}, R) = 0$ for $1 \leqslant i < d$, we have the exact commutative diagram

$$\begin{array}{ccccccccc}
P_0^* & \longrightarrow & P_1^* & \longrightarrow \cdots \longrightarrow & P_{d-1}^* & \longrightarrow & (\Omega^d \mathfrak{m})^* & & \\
\downarrow & & \downarrow & & \downarrow & & \| & & \\
Q_{d-1} & \longrightarrow & Q_{d-2} & \longrightarrow \cdots \longrightarrow & Q_0 & \longrightarrow & (\Omega^d \mathfrak{m})^* & \longrightarrow & 0,
\end{array}$$

which gives rise to the commutative diagram

$$\begin{array}{ccccccccccccc}
0 & \longrightarrow & \Omega^d \mathfrak{m} & \longrightarrow & P_{d-1} & \longrightarrow \cdots \longrightarrow & P_1 & \longrightarrow & P_0 & \longrightarrow & \mathfrak{m} & \longrightarrow & 0 \\
 & & \| & & \uparrow & & \uparrow & & \uparrow & & \uparrow & & \\
0 & \longrightarrow & \Omega^d \mathfrak{m} & \longrightarrow & Q_0^* & \longrightarrow \cdots \longrightarrow & Q_{d-2}^* & \longrightarrow & Q_{d-1}^* & \longrightarrow & E & \longrightarrow & 0.
\end{array}$$

Let $h\colon X \to \mathfrak{m}$, where X is CM. Since $\operatorname{Ext}^i(X, R) = 0$, for $1 \leqslant i < d$, we have natural isomorphisms modulo projectives $\underline{\operatorname{Hom}}(X, \mathfrak{m}) \simeq \underline{\operatorname{Hom}}(\Omega^d X, \Omega^d \mathfrak{m}) \simeq \underline{\operatorname{Hom}}(X, E)$ (see [6]). This shows that E is the nonprojective part of a minimal right almost split map to R.

2. Scrolls of Finite Cohen–Macaulay Type

Consider the matrix

$$\left(\begin{array}{c|c|c} \begin{matrix} Z_0^{(1)} \cdots Z_{m_1-1}^{(1)} \\ Z_1^{(1)} \cdots Z_{m_1}^{(1)} \end{matrix} & \cdots & \begin{matrix} Z_0^{(r)} \cdots Z_{m_r-1}^{(r)} \\ Z_1^{(r)} \cdots Z_{m_r}^{(r)} \end{matrix} \end{array} \right),$$

where the $Z_i^{(j)}$ are indeterminates, and let k be an infinite field. We say that the ring $R = k[[Z_0^{(1)}, ..., Z_{m_1}^{(1)}, ..., Z_0^{(r)}, ..., Z_{m_r}^{(r)}]]/I$, where I is the ideal generated by the determinants of the 2×2 minors of the above matrix, is a scroll of type $(m_1, ..., m_r)$. For basic properties of such rings and their geometric interpretation we refer to [14] and the references given there. A scroll R is known to be an integrally closed CM noetherian domain which is an isolated singularity, and $\dim R = r+1$. It is not hard to see that if $r = 1$, then R is of the form $k[[X, Y]]^G$, and hence of finite CM type. The scroll of type (1, 1) is a hypersurface and is known to be of finite CM type by [18, 22, 23]. The aim of this section is to show the following.

THEOREM 2.1. *The ring* $R = k[[X_0, X_1, X_2, Y_0, Y_1]]/(X_0X_2 - X_1^2, X_0Y_1 - X_1Y_0, X_1Y_1 - X_2Y_0)$ *is of finite CM type.*

Proof. We denote the images of X_i and Y_i in R by x_i and y_i. It is easy to see that the ideals $A = (x_0, x_1)$, $B = (x_0, x_1, y_0)$, $C = (x_0, x_1, x_2) \simeq A^2$ are CM. For consider the exact sequence $0 \to A \to R \to R/A \to 0$. $R/A = k[[X_2, Y_0, Y_1]]/(X_2 Y_0)$ is a complete intersection of dimension 2. Hence depth $R/A = 2$, so that depth $A = 3$ since depth $R = 3$. Similarly $R/B \simeq k[[X_2, Y_1]]$ gives depth $B = 3$ and $R/C \simeq k[[Y_0, Y_1]]$ gives depth $C = 3$. It is known that R, A, B, C are the only indecomposable CM R-modules of rank one, but we do not need to use this fact here. It is also known, or can be computed directly, that A is the dualizing module.

Denote by D the duality from $\mathrm{CM}(R)$ to $\mathrm{CM}(R)$ defined by $D(X) = \mathrm{Hom}_R(X, A)$. We recall that the transpose Tr_L is defined by $\mathrm{Tr}_L X = \Omega^1 X^*$ for X in $\mathrm{CM}(R)$, using that $\dim R = 3$ [2]. We know that D preserves indecomposable CM modules and Tr_L preserves indecomposable nonprojective CM modules, and we have $D^2(X) \simeq X$, and $\mathrm{Tr}_L^2 Y \simeq Y$ when Y is not projective. It is known, or can be computed directly, that $DB \simeq C$, and clearly $DR \simeq A$. We next compute Tr_L for our indecomposable CM modules of rank 1. This computation is needed in the construction of almost split sequences, since they are of the form $0 \to D\,\mathrm{Tr}_L X \to E \to X \to 0$ [2].

We have $B \simeq B' = (1, x_1/x_0, y_0/x_0)$, and $AB' = \mathfrak{m}$, so that $A^* \simeq B$. It follows from this, or can be seen directly, that we have an exact sequence $0 \to B \to R \amalg R \to A \to 0$. Then $\mathrm{Tr}_L B = \Omega^1 B^* = \Omega A = B$ and $\mathrm{Tr}_L A = \Omega^1 A^* = \Omega^1 B = K$, which must be an indecomposable CM module of rank 2. C^* is easily computed to be isomorphic to (x_0, y_0^2). Since it has a minimal set of two generators, $\mathrm{Tr}_L C = \Omega^1 C^* = C$.

We now want to construct candidates for almost split sequences. Considering the exact sequence $0 \to K \to R \amalg R \amalg R \to B \to 0$, we see that K is the submodule of R^3 generated by $(-y_0, 0, x_0)$, $(-y_1, 0, x_1)$, $(0, -y_0, x_1)$, $(0, -y_1, x_2)$, $(-x_1, x_0, 0)$, $(-x_2, x_1, 0)$. It is easy to see that we have maps $f\colon A \to K$ given by $f(x_0) = (-y_0, 0, x_0)$, $f(x_1) = (-y_1, 0, x_1)$, $g\colon A \to K$ given by $g(x_0) = (0, -y_0, x_1)$, $g(x_1) = (0, -y_1, x_2)$, and $h\colon B \to K$ given by $h(x_0) = (-x_1, x_0, 0)$, $h(x_1) = (-x_2, x_1, 0)$, $h(y_0) = (-y_1, y_0, 0)$. These maps induce an epimorphism $u = (f, g, h)\colon A \amalg A \amalg B \to K$. Considering the associated divisors in the class group of R, and using that $\mathrm{Ker}\, u$ has rank 1, we see that $\mathrm{Ker}\, u = R$. (This can also be computed directly.) Hence we have the exact sequence

$$0 \to R \to A \amalg A \amalg B \to K \to 0, \qquad (*)$$

and by dualizing, an exact sequence

$$0 \to DK \to R \amalg R \amalg C \to A \to 0. \qquad (**)$$

Since $D\,\mathrm{Tr}_L A = DK$, $(**)$ is a nonsplit exact sequence of CM modules with the correct end terms for being almost split, it follows as in [8] that it is enough to show $\underline{\mathrm{End}}(A) = k$ to show that $(**)$ is almost split. That $\underline{\mathrm{End}}(A) = k$ is a consequence of the following.

LEMMA 2.2. *Let R be a noetherian domain and X an ideal in R. Let Y be a fractional ideal such that $XY \subset R$. Then a map $f\colon X \to X$ which is multiplication by an element in XY factors through a projective module.*

Proof. Let m be an element in XY, that is, $m = a_1 b_1 + \cdots + a_r b_r$, where $a_1, ..., a_r$ is a set of generators for X, and b_i is in Y. Define $g\colon X \to R^r$ by $g(a) = (ab_1, ..., ab_r)$, and let $h\colon R^r \to X$ be the natural map sending the ith generator

$$(0, ..., \underset{i}{1}, ..., 0)$$

in R^r to a_i. Then hg is multiplication by m, and we are done.

In our case End $A \simeq R$ since R is integrally closed and A is a nonzero ideal. Since $AB' = \mathfrak{m}$, it follows from Lemma 2.2 that $\underline{\mathrm{End}}(A) = k$. Hence $(**)$ is almost split, and by duality $(*)$ is also almost split.

It is now easy to compute the almost split sequences with B and C on the right, by using basic properties of almost split sequences. We have $D\,\mathrm{Tr}_L C = B$, so the middle term of the almost split sequence with C on the right has rank 2. By $(*)$, K is a summand of the middle term, and by $(**)$, DK is. Hence $DK \simeq K$, and we have an almost split sequence

$$0 \to B \to K \to C \to 0.$$

Further we have $D\,\mathrm{Tr}_L B = C$, and by $(*)$ and $(**)$, A and R are both summands of the middle term for an almost split sequence with B on the right. A rank argument again gives an almost split sequence

$$0 \to C \to R \amalg A \to B \to 0.$$

We have now proved that the set $\{R, A, B, C, K\}$ of indecomposable CM modules contains R and is closed under almost split sequences. Since R is not injective, it follows from Theorem 1.1 that these are all indecomposable CM R-modules. This finishes the proof of Theorem 2.1.

3. Scrolls of Infinite Cohen–Macaulay Type

In this section we show that if R is a scroll with $\dim R \geqslant 3$ and not of type (1, 1) or (2, 1), then R is of infinite CM type. We do this by first studying the corresponding graded scrolls $R' = k[Z_0^{(1)}, ..., Z_{m_1}^{(1)}, ..., Z_0^{(r)}, ..., Z_{m_r}^{(r)}]/I$, where I is defined as before. We have chosen to deal with graded modules rather than with sheaves.

To a monomial p in $k[Z_j^{(i)}]$ we define the number $d(p)$ to be the sum of the lower indices for the factors, and $t(p) = (c_1, ..., c_r)$, where c_i is the number of factors of the form $Z_j^{(i)}$ for a fixed i. From the relations for a graded scroll it follows directly that two monomials p and q are equal in R' if and only if $d(p) = d(q)$ and $t(p) = t(q)$. If for the monomials of degree i we choose one monomial for each given $d(p)$ and $t(p)$, we get a k-basis for $\mathfrak{m}^i/\mathfrak{m}^{i+1}$, where $\mathfrak{m} = (Z_0^{(1)}, ..., Z_{m_1}^{(1)}, ..., Z_{m_r}^{(r)})$.

We shall prove the following.

Theorem 3.1. *Let R' be a graded scroll of type $(m_1 = n, m_2 = t, ..., m_r)$, $n \geqslant t \geqslant \cdots \geqslant m_r$. If $r \geqslant 2$ and R' is not of type* (1, 1) *or* (2, 1), *then R' has an infinite number of indecomposable graded CM modules, up to shifts.*

Proof. Write $m = m_1 + \cdots + m_r$, $A = (x_0, x_1)$, and $B = (b_1, ..., b_m)$, where $b_1, ..., b_m$ are the entries in the first row of the matrix. Then it is known that B is CM and A^i is CM for $0 \leqslant i \leqslant m-1$.

We shall first assume $m - n \geqslant 2$, and construct exact sequences in $\operatorname{Ext}^1(A^{n+i}, B)$ when $1 \leqslant i \leqslant m-n-1$. We have $A^{n+i} \simeq (x_0^i x_0 = a_1, x_0^i x_1 = a_2, ..., x_0^i x_n = a_{n+1}, ..., x_1^i x_n = a_{i+n+1})$. For $\lambda \in k$ we define $M_\lambda \subset R \amalg R$ to be generated by $u_j = (x_0^i b_j, 0)$, $1 \leqslant j \leqslant m$, $u_{m+j} = (0, a_j)$; $0 < j \leqslant i+n$, $u_{m+n+i+1} = (x_0^i x_n + \lambda x_0^i y_t, x_1^i x_n)$. Write $s = m+n+i+1$, and $B' = x_0^i B$.

We claim that the sequence $0 \to B' \to^{\beta} M_\lambda \to^{\alpha} A^{n+i} \to 0$ is exact, where β is the natural monomorphism and α is the natural epimorphism. Clearly $\alpha\beta$ is zero. Let $x = \sum_{j=m+1}^{s} r_j u_j$ be an element in M_λ such that $\alpha(x) = 0$. If $r_s = 0$, then clearly $x \in \operatorname{Im} \beta$, so we can assume $r_s \neq 0$. Since $\alpha(u_{m+j}) = a_j$, we have $\sum_{j=1}^{i+n+1} r_j a_j = 0$. Since the a_j are linearly independent in $\mathfrak{m}^{i+1}/\mathfrak{m}^{i+2}$, all r_j must have zero constant term. A monomial p in r_s cannot have only factors of the form $z_{m_k}^{(k)}$, since $d(pa_s)$ would be too large. Hence we must have $p = p' z_j^{(k)}$, $j < m_k$, and $z_j^{(k)}(x_0^i x_n + \lambda x_0^i y_t, 0) = z_{j+1}^{(k)}(x_0^i x_{n-1} + \lambda x_0^i y_{t-1}, 0) \in \beta(B')$. This shows that the sequence is exact, and consequently M_λ is CM.

We now assume that there is an isomorphism of degree zero f: $M_{\lambda'} \to M_\lambda$. A^{n+i} and B' both have all their generators, in a minimal set of generators, in degree $i+1$. A^{n+i} has $n+i+1$ generators, and B has $m \geqslant n+i+1$ generators. Since a nonzero map $B \to A^{n+i}$ must be a monomorphism, it follows that $\mathrm{Hom}_{\mathrm{gr}\,R}(B, A^{n+i})_0$, the graded maps of degree zero, is 0. This gives a commutative diagram

$$\begin{array}{ccccccccc} 0 & \longrightarrow & B' & \longrightarrow & M_{\lambda'} & \longrightarrow & A^{n+i} & \longrightarrow & 0 \\ & & \downarrow{\scriptstyle g} & & \downarrow{\scriptstyle f} & & \downarrow{\scriptstyle h} & & \\ 0 & \longrightarrow & B' & \longrightarrow & M_{\lambda} & \longrightarrow & A^{n+i} & \longrightarrow & 0. \end{array}$$

Since $\mathrm{End}(B') = R'$ and $\mathrm{End}(A^{n+i}) = R'$, we see that $\mathrm{End}(B')_0 = k = \mathrm{End}(A^{n+1})_0$. Hence we can assume that g is the identity and h is the multiplication by some $c \neq 0$. Write $f(u_j) = a_{1j}u_1 + \cdots + a_{sj}u_s$ for $j < s$, $f(u'_s) = a_{1s}u_1 + \cdots + a_{ss}u_s$, where $u_1, \dots, u_{s-1}, u'_s$ are the generators for $M_{\lambda'}$ and $u_1, \dots, u_s$ the generators for M_λ. Since $i \geqslant 1$, we have $x_0 u_{s-1} = x_n u_{s-n-1}$, and this shows $a_{n,s-1} = 0$. Consider the equality $x_0 u'_s = x_1 u_{s-1} + x_1 u_n + \lambda y_t u_1$. Applying f and comparing the coefficients of $x_1 u_n$ we get $a_{s,s} = a_{n,s-1} + 1 = 1$. Applying f and comparing the coefficients of $(x_0^{i+1} y_t, 0)$ gives $\lambda a_{ss} = \lambda'$, so that $\lambda = \lambda'$. This shows that different λ give nonisomorphic M_λ. Since k is an infinite field, we are done in this first case. (We point out that if we let $\lambda = (\lambda_1, \dots, \lambda_r) \in \mathbf{P}^{r-1}(k)$, and in M_λ redefine u_s to be $(\sum_{j=1}^{r} \lambda_j z_{m_j}^{(j)} x_0^i, x_1^i x_n)$, we get similarly an infinite family of nonisomorphic M_λ indexed by $\mathbf{P}^{r-1}(k)$.)

Assume now that the graded scroll R' is of type $(n, 1)$, with $n \geqslant 3$. Let $B' = (x_0^2, x_0 x_1, \dots, x_0 x_{n-1}, x_0 y_0)$ and $A^n = (x_0^2, x_0 x_1, \dots, x_0 x_n)$. For $\lambda \in k$, define $M_\lambda \subset R \amalg R$ to be generated by $u_j = (x_0 x_{j-1}, 0)$, $1 \leqslant j \leqslant n$, $u_{n+1} = (x_0 y_0, 0)$, $u_j = (0, x_0 x_{j-n-2})$; $n+2 \leqslant j \leqslant 2n$, $u_{2n+1} = (x_0 y_1, x_0 x_{n-1})$, $u_{2n+2} = (x_1 y_1 + \lambda x_0 y_1, x_0 x_n)$.

We want to show that we have an exact sequence $0 \to B' \to^{\beta} M_\lambda \to^{\alpha} A^n \to 0$, where β is the natural inclusion and α the natural epimorphism. We clearly have $\alpha\beta$ is zero. Assume that $\alpha(x) = 0$, where $x = \sum_{j=1}^{s} r_j u_j$, $s = 2n+2$. As before, the r_j must have constant term zero, and if r_s and r_{s-1} are both zero, we are done. It is impossible that a monomial p in r_s has all factors of the form x_n and y_1. If p has one factor of the form x_{n-1} or y_0 and the rest of the form x_n and y_1, then a_{s-1} must have a monomial q with $x_0 q = x_1 p$. Then $q(x_0 y_1, 0) = p(x_1 y_1, 0)$ and $p(x_0 y_1, 0) = q(x_0 y_0, 0) \in \beta(B)$. If p does not satisfy any of the above conditions, the $p(x_1 y_1 + \lambda x_0 y_1, 0) \in \beta(B)$. If $r_s = 0$, we argue similarly for r_{s-1}, and get that the sequence is exact, and M_λ is CM.

Assume that $f\colon M_{\lambda'} \to M_\lambda$ is an isomorphism. We get as before a commutative diagram where the induced map $g\colon B' \to B'$ can be assumed to be the identity. We have $x_0 u_s = x_n(0, x_0^2) + y_1(x_0 x_1, 0) + \lambda' y_1(x_0^2, 0)$. Applying f and comparing the coefficients of $(x_1 y_1, 0)$ gives $a_{ss} = 1$. We note that we get no contribution from $x_n f(0, x_0^2)$ since $n > 1 + 1 = 2$. Comparing the coefficients of $(x_0 y_1, 0)$ gives $\lambda a_{ss} = \lambda'$, and hence $\lambda = \lambda'$. This finishes the proof.

We have the following main result of this part.

THEOREM 3.2. *Let R be a scroll of type $(m_1, ..., m_r)$ over an infinite field k. Then R is of finite CM type if and only if R is of type (m), $(1, 1)$, or $(2, 1)$.*

Proof. We have seen that a scroll of type (m), $(1, 1)$, or $(2, 1)$ is of finite CM type. If R is of a different type, we consider $R' \to R$, where R' is the corresponding graded scroll. Let $M = M_0 \amalg M_1 \amalg \cdots$ be a graded R'-module, generated by M_0. R is the completion of R' with respect to the graded maximal ideal $\mathfrak{m}$. $M_i = \mathfrak{m}^{i+1} M_0$. The associated graded module of $R \otimes_{R'} M$ is then M. If $N = N_0 \amalg N_1 \amalg \cdots$ is another graded R'-module, generated by N_0, then an isomorphism $g\colon R \otimes_{R'} M \to R \otimes_{R'} N$ will induce an isomorphism of degree zero between M and N. Since by Theorem 3.1 we have an infinite number of nonisomorphic indecomposable CM modules $M = M_0 \amalg M_1 + \cdots$ generated by their degree zero component, R is of infinite CM type.

PART II: FIXED RINGS

Throughout this part let S be an integrally closed domain, and G a finite nontrivial group acting faithfully on S, such that $|G|$ is invertible in S. Denote by $R = S^G$ the fixed ring, and our main objective is to compare the modules over S and R. It will also be useful to study the modules over the skew group ring SG. When discussing the question of finite CM type, we shall in addition assume that S is complete local CM, with $S/\mathfrak{m} = k$ algebraically closed, where $\mathfrak{m}$ is the maximal ideal of S.

The main result of this part is that when S is the power series ring $k[[X_1, ..., X_n]]$ for $n \geqslant 3$, then $R = S^G$ is of finite Cohen–Macaulay type if and only if $n = 3$ and $G \simeq Z_2$, where the generator of Z_2 acts by sending each variable to its negative.

We start out in Section 1 by studying the relationship between the reflexive modules over SG and R. If the minimal primes in R are unramified in S, we show that the fixed point functor gives an equivalence of categories $\operatorname{Ref} SG \to \operatorname{Ref} R$ of reflexive modules. In the complete case, we use this to show that if R is of finite CM type, then S is. We also reduce the problem of showing that R is of infinite CM type to a similar problem for finite dimensional algebras, which we investigate in Section 2. As a consequence we get most of our results on $R = k[[X_1, ..., X_n]]^G$ $(n \geqslant 3)$ being of infinite CM type. The rest is given in Section 3 using different techniques.

Finally we show in Section 4 that $R = k[[X_1, X_2, X_3]]^{Z_2}$ is of finite CM type with the desired action. We give two different proofs, one in the spirit of the work in this part, and one using the same techniques as in Part 1.

In Section 5 we derive some interesting consequences of $R = S^G$ being of finite CM type and illustrate how our results may be used to show that a given $R = S^G$ is of infinite CM type.

1. Preliminary results

Under our standard assumptions, the fixed ring $R = S^G$ is integrally closed. We want to study the relationship between Ref R and Ref SG. We shall need the following facts on reflexive modules.

LEMMA 1.1. (a) *A module over R or S is reflexive if and only if it is a second syzygy module.*

(b) *S is a finitely generated reflexive R-module.*

(c) *If f: $B \to C$ is a map between reflexive modules in* mod R *or* mod S, *then* Ker f *is reflexive.*

(d) *An SG-module M is SG-reflexive if and only if it is R-reflexive if and only if it is S-reflexive.*

(e) *An SG-module M is reflexive if and only if it is a second syzygy module.*

We have the following direct consequence.

LEMMA 1.2. *The fixed point functor F:* mod $SG \to$ mod R *takes reflexive modules to reflexive modules.*

Proof. If M is a reflexive SG-module, we have an exact sequence $0 \to M \to n(SG) \to m(SG)$, where $n(SG)$ denotes the direct sum of n copies of SG. The sequence $0 \to M^G \to n(SG)^G \to m(SG)^G$ is then an exact sequence of R-modules. There is an isomorphism of R-modules $S \to (SG)^G$ obtained by sending s to $\sum_{\sigma \in G} \sigma(s)\, \sigma$, and since S is a reflexive R-module, we conclude that M^G is a reflexive R-module.

We want to show that if height one primes in R are unramified in S, then F: Ref $SG \to$ Ref R in an equivalence. We recall that this assumption is equivalent to the natural ring map α: $SG \to \mathrm{End}_R(S) = \Gamma$ given by $\alpha(s\sigma)(x) = s\sigma(x)$ being an isomorphism [1]. Since R is integrally closed and S is a reflexive R-module, we know from [11, Chap. 1] that $\mathrm{Hom}_R(S^*, \ )$: mod $R \to$ mod Γ and $\mathrm{Hom}_\Gamma(S, \ )$: mod $\Gamma \to$ mod R induce inverse equivalences of categories between Ref R and Ref Γ, where $S^* = \mathrm{Hom}_R(S, R)$.

Hence we get the following consequence.

PROPOSITION 1.3. *Assume that height one primes in* $R = S^G$ *are unramified in* S. *Then the fixed point functor* F: mod $SG \to$ mod R *induces an equivalence of categories between* Ref SG *and* Ref R.

Proof. Because of the ring isomorphism $SG \to \mathrm{End}_R S = \Gamma$, the equivalence $\mathrm{Hom}_\Gamma(S, \ )$: Ref $\Gamma \to$ Ref R shows that we have an equivalence $\mathrm{Hom}_{SG}(S, \ )$: Ref $SG \to$ Ref R, and $\mathrm{Hom}_{SG}(S, \ )$ is the fixed point functor.

From the standard assumptions on S and $R = S^G$ it follows that R is an R-summand of S, by considering the maps $R \to^i S \to^t R$, where $t(s) = 1/|G| \sum_{\sigma \in G} \sigma(s)$. From this it follows that S is a projective Γ-module. We say that a Γ-module is CM if it is CM over the center of Γ. Since $\mathrm{Hom}_\Gamma(S, \ )$: mod $\Gamma \to$ mod R is an exact functor, it preserves CM modules. Hence we have the following relationship between the CM types.

PROPOSITION 1.4. *Assume that* S *is complete local and* $S/\mathfrak{m}$ *is algebraically closed.*

(a) *If* R *is of finite CM type, then* $\Gamma = \mathrm{End}_R(S)$ *is.*

(b) *If height one primes in* R *are unramified in* S, *and* R *is of finite CM type, then* S *is.*

Proof. Part (a) follows directly from the above comments. (b) Assume R is of finite CM type. Then by our assumption and (a), SG is of finite CM type. If M is an indecomposable CM S-module, then since $|G|$ is a unit in S, M is an S-summand of $SG \otimes_S M$. Since SG is right projective S-module, $SG \otimes_S M$ is a CM SG-module, and hence a CM S-module. This shows that S is of finite CM type.

We note that if we drop the assumption that height one primes in R are unramified in S, then R may be of finite CM type while S is of infinite CM type. For example, let $f(x, y, z)$ be such that the hypersurface $S = k[[x, y, z, u]]/(f + u^2)$ is of infinite CM type (see [18]) and let $G = Z_2$ act on S by sending u to $-u$ and leaving the other variables fixed. Then $R = S^G \simeq k[[x, y, z]]$, which is of finite CM type.

Even if some height one primes in R are ramified in S, there is some relationship between Ref SG and Ref R. We first formulate our result more generally for the situation of a ring map $\Lambda \to \Gamma$.

PROPOSITION 1.5. (a) *Let* $\Lambda \to \Gamma = \mathrm{End}_R(S)$ *be a ring map, such that* Γ *is a reflexive* Λ*-module,* S *is a projective* Λ*-module, and* $\mathrm{Hom}_\Lambda(S, \Gamma) = \mathrm{Hom}_\Gamma(S, \Gamma)$. *Then the restriction induces a fully faithful functor* Ref $\Gamma \to$ Ref Λ.

(b) *The natural ring map* $SG \to \mathrm{End}_R(S) = \Gamma$ *induces by restriction a fully faithful functor* $\mathrm{Ref}\,\Gamma \to \mathrm{Ref}\,SG$, *and hence we get by composition a fully faithful functor* $\mathrm{Ref}\,R \to \mathrm{Ref}\,SG$.

Proof. (a) Since S is projective both as Γ-module and Λ-module, we get, by using $\mathrm{Hom}_\Lambda(S, \Gamma) = \mathrm{Hom}_\Gamma(S, \Gamma)$, natural isomorphisms $\mathrm{Hom}_\Lambda(S, X) \to \mathrm{Hom}_\Gamma(S, X)$ for all X in $\mathrm{mod}\,\Gamma$. $\Gamma = \mathrm{End}_R(S)$ is clearly a finitely generated reflexive R-module and R is integrally closed. Then it is known that a Γ-module M is Γ-reflexive if and only if it is R-reflexive. If Y is in $\mathrm{Ref}\,\Gamma$, then $\mathrm{Hom}_\Lambda(\Gamma, Y)$ is also in $\mathrm{Ref}\,\Gamma$, since it is a reflexive R-module. We have $\mathrm{Hom}_\Gamma(S, \mathrm{Hom}_\Lambda(\Gamma, Y)) \simeq \mathrm{Hom}_\Lambda(S, Y) = \mathrm{Hom}_\Gamma(S, Y)$, and since $\mathrm{Hom}_\Gamma(S, \;)$: $\mathrm{Ref}\,\Gamma \to \mathrm{Ref}\,R$ is an equivalence, we have $Y \simeq \mathrm{Hom}_\Lambda(\Gamma, Y)$. We then have $\mathrm{Hom}_\Lambda(X, Y) \simeq \mathrm{Hom}_\Gamma(X, \mathrm{Hom}_\Lambda(\Gamma, Y)) \simeq \mathrm{Hom}_\Gamma(X, Y)$ for X in $\mathrm{mod}\,\Gamma$. Since if X is a reflexive Γ-module it is a reflexive Λ-module, a fully faithful functor $\mathrm{Ref}\,\Gamma \to \mathrm{Ref}\,\Lambda$ is induced.

(b) Γ is a reflexive R-module, and hence a reflexive SG-module, and S is a projective SG-module since it is a projective S-module. To finish the proof we need only show $\mathrm{Hom}_\Gamma(S, \Gamma) = \mathrm{Hom}_{SG}(S, \Gamma)$.

We have $\mathrm{Hom}_{SG}(S, \Gamma) = \Gamma^G$, where the action of G on Γ is given by $(gf)(s) = gf(s)$ for $f \in \mathrm{End}_R(S)$, $s \in S$. Then clearly $\Gamma^G = \mathrm{Hom}_R(S, S)^G = \mathrm{Hom}_R(S, R) \subset \Gamma$. Consider the map α: $\mathrm{End}_R(S) \to S$ given by $\alpha(f) = f(1)$. Then $\mathrm{Ker}\,\alpha = \{f\colon S \to S; f(1) = 0\} = \{f\colon S \to S; f(R) = 0\}$. $\mathrm{Hom}_\Gamma(S, \Gamma) = \{f\colon S \to S; (\mathrm{Ker}\,\alpha) f = 0\}$. If $\mathrm{Im}\,f \subset R$, then clearly $(\mathrm{Ker}\,\alpha) f = 0$. And if $\mathrm{Im}\,f \not\subset R$, choose $x \in R$, $x \notin \mathrm{Im}\,f$. Since R is an R-summand of S and S is R-reflexive, there is an R-map $h \in \mathrm{Ker}\,\alpha$ with $h(x) \neq 0$. Hence $\mathrm{Hom}_\Gamma(S, \Gamma) = \mathrm{Hom}_R(S, R) \subset \Gamma$, so that $\mathrm{Hom}_{SG}(S, \Gamma) = \mathrm{Hom}_\Gamma(S, \Gamma)$.

We mention without proof that also the following subcategory of $\mathrm{Ref}\,SG$ is equivalent to $\mathrm{Ref}\,R$ via the fixed point functor. The objects are the C such that there is an exact sequence of $S[G]$-modules $0 \to C \to nS \to mS$ such that the induced sequence by taking fixed points and dualizing $(mR)^* \to (nR)^* \to (C^G)^* \to 0$ is exact.

In the complete case, we want to establish a connection between $\mathrm{CM}(R)$ and modules over some finite dimensional algebra, which we shall apply to give necessary conditions for $\mathrm{CM}(R)$ to be of finite representation type. Here f.l. $S[G]$ denotes the SG-modules of finite length. Assume S is complete, $S/\mathfrak{m}$ is algebraically closed, and height one primes in R are unramified in S, and let $\dim S = d \geqslant 3$.

PROPOSITION 1.6. *Let* $\mathfrak{D}$ *be the subcategory of* f.l. SG *consisting of the* A *with* $A^G = 0$. *If* A *is in* $\mathfrak{D}$, *then* $\alpha(A) = (\Omega^2 A)^G$ *is CM.*

Further, A *is indecomposable if and only if* $\alpha(A)$ *is indecomposable, and if* A *and* B *are in* $\mathfrak{D}$, *then* $A \simeq B$ *if and only if* $\alpha(A) \simeq \alpha(B)$.

Proof. Let A be in $\mathfrak{D}$, and let $0 \to \Omega^2 A \to P_1 \to P_0 \to A \to 0$ be the beginning of a minimal projective resolution for A over SG, where $i \geqslant 1$. Then $0 \to (\Omega^2 A)^G \to P_1^G \to P_0^G \to 0$ is an exact sequence of R-modules since $|G|$ is invertible in S. The projective SG-modules P_j are CM since SG is CM. Hence P_j is a CM R-module, and so P_j^G is, since P_j^G is an R-summand of P_j, as is seen by considering $P_j^G \to^i P_j \to^t P_j^G$, where $t(p) = 1/|G| \sum_{\sigma \in G} \sigma(p)$. It follows that $(\Omega^2 A)^G$ is a CM R-module.

Since S is CM and A has finite length, $\mathrm{Ext}_S^i(A, S) = 0$ for $i = 0, 1$. Since SG is a free S-module, $\mathrm{Ext}_S^i(A, SG) = 0$, and hence $\mathrm{Ext}_S^i(A, SG)^G = 0$ for $i = 0, 1$. Since the fixed point functor is exact, $\mathrm{Ext}_{SG}^i(A, SG) \simeq \mathrm{Ext}_S^i(A, SG)^G$, so that $\mathrm{Ext}_{SG}^i(A, SG) = 0$ for $i = 0, 1$. We then get the exact sequence $0 \to A^* \to P_0^* \to P_1^* \to (\Omega^2 A)^* \to 0$, where $X^* = \mathrm{Hom}_{SG}(X, SG)$. Dualizing again gives $0 \to \Omega^2 A \to P_1 \to P_0 \to A \to 0$, so that $P_0^* \to P_1^* \to (\Omega^2 A)^* \to 0$ is a minimal projective presentation, and also $\mathrm{Ext}_{SG}^1((\Omega^2 A)^*, SG) = A$. If $\Omega^2 A$ is indecomposable, A must be, since a module of finite length cannot have projective dimension less then 3. Assume then that A is indecomposable and that $\Omega^2 A \simeq X \amalg Y$, where X and Y are not zero. Then we have $(\Omega^2 A)^* \simeq X^* \amalg Y^*$ with X^* and Y^* not zero. Considering a minimal projective presentation for X^* and Y^*, and dualizing, we get a contradiction to the fact that the presentation $P_1 \to P_0 \to A \to 0$ is minimal. By the above we have $\mathrm{Ext}_{SG}^1((\Omega^2 A)^*, SG) \simeq A$. The rest of our claim now follows by using that the fixed point functor gives an equivalence $\mathrm{Ref}\, SG \to \mathrm{Ref}\, R$.

Let $|G| = n$ and let e be the idempotent $1/n \sum_{\sigma \in G} \sigma$. If A is an SG-module, then $eA \subset A^G$, since $\sigma e = e$ for all $\sigma \in G$. And if $x \in A^G$, then $ex = 1/n \sum \sigma x = x$, so that $A^G \subset eA$. Hence the SG-modules A with $A^G = 0$ are the $SG/(e)$-modules, so that the above category $\mathfrak{D}$ is the category f.l. $SG/(e)$. We further have that $SG/\mathfrak{m}^2 SG = (S/\mathfrak{m}^2)G$, so that $(S/\mathfrak{m}^2)G$ is a factor ring of SG, and hence $(S/\mathfrak{m}^2)\, G/(e)$ is a factor ring of $SG/(e)$. Hence we have the following consequence of the above.

THEOREM 1.7. *Let S be a complete local ring with $S/\mathfrak{m}$ algebraically closed, and assume that height one primes in R are unramified in S. If R is of finite* CM *type, then the artin algebra $(S/\mathfrak{m}^2)\, G/(e)$ is of finite representation type.*

This result motivates investigating when, for an artin algebra Λ with $\mathfrak{r}^2 = 0$, $\Lambda/\mathfrak{r} = k$ algebraically closed, and G a finite group acting, $\Lambda G/(e)$ is of finite representation type. Here ΛG denotes the skew group ring and e the usual idempotent for G. We will deal with this question in the next section.

2. The Representation Type of $\Lambda G/(e)$

Let Λ be an artin algebra with radical $\mathfrak{r}$, such that $\Lambda/\mathfrak{r}=k$ is algebraically closed and $\mathfrak{r}^2=0$. Let G be a finite group acting on Λ such that $|G|$ is invertible in Λ. G induces an action on $\mathfrak{r}=V$, which is then a kG-module. We recall from [3, 10] that the Gabriel quiver for ΛG is the opposite of the McKay quiver for V, where the McKay quiver is defined as follows. The vertices are ine one–one correspondence with the simple kG-modules $V_1=k, ..., V_n$. There are n_{ij} arrows from V_j to V_i if n_{ij} is the multiplicity of V_j in a direct sum decomposition of $V\otimes V_i$. It is easy to see that leaving out the vertex of $V_1=k$ we get (the opposite of) the Gabriel quiver for $\Lambda G/(e)$. It is a well-known theorem of Gabriel that $\Lambda G/(e)$ is of finite (representation) type if and only if the underlying graph of the separated Gabriel quiver is a finite disjoint union of the Dynkin diagrams A_n, D_n, E_6, E_7, and E_8 [16]. The separated quiver is obtained by replacing each vertex v with two vertices v' and v'' and each arrow $v\to w$ with an arrow $v'\to w''$. We will use this to give a characterization of when $\Lambda G/(e)$ is of finite type.

The following lemma reduces our considerations to abelian groups.

Lemma 2.1. *Let Λ be as above. If $\Lambda G/(e)$ is of finite type, then $\Lambda H/(e)$ is of finite type for all subgroups H of G.*

Proof. Let M be an indecomposable ΛH-module with $M^H=0$. We have $\operatorname{Hom}_{\Lambda H}(\Lambda G, M)^G \simeq \operatorname{Hom}_{\Lambda G}(\Lambda, \operatorname{Hom}_{\Lambda H}(\Lambda G, M)) \simeq \operatorname{Hom}_{\Lambda H}(\Lambda, M) = M^H=0$. Since ΛH is a two-sided ΛH-summand of ΛG, M is a summand of the ΛH-module $\operatorname{Hom}_{\Lambda H}(\Lambda G, M)$. This shows that if $\Lambda G/(e)$ is of finite type, then $\Lambda H/(e)$ is.

We now state the main result in this section.

Theorem 2.2. *Let Λ be as before, and assume that* $\dim_k V\geqslant 3$. *Then $\Lambda G/(e)$ is of finite type if and only if $G\simeq Z_2$ and the multiplicity of the nontrivial simple kG-module k_- in V is at most one.*

Proof. Assume $\dim_k V\geqslant 3$, and assume that G has an abelian subgroup H of order at least 3. If some simple summand V_1 of V (as kH-module) occurs with multiplicity $\geqslant 2$, we have $V_1\cdot\rightrightarrows\cdot k$ in the McKay quiver. Let V_2 be a simple kH-module not isomorphic to k or V_1. Then $V_i\otimes_k V_1 \not\simeq k$ for $i=1$ or for $i=2$, and we have $V_i\otimes_k V_1\cdot\rightrightarrows\cdot V_i$ in the MacKay quiver, after removing k. Hence we also have a double arrow in the separated McKay quiver, showing that $\Lambda G/(e)$ is of infinite type. If every summand of V occurs with multiplicity one, there are arrows from at least three vertices to k. Tensoring we see that for any vertex there are arrows from at least three vertices. Hence when we remove k there are arrows from at least two vertices to any given vertex. And similarly there are at least two arrows leaving a given vertex. Hence in the separated quiver of the McKay quiver minus k there are either two arrows leaving a given vertex, or two arrows entering. The underlying graph can then not be a Dynkin diagram, so that we have infinite type.

It follows that if $\Lambda G/(e)$ is of finite type, then every abelian subgroup of G must have order 2. This shows that $G \simeq Z_2$, which we now assume. Let m be the multiplicity of k in V. Then there are m arrows from k to k, hence also from k_- to k_-. The separated McKay quiver minus k is then

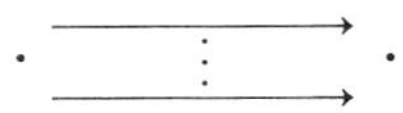

(m arrows)

Hence $\Lambda Z_2/(e)$ is of finite type if and only if $m = 1$. This finishes the proof.

Combining with the results in Section 1 we get the following.

THEOREM 2.3. *Let S and G satisfy the satisfy the standard assumptions of the previous section, and assume also that S is complete local, $S/\mathfrak{m}$ is algebraically closed, height one primes in R are unramified in S, and* dim $S \geq 3$.

If R is of finite CM type, then $G \simeq Z_2$, and the multiplicity of the trivial kG-module k in $\mathfrak{m}/\mathfrak{m}^2$ is at most one.

3. DIMENSION THREE

In this section we assume in addition that S is complete local, $S/\mathfrak{m}$ is algebraically closed, height one primes in R are unramified in S, and dim $R \geq 3$. The aim is to improve Theorem 2.3 by proving the following.

THEOREM 3.1. *If R is of finite CM type, then* dim $R = 3$.

Proof. Assume R is of finite CM type. By Theorem 2.3 we can assume that $G = Z_2$ and k occurs with multiplicity at most one in the kG-module $V = \mathfrak{m}/\mathfrak{m}^2$. We shall need the following.

LEMMA 3.2. *The trivial SG-module k occurs with multiplicity at least* 2 *in $L/\mathfrak{m}L$, where $L = \Omega^2_{SG} k$.*

Proof. The exact sequence $0 \to L \to S \otimes_k V \to \mathfrak{m} \to 0$ gives a projective cover both as SG-modules and S-modules. Tensoring with $\mathfrak{m}$ over S we get $\operatorname{Tor}_1^S(k, \mathfrak{m}) \simeq L/\mathfrak{m}L$. Hence $0 \to \mathfrak{m} \to S \to k \to 0$ gives rise to the exact sequence $0 \to L/\mathfrak{m}L \to \mathfrak{m} \otimes_S \mathfrak{m} \to \mathfrak{m} \to \mathfrak{m}/\mathfrak{m}^2 \to 0$, which again gives $L/\mathfrak{m}L \to \mathfrak{m}/\mathfrak{m}^2 \otimes_k \mathfrak{m}/\mathfrak{m}^2 \to \mathfrak{m}^2/\mathfrak{m}^3 \to 0$.

Considering the commutative diagram

$$
\begin{array}{ccccccccc}
 & & 0 & & & & & & \\
 & & \downarrow & & & & & & \\
0 & \longrightarrow & J & \longrightarrow & V \otimes_k V & \longrightarrow & S_2(V) & \longrightarrow & 0 \\
 & & \downarrow & & \downarrow & & \downarrow & & \\
0 & \longrightarrow & H & \longrightarrow & \mathfrak{m}/\mathfrak{m}^2 \otimes_k \mathfrak{m}/\mathfrak{m}^2 & \longrightarrow & \mathfrak{m}^2/\mathfrak{m}^3 & \longrightarrow & 0,
\end{array}
$$

where $S_2(V)$ is the second symmetric product, it is sufficient to show that k occurs in the kG-module J with multiplicity at least 2.

Assume first that $V = nk_-$, where $n \geqslant \dim R \geqslant 3$. Since $\dim_k V \otimes_k V = n^2$ and $\dim S_2(V) = 1 + \cdots + n = n(n+1)/2$, we have $\dim_k J = n(n-1)/2 \geqslant 3$. Hence k occurs with multiplicity at least 2 in J.

If $V = k \amalg (n-1)\, k_-$, the multiplicity of k in $V \otimes_k V$ is $1 + (n-1)^2$, and in $S_2(V)$ it is $(1 + \cdots + n - 1) + 1 = n(n-1)/2 + 1$. Hence k occurs $(n^2 - 3n + 2)/2 \geqslant 2$ times in J. This finishes the proof of the lemma.

Denote by L_- the SG-module $L \otimes_k k_-$. We know by Proposition 1.6 that $L_- = \Omega^2 k_-$ is an indecomposable SG-module, and hence $\operatorname{End}_{SG}(L_-)$ is local. The SG-modules $\mathfrak{m}L_-$ and $\mathfrak{r}L_-$ are both $\operatorname{End}_{SG}(L_-)$-submodules of L_-, where $\mathfrak{r}$ denotes the radical of $\operatorname{End}_{SG}(L_-)$, and there is the following relationship between them.

Lemma 3.3. *With the above notation,* $\mathfrak{r}L_- \subset \mathfrak{m}L_-$.

Proof. $\operatorname{Ext}^i_{SG}(k_-, SG) \simeq \operatorname{Ext}^i_S(k_-, SG)^G = 0$ for $i = 1, 2$, so that $\underline{\operatorname{End}}_{SG}(L_-) \simeq \underline{\operatorname{End}}_{SG}(k_-) = k$ [6], and hence $\mathfrak{r} = P(L_-, L_-)$, the SG-maps from L_- to L_-, which factor through a projective SG-module. Let $f \in P(L_-, L_-)$, and let $g\colon P \to L_-$ be a projective cover in mod SG. Choose $h\colon L_- \to P$ such that $gh = f$. We must have $\operatorname{Im} h \subset \mathfrak{m}P$, and hence $\operatorname{Im} f \subset \mathfrak{m}L_-$, and this shows $\mathfrak{r}L_- \subset \mathfrak{m}L_-$.

We want to consider modules M_i with $\mathfrak{m}L_- \subset M_i \subset L_-$ and $L_-/M_i \simeq k_-$. By Lemma 3.2 there is an infinite number of choices for M_i, and we show that if $\dim R \geqslant 4$ different M_i give rise to nonisomorphic CM modules.

Lemma 3.4. *Let M_i be as above, and assume* $\dim R \geqslant 4$. *Then $\Omega(M_i)$ is an indecomposable reflexive SG-module, and hence $\Omega(M_i)^G$ is an indecomposable CM R-module.*

Proof. The exact sequence $0 \to M_i \to L_- \to k_- \to 0$ shows that M_i is not projective. Since $\mathrm{Ext}^i_{SG}(k_-, SG) = 0$ for $i = 0, 1, 2$, we have that $\mathrm{Ext}^i_{SG}(L_-, SG) \simeq \mathrm{Ext}^i_{SG}(M_i, SG)$ for $i = 0, 1$, $M_i^{**} \simeq L_-^{**} \simeq L_-$, so that M_i is indecomposable. Since $\dim R \geqslant 4$, we have $\mathrm{Ext}^1_{SG}(L_-, SG) \simeq \mathrm{Ext}^3_{SG}(k_-, SG) = 0$, so that $\mathrm{Ext}^1_{SG}(M_i, SG) = 0$. Since M_i is indecomposable nonprojective, ΩM_i is indecomposable since $\underline{\mathrm{End}}(M_i) \simeq \underline{\mathrm{End}}(\Omega M_i)$ [6]. Since $L_- = \Omega^2 k_-$, where $k_-^G = 0$, L_-^G is a Cohen-Macaulay R-module. The exact sequence $0 \to M_i \to L_- \to k_- \to 0$ shows that $M_i^G \simeq L_-^G$, so that M_i^G is CM, and hence clearly $(\Omega M_i)^G$ is CM.

Since height one primes in R are unramified in S, $\Omega M_i \simeq \Omega M_j$ if and only if $(\Omega M_i)^G \simeq (\Omega M_j)^G$. We have an epimorphism of SG-modules f: $L_- \to 2k_-$. For each one-dimensional subspace U_j of $2k_-$, let $M_j = f^{-1}(U_j)$. We then get a family of subspaces in one–one correspondence with $\mathbf{P}^1(k)$. To finish our proof, we want to show that different M_i give rise to nonisomorphic CM modules $(\Omega M_i)^G$.

LEMMA 3.5. *In the above notation, if $M_1 \neq M_2$, then $\Omega M_1 \not\simeq \Omega M_2$.*

Proof. Assume that $\Omega M_1 \simeq \Omega M_2$. Since $\mathrm{Ext}^1_{SG}(M_i, SG) = 0$ for $i = 1, 2$, we know that we have some isomorphism f: $M_1 \to M_2$. This gives a commutative diagram

$$\begin{array}{ccc} M_1 & \xrightarrow[\sim]{f} & M_2 \\ \cap & & \cap \\ M_1^{**} & \xrightarrow{f^{**}} & M_2^{**} \\ \wr\| & & \wr\| \\ L_- & \xrightarrow[\sim]{g} & L_- \end{array}$$

Since $\mathrm{End}_{SG}(L_-)/\mathfrak{r} = k$, we have $g = z + h$ with $z \in k$ and $h \in \mathfrak{r}$. $M_2 = g(M_1) \subset zM_1 + h(M_1) \subset M_1 + \mathfrak{r}M_1 \subset M_1$, and similarly $M_1 \subset M_2$. This shows that $M_1 = M_2$, and the lemma is proved.

4. A THREE-DIMENSIONAL RING OF FINITE CM TYPE

Let the assumption on S and G be as in the previous section. In this section we give two different proofs of the following.

THEOREM 4.1. *Let $G = Z_2$ act on $S = k[[X_1, X_2, X_3]]$ such that the generator sends x_i to $-x_i$. Then $R = S^G$ has up to isomorphism only the three indecomposables R, S_-^G, $\Omega^2_{SG}(k_-)$, where $S_- = S \otimes_k k_-$.*

For the first proof the following lemma is useful, where $R=S^G$ is assumed to be as in the previous section.

LEMMA 4.2. *Assume that* $\dim R=3$ *and* R *is an isolated singularity, that is,* R_p *is regular local for all nonmaximal prime ideals* p *in* R. *If* A *is in* $\operatorname{Ref} SG$, *then* A^G *is a CM* R*-module if and only if* $\operatorname{Ext}^1_{SG}(A^*, SG)^G=0$.

Proof. Let A be in $\operatorname{Ref} SG$. Then the exact sequence $0\to \Omega A^*=K\to Q\to A^*\to 0$, where Q is a projective SG-module, gives rise to the exact sequence $0\to A\to Q^*\to K^*\to \operatorname{Ext}^1_{SG}(A^*, SG)\to 0$. Since R is an isolated singularity and height one primes in R are unramified in S, S is also an isolated singularity [1]. If p is a nonmaximal prime ideal in R, we then have $\operatorname{gl\,dim} S_pG=\operatorname{gl\,dim} S_p\leqslant 2$, so that A_p^* is a projective S_pG-module, since A^* is a reflexive SG-module. Hence $\operatorname{Ext}^1_{SG}(A^*, SG)$, and consequently $\operatorname{Ext}^1_{SG}(A^*, SG)^G$, has finite length. The exact sequence of R-modules $0\to A^G\to Q^{*G}\to K^{*G}\to \operatorname{Ext}^1_{SG}(A^*, SG)^G\to 0$ then shows that A^G is CM if and only if $\operatorname{Ext}^1_{SG}(A^*, SG)^G=0$, and we are done.

Proof of Theorem 4.1. We first note that $R=S^G$ is known to be an isolated singularity, and the height one primes in R are unramified in S since the action of Z_2 on $V=\mathfrak{m}/\mathfrak{m}^2$ is free. Now let A be an indecomposable nonprojective reflexive SG-module. Since $\operatorname{gl\,dim} SG=3$ and A^* is a reflexive SG-module, we have $\operatorname{pd}_{SG} A^*=1$. This shows that $\operatorname{Ext}^1_{SG}(A^*, SG)\neq 0$. From Section 2 we know that $SG/(e)/\mathfrak{m}^2SG/(e)=k$, so that $SG/(e)=k$. This means that k_- is the only indecomposable SG-module with $k_-^G=0$, so that $\operatorname{Ext}^1_{SG}(A^*, SG)=nk_-$ for some $n>0$. Since A is indecomposable and $\operatorname{pd}_{SG} k_-=3$, nk_- must be indecomposable. It follows that $A\simeq \Omega^2k_-$, so that $A^G\simeq(\Omega^2k_-)^G$. Since S and S_- are the only indecomposable projective SG-modules, we get our desired result from Lemma 4.2.

We now give a proof of Theorem 4.1 using the method discussed in Part I.

Let $V=\mathfrak{m}/\mathfrak{m}^2$, and consider the Koszul complex $0\to S\otimes_k \Lambda^3V\to S\otimes_k \Lambda^2V\to S\otimes_k V\to S\to k\to 0$. This is also an exact sequence of SG-modules, with the natural action of G on S and V. From this we get the exact sequence $0\to S\to S_-^3\to S^3\to S_-\to k_-\to 0$, and hence the exact sequence of R-modules $0\to R\to (S_-^G)^3\to (S^G)^3\to S_-^G\to 0$. Writing $S_-^G=\omega$, we get the two short exact sequences $0\to R\to \omega^3\to L^G\to 0$ and $0\to L^G\to R^3\to \omega\to 0$. Dualizing the first one we get an exact sequence $0\to DL^G\to R^3\to \omega\to 0$, which must be isomorphic to $0\to L^G\to R^3\to \omega$. Since the class group $C(R)$ is isomorphic to $G=Z_2$ (see [9]), $\omega^*\simeq\omega$. We then compute $\operatorname{Tr}_L\omega=\Omega\omega^*=\Omega\omega=L^G$. Hence we have $\operatorname{Tr}_L L^G=\omega$, so that

$D\operatorname{Tr}_L \omega = DL^G \simeq L^G$ and $D\operatorname{Tr}_L L^G \simeq R$. From the exact sequence $S^3 \to S_- \to k \to 0$ we see that any nonisomorphism $g\colon S_- \to S_-$ factors through $S^3 \to S_-$. Hence it follows from the equivalence $\operatorname{Ref} SG \to \operatorname{Ref} R$ that any nonisomorphism $h\colon \omega \to \omega$ factors through $R^3 \to \omega$. Since $D\operatorname{Tr}_L \omega \simeq L^G$, we get that $0 \to L^G \to R^3 \to \omega \to 0$ is almost split [8]. Then the dual sequence $0 \to L^G \to R^3 \to \omega \to 0$ is also almost split. It is now a direct consequence of Theorem 1.1 in Part I that R, ω, L^G are the only indecomposables in CM R.

5. A Consequence of Finite CM Type

Let the assumption on S and G be as in the previous section, and assume in addition that R is an isolated singularity and $\dim R = 3$. We shall derive a consequence of R being of finite CM type by considering the torsionless SG-modules. They have the following relationship with CM R-modules.

Lemma 5.1. *Let $\mathcal{T}$ be the category of torsionless SG-modules T such that* $\operatorname{Ext}^1_{SG}(T, SG) = 0$.

(a) *$\Omega\colon \underline{\mathcal{T}} \to \underline{\operatorname{Ref} SG}$ is an equivalence of categories.*

(b) *If T is in $\mathcal{T}$, then $\Omega(T)^G$ is CM if and only if $(T^{**}/T)^G = 0$.*

Proof. Part (a) follows from [6]. Since $\Omega(T)$ is reflexive, we know that $\Omega(T)^G$ is CM if and only if $\operatorname{Ext}^1_{SG}((\Omega T)^*, SG)^G = 0$. We know that $T^{**}/T \simeq \operatorname{Ext}^2_{SG}(\operatorname{Tr} T, SG)$ (see [6]). Since $\operatorname{Ext}^1_{SG}(T, SG) = 0$, the exact sequence $0 \to \Omega(T) \to P \to T \to 0$ with P projective gives an exact sequence $0 \to T^* \to P^* \to \Omega(T)^* \to 0$, so that $\Omega(T)^* = \Omega(\operatorname{Tr} T)$. Hence $T^{**}/T \simeq \operatorname{Ext}^1_{SG}((\Omega T)^*, SG)$, so that (b) follows.

Since the fixed point functor $\operatorname{Ref} SG \to \operatorname{Ref} R$ is an equivalence, it follows that there is an infinite number of indecomposable CM R-modules if there is an infinite number of indecomposable torsionless T such that $\operatorname{Ex}^1_{SG}(T, SG) = 0$ and $(T^{**}/T)^G = 0$. The following construction of new modules having this property, starting with a reflexive module, will be useful.

Lemma 5.2. *Let B be an indecomposable reflexive SG-module with* $\operatorname{Ext}^1_{SG}(B, SG) = 0$. *If T is a submodule of B with $(B/T)^G = 0$, then T is torsionless,* $\operatorname{Ext}^1_{SG}(T, SG) = 0$, *and* $(T^{**}/T)^G = 0$.

Proof. Since $(B/T)^G = 0$ and S is an isolated singularity, B/T has finite length. The exact sequence $0 \to T \to B \to B/T \to 0$ gives rise to the exact sequence $\operatorname{Ext}^1_{SG}(B, SG) \to \operatorname{Ext}^1_{SG}(T, SG) \to \operatorname{Ext}^2_{SG}(B/T, SG)$. We have that

$\operatorname{Ext}^2_{SG}(B/T, SG)=0$ since B/T has finite length and $\dim S=3$, and by assumption $\operatorname{Ext}^1_{SG}(B, SG)=0$. Hence we get $\operatorname{Ext}^1_{SG}(T, SG)=0$. That B/T has finite length also implies $T^{**}=B$, so that $(T^{**}/T)^G=0$.

To find the reflexive modules B with the desired property we shall need the following.

LEMMA 5.3. *Let A be a reflexive SG-module. Then* $\operatorname{Ext}^1_{SG}(A, SG)=0$ *if and only if A^* is a CM SG-module.*

Proof. Choose $x\neq 0$ in R. The exact sequence of SG-modules $0\to SG\xrightarrow{\cdot x} SG\to (S/xS)\,G\to 0$ gives the exact sequence $0\to \operatorname{Hom}_{SG}(A, SG)\xrightarrow{\cdot x} \operatorname{Hom}_{SG}(A, SG)\to \operatorname{Hom}_{SG}(A, (S/xS)\,G)\to \operatorname{Ext}^1_{SG}(A, SG)\to\cdots$. Since $(S/xS)G$ has an R-sequence of length 2, $\operatorname{Hom}_{SG}(A, (S/xS)\,G)$ also does. Hence $A^*=\operatorname{Hom}_{SG}(A, SG)$ is CM if $\operatorname{Ext}^1_{SG}(A, SG)=0$. Since $\operatorname{Ext}^1_{SG}(A, SG)$ has finite length because S is an isolated singularity, A^* is not Cohen–Macaulay if $\operatorname{Ext}^1_{SG}(A, SG)\neq 0$.

We are now ready to prove the following consequence of finite CM type.

THEOREM 5.4. *In addition to the previous assumptions, assume that $G=Z_2$ and R is of finite CM type. Let A be an indecomposable reflexive SG-module where $G=Z_2$ such that A^* is Cohen–Macaulay. Then* $\dim_k A/(\mathfrak{r}A+\mathfrak{m}A)\leqslant 2$, *and if* $\dim_k A/(\mathfrak{r}A+\mathfrak{m}A)=2$, *then* $A/(\mathfrak{r}A+\mathfrak{m}A)\simeq k\amalg k_-$. *Here* $\mathfrak{m}=\operatorname{rad} S$ and $\mathfrak{r}=\operatorname{rad}\operatorname{End}_{SG}A$.

Proof. Assume first that k_- occurs with multiplicity $m>1$ in $A/(\mathfrak{r}A+\mathfrak{m}A)$. Then we have an infinite family $\{A_i\}$ of SG-submodules of A containing $\mathfrak{r}A+\mathfrak{m}A$. We want to show that if $A_i\neq A_j$, then $A_i\not\simeq A_j$. For assume $f\colon A_i\to A_j$ is an isomorphism. Consider as in Lemma 3.5 the diagram

$$\begin{array}{ccc}
A_i & \xrightarrow[\sim]{f} & A_j\\
\cap & & \cap\\
A_i^{**} & \xrightarrow[\sim]{f^{**}} & A_j^{**}\\
\uparrow\wr & & \uparrow\wr\\
A & \xrightarrow[\sim]{g} & A.
\end{array}$$

$\operatorname{End}_{SG}A$ is local and $\operatorname{End}_{SG}(A)/\mathfrak{r}=k$. We write $g=z+h$ with $z\in k$ and $h\in\mathfrak{r}$. Then we get $A_j=g(A_i)\subset zA_i+h(A_i)\subset A_i$, so that $A_i=A_j$. Hence an infinite number of different A_i gives rise to an infinite number of non-isomorphic reflexives ΩA_i, since $\operatorname{Ext}^1_{AG}(A_i, SG)=0$. Since we know that the

$(\Omega A_i)^G$ are nonisomorphic indecomposable Cohen–Macaulay R-modules, this shows that $m \leqslant 1$.

By considering $A_- = A \otimes_k k_-$, we see that also k occurs with multiplicity at most one in $A/(mA + \mathfrak{r}A)$. This finishes the proof.

As an application of Theorem 5.4 we give the following.

PROPOSITION 5.5. *Let S be the scroll of type* (2, 1), *and let $G = Z_2$ act on $\mathfrak{m}/\mathfrak{m}^2$ by sending each element to its negative. Then $R = S^G$ is of infinite CM type.*

Proof. We recall that $S = k[[X_0, X_1, X_2, Y_0, Y_1]]/I$, where $I = (X_0X_2 - X_1^2, X_0Y_1 - X_1Y_0, X_1Y_1 - X_2Y_0)$. Since the action on S is induced by a free action of G on $U = k[[X_0, X_1, X_2, Y_0, Y_1]]$, nonmaximal primes in U^G are unramified in U, and hence height one primes in $R = S^G$ are unramified in S. Consider the S-module $B = (x_0, x_1, y_0)$, which is also an SG-module, and it is reflexive. $\mathrm{Hom}_{SG}(B, SG) \simeq \mathrm{Hom}_S(B, S) = A = (x_0, x_1)$ is a CM SG-module. We also know that $\underline{\mathrm{End}}(B) = k$, so that $\mathfrak{r}B \subset \mathfrak{m}B$. Hence $\dim_k B/(\mathfrak{r}B + \mathfrak{m}B) = 3$, which shows that R is of infinite CM type.

Theorem 5.4 can also be used to show that $R = \mathbb{C}[[X, Y, Z, T]]/(X^{2m} + Y^2 + Z^2 + T^2)^G$ is of finite CM type, when the action of G is induced by sending each variable to its negative. If m is odd, we can choose the CM ideal $B = (X^m + iT, Y + iZ)$, which is then clearly an SG-module. If m is odd, one can use [18] to find a module B of rank 2 which can be used. We point out, however, that it is already known from [17] that R is of infinite CM type, since it is easy to see that R is a Gorenstein ring which is not an hypersurface.

We end this section with another sufficient condition for infinite CM type, based on using almost split sequences, similar to an argument of Bongartz for finite dimensional algebras.

PROPOSITION 5.6. *Let R be a complete integrally closed local CM domain, with $R/\mathrm{rad}\, R = k$ algebraically closed, which is an isolated singularity. Assume that there is some almost split sequence $0 \to A \to B \amalg B \amalg X \to C \to 0$, where B is indecomposable in CM R.*

(a) *If $rkB < rkC$ and B is not injective, then R is of infinite CM type.*

(b) *If $rkA > rkB$ and B is not projective, then R is of infinite CM type.*

Proof. (a) If B is not injective, we have an almost split sequence $0 \to B \to C \amalg C \amalg Y \to \tau^{-1}B \to 0$, and if $rkB < rkC$, then $rkC < rk(\tau^{-1}B)$. Since $rkC > rkB \geqslant 1$, C is not isomorphic to the dualizing module ω. Hence we have an almost split sequence $0 \to C \to \tau^{-1}B \amalg \tau^{-1}B \amalg Z \to \tau^{-1}C \to 0$. Continuing this way, we get indecomposable CM modules of arbitrarily large rank, and we are done. The proof of (b) is analogous.

We note that in both examples of rings R of finite CM type given in this paper, we have almost split sequences with a repeated term in the middle. For the scroll of type (2, 1) we have, in the notation of Part I, almost split sequences $0 \to R \to A \amalg A \amalg B \to K \to 0$ and $0 \to K \to R \amalg R \amalg C \to A \to 0$. But in the first case A is injective and $rkR = rkA$, and in the second case R is projective and $rkR = rkA$.

For the fixed ring in Section 4 we have the almost split sequences $0 \to L^G \to R^3 \to \omega \to 0$ and $0 \to R \to \omega^3 \to L^G \to 0$. In the first case R is projective, in the second case ω is injective, and $rkR = rk\omega$.

We illustrate Proposition 5.6 by sketching a different proof for the fact mentioned above that $R = \mathbb{C}[[X, Y, Z, T]]/(X^2 + Y^2 + Z^2 + T^2)^{Z_2}$ is of infinite CM type. R can be shown to be isomorphic to the subring $\mathbb{C}[[x^2s^2, xys^2, y^2s^2, x^2st, xyst, y^2st, x^2t^2, xyt^2, y^2t^2]]$ of $\mathbb{C}[[x, y, s, t]]$. Here there is an almost split sequence $0 \to E \to K \amalg K \amalg X \to \operatorname{Tr} E \to 0$, where $E = (xs, xt, ys, yt)$, $K = (xs^2, xst, xt^2, ys^2, yst, yt^2)$, and $rk \operatorname{Tr} E = 3$. Since K is not injective, Proposition 5.6 applies.

References

1. M. Auslander, Purity of branch locus, *Amer. J. Math.* (Jan. 1962), 116–126.
2. M. Auslander, Functors and morphisms determined by objects. Applications of morphisms determined by objects, *in* "Proceedings, Conference on Representation Theory, Philadelphia, 1976," pp. 1–244, 245–327, Dekker, New York, 1976.
3. M. Auslander, Rational singularities and almost split sequences, *Trans. Amer. Math. Soc.* **293**, No. 2 (1986), 511–532.
4. M. Auslander, Isolated singularities and existence of almost split sequences, *in* "Representation Theory II. Groups and Orders" (V. Dlab *et al.*, Eds.), pp. 194–242, Lecture Notes in Mathematics, Vol. 1178, Springer-Verlag, New York/Berlin, 1986.
5. M. Auslander, Almost split sequences and algebraic geometry, *in* "Proceedings Durham Conference on Representation Theory," London Math. Soc. Lecture Note Series 116 (1986).
6. M. Auslander and M. Bridger, "Stable Module Theory," Memoirs Amer. Math. Soc. No. 94, Amer. Math. Soc., Providence, RI, 1969.
7. M. Auslander and I. Reiten, Representation theory of artin algebras. III. Almost split sequences, *Comm. Algebra* **3** (1975), 239–294.
8. M. Auslander and I. Reiten, Representation theory of artin algebras. IV. Invariants given by almost split sequences, *Comm. Algebra* **5** (1977), 519–554.
9. M. Auslander and I. Reiten, Grothendieck groups of algebras and orders, *J. Pure Appl. Algebra* **39** (1986), 1–51.
10. M. Auslander and I. Reiten, McKay quivers and extended Dynkin diagrams, *Trans. Amer. Math. Soc.* **293**, No. 1 (1986), 293–301.
11. M. Auslander and I. Reiten, Almost split sequences in dimension two, *Adv. in Math.* **66** (1987), 88–118.
12. M. Auslander and S. O. Smalø, Lattices over orders: Finitely presented functors and preprojective partitions, *Trans. Amer. Math. Soc.* **273** (1982), 433–445.
13. R. Buchweitz, G. Greuel, and F. Schreyer, Cohen Macaulay modules on hypersurface singularities, II, *Invent. Math.* **88** (1987), 165–182.

14. D. EISENBUD AND J. HARRIS, On varieties of minimal degree, Symposia in Pure and Applied Math.
15. H. ESNAULT, Reflexive modules on quotient surface singularities, *J. Reine Angew. Math.* **362** (1985), 63–71.
16. P. GABRIEL, Unzerlegbare Darstellungen, I, *Manuscripta Math.* **6** (1972), 71–103.
17. J. HERZOG, Ringe mit nur endlich vielen Isomorphieklassen von maximalen unzerlegbaren Cohen–Macaulay-Moduln, *Math. Ann.* **233** (1978), 21–34.
18. H. KNÖRRER, Cohen–Macaulay modules on hypersurface singularities, I, *Invent. Math.* **88** (1987), 153–164.
19. I. REITEN, The use of almost split sequences in the representation theory of artin algebras, *in* "Proceedings, ICRA III, Puebla, 1980," pp. 29–104, Lecture Notes in Mathematics, Vol. 944, Springer-Verlag, New York/Berlin, 1982.
20. I. REITEN, Cohen–Macaulay modules over isolated singularities, Seminaire Malliavin, Springer Lecture Notes **1220**, 25–32.
21. A. WIEDEMANN, Orders with loops in their Auslander–Reiten graph, *Comm. Algebra* **9** (1981), 641–656.
22. R. BUCHWEITZ, D. EISENBUD, AND J. HERZOG, Cohen Macaulay modules on quadrics, *in* "Proceedings, Lambrecht Conference on Singularities Representation of Algebras and Vector Bundles," pp. 58–116, Vol. 1273, Springer-Verlag, New York/Berlin.
23. Ø. SOLBERG, Hypersurface singularities of finite Cohen Macaulay type, *Journal of London Math. Soc.*, (to appear).

Printed by Catherine Press, Ltd., Tempelhof 41, B-8000 Brugge, Belgium

Math. Ann. 277, 345–349 (1987)

Mathematische Annalen

Almost Split Sequences for Cohen-Macaulay-Modules

Maurice Auslander[1,*] and Idun Reiten[2]

[1] Department of Mathematics, Brandeis University, Waltham, MA 02254, USA
[2] University of Trondheim, AVH, N-7055 Dragvoll, Norway

Introduction

Throughout this note R is an equidimensional commutative Gorenstein ring and Λ is an R-algebra which is a finitely generated (maximal) Cohen-Macaulay R-module. We denote by $\mathscr{CM}\Lambda$, the category of finitely generated Λ-modules which are (maximal) Cohen-Macaulay R-modules. Recall that an A in $\mathscr{CM}\Lambda$ is $\mathscr{CM}\Lambda$-injective (or injective in $\mathscr{CM}\Lambda$) if $A \cong \operatorname{Hom}_R(P, R)$ where P is a projective Λ^{op}-module. We can now state our main result.

Suppose, in addition to the above hypothesis, that R is a complete local ring. For a nonprojective indecomposable C in $\mathscr{CM}\Lambda$ the following are equivalent:

a) C_p is Λ_p-projective for all nonmaximal prime ideals p of R;

b) there is an almost split sequence $0 \to A \to B \to C \to 0$ in $\mathscr{CM}\Lambda$.

Dually, for a noninjective indecomposable A in $\mathscr{CM}\Lambda$, the following are equivalent:

a′) A_p is injective in $\mathscr{CM}\Lambda_p$ for all nonmaximal prime ideals p of R;

b′) there is an almost split sequence $0 \to A \to B \to C \to 0$ in $\mathscr{CM}\Lambda$.

Hence if $0 \to A \to B \to C \to 0$ is an almost split sequence in $\mathscr{CM}\Lambda$, then A_p is $\mathscr{CM}(\Lambda_p)$-injective and C_p is Λ_p-projective for all nonmaximal prime ideals p of R.

Suppose now that Λ is an order in the sense of [1]. This means that if X is in $\mathscr{CM}\Lambda$, then X_p is Λ_p-projective for all nonmaximal prime ideals p of Λ if and only if X_p is $\mathscr{CM}\Lambda_p$-injective for all nonmaximal prime ideals p of R. It then follows from the results of this paper that not only does the subcategory of $\mathscr{CM}\Lambda$ consisting of the X such that X_p is Λ_p-projective for all nonmaximal prime ideals p of R have almost split sequences as was first shown in [1], but that these sequences are also almost split sequences in all of $\mathscr{CM}\Lambda$. F. Schreyer first observed this phenomenon when Λ is a certain type of hypersurface singularity. As a consequence of the main theorem we recover the characterization given in [2] of when Λ is an isolated singularity, that is, $\operatorname{gl.dim}\Lambda_p = \dim R_p$ for each nonmaximal prime ideal p of R. Namely Λ is an isolated singularity if and only if $\mathscr{CM}\Lambda$ has almost split sequences.

* Written with partial support of NSF contract No. NCS83-03348

The method of proof in this paper follows closely that of [1] where it was assumed that Λ is an order. The same results about almost split sequences were proven in [4] when R is two-dimensional and Λ is a commutative local ring using a very different approach.

1

Throughout this section R is an arbitrary equidimensional Gorenstein ring of dimension $d \geqq 0$. This section is devoted to proving a generalization of [1, Chap. II, Proposition 8.8] which will be used in the next section to obtain our results concerning almost split sequences under the additional assumption that R is a complete local ring. Before stating and proving the main result of this section, it is convenient to recall some definitions.

We denote by $0 \to R \to I_0 \to \dots \to I_d \to 0$ a minimal injective R-resolution of R. Since R is an equidimensional Gorenstein ring of dimension d, we have that $\mathrm{Hom}_R(A, I_j) = 0$ for all R-modules A of finite length and $0 \leqq j < d$. For each Λ-module X we choose a fixed projective resolution

$$\dots P_k(X) \xrightarrow{d_k} P_1(X) \xrightarrow{d_1} P_0(X) \to X \to 0$$

which we assume to be minimal if R is a complete local ring. We define $\Omega^i X$ to be $\mathrm{Im}\, d_i$ for $i \geqq 1$. Also we define the Λ^{op}-module $\mathrm{Tr}\, X$, the transpose of X, by $\mathrm{Tr}\, X = \mathrm{Coker}(\mathrm{Hom}_\Lambda(P_0(X), \Lambda) \to \mathrm{Hom}_\Lambda(P_1(X), \Lambda))$. Finally, if X and C are Λ-modules, we define $\underline{\mathrm{Hom}}_\Lambda(X, C)$ to be $\mathrm{Coker}(\mathrm{Hom}_\Lambda(X, P_0(C)) \to \mathrm{Hom}_\Lambda(X, C))$.

Proposition 1.1. *Let X be a Λ-module such that X_p is Λ_p-projective for all nonmaximal prime ideals p of R. Then*

a) $\mathrm{Hom}_R(\Omega^d \mathrm{Tr}\, X, R)$ *is in* $\mathscr{CM}\Lambda$ *and* $\mathrm{Hom}(\Omega^d \mathrm{Tr}\, X, R)_p$ *is* $\mathscr{CM}\Lambda_p$*-injective for all nonmaximal prime ideals p of R;*

b) *for each C in $\mathscr{CM}\Lambda$, we have isomorphisms*

$$\mathrm{Ext}^1_\Lambda(C, \mathrm{Hom}_R(\Omega^d \mathrm{Tr}\, X, R) \xrightarrow{\sim} \mathrm{Hom}_R(\underline{\mathrm{Hom}}_\Lambda(X, C), I_d)$$

functorial in C.

Proof. Since C is in $\mathscr{CM}\Lambda$, we have that the exact sequence $0 \to R \to I_0 \to \dots \to I_d \to 0$ gives rise to the exact sequence $0 \to \mathrm{Hom}_R(C, R) \to \mathrm{Hom}_R(C, I_0) \to \dots \to \mathrm{Hom}_R(C, I_d) \to 0$. The fact that each I_j is injective implies that $\mathrm{Ext}^i_\Lambda(\mathrm{Tr}\, X, \mathrm{Hom}_R(C, I_j)) \simeq \mathrm{Hom}_R(\mathrm{Tor}^\Lambda_i(\mathrm{Tr}\, X, C), I_j)$ for all $i \geqq 0$ [5, Chap. VI, Proposition 5.1]. Since X_p is Λ_p-projective for all nonmaximal prime ideals p of R, we have that $\mathrm{Tor}^\Lambda_i(\mathrm{Tr}\, X, C)$ is of finite R-length for all $i \leqq 1$. Therefore for $1 \leqq j < d$ we have that $\mathrm{Ext}^i_\Lambda(\mathrm{Tr}\, X, \mathrm{Hom}_R(C, I_j)) = 0$ for all $i \geqq 1$, and consequently

$$\begin{aligned} \mathrm{Ext}^1_\Lambda(\mathrm{Tr}\, X, \mathrm{Hom}_R(C, I_d)) &\simeq \mathrm{Ext}^{d+1}_\Lambda(\mathrm{Tr}\, X, \mathrm{Hom}_R(C, R)) \\ &\simeq \mathrm{Ext}^1_\Lambda(\Omega^d \mathrm{Tr}\, X, \mathrm{Hom}_R(C, R)). \end{aligned}$$

Since Λ is in $\mathscr{CM}\Lambda$ we have that $\Omega^d \mathrm{Tr}\, X$ is in $\mathscr{CM}\Lambda$. Therefore the duality $\mathrm{Hom}(\ , R): \mathscr{CM}\Lambda^{\mathrm{op}} \to \mathscr{CM}\Lambda$ induces an isomorphism

$$\mathrm{Ext}^1_\Lambda(\Omega^d \mathrm{Tr}\, X, \mathrm{Hom}_R(C, R)) \xrightarrow{\sim} \mathrm{Ext}^1_\Lambda(C, \mathrm{Hom}_R(\Omega^d \mathrm{Tr}\, X, R)).$$

Combining these functorial isomorphisms with the isomorphism

$$\mathrm{Ext}^1_\Lambda(\mathrm{Tr}\,X, \mathrm{Hom}_R(C, I_d)) \cong \mathrm{Hom}_R(\mathrm{Tor}^\Lambda_1(\mathrm{Tr}\,X, C), I_d)$$

we obtain our desired isomorphism

$$\mathrm{Ext}^1_\Lambda(C, \mathrm{Hom}_R(\Omega^d\,\mathrm{Tr}\,X, R)) \xrightarrow{\sim} \mathrm{Hom}_R(\underline{\mathrm{Hom}}_\Lambda(X, C), I_d)$$

using the functorial isomorphism $\mathrm{Tor}^\Lambda_1(\mathrm{Tr}\,X, C) \cong \underline{\mathrm{Hom}}_\Lambda(X, C)$ [3]. The fact that $\mathrm{Tr}\,X_p$ is Λ_p projective for all nonmaximal prime ideals p of R immediately implies that $\mathrm{Hom}_R(\Omega^d\,\mathrm{Tr}\,X, R)_p$ is $\mathscr{CM}\Lambda_p$-injective for all nonmaximal prime ideals p of R, completing the proof of the proposition.

Next we recall that there is a duality $\mathrm{Tr}: \underline{\mathrm{mod}}\,\Lambda \to \underline{\mathrm{mod}}\,\Lambda^{\mathrm{op}}$ given by $X \to \mathrm{Tr}\,X$ for all X in $\mathrm{mod}\,\Lambda$ where $\underline{\mathrm{mod}}\,\Lambda$ denotes the category of finitely generated Λ-modules modulo projectives. Also, there is a functor $\Omega: \underline{\mathrm{mod}}\,\Lambda \to \underline{\mathrm{mod}}\,\Lambda$ given by $X \to \Omega X$. Now let $L_p(\Lambda)$ denote the subcategory of $\mathscr{CM}\Lambda$ consisting of all X such that X_p is Λ_p-projective for all nonmaximal prime ideals p of R. Also define $J(\Lambda)$ to be the subcategory of $\mathrm{mod}\,\Lambda$ consisting of all Λ-modules A such that a) A_p is Λ_p-projective for all nonmaximal prime ideals p of R and b) $\mathrm{Ext}^i_\Lambda(A, \Lambda) = 0$ for $1 \leq i \leq d$. Then an argument identical to that given in [1, Chap. I, Theorems 7.8 and 7.9] shows the following

Proposition 1.2. *The functor* $\Omega: \underline{\mathrm{mod}}\,\Lambda \to \underline{\mathrm{mod}}\,\Lambda$ *induces an equivalence* $\Omega: \underline{J}(\Lambda^{\mathrm{op}}) \to \underline{L}_p(\Lambda^{\mathrm{op}})$ *and the duality* $\mathrm{Tr}: \underline{\mathrm{mod}}\,\Lambda \to \underline{\mathrm{mod}}\,\Lambda^{\mathrm{op}}$ *induces a duality* $\mathrm{Tr}: \underline{L}_p(\Lambda) \to \underline{J}(\Lambda^{\mathrm{op}})$.

We denote the duality $\Omega^d\,\mathrm{Tr}: \underline{L}_p(\Lambda) \to \underline{L}_p(\Lambda^{\mathrm{op}})$ by Tr_L. Also the duality $D = \mathrm{Hom}_R(\ , R): \mathscr{CM}\Lambda \to \mathscr{CM}\Lambda^{\mathrm{op}}$ induces a duality $D: L_p(\Lambda) \to L_I(\Lambda)$ where $L_I(\Lambda)$ is the subcategory of $\mathscr{CM}\Lambda$ consisting of all X such that X_p is $\mathscr{CM}\Lambda_p$-injective for all nonmaximal prime ideals p of R. Then we have the equivalence $D\,\mathrm{Tr}_L: \underline{L}_p(\Lambda) \to \overline{L}_I(\Lambda)$ where $\overline{L}_I(\Lambda)$ is the category $L_I(\Lambda)$ modulo morphisms factoring through $\mathscr{CM}\Lambda$-injective modules.

For ease of reference we state the following special case of Proposition 1.1, in terms of the above terminology.

Proposition 1.3. *Let X be in $L_p(\Lambda)$. Then $D\,\mathrm{Tr}_L X$ is in $L_I(\Lambda)$ and we have for each C in $\mathscr{CM}\Lambda$ an isomorphism*

$$\mathrm{Ext}^1_\Lambda(C, D\,\mathrm{Tr}_L X) \xrightarrow{\sim} \mathrm{Hom}_R(\underline{\mathrm{Hom}}_\Lambda(X, C), I_d)$$

functorial in C.

2

Throughout this section we assume R is a complete local ring of dimension d. We now apply the results in Sect. 1 to obtain the following which is the main result of this paper.

Theorem 2.1. *Let C be an indecomposable nonprojective module in $\mathscr{CM}\Lambda$.*

a) *C is in $L_p(\Lambda)$ if and only if there is an almost split sequence $0 \to A \to B \to C \to 0$ in $\mathscr{CM}\Lambda$.*

b) *If* $0 \to A \to B \to C \to 0$ *is an almost split sequence in* $\mathscr{CM}\Lambda$, *then* $A \cong D\operatorname{Tr}_L C$ *and is therefore in* $L_I(\Lambda)$.

Proof. Suppose there is an almost split sequence $0 \to A \to B \to C \to 0$ in $\mathscr{CM}\Lambda$. Then it follows from [2, Lemma 3.2] and the proof of [2, Proposition 3.1] that C is in $L_p(\Lambda)$.

Suppose now that C is in $L_p(\Lambda)$. Since C is nonprojective and indecomposable, $\operatorname{Tr} C$ in $J(\Lambda^{\mathrm{op}})$ is indecomposable. The fact that $\operatorname{Ext}^i_{\Lambda^{\mathrm{op}}}(\operatorname{Tr} C, \Lambda^{\mathrm{op}}) = 0$ for $1 \leqq i \leqq d$ easily shows that $\Omega^d \operatorname{Tr} C = \operatorname{Tr}_L C$ is indecomposable in $L_p(C)$. Hence $D\operatorname{Tr}_L C$ is indecomposable.

Since C is indecomposable nonprojective in $\mathscr{CM}(\Lambda)$, $\underline{\operatorname{End}}(C)$ is a local ring of finite length. Hence there is a nonzero morphism $f: \underline{\operatorname{Hom}}(C, C) \to I_d$, which is zero on the unique maximal ideal rad $\underline{\operatorname{End}}(C)$ of $\underline{\operatorname{End}}(C)$. Let $0 \to D\operatorname{Tr}_L C \to E \to C \to 0$ be an exact sequence whose corresponding R-morphism if f. Using Proposition 1.3 it follows as in [3, Proposition 4.1] that this sequence is almost split. For the convenience of the reader we outline the argument given in [3].

Let X be indecomposable in $\mathscr{CM}(\Lambda)$ and let $h: X \to C$ be a map which is not an isomorphism. Consider the pullback diagram induced by h

$$\begin{array}{ccccccccc} 0 & \to & D\operatorname{Tr}_L C & \to & F & \to & X & \to & 0 \\ & & \big\downarrow \| & & \big\downarrow & & \big\downarrow h & & \\ 0 & \to & D\operatorname{Tr}_L C & \to & E & \to & C & \to & 0. \end{array}$$

By Proposition 1.3 we have the commutative diagram

$$\begin{array}{ccc} \operatorname{Hom}_R(\underline{\operatorname{Hom}}(C, C), \operatorname{Id}) & \to & \operatorname{Hom}_R(\underline{\operatorname{Hom}}(C, X), \operatorname{Id}) \\ \big| \wr & & \big| \wr \\ \operatorname{Ext}^1_\Lambda(C, D\operatorname{Tr}_L C) & \to & \operatorname{Ext}^1(X, D\operatorname{Tr}_L C). \end{array}$$

Since the induced map $\underline{\operatorname{Hom}}(C, X) \to \underline{\operatorname{Hom}}(C, C)$ has its image in the unique maximal ideal of $\underline{\operatorname{Hom}}(C, C)$ on which $f: \underline{\operatorname{Hom}}(C, C) \to \operatorname{Id}$ vanishes it follows that $0 \to D\operatorname{Tr}_L C \to F \to X \to 0$ splits. This shows that $h: X \to C$ can be lifted to E, so that $0 \to D\operatorname{Tr}_L C \to E \to C \to 0$ is almost split.

We state without proof the dual version of Theorem 2.1.

Theorem 2.2. *Let* A *be an indecomposable module in* $\mathscr{CM}\Lambda$ *which is not* $\mathscr{CM}(\Lambda)$-*injective.*

a) A *is in* $L_I(\Lambda)$ *if and only if there is an almost split sequence* $0 \to A \to B \to C \to 0$ *in* $\mathscr{CM}\Lambda$.

b) *If* $0 \to A \to B \to C \to 0$ *is an almost split sequence, then* $C \cong \operatorname{Tr}_L D(A)$ *and is therefore in* $L_p(\Lambda)$.

As our final result we point out the following consequence of Theorem 2.1.

Proposition 2.3. *Let* $\alpha: \mathscr{CM}\Lambda \to \mathscr{CM}\Lambda$ *be an equivalence of categories. Then* X *in* $\mathscr{CM}\Lambda$ *is in* $L_p(\Lambda)$ *if and only if* $\alpha(X)$ *is in* $L_p(\Lambda)$.

Proof. We first show that an indecomposable module C in $\mathscr{CM}\Lambda$ is in $L_p(\Lambda)$ if and only if there is a right almost split morphism $B \to C$ in $\mathscr{CM}\Lambda$. If C is not projective our desired result is a trivial consequence of Theorem 2.1. If C is projective, then C

is in $L_p(\Lambda)$. But since $\mathscr{C}\mathscr{M}\Lambda$ is functorially finite in $\operatorname{mod}\Lambda$ [2, Proposition 6.6], there is a morphism $B \xrightarrow{f} C$ in $\mathscr{C}\mathscr{M}\Lambda$ such that $\operatorname{Im} f = \underline{r}C$ where $\underline{r} = \operatorname{radical}\Lambda$ and such that if $g: Y \to C$ is a morphism in $\mathscr{C}\mathscr{M}\Lambda$ with $\operatorname{Im} g \subset \underline{r}C$, then g can be lifted to B. It then follows that $f: B \to C$ is right almost split since C is an indecomposable projective Λ-module.

On the other hand, since $\alpha: \mathscr{C}\mathscr{M}\Lambda \to \mathscr{C}\mathscr{M}\Lambda$ is an equivalence of categories it is clear that a C in $\mathscr{C}\mathscr{M}\Lambda$ has a right almost split morphism $B \to C$ in $\mathscr{C}\mathscr{M}\Lambda$ if and only if $\alpha(C)$ has a right almost split morphism $B' \to \alpha(C)$. Thus we have our desired result.

References

1. Auslander, M.: Functors and morphisms determined by objects. Proc. Conf. on Representation Theory, pp. 1–244 (Philadelphia 1976). New York: Dekker 1978
2. Auslander, M.: Isolated singularities and almost split sequences, Proc. ICRA IV. Lect. Notes Math. 1178, 194–241. Berlin, Heidelberg, New York: Springer 1986
3. Auslander, M., Reiten, I.: Representation theory of artin algebras. III. Almost split sequences. Commun. Algebra **3**, 239–294 (1975)
4. Auslander, M., Reiten, I.: Almost split sequences in dimension two. Adv. Math. (to appear)
5. Cartan, H., Eilenberg, S.: Homological algebra. Princeton: Princeton University Press 1956

Received July 2, 1986; in revised form January 1, 1987

Proceedings of the International Congress of Mathematicians
Berkeley, California, USA, 1986

The What, Where, and Why of Almost Split Sequences

MAURICE AUSLANDER

Throughout this paper we assume that R is a noetherian, local Gorenstein ring. We also assume that all R-algebras Λ are finitely generated R-modules. An R-module Λ is said to be complete if R is complete and is said to be Cohen-Macaulay if Λ is a (maximal) Cohen-Macaulay R-module. Recall that if Λ is a complete R-algebra, then $\operatorname{mod}\Lambda$, the category of finitely generated Λ-modules, is a Krull-Schmidt category; i.e., an M in $\operatorname{mod}\Lambda$ is indecomposable if and only if $\operatorname{End}_\Lambda(M)$, the endomorphism ring of M, is a local ring and thus the decomposition of an X in $\operatorname{mod}\Lambda$ into a sum (direct) of indecomposables is unique up to isomorphism. When Λ is a Cohen-Macaulay R-algebra, we denote by $\operatorname{CM}\Lambda$ the full subcategory of $\operatorname{mod}\Lambda$ consisting of the X is $\operatorname{mod}\Lambda$ which are (maximal) Cohen-Macaulay modules when viewed as R-modules. Further, if Λ is a Cohen-Macaulay R-algebra we have the duality $D\colon \operatorname{CM}\Lambda \to \operatorname{CM}\Lambda^{\text{op}}$ given by $D(X) = \operatorname{Hom}_R(X, R)$ for all X in $\operatorname{CM}\Lambda$. Moreover, this duality is exact in the sense that if $0 \to A \to B \to C \to 0$ is exact in $\operatorname{CM}\Lambda$, i.e., it is an exact sequence of Λ-modules with A, B, C in $\operatorname{CM}\Lambda$, then $0 \to D(C) \to D(B) \to D(A) \to 0$ is exact in $\operatorname{CM}\Lambda^{\text{op}}$. We say that A in $\operatorname{CM}\Lambda$ is $\operatorname{CM}\Lambda$-injective if every exact sequence $0 \to A \to B \to C \to 0$ in $\operatorname{CM}\Lambda$ splits. From the exact duality $D\colon \operatorname{CM}\Lambda \to \operatorname{CM}\Lambda^{\text{op}}$ it follows that A is $\operatorname{CM}\Lambda$-injective if and only if $A \cong D(P)$ for some projective Λ^{op}-module.

We will be mainly interested in questions concerning almost split sequences in $\operatorname{CM}\Lambda$ when Λ is a complete Cohen-Macaulay R-algebra. Recall that an almost split sequence in $\operatorname{CM}\Lambda$ is a nonsplit exact sequence $0 \to A \xrightarrow{f} B \xrightarrow{g} C \to 0$ in $\operatorname{CM}\Lambda$ having the following additional properties: (a) A and C are indecomposable; (b) if $h\colon X \to C$ is a nonisomorphism in $\operatorname{CM}\Lambda$ with X indecomposable, then h can be lifted to B; i.e., there is a $t\colon X \to B$ such that $gt = h$; (c) each nonisomorphism $j\colon A \to Y$ in $\operatorname{CM}\Lambda$ with Y indecomposable can be extended to B; i.e., there is an $s\colon B \to Y$ such that $sf = j$. It should be observed that (a) and (b) is equivalent to (a) and (c) and that two almost split sequences

Written with partial support from National Science Foundation Grant MCS 83-03348.

$0 \to A \to B \to C \to 0$ and $0 \to A' \to B' \to C' \to 0$ in $\mathrm{CM}\,\Lambda$ are isomorphic if and only if $A \approx A'$ if and only if $C \cong C'$.

Before stating our main existence theorem for almost split sequences in $\mathrm{CM}\,\Lambda$, it is convenient to make the following definition. We say that a Λ-module X has a certain property outside the maximal ideal if the $\Lambda_{\mathfrak{p}}$-module $X_{\mathfrak{p}}$ has the property for all nonmaximal ideals $\mathfrak{p}$ of R.

THEOREM 1 (AUSLANDER-REITEN [**8**]). *Suppose Λ is a complete Cohen-Macaulay R-algebra.*

(a) *An indecomposable nonprojective C in $\mathrm{CM}\,\Lambda$ has the property that it is projective outside the maximal ideal if and only if there is an almost split sequence $0 \to A \to B \to C \to 0$ in $\mathrm{CM}\,\Lambda$.*

(b) *An indecomposable A in $\mathrm{CM}\,\Lambda$ which is not $\mathrm{CM}\,\Lambda$-injective is $\mathrm{CM}\,\Lambda$-injective outside the maximal ideal if and only if there is an almost split sequence $0 \to A \to B \to C \to 0$ in $\mathrm{CM}\,\Lambda$.*

Let $P_{\mathfrak{m}}(\mathrm{CM}\,\Lambda)$ and $I_{\mathfrak{m}}(\mathrm{CM}\,\Lambda)$ denote the full subcategories of $\mathrm{CM}\,\Lambda$ consisting of the X which are projective ($\mathrm{CM}\,\Lambda$-injective) outside the maximal ideal and have no projective ($\mathrm{CM}\,\Lambda$-injective) indecomposable summands. Combining Theorem 1 with the fact that either end of an almost split sequence determines the sequence (up to isomorphism), we see that we obtain the following bijection between the isomorphism classes of indecomposable objects in $P_{\mathfrak{m}}(\mathrm{CM}\,\Lambda)$ and $I_{\mathfrak{m}}(\mathrm{CM}\,\Lambda)$. We send $[C]$, the isomorphism class of an indecomposable C in $P_{\mathfrak{m}}(\mathrm{CM}\,\Lambda)$, to the uniquely determined $[A]$ with A indecomposable in $I_{\mathfrak{m}}(\mathrm{CM}\,\Lambda)$ such that there is an almost split sequence $0 \to A \to B \to C \to 0$. We now give a description of this bijection without reference to almost split sequences (Auslander [**4**]).

Let $P_1 \xrightarrow{f} P_0 \to C \to 0$ be a minimal projective presentation of C in $P_{\mathfrak{m}}(\mathrm{CM}\,\Lambda)$. As usual, define $\mathrm{Tr}\,C$, the transpose of C, by the exact sequence $P_0^* \xrightarrow{f^*} P_1^* \to \mathrm{Tr}\,C \to 0$ where $X^* = \mathrm{Hom}_\Lambda(X, \Lambda)$. Then define $\mathrm{Tr}_L\,C = \Omega^d\,\mathrm{Tr}\,C$ where Ω^d is the dth syzygy in a miminal projective resolution of $\mathrm{Tr}\,C$ and dimension R is d. Then $\mathrm{Tr}_L\,C$ is in $P_{\mathfrak{m}}(\mathrm{CM}\,\Lambda^{\mathrm{op}})$ and $D\,\mathrm{Tr}_L\,C$ is in $I_{\mathfrak{m}}(\mathrm{CM}\,\Lambda)$. Moreover, if $0 \to A \to B \to C \to 0$ is an almost split sequence in $\mathrm{CM}\,\Lambda$, then $A \cong D\,\mathrm{Tr}_L\,C$ and $C \cong \mathrm{Tr}_L\,DA$.

As a consequence of Theorem 1, we have the following characterization of when a complete Cohen-Macaulay R-algebra Λ is an isolated singularity; i.e., $\mathrm{gl\,dim}\,\Lambda_{\mathfrak{p}} = \dim R_{\mathfrak{p}}$ for all nonmaximal ideals $\mathfrak{p}$ of R.

THEOREM 2 (AUSLANDER [**5**, **4**]). *The following are equivalent for a complete, Cohen-Macaulay R-algebra Λ.*

(a) *Λ is an isolated singularity.*

(b) $P_{\mathfrak{m}}(\Lambda) = \mathrm{CM}\,\Lambda = I_{\mathfrak{m}}(\Lambda)$.

(c) *For each indecomposable, nonprojective C in $\mathrm{CM}\,\Lambda$ there is an almost split sequence $0 \to A \to B \to C \to 0$ in $\mathrm{CM}\,\Lambda$.*

Recall that a complete, Cohen-Macaulay R-algebra is said to be of finite Cohen-Macaulay type if $\operatorname{CM}\Lambda$ has only a finite number of nonisomorphic indecomposable modules. Now it is not difficult to see that if Λ is a complete Cohen-Macaulay R-algebra of finite Cohen-Macaulay type, then every nonprojective C in $\operatorname{CM}\Lambda$ has an almost split sequence $0 \to A \to B \to C \to 0$ in $\operatorname{CM}\Lambda$. Therefore we have the following result as an immediate consequence of Theorem 2.

PROPOSITION 3 (AUSLANDER [**3**, **5**]). *A complete, Cohen-Macaulay R-algebra is an isolated singularity if it is of finite Cohen-Macaulay type.*

An interesting question is to classify the complete Cohen-Macaulay rings of the form $\Lambda = k[[X_1, \dots, X_d]]/\mathfrak{a}$ which are of finite Cohen-Macaulay type, where $\mathfrak{a}$ is an ideal in the formal power series ring $k[[X_1, \dots, X_d]]$ over an algebraically closed field k. This has been accomplished by Knörrer [**24**] and Buchweitz, Greuel, and Schreyer [**17**], when $\mathfrak{a}$ is a principal ideal and characteristic $k \neq 2$. When $\operatorname{ch} k = 0$, one obtains precisely the Ar'nold simple singularities. When $k = \mathbf{C}$, the complex numbers, and dimension $\Lambda = 2$, then Herzog [**23**] has shown that the quotient singularities with respect to finite subgroups of $\operatorname{GL}(2, \mathbf{C})$ are of finite Cohen-Macaulay type while Artin-Verdier (unpublished), Auslander [**6**], and Ésnault [**19**] have shown the converse. Auslander-Reiten [**12**] have found the only known nonhypersurface singularities of finite Cohen-Macaulay type in dimensions greater than 2, both of which have dimension 3. They are the fixed ring $\mathbf{C}[[X_1, X_2, X_3]]^{Z/2Z}$ where the generator σ of $Z/2Z$ operates by $\sigma(X_i) = -X_i$ for all i, which is the only quotient singularity of dimension greater than 2 of finite Cohen-Macaulay type, and the scroll

$$\mathbf{C}[[X_0, X_1, X_2, Y_0, Y_1]]/(X_0X_2 - X_1^2, X_0X_1 - X_1Y_0, X_1Y_1 - X_2Y_0).$$

In this connection, it should be noted that Artin [**1**] has classified an interesting class of noncommutative complete Cohen-Macaulay R-algebras Λ of finite Cohen-Macaulay type. Namely, when $R = k[[X_1, X_2]]$, $\operatorname{Ch} k = 0$ and Λ is a maximal order in a division ring of finite dimension over the field of quotients of R.

Now when $\dim R = 0$, then a complete, Cohen-Macaulay R-algebra Λ is nothing more than an R-algebra Λ and $\operatorname{mod}\Lambda = \operatorname{CM}\Lambda = P_{\mathfrak{m}}(CM\Lambda) = I_{\mathfrak{m}}(\operatorname{CM}\Lambda)$. For example, when R is a field, one is dealing with the theory of finite-dimensional modules over a finite-dimensional algebra. So under these circumstances Theorem 1 says that every indecomposable nonprojective Λ-module C has an almost split sequence $0 \to A \to B \to C \to 0$ in $\operatorname{mod}\Lambda$. While this now appears as a special case of Theorem 1, it was in the zero-dimensional case that almost split sequences were first introduced by Auslander-Reiten [**13**] and where they have been most extensively used, to very good effect. For instance, the first examples of tiltings [**7**] and coverings [**26**] depended in an essential way on almost split sequences. Since then tiltings and coverings as developed by Brenner-Butler [**16**], Happel-Ringel [**22**], and Bongartz-Gabriel [**15**] respectively have become major tools in studying the representation theory of finite-dimensional algebras. The rest of this paper is devoted to showing how some ideas and results first developed in the zero-dimensional case have appeared in higher dimensions.

From now on we assume that Λ is a complete Cohen-Macaulay R-algebra which is a commutative local ring. Since every commutative complete Cohen-Macaulay local ring S is a complete Cohen-Macaulay R-algebra for some R, we are really just talking about commutative complete Cohen-Macaulay rings S. To emphasize this point, we will denote Λ by S and often drop the R. In addition we assume that S is an isolated singularity which simply means that $S_{\mathfrak{p}}$ is regular for all nonmaximal prime ideals of S. Therefore $\mathrm{CM}\,S = P_{\mathfrak{m}}(\mathrm{CM}\,S) = I_{\mathfrak{m}}(\mathrm{CM}\,S)$ so that there is an almost split sequence $0 \to A \xrightarrow{g} B \xrightarrow{f} C \to 0$ in $\mathrm{CM}\,S$ for each indecomposable $C \not\cong S$ and for each $A \not\cong \mathrm{Hom}_R(S, R)$.

Let C be an indecomposable module in $\mathrm{CM}\,S$. Then a morphism $f\colon B \to C$ in $\mathrm{CM}\,S$ is said to be a minimal right almost split morphism if it satisfies the following: (a) f is not a splittable surjection; (b) if $t\colon X \to C$ in $\mathrm{CM}\,S$ is not a splittable surjection, then it can be lifted to B; (c) if $h\colon B \to B$ is such that $fh = f$, then h is an automorphism. It is easily seen that if $f\colon B \to C$ and $f'\colon B' \to C$ are two minimal right almost split morphisms in $\mathrm{CM}\,S$, then there is an isomorphism $h\colon B \to B'$ such that $f = f'h$, so the minimal right almost split morphisms to C are uniquely determined (up to isomorphism) by C. If $C \not\cong S$ and $0 \to A \xrightarrow{g} B \xrightarrow{f} C \to 0$ is an almost split sequence in $\mathrm{CM}\,S$, then $f\colon B \to C$ is easily seen to be a minimal right almost split morphism. If $C \cong S$, then it can be shown that there is also a minimal right almost split morphism $E \to S$ in $\mathrm{CM}\,S$. So for each indecomposable C in $\mathrm{CM}\,S$ there is a uniquely determined (up to isomorphism) minimal right almost split morphism $f\colon B \to C$. Also we have the dual notion of a minimal left almost split morphism $g\colon A \to B$ in $\mathrm{CM}\,S$ with A indecomposable and the dual result that given any indecomposable A in $\mathrm{CM}\,S$, there is a uniquely determined (up to isomorphism) minimal left almost split morphism $A \xrightarrow{g} B$ which is the left-hand side of the almost split sequence $0 \to A \xrightarrow{g} B \xrightarrow{f} C \to 0$ in $\mathrm{CM}\,S$ when $A \not\cong \mathrm{Hom}_R(S, R)$.

A notion intimately connected to that of minimal right (left) almost split morphisms is that of an irreducible morphism first introduced in the case of Artin algebras by Auslander-Reiten [**14**]. A morphism $f\colon X \to Y$ between indecomposable modules in $\mathrm{CM}\,S$ is said to be irreducible in $\mathrm{CM}\,S$ if (a) f is not an isomorphism and (b) if $f = hg$ in $\mathrm{CM}\,S$, then g is a splittable injection or h is a splittable surjection. Let X and Y be indecomposable and suppose $g\colon X \to B$ and $h\colon B' \to Y$ are minimal left and right almost split morphisms respectively. Then for $f\colon X \to Y$ in $\mathrm{CM}\,S$, the following are equivalent: (a) f is irreducible; (b) there exists splittable injection $s\colon X \to B'$ such that $hs = f$; (c) there exists a splittable surjection $t\colon B \to Y$ such that $tg = f$. Thus for each indecomposable module X in $\mathrm{CM}\,S$, there exist only a finite number of nonisomorphic indecomposable Y in $\mathrm{CM}\,S$ which have either an irreducible morphism $Y \to X$ or an irreducible morphism $X \to Y$.

As our first example of how irreducible morphisms have been used in arbitrary dimensions we state the following criterion for determining when S is of finite Cohen-Macaulay type which is essentially the same as a criterion used extensively in the zero-dimensional case [**4**]. This criterion was used to show that the scroll $k[[X_0, X_1, X_2, Y_0, Y_1]]/(X_0X_2 - X_1^2, X_0X_1 - X_1Y_0, X_1Y_1 - X_2Y_0)$, cited earlier, is of finite Cohen-Macaulay type.

PROPOSITION 4 (AUSLANDER-REITEN [**12**]). *Suppose* C *is a subcategory of* CM S *closed under isomorphisms consisting of a finite number of nonisomorphic indecomposable modules including* S*. Further, assume that an indecomposable* X *in* CM S *is in* C *if there exists an irreducible* $C \to X$ *or an irreducible* $X \to C$ *with* C *in* C*. Then* S *is of finite Cohen-Macaulay type and* C *consists of all the indecomposable modules in* CM S.

Assume now that S is a k-algebra where k is an algebraically closed field and $S/\mathfrak{m} = k$ where $\mathfrak{m}$ is the maximal ideal of S. Then if $U \to Y$ and $X \to V$ are minimal right and left almost split morphisms in CM S with Y and X indecomposable, then the multiplicity that Y appears as a summand of V is the same as the multiplicity X appears as a summand of U. This common number is called the number of independent irreducible maps from X to Y. We are now in position to define the AR-quiver of S. This is a directed graph whose vertices are the isomorphism classes $[Z]$ of indecomposable modules Z in CM S and where the number of arrows from $[X]$ to $[Y]$ is the number of independent irreducible morphisms from X to Y. When S is a Gorenstein ring, the AR-quiver of S minus the vertex $[S]$ and the arrows to it and out of it, is called the stable AR-quiver of S. We then have the following striking result which is the consequence of work of Gonzalez-Springberg and Verdier [**21**], Artin-Verdier [**2**], Ésnault-Knörrer [**20**], Auslander-Reiten [**9**].

THEOREM 5. *Suppose* S *is a rational double point over an algebraically closed field of any characteristic. Then the desingularization graph of* S *is naturally isomorphic to the underlying graph of the stable AR-quiver of* S.

In connection with this theorem it is of interest to point out a different but related interpretation of the AR-quiver of two-dimensional quotient singularities in terms of finite group representations. This requires the notion of the McKay-quiver of a representation of a finite group.

Let k be an algebraically closed field and G a finite group whose order is not divisible by Ch k. Let $V_1, \dots, V_s$ be a complete list of nonisomorphic simple kG-modules. Let V be an arbitrary (finite-dimensional) $k[G]$-module. Then the McKay-quiver of V is defined as follows. The V_i are the vertices and one draws n arrows from V_i to V_j if V_i occurs in $V \otimes_k V_j$ with multiplicity n.

Suppose now that V is a faithful two-dimensional kG-module. Associated with this is a linear action of G as a group of k-automorphisms of $k[[X_1, X_2]]$ whose fixed ring we denote by S. Then S is a local complete integrally closed two-dimensional domain and therefore an isolated Cohen-Macaulay singularity.

In addition, assume that the kG-module V has the property that the height one primes of S are unramified in $k[[X_1, X_2]]$ (if $k = \mathbf{C}$, the complex numbers, this means that G has no pseudo-reflections). Then, as noted before, S is of finite Cohen-Macaulay type and we have

PROPOSITION 6 (AUSLANDER [**6**]). *Under the above hypothesis there is a natural isomorphism between the McKay-quiver of V and the AR-quiver of S.*

Since McKay [**25**] had already observed that for G a finite subgroup of $\mathrm{SL}(2, \mathbf{C})$, the McKay-quiver of V, the kG-module given by the inclusion $G \subset \mathrm{SL}(2, \mathbf{C})$, is isomorphic to the disingularization graph of the associated fixed ring, Proposition 6 gave the first connection between almost split sequences and singularity theory.

As our final remark concerning two-dimensional rings, we point out some special features of the structure of almost split sequences for arbitrary S of dimension 2 which played a significant role in the proofs of the above results.

Suppose $\dim S = 2$ and let $\omega = \mathrm{Hom}_R(S, R)$, the dualizing module of S. Then there is a unique (up to isomorphism) exact sequence $0 \to \omega \xrightarrow{g} E \xrightarrow{f} S \to S/\mathfrak{m} \to 0$ with E in $\mathrm{CM}\, S$ called the fundamental sequence of S. It is not difficult to see that $f \colon E \to S$ is the minimal right almost split morphism to S.

PROPOSITION 6 (AUSLANDER [**6**] AND AUSLANDER-REITEN [**10**]). *Suppose $\dim S = 2$ and $0 \to \omega \xrightarrow{g} E \xrightarrow{f} S \to S/\mathfrak{m} \to 0$ is the fundamental sequence and let $M \not\simeq S$ be an indecomposable module in* $\mathrm{CM}\, S$ *and let $M^* = \mathrm{Hom}_S(M, S)$. Then the induced exact sequence $0 \to \mathrm{Hom}_S(M^*, \omega) \to \mathrm{Hom}_S(M^*, E) \to \mathrm{Hom}_S(M^*, S) = M$ has the following properties.*

(a) *$0 \to \mathrm{Hom}_S(M^*, \omega) \to \mathrm{Hom}_S(M^*, E) \to M \to 0$ is exact and is either split or almost split.*

(b) *If* $\mathrm{ch}\, S/\mathfrak{m}$ *does not divide* $\mathrm{rank}\, M$, *then*

$$0 \to \mathrm{Hom}_S(M^*, \omega) \to \mathrm{Hom}_S(M^*, E) \to M \to 0$$

is almost split.

(c) *If $S/\mathfrak{m}$ is algebraically closed, then*

$$0 \to \mathrm{Hom}_S(M^*, \omega) \to \mathrm{Hom}_S(M^*, E) \to M \to 0$$

is almost split if and only if $\mathrm{ch}\, S/\mathfrak{m}$ *does not divide* $\mathrm{rank}\, M$.

We next turn our attention to computing $G_0(S)$, the Grothendieck group of S, which is the free abelian group with basis the isomorphism classes $[X]$ of X in $\mathrm{mod}\, S$ modulo the subgroup generated by $[A] - [B] + C$ whenever there is an exact sequence $0 \to A \to B \to C \to 0$ in $\mathrm{mod}\, S$. Because S is a Cohen-Macaulay ring, it is easily seen that $G_0(S)$ is also the free abelian group with basis $[X]$ with X in $\mathrm{CM}\, S$ modulo the subgroup generated by $[A] - [B] + [C]$ whenever there is an exact sequence in $0 \to A \to B \to C \to 0$ in $\mathrm{CM}\, S$. While the split and almost split sequences in $\mathrm{CM}\, S$ give some of the relations for $G_0(S)$ in general, it is when S is of finite Cohen-Macaulay type that we get a really definitive connection between Grothendieck groups and almost split sequences which is a generalization of Butler's result [**18**] in dimensions at most 1.

PROPOSITION 6 (AUSLANDER-REITEN [**11**]). *Suppose S is of finite Cohen-Macaulay type. Then $G_0(S)$ is the free abelian group with basis the isomorphism classes of X in* $\mathrm{CM}\, S$ *modulo the subgroup generated by $[A]-[B]+[C]$ whenever there is a split or almost split sequence $0 \to A \to B \to C \to 0$.*

As an easy consequence of this result we have

COROLLARY 7. *Let S be of finite Cohen-Macaulay type. Then $G_0(S)$ is completely determined by the AR-quiver of S minus the vertex $[S]$ and all arrows going into and coming out of it.*

We end this discussion of almost split sequences with the following global instance of almost split sequences.

PROPOSITION 8 (AUSLANDER-REITEN [**10**]). *Let X be a connected, Cohen-Macaulay projective curve over an infinite field k. Then an indecomposable Cohen-Macaulay coherent sheaf $\mathcal{H}$ is locally free if and only if there is an almost split sequence $0 \to \mathcal{F} \to \mathcal{G} \to \mathcal{H} \to 0$ in the category of Cohen-Macaulay coherent sheaves.*

An easy consequence of this result is the following which was first shown independently by Schofield [**27**] using sheaf theoretic methods and Auslander-Reiten [**10**] using graded modules.

THEOREM 9. *Let X be a connected nonsingular projective curve over an infinite field. Then the category of coherent sheaves over X has almost split sequences.*

BIBLIOGRAPHY

1. M. Artin, *Maximal orders of global dimension and Krull dimension two*, Invent. Math. (to appear).

2. M. Artin and J.-L. Verdier, *Reflexive modules over rational double points*, Math. Ann. **270** (1985), 71–82.

3. M. Auslander, *Finite type implies isolated singularity* (Proc. Oberwolfach, 1984), Lecture Notes in Math., Vol. 1142, Springer-Verlag, Berlin and New York, 1985, pp. 1–4.

4. ——, *Functors and morphisms determined by objects. Applications of morphisms determined by objects* (Proc. Conf. on Representation Theory, Philadelphia, Pa., 1976), Dekker, New York, 1978, pp. 1–327.

5. ——, *Isolated singularities and almost split sequences*, Representation Theory. II, Groups and Orders (Ottawa, 1984), Lecture Notes in Math., Vol. 1178, Springer-Verlag, Berlin and New York, 1986, pp. 194–242.

6. ——, *Rational singularities and almost split sequences*, Trans. Amer. Math. Soc. **293** (1986), no. 2, 511–532.

7. M. Auslander, M. I. Platzeck, and I. Reiten, *Coxeter functors without diagrams*, Trans. Amer. Math. Soc. **250** (1979), 1–46.

8. M. Auslander and I. Reiten, *Almost split sequences for Cohen-Macaulay modules* (to appear).

9. ——, *Almost split sequences for rational double points*, Trans. Amer. Math. Soc. (to appear).

10. ——, *Almost split sequences in dimension* 2, Adv. in Math. (to appear).

11. ——, *Grothendieck groups of algebras and orders*, J. Pure Appl. Algebra **39** (1986), 1–51.

12. ——, *The Cohen-Macaulay type of Cohen-Macaulay rings* (to appear).

13. ——, *Representation theory of artin algebras.* III, *Almost split sequences*, Comm. Algebra **3** (1975), 239–294.

14. ——, *Representation theory of artin algebras.* IV, *Invariants given by almost split sequences*, Comm. Algebra **5** (1977), 519–554.

15. K. Bongartz and P. Gabriel, *Covering spaces in representation theory*, Invent. Math. **65** (1982), 331-378.

16. S. Brenner and M. C. R. Butler, *Generalizations of the Bernstein-Gelfand-Ponomarev reflection functors*, Representation Theory. II, Lecture Notes in Math., Vol. 832, Springer-Verlag, Berlin and New York, 1980, pp. 103–169.

17. R.-O. Buchweitz, G.-M. Greuel, and F.-O. Schreyer, *Cohen-Macaulay modules on hypersurfaces singularities.* II, Preprint.

18. M. C. R. Butler, *Grothendieck groups and almost split sequences* (Proc. Oberwolfach Conf. on Integral Representations and Applications), Lecture Notes in Math., Vol. 882, Springer-Verlag, Berlin and New York, 1981, pp. 357–368.

19. H. Ésnault, *Reflexive modules on quotient singularities*, J. Reine Angew. Math. (to appear).

20. H. Ésnault and H. Knörrer, *Reflexive modules over rational double points*, Math. Ann. **272** (1985), 545–548.

21. Gonzalez-Springberg and J.-L. Verdier, *Construction geometric de la correspondence de McKay*, Ann. Sci. École Norm. Sup. (4) **16** (1983), 409–449.

22. D. Happel and C. M. Ringel, *Tilted algebras*, Trans. Amer. Math. Soc. **274** (1982), 399–443.

23. J. Herzog, *Ringe mit nur endlich vielen Isomorphieklassen von maximalen, unzerlegbaren Cohen-Macaulay-Moduln*, Math. Ann. **233** (1978), 21–34.

24. H. Knörrer, *Cohen-Macaulay modules on hypersurface singularities.* I, Preprint.

25. J. McKay, *Graphs, singularities and finite groups*, Proc. Sympos. Pure Math., vol. 37, Amer. Math. Soc., Providence, R.I., 1980, pp. 183–186.

26. C. Riedtmann, *Algebren, Dorstellungsköcher, Überlagerungen und Zuräck*, Comment. Math. Helv. **55** (1980), 199–224.

27. A. Schofield, Private communication.

BRANDEIS UNIVERSITY, WALTHAM, MASSACHUSETTS 02254, USA

Cohen-Macaulay Modules for Graded Cohen-Macaulay Rings and their Completions

MAURICE AUSLANDER AND IDUN REITEN

Introduction.

Let $\mathbb{Z}$ denote the integers and $R = \coprod_{i \in \mathbb{Z}} R_i$ a $\mathbb{Z}$-graded commutative ring with $R_0 = k$ a field, $R_i = (0)$ when $i < 0$, and where R is generated as a k-algebra by a finite number of homogenous elements. Denote by $\hat{R}$ the completion at the unique graded maximal ideal $m = \coprod_{i>0} R_i$. When R is Cohen-Macaulay, we shall study the relationship between R and $\hat{R}$, especially with respect to almost split sequences for Cohen-Macaulay modules. Existence theorems for almost split sequences were first proved in the complete case [**1,2,6**], and then in the graded case by imitating the method [**7**]. We here show that there is a more direct connection between the complete case and graded case, by deducing the existence theorem for R from that of $\hat{R}$. An interesting feature of this approach is that it shows at the same time that almost split sequences for R stay almost split under completion.

R is said to be of finite (graded) Cohen-Macaulay type if there is only a finite number of indecomposables up to shift, in the category $\mathrm{CM}(\mathrm{gr}\, R)_0$ of graded Cohen-Macaulay modules with morphisms of degree 0. Similarly $\hat{R}$ is said to be of finite Cohen-Macaulay type if there is only a finite number of indecomposables in the category $\mathrm{CM}(\hat{R})$ of Cohen-Macaulay modules over $\hat{R}$. The known examples indicated that R is of finite Cohen-Macaulay type if and only if $\hat{R}$ is, and that when $\hat{R}$ is of finite Cohen-Macaulay type, then each indecomposable in $\mathrm{CM}(\hat{R})$ is obtained by completion. The motivation for this paper was to show that this is the case, and this is done in Section 1. More generally, we show that if R_p is Gorenstein for each nonmaximal graded prime ideal p in R, then in a connected component of the AR-quiver of $\hat{R}$, if one module is the completion of an R-module, then all are. A corresponding result was proved by Gordon-Green for artin algebras [**15**]. In Section 3 we also show that the AR-quivers for R and $\hat{R}$ are closely related.

A crucial step in our proofs is to investigate the connection between indecomposability for an R-module C and its completion $\hat{C}$, and what it means that $\hat{C} \simeq \hat{C}'$. The results we need are given for artin algebras by Gordon-Green [**14**], and we can use Maranda type reduction to the artin case in the context where we need the result. As suggested to us by M. Van den Bergh, these results are however true more generally, and in Section 2 we give a proof using the method suggested by him.

We would like to thank S. Goto, J. Herzog and M. Van den Bergh for helpful conversations.

1. Finite Cohen-Macaulay type for R and $\hat{R}$.

Throughout this paper $R = \coprod_{i \in \mathbb{Z}} R_i$ is a $\mathbb{Z}$-graded commutative Cohen-Macaulay ring, such that $R_i = (0)$ for $i < 0$, $R_0 = k$ is a field and R is generated as a k-algebra by a finite number of homogenous elements. Then R is noetherian and $m = \coprod_{i>0} R_i$ is the unique graded maximal ideal in R. We denote by $\hat{R}$ the completion of R with respect to the m-adic topology. Then $\hat{R}$ is a complete local Cohen-Macaulay ring with maximal ideal $\hat{m}$. Denote by mod$\hat{R}$ the category of finitely generated $\hat{R}$-modules, by CM($\hat{R}$) the full subcategory whose objects are the Cohen-Macaulay $\hat{R}$-modules and by $L_p(\hat{R})$ the full subcategory of CM($\hat{R}$) whose objects C have the property that C_p is R_p-projective for all nonmaximal prime ideals p in $\hat{R}$. Similarly mod$(\operatorname{gr} R)_0$ denotes the category of finitely generated graded R-modules with degree zero maps, and CM($(\operatorname{gr} R)_0$ and $L_p(\operatorname{gr} R)_0$ subcategories defined similarly to the above. Note that a graded module C has the property that C_p is R_p-projective for every nonmaximal graded prime ideal p if and only if C_p is R_p-projective for every nonmaximal prime ideal p in R. In [**6**] we showed that if C is indecomposable in $L_p(\hat{R})$, there is an almost split sequence $0 \to A \to B \to C \to 0$ in CM($\hat{R}$), and A was expressed in terms of C. Using the same methods, an analogous result was proved for R in [**7**]. In this section we show that the existence theorem in the graded case can be deduced from our results in the complete case. At the same time we prove that the almost split sequences for $\hat{R}$ are obtained by completing those for R. We use this to show that R is of finite Cohen-Macaulay type if and only if $\hat{R}$ is.

We can choose a regular homogenous R-sequence $x_1, \dots, x_d$ in m such that $T = k[x_1, \dots, x_d]$ is a graded polynomial ring with an inclusion map $T \to R$ of degree zero and R is a finitely generated free T-module. We denote the dualizing module $\operatorname{Hom}_T(R, T)$ by ω_R. Then $\omega_{\hat{R}} = \hat{\omega}_R$ is the dualizing module for $\hat{R}$. If C is an R-module, $\hat{C}$ denotes the completion of C. We have the following crucial results, where the assumptions and notation are as above. In addition, for A in $\operatorname{mod}(\operatorname{gr} R)_0$, $A(n)$ is defined by $A(n)_i = A_{n+i}$.

LEMMA 1. *If A is indecomposable in $L_p(\operatorname{gr} R)_0$, then $\hat{A}$ is indecomposable in $L_p(\hat{R})$.*

LEMMA 2. *If A and B are indecomposable in $L_p(\operatorname{gr} R)_0$ and $\hat{A} \simeq \hat{B}$, then there is some n such that $A \simeq B(n)$.*

PROOF: If R is artin, so that $\hat{R} = R$, the lemmas are known by [**14**, Th. 2.4, Th. 3.6]. We can then use a graded version of Maranda type reduction to reduce to the artin case. Namely, we have the following. (See [**13,23**] for details in the complete case.) Since A and B are in $L_p(\operatorname{gr} R)_0$, the annihilator of $\operatorname{Ext}^1_R(A,A) \oplus \operatorname{Ext}^1_R(A,B) \oplus \operatorname{Ext}^1_R(B,B)$ contains a homogenous regular R-sequence $y_1, \ldots, y_d$ such that $A_1 = A/(y_1, \ldots, y_d)A$ and $B_1 = B/(y_1, \ldots, y_d)B$ are indecomposable in $\operatorname{mod}(\operatorname{gr} R)_0$, where $R_1 = R/(y_1, \ldots, y_d)$. And if A_1 and B_1 are isomorphic in $\operatorname{mod}(\operatorname{gr} R)_0$, then A and B are isomorphic in $L_p(\operatorname{gr} R)_0$. Since R_1 is artin, both lemmas follow easily from this.

In Section 2 we prove these lemmas for A and B in $\operatorname{mod}(\operatorname{gr} R)_0$, as suggested by M. Van der Bergh.

We can now prove the following, on the basis of [**6**].

THEOREM 3. *Let C be a nonfree indecomposable in $L_p(\operatorname{gr} R)_0$. Then there exists an almost split sequence $0 \to A \to B \to C \to 0$ in $\operatorname{CM}(\operatorname{gr} R)_0$, and $0 \to \hat{A} \to \hat{B} \to \hat{C} \to 0$ is almost split in $\operatorname{CM}(\hat{R})$.*

PROOF: Since C is in $L_p(\operatorname{gr} R)_0$, the endomorphism ring $\underline{\operatorname{End}}_R(C)$ modulo projectives is of finite length, and hence equal to its completion $\underline{\operatorname{End}}_{\hat{R}}(\hat{C})$. This shows that $\hat{C}$ is in $L_p(\hat{R})$ since $\hat{C}$ is clearly in $\operatorname{CM}(\hat{R})$. Since $\hat{C}$ is indecomposable by Lemma 1 we know that there is an almost split sequence $0 \to D\operatorname{Tr}_L \hat{C} \to E \to \hat{C} \to 0$ in $\operatorname{CM}(\hat{R})$, which is a generator for the simple $\underline{\operatorname{End}}_{\hat{R}}(\hat{C})^{\text{op}}$-socle of $\operatorname{Ext}^1_{\hat{R}}(\hat{C}, D\operatorname{Tr}_L \hat{C})$. (The second statement is a formal consequence of the first one.) We here recall that D denotes the duality $D = \operatorname{Hom}_{\hat{R}}(\ ,\hat{\omega}_R)$ in $\operatorname{CM}(\hat{R})$, and the transpose $\operatorname{Tr}\hat{C}$ is defined as follows. Let $P_1 \to P_0 \to \hat{C} \to 0$ be a minimal projective presentation in $\operatorname{mod}\hat{R}$. Then $\operatorname{Tr}\hat{C}$ is defined by the exact sequence $\operatorname{Hom}_{\hat{R}}(P_0, \hat{R}) \to \operatorname{Hom}_{\hat{R}}(P_1, \hat{R}) \to \operatorname{Tr}\hat{C} \to 0$, and $\operatorname{Tr}_L C$ is defined to be the d-th syzygy module $\Omega^d \operatorname{Tr} C$. The analogous construction can be done in the category $\operatorname{mod}(\operatorname{gr} R)_0$, and we clearly have $\widehat{D \operatorname{Tr}_L C} = D \operatorname{Tr}_L \hat{C}$.

Since C is in $L_p(\operatorname{gr} R)_0$, $\operatorname{Ext}^1_R(C, D \operatorname{Tr}_L C)$ has finite length, and is hence equal to its completion $\operatorname{Ext}^1_{\hat{R}}(\hat{C}, D \operatorname{Tr}_L \hat{C})$. Since we also have $\underline{\operatorname{End}}_R(C) \simeq \underline{\operatorname{End}}_{\hat{R}}(\hat{C})$, $\operatorname{Ext}^1_R(C, D \operatorname{Tr}_L C)$ is a graded $\underline{\operatorname{End}}_R(C)^{\mathrm{op}}$-module of finite length with a simple socle. $\underline{\operatorname{End}}_R(C)^{\mathrm{op}} \simeq \underline{\operatorname{End}}_{\hat{R}}(\hat{C})^{\mathrm{op}}$ is a local ring, and it is easy to see that its radical is $\coprod_{i \neq 0} \underline{\operatorname{End}}_R(C)^{\mathrm{op}} \coprod \operatorname{rad}(\underline{\operatorname{End}}_R(C)^{\mathrm{op}}_0)$. (See [**14**], or Prop. 8.) It then follows that the simple socle of $\operatorname{Ext}^1_R(C, D \operatorname{Tr}_L C)$ is in $\operatorname{Ext}^1_R(C, D \operatorname{Tr}_L C)_i$ for some i. Hence the socle of $\operatorname{Ext}^1_R(C, D \operatorname{Tr}_L C(i))$ is in $\operatorname{Ext}^1_R(C, D \operatorname{Tr}_L C(i))_0$. Let $0 \to D \operatorname{Tr}_L C(i) \to B \to C \to 0$ be a generator for the socle. This is then a nonsplit exact sequence in $\operatorname{CM}(\operatorname{gr} R)_0$, and the completion is almost split in $\operatorname{CM}(\hat{R})$. We want to show that the sequence is almost split in $\operatorname{CM}(\operatorname{gr} R)_0$.

Let $f : X \to C$ be a nonisomorphism in $\operatorname{CM}(\operatorname{gr} R)_0$, with X indecomposable. By considering the exact sequence $0 \to \operatorname{Ker} f \to X \to C \to \operatorname{Coker} f \to 0$, and using that the completion of a nonzero graded module is nonzero, we see that $\hat{f} : \hat{X} \to \hat{C}$ is not an isomorphism. The map $\operatorname{Ext}^1_R(f, \mathrm{id}) : \operatorname{Ext}^1_{\hat{R}}(\hat{C}, D \operatorname{Tr}_L \hat{C}) \to \operatorname{Ext}^1_{\hat{R}}(\hat{X}, D \operatorname{Tr}_L \hat{C})$ sends the almost split sequence $0 \to \widehat{D \operatorname{Tr}_L C(i)} \to \hat{B} \to \hat{C} \to 0$ to 0. $\operatorname{Tr}_L C$ is clearly in $L_p(\operatorname{gr} R)_0$, so that $(D \operatorname{Tr}_L C)_p$ is injective in $\operatorname{CM}(R_p)$ when p is a graded nonmaximal prime ideal in R. Hence $\operatorname{Ext}^1_R(X, D \operatorname{Tr}_L C(i))$ has finite length. The commutative diagram

$$\begin{array}{ccc} \operatorname{Ext}^1_R(C, D \operatorname{Tr}_L C(i)) & \longrightarrow & \operatorname{Ext}^1_R(X, D \operatorname{Tr}_L C(i)) \\ \wr\!\wr & & \wr\!\wr \\ \operatorname{Ext}^1_{\hat{R}}(\hat{C}, D \operatorname{Tr}_L \hat{C}) & \longrightarrow & \operatorname{Ext}^1_{\hat{R}}(\hat{X}, D \operatorname{Tr}_L \hat{C}) \end{array}$$

then shows that $\operatorname{Ext}^1_R(f, \mathrm{id})$ sends $0 \to D \operatorname{Tr}_L C(i) \to B \to C \to 0$ to zero, so that the sequence is almost split.

Remark: Since we know from [**7**, Th. 1.1] that if $0 \to A \to B \to C \to 0$ is almost split in $\operatorname{CM}(\operatorname{gr} R)_0$, then C is in $L_p(\operatorname{gr} R)_0$, we actually have that any almost split sequence in $\operatorname{CM}(\operatorname{gr} R)_0$ remains almost split under completion.

Before we give our application to finite Cohen-Macaulay type, we give a proof of the following $\mathbb{Z}$-graded version of a result in [**2,10**]. (See also [**8**] for a related result.)

Proposition 4. *If R is of finite Cohen-Macaulay type, then R is an isolated singularity.*

PROOF: Assume that R is of finite Cohen-Macaulay type, and let C be indecomposable nonprojective in $\mathrm{CM}(\operatorname{gr} R)_0$. Since for every indecomposable X in $\mathrm{CM}(\operatorname{gr} R)_0$, $\operatorname{Hom}_R(X, C)$ is a finitely generated graded R-module, it is easy to see that we get a right almost split map $g: B \to C$ in $\mathrm{CM}(\operatorname{gr} R)_0$. This means that any $h: Y \to C$ in $\mathrm{CM}(\operatorname{gr} R)_0$, where Y is indecomposable, can be lifted through g. Considering $0 \to \operatorname{Ker} \beta \to B \to C \to 0$ and projecting $\operatorname{Ker} \beta$ onto an appropriate indecomposable summand, we get an almost split sequence $0 \to A \to B' \to C \to 0$. This implies that C is in $L_p(\operatorname{gr} R)_0$ [**7**], so that $\mathrm{CM}(\operatorname{gr} R)_0 = L_p(\operatorname{gr} R)_0$.

Let now C be in $\operatorname{mod}(\operatorname{gr} R)_0$. Since if $d = \dim R$, $\Omega^d C$ is Cohen-Macaulay and hence is in $L_p(\operatorname{gr} R)_0$, we get $pd_{R_p} C_p < \infty$ for any nonmaximal graded prime ideal p in R. This shows that gl. $\dim R_p < \infty$ [**18**, II 8.2], so that R is an isolated singularity.

We shall call the $\hat{R}$-modules of the form $\hat{C}$ gradable $\hat{R}$-modules. We then have the following.

THEOREM 5. *a) R is of finite graded Cohen-Macaulay type if and only if $\hat{R}$ is of finite Cohen-Macaulay type.*

b) If $\hat{R}$ is of finite Cohen-Macaulay type, then every module in $\mathrm{CM}(\hat{R})$ is gradable.

PROOF: a) Assume first that R is of finite Cohen-Macaulay type. By Proposition 4, R is an isolated singularity, hence $\hat{R}$ is the same. Then $\mathrm{CM}(\hat{R})$ has almost split sequences. The set $\mathcal{S}$ of $\hat{R}$-modules of the form $\hat{C}$ where C is indecomposable in $\mathrm{CM}(\operatorname{gr} R)_0$ is a finite set of modules in $\mathrm{CM}(\hat{R})$, which are indecomposable by Lemma 1. For each $\hat{C}$ in $\mathcal{S}$ we have an almost split sequence of the form $0 \to \hat{A} \to \hat{B} \to \hat{C} \to 0$ in $\mathrm{CM}(\hat{R})$, so that $\hat{A}$ and the indecomposable summands of $\hat{B}$ are in $\mathcal{S}$, using Lemma 1. Since $\mathcal{S}$ is finite and contains $\hat{R}$, we know that when $\hat{R}$ is not Gorenstein, $\mathcal{S}$ consists of all indecomposables in $\mathrm{CM}(\hat{R})$ by [**5**, Th. 1.1]. If $\hat{R}$ is Gorenstein, let $g: E \to \hat{R}$ be minimal right almost split. We know that the nonprojective part E' of E has the property that $\Omega^d \hat{m} \simeq \Omega^d E'$ [**5**, Prop. 1.2], and hence E' is clearly gradable. From [**5**, Th. 1.1] we can now conclude that $\mathcal{S}$ consists of all indecomposables also in this case.

Assume conversely that $\hat{R}$ is of finite Cohen-Macaulay type. If for B and C indecomposable in $\mathrm{CM}(\operatorname{gr} R)_0$, $B \not\simeq C(i)$ for all i, we know by Lemmas 1 and 2 that $\hat{B}$ and $\hat{C}$ are nonisomorphic indecomposables in $\mathrm{CM}(\hat{R})$. This shows our claim.

Assume now in addition to our previous assumptions that R_p is Gorenstein whenever p is a nonmaximal graded prime ideal in R. We will then say that R is isolated Gorenstein. Then $\hat{R}$ is also isolated Gorenstein. For $\underline{\mathrm{End}}_R(\omega)$ is a graded R-module which by the assumption on R has finite length, so that also $\underline{\mathrm{End}}_{\hat{R}}(\hat{\omega}_R) \simeq \underline{\mathrm{End}}_R(\omega)$ has finite length. In this case $D\,\mathrm{Tr}_L\, C$ is clearly in $L_p(\mathrm{gr}\, R)_0$ when C is, and similarly for $\hat{R}$, so that $L_p(\hat{R})$ and $L_p(\mathrm{gr}\, R)_0$ have almost split sequences. (This follows from the previous results in this section, or see also [**1**,**6**,**7**].) Also there is a minimal right almost split map $E \to \hat{R}$ (and a minimal left almost split map $\omega_{\hat{R}} \to E'$) in $L_p(\hat{R})$ [**9**, Th. 3.9] [**2**, §6]. And the corresponding result for $L_p(\mathrm{gr}\, R)_0$ follows similarly. (Alternatively one can use results from [**3**], as discussed in Section 3.) We recall that saying that a map $g : E \to \hat{R}$ is right almost split in $L_p(\hat{R})$ means that any nonisomorphism $h : X \to \hat{R}$ with X indecomposable in $L_p(\hat{R})$ can be lifted through g. And g is minimal if the restriction to any nonzero summand is nonzero (see [**4**]). We also recall that a map $g : B \to C$ between indecomposables in $L_p(\hat{R})$ or $L_p(\mathrm{gr}\, R)_0$ is irreducible if g is not an isomorphism, and whenever there is a commutative diagram

$$\begin{array}{ccc} X & = & X \\ {\scriptstyle s}\uparrow & & \downarrow{\scriptstyle t} \\ B & \xrightarrow{g} & C \end{array}$$

then s is split mono or t is split epi [**4**]. The indecomposables in $L_p(\hat{R})$ and $L_p(\mathrm{gr}\, R)_0$ are split up into equivalence classes with respect to the equivalence relation generated by the existence of an irreducible map. The following result shows that if one indecomposable in $L_p(\hat{R})$ is gradable, then all in the same equivalence class are. An analogous result was given by Gordon-Green for artin algebras [**15**].

THEOREM 6. *Let R be isolated Gorenstein, and $g : X \to Y$ an irreducible map between indecomposables in $L_p(\hat{R})$. Then X is gradable if and only if Y is gradable.*

PROOF: Assume that Y is gradable. (The proof is similar when we assume X is gradable.) Choose C in $L_p(\mathrm{gr}\, R)_0$ with $\hat{C} \simeq Y$. If C is not projective, let $0 \to A \to B \to C \to 0$ be almost split in $L_p(\mathrm{gr}\, R)_0$. Then $0 \to \hat{A} \to \hat{B} \to \hat{C} \to 0$ is almost split in $L_p(\hat{R})$ by Proposition 3. Since $X \to \hat{C}$ is irreducible we know that X is a summand of $\hat{B}$ [**4**], and hence gradable, by Lemma 1.

If $Y = R$, assume first that R is not Gorenstein. Then R is not injective in $L_p(\operatorname{gr} R)_0$, so that we have an almost split sequence $0 \to R \to B \to C \to 0$, and hence an almost split sequence $0 \to \hat{R} \to \hat{B} \to \hat{C} \to 0$. Since $\omega_{\hat{R}} \simeq \hat{\omega}_R$ is gradable, we can assume $X \not\simeq \omega_R$. Then we know that $X \simeq D\operatorname{Tr}_L E$ for some indecomposable summand E of $\hat{B}$ (see [4]), and hence X is gradable. If R is Gorenstein, we can by the construction in [**5**, Prop. 1.2], in the graded version used in [**21**], find a map $g : F \coprod R^i \to R$ in $\mathrm{CM}(\operatorname{gr} R)_0$, where F has no nonzero projective summands, $\Omega^d F \simeq \Omega^d m$, and any nonzero map $h : Z \to R$ with Z indecomposable in $\mathrm{CM}(\operatorname{gr} R)_0$ can be lifted through g. We can choose $i \geq 0$ such that g is minimal with respect to this property. By the construction it is clear that $F \coprod R^i$ is in $L_p(\operatorname{gr} R)_0$, so that $g : F \coprod R^i \to R$ is minimal right almost split in $L_p(\operatorname{gr} R)_0$. Since we have a similar construction of the minimal right almost split map $E \coprod \hat{R}^j \to \hat{R}$, we see that $E \simeq \hat{F}$. Since X must be a summand of $E \coprod \hat{R}^j$, X is gradable.

Let $R_1 = k[x,y,z,u,v]/(xz - y^2, xv - yu, yv - zu)$ and $R_2 = k[x,y,z]^{\mathbb{Z}_2}$, where the generator of $\mathbb{Z}_2$ acts by sending each variable to its negative, and assume k is algebraically closed. (For R_2, assume that the characteristic is different from 2, or use the characteristic free description of R_2). Then it was known that R_1, R_2 and $\hat{R}_1, \hat{R}_2$ are of finite Cohen-Macaulay type [**5,7,22**]. It now follows from Theorem 5 that one result can be deduced from the other one.

Also when R is a hypersurface given by an element which is a quadratic form, it was known that R is of finite graded Cohen-Macaulay type if and only if $\hat{R}$ is of finite Cohen-Macaulay type ([**11**], see also [**12**]). For other quasihomogenous hypersurfaces we get the following application, by using known results in the complete case [**17,12,21**].

THEOREM 7. *Let $R = k[x_1, \ldots, x_n]/(f)$, where f is quasihomogenous and k is algebraically closed.*

a) *If char $k \neq 2$, then R is of finite graded Cohen-Macaulay type if and only if $f = g + z_1{}^2 + \cdots + z_t{}^2$, where g is a simple isolated curve singularity in the sense of [**10**].*
b) *If char $k = 2$, R is of finite graded Cohen-Macaulay type if $f = g + x_1 y_1 + \cdots + x_t y_t$, where g is a simple isolated curve singularity in the sense of [**10**] or a rational double point.*

2. Indecomposable modules for R and $\hat{R}$.

In this section we prove the following generalizations of Lemmas 1 and 2.

PROPOSITION 8. *Let A be in* $\text{mod}(\text{gr}\, R)_0$.

a) *If A is indecomposable in* $\text{mod}(\text{gr}\, R)_0$, *then* $\text{End}_R(A)$ *is a graded local ring.*

b) *A is indecomposable in* $\text{mod}(\text{gr}\, R)_0$ *if and only if $\hat{A}$ is an indecomposable $\hat{R}$-module.*

PROPOSITION 9. *Let A and B be indecomposable in* $\text{mod}(\text{gr}\, R)_0$. *Then $\hat{A} \simeq \hat{B}$ if and only if $A \simeq B(n)$ for some $n \in \mathbf{Z}$.*

To prove Proposition 8 we need the following.

LEMMA 10. *Let $\Lambda = \coprod_{i \in \mathbf{Z}} \Lambda_i$ be a $\mathbf{Z}$-graded ring such that Λ_0 is a local ring with radical $\underline{r}$, and there is an integer n such that $\Lambda_i = (0)$ for $i < n$. Then Λ is a graded local ring with unique graded maximal ideal $\underline{n} = \coprod_{i \neq 0} \Lambda_i \coprod \underline{r}$, and this ideal is also a maximal ideal in Λ.*

PROOF: Let x and y be homogenous elements of degrees s and t, where $s \leq t$. It is enough to show that xy is in $\underline{n}$ if either x or y is in $\underline{n}$. This is clear if $s \neq -t$ or if $s = t = 0$. If $s = -t \neq 0$, and xy is not in $\underline{n}$, there is some z is Λ_0 such that $xyz = 1$ since Λ_0 is a local ring. Since $\Lambda_i = (0)$ for $i < n$, x is nilpotent. Choose m minimal with $x^m = 0$. Then $0 = x^{m-1}(xyz) = x^{m-1}$ gives a contradiction. The argument is similar if y has negative degree, and this finishes the proof.

PROOF OF PROPOSITION 8: When A is indecomposable in $\text{mod}(\text{gr}\, R)_0$, $\text{End}_R(A)$ is a graded local ring whose unique graded maximal ideal is a maximal ideal. Then $\hat{R} \otimes_R \text{End}_R(A) \simeq \text{End}_{\hat{R}}(\hat{A})$ is a local ring, so that $\hat{A}$ is an indecomposable $\hat{R}$-module.

PROOF OF PROPOSITION 9: Let A and B be indecomposable in $\text{mod}(\text{gr}\, R)_0$. Hence $\hat{A}$ and $\hat{B}$ are indecomposable $\hat{R}$-modules by Proposition 8. If $\hat{A} \simeq \hat{B}$, the natural $\hat{R}$-map

$$(*) \qquad \text{Hom}_{\hat{R}}(\hat{A}, \hat{B}) \otimes_{\hat{R}} \text{Hom}_{\hat{R}}(\hat{B}, \hat{A}) \xrightarrow{\beta} \text{End}_{\hat{R}}(\hat{A}) \text{ is surjective.}$$

Consider also the natural R-map

$$(**) \qquad \text{Hom}_R(A, B) \otimes_R \text{Hom}_R(B, A) \xrightarrow{\alpha} \text{End}_R(A).$$

Since (*) is the completion of (**) it follows that $\hat{R}\otimes_R \operatorname{Coker}\alpha = \operatorname{Coker}\beta = (0)$, which means that $\operatorname{Coker}\alpha$ is (0) since it is a graded module. Since by Proposition 8 $\operatorname{End}_R(A)$ is a graded local ring, there are maps $g : A \to B$ and $h : B \to A$ of degree j and $-j$ such that $hg = \mathrm{id}_A$. This shows that A is isomorphic to a summand of $B(j)$, and hence isomorphic to $B(j)$.

3. AR-quivers.

In this section we shall refine the results from Section 1 and give relationships between the AR-quivers for $L_p(\operatorname{gr} R)_0$ and $L_p(\hat{R})$, when R is isolated Gorenstein. We recall that the vertices of the AR-quiver are in one-to-one correspondence with the indecomposables in $L_p(\operatorname{gr} R)_0$ or $L_p(\hat{R})$, and there is an arrow between two vertices if there is an irreducible map between the corresponding modules. Associated with each arrow $A \to B$ there is an ordered pair (r_{AB}, s_{AB}) of positive integers, where s_{AB} denotes the multiplicity of B in a minimal left almost split map $A \to E$ and r_{AB} the multiplicity of A in a minimal right almost split map $F \to B$. We say r_{AB} and s_{AB} are zero if there is no arrow $A \to B$.

To be able to compare the AR-quivers for R and $\hat{R}$, we need more information about irreducible maps to R (and dually irreducible maps from ω). We have the following:

PROPOSITION 11. *If $f : E \to R$ is minimal right almost split in $L_p(\operatorname{gr} R)_0$, then $\hat{f} : \hat{E} \to \hat{R}$ is minimal right almost split in $L_p(\hat{R})$.*

PROOF: We use results on Cohen-Macaulay approximations by Auslander-Buchweitz to show that $\hat{f} : \hat{E} \to \hat{R}$ is right almost split. We first show that $f : E \to m \subset R$ is in fact a Cohen-Macaulay approximation, that is, any map $h : X \to m$ with X in $\operatorname{CM}(\operatorname{gr} R)_0$ can be factored through f. For we know by [**7,2**] that there is a minimal Cohen-Macaulay approximation $g : F \to m$, and that $\operatorname{id}_{\operatorname{gr} R} K < \infty$, where $K = \operatorname{Ker} f$ and $\operatorname{id}_{\operatorname{gr} R} K$ denotes the injective dimension in $\operatorname{mod}(\operatorname{gr} R)_0$ [**3**]. If p is a graded nonmaximal prime ideal, we have $m_p \simeq R_p$, and hence $F_p \simeq R_p \coprod K_p$. Since R_p is Gorenstein, $\operatorname{pd} K_p < \infty$, hence K_p is free since it is Cohen-Macaulay. This shows that F is in $L_p(\operatorname{gr} R)_0$, and hence $g : F \to R$ must be isomorphic to $f : E \to R$.

Consider now $0 \to L \to E \xrightarrow{f} m \to 0$ and its completion $0 \to \hat{L} \to \hat{E} \xrightarrow{\hat{f}} \hat{m} \to 0$. Since $\operatorname{id}_{\operatorname{gr} R} L < \infty$, it follows by a graded version of [**19**] that there is an exact sequence $0 \to \omega_t \to \cdots \to \omega_0 \to L \to 0$ in $\mathrm{mod}(\mathrm{gr}\, R)_0$, where each ω_i is isomorphic to a finite sum of shifts of ω. Completing, we see that $\mathrm{id}_{\hat{R}}\, \hat{L} < \infty$, and hence $\hat{f} : \hat{E} \to \hat{m}$ is a Cohen-Macaulay approximation by [**3**]. This means that $\hat{f} : \hat{E} \to \hat{R}$ is right almost split.

We next want to show that the map $\hat{f} : \hat{E} \to \hat{R}$ is minimal.

LEMMA 12. *Let $h : A \to B$ be an indecomposable map in $\mathrm{mod}(\mathrm{gr}\, R)_0$. Then $\hat{h} : \hat{A} \to \hat{B}$ is an indecomposable map in $\mathrm{mod}\hat{R}$.*

PROOF: Denote $h : A \to B$ by X as a module over the lower triangular matrix ring $T = \begin{pmatrix} R & 0 \\ R & R \end{pmatrix}$. Then the completion $\hat{T}$ is clearly $\begin{pmatrix} \hat{R} & 0 \\ \hat{R} & \hat{R} \end{pmatrix}$. $\mathrm{End}(X)$ is a $\mathbb{Z}$-graded ring, where the elements of degree i are the pairs (f_i, g_i) of maps of degree i in $\mathrm{mod}(\mathrm{gr}\, R)_0$, such that $g_i h = h f_i$. $\mathrm{End}(X)_0$ is an artin ring, which must be local since $h : A \to B$ is assumed to be indecomposable in $\mathrm{mod}(\mathrm{gr}\, R)_0$. Since there is some integer n such that $\mathrm{End}(X)_i = (0)$ if $i < n$, it follows from Lemma 10 that $\mathrm{End}(X)$ is a local ring. Since $\widehat{\mathrm{End}(X)} \simeq \mathrm{End}(\hat{X})$, $\hat{X}$ is an indecomposable $\hat{T}$-module. It is easy to see that $\hat{X}$ is isomorphic to the T-module $\hat{h} : \hat{A} \to \hat{B}$.

Combining Proposition 11 with our results in Section 1, we get the following.

THEOREM 13. *Assume R is isolated Gorenstein, and let A and B be indecomposable in $L_p(\mathrm{gr}\, R)_0$. Then there is an irreducible map $f : \hat{A} \to \hat{B}$ in $L_p(\hat{R})$ if and only if there is an irreducible map $g : A \to B(i)$ in $L_p(\mathrm{gr}\, R)_0$ for some i.*

And if there is an irreducible map $f : \hat{A} \to \hat{B}$, then $s_{AB} = \sum_j s_{AB(j)}$ and $r_{AB} = \sum_j r_{AB(j)}$.

This result shows that if $\mathcal{C}$ is a connected component of the AR-quiver for R, then the full subquiver $\hat{\mathcal{C}}$ of the AR-quiver for $\hat{R}$, whose vertices correspond to completions of the modules given by $\mathcal{C}$, is a connected component for $\hat{R}$, and it can be constructed from $\mathcal{C}$. Note however that $\mathcal{C}$ cannot be constructed from $\hat{\mathcal{C}}$, and that there may be more components for R with the same image. $\mathcal{C}$ and $\hat{\mathcal{C}}$ may be isomorphic, for example this is true for all $\mathcal{C}$ when R is the graded ring of an elliptic curve singularity (see [**19**]). But if R is of finite Cohen-Macaulay type, $\mathcal{C}$ and $\hat{\mathcal{C}}$ are clearly not isomorphic. Note that in this case the AR-quiver for R contains more information than the AR-quiver for $\hat{R}$.

References

1. M. Auslander, *Functors and morphisms determined by objects*, in "Proc. Conf. on Representation Theory, Philadelphia 1976," Marcel Dekker, 1978, pp. 1–327.
2. M. Auslander, *Isolated singularities and almost split sequences*, Proc. ICRA IV, Springer Lecture Notes in Math. **1178** (1986), 194–241.
3. M. Auslander and R.-O. Buchweitz, *The homological theory of maximal Cohen-Macaulay approximations*.
4. M. Auslander and I. Reiten, *Representation theory of artin algebras IV: Invariants given by almost split sequences*, Comm. in Algebra **5** (1977), 519–554.
5. M. Auslander and I. Reiten, *The Cohen-Macaulay type of Cohen-Macaulay rings*, Adv. in Math. (to appear).
6. M. Auslander and I. Reiten, *Almost split sequences for Cohen-Macaulay modules*, Math. Annalen **277** (1987), 345–350.
7. M. Auslander and I. Reiten, *Almost split sequences for graded rings*, Proc. Lambrecht Conf. on Vector Bundles, Sing. Theory and Representation Theory, Springer Lecture Notes in Math. **1273** (1987), 232–243.
8. M. Auslander and I. Reiten, *Almost split sequences for abelian group graded rings*, J. Algebra (to appear).
9. M. Auslander and S. Smalø, *Lattices over orders: Finitely presented functors and preprojective partitions*, Trans. AMS **273** (1982), 433–446.
10. W. Barth, C. Peters and A. Van den Ven, "Compact Complex Surfaces," Erg. d. Math. Band 4, Springer, Berlin, 1984.
11. R. O. Buchweitz, D. Eisenbud and J. Herzog, *Cohen-Macaulay modules on quadrics*, Proc. Lambrecht Conf. on Vector Bundles, Sing. Theory and Representation Theory, Springer Lecture Notes in Math. **1273** (1987), 58–133.
12. R. O. Buchweitz, G. M. Greuel and F. O. Schreyer, *Cohen-Macaulay modules on hypersurface singularities II*, Invent. in Math., Fasc 1 **88** (1987), 165–182.
13. E. Dieterich, *Reduction of isolated singularities*, Comm. Math. Helv.
14. R. Gordon and E. L. Green, *Graded artin algebras*, J. Algebra **76** (1982), 111–137.
15. R. Gordon and E. L. Green, *Representation theory of graded artin algebras*, J. Algebra **76** (1982), 138–152.
16. J. Herzog, *The representation type of a graded Cohen-Macaulay ring and its completion*.
17. H. Knörrer, *Cohen-Macaulay modules on hypersurface singularities II*, Invent. Math. **88** (1987), 153–164.
18. C. Natasescu and F. Van Oystayen, "Graded Ring Theory," North Holland, 1982.
19. I. Reiten, *Finite dimensional algebras and singularities*, Proc. Lambrecht Conf. on Vector Bundles, Sing. Theory and Representation Theory, Springer Lecture Notes in Math. **1273** (1987), 35–57.
20. R. Sharp, *Finitely generated modules of finite injective dimension over certain Cohen-Macaulay rings*, Proc. London Math. Soc. (3) **25** (1972), 303–328.
21. Ø. Solberg, *Hypersurface singularities of finite Cohen-Macaulay type*, Proc. Londong Math. Soc. (to appear).
22. Ø. Solberg, *A graded ring of finite Cohen-Macaulay type*, Comm. in Algebra (to appear).
23. Y. Yoshino, *Brauer-Thrall theorems for maximal Cohen-Macaulay modules*, J. Math. Soc. Japan **39, No. 4** (1987), 719–739.

Department of Mathematics, Brandeis University, Waltham MA 02254
Institutt for matematikk og statistikk, Universitetet i Trondheim, AVH, 7055 Dragvoll, NORWAY

TOPICS IN ALGEBRA
BANACH CENTER PUBLICATIONS, VOLUME 26, PART 1
PWN–POLISH SCIENTIFIC PUBLISHERS
WARSAW 1990

GRADED MODULES AND THEIR COMPLETIONS

MAURICE AUSLANDER

Department of Mathematics, Brandeis University
Waltham, Massachusetts, U.S.A.

IDUN REITEN

Department of Mathematics, University of Trondheim
Dragvoll, Norway

Suppose R is a G-graded Cohen–Macaulay $k[x_1, \ldots, x_t]$-algebra where G is a group and k a field. We study the completion functor from graded R-modules to $\hat{R}$-modules where $\hat{R}$ is the completion of R, with special emphasis on Cohen–Macaulay modules. We prove that if the orders of elements of finite order in G are invertible in k and the orders of finite subgroups of G are bounded, then almost split sequences of graded Cohen–Macaulay modules go to direct sums of almost split sequences of Cohen–Macaulay modules over $\hat{R}$. Moreover, we show that R is of finite graded Cohen–Macaulay type if and only if $\hat{R}$ is of finite Cohen–Macaulay type.

Introduction

Before stating the main results in this paper, we need some notation and definitions concerning graded rings and modules for arbitrary groups.

Let G be a group which we write additively even though it may not be commutative. By a *G-graded ring* R we mean a ring (not necessarily commutative) together with a direct sum decomposition $R = \coprod_{g \in G} R_g$ such that $R_h R_g \subset R_{h+g}$ for all h and g in G. A *G-graded R-module* C is an R-module C together with a direct sum decomposition $C = \coprod_{g \in G} C_g$ such that $R_h C_g \subset C_{h+g}$ for all h and g in G. An R-module morphism $\alpha \colon B \to C$ of

This paper is in final form and no version of it will be submitted for publication elsewhere.

G-graded R-modules B and C is said to be of *degree* 0 if $\alpha(B_g) \subset C_g$ for all g in G. We denote by $\mathrm{Mod}(\mathrm{gr}\,R)_0$ the category of G-graded R-modules together with degree 0 morphisms.

Throughout this paper T denotes a polynomial ring $k[X_1, \ldots, X_d]$ with k a field and the X_i indeterminates. We always assume that a G-grading on T has the following properties: (a) $T_0 = k$; (b) each X_i is homogeneous; (c) each T_g is of finite dimension over $k = T_0$. By a *G-graded T-algebra R* we mean throughout this paper that R is a T-algebra by means of the ring morphism f: $T \to R$ and that T and R are G-graded rings such that R is a finitely generated G-graded T-module and f: $T \to R$ is a degree 0 morphism of G-graded T-modules. Hence R is left and right noetherian. Also for each finitely generated R-module C we denote by $\hat{C}$ the completion of C in the $\mathfrak{m}$-adic topology where $\mathfrak{m} = (X_1, \ldots, X_d)$, the ideal in T generated by the indeterminates $X_1, \ldots, X_d$.

Suppose R is a G-graded T-algebra. Then a G-graded R-module C is said to be *Cohen–Macaulay* if it is free when viewed as a T-module. We say that R is a *Cohen–Macaulay G-graded T-algebra* if R is a G-graded Cohen–Macaulay R-module. We say that R *has isolated singularities* if $\mathrm{gl.dim}\, R_\mathfrak{p} = \dim T_\mathfrak{p}$ for all nonmaximal prime ideals $\mathfrak{p}$ of T. Throughout the rest of this introduction we assume that R is a Cohen–Macaulay G-graded T-algebra with isolated singularities.

Our main results are the following theorems connecting the subcategories $\mathrm{CM}(\mathrm{gr}\,R)_0$ and $\mathrm{CM}(\hat{R})$ of Cohen–Macaulay modules in the categories $\mathrm{mod}(\mathrm{gr}\,R)_0$, finitely generated G-graded R-modules with degree zero morphisms, and $\mathrm{mod}\,\hat{R}$, finitely generated $\hat{R}$-modules, respectively.

THEOREM A. (a) *The category* $\mathrm{CM}(\mathrm{gr}\,R)_0$ *has almost split sequences.*

(b) *If* $0 \to A \to B \to C \to 0$ *is an almost split sequence in* $\mathrm{CM}(\mathrm{gr}\,R)_0$, *then* $0 \to \hat{A} \to \hat{B} \to \hat{C} \to 0$ *is a direct sum of almost split sequences in* $\mathrm{CM}(\hat{R})$ *provided the orders of elements of finite order in G are invertible in k.*

This result about almost split sequences is used to prove the following result about when R is of *finite graded Cohen–Macaulay type*, i.e. when there are only a finite number of isomorphism classes of indecomposable modules in $\mathrm{CM}(\mathrm{gr}\,R)_0$ up to shift.

THEOREM B. *Suppose G has the following properties*: (a) *the orders of elements of finite order in G are invertible in k, and* (b) *the orders of finite subgroups of G are bounded. Then R is of finite graded Cohen–Macaulay type if and only if* $\hat{R}$ *is of finite Cohen–Macaulay type.*

These results generalize our results in [4]–[6]. These generalizations were inspired by conversations with H. Lenzing, who pointed out interesting examples from his work with Geigle on two-dimensional graded rings where indecomposability is not preserved under completion. Lenzing has given a different (independent) proof for the first theorem (when G is abelian), using the result for $G = Z$. We take the opportunity to thank Lenzing for his hospitality during our stay in Paderborn. We would also like to thank M. Van den Bergh for helpful conversations about graded rings.

1. Preliminaries

In this section we first give some preliminary results about graded T-algebras. We end the section by giving an existence theorem for almost split sequences.

Denote by $\operatorname{rad} R$ the ordinary radical of R and by $\text{gr-rad}\, R$ the graded radical. For general background on graded theory we refer to [11]. For the first 3 lemmas we assume that if t is the order of an element of finite order in G, then t is invertible in k.

LEMMA 1.1. *Assume that R is a finite-dimensional k-algebra.*

(a) *If R is a graded division ring, then $H = \{g \in G;\ R_g \neq 0\}$ is a finite subgroup of G. Also R is semisimple and is the direct sum of at most $|H|$ indecomposable R-modules, where $|H|$ denotes the order of H. If H is trivial, then R is a division ring.*

(b) $\text{gr-rad}\, R = \operatorname{rad} R$.

Proof. (a) It is well known and easy to see that H is a finite subgroup of G. Since by assumption $|H|$ is invertible in k, $\operatorname{rad} R = (0)$ by [8]. Further $R_0 = D$ is a division ring, and as a left D-module, $R_g \simeq D$ for each $g \in H$ [11]. Hence the last claim follows.

(b) Since R is artin, $\text{gr-rad}\, R$ is nilpotent, so that $\text{gr-rad}\, R \subseteq \operatorname{rad} R$. By a graded version of the Wedderburn theorem

$$R/\text{gr-rad}\, R \simeq M_{n_1}(D_1) \times \ldots \times M_{n_t}(D_t),$$

where each D_i is a graded division ring and $M_{n_i}(D_i)$ denotes the ring of $n_i \times n_i$ matrices over D_i (see [11]). By (a) each D_i is semisimple, so that $\operatorname{rad}(R/\text{gr-rad}\, R) = (0)$, and hence $\text{gr-rad}\, R = \operatorname{rad} R$.

LEMMA 1.2. $(\text{gr-rad}\, R)_{\mathfrak{m}} = \operatorname{rad} R_{\mathfrak{m}}$ *and hence* $(\text{gr-rad}\, R)^{\wedge} = \operatorname{rad} \hat{R}$.

Proof. By the graded version of Nakayama's lemma we have $\mathfrak{m}R \subset \text{gr-rad}\, R$ [11], and we also have $\mathfrak{m}R_{\mathfrak{m}} \subset \operatorname{rad} R_{\mathfrak{m}}$. By Lemma 1.1 we have for $\Lambda = R/\mathfrak{m}R \simeq R_{\mathfrak{m}}/\mathfrak{m}R_{\mathfrak{m}}$ that $\text{gr-rad}\, \Lambda = \operatorname{rad} \Lambda$, since Λ is a finite-dimensional k-algebra. Taking preimages we get $(\text{gr-rad}\, R)_{\mathfrak{m}} = \operatorname{rad} R_{\mathfrak{m}}$. We have $\operatorname{rad} \hat{R}_{\mathfrak{m}} \subseteq \operatorname{rad} \hat{R}$, and since $(R_{\mathfrak{m}}/\operatorname{rad} R_{\mathfrak{m}})^{\wedge} = R_{\mathfrak{m}}/\operatorname{rad} R_{\mathfrak{m}}$ is semisimple, we have $\operatorname{rad} \hat{R}_{\mathfrak{m}} = \operatorname{rad} \hat{R}$, and consequently $(\text{gr-rad}\, R)^{\wedge} = (\operatorname{rad} R_{\mathfrak{m}})^{\wedge} = \operatorname{rad} \hat{R}$.

LEMMA 1.3. *Let A be indecomposable in* $\operatorname{mod}(\operatorname{gr} R)_0$.

(a) *If A is not projective, then* $P(\hat{A}, \hat{A}) \subset \operatorname{rad} \operatorname{End}(\hat{A})$, *where $P(\hat{A}, \hat{A})$ denotes the maps $f\colon \hat{A} \to \hat{A}$ which factor through a projective module.*

(b) *If A is not projective, then $\hat{A}$ has no nonzero projective summands.*

(c) *$\hat{A}$ decomposes into a direct sum of at most n indecomposable summands, where n is the order of the group $H = \{g \in G;\ \Lambda_g \neq (0)\}$, where*

$$\Lambda = \mathrm{End}_R(A)/\text{gr-rad}\,\mathrm{End}_R(A).$$

(d) *If G has no nonzero elements of finite order, then*

$$\mathrm{End}_R(A)/\text{gr-rad}\,\mathrm{End}_R(A) \quad \text{and} \quad \mathrm{End}_{R_{\mathfrak{m}}}(A_{\mathfrak{m}})/\mathrm{rad}\,\mathrm{End}_{R_{\mathfrak{m}}}(A_{\mathfrak{m}})$$

are division rings and $\hat{A}$ is indecomposable in $\mathrm{mod}\,\hat{R}$.

Proof. (a) Since the endomorphism rings of indecomposable modules in $\mathrm{mod}(\mathrm{gr}\,R)_0$ are local, $\mathrm{mod}(\mathrm{gr}\,R)_0$ has projective covers. Let $P \to A$ be a projective cover for A in $\mathrm{mod}(\mathrm{gr}\,R)_0$. We then have an exact sequence $\mathrm{Hom}_R(A, P) \to \mathrm{Hom}_R(A, A) \to \underline{\mathrm{Hom}}_R(A, A) \to 0$, which under completion goes to an exact sequence

$$\mathrm{Hom}_{\hat{R}}(\hat{A}, \hat{P}) \to \mathrm{Hom}_{\hat{R}}(\hat{A}, \hat{A}) \to \underline{\mathrm{Hom}}_{\hat{R}}(\hat{A}, \hat{A}) = (\underline{\mathrm{Hom}}_R(A, A))^{\wedge} \to 0.$$

It follows that $(P(A, A))^{\wedge} = P(\hat{A}, \hat{A})$. Clearly $P(A, A) \subset \text{gr-rad}\,\mathrm{End}_R(A)$, so $P(\hat{A}, \hat{A}) \subset \mathrm{rad}\,\mathrm{End}_R(\hat{A})$ then follows from Lemma 1.2.

(b) This is a direct consequence of (a).

(c) Λ decomposes into at most n indecomposable summands by Lemma 1.1 (b). By Lemma 1.2, $\Lambda = \mathrm{End}_{\hat{R}}(\hat{A})/\mathrm{rad}\,\mathrm{End}_R(\hat{A})$, so that also $\hat{A}$ decomposes into at most n indecomposable summands.

(d) This is a direct consequence of Lemma 1.1 (a), Lemma 1.2 and part (c).

LEMMA 1.4. *Let A and B be indecomposable in* $\mathrm{mod}(\mathrm{gr}\,R)_0$.

(a) $\mathrm{End}_R(A)$ *is a graded local ring.*

(b) *If $\hat{A} \simeq B$, then there is some $g \in G$ such that $A \simeq B(g)$, where $B(g)_h = B_{h+g}$ for all $h \in G$.*

Proof. (a) Let u and v be elements in Γ_g, where $\Gamma = \mathrm{End}_R(A)$, which do not have left inverses. It is then sufficient to show that $u+v$ does not have a left inverse. Assume to the contrary that there is some $w \in \Gamma_{-g}$ such that $w(u+v) = 1 = wu + wv$. Since Γ_0 is a local ring, wu or wv must have a left inverse, so we have a contradiction.

(b) This is proved as in [6, Prop. 9].

LEMMA 1.5. *If P is indecomposable projective in* $\mathrm{mod}(\mathrm{gr}\,R)_0$, *then* $\text{gr-rad}\,P$ *is the unique graded maximal ideal in P.*

Proof. Since the endomorphism rings of indecomposable objects in $\mathrm{mod}(\mathrm{gr}\,R)_0$ are local, the category $\mathrm{mod}(\mathrm{gr}\,R)_0$ has projective covers. Hence P is a projective cover for each nonzero factor module, in particular for $P/M_1 \cap M_2 = P/M_1 \amalg P/M_2$ if M_1 and M_2 are distinct graded maximal submodules of P. Since P is also a projective cover for P/M_1, this easily leads to a contradiction.

To state the next theorem we need to observe that the subgroup H generated by $\{g \in G;\ T_g \neq 0\}$ is commutative since T is a commutative domain.

THEOREM 1.6. *Let R be a Cohen–Macaulay G-graded T-algebra.*

(a) *If C is an indecomposable nonprojective module in $\mathrm{CM}(\mathrm{gr}\,R)_0$, there is an almost split sequence $0 \to A \to B \to C \to 0$ in $\mathrm{CM}(\mathrm{gr}\,R)_0$ if and only if $C_\mathfrak{p}$ is a projective $T_\mathfrak{p}$-module for each nonmaximal prime ideal $\mathfrak{p}$ in T.*

(b) *If A is an indecomposable noninjective object in $\mathrm{CM}(\mathrm{gr}\,R)_0$, there is an almost split sequence $0 \to A \to B \to C \to 0$ in $\mathrm{CM}(\mathrm{gr}\,R)_0$ if and only if $A_\mathfrak{p}$ is an injective object in $\mathrm{CM}(R_\mathfrak{p})$ for each nonmaximal prime ideal $\mathfrak{p}$ in T.*

(c) *If there is an almost split sequence $0 \to A \to B \to C \to 0$ in $\mathrm{CM}(\mathrm{gr}\,R)_0$, then $A = D\mathrm{Tr}_L(-\tilde{d})$, where $\tilde{d} = \deg X_1 + \ldots + \deg X_d$.*

Proof. This is a generalization of [4, Th. 1.1] (and [5]), and the proof is similar to the proof in [4]. We only make a few extra comments. We first recall that D denotes the duality $\mathrm{Hom}_T(\ ,T)$ between $\mathrm{CM}(\mathrm{gr}\,R)_0$ and $\mathrm{CM}(\mathrm{gr}\,R^{\mathrm{op}})_0$, Tr denotes the transpose and $\mathrm{Tr}_L = \Omega^d\mathrm{Tr}$, where Ω^d denotes the dth syzygy.

We note that for each $g \in G$ there is a degree zero isomorphism of graded T-modules $\mathrm{Hom}_R(B, C(g)) \simeq \mathrm{Hom}_R(B, C)(g)$ for all $g \in G$. Also the category $\mathrm{Mod}(\mathrm{gr}\,T)_0$ has enough injectives [11] and injective envelopes. When C is indecomposable in $\mathrm{CM}(\mathrm{gr}\,R)_0$ and $C_\mathfrak{p}$ is $T_\mathfrak{p}$-projective for each nonmaximal prime ideal $\mathfrak{p}$ in R, then $\underline{\mathrm{End}}_R(C)^{\mathrm{op}}$, the endomorphism ring of C modulo projectives, is a finite-dimensional k-algebra, and a crucial point is to show that there is a nonzero degree zero map $f\colon \Gamma \to k$ such that $f(\mathrm{rad}\,\Gamma_0) = 0$. Let $\mathfrak{n} = \text{gr-rad}\,\Gamma$. Then $\mathfrak{n} \cap \Gamma_0 = \mathrm{rad}\,\Gamma_0$, and k is the unique graded simple T-module. Hence we have a nonzero degree zero map $\Gamma/\mathfrak{n} \to k$, and we are done.

We also have the following information on right and left almost split maps.

PROPOSITION 1.7. *Let R be a G-graded T-algebra which is Cohen–Macaulay with isolated singularities.*

(a) *If P is indecomposable projective in $\mathrm{CM}(\mathrm{gr}\,R)_0$, there is a minimal right almost split map $E \to P$.*

(b) *If I is indecomposable injective in $\mathrm{CM}(\mathrm{gr}\,R)_0$, there is a minimal left almost split map $I \to F$.*

Proof. (a) By Lemma 1.5, $\text{gr-rad}\,P$ is a graded maximal submodule of P. By [2] there is a Cohen–Macaulay approximation $g\colon E \to \text{gr-rad}\,P$ in $\mathrm{mod}(\mathrm{gr}\,R)_0$, that is, E is in $\mathrm{CM}(\mathrm{gr}\,R)_0$, and for each map $h\colon C \to \text{gr-rad}\,P$ in $\mathrm{mod}(\mathrm{gr}\,R)_0$, with C in $\mathrm{CM}(\mathrm{gr}\,R)_0$, there is some $t\colon C \to E$ with $gt = h$. Since the endomorphism rings of indecomposable objects in $\operatorname{mod}(\operatorname{gr} R)_0$ are local, it is easy to see that g: $E \to \text{gr-rad}\, P$ can be chosen to be a minimal map. Hence g: $E \to P$ is a minimal right almost split map.

(b) follows from (a) by duality.

2. Completions of almost split sequences

Let R be a Cohen–Macaulay G-graded T-algebra with isolated singularities. In this section we assume in addition that if G has an element of finite order t, then t is invertible in k. In [6] we proved that if R is commutative positively $\mathbf{Z}$-graded, then almost split sequences in $\operatorname{CM}(\operatorname{gr} R)_0$ stay almost split under completion. This is not true in general, but we prove in this section that an almost split sequence goes to a direct sum of almost split sequences under completion. For this we need the following characterization of a direct sum of almost split sequences.

PROPOSITION 2.1. *Assume that C in $\operatorname{CM}(\hat{R})$ has no nonzero projective summands. Then the following are equivalent for an exact sequence*

$$0 \to D\operatorname{Tr}_L C \xrightarrow{f} B \xrightarrow{g} C \to 0 \quad \text{in } \operatorname{CM}(\hat{R}).$$

(a) *The sequence is a direct sum of almost split sequences.*

(b) *A map $h \in \operatorname{End}_R(C)$ factors through g: $B \to C$ if and only if $h \in \operatorname{rad} \operatorname{End}_R(C)$.*

Proof. (a)⇒(b). Let $C = C_1 \amalg \ldots \amalg C_t$ be a decomposition of C into a direct sum of indecomposable modules and denote by

$$0 \to D\operatorname{Tr}_L C_i \xrightarrow{f_i} B_i \xrightarrow{g_i} C_i \to 0$$

the almost split sequences. For h in $\operatorname{End}_R(C)$, denote by h_{ij}: $C_i \to C_j$ the induced maps. Then $h \in \operatorname{rad} \operatorname{End}_R(C)$ if and only if each h_{ij} is not an isomorphism. If $h \in \operatorname{rad} \operatorname{End}_R(C)$, then h_{ij} must factor through g_j: $B_j \to C_j$, so that h factors through g: $B \to C$. If $h \notin \operatorname{rad} \operatorname{End}_R(C)$, then some map h_{ij}: $C_i \to C_j$ is an isomorphism, and hence does not factor through g_j: $B_j \to C_j$. Consequently h does not factor through g: $B \to C$.

(b)⇒(a). The assumption in (b) means that $\operatorname{Im} \operatorname{Hom}(C, g) = \operatorname{rad} \operatorname{End}(C) \subset \operatorname{End}(C)$. If

$$0 \to D\operatorname{Tr}_L C \xrightarrow{f'} B \xrightarrow{g'} C \to 0$$

is a direct sum of almost split sequences, we have $\operatorname{Im}(C, g') = \operatorname{rad} \operatorname{End}(C)$. By the theory of X-determined morphisms an exact sequence of the form

$$0 \to D\operatorname{Tr}_L C \xrightarrow{f''} E \xrightarrow{g''} C \to 0$$

is uniquely determined up to isomorphism by the $\operatorname{End}(C)^{\text{op}}$-submodule

$\operatorname{Im}(C, g'')$ of $\operatorname{End}(C)^{op}$ (see [1, Prop. 9.6]). Hence it follows that our two sequences are isomorphic.

We can now prove the main result of this section.

THEOREM 2.2. *Let* $0 \to A \xrightarrow{f} B \xrightarrow{g} C \to 0$ *be an almost split sequence in* $\mathrm{CM}(\operatorname{gr} R)_0$. *Then* $0 \to \hat{A} \xrightarrow{\hat{f}} \hat{B} \xrightarrow{\hat{g}} \hat{C} \to 0$ *is a direct sum of almost split sequences in* $\mathrm{CM}(\hat{R})$.

Proof. C is indecomposable nonprojective in $\mathrm{CM}(\operatorname{gr} R)_0$. By Lemma 1.3, $\hat{C}$ has no nonzero projective summands. Since $0 \to A \xrightarrow{f} B \xrightarrow{g} C \to 0$ is almost split, we have $\operatorname{Im} \operatorname{Hom}_R(C, g) = \text{gr-rad} \operatorname{End}(C)^{op}$. Since exactness is preserved by completion, we get $\operatorname{Im} \operatorname{Hom}_{\hat{R}}(\hat{C}, \hat{g}) = (\text{gr-rad} \operatorname{End}_R(C)^{op})^\wedge$, which by Lemma 1.2 is $\operatorname{rad} \operatorname{End}_{\hat{R}}(\hat{C})^{op}$. Then we are done by Proposition 2.1.

3. Finite Cohen–Macaulay type

Let R be a G-graded Cohen–Macaulay T-algebra with isolated singularities. In this section we compare the representation type for R and $\hat{R}$. We follow the outline for the corresponding result when R is commutative and $G = \mathbf{Z}$ [6].

In addition to the behavior of almost split sequences given in Theorem 2.2 we need the following result about right and left almost split maps.

PROPOSITION 3.1. (a) *Let P be indecomposable projective in* $\mathrm{CM}(\operatorname{gr} R)_0$ *and* $g\colon E \to P$ *a right almost split map in* $\mathrm{CM}(\operatorname{gr} R)_0$. *Then* $\hat{g}\colon \hat{E} \to \hat{P}$ *is right almost split in* $\mathrm{CM}(\hat{R})$.

(b) *Let I be an indecomposable injective object in* $\mathrm{CM}(\operatorname{gr} R)_0$ *and* $f\colon I \to F$ *a left almost split map in* $\mathrm{CM}(\operatorname{gr} R)_0$. *Then* $\hat{f}\colon \hat{I} \to \hat{F}$ *is left almost split in* $\mathrm{CM}(\hat{R})$.

Proof. (a) We have seen in Section 1 that a right almost split map $g\colon E \to P$ corresponds to a Cohen–Macaulay approximation $g\colon E \to \text{gr-rad}\, P$. From the theory of Cohen–Macaulay approximations [2] it follows that $\hat{g}\colon \hat{E} \to (\text{gr-rad}\, P)^\wedge$ is also a Cohen–Macaulay approximation. From Lemma 1.1 we know that $(\text{gr-rad}\, P)^\wedge = \operatorname{rad} \hat{P}$, and hence we get a right almost split map $\hat{g}\colon \hat{E} \to \hat{P}$.

(b) Similar to (a).

We now get the following main result of this section.

THEOREM 3.2. (a) *If R is of finite graded Cohen–Macaulay type, then $\hat{R}$ is of finite Cohen–Macaulay type, and for every indecomposable module C in* $\mathrm{CM}(\hat{R})$ *there is some indecomposable module B in* $\mathrm{CM}(\operatorname{gr} R)_0$ *such that C is a direct summand of $\hat{B}$.*

(b) *If $\hat{R}$ is of finite Cohen–Macaulay type and there is a bound on the order of the finite subgroups of G, then R is of finite graded Cohen–Macaulay type.*

Proof. (a) Assume that R is of finite graded Cohen–Macaulay type, and let

$B_1, \ldots, B_n$ be a complete set, up to shift, of indecomposable modules in $\mathrm{CM}(\mathrm{gr}\,R)_0$. Let $\mathscr{L}$ be the finite set of indecomposable modules in $\mathrm{CM}(\hat{R})$ which are summands of some $\hat{B}_i$. Then clearly the indecomposable projective $\hat{R}$-modules are in $\mathscr{L}$, and it is easy to see that $\mathscr{L}$ is closed under irreducible maps, by using Theorem 2.2 and Proposition 3.1 (see [6]). Then it follows from [3] that all indecomposable modules in $\mathrm{CM}(\hat{R})$ are in $\mathscr{L}$.

(b) Let A and B be indecomposable in $\mathrm{CM}(\mathrm{gr}\,R)_0$. If $\hat{A} \simeq \hat{B}$, there is by Lemma 1.4 some $g \in G$ such that $A \simeq B(g)$. Denote by N the bound on the order of subgroups of G of finite order. Then it follows from Lemma 1.3 that $\hat{A}$ decomposes into a direct sum of at most N indecomposable summands. From this it follows that if $\hat{R}$ is of finite Cohen–Macaulay type, then R is of finite graded Cohen–Macaulay type.

4. Localizations

Throughout this section we assume that R is a graded T-algebra with grading group G. In the previous sections we have been mainly concerned with relations between R and its completion $\hat{R}$, particularly properties of the completion functor from $\mathrm{gr}(R)_0$ to $\mathrm{mod}\,\hat{R}$. In addition to $\hat{R}$, the ring $R_{\mathfrak{m}}$ is also naturally associated with R. One of the major differences between $\hat{R}$ and $R_{\mathfrak{m}}$ is that while $\mathrm{mod}\,\hat{R}$ is always a Krull–Schmidt category, $\mathrm{mod}\,R_{\mathfrak{m}}$ is rarely a Krull–Schmidt category. However, it follows from Lemma 1.2 that if G has no nonzero elements of finite order, then $(\mathrm{mod}\,R_{\mathfrak{m}})_0$, the full subcategory of $\mathrm{mod}\,R_{\mathfrak{m}}$ consisting of those $R_{\mathfrak{m}}$-modules isomorphic to $A_{\mathfrak{m}}$ for some A in $\mathrm{mod}(\mathrm{gr}\,R)_0$, is a Krull–Schmidt category. For this reason, we assume throughout this section that G has no nonzero elements of finite order.

Our first aim is to study the full subcategory $\mathrm{CM}(R_{\mathfrak{m}})$ of $\mathrm{mod}\,R_{\mathfrak{m}}$ when $\mathrm{CM}(\mathrm{gr}\,R)_0$ is of finite type. To this end we assume now that R is a Cohen–Macaulay T-algebra with isolated singularities.

PROPOSITION 4.1. *Suppose* $\mathrm{CM}(\mathrm{gr}\,R)_0$ *is of finite type. Then the functor* $\mathrm{mod}(\mathrm{gr}\,R)_0 \to \mathrm{mod}\,R_{\mathfrak{m}}$ *given by* $A \mapsto A_{\mathfrak{m}}$ *induces a bijection between the isomorphism classes of indecomposable Cohen–Macaulay modules in* $\mathrm{mod}(\mathrm{gr}\,R)_0$ *and those in* $\mathrm{CM}(R_{\mathfrak{m}})$. *In particular,* $\mathrm{CM}(R_{\mathfrak{m}})$ *is a Krull–Schmidt category of finite type.*

Proof. Suppose A and B are indecomposable modules in $\mathrm{CM}(\mathrm{gr}\,R)_0$ such that $A_{\mathfrak{m}} \simeq B_{\mathfrak{m}}$. Then $\hat{A} \simeq \hat{B}$ and so $A \simeq B$ by [10, Lemma 5.8]. Thus the localization functor $\mathrm{mod}(\mathrm{gr}\,R)_0 \to \mathrm{mod}\,R_{\mathfrak{m}}$ induces an injection from the isomorphism classes of Cohen–Macaulay modules in $\mathrm{mod}(\mathrm{gr}\,R)_0$ to those in $\mathrm{mod}\,R_{\mathfrak{m}}$.

Next suppose A is an indecomposable Cohen–Macaulay module in $\mathrm{mod}\,R_{\mathfrak{m}}$. Then $\hat{A}$ is a Cohen–Macaulay module in $\mathrm{mod}\,\hat{R}$. Since $\mathrm{CM}(\mathrm{gr}\,R)_0$ is of finite type, by Theorem 3.2 there is some B in $\mathrm{CM}(\mathrm{gr}\,R)_0$ such that $\hat{A} \simeq \hat{B}$. Since

$\hat{B} \simeq \hat{B}_{\mathfrak{m}}$ we have $A \simeq B_{\mathfrak{m}}$ because modules in $\operatorname{mod} R_{\mathfrak{m}}$ with isomorphic completions are isomorphic (slight generalization of [10, Lemma 5.8]). Hence B is an indecomposable module in $\mathrm{CM}(\operatorname{gr} R)_0$ such that $B_{\mathfrak{m}} \simeq A$, which finishes the proof of the proposition.

As an immedatie consequence of this result we have the following.

COROLLARY 4.2. *If there is an indecomposable Cohen–Macaulay $R_{\mathfrak{m}}$-module A whose endomorphism ring is not local, then R and $\hat{R}$ are of infinite Cohen–Macaulay type.*

We next want to consider the special case when R is a quasihomogeneous k-algebra, k a field. This means that $R = \coprod_{n \in \mathbf{Z}} R_n$ is a commutative graded ring with grading group $\mathbf{Z}$, the integers, satisfying:

(a) $R_0 = k$.

(b) R_i are finite-dimensional k-vector spaces with $R_i = 0$ for all $i < 0$.

(c) There are a finite number of homogeneous elements $x_1, \ldots, x_n$ such that $R = k[x_1, \ldots x_n]$.

Then R is a graded local ring with graded radical $(x_1, \ldots, x_n)$ which is a maximal ideal in R. Then of course $R_{\mathfrak{m}}$ is a local ring with radical $\mathfrak{m}R_{\mathfrak{m}}$. We now want to investigate the Krull–Schmidt subcategory $(\operatorname{mod} R_{\mathfrak{m}})_0$ of $\operatorname{mod} R_{\mathfrak{m}}$ in this special case.

PROPOSITION 4.3. *Suppose M is in $(\operatorname{mod} R_{\mathfrak{m}})_0$ and $\ldots \xrightarrow{d_2} P_1 \xrightarrow{d_1} P_0 \xrightarrow{d_0} M \to 0$ is a minimal projective resolution of M. Then $\Omega^i(M) = \operatorname{Im} d_i$ is in $(\operatorname{mod} R_{\mathfrak{m}})_0$ for all $i \geqslant 0$.*

Proof. Suppose A in $\operatorname{mod}(\operatorname{gr} R)_0$ is such that $A_{\mathfrak{m}} = M$. Let

$$\ldots \to Q_1 \xrightarrow{\bar{d}_1} Q_0 \xrightarrow{\bar{d}_0} A \to 0$$

be a minimal projective resolution of A in $\operatorname{mod}(\operatorname{gr} R)_0$. Then $\operatorname{Ker} \bar{d}_i \subset \mathfrak{m}Q_i$ for all $i \geqslant 0$, which implies that

$$\ldots \to (Q_1)_{\mathfrak{m}} \xrightarrow{(\bar{d}_1)_{\mathfrak{m}}} (Q_0)_{\mathfrak{m}} \xrightarrow{(\bar{d}_0)_{\mathfrak{m}}} A_{\mathfrak{m}} \to 0$$

is a minimal projective resolution of $A_{\mathfrak{m}} \simeq M$ over $R_{\mathfrak{m}}$. Since $(\Omega^i_R(A))_{\mathfrak{m}} = \Omega^i(M)$ for all $i \geqslant 0$, $\Omega^i(M)$ is in $(\operatorname{mod} R_{\mathfrak{m}})_0$ for all $i \geqslant 0$.

Similar arguments show the following.

PROPOSITION 4.4. *Suppose A and B are in $(\operatorname{mod} R_{\mathfrak{m}})_0$. Then $\operatorname{Ext}^i_{R_{\mathfrak{m}}}(A, B)$ and $\operatorname{Tor}_i^{R_{\mathfrak{m}}}(A, B)$ are in $(\operatorname{mod} R_{\mathfrak{m}})_0$ for all $i \geqslant 0$.*

Thus we see that the Krull–Schmidt subcategory $(\operatorname{mod} R_{\mathfrak{m}})_0$ of $\operatorname{mod} R$ is quite large. It would be interesting to have some description of this subcategory.

Now some of the modules in $(\operatorname{mod} R_{\mathfrak{m}})_0$ can be described without reference to the graded ring R. For example, let $\mathfrak{r}$ be the radical of $R_\mathfrak{m}$. Then $R_\mathfrak{m}/\mathfrak{r}^i$ is in $(\operatorname{mod} R_\mathfrak{m})_0$ for all $i \geqslant 0$. Therefore each indecomposable summand of $\Omega^j(R_\mathfrak{m}/\mathfrak{r}^i)$ has a local endomorphism ring for all j and $i \geqslant 0$. This naturally suggests the following question. If S is a local ring with radical $\mathfrak{r}$, then does every indecomposable summand of $\Omega^j(S/\mathfrak{r}^i)$ have a local endomorphism ring for all j and $i \geqslant 0$? Or can this property be used to test in a fairly effective way whether a local ring is the localization of a quasihomogeneous ring at its graded radical? Clearly the same type of remarks pertain to the class of $R_\mathfrak{m}$-modules in $(\operatorname{mod} R_\mathfrak{m})_0$ which can be described solely in terms of $\operatorname{mod} R_\mathfrak{m}$ without reference to $\operatorname{mod}(\operatorname{gr} R)_0$.

5. Examples

This section is devoted to giving examples illustrating previous results.

A finite-dimensional G-graded k-algebra R is automatically Cohen–Macaulay with isolated singularities, and $\operatorname{CM}(\operatorname{gr} R)_0$ is the same as $\operatorname{mod}(\operatorname{gr} R)_0$. Since further $\hat{R} = R$, we have the following special case of Theorem 2.2.

PROPOSITION 5.1. *Let R be a finite-dimensional G-graded k-algebra, where G is a finite group whose order is invertible in k. If $0 \to A \to B \to C \to 0$ is an almost split sequence in* $\operatorname{mod}(\operatorname{gr} R)_0$, *then it is a direct sum of almost split sequences in* $\operatorname{mod} R$.

This result generalizes the following result about skew group ring constructions from [12].

COROLLARY 5.2. *Let Λ be a finite-dimensional k-algebra and G a group of automorphisms of Λ with the order of G invertible in k. If $0 \to A \to B \to C \to 0$ is an almost split sequence in* $\operatorname{mod} \Lambda$, *then $0 \to \Lambda G \otimes_\Lambda A \to \Lambda G \otimes_\Lambda B \to \Lambda G \otimes_\Lambda C \to 0$ is a direct sum of almost split sequences of ΛG-modules.*

Proof. ΛG is a G-graded algebra, where $(\Lambda G)_g = \Lambda_g$ for $g \in G$. The fuctor $\Lambda G \otimes_\Lambda\colon \operatorname{mod} \Lambda \to \operatorname{mod}(\operatorname{gr} \Lambda G)_0$ is an equivalence of categories [11]. The claim is then a direct consequence of Proposition 5.1.

Interesting commutative examples are provided by the graded algebras studied by Geigle and Lenzing [9]. As Lenzing pointed out to us there are examples where indecomposable graded modules decompose under completion, and this inspired us to generalize our results from [6].

For example, let $R = k[x, y, z]/(x^2 + y^3 + z^4)$, where the characteristic of k is not 2, and let G be the abelian group with generators $\vec{x}$, $\vec{y}$, $\vec{z}$ and relations $\vec{x}^2 = \vec{y}^3 = \vec{z}^4$. Then $G \simeq \mathbf{Z} \oplus \mathbf{Z}_2$. We define degree $x = \vec{x}$, degree $y = \vec{y}$, degree $z = \vec{z}$. $\hat{R}$ is known to be of finite Cohen–Macaulay type, and hence R is of finite graded Cohen–Macaulay type. The AR-quiver for R is the translation quiver $\mathbf{Z}\tilde{\mathbf{E}}_7$ [9], and for $\hat{R}$ it is known to be $\mathbf{Z}\mathbf{E}_6/\langle \tau \rangle$, where τ is the translation $D\operatorname{Tr}$.

Using that almost split sequences in $\mathrm{CM}(\mathrm{gr}\,R)_0$ go to a direct sum of almost split sequences in $\mathrm{CM}(\hat{R})$, that the ranks of the indecomposable modules are known in both cases and that rank is preserved by completion, it is easy to see that up to shift there are exactly two indecomposable modules in $\mathrm{CM}(\mathrm{gr}\,R)_0$ which decompose under completion. One module has rank 4 and the completion decomposes into a direct sum of two nonisomorphic indecomposable modules of rank 2, and the other one has rank 2 and the completion decomposes into a direct sum of two nonisomorphic modules of rank 1.

A similarly defined example from [9] is $R = k[x, y, z]/(x^2+y^2+z^2)$, where the characteristic of k is not 2. Then R is graded by the abelian group G with generators $\bar{x}$, $\bar{y}$, $\bar{z}$ and relations $\bar{x}^2 = \bar{y}^2 = \bar{z}^2$ in a similar way as before, and $G \simeq \mathbf{Z}\oplus\mathbf{Z}_2\oplus\mathbf{Z}_2$. Then R has AR-quiver $\mathbf{Z}\tilde{\mathbf{D}}_4$ [9] and $\hat{R}$ has AR-quiver $\bullet \overset{\rightarrow}{\underset{\leftarrow}{\rightleftarrows}} \bullet$. Here both indecomposable modules in $\mathrm{CM}(\hat{R})$ have rank 1. In $\mathrm{CM}(\mathrm{gr}\,R)_0$ there is only one indecomposable module, up to shift, which has rank greater than 1, and the completion of this module decomposes into a direct sum of two isomorphic modules of rank 1.

As mentioned in the introduction, Lenzing has given a different proof of Theorem 2.2 when G is abelian. He reduces to a skew group construction and the $\mathbf{Z}$-graded case.

If R is a G-graded T-algebra of finite graded Cohen–Macaulay type, it would be interesting to know if there is always a $\mathbf{Z}$-grading on R such that each homogeneous component R_n of R is finite-dimensional over k. It would also be interesting to know if any Cohen–Macaulay $\hat{T}$-algebra of finite Cohen–Macaulay type is of the form $\hat{R}$ for a Cohen–Macaulay $\mathbf{Z}$-graded T-algebra R such that each R_m is finite-dimensional over k, when k is an algebraically closed field of characteristic zero. For finite-dimensional algebras this is known to hold since every k-algebra of finite representation type is standard [7]. (But it is not true if k has characteristic 2.) In the case of two-dimensional tame orders it follows from the work on the classification of such orders of finite representation type [13]. Finally, the only known commutative rings of finite representation type are also completions of graded rings.

References

[1] M. Auslander, *Functors and morphisms determined by objects*, in: Representation Theory of Algebras, Proc. Conf. Philadelphia 1976, Dekker, 1978, 1–327.

[2] M. Auslander and R. O. Buchweitz, *The homological theory of Cohen Macaulay approximations*, Bull. Soc. Math. France, to appear.

[3] M. Auslander and I. Reiten, *The Cohen Macaulay type of Cohen Macaulay rings*, Adv. in Math. 73 (1989), 1–23.

[4] —, —, *Almost split sequences for* **Z**-*graded rings*, in: Vector Bundles, Singularity Theory and Representation Theory, Proc. Conf. Lambrecht, Lecture Notes in Math. 1273, Springer, 1987, 232–243.

[5] —, —, *Almost split sequences for abelian group graded rings*, J. Algebra 113 (1988), 29–39.

[6] —, —, *Cohen–Macaulay modules for graded Cohen–Macaulay rings and their completions*, in: Commutative Algebra, Proc. Berkeley 1987, M. Hochster, C. Huneke and J. D. Sally (eds.), Springer, 1989, 21–31.

[7] R. Bautista, P. Gabriel, A. V. Roiter and L. Salmerón, *Representation-finite algebras and multiplicative bases*, Invent. Math. 81 (1985), 217–285.

[8] M. Cohen and S. Montgomery, *Group graded rings, smash products and group actions*, Trans. Amer. Math. Soc. 282 (1984), 237–258.

[9] W. Geigle and H. Lenzing, *Perpendicular categories with applications to representations and sheaves*, preprint, Paderborn 1988.

[10] J. Herzog und E. Kunz (eds.), *Der kanonische Modul eines Cohen–Macaulay Rings*, Lecture Notes in Math. 238, Springer, 1971.

[11] C. Năstăsescu and F. Van Oystaeyen, *Graded Ring Theory*, North–Holland, 1982.

[12] I. Reiten and C. Riedtmann, *Skew group algebras in the representation theory of artin algebras*, J. Algebra 92 (1985), 224–282.

[13] I. Reiten and M. Van den Bergh, *Two-dimensional tame and maximal orders of finite representation type*, Mem. Amer. Math. Soc. 408 (1989), 72 pp.

CHAPTER IX

Grothendieck groups and Cohen-Macaulay approximations

[73] Grothendieck groups of algebras and orders (with I. Reiten, 1986).
[86] Grothendieck groups of algebras with nilpotent annihilators (with I. Reiten, 1988).
[89] The homological theory of maximal Cohen-Macaulay approximations (with I. Reiten, 1989).
[97] Liftings and weak liftings of modules (with S. Ding, Ø. Solberg, 1993).

Auslander worked on other aspects of higher dimensional rings, in particular, commutative ones, partially inspired by developments for artin algebras.

M. C. R. Butler had shown that for artin algebras of finite representation type the relations given by the almost split sequences generate the relations for the Grothendieck group. He used the principle, emphasized in [22], of expressing the category mod Λ of finitely generated Λ-modules as a quotient category of the finitely presented (that is, coherent) contravariant functors from mod Λ to abelian groups. An important point is that the category of finitely presented functors has global dimension two, as observed in [22]. This principle of viewing the category in question as a quotient category of an abelian category of finite global dimension was further developed and applied in [73]. One application was to lattices over general orders, where the generalization of M. C. R. Butler's result also provided a new method for computing the Grothendieck groups in the case of finite representation type. Another application was to computing Grothendieck groups of invariant rings. These results were generalized in [86], and have been extended further by other authors.

While the work on (maximal) Cohen-Macaulay modules discussed in Chapter VIII dealt with the Cohen-Macaulay modules themselves, the point of view of [89] was to approximate arbitrary modules by Cohen-Macaulay modules and to describe the kernels of such approximations. Formulated in previous terminology, it dealt with the question of proving that the category of Cohen-Macaulay modules over a commutative Cohen-Macaulay ring is contravariantly finite in the category of finitely generated modules. This was inspired by analogous work by Auslander-Smalø on artin algebras, as well as by some results of a similar nature in [27], [55] and [66]. This work has motivated a lot of further research. For example the Cohen-Macaulay approximation gave rise to what is called Auslander's δ-invariant, which counts the number of nonprojective indecomposable summands in a (minimal) approximation. Auslander proved that for a commutative local noetherian Gorenstein ring which is not regular, δ vanishes for the residue field and its syzygies. This concept has been applied and generalized by some of his students and several other authors in commutative ring theory.

In [97] weak liftings of modules are introduced, and the theory of Cohen-Macaulay approximations is applied to liftings and weak liftings of modules. There is ongoing research also in this direction.

Journal of Pure and Applied Algebra 39 (1986) 1-51
North-Holland

GROTHENDIECK GROUPS OF ALGEBRAS AND ORDERS

Maurice AUSLANDER*
Dept. of Mathematics, Brandeis University, Waltham, MA 02254, USA

Idun REITEN*
Dept. of Mathematics, University of Trondheim, Trondheim, Norway

Communicated by H. Bass
Received 12 August 1984
Revised 25 September 1984

Introduction

The aim of this paper is to present a method for finding presentations of Grothendieck groups of abelian categories. This method depends on the concept of a coherent pair $(\mathbf{A}, \mathbf{B})$ of additive categories in which idempotents split over a commutative ring, where the essential properties are that the categories of finitely presented contravariant functors mod $\mathbf{A}$ and mod $\mathbf{B}$ are abelian and that the subcategory $\mathbf{B}$ of $\mathbf{A}$ is such that a functor in mod $\mathbf{A}$ restricts to a functor in mod $\mathbf{B}$. Under these circumstances, the restriction mod $\mathbf{A} \to$ mod $\mathbf{B}$ is the quotient of mod $\mathbf{A}$ by mod $\mathbf{A}/\mathbf{B}$, the Serre subcategory consisting of the functors vanishing on $\mathbf{B}$. We then have the usual exact sequence of Grothendieck groups

$$K_0(\text{mod } \mathbf{A}/\mathbf{b}) \xrightarrow{i} K_0(\text{mod } \mathbf{A}) \to K_0(\text{mod } \mathbf{B}) \to 0.$$

The idea is then, given $\mathbf{B}$, to get a presentation for $K_0(\text{mod } \mathbf{B})$ by making a convenient choice for $\mathbf{A} \supset \mathbf{B}$ such that $(\mathbf{A}, \mathbf{B})$ is a coherent pair. In particular, we would like to have that mod $\mathbf{A}$ is regular, that is, every object has finite projective dimension, for in this case it is well known and easy to see that the Cartan map $c_{\mathbf{A}} : K_0(\mathbf{A}, 0) \to K_0(\text{mod } \mathbf{A})$ is an isomorphism which is given by $A \mapsto (\cdot, A)$ for all A in $\mathbf{A}$. Here $K_0(\mathbf{A}, 0)$ denotes the free group on the isomorphism classes of objects of $\mathbf{A}$, modulo split exact sequences. We would also like $\mathbf{A}$ to be a Krull-Schmidt category, so that $K_0(\mathbf{A}, 0)$ is a free abelian group with the isomorphism classes of indecomposable objects in $\mathbf{A}$ as basis. The next problem is then to be able to compute Im i.

* We would like to thank the N.S.F., the Norwegian Marshall Fund and the Norwegian Research Council for support during the preparation of this paper.

Not only does knowing that the morphism $i: K_0(\text{mod}\,\mathbf{A/B}) \to K_0(\text{mod}\,\mathbf{A})$ is a monomorphism sometimes help in computing the desired relations in $K_0(\text{mod}\,\mathbf{A})$ for $K_0(\text{mod}\,\mathbf{B})$, but it also enables us in some instances to give criteria for when $\mathbf{A}$ is of finite representation type in terms of when the relations in $K_0(\text{mod}\,\mathbf{A})$ for $K_0(\text{mod}\,\mathbf{B})$ are generated by almost split sequences.

We prove that i is a monomorphism in the following situations:

(a) mod $\mathbf{B}$ is regular.

(b) Each object in mod $\mathbf{B}$ has finite length.

(c) mod $\mathbf{A}$ is an R-category such that $\text{Hom}(A_1, A_2)$ is an R-module of finite length for all A_1, A_2 in $\mathbf{A}$ and the objects in mod $\mathbf{A/B}$ are of finite length.

If Λ is an artin algebra and $\mathbf{P}(\Lambda)$ denotes the category of finitely generated projective Λ-modules, we have $K_0(\text{mod}\,\Lambda) = K_0(\text{mod}\,\mathbf{P})$, and in this case $(\text{mod}\,\Lambda, \mathbf{P}(\Lambda))$ is a nice choice for a coherent pair. This gives the following presentation:

$$K_0(\text{mod}(\text{mod}\,\Lambda/\mathbf{P})) \xrightarrow{i} K_0(\text{mod}(\text{mod}\,\Lambda)) \to K_0(\text{mod}\,\Lambda) \to 0$$

and was already studied by Butler in [23]. Here gl.dim $\text{mod}(\text{mod}\,\Lambda) \le 2$ and mod Λ is a Krull–Schmidt category, so $\mathbf{A} = \text{mod}\,\Lambda$ has the desired properties. This sequence gives the theorem of Butler–Auslander that $K_0(\text{mod}\,\Lambda)$ is the free group $K_0(\text{mod}\,\Lambda, 0)$ modulo the relations generated by almost split sequences if and only if Λ is of finite representation type [23]. As a consequence of the more general results stated above we get that i is a monomorphism.

If Λ is a classical order over a complete discrete valuation ring R, a convenient choice for a coherent pair is $(\mathbf{L}, \mathbf{P}(\Lambda))$, where $\mathbf{L}$ is the category of Λ-lattices. Then we get the exact sequence

$$K_0(\text{mod}(\mathbf{L/P})) \xrightarrow{i} K_0(\text{mod}\,\mathbf{L}) \to K_0(\text{mod}\,\Lambda) \to 0,$$

and if Λ is of finite lattice type, this gives again a presentation of $K_0(\text{mod}\,\Lambda)$ as a factor group of $K_0(\mathbf{L}, 0)$ modulo the relations generated by almost split sequences [23]. In this case i is not necessarily a monomorphism, but if it is, we prove the same characterization of finite lattice type as for artin algebras. We prove that i is always a monomorphism when Λ is an order in a simple algebra.

Some of the results mentioned above for artin algebras and classical orders are extended to lattices over the higher-dimensional orders introduced by Auslander in [6], which are regular outside maximal ideals. These include, amongst other rings, the commutative rings which are complete integrally closed two dimensional local domains, and isolated quotient singularities. In both these cases the lattices are the Cohen–Macauley modules. Our main emphasis is giving criteria for when such orders are of finite lattice type.

We next turn our attention to showing how coherent pairs can be used to get information about the Grothendieck groups of quotient singularities, a notion we now recall. Let S be the ring $\mathbb{C}[[X_1, \ldots, X_n]]$ $(n \ge 2)$ of formal power series over $\mathbb{C}$, the complex numbers and let $G \subset \text{GL}(n, \mathbb{C})$ act linearly on S. Then $R = S^G$, the invariant ring, is called a quotient singularity. The skew group ring SG enables us to construct an interesting coherent pair, namely $\mathbf{A}=\mathbf{P}S[G]$, the category of finitely generated projective $S[G]$-modules and $\mathbf{B}$ the additive category generated by the projective $S[G]$-module S. Since $R=\operatorname{End}_{S[G]}(S)$ we have that $\operatorname{mod}\mathbf{B}$ is equivalent to $\operatorname{mod} R$. Further we also know that $\operatorname{gl.dim} S[G]=n<\infty$ and $\operatorname{mod} S[G]$ is Krull–Schmidt. Therefore we obtain the exact sequence of Grothendieck groups

$$K_0(\operatorname{mod}\mathbf{P}S[G]/\mathbf{P}(R))\xrightarrow{i} K_0(\mathbf{P}SG,0)\to K_0(\operatorname{mod}\mathrm{R})\to 0.$$

This implies that $K_0(\operatorname{mod}\mathrm{R})$ is finitely generated since $K_0(\mathbf{P}S[G],0)$ is naturally isomorphic to $K_0(\operatorname{mod}\mathbb{C}[G])$ which is clearly a finitely generated free abelian group. Moreover if we let V be the $\mathbb{C}[G]$-module given by the representation $G\subset \mathrm{GL}(n,\mathbb{C})$, we have that the ideal in the ring $K_0(\operatorname{mod}\mathbb{C}[G])$ generated by the element $\sum_{i=0}^{n}(-1)^i[\bigwedge^i V]$ (when $\bigwedge^i V$ is the i-th exterior product) is contained in $\operatorname{Ker}(K_0(\operatorname{mod}\mathbb{C}[G])\to K_0(\operatorname{mod}\mathrm{R}))$. Moreover, if G acts freely on $V-\{0\}$ (i.e. R is an isolated singularity), we have that

$$K_0(\operatorname{mod}\mathbb{C}[G])\Big/\Big(\sum_{i=0}^{u}(-1)^i[\textstyle\bigwedge^i V]K_0(\operatorname{mod}\mathbb{C}[G])\Big)\cong K_0(\operatorname{mod} R)$$

which implies that $K_0(\operatorname{mod} R)\cong\mathbb{Z}\amalg H$ where H is a finite abelian group.

In the last chapter we study more closely the relationship between Cartan maps and almost split sequences. In particular the kernel and cokernel of a Cartan map can be computed from information given by almost split sequences. We also give a general description of when the Cartan map is mono or when the cokernel of the Cartan map is torsion. Here we only state the following special cases for an artin algebra Λ.

(1) c_Λ is a monomorphism if and only if the following holds. Whenever $0\to A\to B\to C\to 0$ and $0\to A'\to B'\to 0$ are exact sequences in $\operatorname{mod}\Lambda$ such that the nonprojective parts of $A\amalg C\amalg B'$ and $A'\amalg C'\amalg B$ are isomorphic, then they are isomorphic.

(2) $\operatorname{Coker} c_\Lambda$ is torsion if and only if the following holds. For each X in $\operatorname{mod}\Lambda$ there is a pair of exact sequences $0\to A\to B\to C\to 0$ and $0\to A'\to B'\to 0$ and an integer $n>0$, such that the nonprojective parts of $nX\amalg A\amalg C\amalg B'$ and $A'\amalg C'\amalg B$ are isomorphic.

For a two-dimensional integrally closed local domain with algebraically closed residue field we prove that the cokernel of the Cartan map is isomorphic to the class group $C(R)$. Here we do not use the idea of a coherent pair, but consider instead the Serre subcategory $\mathrm{f.l.mod}\, R$ of $\operatorname{mod} R$ consisting of the R-modules of finite length and the quotient category $\operatorname{mod} R/\mathrm{f.l.mod}\, R$. This gives the exact sequence

$$K_0(\mathrm{f.l.mod}\, R)\xrightarrow{i} K_0(\operatorname{mod} R)\to K_0(\operatorname{mod} R/\mathrm{f.l.mod}\, R)\to 0.$$

Here $K_0(\operatorname{mod}\mathrm{R}/\mathrm{f.l.mod}\, R)\simeq Z\amalg C(R)$ and i turns out to be the zero map. When $\operatorname{mod} R$ has only a finite number of reflexive modules, $K_0(\operatorname{mod} R)$ can be computed from knowing the almost split sequences, as pointed out above. Since the class group can be computed from the graph associated with resolutions of rational singularities, the above result ties in nicely with the connection between the singularity graph and almost split sequences established by Auslander [8].

CHAPTER 1

In this chapter we introduce the notion of a coherent pair of categories, and begin a study of the exact sequences of K-groups associated with a coherent pair.

1. Coherent pairs of categories

Throughout this section we assume that R is a fixed commutative ring and $\mathbf{A}$ is an addititive skeletally small R-category. By an $\mathbf{A}$-module we mean an additive contraviariant R-functor $M:\mathbf{A}\to \operatorname{Mod} R$, where $\operatorname{Mod} R$ denotes the R-modules. We denote the category of all $\mathbf{A}$-modules by $\operatorname{Mod}\mathbf{A}$. In analogy with modules over rings, we recall the notions of finitely generated and finitely presented $\mathbf{A}$-modules [3].

An $\mathbf{A}$-module M is finitely generated if there is an epimorphism $(\cdot, A)\to M\to 0$ where $(\cdot, A)$ is the representable functor $\operatorname{Hom}_{\mathbf{A}}(\cdot, A)$. An $\mathbf{A}$-module M is finitely presented if there is an exact sequence $(\cdot, A_1)\to(\cdot, A_0)\to M\to 0$. The full subcategory of $\operatorname{Mod}\mathbf{A}$ consisting of the finitely presented $\mathbf{A}$-modules will be denoted by $\operatorname{mod}\mathbf{A}$. It is the category $\operatorname{mod}\mathbf{A}$ which will be of main interest to us. We recall, without proofs, which are essentially the same as for modules over rings, some of the basic properties of finitely presented $\mathbf{A}$-modules [3]. We begin with the following useful characterization.

Proposition 1.1. *The following are equivalent for an* $\mathbf{A}$*-module* M''.

(a) M'' *is finitely presented.*

(b) M'' *is finitely generated, and given any exact sequence* $0\to M'\to M\to M''\to 0$ *of* $\mathbf{A}$*-modules with* M *finitely generated, then* M' *is finitely generated.*

In addition, we have the following important properties of finitely presented $\mathbf{A}$-modules.

Proposition 1.2. *Let* $0\to M'\to M\to M''\to 0$ *be an exact sequence of* $\mathbf{A}$*-modules.*

(a) *If* M' *and* M'' *are finitely presented, then* M *is also finitely presented.*

(b) *If* M' *is finitely generated and* M *is finitely presented, then* M'' *is finitely presented.*

We now want to describe when $\mathbf{A}$ has the property that $\operatorname{mod}\mathbf{A}$ is abelian and the inclusion $\operatorname{mod}\mathbf{A}\to\operatorname{Mod}\mathbf{A}$ is exact.

Let $g: A_2 \to A_1$ be a morphism in **A**. A pseudokernel of g is a morphism $f: A_3 \to A_2$ in **A** such that

$$(\cdot, A_3) \xrightarrow{(\cdot, f)} (\cdot, A_2) \xrightarrow{(\cdot, g)} (\cdot, A_1)$$

is exact in Mod **A**. A pseudocokernel of f is a morphism $h: A_1 \to A_0$ such that

$$(A_0, \cdot) \xrightarrow{(h, \cdot)} (A_1, \cdot) \xrightarrow{(g, \cdot)} (A_2, \cdot)$$

is exact in Mod $\mathbf{A}^{\mathrm{op}}$. The category **A** is said to be coherent if every morphism in **A** has a pseudokernel. Dually, **A** is said to be cocoherent if every morphism in **A** has a pseudocokernel. We will usually deal with coherent categories, leaving the dual situation for cocoherent categories to the reader. In analogy with the notion of coherent rings we have the following characterizations of when **A** is coherent [3].

Proposition 1.3. *The following are equivalent for* **A**.

(a) **A** *is coherent.*

(b) *If* $0 \to M_1 \to M_2 \to M_3 \to M_4 \to 0$ *is an exact sequence of* **A**-*modules with* M_2 *and* M_3 *finitely presented, then* $M_1, \operatorname{Im} f, M_4$ *are finitely presented.*

(c) mod **A** *is an abelian category and the inclusion* mod **A** $\to$ Mod **A** *is exact.*

By a subcategory **B** of **A** we shall always mean a full additive subcategory of **A** such that if $A_1 \simeq A_2$ and A_1 is in **B**, then A_2 is also in **B**. Suppose **B** is a subcategory of **A**. Then we have the restriction functor Mod **A** $\to$ Mod **B** given by $M \to M|\mathbf{B}$ for all M in Mod **A**. We now describe when M being a finitely presented **A**-module implies that $M|\mathbf{B}$ is a finitely presented **B**-module, at least in the case **A** is coherent.

We recall that **B** is contravariantly finite in **A** or **A** is contravariantly finite over **B** if $(\cdot, A)|\mathbf{B}$ is a finitely generated **B**-module for each A in **A** [17]. Dually, **B** is covariantly finite in **A** or **A** is covariantly finite over **B** if $(A, \cdot)|\mathbf{B}$ is a finitely generated **B**-module for each A in **A**.

It is easy to see that **A** is contravariantly finite over **B** if and only if for each A in **A** there is a morphism $g: B \to A$ with B in **B** such that for each morphism $h: X \to A$ with X in **B** there is a morphism $f: X \to B$ with the property $h = gf$.

Also, it is obvious that **A** is contravariantly finite over **B** if and only if $\mathbf{A}^{\mathrm{op}}$ is covariantly finite over $\mathbf{B}^{\mathrm{op}}$. Therefore, following our usual practice, we will deal with **A** being contravariantly finite over **B** and leave the dual situation of **A** being covariantly finite over **B** to the reader.

Since the restriction Mod **A** $\to$ Mod **B** is exact it follows that **A** is contravariantly finite over **B** if and only if for each finitely generated **A**-module M we have that $M|\mathbf{B}$ is a finitely generated **B**-module. In the case when **A** is coherent we have the following, whose proof is the same as for modules [17, Proposition 2.1].

Proposition 1.4. *Suppose* **A** *is coherent and* **B** is contravariantly finite in **A**. *Then we have the following.*

(a) **B** *is coherent.*

(b) *If M is a finitely presented* **A**-*module, then* $M|\mathbf{B}$ *is a finitely presented* **B**-*module.*

This result suggests the following definitions. A *coherent pair* $(\mathbf{A}, \mathbf{B})$ is a coherent additive R-category $\mathbf{A}$ together with a full additive subcategory $\mathbf{B}$ of $\mathbf{A}$ which is contravariantly finite in $\mathbf{A}$. A *cocoherent* pair $(\mathbf{A}, \mathbf{B})$ is a cocoherent additive R-category $\mathbf{A}$ together with a full additive R-category $\mathbf{B}$ of $\mathbf{A}$ which is covariantly finite in $\mathbf{A}$.

Suppose $(\mathbf{A}, \mathbf{B})$ is a coherent pair. Then by Proposition 1.4 the restriction $\operatorname{Mod}\mathbf{A} \to \operatorname{Mod}\mathbf{B}$ induces an exact functor $\operatorname{mod}\mathbf{A} \to \operatorname{mod}\mathbf{B}$. We denote by $\operatorname{Ker}(\mathbf{A}, \mathbf{B})$ the full subcategory of $\operatorname{mod}\mathbf{A}$ consisting of all M such that $M|\mathbf{B}=0$. Since the restriction is exact, $\operatorname{Ker}(\mathbf{A}, \mathbf{B})$ is a Serre subcategory of $\operatorname{mod}\mathbf{A}$, i.e. if $0 \to M_1 \to M_2 \to M_3 \to 0$ is an exact sequence in $\operatorname{mod}\mathbf{A}$, then M_2 is in $\operatorname{Ker}(\mathbf{A}, \mathbf{B})$ if and only if M_1 and M_3 are in $\operatorname{Ker}(\mathbf{A}, \mathbf{B})$. Therefore there is a quotient category $\operatorname{mod}\mathbf{A} \to \operatorname{mod}\mathbf{A}/\operatorname{Ker}(\mathbf{A}, \mathbf{B})$ [26]. Now we want to show that the exact functor $\operatorname{mod}\mathbf{A} \xrightarrow{\text{res}} \operatorname{mod}\mathbf{B}$ is the quotient $\operatorname{mod}\mathbf{A} \to \operatorname{mod}\mathbf{A}/\operatorname{Ker}(\mathbf{A}, \mathbf{B})$. This is equivalent to showing the following.

Proposition 1.5. *If $F : \operatorname{mod}\mathbf{A} \to \mathbf{X}$ is an exact functor of abelian R-categories such that $F(M) = 0$ for all M in* $\operatorname{Ker}(\mathbf{A}, \mathbf{B})$, *then there is a unique exact functor $G : \operatorname{mod}\mathbf{B} \to \mathbf{X}$ such that $G \operatorname{res} = F$.*

Proof. For each M in $\operatorname{mod}\mathbf{B}$ choose a fixed projective presentation

$$(\cdot, M_1) \xrightarrow{(\cdot, f)} (\cdot, M_0) \to M \to 0$$

subject to the condition that if $M = (\cdot, B)$, then we let $M_0 = B$ and $M_1 = 0$. Define the functor $S : \operatorname{mod}\mathbf{B} \to \operatorname{mod}\mathbf{A}$ by

$$S(M) = \operatorname{Coker}\left(\operatorname{Hom}_{\mathbf{A}}(\cdot, M_1) \xrightarrow{(\cdot, f)} \operatorname{Hom}_A(\cdot, M_0)\right).$$

Then it is not hard to see that S is a right exact fully faithful functor such that $\operatorname{res} S = 1_{\operatorname{mod}\mathbf{B}}$, the identity on $\operatorname{mod}\mathbf{B}$.

It is clear that Proposition 1.5 now follows from the following.

Lemma 1.6. *Suppose $T : \mathbf{C} \to \mathbf{D}$ is an exact functor of abelian R-categories with the property that there is an additive functor $T' : \mathbf{D} \to \mathbf{C}$ such that $TT' = 1_{\mathbf{D}}$. Letting* $\operatorname{Ker} T$ *be the Serre subcategory of* $\mathbf{C}$ *consisting of all C in* $\mathbf{C}$ *such that $T(C) = 0$, we have that given any exact functor $V : \mathbf{C} \to \mathbf{X}$ of abelian R-categories with* $\operatorname{Ker} V \supset \operatorname{Ker} T$, *there is a unique exact functor $W : \mathbf{D} \to X$ such that $V = WT$.*

Proof. Let $0 \to D_1 \to D_2 \to D_3 \to 0$ be an exact sequence in $\mathbf{D}$. Then

$$(*) \qquad 0 \to T'(D_1) \to T'(D_2) \to T'(D_3) \to 0$$

is a complex. Since $(*)$ $0 \to TT'(D_1) \to TT'(D_2) \to TT'(D_3) \to 0$ is exact, we have that $T(\text{homology } *) = 0$. So the homology of $(*)$ is in Ker T. Therefore if $V: \mathbf{C} \to \mathbf{X}$ is exact with Ker $V \supset$ Ker T, then $0 \to VT'(D_1) \to VT'(D_2) \to VT'(D_3) \to 0$ is exact. Hence $VT': \mathbf{D} \to X$ is an exact functor such that $(VT')T = V$ since $T(T'T - 1_{\mathbf{C}}) = 0$ implies that $\operatorname{Im}(T'T - 1_{\mathbf{C}}) \subset \operatorname{Ker} T$ and consequently $V(T'T - 1_{\mathbf{C}}) = 0$.

Now suppose $W: \mathbf{D} \to \mathbf{X}$ is an exact functor such that $V = WT$. Then $VT' = (WT)T' = W$. Therefore the exact functor $VT: \mathbf{D} \to \mathbf{X}$ is our desired exact functor.

We show that there is a coherent additive category $\mathbf{C}$ such that $\ker(\mathbf{A}, \mathbf{B})$ is equivalent to mod $\mathbf{C}$ when $(\mathbf{A}, \mathbf{B})$ is a coherent pair.

For each A in $\mathbf{A}$ define the functor $(\cdot, A)/\mathbf{B}: \mathbf{A} \to \operatorname{Mod} R$ by $(\cdot, A)/\mathbf{B}(X) = \operatorname{Hom}_{\mathbf{A}}(X, A)$ modulo the R-submodule consisting of all $f: X \to A$ which can be factored as $X \to B \to A$ with B in $\mathbf{B}$. Since $\mathbf{B}$ is contravariantly finite in $\mathbf{A}$, we know that for each A in $\mathbf{A}$ there is a morphism $B \to A$ such that $(X, B) \to (X, A) \to 0$ is exact for each X in $\mathbf{B}$. It then follows that $(\cdot, B) \to (\cdot, A) \to (\cdot, A)/\mathbf{B} \to 0$ is exact, which shows that $(\cdot, A)/\mathbf{B}$ is finitely presented for all A in $\mathbf{A}$. Therefore each $(\cdot, A)/\mathbf{B}$ is in $\operatorname{Ker}(\mathbf{A}, \mathbf{B})$ since each $(\cdot, A)/\mathbf{B}$ clearly vanishes on $\mathbf{B}$.

Next we observe from the exact sequence $(\cdot, B) \to (\cdot, A) \to (\cdot, A)/\mathbf{B} \to 0$ such that the epimorphism $(\cdot, A) \to (\cdot, A)/\mathbf{B}$ induces isomorphisms $((\cdot, A)/\mathbf{B}, M) \xrightarrow{\sim} ((\cdot, A), M)$ for all M in $\operatorname{Ker}(\mathbf{A}, \mathbf{B})$. As an easy consequence of this observation we have the following.

Lemma 1.7. *Let* $(\mathbf{A}, \mathbf{B})$ *be a coherent pair. Then we have for each* A *in* $\mathbf{A}$:

(a) $((\cdot, A)/\mathbf{B}, M) = M(A)$ *for all* M *in* $\operatorname{Ker}(\mathbf{A}, \mathbf{B})$.

(b) $(\cdot, A)/\mathbf{B}$ *is a projective object in* $\operatorname{Ker}(\mathbf{A}, \mathbf{B})$.

(c) *For each* M *in* $\operatorname{Ker}(\mathbf{A}, \mathbf{B})$ *there is an epimorphism* $(\cdot, A)/\mathbf{B} \to M \to 0$ *for some* A *in* $\mathbf{A}$.

Define the additive R-category $\mathbf{A}/\mathbf{B}$ as follows. The objects of $\mathbf{A}/\mathbf{B}$ are the objects of $\mathbf{A}$ and $\operatorname{Hom}_{\mathbf{A}/\mathbf{B}}(A_1, A_2) = ((\cdot, A_1)/\mathbf{B}, (\cdot, A_2)/\mathbf{B})$ for all A_1, A_2 in $\mathbf{A}/\mathbf{B}$. Next define the additive functor $\alpha: \operatorname{Ker}(\mathbf{A}, \mathbf{B}) \to \operatorname{Mod} \mathbf{A}/\mathbf{B}$ by $\alpha(M)(A) = M(A) = ((\cdot, A)/\mathbf{B}, M)$ for all A in $\mathbf{A}/\mathbf{B}$. Since $\alpha: \operatorname{Ker}(\mathbf{A}, \mathbf{B}) \to \operatorname{Mod} \mathbf{A}/\mathbf{B}$ is exact, it follows from Lemmas 1.7 that $\alpha(M)$ is in mod $\mathbf{A}/\mathbf{B}$ for all M in $\ker(\mathbf{A}, \mathbf{B})$. Also as a consequence of the exactness of α and Lemma 1.7 we have the following.

Proposition 1.8. *Let* $(\mathbf{A}, \mathbf{B})$ *be a coherent pair. Then*

(a) *The functor* $\operatorname{Ker}(\mathbf{A}, \mathbf{B}) \to \operatorname{mod} \mathbf{A}/\mathbf{B}$ *given by* $M \to \alpha(M)$ (*where* $\alpha(M)(A) = M(A)$) *is an isomorphism of categories.*

(b) $\mathbf{A}/\mathbf{B}$ *is coherent.*

We will usually consider the isomorphism $\alpha: \operatorname{Ker}(\mathbf{A}, \mathbf{B}) \to \operatorname{mod} \mathbf{A}/\mathbf{B}$ an identification and write mod $\mathbf{A}/\mathbf{B}$ for $\operatorname{Ker}(\mathbf{A}, \mathbf{B})$.

We end this section by pointing out the following useful special case of a coherent category. For an object B in a category $\mathbf{B}$ we here denote by add B the full subcategory of $\mathbf{B}$ whose objects are summands of finite direct sums of copies of B.

Proposition 1.9. *Let* $\mathbf{B}$ *be a subcategory of an* R*-category* $\mathbf{A}$*, where* $\mathbf{B} = \operatorname{add} B$ *for some* B *in* $\mathbf{B}$*, and assume that* $\operatorname{Hom}(B, X)$ *is a finitely generated* R*-module for all* X *in* $\mathbf{A}$*. Then* $\mathbf{B}$ *is contravariantly finite in* $\mathbf{A}$*, and* $\mathbf{B}$ *is coherent.*

Proof. Observe that since $\operatorname{Hom}(B, X)$ is a finitely generated R-module, there is some integer n and some morphism $B^n \xrightarrow{h} X$ such that any map from B to X factors through h.

2. K-groups

Let $\mathbf{C}$ be a skeletally small abelian R-category. Then we denote by $K_i(\mathbf{C})$ for $i = 0, 1, \ldots$ the Quillen K-groups of $\mathbf{C}$ relative to short exact sequences in $\mathbf{C}$ [32]. In particular $K_0(\mathbf{C})$ is the free abelian group with basis the set of isomorphism classes $[C]$ of objects in $\mathbf{C}$ modulo the subgroup generated by $[C_1] - [C_2] + [C_3]$ whenever there is an exact sequence $0 \to C_1 \to C_2 \to C_3 \to 0$ in $\mathbf{C}$. In other words, $K_0(\mathbf{C})$ is the Grothendieck group of $\mathbf{C}$.

Suppose $\mathbf{C}'$ is a Serre subcategory of $\mathbf{C}$ with quotient $\mathbf{C}''$, a relationship we denote by $\mathbf{C}'' \hookrightarrow \mathbf{C} \xrightarrow{T} \mathbf{C}''$ where $T : \mathbf{C} \to \mathbf{C}''$ is an exact functor with kernel $\mathbf{C}'$ such that given any exact functor of ableian R-categories $V : \mathbf{C} \to \mathbf{X}$ vanishing on $\mathbf{C}'$ there is a unique exact functor $W : \mathbf{C}'' \to \mathbf{X}$ such that $WT = V$. Then Quillen has defined a long exact sequence of K-groups

$$\cdots \to K_{i+1}(\mathbf{C}'') \to K_i(\mathbf{C}') \to K_i(\mathbf{C}) \to K_i(C'') \to \cdots$$

$$\cdots \to K_1(\mathbf{C}) \to K_1(\mathbf{C}'') \to K_0(\mathbf{C}') \to K_0(\mathbf{C}) \to K_0(\mathbf{C}'') \to 0.$$

In particular, suppose $(\mathbf{A}, \mathbf{B})$ is a coherent pair. Then by the results of Section 1, we have that $\operatorname{mod}(\mathbf{A}/\mathbf{B})$ is a Serre subcategory of $\operatorname{mod} \mathbf{A}$ and $\operatorname{mod}(\mathbf{A}/\mathbf{B}) \hookrightarrow \operatorname{mod} \mathbf{A} \xrightarrow{\text{res}} \operatorname{mod} \mathbf{B}$. Hence we have the exact sequence

$$\cdots \to K_{i+1}(\operatorname{mod} \mathbf{B}) \to K_i(\operatorname{mod} \mathbf{A}/\mathbf{B}) \to K_i(\operatorname{mod} \mathbf{A}) \to K_i(\operatorname{mod} \mathbf{B}) \to \cdots$$

$$\cdots \to K_1(\operatorname{mod} \mathbf{A}) \to K_1(\operatorname{mod} \mathbf{B}) \to K_0(\operatorname{mod} \mathbf{A}/\mathbf{B}) \to K_0(\operatorname{mod} \mathbf{A}) \to K_0(\operatorname{mod} \mathbf{B}) \to 0$$

of K-groups.

We will be primarily interested in when

$$0 \to K_0(\operatorname{mod} \mathbf{A}/\mathbf{B}) \to K_0(\operatorname{mod} \mathbf{A}) \to K_0(\operatorname{mod} \mathbf{B})$$

is exact, or what is the same thing, when $K_1(\operatorname{mod} \mathbf{A}) \to K_1(\operatorname{mod} \mathbf{B}) \to 0$ is exact. In order to state and prove our main result along these lines we need a few preliminary remarks.

Let $\mathbf{A}$ be a skeletally small additive R-category. Then we denote by $K_i(\mathbf{A}, 0)$ the Quillen K-groups of $\mathbf{A}$ relative to split exact sequences. In particular $K_0(\mathbf{A}, 0)$ is the free abelian group with basis $[A]$, the isomorphism classes of objects in $\mathbf{A}$ modulo the subgroup generated by all $[A_1]-[A_2]+[A_3]$ where $A_2 \simeq A_1 \amalg A_3$.

Suppose now that $\mathbf{C}$ is a skeletally small abelian R-category and $\mathbf{C}'$ is a subcategory of $\mathbf{C}$. Then the inclusion $\mathbf{C}' \to \mathbf{C}$ induces maps $K_i(\mathbf{C}', 0) \to K_i(\mathbf{C})$ for all i.

Our aim in this section is to prove the following.

Theorem 2.1. *Let $\mathbf{B}$ be a coherent additive skeletally small R-category and n a non-negative integer. Then the following are equivlent.*

(a) *There is a subcategory* $\mathbf{C}$ *of* $\operatorname{mod} \mathbf{B}$ *such that* $K_n(\mathbf{C}, 0) \to K_n(\operatorname{mod} \mathbf{B})$ *is an epimorphism (splittable epimorphism).*

(b) $K_n(\operatorname{mod} \mathbf{B}, 0) \to K_n(\operatorname{mod} \mathbf{B})$ *is an epimorphism (splittable epimorphism).*

(c) *If* $(\mathbf{A}, \mathbf{B})$ *is a coherent pair, then the restriction* $\operatorname{mod} \mathbf{A} \to \operatorname{mod} \mathbf{B}$ *induces an epimorphism (splittable epimorphism)* $K_n(\operatorname{mod} \mathbf{A}) \to K_n(\operatorname{mod} \mathbf{B})$.

Actually, we prove the following somewhat more general result from which Theorem 2.1 follows trivially.

Theorem 2.2. *Let $\mathbf{D}$ be an abelian R-category and $n \geq 0$. Then the following are equivalent.*

(a) *There is a subcategory* $\mathbf{C}$ *of* $\mathbf{D}$ *such that* $K_n(\mathbf{C}, 0) \to K_n(\mathbf{D})$ *is an epimorphism (splittable epimorphism).*

(b) $K_n(\mathbf{D}, 0) \to K_n(\mathbf{D})$ *is an epimorphism (splittable epimorphism).*

(c) *Suppose* $T: \mathbf{E} \to \mathbf{D}$ *is exact with* $\mathbf{E}$ *abelian such that there is a functor* $S: \mathbf{D} \to \mathbf{E}$ *satisfying* $TS = \mathrm{id}_{\mathbf{D}}$. *Then* $K_n(\mathbf{E}) \to K_n(\mathbf{D})$ *is an epimorphism (splittable epimorphism).*

Proof. (a)$\Rightarrow$(b). Since $K_n(\mathbf{C}, 0) \to K_n(\mathbf{D})$ is the composition $K_n(\mathbf{C}, 0) \to K_n(\mathbf{D}, 0) \to K_n(\mathbf{D})$, it follows that $K_n(\mathbf{D}, 0) \to K_n(\mathbf{D})$ is an epimorphism (splittable epimorphism).

(b)$\Rightarrow$(c). Since $TS = \mathrm{id}_{\mathbf{D}}$ we have that the composition

$$K_n(\mathbf{D}, 0) \xrightarrow{K_n(S)} K_n(\mathbf{E}) \xrightarrow{K_n(T)} K_n(\mathbf{D})$$

is the canonical morphism $K_n(\mathbf{D}, 0) \to K_n(\mathbf{D})$. Hence $K_n(T): K_n(\mathbf{E}) \to K_n(\mathbf{D})$ is an epimorphism (splittable epimorphism).

(c)$\Rightarrow$(a). Since $\mathbf{D}$ is abelian, $\mathbf{D}$ has kernels and so $\operatorname{mod} \mathbf{D}$ is an abelian category with gl.dim.(mod(mod $\mathbf{D}) \leq 2$. Hence $K_i(\mathbf{D}, 0) \to K_i(\operatorname{mod} \mathbf{D})$ is an isomorphism for all $i \geq 0$ [32]. Also the unique exact functor $T: \operatorname{mod} \mathbf{D} \to \mathbf{D}$ such that $T((\cdot, D)) = D$ for all D in $\mathbf{D}$ has the property that $K_n(T): K_n(\operatorname{mod} \mathbf{D}) \to K_n(\mathbf{D})$ is an epimorphism (splittable epimorphism) by hypothesis since the functors $S: \mathbf{D} \to \operatorname{mod} \mathbf{D}$ given by $S(D) = (\cdot, D)$ clearly satisfies $TS = \mathrm{id}_{\mathbf{D}}$. Therefore we have that the composition $K_n(\mathbf{D}, 0) \to K_n(\operatorname{mod} \mathbf{D}) \xrightarrow{K_{n(T)}} K_n(\mathbf{D})$, which is the canonical morphism $K_n(\mathbf{D}, 0) \to K_n(\mathbf{D})$, is an epimorphism (splittable epimorphism). This shows that (c) implies (a).

To see that Theorem 2.2 implies Theorem 2.1, it is only necessary to recall that for a coherent pair $(\mathbf{A}, \mathbf{B})$ there is an $S : \mathrm{mod}\,\mathbf{A}$ such that $\mathrm{res}\, S = \mathrm{id}_{\mathrm{mod}\,\mathbf{B}}$.

As another application of Theorem 2.2 we have the following useful result.

Proposition 2.3. *Suppose* $\mathbf{D}$ *is an abelian R-category satisfying either*

(a) $\mathbf{D}$ *is regular (i.e. every object in* $\mathbf{D}$ *has finite projective dimension) or*

(b) *Every object in* $\mathbf{D}$ *has finite length.*

Let $T : \mathbf{E} \to \mathbf{D}$ *be exact with* $\mathbf{E}$ *abelian such that there is an* $S : \mathbf{D} \to \mathbf{E}$ *with* $TS = \mathrm{id}_{\mathbf{D}}$. *Then* $0 \to K_n(\mathrm{Ker}\, T) \to K_n(\mathbf{E}) \to K_n(\mathbf{D}) \to 0$ *is split exact for all* $n \geq 0$.

Proof. (a) Let $\mathbf{P}$ be the subcategory of $\mathbf{D}$ consisting of the projective objects. Then it is well known that $K_n(\mathbf{P}, 0) \to K_n(\mathbf{D})$ is an isomorphism for all $n \geq 0$. Hence we have our desired result by Theorem 2.2.

(b) Let $\mathbf{C}$ be the subcategory of $\mathbf{D}$ consisting of the semisimple objects in $\mathbf{D}$. Since every object D in $\mathbf{D}$ has a finite filtration $D = D_0 \supset D_1 \supset \cdots \supset D = 0$ with D_i/D_{i+1} in $\mathbf{C}$ for $i = 0, \ldots, n-1$, it follows [32] that $K_n(\mathbf{C}, 0) \to K_n(\mathbf{D})$ is an isomorphism for all $n \geq 0$. Again, we get our desired result by applying Theorem 2.2.

Similarly, applying Theorem 2.1, we obtain the following special case of Proposition 2.3.

Proposition 2.4. *Suppose* $\mathbf{B}$ *is a coherent skeletally small additive R-category satisfying either*

(a) $\mathrm{mod}\,\mathbf{B}$ *is regular or*

(b) *Every object in* $\mathrm{mod}\,\mathbf{B}$ *has finite length.*

Then for each coherent pair $(\mathbf{A}, \mathbf{B})$ *we have that*

$$0 \to K_n(\mathrm{mod}\,\mathbf{A} \mid \mathbf{B}) \to K_n(\mathrm{mod}\,\mathbf{A}) \to K_n(\mathrm{mod}\,\mathbf{B}) \to 0$$

is split exact for all $n \geq 0$.

3. Finite R-categories

Suppose $\mathbf{A}$ is coherent. Then the subcategory f.l.mod $\mathbf{A}$ consisting of the objects of finite length in mod $\mathbf{A}$ is a Serre subcategory of mod $\mathbf{A}$, and so we have the usual short exact sequence

$$K_0(\text{f.l.mod}\,\mathbf{A}) \to K_0(\mathrm{mod}\,\mathbf{A}) \to K_0(\mathrm{mod}\,\mathbf{A}/\text{f.l.mod}\,\mathbf{A}) \to 0.$$

While it is generally not true that $K_0(\text{f.l.mod}\,\mathbf{A}) \to K_0(\mathrm{mod}\,\mathbf{A})$ is a monomorphism, there is a special type of $\mathbf{A}$, the finite R-categories, for which it is true, and this will be of particular interest to us throughout this paper.

Let R be an arbitrary commutative ring. We say that an R-category $\mathbf{A}$ is a finite R-category if idempotents in $\mathbf{A}$ split and $\mathrm{Hom}_{\mathbf{A}}(A_1, A_2)$ is an R-module of finite length for all A_i in $\mathbf{A}$. Since $\operatorname{End}_{\mathbf{A}} A$ is an artin ring for all A in $\mathbf{A}$, it follows that $\mathbf{A}$ has the Krull–Schmidt property, i.e. every A in $\mathbf{A}$ is a finite sum $\coprod_{i=1}^{n} A_i$ of indecomposable A_i in $\mathbf{A}$ and this representation is unique up to order and the isomorphism classes of the A_i. Further, an F in $\operatorname{mod}\mathbf{A}$ is of finite length if and only if its support is finite, i.e. there are only a finite number of nonisomorphic indecomposable A_i in $\mathbf{A}$ such that $F(A_i)\neq 0$ (see [5]).

Keeping these observations in mind, we obtain the following.

Proposition 3.1. *Let R be a commutative ring. Suppose $\mathbf{A}$ is a coherent finite R-category and $\mathbf{C}$ is a Serre subcategory of* $\operatorname{mod}\mathbf{A}$ *contained in* f.l.mod $\mathbf{A}$. *Then the morphism $K_0(\mathbf{C})\to K_0(\operatorname{mod}\mathbf{A})$ is a monomorphism and is an isomorphism if and only if $\mathbf{C}=\operatorname{mod}\mathbf{A}$.*

Proof. Since $\mathbf{A}$ is a Krull–Schmidt category, for each simple functor $S:\mathbf{A}\to\operatorname{mod} R$ there is a unique, up to isomorphism, indecomposable object A_S in $\mathbf{A}$ such that $S(A_S)\neq 0$ and $T(A_S)=0$ for all simple functors $T\not\simeq S$ [5]. Consider the bilinear map $\langle\cdot,\cdot\rangle : K_0(\operatorname{mod}\mathbf{A})\times K_0(\mathbf{A},0)\to Z$ given by $\langle [F],[A]\rangle = l_R F(A)$, where l_R denotes length over R. Now $K_0(\mathbf{C})$ is the free abelian group with basis $[S_i]$ where the S_i are a complete set of nonisomorphic simple objects in $\mathbf{C}$. Therefore, in order to show that $0\to K_0(\mathbf{C})\to K_0(\operatorname{mod}\mathbf{A})$ is exact, it suffices to show that if $\sum n_i[S_i]=0$ in $K_0(\operatorname{mod}\mathbf{A})$ with the S_i nonisomorphic simples in $\mathbf{C}$, then $n_i=0$ for all i. But $0=\langle \sum n_i[S_i], A_{S_j}\rangle = n_j l_R(S_j(A_{S_j}))$, which implies $n_j=0$ for all j since $S_j(A_{S_j})\neq 0$.

Now suppose $K_0(\mathbf{C})\to K_0(\operatorname{mod}\mathbf{A})$ is an isomorphism. Let F be in $\operatorname{mod}\mathbf{A}$. Then $[F]=\sum_{i=1}^{t} n_i[S_i]$ in $K_0(\operatorname{mod}\mathbf{A})$, where the S_i are nonisomorphic simple objects in $\mathbf{C}$. Hence if an indecomposable A in $\mathbf{A}$ is in the support of F, then $A\simeq A_{S_i}$ for some $i=1,\dots,t$. Therefore F has finite length since it has finite support, and all its composition factors are in $\mathbf{C}$. Therefore F is in $\mathbf{C}$, which shows $C=\operatorname{mod}\mathbf{A}$.

CHAPTER 2

In this chapter we study some methods for computing $K_0(\operatorname{mod}\Lambda)$ for certain types of rings Λ.

1. A presentation of $K_0(\operatorname{mod}\Lambda)$

Let Λ be a noetherian ring and $\mathbf{P}=\mathbf{P}(\Lambda)$ the category of finitely generated projective Λ-modules. We shall show how we can use the general results from Chapter 1 to get a method for computing $K_0(\operatorname{mod}\Lambda)=K_0(\operatorname{mod}\mathbf{P})$. The idea is to consider appropriate coherent subcategories $\mathbf{A}$ of $\operatorname{mod}\Lambda$, with $\mathbf{P}\subset\mathbf{A}$. $(\mathbf{A},\mathbf{P})$ will then be a coherent pair, and we have the exact sequence of K-groups

$$K_0(\operatorname{mod}\mathbf{A}/\mathbf{P})\to K_0(\operatorname{mod}\mathbf{A})\to K_0(\operatorname{mod}\Lambda)\to 0.$$

We want to choose $\mathbf{A}$ so that $K_0(\mathrm{mod}\,\mathbf{A})$ is a free abelian group with a specified free basis in such a way that we can also describe $\mathrm{Im}(K_0(\mathrm{mod}\,\mathbf{A/P}) \to K_0(\mathrm{mod}\,\mathbf{A})$. In particular we look for categories $\mathbf{A}$ such that the hypothesis of the following result are satisfied. The proof is obvious, but the result is still quite useful.

Proposition 1.1. *Let Λ be a noetherian ring, $(\mathbf{A},\mathbf{P})$ a coherent pair such that* $\mathrm{gl.dim\,mod}\,\mathbf{A} < \infty$ *and $\mathbf{A}$ a Krull Schmidt category. Assume also that all objects in* $\mathrm{mod}(\mathbf{A/P})$ *have finite length. Then in the exact sequence*

$$K_0(\mathrm{mod}\,\mathbf{A/P}) \xrightarrow{i} K_0(\mathrm{mod}\,\mathbf{A}) \to K_0(\mathrm{mod}\,\Lambda) \to 0,$$

$\mathrm{Im}\, i$ *is the subgroup of the free group* $K_0(\mathrm{mod}\,\mathbf{A}) \simeq K_0(\mathbf{A},0)$ *with the isomorphism classes of indecomposable objects of $\mathbf{A}$ as basis, generated by $[A_1]-[A_2]+[A_3]-\cdots$, where $0 \to (\cdot, A_n) \to \cdots \to (\cdot, A_3) \to (\cdot, A_2) \to (\cdot, A_1) \to S \to 0$ is a projective resolution in* $\mathrm{mod}\,\mathbf{A}$ *of a simple functor S in* $\mathrm{mod}\,\mathbf{A/P}$.

Proof. Since every object in $\mathrm{mod}\,\mathbf{A/P}$ has finite length, $K_0(\mathrm{mod}\,\mathbf{A/P})$ is the free abelian group with basis the isomorphism classes of simples in $\mathrm{mod}\,\mathbf{A/P}$. Considering $i[S]$ for a simple object S in $\mathrm{mod}\,\mathbf{A/P}$ gives our desired result.

We shall apply this result in the next section and also in the next chapter. It will be particularly interesting to have a choice of $\mathbf{A}$ which has almost split sequences, since the description of $\mathrm{Im}\, i$ becomes much simpler in this case.

2. Orders

We shall in this section study the case of orders Λ from the point of view stressed in Section 1. Let R be a commutative noetherian equidimensional Gorentstein ring of Krull dimension d. We recall that an R-order is an R-algebra Λ which is a Cohen–Macaulay R-module and has the property that $\mathrm{Hom}_{R_\mathfrak{p}}(\Lambda_\mathfrak{p}, R_\mathfrak{p})$ is $\Lambda_\mathfrak{p}^{\mathrm{op}}$-projective for all nonmaximal prime ideals $\mathfrak{p}$ in R [6]. We also recall that for an R-order Λ a Λ-lattice is a finitely generated Λ-module L which is a Cohen–Macauley R-module such that $L_\mathfrak{p}$ is $\Lambda_\mathfrak{p}$-projective for all nonmaximal prime ideals $\mathfrak{p}$ of R. We denote the subcategory of $\mathrm{mod}\,\Lambda$ consisting of the Λ-lattices by $\mathbf{L} = \mathbf{L}(\Lambda)$. Before going on we give the following examples of orders and lattices.

(a) $\dim R = 0$. Then the R-orders Λ are artin algebras, and $\mathbf{L}(\Lambda) = \mathrm{mod}\,\Lambda$.

(b) R is a Dedekind domain, Λ is an R-algebra which is a projective R-module and $K \otimes_R \Lambda$ is semisimple, where K is the field of quotients of R. Then the Λ-lattices are the L in $\mathrm{mod}\,\Lambda$ which are projective as R-modules. This is the case of classical orders.

(c) R is an integrally closed equidimensional Gorenstein domain of dimension 2 and S is its integral closure in a separable field extension L of K, the field of quotients of R. Then S is an R-order and the S-lattices are the reflexive S-modules.

The reader is referred to [6] for further examples and the basic facts concerning R-orders and lattices, which will be used rather freely. Also more examples will be given in Chapter 3.

For an R-order Λ we would like to apply Proposition 1.1 to obtain generators and relations for $K_0(\operatorname{mod}\Lambda)$ by choosing $\mathbf{L}$ as our $\mathbf{A}$. Since $\mathbf{L}(\Lambda)/\mathbf{P}$ is always a finite R-category [6], it remains to know when $\mathbf{L}$ is coherent and when gl.dim mod $\mathbf{L}$ is finite. Here we use the following result [9].

Theorem 2.1. *Let Λ be an R-order with R a complete regular local ring. Then we have the following:*

(a) $\mathbf{L}(\Lambda)$ *is contravariantly finite in* mod Λ *if and only if Λ is regular outside the maximal ideal, i.e. $\Lambda_{\mathfrak{p}}$ is regular if $\mathfrak{p}$ is a nonmaximal prime ideal of R.*

(b) *If* $\mathbf{L}(\Lambda)$ *is contravariantly finite in* mod Λ, *for instance if Λ is of finite representation type, then* $\mathbf{L}(\Lambda)$ *is coherent and* gl.dim mod $\mathbf{L}(\Lambda) = \max(2, \dim R) < \infty$.

In view of this result, we assume for the rest of the paper that our R-orders Λ are regular outside maximal ideals. We also assume that R is a complete regular local ring. This will ensure that mod Λ, and hence $\mathbf{L}(\Lambda)$, has the Krull–Schmidt property.

Since $\mathbf{L}(\Lambda)$ has almost split sequences [6], we get the following consequence of Proposition 1.1, generalizing a result of Butler for orders of dimension at most one [23]. We also point out that if Λ is of finite lattice type, then Λ is automatically regular outside maximal ideals [9].

Proposition 2.2. *Let Λ be an R-order (where R is a complete regular local ring). Assume that Λ is of finite lattice type. Then the almost split sequences generate the relations for* $K_0(\operatorname{mod}\Lambda)$.

Proof. We want to apply Proposition 1.1. We have already seen that $(\mathbf{L}(\Lambda), \mathbf{P})$ is a coherent pair and gl.dim mod $\mathbf{L}(\Lambda) < \infty$. Also mod$(\mathbf{L}(\Lambda)/\mathbf{P})$ is a finite R-category since $\mathbf{L}(\Lambda)/\mathbf{P}$ is a finite R-category. Since Λ is of finite lattice type, the support of each F in mod $\mathbf{L}(\Lambda)/\mathbf{P}$ is finite, and hence F has finite length. The result now follows by using that an exact sequence $0 \to L_2 \to L_1 \to L_0 \to 0$ of lattices is an almost split sequence in $\mathbf{L}(\Lambda)$ if and only if

$$0 \to (\cdot, L_2) | \mathbf{L} \to (\cdot, L_1) | \mathbf{L} \to (\cdot, L_0) | \mathbf{L} \to S \to 0$$

is a minimal projective resolution for a simple functor S in mod $\mathbf{L}/\mathbf{P}$.

While it is an open question if the converse of Proposition 2.2 always holds, we have the following, which generalizes to higher dimensions the corresponding result for artin algebras [23], [7].

Proposition 2.3. *Let Λ be an R-order such that* $0 \to K_0(\operatorname{mod}\mathbf{L}/\mathbf{P}) \to K_0(\operatorname{mod}\mathbf{L})$ *is exact, for instance Λ is an artin algebra or* gl.dim $\Lambda < \infty$. *Then Λ is of finite* *representation type if the relations for* $K_0(\text{mod}\,\Lambda)$ *are generated by almost split sequences.*

Proof. As we have seen, the hypothesis that the relations for $K_0(\text{mod}\,\Lambda)$ are generated by almost split sequences is equivalent to

$$K_0(\text{f.l.mod}\,\mathbf{L}/\mathbf{P}) \to K_0(\text{mod}\,\mathbf{L}) \to K_0(\text{mod}\,\mathbf{P}) \to 0$$

being exact. Since mod $\mathbf{L}/\mathbf{P}$ is a finite R-category, we also know by Proposition 3.1 that $0 \to K_0(\text{f.l.mod}\,\mathbf{L}/\mathbf{P}) \to K_0(\text{mod}\,\mathbf{L}/\mathbf{P})$ is exact and hence an isomorphism since $0 \to K_0(\text{mod}\,\mathbf{L}/\mathbf{P}) \to K_0(\text{mod}\,\mathbf{L}) \to K_0(\text{mod}\,\mathbf{P}) \to 0$ is exact by hypothesis. Therefore applying Proposition 3.1 again, we have that f.l.mod $\mathbf{L}/\mathbf{P}$ = mod $\mathbf{L}/\mathbf{P}$. That this implies our desired result that Λ is of finite lattice types is shown in the following.

Lemma 2.4. *Let* Λ *be an R-order. Then* mod $\mathbf{L}/\mathbf{P}$ *has the following properties.*

(a) *There is an* L *in* $\mathbf{L}/\mathbf{P}$ *such that* $(X, L)/\mathbf{P} \neq 0$ *for all nonzero* X *in* $\mathbf{L}/\mathbf{P}$.

(b) Λ *is of finite lattice type if and only if* f.l.mod $\mathbf{L}/\mathbf{P}$ = mod $\mathbf{L}/\mathbf{P}$.

Proof. (a) Let $S_1, \ldots, S_t$ be a complete set of nonisomorphic simple Λ-modules. Since $\mathbf{L}$ is contravariantly finite in mod Λ by Theorem 2.1, we can find nonzero morphisms $L_i \to S_i$ with the L_i in $\mathbf{L}$ such that $(X, L_i) \to (X, S_i) \to 0$ is exact for all X in $\mathbf{L}$. It is now easily seen that $L = \coprod_{i=1}^{t} L_i$ has our desired property.

(b) Suppose f.l.mod $\mathbf{L}/\mathbf{P}$ = mod $\mathbf{L}/\mathbf{P}$ and let L in $\mathbf{L}/\mathbf{P}$ be such that $(\cdot, L)/\mathbf{P}(X) \neq 0$ for each indecomposable X in $\mathbf{L}/\mathbf{P}$. Since the support of $(\cdot, L)/\mathbf{P}$ is finite, this means that there are only a finite number of nonprojective indecomposable lattices. Hence Λ is of finite lattice type. The rest of (b) is trivial.

We end this section by giving a description of when $0 \to K_0(\text{mod}\,\mathbf{L}/\mathbf{P}) \to K_0(\text{mod}\,\mathbf{L})$ is exact for an order Λ of finite lattice type.

Since we are assuming Λ is of finite lattice type, we know that $K_0(\text{mod}\,\mathbf{L})$ is a free abelian group of rank the number of isomorphism classes of indecomposable lattices and $K_0(\text{mod}\,\mathbf{L}/\mathbf{P})$ is a free abelian group of rank the number of isomorphism classes of nonprojective indecomposable lattices. As an easy consequence of these observations, we have the following result, which we state without proof.

Proposition 2.5. *Let* Λ *be an R-order of finite lattice type and let*

$$0 \to U \to K_0(\text{mod}\,\mathbf{L}/\mathbf{P}) \to K_0(\text{mod}\,\mathbf{L}) \to K_0(\text{mod}\,\Lambda) \to 0$$

be exact. Then we have the following.

(a) U *is a free abelian group of finite rank.*

(b) rank $K_0(\text{mod}\,\Lambda)$ = rank U + *number of isomorphism classes of indecomposable projective* Λ*-modules.*

(c) $0 \to K_0(\text{mod}\,\mathbf{L}/\mathbf{P}) \to K_0(\text{mod}\,\mathbf{L})$ *is exact if and only if* rank $K_0(\text{mod}\,\Lambda)$ = *number of nonisomorphic indecomposable projective* Λ*-modules.*

(d) *$U=0$ if and only if the elements $[A]+[C]-[B]$ when $0\to A\to B\to C\to 0$ is an almost split sequence in* **L**, *are linearly independent in* $K_0(\mathbf{L},0)$.

We shall get more information along these lines in the one-dimensional case in the next section.

3. One-dimensional orders

In addition to the exact sequence $K_0(\operatorname{mod}\mathbf{L}/\mathbf{P})\to K_0(\operatorname{mod}\mathbf{L})\to K_0(\operatorname{mod}\mathbf{P})\to 0$ which we have studied for the category of lattices **L** over an order Λ, there is another exact sequence of K-groups which has been studied in the classical one-dimensional case. We here show how these sequences are both parts of a large commutative diagram. We draw some consequences for the case that **L** is of finite type, and our main result in this section is that the sequence $0\to K_0(\operatorname{mod}\mathbf{L}/\mathbf{P})\to K_0(\operatorname{mod}\mathbf{L})$ is exact if Λ is of finite lattice type and $K\otimes_R\Lambda=\Gamma$ is a simple algebra. As we have seen, this means in particular that the almost split sequences are linearly independent in $K_0(\operatorname{mod}\mathbf{L})=K_0(\mathbf{L},0)$.

Throughout this section we assume that Λ is an order over a complete discrete valuation ring R with field of quotients K such that $\Gamma=K\otimes_R\Lambda$ is semisimple. Therefore we have by Theorem 2.1 that **L**, the category of Λ-lattices, is coherent, and gl.dim mod $\mathbf{L}\le 2$ since **L** has kernels. Hence $(\mathbf{L},\mathbf{P})$ is a coherent pair, so that we have the exact sequence

$$(*)\qquad K_0(\operatorname{mod}\mathbf{L}/\mathbf{P})\to K_0(\operatorname{mod}\mathbf{L})\to K_0(\operatorname{mod}\mathbf{P})\to 0.$$

The functor $K\otimes_R\cdot:\operatorname{mod}\Lambda\to\operatorname{mod}\Gamma$ given by $M\mapsto K\otimes_R M$ for all M in mod Λ is an exact functor and moreover Ker $K\otimes_R\cdot$ is the subcategory $t(\operatorname{mod}\Lambda)$ of mod Λ consisting of all Λ-modules which are torsion R-modules. It is well known that $t(\operatorname{mod}\Lambda)$ is a Serre subcategory of mod Λ and

$$t(\operatorname{mod}\Lambda)\hookrightarrow\operatorname{mod}\Lambda\xrightarrow{K\otimes_R\cdot}\operatorname{mod}\Gamma$$

is exact. Hence we have the exact sequence

$$K_0(t(\operatorname{mod}\Lambda))\to K_0(\operatorname{mod}\Lambda)\to K_0(\operatorname{mod}\Gamma)\to 0.$$

We have the following immediate consequence.

Proposition 3.1. *With the above assumptions and notation,* $K_0(t(\operatorname{mod}\Lambda))$ *and* $K_0(\operatorname{mod}\Gamma)$ *are finitely generated free abelian groups, and* $K_0(\operatorname{mod}\Lambda)$ *is finitely generated.*

Proof. Since R is a complete discrete valuation ring, R is noetherian of Krull dimension 1. Hence $t(\operatorname{mod}\Lambda)$ consists of the modules in mod Λ of finite length, and consequently $K_0(t(\operatorname{mod}\Lambda))$ is the free abelian group with basis the isomorphism classes of simple objects in $\operatorname{mod}\Lambda$. Since Γ is semisimple, $K_0(\operatorname{mod}\Gamma)$ is clearly a finitely generated free abelian group. From the exact sequence $K_0(t(\operatorname{mod}\Lambda))\to K_0(\operatorname{mod}\Lambda)\to K_0(\operatorname{mod}\Gamma)\to 0$ it then follows that $K_0(\operatorname{mod}\Lambda)$ is finitely generated.

Our large commutative diagram is gotten from the following more general result, which is proved in a standard way.

Proposition 3.2. *Let* $\mathbf{A}\hookrightarrow\mathbf{B}\xrightarrow{\alpha}\mathbf{C}$ *and* $\mathbf{B}'\hookrightarrow\mathbf{C}\xrightarrow{\beta}\mathbf{D}$ *be exact sequences of abelian categories.*

(a) $\operatorname{Ker}\beta\alpha\hookrightarrow\mathbf{B}\xrightarrow{\beta\alpha}\mathbf{D}$ *is an exact sequence of abelian categories and* $\operatorname{Ker}\beta\alpha$ *contains* $\mathbf{A}$.

(b) *If there is a fully faithful functor* $S:\mathbf{C}\to\mathbf{B}$ *such that* $\alpha S=\mathrm{id}_{\mathbf{C}}$, *then* $\mathbf{A}\hookrightarrow\operatorname{Ker}\beta\alpha\twoheadrightarrow\mathbf{B}'$ *is an exact sequence of abelian categories where* $\operatorname{Ker}\beta\alpha\twoheadrightarrow\mathbf{B}'$ *is the restriction of* $\mathbf{B}\twoheadrightarrow\mathbf{C}$ *to* $\operatorname{Ker}\beta\alpha$.

Making the substitutions $\mathbf{A}=\operatorname{mod}\mathbf{L}/\mathbf{P}$, $\mathbf{B}=\operatorname{mod}\mathbf{L}$, $\mathbf{C}=\operatorname{mod}\Lambda$, $\mathbf{D}=\operatorname{mod}\Gamma$ and $\mathbf{B}'=t\operatorname{mod}(\Lambda)$, it is easy to see that $\ker\beta\alpha$ consists of all F in $\operatorname{mod}\mathbf{L}$ such that $F(\Lambda)$ is a torsion R-module. We denote this subcategory of $\operatorname{mod}\mathbf{L}$ by $t(\operatorname{mod}\mathbf{L})$ and call it the torsion $\mathbf{L}$-modules. The reason for this terminology is that it is not difficult to see that because Γ is semisimple, $t(\operatorname{mod}\mathbf{L})$ consists of those F in $\operatorname{mod}\mathbf{L}$ such that $\Gamma(L)$ is a torsion R-module for all L in $\mathbf{L}$. For writing F in $\operatorname{mod}\mathbf{L}$ as $\operatorname{Coker}((\cdot,L_1)\xrightarrow{(\cdot,f)}(\cdot,L_0))$ with the L_i in $\mathbf{L}$, we see that $F(\Lambda)$ is torsion if and only if $\operatorname{Coker}f$ is torsion. But if $\operatorname{Coker}f$ is torsion, then $K\otimes_R L_1\to K\otimes_R L_0$ is a splittable epimorphism since Γ is semisimple, and so $K\otimes_R F(L)=0$ for all L in $\mathbf{L}$. We then obtain the following.

Proposition 3.3. (a) *Let the assumptions and notation be as above. The following is an exact commutative diagram of abelian categories*

$$\begin{array}{ccccc}
\operatorname{mod}\mathbf{L}/\mathbf{P} & = & \operatorname{mod}\mathbf{L}/\mathbf{P} & & \\
\downarrow & & \downarrow & & \\
t(\operatorname{mod}\mathbf{L}) & \longrightarrow & \operatorname{mod}\mathbf{L} & \twoheadrightarrow & \operatorname{mod}\Gamma \\
\downarrow & & \downarrow & & \| \\
t(\operatorname{mod}\Lambda) & \longrightarrow & \operatorname{mod}\Lambda & \twoheadrightarrow & \operatorname{mod}\Gamma
\end{array}$$

(b) *The following is an exact commutative diagram of abelian groups*

$$
\begin{array}{ccccccccccc}
 & & & & & & 0 & & & & \\
 & & & & & & \downarrow & & & & \\
 & & & & 0 & \longrightarrow & U & & & & \\
 & & & & \downarrow & & \downarrow & & & & \\
 & & & & K_0(\operatorname{mod} \mathbf{L}/\mathbf{P}) & = & K_0(\operatorname{mod} \mathbf{L}/\mathbf{P}) & & & & \\
 & & & & \downarrow & & \downarrow & & & & \\
0 & \longrightarrow & V & \longrightarrow & K_0(t(\operatorname{mod} \mathbf{L})) & \longrightarrow & K_0(\operatorname{mod} \mathbf{L}) & \longrightarrow & K_0(\operatorname{mod} \Gamma) & \longrightarrow & 0 \\
 & & & & \downarrow & & \downarrow & & \| & & \\
0 & \longrightarrow & W & \longrightarrow & K_0(t \operatorname{mod} \Lambda) & \longrightarrow & K_0(\operatorname{mod} \Lambda) & \longrightarrow & K_0(\operatorname{mod} \Gamma) & \longrightarrow & 0 \\
 & & & & \downarrow & & \downarrow & & \downarrow & & \\
 & & & & 0 & & 0 & & 0 & &
\end{array}
$$

where U, V, W are by definition the kernels of the obvious maps.

(c) *There is a morphism $U \to V$ which makes $0 \to U \to V \to W \to 0$ an exact sequence which splits, and U, V:W are free abelian groups.*

(d) $W \neq (0)$.

(e) $0 \to K_0(\operatorname{mod} \mathbf{L}/\mathbf{P}) \to K_0(\operatorname{mod} \mathbf{L})$ *is exact if and only if $V \to W$ is an isomorphism.*

Proof. (a) Since $(\mathbf{L}, \mathbf{P})$ is a coherent pair, we know that there is a fully faithful functor $S : \operatorname{mod} \mathbf{P} \to \operatorname{mod} \mathbf{L}$ such that the composition $\operatorname{mod} \mathbf{P} \xrightarrow{S} \operatorname{mod} L \to \operatorname{mod} \mathbf{P}$ is the identity.

$t(\operatorname{mod} \Lambda)$ consists of the modules in $\operatorname{mod} \Lambda$ of finite length. Since $S(t \operatorname{mod} \Lambda) \subset t(\operatorname{mod} \mathbf{L})$, we know by Proposition 2.4 that

$$0 \to K_0(\operatorname{mod} \mathbf{L}/\mathbf{P}) \to K_0(t(\operatorname{mod} \mathbf{L})) \to K_0(t(\operatorname{mod} \Lambda)) \to 0$$

is exact.

Hence (a) follows directly from Proposition 2.2, and (b) and (c) follow from (a) by diagram chasing. (d) follows directly from (c).

Let L be a lattice and let π generate the maximal ideal of R. Then from the exact sequence $0 \to L \xrightarrow{\pi} L \to L/\pi L \to 0$ we see that $L/\pi L$ is a nonzero torsion module whose image in $K_0(\operatorname{mod} \Lambda)$ is zero. Thus the isomorphism class of $L/\pi L$ in $K_0(t(\operatorname{mod} \Lambda))$ is a nonzero element in W, which proves (e).

We have the following result which is used for our main theorem, but it also is of interest in itself.

Proposition 3.4. *Let Λ be a one-dimensional order as before.*

(a) *Λ is of finite lattice type if and only if $K_0(t(\operatorname{mod} \mathbf{L}))$ is finitely generated.*

(b) *If Λ is of finite lattice type, then*

$$\operatorname{rank} K_0(t(\operatorname{mod} \mathbf{L})) = \operatorname{rank} K_0(\operatorname{mod} \mathbf{L}) \quad \textit{and} \quad \operatorname{rank} K_0(\operatorname{mod} \Gamma) = \operatorname{rank} V,$$

where $V = \operatorname{Ker}(K_0(t(\operatorname{mod} \mathbf{L})) \to K_0(\operatorname{mod} \mathbf{L}))$ *as before.*

Proof. (a) It is easily checked that the simple objects in mod $\mathbf{L}$ are in $t(\operatorname{mod} \mathbf{L})$. Also we know there is a natural bijection between isomorphism classes of indecomposable lattices and isomorphism classes of simple objects in mod $\mathbf{L}$. Hence if Λ is of finite lattices type we know that the subcategory $\mathbf{S}$ of semisimple objects in $t(\operatorname{mod} \mathbf{L})$ has only a finite number of nonisomorphic simple objects. Therefore $K_0(\mathbf{S})$ is finitely generated. But the inclusion $S \to t(\operatorname{mod} \mathbf{L})$ induces an isomorphism $K_0(\mathbf{S}) \to K_0(t(\operatorname{mod} \mathbf{L}))$ since every object in $t(\operatorname{mod} \mathbf{L})$ has finite length, so that $K_0(t(\operatorname{mod} \mathbf{L}))$ is finitely generated.

Assume conversely that $K_0(t(\operatorname{mod} \mathbf{L}))$ is finitely generated. Since also $K_0(\operatorname{mod} \Gamma)$ is finitely generated, it follows from the exact sequence

$$K_0(t(\operatorname{mod} \mathbf{L})) \to K_0(\operatorname{mod} \mathbf{L}) \to K_0(\operatorname{mod} \Gamma) \to 0$$

that $K_0(\operatorname{mod} \mathbf{L})$ is finitely generated. The fact that $K_0(\operatorname{mod} \mathbf{L})$ is the free abelian group with basis the isomorphism classes of indecomposable lattices implies that Λ is of finite lattice type.

(b) $\operatorname{rank}(K_0(t \operatorname{mod} \mathbf{L}))$ and $\operatorname{rank} K_0(\operatorname{mod} \mathbf{L})$ are both equal to the number of indecomposable lattices. The rest then follows from the exact sequence

$$0 \to V \to K_0(t \operatorname{mod} \mathbf{L})) \to K_0(\operatorname{mod} \mathbf{L}) \to K_0(\operatorname{mod} \Gamma) \to 0.$$

We end this section with the following result.

Theorem 3.5. *Let Λ be a one-dimensional order as before, and assume that Λ is of finite lattice type. If $\Gamma = K \otimes_R \Lambda$ is a simple algebra, then $0 \to K_0(\operatorname{mod} \mathbf{L}/\mathbf{P}) \to K_0(\operatorname{mod} \mathbf{L})$ is exact.*

Proof. If Γ is a simple algebra, then $\operatorname{rank} K_0(\operatorname{mod} \Gamma) = 1 = \operatorname{rank} V$. Since $V \to W \to 0$ is split exact, with $V \simeq Z$ and $W \neq (0)$ (see Proposition 3.3), it follows that $V \to W$ is an isomorphism, or equivalently, that $0 \to K_0(\operatorname{mod} \mathbf{L}/\mathbf{P}) \to K_0(\operatorname{mod} \mathbf{L})$ is exact.

CHAPTER 3

In this chapter we use our general method of finding a convenient coherent pair $(\mathbf{A}, \mathbf{P})$ to compute $K_0(\operatorname{mod} \Lambda)$ for certain types of commutative noetherian local rings Λ which we denote by R, including, for instance, quotient singularities. A new feature in this situation is the role of the Cartan map $c_R : K_0(\mathbf{P}, 0) \to K_0(\operatorname{mod} R)$. For instance, if R is a domain, then c_R is a monomorphism which has a canonical splitting, showing that $K_0(\operatorname{mod} \mathrm{R}) \cong Z \amalg \operatorname{Coker} c_R$, reducing the computation of $K_0(\operatorname{mod} R)$ to that of $\operatorname{Coker} c_R$. In the case R is a complete quotient singularity which is an isolated singularity, we obtain a description of $\operatorname{Coker} c_R$ in terms of the integral representation ring of the group defining R, which shows that it is a finite group. We also point out certain module-theoretic consequences for R of the Cartan map being a monomorphism, as well as of $\operatorname{Coker} c_R$ being torsion, the proofs of which will be given in the next chapter. Throughout this chapter we assume R is a commutative noetherian ring.

1. Kernel of the Cartan map

Our purpose in this section is to prove the following easily established result.

Proposition 1.1. *Suppose R is a local ring. Then*

(a) *$K_0(\mathbf{P}, 0) \simeq Z$ with $[R]$ as generator.*

(b) *$c_R : K_0(\mathbf{P}, 0) \to K_0(\operatorname{mod} R)$ is a monomorphism which is a splittable monomorphism if $R_\mathfrak{p}$ is a field for some height zero prime ideal $\mathfrak{p}$.*

Proof. (a) is well known and a trivial consequence of the fact that all projective R-modules are free.

(b) Let $\mathfrak{p}$ be a prime ideal of height zero. Then the exact localization functor $\operatorname{mod} R \to \operatorname{mod} R_\mathfrak{p}$ induces the commutative diagram

$$(*) \qquad \begin{array}{ccc} K_0(\mathbf{P}(R), 0) & \xrightarrow{\alpha} & K_0(\mathbf{P}(R_\mathfrak{p}), 0) \\ \downarrow{\scriptstyle c_R} & & \downarrow{\scriptstyle c_{R_\mathfrak{p}}} \\ K_0(\operatorname{mod} R) & \xrightarrow{\beta} & K_0(\operatorname{mod} R_\mathfrak{p}) \end{array}$$

Since α is obviously an isomorphism, we have that c_R is a monomorphism if $c_{R_\mathfrak{p}}$ is a monomorphism, and is a splittable monomorphism if $c_{R_\mathfrak{p}}$ is an isomorphism. Since $R_\mathfrak{p}$ is an artin local ring, $K_0(\operatorname{mod} R_\mathfrak{p}) \simeq Z$ with $[R_\mathfrak{p}/\mathfrak{p}R_\mathfrak{p}]$ as generator. Therefore $c_{R_\mathfrak{p}} : K_0(\mathbf{P}(R_\mathfrak{p}), 0) \to K_0(\operatorname{mod} \mathrm{R}_\mathfrak{p})$ is the morphism $Z \to Z$ which is multiplication by the length of $R_\mathfrak{p}$. Hence $c_{R_\mathfrak{p}}$ is always a monomorphism and an isomorphism if $R_\mathfrak{p}$ is a field, giving our desired result.

In Chapter 4 we shall show the following consequence of Proposition 1.1.

Proposition 1.2. *Suppose R is a complete local ring. Let $0 \to A_1 \to B_1 \to C_1 \to 0$ and $0 \to A_2 \to B_2 \to C_2 \to 0$ be two exact sequences in* mod R. *Then $A_1 \amalg B_2 \amalg C_1 \simeq A_2 \amalg B_1 \amalg C_2$ if $(A_1 \amalg B_2 \amalg C_1)_\mathbf{P} \simeq (A_2 \amalg B_1 \amalg C_2)_\mathbf{P}$, where $X_\mathbf{P}$ is the unique (up to isomorphism) R-module with no nontrivial summand from* **P** *such that $X \cong X_\mathbf{P} \amalg F$ with F in* **P**.

It should be observed that if R is a domain, the diagram (*) in the proof of part (b) of Proposition 1.1 gives a canonical splitting f for the monomorphism $c_R : K_0(\mathbf{P}(R), 0) \to K_0(\operatorname{mod} R)$. Thus we have a canonical decomposition of $K_0(\operatorname{mod} \mathrm{R})$ as the direct sum $Z \amalg \operatorname{Ker} f$ with $[R]$ as the generator of Z. Identifying $\operatorname{Ker} f$ with $\operatorname{Coker} c_R$ by means of the isomorphism induced by the canonical morphism $K_0(\operatorname{mod} R) \to \operatorname{Coker} c_R$, we obtain the canonical decomposition of $K_0(\operatorname{mod} R)$ as the sum $Z \amalg \operatorname{Coker} c_R$ (with $[R]$ as generator of Z). Thus if R is a domain, $\operatorname{Coker} c_R$ completely determines $K_0(\operatorname{mod} R)$. The rest of this chapter is devoted to examining the cokernel of Cartan maps for certain local rings.

2. dim $R \leq 2$

In this section we consider the cokernel of the Cartan map for various types of low dimensional commutative local rings. In particular, we describe when it is a finite group.

For the sake of completeness, we begin by recalling the following well known result.

Proposition 2.1. *Suppose R is a local ring of dimension zero. Then* $\operatorname{Coker} c_R \simeq Z/nZ$ *where n is the length of R.*

The following general lemma plays a crucial role in our considerations of higher-dimensional local rings.

Lemma 2.2. *Suppose R is a local ring and* $\dim R \geq 1$. *Then the morphism* $\alpha : K_0(\text{f.l.mod } R) \to K_0(\operatorname{mod} R)$ *has the following properties.*

(a) Im α *is a finite cyclic group.*

(b) Im $\alpha = 0$ *if R is complete with algebraically closed residue field $R/\mathfrak{m}$.*

Proof. (a) Since $[R/\mathfrak{m}]$ is a generator for $K_0(\text{f.l.mod } R) \simeq Z$, to prove (a) it suffices to show that $n[R/\mathfrak{m}]$ is zero in $K_0(\operatorname{mod} R)$ for some $n > 0$. Since $\dim R \geq 1$, there is a prime ideal $\mathfrak{p}$ of R such that $\dim R/\mathfrak{p}=1$. Let x be in $\mathfrak{m}-\mathfrak{p}$. From the exact sequence $0\to R/\mathfrak{p}\xrightarrow{x} R/\mathfrak{p}\to R/(\mathfrak{p},x)\to 0$ it follows that $[R/(\mathfrak{p},x)]$ is zero in $K_0(\operatorname{mod} R)$. Since $R/(\mathfrak{p},x)$ is of finite length, we have that $[R/(\mathfrak{p},x)]=n[R/\mathfrak{m}]$ in $K_0(\text{f.l.mod } R)$, where n is the length of $R/(\mathfrak{p},x)$. Hence $n[R/\mathfrak{m}]$ is zero in $K_0(\operatorname{mod} R)$ with $n>0$, proving (a).

(b) Clearly to prove (b) it suffices to show that $[R/\mathfrak{m}]$ is zero in $K_0(\operatorname{mod} R)$. Let S be the integral closure of $R/\mathfrak{p}$ in its field of quotients, where $\mathfrak{p}$ is a prime ideal such that $\dim R/\mathfrak{p}=1$. Since R is complete, $R/\mathfrak{p}$ is complete and so S is a discrete valuation ring which is a finitely generated R-module [29]. Hence $S/\pi S$ is a finite-dimensional field extension of $R/\mathfrak{m}$, where π is a generator for the maximal ideal of S. Therefore $R/\mathfrak{m}=S/\pi S$ since $R/\mathfrak{m}$ is algebraically closed. From the exact sequence $0\to S\xrightarrow{\pi} S\to S/\pi S\to 0$, it follows that $[R/\mathfrak{m}R]=[S/\pi S]$ is zero in $K_0(\operatorname{mod} R)$, proving (b).

We now consider one-dimensional local domains R. Recall from Section 1 that $K_0(\operatorname{mod} R)=Z \amalg \operatorname{Coker} c_R$.

Proposition 2.3. *Suppose R is a one-dimensional local domain. Then we have the following.*

(a) Coker c_R *is a finite cyclic group.*

(b) Coker $c_R=0$ *if R is complete with an algebraically closed residue class field* $R/\mathfrak{m}$.

Proof. Let Q be the field of quotients of R. From the fact that f.l.mod R is a Serre subcategory of mod R with quotient mod Q, we easily deduce the following exact commutative diagram

$$\begin{array}{ccccccc}
 & & 0 & & & & \\
 & & \downarrow & & & & \\
 & & K_0(\mathbf{P}(R),0) & \xrightarrow{\sim} & K_0(\mathbf{P}(Q),0) & & \\
 & & \downarrow{\scriptstyle c_R} & & \downarrow{\scriptstyle c_Q} & & \\
K_0(\text{f.l.mod } R) & \xrightarrow{\alpha} & K_0(\operatorname{mod} R) & \longrightarrow & K_0(\operatorname{mod} Q) & \longrightarrow & 0
\end{array}$$

From this it follows that $K_0(\operatorname{mod} R)=\operatorname{Im} c_R \amalg \operatorname{Im}\alpha$ and $\operatorname{Im}\alpha\simeq\operatorname{Coker} c_R$. Since by Lemma 2.2, $\operatorname{Im}\alpha$ is a finite cyclic group in general and is zero if R is complete with an algebraically closed residue field, we have our desired results.

We now consider integrally closed two-dimensional local domains.

Proposition 2.4. *Suppose R is a two-dimensional integrally closed local domain. Then we have an exact sequence*

$$0 \to Z/nZ \to \operatorname{Coker} c_R \to C(R) \to 0$$

with $n>0$, where $C(R)$ is the ideal class group of R. Moreover, $n=1$, i.e. $\operatorname{Coker} c_R \cong C(R)$, *if R is complete with an algebraically closed residue class field.*

Proof. Suppose $\mathbf{A}$ is the quotient of $\operatorname{mod} R$ by the Serre subcategory f.l.mod R. Then we have the exact sequence

$$K_0(\text{f.l.mod } R) \xrightarrow{\alpha} K_0(\operatorname{mod} R) \to K_0(\mathbf{A}) \to 0.$$

Since R is an integrally closed two-dimensional local domain, we have by [22] that there is a unique isomorphism $K_0(\mathbf{A}) \to Z \amalg C(R)$ such that the composition $K_0(\operatorname{mod} R) \to K_0(\mathbf{A}) \to Z \amalg C(R)$ is induced by $[M] \to (\operatorname{rank} M, c(M))$, where $c(M)$ is the divisor class of the R-module M. In particular, we have $[R] \to (1,0)$. Therefore we have the following exact commutative diagram

$$\begin{array}{ccccccccc}
 & & & & 0 & & 0 & & \\
 & & & & \downarrow & & \downarrow & & \\
 & & & & K_0(\mathbf{P},0) & \xrightarrow{j} & Z & & \\
 & & & & \downarrow{\scriptstyle c_R} & & \downarrow{\scriptstyle i} & & \\
0 & \to & \operatorname{Im} \alpha & \to & K_0(\operatorname{mod} R) & \to & Z \amalg C(R) & \to & 0 \\
 & & \| & & \downarrow & & \downarrow{\scriptstyle p} & & \\
0 & \to & \operatorname{Im} \alpha & \to & \operatorname{Coker} c_R & \to & C(R) & \to & 0 \\
 & & & & \downarrow & & \downarrow & & \\
 & & & & 0 & & 0 & &
\end{array}$$

where $j([R])=1$ and i and p are the canonical inclusion and projection maps. Now by Lemma 2.2 we have $\operatorname{Im} \alpha \simeq Z/nZ$ with $n>0$, and $n=1$ if R is complete with an algebraically closed residue class field, which finishes the proof of the proposition.

As an immediate consequence of this result we have the following.

Corollary 2.5. *Suppose R is a two-dimensional integrally closed local domain. Then* $\operatorname{Coker} c_R$ *is a torsion group if and only if $C(R)$ is a torsion group.*

Our interest in when Coker c_R is a torsion group stems from the following result which will be proven in Chapter 4.

Proposition 2.6. *Let R be an arbitrary complete local ring. Then the following are equivalent.*

(a) Coker c_R *is a torsion group.*

(b) *If M is a non-free indecomposable R-module, then there are exact sequences* $0 \to A_1 \to B_1 \to C_1 \to 0$ *and* $0 \to A_2 \to B_2 \to C_2 \to 0$ *and an integer* $n > 0$ *such that* $nM \amalg (A_1 \amalg B_2 \amalg C_1)_{\mathbf{P}} \cong (A_2 \amalg B_1 \amalg C_1)_{\mathbf{P}}$, *where nM denotes a sum of n copies of M.*

It is well known that if R is the completion of a rational singularity over the complex numbers $\mathbb{C}$, then $C(R)$ is a finite group, [21], [27], and therefore Coker c_R is a finite group. For rational double points we give another proof for the fact that Coker c_R is a finite group in the next chapter. The rest of this chapter is devoted to extending this result to higher dimensional isolated quotient singularities over $\mathbb{C}$ and other algebraically closed fields.

3. Quotient singularities

In the rest of this chapter we will be concerned with the following situation.

Let k be a field, G a finite group whose order is not divisible by the characteristic of k and $\varrho : G \to \mathrm{GL}(n, k)$ a representation, with $n \geq 2$. Then G acts as a group of k-automorphisms on $S = k[[X_1, \ldots, X_n]]$ by means of the obvious action of G on the n-dimensional vector space spanned by the variables $X_1, \ldots, X_n$ given by $\varrho : G \to \mathrm{GL}(n, k)$. We denote by R the invariant subring S^G of S. Then it is well known that R is an integrally closed, local, complete noetherian domain and S is a finitely generated R-module. The rest of this chapter is devoted to studying $K_0(\mathrm{mod}\, R)$. We will here use the general method explained in Chapter 2, and hence for $\mathbf{B} = \mathbf{P}(R)$ we want to choose a convenient $\mathbf{A}$ such that $(\mathbf{A}/\mathbf{P}(R))$ is a coherent pair. This will give us the exact sequence

$$K_0(\mathrm{mod}\, \mathbf{A}/\mathbf{P}(R)) \to K_0(\mathrm{mod}\, \mathbf{A}) \to K_0(\mathrm{mod}\, R) \to 0.$$

The first thing we want to show is that $K_0(\mathrm{mod}\, R)$ is finitely generated. To make a good choice for $\mathbf{A}$ we will need some general properties of the skew group ring $S[G]$ which we now recall (see [16]).

$S]G]$ is the free S-module with the elements of G as a basis. Multiplication is given by $(s_1\sigma_1)(s_2\sigma_2) = s_1\sigma_1(s_2)\sigma_1\sigma_2$ for all s_i in S and σ_i in G. Now it is easily seen that an $S[G]$-module A is and S-module which is also a G-module satisfying $\sigma(sa) = \sigma(s)\sigma(a)$ for all σ in G and s in S. In particular, we shall always consider S an $S[G]$-module by means of the usual action of S on S and the given action of G on S. We now point out some important ways of constructing new $S[G]$-modules from old ones.

Let A and B be two $S[G]$-modules. Then we consider the S-modules $\mathrm{Hom}_S(A, B)$ and $A \otimes_S B$ as G-modules in the following ways. $(\sigma f)(a) = \sigma(f(\sigma^{-1}a))$ for all σ in G and f in $\mathrm{Hom}_S(A, B)$ and $\sigma(a \otimes b) = \sigma(a) \otimes \sigma(b)$ for all σ in G, a in A and b and B. It is easily checked that these G-module structures make $\mathrm{Hom}_S(A, B)$ and $A \otimes_S B$ modules over $S[G]$.

Suppose A and B are $S[G]$-modules. Then it is easily seen that an $S[G]$-morphism $f: A \to B$ is a map which is both an S-morphism and a G-morphism. Thus we have the following easily verified but important observation.

Lemma 3.1. *Suppose A and B are $S[G]$-modules. Then* $\mathrm{Hom}_{S[G]}(A, B) = \mathrm{Hom}_S(A, B)^G$.

As our first application of this lemma we have the following.

Proposition 3.2. *An $S[G]$-module A is projective as an $S[G]$-module if and only if it is free when considered as an S-module.*

Proof. Since $S[G]$ is a free S-module, any projective $S[G]$-module is a projective and hence free S-module.

Suppose A is an $S[G]$-module which is a free S-module. Let $0 \to B_1 \to B_2 \to B_3 \to 0$ be an exact sequence of $S[G]$-modules. Then

$$\mathrm{Hom}_S(A, B_2) \to \mathrm{Hom}_S(A, B_3) \to 0$$

is exact and hence

$$\mathrm{Hom}_S(A, B_2)^G \to \mathrm{Hom}_S(A, B_3)^G \to 0$$

is exact since the fact that the order of G is not divisible by the characteristic of k implies that the group ring $k[G]$ is semisimple. Hence by Lemma 3.1

$$\mathrm{Hom}_{S[G]}(A, B_2) \to \mathrm{Hom}_{S[G]}(A, B_3) \to 0$$

is exact, which shows that A is $S[G]$-projective.

The following are easily verified consequences of the above results.

Corollary 3.3. (a) *S is a projective $S[G]$-module.*

(b) $\mathrm{Hom}_{S[G]}(S, A) \to A^G$, *given by $f \to f(1)$ for all f in* $\mathrm{Hom}_{S[G]}(S, A)$ *is an R-isomorphism which is functorial in A, which we will view as an identification.*

(c) $\mathrm{End}_{S[G]}(S) = R$.

(d) $\mathrm{gl.dim}\, S[G] = \mathrm{gl.dim}\, S = n > \infty$.

We now show how we can use these results on the skew group ring $S[G]$ to make our choice of the category **A**. Since $R = \mathrm{End}_{S[G]} S$, we have that the full subcategory $\mathrm{add}_{S[G]} S$ of $\mathrm{mod}\, S[G]$, consisting of finite sums of the $S[G]$-module S, is equivalent to $\mathbf{P}(R)$. Hence $\text{mod}(\text{add}_{S[G]}S)$ is equivalent to $\text{mod}\, R$. In other words, we have $\mathbf{B}=\mathbf{P}(R)=\text{add}_{S[G]}S$, and we choose $\mathbf{A}=\mathbf{P}(S[G])$. Then gl.dim mod $\mathbf{A}<\infty$ by Corollary 3.3. Since $S[G]$ is an R-algebra which is a finitely generated R-module and R is a complete local noetherian ring, $\mathbf{P}(S[G])$ is a Krull–Schmidt category, with only a finite number of indecomposable objects. The exact sequence

$$K_0(\text{mod}(\mathbf{P}(S[G])/\text{add}_{S[G]}S) \xrightarrow{i} K_0\mathbf{P}(S[G]),0) \to K_0(\text{mod}\, R) \to 0$$

then immediately gives the following, since $K_0(\mathbf{P}(S[G]),0)$ is a finitely generated free abelian group.

Proposition 3.4. *$K_0(\text{mod}\, R)$ is finitely generated.*

The next aim is to get a description of $\text{Im}\, i$ in some cases. As before, the situation will be easier when all objects in $\text{mod}(\mathbf{P}(S[G])/\text{add}_{S[G]}S)$ have finite length. As a preliminary result along these lines, we will show that if A is an $S[G]$-module of finite length, then $[A]$ in $K_0(\text{mod}\, S[G])$ goes to zero under the map $K_0(\text{mod}\, S[G])\to K_0(\text{mod}\, R)$. Later we shall see that these elements generate $\text{Ker}(K_0(\text{mod}\, S[G])\to K_0(\text{mod}\, R))$ when R is an isolated singularity.

To obtain our desired result, it is convenient to have a description of $\text{rad}\, S[G]$, the radical of $S[G]$. Since $\sigma(\text{rad}\, S)\subset \text{rad}\, S$ for all σ in G, it follows easily that $(\text{rad}\, S)S[G]$ is a left and therefore twosided ideal in $S[G]$. Hence $(\text{rad}\, S)S[G]$ is contained in $\text{rad}\, S[G]$ since $S[G]$ is a finitely generated S-module. The fact that $\text{rad}\, S[G]=(\text{rad}\, S)S[G]$ now follows from the fact that $S[G]/(\text{rad}\, S)S[G]$ is semisimple since it is isomorphic to $k[G]$. From this it follows that $(\text{rad}\, S[G])S=\text{rad}\, S$ and so the simple $S[G]$-module $S/(\text{rad}\, S[G])S$ is the simple S-module $k=S/\text{rad}\, S$ with trivial G-action. Hence S is the $S[G]$-projective cover of k. Therefore if A is a simple $S[G]$-module, then $\text{Hom}_{S[G]}(S,A)=0$ if $A\not\simeq k$.

Now the restriction functor $\text{mod}\, S[G]\to\text{mod}(\text{add}\, S)=\text{mod}\, R$ is given by $A\mapsto \text{Hom}_{S[G]}(S,A)=A^G$. Therefore if A is simple and not isomorphic to k, we have that $[A]$ is in $\text{Ker}(K_0(\text{mod}\, S[G])\to K_0(\text{mod}\, R))$. So to get our desired result it suffices to show the following.

Lemma 3.5. *$[k]$ is zero in $K_0(\text{mod}\, R)$.*

Proof. From the exact sequence of R-modules

$$0\to k[[X_1,\dots,X_n]]/(X_1,\dots,X_{n-1})\xrightarrow{X_n} k[[X_1,\dots,X_n]]/(X_1,\dots,X_{n-1})\to k\to 0$$

it follows that $[k]$ is zero in $K_0(\text{mod}\, R)$.

Hence we have established the following.

Proposition 3.6. *The image of the morphism $K_0\text{f.l.mod}\, S[G])\to K_0(\text{mod}\, S[G])$, in-*

duced by the inclusion f.l.mod $S[G] \to \operatorname{mod} S[G]$, *is contained in* $\operatorname{Ker}(K_0(\operatorname{mod} S[G]) \to K_0(\operatorname{mod} R))$.

4. Ring structure

We begin this section by defining a ring structure on $K_0(\operatorname{mod} S[G])$ and showing that $\operatorname{Im}(K_0(\text{f.l.mod } S[G]) \to K_0(\operatorname{mod} S[G])$ is a principal ideal in $K_0(\operatorname{mod} S[G])$.

Let A and B be in $\operatorname{mod} S[G]$. Then as pointed out in Section 3 we consider the S-module $A \otimes_S B$ a G-module by $\sigma(a \otimes b) = \sigma(a) \otimes \sigma(b)$ for all σ in G and a in A and b in B and this makes $A \otimes_S B$ an $S[G]$-module. This also enables us to consider the $\operatorname{Tor}_i^S(A, B)$ as $S[G]$-modules for all $i \geq 0$ as follows. Suppose

$$0 \to P_n \to P_{n-1} \to \cdots \to P_0 \to A \to 0$$

is an $S[G]$-projective resolution of A. Then it is also a projective S-resolution of A which can be used to compute $\operatorname{Tor}^S(A, B)$. Now it is easily verified that

$$0 \to P_d \otimes_S B \to \cdots \to P_0 \otimes_S B \to A \otimes_S B \to 0$$

is a complex of $S[G]$-modules and its homology is also $S[G]$-modules. In this way we obtain that the $\operatorname{Tor}_i^S(A, B)$ are $S[G]$-modules for all $i \geq 0$. We define the product $[A] \cdot [B]$ of $[A]$ and $[B]$ in $K_0(\operatorname{mod} S[G], 0)$ by $[A] \cdot [B] = \sum_{i=0}^d (-1)^i [\operatorname{Tor}_i^S(A, B)]$. This induces a multiplication on $K_0(\operatorname{mod} S[G])$ which makes $K_0(\operatorname{mod} S[G])$ a commutative ring with $[S]$ as identity. Perhaps the easiest way of seeing that this multiplication makes $K_0(\operatorname{mod} S[G])$ a commutative ring is to use the isomorphism $c_{S[G]} K_0(\mathbf{P}S[G], 0) \to K_0(\operatorname{mod} S[G])$. For it is readily checked that if P and Q are in $\mathbf{P}(S[G])$, then $P \otimes_S Q$ is also in $\mathbf{P}(S[G])$ and the multiplication in $K_0(\mathbf{P}(S[G]), 0)$ given by $[P] \cdot [Q] = [P \otimes_S Q]$ makes $K_0(\mathbf{P}(S[G]), 0)$ a commutative ring with identity $[S]$. Since the isomorphism $c_{S[G]} : K_0(\mathbf{P}S[G], 0) \to K_0(\operatorname{mod} S[G])$ is clearly multiplicative, $K_0(\operatorname{mod} S[G])$ is also a commutative ring with identity $[S]$ and the Cartan map is a ring isomorphism.

With these remarks in mind we have the following.

Proposition 4.1. $\operatorname{Im}(K_0(\text{f.l.mod } S[G]) \to K_0(\operatorname{mod} S[G]))$ *is the principal ideal* $[k] \cdot K_0(\operatorname{mod} S[G])$.

Proof. Clearly to show that $\operatorname{Im}(K_0(\text{f.l.mod } S[G]) \to K_0(\operatorname{mod} S[G]))$ is an ideal it suffices to show that if A is an $S[G]$-module of finite length and B is an arbitrary module in $\operatorname{mod} S[G]$, then $[A] \cdot [B]$ is in $\operatorname{Im}(K_0(\text{f.l.mod } S[G]) \to K_0(\operatorname{mod} S[G])$. But this follows from the fact that $[A] \cdot [B] = \sum_{i=0}^d (-1)^i [\operatorname{Tor}_i^S(A, B)]$ and each $\operatorname{Tor}_i^S(A, B)$ has finite length.

Since $\operatorname{Im}(K_0(\text{f.l.mod } S[G]) \to K_0(\operatorname{mod} S[G]))$ contains $[k]$, to show that it is $[k] K_0(\operatorname{mod} S[G])$ it suffices to show that $[A]$ is in $[k] K_0(\operatorname{mod} S[G])$ for each simple

$S[G]$-module A. Suppose P is the projective cover of A. Then

$$k\otimes_S P = S/\text{rad } S\otimes_S P \simeq P/(\text{rad } S)P = A,$$

since $(\text{rad } S)S[G] = \text{rad } S[G]$. Hence $[k]\cdot[P] = [A]$ in $K_0(\text{mod } S[G])$.

Combining Propositions 3.6 and 4.1 we have the following.

Corollary 4.2. *The epimorphism* $K_0(\text{mod } S[G]) \to K_0(\text{mod } R)$ *induces an epimorphism*

$$K_0(\text{mod } S[G])/[k]\cdot K_0(\text{mod } S[G]) \to K_0(\text{mod R}).$$

As is well known $K_0(\text{mod } k[G])$ has a ring structure given by $[A]\cdot[B] = [A\otimes_k B]$ for all A and B in mod $k[G]$. We now describe a ring isomorphism $K_0(\text{mod } S[G]) \to K_0(\text{mod } k[G])$ and the image of $[k]$ under this morphism, thus giving a purely group-theoretic description of the ring $K_0(\text{mod } S[G])/[k]\cdot K_0(\text{mod } S[G])$.

Suppose A is an $S[G]$-module. Then $\text{Tor}_i^S(S/\text{rad } S, A)$ is an $S[G]/\text{rad } S[G] = k[G]$-module for all i. It is then easily seen that there is a unique group homomorphism $K_0(\text{mod } S[G]) \to K_0(\text{mod } k[G])$ such that $[A] \mapsto \sum_{i=0}^d (-1)^i[\text{Tor}_i^S(S/\text{rad } S, A)]$ for all A in mod $S[G]$. To see that this group morphism is actually a ring isomorphism it is convenient to consider the composition

$$K_0(\mathbf{P}S[G], 0) \xrightarrow{c_{S[G]}} K_0(\text{mod } S[G]) \to K_0(\text{mod } k[G])$$

which is obviously given by $[P] \mapsto [P/(\text{rad } S)P]$ for all P in $\mathbf{P}S[G]$. Since $P \to P/(\text{rad } S)P$ is a projective cover and each semisimple $S[G]$-module has a projective cover, it follows that the group morphism $K_0(\mathbf{P}S[G], 0) \to K_0(\text{mod } k[G])$ given by $[P] \mapsto [P/\text{rad } S)P]$ for all P in $\mathbf{P}(S[G])$ is an isomorphism. But from the fact that

$$\begin{aligned} S/\text{rad } S\otimes_S (P\otimes_S Q) &= P/(\text{rad } S)P\otimes_S Q/(\text{rad } S)Q \\ &= P/(\text{rad } S)P\otimes_k Q/(\text{rad } S)Q, \end{aligned}$$

we have that $K_0(\mathbf{P}(S[G]), 0) \to K_0(\text{mod } k[G])$ is multiplicative and therefore a ring isomorphism. Hence $K_0(\text{mod } S[G]) \to K_0(\text{mod } k[G])$ is also a ring isomorphism.

We now turn our attention to describing the image in $K_0(\text{mod } k[G])$ of $[k]$ in $K_0(\text{mod } S[G])$. Let V be the $k[G]$-module given by our given representation $\varrho: G \to \text{GL}(n, k)$ and let $v_1, \ldots, v_n$ be a basis for V. Now given an S-module A and a k-vector space W we consider $A\otimes_k W$ an S-module by $s(a\otimes\omega) = (sa)\otimes\omega$ for all s in S, a in A and w in W. Hence letting $\bigwedge^i V$ be the i-th exterior power we have that $S\otimes_k \bigwedge^i V$ is a free S-module of rank $\dim_k \bigwedge^i V$. Since $\{v_{i_1}\wedge\cdots\wedge v_{i_r} \mid i_1 < i_2 < \cdots < i_r\}$ is a basis for $\bigwedge^r V$ there is a unique S-morphism $d_r: S\otimes_k \bigwedge^r V \to S\otimes_k \bigwedge^{r-1} V$ satisfying

$$d_r(1\otimes v_{i_1}\wedge\cdots\wedge v_{i_r}) = \sum_{j=1}^r (-1)^{j+1} X_{ij}\otimes v_{i_1}\wedge\cdots\wedge v_{i_{j-1}}\wedge\hat{v}_{i_j}\wedge\cdots\wedge v_{ir}$$

where $\hat{v}_{i_j}$ means v_{i_j} has been omitted. Then the sequence of free S-modules

$$0 \to S \otimes_k \overset{n}{\wedge} V \xrightarrow{d_n} S \otimes_k \overset{n-1}{\wedge} V \to \cdots \to S \otimes_k V \xrightarrow{d_1} S \to S/\mathrm{rad}\, S \to 0$$

is the projective S-resolution of $k = S/\mathrm{rad}\, S$ known as the Koszul complex.

Now in general given an $S[G]$-module A and a $k[G]$-module W we consider the S-module $A \otimes_k W$ a G-module by $\sigma(a \otimes w) = \sigma a \otimes \sigma w$, which makes $A \otimes_k W$ an $S[G]$-module. Hence the $S \otimes_k \wedge^r V$ are projective $S[G]$-modules, since they are projective S-modules. Also it is not difficult to check that the morphisms in the Koszul complex are not only S-morphisms but also G-morphisms and hence $S[G]$-morphisms. Therefore the Koszul complex is a projective $S[G]$-resolution of $k = S/\mathrm{rad}\, S$. Hence tensoring this complex over S with $S/\mathrm{rad}\, S$, we obtain that $\mathrm{Tor}_i^S(S/\mathrm{rad}\, S, k)$ is the $k[G]$-module $\wedge^i V$ for all $i \geq 0$. Therefore

$$\sum_{i=0}^{n} (-1)^i [\mathrm{Tor}_i^S(S/\mathrm{rad}\, S, k)] = \sum_{i=0}^{n} (-1)^i \overset{i}{\wedge} V$$

is the image of $[k]$ in $K_0(\mathrm{mod}\, k[G])$.

Summarizing, we have the following result.

Proposition 4.3. *There is a unique ring isomorphism* $K_0(\mathrm{mod}\, S[G]) \to K_0(\mathrm{mod}\, k[G])$ *defined by*

$$[A] \mapsto \sum_{i=0}^{n} (-1)^i [\mathrm{Tor}_i^S(S/\mathrm{rad}\, S, A)$$

for all A in $\mathrm{mod}\, S[G]$. *The element* $[k]$ *in* $K_0(\mathrm{mod}\, S[G])$ *goes to* $\sum_{i=0}^{n} (-1)^i [\wedge^i V]$ *in* $K_0(\mathrm{mod}\, k[G])$ *where V is the $k[G]$-module given by the representation* $\varrho : G \to \mathrm{GL}(n, k)$.

As an immediate consequence of this we have the following.

Corollary 4.4. *There is an epimorphism*

$$K_0(\mathrm{mod}\, k[G]) \Big/ \left(\sum_{i=0}^{n} (-1)^i \left[\overset{i}{\wedge} V \right] \right) K_0(\mathrm{mod}\, k[G]) \to K_0(\mathrm{mod}\, R)$$

where V is the $k[G]$-module given by $\varrho : G \to \mathrm{GL}(n, k)$.

Obviously it is of interest to know when the epimorphism

$$K_0(\mathrm{mod}\, k[G]) \Big/ \left(\sum_{i=0}^{n} (-1)^i \left[\overset{i}{\wedge} V \right] \right) K_0(\mathrm{mod}\, k[G]) \to K_0(\mathrm{mod}\, R)$$

is an isomorphism. This is the question we take up in the next section.

5. Free group action

We continue the same notation as in the previous section, except that we assume that $\varrho: G \to \mathrm{GL}(n, k)$ is an injection which we view as an inclusion.

From the exact sequence

$$K_0(\operatorname{mod} \mathbf{P}S[G]/\operatorname{add}_{S[G]} S) \to K_0(\operatorname{mod} S[G]) \to K_0(\operatorname{mod} R) \to 0$$

we see that if every object in $\operatorname{mod}(\mathbf{P}S[G]/\operatorname{add}_{S[G]} S)$ has finite length, then

$$K_0(\text{f.l.}\operatorname{mod} S[G]) \to K_0(\operatorname{mod} S[G]) \to K_0(\operatorname{mod} R) \to 0$$

is exact, which implies

$$K_0(\operatorname{mod} k[G]) \Big/ \left(\sum_{i=0}^{n} (-1)^i \left[\bigwedge^i V \right] \right) K_0(\operatorname{mod} k[G]) \simeq K_0(\operatorname{mod} R).$$

So we now turn our attention to finding when every object in $\operatorname{mod}(\mathbf{P}S[G]/\operatorname{add}_{S[G]} S)$ has finite length. We begin with some preliminaries.

Straightforward calculations show that the R-submodule $S[G]^G$ of $S[G]$ consists of all $\sum_{\sigma \in G} \sigma(s)\sigma$ with s in S. Thus we obtain the R-isomorphism $S \to S[G]^G$ given by $s \mapsto \sum_{\sigma \in G} \sigma(s)\sigma$ for all s in S, which we will view as an identification. Therefore P^G is in $\operatorname{add}_R S$ for all P in $\mathbf{P}(S[G])$ and so the functor $\operatorname{mod} S[G] \to \operatorname{mod} R$ given by $A \mapsto A^G$ induces a functor $\mathbf{P}(S[G]) \to \operatorname{add}_R S$. We now describe when the functor $\mathbf{P}(S[G]) \to \operatorname{add}_R S$ is an equivalence of categories.

First of all it is easy to see that $\mathbf{P}(S[G]) \to \operatorname{add}_R S$ is an equivalence if and only if the R-algebra morphism $\operatorname{End}_{S[G]}(S[G]) \to \operatorname{End}_R(S)$ given by $f \mapsto f|S$ for all f in $\operatorname{End}_{S[G]}(S[G])$ is an isomorphism. Next we have the R-algebra isomorphism $\beta: S[G]^{\mathrm{op}} \to \operatorname{End}_{S[G]}(S[G])$ given by $\beta(s\sigma)(x) = x \cdot s\sigma$ for all s in S, σ in G and x in $S[G]$. Also we have the R-algebra isomorphism $S[G] \to S[G]^{\mathrm{op}}$ given by $s\sigma \to \sigma^{-1}(s)\sigma^{-1}$ for all s in S and σ in G. Finally, it is straightforward to check that the composition of R-algebra morphisms

$$S[G] \to S[G]^{\mathrm{op}} \xrightarrow{\beta} \operatorname{End}_{S[G]}(S[G]) \to \operatorname{End}_R(S)$$

is the standard R-algebra morphism $\gamma: S[G] \to \operatorname{End}_R(S)$ given $\gamma(\sigma s)(x) = s\sigma(x)$ for all s in S, σ in G and x in S. Combining these remarks, we obtain the first part of the following.

Proposition 5.1. *The following statements are equivalent:*

(a) *The functor $\mathbf{P}(S[G]) \to \operatorname{add}_R S$, given by $P \mapsto P^G$ for all P in $\mathbf{P}(S[G])$, is an equivalence of categories.*

(b) *The R-algebras morphism $\gamma: S[G] \to \operatorname{End}_R(S)$, given by $\sigma(s\sigma)(x) = s\sigma(x)$ for all s in S, σ in G and x in S, is an isomorphism.*

(c) *Each height one prime ideal $\mathfrak{p}$ of R is unramified in S, i.e. $S_\mathfrak{p}$ is a unramified extension of $R_\mathfrak{p}$ for all $\mathfrak{p}$ such that $R_\mathfrak{p}$ is a discrete valuation ring.*

Proof. The equivalence of (a) and (b) follows trivially from remarks preceding that statement of the proposition. The equivalence of (b) and (c) is proven in [2].

Returning to the question of when objects in mod($\mathbf{P}S[G]$)/add$_{S[G]}S$) have finite length, we have the following result.

Proposition 5.2. *Suppose S/R is unramified at height one primes, i.e. each height one prime ideal in R is unramified in S. Then the following are equivalent.*

(a) *The objects in* mod($\mathbf{P}(S[G])/\mathrm{add}_{S[G]}S$) *have finite length.*

(b) *R is an isolated singularity, i.e. $R_\mathfrak{p}$ is a regular local ring for all nonmaximal prime ideals $\mathfrak{p}$ of R.*

Proof. Since S/R is unramified at height one primes, we know that the functor $\mathbf{P}(S[G]) \to \mathrm{add}_R S$, given by $P \mapsto P^G$ for all P in $\mathbf{P}S[G]$, is an equivalence of categories. This induces an equivalence of categories $\mathbf{P}(S[G])/\mathrm{add}_{S[G]}S \to \mathrm{add}_R S/\mathrm{add}_R R$, and hence an equivalence of categories $\mathrm{mod}(\mathbf{P}(S[G])/\mathrm{add}_{S[G]}S) \to \mathrm{mod}(\mathrm{add}_R S/\mathbf{P}(R))$. Therefore every object in $\mathrm{mod}(\mathbf{P}(S[G])/\mathrm{add}_{S[G]}S)$ has finite length if and only if every object in $\mathrm{mod}(\mathrm{add}_R S/\mathbf{P}(R))$ has finite length. Since for every F in $\mathrm{mod}(\mathrm{add}\, S/\mathbf{P}(R))$ there is an epimorphism $(\cdot, A)/\mathbf{P}(R) \to F \to 0$ with A in $\mathrm{add}_R S$, every object in $\mathrm{mod}\, \mathrm{add}_R S/\mathbf{P}(R)$ has finite length if and only if $(\cdot, A)/\mathbf{P}(R)$ has finite length for all A in $\mathrm{add}_R S$, or equivalently, $A_\mathfrak{p}$ is $R_\mathfrak{p}$-free for all A in $\mathrm{add}_R S$ and all nonmaximal prime ideals $\mathfrak{p}$ of R. Therefore we have that every object in $\mathrm{mod}(\mathbf{P}(S[G])/\mathrm{add}_{S[G]}S)$ has finite length if and only if $S_\mathfrak{p}$ is $R_\mathfrak{p}$-free for all nonmaximal prime ideals $\mathfrak{p}$ of R. Hence to finish the proof of the theorem it suffices to show that $S_\mathfrak{p}$ is $R_\mathfrak{p}$-free if and only if $R_\mathfrak{p}$ is regular.

Since S is regular, we have that $S_\mathfrak{p}$ is regular. Hence if $S_\mathfrak{p}$ is $R_\mathfrak{p}$-free, we have that $R_\mathfrak{p}$ is regular. On the other hand S is a Cohen–Macaulay R-module and so $S_\mathfrak{p}$ is a Cohen–Macaulay $R_\mathfrak{p}$-module. Hence if $R_\mathfrak{p}$ is regular, then $S_\mathfrak{p}$ is $R_\mathfrak{p}$-free [29].

Combining our previous remarks in this section we have the following.

Theorem 5.3. *Suppose S/R is unramified at height one primes. If R is an isolated singularity, then*

$$K_0(\mathrm{mod}\, k[G]) \Big/ \left(\sum_{i=0}^{n} (-1)^i \left[\bigwedge^i V \right] \right) K_0(\mathrm{mod}\, k[G]) \simeq K_0(\mathrm{mod}\, R).$$

The rest of this section is devoted to describing in terms of the action of G on V when the hypothesis of Theorem 5.3 is satisfied in case k is algebraically closed. But first it is convenient to make the following general observation which holds even if k is not algebraically closed.

Lemma 5.4. *Suppose S/R is unramified at height one primes. Then R is an isolated singularity if and only if $S_\mathfrak{p}$ is unramified over $R_\mathfrak{p}$ for all nonmaximal prime ideals $\mathfrak{p}$ in R.*

Proof. Let $\mathfrak{p}$ be a nonmaximal prime ideal in R. If $S_\mathfrak{p}$ is unramified over $R_\mathfrak{p}$, then $S_\mathfrak{p}$ is a free $R_\mathfrak{p}$-module since $R_\mathfrak{p}$ is integrally closed. Hence $R_\mathfrak{p}$ is regular since $S_\mathfrak{p}$ is regular [29]. Suppose $R_\mathfrak{p}$ is regular. Then $S_\mathfrak{p}$ is a free $R_\mathfrak{p}$-module, since S being a Cohen–MaCaulay module implies $S_\mathfrak{p}$ is a Cohen–MaCaulay $R_\mathfrak{p}$-module. Since the height one primes in $R_\mathfrak{p}$ are unramified in $S_\mathfrak{p}$, the fact that $S_\mathfrak{p}$ is $R_\mathfrak{p}$-free implies that $S_\mathfrak{p}$ is unramified over $R_\mathfrak{p}$ [29].

Suppose now that k is algebraically closed and $G\subset \mathrm{GL}(n,k)$. Then G acts on $V=k^n$ and on the affine ring $k[X_1,\dots,X_n]$ of V in the obvious ways. Then the inclusion $k[X_1,\dots,X_n]^G\to k[X_1,\dots,X_n]$ gives a surjective morphism $f:V\to V^G$. Since k is algebraically closed, it is well known that v in V is unramified over $f(v)$ in V^G if and only if $f^{-1}(f(x))$ which is the G-orbit of v, has order of G distinct points. Therefore v in V is ramified if and only if v is in $V^{\langle\sigma\rangle}$ for some $\sigma\neq 1$ in G, where $\langle\sigma\rangle$ is the cyclic group generated by σ. Therefore $\bigcup_{G\in G-\{1\}} V^{\langle\sigma\rangle}$ is the ramification locus of $f:V\to V^G$. Hence the height one primes in $k[X_1,\dots,X_n]^G$ are unramified if and only if $\dim_k V^{\langle\sigma\rangle}<n-1$ for each σ in G, or equivalently, no $\sigma\neq 1$ is a pseudoreflection, i.e. 1 is not an eigenvalue of σ of multiplicity $n-1$ for $\sigma\neq 1$. Also we have that (0) in V is the only ramified point if and only if G acts freely on V, i.e. if $v\neq 0$ and $\sigma(v)=v$, then $\sigma=1$. Therefore passing to the completion of $k[X_1,\dots,X_n]$ with respect to $(X_1,\dots,X_n)$ we obtain the following.

Lemma 5.5. *Suppose k is algebraically closed and V is the G-module given by $G\subset \mathrm{GL}(n,k)$.*

(a) *G has no pseudo-reflections if and only if S/R is unramified at height one primes.*

(b) *G acts freely on V if and only if S/R is unramified at all nonmaximal prime ideals.*

As a consequence of Lemmas 5.4 and 5.5, we obtain the following special case of Theorem 5.3.

Theorem 5.6. *Let k be algebraically closed and suppose G acts freely on the G-module V given by $G\subset \mathrm{GL}(n,k)$. Then*

$$K_0(\mathrm{mod}\,k[G])\Big/\left(\sum_{i=0}^{n}(-1)^i\left[\bigwedge^{i} V\right]\right)\cdot K_0(\mathrm{mod}\,k[G])\simeq K_0(\mathrm{mod}\,R).$$

6. Finiteness of Coker c_R

Throughout this section we assume that the field k is algebraically closed. The rest of the hypothesis and notation is the same as in the previous section. Our aim is to prove the following.

Theorem 6.1. *Suppose G acts freely on V. Then the cokernel of the Cartan map $c_R : K_0(\mathbf{P}(R), 0) \to K_0(\mathrm{mod}\, R)$ is finite, so $K_0(\mathrm{mod}\, R) \cong Z \amalg H$, where H is a finite group.*

The proof will proceed in several steps and is based on the isomorphism $K_0(\mathrm{mod}\, k[G])/z\, K_0(\mathrm{mod}\, k[G]) \to K_0(\mathrm{mod}\, R)$ where $z = \sum_{i=0}^n (-1)^i [\bigwedge^i V]$.

Let $V_0, \ldots, V_d$ be a complete set of nonisomorphic simple $k[G]$-modules with $V_0 = k$, the trivial one. Then $\{[V_i]\}_{i=0,\ldots,d}$ is a basis for $K_0(\mathrm{mod}\, k[G])$. Define $\phi : K_0(\mathrm{mod}\, k[G]) \to Z$ by

$$\phi\left(\sum_{i=0}^{d} n_i [V_i]\right) = \sum n_i \dim_k V_i.$$

This is a surjective Z-algebra morphism with $\mathrm{Ker}\, \phi = \{\sum n_i [V_i];\ \sum n_i \dim_k V_i = 0\}$. Thus $\mathrm{Ker}\, \phi$ is free with $\{[V_i] - (\dim V_i)[k]\}_{i=1,\ldots,d}$ as basis, so $K_0(\mathrm{mod}\, k[G]) = z \cdot [k] \amalg \mathrm{Ker}\, \phi$. Since z is in $\mathrm{Ker}\, \phi$ we have

$$K_0(\mathrm{mod}\, k[G])/z \cdot K_0(\mathrm{mod}\, k[G]) = Z \cdot [k] \amalg \mathrm{Ker}\, \phi/z \cdot K_0(\mathrm{mod}\, k[G]).$$

Hence $\mathrm{Coker}\, c_R \cong \mathrm{Ker}\, \phi/z \cdot \mathrm{Ker}(\mathrm{mod}\, k[G])$. Therefore to prove the proposition, it suffices to show that $\ker \phi/z \cdot \mathrm{Ker}\, \phi$ is torsion, or, what is the same thing, $\mathrm{Ker}\, \phi \xrightarrow{z} \mathrm{Ker}\, \phi$ is a monomorphism. We first prove this in the case G is cyclic and then reduce the general case to the cyclic case.

Proposition 6.2. *Let $G = \langle \sigma \rangle$ be a finite cyclic group of order a and suppose G acts freely on an n-dimensional k-vector space V. Then* $\mathrm{Ker}\, \phi/z\, \mathrm{Ker}\, \phi$ *is a finite group of order a^n.*

Proof. Let $V_0, \ldots, V_{a-1}$ be a complete set of nonisomorphic simple G-modules. They are all one-dimensional and form a cyclic group of order a under tensor product, which we denote by C_a. Thus $K_0(\mathrm{mod}\, k[G]) = Z[C_a]$. Since G acts freely on V we have that $V = \coprod_{j=1}^n V_j$ where the order of each $[V_j]$ in C_a is a, i.e. $[V_j]$ is a generator of C_a. Then

$$z = \sum_{i=0}^{n} (-1)^i \left[\bigwedge^i V\right] = \prod_{j=1}^{n} ([V_j] - 1)$$

in $Z[C_a]$. Now let I be the kernel of the usual augmentation $\varepsilon : Z[C_a] \to Z$. Then $I = \mathrm{Ker}\, \phi$, since $\phi = \varepsilon$. Since each $[V_j]$ is a generator for C_a we have $([V_j] - 1) \cdot Z[C_a] = I$ for all j and so $z \cdot K_0(\mathrm{mod}\, k[G]) = I^n$. Hence $\mathrm{Ker}\, \phi/z\, \mathrm{Ker}\, \phi = I/I^{n+1}$. Now it is well known that $I/I^2 \cong C_a$ [24]. Therefore, if τ is a generator for C_a, then $I \xrightarrow{(\tau-1)} I \to I/I^2 \to 0$ is exact and so $I \xrightarrow{\tau - 1} I$ is a monomorphism. Hence $I^j/I^{j+1} \simeq C_a$ for all $j \geq 1$, which implies that I/I^{n+1} is a finite group of order a^n.

Suppose now that G is an arbitrary group which is acting freely on the n-dimensional vector space V. Then for each σ in G we have that $\langle \sigma \rangle$ acts freely on

V. Further the restriction morphism $\operatorname{res}_\sigma : K_0(\operatorname{mod} k[G]) \to K_0(\operatorname{mod} k\langle\sigma\rangle)$ has the property that $\phi_\sigma \operatorname{res}_\sigma = \phi$ where $\phi_\sigma : K_0(\operatorname{mod}(k\langle\sigma\rangle) \to Z$ is the usual morphism. Therefore $\operatorname{res}_\sigma \operatorname{Ker} \phi \subset \operatorname{Ker} \phi_\sigma$ and $\operatorname{res}_j(z) = z_\sigma = \sum_{i=0}^{n} (-1)^i [\Lambda^i V]$ in $K_0(\operatorname{mod}\langle\sigma\rangle)$. Since the induced morphism

$$K_0(\operatorname{mod} k[G]) \xrightarrow{\pi \operatorname{res}_\sigma} \prod_{\sigma \in G} K_0(\operatorname{mod} k\langle\sigma\rangle)$$

is well known to be a monomorphism, we have by Proposition 6.2 that the commutative diagram

$$\begin{array}{ccccc}
 & & & & 0 \\
 & & & & \downarrow \\
0 & \longrightarrow & \operatorname{Ker} \phi & \longrightarrow & \prod_{\sigma \in G} \operatorname{Ker} \phi_\sigma \\
 & & \downarrow z & & \downarrow \prod z_\sigma \\
0 & \longrightarrow & \operatorname{Ker} \phi & \longrightarrow & \prod_{\sigma \in G} \operatorname{Ker} \phi_\sigma
\end{array}$$

is exact. Hence $\operatorname{Ker} \phi \longrightarrow \operatorname{Ker} \phi$ is a monomorphism, which is what we wanted to show. As previously observed, this finishes the proof of Theorem 6.1. It should also be noted that since $\operatorname{Coker} c_R$ is finite, when G acts freely, we have that R satisfies the hypothesis of Proposition 2.6.

7. Class groups

In this section we return to the hypothesis that G is a finite subgroup of $GL(n, k)$ with k an arbitrary field and $n \geq 2$. The rest of the notation is as in the previous sections.

Let m be the order of G. Since char k does not divide m, we have that $1/m$ is in k. Clearly the composition of R-module morphisms

$$R \xrightarrow{i} S \xrightarrow{1/m \sum_{\sigma \in G} \sigma} R$$

is the identity, where

$$\left(\frac{1}{m} \sum_{\sigma \in G} \sigma\right)(s) = \frac{1}{m} \sum_{\sigma \in G} \sigma(s) \quad \text{for all } s \text{ in } S.$$

Therefore the inclusion $R \to S$ is a split monomorphism of R-modules. As a consequence of this observation we have the following.

Lemma 7.1. *Let A be a reflexive R-module. Then we have the following.*

(a) $\mathrm{Hom}_R(\mathrm{Hom}_R(A,R),S)$ *is a reflexive S-module.*

(b) $\mathrm{rank}_R A = \mathrm{rank}_S \mathrm{Hom}_R(\mathrm{Hom}_R(A,R),S)$.

(c) *A is an R-summand of* $\mathrm{Hom}_R(\mathrm{Hom}_R(A,R),S)$ *viewed as an R-module.*

Proof. (a) Since S is integrally closed, we know that an S-module is reflexive if and only if it is of depth at least two. It is easily verified that as long as X is an R-module which is not a torsion module, the S-module $\mathrm{Hom}_R(X,S)$ has depth at least two. Therefore $\mathrm{Hom}_R(A,R),S)$ is S-reflexive.

(b) Obvious.

(c) Since the inclusion $R \to S$ is a split R-module morphism,

$$A = \mathrm{Hom}_R(\mathrm{Hom}_R(\mathrm{Hom}_R(A,R),R) \to \mathrm{Hom}_R(\mathrm{Hom}_R(A,R),S)$$

is a split morphism of R-modules.

As an easy consequence of this lemma we have the following.

Proposition 7.2. (a) *The rank one reflexive R-modules are precisely the rank one R-summands of S.*

(b) *The rank one reflexive R-modules are Cohen–Macaulay modules.*

(c) *$C(R)$, the divisor class group of R, is a finite group.*

Proof. (a) By Lemma 7.1, we know that the rank one reflexive R-modules are precisely the R-summands of rank one reflexive S-modules. Since S is a unique factorization domain, we know that, up to isomorphism, S is the only rank one reflexive S-module. This establishes (a).

(b) Since S is a Cohen–Macaulay S-module and S is a finitely generated R-module, it follows that S is a Cohen–Macaulay R-module. Therefore every R-summand of S is a Cohen–Macaulay R-module. Hence every rank one reflexive R-module is a Cohen–Macaulay R-module by (a).

(c) We recall that $C(R)$ consists of the isomorphism classes $[L]$ of rank one reflexive R-modules with multiplication

$$[L_1]\cdot[L_2] = [\mathrm{Hom}_R(\mathrm{Hom}_R(L_1 \otimes_R L_2, R), R)].$$

That $C(R)$ is finite is then a trivial consequence of (a).

By putting a mild restriction on how G is contained in $\mathrm{GL}(n,k)$, we obtain the following sharpening of part (c) of Proposition 7.2.

Proposition 7.3. *Suppose S/R is unramified at height one primes. Then $C(R) \simeq \mathrm{Hom}(G,k^*)$ where k^* is the multiplicative group of k.*

Proof. Since S/R is unramified at height one primes, we know by Proposition 5.1 that the functor $\mathbf{P}(S[G]) \to \mathrm{add}_R S$ given by $P \mapsto P^G$ is an equivalence of categories.

For the same reason we have that the natural R-algebra map $S[G]\to \mathrm{End}_R(S)$ is an isomorphism. Therefore the projective $S[G]$-modules are the $S[G]$-modules $\mathrm{Hom}_R(A,S)$ with A in $\mathrm{add}_R S$ and $S[G]$ acting on S. Moreover $\mathrm{Hom}_R(A,S)^G=\mathrm{Hom}_R(A,R)$, so $\mathrm{rank}_S \mathrm{Hom}_r(A,S)=\mathrm{rank}_R \mathrm{Hom}_R(A,S)^G$. Hence for each P in $\mathbf{P}(S[G])$ we have $\mathrm{rank}_S P=\mathrm{rank}_R P^G$. Therefore the equivalence $\mathbf{P}(S[G])\to \mathrm{add}_R S$ gives a bijection between the objects in $\mathbf{P}(S[G])$ of rank 1 over S and the isomorphism classes of rank one R-summands of S which, by Proposition 7.2, is the same thing as the isomorphism classes of rank one reflexive R-modules.

It is easily verified that the isomorphism classes of P in $\mathbf{P}S[G]$ with $\mathrm{rank}_S P=1$ form an abelian group $C(S[G])$ under the product $[P]\cdot[Q]=[P\otimes_S Q)$. Therefore the bijection $C(S[G])\to C(R)$ given by $[P]\to[P^G]$ is a group isomorphism provided it is multiplicative. But this follows from the fact that $\mathrm{Hom}_{S[G]}(P,Q)\to \mathrm{Hom}_R(P^G,Q^G)$ is an isomorphism of R-modules for all P and Q in $\mathbf{P}(S[G])$. For suppose P and Q are in $\mathbf{P}(S[G])$. Then they are projective S-modules, so we have that the natural morphism $\mathrm{Hom}_S(P,S)\otimes_S Q\to \mathrm{Hom}_S(P,Q)$ is an S-isomorphism. Moreover it is also a G-morphism. Hence we have

$$(\mathrm{Hom}_S(P,S)\otimes_S Q)^G\simeq \mathrm{Hom}_S(P,Q)^G=\mathrm{Hom}_{S[G]}(P,Q)\xrightarrow{\sim}\mathrm{Hom}_R(P^G,Q^G).$$

Because R is integrally closed and P^G and Q^G are reflexive, we have that the natural morphism $\mathrm{Hom}_R(P^G,R)\otimes_R Q^G\to \mathrm{Hom}_R(P^G,Q^G)$ induces an isomorphism

$$\mathrm{Hom}_R(\mathrm{Hom}_R(\mathrm{Hom}_R(P^G,R)\otimes_R Q^G,R)\to \mathrm{Hom}_R(P^G,Q^G).$$

Combining these R-isomorphisms with the fact that $\mathrm{Hom}_S(P,S)^G=\mathrm{Hom}_R(P^G,R)$, we have that

$$(\mathrm{Hom}_S(P,S)\otimes_S Q)^G\simeq \mathrm{Hom}_R(\mathrm{Hom}_R(\mathrm{Hom}_S(P,S)^G\otimes_R Q^G,R),R)$$

as R-modules. This gives our desired result since every P in $\mathbf{P}(S[G])$ is S-reflexive. Therefore if we show that $C(S[G])\simeq \mathrm{Hom}(G,k^*)$ we will be done since we already have the natural isomorphism $C(S[G])\to C(R)$ given by $[P]\to[P^G]$ for all P in $G(S[R])$.

Now $\mathrm{Hom}(G,k^*)$ is naturally isomorphic to the group of one-dimensional $k[G]$-modules which we now view as an identification. But the ring isomorphism $K_0(\mathbf{P}(S[G]),0)\to K_0(\mathrm{mod}\, k[G])$ given by $P\to P/(\mathrm{rad}\, S)P$ (see Section 4) clearly induces an isomorphism $C(S[G])\to \mathrm{Hom}(G,k^*)$, since $\mathrm{rank}_S P=\dim_k P/(\mathrm{rad}\, S)P$ for all P in $\mathbf{P}(S[G])$. This finishes the proof of Proposition 7.3.

CHAPTER 4. CARTAN MAPS

In this chapter we study methods for investigating when the Cartan map is a monomorphism and when the cokernel of the Cartan map is torsion. In particular, we complete proofs of results stated in the previous chapter. We also study connections with almost split sequences and with contravariantly determined categories.

1. A commutative diagram

The study of Cartan maps is central in the investigation of K-groups. Important properties of Cartan maps are to be monomorphisms or isomorphisms. To investigate such properties we combine previous results to get a commutative diagram which gives a method for computing kernels and cokernels of Cartan maps. Our result is stronger when we have a monomorphism $K_0(\text{mod}\,\mathbf{A/B}) \to K_0(\text{mod}\,\mathbf{A})$. We give some interpretations of the new description of kernels and cokernels in this section and later sections.

Let $(\mathbf{A},\mathbf{B})$ be a coherent pair, where $\mathbf{A}$ is a Krull–Schmidt category, and assume that $c_{\mathbf{A}}$ is an isomorphism so that we can identify $K_0(\mathbf{A},0)$ and $K_0(\text{mod}\,\mathbf{A})$.

Let $\mathbf{B}'$ be the additive subcategory of $\mathbf{A}$ generated by the indecomposable objects of $\mathbf{A}$ which are not in $\mathbf{B}$. We write this as $\mathbf{A}=\mathbf{B}\amalg\mathbf{B}'$. Our standard example is when $\mathbf{A}$ is the category of lattices over an R-order Λ (where as usual R is complete local Gorenstein and Λ is regular outside maximal ideals), and $\mathbf{B}$ is the full subcategory of projective lattices.

The following result is obtained by diagram chasing.

Proposition 1.1. *Let the notation and assumptions be as above.*

(a) *We have the following commutative exact diagram.*

$$
\begin{array}{ccccccc}
 & & & & 0 & & \\
 & & & & \downarrow & & \\
 & & 0 & & \operatorname{Ker} c_{\mathbf{B}} & & \\
 & & \downarrow & & \downarrow & & \\
0 & & K_0(\mathbf{B},0) & \xrightarrow{=} & K_0(\mathbf{B},0) & & \\
\downarrow & & \downarrow & & \downarrow{\scriptstyle c_{\mathbf{B}}} & & \\
K_0(\text{mod}\,\mathbf{A/B}) & \xrightarrow{i} & K_0(\text{mod}\,\mathbf{A}) & \longrightarrow & K_0(\text{mod}\,\mathbf{B}) & \longrightarrow & 0 \\
\| & & \downarrow{\scriptstyle s} & & \downarrow & & \\
K_0(\text{mod}\,\mathbf{A/B}) & \xrightarrow{t} & K_0(\mathbf{B}',0) & \longrightarrow & \operatorname{Coker} c_{\mathbf{B}} & \longrightarrow & 0 \\
\downarrow & & \downarrow & & \downarrow & & \\
0 & & 0 & & 0 & &
\end{array}
$$

(b) *We have an exact sequence*

$$0 \to \operatorname{Ker} c_{\mathbf{B}} \to \operatorname{Im} i \to K_0(\mathbf{B}', 0) \to \operatorname{Coker} c_{\mathbf{B}} \to 0.$$

(c) *If* $K_0(\operatorname{mod} \mathbf{A}/\mathbf{B}) \to K_0(\operatorname{mod} \mathbf{A})$ *is mono, we have an exact sequence*

$$0 \to \operatorname{Ker} c_{\mathbf{B}} \to K_0(\operatorname{mod} \mathbf{A}/\mathbf{B}) \to K_0(\mathbf{B}', 0) \to \operatorname{Coker} c_{\mathbf{B}} \to 0.$$

We see for example that it is a direct consequence of this proposition that if $\mathbf{B}'$ has only a finite number of indecomposable objects, then $\operatorname{Coker} c_{\mathbf{B}}$ is finitely generated. Before we give more interpretations, we introduce the following terminology.

For an additive category $\mathbf{A}$ and a subcategory $\mathbf{B}$ we say that a sequence of maps $A_1 \to A_2 \cdots \to A_i$ is left $\mathbf{B}$-exact if $(B, A_1) \to (B, A_2) \to \cdots \to (B, A_i)$ is exact for all B in $\mathbf{B}$. Then we get the following descriptions of $c_{\mathbf{B}}$ being mono or epi, or having a cokernel which is torsion.

Proposition 1.2. *Let* $(\mathbf{A}, \mathbf{B})$ *be a coherent pair, where* mod $\mathbf{A}$ *is regular and* $\mathbf{A}$ *is a Krull-Schmidt category. Then the following are equivalent.*

(a) $c_{\mathbf{B}}$ *is mono*

(b) *Let* $0 \to A_n \to A_{n-1} \to \cdots \to A_2 \to A_1$ *and* $0 \to A'_n \to A'_{n-1} \to \cdots \to A'_2 \to A'_1$ *be sequences in* $\mathbf{A}$ *which are left* $\mathbf{A}$*-exact, and such that* $0 \to A_n \to A_{n-1} \to \cdots \to A_2 \to A_1 \to 0$ *and* $0 \to A'_n \to A'_{n-1} \to \cdots \to A'_2 \to A'_1 \to 0$ *are left* $\mathbf{B}$*-exact. If* $(A_1 \amalg A'_2 \amalg A_3 \amalg \cdots)_{\mathbf{B}} \simeq (A'_1 \amalg A_2 \amalg A'_3 \amalg \cdots)_{\mathbf{B}}$*, then* $A_1 \amalg A'_2 \amalg A_3 \amalg \cdots \simeq A'_1 \amalg A_2 \amalg A'_3 \amalg \cdots$.

Proof. Assume first that $c_{\mathbf{B}}$ is mono. Let $0 \to A_n \to \cdots \to A_1$ and $0 \to A'_n \to \cdots \to A'_1$ be sequences in $\mathbf{A}$ which are left $\mathbf{A}$-exact, and such that $0 \to A_n \to \cdots \to A_1 \to 0$ and $0 \to A'_n \to \cdots \to A'_1 \to 0$ are left $\mathbf{B}$-exact. Consider the functors F and F' such that $0 \to (\cdot, A_n) \to \cdots \to (\cdot, A_1) \to F \to 0$ and $0 \to (\cdot, A'_n) \to \cdots \to (\cdot, A_n) \to \cdots \to (\cdot, A'_1) \to F' \to 0$ are exact. Then F and F' are in mod $\mathbf{A}/\mathbf{B}$. We have

$$si([F] - [F']) = [(A_1 \amalg A'_2 \amalg A_3 \amalg \cdots)_{\mathbf{B}}] - [(A'_1 \amalg A_2 \amalg A'_3 \amalg \cdots)_{\mathbf{B}}].$$

If $(A_1 \amalg A'_2 \amalg A_3 \amalg \cdots)_{\mathbf{B}} \simeq (A'_1 \amalg A_2 \amalg A'_3 \amalg \cdots)_{\mathbf{B}}$, we have $si([F] - [F']) = 0$, and hence $i([F] - [F']) = 0$ by Proposition 1.1, since $\operatorname{Ker} c_{\mathbf{B}} = 0$. Using the identification of $K_0(\operatorname{mod} \mathbf{A})$ with $K_0(\mathbf{A}, 0)$, we have

$$i([F] - [F']) = [A_1 \amalg A'_2 \amalg A_3 \amalg \cdots) - [A'_1 \amalg A_2 \amalg A'_3 \amalg \cdots],$$

and hence our desired isomorphism.

The converse follows similarly, by using that mod $\mathbf{A}$ is regular.

Proposition 1.3. *Let* $(\mathbf{A}, \mathbf{B})$ *be a coherent pair where* mod $\mathbf{A}$ *is regular and* $\mathbf{A}$ *is a Krull-Schmidt category. Then the following are equivalent.*

(a) Coker $c_{\mathbf{B}}$ *is a torsion group.*

(b) *For each* D *in* $\mathbf{B}'$ *there is a pair of left* $\mathbf{A}$*-exact sequences* $0 \to A_m \to \cdots \to A_1$

and $0\to A'_m\to\cdots\to A'_1$, *where* $0\to A_m\to\cdots\to A_1\to 0$ *and* $0\to A'_m\to\cdots\to A'_1\to 0$ *are left* **B**-*exact, and an integer* $n>0$, such that

$$nD\amalg(A_1\amalg A'_2\amalg A_3\amalg\cdots)_{\mathbf{B}}\simeq(A'_1\amalg A_2\amalg A'_3\amalg\cdots)_{\mathbf{B}}.$$

Proof. Assume that Coker $c_{\mathbf{B}}$ is a torsion group and let D be in **B**. By Proposition 1.1 there is some x in $K_0(\operatorname{mod}\mathbf{A}/\mathbf{B})$ which is sent to $n[D]$ in $K_0(\mathbf{B}',0)$ for some integer $n>0$. We write x in the form $x=[F]-[F']$, where F and F' are in mod **A/B**. Consider the projective resolutions $0\to(\cdot,A_n)\to\cdots\to(\cdot,A_1)\to F\to 0$ and $0\to(\cdot,A'_n)\to\cdots\to(\cdot,A'_1)\to F'\to 0$ in mod **A**. Then the image of x in $K_0(\mathbf{B}',0)$ is

$$[(A_1\amalg A'_2\amalg A_3\amalg\cdots)_{\mathbf{B}}]-[(A'_1\amalg A_2\amalg A'_3\amalg\cdots)_{\mathbf{B}}],$$

so we have our desired isomorphism. The converse is proved similarly.

Note that in the cases when gl.dim mod $\mathbf{A}=2$, it is enough to consider short exact sequences $0\to A_3\to A_2\to A_1$ in Propositions 1.2 and 1.3.

In Chapter 3 we studied conditions for Coker $c_{\mathbf{B}}$ to be a torsion group, and Proposition 1.3 applies to this situation.

The following immediate consequence of Proposition 1.2 will be used later. We here denote by $P(X)$ a projective cover of X, and by ΩA the first *syzygy* module for A.

Corollary 1.4. *let Λ be an artin algebra such that the radical contains no nonzero projective submodule, or let Λ be a selfinjective order with the usual assumptions on an order. Assume that c_Λ is mono. If A and B are Λ-modulos (or Λ-lattices) with no nonzero projective summands, and $A\amalg\Omega A\simeq B\amalg\Omega B$, then $P(A)\simeq P(B)$.*

Proof. Consider the exact sequences $0\to\Omega A\to P(A)\to A\to 0$ and $0\to\Omega B\to P(B)\to B\to 0$, and assume $\Omega A\amalg A\simeq\Omega B\amalg B$. Consider the coherent $(\mathbf{L}(\Lambda),\mathbf{P}(\Lambda))$, where $\mathbf{P}(\Lambda)$ denotes the category of projective Λ-lattices. Since $c_\Lambda$is assumed to be mono, we have by Proposition 1.1 that since $(\Omega A\amalg A\amalg P(B))_{\mathbf{P}(\Lambda)}\simeq(\Omega B\amalg B\amalg P(A))_{\mathbf{P}(\Lambda)}$, then $\Omega A\amalg A\amalg P(B)\simeq\Omega(B)\amalg B\amalg P(A)$, and consequently $P(A)\simeq P(B)$. We here use that ΩA and ΩB have no nonzero projective summands, and that Krull–Schmidt holds, by our assumptions.

We illustrate with the following examples.

Example 1.5. Let $\mathbf{A}=\operatorname{mod}\Lambda$ for an artin algebra Λ, and let **B** consist of the direct sums of one fixed indecomposable M in mod Λ. Then $K_0(\mathbf{B},0)\simeq Z\simeq K_0(\operatorname{mod}\mathbf{B})$, and $K_0(\mathbf{B},0)\to K_0(\operatorname{mod}\mathbf{B})$ is the monomorphism given by multiplication by the length of End M as an End M-module. Let $0\to X_i\to Y_i\to Z_i\to 0$ be exact sequences in mod Λ for $i=1,2$, such that $0\to(M,X_i)\to(M,Y_i)\to(M,Z_i)\to 0$ is exact. If $(X_1\amalg Z_1\amalg Y_2)_{\mathbf{B}}\simeq(X_2\amalg Z_2\amalg Y_1)_{\mathbf{B}}$, then $X_1\amalg Z_1\amalg Y_2\simeq X_2\amalg Z_2\amalg Y_1$.

Example 1.6. Let Λ be a finitely generated R-algebra where R a local commutative noetherian ring, and let $\mathbf{B}$ be the additive category generated by $\Lambda/\underline{r}^i$, $i \geq 0$. When Λ is an artin algebra, we know that gl.dim mod $\mathbf{B} < \infty$ [4], and hence $K_0(\mathbf{B}, 0) \to K_0(\text{mod}\,\mathbf{B})$ is an isomorphism. The general case can be reduced to this case, since showing that $c_{\mathbf{B}}$ is mono or epi only involves considering a finite number of the $\Lambda/\underline{r}^i$ at one time. Let now $0 \to X_i \to Y_i \to Z_i \to Z_i \to 0$ be exact sequences in mod Λ for $i = 1, 2$, such that $0 \to \text{soc}^t X_i \to \text{soc}^t Y_i \to \text{soc}^t Z_i \to 0$ are exact for all t. If $(X_1 \amalg Z_1 \amalg Y_2)_{\mathbf{B}} \simeq (X_2 \amalg Z_2 \amalg Y_1)_{\mathbf{B}}$, then $X_1 \amalg Z_1 \amalg Y_2 \simeq X_2 \amalg Z_2 \amalg Y_1$. And given D in mod Λ, there are exact sequences $0 \to X_i \to Y_i \to Z_i \to 0$, $i = 1, 2$, such that $0 \to \text{soc}^t X_i \to \text{soc}^t Y_i \to \text{soc}^t Z_i \to 0$ are exact for all t, and $D \amalg (X_1 \amalg Z_1 \amalg Y_2)_{\mathbf{B}} \simeq (X_2 \amalg Z_2 \amalg Y_1)_{\mathbf{B}}$.

Example 1.7. Let k be a field and G a finite group. Then it is well known that the Cartan map c_{kG} is a monomorphism. Hence if $0 \to X_i \to Y_i \to Z_i \to 0$ are exact sequences in mod kG, such that $(X_1 \amalg Z_1 \amalg Y_2)_{\mathbf{P}} \simeq (X_2 \amalg Z_2 \amalg Y_1)_{\mathbf{P}}$, then $X_1 \amalg Z_1 \amalg Y_2 \simeq X_2 \amalg Z_2 \amalg Y_1$. Also Coker c_{kG} is torsion, so given any indecomposable non-projective D, there is an integer $n > 0$ and exact sequences $0 \to X_i \to Y_i \to Z_i \to 0$ such that $nD \amalg (X_1 \amalg Z \amalg Y_2)_{\mathbf{P}} \simeq (X_2 \amalg Z_2 \amalg Y_1)_{\mathbf{P}}$.

2. The case when mod A/B is semisimple

Let throughout this section $(\mathbf{A}/\mathbf{B})$ be a coherent pair, where $\mathbf{A} = \mathbf{L}(\Lambda)$ is a category of Λ-lattices over an R-order Λ. In the case when mod$(\mathbf{A}/\mathbf{B})$ is semisimple our results from 4.1 give criteria for when the Cartan map $c_{\mathbf{B}} : K_0(\mathbf{B}, 0) \to K_0(\text{mod}\,\mathbf{B})$ is a monomorphism or an isomorphism. These properties are closely connected with almost split sequences and irreducible maps in $\mathbf{A}$. As usual we get better results when $K_0(\text{mod}(\mathbf{A}/\mathbf{B})) \to K_0(\text{mod}\,\mathbf{A})$ is a monomorphism. From Chapter 2 we know that this is the case when $\mathbf{A} = \text{mod}\,\Lambda$ for an artin algebra Λ, since mod $\mathbf{A}$ is then a finite R-category and the objects in mod$(\mathbf{A}/\mathbf{B})$ are semisimple and hence of finite length.

We start with a description of when mod $\mathbf{A}/\mathbf{B}$ is semisimple. When $\mathbf{B}$ is an additive subcategory of $\mathbf{A}$, we denote by $\mathbf{M} = \{M_i\}_{i \in I}$ the full subcategory of $\mathbf{A}$ whose objects are the indecomposable objects in $\mathbf{A}$ which are not in $\mathbf{B}$.

Proposition 2.1. *Let* $\mathbf{A} = \mathbf{L}(\Lambda)$, $\mathbf{B}$ *a subcategory of* $\mathbf{A}$, *and let* $\mathbf{M}$ *be as defined above.*

$(\mathbf{A}, \mathbf{B})$ *is a coherent pair with* mod$(\mathbf{A}/\mathbf{B})$ *semisimple if and only if there are no irreducible maps between any of the* M_i. *And when* mod$(\mathbf{A}/\mathbf{B})$ *is semisimple, then the simple objects are the* $(\cdot, M_i)/\mathbf{B}$.

Proof. Assume first that there are no irreducible maps between the M_i. For M_i in $\mathbf{M}$, consider the right almost split map $E_i \to M_i$. By assumption, E_i is in $\mathbf{B}$, and $(\cdot, E_i)|_{\mathbf{B}} \to (\cdot, M_i)|_{\mathbf{B}} \to 0$ is exact, so that $\mathbf{B}$ is contravariantly finite in $\mathbf{A}$. It also follows that $(\cdot, M_i)/\mathbf{B}$ is simple. If there is an irreducible map $M_i \to M_j$, then

$(\cdot, M_j)/\mathbf{B}$ is indecomposable and is not simple since $M_i \to M_j$ does not factor through any object in $\mathbf{B}$.

We recall that for an indecomposable non-projective lattice C, $\mathrm{DTr}\, C$ denotes the left hand term for the almost split sequence whose right hand term is C. We have the following.

Theorem 2.2. *Let* $(\mathbf{A}, \mathbf{B})$ *be a coherent pair with* $\mathrm{mod}(\mathbf{A}/\mathbf{B})$ *semisimple, where* $\mathbf{A} = \mathbf{L}(\Lambda)$ *is a category of lattices with the usual assumptions and let* $\mathbf{M}$ *denote the indecomposable objects in* $\mathbf{A}$ *not in* $\mathbf{B}$ *as before. If* $\dim R \geq 2$ *we assume that* $\mathbf{M}$ *contains no projectives. Then we have the following.*

(a) $\mathrm{mod}\, \mathbf{B}$ *is regular if and only if for each* M *in* $\mathbf{M}$ *there is some* $n > 0$ *such that* $\mathrm{DTr}^n M$ *is zero or in* $\mathbf{B}$ *if and only if* $c_{\mathbf{B}}$ *is an isomorphism.*

(b) *If there is no* M *in* $\mathbf{M}$ *of even period with all* $\mathrm{DTr}^j M$ *in* $\mathbf{M}$, *then* $c_{\mathbf{B}}$ *is a monomorphism. If* $\dim R = 0$, *then converse also holds.*

Proof. Recall that $\mathrm{gl.dim}\, \mathrm{mod}\, \mathbf{A} < \infty$. We observe that $\mathrm{mod}\, \mathbf{B}$ is regular if and only if $\mathrm{pd}(\cdot, M_i)|_{\mathbf{B}} < \infty$ for all M_i in $\mathbf{M}$: For if F is in $\mathrm{mod}\, \mathbf{B}$ we have an exact sequence

$$0 \to (\cdot, X) \to (\cdot, B_n) \to \cdots \to (\cdot, B_1) \to (\cdot, B_0) \to F \to 0$$

in $\mathrm{mod}\, \mathbf{A}$, with the B_i in $\mathbf{B}$. We see that $\mathrm{pd}_{\mathbf{B}} F < \infty$ if and only if $\mathrm{pd}(\cdot, X)|_{\mathbf{B}} < \infty$, which proves the claim.

Assume that for each M in $\mathbf{M}$ there is some $n > 0$ such that $\mathrm{DTr}^n M$ is zero or in $\mathbf{B}$. We want to show that this implies $\mathrm{pd}(\cdot, M)|\mathbf{B} < \infty$ for all M in $\mathbf{M}$. If M is projective, then $\dim R \leq 1$ by our assumption. We then have a minimal right almost split $\mathrm{rad}\, M \to M$ [6]. Since $\mathrm{rad}\, M$ is in $\mathbf{B}$ and $(\cdot, \mathrm{rad}\, M)|\mathbf{B} \to (\cdot, M)|\mathbf{B}$ is an isomorphism, $(\cdot, M)|\mathbf{B}$ is projective in $\mathrm{mod}\, \mathbf{B}$. If M is not projective, the almost split sequence $0 \to \mathrm{DTr}\, M \to E \to M \to 0$ induces an exact sequence

$$0 \to (\cdot, \mathrm{DTr}\, M)|\mathbf{B} \to (\cdot, E)|\mathbf{B} \to (\cdot, M)|\mathbf{B} \to 0.$$

Since E is in $\mathbf{B}$, $(\cdot, E)|\mathbf{B}$ is projective. If $\mathrm{DTr}\, M$ is projective or in $\mathbf{B}$, $(\cdot, \mathrm{DTr}\, M)|\mathbf{B}$ is projective, and hence $\mathrm{pd}(\cdot, M)|\mathbf{B} < \infty$. Otherwise we consider the almost split sequence whose right hand term is $\mathrm{DTr}\, M$, and so on. This shows that $\mathrm{mod}\, \mathbf{B}$ is regular.

Assume now that $\mathrm{mod}\, \mathbf{B}$ is regular, and assume there is some M in $\mathbf{M}$ such that for $i \geq 0$, no $\mathrm{DTr}^i M$ is zero or in $\mathbf{B}$. Considering resolutions as above, we get that $(\cdot, \mathrm{DTr}^i M)|\mathbf{B}$ is projective for $i = \mathrm{pd}(\cdot, M)|\mathbf{B}$. Since $\mathrm{DTr}^i M$ and $\mathrm{DTr}^{i+1} M$ are not projective, we have almost split sequences $0 \to \mathrm{DTr}^{i+1} M \to E \to \mathrm{DTr}^i M \to 0$ and $0 \to \mathrm{DTr}^{i+2} M \to E' \to \mathrm{DTr}^{i+1} M \to 0$, and hence the exact sequence $(\cdot, E')|\mathbf{B} \to (\cdot, E)|\mathbf{B} \to (\cdot, \mathrm{DTr}^i M)|\mathbf{B} \to 0$. If X is an indecomposable summand of E, then the induced map $X \to Y$ is the composite of two irreducible maps, and hence not an isomorphism. This shows that $(\cdot, E)|\mathbf{B} \to (\cdot, \mathrm{DTr}^i M)|\mathbf{B}$ is a projective cover, and consequently $(\cdot, \mathrm{DTr}^{i+1} M)\,|\,\mathbf{B}$ is zero. This is a contradiction, since $(\cdot, Z)\,|\,\mathbf{B}$ is zero only if $\dim R = 0$ and Z is simple projective.

The rest of the proof of the theorem is a consequence of the following description of when the composite map

$$K_0(\mathrm{mod}\ \mathbf{A/B}) \xrightarrow{i} K_0(\mathrm{mod}\ \mathbf{A}) \xrightarrow[\sim]{s} K_0(\mathbf{A}, 0) \xrightarrow{t} K_0(\mathbf{M}, 0),$$

is a monomorphism or an epimorphism or an isomorphism, where s denotes the inverse of the Cartan map and t the natural projection map. Since $\mathrm{mod}\ \mathbf{A/B}$ is semisimple, the $(\cdot, M)/\mathbf{B}$ with M in $\mathbf{M}$ give rise to a basis for $K_0(\mathrm{mod}\ \mathbf{A/B})$. If M is in $\mathbf{M}$ and is not projective, we consider the almost split sequence $0 \to \mathrm{DTr}\, M \to E \to M \to 0$ and the projective resolution $0 \to (\cdot, \mathrm{DTr}\, M) \to (\cdot, E) \to (\cdot, M) \to (\cdot, M)/\mathbf{B} \to 0$. This shows that the image of $[(\cdot, M)/\mathbf{B}]$ under the above map is $[M] + [\mathrm{DTr}\, M]$ if $\mathrm{DTr}\, M$ is in $\mathbf{M}$ and $[M]$ if $\mathrm{DTr}\, M$ is not in $\mathbf{M}$. If M is in $\mathbf{M}$ and is projective, the image of $[(\cdot, M)/\mathbf{B}]$ is $[M]$.

For each M in $\mathbf{M}$ let $\mathscr{S}_M = \{\mathrm{DTr}^i M \mid \mathrm{DTr}^k M \in \mathbf{M}$ for all k between and including 0 and $i\}$. Here $\mathrm{DTr}^i M = \mathrm{TrD}^{-i} M$. Two such subsets are either disjoint or they coincide, so that we get a partition of $\mathbf{M}$. For each such $\mathscr{S} = \mathscr{S}_M$ we consider the free subgroup of $K_0(\mathrm{mod}\ \mathbf{A/B})$ with the $[(\cdot, M)/\mathbf{B}]$ for M in $\mathscr{S}$ as basis and the free subgroup of $K_0(\mathbf{M}, 0)$ with the $[M]$ for M in $\mathscr{S}$ as basis. The map $K_0(\mathrm{mod}\ \mathbf{A/B}) \to K_0(\mathbf{M}, 0)$ restricts to a map between the free subgroups determined by $\mathscr{S}$, and is hence a direct sum of such maps. So to decide whether $K_0(\mathrm{mod}\ \mathbf{A/B}) \to K_0(\mathbf{M}, 0)$ is a monomorphism, an epimorphism or an isomorphism it is enough to consider these restriction maps. There are the following four types of subsets $\mathscr{S}$.

(1) For some M in $\mathscr{S}$ and $i \geq 0$ we have that $\mathrm{DTr}^i M$ is projective or is not in $\mathbf{M}$. Then clearly the corresponding restriction map is an isomorphism.

(2) $\mathscr{S}$ contains M of odd DTr-period, with all $\mathrm{DTr}^i M$ in $\mathbf{M}$. Then the corresponding restriction map is a monomorphism but not an isomorphism. For if $r_0[(\cdot, M)/\mathbf{B}] + r_1[(\cdot, \mathrm{DTr}\, M)/\mathbf{B}] + \cdots + r_{2n}[(\cdot, \mathrm{DTr}^{2n} M)/\mathbf{B}]$, where the r_i are nonnegative integers, is mapped to zero, that is $r_0([M] + [\mathrm{DTr}\, M]) + \cdots + r_{2n}([\mathrm{DTr}^{2n} M] + [M])$ is zero, then $r_0 + r_{2n} = 0$, $r_i + r_{i+1} = 0$; $0 \leq i \leq 2n-1$, so that $r_i = 0$ for all i. Clearly $[M]$ is not in the image.

(3) For all M in $\mathscr{S}$, $\mathrm{DTr}^i M$ is not zero and in $\mathbf{M}$ for all $i > 0$ and $\mathscr{S}$ is infinite. Then it is easy to see that the corresponding restriction map is also a monomorphism which is not an isomorphism.

(4) $\mathscr{S}$ contains M of even DTr-period $2n$, with all $\mathrm{DTr}^i M$ in $\mathbf{M}$
Then $\sum_{k=0}^{2n-1} (-1)^k [(\cdot, \mathrm{DTr}^k M)/\mathbf{B}]$ is a nonzero element in $K_0(\mathrm{mod}\ \mathbf{A/B})$ which is mapped to zero. Also $[M]$ is not in the image, so that the restriction map is neither a monomorphism nor an epimorphism.

Using Proposition 1.1, we conclude that $c_{\mathbf{B}}$ is a monomorphism if $\mathbf{M}$ contains no M of even DTr-period with all $\mathrm{DTr}^i M$ in $\mathbf{M}$, and that the converse holds when $K_0(\mathrm{mod}\ \mathbf{A/B}) \to K_0(\mathrm{mod}\ \mathbf{A})$ is a monomorphism, in particular when $\dim R = 0$. Since the restriction map is not an epimorphism in cases (2), (3), (4), $c_{\mathbf{B}}$ is not an epimorphism by Proposition 1.1 and hence not an isomorphism if any of these cases occur. This shows that mod **B** is regular if and only if $c_{\mathbf{B}}$ is an isomorphism.

We point out that it is essential that **M** contains no projectives when $\dim R \geq 2$, as is shown by the following.

Example 2.3. Let R be a complete local, noetherian, integrally closed Gorenstein domain of dimension 2 and let $\mathbf{M}=\{R\}$. Then $\mathrm{pd}(\cdot,R)|\mathbf{B}=\infty$, so that mod **B** is not regular. Now it is well known that $\mathrm{Ext}_R^2(R/\mathfrak{m},R)\simeq R/\mathfrak{m}$ where $\mathfrak{m}$ is the maximal ideal. Let $0\to R\to E\to R\to R/\mathfrak{m}\to 0$ represent some nonzero element of $\mathrm{Ext}_R^2(R/\mathfrak{m},R)$. Then E is a lattice and $E\to R$ is a minimal right almost split morphism of lattices (see [6]). Hence $0\to(\cdot,R)|\mathbf{B}\to(\cdot,E)|\mathbf{B}\to(\cdot,R)|\mathbf{B}\to 0$ is exact with E in **B**. Therefore $\mathrm{pd}_{\mathbf{B}}(\cdot,R)|\mathbf{B}=0$ or ∞. But if $(\cdot,R)|\mathbf{B}$ is projective, then there would be a summand E' of E such that $(\cdot,E')|\mathbf{B}=(\cdot,R)|\mathbf{B}$ which would contradict the fact that $E\to R$ is minimal right almost split. Hence we have that $\mathrm{pd}_{\mathbf{B}}(\cdot,R)|\mathbf{B}=\infty$.

Specializing to the case when M has only one object, we get the following.

Corollary 2.4. *Let* $\mathbf{A}=\mathbf{L}(\Lambda)$ *be a category of lattices as before, and* **B** *be a subcategory such that* **M** *has only one object* M. *Then we have the following.*

(a) $c_{\mathbf{B}}$ *is a monomorphism.*

(b) $c_{\mathbf{B}}$ *is an isomorphism if and only if* mod **B** *is regular if and only if* $\mathrm{DTr}\, M \not\simeq M$.

We also note that Theoem 2.2 gives interesting examples of categories **B** such that gl.dim mod $\mathbf{B}<\infty$ and gl.dim mod $\mathbf{B}^{\mathrm{op}}=\infty$. Let for example Λ be an hereditary artin algebra of infinite representation type, $\mathbf{A}=\mathrm{mod}\,\Lambda$, and choose **M** to be the indecomposable preprojective Λ-modules. Then for each M in **M**, $\mathrm{DTr}^i M$ is projective for some i, so that mod **B** is regular. But since all $\mathrm{DTr}^i M$, $i\geq 0$, are in **M** and are not injective, we have that mod $\mathbf{B}^{\mathrm{op}}$ is not regular.

In this case we also note that $c_{\mathbf{B}^{\mathrm{op}}}$ is not an isomorphism, by Theorem 2.2, even though mod **B** is regular.

3. Cartan maps and almost split sequences

In this section we study the connection between Cartan maps on one hand and almost split sequences and AR-quivers (Auslander–Reiten quivers) on the other hand, from the matrix point of view. We have already seen that whether $K_0(\mathrm{mod}(\mathbf{A}/\mathbf{B})\to K_0(\mathrm{mod}\,\mathbf{A})$ is mono often has to do with properties of almost split sequences. We here study this aspect more systematically.

Let $\mathbf{L}(\Lambda)$ be a category of lattices of finite type for an R-order Λ, with the usual assumptions that R is a complete local Gorenstein ring and Λ is regular outside maximal ideals. Then we denote by $T(\Lambda)$ the following associated maxtrix of almost split sequences and right almost split maps: For each of the indecomposable objects $L_1, \ldots, L_n$ in $\mathbf{L}(\Lambda)$ (where the ordering is such that all indecomposable projectives appear at the end), we consider the almost split sequence $0 \to \mathrm{DTr}\, L_i \to E_i \to L_i \to 0$ if L_i is not projective, and the minimal right almost split map $E_i \to L_i$ if L_i is projective. For $[L_i]$ in $K_0(\mathbf{L}(\Lambda), 0)$ we consider $[L_i] + [\mathrm{DTr}\, L_i] - [E_i]$ if L_i is not projective, and $[L_i] - [E_i]$ if L_i is projective. These elements give the rows of our desired matrix, with respect to the basis $\{[L_1], \ldots, [L_n]\}$ of $K_0(\mathbf{L}, 0)$. We are also interested in certain submatrices of $T(\Lambda)$, and denote by $T(\Lambda)_p$ the matrix obtained from $T(\Lambda)$ when leaving out the rows corresponding to the projectives and by $T(\Lambda)_p^p$ the matrix obtained by also leaving out the columns corresponding to projectives. We note that these matrices are already determined by the (valued) AR-quiver Γ_Λ of Λ. An AR-quiver is a special case of a (valued) translation quiver [30], and an associated matrix can be similarly defined for a translation quiver. Clearly a translation quiver Γ and its associated matrix $T(\Gamma)$ determine each other uniquely, up to simultaneous change of order of rows and columns. In particular, being given the AR-quiver Γ for a lattice category $\mathbf{L}(\Lambda)$ is equivalent to being given the matrix $T(\Lambda)$ of almost split sequences.

The Cartan matrix C_Λ associated with an artin algebra Λ determines the kernel and cokernel of the Cartan map c_Λ and we also write them as $\mathrm{Ker}\, C_\Lambda$ and $\mathrm{Coker}\, C_\Lambda$. When Λ is of finite type, there is the following relationship between the matrix $T(\Lambda)$, and the Cartan matrix for the corresponding Auslander algebra.

Proposition 3.1. *Let Λ be an artin algebra of finite type and Σ the corresponding Auslander algebra. Then $C_\Sigma = T(\Lambda)^{-1}$.*

Proof. $\mathrm{mod}\,\Lambda$ is equivalent to the finitely generated projective Γ-modules, so $c_\Sigma = c_{\mathrm{mod}\,\Lambda}$. Since $\mathrm{gl.dim}\,\mathrm{mod}(\mathrm{mod}\,\Lambda) \leq 2$, $c_{\mathrm{mod}\,\Lambda} : K_0(\mathrm{mod}\,\Lambda, 0) \to K_0(\mathrm{mod}(\mathrm{mod}\,\Lambda))$ is an isomorphism, and the inverse is well known and easily seen to be defined as follows.

Let $0 \to (\cdot, A) \to (\cdot, B) \to (\cdot, C) \to F \to 0$ be a minimal projective resolution of a simple functor F in $\mathrm{mod}(\mathrm{mod}\,\Lambda)$. Then $[F]$ is sent to $[A] + [C] - [B]$ in $K_0(\mathrm{mod}\,\Lambda, 0)$. Since $B \to C$ is a minimal right almost split map (see [12]) and $0 \to A \to B \to C \to 0$ is an almost split sequence if A is not zero, we are done.

Clearly an artin algebra has the same Cartan matrix as its associated graded algebra, but not much is known in general about connection between algebras having the same Cartan matrices. However, as a direct consequence of the above and the fact that if two basic Auslander algebras over an algebraically closed field k have the same AR-quiver, then their graded algebras are isomorphic [20], we have the following result.

Theorem 3.2. *Let Σ and Σ' be two basic Auslander algebras over an algebraically* *closed field k. Then* $c_\Sigma = c_{\Sigma'}$ *if and only if the associated graded algebras are isomorphic.*

Actually, recent results of Bautista–Gabriel–Roiter–Salmeron say that if Σ and Σ' have the same AR-quiver and char $k \neq 2$, then $\Sigma \simeq \Sigma'$ [18]. So if Λ and Λ' are basic k-algebras, with char $k \neq 2$, such that their AR-quivers are isomorphic, then $\Lambda \simeq \Lambda'$. We shall see that for an arbitrary lattice category $\mathbf{L}(\Lambda)$ of finite type the matrices $T(\Lambda)$, $T(\Lambda)_p$, and $T(\Lambda)_p^p$, determine various properties connected with Cartan maps. Some of these properties are restatements of results in Chapter 2, and for the others Proposition 1.1 is useful. We note that for Λ artin corank $T(\Lambda)_p = 0$, since $T(\Lambda)$ is invertible.

Proposition 3.3. *Let* $\mathbf{L}(\Lambda)$ *be the category of lattices for an R-order* Λ *of finite type. Then we have the following.*

(a) rank(Ker $K_0(\text{mod } \mathbf{L}/\mathbf{P}) \to K_0(\text{mod } \mathbf{L})$) = corank $T(\Lambda)_p$.

(b) rank(Ker c_Λ) = corank $T(\Lambda)_p^p$ − corank $T(\Lambda)_p$, *and for artin algebras,* Ker c_Λ = Ker $T(\Lambda)_p^p$.

(c) Coker c_Λ = Coker $T(\Lambda)_p^p$.

(d) $K_0(\text{mod } \Lambda)$ *is a factor of the free group* $K_0(\mathbf{L}, 0)$ *modulo relations determined by* $T(\Lambda)_p$. *In particular,* rank $K_0(\text{mod } \Lambda)$ = corank $T(\Lambda)_p$ + *the number of indecomposable projective* Λ*-modules.*

(e) *If* dim $R = 1$, *and* Λ *is an order in a semisimple algebra* A, *then the number of simple* A*-modules is equal to* corank $T(\Lambda)$.

Proof. (a) and (d) follow from the exact sequence

$$K_0(\text{mod } \mathbf{L}/\mathbf{P}) \xrightarrow{f} K_0(\text{mod } \mathbf{L}) \simeq K_0(\mathbf{L}, 0) \to K_0(\text{mod } \Lambda) \to 0,$$

and the fact that all objects in mod $\mathbf{L}/\mathbf{P}$ are of finite length, so that $T(\Lambda)_p$ is an associated matrix for f.

(e) follows from the exact sequence

$$0 \to V \to K_0(t(\text{mod } \mathbf{L})) \xrightarrow{f} K_0(\text{mod } \mathbf{L}) \to K_0(\text{mod } A) \to 0,$$

since $t(\text{mod } \mathbf{L})$ are the objects of finite length in mod $\mathbf{L}$, and $T(\Lambda)$ is a matrix associated with g, and rank $K_0(\text{mod } \mathbf{A})$ = rank V. Similarly, (b) and (c) follow by using Proposition 1.1.

Since the matrix $T(\Lambda)_p^p$ only depends on the subquiver Γ_p of the corresponding AR-quiver, obtained by leaving out the projective vertices, we have the following consequence.

Corollary 3.4. *Let* Λ *be an artin algebra of finite representation type. Then the translation quiver* Γ_p *determines* Ker c_Λ, Coker c_Λ, *and hence* $|\det C_\Lambda|$.

Proof. det $C_\Lambda \neq 0$ if and only if c_Λ is mono, and if c_Λ is mono, it is easy to see that $|\det C_\Lambda|$ is the order of Coker c_Λ [22]. The rest follows from Proposition 3.3.

Since stably equivalent algebras often have the same AR-quivers after removing the projectives, we have the following consequence.

Corollary 3.5. *Let Λ and Λ' be stably equivalent algebras, of finite representation type, and assume that we have a stable equivalence between Λ and Λ', which commutes with* DTr. *Then* $\operatorname{Ker} c_\Lambda \simeq \operatorname{Ker} c_\Lambda$, $\operatorname{Coker} c_\Lambda \simeq \operatorname{Coker} c_{\Lambda'}$, *and* $|\det c_\Lambda| = |\det C_{\Lambda'}|$.

In particular, the assumptions of the above corollary hold when Λ and Λ' are stably equivalent selfinjective algebras of finite type (with no ring summand whose radical square is zero) [14]. Actually, for selfinjective algebras we have a more general result along these lines.

Theorem 3.6. *Let Λ and Λ' be stably equivalent selfinjective algebras of finite representation type. Then* $\operatorname{Ker} c_\Lambda \simeq \operatorname{Ker} c_{\Lambda'}$, $\operatorname{Coker} c_\Lambda \simeq \operatorname{Coker} c_{\Lambda'}$, *and* $|\det C\Lambda| = |\det C_{\Lambda'}|$.

Proof. In order to apply Proposition 1.1 we need to investigate what happens to exact sequences under stable equivalence.

Let $\alpha : \underline{\operatorname{mod}\Lambda} \to \underline{\operatorname{mod}\Lambda'}$ be a stable equivalence, and denote also by α the induced correspondence between the modules with no nonzero projective summands. Let $0 \to A \xrightarrow{f} B \xrightarrow{g} C \to 0$ be an exact sequence with no split exact summands. We know that there is a related sequence $0 \to K \xrightarrow{f'} \alpha(B) \xrightarrow{g'} \alpha(C) \to 0$ in mod Λ', such that $\alpha(\underline{g}) = \underline{g}'$ [14]. Assume that $0 \to A \xrightarrow{f} B \xrightarrow{g} C \to 0$ is indecomposable and B not projective. We want to show that also $K \simeq \alpha(a)$ and $\alpha(\underline{f}) = \underline{f}'$. So consider the exact sequence $0 \to DC \to DB \to DA \to 0$ in mod Λ^{op}, which also must be indecomposable. Consider the next exact sequence $(\cdot, DB) \to (\cdot, DA) \to \tilde{F} \to 0$. Here $\tilde{F}$ must be indecomposable and the presentation minimal, hence $(\cdot, \underline{DB}) \to (\cdot, \underline{DA}) \to \tilde{F} \to 0$ is also a minimal projective presentation in mod($\underline{\operatorname{mod}\Lambda^{\mathrm{op}}}$) [10]. F is not projective since DB by assumption is not projective. Then $\underline{DB} \to \underline{DA}$ can have no summand of the form $0 \to \underline{Y}$, and hence $\underline{A} \to \underline{B}$ has no summand of the form $\underline{X} \to 0$. This shows that $(\cdot, \underline{A}) \to (\cdot, \underline{B}) \to (\cdot, \underline{C}) \to F \to 0$ is part of a minimal projective resolution, giving our desired claim that we have an exact sequence $0 \to \alpha A \xrightarrow{f'} \alpha B \xrightarrow{g'} \alpha C \to 0$ with $\alpha(\underline{f}) = \underline{f}'$.

Let F be in mod($\underline{\operatorname{mod}\Lambda}$), and $[F]$ the corresponding element in $K_0(\operatorname{mod}(\underline{\operatorname{mod}\Lambda}))$. Under the map $K_0(\operatorname{mod}(\underline{\operatorname{mod}\Lambda})) \to K_0(\operatorname{mod}\Lambda \setminus \mathbf{P}, 0)$, $[F]$ is sent to $[A] + [C] - [B]$ if $0 \to (\cdot, A) \to (\cdot, B) \to (\cdot, C) \to F \to 0$ is a minimal projective resolution in mod(mod Λ). We then get our desired result by applying Proposition 1.1.

As a special case we have that if Λ and Λ' are algebras over an algebraically closed field k, both given by a Brauer tree with e edges an multiplicity m of the exceptional vertex, then $|\det C_\Lambda| = |\det C_{\Lambda'}|$, since Λ and Λ' are stably equivalent selfinjective algebras in this case [25]. It is trivial to compute that $\det C_\Lambda = e^2 m + 1$ if the Brauer tree is a star with the exceptional vertex in the middle. We point out that Alperin has shown that $\det C_\Lambda = e^2 m + 1$ for any Brauer tree.

For lattices over classical orders we illustrate some of the above through the following example.

Example 3.6. Consider the translation quiver Γ:

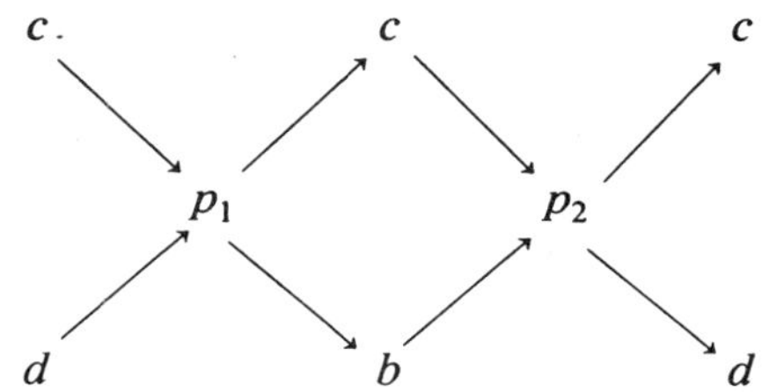

where p_1 and p_2 are projective vertices, $\tau c = a$, $\tau a = c$, $\tau b = d$, $\tau d = b$.

Let Λ be a classical R-order, K the quotient field of R, and $A = K \otimes_R \Lambda$ a semisimple algebra, and assume that the above Γ is the associated AR-quiver. We point out that Γ is realized by a Bäckström order [31]. If the points are ordered as $\{a, b, c, d, p_1, p_2\}$ we have the following matrix $T(\Gamma)$ (with submatrices $T(\Gamma)_p$ and $T(\Gamma)_p^p$).

$$T(\Gamma) = \left(\begin{array}{cccc|cc} 1 & 0 & 1 & 0 & -1 & 0 \\ 0 & 1 & 0 & 1 & -1 & 0 \\ 1 & 0 & 1 & 0 & 0 & -1 \\ 0 & 1 & 0 & 1 & 0 & -1 \\ \hline 0 & 0 & -1 & -1 & 1 & 0 \\ -1 & -1 & 0 & 0 & 0 & 1 \end{array}\right).$$

We see that rank $T(\Gamma) = 4$, rank $T(\Gamma)_p = 3$, rank $T(\Gamma)_p^p = 2$. $K_0(\operatorname{mod} \Lambda)$ is the free abelian group with basis $\{a, b, c, d, p_1, p_2\}$, modulo the relations generated by $a + c - p_1$, $b + d - p_1$, $a + c - p_2$, $b + d - p_2$. It follows that $K_0(\operatorname{mod} \Lambda) \simeq Z \times Z \times Z$. We further have rank $\operatorname{Ker}(K_0(\operatorname{mod} \mathbf{L}/\mathbf{P}) \to K_0(\operatorname{mod} \mathbf{L})) = 4 - 3 = 1$, $\operatorname{rank}(\operatorname{Ker} c_\Lambda) = 3 - 2 = 1$, and the number of simple A-modules is $6 - 4 = 2$.

We also consider some applications to the two-dimensional case. Let $R = C[[X, Y]]^G$, where G is a finite subgroup of $\mathrm{GL}(2, \mathbb{C})$ acting naturaly on $C[[X, Y]]$, such that no $g \neq 1$ has 1 as eigenvalue. Then R is an order of dimension 2, and $\mathbf{L}(R)$ is the reflexive modules. $\mathbf{L}(R)$ is of finite type in this case [1], [8]. By Proposition 3.3 we know that we can use the AR-quiver to compute $K_0(\operatorname{mod} R)$. And in the case when G is a subgroup of $\mathrm{SL}(2, \mathbb{C})$, Auslander has proved that the AR-quiver for $\mathbf{L}(R)$, after leaving out the projective module, is the same as the graph occuring in the resolution of singularities. Here it is understood that $\cdot \rightleftarrows \cdot$ is replaced by a single edge, and the translation τ in the AR-quiver is the identity [8]. The graphs occuring this way are known to be Dynkin, so that the various $K_0(\operatorname{mod} R)$ can be computed, as has been done with this method by Rohnes.

On the other hand, we have shown in Chapter 3 that $K_0(\operatorname{mod} R) = Z \amalg C(R)$ where $C(R)$ denotes the class group. And it is known that $C(R)$ can be computed from the graph associated with the resolution of singularities in the following way. Let $1, \ldots, n$ be a numbering of the curves, or equivalently the vertices. Let $A = (a_{ij})$, where a_{ij} is the intersection number between the ith and jth curve. It is known that $a_{ij} = 1$ when $i \neq j$. Then it is known that $\operatorname{Coker} A = C(R)$, in particular $|\det A|$ is the order of $C(R)$ [27]. For rational double points, that is when $G \subset \mathrm{SL}(2, \mathbb{C})$, it is known that $a_{ii} = -2$ for all i [21]. In this case we have for the matrix $T(R)_p^p$ that we have 2 on the diagonal, since τ is the identity, and we have -1 whenever we have an irreducible map. We clearly have $T(R)_p^p = -A$ in this case, which agrees nicely with the fact that Coker $T(R)_p^p = -A$ in this case, which agrees nicely with the fact that Coker $T(R)_p^p = \operatorname{Coker} c_R$. This gives for example another proof of the fact that $C(R) \simeq \operatorname{Coker} c_R$ for rational double points, or if we use this fact, another proof that $C(R)$ is given by the singularity matrix A.

For arbitrary rational singularities the singularity graph and the AR-quiver (minus the projective vertex) are not the same, or, in other words, the matrices A and $T(R)_p^p$ are not the same. But the fact $\operatorname{Coker} A \simeq \operatorname{Coker} T(R)_p^p$ gives an interesting connection between the graphs in general. In particular, $|\det A| = |\det C_R|$. So knowing the singularity graph gives some restriction on what the corresponding AR-quiver looks like.

We illustrate with the following example. Let ε be a primitive nth root of 1, and $G = \langle (\begin{smallmatrix} \xi & 0 \\ 0 & \xi \end{smallmatrix}) \rangle \subset \mathrm{GL}(2, \mathbb{C})$. By considering the corresponding group representations [8] it is easy to see that the AR-quiver is

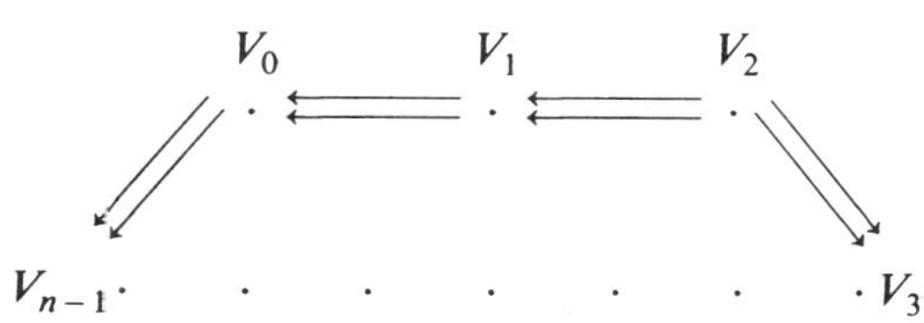

where V_0 is the projective module and $\tau(v_i) = V_{i+2}$ (addition modulo n) for $i \neq 0$. The singularity graph is $\overset{-n}{\cdot}$ in this case. For example for $n = 3$ we have the matrices $A = (-3)$ and $T(R)_p^p = (\begin{smallmatrix} 1 & -2 \\ 1 & 1 \end{smallmatrix})$.

4. Contravariantly determined categories

Let $\mathbf{A}$ be a finite R-category over a commutative artin ring R. $\mathbf{A}$ is said to be contravariantly determined if for A, B in $\mathbf{A}$, $\langle X, A \rangle = \langle X, B \rangle$ for all X in $\mathbf{A}$ implies $A \simeq B$. Here $\langle X, A \rangle$ denotes the length of (X, A) over R. Dually $\mathbf{A}$ is said to be covariantly determined if $\mathbf{A}^{\mathrm{op}}$ is contravariantly determined. These concepts were introduced in [15] and were useful for investigating when indecomposable Λ-modules are determined by their composition factors. Our interest in the property contravariantly determined in this paper comes from the connection with the property that the Cartan map $c_{\mathbf{A}}: K_0(\mathbf{A}, 0) \to K_0(\operatorname{mod} \mathbf{A})$ is mono. For if $\mathbf{A}$ is the category of finitely generated projectives over an artin algebra Λ, then $c_{\mathbf{A}} = c_{\Lambda}$ is mono if and only if $\mathbf{A}$ is contravariantly determined. This follows from the fact that if M and N are in $\operatorname{mod} \Lambda$, then $[M] = [N]$ in $K_0(\operatorname{mod} \Lambda)$ if and only if M and N have the same composition factors. For an arbitrary coherent finite R-category we do not, however, know if the concepts are equivalent. Here the following property comes up. We say that $\operatorname{mod} \mathbf{A}$ is determined by composition factors if for F and G in $\operatorname{mod} \mathbf{A}$, $\langle (\cdot, A_i), F\rangle = \langle (\cdot, A_i), G\rangle$ for all indecomposable A_i in $\mathbf{A}$ implies that $[F] = [G]$ in $K_0(\operatorname{mod} \mathbf{A})$. The following is then an easy consequence of the definitions.

Proposition 4.1. *Let* $\mathbf{A}$ *be a coherent finite R-category. If* $\mathbf{A}$ *is contravariantly determined, then* $c_{\mathbf{A}}$ *is mono, and if* $\operatorname{mod} \mathbf{A}$ *is determined by composition factors and* $c_{\mathbf{A}}$ *is mono, then* $\mathbf{A}$ *is contravariantly determined.*

A large class of contravariantly finite categories including $\operatorname{mod} \Lambda$, is obtained from the following result.

Proposition 4.2. *Let* $\mathbf{A}$ *be a finite R-category.*

(a) *If the simple functors in* $\operatorname{Mod} \mathbf{A}$ *have finite projective resolutions in* $\operatorname{Mod} \mathbf{A}$ *with projectives from* $\operatorname{mod} \mathbf{A}$, *then* $\mathbf{A}$ *is contravariantly determined.*

(b) *If the simple functors in* $\operatorname{Mod} \mathbf{A}^{\mathrm{op}}$ *have finite projective resolutions in* $\operatorname{Mod} \mathbf{A}^{\mathrm{op}}$ *with projectives from mod* $\mathbf{A}^{\mathrm{op}}$, *then* $\mathbf{A}$ *is covariantly determined.*

Proof. Let X and Y be in $\mathbf{A}$, and assume that $\langle X, A\rangle = \langle Y, A\rangle$ for all A in $\mathbf{A}$, that is, $\langle (\cdot, X), (\cdot, A)\rangle = \langle (\cdot, Y), (\cdot, A)\rangle$. By the assumption on the resolutions of simple functors, we get $\langle (\cdot, X), S\rangle = \langle (\cdot, Y), S\rangle$ for all simple functors S in $\operatorname{Mod} \mathbf{A}$. Since $\operatorname{mod} \mathbf{A}$ is a Krull-Schmidt category, each S is of the form $(\cdot, Z)/\operatorname{rad}(\cdot, Z)$, and further $(\cdot, Z)$ is determined by $(\cdot, Z)/\operatorname{rad}(\cdot, Z)$. It follows that $(\cdot, X) \simeq (\cdot, Y)$, and hence $X \simeq Y$. This proves (a), and (b) is dual.

$\operatorname{mod} \Lambda$ satisfies the above condition since $\operatorname{mod} \Lambda$ has kernels, and hence $\operatorname{gl.dim} \operatorname{mod}(\operatorname{mod} \Lambda) \le 2$. By using almost split sequences to get resolutions, it follows that more generally $\mathbf{A} = \operatorname{add} \mathbf{C}$ is contravariantly and covariantly determined, when $\mathbf{C}$ is a subcategory of $\operatorname{ind} \Lambda$ closed under irreducible maps (see [7], [19], [15]).

Another special case is the following. Let Λ be a k-algebra, where k is an algebraically closed field, $\mathbf{C}$ a subcategory of $\operatorname{ind} \Lambda$ closed under irreducible maps and $\mathbf{C}'$ is $\mathbf{C}$ modulo the maps in the infinite radical (see [14]). If $\mathbf{A} = \operatorname{add} \mathbf{C}'$, it is proved in [28] that for each simple functor S in $\operatorname{Mod} \mathbf{A}$ there is a finite resolution with projectives from $\operatorname{mod} \mathbf{A}$, in fact such that $\operatorname{pd} S \le 2$.

Unlike $\text{mod}\,\Lambda$, the category $\underline{\text{mod}}\,\Lambda$ is not contravariantly determined in general, and we shall devote the rest of this section to get criteria for when it is. We first point out that this is equivalent to investigating when the Cartan map is mono, as is a consequence of the following result and Proposition 4.1.

Proposition 4.3. (a) *Let* $\mathbf{A}$ *be a coherent finite R-category. If* $c_{\mathbf{A}}$ *is surjective and* $\mathbf{A}$ *is contravariantly determined, then* $\text{mod}\,\mathbf{A}$ *is determined by composition factors.*
(b) $\underline{\text{mod}}\,\Lambda$ *is determined by composition factors.*

Proof. (a) Since $c_{\mathbf{A}}$ is surjective, any element in $K_0(\text{mod}\,\mathbf{A})$ is of the form $[(\cdot, A)] - [(\cdot, B)]$ for A and B in $\mathbf{A}$. If $\langle(\cdot, X), (\cdot, A)\rangle = \langle(\cdot, X), (\cdot, B)\rangle$ for all X in $\mathbf{A}$, then $A \simeq B$ since $\mathbf{A}$ is contravariantly determined, and hence $[(\cdot, A)] = [(\cdot, B)]$. This shows that $\text{mod}\,\mathbf{A}$ is determined by composition factors.
(b) This follows from (a) and the fact that $K_0(\text{mod}(\underline{\text{mod}}\,\Lambda)) \to K_0(\text{mod}(\text{mod}\,\Lambda))$ is a monomorphism.

We now go on to investigate to which extent $\underline{\text{mod}}\,\Lambda$ is contravariantly determined. For X in $\text{mod}\,\Lambda$ the corresponding object in $\underline{\text{mod}}\,\Lambda$ is denoted by $\underline{X}$.

Proposition 4.4. *Let* Λ *be an artin algebra, and assume that* A *and* B *in* $\text{mod}\,\Lambda$ *have no nonzero projective summands. Then* $\langle \underline{X}, \underline{A}\rangle = \langle \underline{X}, \underline{B}\rangle$ *for all* X *in* $\text{mod}\,\Lambda$ *if and only if* $A \amalg \Omega A \amalg P(B) \simeq B \amalg \Omega B \amalg P(A)$.

Proof. Consider the projective resolutions

$$0 \to (\cdot, \Omega A) \to (\cdot, P(A)) \to (\cdot, A) \to (\cdot, \underline{A}) \to 0$$

and

$$0 \to (\cdot, \Omega B) \to (\cdot, P(B)) \to (\cdot, B) \to (\cdot, \underline{B}) \to 0$$

in $\text{mod}(\text{mod}\,\Lambda)$. Hence $\langle \underline{X}, \underline{A}\rangle = \langle \underline{X}, \underline{B}\rangle$ for all X in $\text{mod}\,\Lambda$ if and only if $\langle X, A \amalg \Omega A \amalg P(B)\rangle = \langle X, B \amalg \Omega B \amalg P(A)\rangle$ for all X in $\text{mod}\,\Lambda$. Since $\text{mod}\,\Lambda$ is contravariantly determined, this is equivalent to $A \amalg \Omega A \amalg P(B) \simeq B \amalg \Omega B \amalg P(A)$.

For a class of algebras containing the selfinjective algebras we can improve this result as follows.

Corollary 4.5. *Let* Λ *be an artin algebra whose radical contains no nonzero projective summands, and assume that* A *and* B *are in* $\text{mod}\,\Lambda$ *and have no nonzero projective summands.*
(a) $\langle \cdot, \underline{A}\rangle = \langle \cdot, \underline{B}\rangle$ *if and only if* $A \amalg \Omega A \simeq B \amalg \Omega B$ *and* $A/\underline{r}A \simeq B/\underline{r}B$.
(b) *If* c_Λ *is mono, then* $\langle \cdot, \underline{A}\rangle = \langle \cdot, \underline{B}\rangle$ *if and only if* $A \amalg \Omega A \simeq B \amalg \Omega B$.

Proof. We observe that ΩA and ΩB have no nonzero projective summands, and that $P(A) \simeq P(B)$ if and only if $A/\boldsymbol{r}A \simeq B/\boldsymbol{r}B$, where $\boldsymbol{r}$ is the radical of Λ. The last part follows by combining with Corollary 1.4.

The following example shows that the condition $A/rA \simeq B/rB$ can not always be left out.

Example 4.6. Let Λ be a symmetric algebra of Loewy length 3, having two simple modules T_1 and T_2, such that $rP(T_1)/\text{soc}\, P(T_1) \simeq T_2 \amalg T_2$ and $rP(T_2)/\text{soc}\, P(T_2) \simeq T_1 \amalg T_1$. If M is a Λ-module of length 2, it can be shown that $\Omega^2 M \simeq M$, but $M/\underline{r}N \not\simeq \Omega M/\underline{r}\Omega M$.

We also give an example to show that $\underline{\text{mod}\,\Lambda}$ is not always contravariantly determined.

4.7. Let k be a field and $\Lambda = k[x]/(x^5)$, $A = k[x]/(x^2)$, $B = k[x]/(x^3)$. Then $\Omega A \amalg A \simeq \Omega B \amalg B$ and $A/rA \simeq B/rB$, showing that $\underline{\text{mod}\,\Lambda}$ is not contravariantly determined.

We end with the following criterion for when $\underline{\text{mod}\,\Lambda}$ is contravariantly determined for a class of selfinjective algebras.

Proposition 4.8. *Assume that Λ is selfinjective and c_Λ is mono.*

(a) Ker $c_{\underline{\text{mod}\,\Lambda}}$ *is generated by elements of the form* $[X \amalg \Omega^2 X \amalg \cdots \amalg \Omega^{2n-2} X] - [\Omega X \amalg \cdots \amalg \Omega^{2n-1} X]$, *for X indecomposable of even period $2n$, and these elements are all nonzero.*

(b) $\underline{\text{mod}\,\Lambda}$ *is contravariantly determined if and only if there exists no indecomposable Λ-module of even period.*

Proof. (a) Assume that X is indecomposable of period $2n$, and let $A = X \amalg \Omega^2 X \amalg \cdots \amalg \Omega^{2n-2} X$. Then $\Omega^2 A \simeq A$, and $\Omega A \not\simeq A$, so that $[A] - [\Omega A]$ is a nonzero element in Ker $c_{\underline{\text{mod}\,\Lambda}}$, by Corollary 4.5.

Assume conversely that $[A] - [B]$ is a nonzero element in Ker $c_{\underline{\text{mod}\,\Lambda}}$. We then have $A \not\simeq B$ and $A \amalg \Omega A \simeq B \amalg \Omega B$. We can clearly assume that A and B have no common indecomposable summands, hence the same is the case for ΩA and ΩB. Let $A = C_1 \amalg \cdots \amalg C_t$ and $B = D_1 \amalg \cdots \amalg D_s$, where the C_i and D_i are indecomposable. If $\Omega C_1 \simeq C_j$ for some j, then since $C_j \simeq \Omega D_t$ for some t, we would get $C_1 \simeq D_t$, a contradiction. It follows that $\Omega C_1 \simeq D_t$ for some t. We then see that C_1 must have even period.

References

[1] M. Artin and J. Verdier, Reflexive modules over rational double ponts.

[2] M. Auslander, On the purity of the branch locus. Amer. J. Math. 84 (1) (1962) 116–125.

[3] M. Auslander, Coherent Functors, Proc. Conf. Categorical Algebra (Springer, New York, 1966).

[4] M. Auslander, Representation Dimension of Artin Algebras, Queen Mary College Math. Notes (London, 1971).

[5] M. Auslander, Representation theory of artin algebras II, Comm. Algebra 1 (1974) 269–310.
[6] M. Auslander, Functors and morphisms determined by objects, Proc. Conf. Repr. Theory (Philadelphia 1976) (Marcel Dekker, New York, 1978) 1–244.
[7] M. Auslander, Relations for Grothendieck groups of artin algebras, Proc. Amer. Math. Soc., to appear.
[8] M. Auslander, Almost split sequences and rational singularities.
[9] M. Auslander, to appear.
[10] M. Auslander and I. Reiten, Stable equivalence of artin algebras, Proc. Conf. on Orders, Group Rings and Related Topics, Lecture Notes in Math. 353 (Springer, Berlin, 1972) 8–71.
[11] M. Auslander and I. Reiten, Representation theory of artin algebras III: Almost split sequences, Comm. Algebra 3 (1975) 239–294.
[12] M. Auslander and I. Reiten, Representation theory of artin algebras IV: Invariants given by almost split sequences, Comm. Algebra 5 (1977) 443–518.
[13] M. Auslander and I. Reiten, Representation theory of artin algebras V: Methods for computing almost split sequences and irreducible morphisms, Comm. Algebra 5 (1977) 519–554.
[14] M. Auslander and I. Reiten, Representation theory of artin algebras VI: A functorial approach to almost split sequences, Comm. Algebra 11 (1977).
[15] M. Auslander and I. Reiten, Modules determined by their composition factors, Illinois J. Math.
[16] M. Auslander and D. Rim, Ramification index and multiplicity, Illinois J. Math. 7 (4) (1963) 566–581.
[17] M. Auslander and S. Smalø, Almost split sequences in subcategories, J. Algebra 69 (2) (1981) 426–454.
[18] R. Bautista, P. Gabriel, A. Roiten and L. Salmeron, Representation finite algebras and multiplicative basis.
[19] R. Benson and D. Parker, The Green ring of a finite group, J. Algebra, to appear.
[20] K. Bongartz and P. Gabriel, Coverings in representation theory, Invent. Math. 65 (1982) 331–378.
[21] E. Brieskorn, Rationale Singularitäten komplexer Flächen, Invent. Math. 4 (1968) 336–358.
[22] N. Bourbaki, Commutative Algebra, Chapter 7.
[23] M. Butler, Grothendieck groups and almost split sequences, Proc. Oberwolfach Conf. on Integral Representations and Applications, Lecture Notes in Math. 882 (1981) 357–368.
[24] H. Cartan and S. Eilenberg, Homological Algebra (Princeton Univ. Press, Princeton, 1956).
[25] P. Gabriel and C. Riedtmann, Group representations without groups, Comment. Math. Helv. 54 (1979) 240–287.
[26] P. Gabriel, Des categories abeliennes, Bull. Soc. Math. France 90 (1962) 323–448.
[27] F. Hirzebruch, The topology of normal singularities of an algebraic surface, Sem. Bourbaki 1962/63 no. 250, 01–09.
[28] K. Igusa and G. Todorov, Radical layers of representable functors.
[29] M. Nagata, Local Rings (Interscience, New York, 1960).
[30] C. Riedtmann, Algebren, Darstellungskörper, Überlangerungen und Zurück, Comment. Math. Helv. 55 (1980) 199–224.
[31] K. Roggenkamp, The lattice type of orders II: Auslander-Reiten quivers, Proc. Oberwolfach Conf. on Integeral Representations and Applications, Lecture Notes in Math. 882 (Springer, Berlin, 1981) 430–477.
[32] D. Quillen, Higher algebraic *K*-theory I, Proc. Conf. on Algebraic *K*-theory, Seattle, Vol. 1, Lecture Notes in Math. 341 (Springer, Berlin, 1973) 77–139.

PROCEEDINGS OF THE
AMERICAN MATHEMATICAL SOCIETY
Volume 103, Number 4, August 1988

GROTHENDIECK GROUPS OF ALGEBRAS WITH NILPOTENT ANNIHILATORS

MAURICE AUSLANDER AND IDUN REITEN

(Communicated by Donald S. Passman)

ABSTRACT. Let R be a commutative noetherian ring and $i: R \to \Lambda$ an R-algebra such that Λ is a finitely generated R-module. Then the annihilator of Λ in R is nilpotent if and only if the cokernel of the induced map of Grothendieck groups $i^*: K_0(\text{mod } \Lambda) \to K_0(\text{mod } R)$ is a torsion group.

Let k be an algebraically closed field, and let $S = k[[X_1, \ldots, X_n]]$ be the formal power series ring in n variables over k. Let G be a finite subgroup of $\mathrm{GL}(n, k)$. Hence G acts as a group of k-automorphisms of S, and we denote by R the fixed ring $R = S^G$. Denote by $K_0(\text{mod } R)$ the Grothendieck group of the category of finitely generated R-modules mod R modulo exact sequences. In [**1**] we proved that $K_0(\text{mod } R)$ is finitely generated. Under the additional assumption that G acts freely we proved that $K_0(\text{mod } R)$ is isomorphic to $Z[R] \amalg H$, where H is a finite group and $[R]$ denotes the image of R in $K_0(\text{mod } R)$. The motivation for this paper was to show that the assumption that G acts freely is not necessary. The proof we give here is based on the proof given in the Bielefeld May 1985 conference on representation theory and singularity theory. A different proof in the case that G is abelian has been given in [**3**]. Our desired result is an easy consequence of the following general result which is also of independent interest. For the rest of the paper R denotes a commutative noetherian ring and Λ an R-algebra via a fixed map $i: R \to \Lambda$ such that Λ is a finitely generated R-module.

THEOREM. *The annihilator of* Λ, $\operatorname{ann}_R \Lambda$, *is a nilpotent ideal in* R *if and only if* $\operatorname{Coker}(K_0(\text{mod } \Lambda) \xrightarrow{i^*} K_0(\text{mod } R))$ *is a torsion group.*

This result is a direct consequence of the following two propositions.

PROPOSITION 1. *If* $\operatorname{ann}_R \Lambda$ *is nilpotent, then*

$$\operatorname{Coker}(K_0(\text{mod } \Lambda) \xrightarrow{i^*} K_0(\text{mod } R))$$

is torsion.

PROOF. Now, $\operatorname{ann}_R \Lambda$ is nilpotent if and only if $\Lambda_p \neq 0$ for all prime ideals p in R. From this characterization it follows easily that if $\mathfrak{A}$ is an ideal in R and we consider the $R/\mathfrak{A}$-algebra $R/\mathfrak{A} \to \Lambda/\mathfrak{A}\Lambda$, then $\operatorname{ann}_{R/\mathfrak{A}} \Lambda/\mathfrak{A}\Lambda$ is nilpotent. Assume that the proposition is false. Then because R is noetherian there is an ideal $\mathfrak{A}$ in R such that $\operatorname{Coker}(K_0(\text{mod } \Lambda/\mathfrak{A}\Lambda) \to K_0(\text{mod } R/\mathfrak{A}))$ is not torsion, and

Received by the editors December 1, 1986 and, in revised form, March 30, 1987.
1980 *Mathematics Subject Classification* (1985 *Revision*). Primary 13D15; Secondary 19A49.
Written with partial support from NSF.

Coker(K_0(mod $\Lambda/b\Lambda$) $\to$ K_0(mod R/b)) is torsion for any ideal $\mathfrak{B}$ in R properly containing $\mathfrak{A}$. We can clearly assume that $\mathfrak{A}$ is zero.

Consider for a minimal prime ideal p in R the commutative exact diagram [**2**, p. 642]

$$\begin{array}{ccccccc} \lim_{t\notin p} K_0(\text{mod } \Lambda/t\Lambda) & \longrightarrow & K_0(\text{mod } \Lambda) & \longrightarrow & K_0(\text{mod } \Lambda_p) & \longrightarrow & 0 \\ \downarrow{\scriptstyle \alpha} & & \downarrow{\scriptstyle i} & & \downarrow{\scriptstyle \beta} & & \\ \lim_{t\notin p} K_0(\text{mod } R/tR) & \longrightarrow & K_0(\text{mod } R) & \longrightarrow & K_0(\text{mod } R_p) & \longrightarrow & 0 \end{array}$$

The snake lemma gives the exact sequence $\operatorname{Coker}\alpha \to \operatorname{Coker} i^* \to \operatorname{Coker}\beta \to 0$. Since R_p is a local artin ring, $K_0(\text{mod } R_p) \simeq Z$, and since $\Lambda_p \neq 0$, β is not zero, so that $\operatorname{Coker}\beta$ is torsion. By our assumption Coker(K_0(mod $\Lambda/t\Lambda$) $\to$ K_0(mod R/tR)) is torsion since $t \neq 0$, and hence $\operatorname{Coker}\alpha$ is torsion. From this it follows that $\operatorname{Coker} i^*$ is torsion, which contradicts our assumption, and the proof is done.

PROPOSITION 2. *We have* $\operatorname{rank}(\operatorname{Coker} K_0(\text{mod } \Lambda \xrightarrow{i^*} K_0(\text{mod } R)) \geq n$, *where n denotes the number of minimal primes in R not containing* $\mathfrak{A} = \operatorname{ann}_R \Lambda$.

PROOF. Let $p_1, \dots, p_n, \dots, p_r$ be the minimal primes in R, where $\mathfrak{A} \subsetneq p_i$ for $1 \leq i \leq n$ and $\mathfrak{A} \subset p_i$ for $n < i \leq r$. Let $T = R \backslash \bigcup_{i=1}^r p_i$, and consider the diagram

$$\begin{array}{ccc} K_0(\text{mod } R/\mathfrak{A}) & \longrightarrow & K_0(\text{mod}(R/\mathfrak{A})_T) \\ \downarrow{\scriptstyle \delta} & & \downarrow{\scriptstyle \gamma} \\ K_0(\text{mod } R) & \longrightarrow & K_0(\text{mod } R_T) \end{array}$$

Since R_T is artin, $K_0(\text{mod } R_T)$ is a free group with basis $\{[(R/p_i)_T]; 1 \leq i \leq r\}$ and since $(R/\mathfrak{A})_T$ is artin $K_0(\text{mod}(R/\mathfrak{A})_T)$ is a free group with basis $\{[(R/p_i)_T]; r \leq i \leq r\}$. Hence we have rank $\operatorname{Coker}\delta \geq$ rank $\operatorname{Coker}\gamma \geq n$. Since $i\colon R \to \Lambda$ has the factorization $R \to R/\mathfrak{A} \to \Lambda$, we get rank $\operatorname{Coker} i^* \geq$ rank $\operatorname{Coker}\delta \geq n$.

We end this note with the following consequences of Proposition 1.

COROLLARY 3. *If Λ is a commutative semilocal regular domain and $R \subset \Lambda$, then* $\operatorname{rank} K_0(\text{mod } R) = 1$.

PROOF. Since Λ is regular, $K_0(\text{mod } \Lambda) \simeq K_0(\mathscr{P}(\Lambda), 0)$, where $\mathscr{P}$ denotes the category of finite generated projective Λ-modules and $K_0(\mathscr{P}, 0)$ denotes the free group on the isomorphism classes of objects in $\mathscr{P}$ modulo split exact sequences. Then $K_0(\text{mod } \Lambda) \simeq Z$ since every finitely generated projective Λ-module is free. Since $R \subset \Lambda$, R is also a domain. If L denotes the quotient field of R, we have a surjection $K_0(\text{mod } R) \to K_0(\text{mod } L)$, so that $\operatorname{rank} K_0(\text{mod } R) \geq 1$. It then follows from Proposition 1 that $\operatorname{rank} K_0(\text{mod } R) = 1$.

COROLLARY 4. *Let Λ be a commutative complete regular local domain and G a finite group acting on Λ as ring automorphisms, such that the order of G is invertible in Λ. Then* $K_0(\text{mod } \Lambda^G) \simeq Z \oplus H$, *where H is a finite group.*

PROOF. Λ is a finitely generated Λ^G-module [**4**, Corollary 5.9] and $R = \Lambda^G$ is noetherian [**4**, Corollary 1.12]. Then $\operatorname{rank} K_0(\text{mod } \Lambda^G) = 1$ follows from Corollary 3, and it only remains to show that $K_0(\text{mod } \Lambda^G)$ is finitely generated. This follows as in [**1**, Proposition 3.4]: Since $\mathrm{End}_{\Lambda G}(\Lambda) \simeq \Lambda^G$, there is a surjection $K_0(\mathrm{mod}\ \Lambda G) \to K_0(\mathrm{mod}\ \Lambda^G)$. Further ΛG has finite global dimension since the order of G is invertible in Λ, and hence $K_0(\mathrm{mod}\ \Lambda G) \simeq K_0(\mathscr{P}(\Lambda G), 0)$. Since by our assumption on Λ the Krull-Schmidt property holds for ΛG, $K_0(\mathscr{P}(\Lambda G), 0)$ is finitely generated.

REFERENCES

1. M. Auslander and I. Reiten, *Grothendieck groups of algebras and orders*, J. Pure Appl. Algebra **29** (1986), 1–51.
2. H. Bass, *Algebraic K-theory*, Benjamin, New York, 1968.
3. J. Herzog and H. Sanders, *The Grothendieck group of invariant rings and of simple hypersurface singularities*,
4. S. Montgomery, *Fixed rings of finite automorphism groups of associative rings*, Lecture Notes in Math., vol. 818, Springer-Verlag, Berlin and New York, 1980.

DEPARTMENT OF MATHEMATICS, VIRGINIA POLYTECHNIC INSTITUTE AND STATE UNIVERSITY, BLACKSBURG, VIRGINIA 24061-4097

Current address (Maurice Auslander): Department of Mathematics, Brandeis University, Waltham, Massachusetts 02254

Current address (Idun Reiten): Department of Mathematics, University of Trondheim, 7055 Dragvoll, Norway

Société Mathématique de France
Mémoire n° 38, 1989, p.5-37.

THE HOMOLOGICAL THEORY OF MAXIMAL COHEN-MACAULAY APPROXIMATIONS

by

Maurice Auslander (Brandeis) and Ragnar-Olaf Buchweitz (Toronto)

Summary: Let R be a commutative noetherian Cohen-Macaulay ring which admits a dualizing module. We show that for any finitely generated R-module N there exists a maximal Cohen-Macaulay R-module M which surjects onto N and such that any other surjection from a maximal Cohen-Macaulay module onto N factors over it. Dually, there is a finitely generated R-module I of finite injective dimension into which N embeds, universal for such embeddings. We prove and investigate these results in the broader context of abelian categories with a suitable subcategory of "maximal Cohen-Macaulay objects" extracting for this purpose those ingredients of Grothendieck-Serre duality theory which are needed.

Résumé: Soit R un anneau commutatif, noethérien et de Cohen-Macaulay, tel que un module dualisant existe pour R. On démontre que pour chaque R-module N de type fini il existe un R-module M de profondeur maximale et un homomorphisme surjectif de M sur N, tel que toute autre surjection d'un tel module sur N s'en factorise. De manière duale, il existe aussi un plongement de N dans un R-module I de type fini et de dimension injective finie, universelle pour telles plongements. Nous démontrons et examinons ces résultats dans le cadre des catégories abéliennes avec une sous-catégorie convenable des "objets de Cohen-Macaulay maximaux", à cet effet mettant en évidence les propriétés de la théorie de dualité de Grothendieck-Serre dont on a besoin.

§0. *A Commutative Introduction*

The aim of this work is to analyze the framework in which the theory of *maximal Cohen-Macaulay approximations* can be developed. Instead of outlining right away the abstract results, we want to start by describing the situation in the classical case of a commutative local noetherian ring R with maximal ideal $\mathbf{m}$ and residue class field $k = R/\mathbf{m}$.

Assume that R admits a *dualizing module* ω. Then R is Cohen-Macaulay, and the finitely generated R-modules M which are maximal Cohen-Macaulay in the sense that $\operatorname{depth}_{\mathbf{m}} M = \dim R$ can be characterized homologically as those modules for which $\operatorname{Ext}^i_R(M, \omega) = 0$ for $i \neq 0$.

Our main result can then be paraphrased as saying that **R-mod**, the category of finitely generated R-modules, is obtained by glueing together the orthogonal subcategories of *modules of finite injective dimension* over R and the category of *maximal Cohen-Macaulay modules* along their common intersection which is spanned by ω.

More precisely, let us recall that ω is *dualizing* for the local ring $(R,\mathbf{m},k)$ if and only if it satisfies the following three conditions:

(i) ω is finitely generated and of finite injective dimension over R.

(ii) The natural ring homomorphism which is given by multiplication with scalars from R on ω, $R \to \mathrm{Hom}_R(\omega,\omega)$ is an isomorphism.

(iii) For any integer $i \neq 0$, one has $\mathrm{Ext}^i_R(\omega,\omega) = 0$.

Now our main results in this context are

<u>Theorem A:</u> *(Existence of the decomposition).* Let $(R,\mathbf{m},k)$ be a commutative, local noetherian ring with dualizing module ω. For any finitely generated R-module N there exist finitely generated R-modules M_N and I^N together with an R-linear map

$$d_N : M_N \to I^N$$

such that

(a) The image of d_N is isomorphic to N.

(b) M_N is maximal Cohen-Macaulay and $I_N = \mathrm{Ker} d_N$ is an R-module of finite injective dimension.

(c) I^N is of finite injective dimension and $M^N = \mathrm{Cok} d_N$ is maximal Cohen-Macaulay.

(d) There exists an integer $n \geq 0$ such that d_N can be factored into an injection $j : M_N \to \omega^{\oplus n}$ and a surjection $p : \omega^{\oplus n} \to I^N$. ∎

If $d_N = \iota^N \circ \pi_N$ denotes the factorization of d_N over its image N, we can arrange the data given in the theorem into the following exact commutative diagram of R-modules:

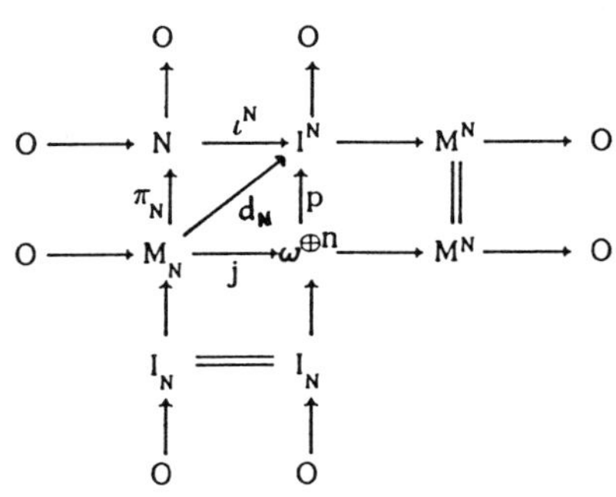

Theorem B: (Essential Uniqueness)

(a) Assume given a second homomorphism $d'_N : M'_N \to {}'I^N$ satisfying Theorem A for the same module N. If the image factorization of d'_N is given as

$$M'_N \xrightarrow{\pi'_N} N \xrightarrow{\iota'^N} {}'I^N,$$

there exist modules P, P' and Q, 'Q which are each finite direct sums of copies of ω, and R-module isomorphisms μ, κ so that the following diagram commutes:

$$\begin{array}{ccccc} M'_N \oplus P & \xrightarrow{\pi'_N \oplus O} & N & \xrightarrow{\iota'^N \oplus O} & {}'I^N \oplus Q \\ \downarrow \mu & & \| & & \downarrow \kappa \\ M_N \oplus P' & \xrightarrow{\pi_N \oplus O} & N & \xrightarrow{\iota^N \oplus O} & I^N \oplus {}'Q \end{array}$$

(b) If $f: M \to N$ is any homomorphism from a maximal Cohen-Macaulay R-module M into N, it factors over π_N. If $g: N \to J$ is any homomorphism from N into an R-module J of finite injective dimension, it factors over ι^N. ■

These results suggest to call $0 \to I_N \to M_N \xrightarrow{\pi_N} N \to 0$ a *maximal Cohen-Macaulay approximation of* N and $0 \to N \xrightarrow{\iota^N} I^N \to M^N \to 0$ a *hull of finite injective dimension* for N.

To give a simple illustration, consider the case where N itself is a Cohen-Macaulay R-module, hence satisfying $\text{depth}_{\mathfrak{m}} N = \dim N$.
Set $n = \text{codepth}_{\mathfrak{m}} N = \dim R - \dim N$. Then local duality theory implies:

(i) $\text{Ext}^i_R(N, \omega) = 0$ for $i \neq n$.

(ii) $N^\vee = \text{Ext}^n_R(N, \omega)$ is again Cohen-Macaulay of codepth n.

(iii) $N = \text{Ext}^r_R(N^\vee, \omega) = N^{\vee\vee}$.

Using this information, let

$$0 \to \Omega_n(N) \to R^{\oplus b_{n-1}} \xrightarrow{d_{n-2}} \cdots \xrightarrow{d_0} R^{\oplus b_0} \longrightarrow N^\vee \to 0$$

be an exact sequence obtained by truncating a free resolution of $N^\vee$. It follows that $\Omega_n(N)$ is maximal Cohen-Macaulay and that dualizing with respect to ω results in an exact sequence

$$0 \longrightarrow \omega^{\oplus b_0} \longrightarrow \cdots \xrightarrow{d^\vee_{n-2}} \omega^{\oplus b_{n-1}} \longrightarrow \text{Hom}_R(\Omega_n(N), \omega) \xrightarrow{\pi} N^{\vee\vee} = N \longrightarrow 0.$$

Then $M_n = \text{Hom}_R(\Omega_n(N), \omega) \xrightarrow{\pi} N$ is a desired maximal Cohen-Macaulay approximation of N, and $I_N = \text{Cok}\, d^\vee_{n-2}$ admits a finite resolution "by ω", which shows that I_N is of finite injective dimension. The hull of finite injective dimension I^N is then simply the cokernel of the ω-dual of the next differential in the resolution of $N^\vee$, namely
$I^N = \text{Cok}\, \text{Hom}_R(d_{n-1}, \omega)$.

If R is a domain, for example, we get even more precise information:

(i) The rank of M_N equals the alternating sum

$$\sum_{i=1}^{n}(-1)^{i+1}b_{n-i} + (-1)^n \text{ rk } N,$$

(ii) $M_N^{\vee} = \mathrm{Hom}_R(M_N, \omega) = \Omega_n(N)$ embeds into $R^{\oplus b_{n-1}}$,

(iii) M_N contains no copy of ω as a direct summand if and only if $\Omega_n(N)$, the n-th syzygy module of N, contains no free summand.

It follows that one can attach new numerical invariants to an R-module N in this way. The minimum number of copies of ω necessarily contained in M_N or I^N, the rank of the ω-free summand of either M_N or I^N, their minimum number of generators and so forth.

Here, we are not concerned with these more detailed consequences of the theory but rather with its general framework.

The first author first proved an essentially equivalent version of Theorem A but for the category of additive functors on **R-mod**, see [Ausl], where the result was phrased by saying that the category of maximal Cohen-Macaulay modules is "coherently (co-)finite". The essential step then was to establish the representability of the functors involved.

This background illuminates our approach here. Although the primary applications of the theory might be within the classical theory of rings and algebras, to a large extent it can be developed in any abelian category **C** which admits a suitable subcategory **X** of "maximal Cohen-Macaulay objects".

Here we establish sufficient conditions on **X** to guarantee the categorical analogues of Theorems A and B. Section 1 deals with the decomposition theorem and section 2 addresses the uniqueness question. Sections 3 and 4 investigate the circumstances under which - in the terminology of the above example - the category of modules with "finite ω-resolution" are *all* the modules of finite injective dimension. Section 5 assembles a few remarks on finiteness conditions and section 6 contains more examples, among other purposes highlighting the differences in the theory when applied to either commutative or non-commutative rings.

§1. *The Basic Decomposition Theorem*

In this section we prove the basic decomposition theorem on which this paper rests. Before stating the result, we give some definitions and notations.

Throughout, **C** will be an *abelian* category. By a *subcategory* **A** of **C** we will always mean a *full, additive* and *essential* subcategory of **C**, so that **A** is closed under finite direct sums in **C** and such that any object C in **C** which is isomorphic to an object in **A** is already an object in **A**.

A subcategory of **C** is said to be *additively closed* (or *karoubian* in the terminology of [SGA IV] or [Qu]), if it is closed under direct summands in **C**, or, equivalently, if any projector ($p = p^2$) in the subcategory admits an image in that subcategory. Any subcategory **A** of **C** admits an *additive closure* **add A** in **C**, consisting of all those objects C in **C** which are isomorphic to a direct summand (in **C**) of an object in **A**. Clearly **A** is additively closed in **C** if and only if **A** = **add A**.

More generally, given any collection $\{C_\iota\}_{\iota\epsilon I}$ of objects in **C**, there is a unique smallest additively closed subcategory **add** $\{C_\iota\}_{\iota\epsilon I}$ containing each object C_ι, $\iota\epsilon I$. It can be described by the following "universal mapping property": If $F:\mathbf{C} \to \mathbf{D}$ is any additive functor from **C** into another additive category **D** such that $F(C_\iota)$ is a zero-object in **D** for each $\iota\epsilon I$, then F(**add** $\{C_\iota\}_{\iota\epsilon I}$) consists entirely of zero-objects.

In particular, (cf. also [He]), there exists the *additive quotient category* $\pi:\mathbf{C} \to \mathbf{C}/\mathbf{add}\ \{C_\iota\}_{\iota\epsilon I}$, where $\mathbf{C}/\mathbf{add}\ \{C_\iota\}_{\iota\epsilon I}$ has the same objects as **C** and π is a full, additive functor which is the identity on objects.

The projection functor π is characterized by the property that any additive functor F as before factors uniquely over π. Of course, even if **C** is assumed to be abelian, as here, $\mathbf{C}/\mathbf{add}\ \{C_\iota\}_{\iota\epsilon I}$ need not to be so.

If **A** is an additively closed subcategory of **C**, the morphism groups in **C/A** are given by

$$\mathrm{Hom}_{\mathbf{C}/\mathbf{A}}(C_1, C_2) = \frac{\mathrm{Hom}_{\mathbf{C}}(C_1, C_2)}{\{\phi:C_1 \to C_2 \mid \phi \text{ factors over an object in } \mathbf{A}\}}$$

Now suppose again that **A** is any subcategory of **C** in the sense fixed above. We say that a sequence of morphisms $\dots \to A_{i+1} \to A_i \to A_{i-1} \to \dots$ in **A** is *exact*, if when viewed as a sequence in **C** it is exact.

Suppose C is an object in **C**. We define **A**–resol.dimC, the **A**-*resolution dimension* of C, to be the smallest nonnegative integer n such that there exists an exact sequence $0 \to A_n \to A_{n-1} \to \dots \to A_0 \to C \to 0$, with each A_i in **A**, if such an integer exists. We say that **A**–resol.dim C $< \infty$ if **A**–resol.dimC $= n$ for some non-negative integer n. The subcategory of **C** consisting of all C in **C** such that **A**–resol.dimC $< \infty$ will be denoted $\hat{\mathbf{A}}$.

Finally, we say a subcategory **B** of **A** is a *cogenerator* for **A** if for each object A in **A** there is an exact sequence $0 \to A \to B \to A' \to 0$ in **A** with B in **B**.

With these notations, we fix throughout the rest of this paper an additively closed subcategory **X** of **C** which is furthermore *closed under extensions*, i.e. if $0 \to C_1 \to C_2 \to C_3 \to 0$ is exact in **C** with C_1 and C_3 in **X**, then also C_2 is in **X**. (In the terminology of [Qu], for example, **X** is a *karoubian exact* subcategory of **C**.) Also we assume given an additively closed subcategory ω of **X** which is a cogenerator of **X**.

The paper is now devoted to studying how the categories **X**, ω, $\hat{\mathbf{X}}$ and $\hat{\omega}$ are related.

All of our results depend on the following

Theorem 1.1. For each C in $\hat{\mathbf{X}}$ there are exact sequences

$$0 \to Y_C \to X_C \to C \to 0 \quad \text{and}$$
$$0 \to C \to Y^C \to X^C \to 0$$

with Y_C and Y^C in $\hat{\omega}$ and X_C and X^C in $\mathbf{X}$.

Proof. The proof proceeds by induction on $\mathbf{X}$–resol.dimC and is based on the following two easily proven observations.

Lemma 1.2. Suppose given exact sequences $0 \to K \to X \to C \to 0$ and $0 \to K \to Y^K \to X^K \to 0$ with X and X^K in $\mathbf{X}$ and Y^K in $\hat{\omega}$. Then in the pushout diagram

$$\begin{array}{ccccccccc}
 & & 0 & & 0 & & & & \\
 & & \downarrow & & \downarrow & & & & \\
0 & \longrightarrow & K & \longrightarrow & X & \longrightarrow & C & \longrightarrow & 0 \\
 & & \downarrow & & \downarrow & & \| & & \\
0 & \longrightarrow & Y^K & \longrightarrow & U & \longrightarrow & C & \longrightarrow & 0 \\
 & & \downarrow & & \downarrow & & & & \\
 & & X^K & = & X^K & & & & \\
 & & \downarrow & & \downarrow & & & & \\
 & & 0 & & 0 & & & &
\end{array}$$

the exact sequence

$$0 \to Y^K \to U \to C \to 0$$

has the property that Y^K is in $\hat{\omega}$ and U is in $\mathbf{X}$.

Proof. As Y^K is in $\hat{\omega}$ by assumption, it remains to be seen that U is in $\mathbf{X}$. This follows from the fact that both X and X^K are in $\mathbf{X}$ and $\mathbf{X}$ is closed under extensions. ∎

The other observation we need is the following.

Lemma 1.3. Suppose that we have an exact sequence $0 \to Y_C \to X_C \to C \to 0$ with Y_C in $\hat{\omega}$ and X_C in $\mathbf{X}$. Let $0 \to X_C \to W \to X \to 0$ be exact with X in $\mathbf{X}$ and W in ω. Then in the pushout diagram

$$\begin{array}{ccccccccc}
 & & 0 & & 0 & & & & \\
 & & \downarrow & & \downarrow & & & & \\
 & & Y_C & = & Y_C & & & & \\
 & & \downarrow & & \downarrow & & & & \\
0 & \longrightarrow & X_C & \longrightarrow & W & \longrightarrow & X & \longrightarrow & 0 \\
 & & \downarrow & & \downarrow & & \| & & \\
0 & \longrightarrow & C & \longrightarrow & Z & \longrightarrow & X & \longrightarrow & 0 \\
 & & \downarrow & & \downarrow & & & & \\
 & & 0 & & 0 & & & &
\end{array}$$

the exact sequence

$$0 \to C \to Z \to X \to 0$$

has the property that Z is in $\hat{\omega}$ and X is in **X**.

<u>*Proof.*</u> As again X is in **X** by assumption, it is only required to prove that Z is in $\hat{\omega}$. But in the exact sequence $0 \to Y_C \to W \to Z \to 0$, we have Y_C in $\hat{\omega}$ and W in ω, so that Z is in $\hat{\omega}$ by definition of that category. ■

<u>*The proof of theorem 1.1.*</u> follows now easily from these lemmas.

Suppose **X**$-$resol.dim$C = n$ and let $0 \to X_n \to \dots \to X_1 \xrightarrow{d_0} X_0 \to C \to 0$ be exact with each X_i in **X**. If $n = 0$, we have that C is already in **X**. Since ω is a cogenerator for **X**, there is an exact sequence $0 \to C \to W \to X \to 0$ in **X** with W in ω which is one of our desired exact sequences. The other one is $0 \to 0 \to C \xrightarrow{=} C \to 0$. Now suppose that $n > 0$ and set $K = \operatorname{Im} d_0$, so that we have exact sequences $0 \to K \to X_0 \to C \to 0$ and $0 \to X_n \to \dots \to X_1 \to K \to 0$ with each X_i in **X**. By the inductive hypothesis we know there is an exact sequence $0 \to K \to Y^K \to X^K \to 0$ with Y^K in $\hat{\omega}$ and X^K in **X**. Therefore, by Lemma 1.2, the pushout diagram

$$\begin{array}{ccccccccc}
0 & \longrightarrow & K & \longrightarrow & X_0 & \longrightarrow & C & \longrightarrow & 0 \\
 & & \downarrow & & \downarrow & & \| & & \\
0 & \longrightarrow & Y^K & \longrightarrow & U & \longrightarrow & C & \longrightarrow & 0
\end{array}$$

has the property that U is in **X**. Hence we may choose $0 \to Y^K \to U \to C \to 0$ as one of our desired sequences for C. From the existence of this exact sequence, it follows by Lemma 1.3 that we also have an exact sequence $0 \to C \to Y^C \to X^C \to 0$ with Y^C in $\hat{\omega}$ and X^C in **X**. This finishes the proof of theorem 1.1. ■

For ease of reference, we call an exact sequence $0 \to Y_C \to X_C \xrightarrow{\pi_C} C \to 0$ with X_C in **X** and Y_C in $\hat{\omega}$ an ***X**-approximation* of C. Dually, we call an exact sequence $0 \to C \xrightarrow{\iota^C} Y^C \to X^C \to 0$ with Y^C in $\hat{\omega}$ and X^C in **X** an *$\hat{\omega}$-hull* of C.

From now on, we assume that **X** has the property that if $0 \to X_0 \to X_1 \to X_2 \to 0$ is an exact sequence with X_1 and X_2 in **X**, then X_0 is also in

X, in addition to **X** being an additively closed subcategory of **C** which is closed under extensions. (In D.Quillen's terminology, (loc. cit.), all epimorphisms from **C** in **X** are *admissible.*)

It should be noted that in all our examples the categories **X** satisfy this additional condition. As a consequence of this further hypothesis on **X**, we get the following

Lemma 1.4. Suppose C in **C** has an $\hat{\omega}$-hull $0 \to C \to Y^C \to X^C \to 0$. Then it also admits an **X**-approximation $0 \to Y_C \to X_C \to C \to 0$. Furthermore, Y_C can be chosen such that ω–resol.dimY_C < ω–resol.dim Y^C if Y^C is not already in ω.

Proof. Let $0 \to W_n \to W_{n-1} \to \dots \xrightarrow{d_0} W_0 \to Y^C \to 0$ be exact with the W_i in ω. Then we obtain the following pullback diagram

$$\begin{array}{ccccccccc} & & 0 & & 0 & & & & \\ & & \downarrow & & \downarrow & & & & \\ & & K & = & K & & & & \\ & & \downarrow & & \downarrow & & & & \\ 0 & \to & L & \to & W_0 & \to & X^C & \to & 0 \\ & & \downarrow & & \downarrow & & \| & & \\ 0 & \to & C & \to & Y^C & \to & X^C & \to & 0 \\ & & \downarrow & & \downarrow & & & & \\ & & 0 & & 0 & & & & \end{array}$$

where $K = \operatorname{Im} d_0$. Since X^C and W_0 are in **X**, the additional assumption yields that L is in **X** too. By definition, K is in $\hat{\omega}$ and so $0 \to K \to L \to C \to 0$ is an **X**-approximation of C. Now set $Y_C = K$ and $X_C = L$. ∎

As a consequence of this lemma, we obtain the following characterization of the objects in $\hat{\mathbf{X}}$.

Proposition 1.5. Let **X** be an additively closed and exact subcategory of **C** in which every epimorphism is admissible. If ω is a cogenerator of **X**, the following are equivalent for an object C in **C**:

(a) C is in $\hat{\mathbf{X}}$.

(b) There exists an **X**-approximation $0 \to Y_C \to X_C \to C \to 0$ of C.

(c) There is an $\hat{\omega}$-hull $0 \to C \to Y^C \to X^C \to 0$ of C.

Proof. Since (a) implies (b) and (c) by theorem 1.1, it is only required to show that (b) implies (a) and (c) implies (b).

(b) ⇒ (a): Since Y_C is in $\hat{\omega}$ by assumption, there is an exact sequence $0 \to W_n \to \dots \to W_0 \to Y_C \to 0$ with each W_i in ω. Since ω is a subcategory of **X**, it follows from the exact sequence $0 \to W_n \to \dots \to W_0 \to X_C \to C \to 0$ that C is in $\hat{\mathbf{X}}$.

(c) $\Rightarrow$ (b): This is just a restatement of Lemma 1.4. ∎

We end this section with three examples, illustrating the theory developed so far.

Example 1. Let $X \to \operatorname{Spec} k$ be a scheme of finite type over a field k. Assume that X is equidimensional of dimension d and locally Cohen-Macaulay in the sense that $\mathcal{O}_{X,x}$ is a local Cohen-Macaulay ring for each x in X. Let **C** be the category of coherent sheaves of $\mathcal{O}_X$*-modules* and define **X** to be the subcategory of maximal Cohen-Macaulay coherent sheaves, where a coherent $\mathcal{O}_X$-module $\mathcal{M}$ is said to be maximal Cohen-Macaulay if for every $x \epsilon X$ one has $\operatorname{depth}_{\mathbf{m}_x} \mathcal{M}_x = \dim \mathcal{O}_{X,x}$; $\mathbf{m}_x$ the unique maximal ideal of $\mathcal{O}_{X,x}$.

It is then clear that if $0 \to \mathcal{M}_1 \to \mathcal{M}_2 \to \mathcal{M}_3 \to 0$ is an exact sequence in **C**, then

(a) $\mathcal{M}_2$ is in **X** if $\mathcal{M}_1$ and $\mathcal{M}_3$ are in **X**, and

(b) $\mathcal{M}_1$ is in **X** if $\mathcal{M}_2$ and $\mathcal{M}_3$ are in **X**.

Remark also that, by hypothesis, the structure sheaf $\mathcal{O}_X$ is in **X** and that consequently **X** contains all locally free sheaves of $\mathcal{O}_X$-modules. Conversely, a maximal Cohen-Macaulay $\mathcal{O}_X$-module is locally free on the regular locus $X_{reg} \subseteq X$. Moreover, $\mathbf{C} = \hat{\mathbf{X}}$, and if $C \neq 0$ is in **C**, then **X**$-\text{resol.dim}\, C = n$ if and only if n is the largest integer such that $\mathcal{E}xt^n_{\mathcal{O}_X}(C, \omega_X) \neq 0$, where ω_X is a dualizing sheaf for X.

Now assume that either X admits a very ample invertible sheaf $\mathcal{L}$ or that X is affine (in which case $\mathcal{L} = \mathcal{O}_X$ in the following). Then X can be embedded into a projective space over k, say $i: X \to \mathbf{P}^N_k$, with $\mathcal{L} = i^* \mathcal{O}_{\mathbf{P}^N}(1)$.

Denoting by $\omega_{\mathcal{L}}$ the smallest additively closed subcategory which contains the family of objects $\{\omega_X \otimes \mathcal{L}^{\otimes n}\}_{n \epsilon \mathbf{Z}}$, it follows easily from Grothendieck-Serre duality theory that $\omega_{\mathcal{L}}$ is a cogenerator for **X**.

Proposition 1.6. For each coherent sheaf C of $\mathcal{O}_X$-modules we have both an **X**-approximation with respect to $\omega_{\mathcal{L}}$ of the form $0 \to Y_C \to X_C \to C \to 0$ and an $\hat{\omega}_{\mathcal{L}}$-hull $0 \to C \to Y^C \to X^C \to 0$. ∎

Remark that in this example the category **X** depends only on the scheme X whereas its cogenerator depends on the choice of both a dualizing module ω_X and a very ample sheaf $\mathcal{L}$. Also the **X**-approximations and $\hat{\omega}_{\mathcal{L}}$-hulls will vary with these choices.

Next consider the following modified version of Example 1.

Example 2. As in Example 1, we let $X \to \operatorname{Spec} k$ be an equidimensional Cohen-Macaulay scheme over a field k. Let $X' \subset X$ be the *Gorenstein locus* of X, which is the set of all points x in X for which $\mathcal{O}_{X,x}$ is a Gorenstein local ring. Let **X'** be the subcategory of **C**, the category of coherent sheaves of $\mathcal{O}_X$-modules, consisting of those Cohen-Macaulay sheaves $\mathcal{M}$ such that $\mathcal{M}_x$ is $\mathcal{O}_{X,x}$-free for all x in X'. It is clear again that **X'** is an exact subcategory of **C** in which every epimorphism is admissible. Also $\hat{\mathbf{X}}'$ consists of all those

$\mathcal{M}$ in **C** for which $\mathcal{M}_x$ is of finite projective dimension over $\mathcal{O}_{X,x}$ for each x in X'. This implies that an exact sequence $0 \to A \to B \to C \to 0$ in **C** is in $\hat{\mathbf{X}}'$ if any two of A,B or C are in $\hat{\mathbf{X}}'$.

Again, any invertible sheaf of $\mathcal{O}_X$–modules is in **X'**, and in particular for any dualizing sheaf ω_X and each very ample invertible sheaf $\mathcal{L}$ on X, the category $\omega_{\mathcal{L}}$ defined above is a cogenerator of **X'**. We leave it to the reader to give in this case the analogue of Proposition 1.6.

Our final example in this section treats a not necessarily commutative version of Gorenstein rings of finite Krull dimension.

Example 3. Let R be a ring with unit which is noetherian on both sides and such that the *injective dimension of* R *as a right module over itself is finite,* say equal to d.

Take **C** = **R-mod**, the category of finitely generated left R-modules, and let **X** be the subcategory consisting of all modules M in **R-mod** which satisfy $\mathrm{Ext}^i_R(M,R) = 0$ for $i \neq 0$.

Then **X** is certainly additively closed and has the property that an exact sequence $0 \to M_1 \to M_2 \to M_3 \to 0$ is in **X** as soon as either M_1 and M_3 or M_2 and M_3 are in **X**. Hence **X** satisfies our general assumptions.

For ω, take the subcategory of all finitely generated *projective* left R-modules. Then ω is by definition a subcategory of **X** which is additively closed.

For our theory to apply, we have hence to show that ω constitutes a cogenerator for **X**. To obtain this result we need our assumption on R. Namely, let $\dots \xrightarrow{d_j} P_j \longrightarrow \dots \longrightarrow P_1 \xrightarrow{d_0} P_0 \longrightarrow M \longrightarrow 0$ be a projective resolution of a module M in **X**. By definition of **X**, the dualized complex

$$0 \longrightarrow M^* \longrightarrow P_0^* \xrightarrow{d_0^*} P_1^* \longrightarrow \dots \longrightarrow P_j^* \xrightarrow{d_j^*} \dots$$

is acyclic. But then our hypothesis furnishes the following more precise information.

Lemma 1.7. With notations and assumptions as above, for every module M in **X** one has

(a) For all integers $j \geq 0$, the right R-modules $K_j = \mathrm{Ker}\, d_j^*$ satisfy $\mathrm{Ext}^i_R(K_j,R) = 0$ for $i \neq 0$.

(b) M is *reflexive,* that is, the natural morphism of left R-modules $M \to M^{**}$ is an isomorphism.

(c) If $0 \longrightarrow L \longrightarrow Q \xrightarrow{p} M^* \longrightarrow 0$ is an exact sequence of right R-modules with Q finitely generated projective, then L^* satisfies $\mathrm{Ext}^i_R(L^*,R) = 0$ for $i \neq 0$.

Proof. (a) As all the modules P_j^* are finitely generated projective right R-modules, they satisfy necessarily $\mathrm{Ext}^i_R(P_j^*,R) = 0$ for $i \neq 0$. But this implies that for any integer $n \geq 0$ one has natural isomorphisms $\mathrm{Ext}^i_R(K_{j-n},R) \xrightarrow{\sim} \mathrm{Ext}^{i+n}_R(K_j,R)$ for any $i > 0$. Since by

assumption $\mathrm{Ext}_R^k(-,R) = 0$ as soon as $k > d$, it suffices to take $n \geq d$ above to conclude $\mathrm{Ext}_R^i(K_j,R) = 0$ for all $i > 0$ and $j \geq 0$.
(b) This is a consequence of (a). By [A-B; 2.1.], for any left R-module M the natural morphism $M \to M^{**}$ fits into an exact sequence

$$0 \longrightarrow \mathrm{Ext}_R^1(D(M),R) \longrightarrow M \longrightarrow M^{**} \longrightarrow \mathrm{Ext}_R^2(D(M),R) \longrightarrow 0,$$

where $D(M) = \mathrm{Cok}\, d_0^*$. But if M is in **X**, we have $\mathrm{Cok}\, d_0^* = \mathrm{Ker}\, d_2^*$ and (a) shows that the extreme terms of this exact sequence vanish, establishing (b).
(c) As $M^* = K_0$, part (a) implies that the sequence

$$(*)\ 0 \longrightarrow M^{**} \xrightarrow{p^*} Q^* \longrightarrow L^* \longrightarrow 0$$

is exact. From (b) we have $M \cong M^{**}$ and as M is in **X**, it follows already that $\mathrm{Ext}_R^i(L^*,R) = 0$ for $i > 1$. It hence only remains to be seen that $\mathrm{Ext}_R^1(L^*,R) = 0$, or, equivalently, that the dual sequence of (*):

$$0 \longrightarrow L^{**} \longrightarrow Q^{**} \xrightarrow{p^{**}} M^{***} \longrightarrow 0$$

is again exact. But this is obvious as both Q and M^* are reflexive right R-modules and $p^{**} = p$. ∎

Combining (b) and (c) of this lemma, we have now that any module M in **X** embeds into the finitely generated projective module $\mathrm{Hom}_R(Q,R)$ and that the cokernel, isomorphic to L^*, is again in **X**. This shows that ω is indeed a cogenerator for **X**.

Finally observe that $\hat{\mathbf{X}}$ consists of all left R-modules N in **C** satisfying $\mathrm{Ext}_R^i(N,R) = 0$ for all sufficiently large i, and that $\hat{\omega}$ is the category of all finitely generated left R-modules *of finite projective dimension.*

Now Theorem 1.1 yields in this context the following.

Theorem 1.8. Let R be a ring which is noetherian on both sides and of finite injective dimension as a right module over itself. Then for any finitely generated left R-module N satisfying $\mathrm{Ext}_R^i(N,R) = 0$ for all sufficiently large i, there are modules Y_N and Y^N in **R-mod** of finite projective dimension and modules X_N and X^N in **R-mod** with $\mathrm{Ext}_R^i(X_N,R) = \mathrm{Ext}_R^i(X^N,R) = 0$ for $i \neq 0$ which fit into exact sequences

$$0 \longrightarrow Y_N \longrightarrow X_N \longrightarrow N \longrightarrow 0 \quad \text{and}$$

$$0 \longrightarrow N \longrightarrow Y^N \longrightarrow X^N \longrightarrow 0.$$ ∎

§2. *Injective Cogenerators*

Having established the *existence* of **X**-approximations and $\hat{\omega}$–hulls for a pair $(\mathbf{X},\omega)$ of subcategories as in the preceding section, the important question which remains is their *uniqueness*.

To see which conditions ought to be imposed, assume given two $\mathbf{X}$-approximations for the same object C in $\hat{\mathbf{X}}$, say $0 \to Y_C \to X_C \to C \to 0$ and $0 \to Y'_C \to X'_C \to C \to 0$. Then the least one should ask for is that these $\mathbf{X}$-approximations can be *compared* in the sense that there exists a morphism $\phi : X_C \to X'_C$ making the following diagram commutative

$$\begin{array}{ccccccccc} 0 & \longrightarrow & Y_C & \longrightarrow & X_C & \longrightarrow & C & \longrightarrow & 0 \\ & & \downarrow & & \downarrow{\scriptstyle\phi} & & \downarrow & & \\ 0 & \longrightarrow & Y'_C & \longrightarrow & X'_C & \longrightarrow & C & \longrightarrow & 0 \end{array}$$

Apparently, the existence of such a comparison morphism is guaranteed as soon as $\mathrm{Ext}^1_{\mathbf{C}}(X_C, Y'_C) = 0$.

Hence, for comparisons to exist and to yield an equivalence relation, it certainly suffices to have that $\mathrm{Ext}^1_{\mathbf{C}}(X,Y) = 0$ for *all* X in $\mathbf{X}$ and Y in $\hat{\omega}$. This section is devoted to a study of this condition and its consequences. First, once again, some general remarks and notations.

Let A and C be objects in $\mathbf{C}$. As $\mathbf{C}$ is supposed to be abelian, the groups $\mathrm{Ext}^i_{\mathbf{C}}(A,C)$ are defined for all $i \geq 0$. If there is an integer n such that $\mathrm{Ext}^i_{\mathbf{C}}(A,C) = 0$ for all $i > n$, then the smallest nonnegative such integer n is called the *A-injective dimension of* C, (notation: A–inj.dimC), or the *C-projective dimension of* A, (notation: C–proj.dimA).

Otherwise we set A–inj.dimC $= \infty =$ C–proj.dimA . If $\mathbf{B}$ is a subcategory of $\mathbf{C}$, for each A in $\mathbf{C}$ we define A–inj.dim$\mathbf{B}$ to be the maximum (in $\mathbf{Z} \cup \{\infty\}$) of A–inj.dimB for all B in $\mathbf{B}$. Dually, for each C in $\mathbf{C}$, we define C–proj.dim$\mathbf{B}$ to be the maximum of C–proj.dimB for all B in $\mathbf{B}$.

Clearly A–inj.dim$\mathbf{B}$ = $\mathbf{B}$–proj.dimA.

Suppose now that $\mathbf{A}$ and $\mathbf{B}$ are subcategories of $\mathbf{C}$. Then define $\mathbf{A}$–proj.dim$\mathbf{B}$ to be the maximum of A–proj.dimB for all A in $\mathbf{A}$ and B in $\mathbf{B}$. We define dually $\mathbf{A}$–inj.dim$\mathbf{B}$ to be the maximum of A–inj.dimB for all A in $\mathbf{A}$ and B in $\mathbf{B}$. Again, one has clearly $\mathbf{A}$–inj.dim$\mathbf{B}$ = $\mathbf{B}$–proj.dim$\mathbf{A}$.

If for two such subcategories $\mathbf{A}$–inj.dim$\mathbf{B}$ = 0 = $\mathbf{B}$–proj.dim$\mathbf{A}$, we follow J.L. Verdier, [SGA 4½, C.D.; I.2.6.1.], and say that $\mathbf{A}$ is *left orthogonal* to $\mathbf{B}$ and $\mathbf{B}$ is *right orthogonal* to $\mathbf{A}$ - with respect to the "augmented" bilinear $\mathbf{Z}$-graded pairing induced by $(\mathrm{Ext}^i_{\mathbf{C}}(-,-))_{i>0}$ on the monoid of isomorphism classes of objects of $\mathbf{C}$.

Consequently, if $\mathbf{A}$ consists precisely of those objects A in $\mathbf{C}$ for which A–inj.dim$\mathbf{B}$ = 0, we call $\mathbf{A}$ the *left orthogonal complement* of $\mathbf{B}$ in $\mathbf{C}$, denoted $\mathbf{A} = {}^{\perp}\mathbf{B}$. Dually again, $\mathbf{A}^{\perp}$, the *right orthogonal complement* of $\mathbf{A}$ in $\mathbf{C}$, is the subcategory $\mathbf{B}$ consisting of all objects B in $\mathbf{C}$ for which $\mathbf{A}$–inj.dimB = 0.

One has obviously $\mathbf{A} \subseteq {}^{\perp}(\mathbf{A}^{\perp})$ and $\mathbf{A} \subseteq ({}^{\perp}\mathbf{A})^{\perp}$, but not necessarily ${}^{\perp}(\mathbf{A}^{\perp}) = ({}^{\perp}\mathbf{A})^{\perp}$. If $\mathbf{B}'$ is a subcategory of $\mathbf{B}$ in $\mathbf{C}$, then ${}^{\perp}\mathbf{B}$ is contained in ${}^{\perp}\mathbf{B}'$ and similary for right orthogonal complements. Remark also that by definition ${}^{\perp}\mathbf{C}$, the *left radical* of $\mathbf{C}$ with respect to the pairing $(\mathrm{Ext}^i_{\mathbf{C}}(-,-))_{i>0}$, consists precisely of all *projective* objects of $\mathbf{C}$, whereas $\mathbf{C}^{\perp}$, the *right radical* of $\mathbf{C}$, is given by all *injective* objects of $\mathbf{C}$.

Furthermore, it is obvious that orthogonal complements are additively closed and exact subcategories of $\mathbf{C}$ and that in a left orthogonal complement ${}^{\perp}\mathbf{B}$ all epimorphisms are admissible, whereas in a right orthogonal complement $\mathbf{A}^{\perp}$ all monomorphisms are admissible.

Returning to our subcategories $\mathbf{X}$ and ω of $\mathbf{C}$ from the previous section, we say that ω is an *injective cogenerator* for $\mathbf{X}$ if $\mathbf{X}$–inj.dim$\omega = 0$, that is, $\omega \subseteq \mathbf{X}^{\perp}$. If there is a cogenerator for $\mathbf{X}$ in $\mathbf{X} \cap \mathbf{X}^{\perp}$, we say also that the exact category $\mathbf{X}$ has *enough relatively injective objects.*

Unless stated to the contrary, we assume from now on that ω is an injective cogenerator for $\mathbf{X}$. Our next aim is to explore some important properties of $\mathbf{X}$-approximations and $\hat{\omega}$–hulls implied by this additional assumption.

We begin with the following relations between some of the dimensions we have just introduced for an object C in $\hat{\mathbf{X}}$. These relations do not require that any epimorphism in $\mathbf{X}$ is admissible.

Proposition 2.1. Given an object C in $\hat{\mathbf{X}}$, where $\mathbf{X}$ is an additively closed exact subcategory of $\mathbf{C}$ and ω is an injective cogenerator for $\mathbf{X}$, the following are equivalent for any integer $n \geq 0$.

(a) $\mathbf{X}$–resol.dim$C = n$,

(b) C–inj.dim$\omega = n$,

(c) C–inj.dim$\hat{\omega} = n$,

(d) $\mathrm{Ext}^{n+1}_{\mathbf{C}}(C, Y) = 0$ for all Y in $\hat{\omega}$.

Proof: Proceed by induction on $n = \mathbf{X}$–resol.dimC, the case $n = 0$ being settled as follows.

(a) $\Rightarrow$ (b) is true because ω is contained in $\mathbf{X}^{\perp}$ by assumption.

(b) $\Rightarrow$ (c) follows from the usual dimension shift argument.

(c) $\Rightarrow$ (d) is the definition of C–inj.dim$\hat{\omega}$.

(d) $\Rightarrow$ (a): Since C is in $\hat{\mathbf{X}}$ by the general hypothesis, there is an $\mathbf{X}$-approximation $0 \to Y_C \to X_C \to C \to 0$ which splits by (d). Hence C is a direct summand of X_C in $\mathbf{X}$ and so C is in $\mathbf{X}$.

The proof of the inductive step follows easily from what we have just shown and is left to the reader. ■

As an obvious consequence of this proposition we have

Corollary 2.2. $\mathbf{X}$–inj.dim$\hat{\omega} = 0$. ■

This corollary yields the following important properties of $\mathbf{X}$-approximations and $\hat{\omega}$-hulls.

Theorem 2.3. Let $0 \to Y_C \to X_C \xrightarrow{\pi_C} C \to 0$ be an $\mathbf{X}$-approximation for C in $\hat{\mathbf{X}}$. Then for each X in $\mathbf{X}$ we have

(a) $0 \to \mathrm{Hom}_{\mathbf{C}}(X, Y_C) \to \mathrm{Hom}_{\mathbf{C}}(X, X_C) \to \mathrm{Hom}_{\mathbf{C}}(X, C) \to 0$ is exact,

(b) π_C induces isomorphisms $\mathrm{Ext}^i_{\mathbf{C}}(X, X_C) \to \mathrm{Ext}^i_{\mathbf{C}}(X, C)$ for all $i > 0$.

Proof: As $\mathbf{X}$–inj.dim$\hat{\omega} = 0$, one has $\mathrm{Ext}^i_{\mathbf{C}}(X, Y_C) = 0$ for all $i > 0$. ■

The exact sequence $0 \to Y_C \to X_C \to C \to 0$ is called an $\mathbf{X}$-approximation precisely because $\mathrm{Hom}_{\mathbf{C}}(X, X_C) \to \mathrm{Hom}_{\mathbf{C}}(X, C) \to 0$ is exact for all X in $\mathbf{X}$. This property of $\mathbf{X}$-approximations of C gives rise to a weak sort of uniqueness for such approximations as we now explain.

Let us call two morphisms $f:B \to C$ and $f':B' \to C$ in $\mathbf{C}$ *equivalent* if there are morphisms $g:B \to B'$ and $h:B' \to B$ such that $f = f'g$ and $f' = fh$. Also, we say that two exact sequences $0 \to A \to B \xrightarrow{f} C$ and $0 \to A' \to B' \xrightarrow{f'} C$ are *(right) equivalent*, if f and f' are equivalent, which amounts to the same as saying that there is a commutative diagram

$$\begin{array}{ccccccc}
0 & \longrightarrow & A' & \longrightarrow & B' & \xrightarrow{f'} & C \\
 & & \downarrow & & \downarrow h & & \| \\
0 & \longrightarrow & A & \longrightarrow & B & \xrightarrow{f} & C \\
 & & \downarrow & & \downarrow g & & \| \\
0 & \longrightarrow & A' & \longrightarrow & B' & \xrightarrow{f'} & C
\end{array}$$

In particular, $\mathrm{id}_B - hg$ factors over A and $\mathrm{id}_{B'} - gh$ factors over A', so that h, g become inverse isomorphisms in $\mathbf{C}/\mathbf{add}\{A, A'\}$.

As an immediate consequence of theorem 2.3 we obtain the following uniqueness result.

Corollary 2.4. $\mathbf{X}$-approximations for an object C in $\hat{\mathbf{X}}$ are unique up to equivalence, that is, any two $\mathbf{X}$-approximations for C are (right) equivalent exact sequences. ■

There are also similar results for $\hat{\omega}$-hulls of an object C in $\hat{\mathbf{X}}$ as we now point out.

Theorem 2.5. Let $0 \to C \xrightarrow{\iota^C} Y^C \to X^C \to 0$ be an $\hat{\omega}$-hull for C in $\hat{\mathbf{X}}$. Then for each Y in $\hat{\omega}$ we have the following

(a) $0 \to \mathrm{Hom}_{\mathbf{C}}(X^C, Y) \to \mathrm{Hom}_{\mathbf{C}}(Y^C, Y) \to \mathrm{Hom}_{\mathbf{C}}(C, Y) \to 0$ is exact,

(b) ι^C induces isomorphisms $\mathrm{Ext}^i_{\mathbf{C}}(Y^C, Y) \to \mathrm{Ext}^i_{\mathbf{C}}(C, Y)$ for all $i > 0$.

Proof: This follows again from the fact that $\mathbf{X}$–inj.dim$\hat{\omega} = 0$. ■

The exact sequence $0 \to C \xrightarrow{\iota^C} Y^C \to X^C \to 0$ is called an $\hat{\omega}$-hull precisely because $\mathrm{Hom}_{\mathbf{C}}(Y^C, Y) \to \mathrm{Hom}_{\mathbf{C}}(C, Y) \to 0$ is exact for all Y in $\hat{\omega}$. Again, this property gives rise to a weak sort of uniqueness for $\hat{\omega}$-hulls, similar to that already discussed for **X**-approximations, as we now explain.

Dually to the above, we say two morphisms $f:C \to D$ and $f':C \to D'$ are equivalent if there are morphisms $g:D \to D'$ and $h:D' \to D$ such that $f' = gf$ and $f = hf'$. Also, we say that two exact sequences $C \xrightarrow{f} D \to E \to 0$ and $C' \xrightarrow{f'} D' \to E' \to 0$ are (*left*) *equivalent* if f and f' are equivalent, which is the same thing as saying that there is a commutative diagram

$$\begin{array}{ccccccc} C & \xrightarrow{f} & D & \longrightarrow & E & \longrightarrow & 0 \\ \| & & \downarrow g & & \downarrow & & \\ C & \xrightarrow{f'} & D' & \longrightarrow & E' & \longrightarrow & 0 \\ \| & & \downarrow h & & \downarrow & & \\ C & \xrightarrow{f} & D & \longrightarrow & E & \longrightarrow & 0 \end{array}$$

In particular, $id_D - hg$ factors over E and $id_{D'} - gh$ factors over E', so that h and g become inverse isomorphisms in $\mathbf{C}/\mathbf{add}\{E, E'\}$.

As an immediate consequence of theorem 2.5 we have the following uniqueness theorem.

Corollary 2.6. $\hat{\omega}$-hulls for an object C in $\hat{\mathbf{X}}$ are unique up to equivalence, that is, any two $\hat{\omega}$-hulls are (left) equivalent exact sequences. ■

We may reformulate and sharpen these uniqueness results slightly by considering the situation "modulo ω". This depends on the following simple observation.

Lemma 2.7. Let $f:X \to C$ be a morphism in **C** with X in **X** and C in $\hat{\mathbf{X}}$. Then the following conditions on f are equivalent.

(a) f factors through an object in $\hat{\omega}$.

(b) f factors through an object in ω.

Proof: As (b) is a priori a special case of (a), we need only to show that in fact (a) implies (b). Hence assume that $f = gh$ where $h:X \to Y$ and $g:Y \to C$ are morphisms in **C** and Y is in $\hat{\omega}$. By definition of $\hat{\omega}$, there is an exact sequence $0 \to K \to W \to Y \to 0$ with W in ω and K again in $\hat{\omega}$. By corollary 2.2, $\mathbf{X}$–inj.dim $\hat{\omega} = 0$ and so $\mathrm{Ext}^1_{\mathbf{C}}(X, K) = 0$. This shows that h, and then also f, factor over W in ω. ■

Now choose for any object C in $\hat{\mathbf{X}}$ an **X**-approximation $0 \to Y_C \to X_C \xrightarrow{\pi_C} C \to 0$ and an $\hat{\omega}$-hull $0 \to C \xrightarrow{\iota^C} Y^C \to X^C \to 0$, as well as for any morphism $f:C \to D$ in $\hat{\mathbf{X}}$ liftings $f_*:X_C \to X_D$ and $f^*:Y^C \to Y^D$ which exist by the above.

By the uniqueness results just established, it follows that given a second morphism $g:D \to E$ in $\hat{\mathbf{X}}$, the differences $g^*f^* - (gf)^*$ and $g_*f_* - (gf)_*$ factor over objects in ω, hence become zero-morphisms in $\hat{\mathbf{X}}/\omega$, the full subcategory spanned by $\hat{\mathbf{X}}$ in $\mathbf{C}/\omega$.

From this we obtain immediately the following

Theorem 2.8. Denote $i: \hat{\omega} \to \hat{\mathbf{X}}$ and $j:\mathbf{X} \to \hat{\mathbf{X}}$ the natural inclusion functors. Then

(a) The induced functor $j_!:\mathbf{X}/\omega \to \hat{\mathbf{X}}/\omega$ is fully faithful and admits a *right adjoint* $j^!:\hat{\mathbf{X}}/\omega \to \mathbf{X}/\omega$ which associates to an object C in $\hat{\mathbf{X}}$ the chosen $\mathbf{X}$-approximation X_C. The adjunction morphism $j_!j^!C \to C$ is given by the class of $\pi_C:X_C \to C$ in $\mathrm{Hom}_{\hat{\mathbf{X}}/\omega}(X_C, C)$.

(b) The induced functor $i_*: \hat{\omega}/\omega \to \hat{\mathbf{X}}/\omega$ is fully faithful and admits a *left adjoint* $i^*: \hat{\mathbf{X}}/\omega \to \hat{\omega}/\omega$ which associates to an object C in $\hat{\mathbf{X}}$ the chosen $\hat{\omega}$-hull Y^C. The adjunction morphism $C \to i_*i^*C$ is given by the class of $\iota^C:C \to Y^C$ in $\mathrm{Hom}_{\hat{\mathbf{X}}/\omega}(C, Y^C)$.

(c) One has $j^!i_* = 0$ and $i^*j_! = 0$.

(d) The composition of the adjunction morphisms

$$j_!j^! \xrightarrow{\pi} \mathrm{id}_{\hat{\mathbf{X}}/\omega} \xrightarrow{\iota} i_*i^*$$

is zero in $\hat{\mathbf{X}}/\omega$.

Proof: The remarks preceding the theorem show that X_- and Y^- define functors from $\hat{\mathbf{X}}$ into $\mathbf{X}/\omega$ and $\hat{\omega}/\omega$ respectively. By the universal property of quotient categories these functors factor over $\hat{\mathbf{X}}/\omega$, yielding $j_!$ and i^*. To prove that $j_!$ is indeed right adjoint to the inclusion functor $j^!:\mathbf{X}/\omega \to \hat{\mathbf{X}}/\omega$, it suffices to give the natural isomorphisms $\phi_{X,C}:\mathrm{Hom}_{\mathbf{X}/\omega}(X, j_!C) \xrightarrow{\sim} \mathrm{Hom}_{\hat{\mathbf{X}}/\omega}(j^!X, C)$. Now composition with $\pi_C:X_C = j_!C \to C$ defines the natural map $\mathrm{Hom}_{\mathbf{X}}(X, X_C) \to \mathrm{Hom}_{\hat{\mathbf{X}}}(X, C)$ which is surjective by theorem 2.3.(a). Let $\phi_{X,C}$ be the induced map on the quotient groups, which is hence still surjective. To prove that it is injective, let f in $\mathrm{Hom}_{\mathbf{X}}(X, X_C)$ be a morphism such that $\pi_C f:X \to C$ factors over some object W in ω. This means that there is a commutative diagram in $\mathbf{C}$

$$\begin{array}{ccccccccc} 0 & \longrightarrow & Y_C & \longrightarrow & X_C & \xrightarrow{\pi_C} & C & \longrightarrow & 0 \\ & & & & f\big\uparrow & & \big\uparrow g & & \\ & & & & X & \xrightarrow[h]{} & W & & \end{array}$$

As W is a priori in $\mathbf{X}$ and Y_C is in $\hat{\omega}$, corollary 2.2 applies once again to yield $\mathrm{Ext}^1_{\mathbf{C}}(W, Y_C) = 0$ and hence to establish the existence of a morphism $g':W \to X_C$ such that $\pi_C g' = g$. But then $f-g'h$ satisfies $\pi_C(f-g'h) = \pi_C f-(\pi_C g')h = gh-gh = 0$, so that $f-g'h$ factors over Y_C. Then lemma 2.7 shows that $f-g'h$ factors already over some object W' in ω and hence the class of $f-g'h$ in $\mathrm{Hom}_{\mathbf{X}/\omega}(X, X_C)$ is the zero-morphism. As $g'h$ factors over W in ω, its class is zero as well, which shows that f and $f-g'h$ define the same morphism in $\mathrm{Hom}_{\mathbf{X}/\omega}(X, X_C) = \mathrm{Hom}_{\mathbf{X}/\omega}(X, j_!C)$. Hence already the class of f is the zero-morphism and $\phi_{X,C}$ is injective as claimed.

The definition of $\phi_{X,C}$ is natural in both arguments, so that the adjointness of $j_!$ and $j^!$ is established. Furthermore, the construction of $\phi_{X,C}$ shows that π_C induces the adjunction morphism $j^!j_!C \to C$.

This proves (a).
As the proof of (b) is completely analogous, it is left to the reader.
For (c), just remark again that by definition of $\hat{\omega}$, any object Y in $\hat{\omega}$ appears in an exact sequence $0 \to K \to W \to Y \to 0$ with W in ω and K in $\hat{\omega}$. But this sequence serves as an $\mathbf{X}$-approximation for Y whence $j^! i_* Y$, the chosen $\mathbf{X}$-approximation of Y, is isomorphic to W in $\mathbf{X}/\omega$, i.e. it is a zero object. This shows $j^! i_* = 0$ and $i^* j_! = 0$ follows then by adjunction.
(d) follows now from (c), as one has by naturality the commutative diagram of morphisms of functors

$$\begin{array}{ccc} j_! j^! & \xrightarrow{\pi} & \mathrm{id}_{\hat{\mathbf{X}}/\omega} \\ {\scriptstyle \iota*(j_! j^!)}\downarrow & & \downarrow{\scriptstyle \iota} \\ i_* i^* j_! j^! & \xrightarrow[(i_* i^*)*\pi]{} & i_* i^* \end{array}$$

in which the lower left corner is zero by (c). (In more concrete terms, (d) says that for any object C in $\hat{\mathbf{X}}$ there is a commutative diagram

$$\begin{array}{ccc} X_C & \xrightarrow{\pi_C} & C \\ {\scriptstyle j}\downarrow & & \downarrow{\scriptstyle \iota^C} \\ W & \longrightarrow & Y^C \end{array}$$

with W in ω, and we have seen indeed in lemma 1.3 and the proof of theorem 1.1 that ι^C can be obtained as the push-out of such a morphism j along π_C.) This finishes the proof of theorem 2.8. ∎

The reader puzzled by the notations used in the preceding theorem should compare it with the treatment of the "glueing of categories" in [BBD; 1.4]. It shows that in our situation one should think of $\hat{\mathbf{X}}$ as being obtained by "glueing together the open subcategory $\mathbf{X}$ and the closed subcategory $\hat{\omega}$ along ω ". What is missing for a complete glueing in the sense of (loc. cit.) is the existence of the other adjoints j_* and $i^!$.

The statements (c) and (d) in theorem 2.8 also explain why we think of theorem 1.1 as a "decomposition theorem": an object C in $\hat{\mathbf{X}}$ is *decomposed* - at least in $\hat{\mathbf{X}}/\omega$ - into its $\mathbf{X}$-approximation X_C and its $\hat{\omega}$-hull Y^C, which belong to "orthogonal" subcategories of $\hat{\mathbf{X}}/\omega$.

The property which is desirable but missing yet is that $\mathbf{X}$ and $\hat{\omega}$ should have ω as their common intersection. This will be addressed later on in Proposition 3.6.

For now, we return to the examples 1 and 2 of the previous section. As soon as $X \to \mathrm{Spec}\ k$ is *projective*, $\omega_{\mathcal{L}}$ is not an injective generator in either $\mathbf{X}$ or $\mathbf{X}'$, as $\mathrm{Ext}^d_{\mathcal{O}_X}(\mathcal{O}_X, \omega_X \otimes \mathcal{L}^{\otimes n}) = H^d(X, \omega_X \otimes \mathcal{L}^{\otimes n})$ does not vanish for all integers n. None the less, the following analogues of the results for injective cogenerators are valid for these examples if one substitutes $\mathcal{E}xt^i_{\mathcal{O}_X}(A, B)$ for $\mathrm{Ext}^i_{\mathcal{O}_X}(A, B)$.

Lemma 2.9. With notations as in examples 1 and 2, the following are equivalent for a sheaf $\mathcal{M}$ in $\mathbf{C}$.

(a) $\mathcal{M}$ is in $\mathbf{X}$.

(b) $\mathcal{E}xt^i_{\mathcal{O}X}(\mathcal{M}, \omega_X) = 0$ for all $i > 0$.

(c) $\mathcal{E}xt^i_{\mathcal{O}X}(\mathcal{M}, \omega_X \otimes \mathcal{L}^{\otimes n}) = 0$ for all $i > 0$ and all n in $\mathbf{Z}$.

(d) $\mathcal{E}xt^i_{\mathcal{O}X}(\mathcal{M}, Y) = 0$ for all $i > 0$ and Y in $\omega_{\mathcal{L}}$.

Proof: Easy consequence of the fact that the corresponding statements hold for Cohen-Macaulay local rings with a dualizing module. ∎

Proposition 2.10. With the same assumptions and notations as above, let $0 \to Y_C \to X_C \to C \to 0$ be an $\mathbf{X}$-approximation for C in $\mathbf{C}$.
Then we have for any $\mathcal{M}$ in $\mathbf{X}$:

(a) $0 \to Hom_{\mathcal{O}X}(\mathcal{M}, Y_C) \to Hom_{\mathcal{O}X}(\mathcal{M}, X_C) \to Hom_{\mathcal{O}X}(\mathcal{M}, C) \to 0$ is exact.

(b) The induced morphisms $\mathcal{E}xt^i_{\mathcal{O}X}(\mathcal{M}, X_C) \to \mathcal{E}xt^i_{\mathcal{O}X}(\mathcal{M}, C)$ are isomorphisms for $i > 0$.

Proof: Immediate consequence of lemma 2.7. ∎

Remark that in Example 3 the category ω is in fact an injective cogenerator as by definition there $\mathbf{X} = {}^{\perp}\omega$. Furthermore, in that example $\mathbf{X}/\omega$ is the category of left *maximal Cohen-Macaulay R-modules* - in the sense that $\mathrm{Ext}^i_R(M, R) = 0$ for $i \neq 0$ - *modulo stable equivalence*: two modules M and M' from $\mathbf{X}$ become isomorphic in $\mathbf{X}/\omega$ if and only if there are finitely generated projective left R-modules P and P' such that $M \oplus P'$ is isomorphic to $M' \oplus P$ in **R-mod**.

We end this section with two more illustrations of situations where ω is an injective cogenerator for $\mathbf{X}$.

Example 4. Suppose R is a commutative noetherian Cohen-Macaulay ring in the sense that all its localizations $R_{\mathbf{p}}$ at primes $\mathbf{p}$ are local Cohen-Macaulay rings. We say that a finitely generated R-module M is *maximal Cohen-Macaulay* (MCM for short), if $M_{\mathbf{p}}$ satisfies depth $M_{\mathbf{p}} = \dim R_{\mathbf{p}}$ for all primes $\mathbf{p}$.

Now suppose that R is a *Gorenstein* ring and that S is a commutative R-algebra which is MCM as an R-module. Let $\mathbf{C} =$ **S-mod** be the category of finitely generated S-modules and let $\mathbf{X}$ be the subcategory of $\mathbf{C}$ consisting of those S-modules M which are maximal Cohen-Macaulay as R-modules. Then $\mathbf{X}$ satisfies the usual properties. Set $\omega_{S/R} = \mathrm{Hom}_R(S, R)$, which is a relative dualizing module for the algebra $R \to S$. Then $\omega = \mathbf{add}\{\omega_{S/R}\}$ consists of all S-modules of the form $\mathrm{Hom}_R(P, R)$ with P finitely generated projective over S. It is easily seen - and well-known - that ω is an injective cogenerator for $\mathbf{X}$. Also, if the Krull dimension of R is finite, then $\hat{\mathbf{X}} = \mathbf{C}$.

To acknowledge the scope of this example and to emphasize its relevance for Grothendieck duality theory, we quote the following from [FGR; Cor. 5.9.].

Theorem. Suppose S is a commutative ring with finite Krull dimension and with connected prime spectrum. Then S *admits a canonical module* if and only if S is a homomorphic image of a Gorenstein ring R such that S is maximal Cohen-Macaulay as an R-module. ∎

Our final illustration of this section is the following variant of Examples 4 and 2.

Example 5. Maintain the hypotheses on S and R from Example 4. Let $X \subset \operatorname{Spec} R$ have the property that if $\mathbf{p}$ is in X, then $S_{\mathbf{p}}$ is a Gorenstein ring, or equivalently, $(\omega_{S/R})_{\mathbf{p}}$ is $S_{\mathbf{p}}$-free.

Set again $\mathbf{C} = \mathbf{S\text{-}mod}$ and let $\mathbf{X}'$ consist of those S-modules M which are MCM over R and satisfy furthermore that $M_{\mathbf{p}}$ is $S_{\mathbf{p}}$-projective for all $\mathbf{p}$ in X. Then $\mathbf{X}'$ satisfies the usual properties and contains $\omega = \mathbf{add}\{\omega_{S/R}\}$. Again, ω is an injective cogenerator for $\mathbf{X}'$ and $\hat{\mathbf{X}}'$ consists of all S-modules C such that $\operatorname{proj.dim}_{S_{\mathbf{p}}} C_{\mathbf{p}} < \infty$ for all $\mathbf{p}$ in X.

§3. *Exactness properties of* $\hat{\mathbf{X}}$ *and* $\hat{\omega}$.

We maintain our general assumption that $\mathbf{X}$ is an additively closed and exact subcategory of $\mathbf{C}$ in which every epimorphism is admissible, and that ω is an injective cogenerator for $\mathbf{X}$.

In this situation, we show that $\hat{\mathbf{X}}$ is an additively closed subcategory of $\mathbf{C}$ which has the property that an exact sequence $0 \to A \to B \to C \to 0$ is in $\hat{\mathbf{X}}$ whenever two of A, B and C are in $\hat{\mathbf{X}}$. This result is then used to prove that $\hat{\omega}$ is an additively closed subcategory of $\mathbf{C}$ having the property that an exact sequence $0 \to A \to B \to C \to 0$ in $\mathbf{C}$ is already in $\hat{\omega}$ if either A and C are in $\hat{\omega}$ or A and B are in $\hat{\omega}$. Hence $\hat{\omega}$ is seen to be an *additively closed exact subcategory* of $\mathbf{C}$ in which *every monomorphism is admissible.*

We begin with the following

Lemma 3.1. The category $\hat{\mathbf{X}}$ is closed under extensions.

Proof: Suppose $0 \to A \to B \to C \to 0$ is an exact sequence in $\mathbf{C}$ with A and C in $\hat{\mathbf{X}}$. Proceed by induction on $n = \mathbf{X}\text{-resol.dim} C$. Suppose $n = 0$, which means that C is in $\mathbf{X}$. As A is in $\hat{\mathbf{X}}$, there is an $\mathbf{X}$-approximation $0 \to Y_A \to X_A \to A \to 0$ of A. Since C is in $\mathbf{X}$, we know by theorem 2.3, that the induced map $\operatorname{Ext}^1_{\mathbf{C}}(C, X_A) \to \operatorname{Ext}^1_{\mathbf{C}}(C, A)$ is an isomorphism. Hence there exists an exact commutative diagram

$$\begin{array}{ccccccccc}
 & & 0 & & 0 & & & & \\
 & & \downarrow & & \downarrow & & & & \\
 & & Y_A & = & Y_A & & & & \\
 & & \downarrow & & \downarrow & & & & \\
0 & \to & X_A & \to & Z & \to & C & \to & 0 \\
 & & \downarrow & & \downarrow & & \| & & \\
0 & \to & A & \to & B & \to & C & \to & 0 \\
 & & \downarrow & & \downarrow & & & & \\
 & & 0 & & 0 & & & &
\end{array}$$

Since X_A and C are in **X**, the object Z is also in **X**, as that category is closed under extensions. Now Y_A is in $\hat{\omega}$, hence in $\hat{\mathbf{X}}$, and it follows that B is in $\hat{\mathbf{X}}$ as required.

Suppose now that $n > 0$ and let $0 \to L \to X_0 \to C \to 0$ be exact with **X**–resol.dim$L = n\text{-}1$. Since X_0 is in **X**, we have that $\mathrm{Ext}^1_{\mathbf{C}}(X_0, X_A) \to \mathrm{Ext}^1_{\mathbf{C}}(X_0, A)$ is an isomorphism by theorem 2.3, and so there exists an exact commutative diagram in **C**

$$\begin{array}{ccccccccc}
 & & 0 & & 0 & & 0 & & \\
 & & \downarrow & & \downarrow & & \downarrow & & \\
0 & \to & Y_A & \to & U & \to & L & \to & 0 \\
 & & \downarrow & & \downarrow & & \downarrow & & \\
0 & \to & X_A & \to & V & \to & X_0 & \to & 0 \\
 & & \downarrow & & \downarrow & & \downarrow & & \\
0 & \to & A & \to & B & \to & C & \to & 0 \\
 & & \downarrow & & \downarrow & & \downarrow & & \\
 & & 0 & & 0 & & 0 & &
\end{array}$$

This shows that B is in $\hat{\mathbf{X}}$ since V is necessarily in **X** and U is in $\hat{\mathbf{X}}$ by the inductive hypothesis. ∎

We now use the fact that $\hat{\mathbf{X}}$ is closed under extensions to prove the following

Lemma 3.2. Let $0 \to K \to X \to C \to 0$ be an exact sequence in **C** with X in **X**. Then C is in $\hat{\mathbf{X}}$ if and only if K is in $\hat{\mathbf{X}}$.

Proof: By definition, if K is in $\hat{\mathbf{X}}$ then also C is in $\hat{\mathbf{X}}$. Hence assume that C is in $\hat{\mathbf{X}}$ and let $0 \to Y_C \to X_C \to C \to 0$ be an **X**-approximation of C. Since X is in **X**, we have by theorem 2.3 that $\mathrm{Hom}_{\mathbf{C}}(X, X_C) \to \mathrm{Hom}_{\mathbf{C}}(X, C)$ is surjective. Therefore we obtain a commutative exact diagram

$$\begin{array}{ccccccccc} 0 & \longrightarrow & K & \longrightarrow & X & \longrightarrow & C & \longrightarrow & 0\\ & & \downarrow & & \downarrow & & \| & & \\ 0 & \longrightarrow & Y_C & \longrightarrow & X_C & \longrightarrow & C & \longrightarrow & 0\end{array}$$

Since Y_C is in $\hat{\omega}$, we know there is an epimorphism $W \to Y_C$ with W in ω. Adding this epimorphism to the foregoing diagram, we obtain the following commutative exact diagram

$$\begin{array}{ccccccccc} & & 0 & & 0 & & & & \\ & & \downarrow & & \downarrow & & & & \\ & & V & = & V & & & & \\ & & \downarrow & & \downarrow & & & & \\ 0 & \longrightarrow & K\oplus W & \longrightarrow & X\oplus W & \longrightarrow & C & \longrightarrow & 0\\ & & \downarrow & & \downarrow & & \| & & \\ 0 & \longrightarrow & Y_C & \longrightarrow & X_C & \longrightarrow & C & \longrightarrow & 0\\ & & \downarrow & & \downarrow & & & & \\ & & 0 & & 0 & & & & \end{array}$$

where $K\oplus W \to X\oplus W$ is the sum of $K \to X$ and the identity on W. Since W and X are both in $\mathbf{X}$, we have that V is in $\mathbf{X}$, as any epimorphism of $\mathbf{C}$ in $\mathbf{X}$ is admissible by assumption. Therefore $K\oplus W$ is in $\hat{\mathbf{X}}$, since Y_C and V are in $\hat{\mathbf{X}}$ and $\hat{\mathbf{X}}$ is closed under extensions. We now show that this implies that K is in $\hat{\mathbf{X}}$. Since $K\oplus W$ is in $\hat{\mathbf{X}}$, we obtain the following exact commutative diagram from an $\mathbf{X}$-approximation of $K\oplus W$

$$\begin{array}{ccccccccc} & & & & 0 & & 0 & & \\ & & & & \downarrow & & \downarrow & & \\ & & 0 & \longrightarrow & Y_{K\oplus W} & \longrightarrow & Z & & \\ & & & & \downarrow & & \downarrow & & \\ & & & & X_{K\oplus W} & = & X_{K\oplus W} & & \\ & & & & \downarrow & & \downarrow & & \\ 0 & \longrightarrow & W & \longrightarrow & K\oplus W & \longrightarrow & K & \longrightarrow & 0\\ & & & & \downarrow & & \downarrow & & \\ & & & & 0 & & 0 & & \end{array}$$

Hence we have the exact sequence $0 \to Y_{K\oplus W} \to Z \to W \to 0$. Since W and $Y_{K\oplus W}$ are in $\hat{\mathbf{X}}$, (in fact already in $\hat{\omega}$), and $\hat{\mathbf{X}}$ is closed under extensions, we have that Z is in $\hat{\mathbf{X}}$ as well. This implies that K is in $\hat{\mathbf{X}}$ since $X_{K\oplus W}$ is in $\mathbf{X}$. This completes the proof of the lemma. ∎

We now apply the foregoing lemma to prove

Proposition 3.3. Suppose C is an object in $\hat{\mathbf{X}}$. Then the following are equivalent for any integer $n \geq 0$:

(a) $\mathbf{X}$–resol.dim$C \leq n$,

(b) If $0 \to U \to X_{n-1} \to \ldots \to X_0 \to C \to 0$ is exact with X_i in $\mathbf{X}$ for $i = 0,\ldots,n\text{-}1$, then U is in $\mathbf{X}$.

Proof: For $n = 0$ there is nothing to prove. So suppose $n > 0$. Assuming (a), repeated application of lemma 3.2 shows that U is in $\hat{\mathbf{X}}$. Also we have $\mathrm{Ext}^1_{\mathbf{C}}(U,W) = \mathrm{Ext}^{n+1}_{\mathbf{C}}(C,W) = 0$ for all W in ω since $\mathbf{X}$–inj.dim$\omega = 0$ and $\mathbf{X}$–resol.dim$C \leq n$. Therefore by proposition 2.1, it follows that U is in $\mathbf{X}$ proving that (a) implies (b). As (a) follows from (b) by definition of $\mathbf{X}$–resol.dimC, we are done. ∎

As a first application of this proposition we prove the following

Proposition 3.4. $\hat{\mathbf{X}}$ is an additively closed subcategory of $\mathbf{C}$, that is $\hat{\mathbf{X}} = \mathbf{add}\,\hat{\mathbf{X}}$.

Proof: Suppose $C_1 \oplus C_2$ is in $\hat{\mathbf{X}}$ for two objects C_1 and C_2 in $\mathbf{C}$. Proceed by induction on $n = \mathbf{X}$–resol.dim $(C_1 \oplus C_2)$. If $n = 0$, the summands C_1 and C_2 are in $\mathbf{X}$ as $\mathbf{X}$ is an additively closed subcategory of $\mathbf{C}$. Suppose $n > 0$. Since $C_1 \oplus C_2$ is in $\hat{\mathbf{X}}$, there is an epimorphism $X \to C_1 \oplus C_2 \to 0$ with X in $\mathbf{X}$. Therefore we obtain exact sequences $0 \to L_i \to X \to C_i \to 0$ for $i = 1,2$ which yield the exact sequence $0 \to L_1 \oplus L_2 \to X \oplus X \to C_1 \oplus C_2 \to 0$. Now by Lemma 3.2, we know that $L_1 \oplus L_2$ is in $\hat{\mathbf{X}}$ and proposition 3.3 shows that $\mathbf{X}$–resol.dim$(L_1 \oplus L_2) \leq n-1$. By the inductive hypothesis L_1 and L_2 are in $\hat{\mathbf{X}}$ and another application of lemma 3.2 shows that then also C_1 and C_2 are in $\hat{\mathbf{X}}$. ∎

We are now in position to establish one of the results promised in the beginning of this section.

Proposition 3.5. An exact sequence $0 \to A \to B \to C \to 0$ from $\mathbf{C}$ is in $\hat{\mathbf{X}}$ if any two of A, B and C are in $\hat{\mathbf{X}}$.

Proof: Since we already know that $\hat{\mathbf{X}}$ is closed under extensions by lemma 3.1, it suffices to show that if B is in $\hat{\mathbf{X}}$ then A is in $\hat{\mathbf{X}}$ if and only if C is in $\hat{\mathbf{X}}$. We first show that if A and B are in $\hat{\mathbf{X}}$ then C is in $\hat{\mathbf{X}}$. Choose an $\mathbf{X}$-approximation $0 \to Y_B \to X_B \to B \to 0$ for B. It gives rise to an exact commutative diagram

$$
(*)\qquad
\begin{array}{ccccccccc}
 & & & & 0 & & 0 & & \\
 & & & & \downarrow & & \downarrow & & \\
 & & 0 & \longrightarrow & Y_B & \longrightarrow & L & & \\
 & & & & \downarrow & & \downarrow & & \\
 & & & & X_B & = & X_B & & \\
 & & & & \downarrow & & \downarrow & & \\
0 & \longrightarrow & A & \longrightarrow & B & \longrightarrow & C & \longrightarrow & 0\\
 & & & & \downarrow & & \downarrow & & \\
 & & & & 0 & & 0 & &
\end{array}
$$

from which we get an exact sequence $0 \to Y_B \to L \to A \to 0$. It follows that L is in $\hat{\mathbf{X}}$ since Y_B and A are in $\hat{\mathbf{X}}$ and $\hat{\mathbf{X}}$ is closed under extensions. Therefore C is in $\hat{\mathbf{X}}$ since X_B is in $\mathbf{X}$.

Suppose now that B and C are in $\hat{\mathbf{X}}$. Using lemma 3.2, the exact sequence $0 \to L \to X_B \to C \to 0$ from (*) shows that L is in $\hat{\mathbf{X}}$. Applying the just established result to the exact sequence $0 \to Y_B \to L \to A \to 0$, it follows that A is in $\hat{\mathbf{X}}$. This completes the proof of the proposition. ∎

We now turn our attention to $\hat{\omega}$. We begin with the characterization of $\hat{\omega}$ as a subcategory of $\hat{\mathbf{X}}$, proving that $\hat{\omega} = \mathbf{X}^{\perp} \cap \hat{\mathbf{X}}$ in $\mathbf{C}$.

Proposition 3.6. The following statements are equivalent for an object C in $\hat{\mathbf{X}}$.

(a) C is in $\hat{\omega}$.

(b) $\mathbf{X}$-inj.dim C = 0, that is: C is in $\mathbf{X}^{\perp} \cap \hat{\mathbf{X}}$.

(c) If $0 \to Y_C \to X_C \to C \to 0$ is any $\mathbf{X}$-approximation of C, then X_C is in ω.

Proof: That (a) implies (b) was seen in corollary 2.2, and it is obvious that (c) implies (a). Hence we only need to show that (b) implies (c).
Since $\mathbf{X}$-inj.dim C = 0 = $\mathbf{X}$-inj.dim Y_C, it follows that $\mathbf{X}$-inj.dim X_C = 0. Our desired result is therefore a trivial consequence of the following, which proves $\omega = \mathbf{X} \cap \hat{\omega}$.

Lemma 3.7. The following are equivalent for an object X in $\mathbf{X}$.

(a) X is in ω.

(b) X is in $\hat{\omega}$.

(c) $\mathbf{X}$-inj.dim X = 0.

Proof: Again it is obvious that (a) implies (b), and Corollary 2.2 shows that (b) implies (c). It remains to prove
(c) ⇒ (a): Let $0 \to X \to W \to X' \to 0$ be an exact sequence in $\mathbf{X}$ with W in ω which exists as ω is a cogenerator for $\mathbf{X}$. Then by (c) this sequence splits. Therefore X is a direct summand of W which implies that X is in ω as that category is assumed to be additively closed.
This establishes lemma 3.7 and finishes the proof of proposition 3.6. ∎

These results prove the following fact, already announced in the introduction to this section.

Proposition 3.8. $\hat{\omega}$ is an additively closed and exact subcategory of $\mathbf{C}$ in which any monomorphism is admissible. In more detail, let $0 \to A \to B \to C \to 0$ be an exact sequence in $\mathbf{C}$. Then

(a) B is in $\hat{\omega}$ if A and C are in $\hat{\omega}$, and

(b) C is in $\hat{\omega}$ if A and B are in $\hat{\omega}$.

Proof: We know by now that $\hat{\omega} = \mathbf{X}^{\perp} \cap \hat{\mathbf{X}}$ in $\mathbf{C}$. But the statements hold for $\hat{\mathbf{X}}$ by propositions 3.4 and 3.5, and as $\mathbf{X}^{\perp}$ is a right orthogonal complement, it also is an additively closed and exact subcategory of $\mathbf{C}$ in which every monomorphism is admissible. As all these properties are stable under intersection in $\mathbf{C}$, the result for $\hat{\omega}$ follows. ■

We sum up the foregoing results as

Theorem 3.9. Let $\mathbf{X}$ be an additively closed and exact subcategory of an abelian category $\mathbf{C}$. Assume that

(i) all epimorphisms from $\mathbf{C}$ in $\mathbf{X}$ are admissible, and

(ii) $\mathbf{X}$ has enough relatively injective objects.

Let ω be an injective cogenerator for $\mathbf{X}$. Then there results a diagram of additively closed and exact subcategories of $\mathbf{C}$

$$\begin{array}{ccccc}
\mathbf{X} & \longrightarrow & \hat{\mathbf{X}} & \longrightarrow & \mathbf{C} \\
\uparrow & & \uparrow & & \uparrow \\
\omega & \longrightarrow & \hat{\omega} & \longrightarrow & \mathbf{X}^{\perp}
\end{array}$$

such that

(a) each square is *cartesian*, i.e.: $\hat{\omega} = \hat{\mathbf{X}} \cap \mathbf{X}^{\perp}$ and $\omega = \mathbf{X} \cap \mathbf{X}^{\perp}$,

(b) in $\hat{\mathbf{X}}$ all *mono- or epimorphisms* from $\mathbf{C}$ *are admissible*,

(c) in $\mathbf{X}^{\perp}$ and $\hat{\omega}$ *all monomorphisms* from $\mathbf{C}$ *are admissible*.

In particular, there is a *unique* injective cogenerator ω for $\mathbf{X}$ in $\mathbf{C}$, namely $\omega = \mathbf{X} \cap \mathbf{X}^{\perp}$. ■

To reformulate it once again modulo $\omega = \mathbf{X} \cap \mathbf{X}^{\perp}$, let us say that a sequence $0 \to \mathbf{A} \to \mathbf{B} \xrightarrow{p} \mathbf{C} \to 0$ of additive functors between additive categories is *exact* if and only if $\mathbf{A}$ is a full, essential and additively closed subcategory of $\mathbf{B}$ and p is equivalent to the projection functor $\pi : \mathbf{B} \to \mathbf{B}/\mathbf{A}$.

With the notations of theorem 2.8 we have then the following

Corollary 3.10. The adjoint pairs of functors (i^*, i_*) and $(j_!, j^!)$ fit into the commutative diagram of exact sequences of additive categories

$$\begin{array}{ccccccccc} 0 & \longrightarrow & \hat{\omega}/\omega & \xrightarrow{i_*} & \hat{X}/\omega & \xrightarrow{j^!} & X/\omega & \longrightarrow & 0 \\ & & \Downarrow & & \| & & \Uparrow & & \\ 0 & \longleftarrow & \hat{\omega}/\omega & \xleftarrow{i^*} & \hat{X}/\omega & \xleftarrow{j_!} & X/\omega & \longleftarrow & 0 \end{array}$$

§4. *The category $\hat{\omega}$.*

Our aim in this section is to describe under which assumptions $\hat{\omega}$ has the further property that an exact sequence $0 \to A \to B \to C \to 0$ is in $\hat{\omega}$ if B and C are in $\hat{\omega}$.

To investigate this problem, we first define $\tilde{\omega}$ to be the subcategory of **C** consisting of all objects C in **C** which appear in an exact sequence $0 \to C \to Y_0 \to \dots \to Y_n \to 0$ with each Y_i in $\hat{\omega}$. Clearly such an object C is in $\hat{\mathbf{X}}$ since the Y_i are in $\hat{\omega} \subseteq \hat{\mathbf{X}}$ and the kernel of an epimorphism in $\hat{\mathbf{X}}$ is again in $\hat{\mathbf{X}}$ by proposition 3.5. Also it is obvious that $\hat{\omega}$ is a subcategory of $\tilde{\omega}$.

Lemma 4.1. The following statements are equivalent:

(a) An exact sequence $0 \to A \to B \to C \to 0$ is in $\hat{\omega}$ if B and C are in $\hat{\omega}$.

(b) $\hat{\omega} = \tilde{\omega}$.

The proof is trivial. ∎

This simple observation explains the relevance of the category $\tilde{\omega}$ to our problem about $\hat{\omega}$. The following description of $\tilde{\omega}$ is basic to the results in this paragraph.

Proposition 4.2. For an object C in $\hat{\mathbf{X}}$ the following are equivalent:

(a) C is in $\tilde{\omega}$,

(b) **X**-inj.dim $C < \infty$.

Proof: (a) $\Rightarrow$ (b). Let $0 \to C \to Y_0 \to \dots \to Y_n \to 0$ be exact with each Y_i in $\hat{\omega}$. Since **X**-inj.dim $\hat{\omega} = 0$, it follows by induction on n that **X**-inj.dim $C \leq n < \infty$.

(b) $\Rightarrow$ (a): Since C is in $\hat{\mathbf{X}}$, it admits an $\hat{\omega}$-hull $0 \to C \to Y^C \to X^C \to 0$. The assumption that **X**-inj.dim $C < \infty$ and the fact that **X**-inj.dim $Y^C = 0$ imply that **X**-inj.dim $X^C < \infty$. Therefore if we show that an object X from **X** which satisfies **X**-inj.dim $X < \infty$ is necessarily in $\tilde{\omega}$, we will be done. Indeed we have the following more specific result.

Lemma 4.3. Let X be an object in **X**, n a nonnegative integer. Then **X**-inj.dim $X \leq n$ if and only if there is an exact sequence $0 \to X \to W_0 \to W_1 \to \dots \to W_n \to 0$ with W_i in ω for $i = 0, \dots, n$.

Proof: The if-part follows as before from **X**-inj.dim $\omega = 0$. Hence suppose that **X**-inj.dim $X = n$. Since ω is a cogenerator for **X**, we can construct an exact sequence $0 \to X \to W_0 \to \dots \to W_{n-1} \to X' \to 0$ in **X** such that each W_i is in ω for $i = 0,\dots,n-1$. As **X**-inj.dim $\omega = 0$ and **X**-inj.dim $X \leq n$ by assumption, it follows that for any integer $i > 0$ and all objects Z in **X** one has $\mathrm{Ext}^i_{\mathbf{C}}(Z,X') \cong \mathrm{Ext}^{n+i}_{\mathbf{C}}(Z,X) = 0$. But by lemma 3.7 this shows that X' is already in ω as desired.
This concludes the proof of lemma 4.3 and proposition 4.2. ∎

As a first application we get the following.

Corollary 4.4. $\check{\omega}$ is an additively closed subcategory of **C** with the property that an exact sequence $0 \to A \to B \to C \to 0$ is in $\check{\omega}$ if any two of A, B and C are in $\check{\omega}$.

Proof: Since $\hat{\mathbf{X}}$ is additively closed, it contains with $\check{\omega}$ also **add** $\check{\omega}$. It then follows from proposition 4.2 that $\check{\omega} =$ **add** $\check{\omega}$. Also if $0 \to A \to B \to C \to 0$ is an exact sequence in **C** with two of A, B and C in $\check{\omega}$, then all of A, B and C are in $\hat{\mathbf{X}}$ by Proposition 3.5. It then follows from proposition 4.2 that they are all in $\check{\omega}$. ∎

As another immediate consequence of proposition 4.2 we get the following.

Corollary 4.5. The following are equivalent:

(a) $\hat{\omega} = \check{\omega}$,

(b) If C is an object in $\hat{\mathbf{X}}$ with **X**-inj.dim $C < \infty$, then **X**-inj.dim $C = 0$.

Proof: Let C be in $\hat{\mathbf{X}}$. By proposition 4.2 we have that C is in $\check{\omega}$ if and only if **X**-inj.dim $C < \infty$. By Proposition 3.6 we have that C is in $\hat{\omega}$ if and only if **X**-inj.dim $C = 0$.
Hence the equivalence of (a) and (b). ∎

We now give criteria for the property $\hat{\omega} = \check{\omega}$ in terms of the categories ω and **X** themselves.

Proposition 4.6. The following are equivalent:

(a) $\hat{\omega} = \check{\omega}$.

(b) If $0 \to C \to W_0 \to W_1 \to 0$ is exact in **C** with W_0 and W_1 in ω, then C is in ω.

(c) If $0 \to C \to W_0 \to W_1 \to \dots \to W_n \to 0$ is exact with each W_i in ω for $i = 0,\dots,n$, then C is in ω.

(d) If X is in **X** and **X**-inj.dim $X < \infty$, then X is in ω.

Proof: (a) ⇒ (b). Since W_0 and W_1 are objects in ω, they are in **X**, so C is in **X**. Clearly C is in $\check{\omega}$ which means by the assumption that it is in $\hat{\omega}$. Therefore C is in $\mathbf{X} \cap \hat{\omega}$ which category equals ω by Lemma 3.7.
(b) ⇒ (c) by induction on n.
(c) ⇒ (d). Suppose X is in **X** with **X**-inj.dim $X < \infty$. Then by Lemma 4.3, we know there

is an exact sequence $0 \to X \to W_0 \to \ldots \to W_n \to 0$ with W_i in ω for $i = 0, \ldots, n$. Therefore by (c) we have that X is in ω.
(d) $\Rightarrow$ (a). Let C be an object in $\check{\omega}$. Then we can choose an $\mathbf{X}$-approximation $0 \to Y_C \to X_C \to C \to 0$ for C and proposition 4.2 shows that $\mathbf{X}$-inj.dim $C < \infty$. But $\mathbf{X}$-inj.dim $Y_C = 0$ so that $\mathbf{X}$-inj.dim $X_C < \infty$. Therefore X_C is in ω by the hypothesis (d), which shows that C is in $\hat{\omega}$. ∎

Now we establish the following.

Proposition 4.7. Let C be an object in $\check{\omega}$. Then ω-inj.dim C = $\mathbf{X}$-inj.dim C.

Proof: Since ω is a subcategory of $\mathbf{X}$, we have that always ω-inj.dim C $\leq$ $\mathbf{X}$-inj.dim C . So it suffices to show that ω-inj.dim C $\geq$ $\mathbf{X}$-inj.dim C. As C is in $\check{\omega}$ by assumption, we also know from proposition 4.2 that $\mathbf{X}$-inj.dim C is finite.

To begin with, we prove the proposition when C = X is an object in $\mathbf{X} \cap \check{\omega}$. By lemma 4.3, we have that then there is an exact sequence $0 \to X \to W_0 \to \ldots \to W_n \to 0$ with each W_i in ω for $i = 0, \ldots, n$. Assume that ω-inj.dim X = 0. Since ω-inj.dim ω = 0, it follows by induction on n that the exact sequence $0 \to X \to W_0 \to \ldots \to W_n \to 0$ splits. Hence $\mathbf{X}$ is already in ω which implies $\mathbf{X}$-inj.dim X = 0. This result shows furthermore that ω-inj.dim X $\leq$ n if and only if there is an exact sequence $0 \to X \to W_0 \to \ldots \to W_n \to 0$ with each W_i in ω. But we have seen in lemma 4.3 that the existence of such an exact sequence is equivalent to $\mathbf{X}$-inj.dim X $\leq$ n. Hence we have shown that ω-inj.dim C = $\mathbf{X}$-inj.dim C when C is in $\mathbf{X} \cap \check{\omega}$.

Assume now that C is an arbitrary object in $\check{\omega}$. Let $0 \to C \to Y^C \to X^C \to 0$ be an $\hat{\omega}$-hull of C. Since C and Y^C are in $\check{\omega}$ by assumption, we get that X^C is in $\check{\omega}$ by corollary 4.4. Suppose now ω-inj.dim C = 0. Then also ω-inj.dim X^C = 0 which implies that X^C is in ω by our previous result. But our current hypothesis then implies that $\mathrm{Ext}^1_{\mathbf{C}}(X^C, C) = 0$, which means that the chosen $\hat{\omega}$-hull of C splits. So C is a direct summand of Y^C in $\hat{\omega}$ and is hence itself in $\hat{\omega}$, as $\hat{\omega}$ is additively closed by proposition 3.8. Therefore $\mathbf{X}$-inj.dim C = 0 by corollary 2.2 and we are done in this case.

Finally suppose ω-inj.dim C = n > 0. Since ω-inj.dim Y^C = 0, it follows that ω-inj.dim X^C = n−1. Therefore $\mathbf{X}$-inj.dim X^C = n−1 by our first result, which implies that $\mathbf{X}$-inj.dim C $\leq$ n. Hence ω-inj.dim C $\geq$ $\mathbf{X}$-inj.dim C for all C in $\check{\omega}$, which completes the proof of the proposition. ∎

The following is an immediate consequence of our earlier results and summarizes sufficient conditions for $\hat{\omega} = \check{\omega}$ to hold.

Corollary 4.8. Consider the following conditions:
(a) ω-inj.dim $\mathbf{X}$ = 0,
(b) ω-inj.dim $\hat{\mathbf{X}}$ = 0,
(c) ω-inj.dim $\check{\omega}$ = 0,
(d) Every epimorphism $W' \to W \to 0$ in $\mathbf{C}$ with W and W' in ω admits a section.
(e) $\hat{\omega} = \check{\omega}$.

Then one has (a) $\Leftrightarrow$ (b) $\Rightarrow$ (c) $\Rightarrow$ (d) $\Rightarrow$ (e).

Proof. As $\mathbf{X}$ and $\tilde{\omega}$ are subcategories of $\hat{\mathbf{X}}$, the implications (b) $\Rightarrow$ (a) and (b) $\Rightarrow$ (c) are trivial. That (a) $\Rightarrow$ (b) follows from the existence of an $\mathbf{X}$-approximation $0 \to Y_C \to X_C \to C \to 0$ for any object C in $\hat{\mathbf{X}}$ and the fact that ω-inj.dim $\hat{\omega}$ = $\mathbf{X}$-inj.dim $\hat{\omega} = 0$. That (c) $\Rightarrow$ (d) follows from the fact that the kernel K of any epimorphism $W' \to W \to 0$ between objects from ω is by definition an object in $\tilde{\omega}$. But (c) implies $\mathrm{Ext}^1_{\mathbf{C}}(W,K) = 0$, whence the exact sequence $0 \to K \to W' \to W \to 0$ splits. The remaining implication (d) $\Rightarrow$ (e) is a special case of proposition 4.2. ∎

Example 6. A special case in which $\hat{\omega} = \tilde{\omega}$ holds, has been investigated already by A.Heller [He]. Following him let us say that $\mathbf{X}$ in $\mathbf{C}$ is a *Frobenius category* if it satisfies our usual assumptions of being additively closed and exact in $\mathbf{C}$ with every epimorphism from $\mathbf{C}$ in $\mathbf{X}$ being admissible and if furthermore $\omega = \mathbf{X} \cap \mathbf{X}^{\perp}$ is also a *projective generator* for $\mathbf{X}$, which is equivalent to ω^{op} being an injective cogenerator of $\mathbf{X}^{op}$. This means hence that

(i) $\mathbf{X}$-inj.dim ω = $\mathbf{X}$-proj.dim ω = 0 and

(ii) for every object X in $\mathbf{X}$ there exists both a monomorphism $i : X \to W$ as well as an epimorphism $p : W' \to X$ with W, W' in ω such that the objects Ker p and Coki, calculated in $\mathbf{C}$, are again objects in $\mathbf{X}$.

A.Heller himself gave already some examples of such categories and further such categories are discussed in [Ha]. Also, it is clear from the definitions that in Example 3 the category $\mathbf{X}$ of maximal Cohen-Macaulay R-modules is Frobenius.

§5. *Some remarks on the $\mathbf{X}$-resolution dimension of $\hat{\mathbf{X}}$.*

We define $\mathbf{X}$-resol.dim $\hat{\mathbf{X}}$ to be the maximum (including ∞) of $\mathbf{X}$-resol.dim C for all objects C in $\hat{\mathbf{X}}$. This paragraph is devoted to interpreting some of our previous results when $\mathbf{X}$-resol.dim $\hat{\mathbf{X}}$ is *finite*. So for the remainder of this section we assume $\mathbf{X}$-resol.dim $\hat{\mathbf{X}} = d < \infty$.

Our remarks are based on the following observation.

Lemma 5.1. The following statements are equivalent for an object C in $\mathbf{C}$.

(a) $\mathbf{X}$-inj.dim C $< \infty$,

(b) $\hat{\mathbf{X}}$-inj.dim C $< \infty$.

Moreover, if $\mathbf{X}$-inj.dim C $= m < \infty$, then $\hat{\mathbf{X}}$–inj.dim C $\leq d+m$.

Proof: Usual dimension shift argument. ∎

This lemma implies immediately the following.

Proposition 5.2. Suppose C is an object in $\hat{\mathbf{X}}$.
(a) $\hat{\mathbf{X}}$-inj.dimC $< \infty$ if and only if C is in $\tilde{\omega}$.
(b) If C is an object in $\hat{\omega}$, then $\hat{\mathbf{X}}$-inj.dimC $\leq$ d .

Proof: (a): By lemma 5.1, we know that $\hat{\mathbf{X}}$-inj.dimC $< \infty$ if and only if $\mathbf{X}$-inj.dimC $< \infty$. But by proposition 4.2, we know that $\mathbf{X}$-inj.dimC $< \infty$ if and only if C is in $\tilde{\omega}$.
(b): Since $\mathbf{X}$-inj.dim $\hat{\omega} = 0$ by corollary 2.2, the result follows from lemma 5.1. ■

As a special case we obtain the following consequence.

Corollary 5.3. If $\hat{\mathbf{X}} = \mathbf{C}$, then we have:
(a) C is in $\tilde{\omega}$ if and only if inj.dimC $< \infty$.
(b) If C is in $\hat{\omega}$, then inj.dimC $\leq$ d. ■

Remark that the injective dimension of an object C in $\mathbf{C}$ is defined here in terms of vanishing of the functors $\text{Ext}^*_{\mathbf{C}}(-,C)$. As soon as $\mathbf{C}$ itself has enough injective objects it coincides with the notion obtained from the length of a shortest injective resolution.

Applying the foregoing result to our decomposition into $\mathbf{X}$-approximations and $\hat{\omega}$-hulls we have the following.

Corollary 5.4. Suppose again $\hat{\mathbf{X}} = \mathbf{C}$ and let $0 \to Y_C \to X_C \to C \to 0$ and $0 \to C \to Y^C \to X^C \to 0$ be an $\mathbf{X}$-approximation and an $\hat{\omega}$-hull of an object C in $\mathbf{C}$ respectively. Then inj.dim $Y_C <$ inj.dim $Y^C \leq$ d or both Y_C and Y^C are already injective. ■

Finally, consider the case where $\hat{\omega} = \tilde{\omega}$. Then we have first the following consequence of lemma 5.1.

Proposition 5.5. Suppose $\hat{\omega} = \tilde{\omega}$. Then the following statements are equivalent for an object C in $\hat{\mathrm{X}}$.
(a) C is in $\hat{\omega}$.
(b) $\hat{\mathbf{X}}$-inj.dim C $\leq$ d.
(c) $\hat{\mathbf{X}}$-inj.dim C $< \infty$.

Proof: (a) $\Rightarrow$ (b) by proposition 5.2.
(b) $\Rightarrow$ (c) is trivial.
(c) $\Rightarrow$ (a): Since $\hat{\mathbf{X}}$-inj.dim C $< \infty$, we have that $\mathbf{X}$-inj.dim C $< \infty$.
Therefore C is in $\tilde{\omega}$ by proposition 4.2. Hence C is in $\hat{\omega}$ since $\tilde{\omega} = \hat{\omega}$ by assumption. ■

As an obvious consequence of this proposition we have the following.

Corollary 5.6. Suppose $\hat{\mathbf{X}} = \mathbf{C}$ and $\check{\omega} = \hat{\omega}$. Then the following conditions are equivalent for an object C in **C**.

(a) C is in $\hat{\omega}$,

(b) inj.dim $C \leq d$,

(c) inj.dim $C < \infty$. ■

§6. *More Examples*

In this section we describe various situations where the theory we have developed is applicable.

First we consider a generalization of Example 4.

Example 7. Let R be a commutative noetherian Gorenstein ring of finite dimension d. Let Λ be an R-algebra, not necessarily commutative, which is a *maximal Cohen-Macaulay R-module.* Set $\mathbf{C} = \mathbf{mod}\text{-}\Lambda$, the category of finitely generated right Λ-modules, and let **X** be the full subcategory of **C** whose objects are the Λ-modules which are MCM if considered as R-modules. Then **X** is again additively closed, exact and has all its epimorphisms admissible. Also we have that $\hat{\mathbf{X}} = \mathbf{mod}\text{-}\Lambda$ and that **X**-resol.dim $\hat{\mathbf{X}} = d < \infty$.

As in Example 4, we let ω consist of all Λ-modules isomorphic to $\mathrm{Hom}_R(P,R)$ for some finitely generated projective Λ^{op}–module P. Again, ω is just the additive closure of $\omega_{\Lambda/R} = \mathrm{Hom}_R(\Lambda,R)$, and it is an injective cogenerator for **X**.

Applying the results in section 5, we have the following.

Proposition 6.1. Let C be in $\mathbf{mod}\text{-}\Lambda$.

(a) inj.dim $C < \infty$ if and only if C is in $\check{\omega}$.

(b) If C is in $\hat{\omega}$ then inj.dim $C \leq d$.

Proof: See Corollary 5.3. ■

As a consequence of this we obtain hence the following.

Corollary 6.2. Let C be in $\mathbf{mod}\text{-}\Lambda$. Then $0 \to Y_C \to X_C \to C \to 0$, the **X**–approximation of C, and $0 \to C \to Y^C \to X^C \to 0$, the $\hat{\omega}$-hull of C, have the property that X_C and X^C are maximal Cohen-Macaulay R-modules and that inj.dim $Y_C \leq d-1$ and inj.dim $Y^C \leq d$. ■

We now turn our attention to the question of when $\hat{\omega} = \check{\omega}$ in this context.

Proposition 6.3. The following statements are equivalent for Λ.

(a) $\hat{\omega} = \check{\omega}$.

(b) If X is a Λ^{op}–module which is MCM as an R-module and such that $\mathrm{proj.dim}_{\Lambda^{op}} X < \infty$, then X is a projective Λ^{op}-module.

Proof: We know by Proposition 4.6 that $\check{\omega} = \hat{\omega}$ if and only if an exact sequence $0 \to C \to W_0 \to \dots \to W_n \to 0$ is in ω as soon as each W_i is in ω for $i = 0,\dots n$.
(a) $\Rightarrow$ (b): Suppose $0 \to P_m \to \dots \to P_0 \to X \to 0$ is a projective resolution for a given Λ^{op}-module X which is MCM over R. Then

$$0 \to \mathrm{Hom}_R(X,R) \to \mathrm{Hom}_R(P_0,R) \to \dots \to \mathrm{Hom}_R(P_m,R) \to 0$$

is exact in **mod**-Λ with each $\mathrm{Hom}_R(P_i,R)$ in ω for $i = 0,\dots,m$. By (a) we have that $\mathrm{Hom}_R(X,R)$ is necessarily in ω. Therefore $\mathrm{Hom}_R(X,R) \cong \mathrm{Hom}_R(P,R)$ for some projective Λ^{op}-module P, which then yields $X \cong P$.
(b) $\Rightarrow$ (a): Suppose that X in **mod**-Λ is MCM over R and that $0 \to X \to W_0 \to \dots \to W_m \to 0$ is an exact sequence with W_i in ω for all $i = 0,\dots,m$. Then $0 \to \mathrm{Hom}_R(W_m,R) \to \dots \to \mathrm{Hom}_R(W_0,R) \to \mathrm{Hom}_R(X,R) \to 0$ is exact and $\mathrm{Hom}_R(W_i,R)$ is a projective Λ^{op}-module for each i. The Λ^{op}-module $\mathrm{Hom}_R(X,R)$ is still MCM as an R-module and hence $\mathrm{Hom}_R(X,R) \cong P$ for some projective Λ^{op}-module by our assumption. As MCM's are reflexive, $X \cong \mathrm{Hom}_R(P,R)$ and X is therefore in ω. ■

This proposition gives the following generalization of a result of R. Sharp [Sh].

Corollary 6.4. Suppose Λ is a commutative ring. Then the following are equivalent for a finitely generated Λ-module M.

(a) $\mathrm{inj.dim}_\Lambda\ M < \infty$.

(b) There is an exact sequence $0 \to W_m \to \dots \to W_0 \to M \to 0$ with W_i in ω for all $i = 0,\dots,m$.

Proof: The equivalence of (a) and (b) is nothing more than the statement that $\check{\omega} = \hat{\omega}$. But this follows from Proposition 6.3 since it is well-known for commutative rings, that a maximal Cohen-Macaulay module of finite projective dimension is projective. ■

However, if Λ is not commutative, it is not necessarily true that a Λ^{op}-module which is MCM over R and of finite projective dimension over Λ^{op} is necessarily projective. For example, let R be a regular local ring of dimension $d > 0$ and let Λ be the algebra of lower triangular $n \times n$ matrices over R with $n \geq 2$. Then Λ is a free and finitely generated R-module and $\mathrm{gl.dim}\ \Lambda^{op} = d+1$. Let

$$0 \to P_{d+1} \to P_d \to \dots \to P_0 \to M \to 0$$

be a projective Λ^{op}-resolution of a Λ^{op}-module M with $\mathrm{proj.dim}_{\Lambda^{op}}\ M = d+1$. Then $\mathrm{Im}(P_d \to P_{d-1})$ is an MCM over R which is of projective dimension one over Λ^{op} and is hence not Λ^{op}-projective.

Example 8. Let k be a field and $P = k[x_0,\dots,x_n]$ a polynomial ring over k in n+1 variables which we grade by assigning arbitrary positive integral weights to the variables. Let I be a homogeneous ideal in P and set $S = P/I$ which is hence a positively graded k-algebra.

We assume that S is a Cohen-Macaulay ring. It is known then that there exists a sequence $y_1, \dots, y_m$ of homogeneous elements of strictly positive degrees in I which is a regular $k[x_0, \dots, x_n]$-sequence, and such that $R = P/(y_1, \dots, y_m)$ has the same dimension as S. As R is a complete intersection, it is a Gorenstein ring and by construction the natural surjection $R \to S \to 0$ is a degree-preserving homomorphism of rings. Let **C** = **S-grmod** be the category of finitely generated graded S-modules with degree zero graded maps as morphisms. Also let **X** be the subcategory of **C** consisting of all maximal Cohen-Macaulay modules. In addition to the usual properties, **X** also satisfies $\hat{\mathbf{X}} = \mathbf{C}$ and $\mathbf{X}$-resol.dim $\hat{\mathbf{X}} = n+1-m = d$, the dimension of S.

Set $\omega_{S/R} = \mathrm{Hom}_R(S, R)$, which is a dualizing module of S over R, and define ω to be the subcategory of **C** consisting of all $\omega_{S/R}(n)$ for n in **Z**. Then ω is an injective cogenerator for **X**. Moreover we know that X in **X** is of finite projective dimension if and only if it is isomorphic to a direct sum $\bigoplus_i S(a_i)$.

As in the previous example, this implies $\hat{\omega} = \check{\omega}$. We leave it to the reader to write down in detail what this means for **X**-approximations, $\hat{\omega}$-hulls and modules of finite injective dimension.

We now give our last example.

Example 9. Let $\Lambda \to \Gamma$ be a ring homomorphism with Λ both left and right noetherian and Γ a finitely generated projective Λ-module on both left and right. Let **C** = Γ-**mod** be the category of all finitely generated left Γ-modules and let **X** consist of all M in **C** such that M is a projective Λ-module. In addition to the usual properties, we have that $\hat{\mathbf{X}}$ consists of all N in **C** such that $\mathrm{proj.dim}_\Lambda\, N < \infty$.

Define ω to be the category of all modules isomorphic to $\mathrm{Hom}_\Lambda(P, \Lambda)$ for some finitely generated projective Γ^{op}-module P. Then ω is an injective cogenerator for **X**. In general $\hat{\omega} \neq \check{\omega}$, but if all the modules in ω are projective Γ-modules, then we do have $\hat{\omega} = \check{\omega}$ by Corollary 4.8 and in fact **X** becomes a Frobenius category, see Example 6.

References

[Ausl] Auslander,M.: Isolated singularities and existence of almost split sequences. In: Representation Theory II; Lecture Notes **1178**, p. 194-241, Springer Verlag, New York, 1986.

[A-B] Auslander,M.; Bridger,M.: "Stable Module Theory", Mem. of the AMS **94**, AMS Providence, 1969.

[BBD] Beilinson,A.A.; Bernstein,J.; Deligne,D.: "Faisceaux pervers", astèrisque **100**, (1984).

[FGR] Fossum,R.; Griffith,P.A.; Reiten,I.: "Trivial Extensions of Abelian Categories", Springer Lecture Notes in Math. **456**, Springer Verlag, New York 1975.

[Ha] Happel,D.: "Triangulated Categories in Representation Theory of Finite Dimensional Algebras", Habilitationsschrift, Bielefeld 1985.

[He] Heller,A.: The loop-space functor in homological algebra, Trans. AMS **96**, p. 382-394, (1960).

[Qu] Quillen,D.: Higher algebraic K-theory I, in: Algebraic K-theory I, p. 85-147, Lecture Notes **341**, Springer Verlag, New York, 1973.

[Sh] Sharp,R.Y.: Finitely generated modules of finite injective dimension over certain Cohen-Macaulay rings, Proc. London Math. Soc. **25**, p. 303-328, (1972).

[SGA] Séminaire de Géométrie Algébrique IV and 4½; Springer Lecture Notes in Math. **269**, **270**, **305** and **569**, Springer Verlag, New York, 1970 ff.

Reprinted from JOURNAL OF ALGEBRA
All Rights Reserved by Academic Press, New York and London

Vol. 156, No. 2, April 15, 1993
Printed in Belgium

Liftings and Weak Liftings of Modules

MAURICE AUSLANDER, SONGQING DING, AND ØYVIND SOLBERG

Department of Mathematics, Brandeis University, Waltham, Massachusetts 02254

Communicated by Melvin Hochster

Received April 9, 1990

INTRODUCTION

Throughout this paper we assume that all modules are finitely generated (left) modules over a noetherian R-algebra Λ for a commutative local ring R. By a noetherian R-algebra Λ, we mean that Λ is finitely generated as an R-module and that R is a noetherian ring. We denote by mod Λ the category of all finitely generated Λ-modules.

This paper is devoted to studying liftings and weak liftings of modules, which are defined in the following way. Let $\Lambda \to \Gamma$ be a homomorphism of rings and let M be in mod Γ. Then L in mod Λ is called a *lifting* of M to Λ if (a) $M \simeq \Gamma \otimes_\Lambda L$ and (b) $\mathrm{Tor}_i^\Lambda(\Gamma, L) = (0)$ for all $i \geq 1$. The Γ-module M is said to be *liftable* to Λ, when such a Λ-module L exists. If M is only a direct summand of $\Gamma \otimes_\Lambda L$ for some Λ-module L satisfying (b), then L is a *weak lifting of* M to Λ and M is said to be *weakly liftable* to Λ. The notion of a lifting of a module has existed for some time, but the notion of a weak lifting of a module is introduced in this paper. A notion of a lifting has been studied in modular representation theory. Let $\Lambda = RG$ where R is a complete discrete valuation ring with maximal ideal generated by a prime number p and G is a finite group such that p divides the order of G. Assume that the characteristic of R is zero and that the characteristic of the residue class field $R/(p)$ is p. Then a lifting of a module M in mod $\Lambda/p\Lambda$ is a Λ-lattice L such that $M \simeq L/pL$. It is not hard to see that this notion of a lifting is a special case of the general definition given above and many of the results in this paper are in fact inspired by the work of J. A. Green in [9]. An important part of this paper is to introduce and study the properties of weak liftings. The notion of a weak lifting of a module not only is a natural generalization of the notion of a lifting of a module, but we show that notion of weak liftability of a module corresponds to that the module has an infinitesimal deformation, a notion occurring in deformation theory.

0021-8693/93 $5.00

Let Λ be a noetherina R-algebra where R is a commutative complete local ring with maximal ideal $\underline{m}$. Let $\{x_1, x_2, ..., x_t\}$ be a Λ-regular sequence in $\underline{m}$ and $\Gamma = \Lambda/(x_1, x_2, ..., x_t)\Lambda$. The main aim is to prove that if $\mathrm{Ext}^2_\Gamma(M, M) = (0)$ for M in mod Γ, then M is liftable to Λ. This is done by first considering the lifting problem in the situations $\Lambda \to \Lambda/(x) = \Lambda_1$ and $\Lambda_i = \Lambda/(x^i) \to \Lambda_1$, where x is a Λ-regular element in $\underline{m}$. In this situation we characterize Λ_1-modules liftable to Λ in terms of a lifting sequence and obtain homological obstructions for lifting a module. In particular, we show what seems to be folklore to deformation theorists, namely, that if $\mathrm{Ext}^2_{\Lambda_1}(M, M) = (0)$, then M is liftable to Λ (see [11]). We end this section by giving some applications of our main result to module theory. For these applications assume that Λ has finite global dimension. Then we show that if $\mathrm{Ext}^2_\Gamma(M, M) = (0)$ for a Γ-module M, then $pd_\Gamma M < \infty$. Also, if $\mathrm{Ext}^i_\Gamma(M \oplus \Gamma,\ M \oplus \Gamma) = (0)$ for all $i \geqslant 1$ and M in mod Γ, then M is projective. Examples of such rings Γ where these results apply are complete intersections and therefore also group rings of finite abelian groups. Since we consider liftings in the situations $\Lambda \to \Lambda_1$, $\Lambda_i \to \Lambda_1$, and $\Lambda \to \Lambda/(x_1, x_2, ..., x_t)\Lambda$, this justifies the generality in the definition of a lifting of a module.

The setting in Section 2 is the same as in Section 1. Here, the main object is to study realizations of liftings of a Λ_1-module M and when it is unique up to isomorphism. In particular, we show that every lifting of M to Λ is a factor module of $\Omega_\Lambda(M)$, where $\Omega_\Lambda(M)$ denotes the first syzygy of M over Λ given by the projective cover of M over Λ. Further, we prove the following result which seems to be folklore to deformation theorists. If $\mathrm{Ext}^1_{\Lambda_1}(M, M) = (0)$ for a Λ_1-module M liftable to Λ, then the lifting to Λ of M is unique up to isomorphism (see [11]). This result is also generalized to the situation $\Lambda \to \Gamma$, where $\Gamma = \Lambda/(x_1, x_2, ..., x_t)\Lambda$ for a Λ-regular sequence $\{x_1, x_2, ..., x_t\}$ in $\underline{m}$. Finally, we show that every Λ_1-module liftable to Λ has a unique lifting to Λ up to isomorphism if and only if $x \cdot \mathrm{Ext}^1_\Lambda(L, L') = (0)$ for all Λ-modules L and L' for which x is regular and every weakly Λ-liftable Λ_2-lifting of a Λ_1-module liftable to Λ is unique up to isomorphism.

In Section 3 we introduce the notion of a weak lifting of a module. Here we assume that Λ is a noetherian R-algebra over a commutative local ring R (not necessarily complete) with maximal ideal $\underline{m}$. Let $\{x_1, x_2, ..., x_t\}$ be a Λ-regular sequence of central elements in Λ and denote by I the ideal $(x_1, x_2, ..., x_t)\Lambda$ in Λ. Let $\Gamma = \Lambda/I\Lambda$ and $\Gamma_2 = \Lambda/I^2\Lambda$. Then the aim is to characterize Γ-modules weakly liftable to Λ. In fact, we show that the following are equivalent for a Γ-module M, (a) the module M is weakly liftable to Λ, (b) the module M is liftable to Γ_2, and (c) the module M is isomorphic to a direct summand of $\Omega'_\Lambda(M)/I\Omega'_\Lambda(M) \oplus Q/IQ$ for some projective Λ-module Q and for any given projective resolution defining a

tth syzygy $\Omega^t_\Lambda(M)$. Since an infinitesimal deformation of a Γ-module is a lifting to Γ_2 (see [11]), weak liftability of a Γ-module corresponds to that the module has an infinitesimal deformation. Hence, if M in mod Γ has an infinitesimal deformation and gl.dim $\Lambda < \infty$, then $\mathrm{pd}_\Gamma M < \infty$. We also show if M in mod Γ has an infinitesimal deformation, then the completion of M with respect to the topology induced by the $\underline{m}$-adic topology also has an infinitesimal deformation as a module over the completion of Γ. It is not obvious that the notions of weakly liftable and liftable are different, but we give some examples showing that they are indeed different.

Section 4 is devoted to studying liftings and weak liftings in the situation $R \to R/(x) = \bar{R}$, where R is a commutative local Gorenstein ring and x an R-regular element in the maximal ideal of R. Let CM(R) denote the category of all finitely generated maximal Cohen–Macaulay modules over R. Here, we only consider (weak) liftability of maximal Cohen–Macaulay modules over $\bar{R}$. One of the main aims in this section is to characterize liftable and weakly liftable modules in CM($\bar{R}$). We show that every lifting of a module C in $CM(\bar{R})$ is a submodule of the minimal Cohen–Macaulay approximation X_C of C over R and use this to characterize the liftable modules. Further, we prove that a module C in CM($\bar{R}$) is weakly liftable to R if and only if $X_c/xX_C \simeq C \oplus \Omega^{-1}_{\bar{R}}(C)$. We prove that the category of liftable and also the category of weakly liftable modules to R in CM($\bar{R}$) are closed under syzygies, cosyzygies, and taking duals. Finally, we show that the category of weakly liftable modules to R in CM($\bar{R}$) is functorially finite in mod $\bar{R}$ (see [4] or Section 4 for definition of a functorially finite subcategory).

Also in the last section we study weak liftings over a commutative local Gorenstein ring R, but now we consider all the finitely generated modules over $\bar{R} = R/(x)$ for an R-regular element x in the maximal ideal of R. We use the definition of a weak lifting of a module given in Section 3 and this section is mainly devoted to studying the properties of weakly liftable $\bar{R}$-modules in terms of their minimal Cohen–Macaulay approximation over R.

1. Lifting

Throughout this section let Λ be a noetherian R-algebra over a commutative complete local ring R and let $\underline{m}$ denote the maximal ideal in R. Let $\{x_1, x_2, ..., x_t\}$ be a Λ-regular sequence in $\underline{m}$ and $\Gamma = \Lambda/(x_1, x_2, ..., x_t)\Lambda$. This section is devoted to developing a obstruction theory for lifting modules over Γ to Λ.

We start by recalling the definition of a lifting of a module over arbitary rings.

DEFINITION. Let $\Delta \to \Sigma$ be a homomorphism of rings and let M be in mod Σ. Then L in mod Δ is called a *lifting* of M to Δ if the following two conditions are satisfied

(a) $M \simeq \Sigma \otimes_\Delta L$;

(b) $\operatorname{Tor}_i^\Delta(\Sigma, L) = (0)$ for all $i > 0$.

The Σ-module M is said to be *liftable* to Δ, when such a Δ-module L exists.

Remark. Let Λ be an R-order, where R is a discrete valuation ring and $0 \neq x \in \underline{m}$. If $\Gamma = \Lambda/x\Lambda$, then it is easy to see that $\operatorname{Tor}_i^\Lambda(\Gamma, L) = (0)$ for $i > 0$ for a Λ-module L if and only if x is L-regular. Since x is L-regular if and only if L is a free R-module (L is a Λ-lattice), the generalized notion of lifting specializes to the usual notion of lifting.

From now on let Λ be a noetherian R-algebra, where R is a commutative complete local ring and x a Λ-regular element, although some of the results hold true in a more general setting. Since Λ is a finitely generated R-module, Λ is complete with respect to the $\underline{m}$-adic topology and therefore also complete with respect to the topology induced by any proper ideal of R. Let $\Lambda_i = \Lambda/x^i\Lambda$ for all $i \geqslant 1$. We will now restrict ourselves to studying the lifting problem in the following cases, $\Lambda \to \Lambda_1$ and $\Lambda_i \to \Lambda_1$.

Our first aim is to show that if $\operatorname{Ext}^2_{\Lambda_1}(M, M) = (0)$ for a Λ_1-module M, then M is liftable to Λ. The proof consists of several steps. First we characterize liftable Λ_1-modules in terms of a lifting sequence and then give obstructions for lifting in terms of elements in $\operatorname{Ext}^2_{\Lambda_1}(M, M)$.

In order to prove the characterization of liftable Λ_1-modules in terms of a lifting sequence we need the following lemma.

LEMMA 1.1. (a) *For M in* mod Λ *we have*

$$\operatorname{Tor}_k^\Lambda(\Lambda_1, M) = \begin{cases} \{m \in M \mid xm = 0\}, & k = 1 \\ 0, & k > 1 \end{cases}.$$

(b) *For M in* mod Λ_i *we have*

$$\operatorname{Tor}_k^{\Lambda_i}(\Lambda_1, M) = \begin{cases} m \in M \mid xm = 0\}/x^{i-1} M, & k \textit{ odd} \\ \{m \in M \mid x^{i-1}m = 0\}/xM, & k \textit{ even} \end{cases}.$$

Proof. Using that $0 \to \Lambda \xrightarrow{x} \Lambda \to \Lambda_1 \to 0$ is free resolution of Λ_1 over Λ the proof of (a) is easy. Similarly, observing that

$$\cdots \Lambda_i \xrightarrow{x^{i-1}} \Lambda_i \xrightarrow{x} \Lambda_i \xrightarrow{x^{i-1}} \Lambda_i \xrightarrow{x} \Lambda_i \longrightarrow \Lambda_1 \longrightarrow 0$$

is a free resolution of Λ_1 over Λ_i the statement in (b) follows easily.

Let M in mod Λ_1 be liftable to Λ with lifting L and let $L_n = L/x^nL$. It is easy to see that L_n in mod Λ_n is a lifting of L_{n-1} to Λ_n for all $n \geqslant 2$. In fact, such a sequence $L_1 = M, L_2, L_3, \ldots$ characterizes liftable Λ-modules.

THEOREM 1.2. *The following are equivalent for M in* mod Λ_1.

(a) *M is liftable to Λ.*

(b) *There exists a sequence $L_1 \simeq M, L_2, L_3, \ldots$ of Λ-modules such that L_{n+1} in* mod Λ_{n+1} *is a lifting of L_n in* mod Λ_n *to Λ_{n+1} for all $n \geqslant 1$.*

Proof. The proof of the fact that (a) implies (b) is already given and this implication is true even in a more general setting.

Assume that there is a sequence $L_1 \simeq M, L_2, L_3, \ldots$ of Λ-modules such that L_{n+1} in mod Λ_{n+1} is a lifting of L_n in mod Λ_n to Λ_{n+1} for all $n \geqslant 1$. Since L_{n+1} is a lifting of L_n to Λ_{n+1}, it is not hard to see that L_{n+1} is a lifting of M to Λ_{n+1}. Then it follows from Lemma 1.1(b) that we have the commutative diagram

$$\begin{array}{ccccccccc}
0 & \longrightarrow & L_{n+1}/x^nL_{n+1} & \xrightarrow{\bar{x}} & L_{n+1} & \longrightarrow & M & \longrightarrow & 0 \\
 & & \downarrow{\scriptstyle \bar{p}_n^{n+1}} & & \downarrow{\scriptstyle p_n^{n+1}} & & \| & & , \\
0 & \longrightarrow & L_n/x^{n-1}L_n & \xrightarrow{\bar{x}} & L_n & \longrightarrow & M & \longrightarrow & 0(*)
\end{array}$$

where the maps p_n^{n+1} and $\bar{p}_n^{n+1}$ are the epimorphisms induced by the epimorphism $L_{n+1} \to \Lambda_n \otimes_{\Lambda_{n+1}} L_{n+1} \simeq L_{n+1}/x^nL_{n+1} \simeq L_n$. Then the modules L_n and the maps $p_n^{n+1}\colon L_{n+1} \to L_n$ define an inverse system. Similarly we have that $L_n/x^{n-1}L_n$ and $\bar{p}_n^{n+1}\colon L_{n+1}/x^nL_{n+1} \to L_n/x^{n-1}L_n$ define an inverse system and it is not hard to see that they are isomorphic. Since $\{L_n, p_n^m\}$ is a surjective inverse system, the sequence $(*)$ remains exact when passing to the inverse limit by [5, Prop. 10.2, p. 104]. Let L denote $\varprojlim_n L_n$. We obtain the exact sequence of Λ-modules

$$0 \longrightarrow L \xrightarrow{x} L \longrightarrow M \longrightarrow 0,$$

since evidently multiplication by x is the unique Λ-homomorphism commuting with the maps $L_{n-1} \xrightarrow{x} L_n$. According to the definition of a lifting, L is a lifting of M to Λ if L is a finitely generated Λ-module, which we will prove next.

We could use the same method of proof as above replacing M with L_n and x with x^n and obtain the following exact sequence $0 \to L \xrightarrow{x^n} L \to L_n \to 0$. Hence, we have $L/x^nL \simeq L_n$. Since M is a finitely generated Λ_1-module, there exists an epimorphism $\pi\colon \Lambda^t \to M$ for a finite t. We then have the following commutative diagram:

$$
\begin{array}{ccccc}
 & & \Lambda^t & & \\
 & \swarrow^{\varphi} & \downarrow{\scriptstyle \pi} & & \\
L & \longrightarrow & M & \longrightarrow & 0. \\
\downarrow{\scriptstyle \alpha_n} & & \| & & \\
L_n & \longrightarrow & M & \longrightarrow & 0
\end{array}
$$

Using that L_n is a finitely generated R-module and the Nakayama Lemma, it follows that $\varphi_n = \alpha_n \circ \varphi$ is an epimorphism for every n. Therefore the induced map $\bar{\varphi}_n \colon \Lambda^t/x^n\Lambda^t \to L_n$ is also an epimorphism for every n. These maps induce the following commutative diagram:

$$
\begin{array}{ccccccccc}
 & & & & 0 & & 0 & & \\
 & & & & \downarrow & & \downarrow & & \\
 & & & & x^{n-1}\Lambda^t/x^n\Lambda^t & \xrightarrow{s_n} & x^{n-1}L_n & & \\
 & & & & \downarrow & & \downarrow & & \\
0 & \longrightarrow & K_n & \longrightarrow & \Lambda^t/x^n\Lambda^t & \xrightarrow{\bar{\varphi}_n} & L_n & \longrightarrow & 0. \\
 & & \downarrow{\scriptstyle r^n_{n-1}} & & \downarrow{\scriptstyle t^n_{n-1}} & & \downarrow{\scriptstyle p^n_{n-1}} & & \\
0 & \longrightarrow & K_{n-1} & \longrightarrow & \Lambda^t/x^{n-1}\Lambda^t & \xrightarrow{\bar{\varphi}_{n-1}} & L_{n-1} & \longrightarrow & 0 \\
 & & & & \downarrow & & \downarrow & & \\
 & & & & 0 & & 0 & &
\end{array}
$$

Since $\bar{\varphi}_n$ is an epimorphism, it follows easily that s_n is an epimorphism and therefore by the Snake Lemma r^n_{n-1} is also an epimorphism. The exact sequences above define an exact sequence of inverse systems $0 \to \{K_n, r^m_n\} \to \{\Lambda^t/x^n\Lambda^t, t^m_n\} \to \{L_n, p^m_n\} \to 0$, where $\{K_n, r^m_n\}$ is a surjective inverse system. Again, by [5, Prop. 10.2, p. 104] the sequence remains exact passing to the inverse limit and we get an epimorphism $\varprojlim_n \Lambda^t/x^n\Lambda^t = \Lambda^t \to L$, since Λ is complete with respect to the x-adic topology. Hence, L is a finitely generated Λ-module and therefore L is a lifting of M to Λ.

Theorem 1.2 gives us a possible way to construct liftings of M in mod Λ_1 by constructing a lifting sequence $L_1 \simeq M$, $L_2, L_3, \ldots$. When considering such a lifting sequence the following observation is useful. A module L in mod Λ_i is a lifting of M in mod Λ_1 to Λ_i by Lemma 1.1(b) if and only if (a) $M \simeq L/xL$, (b) $\{l \in L \mid xl = 0\} = x^{i-1}L$ and (c) $\{l \in L \mid x^{i-1}l = 0\} = xL$. It is easy to verify that the following lemma is true.

LEMMA 1.3. *For L in* mod Λ_i, *the set* $\{l \in L \mid x^{i-1}l = 0\}$ *equals* xL *if and only if* $\{l \in L \mid xl = 0\}$ *equals* $x^{i-1}L$.

Now we will describe obstructions for constructing such a lifting sequence mentioned above in terms of elements in $\mathrm{Ext}^2_{\Lambda_1}(M, M)$. Assume that L in mod Λ_i is a lifting of M in mod Λ_1 to Λ_i. The exact sequence $0 \to \Omega_\Lambda(L) \to P \to L \to 0$ induces the following exact sequence by tensoring it by $\Lambda_1 \otimes_\Lambda -$

$$0 \to \mathrm{Tor}_1^\Lambda(\Lambda_1, L) \to \Omega_\Lambda(L)/x\Omega_\Lambda(L) \to P/xP \to M \to 0.$$

Since L is a lifting of M to Λ_i, then $\{l \in L \mid x^{i-1}l = 0\} = xL$ and $\{l \in L \mid x^l = 0\} = x^{i-1}L$ by Lemma 1.1. Since $\mathrm{Tor}_1^\Lambda(\Lambda_1, L) = \{l \in L \mid xl = 0\}$ by Lemma 1.1 and L is a Λ_i-lifting of M, it follows easily that $\mathrm{Tor}_1^\Lambda(\Lambda_1, L) = x^{i-1}L \simeq M$. Hence, we have the element in $\mathrm{Ext}^2_{\Lambda_1}(M, M)$,

$$0 \to M \to \Omega_\Lambda(L)/\Omega_\Lambda(L) \to P/xP \to M \to 0,$$

which we will denote by θ_L. Since $\mathrm{Ext}^2_{\Lambda_1}(M, M) \simeq \mathrm{Ext}^1_{\Lambda_1}(\Omega_{\Lambda_1}(M), M)$ and $P/xP \twoheadrightarrow M$ is a projective cover, the element θ_L corresponds to the element, θ'_L, $0 \to M \to \Omega_\Lambda(L)/x\Omega_\Lambda(L) \to \Omega_{\Lambda_1}(M) \to 0$ in $\mathrm{Ext}^1_{\Lambda_1}(\Omega_{\Lambda_1}(M), M)$. Whether θ_L (or θ'_L) is zero or not will give an obstruction for lifting L to Λ_{i+1}. But to prove this we need the following lemma.

LEMMA 1.4. *Let L' in* mod Λ_i *be a lifting of M to Λ_i. A module L in* mod Λ_{i+1} *is lifting of L' to Λ_{i+1} if and only if there is an exact sequence*

$$0 \to M \to L \to L' \to 0,$$

where $L/xL \simeq M$.

Proof. Assume that L in mod Λ_{i+1} is a lifting of L' in mod Λ_i to Λ_{i+1}. Since $\mathrm{Tor}^{\Lambda_{i+1}}_{2n+1}(\Lambda_i, L) = \{l \in L \mid x^i l = 0\}/xL$ and $\mathrm{Tor}^{\Lambda_{i+1}}_{2n}(\Lambda_i, L) = \{l \in L \mid xl = 0\}/x^iL$ and L is a Λ_{i+1}-lifting of L', we have the following exact sequences $0 \to xL \to L \xrightarrow{x^i} x^iL \to 0$ and $0 \to x^iL \to L \xrightarrow{x} xL \to 0$. Hence, $xL \simeq L/x^iL \simeq L'$ and $x^iL \simeq L/xL \simeq L'/xL' \simeq M$, so we have an exact sequence

$$0 \to M \to L \to L' \to 0,$$

where $L/xL \simeq M$.

Assume that there is an exact sequence $0 \to M \xrightarrow{g} L \xrightarrow{f} L' \to 0$ where $L/xL \simeq M$ and L' in mod Λ_i is a lifting of M to Λ_i. Then we have the following commutative diagram:

$$
\begin{array}{ccccccccc}
 & & 0 & & 0 & & & & \\
 & & \downarrow & & \downarrow & & & & \\
0 & \longrightarrow & \ker f\,|_{xL} & \longrightarrow & M & & & & \\
 & & \downarrow & & \downarrow{\scriptstyle g} & & & & \\
0 & \longrightarrow & xL & \longrightarrow & L & \longrightarrow & L/xL & \longrightarrow & 0. \\
 & & \downarrow{\scriptstyle f} & & \downarrow{\scriptstyle f} & & \downarrow{\scriptstyle \bar{f}} & & \\
0 & \longrightarrow & xL' & \longrightarrow & L' & \longrightarrow & L'/xL' & \longrightarrow & 0 \\
 & & \downarrow & & \downarrow & & \downarrow & & \\
 & & 0 & & 0 & & 0 & &
\end{array}
$$

Since $\bar{f}\colon M \simeq L/xL \to L'/xL' \simeq M$ is a surjective endomorphism of a noetherian module, $\bar{f}$ is an isomorphism. We want to show that $x^iL = \{l \in L \mid xl = 0\}$. Assume that $f(xl) = xf(l) = 0$. Since L' is a Λ_i-lifting of M, $f(l) = x^{i-1}l_1$ for some $l_1$in L'. There exists l' in L such that $f(l') = l_1$. Then $f(l - x^{i-1}l') = 0$ and therefore $l - x^{i-1}l' = g(m)$ for some $m \in M$. By multiplying this equality with x, we have $xl = x^il'$ and therefore ker $f\,|_{xL} \subset x^iL$. Since $x^iL \subset \ker f\,|_{xL}$, we have $\ker f\,|_{xL} = x^iL = \operatorname{Im} g \simeq M$. Now, consider the following commutative diagram:

$$
\begin{array}{ccccccccc}
 & & 0 & & 0 & & 0 & & \\
 & & \downarrow & & \downarrow & & \downarrow & & \\
0 & \longrightarrow & M & \longrightarrow & \{l \in L \mid xl = 0\} & \longrightarrow & \{l \in L' \mid xl' = 0\} & & \\
 & & \| & & \downarrow & & \downarrow & & \\
0 & \longrightarrow & M & \xrightarrow{\;g\;} & L & \xrightarrow{\;f\;} & L' & \longrightarrow & 0 \\
 & & \downarrow{\scriptstyle 0} & & \downarrow{\scriptstyle x} & & \downarrow{\scriptstyle x} & & \\
0 & \longrightarrow & M & \longrightarrow & xL & \xrightarrow{\;f\;} & xL' & \longrightarrow & 0 \\
 & & \| & & \downarrow & & \downarrow & & \\
 & & M & \longrightarrow & 0 & & 0 & & \\
 & & \downarrow & & & & & & \\
 & & 0 & & & & & &
\end{array}
$$

Since L' is a Λ_i-lifting of M, we have that $\{l \in L' \mid xl = 0\} = x^{i-1}L' \simeq M$. By the Snake Lemma the connecting homomorphism $\partial\colon M \simeq \{l \in L' \mid xl = 0\}$

$\to M$ is a surjective endomorphism of the noetherian module M, so ∂ is an isomorphism. Hence, $\{l \in L \mid xl = 0\} = \operatorname{Im} g = x^i L$ and by Lemma 1.3 $\operatorname{Tor}_k^{\Lambda_{i+1}}(\Lambda_i, L) = (0)$ for all $k > 0$. Since $\operatorname{Im} g = x^i L$, we have $L/x^i L \simeq L'$ and L is a lifting of L' to Λ_{i+1}.

Now we are ready to prove that whether θ_L is zero or not for a Λ_i-lifting L of M gives an obstruction for lifting L to Λ_{i+1}.

PROPOSITION 1.5. *Let M be in* mod Λ_1 *and assume that L in* mod Λ_i *is a lifting of M to Λ_i. Then the following are equivalent*:

(a) $\theta_L = 0$ *in* $\operatorname{Ext}^2_{\Lambda_1}(M, M)$.

(b) $\Omega_\Lambda(L)/x\Omega_\Lambda(L) \simeq M \oplus \Omega_{\Lambda_1}(M)$.

(c) L *is liftable to* Λ_{i+1}.

Proof. We have already remarked that the element θ_L in $\operatorname{Ext}^2_{\Lambda_1}(M, M)$ corresponds to the element $\theta'_L : 0 \to M \to \Omega_\Lambda(L)/x\Omega_\Lambda(L) \to \Omega_{\Lambda_1}(M) \to 0$ in $\operatorname{Ext}^1_{\Lambda_1}(\Omega_{\Lambda_1}(M), M)$. The element $\theta_L = 0$ if and only if $\theta'_L = 0$, that is, if and only if $0 \to M \xrightarrow{\alpha} \Omega_\Lambda(L)/x\Omega_\Lambda(L) \to \Omega_{\Lambda_1}(M) \to 0$ is a split exact sequence. Since α is a split monomorphism if and only if $\Omega_\Lambda(L)/x\Omega_\Lambda(L) \simeq M \oplus \Omega_{\Lambda_1}(M)$ by [12], we have shown that (a) and (b) are equivalent.

(b)$\Rightarrow$(c). Assume that the non-zero map $\alpha : M \to \Omega_\Lambda(L)/x\Omega_\Lambda(L)$ in θ'_L is a split monomorphism with splitting $\beta : \Omega_\Lambda(L)/x\Omega_\Lambda(L) \to M$. Let γ be the composition $\Omega_\Lambda(L) \to \Omega_\Lambda(L)/x\Omega_\Lambda(L) \xrightarrow{\beta} M$. Then we have the commutative diagram

$$\begin{array}{ccccccccc} & 0 & \longrightarrow & \Omega_\Lambda(L) & \longrightarrow & P & \longrightarrow & L & \longrightarrow & 0 \\ & & & \downarrow{\scriptstyle\gamma} & & \downarrow & & \| & & \\ \eta: & 0 & \longrightarrow & M & \longrightarrow & E & \longrightarrow & L & \longrightarrow & 0 \end{array}$$

and applying the functor $\Lambda_1 \otimes_\Lambda -$ the following diagram is induced:

$$\begin{array}{ccccccccccc} \theta_L: & 0 & \longrightarrow & M & \xrightarrow{\alpha} & \Omega_\Lambda(L)/x\Omega_\Lambda(L) & \longrightarrow & P/xP & \longrightarrow & M & \longrightarrow & 0 \\ & & & \downarrow{\scriptstyle s} & & \downarrow{\scriptstyle\beta} & & \downarrow & & \| & & \\ & & & M & \xrightarrow{t} & M & \longrightarrow & E/xE & \longrightarrow & M & \longrightarrow & 0 \end{array}.$$

Since $\beta \circ \alpha = \mathrm{id}_M = t \circ s$, the map t must be an epimorphism and $E/xE \simeq M$. Since the exact sequence η satisfies all the requirements in Lemma 1.4, we have that E in mod Λ_{i+1} is a lifting of L to Λ_{i+1}.

(c)$\Rightarrow$(*a*). Assume that E in mod Λ_{i+1} is a lifting of L to Λ_{i+1}. By Lemma 1.4 we have the commutative diagram

$$\begin{array}{ccccccccc} 0 & \longrightarrow & \Omega_\Lambda(L) & \longrightarrow & P & \longrightarrow & L & \longrightarrow & 0 \\ & & \downarrow & & \downarrow & & \| & & \\ 0 & \longrightarrow & M & \longrightarrow & E & \xrightarrow{f} & L & \longrightarrow & 0 \end{array},$$

where $E/xE \simeq M$. Applying the functor $\Lambda_1 \otimes_\Lambda -$ to the diagram above induces the following commutative diagram:

$$\begin{array}{ccccccccc} \theta_L\colon 0 \longrightarrow & M & \xrightarrow{\alpha} & \Omega_\Lambda(L)/x\Omega_\Lambda(L) & \longrightarrow & P/xP & \longrightarrow & M & \longrightarrow 0 \\ & \| & & \downarrow{\scriptstyle\beta} & & \downarrow & & \| & \\ & M & \xrightarrow{\delta} & M & \xrightarrow{0} & E/xE & \xrightarrow{\bar{f}} & M & \longrightarrow 0 \end{array}.$$

The map $\bar{f}\colon E/xE \to M$ induced by f is a surjective endomorphism of a noetherian module. Therefore $\bar{f}$ is an isomorphism and for the same reason the map δ is an isomorphism. Since $\delta = \beta \circ \alpha$, the map α is a split monomorphism and therefore $\theta_L = 0$ in $\mathrm{Ext}^2_{\Lambda_1}(M, M)$.

Combining Theorem 1.2 and Proposition 1.5 we reach our first aim, namely, the following sufficient condition for liftability (see also [11]).

PROPOSITION 1.6. *Let M be in* mod Λ_1. *If* $\mathrm{Ext}^2_{\Lambda_1}(M, M) = (0)$, *then M is liftable to Λ.*

One immediate consequence of this result is that every Λ_1-module of projective dimension less or equal to 1 is liftable to Λ. Another observation is that the converse of this result is *not* true, due to the following counterexample.

Let G be a cyclic group of prime order p and R a discrete valuation ring with maximal ideal generated by p. If $\Lambda = RG$ and $x = p$, then x is a Λ-regular element, the ring Λ is isomorphic to $R[t]/(t^p - 1)$ and $\Lambda_1 = \Lambda/x\Lambda = R/pR[t]/(t-1)^p$. Let $S_i = R/pR[t]/(t-1)^i$ for $i = 1, 2, ..., p$, which are all the indecomposable Λ_1-modules. It is shown in [15; see Section 2] that S_1, S_{p-1} and S_p are liftable to Λ. Since $\Omega_{\Lambda_1}(S_1) = S_{p-1}$ and $0 \to S_1 \to \Lambda_1 \to \Omega_{\Lambda_1}(S_1) \to 0$ is a non-split exact sequence, $\mathrm{Ext}^2_{\Lambda_1}(S_1, S_1) \neq (0)$ but S_1 is liftable to Λ.

Our next aim is to generalize Proposition 1.6 to the situation $\Lambda \to \Lambda/(x_1, x_2, ..., x_t)\,\Lambda = \Gamma$, where $\{x_1, x_2, ..., x_t\}$ is a Λ-regular sequence.

PROPOSITION 1.7. *Let M be in* mod Γ. *If* $\mathrm{Ext}^2_\Gamma(M, M) = (0)$, *then M is liftable to Λ.*

Proof. Let $\Gamma_i = \Lambda/(x_1, x_2, ..., x_i)\,\Lambda$ for $i = 1, ..., t$ and let $\Gamma_0 = \Lambda$. Since $\mathrm{Ext}^2_{\Gamma_t}(M, M) = (0)$ and x_t is regular on Γ_{t-1}, there exists a lifting L_{t-1} of

M to Γ_{t-1} by Proposition 1.6. Applying $\operatorname{Hom}_{\Gamma_{t-1}}(L_{t-1},\ )$ to the sequence $0 \to L_{t-1} \xrightarrow{x_t} L_{t-1} \to M \to 0$, it induces the following long exact sequence:

$$\cdots \longrightarrow \operatorname{Ext}^2_{\Gamma_{t-1}}(L_{t-1}, L_{t-1}) \xrightarrow{x_t} \operatorname{Ext}^2_{\Gamma_{t-1}}(L_{t-1}, L_{t-1})$$
$$\longrightarrow \operatorname{Ext}^2_{\Gamma_{t-1}}(L, M) \to \cdots .$$

Since L_{t-1} is a lifting of M, we have that $\operatorname{Ext}^i_{\Gamma_t}(M, M) \simeq \operatorname{Ext}^i_{\Gamma_{t-1}}(L_{t-1}, M)$, so $\operatorname{Ext}^2_{\Gamma_{t-1}}(L_{t-1}, M) = (0)$ and $x_t \cdot \operatorname{Ext}^2_{\Gamma_{t-1}}(L_{t-1}, L_{t-1}) = \operatorname{Ext}^2_{\Gamma_{t-1}}(L_{t-1}, L_{t-1})$. Since $\operatorname{Ext}^2_{\Gamma_{t-1}}(L_{t-1}, L_{t-1})$ is a finitely generated R-module and $x_t \in \underline{m}$, we have $\operatorname{Ext}^2_{\Gamma_{t-1}}(L_{t-1}, L_{t-1}) = (0)$ by the Nakayama Lemma. Hence, the module L_{t-1} can be lifted to Γ_{t-2}. Continuing this process, we construct a sequence of modules $L_t = M, L_{t-1}, L_{t-2}, \ldots, L_0$ where L_i is a lifting of L_{i+1} to Γ_i for $i = t-1, t-2, \ldots, 0$. It follows that $\Gamma_t \otimes_\Lambda L_0 \simeq M$, and L_0 is a lifting of M to Λ if $\operatorname{Tor}^\Lambda_i(\Gamma_t, L_0) = (0)$ for all $i \geqslant 1$. Assume that $\operatorname{Tor}^{\Gamma_j}_i(\Gamma_t, L_j) = (0)$ for all $i \geqslant 1$ and $j \geqslant r+1$. Let $\mathbf{P}^\bullet$ be a projective resolution of L_r over Γ_r. Since L_r is a lifting of L_{r+1} to Γ_r, we have that $\Gamma_{r+1} \otimes_{\Gamma_r} \mathbf{P}^\bullet$ is a projective resolution of L_{r+1} over Γ_{r+1}. Further, since $\Gamma_t \otimes_{\Gamma_r} \mathbf{P}^\bullet \simeq \Gamma_t \otimes_{\Gamma_{r+1}} (\Gamma_{r+1} \otimes_{\Gamma_r} \mathbf{P}^\bullet)$, it follows that $\operatorname{Tor}^{\Gamma_r}_i(\Gamma_t, L_r) \simeq \operatorname{Tor}^{\Gamma_{r+1}}_i(\Gamma_t, L_{r+1}) = (0)$ for $i \geqslant 1$. By induction on r we have $\operatorname{Tor}^\Lambda_i(\Gamma_t, L_0) = (0)$ for all $i \geqslant 1$, and L_0 is a lifting of M to Λ.

We end this section by proving some applications of Proposition 1.7 in module theory. Let Λ and Γ be as above, then the following remark follows immediately from the definition of a lifting of a module. If L in mod Λ is a lifting of M in mod Γ and $\mathbf{P}^\bullet$ is a Λ-projective resolution of L, then $\Gamma \otimes_\Lambda \mathbf{P}^\bullet$ is a Γ-projective resolution of M. Then, the following proposition is an immediate consequence of this remark and Proposition 1.7.

PROPOSITION 1.8. *Assume that* gl. dim $\Lambda < \infty$. *If* $\operatorname{Ext}^2_\Gamma(M, M) = (0)$ *for* M *in* mod Γ, *then* $pd_\Gamma M < \infty$.

The second and last application is connected to the Nakayama Conjecture: If Λ is a finite dimensional algebra over a field k and dom.dim $\Lambda = \infty$, that is, in a minimal injective resolution

$$0 \to \Lambda \to E_0 \to E_1 \to \cdots \to E_i \to \cdots$$

for Λ, all the E_i are projective, then Λ is selfinjective. In [3], Auslander and Reiten made the following conjecture: If Λ is an artin algebra and M is a Λ-generator with $\operatorname{Ext}^i_\Lambda(M, M) = (0)$ for all $i \geqslant 1$, then M is projective. Then they showed that the Nakayama Conjecture holds if their conjecture holds for all artin algebras. The conjecture of Auslander and Reiten mentioned above makes sense for any ring. We consider this conjecture in our setting and it is not hard to see that it is equivalent to the following.

If Λ is a noetherian R-algebra over a complete commutative local ring R and $\operatorname{Ext}^i_\Lambda(M \oplus \Lambda, M \oplus \Lambda) = (0)$ for all $i \geqslant 1$, then M is projective. Then, the following proposition is an easy consequence of Proposition 1.8.

PROPOSITION 1.9. *Let Λ be a noetherian R-algebra of finite global dimension over a complete commutative local ring R and let $\Gamma = \Lambda/(x_1, x_2, ..., x_t)\Lambda$, where $\{x_1, x_2, ..., x_t\}$ is a Λ-sequence in the maximal ideal of R. If $\operatorname{Ext}^i_\Gamma(M \oplus \Gamma, M \oplus \Gamma) = (0)$ for M in $\operatorname{mod} \Gamma$ for all $i \geqslant 1$, then M is projective.*

Proof. Assume that $\operatorname{Ext}^i_\Gamma(M \oplus \Gamma, M \oplus \Gamma) = (0)$ for M in $\operatorname{mod} \Gamma$ for all $i \geqslant 1$. By Proposition 1.8 $\operatorname{pd}_\Gamma M < \infty$, since $\operatorname{gl.dim} \Lambda < \infty$ and $\operatorname{Ext}^2_\Gamma(M \oplus \Gamma, M \oplus \Gamma) = (0)$. Since $\operatorname{Ext}^i_\Gamma(M, \Gamma) = (0)$ for all $i \geqslant 1$, it follows that M is projective.

This raises the question whether all artin algebras Γ are a factor of a noetherian R-algebra Λ of finite global dimension over a complete commutative local ring R of the form $\Gamma = \Lambda/(x_1, x_2, ..., x_t)\Lambda$ where $\{x_1, x_2, ..., x_t\}$ is a Λ-regular sequence in the maximal ideal of R. All complete intersections and therefore every group ring kG where G is a finite abelian group, k a field and the characteristic of k divides the order of G are of this form. But the following example suggested by J. Alperin shows that not all group rings kG are such a factor. Let k be a field of characteristic two and let G be the semi-direct product of a (normal) elementary abelian 2-group of order eight extended by a group of order seven which acts faithfully on the abelian 2-group. Let $\Gamma = kG$ and k the trivial Γ-module. Then it is known that $\operatorname{Ext}^2_\Gamma(k, k) = (0)$ and that $\operatorname{pd}_\Gamma k = \infty$. By Proposition 1.8 it follows that Γ is not of the form $\Lambda/(x_1, x_2, ..., x_t)\Lambda$ for any noetherian R-algebra Λ of finite global dimension where R is a complete commutative local ring and $\{x_1, x_2, ..., x_t\}$ is a Λ-regular sequence in the maximal ideal of R.

2. REALIZATION AND UNIQUENESS OF LIFTINGS

Throughout this section Λ will be a noetherian R-algebra over a commutative complete local ring R with maximal ideal $\underline{m}$. Let x be a Λ-regular element in $\underline{m}$ and $\{x_1, x_2, ..., x_t\}$ a Λ-regular sequence in $\underline{m}$. We will also in this section restrict our attention to studying the lifting problem in the following cases, $\Lambda \to \Lambda/x\Lambda = \Lambda_1$, $\Lambda_i = \Lambda/x^i\Lambda \to \Lambda_1$, and $\Lambda \to \Gamma$, where $\Gamma = \Lambda/(x_1, x_2, ..., x_t)\Lambda$. Throughout this section a regular element or a regular sequence on a Λ-module is always assumed to be elements in $\underline{m}$.

This section is devoted to showing that every lifting of a liftable Λ_1-module M is a factor module of $\Omega_\Lambda(M)$ and to discussing which properties the category of Λ_1-modules liftable to Λ or Λ_i has. Here, $\Omega_\Lambda(M)$ denotes the first syzygy of M over Λ given by the projective cover of M over Λ. We also show that if M in mod Γ is liftable to Λ and $\operatorname{Ext}^1_\Gamma(M, M) = (0)$, then the lifting is unique up to isomorphism. This is done by first showing that if M in mod Λ_1 is liftable and $\operatorname{Ext}^1_{\Lambda_1}(M, M) = (0)$, then the lifting is unique up to isomorphism (see [11]). Finally, we discuss when every liftable Λ_1-module has a unique lifting up to isomorphism. In the following a module M is said to have a unique lifting or to be uniquely liftable if M has a unique lifting up to isomorphism.

The aim now is to show that every lifting of a liftable Λ_1-module M is a factor module of $\Omega_\Lambda(M)$. This will lead us to a characterization of a liftable Λ_1-module M in terms of existence of special submodules of $\Omega_\Lambda(M)$.

PROPOSITION 2.1. *If M in* mod Λ_1 *is liftable to Λ, then every lifting of M is a factor module of $\Omega_\Lambda(M)$.*

Proof. Assume that M in mod Λ_1 is liftable to Λ with lifting L. Then we have the following commutative diagram:

$$\begin{array}{ccccccccc}
 & & & & & & 0 & & \\
 & & & & & & \downarrow & & \\
 & & 0 & & 0 & & L & & \\
 & & \downarrow & & \downarrow & & \downarrow{\scriptstyle x} & & \\
0 & \longrightarrow & \Omega_\Lambda(L) & \longrightarrow & P & \longrightarrow & L & \longrightarrow & 0\\
 & & \downarrow & & \| & & \downarrow & & \\
0 & \longrightarrow & \Omega_\Lambda(M) & \longrightarrow & P & \longrightarrow & M & \longrightarrow & 0\\
 & & \downarrow & & \downarrow & & \downarrow & & \\
 & & L & & 0 & & 0 & & \\
 & & \downarrow & & & & & & \\
 & & 0 & & & & & &
\end{array}.$$

It follows immediately from the Snake Lemma that we have an exact sequence $0 \to \Omega_\Lambda(L) \to \Omega_\Lambda(M) \to L \to 0$, hence L is a factor module of $\Omega_\Lambda(M)$.

Before we characterize a liftable Λ_1-module M in terms of existence of special submodules of $\Omega_\Lambda(M)$, we need to recall some notions from

Section 1. The element $\theta_M: 0 \to M \to \Omega_\Lambda(M)/x\Omega_\Lambda(M) \to P/xP \to M \to 0$ in $\mathrm{Ext}^2_{\Lambda_1}(M, M)$ is induced by tensoring the exact sequence $0 \to \Omega_\Lambda(M) \to P \to M \to 0$ by $\Lambda_1 \otimes_\Lambda -$. The element $\theta'_M: 0 \to M \to \Omega_\Lambda(M)/x\Omega_\Lambda(M) \to \Omega_{\Lambda_1}(M) \to 0$ is the element corresponding to the element θ_M by the isomorphism $\mathrm{Ext}^2_{\Lambda_1}(M, M) \simeq \mathrm{Ext}^1_{\Lambda_1}(\Omega_{\Lambda_1}, (M), M)$.

Proposition 2.2. *A module M in* $\mathrm{mod}\,\Lambda_1$ *is liftable to Λ if and only if there exists $N \subset \Omega_\Lambda(M)$ such that x is a $\Omega_\Lambda(M)/N$-regular element and $\pi(N) \hookrightarrow \Omega_\Lambda(M)/x\Omega_\Lambda(M) \xrightarrow{\beta} \Omega_{\Lambda_i}(M)$ is an isomorphism, where $\pi: \Omega_\Lambda(M) \to \Omega_\Lambda(M)/x\Omega_\Lambda(M)$ is the natural epimorphism and β is the map $\Omega_\Lambda(M)/x\Omega_\Lambda(M) \to \Omega_{\Lambda_1}(M)$ in the exact sequence θ'_M.*

Proof. We will now only prove the "only if"-part, because the "if"-part will be stated as Lemma 2.3 for later use.

Assume that M is liftable to Λ with lifting L. Then we have seen that we have the following commutative diagram:

$$
\begin{array}{ccccccccc}
 & & & & & & 0 & & \\
 & & & & & & \downarrow & & \\
 & & 0 & & 0 & & L & & \\
 & & \downarrow & & \downarrow & & \downarrow{\scriptstyle x} & & \\
0 & \longrightarrow & \Omega_\Lambda(L) & \longrightarrow & P & \longrightarrow & L & \longrightarrow & 0 \\
 & & \downarrow & & \| & & \downarrow & & \\
0 & \longrightarrow & \Omega_\Lambda(M) & \longrightarrow & P & \longrightarrow & M & \longrightarrow & 0 \\
 & & \downarrow & & \downarrow & & \downarrow & & \\
 & & L & & 0 & & 0 & & \\
 & & \downarrow & & & & & & \\
 & & 0 & & & & & &
\end{array}
$$

Let $N = \Omega_\Lambda(L)$, then x is a $\Omega_\Lambda(M)/N$-regular element. Applying the functor $\Lambda_1 \otimes_\Lambda -$ to the above diagram the following commutative diagram is induced:

$$
\begin{array}{cccccccccc}
 & 0 & \longrightarrow & & N/xN & \longrightarrow & P/xP & \longrightarrow & L/xL & \longrightarrow 0 \\
 & & & & \downarrow{\scriptstyle t} & & \| & & \| & \\
\theta_M: & 0 & \longrightarrow & M \longrightarrow & \Omega_\Lambda(M)/x\Omega_\Lambda(M) & \longrightarrow & P/xP & \longrightarrow & M & \longrightarrow 0
\end{array}
$$

Hence, $N/xN \simeq \Omega_{\Lambda_1}(M)$ and it is easy to see that $\beta \circ t$ is an isomorphism. From the commutative diagram

$$\begin{array}{ccccc} & & N & \hookrightarrow & \Omega_\Lambda(M) \\ & & \downarrow & & \downarrow{\scriptstyle \pi} \\ 0 & \longrightarrow & N/xN & \xrightarrow{t} & \Omega_\Lambda(M)/x\Omega_\Lambda(M) \end{array}$$

we see that $\pi(N) = t(N/xN)$ and therefore the composition

$$\pi(N) \hookrightarrow \Omega_\Lambda(M)/x\Omega_\Lambda(M) \xrightarrow{\beta} \Omega_{\Lambda_1}(M)$$

is an isomorphism.

The following lemma completes the proof of Proposition 2.2 and the lemma will be used in Section 4.

LEMMA 2.3. *Let E in* mod Λ *be such that x is an E-regular element and let $0 \to A \xrightarrow{\alpha} E/xE \xrightarrow{\beta} B \to 0$ be an exact sequence in* mod Λ_1. *Let $\pi: E \to E/xE$ be the natural epimorphism. If there exists $N \subset E$, such that x is an E/N-regular element and the composition $\pi(N) \hookrightarrow E/xE \xrightarrow{\beta} B$ is an isomorphism, then N is a lifting of B and E/N is a lifting of A.*

Proof. Applying the functor $\Lambda_1 \otimes_\Lambda -$ to the sequence $0 \to N \xrightarrow{i} E \to L \to 0$ induces the following commutative diagram where $L = E/N$:

$$\begin{array}{ccccccccc} & & N & \xrightarrow{i} & E & & & & \\ & & \downarrow & & \downarrow{\scriptstyle \pi} & & & & \\ 0 & \longrightarrow & N/xN & \xrightarrow{\bar{i}} & E/xE & \longrightarrow & L/xL & \longrightarrow & 0 \end{array}.$$

Since $N/xN \simeq \operatorname{Im} \bar{i} = \operatorname{Im}(\pi \circ i) = \pi(N)$ and the composition $\pi(N) \hookrightarrow E/xE \xrightarrow{\beta} B$ is an isomorphism, $N/xN \simeq B$ and the map $\bar{i}$ is a split monomorphism. Hence, $E/xE \simeq N/xN \oplus L/xL \simeq B \oplus A$. Since Λ_1 is a Krull–Schmidt ring and $N/xN \simeq B$, it follows that $L/xL \simeq A$. This shows that N is a lifting of B and L is a lifting of A.

Our next aim is to study the category of Λ_1-modules liftable to Λ or Λ_i. First, we show that the category of Λ_1-modules liftable to Λ or Λ_i is closed under syzygies.

PROPOSITION 2.4. *If M in* mod Λ_1 *is liftable to $\Lambda(\Lambda_i)$, then $\Omega_{\Lambda_1}(M)$ is also liftable to Λ (Λ_i). Moreover, if L is a lifting of M to Λ (Λ_i), then $\Omega_\Lambda(L)$ $(\Omega_{\Lambda_i}(L))$ is a lifting of $\Omega_{\Lambda_1}(M)$ to Λ (Λ_i).*

Proof. We only prove the result for Λ, since the proof for Λ_i is similar. Assume that M is liftable to Λ with lifting L. Tensoring the commutative diagram

$$\begin{array}{ccccccccc} 0 & \longrightarrow & \Omega_\Lambda(L) & \longrightarrow & P & \longrightarrow & L & \longrightarrow & 0 \\ & & \downarrow & & \| & & \downarrow & & \\ 0 & \longrightarrow & \Omega_\Lambda(M) & \longrightarrow & P & \longrightarrow & M & \longrightarrow & 0 \end{array}$$

with $\Lambda_1 \otimes_\Lambda -$ the following commutative diagram is induced:

$$\begin{array}{ccccccccc} 0 & \longrightarrow & \Omega_\Lambda(L)/x\Omega_\Lambda(L) & \longrightarrow & P/xP & \longrightarrow & L/xL & \longrightarrow & 0 \\ & & \downarrow & & \| & & \| & & \\ 0 & \longrightarrow & \Omega_{\Lambda_1}(M) & \longrightarrow & P/xP & \longrightarrow & M & \longrightarrow & 0 \end{array} .$$

Hence $\Omega_\Lambda(L)/x\Omega_\Lambda(L) \simeq \Omega_{\Lambda_1}(M)$ and since $\Omega_\Lambda(L)$ is a submodule of P, the module $\Omega_\Lambda(L)$ is a lifting of $\Omega_{\Lambda_1}(M)$ to Λ.

As mentioned in Section 1 the lifting problem has been studied for group rings over a ring of p-adic integers in [15]. Because of Proposition 2.4 it may be tempting to think that the category of Λ_1-module liftable to Λ (or Λ_i) is closed under kernels of epimorphisms, cokernels of monomorphisms and summands, but using the results in [15] it is easy to see the following.

(1) The category of Λ_1-modules liftable to Λ (or Λ_i) is not closed under kernels of epimorphisms or cokernels of monomorphisms.

(2) The category of Λ_1-modules liftable to Λ (or Λ_i) is not closed under extensions.

(3) The category of Λ_1-modules liftable to Λ is not closed under summands.

In the rest of this section we will study uniqueness of liftings. Let $\Gamma = \Lambda/(x_1, x_2, ..., x_t)\Lambda$ where $\{x_1, x_2, ..., x_t\}$ is a Λ-regular sequence. The first aim is to show that if M in $\operatorname{mod}\Gamma$ is liftable to Λ and $\operatorname{Ext}^1_\Gamma(M, M) = (0)$, then the lifting of M is unique. In order to prove this sufficient conditions for unique liftability we first consider the problem in the situation $\Lambda \to \Lambda_1$.

PROPOSITION 2.5. *If M in* $\operatorname{mod}\Lambda_1$ *is liftable and* $\operatorname{Ext}^1_{\Lambda_1}(M, M) = (0)$, *then the lifting of M to Λ is unique.*

Proof. Let L and L' be two liftings of M to Λ. The exact sequence $0 \to L \xrightarrow{x} L \to M \to 0$ induces the following long exact sequence:

$$\begin{aligned} 0 \longrightarrow \operatorname{Hom}_\Lambda(L', L) &\xrightarrow{x} \operatorname{Hom}_\Lambda(L', L) \longrightarrow \operatorname{Hom}_\Lambda(L', M) \\ \longrightarrow \operatorname{Ext}^1_\Lambda(L', L) &\xrightarrow{x} \operatorname{Ext}^1_\Lambda(L', L) \longrightarrow \operatorname{Ext}^1_\Lambda(L', M) \end{aligned}$$

Since L' is a lifting of M to Λ, we have $\operatorname{Ext}^1_\Lambda(L', M) \simeq \operatorname{Ext}^1_{\Lambda_1}(M, M) = (0)$ and therefore $\operatorname{Ext}^1_\Lambda(M', L) = x \cdot \operatorname{Ext}^1_\Lambda(L', L)$. Since $\operatorname{Ext}^1_\Lambda(L', L)$ is a finitely generated R-module and $x \in \underline{m}$, the Nakayama Lemma implies that $\operatorname{Ext}^1_\Lambda(L', L) = (0)$. Hence, the following sequence is exact:

$$0 \longrightarrow \operatorname{Hom}_\Lambda(L', L) \xrightarrow{x} \operatorname{Hom}_\Lambda(L', L) \longrightarrow \operatorname{Hom}_\Lambda(L', M) \longrightarrow 0.$$

This implies that we have the commutative diagram

$$\begin{array}{ccccccccc} 0 & \longrightarrow & L' & \xrightarrow{x} & L' & \longrightarrow & M & \longrightarrow & 0 \\ & & \downarrow & & \downarrow{\scriptstyle f} & & \| & & \\ 0 & \longrightarrow & L & \xrightarrow{x} & L & \longrightarrow & M & \longrightarrow & 0, \\ & & \downarrow & & \downarrow{\scriptstyle g} & & \| & & \\ 0 & \longrightarrow & L' & \xrightarrow{x} & L' & \longrightarrow & M & \longrightarrow & 0 \end{array}$$

where $\Lambda_1 \otimes_\Lambda g \circ f \colon L'/xL' \to L'/xL'$ is an isomorphism. Let $h = g \circ f$. Then the composition $L' \xrightarrow{h} L' \xrightarrow{\pi} L'/xL'$ is an epimorphism and therefore $xL'/\operatorname{Im} h = L'/\operatorname{Im} h$. Hence, h is an epimorphism by the Nakayama Lemma and h is an isomorphism since it is a surjective endomorphism of L'. Similarly, we can show that there are maps $f' \colon L \to L'$ and $g' \colon L' \to L$ such that the composition $g' \circ f'$ is an isomorphism. Since Λ has the Krull–Schmidt property, we have shown that L and L' are isomorphic.

Remark. The converse of this statement is not true. Let $R = \mathbf{Z}_{(3)}$ be the ring of 3-adic integers, G a cyclic group of order 3, $x = 3$ and $\Lambda = RG$. Then $\Lambda \simeq R[t]/(t^3 - 1)$ and $\Lambda_1 = (R/3R[t])/(t-1)^3 = \mathbf{Z}_3[t]/(t-1)^3$. Let $M = \mathbf{Z}_3[t]/(t-1)$, which is liftable to Λ and a lifting is $L = R[t]/(t-1)$. By [15, Lemma 2.1] every lifting of M is isomorphic to $R[t]/(Q)$, where Q is a monic polynomial of degree 1 dividing $t^3 - 1$ over R. Since $R[t]$ is a UFD, $t^3 - 1 = (t-1) \cdot (t^2 + t + 1)$ and $t - 1$ and $t^2 + t + 1$ are irreducible elements in $R[t]$, every lifting of M to Λ is isomorphic to L. But even though the lifting of M to Λ is unique $\operatorname{Ext}^1_{\Lambda_1}(M, M) \neq (0)$, since $0 \to M \to \mathbf{Z}_3[t]/(t-1)^2 \to M \to 0$ is a non-split exact sequence.

Let Γ denote $\Lambda/(x_1, x_2, ..., x_t)\Lambda$, where $\{x_1, x_2, ..., x_t\}$ is a Λ-regular sequence.

PROPOSITION 2.6. *If M in* mod Γ *is liftable and* $\operatorname{Ext}^1_\Gamma(M, M) = (0)$, *then the lifting of M to Λ is unique.*

Proof. Let L and L' in mod Λ be two liftings of M in mod Γ. Denote by Γ_i the R-algebra $\Lambda/(x_1, x_2, ..., x_i)\Lambda$ for all $i = 1, 2, ..., t$, by Γ_0 the R-algebra Λ and by L_r the Γ_r-module $\Gamma_r \otimes_\Lambda L$. We want to show that it is enough to show that L_r is a lifting of L_{r+1} to Γ_r for all $r = 0, 1, ..., t-1$ and similarly for L'. Then, since L_{t-1} and L'_{t-1} both are liftings of M to Γ_{t-1}, we have that $L_{t-1} \simeq L'_{t-1}$ by Proposition 2.5. Using the same method of proof as in Proposition 1.7 we can show that $\operatorname{Ext}^1_{\Gamma_{t-1}}(L_{t-1}, L_{t-1}) = (0)$ and similarly for L'_{t-1}. Continuing this process it follows that $L \simeq L'$.

It is clear that $\Gamma_{r+1} \otimes_{\Gamma_r} L_r \simeq L_{r+1}$, so we need to prove that $\operatorname{Tor}_i^{\Gamma_r}(\Gamma_{r+1}, L_r) = (0)$ for all $i \geqslant 1$. Applying the functor $-\otimes_\Lambda L$ to the exact sequence $0 \to \Gamma_{t-1} \xrightarrow{x_t} \Gamma_{t-1} \to \Gamma_t = \Gamma \to 0$, we get the following long exact sequence:

$$\cdots \longrightarrow \operatorname{Tor}_i^\Lambda(\Gamma_{t-1}, L) \xrightarrow{x_t} \operatorname{Tor}_i^\Lambda(_{t-1}, L) \longrightarrow \operatorname{Tor}_i^\Lambda(\Gamma_t, L) \longrightarrow \cdots .$$

Since L is a lifting of M to Λ, we have that $\operatorname{Tor}_i^\Lambda(\Gamma_t, L) = (0)$ for all $i \geqslant 1$. Hence, it follows that $\operatorname{Tor}_i^\Lambda(\Gamma_{t-1}, L) = x_t \cdot \operatorname{Tor}_i^\Lambda(\Gamma_{t-1}, L)$ and therefore $\operatorname{Tor}_i^\Lambda(\Gamma_{t-1}, L) = (0)$ for all $i \geqslant 1$ by the Nakayama Lemma since $x_t \in \underline{m}$. By induction we can show that $\operatorname{Tor}_i^\Lambda(\Gamma_r, L) = (0)$ for all $i \geqslant 1$ and for all $r = 1, 2, ..., t$. If $\mathbf{P}^\cdot$ is a projective resolution of L over Λ, then $\Gamma_r \otimes_\Lambda \mathbf{P}^\cdot$ is a projective resolution of L_r over Γ_r, since $\operatorname{Tor}_i^\Lambda(\Gamma_r, L) = (0)$ for all $i \geqslant 1$. Since $\Gamma_t \otimes_\Lambda \mathbf{P}^\cdot \simeq \Gamma_t \otimes_{\Gamma_r} (\Gamma_r \otimes_\Lambda \mathbf{P}^\cdot)$, we have that $\operatorname{Tor}_i^{\Gamma_r}(\Gamma_t, L_r) \simeq \operatorname{Tor}_i^\Lambda(\Gamma_t, L) = (0)$ for all $i \geqslant 1$. Similar as for L we show that $\operatorname{Tor}_i^{\Gamma_r}(\Gamma_{r+1}, L_r) = (0)$ for all $i \geqslant 1$ and for all $r = 0, 1, ..., t-1$. Hence, we have shown that L_r is a lifting of L_{r+1} to Γ_r.

If Λ_1 is a semisimple ring, then Proposition 2.5 immediately implies that every lifting to Λ is unique. We end this section by characterizing when every liftable Λ_1-module has an unique lifting. First, we give a necessary condition.

PROPOSITION 2.7. *If every liftable Λ_1-module has a unique lifting to Λ, then* $x \cdot \operatorname{Ext}^1_\Lambda(A, \ \ ) = (0)$ *for every A in* mod Λ *for which x is an A-regular element.*

Proof. Assume that every liftable Λ_1-module has a unique lifting to Λ. Let L be in mod Λ where x is a L-regular element and let $M = L/xL$. Since M is liftable to Λ and in particular liftable to Λ_2, we have that $\Omega_\Lambda(M)/x\Omega_\Lambda(M) \simeq M \oplus \Omega_{\Lambda_1}(M)$ by Proposition 1.5 and by Proposition 2.4 we have that $L \oplus \Omega_\Lambda(L)$ is a lifting of $\Omega_\Lambda(M)/x\Omega_\Lambda(M)$ to Λ. Since $\Omega_\Lambda(M)$ is clearly also a lifting of $\Omega_\Lambda(M)/x\Omega_\Lambda(M)$ and every lifting is unique, $\Omega_\Lambda(M) \simeq L \oplus \Omega_\Lambda(L)$. The sequence $0 \to L \xrightarrow{x} L \xrightarrow{\pi} M \to 0$ induces the long exact sequence

$$\begin{array}{lccccc}
 & \cdots \longrightarrow \operatorname{Ext}^1_\Lambda(L,\ ) & \xrightarrow{x} & \operatorname{Ext}^1_\Lambda(L,\ ) & \xrightarrow{\partial} & \operatorname{Ext}^2_\Lambda(M,\ ) \\
 & & & \downarrow{\scriptstyle\gamma} & & \downarrow{\scriptstyle\varphi_M} \\
\theta: & 0 & \longrightarrow & \operatorname{Ext}^1_\Lambda(L,\ ) & \longrightarrow & \operatorname{Ext}^1_\Lambda(\Omega_\Lambda(M),\ ) \\
 & & & & \xrightarrow{\operatorname{Ext}^2_\Lambda(\pi,\ )} & \operatorname{Ext}^2_\Lambda(L,\ ) \longrightarrow \cdots \\
 & & & & & \downarrow{\scriptstyle\varphi_L} \\
 & & & & \xrightarrow{\operatorname{Ext}^1_\Lambda(\Omega_\Lambda(\pi),\ )} & \operatorname{Ext}^1(\Omega_\Lambda(L),\ ) \longrightarrow 0
\end{array},$$

where the natural isomorphisms φ_M and φ_L are induced from the following commutative diagram:

$$\begin{array}{ccccccccc}
 & & & & 0 & & 0 & & \\
 & & & & \downarrow & & \downarrow & & \\
\delta: 0 & \longrightarrow & \Omega_\Lambda(L) & \xrightarrow{\Omega_\Lambda(\pi)} & \Omega_\Lambda(M) & \longrightarrow & L & \longrightarrow & 0 \\
 & & \| & & \downarrow & & \downarrow{\scriptstyle x} & & \\
\eta: 0 & \longrightarrow & \Omega_\Lambda(L) & \longrightarrow & P & \longrightarrow & L & \longrightarrow & 0. \\
 & & & & \downarrow & & \downarrow{\scriptstyle\pi} & & \\
 & & & & M & = & M & & \\
 & & & & \downarrow & & \downarrow & & \\
 & & & & 0 & & 0 & &
\end{array}$$

Since $\Omega_\Lambda(M) \simeq L \oplus \Omega_\Lambda(L)$, the sequence δ is split exact by [12] and therefore the sequence θ is also a split exact sequence. Since $\operatorname{Ext}^2_\Lambda(\pi,\ )$ is an epimorphism, the induced map $\gamma\colon \operatorname{Ext}^1_\Lambda(L,\ ) \to \operatorname{Ext}^1_\Lambda(L,\ )$ is an epimorphism by the Snake Lemma. Since γ is an surjective endomorphism of a noetherian R-module, γ is an isomorphism and therefore the connecting homomorphism $\partial\colon \operatorname{Ext}^1_\Lambda(L,\ ) \to \operatorname{Ext}^2_\Lambda(M,\ )$ is a monomorphism. Hence, it follows that $x \cdot \operatorname{Ext}^1_\Lambda(L,\ ) = (0)$.

Remarks. (1) The converse statement is not true, due to the following counterexample. Let $R = \mathbf{Z}_{(2)}$ be the ring of 2-adic integers, G a cyclic group of order 2, $x=2$ and $\Lambda = RG$. Then $\Lambda \simeq R[t]/(t^2-1)$, the factor ring $\Lambda_1 = R/2R[t]/(t-1)^2 = \mathbf{Z}_2[t]/(t-1)^2$ and $x \cdot \operatorname{Ext}^1_\Lambda(A, \ ) = (0)$ for every A in mod Λ where x is an Λ-regular element. Let $M = \mathbf{Z}_2[t]/(t-1)$, which is an indecomposable Λ_1-module. Since $E = R[t]/(t+1) = \Lambda(t-1)$ and $D = R[t]/(t-1) = \Lambda(t+1)$ are liftings of M, and E and D are not isomorphic, every lifting with respect to x is not unique.

(2) The condition $x \cdot \operatorname{Ext}^1_\Lambda(A, \ ) = (0)$ for A in mod Λ where x is an Λ-regular element, can be reformulated in the following way (see [10]). The condition that $x \cdot \operatorname{Ext}^1_\Lambda(A, \ ) = (0)$ for A in mod Λ is equivalent to the fact that the map given by multiplication by x, $A \xrightarrow{x} A$, factors through a projective Λ-module.

(3) If we consider the proof of Proposition 2.7 again, we see that if the lifting of a liftable Λ_1-module M and the lifting of $\Omega_\Lambda(M)/x\Omega_\Lambda(M) \simeq M \oplus \Omega_{\Lambda_1}(M)$ to Λ are unique, then $x \cdot \operatorname{Ext}^1_\Lambda(L, \ ) = (0)$ for the lifting L of M to Λ. But, if the lifting of M to Λ is unique it does not imply that $x \cdot \operatorname{Ext}^1_\Lambda(L, \ ) = (0)$ for every lifting L of M to Λ. The following example shows this. Let R be a discrete valuation ring with maximal ideal generated by an element x and let k denote the residue class field $R/(x)$. Let Λ be the ring $\left(\begin{smallmatrix} R & 0 \\ R & R \end{smallmatrix}\right)$, then the element x is a Λ-regular element and $\Lambda_1 = \left(\begin{smallmatrix} k & 0 \\ k & k \end{smallmatrix}\right)$. Let M be the Λ_1-module $\left(\begin{smallmatrix} k & 0 \\ k & 0 \end{smallmatrix}\right)/\left(\begin{smallmatrix} 0 & 0 \\ k & 0 \end{smallmatrix}\right)$ which is a simple injective module. Therefore $\operatorname{Ext}^i_{\Lambda_1}(M, M) = (0)$ for $i = 1, 2$ and M is uniquely liftable to Λ. The Λ-module L given by $\left(\begin{smallmatrix} R & 0 \\ R & 0 \end{smallmatrix}\right)/\left(\begin{smallmatrix} 0 & 0 \\ R & 0 \end{smallmatrix}\right)$ is the lifting of M. Since the map given by multiplication by x, $L \xrightarrow{x} L$, does not factor through a projective Λ-module, $x \cdot \operatorname{Ext}^1_\Lambda(L, \ ) \neq (0)$. So, even if M is uniquely liftable, $x \cdot \operatorname{Ext}^1_\Lambda(L, \ )$ is not equal to (0) for the lifting L of M to Λ. Therefore we can also conclude that even if M in mod Λ is uniquely liftable, then $\Omega_\Lambda(M)/x\Omega_\Lambda(M) \simeq M \oplus \Omega_{\Lambda_1}(M)$ is not necessarily uniquely liftable to Λ. The Λ_1-module $M' = \Omega_{\Lambda_1}(M)$ given by $\left(\begin{smallmatrix} 0 & 0 \\ 0 & k \end{smallmatrix}\right)$ is a simple projective module. Then, $\operatorname{Ext}^i_{\Lambda_i}(M', M') = (0)$ for $i = 1, 2$ and therefore M' is uniquely liftable to Λ. So, even if M and M' are uniquely liftable to Λ, the direct sum $M \oplus M'$ is not uniquely liftable to Λ.

Our characterization of when every liftable Λ_1-module has a unique lifting uses a generalization of the Maranda Theorem [6]. For the convenience of the reader we recall the Maranda Theorem from [6]. Let Σ be an S-order in a finite dimensional separable K-algebra A, where S is a discrete valuation ring with prime element π and quotient ring K. Then there exists a nonnegative integer i_0 such that $\pi^{i_0} \cdot \operatorname{Ext}^1_\Sigma(M, N) = (0)$ for all Σ-lattices M and N [6, Coro. 29.5]. The Maranda Theorem reads as follows.

THEOREM 2.8 [6, Thm. 30.14]. *Let M and N be Σ-lattices and let i be a nonnegative integer. Then the following are true*:

(a) *If* $M \simeq N$ *as* Σ*-modules, then* $M/\pi^i M \simeq N/\pi^i N$ *as* $\Sigma/\pi^i\Sigma$*-modules for each* $i \geqslant 0$.

(b) *If* $M/\pi^i M \simeq N/\pi^i N$ *for some* $i \geqslant i_0 + 1$, *then* $M \simeq N$.

The Maranda Theorem gives examples when every liftable Λ_1-module has a unique lifting to Λ (when Λ is an order over a discrete valuation ring). The following generalization of the Maranda Theorem plays an important role in our characterization of when every liftable Λ_1-module has a unique lifting to Λ.

THEOREM 2.9. *Assume that* $x \cdot \operatorname{Ext}^1_\Lambda(E, E') = (0)$ *for all* Λ*-modules* E *and* E' *where* x *is regular on* E *and* E'. *Let* L *and* L' *be in* mod Λ *for which* x *is regular. If* $L/x^2L \simeq L'/x^2L'$, *then* $L \simeq L'$.

Proof. The proof follows along the lines of the proof for orders in [14, Theorem 1.1]. Let L and L' be Λ-modules where x is regular on L and L', such that $L/x^2L \simeq L'/x^2L'$. Let $\varphi\colon L/x^2L \to L'/x^2L'$ be an isomorphism and $f\colon P \to L'/x^2L'$ a projective cover of L'/x^2L' over Λ. By the Nakayama Lemma and since P is projective, there exists an epimorphism $g\colon P \to L$ such that $\pi \circ g = \varphi^{-1} \circ f$, where $\pi\colon L \to L/x^2L$ is the natural epimorphism. By [10] $x \cdot \operatorname{Ext}^1_\Lambda(L, \ \ ) = (0)$ if and only if $L \xrightarrow{x} L$ factors through a projective Λ-module. Since $L \xrightarrow{x} L$ factors through a projective Λ-module if and only if it factors through a projective cover of L, we have the following commutative diagram:

$$\begin{array}{ccccccc} & & P & = & P & = & P \\ & \overset{s}{\nearrow} & \downarrow & & \downarrow & & \downarrow \\ L & \xrightarrow{x} & L & \xrightarrow{\pi} & L/x^2L & \xrightarrow{\varphi} & L'/x^2L' \end{array}.$$

It follows from this diagram that the map $\psi = \varphi \circ \pi \circ x\colon L \to L'/x^2L'$ factors through a projective, hence there exists a map $\sigma\colon L \to L'$ such that $\rho \circ \sigma = \psi$ where $\rho\colon L' \to L'/x^2L'$ is the natural epimorphism. Since $\operatorname{Im}\psi = xL'/x^2L'$ and $\rho^{-1}(xL'/x^2L') = xL'$, the image $\operatorname{Im}\sigma \subset xL'$. By the Nakayama Lemma, $\operatorname{Im}\sigma = xL'$, since for every $xl' \in xL'$ there exists an $l \in L$ such that $xl' = \sigma(l) - x^2l''$ for some $l'' \in L'$. Hence, $\sigma\colon L \to xL' \simeq L'$ is an epimorphism, since x is regular on L'. Assume that $\sigma(l) = 0$ for some $l \in L$. Since $\psi(l) = \rho \circ \sigma(l) = \varphi(\pi(xl))$ and φ is an isomorphism, $xl = x^2 l_1$ for some $l_1 \in L$. Since x is regular on L, we have $l = xl_1$ and $\sigma(l_1) = 0$. So by continuing this process the element $l \in \bigcap_{i=1}^{\infty} x^i L = (0)$, hence σ is a monomorphism and $L \simeq L'$.

Now the characterization of when every liftable Λ_1-module has a unique lifting to Λ follows easily from Proposition 2.7, Theorem 2.9 and [8, Thm. 2.2].

PROPOSITION 2.10. *Every liftable Λ_1-module has a unique lifting to Λ if and only if the following two conditions are satisfied*:

(a) $x \cdot \operatorname{Ext}^1_\Lambda(L, L') = (0)$ *for all Λ-modules L and L', where x is regular on L and L'.*

(b) *Every weakly Λ-liftable Λ_2-lifting of a Λ_1-module liftable to Λ is unique up to isomorphism.*

Proof. Assume that every liftable Λ_1-module has a unique lifting to Λ. Then by Proposition 2.7 we have $x \cdot \operatorname{Ext}^1_\Lambda(L, \ ) = (0)$ for every Λ-module L where x is regular on L.

Let M be liftable to Λ and let L_1 and L_2 be two Λ_2-liftings of M which are weakly liftable to Λ. Since weak liftability and liftability are the same for Λ_2-modules by [8, Thm. 2.2], the modules L_1 and L_2 are liftable to Λ. Then their liftings to Λ are also liftings of M to Λ and therefore L_1 and L_2 are isomorphic.

Assume that the conditions (a) and (b) are true. The condition (b) implies that if L and L' are two liftings of a Λ_1-module M, then L/x^2L and L'/x^2L' are two Λ_2-liftings of M and therefore $L/x^2L \simeq L'/x^2L'$. By condition (a) and the Maranda Theorem we have that $L \simeq L'$, hence every Λ_1-module liftable to Λ has a unique lifting.

3. WEAK LIFTING

Throughout this section let Λ be a noetherian R-algebra over a commutative local ring R with maximal ideal $\underline{m}$. Let $\{x_1, x_2, ..., x_t\}$ be a Λ-regular sequence of central elements in Λ and denote by Γ the factor ring $\Lambda/(x_1, x_2, ..., x_t)\,\Lambda$ and I the ideal $(x_1, x_2, ..., x_t)\,\Lambda$.

This section is mainly devoted to introducing and characterizing the notion of a weak lifting of a Γ-module. We show that the following are equivalent for M in mod Γ, (a) the module M is weakly liftable to Λ, (b) the module M is liftable to $\Lambda/I^2\Lambda$ and (c) the module M is isomorphic to a direct summand of $\Omega^t_\Lambda(M)/I\Omega^t_\Lambda(M) \oplus Q/IQ$ for a projective Λ-module Q and for any given projective resolution of M defining a tth syzygy $\Omega^t_\Lambda(M)$. We prove this result by first considering the case $t = 1$, which follows quite easily from Proposition 1.5 observing that we never used that R was complete and that the Λ-regular element x was an element in the maximal ideal of R.

Let $\Gamma_i = \Lambda/I^i\Lambda$ for all $i \geqslant 1$. In deformation theory a module L in mod Γ_2 is called an infinitesimal deformation of M in mod Γ if L is a Γ_2-lifting of

M (see [11]). By the result mentioned above a Γ-module has an infinitesimal deformation if and only if M has a weak lifting to Λ. This connection with infinitesimal deformations is one of the reasons for introducing the notion of a weak lifting of a module, but it also seems to be the right class of modules to consider in view of some of the problems studied in [15]. We define the notion of a weak lifting of a module in the same general context as for a lifting of a module in the following way.

DEFINITION. Let $\Delta \to \Sigma$ be a homomorphism rings and let M be in mod Σ. Then, L in mod Δ is called a *weak lifting* of M to Δ if the following two conditions are satisfied

(a) M is a direct summand of $\Sigma \otimes_\Delta L$;

(b) $\operatorname{Tor}_i^\Delta(\Sigma, L) = (0)$ for all $i > 0$.

The Σ-module M is said to be *weakly liftable* to Δ, if it has a weak lifting.

Every Γ-module liftable to Λ is obviously also weakly liftable to Λ. It is not obvious that these two notions of liftability are different, but we see later they indeed are different. An immediate consequence of the definition of weak liftability and our main result mentioned above is the following result.

PROPOSITION 3.1. *Assume that* $\operatorname{gl.dim} \Lambda < \infty$. *If* M *in* $\operatorname{mod} \Gamma$ *has an infinitesimal deformation, then* $\operatorname{pd}_\Gamma M < \infty$.

Examples of algebras Γ where this result applies are complete intersections and therefore also every group ring kG where G is a finite abelian group, k a field and the characteristic of k divides the order of G.

Our first step towards characterizing the weakly liftable Γ-modules is to consider the case $t = 1$. Let x be a central Λ-regular element in Λ and let $\Lambda_i = \Lambda/x^i\Lambda$ for all $i \geqslant 1$. Since we in this section do not assume that R is complete, we do not necessarily have projective covers. For M in mod Λ_1 let $P \to M$ be an epimorphism where P is a projective Λ-module and denote the kernel by $\Omega_\Lambda(M)$. Tensoring the exact sequence $0 \to \Omega_\Lambda(M) \to P \to M \to 0$ with $\Lambda_1 \otimes_\Lambda -$ the following exact sequence is induced:

$$0 \to M \to \Omega_\Lambda(M)/x_1\Omega_\Lambda(M) \to P/x_1P \to M \to 0.$$

This sequence represents an element in $\operatorname{Ext}^2 \Lambda_1(M, M)$ and observing that this element is independent of the choice of an epimorphism $P \to M$, we denote this element by θ_M as in Section 1. Then we have the following characterization of a Λ_1-module M weakly liftable to Λ.

PROPOSITION 3.2. *For M in* mod Λ_1 *the following are equivalent*

(a) *M is weakly liftable to Λ.*

(b) *$\Omega_\Lambda(M)/x\Omega_\Lambda(M) \simeq M \oplus \Omega_{\Lambda_1}(M)$, where $\Omega_{\Lambda_1}(M)$ is induced from the projective resolution defining $\Omega_\Lambda(M)$.*

(c) *M is liftable to Λ_2.*

Proof. Observing that we never used that R was complete and that x was an element in $\underline{m}$ in Proposition 1.5, the statements in (b) and (c) are equivalent by Proposition 1.5. Since (b) obviously implies (a), we only need to prove that (a) implies (b) or (c).

Assume that E is a weak lifting of M to Λ. Then we have the following commutative diagram:

$$\begin{array}{ccccccccc} 0 & \longrightarrow & \Omega_\Lambda(M) & \longrightarrow & P & \longrightarrow & M & \longrightarrow & 0 \\ & & \downarrow & & \downarrow & & \| & & \\ 0 & \longrightarrow & K & \longrightarrow & E & \longrightarrow & M & \longrightarrow & 0 \end{array}.$$

Applying the functor $\Lambda_1 \otimes_\Lambda -$, we obtain the following commutative diagram:

$$\begin{array}{ccccccccccc} \theta_M\colon 0 & \longrightarrow & M & \longrightarrow & \Omega_\Lambda(M)/x\Omega_\Lambda(M) & \longrightarrow & P/xP & \longrightarrow & M & \longrightarrow & 0 \\ & & \| & & \downarrow & & \downarrow & & \| & & \\ \eta\colon 0 & \longrightarrow & M & \longrightarrow & K/xK & \longrightarrow & E/xE & \xrightarrow{s} & M & \longrightarrow & 0 \end{array}.$$

This shows that θ_M and η represent the same element in $\operatorname{Ext}^2_{\Lambda_1}(M, M)$ and since s is a split epimorphism, θ_M is zero in $\operatorname{Ext}^2_{\Lambda_1}(M, M)$. The element θ_M is zero if and only if (b) or (c) is true by Proposition 1.5, hence we have shown that (a) implies (b).

Using this characterization of weakly liftable Λ_1-modules it is not hard to find examples of Λ_1-modules which are weakly liftable but not liftable.

(1) Let $R = \mathbf{Z}_{(3)}$ be the ring or p-adic integers, G a cyclic group of order 27, $x = 3$ and $\Lambda = RG$. Then $\Lambda \simeq R[t]/(t^{27} - 1)$ and $\Lambda_1 = \Lambda/x\Lambda = \mathbf{Z}_3[t]/(t-1)^{27}$. Then it is shown in [15] that the Λ_1-modules $\mathbf{Z}_3[t]/(t-1)^i$ for $i = 4, 5, 22, 23$ are liftable to Λ_2 but not liftable to Λ.

(2) Let S be a commutative local domain of dimension 1 with maximal ideal $\underline{m}$ and not a discrete valuation ring. We want to prove that the residue field $k \simeq S/\underline{m}$ is weakly liftable but not liftable with respect to any element in $\underline{m}\backslash\underline{m}^2$. Let x be in $\underline{m}\backslash\underline{m}^2$ and $S_1 = S/(x)$. Then x is S-regular and we have the exact sequence $0 \to k \xrightarrow{f} \underline{m}/x\underline{m} \to \underline{m}/(x) \to 0$,

where $f(1)=x$, $\underline{m}=\Omega_S(S/\underline{m})$ and $\underline{m}/(x)=\Omega_{S_1}(k)$. Since the composition $k \xrightarrow{f} \underline{m}/x\underline{m} \xrightarrow{\pi} \underline{m}/\underline{m}^2$ is nonzero, there is a map $g: \underline{m}/\underline{m}^2 \to k$ such that $g \circ \pi \circ f = \mathrm{id}_k$. Therefore f is a split monomorphism, so that $\underline{m}/x\underline{m} \simeq k \oplus \underline{m}/(x)$. Hence, k as an S_1-module is weakly liftable to S. It is not hard to see that that k is liftable if and only if S is a discrete valuation ring.

The aim now is to generalize Proposition 3.2 to the situation $\Lambda \to \Lambda/(x_1, x_2, ..., x_t)\,\Lambda = \Gamma$, where $\{x_1, x_2, ..., x_t\}$ is a Λ-regular sequence of central elements in Λ. Denote by I the ideal $(x_1, x_2, ..., x_t)\,\Lambda$ in Λ and by Γ_i the factor ring $\Lambda/I^i\Lambda$ for all $i \geqslant 1$. We want to show that M in mod Γ is weakly liftable to Λ if and only if M is liftable to Γ_2. When considering liftings of M to Γ_2 we must compute $\mathrm{Tor}_i^{\Gamma_i}(\Gamma, E)$ for $i \geqslant 1$ and for Γ_2-modules E. These Tor-groups are found in the following lemma.

LEMMA 3.3. *Let E be in* mod Γ_2. *Then* $\mathrm{Tor}_1^{\Gamma_2}(\Gamma, E) = \{(e_i) \in E^t \mid \sum_{i=1}^t x_i e_i = 0\}/IE^t$ *and* $\mathrm{Tor}_i^{\Gamma_2}(\Gamma, E) = (0)$ *for all* $i \geqslant 1$ *if and only if* $\mathrm{Tor}_1^{\Gamma_2}(\Gamma, E) = (0)$.

Proof. Using the Koszul complex as a free resolution of Γ over Λ and that the maps in the complex are given by multiplication by elements in I, it follows that $\mathrm{Tor}_1^{\Lambda}(\Gamma, \Gamma) = \Gamma^t$. Applying the functor $\Gamma \otimes_\Lambda -$ to the exact sequence $0 \to I \to \Lambda \to \Gamma \to 0$ it follows easily that $I/I^2 \simeq \Gamma^t$. Hence, Γ has the following free resolution $\mathbf{P}^\bullet$ over Γ_2

$$\mathbf{P}^\bullet : \cdots \longrightarrow \Gamma_2^{t^2} \xrightarrow{\oplus_{i=1}^t d} \Gamma_2^t \xrightarrow{(x_1, x_2, ..., x_t) = d} \Gamma_2 \longrightarrow \Gamma \longrightarrow 0.$$

Using this free resolution of Γ over Γ_2 it follows that $\mathrm{Tor}_1^{\Gamma_2}(\Gamma, E) = \{(e_i) \in E^t \mid \sum_{i=1}^t x_i e_i = 0\}/IE^t$. Since $\Omega_{\Gamma_2}^i(\Gamma) \simeq {}^{t^i}$, we have that $\mathrm{Tor}_i^{\Gamma_i}(\Gamma, E) = (0)$ for all $i \geqslant 1$ if and only if $\mathrm{Tor}_1^{\Gamma_2}(\Gamma, E) = (0)$ and the desired results follow directly.

Similar as for the case $t = 1$, we can give the following characterization of a lifting of M in mod Γ to Γ_2.

LEMMA 3.4. *A module E in* mod Γ_2 *is a lifting of M in* mod Γ *to* Γ_2 *if and only if there exists an exact sequence*

$$0 \longrightarrow M^t \longrightarrow E \xrightarrow{f} M \longrightarrow 0,$$

where E is in mod Γ_2 *and the map f induces an isomorphism $E/IE \to M$.*

Proof. Assume that E in mod Γ_2 is a lifting of M to Γ_2. Then we have the following exact sequence $0 \to IE \to E \to M \to 0$ and $\mathrm{Tor}_i^{\Gamma_2}(\Gamma, E) = (0)$

for all $i \geqslant 1$. Applying the functor $\Gamma \otimes_{\Gamma_2} -$ to this sequence induces the exact sequence

$$0 \longrightarrow \operatorname{Tor}_1^{\Gamma_2}(\Gamma, M) \longrightarrow IE \xrightarrow{0} E/IE \longrightarrow M \longrightarrow 0,$$

since $I^2E = (0)$. By Lemma 3.3 we have that $\operatorname{Tor}_1^{\Gamma_2}(\Gamma, M) = M^t$ and hence $IE \simeq M^t$. Therefore there exists an exact sequence $0 \to M^t \to E \xrightarrow{f} M \to 0$, where E is in mod Γ_2 and the map f induces an isomorphism $E/IE \to M$.

Assume that there exists an exact sequence $0 \to M^t \to E \xrightarrow{f} M \to 0$, where E is in mod Γ_2 and the map f induces an isomorphism $E/IE \to M$. In order to show that E is a lifting of M to Γ_2, it is sufficient to show that $\operatorname{Tor}_1^{\Gamma_2}(\Gamma, E) = (0)$ by Lemma 3.3. Consider the commutative diagram

$$\begin{array}{ccccccccc}
 & & 0 & & 0 & & 0 & & \\
 & & \downarrow & & \downarrow & & \downarrow & & \\
0 & \longrightarrow & IE^t & \longrightarrow & K & \xrightarrow{0} & IE & & \\
 & & \| & & \downarrow & & \downarrow & & \\
0 & \longrightarrow & IE^t & \longrightarrow & E^t & \longrightarrow & M^t & \longrightarrow & 0 \\
 & & \downarrow{\scriptstyle 0} & & \downarrow{\scriptstyle d} & & \downarrow{\scriptstyle 0} & & \\
0 & \longrightarrow & IE & \longrightarrow & E & \longrightarrow & M & \longrightarrow & 0 \\
 & & \| & & \downarrow & & \| & & \\
 & & IE & \longrightarrow & M & = & M & \longrightarrow & 0 \\
 & & \downarrow & & \downarrow & & \downarrow & & \\
 & & 0 & & 0 & & 0 & &
\end{array},$$

where $d = (x_1, x_2, ..., x_t)$. By the Snake Lemma we have that $\ker d = IE^t$. Since Γ has the following free resolution $\mathbf{P}^\bullet$ over Γ_2,

$$\mathbf{P}^\bullet : \cdots \longrightarrow \Gamma_2^{t^2} \xrightarrow{\oplus_{i=1}^t d} \Gamma_2^t \xrightarrow{(x_1, x_2, ..., x_t) = d} \Gamma_2 \longrightarrow \Gamma \longrightarrow 0,$$

the middle vertical sequence in the diagram above is the start of the complex $\mathbf{P}^\bullet \otimes_{\Gamma_2} E$. Since $IE^t = \operatorname{Im}((\oplus_{i=1}^t d) \otimes_{\Gamma_2} E) \subseteq \ker d$, we have that $\operatorname{Im}((\oplus_{i=1}^t d) \otimes_{\Gamma_2} E) = \ker d$ and therefore $\operatorname{Tor}_1^{\Gamma_2}(\Gamma, E) = (0)$ and E is a lifting of M to Γ_2.

Let M be in mod Γ. Similar to the case $t = 1$, tensoring the exact sequence $0 \to \Omega_\Lambda(M) \to P \to M \to 0$ with $\Gamma \otimes_\Lambda -$, the following exact sequence is induced:

$$0 \to \operatorname{Tor}_1^\Lambda(\Gamma, M) \to \Omega_\Lambda(M)/I\Omega_\Lambda(M) \to P/IP \to M \to 0.$$

Using the Koszul complex as a free resolution of Γ over Λ and that the maps in the complex are given by multiplication by elements in I, it follows that $\operatorname{Tor}_i^\Lambda(\Gamma, M) = M^{\binom{t}{i}}$ for $i = 1, 2, \ldots, t$ and (0) for $i > t$. Hence, the exact sequence above is

$$0 \to M' \to \Omega_\Lambda(M)/I\Omega_\Lambda(M) \to P/IP \to M \to 0,$$

which represents an element in $\operatorname{Ext}_\Gamma^2(M, M')$. Observing that this element is independent of the choice of an epimorphism $P \to M$, we denote it by θ_M. Now, we show that M in mod Γ is liftable to Γ_2 if and only if $\theta_M = 0$ in $\operatorname{Ext}_\Gamma^2(M, M')$.

Proposition 3.5. *The following are equivalent for M in* mod Γ.

(a) $\theta_M = 0$ *in* $\operatorname{Ext}_\Gamma^2(M, M')$.

(b) *M is liftable to Γ_2.*

Proof. (a)$\Rightarrow$(b). Assume that $\theta_M = 0$ in $\operatorname{Ext}_\Gamma^2(M, M')$. Let $0 \to \Omega_\Lambda(M) \to P \to M \to 0$ be an exact sequence with P a projective Λ-module and denote the kernel of the induced epimorphism $P/IP \to M$ by $\Omega_\Gamma(M)$. Since $\operatorname{Ext}_\Gamma^2(M, M') \simeq \operatorname{Ext}_\Gamma^1(\Omega_\Gamma(M), M')$, the element $\theta_M = 0$ if and only if $\theta'_M : 0 \to M' \xrightarrow{\alpha} \Omega_\Lambda(M)/I\Omega_\Lambda(M) \to \Omega_\Gamma(M) \to 0$ is a split exact sequence. Let β be a splitting of α and let γ be the composition $\Omega_\Lambda(M) \to \Omega_\Lambda(M)/I\Omega_\Lambda(M) \xrightarrow{\beta} M'$. Then we have the commutative diagram

$$\begin{array}{ccccccccc}
0 & \longrightarrow & \Omega_\Lambda(M) & \longrightarrow & P & \longrightarrow & M & \longrightarrow & 0 \\
 & & \downarrow & & \downarrow & & \| & & \\
0 & \longrightarrow & M' & \longrightarrow & E & \longrightarrow & M & \longrightarrow & 0
\end{array}$$

and applying the functor $\Gamma \otimes_\Lambda -$, the following commutative diagram is induced:

$$\begin{array}{ccccccccc}
\theta_M : 0 \longrightarrow & M' & \xrightarrow{\alpha} & \Omega_\Lambda(M)/I\Omega_\Lambda(M) & \longrightarrow & P/IP & \longrightarrow & M & \longrightarrow 0 \\
 & \| & & \downarrow{\scriptstyle \beta} & & \downarrow & & \| & . \\
 & M' & \xrightarrow{r} & M' & \longrightarrow & E/IE & \longrightarrow & M & \longrightarrow 0
\end{array}$$

Since $\beta \circ \alpha = \mathrm{id}_M = r$, we have that $E/IE \simeq M$. Since E is in mod Γ_2, the module E is a lifting of M to Γ_2 by Lemma 3.4.

(b) ⇒ (a). Assume that E is a lifting of M to Γ_2. Applying the functor $\Gamma \otimes_\Lambda -$ to the commutative diagram

$$\begin{array}{ccccccccc}
0 & \longrightarrow & \Omega_\Lambda(M) & \longrightarrow & P & \longrightarrow & M & \longrightarrow & 0 \\
 & & \downarrow & & \downarrow & & \| & & \\
0 & \longrightarrow & M' & \longrightarrow & E & \longrightarrow & M & \longrightarrow & 0
\end{array},$$

the following commutative diagram is induced:

$$\begin{array}{ccccccccccc}
\theta_M\colon 0 & \longrightarrow & M' & \xrightarrow{\alpha} & \Omega_\Lambda(M)/I\Omega_\Lambda(M) & \longrightarrow & P/IP & \longrightarrow & M & \longrightarrow & 0 \\
 & & \| & & \downarrow{\scriptstyle\beta} & & \downarrow & & \| & & \\
 & & M' & \xrightarrow{\delta} & M' & \xrightarrow{0} & E/IE & \xrightarrow{\sim} & M & \longrightarrow & 0
\end{array}.$$

The map $\delta\colon M' \to M'$ is a surjective endomorphism of a noetherian module, so δ is an isomorphism. Since $\delta = \beta \circ \alpha$ is an isomorhism, α is a split monomorphism and therefore $\theta_M = 0$ in $\operatorname{Ext}^2_\Gamma(M, M')$.

For M in mod Γ the obstructions to lifting M to Γ_2 are t elements in $\operatorname{Ext}^2_\Gamma(M, M)$. Similar obstructions have been studied by Nastold in [13]. It follows from the above proof that if M is liftable to Γ_2, then M' and therefore M are direct summands of $\Omega_\Lambda(M)/I\Omega_\Lambda(M)$. But since $\operatorname{Tor}_i^\Lambda(\Gamma, \Omega_\Lambda(M))$ is not necessarily zero for all $i \geqslant 1$ when $t > 1$, we can not conclude from this that M is weakly liftable to Λ as in the case $t = 1$. The next aim is to show that liftability to Γ_2 is equivalent to weak liftability to Λ.

PROPOSITION 3.6. *The following are equivalent for M in* mod Γ.

(a) *M is weakly liftable to Λ.*

(b) *M is isomorphic to a direct summand of $\Omega'_\Lambda(M)/I\Omega'_\Lambda(M) \oplus Q/IQ$ for a projective Λ-module Q and for any given projective resolution defining a tth syzygy $\Omega'_\Lambda(M)$.*

(c) *M is liftable to Γ_2.*

Proof. (a) ⇒ (c). Let L in mod Λ be a weak lifting of M to Λ. Applying the functor $\Gamma \otimes_\Lambda -$ to the commutative diagram

$$\begin{array}{ccccccccc}
0 & \longrightarrow & \Omega_\Lambda(M) & \longrightarrow & P & \longrightarrow & M & \longrightarrow & 0 \\
 & & \downarrow & & \downarrow & & \downarrow & & \\
0 & \longrightarrow & K & \longrightarrow & L & \xrightarrow{f} & M & \longrightarrow & 0
\end{array}$$

induces the following commutative diagram:

$$\begin{array}{ccccccccccc}
\theta_m\colon 0 & \longrightarrow & M' & \longrightarrow & \Omega_\Lambda(M)/I\Omega_\Lambda(M) & \longrightarrow & P/IP & \longrightarrow & M & \longrightarrow & 0 \\
 & & \| & & \downarrow & & \downarrow & & \| & & \\
\eta\colon 0 & \longrightarrow & M' & \longrightarrow & K/IK & \longrightarrow & L/IL & \xrightarrow{\bar f} & M & \longrightarrow & 0
\end{array}.$$

This shows that θ_M and η represent the same element in $\operatorname{Ext}^2_\Gamma(M, M')$ and since $\bar f$ is a split epimorphism, $\theta_M = 0$ in $\operatorname{Ext}^2_\Gamma(M, M')$. By Proposition 3.5 we have that M is liftable to Γ_2.

(c)$\Rightarrow$(b). Let E in $\operatorname{mod}\Gamma_2$ be a lifting of M to Γ_2 and let $E' = E/(x_2, ..., x_t)E$. Then $\Lambda/x_1\Lambda \otimes_{\Lambda/x_1^2\Lambda} E' \simeq \Gamma \otimes_\Lambda E$, so E' is a lifting of M viewed as a $\Lambda/x_1\Lambda$-module to $\Lambda/x_1^2\Lambda$, if $\operatorname{Tor}_i^{\Lambda/x_1^2\Lambda}(\Lambda/x_1\Lambda, E') = (0)$ for all $i \geqslant 1$. Since $\cdots \to \Lambda/x_1^2\Lambda \xrightarrow{x_1} \Lambda/x_1^2 \xrightarrow{x_1} \Lambda/x_1^2\Lambda \to \Lambda/x_1\Lambda \to 0$ is a free resolution of $\Lambda/x_1\Lambda$ over $\Lambda/x_1^2\Lambda$, we have that $\operatorname{Tor}^{\Lambda/x_1^2\Lambda}(\Lambda/x_1\Lambda, E') = \{e \in E' \mid x_1 e = 0\}/x_1 E'$. Let $e \in E$ and assume that $x_1 e = \sum_{i=2}^t x_i e_i$ for some $e_i \in E$. Since $\operatorname{Tor}_i^{\Gamma_2}(\Gamma, E) = (0)$ for all $i \geqslant 1$, we have that $\{(e_i) \in E^t \mid \sum_{i=1}^t x_i e_i = 0\} = IE^t$ by Lemma 3.3. Since $x_1 e - \sum_{i=2}^t x_i e_i = 0$, it follows that $e \in IE$ and therefore if $x_1 e' = 0$ for some $e' \in E'$, then $e' = x_1 e''$ for some $e'' \in E'$. This shows that $\{e \in E' \mid x_1 e = 0\} \subseteq x_1 E'$ and since the opposite inclusion always holds, we have shown that $\operatorname{Tor}_i^{\Lambda/x_1^2\Lambda}(\Lambda/x_1\Lambda, E') = (0)$ for all $i \geqslant 1$ and E' is a lifting of M to $\Lambda/x_1^2\Lambda$. By Proposition 3.2 we have that $\Omega_\Lambda(M)/x_1\Omega_\Lambda(M) \simeq M \oplus \Omega_{\Lambda/x_1\Lambda}(M)$. Since (c) implies (b) for $t = 1$ by Proposition 3.2, we want to finish the proof by induction on t.

Let $\cdots \to P_2 \to P_1 \to P_0 \to M \to 0$ be a projective resolution of M over Λ, then $\cdots \to P_2 \to P_1 \to \Omega_\Lambda(M) \to 0$ is a projective resolution of $\Omega_\Lambda(M)$ over Λ. Since x_1 is regular on $\Omega_\Lambda(M)$, we have that $\cdots \to P_2/x_1P_2 \to P_1/x_1P_1 \to \Omega_\Lambda(M)/x_1\Omega_\Lambda(M) \to 0$ is a projective resolution of $\Omega_\Lambda(M)/x_1\Omega_\Lambda(M)$ over $\Lambda/x_1\Lambda$. Since $\Omega_\Lambda(M)/x_1\Omega(M) \simeq M \oplus \Omega_{\Lambda/x_1\Lambda}(M)$ by Proposition 3.2, we have that $\Omega^{t-1}_{\Lambda/x_1\Lambda}(M)$ is a summand of $\Omega^t_\Lambda(M)/x_1\Omega^t_\Lambda(M) \oplus U$, where U is a projective $\Lambda/x_1\Lambda$-module by Schanuel's Lemma. Let $\Lambda' = \Lambda/x_1\Lambda$, then $\{x_2, ..., x_t\}$ is a Λ'-regular sequence and denote by Γ'_2 the factor ring $\Gamma_2/x_1\Gamma_2 = \Lambda'/(x_2, ..., x_t)^2\Lambda'$. Using that

$$\cdots \longrightarrow \Gamma_2^{t^2} \xrightarrow{\oplus_{i=1}^t d} \Gamma_2^t \xrightarrow{(x_1, x_2, ..., x_t) = d} \Gamma_2\Gamma_1 \xrightarrow{x_1} \Gamma_2 \longrightarrow \Gamma'_2 \longrightarrow 0$$

is a free resolution of Γ'_2 over Γ_2 and that $\operatorname{Tor}_i^{\Gamma_2}(\Gamma, E) = (0)$ for all $i \geqslant 1$, it is not hard to see that $\operatorname{Tor}_i^{\Gamma_2}(\Gamma'_2, E) = (0)$ for all $i \geqslant 1$. Therefore, if $\mathbf{P}^\cdot$ is a projective resolution of E over Γ_2, then $\Gamma'_2 \otimes_{\Gamma_2} \mathbf{P}^\cdot$ is a projective resolution of $\Gamma'_2 \otimes_{\Gamma_2} E \simeq E/x_1E$ over Γ'_2. Since $\Gamma \otimes_{\Gamma_2} \mathbf{P}^\cdot \simeq \Gamma \otimes_{\Gamma'_2} (\Gamma'_2 \otimes_{\Gamma_2} \mathbf{P}^\cdot)$ and $\Gamma \otimes_{\Gamma'_2} E/x_1E \simeq E/IE$, we have that $\operatorname{Tor}_i^{\Gamma'_2}(\Gamma, E/x_1E) \simeq \operatorname{Tor}_i^{\Gamma_2}(\Gamma, E) = (0)$ for all $i \geqslant 1$. Hence, E/x_1E is a lifting of M to Γ'_2 and by induction on t we have that M is a summand of $\Omega^{t-1}_{\Lambda'}(M)/(x_2, ..., x_t)\Omega^{t-1}_{\Lambda'}(M) \oplus Q'$, where

Q' is a projective Γ-module. Since $\Omega_{\Lambda'}^{t-1}(M)$ is a summand of $\Omega_{\Lambda}^{t}(M)/x_1\Omega_{\Lambda}^{t}(M)\oplus U$, we have that $\Omega_{\Lambda'}^{t-1}(M)/(x_2, ..., x_t)\,\Omega_{\Lambda'}^{t-1}(M)\oplus Q'$ is a direct summand of

$$\Omega_{\Lambda}^{t}(M)/x_1\Omega_{\Lambda}^{t}(M)/((x_2, ..., x_t)\,\Omega_{\Lambda}^{t}(M)/x_1\Omega_{\Lambda}^{t}(M))$$
$$\oplus\, U/(x_2, ..., x_t)\,U\oplus Q',$$

which is isomorphic to $\Omega_{\Lambda}^{t}(M)/I\Omega_{\Lambda}^{t}(M)\oplus U/(x_2, ..., x_t)\,U\oplus Q'$. Since $U/(x_2, ..., x_t)\,U\oplus Q'$ is a direct summand of Q/IQ for some projective Λ-module Q, the module M is isomorphic to a direct summand of $\Omega_{\Lambda}(M)/I\Omega_{\Lambda}(M)\oplus Q/IQ$ for some projective Λ-module Q.

(b) $\Rightarrow$ (a). Since $\mathrm{Tor}_i^{\Lambda}(\Gamma, \Omega_{\Lambda}^{t}(M)\oplus Q)=\mathrm{Tor}_{i+t}^{\Lambda}(\Gamma, M)=(0)$ for $i\geqslant 1$ and M is a direct summand of $\Omega_{\Lambda}^{t}(M)/I\Omega_{\Lambda}^{t}(M)\oplus Q/IQ$, it follows directly that M is weakly liftable to Λ.

Remark. If we assume that R is complete, the condition (b) in the Proposition 3.6 can be replaced with that M is a direct summand of $\Omega_{\Lambda}^{t}(M)/I\Omega_{\Lambda}^{t}(M)$, where the syzygies $\Omega_{\Lambda}^{i}(M)$ are defined by the minimal projective resolution of M over Λ.

Let $\hat{\Lambda}$ denote the completion of Λ with respect to the $\underline{m}$-adic topology. Then $\{x_1, x_2, ..., x_t\}$ is a $\hat{\Lambda}$-regular sequence of central elements in $\hat{\Lambda}$. Since the functor given by the completion with respect to the $\underline{m}$-adic topology is a full and faithful exact functor on mod Γ, the following result follows immediately from Proposition 3.5.

PROPOSITION 3.7. *A module M in* mod Γ *has an infinitesimal deformation if and only if $\hat{M}$ in* mod $\hat{\Gamma}$ *has.*

We end this section by studying some properties of the category of Γ-modules weakly liftable to Λ. Again, let x be a regular on Λ in $\underline{m}$ and $\Lambda_i=\Lambda/x^i\Lambda$ for all $i\geqslant 1$. Similar to when we consider the category of Λ_1-modules liftable to Λ, we can use the results in [15; see Section 2] to show that the category of Λ_1-modules weakly liftable to Λ is not closed under (a) kernels of epimorphisms, (b) cokernels of monomorphisms, and (c) extensions. But, contrary to the category of Λ_1-modules liftable to Λ the category of Λ_1-modules or Γ-modules weakly liftable to Λ is closed under summands, which is obvious from the definition a weakly liftable module. The category of Λ_1-modules or Γ-modules weakly liftable to Λ is as the category of Λ_1-modules liftable to Λ closed under syzygies, which we want to prove next. For the complete case and for the category of Λ_1-modules weakly liftable to Λ, this follows in two different ways from what we already have done. If M in mod Λ_1 is weakly liftable, then $\Omega_{\Lambda}(M)/x\Omega_{\Lambda}(M)\simeq M\oplus\Omega_{\Lambda_1}(M)$ by Proposition 3.2, hence $\Omega_{\Lambda_1}(M)$ is weakly liftable. Since M in mod Λ_1 is weakly liftable if and only if M is liftable to Λ_2, then $\Omega_{\Lambda_1}(M)$ is liftable to Λ_2 by Proposition 2.4 and therefore $\Omega_{\Lambda_1}(M)$ is weakly liftable to Λ if M is.

PROPOSITION 3.8. *If M in* mod *Γ is weakly liftable to Λ, then $\Omega_\Gamma(M)$ is also weakly liftable to Λ.*

Proof. Let L in mod Λ be a weak lifting of M to Λ. Applying the functor $\Gamma \otimes_\Lambda -$ to the exact sequence $0 \to \Omega_\Lambda(L) \to P \to L \to 0$ induces the exact sequence $0 \to \Omega_\Lambda(L)/I\Omega_\Lambda(L) \to P/IP \to L/IL \to 0$. Since M is a direct summand of L/IL, we have that $\Omega_\Gamma(M)$ is a direct summand of $\Omega_\Lambda(L)/I\Omega_\Lambda(L) \oplus U$, where U is a projective Γ-module by Schanuel's Lemma. Since U is a direct summand of Q/IQ for some projective Λ-module Q, it follows that $\Omega_\Gamma(M)$ is weakly liftable to Λ.

4. LIFTING, WEAK LIFTING, AND COHEN–MACAULAY MODULES

Throughout the rest of the paper R is a commutative local Gorenstein ring of dimension d and x is an R-regular element in the maximal ideal $\underline{m}$ of R. We denote by $CM(R)$ the category of all finitely generated maximal Cohen–Macaulay modules over R. We want to study the lifting problems for the situation $R \to R/(x) = \bar{R}$ and the modules we consider over $\bar{R}$ will be maximal Cohen–Macaulay modules. But before we briefly describe the results in this section, we will recall for the convenience of the reader some of the notions involved in this section.

The notion of Cohen–Macaulay approximations was introduced by M. Auslander and R.-O. Buchweitz in [2]. Given an R-module C, an exact sequence $0 \to Y_C \to X_C \xrightarrow{f} C \to 0$ is called a right *Cohen–Macaulay approximation* of C if X_C is in $\mathrm{CM}(R)$ and $\mathrm{pd}_R Y_C < \infty$. An important property of these approximations is that every morphism $X \to C$ with X in $\mathrm{CM}(R)$ factors through f. The Cohen–Macaulay approximation is called *minimal* if f is a right minimal morphism, that is, an endomorphism $g: X_C \to X_C$ is an automorphism whenever $f = f \circ g$. The existence of Cohen–Macaulay and minimal Cohen–Macaulay approximations was established in [1, 2]. It was also shown that the minimal Cohen–Macaulay approximations are unique up to isomorphism.

Now, we define a lifting and a weak lifting in the setting in this section. A module L in $\mathrm{CM}(R)$ is said to be a *lifting* of M in $\mathrm{CM}(\bar{R})$ to R if $L/xL \simeq M$. If M is only isomorphic to a direct summand of L/xL, then L is said to be a weak lifting of M to R. The module M is then said to be *liftable* or *weakly liftable* to Λ, respectively. Since every module L in $\mathrm{CM}(R)$ is a submodule of a free R-module, every R-regular element is also

L-regular. So, the definition of a lifting above agrees with the definition given in Section 1. The definition of a weak lifting above does not agree with the definition in Section 1, but it follows easily from Proposition 3.2 that these two definitions are equivalent.

One of the main aims in this section is to characterize liftable and weakly liftable modules C in $\mathrm{CM}(\bar{R})$ in terms of their minimal Cohen–Macaulay approximation X_C over R and show that every lifting of a liftable module C in $\mathrm{CM}(\bar{R})$ is a submodule of X_C. We also study the category of modules in $\mathrm{CM}(\bar{R})$ liftable to R and the category of modules in $\mathrm{CM}(\bar{R})$ weakly liftable to R and show that they are closed under syzygies, cosyzgies, and taking duals. In addition, we show that the category of modules in $\mathrm{CM}(\bar{R})$ weakly liftable to R is functorially finite in mod $\bar{R}$, a notion we recall later in this section from [4]. Since we only deal with the lifting problems in the setting $R \to \bar{R}$, we denote by $\bar{L}$ the $\bar{R}$-module L/xL for any R-module L.

Our first aim is to show that every lifting of a liftable Cohen–Macaulay $\bar{R}$-module C is a submodule of the minimal Cohen–Macaulay approximation X_C in $\mathrm{CM}(R)$. This leads us to a characterization of liftability of C in $\mathrm{CM}(\bar{R})$ in terms of existence of special submodules of X_C.

PROPOSITION 4.1. *If C in $\mathrm{CM}(\bar{R})$ is liftable, then every lifting of C is a submodule of the minimal Cohen–Macaulay approximation of C of R.*

Proof. Assume that L in $\mathrm{CM}(R)$ is q lifting of C in $\mathrm{CM}(\bar{R})$. Then we have the following commutative diagram:

$$\begin{array}{ccccccccc}
 & & & & 0 & & 0 & & \\
 & & & & \downarrow & & \downarrow & & \\
 & & 0 & \longrightarrow & \ker t & \longrightarrow & L & & \\
 & & & & \downarrow & & \downarrow{\scriptstyle x} & & \\
 & & & & L & = & L & & \\
 & & & & \downarrow{\scriptstyle t} & & \downarrow & & \\
0 & \longrightarrow & Y_C & \longrightarrow & X_C & \longrightarrow & C & \longrightarrow & 0.
\end{array}$$

We want to show that $\ker t = (0)$. If $t(l) = 0$ for an $l \in L$, then it is easy to see that $l = xl_1$ for some $l_1 \in L$. Hence, $f(l) = xf(l_1) = 0$ and using that x is a X_C-regular element, $f(l_1) = 0$. Then we can continue this process and show that $l \in \bigcap_{i=1}^{\infty} x^i L$. Since L is a finitely generated R-module and $x \in \operatorname{rad} R$, we have that $\bigcup_{i=1}^{\infty} x^i L = (0)$ and therefore $\ker t = (0)$.

Next, we use the result above to characterize liftability of C in $\mathrm{CM}(\bar{R})$ in terms of existence of special submodules of X_C. But first we need to introduce some notation. Let $0 \to Y_C \to X_C \xrightarrow{\alpha_C} C \to 0$ be the minimal Cohen–Macaulay approximation of C in $\mathrm{CM}(\bar{R})$. Denote by $\pi: X_C \to \bar{X}_C$ the natural epimorphism and by $\bar{\alpha}_C: \bar{X}_C \to C$ the map $\bar{R} \otimes_R \alpha_C$ for $\alpha_C: X_C \to C$.

Proposition 4.2. *A module C in $\mathrm{CM}(\bar{R})$ is liftable if and only if there exists a submodule L of X_C, such that L is in $\mathrm{CM}(R)$, x is a X_C/L-regular element and the composition $\pi(L) \hookrightarrow \bar{X}_C \xrightarrow{\bar{\alpha}_C} C$ is an isomorphism.*

Proof. Assume that L in $\mathrm{CM}(R)$ is a lifting of C in $\mathrm{CM}(\bar{R})$. By Proposition 4.1 we have the commutative diagram

$$\begin{array}{ccccccccc}
 & & & & 0 & & & & \\
 & & & & \downarrow & & & & \\
 & & & & L & = & L & & \\
 & & & & \downarrow{\scriptstyle \iota} & & \downarrow & & \\
0 & \longrightarrow & Y_C & \longrightarrow & X_C & \xrightarrow{\alpha_C} & C & \longrightarrow & 0,
\end{array}$$

and applying the functor $\bar{R} \otimes_R -$, the following commutative diagram is induced:

$$\begin{array}{ccccc}
0 & & & & \\
\downarrow & & & & \\
\mathrm{Tor}_1^R(\bar{R}, X_C/L) & & & & \\
\downarrow & & & & \\
\bar{L} & = & \bar{L} & & \\
\downarrow{\scriptstyle \bar{\iota}} & & \downarrow \wr & & \\
\bar{X}_C & \xrightarrow{\bar{\alpha}_C} & C & \longrightarrow & 0.
\end{array}$$

Since $\bar{\alpha}_C \circ \bar{\iota}$ is an isomorphism, $\mathrm{Tor}_1^R(\bar{R}, X_C/L) = (0)$ and therefore x is a X_C/L-regular element. Since

$$\begin{array}{ccccc}
0 & \longrightarrow & L & \xrightarrow{\iota} & X_C \\
 & & \downarrow & & \downarrow{\scriptstyle \pi} \\
0 & \longrightarrow & L/xL & \xrightarrow{\bar{\iota}} & \bar{X}_C \\
 & & \downarrow & & \\
 & & 0 & &
\end{array}$$

is a commutative diagram, $\pi(L) = \pi(L) = \pi \circ t(L) = \bar{t}(L)$ and therefore the composition $\pi(L) \hookrightarrow \bar{X}_C \xrightarrow{\bar{\alpha}_C} C$ is an isomorphism.

The proof of the "if"-part of the statement follows from Lemma 2.3.

Now, we combine the characterization of liftable modules above with the characterization we found for liftable modules in Proposition 2.2 to get the following characterization of liftable Cohen–Macaulay modules over $\bar{R}$.

PROPOSITION 4.3. *Consider the following commutative diagram with C in* $\mathrm{CM}(\bar{R})$:

$$
\begin{array}{ccccccccc}
0 & \longrightarrow & \Omega_R(C) & \longrightarrow & P & \xrightarrow{\varphi} & C & \longrightarrow & 0 \\
 & & \downarrow & & \downarrow{\scriptstyle t} & & \| & & \\
0 & \longrightarrow & Y_C & \longrightarrow & X_C & \xrightarrow{\alpha_C} & C & \longrightarrow & 0
\end{array}
$$

The module C is liftable to R if and only if there exists a map $t\colon P \to X_C$ such that $\alpha_C \circ t = \varphi$, the composition $(\pi \circ t)(P) \hookrightarrow \bar{X}_C \xrightarrow{\alpha_C} C$ is an isomorphism and x is a $X_C/\mathrm{Im}\ t$-regular element.

Proof. The "if"-part of the statement follows from Proposition 4.2.

Assume that L in $\mathrm{CM}(R)$ is a lifting of C to R. By Proposition 2.1 we have the following commutative diagram:

$$
\begin{array}{ccccccccc}
 & & 0 & & 0 & & & & \\
 & & \downarrow & & \downarrow & & & & \\
 & & \Omega_R(L) & = & \Omega_R(L) & & & & \\
 & & \downarrow & & \downarrow & & & & \\
0 & \longrightarrow & \Omega_R(C) & \longrightarrow & P & \longrightarrow & C & \longrightarrow & 0 \\
 & & \downarrow & & \downarrow{\scriptstyle r} & & \| & & \\
0 & \longrightarrow & \Omega_R(L) & \longrightarrow & L & \longrightarrow & C & \longrightarrow & 0 \\
 & & \downarrow & & \downarrow & & & & \\
 & & 0 & & 0 & & & &
\end{array}
$$

By Proposition 4.1 we have the following commutative diagram:

$$\begin{array}{ccccccccc}
 & & 0 & & 0 & & & & \\
 & & \downarrow & & \downarrow & & & & \\
0 & \longrightarrow & L & \xrightarrow{x} & L & \longrightarrow & C & \longrightarrow & 0 \\
 & & \downarrow & & \downarrow{\scriptstyle s} & & \| & & \\
0 & \longrightarrow & Y_C & \longrightarrow & X_C & \longrightarrow & C & \longrightarrow & 0 \\
 & & \downarrow & & \downarrow & & & & \\
 & & X_C/L & = & X_C/L & & & & \\
 & & \downarrow & & \downarrow & & & & \\
 & & 0 & & 0 & & & &
\end{array}$$

If we let $t = s \circ r$, then t has the desired properties.

The preceeding result shows the close connection between an $\bar{R}$-module C in $\mathrm{CM}(\bar{R})$ and its minimal Cohen–Macaulay approximation X_C over R. Our next result shows that the category of liftable Cohen–Macaulay $\bar{R}$-modules is closed under syzygies, cosyzygies and taking dual. Every module C in $\mathrm{CM}(\bar{R})$ has a coresolution of the form

$$0 \longrightarrow C \longrightarrow \bar{R}^{n_0} \xrightarrow{d^1} \bar{R}^{n_1} \xrightarrow{d^2} \bar{R}^{n_2} \xrightarrow{d^3} \cdots,$$

where $\operatorname{Im} d^i$ is in $\mathrm{CM}(\bar{R})$ for all $i \geqslant 1$ and has no nonzero free summands. Then we call the module $\operatorname{Im} d^i$ an ith cosyzygy of C in $\mathrm{CM}(\bar{R})$ and denote it by $\Omega_{\bar{R}}^{-i}(C)$. The dual of a module C in $\mathrm{CM}(\bar{R})$ is defined by $D(C) = \operatorname{Hom}_{\bar{R}}(C, \bar{R})$.

Proposition 4.4. *Let C be in $\mathrm{CM}(\bar{R})$. If C is liftable to R, then the syzygies $\Omega^i_{\bar{R}}(C)$ for any i in $\mathbf{Z}$ and the dual $\operatorname{Hom}_{\bar{R}}(C, \bar{R})$ of C are liftable to R.*

Proof. Let L in $CM(R)$ be a lifting of C in $\mathrm{CM}(\bar{R})$. Then $\overline{\Omega^i_R(L)} \simeq \Omega^i_{\bar{R}}(C)$ for all positive integers i, since tensoring a projective resolution of L over R with $\bar{R} \otimes_R -$ induces a projective resolution of $C \simeq \bar{R} \otimes_R L$. Therefore all the syzygies of C are liftable.

Applying $\operatorname{Hom}_R(L,\ )$ to the exact sequence $0 \to R \xrightarrow{x} R \to \bar{R} \to 0$ it induces the exact sequence

$$0 \longrightarrow \operatorname{Hom}_R(L, R) \xrightarrow{x} \operatorname{Hom}_R(L, R) \longrightarrow \operatorname{Hom}_{\bar{R}}(\bar{L}, \bar{R}) \longrightarrow 0,$$

since R is an injective module in $\mathrm{CM}(R)$. Hence, the dual of C given by

$D(C) \simeq D(\bar{L})$ is liftable, since the dual of L given by $D_R(L) = \operatorname{Hom}_R(L, R)$ is in $\mathrm{CM}(R)$. This also shows that taking dual and reduction modulo x commutes, that is, $\operatorname{Hom}_{\bar{R}}(\bar{L}, \bar{R}) \simeq \bar{R} \otimes_R \operatorname{Hom}_R(L, R)$.

Using the duality $\operatorname{Hom}_{\bar{R}}(\ , \bar{R}) : \mathrm{CM}(\bar{R}) \to \mathrm{CM}(\bar{R})$ we have that

$$\Omega_{\bar{R}}^{-i}(C) \simeq \operatorname{Hom}_{\bar{R}}(\Omega_{\bar{R}}^{i}(\operatorname{Hom}_{\bar{R}}(C, \bar{R})), \bar{R}).$$

From this it follows immediately that the cosyzygies of C are liftable.

In the rest of this section we study the properties of the category of weakly liftable Cohen–Macaulay $\bar{R}$-modules which we denote by w.l.$(\bar{R})$. Our aim is to show that w.l.$(\bar{R})$ is a functorially finite subcategory in mod $\bar{R}$. First we give a characterization of a weakly liftable Cohen–Macaulay $\bar{R}$-module C in terms of its minimal Cohen–Macaulay approximation X_C over R. Before we prove this we need to introduce some notation.

Let C be in $\mathrm{CM}(\bar{R})$. Then the minimal Cohen–Macaulay approximation of C over R can be constructed in the following way. Let $0 \to \Omega_R^1(D(C)) \to R^n \to D(C) \to 0$ be a projective cover of $D(C) = \operatorname{Hom}_{\bar{R}}(C, \bar{R})$ over R. It follows that $\Omega_R^1(D(C))$ is in $\mathrm{CM}(R)$ and that dualizing with respect to R induces the following exact sequence:

$$0 \to R^n \to \operatorname{Hom}_R(\Omega_R^1(D(C)), R) \to \operatorname{Ext}_R^1(D(C), R) \to 0. \ (*)$$

It is easy to check that $\operatorname{Ext}_R^1(D(C), R) \simeq \operatorname{Hom}_{\bar{R}}(D(C), \bar{R}) \simeq C$ and that the sequence $(*)$ is a minimal Cohen–Macaulay approximation of C over R. In the following we denote by $0 \to R^n \to X_C \xrightarrow{g} C \to 0$ the minimal Cohen–Macaulay approximation of C over R where $n = \mu(\operatorname{Ext}_R^1(C, R))$, the minimal number of generators of $D(C)$. Applying the functor $\bar{R} \otimes_R -$, we obtain an exact sequence

$$\xi_C : 0 \longrightarrow C \longrightarrow \bar{R}^n \longrightarrow \bar{X}_C \xrightarrow{\bar{g}} C \longrightarrow 0.$$

Then ξ_C represents an element in $\operatorname{Ext}_{\bar{R}}^2(C, C)$ and $\xi_C = \theta_C$, where

$$\theta_C : 0 \to C \to \Omega_R(C)/x\Omega_R(C) \to P/xP \to C \to 0$$

is induced from the exact sequence $0 \to \Omega_R(C) \to P \to C \to 0$ by tensoring with $\bar{R} \otimes_R -$ as defined in Section 1. Because, applying the functor $\bar{R} \otimes_R -$ to the commutative diagram

$$\begin{array}{ccccccccc} 0 & \longrightarrow & \Omega_R(C) & \longrightarrow & P & \longrightarrow & C & \longrightarrow & 0 \\ & & \downarrow & & \downarrow & & \| & & \\ 0 & \longrightarrow & R^n & \longrightarrow & X_C & \xrightarrow{g} & C & \longrightarrow & 0 \end{array},$$

the following commutative diagram is induced:

$$\begin{array}{lcccccccccc}
\theta_C\colon & 0 & \longrightarrow & C & \longrightarrow & \overline{\Omega_R(C)} & \longrightarrow & \bar{P} & \longrightarrow & C & \longrightarrow 0 \\
 & & & \| & & \downarrow & & \downarrow & & \| & \\
\xi_C\colon & 0 & \longrightarrow & C & \longrightarrow & \bar{R}^n & \longrightarrow & \bar{X}_C & \xrightarrow{\bar{g}} & C & \longrightarrow 0
\end{array}.$$

This shows that θ_C and ξ_C represent the same element in $\operatorname{Ext}^2_{\bar{R}}(C, C)$.

As a consequence of Proposition 3.2 we have the following.

PROPOSITION 4.5. *The following are equivalent for C in* $\mathrm{CM}(\bar{R})$.

(a) C *is in* $\mathrm{w.l.}(\bar{R})$.

(b) $\xi_C = 0$ *in* $\operatorname{Ext}^2_{\bar{R}}(C, C)$.

(c) $\bar{g}$ *is a split epimorphism.*

(d) $\bar{X}_C \simeq C \oplus \Omega^{-1}_{\bar{R}}(C)$.

Proof. (a)$\Leftrightarrow$(b). Follows from Proposition 3.2 and above argument.

(a)$\Rightarrow$(c). Suppose C is in $\mathrm{w.l.}(\bar{R})$. Then there is an exact sequence $0 \to K \to L \xrightarrow{\varphi} C \to 0$ such that L is in $\mathrm{CM}(R)$ and φ induces a split epimorphism $\bar{L} \xrightarrow{\bar{\varphi}} C \to 0$. Then we have the commutative diagram

$$\begin{array}{ccccccccc}
0 & \longrightarrow & K & \longrightarrow & L & \xrightarrow{\varphi} & C & \longrightarrow & 0 \\
 & & \downarrow & & \downarrow & & \| & & \\
0 & \longrightarrow & R^n & \longrightarrow & X_C & \xrightarrow{g} & C & \longrightarrow & 0
\end{array},$$

and applying the functor $\bar{R} \otimes_R -$ the following commutative diagram is induced:

$$\begin{array}{lcccccccccc}
 & 0 & \longrightarrow & C & \longrightarrow & \bar{K} & \longrightarrow & \bar{L} & \xrightarrow{\bar{\varphi}} & C & \longrightarrow 0 \\
 & & & \| & & \downarrow & & \downarrow & & \| & \\
\xi_C\colon & 0 & \longrightarrow & C & \longrightarrow & \bar{R}^n & \longrightarrow & \bar{X}_C & \xrightarrow{\bar{g}} & C & \longrightarrow 0
\end{array}.$$

Since $\bar{\varphi}$ is a split epimorphism, $\bar{g}$ is also a split epimorphism.

(c)$\Rightarrow$(d). Suppose $\bar{g}$ is a split epimorphism. We want to show that $\ker \bar{g} \simeq \Omega^{-1}_{\bar{R}}(C)$. The exact sequence $0 \to R \xrightarrow{x} R \to \bar{R} \to 0$ induces the following long exact sequence:

$$0 \to \operatorname{Hom}_R(C, R) \xrightarrow{x} \operatorname{Hom}_R(C, R) \to \operatorname{Hom}_R(C, \bar{R}) \to \operatorname{Ext}^1_R(C, R) \xrightarrow{x} \cdots .$$

Since $x \in \operatorname{Ann}_R(C)$, we have $\operatorname{Hom}_{\bar{R}}(C, \bar{R}) \simeq \operatorname{Ext}^1_R(C, R)$. Hence we have $n = \mu(\operatorname{Hom}_{\bar{R}}(C, \bar{R}))$ and $\ker \bar{g} \simeq \Omega^{-1}_{\bar{R}}(C)$.

(d) $\Rightarrow$ (a). Follows from the definition.

As consequences of this proposition, we haven the following corollaries which will be used later.

COROLLARY 4.6. *Let C be indecomposable in* $\mathrm{CM}(\bar{R})$ *and suppose that* $C \not\simeq \bar{R}$. *If C is weakly liftable to R, then X_C has no free summands.*

Proof. Since C is weakly liftable to R, we have $\bar{X}_C \simeq C \oplus \Omega_{\bar{R}}^{-1}(C)$ by (d) of Proposition 4.5. Since C is indecomposable in $\mathrm{CM}(\bar{R})$ and $C \not\simeq \bar{R}$, we know that $\Omega_{\bar{R}}^{-1}(C)$ is also indecomposable and $\Omega_{\bar{R}}^{-1}(C) \not\simeq \bar{R}$. So the module $\bar{X}_C$ has no free summands. Therefore X_C has no free summands. Since if F is a free summand of X_C then $\bar{F}$ is a free summand of $\bar{X}_C$.

COROLLARY 4.7. *Let C be in* $\mathrm{CM}(\bar{R})$ *and suppose C is indecomposable and not liftable. Assume* $\mathrm{CM}(\bar{R})$ *is a Krull-Schmidt category (for instance, when R is complete). If C is weakly liftable, then X_C is indecomposable.*

Proof. By (d) of Proposition 4.5 we have $\bar{X}_C \simeq C \oplus \Omega_{\bar{R}}^{-1}(C)$. Since C is indecomposable, $\Omega_{\bar{R}}^{-1}(C)$ is also indecomposable. Then if X_C is decomposable, we have that C is liftable which contradicts our assumption.

We end this section by showing that $\mathrm{w.l.}(\bar{R})$ is a functorially finite subcategory in $\mathrm{mod}\,\bar{R}$, a notion we now recall from [4]. Let $\mathscr{A}$ be a category and $\mathscr{C}$ a subcategory of $\mathscr{A}$. The subcategory $\mathscr{C}$ is said to be *contravariantly finite* in $\mathscr{A}$ if for any object A in $\mathscr{A}$, there is a C in $\mathscr{C}$ and a morphism $f: C \to A$ such that any morphism $g: C' \to A$ with C' in $\mathscr{C}$ factors through f. If the dual condition is satisfied for the subcategory $\mathscr{C}$, then $\mathscr{C}$ is said to be *covariantly* finite in $\mathscr{A}$. If $\mathscr{C}$ is both covariantly and contravariantly finite in $\mathscr{A}$, then $\mathscr{C}$ is said to be *functorially finite* in $\mathscr{A}$.

PROPOSITION 4.8. *The subcategory* $\mathrm{w.l.}(\bar{R})$ *is functorially finite in* $\mathrm{mod}\,\bar{R}$.

Proof. Given an $\bar{R}$-module C in $\mathrm{mod}\,\bar{R}$, we have an epimorphism $\bar{f}: \bar{X}_C \to C \to 0$. We want to show that for any D in $\mathrm{w.l.}(\bar{R})$ and any morphism $h: D \to C$, the morphism h factors through $\bar{f}$. Let L be in $\mathrm{CM}(R)$ such that there is a map $\varphi: L \to D$ which induces a split epimorphism $\bar{\varphi}: \bar{L} \to D$. Then the commutative diagram

$$\begin{array}{ccccccccc} 0 & \longrightarrow & K & \longrightarrow & L & \xrightarrow{\varphi} & D & \longrightarrow & 0 \\ & & \downarrow & & \downarrow & & \downarrow{\scriptstyle h} & & \\ 0 & \longrightarrow & Y_C & \longrightarrow & X_C & \xrightarrow{f} & C & \longrightarrow & 0 \end{array}$$

induces the following commutative diagram by applying the functor $\bar{R}\otimes_R -$:

$$\begin{array}{ccccccccc} 0 & \longrightarrow & D & \longrightarrow & \bar{K} & \longrightarrow & \bar{L} & \xrightarrow{\bar{\varphi}} & D & \longrightarrow & 0 \\ & & \downarrow & & \downarrow & & \downarrow & & \downarrow{\scriptstyle h} & & \\ 0 & \longrightarrow & C & \longrightarrow & \bar{Y}_C & \longrightarrow & \bar{X}_C & \xrightarrow{\bar{f}} & C & \longrightarrow & 0 \end{array} .$$

Since $\bar{\varphi}$ is a split epimorphism, h factors through $\bar{f}$. This shows that w.l.$(\bar{R})$ is contravariantly finite in mod $\bar{R}$.

Now we want to show that w.l.$(\bar{R})$ is covariantly finite in mod $\bar{R}$. For C in mod $\bar{R}$ let C^* denote the $\bar{R}$-module $\operatorname{Hom}_{\bar{R}}(C, \bar{R})$. Then we have a natural morphism $\alpha_C\colon C \to C^{**}$ for every C in mod $\bar{R}$. Since w.l.$(\bar{R})$ is contravariantly finite in mod $\bar{R}$, we have a morphism $\beta\colon X \to C^*$ such that X is in w.l.$(\bar{R})$ and any morphism $h\colon D \to C^*$ with D in w.l.$(\bar{R})$ factors through β. Applying the functor $\operatorname{Hom}_{\bar{R}}(\ , \bar{R})$, we get $\beta^*\colon C^{**} \to X^*$. Let $g\colon C \to X^*$ be the composition $C \xrightarrow{\alpha_C} C^{**} \xrightarrow{\beta^*} X^*$. We claim that any morphism $h\colon C \to E$ with E in w.l.$(\bar{R})$ can be extended to X^*. Since E is reflexive, we have the following commutative diagram:

$$\begin{array}{ccc} C & \xrightarrow{\alpha_C} & C^{**} \\ \downarrow{\scriptstyle h} & & \downarrow{\scriptstyle h^{**}} \\ E & \longrightarrow & E^{**} \end{array} .$$

Since E^*, C^*, and X are reflexive $\bar{R}$-modules, applying the functor $\operatorname{Hom}_{\bar{R}}(\ , \bar{R})$ to the diagram

$$\begin{array}{ccc} C^{**} & \xrightarrow{\beta^*} & X^* \\ \downarrow{\scriptstyle h^{**}} & & \\ E^{**} & & \end{array} ,$$

it induces the following commutative diagram:

$$\begin{array}{ccc} & & E^* \\ & \swarrow{\scriptstyle s} & \downarrow{\scriptstyle h^*} \\ X & \xrightarrow{\beta} & C^* \end{array} .$$

Since w.l.$(\bar{R})$ is contravariantly finite in mod $\bar{R}$, the map h^* factors through β. Applying the functor $\operatorname{Hom}_{\bar{R}}(\ , \bar{R})$ again, we have the commutative diagram

$$\begin{array}{ccc} C^{**} & \xrightarrow{\beta^*} & X^* \\ \downarrow{\scriptstyle h^{**}} & \swarrow{\scriptstyle s^*} & \\ E^{**} & & \end{array}.$$

This shows $h: C \to E$ can be extended to X^* and w.l.$(\bar{R})$ is also covariantly finite in mod $\bar{R}$, and therefore w.l.$(\bar{R})$ is functorially finite in mod $\bar{R}$.

5. Weak Lifting over Gorenstein Rings

In this section, instead of studying the weak liftability of modules in $\mathrm{CM}(\bar{R})$, we study the weak liftability of modules in mod $\bar{R}$. A finitely generated $\bar{R}$-modules C is said to be weakly liftable to R if there exists an R-module L such that x is an L-regular element and C is isomorphic to a direct summand of L/xL. This section is mainly devoted to studying the properties of weakly liftable $\bar{R}$-modules in terms of their minimal Cohen–Macaulay approximations over R.

All the results in this section are based on the following lemma.

Lemma 5.1. *Let L be in* mod R *such that x is L-regular and let $0 \to Y_L \to X_L \longrightarrow L \to 0$ be the minimal Cohen–Macaulay approximation of L over R. Then the reduction $0 \to \bar{Y}_L \to \bar{X}_L \xrightarrow{\bar{\varphi}} \bar{L} \to 0$ is the minimal Cohen–Macaulay approximation of $\bar{L}$ over $\bar{R}$.*

Proof. Since x is L-regular, the sequence $0 \to \bar{Y}_L \to \bar{X}_L \xrightarrow{\bar{\varphi}} \bar{L} \to 0$ is exact. We know that x is regular on X_L and Y_L, so $\bar{X}_L$ is in $\mathrm{CM}(\bar{R})$ and $\bar{Y}_L$ is of finite projective dimension as an $\bar{R}$-module. Therefore $0 \to \bar{Y}_L \to \bar{X}_L \xrightarrow{\bar{\varphi}} \bar{L} \to 0$ is a Cohen–Macaulay approximation of $\bar{L}$ over $\bar{R}$. It remains to show that it is minimal. This is done by using a general criterion of minimal Cohen–Macaulay approximation proved in [1]. It says that a Cohen–Macaulay approximation $0 \to Y \to X \xrightarrow{f} C \to 0$ of an R-module C is minimal if and only if X has a decomposition $F \oplus U$ with U no non-zero free summands and F a free module such that the induced map $F \to C/f(U)$ is a projective cover. Now suppose $X_L = F \oplus U$ is such a decomposition for X_L. Then we have $\bar{X}_L = \bar{F} \oplus \bar{U}$. It follows that $\bar{F}$ is a free

$\bar{R}$-module with rank $\bar{F} = \operatorname{rank} F$ and $\bar{U}$ has no free summands as an $\bar{R}$-module [1]. We now show that the induced map $\bar{F} - \bar{L}/\bar{\varphi}(\bar{U}) \to 0$ is a projective cover over $\bar{R}$. Since $\bar{L}/\bar{\varphi}(\bar{U}) \simeq L/(\varphi(U) + xL)$ and x is in the maximal ideal of R, we have that $\mu(\bar{L}/\bar{\varphi}(\bar{U})) = \mu(L/\varphi(U))$ and therefore $\bar{F} \to \bar{L}/\bar{\varphi}(\bar{U}) \to 0$ is a projective cover of $\bar{L}/\bar{\varphi}(\bar{U})$ over $\bar{R}$. By the criterion cited above we have showed that $0 \to \bar{Y}_L \to \bar{X}_L \xrightarrow{\bar{\varphi}} \bar{L} \to 0$ is a minimal Cohen–Macaulay approximation of $\bar{L}$.

Let C be an arbitrary $\bar{R}$-module. We denote by $X_C^{\bar{R}}$ the minimal Cohen–Macaulay approximation of C over $\bar{R}$. As an immediate consequence of Lemma 5.1, we have the following result which shows the connection between the (weak) liftability of C and $X_C^{\bar{R}}$.

PROPOSITION 5.2. *Let C be a finitely generated $\bar{R}$-module. If C is* (weakly) *liftable to R, then $X_C^{\bar{R}}$ and $Y_C^{\bar{R}}$ are* (weakly) *liftable to R.*

When $\dim R \leqslant 2$, the converse of Proposition 5.2 is also true. We state it in the following proposition.

PROPOSITION 5.3. *Let* $\dim R \leqslant 2$ *and C be a finitely generated $\bar{R}$-module. Then C is (weakly) liftable to R if and only if $X_C^{\bar{R}}$ is (weakly) liftable to R.*

Proof. If $\dim R = 1$, then $\dim \bar{R} = 0$. So $X_C^{\bar{R}} = C$ and there is nothing to prove. Suppose $\dim R = 2$. Let $0 \to \bar{R}^n \to X_C^{\bar{R}} \to C \to 0$ be the minimal Cohen–Macaulay approximation of C over $\bar{R}$. Suppose $X_C^{\bar{R}}$ is liftable to R and suppose $0 \to X \xrightarrow{x} X \to X_C^{\bar{R}} \to 0$ is an exact sequence with X in $\mathrm{CM}(R)$. Then we have the following commutative exact diagram:

$$
\begin{array}{ccccccccc}
 & & & & & & 0 & & \\
 & & & & & & \downarrow & & \\
0 & \longrightarrow & R^n & \xrightarrow{x} & R^n & \longrightarrow & \bar{R}^n & \longrightarrow & 0 \\
 & & \downarrow{\scriptstyle f} & & \downarrow{\scriptstyle f} & & \downarrow & & \\
0 & \longrightarrow & X & \xrightarrow{x} & X & \longrightarrow & X_C^{\bar{R}} & \longrightarrow & 0. \\
 & & & & & & \downarrow & & \\
 & & & & & & C & \longrightarrow & 0 \\
 & & & & & & \downarrow & & \\
 & & & & & & 0 & &
\end{array}
$$

By the Snake Lemma, we obtain an exact sequence

$$0 \to \operatorname{coker} f \xrightarrow{x} \operatorname{coker} f \to C \to 0.$$

Since x is regular on coker f, we have that C is liftable to R.

We have shown in Section 4 for C in $\mathrm{CM}(\bar{R})$, that if C is indecomposable and $C \not\simeq \bar{R}$, then C weakly liftable to R implies that X_C has no free summands. Now we generalize this result in the following form. Given an R-module C, we denote by $\delta_R(C)$ the maximal rank of free summands in X_C, the minimal Cohen–Macaulay approximation of C over R. We define $\delta^i_R(C) = \delta(\Omega^i_R(C))$. Since $\Omega^i_R(C)$ is in $\mathrm{CM}(R)$ and has no free summands for $i > \dim R$, we have $\delta^i_R(C) = 0$ for $i > \dim R$, so that the sum $\sum_{i \geqslant 0} (-1)^i \delta^i_R(C)$ makes sense.

Proposition 5.4. *Let C be in* mod R. *If there is an R-regular element $x \in \operatorname{Ann}_R(C)$ such that, as an $\bar{R} = R/(x)$-module, C is weakly liftable to R, then $\sum_{i \geqslant 0}(-1)^i \delta^i_R(C) = 0$.*

Proof. Since C is weakly liftable to R, we have that $\overline{\Omega^1_R(C)} \simeq C \oplus \Omega^1_R(C)$ by Proposition 3.2. Let $0 \to Y \to X \to \Omega^1_R(C) \to 0$ be the minimal Cohen–Macaulay approximation of $\Omega^1_R(C)$ over R. Then $0 \to \bar{Y} \to \bar{X} \to \overline{\Omega^1_R(C)} \to 0$ is the minimal Cohen–Macaulay approximation of $\overline{\Omega^1_R(C)} \simeq C \oplus \Omega^1_{\bar{R}}(C)$ over $\bar{R}$. Therefore we have $\bar{X} \simeq X^{\bar{R}}_C \oplus X^{\bar{R}}_{\Omega^1_{\bar{R}}(C)}$ by Lemma 5.1 and hence $\delta_R(\Omega^1_R(C)) = \delta_{\bar{R}}(C) + \delta_{\bar{R}}(\Omega^1_{\bar{R}}(C))$, in other words, we have the relation $\delta^1_R(C) = \delta^0_{\bar{R}}(C) + \delta^1_{\bar{R}}(C)$.

Let $\cdots \to P_1 \to P_0 \to \Omega^1_R(C) \to 0$ be an R-free resolution of $\Omega^1_R(C)$. Since x is a regular element on $\Omega^1_R(C)$, we have that $\cdots \to \bar{P}_1 \to \bar{P}_0 \to \overline{\Omega^1_R(C)} \to 0$ is an $\bar{R}$-free resolution of $\overline{\Omega^1_R(C)}$. Therefore we have $\overline{\Omega^{i+1}_R(C)} \simeq \Omega^1_{\bar{R}}(\overline{\Omega^1_R(C)})$ for all $i \geqslant 1$. Since C is weakly liftable to R as an $\bar{R}$-module, the syzygies $\Omega^i_{\bar{R}}(C)$ are also weakly liftable to R for all $i > 0$. Therefore by the above argument we have $\bar{X}_{\Omega^{i+1}_R(C)} \simeq X^{\bar{R}}_{\Omega^i_{\bar{R}}(C)} \oplus X^{\bar{R}}_{\Omega^{i+1}_{\bar{R}}(C)}$ for all $i \geqslant 0$ and hence $\delta^i_{\bar{R}}(C) + \delta^{i+1}_{\bar{R}}(C) = \delta^{i+1}_R(C)$ for all $i \geqslant 0$. Therefore the alternating sum of $\delta^{i+1}_R(C)$ for $i \geqslant 0$ is $\delta^0_{\bar{R}}(C)$. Hence we have $\sum_{i \geqslant 0} (-1)^i \delta^i_R(C) = \delta^0_R(C) - \delta^0_{\bar{R}}(C)$. Now we must show that $\delta_R(C) = \delta_{\bar{R}}(C)$. Since C is weakly liftable to R as an $\bar{R}$-module, $X^{\bar{R}}_C$, the minimal Cohen–Macaulay approximation of C over $\bar{R}$, is also weakly liftable to R by Proposition 5.2. Then our result follows from the following fact which we state as a lemma.

Lemma 5.5. *Suppose C is an $\bar{R} = R/(x)$-module such that $X^{\bar{R}}_C$ is weakly liftable. Then $\delta_R(C) = \delta_{\bar{R}}(C)$.*

Proof. We write $X_C^{\bar{R}} = U \oplus \bar{R}^n$ where U is an $\bar{R}$-module without non-zero free summands. Let $X_U \to U$ be the minimal Cohen–Macaulay approximation of U over R. Then since U is in $\mathrm{CM}(\bar{R})$ and weakly liftable to R, we have $\delta_R(U) = 0$, that is, X_U has no free summands by Corollary 4.6. Now consider the following commutative exact diagram:

$$
\begin{array}{ccccccccc}
 & & 0 & & 0 & & & & \\
 & & \downarrow & & \downarrow & & & & \\
 & & R^m \oplus R^n & = & R^m \oplus R^n & & & & \\
 & & \downarrow & & \downarrow & & & & \\
0 & \longrightarrow & Z & \longrightarrow & X_U \oplus R^n & \longrightarrow & C & \longrightarrow & 0. \\
 & & \downarrow & & \downarrow & & \| & & \\
0 & \longrightarrow & Y_C^{\bar{R}} & \longrightarrow & U \oplus \bar{R}^n & \longrightarrow & C & \longrightarrow & 0 \\
 & & \downarrow & & \downarrow & & & & \\
 & & 0 & & 0 & & & &
\end{array}
$$

Since $\mathrm{pd}_{\bar{R}} Y_C^{\bar{R}} < \infty$, we have that $\mathrm{pd}_R Y_C^{\bar{R}} < \infty$. Therefore $\mathrm{pd}_R Z < \infty$ and $0 \to Z \to X \oplus R^n \to C \to 0$ is a Cohen-Macaulay approximation of C over R and we have $\delta_R(C) \leqslant n = \delta_{\bar{R}}(C)$. Since $\delta_{\bar{R}}(C) \leqslant \delta_R(C)$ for all $\bar{R}$-modules C, we obtain that $\delta_R(C) = \delta_{\bar{R}}(C)$ whenever $X_C^{\bar{R}}$ is weakly liftable to R.

Combining this proposition and Corollary 4.6, we have the following.

COROLLARY 5.6. *Let C be in $\mathrm{CM}(\bar{R})$ and $\bar{R}$ is not a summand of C. If C is weakly liftable to R, then $\delta_R^i(C) = 0$ for all $i \geqslant 0$.*

Let k denote the residue field $R/\underline{m}$. If k contains infinitely many elements, there exists an R-regular element $x \in \underline{m} \setminus \underline{m}^2$. As in Section 3, it follows that k as an $\bar{R} = R/(x)$-module is weakly liftable to R. Therefore we have $\sum_{i \geqslant 0} (-1)^i \delta_R^i(k) = 0$. In fact, more has been shown in [1]. We have the following.

PROPOSITION 5.7 [1]. *Let $k = R/\underline{m}$ be the residue field of R. Then $\delta_R^i(k) = 0$ for all $i \geqslant 0$.*

In view of these results we naturally ask whether the converse of Proposition 5.4 is true, that is, given an R-module C such that $\sum_{i \geqslant 0} (-1)^i \delta_R^i(C) = 0$ (or even $\delta_R^i(C) = 0$ for all $i \geqslant 0$) and such that $\mathrm{Ann}_R(C)$ contains R-regular elements. Then is there an R-regular element

$x \in \mathrm{Ann}_R(C)$ such that as an $\bar{R} = R/(x)$-module, C is weakly liftable to R. The answer for this question in general is negative as showed by the following example.

EXAMPLE. Let $R = \mathbf{C}[[t^3, t^4]]$ and $C = R/\underline{a}$, where $\underline{a} = (t^8, t^9)$. Then we know that $\delta^i_R(C) = 0$ for all $i \geq 0$ [7]. If there is an element $f \in \underline{a}$ such that C as an $\bar{R} = R/(f)$-module is weakly liftable to R, then we would have $(t^8, t^9) \simeq C \oplus \Omega^1_{\bar{R}}(C)$. Now $\underline{a}$ has the presentation over R

$$R^2 \xrightarrow{\begin{pmatrix} t^8 & -t^9 \\ t^3 & -t^4 \end{pmatrix}} R^2 - \underline{a} \to 0,$$

and reducing this modulo f we obtain the following exact sequence:

$$\bar{R}^2 \xrightarrow{\begin{pmatrix} \bar{t}^8 & -\bar{t}^9 \\ \bar{t}^3 & -\bar{t}^4 \end{pmatrix}} \bar{R}^2 \to \bar{\underline{a}} \to 0.$$

There exist invertible matrices P and Q in $\mathrm{End}_{\bar{R}}(\bar{R}^2)$ such that

$$Q \begin{pmatrix} \bar{t}^8 & -\bar{t}^9 \\ \bar{t}^3 & -\bar{t}^4 \end{pmatrix} P = \begin{pmatrix} \bar{g} & 0 \\ 0 & \bar{z} \end{pmatrix}$$

with $\bar{g} \in \bar{\underline{a}}$. $(\bar{t}^8, \bar{t}^9, \bar{t}^3, \bar{t}^4) = \bar{\underline{m}} \neq (\bar{g}, \bar{z})$ as ideals in $\bar{R}$. This contradiction shows that $R/\underline{a}$ is not weakly liftable to R with respect to any element in $\underline{a}$.

REFERENCES

1. M. AUSLANDER, Minimal Cohen–Macaulay approximations, in preparation.
2. M. AUSLANDER AND R.-O. BUCHWEITZ, The homological theory of maximal Cohen–Macaulay approximations, preprint, Dept. Math., Univ. of Toronto.
3. M. AUSLANDER AND I. REITEN, On a generalized version of the Nakayama Conjecture, *Proc. Amer. Math. Soc.* **52** (1975), 69–74.
4. M. AUSLANDER AND S. O. SMALØ, Preprojective modules over artin algebras, *J. Algebra* **66** (1980), 61–122.
5. M. F. ATIYAH AND I. G. MACDONALD, "Introduction to Commutative Algebra," Addison–Wesley, Reading, MA, 1969.
6. C. W. CURTIS AND I. REINER, "Methods of Representation Theory," Wiley, New York, 1981.
7. S. DING, Cohen–Macaulay approximations over a local Gorenstein ring, preprint, Brandeis University.
8. S. DING AND Ø. SOLBERG, The Maranda Theorem and liftings of modules, preprint, Mathematics No. 8/1991, Departments of Mathematics, The University of Trondheim, Norway.
9. J. A. GREEN, A lifting theorem for modular representations, *Proc. Roy. Soc. London, Ser. A* (1959), 135–142.
10. P. HILTON AND D. REES, Natural maps of extension functors and a theorem of R. G. Swan, *Proc. Cambridge Philos. Soc. Math. Phys. Sci.* **51** (1961).

11. O. A. LAUDAL, Matric Massey products and formal moduli I, *in* "Lecture Notes in Math.," Vol. 1183, pp. 218–240, Springer, New York, 1986.
12. T. MIYATA, Note on direct summands of modules, *J. Math. Kyoto Univ.* **7**, Vol. 1 (1967), 65–69.
13. H.-J. NASTOLD, Zur Serreschen Multiplizitätstheorie in der arithmetischen Geometrie, *Math. Ann.* **143** (1961), 333–343.
14. K. W. ROGGENKAMP, Lattices over orders II, *in* "Lecture Notes in Math.," p. 142, Springer, New York, 1970.
15. J. THÉVENAZ, Representations of finite groups in characteristic p^r, *J. Algebra* **72** (1981), 478–500.

Printed by Catherine Press, Ltd., Tempelhof 41, B-8000 Brugge, Belgium

CHAPTER X

Relative theory and syzygy modules

[98] Relative homology and representation theory I, Relative homology and homologically finite subcategories (with Ø. Solberg, 1993).
[99] Relative homology and representation theory II, Relative cotilting theory (with Ø. Solberg, 1993).
[102] k-Gorenstein algebras and syzygy modules (with I. Reiten, 1994).
[108] Syzygy modules for noetherian rings (with I. Reiten, 1996).

The theory of contravariantly and covariantly finite subcategories and their relationship with tilting theory was important both for Auslander's work on relative homology and his later work on syzygy modules. In both cases, the theory provided a new point of view on old topics.

Relative homological algebra goes back to work of G. Hochschild and Butler-Horrocks in the fifties. In the papers [98, 99], and also [100, 101], a new twist is given on the original theory by taking advantage of developments in the representation theory of artin algebras. An interesting relationship between relative homology and contravariantly finite subcategories is established, and almost split sequences are used to express the relationship between the relative projective and injective modules in terms of the operation $D\,\mathrm{Tr}$.

More general methods for constructing new contravariantly or covariantly finite subcategories from old ones are found by extending results from [92] to the relative setting. The ordinary theory of tilting and cotilting modules is also generalized to the relative setting, thus giving new interesting relationships between algebras. Recently connections between relative homology and subspace categories of vector space categories have been established and applied.

The starting point of [102] was to investigate the category of dth syzygy modules, for a given positive integer d, illustrating the connection between contravariantly finite resolving subcategories and cotilting modules established in [92]. The crucial question turned out to be whether the category of dth syzygy modules is closed under extensions. This is investigated thoroughly in [102, 108], also in the context of noetherian rings.

The papers [102] and [108] were strongly influenced by Auslander's work from the late sixties, in particular by [27]. With the idea of providing a fruitful noncommutative generalization of the commutative noetherian Gorenstein rings, Auslander introduced the k-Gorenstein rings for a nonnegative integer k. They are defined in terms of flat dimension of the injective modules appearing in the minimal injective resolution of a noetherian ring. The noetherian rings which are k-Gorenstein for all k are now called Auslander rings, or rings satisfying the Auslander conditions, and are quite popular in the theory of noncommutative noetherian rings. The terms Auslander regular and Auslander-Gorenstein are used when, in addition, the ring has finite global dimension or finite injective dimension respectively.

That a noetherian ring is k-Gorenstein is shown to be a sufficient condition for the dth syzygy modules to be closed under extensions for $d \leq k$ in [102]. Necessary and sufficient conditions for the dth syzygy modules to be closed under extensions for $d \leq k$ are established in [108], and have strong relationships to grade conditions investigated in [27].

Relative homology and representation theory I
Relative homology and homologically finite subcategories[1]

M. Auslander[2] and Ø. Solberg[3]

Introduction

While homological algebra has been used extensively in the representation theory of artin algebras, relative homological algebra has been used only sporadically. It is our hope that this series of three papers devoted to applying relative homological algebra to the representation theory of artin algebras will encourage a more systematic use of this theory.

This, the first of the three papers, is devoted to developing certain foundational aspects of the theory of relative homological algebra particular to the category $\operatorname{mod}\Lambda$ of finitely generated modules over an artin algebra Λ. These results are then used to study homologically finite subcategories of $\operatorname{mod}\Lambda$. As an application of the general theory developed in this paper, we develop in the second paper in the series a theory of relative cotilting modules which while analogous to the standard cotilting theory has some particular features of its own. The final paper in the series is devoted to studying particular types of relative cotilting modules. Connections with the theory of Wedderburn correspondence as defined in [**1**] are developed. In addition we show how to construct as endomorphism rings of particular types of cotilting modules artin algebra Γ satisfying (i) the injective dimension of Γ as both a left and a right Γ-module is two and (ii) dominant dimension of Γ is also two. Having given a brief survey of the contents of all the papers in this series, we now turn our attention to giving a more specific description of the content of this paper.

Section one is devoted to studying additive subfunctors F of the additive bifunctor $\operatorname{Ext}^1_\Lambda(\ ,\ )\colon (\operatorname{mod}\Lambda)^{\mathrm{op}} \times \operatorname{mod}\Lambda \to \mathrm{Ab}$. Given such a functor F we say that an exact sequence $\eta\colon 0 \to A \to B \to C \to 0$ in $\operatorname{mod}\Lambda$ is F-exact if η is in $F(C,A)$. A Λ-module P is said to be F-projective if for each F-exact sequence $0 \to A \to B \to C \to 0$, the sequence $0 \to (P,A) \to (P,B) \to (P,C) \to 0$ is exact. The full subcategory of $\operatorname{mod}\Lambda$ consisting of all F-projective modules is denoted by $\mathcal{P}(F)$. Clearly $\mathcal{P}(F)$ contains $\mathcal{P}(\Lambda)$, the full subcategory of $\operatorname{mod}\Lambda$ consisting of all projective Λ-modules. Analogously, a Λ-module I is said to be F-injective if for each F-exact sequence $0 \to A \to B \to C \to 0$, the sequence $0 \to (C,I) \to (B,I) \to (A,I) \to 0$ is exact. The subcategory of $\operatorname{mod}\Lambda$ consisting of all F-injective modules is denoted by $\mathcal{I}(F)$. Clearly $\mathcal{I}(F)$ contains $\mathcal{I}(\Lambda)$,

[1] COMMUNICATIONS IN ALGEBRA, 21(9), 2995–3031 (1993)
[2] Partially supported by NSF Grant No. DMS–8904594.
[3] Supported by the Norwegian Research Council during the preparation of this paper.

the full subcategory of $\operatorname{mod}\Lambda$ consisting of all injective Λ-modules. It is shown that $\mathcal{P}(F) = \operatorname{Tr} D(\mathcal{I}(F)) \cup \mathcal{P}(\Lambda)$ and $\mathcal{I}(F) = D\operatorname{Tr}(\mathcal{P}(F)) \cup \mathcal{I}(\Lambda)$.

Of particular concern to us are the additive subfunctors F of $\operatorname{Ext}^1_\Lambda(\ ,\ )$ which have the property that given any module C in $\operatorname{mod}\Lambda$ there is an F-exact sequence $0 \to A \to P \to C \to 0$ with P in $\mathcal{P}(F)$. Such functors F are said to have enough projectives. Dually, F is said to have enough injectives if for each module A in $\operatorname{mod}\Lambda$ there is an F-exact sequence $0 \to A \to I \to C \to 0$ with I in $\mathcal{I}(F)$. We show that the map $F \mapsto \mathcal{P}(F)$ gives a bijection between additive subfunctors F of $\operatorname{Ext}^1_\Lambda(\ ,\ )$ with enough projectives and contravariantly finite subcategories of $\operatorname{mod}\Lambda$ containing $\mathcal{P}(\Lambda)$. Dually, the map $F \mapsto \mathcal{I}(F)$ gives a bijection between F with enough injectives and covariantly subcategories of $\operatorname{mod}\Lambda$ containing $\mathcal{I}(\Lambda)$. From this it follows that F has enough projectives and injectives if and only if $\mathcal{P}(F)$ is a functorially finite subcategory of $\operatorname{mod}\Lambda$ and an exact sequence $0 \to A \to B \to C \to 0$ is F-exact if $(P, B) \to (P, C) \to 0$ is exact for all modules P in $\mathcal{P}(F)$.

Suppose F has enough projectives. In section 2 we recall the notion of F-exact projective resolution of Λ-modules and the derived functors $\operatorname{Ext}^i_F(C,\ )$ and $\operatorname{Tor}^F_i(B,\ )$ of the functors $\operatorname{Hom}_\Lambda(C,\ )$ and $B \otimes_\Lambda -$ respectively. Since the functors $F(C,\ )$ and $\operatorname{Ext}^1_F(C,\ )$ are naturally isomorphic, the derived functors $\operatorname{Ext}^i_F(C,\ )$ and $\operatorname{Tor}^F_i(B,\ )$ give ways of studying amongst other things the functor F. First we show that there are isomorphisms $\phi^i\colon \operatorname{Ext}^i_F(A, DB) \to D\operatorname{Tor}^F_i(B, A)$ for all $i > 0$ which are functorial in both variables. This is then applied to prove that there is an isomorphism $\operatorname{Ext}^1_F(C, D\operatorname{Tr} A) \simeq D\,\underline{\operatorname{Hom}}_{\mathcal{P}(F)}(A, C)$ functorial in both variables where $\underline{\operatorname{Hom}}_{\mathcal{P}(F)}(A, C)$ is $\operatorname{Hom}_\Lambda(A, C)$ modulo $\mathcal{P}(F)(A, C)$, the subgroup of $\operatorname{Hom}_\Lambda(A, C)$ consisting of all morphisms factoring through modules in $\mathcal{P}(F)$. The reader should observe that these results are relative versions of well known homological facts.

The rest of the paper is devoted to further elaborating the connections between relative homological algebra and homologically finite subcategories of $\operatorname{mod}\Lambda$. The results given here are relative versions of results already obtained in [**4**] and [**3**]. In section three we prove the following. Let $\mathcal{X}$ be a functorially finite subcategory of $\operatorname{mod}\Lambda$ containing $\mathcal{P}(\Lambda)$. For each C in $\operatorname{mod}\Lambda$ let $\cdots \to X_1(C) \xrightarrow{f_1(C)} X_0(C) \to C \to 0$ be exact with each $f_i(C)\colon X_i(C) \to \operatorname{Im} f_i(C)$ being a minimal right $\mathcal{X}$-approximation. For each C in $\operatorname{mod}\Lambda$ and $n > 0$, let $\Omega^n_{\mathcal{X}}(C)$ be $\operatorname{Im} f_n(C)$. Then $\mathcal{X} \cup \Omega^n_{\mathcal{X}}(\operatorname{mod}\Lambda)$ is functorially finite in $\operatorname{mod}\Lambda$ where $\Omega^n_{\mathcal{X}}(\operatorname{mod}\Lambda) = \{\Omega^n_{\mathcal{X}}(C) \mid C \in \operatorname{mod}\Lambda\}$.

Section four is devoted to giving relative versions of the Wakamatsu, Auslander-Reiten and Ringel lemmas given in [**4**].

Throughout this paper Λ denotes an artin algebra and $\operatorname{mod}\Lambda$ denotes the category of finitely generated left Λ-modules. Moreover, we assume that all subcategories are full subcategories.

1. Subfunctors of $\operatorname{Ext}^1_\Lambda(\ ,\ )$

We denote by $(\operatorname{mod}\Lambda)^{\mathrm{op}}$ the opposite category of $\operatorname{mod}\Lambda$. The objects in the product category $(\operatorname{mod}\Lambda)^{\mathrm{op}} \times \operatorname{mod}\Lambda$ are pairs (A, B) with A and B a pair of modules in $\operatorname{mod}\Lambda$. A morphism $(\alpha, \beta)\colon (A, B) \to (A', B')$ between two objects (A, B) and (A', B') in $(\operatorname{mod}\Lambda)^{\mathrm{op}} \times \operatorname{mod}\Lambda$ is a pair of Λ-homomorphisms $\alpha\colon A' \to A$ and $\beta\colon B \to B'$, where composition and addition of morphisms are componentwise. Then a functor $F\colon (\operatorname{mod}\Lambda)^{\mathrm{op}} \times \operatorname{mod}\Lambda \to \mathrm{Ab}$ is called a *bifunctor*, where Ab denotes the category of abelian groups. A bifunctor has the property that $F(C,\ )\colon \operatorname{mod}\Lambda \to \mathrm{Ab}$ is a covariant functor for all modules C in $\operatorname{mod}\Lambda$ and $F(\ , A)\colon \operatorname{mod}\Lambda \to \mathrm{Ab}$ is a contravariant functor for all modules A in $\operatorname{mod}\Lambda$. We say that a bifunctor F is *additive* if these induced functors $F(C,\ )$ and $F(\ ,A)$ are additive for all modules A and C in $\operatorname{mod}\Lambda$. Let $F\colon (\operatorname{mod}\Lambda)^{\mathrm{op}}\times\operatorname{mod}\Lambda\to\mathrm{Ab}$ be a bifunctor. A subfunctor $G\colon (\operatorname{mod}\Lambda)^{\mathrm{op}}\times\operatorname{mod}\Lambda\to\mathrm{Ab}$ of F is said to be an *additive* subfunctor if G is an additive bifunctor.

The functor $\operatorname{Ext}^1_\Lambda(\ ,\ )$ is an additive bifunctor from $(\operatorname{mod}\Lambda)^{\mathrm{op}}\times\operatorname{mod}\Lambda$ to Ab. Let F be an additive subfunctor of $\operatorname{Ext}^1_\Lambda(\ ,\ )$. A short exact sequence

$$\eta\colon 0\to A\to B\to C\to 0$$

is said to be *F-exact* if η is in $F(C,A)$. Then for all pairs of objects (C,A) and (C',A') in $(\operatorname{mod}\Lambda)^{\mathrm{op}}\times\operatorname{mod}\Lambda$ and morphisms $(\alpha,\beta)\colon (C,A)\to(C',A')$ the following diagram is commutative

$$\begin{array}{ccc} F(C,A) & \hookrightarrow & \operatorname{Ext}^1_\Lambda(C,A) \\ \big\downarrow F(\alpha,\beta) & & \big\downarrow \operatorname{Ext}^1_\Lambda(\alpha,\beta) \\ F(C',A') & \hookrightarrow & \operatorname{Ext}^1_\Lambda(C',A'), \end{array}$$

where the maps $\operatorname{Ext}^1_\Lambda(\alpha,\beta)$ and $F(\alpha,\beta)$ are taking the pullback by the map $\alpha\colon C'\to C$ and then taking the pushout by the map $\beta\colon A\to A'$. This implies that every pullback and pushout of an F-exact sequence $0\to A\to B\to C\to 0$ given by maps $C'\to C$ and $A\to A'$ are again F-exact. In particular, a subfunctor F of $\operatorname{Ext}^1_\Lambda(\ ,\ )$ determines a collection of short exact sequences which is closed under pushouts, pullbacks and Baer sums. Conversely, given a collection of short exact sequences which is closed under pushouts, pullbacks and Baer sums, gives rise to a subfunctor of $\operatorname{Ext}^1_\Lambda(\ ,\ )$ in the obvious way.

When constructing subfunctors F of $\operatorname{Ext}^1_\Lambda(\ ,\ )\colon (\operatorname{mod}\Lambda)^{\mathrm{op}}\times\operatorname{mod}\Lambda\to\mathrm{Ab}$, one must in particular show that $F(C,A)$ is a subgroup of $\operatorname{Ext}^1_\Lambda(C,A)$ for all pairs of modules A and C in $\operatorname{mod}\Lambda$. The following result is useful to show this property.

LEMMA 1.1. *Let F be a subfunctor of* $\operatorname{Ext}^1_\Lambda(\ ,\ )\colon (\operatorname{mod}\Lambda)^{\mathrm{op}}\times\operatorname{mod}\Lambda\to\mathrm{Sets}$. *Then F is an additive subfunctor of* $\operatorname{Ext}^1_\Lambda(\ ,\ )\colon (\operatorname{mod}\Lambda)^{\mathrm{op}}\times\operatorname{mod}\Lambda\to\mathrm{Ab}$ *if and only if F is closed under direct sums of F-exact sequences.*

PROOF. Let F be a subfunctor of $\operatorname{Ext}^1_\Lambda(\ ,\ )\colon (\operatorname{mod}\Lambda)^{\mathrm{op}}\times\operatorname{mod}\Lambda\to\mathrm{Sets}$. Then for all modules A and C in $\operatorname{mod}\Lambda$ we have that 0 is in $F(C,A)$ and if η is in $F(C,A)$, then $-\eta$ is in $F(C,A)$. The proof of these facts goes as follows. Let $\eta\colon 0\to A\to B\to C\to 0$ be F-exact. Because $\operatorname{Ext}^1_\Lambda(\ ,\ )$ is an additive bifunctor, the zero element in $\operatorname{Ext}^1_\Lambda(C,A)$ is $\operatorname{Ext}^1_\Lambda(0,\mathrm{id}_A)(\eta)$ and the inverse of η is $\operatorname{Ext}^1_\Lambda(-\mathrm{id}_C,\mathrm{id}_A)(\eta)$. Since F is a subfunctor of $\operatorname{Ext}^1_\Lambda(\ ,\ )$ and η is F-exact, the sequences $\operatorname{Ext}^1_\Lambda(0,\mathrm{id}_A)(\eta)$ and $\operatorname{Ext}^1_\Lambda(-\mathrm{id}_C,\mathrm{id}_A)(\eta)$ are again F-exact. So, $F(C,A)$ is a subgroup of $\operatorname{Ext}^1_\Lambda(C,A)$ for all pairs of modules A and C in $\operatorname{mod}\Lambda$ if and only if $F(C,A)$ is closed under Baer sum.

Assume that F is closed under direct sums of F-exact sequences. We want to show that F is an additive subfunctor of $\operatorname{Ext}^1_\Lambda(\ ,\ )\colon (\operatorname{mod}\Lambda)^{\mathrm{op}}\times\operatorname{mod}\Lambda\to\mathrm{Ab}$. It is enough to show that $F(C,A)$ is a subgroup of $\operatorname{Ext}^1_\Lambda(C,A)$ for a pair of modules A and C in $\operatorname{mod}\Lambda$ and that F is an additive bifunctor. Let $\eta\colon 0\to A\to B\to C\to 0$ and $\eta'\colon 0\to A\to B'\to C\to 0$ be in $F(C,A)$. Let $\Delta\colon C\to C\oplus C$ be given by $c\mapsto(c,c)$ and $\nabla\colon A\oplus A\to A$ be given by $(a,a')\mapsto a+a'$. Then the Baer sum $\eta+\eta'$ is given by $\operatorname{Ext}^1_\Lambda(\Delta,\nabla)(\eta\oplus\eta')$. Since $\eta\oplus\eta'$ is in $F(C\oplus C,A\oplus A)$ and F is a subfunctor of $\operatorname{Ext}^1_\Lambda(\ ,\ )$ by assumption, it follows that $\eta+\eta'$ is in $F(C,A)$ and $F(C,A)$ is a subgroup of $\operatorname{Ext}^1_\Lambda(C,A)$.

Next we want to show that F is an additive subfunctor of $\operatorname{Ext}^1_\Lambda(\ ,\ )$. Given an F-exact sequence $\eta\colon 0\to A\oplus A'\to B\to C\to 0$ we can take the pushouts by the maps $(1,0)\colon A\oplus A'\to A$ and $(0,1)\colon A\oplus A'\to A'$. On the other hand given two F-exact sequences $\eta\colon 0\to A\to B\to C\to 0$ and $\eta'\colon 0\to A'\to B'\to C\to 0$, the direct sum $\eta\oplus\eta'$

is F-exact and therefore the pullback by the diagonal map $\Delta\colon C\oplus C \to C$ is F-exact. This gives maps $F(C, A\oplus A') \to F(C, A)\oplus F(C, A')$ and $F(C, A)\oplus F(C, A') \to F(C, A\oplus A')$. It is easy to see that these maps are inverse isomorphisms. Dually we can show that $F(\ ,A)\colon \operatorname{mod}\Lambda \to \mathrm{Ab}$ is an additive contravariant functor. This completes the proof of the "if" part of the lemma.

Conversely, assume that F is an additive subfunctor of $\operatorname{Ext}^1_\Lambda(\ ,\ )\colon (\operatorname{mod}\Lambda)^{\mathrm{op}} \times \operatorname{mod}\Lambda \to \mathrm{Ab}$. In particular, $F(C, A)$ is a subgroup of $\operatorname{Ext}^1_\Lambda(C, A)$ for all pair of modules A and C in $\operatorname{mod}\Lambda$. Let $\eta\colon 0 \to A \to B \to C \to 0$ be in $F(C, A)$ and $\eta'\colon 0 \to A' \to B' \to C' \to 0$ be in $F(C', A')$. We want to show that $\eta\oplus\eta'$ is in $F(C\oplus C', A\oplus A')$. Let $((1,0),\binom{1}{0})\colon (C, A) \to (C\oplus C', A\oplus A')$ and $((0,1),\binom{0}{1})\colon (C', A') \to (C\oplus C', A\oplus A')$. Denote by η_1 and η'_1 the sequences $\operatorname{Ext}^1_\Lambda((1,0),\binom{1}{0})(\eta)$ and $\operatorname{Ext}^1_\Lambda((0,1),\binom{0}{1})(\eta')$, respectively. Since $F(C\oplus C', A\oplus A')$ is a subgroup of $\operatorname{Ext}^1_\Lambda(C\oplus C', A\oplus A')$, the Baer sum $\eta_1 + \eta'_1$ is again in $F(C\oplus C', A\oplus A')$. To complete the proof we need to show that $\eta_1 + \eta'_1 = \eta\oplus\eta'$. Because $\operatorname{Ext}^1_\Lambda(\ ,\ )$ is an additive bifunctor, there is a group isomorphism

$$\operatorname{Ext}^1_\Lambda(C\oplus C', A\oplus A') \simeq \operatorname{Ext}^1_\Lambda(C, A)\oplus \operatorname{Ext}^1_\Lambda(C, A')\oplus \operatorname{Ext}^1_\Lambda(C', A)\oplus \operatorname{Ext}^1_\Lambda(C', A'),$$

where the isomorphism is given by taking pullbacks to the summands C and C' and then taking pushouts to the summands A and A'. It is easy to see that η_1 and η'_1 correspond to the elements $(\eta, 0, 0, 0)$ and $(0, 0, 0, \eta')$ on the right hand side, respectively. Therefore $\eta_1 + \eta'_1$ corresponds to the element $(\eta, 0, 0, \eta')$. It is not hard to see that the direct sum $\eta\oplus\eta'$ also corresponds to $(\eta, 0, 0, \eta')$. It follows that $\eta_1 + \eta'_1$ is $\eta\oplus\eta'$ and this completes the proof. □

Let F be an additive subfunctor of $\operatorname{Ext}^1_\Lambda(\ ,\ )$. By the above result the direct sum of two F-exact sequences is again F-exact. Conversely, given an F-exact sequence which is a direct sum of two short exact sequences, it is easy to see taking pushouts and pullbacks that each direct summand must be F-exact. We record this property in the following lemma.

LEMMA 1.2. *Let $\eta\colon 0 \to A \to B \to C \to 0$ and $\eta'\colon 0 \to A' \to B' \to C' \to 0$ be two short exact sequences. Then $\eta\oplus\eta'$ is F-exact if and only if both η and η' are F-exact.*

Let F be an additive subfunctor of $\operatorname{Ext}^1_\Lambda(\ ,\ )\colon (\operatorname{mod}\Lambda)^{\mathrm{op}} \times \operatorname{mod}\Lambda \to \mathrm{Ab}$. To every F we associate the subcategory $\mathcal{P}(F)$ of $\operatorname{mod}\Lambda$ consisting of all modules P such that if $0 \to A \to B \to C \to 0$ is F-exact, the sequence $0 \to (P, A) \to (P, B) \to (P, C) \to 0$ is exact. This subcategory is called the *projective modules* of the subfunctor or the *F-projectives.* Dually we associate to every F the subcategory $\mathcal{I}(F)$ of $\operatorname{mod}\Lambda$ consisting of all modules I such that if $0 \to A \to B \to C \to 0$ is F-exact, the sequence $0 \to (C, I) \to (B, I) \to (A, I) \to 0$ is exact. The subcategory $\mathcal{I}(F)$ is called the *injective modules* of the subfunctor or the *F-injectives.* Let $\mathcal{P}(\Lambda)$ denote the subcategory of projective modules in $\operatorname{mod}\Lambda$ and $\mathcal{I}(\Lambda)$ the subcategory of injective modules in $\operatorname{mod}\Lambda$. Clearly, we have that $\mathcal{P}(\Lambda)$ is contained in $\mathcal{P}(F)$ and that $\mathcal{I}(\Lambda)$ is contained in $\mathcal{I}(F)$. An additive subfunctor F is said to have *enough projectives* if for every A in $\operatorname{mod}\Lambda$ there exists an F-exact sequence $0 \to A' \to P \to A \to 0$ with P in $\mathcal{P}(F)$. To have *enough injectives* is defined dually.

The rest of this section is devoted to showing basic facts about additive subfunctors F of $\operatorname{Ext}^1_\Lambda(\ ,\ )$ and the corresponding subcategories $\mathcal{P}(F)$ and $\mathcal{I}(F)$. We show that the map $F \mapsto \mathcal{P}(F)$ is a bijection between additive subfunctors of $\operatorname{Ext}^1_\Lambda(\ ,\ )$ with enough projectives and contravariantly finite subcategories of $\operatorname{mod}\Lambda$ containing $\mathcal{P}(\Lambda)$. Dually, we have that the map $F \mapsto \mathcal{I}(F)$ is a bijection between additive subfunctors of $\operatorname{Ext}^1_\Lambda(\ ,\ )$ with enough injectives and covariantly finite subcategories of $\operatorname{mod}\Lambda$ containing $\mathcal{I}(\Lambda)$. We also characterize the projective and injective modules of an additive subfunctor and show that we have the following connection between injectives and projectives of an additive subfunctor F, namely $\mathcal{I}(F) = D\operatorname{Tr}(\mathcal{P}(F)) \cup \mathcal{I}(\Lambda)$.

From now on we assume that all subfunctors of $\operatorname{Ext}^1_\Lambda(\ ,\ )\colon (\operatorname{mod}\Lambda)^{\mathrm{op}} \times \operatorname{mod}\Lambda \to \mathrm{Ab}$ are additive subfunctors. We start by characterizing the projective and injective modules of a subfunctor of $\operatorname{Ext}^1_\Lambda(\ ,\ )$. The first step in doing this is the following easily proven proposition.

PROPOSITION 1.3. *Let $\eta\colon 0 \to A \to B \to C \to 0$ be an F-exact sequence. For all M in* $\operatorname{mod}\Lambda$ *the sequences*

$$0 \to \operatorname{Hom}_\Lambda(C, M) \to \operatorname{Hom}_\Lambda(B, M) \to \operatorname{Hom}_\Lambda(A, M) \xrightarrow{\partial} F(C, M)$$

and

$$0 \to \operatorname{Hom}_\Lambda(M, A) \to \operatorname{Hom}_\Lambda(M, B) \to \operatorname{Hom}_\Lambda(M, C) \xrightarrow{\partial} F(M, A)$$

are exact.

PROOF. Since η is an F-exact sequence and F is a subfunctor of $\operatorname{Ext}^1_\Lambda(\ ,\ )$, the image of the standard connecting homomorphism is in $F(C, M)$ or $F(M, A)$. Then it follows from the exactness of the standard connecting homomorphism that the sequences are exact. □

As an immediate consequence of this we have the following characterization of the projective and injective modules for a subfunctor F of $\operatorname{Ext}^1_\Lambda(\ ,\ )$.

COROLLARY 1.4. (a) *The following are equivalent.*
(i) *P is in $\mathcal{P}(F)$.*
(ii) *$F(P, A) = 0$ for all modules A in* $\operatorname{mod}\Lambda$.
(b) *The following are equivalent.*
(i) *I is in $\mathcal{I}(F)$.*
(ii) *$F(C, I) = 0$ for all modules C in* $\operatorname{mod}\Lambda$.

The following characterization of F-exact sequences when F has enough projectives or injectives will be useful.

PROPOSITION 1.5. (a) *If F has enough projectives, then an exact sequence $0 \to A \to B \to C \to 0$ is F-exact if and only if $0 \to \operatorname{Hom}_\Lambda(P, A) \to \operatorname{Hom}_\Lambda(P, B) \to \operatorname{Hom}_\Lambda(P, C) \to 0$ is exact for all P in $\mathcal{P}(F)$.*

(b) *If F has enough injectives, then an exact sequence $0 \to A \to B \to C \to 0$ is F-exact if and only if $0 \to \operatorname{Hom}_\Lambda(C, I) \to \operatorname{Hom}_\Lambda(B, I) \to \operatorname{Hom}_\Lambda(A, I) \to 0$ is exact for all I in $\mathcal{I}(F)$.*

PROOF. (a) Let $\eta\colon 0 \to A \to B \to C \to 0$ be F-exact. Then the sequence obtained by applying $\operatorname{Hom}_\Lambda(P,\ )$ for P in $\mathcal{P}(F)$ is short exact by definition.

Let $\eta\colon 0 \to A \to B \xrightarrow{f} C \to 0$ be exact and assume that $0 \to \operatorname{Hom}_\Lambda(P, A) \to \operatorname{Hom}_\Lambda(P, B) \to \operatorname{Hom}_\Lambda(P, C) \to 0$ is exact for all P in $\mathcal{P}(F)$. If F has enough projectives, there exists an F-exact sequence $\theta\colon 0 \to C' \to P \xrightarrow{s} C \to 0$ with P in $\mathcal{P}(F)$. Then there exists a homomorphism $t\colon P \to B$ such that $f \circ t = s$. Therefore η is a pushout of θ and η is F-exact.

The proof when F has enough injectives is similar just using pullbacks instead of pushouts. □

Let $0 \to A \to B \to C \to 0$ be an exact sequence in $\operatorname{mod}\Lambda$. Then by [**1**, III, Corollary 4.2] the sequence $\operatorname{Hom}_\Lambda(X,B) \to \operatorname{Hom}_\Lambda(X,C) \to 0$ is exact if and only if $\operatorname{Hom}_\Lambda(B, D\operatorname{Tr}X) \to \operatorname{Hom}_\Lambda(A, D\operatorname{Tr}X) \to 0$ is exact. Using this fact we have the following consequence.

COROLLARY 1.6. (a) *The subcategory* $\mathcal{I}(F)$ *is equal to* $D\operatorname{Tr}(\mathcal{P}(F)) \cup \mathcal{I}(\Lambda)$.

(b) *The subcategory* $\mathcal{P}(F)$ *is equal to* $\operatorname{Tr}D(\mathcal{I}(F)) \cup \mathcal{P}(\Lambda)$.

(c) *If the subcategory* $\mathcal{P}(F)$ *or equivalently* $\mathcal{I}(F)$ *is of finite type, then the number of nonisomorphic indecomposable modules in* $\mathcal{P}(F)$ *and* $\mathcal{I}(F)$ *are the same.*

Let $\mathcal{X}$ be a subcategory of $\operatorname{mod}\Lambda$. Let $\mathcal{X}$ be a subcategory of $\operatorname{mod}\Lambda$ and $H\colon \operatorname{mod}\Lambda \to \mathrm{Ab}$ a functor. Denote by $H(\mathcal{X})$ the functor $H|_{\mathcal{X}}\colon \mathcal{X} \to \mathrm{Ab}$. The previous result shows that when for example F has enough projectives, the F-exact sequences are exactly those short exact sequences $0 \to A \to B \to C \to 0$ where $0 \to (\mathcal{P}(F), A) \to (\mathcal{P}(F), B) \to (\mathcal{P}(F), C) \to 0$ is exact. But we can do this construction for any subcategory in $\operatorname{mod}\Lambda$ and construct a subfunctor of $\operatorname{Ext}^1_\Lambda(\ ,\ )$. Namely, let $\mathcal{X}$ be a subcategory of $\operatorname{mod}\Lambda$ and for each pair of modules A and C in $\operatorname{mod}\Lambda$ define

$$F_{\mathcal{X}}(C,A) = \{0 \to A \to B \to C \to 0 \mid (\mathcal{X}, B) \to (\mathcal{X}, C) \to 0 \text{ is exact}\}.$$

Dually we define for each pair of modules A and C in $\operatorname{mod}\Lambda$

$$F^{\mathcal{X}}(C,A) = \{0 \to A \to B \to C \to 0 \mid (B, \mathcal{X}) \to (A, \mathcal{X}) \to 0 \text{ is exact}\}.$$

It is not obvious that these constructicns give additive subfunctors of $\operatorname{Ext}^1_\Lambda(\ ,\ )$, a fact we prove next.

PROPOSITION 1.7. *Let* $\mathcal{X}$ *be any subcategory in* $\operatorname{mod}\Lambda$, *then* $F_{\mathcal{X}}$ *and* $F^{\mathcal{X}}$ *are additive subfunctors of* $\operatorname{Ext}^1_\Lambda(\ ,\ )$.

PROOF. We only prove that $F_{\mathcal{X}}$ is an additive subfunctor of $\operatorname{Ext}^1_\Lambda(\ ,\ )$, since the proof for $F^{\mathcal{X}}$ is given by the dual argument.

By Lemma 1.1 it is enough to prove that $F_{\mathcal{X}}$ is closed under direct sums of F-exact sequences and that $F_{\mathcal{X}}$ is a subfunctor of $\operatorname{Ext}^1_\Lambda(\ ,\ )\colon (\operatorname{mod}\Lambda)^{\mathrm{op}} \times \operatorname{mod}\Lambda \to \mathrm{Ab}$.

We first show that $F_{\mathcal{X}}$ is a subfunctor. Let $\eta\colon 0 \to A \to B \to C \to 0$ be in $F_{\mathcal{X}}(C,A)$ and let $t\colon C' \to C$ be a Λ-homomorphism. Then by the definition of a pullback and since η is in $F_{\mathcal{X}}(C,A)$, the pullback ηt,

$$\begin{array}{lccccccccc}
\eta t\colon & 0 & \to & A & \to & E & \to & C' & \to & 0 \\
 & & & \| & & \downarrow & & \downarrow{\scriptstyle t} & & \\
\eta\colon & 0 & \to & A & \to & B & \to & C & \to & 0,
\end{array}$$

is also in $F_{\mathcal{X}}(C',A)$. Let $s\colon A \to A'$ be a Λ-homomorphism. Then since η is in $F_{\mathcal{X}}(C,A)$, the pushout $s\eta$,

$$\begin{array}{lccccccccc}
\eta\colon & 0 & \to & A & \to & B & \to & C & \to & 0 \\
 & & & \downarrow{\scriptstyle s} & & \downarrow & & \| & & \\
s\eta\colon & 0 & \to & A' & \to & E & \to & C & \to & 0,
\end{array}$$

is also in $F_{\mathcal{X}}(C,A')$. It is easy to check that $F_{\mathcal{X}}\colon (\operatorname{mod}\Lambda)^{\mathrm{op}} \times \operatorname{mod}\Lambda \to \mathrm{Sets}$ is a bifunctor, hence $F_{\mathcal{X}}$ is a subfunctor of $\operatorname{Ext}^1_\Lambda(\ ,\ )\colon (\operatorname{mod}\Lambda)^{\mathrm{op}} \times \operatorname{mod}\Lambda \to \mathrm{Sets}$.

Since the subfunctor $F_{\mathcal{X}}$ clearly is closed under direct sums of $F_{\mathcal{X}}$-exact sequences, the proof is complete. □

Again using [**1**, III, Corollary 4.2] we have the following connection between the two different methods of constructing subfunctors of $\operatorname{Ext}^1_\Lambda(\ ,\ )$.

PROPOSITION 1.8. *Let $\mathcal{X}$ be any subcategory of* $\operatorname{mod}\Lambda$, *then* $F_{\mathcal{X}} = F^{D\operatorname{Tr}\mathcal{X}}$.

Our next aim is to characterize when subfunctors F of $\operatorname{Ext}^1_\Lambda(\ ,\ )$ have enough projectives or injectives, since these are the subfunctors of main interest to us in the rest of the paper. Suppose F has enough projectives. By Proposition 1.5 (a) we have that an exact sequence $0 \to A \to B \to C \to 0$ is F-exact if and only if $(\mathcal{P}(F), B) \to (\mathcal{P}(F), C) \to 0$ is exact. From this it follows that $F = F_{\mathcal{P}(F)}$. Before characterizing the subfunctors with this property the following characterization of the projective and injective modules of a subfunctor in terms of almost split sequences is useful. This characterization was pointed out to us by J. Carlson.

PROPOSITION 1.9. (a) *An indecomposable nonprojective module P is in $\mathcal{P}(F)$ if and only if the almost split sequence $0 \to D\operatorname{Tr}P \to E \to P \to 0$ is not F-exact.*

(b) *An indecomposable noninjective module I is in $\mathcal{I}(F)$ if and only if the almost split sequence $0 \to I \to E \to \operatorname{Tr}DI \to 0$ is not F-exact.*

PROOF. We only prove (a), since the proof of (b) is the dual of (a). Assume that P is an indecomposable nonprojective module in $\mathcal{P}(F)$ and that the almost split sequence η ending in P is F-exact. Then η must be split exact, but this is a contradiction, hence η is not F-exact.

Conversely, assume that the almost split sequence $\eta\colon 0 \to D\operatorname{Tr}P \to E \to P \to 0$ is not F-exact. Suppose there is a nonsplit F-exact sequence $0 \to A \to B \xrightarrow{f} P \to 0$. Then, since η is an almost split sequence and f is a nonsplit epimorphism, we have the following commutative exact diagram

$$\begin{array}{ccccccccc} 0 & \to & A & \to & B & \xrightarrow{f} & P & \to & 0 \\ & & \downarrow & & \downarrow & & \| & & \\ 0 & \to & D\operatorname{Tr}P & \to & E & \to & P & \to & 0, \end{array}$$

which shows that η is F-exact. This contradicts our assumption, therefore every F-exact sequence ending in P must split and P is in $\mathcal{P}(F)$. □

The next result is the key to our characterization of the subfunctors F with the property that $F = F_{\mathcal{P}(F)}$.

PROPOSITION 1.10. *Let $\mathcal{X}$ be any additive subcategory of* $\operatorname{mod}\Lambda$. *Then $\mathcal{X} \cup \mathcal{P}(\Lambda) = \mathcal{P}(F_{\mathcal{X}})$.*

PROOF. It follows easily from the definitions involved, that $\mathcal{X} \cup \mathcal{P}(\Lambda)$ is contained in $\mathcal{P}(F_{\mathcal{X}})$. Let M be an indecomposable nonprojective module not in $\mathcal{X}$. We want to show that M is not in $\mathcal{P}(F_{\mathcal{X}})$. Let $\eta\colon 0 \to D\operatorname{Tr}M \to E \to M \to 0$ be the almost split sequence ending in M. Since M is not in $\mathcal{X}$, every homomorphism $f\colon X \to M$ with X in $\mathcal{X}$ is not a splittable epimorphism. Therefore, every homomorphism $f\colon X \to M$ with X in $\mathcal{X}$ can be lifted to E, hence η is an $F_{\mathcal{X}}$-exact sequence. By Proposition 1.9 the module M is not in $\mathcal{P}(F_{\mathcal{X}})$ and it follows that $\mathcal{X} \cup \mathcal{P}(\Lambda) = \mathcal{P}(F_{\mathcal{X}})$. □

The similar result for subfunctors F of $\operatorname{Ext}^1_\Lambda(\ ,\ )$ is not true in general, namely that $F = F_{\mathcal{P}(F)}$. The promised characterization of the subfunctors F with the property that $F = F_{\mathcal{P}(F)}$ follows essentially from Proposition 1.10. We leave it to the reader to fill in the details.

PROPOSITION 1.11. *Let F be a subfunctor of* $\operatorname{Ext}^1_\Lambda(\ ,\ )$. *Then we have the following.*
(a) *The subfunctor F is contained in $F_{\mathcal{P}(F)}$.*

(b) *The subfunctor* $F = F_{\mathcal{P}(F)}$ *if and only if* $F = F_{\mathcal{X}}$ *for some subcategory* $\mathcal{X}$ *in* $\operatorname{mod}\Lambda$.

We leave it to the reader to state and prove dual statements involving $\mathcal{I}(F)$ and $F^{\mathcal{Y}}$. The characterization of when subfunctors of $\operatorname{Ext}^1_\Lambda(\ ,\ )$ has enough projectives or injectives involves the notions of covariantly and contravariantly finite subcategories of $\operatorname{mod}\Lambda$. We recall these notions from [**6**].

Let $\mathcal{X}$ be a subcategory of $\operatorname{mod}\Lambda$. A *right $\mathcal{X}$-approximation* of a module C in $\operatorname{mod}\Lambda$ is a Λ-homomorphism $f\colon X \to C$ with X in $\mathcal{X}$ such that $\operatorname{Hom}_\Lambda(\mathcal{X}, X) \xrightarrow{\operatorname{Hom}_\Lambda(\mathcal{X}, f)} \operatorname{Hom}_\Lambda(\mathcal{X}, C) \to 0$ is exact. The approximation is called *minimal* if f is a right minimal homomorphism. A *left $\mathcal{X}$-approximation* of a module A in $\operatorname{mod}\Lambda$ is a Λ-homomorphism $g\colon A \to X$ with X in $\mathcal{X}$ such that $\operatorname{Hom}_\Lambda(X, \mathcal{X}) \xrightarrow{\operatorname{Hom}_\Lambda(g, \mathcal{X})} \operatorname{Hom}_\Lambda(A, \mathcal{X}) \to 0$ is exact. The approximation is called *minimal* if g is a left minimal homomorphism. A subcategory $\mathcal{X}$ is called *contravariantly finite* in $\operatorname{mod}\Lambda$ if every module in $\operatorname{mod}\Lambda$ has a right $\mathcal{X}$-approximation and a subcategory $\mathcal{Y}$ is called *covariantly finite* in $\operatorname{mod}\Lambda$ if every module in $\operatorname{mod}\Lambda$ has a left $\mathcal{Y}$-approximation. A subcategory $\mathcal{Z}$ is called *functorially finite* in $\operatorname{mod}\Lambda$ if it is both contravariantly and covariantly finite in $\operatorname{mod}\Lambda$. A subcategory $\mathcal{X}$ is said to be *homologically finite* in $\operatorname{mod}\Lambda$ if $\mathcal{X}$ is either a covariantly, contravariantly or functorially finite subcategory of $\operatorname{mod}\Lambda$.

Using the above notions, we have the following characterization of when subfunctors of $\operatorname{Ext}^1_\Lambda(\ ,\ )$ have enough projectives or injectives.

THEOREM 1.12. (a) *A subfunctor F of* $\operatorname{Ext}^1_\Lambda(\ ,\ )$ *has enough projectives if and only if $\mathcal{P}(F)$ is contravariantly finite in* $\operatorname{mod}\Lambda$ *and* $F = F_{\mathcal{P}(F)}$.

(b) *A subfunctor F of* $\operatorname{Ext}^1_\Lambda(\ ,\ )$ *has enough injectives if and only if $\mathcal{I}(F)$ is covariantly finite in* $\operatorname{mod}\Lambda$ *and* $F = F^{\mathcal{I}(F)}$.

PROOF. (a) Suppose that F has enough projectives. It is clear that $\mathcal{P}(F)$ is contravariantly finite in $\operatorname{mod}\Lambda$ and $F = F_{\mathcal{P}(F)}$ by Proposition 1.5 (a).

Suppose that $\mathcal{P}(F)$ is contravariantly finite in $\operatorname{mod}\Lambda$ and that $F = F_{\mathcal{P}(F)}$. Since $\mathcal{P}(F)$ is contravariantly finite in $\operatorname{mod}\Lambda$ and contains $\mathcal{P}(\Lambda)$, we know there is an exact sequence $\eta_C\colon 0 \to C' \to P \to C \to 0$ for all C in $\operatorname{mod}\Lambda$, such that $\operatorname{Hom}_\Lambda(\mathcal{P}(F), P) \to \operatorname{Hom}_\Lambda(\mathcal{P}(F), C) \to 0$ is exact. Hence η_C is in $F_{\mathcal{P}(F)}(C, C') = F(C, C')$ and therefore F has enough projectives.

The proof of (b) is the dual of (a). □

If F has enough injectives, then $\mathcal{I}(F)$ is covariantly finite in $\operatorname{mod}\Lambda$ by Theorem 1.12. By Corollary 1.6 (b) the subcategory $\mathcal{P}(F) = \operatorname{Tr} D(\mathcal{I}(F)) \cup \mathcal{P}(\Lambda)$. We claim that $\mathcal{P}(F)$ is covariantly finite in $\operatorname{mod}\Lambda$. A subcategory $\mathcal{X}$ of $\operatorname{mod}\Lambda$ is covariantly (contravariantly) finite in $\operatorname{mod}\Lambda$ if and only if the subcategory $\operatorname{Tr}\mathcal{X}$ is contravariantly (covariantly) finite in $\operatorname{mod}\Lambda$ by [**6**, Corollary 7.3]. The duality $D\colon \operatorname{mod}\Lambda \to \operatorname{mod}\Lambda^{\mathrm{op}}$ maps contravariantly finite subcategories to covariantly finite subcategories and vice versa. Since $\mathcal{I}(F)$ is covariantly finite in $\operatorname{mod}\Lambda$, it follows that $\operatorname{Tr} D(\mathcal{I}(F))$ is covariantly finite in $\operatorname{mod}\Lambda$. Since a union of two covariantly finite subcategories in $\operatorname{mod}\Lambda$ is again covariantly finite in $\operatorname{mod}\Lambda$, it follows that $\mathcal{P}(\Lambda) \cup \operatorname{Tr} D(\mathcal{I}(F)) = \mathcal{P}(F)$ is covariantly finite in $\operatorname{mod}\Lambda$. So if F also have enough projective the subcategory $\mathcal{P}(F)$ is functorially finite and $F = F_{\mathcal{P}(F)}$. Using this and similar arguments we have the following corollary of Theorem 1.12.

COROLLARY 1.13. *The following are equivalent for a subfunctor F of* $\operatorname{Ext}^1_\Lambda(\ ,\ )$.

(a) *F has enough projectives and injectives.*

(b) *The subcategory $\mathcal{P}(F)$ is functorially finite in* $\operatorname{mod}\Lambda$ *and* $F = F_{\mathcal{P}(F)}$.

(c) *The subcategory* $\mathcal{I}(F)$ *is functorially finite in* $\operatorname{mod}\Lambda$ *and* $F = F^{\mathcal{I}(F)}$.

In the setting of subcategories the above results read as follows.

THEOREM 1.14. (a) *A subcategory* $\mathcal{X}$ *containing* $\mathcal{P}(\Lambda)$ *is a contravariantly finite subcategory of* $\operatorname{mod}\Lambda$ *if and only if* $F_{\mathcal{X}}$ *has enough projectives.*

(b) *A subcategory* $\mathcal{Y}$ *containing* $\mathcal{I}(\Lambda)$ *is a covariantly finite subcategory of* $\operatorname{mod}\Lambda$ *if and only if* $F^{\mathcal{Y}}$ *has enough injectives.*

(c) *Let* $\mathcal{X}$ *be a subcategory of* $\operatorname{mod}\Lambda$ *containing* $\mathcal{P}(\Lambda)$. *Then* $F_{\mathcal{X}} = F^{\mathcal{I}(\Lambda)\cup D\operatorname{Tr}(\mathcal{X})}$ *has enough projectives and injectives if and only if* $\mathcal{X}$ *is a functorially finite subcategory of* $\operatorname{mod}\Lambda$.

(d) *Let* $\mathcal{Y}$ *be a subcategory of* $\operatorname{mod}\Lambda$ *containing* $\mathcal{I}(\Lambda)$. *Then* $F^{\mathcal{Y}} = F_{\mathcal{P}(\Lambda)\cup\operatorname{Tr}D(\mathcal{Y})}$ *has enough projectives and injectives if and only if* $\mathcal{Y}$ *is a functorially finite subcategory of* $\operatorname{mod}\Lambda$.

Let M be a module in $\operatorname{mod}\Lambda$. We denote by $\operatorname{add} M$ the full subcategory of $\operatorname{mod}\Lambda$ consisting of all direct summands of all finite direct sums of copies of M. Let $\mathcal{X}$ be a subcategory of $\operatorname{mod}\Lambda$. The subcategory $\mathcal{X}$ is said to be of *finite type* if $\mathcal{X} \subset \operatorname{add} M$ for some module M in $\operatorname{mod}\Lambda$.

An important class of functorially finite subcategories of $\operatorname{mod}\Lambda$ are the subcategories of $\operatorname{mod}\Lambda$ of finite type, since all subcategories of finite type of $\operatorname{mod}\Lambda$ are functorially finite by [**6**, Proposition 4.2]. So if $\mathcal{X}$ is a subcategory of finite type of $\operatorname{mod}\Lambda$ containing $\mathcal{P}(\Lambda)$, then $F_{\mathcal{X}}$ has enough projectives and injectives by Theorem 1.14.

Combining our previous results we have the following connection between subfunctors of $\operatorname{Ext}^1_\Lambda(\ ,\ )$ with enough projectives or injective and homologically finite subcategories of $\operatorname{mod}\Lambda$, where we make the following definitions. Let $\mathcal{X}$ be a subcategory of $\operatorname{mod}\Lambda$. If $\mathcal{P}(\Lambda)$ is contained in $\mathcal{X}$, the subcategory $\mathcal{X}$ is said to be a *generator*. Dually, when $\mathcal{I}(\Lambda)$ is contained in $\mathcal{X}$, the subcategory $\mathcal{X}$ is called a *cogenerator*.

THEOREM 1.15. (a) *The maps* $F \mapsto \mathcal{P}(F)$ *and* $\mathcal{X} \mapsto F_{\mathcal{X}}$ *induce inverse bijections between additive subfunctors* F *of* $\operatorname{Ext}^1_\Lambda(\ ,\ )$ *with enough projectives and contravariantly finite generators of* $\operatorname{mod}\Lambda$.

(b) *The maps* $F \mapsto \mathcal{I}(F)$ *and* $\mathcal{Y} \mapsto F^{\mathcal{Y}}$ *induce inverse bijections between additive subfunctors* F *of* $\operatorname{Ext}^1_\Lambda(\ ,\ )$ *with enough injectives and covariantly finite cogenerators of* $\operatorname{mod}\Lambda$.

2. Relative homology

Let F be a subfunctor of $\operatorname{Ext}^1_\Lambda(\ ,\ )$. In this section we recall the notion of F-exact projective and injective resolutions of Λ-modules and the derived functors of $\operatorname{Hom}_\Lambda(C,\ )$, $\operatorname{Hom}_\Lambda(\ ,A)$, $B\otimes_\Lambda -$ and $-\otimes_\Lambda A$ for all modules A and C in $\operatorname{mod}\Lambda$ and B in $\operatorname{mod}\Lambda^{\operatorname{op}}$. Denote the corresponding derived functors with $\operatorname{Ext}^i_F(\ ,\ )$ and $\operatorname{Tor}^F_i(\ ,\ )$, respectively. Then we show that there are isomorphisms $\phi^i\colon \operatorname{Ext}^i_F(A,DB) \to D\operatorname{Tor}^F_i(B,A)$ functorial in both variables for all $i > 0$. This is used to prove that there is an isomorphism $\operatorname{Ext}^1_F(C, D\operatorname{Tr}A) \to D\,\underline{\operatorname{Hom}}_F(A,C)$ functorial in both variables for all modules A and C in $\operatorname{mod}\Lambda$, where $\underline{\operatorname{Hom}}_F(A,C)$ denotes homomorphism group $\operatorname{Hom}_\Lambda(A,C)$ modulo the morphisms that factor through modules in $\mathcal{P}(F)$.

Let F be a subfunctor of $\operatorname{Ext}^1_\Lambda(\ ,\ )$. Suppose F has enough projectives. Then for any A in $\operatorname{mod}\Lambda$ there is an exact sequence of Λ-modules of the form

$$\cdots \to P_n \xrightarrow{d_n} P_{n-1} \xrightarrow{d_{n-1}} \cdots \to P_1 \xrightarrow{d_1} P_0 \xrightarrow{d_0} A \to 0,$$

where P_i is in $\mathcal{P}(F)$ for $i \geq 0$ and $0 \to \operatorname{Im} d_{i+1} \to P_i \to \operatorname{Im} d_i \to 0$ is F-exact for all $i \geq 0$. Such a sequence is called an *F-exact projective resolution* of A. If F has enough injectives, there exists an exact sequence of Λ-modules for any module A of the form

$$0 \to A \to I_0 \xrightarrow{d_0} I_1 \xrightarrow{d_1} \cdots \to I_{n-1} \xrightarrow{d_{n-1}} I_n \xrightarrow{d_n} \cdots,$$

where I_i is in $\mathcal{I}(F)$ for $i \geq 0$ and $0 \to \ker d_i \to I_i \to \ker d_{i+1} \to 0$ is F-exact for all $i \geq 0$. Such a sequence is called an *F-exact injective resolution* of A. Then as for the case when $\mathcal{P}(F) = \mathcal{P}(\Lambda)$ one can define the right derived functors of $\operatorname{Hom}_\Lambda(C, \ )$ and $\operatorname{Hom}_\Lambda(\ , A)$ for any Λ-modules A and C when F has enough injectives and enough projectives respectively. Also if F has enough projectives one can define the left derived functors of $B \otimes_\Lambda -$ for any Λ^{op}-module B. Let F^{op} be the subfunctor of $\operatorname{Ext}^1_{\Lambda^{\mathrm{op}}}(\ ,\ )$ given by

$$F^{\mathrm{op}}(C, A) = \{\eta \colon 0 \to A \to B \to C \to 0 \mid D(\eta) \in F(DA, DC)\}$$

for all modules A and C in $\operatorname{mod} \Lambda^{\mathrm{op}}$. Then $\mathcal{P}(F^{\mathrm{op}}) = D\mathcal{I}(F)$ and $\mathcal{I}(F^{\mathrm{op}}) = D\mathcal{P}(F)$. If F^{op} has enough projectives or equivalently F has enough injectives, one can define the left derived functors of $- \otimes_\Lambda A$ for any module A in $\operatorname{mod} \Lambda$. These constructions were introduced in **[7]** and **[8]** and we refer to these papers for details.

Similarly as in the standard case it is shown that all these constructions are independent of the choice of resolutions. Moreover, if F has enough projectives and injectives, the right derived functors of $\operatorname{Hom}_\Lambda(C, \ )$ and $\operatorname{Hom}_\Lambda(\ , A)$ using injective and projective resolutions, respectively, coincide. We denote by $\operatorname{Ext}^i_F(C, \ )$ the right derived functors of $\operatorname{Hom}_\Lambda(C, \ )$ and by $\operatorname{Ext}^i_F(\ , A)$ the right derived functors of $\operatorname{Hom}_\Lambda(\ , A)$ for all modules A and C in $\operatorname{mod} \Lambda$. Similar to the proof that $\operatorname{Ext}^1_\Lambda(C, A)$ corresponds to equivalence classes of short exact sequences, one can show that if F has enough projectives or injectives, then $\operatorname{Ext}^1_F(C, A) = F(C, A)$. If F has enough projectives and injectives, the left derived functors of $B \otimes_\Lambda -$ and $- \otimes_\Lambda A$ using F-exact projective resolutions in $\operatorname{mod} \Lambda$ and F^{op}-exact projective resolutions in $\operatorname{mod} \Lambda^{\mathrm{op}}$, respectively, coincide. We denote by $\operatorname{Tor}^F_i(B, \ )$ the left derived functors of $B \otimes_\Lambda -$ for any module B in $\operatorname{mod} \Lambda^{\mathrm{op}}$ and by $\operatorname{Tor}^F_i(\ , A)$ the left derived functors of $- \otimes_\Lambda A$ for any module A in $\operatorname{mod} \Lambda$.

Moreover, as for the standard case, if F has enough projectives or injectives, short F-exact sequences in $\operatorname{mod} \Lambda$ or short F^{op}-exact sequences in $\operatorname{mod} \Lambda^{\mathrm{op}}$ induce long exact sequences in $\operatorname{Ext}^i_F(C, \ )$, $\operatorname{Ext}^i_F(\ , A)$, $\operatorname{Tor}^F_i(B, \ )$ and $\operatorname{Tor}^F_i(\ , A)$ for all modules A and C in $\operatorname{mod} \Lambda$ and any module B in $\operatorname{mod} \Lambda^{\mathrm{op}}$.

Let $\mathcal{P}(F)(A, B)$ denote all the Λ-homomorphisms from A to B which factor through an object in $\mathcal{P}(F)$. Denote by $\underline{\operatorname{Hom}}_{\mathcal{P}(F)}(A, B)$ the factor group $\operatorname{Hom}_\Lambda(A, B)/\mathcal{P}(F)(A, B)$. The rest of this section is devoted to giving the relative version of the Auslander-Reiten isomorphism $\operatorname{Ext}^1_\Lambda(C, D \operatorname{Tr} A) \simeq D \underline{\operatorname{Hom}}_\Lambda(A, B)$.

Assume that F is a subfunctor of $\operatorname{Ext}^1_\Lambda(\ ,\ )$ with enough projectives. Then for each C in $\operatorname{mod} \Lambda$ there is an F-exact sequence $0 \to C' \to P \to C \to 0$ with P in $\mathcal{P}(F)$. It is easily seen that the sequence

$$\operatorname{Hom}_\Lambda(A, P) \to \operatorname{Hom}_\Lambda(A, C) \to \underline{\operatorname{Hom}}_{\mathcal{P}(F)}(A, C) \to 0$$

is exact for all A in $\operatorname{mod} \Lambda$. The following lemma is the first of two steps to prove the relative Auslander-Reiten isomorphism.

LEMMA 2.1. *For all A in $\operatorname{mod} \Lambda$ and B in $\operatorname{mod} \Lambda^{\mathrm{op}}$ we have isomorphisms*

$$\phi^i \colon \operatorname{Ext}^i_F(A, DB) \to D \operatorname{Tor}^F_i(B, A)$$

functorial in both variables for all $i > 0$.

PROOF. Let A be in $\operatorname{mod}\Lambda$ and let $\cdots \to P_n \to P_{n-1} \to \cdots \to P_0 \to A \to 0$ be an F-exact projective resolution of A. Then by the adjointness isomorphism we have the following commutative diagram for B in $\operatorname{mod}\Lambda^{\mathrm{op}}$

$$\begin{array}{ccccccccc} 0 & \to & (A, DB) & \to & (P_0, DB) & \to & (P_1, DB) & \to & \cdots \\ & & \downarrow\wr & & \downarrow\wr & & \downarrow\wr & & \\ 0 & \to & D(B \otimes_\Lambda A) & \to & D(B \otimes_\Lambda P_0) & \to & D(B \otimes_\Lambda P_1) & \to & \cdots, \end{array}$$

where the cohomology of the upper sequence is $\operatorname{Ext}^i_F(A, DB)$ and the lower sequence is $D\operatorname{Tor}^F_i(B, A)$, hence our desired result. Because the adjointness isomorphism is functorial in both variables, it follows immediately that ϕ^i is functorial in both variables. □

LEMMA 2.2. *For all A and B in $\operatorname{mod}\Lambda$ we have an isomorphism*

$$\psi\colon \operatorname{Tor}^F_1(\operatorname{Tr} A, B) \simeq \underline{\operatorname{Hom}}_{\mathcal{P}(F)}(A, B)$$

functorial in both variables.

PROOF. Let $Q_1 \to Q_0 \to A \to 0$ be the minimal projective resolution of A and let $P_2 \to P_1 \to P_0 \to B \to 0$ an F-exact projective resolution of B, where A and B are in $\operatorname{mod}\Lambda$. Let M^* denote the Λ^{op}-module $\operatorname{Hom}_\Lambda(M, \Lambda)$ for M in $\operatorname{mod}\Lambda$. Then we have the following commutative diagram

$$\begin{array}{ccccccccc} Q_0^* \otimes_\Lambda P_2 & \to & Q_0^* \otimes_\Lambda P_1 & \to & Q_0^* \otimes_\Lambda P_0 & \to & Q_0^* \otimes_\Lambda B & \to & 0 \\ \downarrow & & \downarrow & & \downarrow & & \downarrow & & \\ Q_1^* \otimes_\Lambda P_2 & \to & Q_1^* \otimes_\Lambda P_1 & \to & Q_1^* \otimes_\Lambda P_0 & \to & Q_1^* \otimes_\Lambda B & \to & 0 \\ \downarrow & & \downarrow & & \downarrow & & \downarrow & & \\ \operatorname{Tr} A \otimes_\Lambda P_2 & \to & \operatorname{Tr} A \otimes_\Lambda P_1 & \to & \operatorname{Tr} A \otimes_\Lambda P_0 & \to & \operatorname{Tr} A \otimes_\Lambda B & \to & 0. \\ \downarrow & & \downarrow & & \downarrow & & \downarrow & & \\ 0 & & 0 & & 0 & & 0 & & \end{array}$$

Denote $\underline{\operatorname{Hom}}_{\mathcal{P}(F)}(A, B)$ by X. Since $P^* \otimes_\Lambda M \simeq \operatorname{Hom}_\Lambda(P, M)$ for all M in $\operatorname{mod}\Lambda$ and P in $\mathcal{P}(\Lambda)$, it follows that we have the following commutative diagram

$$\begin{array}{ccccccccc} 0 & & 0 & & 0 & & 0 & & \\ \downarrow & & \downarrow & & \downarrow & & \downarrow & & \\ (A, P_2) & \to & (A, P_1) & \to & (A, P_0) & \to & (A, B) & \to & X \to 0 \\ \downarrow & & \downarrow & & \downarrow & & \downarrow & & \\ (Q_0, P_2) & \to & (Q_0, P_1) & \to & (Q_0, P_0) & \to & (Q_0, B) & \to & 0 \\ \downarrow & & \downarrow & & \downarrow & & \downarrow & & \\ (Q_1, P_2) & \to & (Q_1, P_1) & \to & (Q_1, P_0) & \to & (Q_1, B) & \to & 0 \\ \downarrow & & \downarrow & & \downarrow & & \downarrow & & \\ \operatorname{Tr} A \otimes_\Lambda P_2 & \to & \operatorname{Tr} A \otimes_\Lambda P_1 & \to & \operatorname{Tr} A \otimes_\Lambda P_0 & \to & \operatorname{Tr} A \otimes_\Lambda B & \to & 0, \\ \downarrow & & \downarrow & & \downarrow & & \downarrow & & \\ 0 & & 0 & & 0 & & 0 & & \end{array}$$

where all the vertical sequences are exact and the middle two horizontal rows are exact. By diagram chasing we have inverse isomorphisms $\operatorname{Tor}^F_1(\operatorname{Tr} A, B) \to \underline{\operatorname{Hom}}_{\mathcal{P}(F)}(A, B)$ and $\underline{\operatorname{Hom}}_{\mathcal{P}(F)}(A, B) \to \operatorname{Tor}^F_1(\operatorname{Tr} A, B)$, hence our desired result. Because the isomorphism $P^* \otimes_\Lambda M \simeq \operatorname{Hom}_\Lambda(P, M)$ is functorial in both variables and $\operatorname{Hom}_\Lambda(\ ,\ )$ is a bifunctor, we have that the induced isomorphism ψ between $\operatorname{Tor}_1^F(\operatorname{Tr} A, B)$ and $\underline{\operatorname{Hom}}_{\mathcal{P}(F)}(A, B)$ is also functorial in both variables. □

Combining the two lemmas above we have the relative version of the Auslander-Reiten isomorphism.

PROPOSITION 2.3. *Let F be a subfunctor of $\operatorname{Ext}^1_\Lambda(\ ,\ )$ with enough projectives. Then for all modules A and C in* $\operatorname{mod}\Lambda$ *we have an isomorphism*

$$\operatorname{Ext}^1_F(C, D\operatorname{Tr} A) \simeq D\,\underline{\operatorname{Hom}}_{\mathcal{P}(F)}(A, C)$$

functorial in both variables.

Remark: When F has enough projectives, this result can be used to prove $\mathcal{P}(F) = \mathcal{P}(\Lambda) \cup \operatorname{Tr} D\mathcal{I}(F)$ and that $\mathcal{I}(F) = \mathcal{I}(\Lambda) \cup D\operatorname{Tr}\mathcal{P}(F)$.

3. Homologically finite subcategories

Let $\Omega^n_\Lambda(\operatorname{mod}\Lambda)$ denote the full subcategory of $\operatorname{mod}\Lambda$ consisting of all the direct summands of all the n-th syzygies of all the modules in $\operatorname{mod}\Lambda$. It is proved in **[3, 5]** that $\Omega^n_\Lambda(\operatorname{mod}\Lambda)$ is functorially finite for all positive n using a construction due to Smalø. This section is devoted to proving the following generalization of this result. Let $\mathcal{X}$ be a functorially finite subcategory of $\operatorname{mod}\Lambda$ containing $\mathcal{P}(\Lambda)$. For each C in $\operatorname{mod}\Lambda$ let $\cdots \to X_1(C) \xrightarrow{f_1(C)} X_0(C) \to C \to 0$ be exact with each $f_i(C)\colon X_i(C) \to \operatorname{Im} f_i(C)$ being a minimal right $\mathcal{X}$-approximation. For each C in $\operatorname{mod}\Lambda$ and $n > 0$, let $\Omega^n_{\mathcal{X}}(C)$ be $\operatorname{Im} f_n(C)$. Then we show that $\mathcal{X} \cup \Omega^n_{\mathcal{X}}(\operatorname{mod}\Lambda)$ is functorially finite in $\operatorname{mod}\Lambda$, where $\Omega^n_{\mathcal{X}}(\operatorname{mod}\Lambda) = \{\Omega^n_{\mathcal{X}}(C) \mid C \in \operatorname{mod}\Lambda\}$.

The subcategory $\Omega^n_\Lambda(\operatorname{mod}\Lambda)$ is functorially finite in $\operatorname{mod}\Lambda$ for all positive n. This result can be formulated in the following way. For each module C in $\operatorname{mod}\Lambda$ let $\cdots \to P_1(C) \xrightarrow{f_1(C)} P_0(C) \to C \to 0$ be exact with $f_i(C)\colon P_i(C) \to \operatorname{Im} f_i(C)$ a projective cover. For each $n > 0$ denote by $\Omega^n_{\mathcal{P}(\Lambda)}(C)$ the module $\operatorname{Im} f_i(C)$. Then the above result states that $\mathcal{P}(\Lambda) \cup \Omega^n_{\mathcal{P}(\Lambda)}(\operatorname{mod}\Lambda)$ is functorially finite in $\operatorname{mod}\Lambda$ where $\Omega^n_{\mathcal{P}(\Lambda)}(\operatorname{mod}\Lambda) = \{\Omega^n_{\mathcal{P}(\Lambda)}(C) \mid C \in \operatorname{mod}\Lambda\}$. The subcategory $\mathcal{P}(\Lambda)$ is functorially finite in $\operatorname{mod}\Lambda$ and contains Λ. This turns out to be the essential assumption in this result. Let $\mathcal{X}$ be a functorially finite generator subcategory of $\operatorname{mod}\Lambda$, where we use the same notation as given in the introduction to this section. Our aim in this section is to prove the following result.

PROPOSITION 3.1. *The subcategory $\mathcal{X} \cup \Omega^n_{\mathcal{X}}(\operatorname{mod}\Lambda)$ is functorially finite in* $\operatorname{mod}\Lambda$ *for all positive n.*

Remark: (1) Note that $\Omega_{\mathcal{X}}(\operatorname{mod}\Lambda)$ is the same as the kernels of all the minimal right $\mathcal{X}$-approximations.

(2) The above result is one of the few results where one is able to prove something about homologically finite subcategories $\mathcal{X}$ of $\operatorname{mod}\Lambda$ without assuming that $\mathcal{X}$ is closed under extensions.

This result was inspired by the standard result by viewing it in a relative setting. Let $F = F_{\mathcal{X}}$, where $\mathcal{X}$ is as above. Then $\mathcal{P}(F) = \mathcal{X}$ by Proposition 1.10 and F has enough projectives and injectives by Corollary 1.13. For any module M in $\operatorname{mod}\Lambda$ let $\Omega^n_F(M)$ and $\Omega^{-n}_F(M)$ denote the n-th syzygy and the n-th cosyzygy in a minimal F-exact projective resolution and in a minimal F-exact injective resolution, respectively. Then we have that $\Omega^n_{\mathcal{X}}(C) = \Omega^n_F(C)$ for all modules C in $\operatorname{mod}\Lambda$ and $n > 0$. We split the proof of Proposition 3.1 in two parts, where we first prove that $\mathcal{X} \cup \Omega^n_{\mathcal{X}}(\operatorname{mod}\Lambda)$ is contravariantly finite in $\operatorname{mod}\Lambda$.

PROPOSITION 3.2. *Let $\mathcal{X}$ be a functorially finite generator of* $\operatorname{mod}\Lambda$. *Then $\mathcal{X} \cup \Omega^n_{\mathcal{X}}(\operatorname{mod}\Lambda)$ is contravariantly finite in* $\operatorname{mod}\Lambda$ *for all positive n. For all M in* $\operatorname{mod}\Lambda$ *a right $(\mathcal{X} \cup \Omega^n_F(\operatorname{mod}\Lambda))$-approximation of M is given by $\Omega^n_F(\Omega^{-n}_F(M)) \oplus X_M \to M$ for some X_M is in $\mathcal{X}$.*

PROOF. The proof goes by induction on n. First let $n = 1$ and let M be an arbitrary module in $\operatorname{mod}\Lambda$. Then we can construct the following commutative exact diagram

$$\begin{array}{cccccccccc}
 & & & 0 & & 0 & & & & \\
 & & & \downarrow & & \downarrow & & & & \\
 & & & \Omega^1_F(I) & = & \Omega^1_F(I) & & & & \\
 & & & \downarrow & & \downarrow & & & & \\
\eta' \colon & 0 & \to & Z & \xrightarrow{s} & P_I & \xrightarrow{t} & \Omega^{-1}_F(M) & \to & 0 \\
 & & & \downarrow{\scriptstyle f} & & \downarrow{\scriptstyle f'} & & \| & & \\
\eta \colon & 0 & \to & M & \xrightarrow{\alpha} & I & \xrightarrow{\beta} & \Omega^{-1}_F(M) & \to & 0, \\
 & & & \downarrow & & \downarrow & & \downarrow & & \\
 & & & 0 & & 0 & & 0 & &
\end{array}$$

where the diagram is induced by starting with the minimal right $\mathcal{X}$-approximation $0 \to \Omega^1_F(I) \to P_I \to I \to 0$. We want to show that the leftmost vertical sequence θ is a right $(\mathcal{X} \cup \Omega^1_F(\operatorname{mod}\Lambda))$-approximation of M. It is straightforward to see that η' is F-exact. Since P_I is in $\mathcal{P}(F)$ and the map t is not necessarily right minimal, the module Z is in $\mathcal{X} \cup \Omega^1_F(\operatorname{mod}\Lambda)$ and $Z \simeq \Omega^1_F(\Omega^{-1}_F(M)) \oplus X$ for some X in $\mathcal{X}$. The sequence θ is F-exact, because it is a pullback of an F-exact sequence. In order to show that θ is a right $(\mathcal{X} \cup \Omega^1_F(\operatorname{mod}\Lambda))$-approximation it is sufficient to prove that every homomorphism $g \colon B \to M$ with B in $\Omega^1_F(\operatorname{mod}\Lambda)$ factors through f. Let $g \colon B \to M$ with B in $\Omega^1_F(\operatorname{mod}\Lambda)$. Then B is a direct summand of a module of the form $\Omega^1_F(A)$ for some module A in $\operatorname{mod}\Lambda$. We can assume without loss of generality that $g \colon B = \Omega^1_F(A) \to M$ for some module A in $\operatorname{mod}\Lambda$. Then we have the following commutative exact diagram

$$\begin{array}{ccccccccc}
0 & \to & Z & \xrightarrow{s} & P_I & \xrightarrow{t} & \Omega^{-1}_F(M) & \to & 0 \\
 & & \downarrow{\scriptstyle f} & & \downarrow{\scriptstyle f'} & & \| & & \\
0 & \to & M & \xrightarrow{\alpha} & I & \xrightarrow{\beta} & \Omega^{-1}_F(M) & \to & 0 \\
 & & \uparrow{\scriptstyle g} & & \uparrow{\scriptstyle g'} & & \uparrow{\scriptstyle g''} & & \\
0 & \to & \Omega^1_F(A) & \xrightarrow{u} & P_A & \xrightarrow{v} & A & \to & 0,
\end{array}$$

where the maps g' and g'' exist by the following arguments. Since the lower sequence is F-exact and I is in $\mathcal{I}(F)$, there exists a map $g' \colon P_A \to I$ such that $\alpha \circ g = g' \circ u$. The map $g'' \colon A \to \Omega^{-1}_F(M)$ is induced from the first commutative square. Since $0 \to \Omega^1_F(I) \to P_I \to I \to 0$ is F-exact with P_I in $\mathcal{P}(F)$, there is a homomorphism $h' \colon P_A \to P_I$ such that $f' \circ h' = g'$. Therefore we have that $t \circ h' = g'' \circ v$ and this induces a map $h \colon \Omega^1_F(A) \to M$ such that $h' \circ u = s \circ h$. It is easily seen that $\alpha \circ f \circ h = \alpha \circ g$ and since α is a monomorphism we have that $g = f \circ h$. Therefore $Z \xrightarrow{f} M$ is a right $(\mathcal{X} \cup \Omega^1_F(\operatorname{mod}\Lambda))$-approximation of M.

Now let $n > 1$ and let M be an arbitrary module in $\operatorname{mod}\Lambda$. By induction we have that $\Omega_F^{n-1}(\Omega_F^{-n}(M)) \oplus X \xrightarrow{f''} \Omega_F^{-1}(M)$ is a right $\mathcal{X} \cup \Omega_F^{n-1}(\operatorname{mod}\Lambda)$-approximation of $\Omega_F^{-1}(M)$ for some X in $\mathcal{X}$. Then we can construct the following commutative exact diagram

$$\begin{array}{ccccccccc}
 & & & & 0 & & 0 & & \\
 & & & & \downarrow & & \downarrow & & \\
 & & & & K' & & K & & \\
 & & & & \downarrow & & \downarrow & & \\
0 & \to & Z & \xrightarrow{s} & Q & \xrightarrow{t} & \Omega_F^{n-1}(\Omega_F^{-n}(M)) \oplus X & \to & 0 \\
 & & \downarrow{\scriptstyle f} & & \downarrow{\scriptstyle f'} & & \downarrow{\scriptstyle f''} & & \\
0 & \to & M & \xrightarrow{\alpha} & I & \xrightarrow{\beta} & \Omega_F^{-1}(M) & \to & 0, \\
 & & & & \downarrow & & \downarrow & & \\
 & & & & 0 & & 0 & &
\end{array}$$

where Q in $\mathcal{X}$ can be chosen such that all the indicated short exact sequences are F-exact (e.g. $Q = P_I \oplus P_{\Omega_F^{n-1}(\Omega_F^{-n}(M))} \oplus X$). Since t is not necessarily a right minimal map and Q is in $\mathcal{P}(F)$, the module Z is in $\mathcal{X} \cup \Omega_F^n(\operatorname{mod}\Lambda)$ and $Z \simeq \Omega_F^n(\Omega_F^{-n}(M)) \oplus X'$ for some X' in $\mathcal{X}$. Let $0 \to \Omega_F^1(M) \to P_M \xrightarrow{\pi} M \to 0$ be F-exact with π right minimal and P_M in $\mathcal{P}(F)$. We want to show that $(f,\pi)\colon Z \oplus P_M \to M$ is a right $(\mathcal{X} \cup \Omega_F^n(\operatorname{mod}\Lambda))$-approximation of M. Every map $g\colon Y \to M$ with Y in $\mathcal{X}$ clearly factors through (f,π), so we only need to show that every map $g\colon B \to M$ with B in $\Omega_F^n(\operatorname{mod}\Lambda)$ factors through (f,π). As before we can assume without loss of generality that $g\colon B = \Omega_F^n(A) \to M$ for some module A in $\operatorname{mod}\Lambda$. Then we have the following commutative exact diagram

$$\begin{array}{ccccccccc}
0 & \to & Z & \xrightarrow{s} & Q & \xrightarrow{t} & \Omega_F^{n-1}(\Omega_F^{-n}(M)) \oplus X & \to & 0 \\
 & & \downarrow{\scriptstyle f} & & \downarrow{\scriptstyle f'} & & \downarrow{\scriptstyle f''} & & \\
0 & \to & M & \xrightarrow{\alpha} & I & \xrightarrow{\beta} & \Omega_F^{-1}(M) & \to & 0 \\
 & & \uparrow{\scriptstyle g} & & \uparrow{\scriptstyle g'} & & \uparrow{\scriptstyle g''} & & \\
0 & \to & \Omega_F^n(A) & \xrightarrow{u} & P & \xrightarrow{v} & \Omega_F^{n-1}(A) & \to & 0,
\end{array}$$

since the bottom row is F-exact and I is in $\mathcal{I}(F)$. Because f'' is a right $(\mathcal{X} \cup \Omega_F^{n-1}(\operatorname{mod}\Lambda))$-approximation, there exists a map $h''\colon \Omega_F^{n-1}(A) \to \Omega_F^{n-1}(\Omega_F^{-n}(M)) \oplus X$ such that $g'' = f''{\circ}h''$. Moreover, P is in $\mathcal{P}(F)$ and the upper sequence is F-exact, so that there exists a map $h'\colon P \to Q$ such that $t{\circ}h' = h''{\circ}v$. Then this induces a homomorphism $h\colon \Omega_F^n(A) \to Z$, such that $s{\circ}h = h'{\circ}u$. It is easy to see that $\beta{\circ}f'{\circ}h' = \beta{\circ}g'$, so that there exists a map $r\colon P \to M$ such that $f'{\circ}h' - g' = \alpha{\circ}r$. Further, it is not hard to show that $f{\circ}h - r{\circ}u = g$. Since P is in $\mathcal{P}(F)$ and $\eta\colon 0 \to C \to Z \oplus P_M \xrightarrow{(f,\pi)} M \to 0$ is F-exact, there is a map $h_0\colon P \to Z \oplus P_M$ such that $r = (f,\pi){\circ}h_0$. Hence it follows that $(f,\pi){\circ}(h_0{\circ}u + (h,0)) = g$. Therefore the sequence η is a right $(\mathcal{X} \cup \Omega_F^n(\operatorname{mod}\Lambda))$-approximation of M.

□

In order to show that $\mathcal{X} \cup \Omega_{\mathcal{X}}^n(\operatorname{mod}\Lambda)$ is covariantly finite in $\operatorname{mod}\Lambda$ we use the following result from [**4**]. In an arbitrary category $\mathcal{C}$ a covariantly or contravariantly finite subcategory is define by the natural generalization of the definitions given in section 1.

PROPOSITION 3.3 ([**4**, Proposition 1.2 (b)]). *Suppose $\mathcal{C}$ and $\mathcal{D}$ are categories and $G\colon \mathcal{C} \to \mathcal{D}$ and $H\colon \mathcal{D} \to \mathcal{C}$ is an adjoint pair of functors with G a left adjoint and H*

a right adjoint. Then $\operatorname{Im} H$ *is covariantly finite in* $\mathcal{C}$ *and the homomorphism* $I \to HG$ *given by the adjunction gives a left* $\operatorname{Im} H$*-approximation* $C \to HG(C)$ *for each* C *in* $\mathcal{C}$.

Let $\underline{\operatorname{mod}}_F \Lambda$ denote the category $\operatorname{mod} \Lambda$ modulo the subcategory $\mathcal{P}(F)$. Recall from section 2 that F^{op} denotes the subfunctor of $\operatorname{Ext}^1_{\Lambda^{\mathrm{op}}}(\ ,\ )$ given by

$$F^{\mathrm{op}}(C, A) = \{\eta \colon 0 \to A \to B \to C \to 0 \mid D(\eta) \in F(DA, DC)\}$$

for all modules A and C in $\operatorname{mod} \Lambda^{\mathrm{op}}$. Then $\Omega^n_F = \Omega^n_{\mathcal{X}}$ is a functor $\Omega^n_F \colon \underline{\operatorname{mod}}_F \Lambda \to \underline{\operatorname{mod}}_{F^{\mathrm{op}}} \Lambda^{\mathrm{op}}$. Our aim is to show that Ω^n_F has a left adjoint and therefore prove that $\Omega^n_{\mathcal{X}}(\operatorname{mod} \Lambda)$ is covariantly finite in $\underline{\operatorname{mod}}_F \Lambda$ by Proposition 3.3. This means that for every module A in $\operatorname{mod} \Lambda$ there exists a Λ-homomorphism $g \colon A \to Z^A$ with Z^A in $\Omega^n_{\mathcal{X}}(\operatorname{mod} \Lambda)$ such that for every Λ-homomorphism $h \colon A \to U$ with U in $\Omega^n_{\mathcal{X}}(\operatorname{mod} \Lambda)$ there exists a Λ-homomorphism $f \colon Z \to U$ such that $h - f \circ g$ factors through a module in $\mathcal{X}$. If $g' \colon A \to X^A$ is a left $\mathcal{X}$-approximation of A, it follows that $(g, g') \colon A \to Z^A \oplus X^A$ is a left $(\mathcal{X} \cup \Omega^n_{\mathcal{X}}(\operatorname{mod} \Lambda))$-approximation of A. This implies that $\mathcal{X} \cup \Omega^n_{\mathcal{X}}(\operatorname{mod} \Lambda)$ is covariantly finite in $\operatorname{mod} \Lambda$. We observed in section 2 that $\mathcal{P}(F^{\mathrm{op}}) = D\mathcal{I}(F)$. From this it is easy to see that the transpose is a functor from $\underline{\operatorname{mod}}_F \Lambda$ to $\underline{\operatorname{mod}}_{F^{\mathrm{op}}} \Lambda^{\mathrm{op}}$. Then using the proof for the standard case in [**3**, Proposition 1.7] verbatim, one can show that $\operatorname{Tr} \Omega^n_{F^{\mathrm{op}}} \operatorname{Tr} \colon \underline{\operatorname{mod}}_F \Lambda \to \underline{\operatorname{mod}}_F \Lambda$ is a left adjoint for $\Omega^n_{\mathcal{X}} \colon \underline{\operatorname{mod}}_F \Lambda \to \underline{\operatorname{mod}}_F \Lambda$. By the above remarks and Proposition 3.3 we have the following result, which completes the proof of Proposition 3.1.

PROPOSITION 3.4. *Let* $\mathcal{X}$ *be a functorially finite generator in* $\operatorname{mod} \Lambda$. *Then* $\mathcal{X} \cup \Omega^n_{\mathcal{X}}(\operatorname{mod} \Lambda)$ *is covariantly finite in* $\operatorname{mod} \Lambda$ *for all positive* n. *For all* M *in* $\operatorname{mod} \Lambda$ *a left* $(\mathcal{X} \cup \Omega^n_F(\operatorname{mod} \Lambda))$*-approximation of* M *is given by* $M \to \Omega^n_F(\operatorname{Tr} \Omega^n_{F^{\mathrm{op}}} \operatorname{Tr}(M)) \oplus X_M$ *for some* X_M *is in* $\mathcal{X}$.

4. Construction of homologically finite subcategories

Throughout this section let F be an additive subfunctor of $\operatorname{Ext}^1_\Lambda(\ ,\ )$. In view of the definitions of covariantly and contravariantly finite subcategories of $\operatorname{mod} \Lambda$ we recalled in section 1, it is natural to make the following definitions in the relative case. Let $\mathcal{X}$ be a subcategory of $\operatorname{mod} \Lambda$. A *right* F*-*$\mathcal{X}$*-approximation* of a module C in $\operatorname{mod} \Lambda$ is an F-exact sequence $0 \to Y \to X \xrightarrow{f} C \to 0$ with X in $\mathcal{X}$ such that $\operatorname{Hom}_\Lambda(\mathcal{X}, X) \to \operatorname{Hom}_\Lambda(\mathcal{X}, C) \to 0$ is exact. The approximation is called *minimal* if f is a right minimal homomorphism. A *left* F*-*$\mathcal{X}$*-approximation* of a module A in $\operatorname{mod} \Lambda$ is an F-exact sequence $0 \to A \xrightarrow{g} X \to Z \to 0$ with X in $\mathcal{X}$ such that $\operatorname{Hom}_\Lambda(X, \mathcal{X}) \to \operatorname{Hom}_\Lambda(A, \mathcal{X}) \to 0$ is exact. The approximation is called *minimal* if g is a left minimal homomorphism. A subcategory $\mathcal{X}$ is called F*-contravariantly finite* in $\operatorname{mod} \Lambda$ if every module in $\operatorname{mod} \Lambda$ has a right F-$\mathcal{X}$-approximation and a subcategory $\mathcal{Y}$ is called F*-covariantly finite* in $\operatorname{mod} \Lambda$ if every module in $\operatorname{mod} \Lambda$ has a left F-$\mathcal{Y}$-approximation. A subcategory $\mathcal{Z}$ is called F*-functorially finite* in $\operatorname{mod} \Lambda$ if it is both F-contravariantly and F-covariantly finite in $\operatorname{mod} \Lambda$. We say that a subcategory $\mathcal{X}$ of $\operatorname{mod} \Lambda$ is F*-homologically finite* in $\operatorname{mod} \Lambda$ if it is either F-covariantly, F-contravariantly or F-functorially finite.

It is clear from the definition of an F-contravariantly finite subcategory $\mathcal{X}$ of $\operatorname{mod} \Lambda$, that $\mathcal{X}$ contains $\mathcal{P}(F)$ and it is contravariantly in $\operatorname{mod} \Lambda$ in the standard sense. A subcategory $\mathcal{X}$ is called an F*-generator* if $\mathcal{X}$ contains $\mathcal{P}(F)$. Dually, a subcategory $\mathcal{Y}$ is called an F*-cogenerator* if $\mathcal{Y}$ contains $\mathcal{I}(F)$. We record the above observation in the following result.

PROPOSITION 4.1. (a) *If a subcategory* $\mathcal{X}$ *is* F*-contravariantly finite in* $\operatorname{mod}\Lambda$, *then* $\mathcal{X}$ *is a contravariantly finite* F*-generator in* $\operatorname{mod}\Lambda$.

(b) *If a subcategory* $\mathcal{Y}$ *is* F*-covariantly finite in* $\operatorname{mod}\Lambda$, *then* $\mathcal{Y}$ *is a covariantly finite* F*-cogenerator in* $\operatorname{mod}\Lambda$.

The converse of the above proposition is not necessarily true in general, but with one additional assumption we have the following equivalence.

PROPOSITION 4.2. *Assume that* F *is a subfunctor of* $\operatorname{Ext}^1_\Lambda(\ ,\ )$ *with the property that* $F = F_{\mathcal{P}(F)}$ *or equivalently that* $F = F^{\mathcal{I}(F)}$.

(a) *A subcategory* $\mathcal{X}$ *is* F*-contravariantly finite in* $\operatorname{mod}\Lambda$ *if and only if* $\mathcal{X}$ *is a contravariantly finite* F*-generator in* $\operatorname{mod}\Lambda$.

(b) *A subcategory* $\mathcal{Y}$ *is* F*-covariantly finite in* $\operatorname{mod}\Lambda$ *if and only if* $\mathcal{Y}$ *is a covariantly finite* F*-cogenerator in* $\operatorname{mod}\Lambda$.

PROOF. We only prove (a) since the proof of (b) is dual. The "only if" part is the previous proposition. Assume that $\mathcal{X}$ is a contravariantly finite F-generator in $\operatorname{mod}\Lambda$. Let $0 \to Y \to X \to C \to 0$ be a right $\mathcal{X}$-approximation of a module C in $\operatorname{mod}\Lambda$. Since $\mathcal{P}(F)$ is contained in $\mathcal{X}$, the sequence $\operatorname{Hom}_\Lambda(\mathcal{P}(F), X) \to \operatorname{Hom}_\Lambda(\mathcal{P}(F), C) \to 0$ is exact. This shows that $0 \to Y \to X \to C \to 0$ is F-exact since $F = F_{\mathcal{P}(F)}$, so it is a right F-$\mathcal{X}$-approximation of C. Therefore $\mathcal{X}$ is F-contravariantly finite in $\operatorname{mod}\Lambda$. □

Remark: (1) The F-homologically finite subcategories are homologically finite in the standard sense, so that this gives a new way of constructing homologically finite subcategories.

(2) Let $F = F_{\mathcal{P}(F)}$ and let $\mathcal{X}$ be an F-generator in $\operatorname{mod}\Lambda$. Then we saw in the proof above that a right F-$\mathcal{X}$-approximation of a module M in $\operatorname{mod}\Lambda$ is the same as a right $\mathcal{X}$-approximation of M. However, in general standard left or right $\mathcal{X}$-approximations are not necessarily left of right F-$\mathcal{X}$-approximations. A direct sum of two short exact sequences is F-exact if and only if each summand is F-exact by Lemma 1.2. Therefore it follows that if a module M in $\operatorname{mod}\Lambda$ has a left or right F-$\mathcal{Y}$-approximation, then it also has a minimal approximation.

This section is mostly devoted to illustrating that some of the results from the standard theory have relative versions. Generally speaking, as we saw in section 3, if the proof of a standard result only depends on arguments involving pullbacks and pushouts, the result has a relative version. Moreover, the proof of the relative result is the same as the standard proof once one observes that whenever one starts with F-exact sequences the sequences constructed are again F-exact. Due to this fact we leave most of the proofs to the reader after giving the references to the standard proofs.

This section consists mainly of generalizations of results from the standard theory where one assumes that certain subcategories are closed under extensions. In the generalizations we assume that the subcategories are closed under F-extensions instead of all extensions. For example let $\mathcal{X} = \operatorname{Sub}(\Lambda)$. Then by [**4**, Proposition 3.5] $\mathcal{X}$ is extension closed if and only if $\operatorname{pd}_{\Lambda^{\mathrm{op}}} I(\Lambda^{\mathrm{op}}) \leq 1$, where $I(\Lambda^{\mathrm{op}})$ is the Λ^{op}-injective envelope of Λ^{op}. Let $F = F^{\mathcal{P}(\Lambda)\cup\mathcal{I}(\Lambda)}$, then it is not hard to see that $\mathcal{X}$ is always closed under F-extensions.

The first result we generalize from the standard theory is Wakamatsu's Lemma, where we include the proof to illustrate the general philosophy mentioned above. Recall that if $\mathcal{X}$ is a subcategory of $\operatorname{mod}\Lambda$ and $H\colon \operatorname{mod}\Lambda \to \operatorname{Ab}$ is a functor, then $H(\mathcal{X})$ denotes the functor $H|_{\mathcal{X}}\colon \mathcal{X} \to \operatorname{Ab}$. For notational convenience we let $\operatorname{Ext}^1_F(\ ,\ )$ denote F even though F has not enough projectives or injectives.

LEMMA 4.3 ([**10**]). *Suppose $\mathcal{X}$ is a subcategory of* $\operatorname{mod}\Lambda$ *which is F-extension closed. Let C be an arbitrary module in* $\operatorname{mod}\Lambda$. *Then we have the following.*

(a) *If $0 \to Y_C \to X_C \to C \to 0$ is a minimal right F-$\mathcal{X}$-approximation of C, then* $\operatorname{Ext}^1_F(\mathcal{X}, Y_C) = 0$.

(b) *If $0 \to C \to X^C \to Y^C \to 0$ is a minimal left F-$\mathcal{X}$-approximation of C, then* $\operatorname{Ext}^1_F(Y^C, \mathcal{X}) = 0$.

PROOF. (a) Let $0 \to Y_C \to X_C \xrightarrow{f} C \to 0$ be a minimal right F-$\mathcal{X}$-approximation of C and let $0 \to Y_C \to B \to X \to 0$ be F-exact with X in $\mathcal{X}$. Then we have the commutative exact pushout diagram

$$\begin{array}{ccccccccc} & & 0 & & 0 & & & & \\ & & \downarrow & & \downarrow & & & & \\ 0 & \to & Y_C & \to & X_C & \to & C & \to & 0 \\ & & \downarrow & & \downarrow & & \| & & \\ 0 & \to & B & \to & E & \to & C & \to & 0. \\ & & \downarrow & & \downarrow & & & & \\ & & X & = & X & & & & \\ & & \downarrow & & \downarrow & & & & \\ & & 0 & & 0 & & & & \end{array}$$

Since $0 \to Y_C \to B \to X \to 0$ is F-exact and $\eta\colon 0 \to X_C \to E \to X \to 0$ is the pushout of this sequence, the sequence η is F-exact. By assumption $\mathcal{X}$ is F-extension closed, so that E must be in $\mathcal{X}$. Since $0 \to Y_C \to X_C \to C \to 0$ is a right F-$\mathcal{X}$-approximation, it follows that we have the following commutative exact diagram

$$\begin{array}{ccccccccc} 0 & \to & Y_C & \to & X_C & \xrightarrow{f} & C & \to & 0 \\ & & \downarrow & & \downarrow & & \| & & \\ 0 & \to & B & \to & E & \to & C & \to & 0 \\ & & \downarrow & & \downarrow & & \| & & \\ 0 & \to & Y_C & \to & X_C & \xrightarrow{f} & C & \to & 0. \end{array}$$

This shows that $0 \to Y_C \to B \to X \to 0$ splits, because f is a minimal right homomorphism and therefore $F(\mathcal{X}, Y_C) = 0$.

The proof of (b) is the dual (a). □

Next we give a characterization of when a contravariantly finite F-generator or a covariantly finite F-cogenerator in $\operatorname{mod}\Lambda$ is F-extension closed inspired by Wakamatsu's Lemma under the assumption that $F = F_{\mathcal{P}(F)}$. Then, as in the standard case, we make the following definitions.

Let $\mathcal{C}$ be any subcategory of $\operatorname{mod}\Lambda$. Then let $\operatorname{Ext}_F(\mathcal{C})$ denote the intersection of all F-extension closed subcategories of $\operatorname{mod}\Lambda$ containing $\mathcal{C}$. The subcategory $\operatorname{Ext}_F(\mathcal{C})$ is obviously F-extension closed and we call it the *F-extension closure* of $\mathcal{C}$ in $\operatorname{mod}\Lambda$. Moreover, for a module M in $\operatorname{mod}\Lambda$ a sequence $M = M_0 \supset M_1 \supset \cdots \supset M_t = 0$ is called an *F-filtration* of M if the induced sequences $0 \to M_{i+1} \to M_i \to M_i/M_{i+1} \to 0$ are F-exact for all $i = 0, \dots, t-1$. Again, let $\mathcal{C}$ be any subcategory of $\operatorname{mod}\Lambda$. Then a module M in $\operatorname{mod}\Lambda$ is said to have an *F-filtration in $\mathcal{C}$* if there is an F-filtration $M = M_0 \supset M_1 \supset \cdots \supset M_t = 0$, where M_i/M_{i+1} is in $\mathcal{C}$ for all $i = 0, \dots, t-1$. Then we can describe $\operatorname{Ext}_F(\mathcal{C})$ in the following way. We leave the proof to the reader.

PROPOSITION 4.4. *Let $\mathcal{C}$ be any additive subcategory of* $\operatorname{mod}\Lambda$ *and assume that* $F = F_{\mathcal{P}(F)}$. *Then* $\operatorname{Ext}_F(\mathcal{C})$ *consists of all the modules in* $\operatorname{mod}\Lambda$ *which are isomorphic to summands of modules which have an* F*-filtration in* $\mathcal{C}$.

For any additive subcategory $\mathcal{C}$ of $\operatorname{mod}\Lambda$ define $\ker_F \mathcal{C}$ to be the F-extension closure of the subcategory of $\operatorname{mod}\Lambda$ consisting of all the modules of the form $\ker f$, where $f\colon C_M \to M$ is a minimal right F-$\mathcal{C}$-approximation. Dually, we define $\operatorname{coker}_F \mathcal{C}$ to be the F-extension closure of the subcategory of $\operatorname{mod}\Lambda$ consisting all the modules of the form $\operatorname{coker} g$, where $g\colon M \to C^M$ is a minimal left F-$\mathcal{C}$-approximation. Also, let $\mathcal{X}_F(\mathcal{C})$ denote the subcategory $\{X \in \operatorname{mod}\Lambda \mid F(X,\mathcal{C}) = 0\}$ and $\mathcal{Y}_F(\mathcal{C})$ the subcategory $\{Y \in \operatorname{mod}\Lambda \mid F(\mathcal{C},Y) = 0\}$. In this language Wakamatsu's Lemma says that if $\mathcal{C}$ is F-extension closed, then $\ker_F \mathcal{C}$ is contained in $\mathcal{Y}_F(\mathcal{C})$ and $\operatorname{coker}_F \mathcal{C}$ is contained in $\mathcal{X}_F(\mathcal{C})$.

We now have the necessary preliminaries to characterize when a contravariantly finite F-generator or a covariantly finite F-cogenerator in $\operatorname{mod}\Lambda$ is F-extension closed, assuming that $F = F_{\mathcal{P}(F)}$. The proof is the same as in the standard case.

PROPOSITION 4.5 ([**4**, Proposition 3.2]). *Assume that* $F = F_{\mathcal{P}(F)}$. *Then we have the following.*

(a) *Suppose* $\mathcal{C}$ *is an* F*-contravariantly finite* F*-generator in* $\operatorname{mod}\Lambda$. *Then* $\mathcal{C}$ *is* F*-extension closed if and only if* $\ker_F \mathcal{C}$ *is contained in* $\mathcal{Y}_F(\mathcal{C})$.

(b) *Suppose* $\mathcal{C}$ *is an* F*-covariantly finite* F*-cogenerator in* $\operatorname{mod}\Lambda$. *Then* $\mathcal{C}$ *is* F*-extension closed if and only if* $\operatorname{coker}_F \mathcal{C}$ *is contained in* $\mathcal{X}_F(\mathcal{C})$.

The following results are key results for constructing homologically finite subcategories from other homologically finite subcategories. We leave the proofs to the reader.

PROPOSITION 4.6 ([**3**, Corollary 1.5]). *Assume that* $F = F_{\mathcal{P}(F)}$. *Suppose* $\mathcal{Y}$ *is a subcategory of* $\operatorname{mod}\Lambda$ *closed under* F*-extensions and let* $\mathcal{X} = \mathcal{X}_F(\mathcal{Y}) = \{X \in \operatorname{mod}\Lambda \mid \operatorname{Ext}^1_F(X,\mathcal{Y}) = 0\}$. *Then the following are equivalent.*

(a) *The functor* $\operatorname{Ext}^1_F(C,\ )|_{\mathcal{Y}}\colon \mathcal{Y} \to \mathrm{Ab}$ *is finitely generated for all modules* C *in* $\operatorname{mod}\Lambda$.

(b) *The subcategory* $\mathcal{X}$ *is a contravariantly finite* F*-generator in* $\operatorname{mod}\Lambda$ *and if* $0 \to Y \to X \to C \to 0$ *is a minimal right* $\mathcal{X}$*-approximation of* C *in* $\operatorname{mod}\Lambda$, *then* Y *is in* $\mathcal{Y}$.

PROPOSITION 4.7 ([**3**, Corollary 1.7]). *Assume that* $F = F_{\mathcal{P}(F)}$. *Suppose* $\mathcal{X}$ *is a subcategory of* $\operatorname{mod}\Lambda$ *closed under* F*-extensions and let* $\mathcal{Y} = \mathcal{Y}_F(\mathcal{X}) = \{Y \in \operatorname{mod}\Lambda \mid \operatorname{Ext}^1_F(\mathcal{X},Y) = 0\}$. *Then the following are equivalent.*

(a) *The functor* $\operatorname{Ext}^1_F(\ ,C)|_{\mathcal{X}}\colon \mathcal{X} \to \mathrm{Ab}$ *is finitely generated for all modules* C *in* $\operatorname{mod}\Lambda$.

(b) *The subcategory* $\mathcal{Y}$ *is a covariantly finite* F*-cogenerator in* $\operatorname{mod}\Lambda$ *and if* $0 \to C \to Y \to X \to 0$ *is a minimal left* $\mathcal{Y}$*-approximation of* C *in* $\operatorname{mod}\Lambda$, *then* X *is in* $\mathcal{X}$.

As an application of these results we have the generalization of the Auslander-Reiten Lemma, which enables us to construct homologically finite subcategories from other homologically finite subcategories.

LEMMA 4.8. *Let* $\mathcal{Z}$ *be an* F*-extension closed subcategory of* $\operatorname{mod}\Lambda$.

(a) *If* $\mathcal{Z}$ *is a covariantly finite subcategory of* $\operatorname{mod}\Lambda$ *and* F *has enough projectives, then the subcategory* $\mathcal{X}_F(\mathcal{Z}) = \{X \in \operatorname{mod}\Lambda \mid \operatorname{Ext}^1_F(X,\mathcal{Z}) = 0\}$ *is an* F*-extension closed contravariantly finite* F*-generator in* $\operatorname{mod}\Lambda$ *and* $\ker_F \mathcal{X}_F(\mathcal{Z})$ *is contained in* $\mathcal{Z}$.

(b) *If* $\mathcal{Z}$ *is a contravariantly finite subcategory of* $\operatorname{mod}\Lambda$ *and* F *has enough injectives, then the subcategory* $\mathcal{Y}_F(\mathcal{Z}) = \{Y \in \operatorname{mod}\Lambda \mid \operatorname{Ext}^1_F(\mathcal{Z},Y) = 0\}$ *is an* F*-extension closed covariantly finite* F*-cogenerator in* $\operatorname{mod}\Lambda$ *and* $\operatorname{coker}_F \mathcal{Y}_F(\mathcal{Z})$ *is contained in* $\mathcal{Z}$.

PROOF. (a) By Proposition 4.6 it is enough to show that $\operatorname{Ext}^1_F(C,\ )|_{\mathcal{Z}}\colon \mathcal{Z} \to \mathrm{Ab}$ is a finitely generated functor for all modules C in $\operatorname{mod}\Lambda$. Let $0 \to A \to P \to C \to 0$ be F-exact with P in $\mathcal{P}(F)$. This induces the exact sequence $(P,\ ) \to (A,\ ) \to \operatorname{Ext}^1_F(C,\ ) \to 0$. Since $\mathcal{Z}$ is covariantly finite in $\operatorname{mod}\Lambda$, the module A has a minimal left $\mathcal{Z}$-approximation $A \to Z^A \to U^A \to 0$ and therefore $(Z^A, \mathcal{Z}) \to (A, \mathcal{Z}) \to 0$ is exact. Combining the two sequences above we get that $(Z^A,\ )|_{\mathcal{Z}} \to \operatorname{Ext}^1_F(C,\ )|_{\mathcal{Z}} \to 0$ is exact and our desired results follow.

(b) The proof of (b) is the dual of (a) using Proposition 4.7. □

Let $\mathcal{Z}$ be a subcategory of $\operatorname{mod}\Lambda$. Since F is a subfunctor of $\operatorname{Ext}^1_\Lambda(\ ,\ )$, the subcategory $\mathcal{X}_F(\mathcal{Z}) = \{X \in \operatorname{mod}\Lambda \mid \operatorname{Ext}^1_F(X, \mathcal{Z}) = 0\}$ contains the subcategory $\mathcal{X}_{\mathrm{std}}(\mathcal{Z}) = \{X \in \operatorname{mod}\Lambda \mid \operatorname{Ext}^1_\Lambda(X, \mathcal{Z}) = 0\}$ in addition to $\mathcal{P}(F)$. Then it is natural to ask when is $\mathcal{X}_F(\mathcal{Z}) = \mathcal{X}_{\mathrm{std}}(\mathcal{Z})$. We answer this question when $F = F_{\mathcal{P}(F)}$ in Proposition 4.11 after some preliminary results.

PROPOSITION 4.9. *Let $\mathcal{Z}$ be a subcategory of* $\operatorname{mod}\Lambda$ *and assume that $F = F_{\mathcal{P}(F)}$. Then $\mathcal{X}_F(\mathcal{Z}) = \{X \in \operatorname{mod}\Lambda \mid \operatorname{Ext}^1_F(X, \mathcal{Z}) = 0\}$ is contained in any subcategory $\mathcal{C}$ of* $\operatorname{mod}\Lambda$ *satisfying the following two conditions*

(a) *The subcategory $\mathcal{C}$ is contravariantly finite F-generator in* $\operatorname{mod}\Lambda$.

(b) *If $0 \to U_M \to C_M \to M \to 0$ is the minimal right $\mathcal{C}$-approximation of M, then U_M is in $\mathcal{Z}$.*

PROOF. The property (a) implies that $\mathcal{C}$ is F-contravariantly finite in $\operatorname{mod}\Lambda$ by Proposition 4.2. Let X be in $\mathcal{X}_F(\mathcal{Z})$ and let $0 \to U_X \to C_X \to X \to 0$ be the minimal right F-$\mathcal{C}$-approximation of X. Then by condition (b) U_X is in $\mathcal{Z}$ and therefore the sequence splits and X is in $\mathcal{C}$. Hence $\mathcal{X}_F(\mathcal{Z})$ is contained in $\mathcal{C}$. □

An immediate consequence is the following.

COROLLARY 4.10. *Let $\mathcal{Z}$ be a subcategory of* $\operatorname{mod}\Lambda$ *and assume that $F = F_{\mathcal{P}(F)}$. Then $\mathcal{X}_F(\mathcal{Z}) = \{X \in \operatorname{mod}\Lambda \mid \operatorname{Ext}^1_F(X, \mathcal{Z}) = 0\}$ is the unique minimal subcategory $\mathcal{C}$ of* $\operatorname{mod}\Lambda$ *satisfying the following two properties*

(a) *The subcategory $\mathcal{C}$ is contravariantly finite F-generator in* $\operatorname{mod}\Lambda$.

(b) *If $0 \to U_M \to C_M \to M \to 0$ is the minimal right $\mathcal{C}$-approximation of M, then U_M is in $\mathcal{Z}$.*

We now show when $\mathcal{X}_F(\mathcal{Z}) = \mathcal{X}_{\mathrm{std}}(\mathcal{Z})$.

PROPOSITION 4.11. *Let $\mathcal{Z}$ be a subcategory of* $\operatorname{mod}\Lambda$ *and assume that $F = F_{\mathcal{P}(F)}$. Then $\mathcal{X}_F(\mathcal{Z}) = \mathcal{X}_{\mathrm{std}}(\mathcal{Z})$ if and only if the following two conditions are satisfied.*

(a) *The subcategory $\mathcal{X}_{\mathrm{std}}(Z)$ is contravariantly finite F-generator in* $\operatorname{mod}\Lambda$.

(b) *If $0 \to U_M \to C_M \to M \to 0$ is the minimal right $\mathcal{X}_{\mathrm{std}}(\mathcal{Z})$-approximation of M, then U_M is in $\mathcal{Z}$.*

PROOF. If $\mathcal{X}_F(\mathcal{Z}) = \mathcal{X}_{\mathrm{std}}(\mathcal{Z})$, then the properties (a) and (b) are satisfied by Proposition 4.9. On the other hand, if $\mathcal{X}_{\mathrm{std}}(\mathcal{Z})$ satisfies the conditions (a) and (b), then by Corollary 4.10 we have that $\mathcal{X}_F(\mathcal{Z})$ is contained in $\mathcal{X}_{\mathrm{std}}(\mathcal{Z})$. Since we always have the opposite inclusion, it follows that $\mathcal{X}_F(\mathcal{Z}) = \mathcal{X}_{\mathrm{std}}(\mathcal{Z})$. □

Suppose $\mathcal{Z}$ is a covariantly finite subcategory of $\operatorname{mod}\Lambda$. Now if $\mathcal{Z}$ is closed under extensions, then we know that the subcategory $\mathcal{X}_{\mathrm{std}}(\mathcal{Z}) = \{X \in \operatorname{mod}\Lambda \mid \operatorname{Ext}^1_\Lambda(X, \mathcal{Z}) = 0\}$ is contravariantly finite in $\operatorname{mod}\Lambda$ and closed under extensions. However it might happen that although the covariantly finite subcategory $\mathcal{Z}$ is not closed under extensions, there is some subfunctor F of $\operatorname{Ext}^1_\Lambda(\ ,\ )$ with enough projectives such that $\mathcal{Z}$ is F-extension closed. Therefore $\mathcal{X}_F(\mathcal{Z}) = \{X \in \operatorname{mod}\Lambda \mid \operatorname{Ext}^1_F(X,\mathcal{Z}) = 0\}$ is contravariantly finite, thus giving a new way of constructing homologically finite subcategories from others. To illustrate this point let $F = F^{\mathcal{P}(\Lambda)\cup\mathcal{I}(\Lambda)}$. We have stated already that $\operatorname{Sub}(\Lambda)$ is not closed under extensions in general, but it is always closed under F-extensions. It is shown in [**6**, Corollary 4.7] that $\operatorname{Sub}(\Lambda)$ is functorially finite in $\operatorname{mod}\Lambda$. Therefore $\mathcal{X}_F(\mathcal{Z}) = \{X \in \operatorname{mod}\Lambda \mid \operatorname{Ext}^1_F(X,\mathcal{Z}) = 0\}$ is contravariantly finite in $\operatorname{mod}\Lambda$. This example will be discussed further at the end of this section.

Let $\mathcal{C}$ be a subcategory of $\operatorname{mod}\Lambda$. Let F-$\operatorname{Rapp}\mathcal{C}$ denote the modules in $\operatorname{mod}\Lambda$ which have a minimal right F-$\mathcal{C}$-approximation and let F-$\operatorname{Lapp}\mathcal{C}$ denote the modules in $\operatorname{mod}\Lambda$ which have a minimal left F-$\mathcal{C}$-approximation. Moreover, let F-$\operatorname{Fac}\mathcal{C}$ denote all the factor modules M of modules in $\mathcal{C}$ such that there exists an F-exact sequence $0 \to C' \to C \to M \to 0$ with C in $\mathcal{C}$. Dually, let F-$\operatorname{Sub}\mathcal{C}$ denote all submodules M of modules in $\mathcal{C}$ such that there exists an F-exact sequence $0 \to M \to C \to N \to 0$ with C in $\mathcal{C}$. Then the following result is another way to find homologically finite subcategories. This result is the analog of [**4**, Proposition 3.10] and the proof is the same as the standard one.

LEMMA 4.12. *Let $\mathcal{C}$ and $\mathcal{Z}$ be two F-extension closed subcategories of* $\operatorname{mod}\Lambda$.

(a) *Assume that F-$\operatorname{Sub}(\mathcal{Z})$ is covariantly finite in* $\operatorname{mod}\Lambda$. *If $\mathcal{Z}$ satisfies the two conditions* (i) $\ker_F\mathcal{C} \subset \mathcal{Z} \subset \mathcal{Y}_F(\mathcal{C})$ *and* (ii) F-$\operatorname{Fac}\mathcal{Z} \subset F$-$\operatorname{Rapp}\mathcal{C}$, *then $\mathcal{Z}$ is covariantly finite in* $\operatorname{mod}\Lambda$.

(b) *Assume that F-$\operatorname{Fac}(\mathcal{Z})$ is contravariantly finite in* $\operatorname{mod}\Lambda$. *If $\mathcal{Z}$ satisfies the two conditions* (i) $\operatorname{coker}_F\mathcal{C} \subset \mathcal{Z} \subset \mathcal{X}_F(\mathcal{C})$ *and* (ii) F-$\operatorname{Sub}\mathcal{Z} \subset F$-$\operatorname{Lapp}\mathcal{C}$, *then $\mathcal{Z}$ is contravariantly finite in* $\operatorname{mod}\Lambda$.

If the subcategory $\mathcal{C}$ is F-contravariantly finite in $\operatorname{mod}\Lambda$, then F-$\operatorname{Rapp}\mathcal{C} = \operatorname{mod}\Lambda$. Specializing this result using this fact we have the following relative version of Ringel's Lemma [**4**, Proposition 3.9].

LEMMA 4.13 ([**9**, Lemma 2]). *Let $\mathcal{C}$ be an F-extension closed F-contravariantly finite subcategory of* $\operatorname{mod}\Lambda$ *and $\mathcal{Z}$ an F-extension closed subcategory of* $\operatorname{mod}\Lambda$. *Assume that F-$\operatorname{Sub}(\mathcal{Z})$ is covariantly finite in* $\operatorname{mod}\Lambda$. *If* $\ker_F\mathcal{C} \subset \mathcal{Z} \subset \mathcal{Y}_F(\mathcal{C})$, *then $\mathcal{Z}$ is covariantly finite in* $\operatorname{mod}\Lambda$.

Assume that F has enough injectives. If a subcategory $\mathcal{Z}$ of $\operatorname{mod}\Lambda$ contains $\mathcal{I}(F)$, then F-$\operatorname{Sub}(\mathcal{Z})$ is equal to $\operatorname{mod}\Lambda$ and clearly covariantly finite in $\operatorname{mod}\Lambda$. Using this fact and its dual we have the following special case of Lemma 4.12.

LEMMA 4.14. *Let $\mathcal{C}$ be a F-extension closed subcategory of* $\operatorname{mod}\Lambda$.

(a) *Assume that F has enough injectives. If $\mathcal{Z}$ is a F-extension closed F-cogenerator satisfying the two conditions* (i) $\ker_F\mathcal{C} \subset \mathcal{Z} \subset \mathcal{Y}_F(\mathcal{C})$ *and* (ii) F-$\operatorname{Fac}\mathcal{Z} \subset F$-$\operatorname{Rapp}\mathcal{C}$, *then $\mathcal{Z}$ is a covariantly finite F-cogenerator in* $\operatorname{mod}\Lambda$.

(b) *Assume that F has enough projectives. If $\mathcal{Z}$ is a F-extension closed F-generator satisfying the two conditions* (i) $\operatorname{coker}_F\mathcal{C} \subset \mathcal{Z} \subset \mathcal{X}_F(\mathcal{C})$ *and* (ii) F-$\operatorname{Sub}\mathcal{Z} \subset F$-$\operatorname{Lapp}\mathcal{C}$, *then $\mathcal{Z}$ is a contravariantly finite F-generator in* $\operatorname{mod}\Lambda$.

As a consequence of Wakamatsu's Lemma and Lemma 4.14 we have the following, whose proof we leave to the reader.

PROPOSITION 4.15 ([**4**, Proposition 3.11]). *Assume that F has enough injectives. Let $\mathcal{C}$ be an F-extension closed contravariantly finite F-generator in* $\operatorname{mod}\Lambda$ *and let $\mathcal{Z}$ be an F-extension closed F-cogenerator in* $\operatorname{mod}\Lambda$. *Then $\mathcal{Z}$ is a covariantly finite F-cogenerator in* $\operatorname{mod}\Lambda$ *with the property that $\mathcal{X}_F(\mathcal{Z}) = \mathcal{C}$ if and only if* $\ker_F\mathcal{C} \subset \mathcal{Z} \subset \mathcal{Y}_F(\mathcal{C})$.

The assumption in Lemma 4.8 (a) is that $\mathcal{Z}$ is a covariantly finite subcategory in $\operatorname{mod}\Lambda$. From the proof of Lemma 4.8 we see that if F has enough projectives, then $\operatorname{Ext}^1_F(C, \ )|_{\mathcal{Z}}$ is finitely generated. On the other hand if we start with a subcategory $\mathcal{Z}$ satisfying the condition that $\operatorname{Ext}^1_F(C, \ )|_{\mathcal{Z}}$ is finitely generated, then we have the following result.

PROPOSITION 4.16 ([**3**, Proposition 1.8]). *Assume that* $F = F_{\mathcal{P}(F)}$. *Suppose that* $\mathcal{Z}$ *is an* F*-extension closed subcategory of* $\operatorname{mod}\Lambda$ *such that the functor* $\operatorname{Ext}^1_F(C, \ )|_{\mathcal{Z}} \colon \mathcal{Z} \to \mathrm{Ab}$ *is finitely generated for all* C *in* $\operatorname{mod}\Lambda$. *Then we have the following.*

(a) *The subcategory* $\mathcal{X}_F(\mathcal{Z}) = \{X \in \operatorname{mod}\Lambda \mid \operatorname{Ext}^1_F(X, \mathcal{Z}) = 0\}$ *is an* F*-extension closed contravariantly finite* F*-generator in* $\operatorname{mod}\Lambda$.

(b) *If* $\mathcal{Y}_F(\mathcal{X}_F(\mathcal{Z})) = \{Y \in \operatorname{mod}\Lambda \mid \operatorname{Ext}^1_F(\mathcal{X}_F, Y) = 0\}$, *then* $\mathcal{X}_F(\mathcal{Z}) = \{X \in \operatorname{mod}\Lambda \mid \operatorname{Ext}^1_F(X, \mathcal{Y}_F(\mathcal{X}_F(\mathcal{Z})) = 0\}$.

(c) *Every module* C *in* $\operatorname{mod}\Lambda$ *has a minimal* $\mathcal{X}_F(\mathcal{Z})$*-approximation* $0 \to Z_C \to X_C \to C \to 0$ *where* Z_C *is in* $\mathcal{Z}$.

If F *has enough injectives, then we have the following.*

(d) *The subcategory* $\mathcal{Y}_F(\mathcal{X}_F(\mathcal{Z})) = \{Y \in \operatorname{mod}\Lambda \mid \operatorname{Ext}^1_F(\mathcal{X}_F(\mathcal{Z}), Y) = 0\}$ *is an* F*-extension closed covariantly finite* F*-cogenerator in* $\operatorname{mod}\Lambda$ *containing* $\mathcal{Z}$.

(e) *If* $0 \to C \to Y^C \to X^C \to 0$ *is a minimal left* $\mathcal{Y}_F(\mathcal{X}_F(\mathcal{Z}))$*-approximation of a module* C *in* $\operatorname{mod}\Lambda$, *then* X^C *is in* $\mathcal{X}_F(\mathcal{Z})$.

The next result is an easy consequence of earlier results in this section and the definitions involved.

PROPOSITION 4.17 ([**3**, Proposition 1.9]). *Assume that* F *has enough projectives and injectives. Let* $\mathcal{X}$ *and* $\mathcal{Y}$ *be subcategories of* $\operatorname{mod}\Lambda$ *satisfying that* X *is in* $\mathcal{X}$ *if and only if* $\operatorname{Ext}^1_F(X, \mathcal{Y}) = 0$ *and* Y *is in* $\mathcal{Y}$ *if and only if* $\operatorname{Ext}^1_F(\mathcal{X}, Y) = 0$. *Let* $\mathcal{W} = \mathcal{X} \cap \mathcal{Y}$. *Then we have the following.*

(a) *The subcategory* $\mathcal{X}$ *is a contravariantly finite* F*-generator in* $\operatorname{mod}\Lambda$ *if and only if the subcategory* $\mathcal{Y}$ *is a covariantly finite* F*-cogenerator in* $\operatorname{mod}\Lambda$.

(b) *If the subcategory* $\mathcal{X}$ *is a contravariantly finite* F*-generator in* $\operatorname{mod}\Lambda$, *then each module* C *in* $\operatorname{mod}\Lambda$ *has a minimal right* $\mathcal{X}$*-approximation* $0 \to Y_C \to X_C \to C \to 0$ *and a minimal left* $\mathcal{Y}$*-approximation* $0 \to C \to Y^C \to X^C \to 0$, *where* Y_C *is in* $\mathcal{Y}$ *and* X^C *is in* $\mathcal{X}$.

(c) $\operatorname{Ext}^1_F(\mathcal{W}, \mathcal{W}) = 0$.

(d) *Suppose* $\mathcal{X}$ *is a contravariantly finite* F*-generator in* $\operatorname{mod}\Lambda$. *Then*

(i) *For each* X *in* $\mathcal{X}$ *there is an* F*-exact sequence* $0 \to X \to W^X \to X' \to 0$ *in* $\mathcal{X}$ *with* W^X *in* $\mathcal{W}$.

(ii) *For each* Y *in* $\mathcal{Y}$ *there is an* F*-exact sequence* $0 \to Y' \to W_Y \to Y \to 0$ *in* $\mathcal{Y}$ *with* W_Y *in* $\mathcal{W}$.

In the following result we describe a situation where the subcategories $\mathcal{X}$ and $\mathcal{Y}$ satisfy the hypothesis of the previous result.

PROPOSITION 4.18 ([**3**, Proposition 1.10]). *Suppose* $\mathcal{X}$ *is an* F*contravariantly finite subcategory of* $\operatorname{mod}\Lambda$ *which is closed under* F*-extensions. Let* $\mathcal{Y} = \{Y \in \operatorname{mod}\Lambda \mid \operatorname{Ext}^1_F(\mathcal{X}, Y) = 0\}$. *Then* $\mathcal{X} = \{X \in \operatorname{mod}\Lambda \mid \operatorname{Ext}^1_F(X, \mathcal{Y}) = 0\}$.

As a consequence of Proposition 4.17 and Proposition 4.18 and their duals we have the following result.

PROPOSITION 4.19 ([**4**, Proposition 3.4]). *Assume that* F *has enough injectives and projectives. Let* $\mathcal{S}$ *be the set of all* F*-extension closed contravariantly finite* F*-generators* *in* $\operatorname{mod}\Lambda$ *and let* $\mathcal{T}$ *be the set of all* F*-extension closed covariantly finite* F*-cogenerators in* $\operatorname{mod}\Lambda$. *Then the maps* $\mathcal{S} \mapsto \mathcal{T}$ *and* $\mathcal{T} \mapsto \mathcal{S}$ *given by* $\mathcal{C} \mapsto \mathcal{Y}_F(\mathcal{C})$ *and* $\mathcal{Z} \mapsto \mathcal{X}_F(\mathcal{Z})$ *are inverse bijections.*

We conclude this section by using the Auslander-Reiten Lemma to construct a homologically finite subcategory of $\operatorname{mod}\Lambda$.

Let M be any module in $\operatorname{mod}\Lambda$ and let $\Gamma = \operatorname{End}_\Lambda(M)$. It is not hard to see that a module A in $\operatorname{mod}\Lambda$ is in $\operatorname{Sub}(M)$ if and only if the natural Λ-homomorphism $A \to \operatorname{Hom}_\Gamma(\operatorname{Hom}_\Lambda(A, M), {}_\Gamma M)$ is a monomorphism. Let $F = F^{\operatorname{add} M \cup \mathcal{I}(\Lambda)}$, then F has enough projective and injectives by Theorem 1.14, since $\operatorname{add} M \cup \mathcal{I}(\Lambda)$ is functorially finite in $\operatorname{mod}\Lambda$. Using the above characterization of the modules in $\operatorname{Sub}(M)$, it follows that $\operatorname{Sub}(M)$ is F-extension closed. Denote by $M(X, Y)$ the subgroup of $\operatorname{Hom}_\Lambda(X, Y)$ consisting of all maps that factor through a module in $\operatorname{add} M$. Then we have the following result.

PROPOSITION 4.20. *Let* M *and* F *be as above. Then*

(a) *The subcategory* $\mathcal{X}_F(\operatorname{Sub}(M)) = \{X \in \operatorname{mod}\Lambda \mid \operatorname{Ext}^1_F(X, \operatorname{Sub}(M)) = 0\}$ *is a contravariantly finite* F*-extension closed subcategory of* $\operatorname{mod}\Lambda$.

(b) *If* $\operatorname{Ext}^1_\Lambda(M, M) = 0$, *then*

$$\mathcal{X}_F(\operatorname{Sub}(M)) = \{X \in \operatorname{mod}\Lambda \mid \operatorname{Hom}_\Lambda(X, \mathcal{A}_M) = M(X, \mathcal{A}_M)\},$$

where $\mathcal{A}_M = \{A \in \operatorname{mod}\Lambda \mid \operatorname{Ext}^1_\Lambda(A, M) = 0\} \cap \operatorname{Fac} M$.

PROOF. (a) Since $\operatorname{Sub}(M)$ is a covariantly finite F-extension closed subcategory of $\operatorname{mod}\Lambda$ by [**6**, Corollary 4.7] and F has enough projectives, the subcategory $\mathcal{X}_F(\operatorname{Sub}(M))$ is a contravariantly finite F-extension closed subcategory of $\operatorname{mod}\Lambda$ by the Auslander-Reiten Lemma (Lemma 4.8).

(b) The subcategory $\operatorname{add} M$ is covariantly finite in $\operatorname{mod}\Lambda$. If Y is in $\operatorname{Sub}(M)$, then the minimal left $\operatorname{add} M$-approximation $\eta\colon 0 \to Y \to M^Y \to A^Y \to 0$ is a monomorphism, F-exact and A^Y is in $\mathcal{A}_M$ by Wakamatsu's Lemma. On the other hand given A in $\mathcal{A}_M$, there is an epimorphism $g\colon M_A \to A$ such that $0 \to Y_A \to M_A \xrightarrow{g} A \to 0$ is F-exact and Y_A is in $\operatorname{Sub}(M)$. Applying $\operatorname{Hom}_\Lambda(X,\ )$ for a module X in $\operatorname{mod}\Lambda$ to η we get the exact sequence

$$0 \to (X, Y) \to (X, M^Y) \to (X, A^Y) \to \operatorname{Ext}^1_F(X, Y) \to \operatorname{Ext}^1_F(X, M^Y) = 0,$$

since M^Y is in $\mathcal{I}(F)$. Therefore X is in $\mathcal{X}_F(\operatorname{Sub}(M))$ if and only if $\operatorname{Hom}_\Lambda(X, \mathcal{A}_M) = M(X, \mathcal{A}_M)$.

□

Now we apply the above to the special case when $M = \Lambda$. Here we are able to give a more explicit description of the subcategory $\mathcal{X}_F(\operatorname{Sub}(\Lambda))$.

Since $\operatorname{Fac}\Lambda = \operatorname{mod}\Lambda$, the subcategory $\mathcal{A}_\Lambda = \{A \in \operatorname{mod}\Lambda \mid \operatorname{Ext}^1_\Lambda(A, \Lambda) = 0\}$. By [**6**, Corollary 7.3] a subcategory $\mathcal{C}$ is covariantly finite if and only if $\operatorname{Tr}\mathcal{C}$ is contravariantly finite. Therefore $\mathcal{A}_\Lambda$ is covariantly finite in $\operatorname{mod}\Lambda$, since $\operatorname{Tr}\colon \mathcal{A}_\Lambda \to \operatorname{Sub}(\Lambda^{\operatorname{op}})$ is a duality modulo projectives. Using these remarks, we have the following special case of Proposition 4.20, where the second part of the proposition is an easy consequence of the first part and is left to the reader.

PROPOSITION 4.21. (a) *The subcategory*

$$\mathcal{X}_F(\operatorname{Sub}(\Lambda)) = \{X \in \operatorname{mod}\Lambda \mid \underline{\operatorname{Hom}}_\Lambda(X, \mathcal{A}_\Lambda) = 0\}$$

is a contravariantly finite F-extension closed subcategory of $\operatorname{mod}\Lambda$, *where* $\mathcal{A}_\Lambda = \{A \in \operatorname{mod}\Lambda \mid \operatorname{Ext}^1_\Lambda(A,\Lambda) = 0\}$.

(b) *A module* X *is in* $\mathcal{X}_F(\operatorname{Sub}(\Lambda))$ *if and only if* A^X *is projective, where* $X \to A^X \to B^X \to 0$ *is the minimal left* $\mathcal{A}_\Lambda$*-approximation.*

The subcategory $\mathcal{X}_F(\operatorname{Sub}(\Lambda))$ is not only contravariantly finite in $\operatorname{mod}\Lambda$, but also functorially finite in $\operatorname{mod}\Lambda$, as we now show.

PROPOSITION 4.22. *The subcategory* $\mathcal{X}_F(\operatorname{Sub}(\Lambda))$ *is functorially finite in* $\operatorname{mod}\Lambda$ *and* $\mathcal{X}_F(\operatorname{Sub}(\Lambda)) = \operatorname{Tr} D(\mathcal{Y}_{\rm std}(\mathcal{A}_\Lambda)) \cup \mathcal{P}(\Lambda)$, *where the subcategory* $\mathcal{Y}_{\rm std}(\mathcal{A}_\Lambda)$ *is given by* $\{Y \in \operatorname{mod}\Lambda \mid \operatorname{Ext}^1_\Lambda(\mathcal{A}_\Lambda, Y) = 0\}$.

PROOF. By the Auslander-Reiten Lemma the subcategory $\mathcal{A}_\Lambda = \{A \in \operatorname{mod}\Lambda \mid \operatorname{Ext}^1_\Lambda(A,\Lambda) = 0\}$ is contravariantly finite in $\operatorname{mod}\Lambda$ and closed under extensions. Applying the Auslander-Reiten Lemma again we have that the subcategory $\mathcal{Y}_{\rm std}(\mathcal{A}_\Lambda) = \{Y \in \operatorname{mod}\Lambda \mid \operatorname{Ext}^1_\Lambda(\mathcal{A}_\Lambda, Y) = 0\}$ is covariantly finite in $\operatorname{mod}\Lambda$. Since $\mathcal{X}_F(\operatorname{Sub}(\Lambda)) = \{X \in \operatorname{mod}\Lambda \mid \operatorname{Ext}^1_\Lambda(\mathcal{A}_\Lambda, D\operatorname{Tr}X) = 0\}$, we have that $\mathcal{X}_F(\operatorname{Sub}(\Lambda)) = \operatorname{Tr} D(\mathcal{Y}_{\rm std}(\mathcal{A}_\Lambda)) \cup \mathcal{P}(\Lambda)$, which is covariantly finite in $\operatorname{mod}\Lambda$ by [**6**, Corollary 7.3]. Therefore $\mathcal{X}_F(\operatorname{Sub}(\Lambda))$ is functorially finite in $\operatorname{mod}\Lambda$. □

We have already remarked that $\operatorname{Tr}\colon \mathcal{A}_\Lambda \to \operatorname{Sub}(\Lambda^{\rm op})$ is a duality modulo projectives. Using this and Proposition 4.21 we have that $\mathcal{X}_F(\operatorname{Sub}(\Lambda))$ is given by $\{X \in \operatorname{mod}\Lambda \mid \underline{\operatorname{Hom}}_{\Lambda^{\rm op}}(\operatorname{Sub}(\Lambda^{\rm op}), \operatorname{Tr}X) = 0\}$. Let $\mathcal{C} = \{C \in \operatorname{mod}\Lambda^{\rm op} \mid \underline{\operatorname{Hom}}_{\Lambda^{\rm op}}(\operatorname{Sub}(\Lambda^{\rm op}), C) = 0\}$, then $\mathcal{X}(\operatorname{Sub}(\Lambda)) = \operatorname{Tr}\mathcal{C} \cup \mathcal{P}(\Lambda)$. Next we give another description of $\mathcal{C}$. The subcategory $\operatorname{Sub}(\Lambda^{\rm op})$ is contravariantly finite in $\operatorname{mod}\Lambda^{\rm op}$ and let $0 \to K_M \to X_M \to M \to 0$ denote the minimal right $\operatorname{Sub}(\Lambda^{\rm op})$-approximation of a module M in $\operatorname{mod}\Lambda^{\rm op}$.

PROPOSITION 4.23. *A module* C *in* $\operatorname{mod}\Lambda^{\rm op}$ *is in* $\mathcal{C}$ *if and only if the projective cover* $0 \to \Omega_{\Lambda^{\rm op}}(C) \to P_C \to C \to 0$ *is a minimal right* $\operatorname{Sub}(\Lambda^{\rm op})$*-approximation. Moreover,* $\mathcal{I}(\Lambda^{\rm op})$ *is contained in* $\mathcal{C}$.

PROOF. Let C be a module in $\operatorname{mod}\Lambda^{\rm op}$. Clearly, if the projective cover $0 \to \Omega_{\Lambda^{\rm op}}(C) \to P_C \to C \to 0$ is a minimal right $\operatorname{Sub}(\Lambda^{\rm op})$-approximation, then $\underline{\operatorname{Hom}}_{\Lambda^{\rm op}}(\operatorname{Sub}(\Lambda^{\rm op}), C) = 0$ and C is in $\mathcal{C}$.

Conversely, assume that $\underline{\operatorname{Hom}}_{\Lambda^{\rm op}}(\operatorname{Sub}(\Lambda^{\rm op}), C) = 0$ and let $0 \to \Omega_{\Lambda^{\rm op}}(C) \to P_C \xrightarrow{f} C \to 0$ be the projective cover of C in $\operatorname{mod}\Lambda^{\rm op}$. Then we have the following commutative diagram

$$\begin{array}{ccccccccc}
0 & \to & \Omega_{\Lambda^{\rm op}}(C) & \to & P_C & \xrightarrow{f} & C & \to & 0 \\
 & & \downarrow & & \downarrow{\scriptstyle r} & & \| & & \\
0 & \to & K_C & \to & X_C & \xrightarrow{g} & C & \to & 0 \\
 & & \downarrow & & \downarrow{\scriptstyle s} & & \| & & \\
0 & \to & \Omega_{\Lambda^{\rm op}}(C) & \to & P_C & \xrightarrow{f} & C & \to & 0.
\end{array}$$

Since both f and g is a right minimal homomorphism, the compositions $r \circ s$ and $s \circ r$ must be isomorphisms and therefore $X_C \simeq P_C$.

Let M be in $\operatorname{Sub}(\Lambda^{\rm op})$, then we have a monomorphism $0 \to M \to (\Lambda^{\rm op})^n$ for some positive integer n. The claim that $\mathcal{I}(\Lambda^{\rm op})$ is contained in $\mathcal{C}$ follows directly from this. □

We now apply this proposition to obtain a new criterion for $\operatorname{Sub}(\Lambda)$ to be closed under extensions.

Proposition 4.24. *Let* $0 \to \Omega_\Lambda(D(\Lambda^{\mathrm{op}})) \to P_{D(\Lambda^{\mathrm{op}})} \to D(\Lambda^{\mathrm{op}}) \to 0$ *be the projective cover of* $D(\Lambda^{\mathrm{op}})$ *in* $\operatorname{mod}\Lambda$. *The subcategory* $\operatorname{Sub}(\Lambda)$ *is extension closed if and only if* $P_{D(\Lambda^{\mathrm{op}})} \oplus \Omega_\Lambda(D(\Lambda^{\mathrm{op}}))$ *is a cotilting module.*

Proof. Assume that $\operatorname{Sub}(\Lambda)$ is extension closed. Let M be in $\operatorname{mod}\Lambda$ and let $0 \to K_M \to X_M \to M \to 0$ be the minimal right $\operatorname{Sub}(\Lambda)$-approximation of M in $\operatorname{mod}\Lambda$. Then by Wakamatsu's Lemma $\operatorname{Ext}^1_\Lambda(\operatorname{Sub}(\Lambda), K_M) = 0$. Since $\operatorname{Sub}(\Lambda)$ is closed under syzygies, $\operatorname{Ext}^i_\Lambda(\operatorname{Sub}(\Lambda), K_M) = 0$ for all $i > 0$. Therefore $\operatorname{Ext}^i_\Lambda(\ , K_M) = 0$ for all $i > 1$ and $\operatorname{id}_\Lambda K_M \leq 1$. Let $\mathcal{C} = \{C \in \operatorname{mod}\Lambda \mid \underline{\operatorname{Hom}}_\Lambda(\operatorname{Sub}(\Lambda), C) = 0\}$. Since $\mathcal{I}(\Lambda) = \operatorname{add} D(\Lambda^{\mathrm{op}})$ is contained in $\mathcal{C}$ by Proposition 4.23, the projective cover of $D(\Lambda^{\mathrm{op}})$ is the minimal right $\operatorname{Sub}(\Lambda)$-approximation of $D(\Lambda^{\mathrm{op}})$ and therefore $\operatorname{id}_\Lambda \Omega_\Lambda(D(\Lambda^{\mathrm{op}})) \leq 1$. By assumption and [**4**, Proposition 3.5] $\operatorname{id}_\Lambda P_{D(\Lambda^{\mathrm{op}})} \leq 1$. Let $T = P_{D(\Lambda^{\mathrm{op}})} \oplus \Omega_\Lambda(D(\Lambda^{\mathrm{op}}))$, then we have shown that $\operatorname{id}_\Lambda T \leq 1$ and that $\mathcal{I}(\Lambda)$ has a finite resolution in $\operatorname{add} T$. We have that $\operatorname{Ext}^i_\Lambda(T, T) = \operatorname{Ext}^{i+1}_\Lambda(D(\Lambda^{\mathrm{op}}), T)$ for all $i > 0$, which is zero since $\operatorname{id}_\Lambda T \leq 1$. Hence, we have shown that T is a cotilting module.

Conversely, assume that $T = P_{D(\Lambda^{\mathrm{op}})} \oplus \Omega_\Lambda(D(\Lambda^{\mathrm{op}}))$ is a cotilting module with $\operatorname{id}_\Lambda T = n$. All we have to prove according to [**4**, Proposition 3.5] is that $\operatorname{id}_\Lambda P_{D(\Lambda^{\mathrm{op}})} \leq 1$. Since T is a cotilting module,

$$\operatorname{Ext}^i_\Lambda(\Omega_\Lambda(D(\Lambda^{\mathrm{op}})), P_{D(\Lambda^{\mathrm{op}})}) = \operatorname{Ext}^{i+1}_\Lambda(D(\Lambda^{\mathrm{op}}), P_{D(\Lambda^{\mathrm{op}})}) = 0$$

for all $i > 0$ and therefore $\operatorname{Ext}^i_\Lambda(I, P_{D(\Lambda^{\mathrm{op}})}) = 0$ for all $i > 1$ and all injectives I in $\operatorname{mod}\Lambda$. Let C be a module in $\operatorname{mod}\Lambda$. Then it is easy to see using dimension shift that $\operatorname{Ext}^i_\Lambda(C, P_{D(\Lambda^{\mathrm{op}})}) \simeq \operatorname{Ext}^{i+n+1}_\Lambda(\Omega^{-n-1}_\Lambda(C), P_{D(\Lambda^{\mathrm{op}})}) = 0$ which in turn is isomorphic to $\operatorname{Ext}^{i+n}_\Lambda(\Omega^{-n-1}_\Lambda(C), \Omega^{-1}_\Lambda(P_{D(\Lambda^{\mathrm{op}})}))$ for all $i > 1$. Hence $\operatorname{id}_\Lambda P_{D(\Lambda^{\mathrm{op}})} \leq 1$ and we are done. □

References

[1] M. Auslander, *Functors and morphisms determined by objects*, in Representation theory of artin algebras, Proceedings of the Philadelphia conference, Lecture Notes in Pure and Appl. Math., vol. 37, Dekker, New York, 1978, 1–244.

[2] M. Auslander, R.-O. Buchweitz, *The homological theory of maximal Cohen-Macaulay approximations*, Soc. Math. de France, Mem. No. 38 (1989) 5–37.

[3] M. Auslander, I. Reiten, *Applications of homologically finite subcategories*, Adv. in Math., 86 (1991) 111–152.

[4] M. Auslander, I. Reiten, *Homologically finite subcategories*, Proceedings of the ICRA V, Tsukuba, Japan.

[5] M. Auslander, I. Reiten, *Syzygy modules over artin algebras*, in preparation.

[6] M. Auslander, S. O. Smalø, *Preprojective modules over artin algebras*, J. Algebra 66 (1980) 61–122.

[7] M. C. R. Butler, G. Horrocks, *Classes of extensions and resolutions*, Phil. Trans. Royal Soc., London, Ser. A, 254 (1961) 155–222.

[8] G. Hochschild, Relative homological algebra, Trans. Amer. Math. Soc. 82 (1956) 246–269.

[9] C. M. Ringel, The category of good modules over a quasihereditary algebra, Math. Zeitschrift 208 (1991) 209–223.

[10] T. Wakamatsu, *Stable equivalence of selfinjective algebras and a generalization of tilting modules*, J. Algebra No. 2, 134 (1990) 298–325.

Department of Mathematics, Brandeis University, Waltham, Mass. 02254–9110, USA

Department of Mathematics, Brandeis University, Waltham, Mass. 02254–9110, USA

Relative homology and representation theory II
Relative cotilting theory[1]

M. Auslander[2] and Ø. Solberg[3]

Introduction

This paper, the second in a series of three papers, studying the use of relative homological algebra in the representation theory of artin algebras, is devoted to developing a general theory of relative cotilting modules for artin algebras. The reader is referred to the first paper in the series [**5**] for basic definitions and results, as well as notation, concerning the relative homological algebra used in this paper.

Let $\operatorname{mod}\Lambda$ be the category of finitely generated left modules over an artin algebra Λ. Suppose F is an additive subfunctor of the additive bifunctor $\operatorname{Ext}^1_\Lambda(\ ,\ )\colon (\operatorname{mod}\Lambda)^{\mathrm{op}} \times \operatorname{mod}\Lambda \to \mathrm{Ab}$. We assume that F has enough projectives and injectives. In [**5**] F-exact sequences and the functors $\operatorname{Ext}^i_F(\ ,\ )$ for all $i \geq 0$ were discussed. In analogy with the standard situation relative F-cotilting modules are defined as follows. A Λ-module T is an F-cotilting module if it satisfies (a) $\operatorname{Ext}^i_F(T,T) = 0$ for all $i > 0$, (b) $\operatorname{id}_F T < \infty$ and (c) for each F-injective module I there is an F-exact sequence $0 \to T_n \to \cdots \to T_0 \to I \to 0$ with the T_i in $\operatorname{add} T$. It is shown that if there is an F-cotilting module T, then there is a module G in $\mathcal{P}(F)$, the category of all F-projective modules, such that $\mathcal{P}(F) = \operatorname{add} G$. So we assume from now on that F has the property that $\mathcal{P}(F) = \operatorname{add} G$.

Suppose now that T is an F-cotilting module. As in the standard theory the algebra $\Gamma = \operatorname{End}_\Lambda(T)$, the Γ-Λ-bimodule T and the functors $\operatorname{Hom}_\Lambda(\ ,{}_\Lambda T)\colon \operatorname{mod}\Lambda \to \operatorname{mod}\Gamma$ and $\operatorname{Hom}_\Gamma(\ ,{}_\Gamma T)\colon \operatorname{mod}\Gamma \to \operatorname{mod}\Lambda$ play a critical role in studying the F-cotilting module T. It is shown that $T_0 = \operatorname{Hom}_\Lambda(G,T)$ is a standard cotilting Γ-module satisfying (a) $\operatorname{id}_\Gamma T_0 \leq \operatorname{id}_F T + 2$ and $\operatorname{id}_\Gamma({}_\Gamma T) \leq \operatorname{id}_F({}_\Lambda T)$, where id_Γ and id_F are the injective dimensions of Γ-modules and the F-injective dimension of Λ-modules respectively. Further, it is shown that the functors $\operatorname{Hom}_\Lambda(\ ,{}_\Lambda T)\colon \operatorname{mod}\Lambda \to \operatorname{mod}\Gamma$ and $\operatorname{Hom}_\Gamma(\ ,{}_\Gamma T)\colon \operatorname{mod}\Gamma \to \operatorname{mod}\Lambda$ induce inverse dualities between the subcategory ${}^{\perp}_{\Lambda}T$ of $\operatorname{mod}\Lambda$ consisting of all X in $\operatorname{mod}\Lambda$ such that $\operatorname{Ext}^i_F(X,T) = 0$ for all $i > 0$ and ${}^{\perp}\operatorname{Hom}_\Gamma(G,T)$. Amongst other things these results are based on the fact that the natural algebra morphism $\Lambda \to \operatorname{End}_\Gamma({}_\Gamma T)$ is an isomorphism.

In standard cotilting theory, one knows that all cotilting Λ-modules have the same number of nonisomorphic indecomposable summands, namely the number of nonisomorphic simple Λ-modules. We show that all F-cotilting Λ-modules also have the same

[1] Communications in algebra, 21(9), 3033–3079 (1993)
[2] Partially supported by NSF Grant No. DMS–8904594.
[3] Supported by the Norwegian Research Council during the preparation of this paper.

number of nonisomorphic indecomposable summands, namely, the number of nonisomorphic indecomposable summands of G where $\operatorname{add} G = \mathcal{P}(F)$. Since Λ is in $\mathcal{P}(F)$, the number of nonisomorphic summands of an F-cotilting module is greater than or equal to the number of nonisomorphic simple Λ-modules. More particularly, the number of nonisomorphic indecomposable summands of an F-cotilting module T is equal to the number of nonisomorphic simple Λ-modules if and only if T is a standard cotilting Λ-module. The proofs of these results are based on the fact that the standard Grothendieck group of Γ and the F-Grothendieck group of Λ are isomorphic where in the F-Grothendieck group of Λ the relations are given only by the F-exact sequences rather than all exact sequences.

From this discussion, it is clear that while the relative cotilting and standard cotilting theories are intimately related, the relative cotilting theory has some features of its own. One of the limitations of the standard theory is that there are no nontrivial standard cotilting modules for selfinjective algebras. This is not the case for the relative theory, as we show by examples in this paper. This feature of the relative cotilting theory will be exploited in a forthcoming paper (not in this series) showing how to construct certain types of algebras Γ of dominant dimension 2 and $\operatorname{id}_\Gamma \Gamma = 2$ as the endomorphism rings of special types of relative cotilting modules over selfinjective algebras.

Having introduced $\operatorname{pd}_F C$ and $\operatorname{id}_F C$ for Λ-modules, we define $\operatorname{gl.dim}_F \Lambda$, the F-global dimension of Λ, to the maximum (including infinity) of $\operatorname{pd}_F C$ for all C in $\operatorname{mod} \Lambda$, which is the same as the maximum of $\operatorname{id}_F C$ for all C in $\operatorname{mod} \Lambda$. Then $\operatorname{gl.dim}_F \Lambda$ and $\operatorname{gl.dim} \Gamma$ are related by $\operatorname{gl.dim}_F \Lambda - \operatorname{id}_F T \leq \operatorname{gl.dim} \Gamma \leq \operatorname{id}_F T + 2 + \operatorname{gl.dim}_F \Lambda$, a formula reminiscent but not the same as the relationship between $\operatorname{gl.dim} \Lambda$ and $\operatorname{gl.dim} \Gamma$ in the standard case. There are even more striking differences in behavior since, as we show by an example, it can happen that $\operatorname{gl.dim} \Gamma$ is finite and $\operatorname{gl.dim} \Lambda$ is infinite. More precisely, we show that if $\operatorname{gl.dim} \Gamma$ is finite, then $\operatorname{gl.dim} \Lambda < \infty$ if and only if either $\operatorname{pd}_\Lambda T < \infty$ or $\operatorname{id}_\Lambda T < \infty$. We also have the following formula: $\operatorname{gl.dim} \Lambda \leq \min\{\operatorname{id}_\Lambda T + \operatorname{gl.dim} \Gamma + 1, \operatorname{pd}_\Lambda T + \operatorname{id}_F T\}$.

The paper ends with a discussion of the relative finitistic dimension, $\operatorname{rel.fin.dim} \Lambda$, of Λ, which is defined as follows. Let C in $\operatorname{mod} \Lambda$ be a generator for $\operatorname{mod} \Lambda$ and let $F = F_{\mathcal{C}}$ where $\mathcal{C} = \operatorname{add} C$. Define the F-finitistic dimension, $\operatorname{fin.dim}_F \Lambda$, to be $\sup\{\operatorname{pd}_F A \mid \operatorname{pd}_F A < \infty, A \in \operatorname{mod} \Lambda\}$. Then $\operatorname{rel.fin.dim} \Lambda$ is defined to be

$$\sup\{\operatorname{fin.dim}_{F_{\mathcal{C}}} \Lambda \mid \mathcal{C} = \operatorname{add} C \text{ for all generators } C \text{ of } \operatorname{mod} \Lambda\}.$$

We say that Λ satisfies the relative finitistic dimension conjecture if $\operatorname{rel.fin.dim} \Lambda$ is finite. Our main result is that the relative finitistic dimension conjecture is true for all artin algebras if and only if the standard finitistic dimension conjecture is true for all artin algebras.

1. Preliminaries

In this section we recall definitions and results we need from [**5**].

Let F be an additive subfunctor of $\operatorname{Ext}^1_\Lambda(\ ,\ )\colon (\operatorname{mod} \Lambda)^{\mathrm{op}} \times \operatorname{mod} \Lambda \to \mathrm{Ab}$, where $(\operatorname{mod} \Lambda)^{\mathrm{op}}$ denotes the opposite category of $\operatorname{mod} \Lambda$. This means that F is a subfunctor of $\operatorname{Ext}^1_\Lambda(\ ,\ )$ and that the induced functors $F(C,\ )\colon \operatorname{mod} \Lambda \to \mathrm{Ab}$ and $F(\ ,A)\colon \operatorname{mod} \Lambda \to \mathrm{Ab}$ are additive covariant and contravariant functors, respectively. A short exact sequence $\eta\colon 0 \to A \to B \to C \to 0$ is said to be *F-exact* if η is in $F(C, A)$. Moreover, we recall the following result from [**5**].

LEMMA 1.1 ([**5**, Lemma 1.2]). *Let $\eta\colon 0 \to A \to B \to C \to 0$ and $\eta'\colon 0 \to A' \to B' \to C' \to 0$ be two short exact sequences. Then $\eta \oplus \eta'$ is F-exact if and only if both η and η' are F-exact.*

Since F is a subfunctor, it follows that any pullback, pushout and Baer sum of F-exact sequences are again F-exact. Let $\mathcal{P}(F)$ be the subcategory of $\operatorname{mod}\Lambda$ consisting of all modules P such that if $0 \to A \to B \to C \to 0$ is F-exact, then $0 \to (P, A) \to (P, B) \to (P, C) \to 0$ is exact. The subcategory $\mathcal{P}(F)$ is called the *projective modules* of the subfunctor F. If $\mathcal{P}(\Lambda)$ denotes the subcategory of projective modules in $\operatorname{mod}\Lambda$, then $\mathcal{P}(\Lambda)$ is contained in $\mathcal{P}(F)$. Dually, denote by $\mathcal{I}(F)$ the subcategory of $\operatorname{mod}\Lambda$ consisting of all modules I such that if $0 \to A \to B \to C \to 0$ is F-exact, then $0 \to (C, I) \to (B, I) \to (A, I) \to 0$ is exact. Then $\mathcal{I}(F)$ is said to be the *injective modules* of F. If $\mathcal{I}(\Lambda)$ denotes the subcategory of injective modules in $\operatorname{mod}\Lambda$, then $\mathcal{I}(\Lambda)$ is contained in $\mathcal{I}(F)$. A subfunctor F is said to have *enough projectives* if for every module C in $\operatorname{mod}\Lambda$, there is an F-exact sequence $0 \to C' \to P \to C \to 0$ with P in $\mathcal{P}(F)$. To have *enough injectives* is defined dually. We have the following connection between $\mathcal{P}(F)$ and $\mathcal{I}(F)$.

LEMMA 1.2 ([**5**, Corollary 1.6]). (a) *The subcategory $\mathcal{I}(F)$ is equal to $D\operatorname{Tr}(\mathcal{P}(F)) \cup \mathcal{I}(\Lambda)$.*

(b) *The subcategory $\mathcal{P}(F)$ is equal to $\operatorname{Tr}D(\mathcal{I}(F)) \cup \mathcal{P}(\Lambda)$.*

(c) *If the subcategory $\mathcal{P}(F)$ or equivalently $\mathcal{I}(F)$ is of finite type, then the number of nonisomorphic indecomposable modules in $\mathcal{P}(F)$ and $\mathcal{I}(F)$ are the same.*

Let $\mathcal{X}$ be a subcategory of $\operatorname{mod}\Lambda$ and $H\colon \operatorname{mod}\Lambda \to \mathrm{Ab}$ a functor. Then $H(\mathcal{X})$ denotes the functor $H|_{\mathcal{X}}\colon \mathcal{X} \to \mathrm{Ab}$.

Let $\mathcal{X}$ be a subcategory of $\operatorname{mod}\Lambda$. For each pair of modules A and C in $\operatorname{mod}\Lambda$ define

$$F_{\mathcal{X}}(C, A) = \{0 \to A \to B \to C \to 0 \mid (\mathcal{X}, B) \to (\mathcal{X}, C) \to 0 \text{ is exact}\}.$$

Dually we define for each pair of modules A and C in $\operatorname{mod}\Lambda$

$$F^{\mathcal{X}}(C, A) = \{0 \to A \to B \to C \to 0 \mid (B, \mathcal{X}) \to (A, \mathcal{X}) \to 0 \text{ is exact}\}.$$

In [**5**, Proposition 1.7] we showed that these constructions give additive subfunctors of $\operatorname{Ext}^1_\Lambda(\ ,\ )$. Further, we proved the following result.

PROPOSITION 1.3 ([**5**, Proposition 1.8 and 1.10]). *Let $\mathcal{X}$ be an additive subcategory of* $\operatorname{mod}\Lambda$. *Then*

(a) $F_{\mathcal{X}} = F^{D\operatorname{Tr}\mathcal{X}}$.

(b) $\mathcal{P}(F_{\mathcal{X}}) = \mathcal{X} \cup \mathcal{P}(\Lambda)$.

Subfunctors of the form $F_{\mathcal{X}}$ for a subcategory $\mathcal{X}$ in $\operatorname{mod}\Lambda$ are the subfunctors of $\operatorname{Ext}^1_\Lambda(\ ,\ )$ of main interest to us in this paper. This is because of the following reasons. We either assume that F has enough projective or injectives. In this case we showed in [**5**, Theorem 1.12] that $F = F_{\mathcal{P}(F)}$ or equivalently $F = F^{\mathcal{I}(F)}$. Moreover, we characterized when a subfunctor F has enough projectives or injectives. This characterization involves the notions of contravariantly and covariantly finite subcategories of $\operatorname{mod}\Lambda$, notions we recall from [**4**] next.

Let $\mathcal{X}$ be a subcategory of $\operatorname{mod}\Lambda$. A *right $\mathcal{X}$-approximation* of a module C in $\operatorname{mod}\Lambda$ is a Λ-homomorphism $f\colon X \to C$ with X in $\mathcal{X}$ such that $\operatorname{Hom}_\Lambda(\mathcal{X}, X) \xrightarrow{\operatorname{Hom}_\Lambda(\mathcal{X}, f)} \operatorname{Hom}_\Lambda(\mathcal{X}, C) \to 0$ is exact. The approximation is called *minimal* if f is a right minimal homomorphism. A *left $\mathcal{X}$-approximation* of a module A in $\operatorname{mod}\Lambda$ is a Λ-homomorphism $g\colon A \to X$ with X in $\mathcal{X}$ such that $\operatorname{Hom}_\Lambda(X, \mathcal{X}) \xrightarrow{\operatorname{Hom}_\Lambda(g, \mathcal{X})} \operatorname{Hom}_\Lambda(A, \mathcal{X}) \to 0$ is exact. The approximation is called *minimal* if g is a left minimal homomorphism. A subcategory $\mathcal{X}$ is called *contravariantly finite* in $\operatorname{mod}\Lambda$ if every module in $\operatorname{mod}\Lambda$ has a right $\mathcal{X}$-approximation and a subcategory $\mathcal{Y}$ is called *covariantly finite* in $\operatorname{mod}\Lambda$ if every module in $\operatorname{mod}\Lambda$ has a left $\mathcal{Y}$-approximation. A subcategory $\mathcal{Z}$ is called *functorially finite* in $\operatorname{mod}\Lambda$ if it is both contravariantly and covariantly finite in $\operatorname{mod}\Lambda$. A subcategory $\mathcal{X}$ is said to be *homologically finite* in $\operatorname{mod}\Lambda$ if $\mathcal{X}$ is either a covariantly, contravariantly or functorially finite subcategory of $\operatorname{mod}\Lambda$.

Then the characterization of when F has enough projectives or injectives is the following.

THEOREM 1.4 ([**5**, Theorem 1.12]). (a) *A subfunctor F of* $\operatorname{Ext}^1_\Lambda(\ ,\ )$ *has enough projectives if and only if $\mathcal{P}(F)$ is contravariantly finite in* $\operatorname{mod}\Lambda$ *and* $F = F_{\mathcal{P}(F)}$.

(b) *A subfunctor F of* $\operatorname{Ext}^1_\Lambda(\ ,\ )$ *has enough injectives if and only if $\mathcal{I}(F)$ is covariantly finite in* $\operatorname{mod}\Lambda$ *and* $F = F^{\mathcal{I}(F)}$.

Normally we start with a subcategory $\mathcal{X}$ of $\operatorname{mod}\Lambda$ and construct $F = F_{\mathcal{X}}$ or $F = F^{\mathcal{X}}$. Then the next result is a convenient reformulation of this result.

THEOREM 1.5 ([**5**, Theorem 1.14]). (a) *A subcategory $\mathcal{X}$ containing $\mathcal{P}(\Lambda)$ is a contravariantly finite subcategory of* $\operatorname{mod}\Lambda$ *if and only if $F_{\mathcal{X}}$ has enough projectives.*

(b) *A subcategory $\mathcal{Y}$ containing $\mathcal{I}(\Lambda)$ is a covariantly finite subcategory of* $\operatorname{mod}\Lambda$ *if and only if $F^{\mathcal{Y}}$ has enough injectives.*

(c) *Let $\mathcal{X}$ be a subcategory of* $\operatorname{mod}\Lambda$ *containing $\mathcal{P}(\Lambda)$. Then $F = F_{\mathcal{X}} = F^{\mathcal{I}(\Lambda)\cup D\operatorname{Tr}(\mathcal{X})}$ has enough projectives and injectives if and only if $\mathcal{X}$ is a functorially finite subcategory of* $\operatorname{mod}\Lambda$.

(d) *Let $\mathcal{Y}$ be a subcategory of* $\operatorname{mod}\Lambda$ *containing $\mathcal{I}(\Lambda)$. Then $F^{\mathcal{Y}} = F_{\mathcal{P}(\Lambda)\cup\operatorname{Tr}D(\mathcal{Y})}$ has enough projectives and injectives if and only if $\mathcal{Y}$ is a functorially finite subcategory of* $\operatorname{mod}\Lambda$.

Suppose that F is a subfunctor of $\operatorname{Ext}^1_\Lambda(\ ,\ )$ with enough projectives. Then for any module C in $\operatorname{mod}\Lambda$ there is an exact sequence of Λ-modules of the form

$$\cdots \to P_n \xrightarrow{d_n} P_{n-1} \xrightarrow{d_{n-1}} \cdots \to P_1 \xrightarrow{d_1} P_0 \xrightarrow{d_0} C \to 0,$$

where P_i is in $\mathcal{P}(F)$ for $i \geq 0$ and $0 \to \operatorname{Im} d_{i+1} \to P_i \to \operatorname{Im} d_i \to 0$ is F-exact for all $i \geq 0$. Such a sequence is called an *F-exact projective resolution* of C. If F has enough injectives, there exists an exact sequence of Λ-modules for any module A of the form

$$0 \to A \to I_0 \xrightarrow{d_0} I_1 \xrightarrow{d_1} \cdots \to I_{n-1} \xrightarrow{d_{n-1}} I_n \xrightarrow{d_n} \cdots,$$

where I_i is in $\mathcal{I}(F)$ for $i \geq 0$ and $0 \to \ker d_i \to I_i \to \ker d_{i+1} \to 0$ is F-exact for all $i \geq 0$. Such a sequence is called an *F-exact injective resolution* of A. Then as for the case when $\mathcal{P}(F) = \mathcal{P}(\Lambda)$ one can define the right derived functors $\operatorname{Ext}^i_F(C,\ )$ of $\operatorname{Hom}_\Lambda(C,\ )$ and $\operatorname{Ext}^i_F(\ ,A)$ of $\operatorname{Hom}_\Lambda(\ ,A)$ for any Λ-modules A and C when F has enough injectives and enough projectives respectively. Moreover, if F has enough projectives and injectives, we have the $\operatorname{Ext}^i_F(C,\ )(A)$ is isomorphic to $\operatorname{Ext}^i_F(\ ,A)(C)$. Similar to the proof that $\operatorname{Ext}^1_F(C,A)$ corresponds to equivalence classes of short exact sequences, one can show that if F has enough projectives or injectives then $\operatorname{Ext}^1_F(C,A) = F(C,A)$. Also if F has enough projectives one can define the left derived functors $\operatorname{Tor}^F_i(B,\ )$ of $B \otimes_\Lambda -$ for any Λ^{op}-module B. These construction were considered in [**6**] and [**8**] and we refer to these papers for details. Moreover, as in the standard case short F-exact sequences induce long exact sequences in $\operatorname{Ext}^i_F(C,\ )$, $\operatorname{Ext}^i_F(\ ,A)$ and $\operatorname{Tor}^F_i(B,\ )$ for all modules A and C in $\operatorname{mod}\Lambda$ and all modules B in $\operatorname{mod}\Lambda^{\mathrm{op}}$.

Let $\mathcal{P}(F)(A,B)$ denote all the Λ-homomorphisms from A to B which factors through an object in $\mathcal{P}(F)$. Denote by $\underline{\operatorname{Hom}}_{\mathcal{P}(F)}(A,B)$ the factor group $\operatorname{Hom}_\Lambda(A,B)/\mathcal{P}(F)(A,B)$. Then we have the following relative version of the Auslander-Reiten isomorphism from [**5**].

PROPOSITION 1.6 ([**5**, Proposition 2.3]). *Let F be an additive subfunctor of* $\operatorname{Ext}_\Lambda(\ ,\ )$ *with enough projectives. Then for all A and C in* $\operatorname{mod}\Lambda$ *we have an isomorphism*

$$\operatorname{Ext}^1_F(C, D\operatorname{Tr} A) \simeq D\,\underline{\operatorname{Hom}}_{\mathcal{P}(F)}(A, C),$$

which is functorial in both variables.

Let $\mathcal{X}$ be a subcategory of $\operatorname{mod}\Lambda$. If $\mathcal{X}$ contains the subcategory $\mathcal{P}(F)$, then $\mathcal{X}$ is called an *F-generator* in $\operatorname{mod}\Lambda$. Dually we say that $\mathcal{X}$ is an *F-cogenerator* in $\operatorname{mod}\Lambda$ if $\mathcal{X}$ contains $\mathcal{I}(F)$.

PROPOSITION 1.7 ([**5**, Proposition 4.17]). *Assume that F has enough projectives and injectives. Let $\mathcal{X}$ and $\mathcal{Y}$ be subcategories of* $\operatorname{mod}\Lambda$ *satisfying that X is in $\mathcal{X}$ if and only if $\operatorname{Ext}^1_F(X, \mathcal{Y}) = 0$ and Y is in $\mathcal{Y}$ if and only if $\operatorname{Ext}^1_F(\mathcal{X}, Y) = 0$. Let $\mathcal{W} = \mathcal{X} \cap \mathcal{Y}$. Then we have the following.*

(a) *The subcategory $\mathcal{X}$ is a contravariantly finite F-generator in* $\operatorname{mod}\Lambda$ *if and only if the subcategory $\mathcal{Y}$ is a covariantly finite F-cogenerator in* $\operatorname{mod}\Lambda$.

(b) *If the subcategory $\mathcal{X}$ is a contravariantly finite F-generator in* $\operatorname{mod}\Lambda$, *then each module C in* $\operatorname{mod}\Lambda$ *has a minimal right $\mathcal{X}$-approximation $0 \to Y_C \to X_C \to C \to 0$ and a minimal left $\mathcal{Y}$-approximation $0 \to C \to Y^C \to X^C \to 0$, where Y_C is in $\mathcal{Y}$ and X^C is in $\mathcal{X}$.*

(c) $\operatorname{Ext}^1_F(\mathcal{W}, \mathcal{W}) = 0$.

(d) *Suppose $\mathcal{X}$ is a contravariantly finite F-generator in* $\operatorname{mod}\Lambda$. *Then*

(i) *For each X in $\mathcal{X}$ there is an F-exact sequence $0 \to X \to W^X \to X' \to 0$ in $\mathcal{X}$ with W^X in $\mathcal{W}$.*

(ii) *For each Y in $\mathcal{Y}$ there is an F-exact sequence $0 \to Y' \to W_Y \to Y \to 0$ in $\mathcal{Y}$ with W_Y in $\mathcal{W}$.*

A criterion for when the hypothesis in this result is true, is given in the following.

PROPOSITION 1.8 ([**5**, Proposition 4.18]). *Assume that $F = F_{\mathcal{P}(F)}$. Suppose $\mathcal{X}$ is an F-extension closed contravariantly finite F-generator in* $\operatorname{mod}\Lambda$. *Let $\mathcal{Y} = \{Y \in \operatorname{mod}\Lambda \mid \operatorname{Ext}^1_F(\mathcal{X}, Y) = 0\}$. Then $\mathcal{X} = \{X \in \operatorname{mod}\Lambda \mid \operatorname{Ext}^1_F(X, \mathcal{Y}) = 0\}$.*

Let $\mathcal{C}$ be any subcategory of $\operatorname{mod}\Lambda$. Then let $\operatorname{Ext}_F(\mathcal{C})$ denote the intersection of all F-extension closed subcategories of $\operatorname{mod}\Lambda$ containing $\mathcal{C}$. The subcategory $\operatorname{Ext}_F(\mathcal{C})$ is obviously F-extension closed and we call it the *F-extension closure* of $\mathcal{C}$ in $\operatorname{mod}\Lambda$. Moreover, for a module M in $\operatorname{mod}\Lambda$ a sequence $M = M_0 \supset M_1 \supset \cdots \supset M_t = 0$ is called an *F-filtration* of M if the induced sequences $0 \to M_{i+1} \to M_i \to M_i/M_{i+1} \to 0$ are F-exact for all $i = 0, \dots, t-1$. Again, let $\mathcal{C}$ be any subcategory of $\operatorname{mod}\Lambda$. Then a module M in $\operatorname{mod}\Lambda$ is said to have an *F-filtration in $\mathcal{C}$* if there is an F-filtration $M = M_0 \supset M_1 \supset \cdots \supset M_t = 0$, where M_i/M_{i+1} is in $\mathcal{C}$ for all $i = 0, \dots, t-1$. Then $\operatorname{Ext}_F(\mathcal{C})$ can be described in the following way.

PROPOSITION 1.9 ([**5**, Proposition 4.4]). *Let $\mathcal{C}$ be any additive subcategory of* $\operatorname{mod}\Lambda$ *and assume that F has enough projectives or injectives. Then $\operatorname{Ext}_F(\mathcal{C})$ consists of all the modules in* $\operatorname{mod}\Lambda$ *which are isomorphic to summands of modules which have an F-filtration in $\mathcal{C}$.*

The final result we recall from [**5**] is Wakamatsu's Lemma.

LEMMA 1.10 ([**5**, Lemma 4.3]). *Assume that $F = F_{\mathcal{P}(F)}$. Suppose $\mathcal{X}$ is an F-extension closed F-generator in* $\operatorname{mod}\Lambda$. *Let C be an arbitrary module in* $\operatorname{mod}\Lambda$. *Then we have the following.*

(a) *If* $0 \to Y_C \to X_C \to C \to 0$ *is a minimal right* $\mathcal{X}$*-approximation of* C*, then* $\operatorname{Ext}^1_F(\mathcal{X}, Y_C) = 0$.

(b) *If* $0 \to C \to X^C \to Y^C \to 0$ *is a minimal left* $\mathcal{X}$*-approximation of* C*, then* $\operatorname{Ext}^1_F(Y^C, \mathcal{X}) = 0$.

2. Relative resolving subcategories

Throughout this section let F be an additive subfunctor of $\operatorname{Ext}^1_\Lambda(\ ,\ )$ with enough projectives and injectives. In [**5**] we mainly considered subcategories closed under F-extensions. In this section we study relative resolving and coresolving subcategories, which in addition to being F-extension closed satisfy two additional properties. This section gives some of the preliminary results we need in order to study relative cotilting and tilting modules. It is mostly devoted to proving that results about resolving and coresolving subcategories given in [**2**] carry over to relative resolving and coresolving subcategories, which we define next.

A subcategory $\mathcal{X}$ of $\operatorname{mod}\Lambda$ is said to be an F*-resolving* subcategory if it satisfies the following three conditions (a) it is closed under F-extensions, (b) if $0 \to A \to B \to C \to 0$ is F-exact and B and C are in $\mathcal{X}$, then A is in $\mathcal{X}$ and (c) it contains $\mathcal{P}(F)$.

A subcategory $\mathcal{Y}$ of $\operatorname{mod}\Lambda$ is said to be an F*-coresolving* subcategory if it satisfies the following three conditions (a) it is closed under F-extensions, (b) if $0 \to A \to B \to C \to 0$ is F-exact and A and B are in $\mathcal{Y}$, then C is in $\mathcal{Y}$ and (c) it contains $\mathcal{I}(F)$.

Let $\mathcal{C}$ be any subcategory of $\operatorname{mod}\Lambda$. We denote by ${}^\perp\mathcal{C}$ the subcategory of $\operatorname{mod}\Lambda$ given by $\{X \in \operatorname{mod}\Lambda \mid \operatorname{Ext}^i_F(X, \mathcal{C}) = 0, \text{for all } i > 0\}$ and by $\mathcal{C}^\perp$ the subcategory of $\operatorname{mod}\Lambda$ given by $\{Y \in \operatorname{mod}\Lambda \mid \operatorname{Ext}^i_F(\mathcal{C}, Y) = 0, \text{for all } i > 0\}$. It follows from the long homological exact sequences that ${}^\perp\mathcal{C}$ is an F-resolving subcategory and that $\mathcal{C}^\perp$ is an F-coresolving subcategory of $\operatorname{mod}\Lambda$ for all subcategories $\mathcal{C}$ of $\operatorname{mod}\Lambda$.

The following result shows that the theory developed in [**5**] and in particular Proposition 1.7 applies to relative resolving and coresolving subcategories.

LEMMA 2.1 ([**2**, Lemma 3.1 and 3.2]). (a) *Suppose* $\mathcal{Y}$ *is an* F*-coresolving subcategory of* $\operatorname{mod}\Lambda$. *Let* $\mathcal{X} = \{X \in \operatorname{mod}\Lambda \mid \operatorname{Ext}^1_F(X, \mathcal{Y}) = 0\}$. *Then*

(i) $\mathcal{X} = {}^\perp\mathcal{Y}$.

(ii) *The subcategory* $\mathcal{X}$ *is an* F*-resolving subcategory of* $\operatorname{mod}\Lambda$.

(b) *Suppose* $\mathcal{X}$ *is an* F*-resolving subcategory of* $\operatorname{mod}\Lambda$. *Let* $\mathcal{Y} = \{Y \in \operatorname{mod}\Lambda \mid \operatorname{Ext}^1_F(\mathcal{X}, Y) = 0\}$. *Then*

(i) $\mathcal{Y} = \mathcal{X}^\perp$.

(ii) *The subcategory* $\mathcal{Y}$ *is an* F*-coresolving subcategory of* $\operatorname{mod}\Lambda$.

PROOF. (a) Since ${}^\perp\mathcal{Y}$ is contained in $\mathcal{X}$, we only have to show that if $\operatorname{Ext}^1_F(X, \mathcal{Y}) = 0$, then $\operatorname{Ext}^i_F(X, \mathcal{Y}) = 0$ for all $i > 0$. Suppose Y is in $\mathcal{Y}$ and let $0 \to Y \to I_0 \xrightarrow{d_0} I_1 \xrightarrow{d_1} \cdots$ be an F-injective resolution of Y. Since $\mathcal{Y}$ is F-coresolving, $\operatorname{Im} d_i$ is in $\mathcal{Y}$ for all $i \geq 0$. Therefore we have by dimension shift that $\operatorname{Ext}^i_F(X, Y)$ is isomorphic to $\operatorname{Ext}^1_F(X, \operatorname{Im} d_{i-2})$ and the claim in (i) follows.

The statement (ii) follows from (a) and the remark above. The proof of (b) is the dual of (a). □

If $\mathcal{X}$ is an F-resolving subcategory of $\operatorname{mod}\Lambda$, the subcategory $\mathcal{P}(F)$ is contained in $\mathcal{X}$. So, if $\mathcal{X}$ is an F-resolving contravariantly finite subcategory of $\operatorname{mod}\Lambda$, then $\mathcal{X}$ is in particular a contravariantly finite F-generator in $\operatorname{mod}\Lambda$. With this remark and the previous result in mind, we have the following result. This is essentially is a restatement of Proposition 1.7 and Proposition 1.8 and we leave the proof to the reader.

PROPOSITION 2.2 ([**2**, Proposition 3.3]). *Let $\mathcal{X}$ be an F-resolving contravariantly finite subcategory of* $\operatorname{mod}\Lambda$. *Then we have the following.*

(a) *The subcategory $\mathcal{Y} = \mathcal{X}^{\perp}$ is an F-coresolving covariantly finite subcategory of* $\operatorname{mod}\Lambda$.

(b) $\mathcal{X} = {}^{\perp}\mathcal{Y} = {}^{\perp}(\mathcal{X}^{\perp})$.

(c) *Each C in* $\operatorname{mod}\Lambda$ *has a unique up to isomorphism minimal right $\mathcal{X}$-approximation $0 \to Y_C \to X_C \xrightarrow{f} C \to 0$ where Y_C is in $\mathcal{Y}$ and the homomorphism f induces isomorphisms $\operatorname{Ext}^i_F(X, X_C) \simeq \operatorname{Ext}^i_F(X, C)$ for all X in $\mathcal{X}$ and all $i > 0$.*

(d) *Each A in* $\operatorname{mod}\Lambda$ *has a unique up to isomorphism minimal left $\mathcal{Y}$-approximation $0 \to A \xrightarrow{g} Y^A \to X^A \to 0$ where X^A is in $\mathcal{X}$ and the homomorphism g induces isomorphisms $\operatorname{Ext}^i_F(Y^C, Y) \simeq \operatorname{Ext}^i_F(A, Y)$ for all Y in $\mathcal{Y}$ and all $i > 0$.*

Given an F-resolving contravariantly finite subcategory $\mathcal{X}$ of $\operatorname{mod}\Lambda$, we will usually denote $\mathcal{X}^{\perp}$ by $\mathcal{Y}$ and denote $\mathcal{X} \cap \mathcal{Y}$ by $\mathcal{W}$. A module C is said to be an *F-selforthogonal* module if $\operatorname{Ext}^i_F(C, C) = 0$ for all $i > 0$. If $\mathcal{C}$ is a subcategory of $\operatorname{mod}\Lambda$ and $\operatorname{Ext}^i_F(C, C') = 0$ for all $i > 0$ and all pairs of modules C and C' in $\mathcal{C}$, the subcategory $\mathcal{C}$ is called an *F-selforthogonal* subcategory. Then we have the following result, which is an easy consequence of the definitions involved and Proposition 2.2.

PROPOSITION 2.3 ([**2**, Proposition 3.4]). *Let $\mathcal{X}$ be an F-resolving F-contravariantly finite subcategory of* $\operatorname{mod}\Lambda$. *Then $\mathcal{W} = \mathcal{X} \cap \mathcal{Y}$ has the following properties.*

(a) *The subcategory $\mathcal{W}$ is F-selforthogonal.*

(b) *For each X in $\mathcal{X}$ there is an F-exact sequence $0 \to X \to W^X \to X' \to 0$ with W^X in $\mathcal{W}$ and X' in $\mathcal{X}$.*

(c) *For each Y in $\mathcal{Y}$ there is an F-exact sequence $0 \to Y' \to W_Y \to Y \to 0$ with W_Y in $\mathcal{W}$ and Y' in $\mathcal{Y}$.*

The subcategory $\mathcal{W}$ in Proposition 2.3 is an Ext_F-injective cogenerator for $\mathcal{X}$. The precise definition is the following. A subcategory $\mathcal{W}$ of $\operatorname{mod}\Lambda$ is called an *Ext_F-injective cogenerator* for a subcategory $\mathcal{X}$ of $\operatorname{mod}\Lambda$ if (i) $\mathcal{W}$ is contained in $\mathcal{X}$, (ii) $\mathcal{W}$ is contained in $\mathcal{X}^{\perp}$ and (iii) for each X in $\mathcal{X}$ there exists an F-exact sequence $0 \to X \to W \to X' \to 0$ in $\mathcal{X}$ with W in $\mathcal{W}$.

In stating the next key result in studying relative cotilting theory, it is convenient to give the following definitions.

For a subcategory $\mathcal{C}$ in $\operatorname{mod}\Lambda$ we denote by $\widehat{\mathcal{C}}$ the subcategory of $\operatorname{mod}\Lambda$ whose objects are the Λ-modules M for which there is an F-exact sequence

$$0 \to C_n \to C_{n-1} \to \cdots \to C_0 \to M \to 0$$

with C_i in $\mathcal{C}$. Dually we denote by $\check{\mathcal{C}}$ the subcategory of $\operatorname{mod}\Lambda$ whose objects are the Λ-modules which have a finite F-exact coresolution in $\mathcal{C}$.

If F has enough projectives, we say that the relative projective dimension $\operatorname{pd}_F M$ of a Λ-module M is the minimum of n including infinity such that there exists an F-exact sequence

$$0 \to P_n \xrightarrow{f_n} P_{n-1} \xrightarrow{f_{n-1}} \cdots \to P_1 \xrightarrow{f_1} P_0 \xrightarrow{f_0} M \to 0,$$

where P_i is in $\mathcal{P}(F)$. If all the F-exact sequences $0 \to \ker f_i \to P_i \xrightarrow{f_i} \operatorname{Im} f_i \to 0$ have the property that f_i is a right minimal homomorphism, then we denote by $\Omega^i_F(M)$ the i-th syzygy $\ker f_{i-1}$. If F has enough injectives, we define the relative injective dimension $\operatorname{id}_F M$ of a Λ-module M dually by the minimum length of coresolutions of M in $\mathcal{I}(F)$. Similarly we let $\Omega^{-i}_F(M)$ denote the i-th cosyzygy in a minimal F-exact injective resolution of M.

Let $\mathcal{X}$ be a contravariantly finite F-generator in $\operatorname{mod}\Lambda$. Then the resolution dimension of a Λ-module M with respect to $\mathcal{X}$ is defined to be the minimum of n including infinity such that there exists an F-exact sequence

$$0 \to X_n \to X_{n-1} \to \cdots \to X_1 \to X_0 \to M \to 0,$$

where X_i is in $\mathcal{X}$. We denote this dimension by $\mathcal{X}\text{-resdim}_F M$. If $\mathcal{Y}$ is a subcategory of $\operatorname{mod}\Lambda$ then $\mathcal{X}\text{-resdim}_F \mathcal{Y}$ is defined to be $\sup\{\mathcal{X}\text{-}\operatorname{resdim}_F Y \mid Y \in \mathcal{Y}\}$.

The following result is a key result in studying relative cotilting modules. The proof is the same as in the standard case and is left to the reader to check that it works.

THEOREM 2.4 ([**1**, Theorem 2.3]). *Let $\mathcal{X}$ be an F-resolving subcategory of* $\operatorname{mod}\Lambda$ *with an* Ext_F*-injective cogenerator $\mathcal{W}$. If $\widehat{\mathcal{X}} = \operatorname{mod}\Lambda$, we have the following.*

(a) *The subcategory $\mathcal{X}$ is contravariantly finite in* $\operatorname{mod}\Lambda$.

(b) $\mathcal{Y} = \mathcal{X}^{\perp} = \widehat{\mathcal{W}}$.

If $\mathcal{X}$ is an F-resolving contravariantly finite subcategory in $\operatorname{mod}\Lambda$, then $\mathcal{X}$-$\operatorname{resdim}_F(\operatorname{mod}\Lambda)$ being finite has the following characterization in terms of $\mathcal{Y} = \mathcal{X}^{\perp}$, where $\mathcal{I}^n(F)$ denotes the subcategory of $\operatorname{mod}\Lambda$ consisting of all modules M with $\operatorname{id}_F M \leq n$.

THEOREM 2.5 ([**2**, Theorem 5.2]). *Let $\mathcal{X}$ be an F-resolving contravariantly finite subcategory of* $\operatorname{mod}\Lambda$ *and $\mathcal{Y} = \mathcal{X}^{\perp}$ the associated F-coresolving covariantly finite subcategory of* $\operatorname{mod}\Lambda$. *Then* $\mathcal{X}\text{-resdim}_F(\operatorname{mod}\Lambda)$ *is finite if and only if $\mathcal{Y}$ is contained in $\mathcal{I}^n(F)$ for some finite n.*

PROOF. Assume that $\mathcal{X}\text{-resdim}_F(\operatorname{mod}\Lambda) = n$ is finite. For each Λ-module C there is an F-exact sequence $0 \to X_n \to X_{n-1} \to \cdots \to X_0 \to C \to 0$ with X_i in $\mathcal{X}$ for all i. Since $\operatorname{Ext}^i_F(\mathcal{X}, \mathcal{Y}) = 0$ for all $i > 0$, we get that $\operatorname{Ext}^i_F(C, \mathcal{Y}) = 0$ for all $i > n$. This shows that $\operatorname{Ext}^i_F(\ ,\mathcal{Y}) = 0$ for all $i > n$, so that $\mathcal{Y}$ is contained in $\mathcal{I}^n(F)$.

Assume that $\mathcal{Y}$ is contained in $\mathcal{I}^n(F)$ for a finite n. Then $\operatorname{Ext}^i_F(\ ,\mathcal{Y}) = 0$ for all $i > n$. Since $\mathcal{X}$ is an F-resolving contravariantly finite subcategory of $\operatorname{mod}\Lambda$, we have that $\mathcal{X} = {}^{\perp}\mathcal{Y}$ by Proposition 2.2 (b). By assumption $\operatorname{Ext}^i_F(\Omega^n_F(C), \mathcal{Y}) = 0$ for all $i > 0$ and all C in $\operatorname{mod}\Lambda$, so that $\Omega^n_F(C)$ is in $\mathcal{X}$ for all C in $\operatorname{mod}\Lambda$. Because $\mathcal{X}$ contains $\mathcal{P}(F)$, this shows that $\mathcal{X}\text{-resdim}_F(\operatorname{mod}\Lambda) \leq n$. □

The following observation is a restatement of Wakamatsu's Lemma using Lemma 2.1 (b).

LEMMA 2.6. *Suppose $\mathcal{X}$ is an F-resolving subcategory of* $\operatorname{mod}\Lambda$ *and $0 \to Y_C \to X_C \to C \to 0$ is a minimal right $\mathcal{X}$-approximation of a module C in* $\operatorname{mod}\Lambda$. *Then Y_C is in $\mathcal{X}^{\perp}$.*

Let $\mathcal{X}$ be an F-resolving subcategory of $\operatorname{mod}\Lambda$. Next we point out that the modules having a right $\mathcal{X}$-approximation is closed under F-extensions. This is essentially the content of the following result, where we leave to the reader to check that the standard proof works.

PROPOSITION 2.7 ([**2**, Proposition 3.6]). *Suppose $\mathcal{X}$ is an F-resolving subcategory of* $\operatorname{mod}\Lambda$. *Let*

$$\begin{array}{ccccccccc}
 & & 0 & & & & 0 & & \\
 & & \downarrow & & & & \downarrow & & \\
 & & Y_1 & & & & Y_3 & & \\
 & & \downarrow & & & & \downarrow & & \\
 & & X_1 & & & & X_3 & & \\
 & & \downarrow{\scriptstyle f_1} & & & & \downarrow{\scriptstyle f_3} & & \\
0 & \to & C_1 & \to & C_2 & \to & C_3 & \to & 0 \\
 & & \downarrow & & & & \downarrow & & \\
 & & 0 & & & & 0 & &
\end{array}$$

be an F-exact diagram of Λ-modules with X_i in $\mathcal{X}$ and Y_i in $\mathcal{Y} = \mathcal{X}^{\perp}$ for $i = 1, 3$. Then there is an F-exact commutative diagram

$$\begin{array}{ccccccccc}
 & & 0 & & 0 & & 0 & & \\
 & & \downarrow & & \downarrow & & \downarrow & & \\
0 & \to & Y_1 & \to & Y_2 & \to & Y_3 & \to & 0 \\
 & & \downarrow & & \downarrow & & \downarrow & & \\
0 & \to & X_1 & \to & X_2 & \to & X_3 & \to & 0 \\
 & & \downarrow{\scriptstyle f_1} & & \downarrow{\scriptstyle f_2} & & \downarrow{\scriptstyle f_3} & & \\
0 & \to & C_1 & \to & C_2 & \to & C_3 & \to & 0 \\
 & & \downarrow & & \downarrow & & \downarrow & & \\
 & & 0 & & 0 & & 0 & &
\end{array}$$

with X_2 in $\mathcal{X}$ and Y_2 in $\mathcal{Y}$.

Recall from section 1 that if $\mathcal{C}$ is a subcategory of $\operatorname{mod}\Lambda$, then $\operatorname{Ext}_F(\mathcal{C})$ denotes the intersection of all F-extension closed subcategories in $\operatorname{mod}\Lambda$ containing $\mathcal{C}$. The proof is an easy combination of Proposition 1.9, Lemma 2.6 and Lemma 2.1 (b).

PROPOSITION 2.8 ([**2**, Proposition 3.7]). *Suppose $\mathcal{X}$ is an F-resolving subcategory of* $\operatorname{mod}\Lambda$. *Then we have the following.*

(a) *The subcategory of all modules having right $\mathcal{X}$-approximations is closed under F-extensions.*

(b) *Assume that* $\operatorname{mod}\Lambda = \operatorname{Ext}_F(\{M_1, \dots, M_n\})$. *Then $\mathcal{X}$ is contravariantly finite in* $\operatorname{mod}\Lambda$ *if and only if all the M_i have right $\mathcal{X}$-approximations.*

Another easy consequence of Proposition 2.7 is the following.

PROPOSITION 2.9 ([**2**, Proposition 3.8]). *Suppose $\mathcal{X}$ is F-resolving contravariantly finite subcategory of* $\operatorname{mod}\Lambda$ *and assume that* $\operatorname{mod}\Lambda$ *is equal to* $\operatorname{Ext}_F(\{M_1, \dots, M_n\})$. *Let $f_i \colon X_i \to M_i$ be the minimal right $\mathcal{X}$-approximation of M_i for $i = 1, \dots, n$. Then $\mathcal{X} = \operatorname{Ext}_F(\{X_1, \dots, X_n\})$.*

PROOF. Since $\mathcal{X}$ is F-extension closed, $\operatorname{Ext}_F(\{X_1, \dots, X_n\})$ is contained in $\mathcal{X}$.

Let C be arbitrary in $\operatorname{mod}\Lambda$. Then C is a summand of a module N with an F-filtration $N = N_0 \supset N_1 \supset \cdots \supset N_t = 0$ such that for each $i = 0, \dots, t-1$ we have $N_i/N_{i+1} \simeq M_j$ for some $j = 1, \dots, n$ by Proposition 1.9. Then it follows from Proposition 2.7 that there is an F-exact sequence $0 \to Y \to X \to C \to 0$ with Y in $\mathcal{Y} = \mathcal{X}^{\perp}$

and X in $\operatorname{Ext}_F(\{X_1,\dots,X_n\})$. Suppose now that C is in $\mathcal{X}$. Then the F-exact sequence $0 \to Y \to X \to C \to 0$ splits since Y is in $\mathcal{Y}$. Therefore $\mathcal{X}$ is contained in $\operatorname{Ext}_F(\{X_1,\dots,X_n\})$ and we have $\mathcal{X} = \operatorname{Ext}_F(\{X_1,\dots,X_n\})$. □

The following corollary illustrates that even if one is not interested in relative homologically finite subcategories or relative homology, this relative theory can produce results concerning the standard theory.

COROLLARY 2.10 ([**2**, Corollary 3.9]). *Let $\mathcal{X}$ be an F-resolving contravariantly finite subcategory of* $\operatorname{mod}\Lambda$. *Assume that* $\operatorname{mod}\Lambda = \operatorname{Ext}_F(\{M_1,\dots,M_n\})$ *and let $f_i\colon X_i \to M_i$ be the minimal right $\mathcal{X}$-approximation for M_i for $i = 1,\dots,n$. Then the following are equivalent.*

(a) $k = \sup\{\operatorname{pd}_\Lambda X_i \mid i = i,\dots,n\} < \infty$.

(b) $\sup\{\operatorname{pd}_\Lambda X \mid X \in \mathcal{X}\} = k < \infty$.

(c) $\operatorname{pd}_\Lambda X < \infty$ *for each X in $\mathcal{X}$.*

PROOF. (a) *implies* (b). Let C be in $\mathcal{X}$. We want to show that $\operatorname{pd}_\Lambda C \leq k$. By Proposition 2.9 the module C is in $\operatorname{Ext}_F(\{X_1,\dots,X_n\})$. Therefore C is a summand of a module N which has an F-filtration $N = N_0 \supset N_1 \supset \cdots \supset N_t = 0$ such that each $N_i/N_{i+1} \simeq X_j$ for some $i = 1,\dots,n$ by Proposition 1.9. Then it follows that $\operatorname{pd}_\Lambda N \leq k$, which implies that $\operatorname{pd}_\Lambda C \leq k$.

The implications (b) *implies* (c) and (c) *implies* (a) are obvious. □

3. Relative tilting and cotilting modules

In this section we introduce the notion of relative tilting and cotilting modules. This is a generalization of the notion of tilting and cotilting modules in [**10**] and the approach will be the same as in [**3**].

Throughout this section let F be an additive subfunctor of $\operatorname{Ext}^1_\Lambda(\ ,\ )$ with enough projectives and injectives. Then we define relative tilting and cotilting modules as follows. An F-selforthogonal module T is called an *F-tilting module* if $\operatorname{pd}_F T < \infty$ and $\mathcal{P}(F)$ is contained in $\operatorname{add} T$. An F-selforthogonal module T is called an *F-cotilting module* if $\operatorname{id}_F T < \infty$ and $\mathcal{I}(F)$ is contained in $\widehat{\operatorname{add} T}$. Then a standard tilting or cotilting module T is a module satisfying one of the above definitions with $F = \operatorname{Ext}^1_\Lambda(\ ,\ )$.

One main motivation for studying relative cotilting and tilting modules is of course that standard cotilting and tilting modules have appeared in many different contexts and have had many applications. Some additional inspiration for considering relative cotilting and tilting modules is that these constructions are connected to the Wedderburn correspondence and standard cotilting modules. Moreover, studying relative cotilting modules over selfinjective artin algebras give nontrivial new results. For details on this we refer to the last paper in this series and a forthcoming paper.

We only consider relative cotilting modules in the rest of this paper and we leave it to the reader to state and prove the dual results for relative tilting modules.

Let T be an F-cotilting module. We show that ${}^\perp T$ is an F-resolving functorially finite subcategory of $\operatorname{mod}\Lambda$ with ${}^\perp T$-$\operatorname{resdim}_F(\operatorname{mod}\Lambda)$ finite, which is also true for standard cotilting modules. We also have that ${}^\perp T \cap ({}^\perp T)^\perp = \operatorname{add} T$. Every F-cotilting module is shown to have the same number of nonisomorphic indecomposable summands as there are nonisomorphic indecomposable modules in $\mathcal{P}(F)$. So, relative cotilting modules only exist when $\mathcal{P}(F)$ or equivalently $\mathcal{I}(F)$ is of finite type. If $\mathcal{P}(F)$ is of finite type, we show that the maps $T \mapsto {}^\perp T$ and $\mathcal{X} \mapsto \mathcal{X} \cap \mathcal{X}^\perp$ give inverse bijections between isomorphism classes of basic F-cotilting modules in $\operatorname{mod}\Lambda$ and the family of F-resolving contravariantly finite subcategories of $\mathcal{X}$ of $\operatorname{mod}\Lambda$ with $\mathcal{X}$-$\operatorname{resdim}_F(\operatorname{mod}\Lambda)$ finite. Recall that a module M is said to be *basic* if it is a finite direct sum of nonisomorphic modules.

Denote by Γ the artin algebra $\operatorname{End}_\Lambda(T)$. As in the standard case we have that $\Lambda \simeq \operatorname{End}_\Gamma({}_\Gamma T)$. But ${}_\Gamma T$ is not a cotilting module over Γ only a direct summand of a cotilting module T_0. If $\operatorname{add} P = \mathcal{P}(F)$, then T_0 is given by $\operatorname{Hom}_\Lambda(P,T)$, where $\operatorname{id}_\Gamma T_0 \leq \operatorname{id}_F T + 2$ and $\operatorname{id}_\Gamma T \leq \operatorname{id}_F T$. The functors $\operatorname{Hom}_\Lambda(\ ,T)\colon \operatorname{mod}\Lambda \to \operatorname{mod}\Gamma$ and $\operatorname{Hom}_\Gamma(\ ,{}_\Gamma T) \operatorname{mod}\Gamma \to \operatorname{mod}\Lambda$ induce inverse dualities between ${}^\perp T$ and $\operatorname{Hom}_\Lambda({}^\perp T, T)$. Moreover, $\operatorname{Hom}_\Lambda({}^\perp T,T) = {}^\perp T_0$, where T_0 is the cotilting module given above. The module ${}_\Gamma T$ has the property that the natural homomorphism $X \to \operatorname{Hom}_\Lambda(\operatorname{Hom}_\Gamma(X,T),T)$ is an isomorphism for all modules X in ${}^\perp T_0$. We end this section by considering some examples of and methods of constructing relative cotilting modules.

We start by defining the subcategory $\mathcal{X}_T$ corresponding to an F-cotilting module T, which will be one of our main tools in studying relative cotilting modules. For an F-selforthogonal Λ-module T denote by $\mathcal{X}_T$ the subcategory of ${}^\perp T$ whose objects are the Λ-modules C such that there is an F-exact sequence

$$0 \to C \to T_0 \xrightarrow{f_0} T_1 \to \cdots \to T_n \xrightarrow{f_n} T_{n+1} \to \cdots$$

with T_i in $\operatorname{add} T$ and $\operatorname{Im} f_i$ in ${}^\perp T$ for all $i \geq 0$. For an F-selforthogonal subcategory $\mathcal{W}$ of $\operatorname{mod}\Lambda$ the subcategory $\mathcal{X}_\mathcal{W}$ is defined similarly.

Our first aim is to prove that if T is an F-cotilting module, then $\mathcal{X}_T$ is an F-resolving contravariantly finite subcategory of $\operatorname{mod}\Lambda$ with $\mathcal{X}_T$-$\operatorname{resdim}_F(\operatorname{mod}\Lambda)$ finite, as in the standard case. The following result is the first step in that direction.

PROPOSITION 3.1 ([**2**, Proposition 5.1]). *For an F-selforthogonal Λ-module T the subcategory $\mathcal{X}_T$ is closed under* (a) *F-extensions,* (b) *direct summands and* (c) *if $0 \to A \to B \to C \to 0$ is F-exact with B and C in $\mathcal{X}_T$, then A is in $\mathcal{X}_T$.*

PROOF. (a) Let $0 \to A \to B \to C \to 0$ be an F-exact sequence with A and C in $\mathcal{X}_T$. We want to show that B is in $\mathcal{X}_T$. Then we have F-exact sequences $0 \to A \to T_0 \to L \to 0$ and $0 \to C \xrightarrow{t} T_0' \to K \to 0$ with T_0 and T_0' in $\operatorname{add} T$ and L and K in $\mathcal{X}_T$. Consider the commutative exact pushout diagram

$$\begin{array}{ccccccccc}
 & & 0 & & 0 & & & & \\
 & & \downarrow & & \downarrow & & & & \\
0 & \to & A & \to & B & \to & C & \to & 0 \\
 & & \downarrow & & \downarrow & & \| & & \\
0 & \to & T_0 & \to & U & \to & C & \to & 0. \\
 & & \downarrow & & \downarrow & & & & \\
 & & L & = & L & & & & \\
 & & \downarrow & & \downarrow & & & & \\
 & & 0 & & 0 & & & &
\end{array}$$

Since C is in $\mathcal{X}_T \subset {}^\perp T$, we have that $\operatorname{Ext}^1_F(C,T_0) = 0$. The sequence $0 \to T_0 \to U \to C \to 0$ is F-exact, so that it splits and therefore $U \simeq T_0 \oplus C$. Then we have the following commutative exact diagram

$$\begin{array}{ccccccccc} & & & & 0 & & 0 & & \\ & & & & \downarrow & & \downarrow & & \\ 0 & \to & B & \to & U \simeq T_0 \oplus C & \to & L & \to & 0 \\ & & \| & & \downarrow{\scriptstyle \mathrm{id} \oplus t} & & \downarrow & & \\ 0 & \to & B & \to & T_0 \oplus T_0' & \to & V & \to & 0 \\ & & & & \downarrow & & \downarrow & & \\ & & & & K & = & K. & & \\ & & & & \downarrow & & \downarrow & & \\ & & & & 0 & & 0 & & \end{array}$$

We want to show that the sequence $\eta\colon 0 \to B \to T_0 \oplus T_0' \to V \to 0$ is F-exact, V is in ${}^{\perp}T$ and B is in ${}^{\perp}T$. The middle vertical sequence in the diagram above is a direct sum of two F-exact sequences and therefore it is F-exact by Lemma 1.1. The sequence $0 \to L \to V \to K \to 0$ is F-exact, since it is a pushout of the F-exact middle vertical sequence. Because $F = F_{\mathcal{P}(F)}$ by Theorem 1.4, it follows that the sequence η is F-exact. Since K and L are in ${}^{\perp}T$ and ${}^{\perp}T$ is closed under F-extensions, V is in ${}^{\perp}T$. Clearly, for the same reason B in ${}^{\perp}T$. Using that $0 \to L \to V \to K \to 0$ is F-exact and that K and L are in $\mathcal{X}_T$, we have by induction that B is in $\mathcal{X}_T$.

(b) and (c). Let $0 \to A \to B \to C \to 0$ be F-exact with B in $\mathcal{X}_T$. Then we have an F-exact sequence $0 \to B \to T_0 \to E \to 0$ with T_0 in $\operatorname{add} T$ and E in $\mathcal{X}_T$. We have the following commutative exact diagram

$$\begin{array}{ccccccccc} & & & & 0 & & 0 & & \\ & & & & \downarrow & & \downarrow & & \\ 0 & \to & A & \to & B & \to & C & \to & 0 \\ & & \| & & \downarrow & & \downarrow{\scriptstyle r} & & \\ 0 & \to & A & \to & T_0 & \to & K & \to & 0 \\ & & & & \downarrow & & \downarrow & & \\ & & & & E & = & E. & & \\ & & & & \downarrow & & \downarrow & & \\ & & & & 0 & & 0 & & \end{array}$$

The sequence $\gamma\colon 0 \to C \to K \to E \to 0$ is F-exact, since it is a pushout of an F-exact sequence. Again, since $F = F_{\mathcal{P}(F)}$ by Theorem 1.4, we have that the sequence $\theta\colon 0 \to A \to T_0 \to K \to 0$ is F-exact.

(c) Assume that C is in $\mathcal{X}_T$. We want to show that A is in $\mathcal{X}_T$. Since $\mathcal{X}_T$ is closed under F-extensions and γ is F-exact, the module K is in $\mathcal{X}_T$. The sequence θ is F-exact and K is in $\mathcal{X}_T$, hence the module A is in $\mathcal{X}_T$. Therefore $\mathcal{X}_T$ is closed under kernels of epimorphisms in F-exact sequences.

(b) Assume that $B \simeq A \oplus C$. We want to show that A is in $\mathcal{X}_T$. The sequence $0 \to B \simeq A \oplus C \xrightarrow{(\mathrm{id}, r)} A \oplus K \to E \to 0$ is a direct sum of two F-exact sequences, hence it is F-exact by Lemma 1.1. Since B and E are in $\mathcal{X}_T$, the direct sum $A \oplus K$ is in $\mathcal{X}_T$ and therefore K is in ${}^{\perp}T$. By repeating the process with $0 \to K \to A \oplus K \to A \to 0$, we have by induction that A is in $\mathcal{X}_T$. Therefore $\mathcal{X}_T$ is closed under direct summands. $\square$

Now we show that $\mathcal{X}_T$ is an F-resolving contravariantly finite subcategory with $\mathcal{X}_T$-resdim$_F(\operatorname{mod}\Lambda)$ finite.

THEOREM 3.2. *Let T be an F-cotilting module. Then we have the following.*

(a) *The subcategory $\mathcal{X}_T = {}^{\perp}T$ is an F-resolving contravariantly finite subcategory of* $\operatorname{mod}\Lambda$ *with $\mathcal{X}_T$-resdim$_F(\operatorname{mod}\Lambda)$ finite.*

(b) *The subcategory $\widehat{\operatorname{add} T} = (\mathcal{X}_T)^{\perp}$ is an F-coresolving covariantly finite subcategory of* $\operatorname{mod}\Lambda$ *with $\operatorname{id}_F \widehat{\operatorname{add} T}$ finite.*

PROOF. Let T be an F-cotilting module, then $\operatorname{id}_F T = n$ for some finite n. We want to show that $\mathcal{X}_T = {}^{\perp}T$. Let C be a module in ${}^{\perp}T$ and let $0 \to C \to I \to C_0 \to 0$ be F-exact with I in $\mathcal{I}(F)$. Since I is in $\mathcal{I}(F) \subset \widehat{\operatorname{add} T}$, there is finite F-exact resolution $0 \to T_n \to \cdots T_1 \to T_0 \to I \to 0$ with T_i in $\operatorname{add} T$. Let $0 \to L \to T_0 \to I \to 0$ be the first part of this finite F-exact resolution of I. Then we have the following exact diagram

$$\begin{array}{ccccccccc}
 & & & & 0 & & 0 & & \\
 & & & & \downarrow & & \downarrow & & \\
0 & \to & L & \to & E & \to & C & \to & 0 \\
 & & \| & & \downarrow & & \downarrow & & \\
0 & \to & L & \to & T_0 & \to & I & \to & 0 \\
 & & & & \downarrow & & \downarrow & & \\
 & & & & C_0 & = & C_0. & & \\
 & & & & \downarrow & & \downarrow & & \\
 & & & & 0 & & 0 & &
\end{array}$$

By dimension shift we have that $\operatorname{Ext}^1_F({}^{\perp}T, L) \simeq \operatorname{Ext}^n_F({}^{\perp}T, T_n) = 0$. This implies that the F-exact sequence $0 \to L \to E \to C \to 0$ splits. Hence we get an exact sequence $0 \to C \to T_0 \to A \to 0$. Since $\operatorname{add} T$ is covariantly finite in $\operatorname{mod}\Lambda$, there is a map $C \to T_0'$ such that any map $C \to T$ factors through $C \to T_0 \oplus T_0'$, where T_0' is in $\operatorname{add} T$. We want to show that the sequence $\eta\colon 0 \to C \to T_0 \oplus T_0' \to K \to 0$ is F-exact. We have the following commutative exact diagram

$$\begin{array}{ccccccccc}
0 & \to & C & \to & T_0 & \to & A & \to & 0 \\
 & & \| & & \downarrow & & \downarrow & & \\
0 & \to & C & \to & I & \to & C_0 & \to & 0,
\end{array}$$

hence $0 \to C \to T_0 \to A \to 0$ is F-exact. We also have the following commutative exact diagram

$$\begin{array}{ccccccccc}
0 & \to & C & \to & T_0 \oplus T_0' & \to & K & \to & 0 \\
 & & \| & & \downarrow{\scriptstyle (\mathrm{id},0)} & & \downarrow & & \\
0 & \to & C & \to & T_0 & \to & A & \to & 0
\end{array}$$

and therefore η is F-exact. Applying $\operatorname{Hom}_\Lambda(\ ,T)$ to η, it follows that $\operatorname{Ext}^i_F(K,T) = 0$ for all $i > 0$, so that K is in ${}^{\perp}T$. Continuing this process with K, we have by induction that C is in $\mathcal{X}_T$. Therefore ${}^{\perp}T = \mathcal{X}_T$, since $\mathcal{X}_T$ is contained in ${}^{\perp}T$. This also shows that $\mathcal{X}_T$ is an F-resolving subcategory of $\operatorname{mod}\Lambda$.

Next we want to prove that $\mathcal{X}_T$-resdim$_F(\operatorname{mod}\Lambda)$ is finite. Since $\operatorname{id}_F T = n$, we have that $\operatorname{Ext}^i_F(C,T) = 0$ for all $i > n$ and for all C in $\operatorname{mod}\Lambda$. This shows that $\Omega^n_F(C)$ is in ${}^{\perp}T = \mathcal{X}_T$ for all C in $\operatorname{mod}\Lambda$, so that $\mathcal{X}_T$-resdim$_F(\operatorname{mod}\Lambda) \leq n$, because $\mathcal{P}(F)$ is contained in $\mathcal{X}_T$.

Since $\mathcal{X}_T = {}^{\perp}T$, the subcategory $\operatorname{add} T$ is an Ext_F-injective cogenerator for $\mathcal{X}_T$. Therefore by Theorem 2.4 the subcategory $\mathcal{X}_T$ is contravariantly finite, which completes the proof of (a). Moreover, Theorem 2.4 also says that $\widehat{\operatorname{add} T} = (\mathcal{X}_T)^{\perp}$. Because $\mathcal{X}_T$ is F-resolving contravariantly finite by (a), the subcategory $\widehat{\operatorname{add} T}$ is F-coresolving covariantly finite in $\operatorname{mod} \Lambda$ by Proposition 2.2 (a). By (a) $\mathcal{X}_T\text{-}\operatorname{resdim}_F(\operatorname{mod} \Lambda) \leq n$, therefore the subcategory $\widehat{\operatorname{add} T}$ is contained in $\mathcal{I}^n(F)$ by Theorem 2.5. This completes the proof of the theorem. □

In the standard case we also have the converse, namely that if $\mathcal{X}$ is a resolving contravariantly finite subcategory of $\operatorname{mod} \Lambda$ with $\mathcal{X}\text{-}\operatorname{resdim}(\operatorname{mod} \Lambda)$ finite, then $\mathcal{X} \cap \mathcal{X}^{\perp} = \operatorname{add} T$ for some cotilting module T over Λ [**3**, Theorem 5.4]. Our main aim is to show that we have the same correspondence between F-resolving contravariantly finite subcategories $\mathcal{X}$ with $\mathcal{X}\text{-}\operatorname{resdim}_F(\operatorname{mod} \Lambda)$ finite and F-cotilting modules. This will take several steps. In the following let T be an F-cotilting module and let $\Gamma = \operatorname{End}_\Lambda(T)$.

LEMMA 3.3. *Let T be in $\operatorname{mod} \Lambda$ and let $\Gamma = \operatorname{End}_\Lambda(T)$.*

(a) *Then*

$$G = \operatorname{Hom}_\Lambda(\ , {}_\Lambda T_{\Gamma^{\mathrm{op}}})\colon \operatorname{mod} \Lambda \to \operatorname{mod} \Gamma$$

and

$$H = \operatorname{Hom}_\Gamma(\ , {}_\Gamma T_{\Lambda^{\mathrm{op}}})\colon \operatorname{mod} \Gamma \to \operatorname{mod} \Lambda$$

are an adjoint pair of contravariant functors, that is, we have the isomorphism

$$\phi\colon \operatorname{Hom}_\Lambda(A, \operatorname{Hom}_\Gamma(C, {}_\Gamma T_{\Lambda^{\mathrm{op}}})) \to \operatorname{Hom}_\Gamma(C, \operatorname{Hom}_\Lambda(A, {}_\Lambda T_{\Gamma^{\mathrm{op}}}))$$

functorial in both variables for all A in $\operatorname{mod} \Lambda$ and C in $\operatorname{mod} \Gamma$ given by $\phi(f)(c)(a) = f(a)(c)$ for f in $\operatorname{Hom}_\Lambda(A, \operatorname{Hom}_\Gamma(C, {}_\Gamma T_{\Lambda^{\mathrm{op}}}))$.

(b) *If T is an F-cotilting module, then the functor*

$$e_T = \operatorname{Hom}_\Lambda(\ , {}_\Lambda T_{\Gamma^{\mathrm{op}}})\colon \mathcal{X}_T = {}^{\perp}T \to \operatorname{mod} \Gamma$$

is an F-exact fully faithful contravariant functor.

PROOF. (a) Define $\psi\colon \operatorname{Hom}_\Gamma(C, \operatorname{Hom}_\Lambda(A, {}_\Lambda T_{\Gamma^{\mathrm{op}}})) \to \operatorname{Hom}_\Lambda(A, \operatorname{Hom}_\Gamma(C, {}_\Gamma T_{\Lambda^{\mathrm{op}}}))$ by $\psi(g)(a)(c) = g(c)(a)$ for g in $\operatorname{Hom}_\Gamma(C, \operatorname{Hom}_\Lambda(A, {}_\Lambda T_{\Gamma^{\mathrm{op}}}))$. It is easy to see that ϕ and ψ are inverse isomorphisms functorial both in A and C.

(b) It is easy to see that $e_T = \operatorname{Hom}_\Lambda(\ , T)\colon \operatorname{Hom}_\Lambda(X, Z) \to \operatorname{Hom}_\Gamma(e_T(Z), e_T(X))$ is an isomorphism for any X in $\operatorname{mod} \Lambda$ and Z in $\operatorname{add} T$.

Let C be in $\mathcal{X}_T = {}^{\perp}T$. Then there is an F-exact sequence $0 \to C \to T_0 \xrightarrow{d} T_1$, where $\operatorname{Im} d$ and $\operatorname{Coker} d$ are in ${}^{\perp}T$. Therefore the sequence

$$\operatorname{Hom}_\Lambda(T_1, T) \to \operatorname{Hom}_\Lambda(T_0, T) \to \operatorname{Hom}_\Lambda(C, T) \to 0$$

is exact. Applying $\operatorname{Hom}_\Gamma(\ , \operatorname{Hom}_\Lambda(X, T))$ to this sequence we get the following commutative exact diagram

$$\begin{array}{ccccccc}
0 & \to & ((C,T),(X,T)) & \to & ((T_0,T),(X,T)) & \to & ((T_1,T),(X,T)) \\
 & & \uparrow & & \uparrow \wr & & \uparrow \wr \\
0 & \to & \operatorname{Hom}_\Lambda(X,C) & \to & \operatorname{Hom}_\Lambda(X,T_0) & \to & \operatorname{Hom}_\Lambda(X,T_1).
\end{array}$$

Hence, $\operatorname{Hom}_\Lambda(X, C) \simeq \operatorname{Hom}_\Gamma((C,T),(X,T))$ for all X in $\operatorname{mod} \Lambda$ and all C in $\mathcal{X}_T = {}^{\perp}T$.

Let $0 \to A \to B \to C \to 0$ be F-exact with C in $\mathcal{X}_T$. Then

$$0 \to e_T(C) \to e_T(B) \to e_T(A) \to \operatorname{Ext}^1_F(C, T) = 0$$

is exact and therefore e_T is F-exact. □

As in the standard case we recover the algebra we started with by taking the endomorphism ring of the module T over Γ. We record this in the following corollary.

COROLLARY 3.4. *If T is an F-cotilting module and $\Gamma = \operatorname{End}_\Lambda(T)$, then*

$$\Lambda \simeq \operatorname{End}_\Gamma({}_\Gamma T)$$

via the homomorphism $\lambda \mapsto f_\lambda$, where $f_\lambda(t) = \lambda \cdot t$ for all t in T.

PROOF. The module Λ is in $\mathcal{X}_T$. By Lemma 3.3 (b) $\Lambda \simeq \operatorname{Hom}_\Lambda(\Lambda, \Lambda) \simeq \operatorname{Hom}_\Gamma((\Lambda, T), (\Lambda, T)) \simeq \operatorname{Hom}_\Gamma({}_\Gamma T, {}_\Gamma T)$ as Λ-modules, since $\operatorname{Hom}_\Lambda(\Lambda, {}_\Lambda T_{\Gamma^{\mathrm{op}}}) \simeq {}_\Gamma T$. Every element in $\operatorname{Hom}_\Lambda(\Lambda, \Lambda)$ is of the form g_a where $g_a(\lambda) = \lambda \cdot a$ for some a in Λ and all λ in Λ. Let g_a be in $\operatorname{Hom}_\Lambda(\Lambda, \Lambda)$. From the diagram

$$\begin{array}{ccc}
\operatorname{Hom}_\Lambda(\Lambda, T) & \xrightarrow{e_T(g_a)} & \operatorname{Hom}_\Lambda(\Lambda, T) \\
\downarrow \wr & & \downarrow \wr \\
{}_\Gamma T & \longrightarrow & {}_\Gamma T,
\end{array}$$

where the vertical maps are given by $h \mapsto h(1)$ for h in $\operatorname{Hom}_\Lambda(\Lambda, T)$, it follows that the homomorphism induced by $e_T(g_a)\colon {}_\Gamma T \to {}_\Gamma T$ is given by $t \mapsto a \cdot t$. So the map $\Lambda \to \operatorname{End}_\Gamma({}_\Gamma T)$ given by $\lambda \mapsto f_\lambda$, where $f_\lambda(t) = \lambda \cdot t$ for t in T, is a ring isomorphism. $\square$

We have seen that $\mathcal{X}_T = {}^\perp T$, so that in particular $\mathcal{P}(\Lambda)$ is contained in $\mathcal{X}_T$. Since $e_T = \operatorname{Hom}_\Lambda(\ , T)\colon \mathcal{X}_T \to \operatorname{mod}\Gamma$ is a fully faithful functor by Lemma 3.3 (b), we have that

$$X \simeq \operatorname{Hom}_\Lambda(\Lambda, X) \simeq \operatorname{Hom}_\Gamma((X, T), (\Lambda, T)) \simeq \operatorname{Hom}_\Gamma((X, T), T)$$

for all X in $\mathcal{X}_T$. This shows that $\mathcal{X}_T$ is dual to $(\mathcal{X}_T, T)$ in $\operatorname{mod}\Gamma$ given by the functor e_T. Using the following general result we can show that the subcategory $(\mathcal{X}_T, T)$ is covariantly finite in $\operatorname{mod}\Gamma$. The result is the dual of [**2**, Proposition 5.8] and we leave the proof to the reader.

PROPOSITION 3.5. *Let (G, H) be an adjoint pair of contravariant functors, where $G\colon \operatorname{mod}\Lambda \to \operatorname{mod}\Gamma$ and $H\colon \operatorname{mod}\Gamma \to \operatorname{mod}\Lambda$. Let $\mathcal{Y}$ be a subcategory of* $\operatorname{mod}\Lambda$ *and $\mathcal{Z}$ a subcategory of* $\operatorname{mod}\Gamma$ *such that the natural morphisms $\beta\colon I \to HG$ and $\alpha\colon J \to GH$ are isomorphisms on objects in $\mathcal{Y}$ and $\mathcal{Z}$, where I and J are the identity functors on* $\operatorname{mod}\Lambda$ *and* $\operatorname{mod}\Gamma$, *respectively. Assume that $G(\mathcal{Y}) \subset \mathcal{Z}$ and that $H(\mathcal{Z}) \subset \mathcal{Y}$.*

(a) *If $\mathcal{Y}$ is contravariantly finite in* $\operatorname{mod}\Lambda$, *then $\mathcal{Z}$ is covariantly finite in* $\operatorname{mod}\Gamma$.

(b) *If $\mathcal{Z}$ is contravariantly finite in* $\operatorname{mod}\Gamma$, *then $\mathcal{Y}$ is covariantly finite in* $\operatorname{mod}\Lambda$.

From Lemma 3.3 the functors $G = \operatorname{Hom}_\Lambda(\ , {}_\Lambda T_{\Gamma^{\mathrm{op}}})\colon \operatorname{mod}\Lambda \to \operatorname{mod}\Gamma$ and $H = \operatorname{Hom}_\Gamma(\ , {}_\Gamma T_{\Lambda^{\mathrm{op}}})\colon \operatorname{mod}\Gamma \to \operatorname{mod}\Lambda$ are an adjoint pair of contravariant functors satisfying the assumptions in Proposition 3.5 when $\mathcal{Y} = \mathcal{X}_T$ and $\mathcal{Z} = (\mathcal{X}_T, T)$. Then we have the following corollary of Proposition 3.5.

COROLLARY 3.6. (a) *The subcategory $(\mathcal{X}_T, T)$ is covariantly finite in* $\operatorname{mod}\Gamma$. *The functors* $\operatorname{Hom}_\Lambda(\ , {}_\Lambda T_{\Gamma^{\mathrm{op}}})\colon \operatorname{mod}\Lambda \to \operatorname{mod}\Gamma$ *and* $\operatorname{Hom}_\Gamma(\ , {}_\Gamma T_{\Lambda^{\mathrm{op}}})\colon \operatorname{mod}\Gamma \to \operatorname{mod}\Lambda$ *induce inverse dualities between $\mathcal{X}_T$ and $(\mathcal{X}_T, T)$.*

(b) *If F has enough projectives, then the subcategory $(\mathcal{P}(F), T)$ is covariantly finite in* $\operatorname{mod}\Gamma$. *The functors* $\operatorname{Hom}_\Lambda(\ , {}_\Lambda T_{\Gamma^{\mathrm{op}}})\colon \operatorname{mod}\Lambda \to \operatorname{mod}\Gamma$ *and* $\operatorname{Hom}_\Gamma(\ , {}_\Gamma T_{\Lambda^{\mathrm{op}}})\colon \operatorname{mod}\Gamma \to \operatorname{mod}\Lambda$ *induce inverse dualities between $\mathcal{P}(F)$ and $(\mathcal{P}(F), T)$.*

The isomorphism between the homomorphism groups over Λ and Γ extends to the Ext-groups. We get the following connection between the relative cohomology and the standard cohomology over Λ and Γ, respectively.

PROPOSITION 3.7. *If T is an F-cotilting module and $\Gamma = \operatorname{End}_\Lambda(T)$, then*

$$\operatorname{Hom}_\Lambda(\ ,T)\colon \operatorname{Ext}^i_F(C,A) \to \operatorname{Ext}^i_\Gamma((A,T),(C,T))$$

is an isomorphism for all A and C in $\mathcal{X}_T$ functorial in both variables.

PROOF. Let A and C be in $\mathcal{X}_T$. Since A is in $\mathcal{X}_T$, there is an F-exact sequence $0 \to A \to T_0 \xrightarrow{d_0} T_1 \xrightarrow{d_1} T_2 \to \cdots$, where $\operatorname{Im} d_i$ is in $\mathcal{X}_T$ for all i. Then the exact sequence

$$\cdots \to (T_2,T) \to (T_1,T) \to (T_0,T) \to (A,T) \to 0$$

is a Γ-projective resolution of (A,T). Applying $\operatorname{Hom}_\Gamma(\ ,(C,T))$ to this sequence we get the following commutative diagram by Lemma 3.3

$$\begin{array}{ccccccccc} 0 & \to & ((A,T),(C,T)) & \to & ((T_0,T),(C,T)) & \to & ((T_1,T),(C,T)) & \to & \cdots \\ & & \uparrow\wr & & \uparrow\wr & & \uparrow\wr & & \\ 0 & \to & (C,A) & \to & (C,T_0) & \to & (C,T_1) & \to & \cdots, \end{array}$$

where the cohomology of the upper row is $\operatorname{Ext}^i_\Gamma((A,T),(C,T))$ and the cohomology of the lower row is $\operatorname{Ext}^i_F(C,A)$, hence our desired result. The isomorphism is functorial in both variables, since the isomorphism between the homomorphism groups is functorial in both variables. $\square$

Next we show that the subcategory $(\mathcal{X}_T,T)$ of $\operatorname{mod}\Gamma$ is a resolving subcategory.

PROPOSITION 3.8. *Let T be an F-cotilting module with $\operatorname{id}_F T = r$ and let $\Gamma = \operatorname{End}_\Lambda(T)$.*

(a) *If $\operatorname{Ext}^i_\Gamma(C,(\mathcal{P}(F),T)) = 0$ for $0 < i < r+3$, then C is in $(\mathcal{X}_T,T)$.*

(b) *The subcategory $(\mathcal{X}_T,T)$ is a resolving subcategory of $\operatorname{mod}\Gamma$ and we have ${}^\perp(\mathcal{P}(F),T) = (\mathcal{X}_T,T)$.*

PROOF. (a) Let C be a module in $\operatorname{mod}\Gamma$ such that $\operatorname{Ext}^i_\Gamma(C,(\mathcal{P}(F),T)) = 0$ for $0 < i < r+3$. A start of a minimal projective resolution of C is of the form $(T_1,T) \xrightarrow{(f_0,T)} (T_0,T) \to C \to 0$. By Lemma 3.3 (b) the sequence is induced from a map $f_0\colon T_0 \to T_1$. If C_1 denotes $\operatorname{Coker} f_0$, then $\Omega^2_\Gamma(C) = (C_1,T)$. Let $C_1 \to T_2 \to C_2 \to 0$, $C_2 \to T_3 \to C_3 \to 0$, $C_3 \to T_4 \to C_4 \to 0 \ldots$, be a succession of minimal right $\operatorname{add} T$-approximations. This gives rise to a complex

$$C_1 \to T_2 \xrightarrow{f_2} T_3 \xrightarrow{f_3} T_4 \xrightarrow{f_4} T_5 \to \cdots$$

and

$$\cdots \to (T_n,T) \to (T_{n-1},T) \to \cdots \to (T_0,T) \to C \to 0$$

is a Γ-projective resolution of C. Denote $\operatorname{Im} f_0$ by C_0. Applying $\operatorname{Hom}_\Gamma(\ ,(P,T))$ with P in $\mathcal{P}(F)$ to this resolution we get the following commutative exact diagram by Lemma 3.3 (a)

$$\begin{array}{ccccccccc} 0 & \to & (C,(P,T)) & \to & ((T_0,T),(P,T)) & \to & ((T_1,T),(P,T)) & \to & \cdots \\ & & \uparrow\wr & & \uparrow\wr & & \uparrow\wr & & \\ 0 & \to & (P,(C,T)) & \to & (P,T_0) & \to & (P,T_1) & \to & \cdots. \end{array}$$

The cohomology of the upper row is $\operatorname{Ext}^i_\Gamma(C,(P,T)) = 0$ for $0 < i < r+3$, hence the lower sequence is exact up to $i = r+3$. Since this is true for all P in $\mathcal{P}(F)$ and Λ is in $\mathcal{P}(F)$, it follows that the sequence

$$0 \to (C,T) \to T_0 \xrightarrow{f_0} T_1 \to \cdots \to T_{r+1} \to C_{r+1} \to 0$$

is F-exact. Denote $\operatorname{Ext}_F^1(C_0, T)$ by U. Applying $\operatorname{Hom}_\Lambda(\ ,T)$ to this sequence induces the following commutative diagram

$$\begin{array}{ccccccccc}
\cdots\to & (T_2,T) & \xrightarrow{(f_1,T)} & (T_1,T) & \xrightarrow{(f_0,T)} & (T_0,T) & \to & ((C,T),T) & \to & U\to 0\\
& \| & & \| & & \| & & \uparrow & & \\
\cdots\to & (T_2,T) & \xrightarrow{(f_1,T)} & (T_1,T) & \xrightarrow{(f_0,T)} & (T_0,T) & \to & C & \to & 0.
\end{array}$$

Since the lower row is exact, it follows that $\operatorname{Ext}_F^1(C_i, T) = 0$ for $2 \le i \le r+1$. From the exact sequences $0 \to C_i \to T_{i+1} \to C_{i+1} \to 0$, we get that $\operatorname{Ext}_F^j(C_i, T) \simeq \operatorname{Ext}_F^{j+1}(C_{i+1}, T)$ for $j \ge 1$. Therefore $\operatorname{Ext}_F^i(C_{r+1}, T) = 0$ for $0 < i < r+1$. Since $\operatorname{id}_F T = r$, we have that C_{r+1} is in $\mathcal{X}_T$. The subcategory $\mathcal{X}_T$ is F-resolving by Theorem 3.2 (a), so that (C,T) and C_i for $i = 0, \dots, r$ are also in $\mathcal{X}_T$. Since $\operatorname{Ext}_F^1(C_0, T) = 0$, it follows from the above diagram that $C \simeq ((C,T),T)$ and therefore C is in $(\mathcal{X}_T, T)$.

(b) It follows immediately from the above and Proposition 3.7 that ${}^{\perp}(\mathcal{P}(F), T) = (\mathcal{X}_T, T)$. Therefore (X_T, T) is a resolving subcategory of $\operatorname{mod}\Gamma$.

□

Our next aim is to show that $(\mathcal{X}_T, T)$-resdim$(\operatorname{mod}\Gamma)$ is finite. Before we prove this, we need the following two results.

LEMMA 3.9. *Let T be an F-cotilting module and let $\Gamma = \operatorname{End}_\Lambda(T)$. Then the map*

$$\phi\colon \operatorname{Hom}_\Lambda(T,\omega) \otimes_\Gamma \operatorname{Hom}_\Lambda(X,T) \to \operatorname{Hom}_\Lambda(X,\omega)$$

given by $\phi(f \otimes g) = f \circ g$ is an isomorphism for all ω in $\widehat{\operatorname{add} T}$ and X in $\mathcal{X}_T$ and is functorial in both variables.

PROOF. It is clear that ϕ is an isomorphism for $\omega = T$ and therefore also for ω in $\operatorname{add} T$. Let ω be in $\widehat{\operatorname{add} T}$ and X in $\mathcal{X}_T$. Then we have an F-exact sequence $T_1 \to T_0 \to \omega \to 0$, which remains exact when applying $\operatorname{Hom}_\Lambda(T,\ )$. This gives the exact sequence

$$(T,T_1) \to (T,T_0) \to (T,\omega) \to 0.$$

Applying $-\otimes_\Gamma (X,T)$ to this sequence we obtain the following commutative diagram

$$\begin{array}{ccccccc}
(T,T_1)\otimes_\Gamma (X,T) & \to & (T,T_0)\otimes_\Gamma (X,T) & \to & (T,\omega)\otimes_\Gamma (X,T) & \to & 0\\
\downarrow\wr & & \downarrow\wr & & \downarrow & & \\
(X,T_1) & \to & (X,T_0) & \to & (X,\omega) & \to & 0.
\end{array}$$

Since $\mathcal{X}_T^{\perp} = \widehat{\operatorname{add} T}$ by Theorem 3.2 (b), the lower sequence is exact. Therefore $\phi\colon (T,\omega)\otimes_\Gamma (X,T) \to (X,\omega)$ is an isomorphism. It is easy to see that the isomorphism is functorial in both variables. □

LEMMA 3.10. *If T is an F-cotilting module and $\Gamma = \operatorname{End}_\Lambda(T)$, then*

$$\operatorname{id}_\Gamma D(T, \widehat{\operatorname{add} T}) \le r.$$

In particular $\operatorname{id}_\Gamma T \le r$.

PROOF. Let $\operatorname{id}_F T = r$. Every module ω in $\widehat{\operatorname{add} T}$ has an F-exact resolution in $\operatorname{add} T$ of the form $0 \to T_n \to T_{n-1} \to \cdots \to T_0 \to \omega \to 0$ with $n \le r$. The module T being F-selforthogonal, implies that $\operatorname{Ext}_F^i(T,\omega) = 0$ for all $i > 0$. Therefore the sequence above remains exact when we apply $\operatorname{Hom}_\Lambda(T,\ )$. This shows that $\operatorname{pd}_{\Gamma^{\mathrm{op}}}(T,\omega) \le r$ and consequently $\operatorname{id}_\Gamma D(T, \widehat{\operatorname{add} T}) \le r$.

Since $D(\Lambda^{\mathrm{op}})$ is in $\widehat{\operatorname{add} T}$, we have that $\operatorname{pd}_{\Gamma^{\mathrm{op}}}(T, D(\Lambda^{\mathrm{op}})) \leq r$. Combining this with the isomorphism $\operatorname{Hom}_\Lambda(T, D(\Lambda^{\mathrm{op}})) \simeq \operatorname{Hom}_{\Lambda^{\mathrm{op}}}(\Lambda^{\mathrm{op}}, DT) \simeq DT$, it follows that $\operatorname{id}_\Gamma T \leq r$. □

Now we are ready to prove that $(\mathcal{X}_T, T)$-resdim$(\operatorname{mod}\Gamma)$ is finite.

PROPOSITION 3.11. *Let T be an F-cotilting module and let $\Gamma = \operatorname{End}_\Lambda(T)$. Then* $\widehat{(\mathcal{X}_T, T)} = \operatorname{mod}\Gamma$ *and* $(\mathcal{X}_T, T)$-resdim$(\operatorname{mod}\Gamma) \leq \operatorname{id}_F T + 2$.

PROOF. Let B be a module in $\operatorname{mod}\Gamma$ and let $r = \operatorname{id}_F T$. A start of a minimal projective resolution of B is of the form $(T'', T) \to (T', T) \to B \to 0$. By Lemma 3.3 (b) the sequence is induced from a map $f\colon T' \to T''$. If C denotes $\operatorname{Coker} f$, then $\Omega^2_\Gamma(B) = (C, T)$. If $r = 0$, then $\mathcal{X}_T = \operatorname{mod}\Lambda$ and the proposition is proved in this case.

Let $C \to T_0 \to C_1 \to 0$, $C_1 \to T_1 \to C_2 \to 0$, $C_2 \to T_2 \to C_3 \to 0 \ldots$, be a succession of minimal left $\operatorname{add} T$-approximations. This gives rise to a complex

$$C \to T_0 \xrightarrow{f_0} T_1 \xrightarrow{f_1} T_2 \xrightarrow{f_2} T_3 \to \cdots$$

and

$$\cdots \to (T_n, T) \to (T_{n-1}, T) \to \cdots \to (T_0, T) \to (C, T) \to 0$$

is a Γ-projective resolution of (C, T).

Assume that $r \geq 1$. We want to show that $\Omega^{r+2}_\Gamma(B) \simeq (C_r, T)$ is in $(\mathcal{X}_T, T)$. Since $\operatorname{id}_\Gamma D(T, \widehat{\operatorname{add} T}) \leq r$ by Lemma 3.10, we have that $\operatorname{Ext}^j_\Gamma(B, D(T, \widehat{\operatorname{add} T})) = 0$ for $j > r$. So if we apply $\operatorname{Hom}_\Gamma(\ , D(T, \widehat{\operatorname{add} T}))$ to the sequence

$$\eta\colon \cdots \to (T_n, T) \to (T_{n-1}, T) \to \cdots \to (T_r, T) \to (C_r, T) \to 0$$

it remains exact. Let ω be in $\widehat{\operatorname{add} T}$, then we have the following commutative exact diagram by the adjointness isomorphism and Lemma 3.9

$$\begin{array}{ccccccc}
0 & \to & ((C_r, T), D(T, \omega)) & \to & ((T_r, T), D(T, \omega)) & \to & \cdots \\
 & & \downarrow \wr & & \downarrow \wr & & \\
0 & \to & D((T, \omega) \otimes_\Gamma (C_r, T)) & \to & D((T, \omega) \otimes_\Gamma (T_r, T)) & \to & \cdots \\
 & & \uparrow & & \uparrow \wr & & \\
0 & \to & D((C_r, \omega)) & \to & D((T_r, \omega)) & \to & \cdots .
\end{array}$$

It follows that $0 \to (C_{i+1}, \omega) \to (T_i, \omega) \to (C_i, \omega) \to 0$ is exact for $i \geq r$ and for all ω in $\widehat{\operatorname{add} T}$. Since $\mathcal{I}(F)$ is contained in $\widehat{\operatorname{add} T}$, the sequences $0 \to C_i \to T_i \to C_{i+1} \to 0$ for $i \geq r$ are F-exact, because $F = F^{\mathcal{I}(F)}$ by Theorem 1.4. Therefore $\operatorname{Ext}^1_F(C_i, \widehat{\operatorname{add} T}) = 0$ for all $i \geq r+1$. Using the F-exact sequences $0 \to C_i \to T_i \to C_{i+1} \to 0$ for $i \geq r$ we have that $\operatorname{Ext}^j_F(C_i, \widehat{\operatorname{add} T}) \simeq \operatorname{Ext}^{j+1}_F(C_{i+1}, \widehat{\operatorname{add} T})$ for $j \geq 1$ and $i \geq r$. Hence we have that $\operatorname{Ext}^j_F(C_{2r}, \widehat{\operatorname{add} T}) = 0$ for $0 < j < r+1$. Therefore C_{2r} is in ${}^\perp\widehat{\operatorname{add} T} = \mathcal{X}_T$, since $\widehat{\operatorname{add} T}$ is contained in $\mathcal{I}^r(F)$. But, since $\mathcal{X}_T \supset \operatorname{add} T$ is F-resolving and $0 \to C_i \to T_i \to C_{i+1} \to 0$ is F-exact for $i \geq r$, then $(C_r, T) \simeq \Omega^{r+2}_\Gamma(B)$ is in $(\mathcal{X}_T, T)$. This implies that $(\mathcal{X}_T, T)$-resdim$(\operatorname{mod}\Gamma) \leq r + 2$. Therefore $\widehat{(\mathcal{X}_T, T)} = \operatorname{mod}\Gamma$. This completes the proof of the proposition.

□

Let B be a module in $\operatorname{mod}\Gamma$ and let $(T_1, T) \to (T_0, T) \to B \to 0$ be the start of a minimal Γ-projective resolution. By Lemma 3.3 (a) the sequence is induced from a map $f\colon T_0 \to T_1$. Denote by C the cokernel of f, then $\Omega^2_\Gamma(B) \simeq (C, T)$. Taking a minimal

Γ-projective resolution of (C,T) amounts to taking a succession of minimal right $\operatorname{add} T$-approximations, $C \to T^C \to C_1 \to 0$, $C_1 \to T^{C_1} \to C_2 \to 0$, $C_2 \to T^{C_2} \to C_3 \to 0$, and so on. Then we actually proved the following result in the proof of Proposition 3.11.

LEMMA 3.12. *Let T be an F-cotilting module with $\operatorname{id}_F T = r$ and C a module in $\operatorname{mod}\Lambda$ given by an exact sequence $T_0 \to T_1 \to C \to 0$ with T_i in $\operatorname{add} T$. Let $C \to T^C \to C_1 \to 0$, $C_1 \to T^{C_1} \to C_2 \to 0$, $C_2 \to T^{C_2} \to C_3 \to 0$, ... be a sequence of minimal right $\operatorname{add} T$-approximations. Then C_r is in $\mathcal{X}_T$, where $C_0 = C$.*

Now we have the preliminaries needed for showing one of our main result in this section.

THEOREM 3.13. *Let T be an F-cotilting module in $\operatorname{mod}\Lambda$ and let $\Gamma = \operatorname{End}_\Lambda(T)$. Then we have the following.*

(a) *The subcategory $(\mathcal{X}_T,T) = {}^\perp(\mathcal{P}(F),T)$ is a resolving functorially finite subcategory of $\operatorname{mod}\Gamma$ with $(\mathcal{X}_T,T)$-$\operatorname{resdim}(\operatorname{mod}\Gamma) \le \operatorname{id}_F T + 2$.*

(b) *The subcategory $\widehat{(\mathcal{P}(F),T)} = (\mathcal{X}_T,T)^\perp$ is a coresolving covariantly finite subcategory of $\operatorname{mod}\Gamma$ with $\operatorname{id}_\Gamma(\mathcal{P}(F),T) \le \operatorname{id}_F T + 2$.*

(c) *$(\mathcal{X}_T,T) \cap (\mathcal{X}_T,T)^\perp = (\mathcal{P}(F),T)$.*

(d) *The subcategory $(\mathcal{P}(F),T) = \operatorname{add} T_0$ for a cotilting module T_0 over Γ with $\operatorname{id}_\Gamma T_0 \le \operatorname{id}_F T + 2$. The subcategory $(\mathcal{X}_T,T) = \mathcal{X}_{T_0} = {}^\perp T_0$.*

PROOF. (a) The subcategory $(\mathcal{X}_T,T) = {}^\perp(\mathcal{P}(F),T)$ is a resolving subcategory of $\operatorname{mod}\Gamma$ with $(\mathcal{X}_T,T)$-$\operatorname{resdim}(\operatorname{mod}\Gamma) \le \operatorname{id}_F T + 2$ by Proposition 3.8 and Proposition 3.11. Since $(\mathcal{P}(F),T)$ is an $\operatorname{Ext}_\Gamma$-injective cogenerator for $(\mathcal{X}_T,T)$, the subcategory $(\mathcal{X}_T,T)$ is contravariantly finite in $\operatorname{mod}\Gamma$ by Theorem 2.4. By Corollary 3.6 the subcategory $(\mathcal{X}_T,T)$ is covariantly finite in $\operatorname{mod}\Gamma$, which completes the proof of (a).

(b) By (a) and Theorem 2.4 we have that $\widehat{(\mathcal{P}(F),T)} = (\mathcal{X}_T,T)^\perp$. Since $(\mathcal{X}_T,T)$ is resolving contravariantly finite in $\operatorname{mod}\Gamma$, it follows by Proposition 2.2 that $\widehat{(\mathcal{P}(F),T)}$ is coresolving covariantly finite in $\operatorname{mod}\Gamma$. By (a) $(\mathcal{X}_T,T)$-$\operatorname{resdim}(\operatorname{mod}\Gamma) \le \operatorname{id}_F T + 2$, therefore we have by Theorem 2.5 that $\operatorname{id}_\Gamma(\widehat{(\mathcal{P}(F),T)}) \le \operatorname{id}_F T + 2$.

(c) The subcategory $(\mathcal{X}_T,T) \cap (\mathcal{X}_T,T)^\perp$ is equal to $(\mathcal{X}_T,T) \cap \widehat{(\mathcal{P}(F),T)}$. Hence $(\mathcal{P}(F),T)$ is contained in $(\mathcal{X}_T,T) \cap (\mathcal{X}_T,T)^\perp$. Let (X,T) be in $(\mathcal{X}_T,T) \cap (\mathcal{X}_T,T)^\perp$. We want to show that (X,T) is in $(\mathcal{P}(F),T)$. Then there is an exact sequence $0 \to (P_n,T) \xrightarrow{(f_n,T)} (P_{n-1},T) \xrightarrow{(f_{n-1},T)} \cdots \to (P_0,T) \xrightarrow{(f_0,T)} (X,T) \to 0$, where P_i is in $\mathcal{P}(F)$. Since $(\mathcal{X}_T,T)$ is resolving, $\operatorname{Coker}(f_i,T) \simeq (X_{i-2},T)$ with X_{i-2} in $\mathcal{X}_T$ for all i. By Proposition 3.7 the exact sequence $0 \to (P_n,T) \to (P_{n-1},T) \to (X_{n-2},T) \to 0$ comes from an F-exact sequence $0 \to X_{n-2} \to P_{n-1} \to P_n \to 0$. Since P_n is in $\mathcal{P}(F)$, the sequence splits and X_{n-2} must be in $\mathcal{P}(F)$. By induction X must be in $\mathcal{P}(F)$ and therefore $(\mathcal{X}_T,T) \cap (\mathcal{X}_T,T)^\perp = (\mathcal{P}(F),T)$.

(d) Since $(\mathcal{X}_T,T)$ is a resolving contravariantly finite subcategory of $\operatorname{mod}\Gamma$ with $(\mathcal{X}_T,T)$-$\operatorname{resdim}(\operatorname{mod}\Gamma)$ finite, the subcategory $(\mathcal{X}_T,T) \cap (\mathcal{X}_T,T)^\perp$ is $\operatorname{add} T_0$ for a cotilting module T_0 over Γ by [**3**, Theorem 5.4 (a)]. By (c) the intersection $(\mathcal{X}_T,T) \cap (\mathcal{X}_T,T)^\perp = (\mathcal{P}(F),T)$, which completes the proof of the theorem. □

As immediate consequence of this we have the following.

COROLLARY 3.14. *The subcategories $\mathcal{P}(F)$ and $\mathcal{I}(F)$ of $\operatorname{mod}\Lambda$ are of finite type.*

PROOF. Since $\mathcal{P}(F)$ is dual to $(\mathcal{P}(F),T)$ by Corollary 3.6 (b) and $(\mathcal{P}(F),T)$ is of finite type by Theorem 3.13 (d), the subcategory $\mathcal{P}(F)$ is of finite type. By Lemma 1.2 (c) $\mathcal{I}(F)$ is also of finite type. □

It follows from this result, that if $\mathcal{P}(F)$ or equivalently $\mathcal{I}(F)$ is of infinite representation type, then $\operatorname{mod}\Lambda$ has no F-cotilting modules.

Let T be a standard cotilting module and let $\Gamma = \operatorname{End}_\Lambda(T)$. It is shown in [**10**] that ${}_\Gamma T$ is a standard cotilting module over Γ. This is not the case for relative cotilting modules for proper subfunctors F of $\operatorname{Ext}^1_\Lambda(\ ,\ )$, since we have the following result.

PROPOSITION 3.15. *Let T be an F-cotilting module in* $\operatorname{mod}\Lambda$ *and let* $\Gamma = \operatorname{End}_\Lambda(T)$. *The module ${}_\Gamma T$ is a direct summand of a cotilting module T_0 over Γ with* $\operatorname{add} T_0 = (\mathcal{P}(F), T)$, $\operatorname{id}_\Gamma T_0 \leq \operatorname{id}_F T + 2$ *and* $\operatorname{id}_\Gamma T \leq r$. *Moreover the natural homomorphism* $X \to \operatorname{Hom}_\Lambda(\operatorname{Hom}_\Gamma(X,T),T)$ *is an isomorphism for all X in* $\mathcal{X}_{T_0} = {}^\perp T_0$.

PROOF. By Theorem 3.13 (d) the subcategory $(\mathcal{P}(F),T) = \operatorname{add} T_0$ for a cotilting module T_0 over Γ with $\operatorname{id}_\Gamma T_0 \leq \operatorname{id}_F T + 2$. Since ${}_\Gamma T$ is in $(\mathcal{P}(F),T)$, we can choose the cotilting Γ-module T_0 such that the module ${}_\Gamma T$ is a direct summand. By Lemma 3.10 $\operatorname{id}_\Gamma T \leq \operatorname{id}_F T$. Statement (d) of Theorem 3.13 implies that $\mathcal{X}_{T_0} = (\mathcal{X}_T, T)$. Therefore the natural homomorphism $X \to \operatorname{Hom}_\Lambda(\operatorname{Hom}_\Gamma(X,T),T)$ is an isomorphism for all X in $\mathcal{X}_{T_0}$, since the natural homomorphism $Z \to \operatorname{Hom}_\Gamma(\operatorname{Hom}_\Lambda(Z,T),T)$ is an isomorphism for all Z in $\mathcal{X}_T$. □

The connection between the global dimensions of Λ and Γ is discussed in the next section. Here we give the following upper bound on $\operatorname{gl.dim}\Gamma$.

PROPOSITION 3.16. *Let T be an F-cotilting module in* $\operatorname{mod}\Lambda$ *and let* $\Gamma = \operatorname{End}_\Lambda(T)$. *Then* $\operatorname{gl.dim}\Gamma = \operatorname{id}_F T + 2 + \max\{\operatorname{pd}_\Gamma(X_i,T) \mid (X_i,T) \to S_i\}$, *where $(X_i,T) \to S_i$ is the minimal right $(\mathcal{X}_T,T)$-approximation of the simple Γ-modules $\{S_1,\dots,S_t\}$.*

PROOF. By Theorem 3.13 (a) $(\mathcal{X}_T,T)\text{-resdim}(\operatorname{mod}\Gamma) = \operatorname{id}_F T + 2$. Therefore $\Omega_\Gamma^{\operatorname{id}_F T+2}(B)$ is in $(\mathcal{X}_T,T)$ for any Γ-module B. By the proof of Corollary 2.10 $\max\{\operatorname{pd}_\Gamma(X,T) \mid X \in \mathcal{X}_T\} = \max\{\operatorname{pd}_\Gamma(X_i,T) \mid (X_i,T) \to S_i\}$, where $(X_i,T) \to S_i$ is the minimal right $(\mathcal{X}_T,T)$-approximations of the simple Γ-modules $\{S_1,\dots,S_t\}$. The claim follows from this. □

By combining some of our results in this section we have the following corollary.

COROLLARY 3.17. *If T is an F-cotilting module, then the subcategory $\mathcal{X}_T$ of* $\operatorname{mod}\Lambda$ *is F-resolving functorially finite in* $\operatorname{mod}\Lambda$ *with* $\mathcal{X}_T\text{-resdim}_F(\operatorname{mod}\Lambda)$ *finite.*

PROOF. The subcategory $\mathcal{X}_T$ is F-resolving and contravariantly finite in $\operatorname{mod}\Lambda$ with $\mathcal{X}_T\text{-resdim}_F(\operatorname{mod}\Lambda)$ finite by Theorem 3.2 (a). It follows from Proposition 3.5 (b) and Corollary 3.6 (a) that $\mathcal{X}_T$ is covariantly finite in $\operatorname{mod}\Lambda$. □

In the standard cotilting theory we have that a resolving contravariantly finite subcategory $\mathcal{X}$ of $\operatorname{mod}\Lambda$ with $\mathcal{X}\text{-resdim}(\operatorname{mod}\Lambda)$ finite corresponds to a cotilting module via the map $\mathcal{X} \mapsto \mathcal{X}\cap\mathcal{X}^\perp$. Now we investigate in what generality we have a similar correspondence between F-cotilting modules and F-resolving contravariantly finite subcategories $\mathcal{X}$ of $\operatorname{mod}\Lambda$ with $\mathcal{X}\text{-resdim}_F(\operatorname{mod}\Lambda)$ finite. In Theorem 3.2 (a) we saw that if T is an F-cotilting module, then $\mathcal{X}_T = {}^\perp T$ is an F-resolving contravariantly finite subcategory of $\operatorname{mod}\Lambda$ with $\mathcal{X}_T\text{-resdim}_F(\operatorname{mod}\Lambda)$ finite. By Corollary 3.14 the subcategory $\mathcal{P}(F)$ being of finite type is a necessary condition for the existence of F-cotilting modules. So in order to have a similar correspondence as in the standard, it is necessary to assume that $\mathcal{P}(F)$ is of finite type. This turns out also to be sufficient.

As for the standard cotilting theory it is useful to study the Grothendieck groups in order to determined the number of indecomposable nonisomorphic summands a relative cotilting module has. Contrary to the standard case we can not use the standard

Grothendieck groups for both $\operatorname{mod}\Lambda$ and $\operatorname{mod}\Gamma$, but we must introduce the following group.

Let $\mathbf{Z}(\operatorname{mod}\Lambda)$ denote the free abelian group with the isomorphism classes $[A]$ of modules A in $\operatorname{mod}\Lambda$ as basis. We define the *F-Grothendieck group* of $\operatorname{mod}\Lambda$, $F\text{-}K_0(\operatorname{mod}\Lambda)$, to be $\mathbf{Z}(\operatorname{mod}\Lambda)/F\text{-}R(\operatorname{mod}\Lambda)$, where $F\text{-}R(\operatorname{mod}\Lambda)$ is the subgroup of $\mathbf{Z}(\operatorname{mod}\Lambda)$ generated by the elements $[A]+[C]-[B]$ whenever there is an F-exact sequence of modules $0 \to A \to B \to C \to 0$.

Let T be an F-cotilting module and let $\operatorname{id}_F T = r$. Let $\Gamma = \operatorname{End}_\Lambda(T)$ and define $\phi\colon F\text{-}K_0(\operatorname{mod}\Lambda) \to K_0(\operatorname{mod}\Gamma)$ by

$$\phi([C]) = \sum_{i=0}^{r}(-1)^i[\operatorname{Ext}^i_F(C,T)]$$

for any module C in $\operatorname{mod}\Lambda$. This map is well-defined, since we have long exact sequences and $\operatorname{Ext}^i_F(\ ,T) = 0$ for all $i \geq r+1$. Then we have the following connection between $F\text{-}K_0(\operatorname{mod}\Lambda)$ and $K_0(\operatorname{mod}\Gamma)$.

PROPOSITION 3.18. *If T is an F-cotilting module and $\Gamma = \operatorname{End}_\Lambda(T)$, then*

$$F{-}K_0(\operatorname{mod}\Lambda) \simeq K_0(\operatorname{mod}\Gamma).$$

PROOF. Define $\psi\colon K_0(\operatorname{mod}\Gamma) \to F{-}K_0(\operatorname{mod}\Lambda)$ by

$$\psi([B]) = \sum_{i=0}^{n-1}(-1)^i[T_i] + (-1)^n[X_B]$$

for any module B in $\operatorname{mod}\Gamma$, where

$$0 \to (X_B,T) \to (T_{n-1},T) \to (T_{n-2},T) \to \cdots \to (T_0,T) \to B \to 0$$

is a Γ-projective resolution of B and $n = \operatorname{id}_F T + 2$. We want to show that ψ is an inverse of ϕ. First we need to show that ψ is a well-defined map.

Since $\operatorname{id}_\Gamma(\mathcal{P}(F),T) \leq n$ and $(\mathcal{X}_T,T) = {}^\perp(\mathcal{P}(F),T)$ by Theorem 3.13 (b), we have that X_B is in $\mathcal{X}_T$ for all B in $\operatorname{mod}\Gamma$. For X in $\mathcal{X}_T$ direct summands of (X,T) corresponds to direct summands of X and vice versa by Lemma 3.3 (b). It follows from this that the definition of $\psi([B])$ is independent of the choice of Γ-projective resolution of B. Since $\operatorname{Ext}^1_\Gamma((X,T),(X',T)) \simeq \operatorname{Ext}^1_F(X',X)$ for all X and X' in $\mathcal{X}_T$, we have that ψ maps a short exact sequence to a sum of short F-exact sequences. Therefore ψ is a well-defined homomorphism. The subcategory $\mathcal{X}_T$ generates $F\text{-}K_0(\operatorname{mod}\Lambda)$ by Theorem 3.2 (a). Since $(\mathcal{X}_T,T)$-resdim$(\operatorname{mod}\Gamma)$ is finite, $(\mathcal{X}_T,T)$ generates $K_0(\operatorname{mod}\Gamma)$. It is easy to see that $\psi\circ\phi$ and $\phi\circ\psi$ are the identities on these generators, so that ϕ is an isomorphism. □

As a consequence of Proposition 3.18 we have the following.

PROPOSITION 3.19. *Let T be an F-cotilting module and $\Gamma = \operatorname{End}_\Lambda(T)$. Then the following three numbers are the same.*

(i) *the number of nonisomorphic indecomposable summands of ${}_\Lambda T$.*
(ii) *the number of nonisomorphic simple Γ-modules.*
(iii) *the number of nonisomorphic indecomposable modules in $\mathcal{P}(F)$.*

PROOF. Since $\Gamma = \operatorname{End}_\Lambda(T)$, the number of indecomposable nonisomorphic summands of T is the same as the number of nonisomorphic simple Γ-modules. The subcategory $(\mathcal{P}(F),T)$ of $\operatorname{mod}\Gamma$ is $\operatorname{add} T_0$ for some cotilting module T_0 in $\operatorname{mod}\Gamma$ by Theorem 3.13 (d). So the number of nonisomorphic indecomposable modules in $\mathcal{P}(F)$ is the same as the number of nonisomorphic simple Γ-modules. □

Since the last number in this result is independent of the cotilting module T, the following consequence is immediate.

Corollary 3.20. *All F-cotilting modules have the same number of nonisomorphic indecomposable summands and it is the same as the number of nonisomorphic indecomposable modules in $\mathcal{P}(F)$.*

Since the number of nonisomorphic indecomposable modules in $\mathcal{P}(F)$ can be greater than the number of nonisomorphic simple Λ-modules, the number of nonisomorphic indecomposable summands of an F-cotilting module can be greater than the number of nonisomorphic simple Λ-modules. More specifically we have the following.

Corollary 3.21. *Let T be an F-cotilting module in* $\operatorname{mod}\Lambda$. *Denote* $\operatorname{End}_\Lambda(T)$ *by* Γ. *Then ${}_\Lambda T$ is a standard cotilting module if and only if the number nonisomorphic indecomposable summands of ${}_\Lambda T$ is the same as the number of nonisomorphic simple Λ-modules.*

Proof. If ${}_\Lambda T$ is a standard cotilting module, then by Proposition 3.18 the module ${}_\Lambda T$ has the same number of nonisomorphic indecomposable summands as nonisomorphic simple Λ-modules.

If ${}_\Lambda T$ has the same number of nonisomorphic indecomposable summands as nonisomorphic simple Λ-modules, the subcategory $\mathcal{P}(F) = \mathcal{P}(\Lambda)$ by Corollary 3.20. Therefore ${}_\Lambda T$ is a standard cotilting module. □

This is one aspect of relative cotilting theory which is different from the standard cotilting theory.

If there exists a relative cotilting module T, then we know that $\mathcal{P}(F)$ and $\mathcal{I}(F)$ are of finite type and that all relative cotilting modules have the same number of nonisomorphic indecomposable summands. This enables us to prove the converse of Corollary 3.17.

Proposition 3.22. *Assume that the subcategory $\mathcal{P}(F)$ is of finite type. Let $\mathcal{X}$ be an F-resolving contravariantly finite subcategory of* $\operatorname{mod}\Lambda$ *with $\mathcal{X}$-$\operatorname{resdim}_F(\operatorname{mod}\Lambda)$ finite. Let $\mathcal{W} = \mathcal{X} \cap \mathcal{X}^\perp$. Then $\mathcal{W} = \operatorname{add} T$ for some F-cotilting module T.*

Proof. Let $\mathcal{W} = \mathcal{X} \cap \mathcal{X}^\perp$. Then by Proposition 2.3 the subcategory $\mathcal{W}$ is an Ext_F-injective cogenerator for $\mathcal{X}$. Theorem 2.4 (a) implies that $\mathcal{X}^\perp = \widehat{\mathcal{W}}$ and furthermore by Proposition 2.5 we have that $\mathcal{I}(F) \subset \widehat{\mathcal{W}} \subset \mathcal{I}^n(F)$ for some finite n. Since $\mathcal{P}(F)$ is of finite type, then $\mathcal{I}(F)$ is also of finite type. Therefore let $\mathcal{I}(F) = \operatorname{add} I$ for some Λ-module I and let $0 \to \omega_m \to \omega_{m-1} \to \cdots \to \omega_0 \to I \to 0$ be an F-exact resolution of I in $\mathcal{W}$. Then $T = \oplus_{i=0}^m \omega_i$ is an F-cotilting module. Let ω be in $\mathcal{W}$, then $T \oplus \omega$ is also an F-cotilting module. But since the number of nonisomorphic indecomposable summands in an F-cotilting module is the same by Corollary 3.20, we have that $\mathcal{W} = \operatorname{add} T$. □

As a restatement of Proposition 3.22 and Theorem 3.2 (a) we have the following results.

Theorem 3.23. *Let $\mathcal{W}$ be an F-selforthogonal additive subcategory of* $\operatorname{mod}\Lambda$ *and assume that the subcategory $\mathcal{P}(F)$ is of finite type. Then $\mathcal{W} = \operatorname{add} T$ for some F-cotilting module if and only if $\mathcal{X}_\mathcal{W}$ is an F-resolving contravariantly finite subcategory of* $\operatorname{mod}\Lambda$ *with $\mathcal{X}_\mathcal{W}$-$\operatorname{resdim}_F(\operatorname{mod}\Lambda)$ finite.*

Theorem 3.24. *Assume that the subcategory $\mathcal{P}(F)$ is of finite type. Then the maps $T \mapsto \mathcal{X}_T$ and $\mathcal{X} \mapsto \mathcal{X} \cap \mathcal{X}^\perp$ induce inverse bijections between the isomorphism classes of basic F-cotilting modules in* $\operatorname{mod}\Lambda$ *and the family of F-resolving contravariantly finite subcategories $\mathcal{X}$ of* $\operatorname{mod}\Lambda$ *with $\mathcal{X}$-$\operatorname{resdim}_F(\operatorname{mod}\Lambda)$ finite.*

Tilting and cotilting modules of projective and injective dimension at most one, respectively, have been studied extensively. Let T be a module in $\operatorname{mod}\Lambda$ with $\operatorname{Ext}^i_\Lambda(T,T) = 0$ for all $i > 0$. If $\operatorname{id}_\Lambda T \leq 1$, then there is a module X in $\operatorname{mod}\Lambda$ such that $T \oplus X$ is cotilting module of injective dimension at most one. If $\operatorname{id}_\Lambda T > 1$, then is known that T is not necessarily a summand of a cotilting module [**9**]. The fact that a selforthogonal module T with $\operatorname{id}_\Lambda T \leq 1$ can be extended to a cotilting module, can also be generalized to relative cotilting modules. The proof is the same as in the standard case, so it is left to the reader. Let F be a subfunctor of $\operatorname{Ext}^1_\Lambda(\ ,\ )$ with enough projectives and injectives, where $\mathcal{P}(F)$ is of finite type.

PROPOSITION 3.25. *Let T be a module in* $\operatorname{mod}\Lambda$ *with* $\operatorname{Ext}^1_F(T,T) = 0$ *and* $\operatorname{id}_F T \leq 1$.
(a) *There is some module X in* $\operatorname{mod}\Lambda$ *such that $T \oplus X$ is an F-cotilting module with* $\operatorname{id}_F T \oplus X \leq 1$ *and such that there is an F-exact sequence $0 \to T_0 \to X \to I \to 0$ with T_0 in* $\operatorname{add} T$ *and* $\operatorname{add} I = \mathcal{I}(F)$.
(b) *The module T is an F-cotilting module if and only if the number of nonisomorphic indecomposable summands of T is equal to the number of nonisomorphic indecomposable modules in $\mathcal{P}(F)$.*

We end this section by considering some examples of cotilting modules.

Example I.. Let F be a subfunctor of $\operatorname{Ext}^1_\Lambda(\ ,\ )$ with enough projectives and assume that $\mathcal{P}(F)$ is of finite type. Then the easiest example of an F-cotilting module T is a module T such that $\operatorname{add} T = \mathcal{I}(F)$. If G is a Λ-module such that $\operatorname{add} G = \mathcal{P}(F)$, then $\operatorname{Hom}_\Lambda(G,T)$ is a cotilting module over $\Gamma = \operatorname{End}_\Lambda(T)$ with $\operatorname{id}_\Gamma(G,T) \leq 2$. Moreover we have the following stronger result.

PROPOSITION 3.26. *Assume that F is a proper subfunctor of* $\operatorname{Ext}^1_\Lambda(\ ,\ )$. *Let T be an F-cotilting module with* $\operatorname{id}_F T = 0$ *and let* $\Gamma = \operatorname{End}_\Lambda(T)$. *If G is a Λ-module such that* $\operatorname{add} G = \mathcal{P}(F)$, *then* $\operatorname{Hom}_\Lambda(G,T)$ *is a cotilting module over Γ with* $\operatorname{id}_\Gamma(G,T) = 2$.

PROOF. By Theorem 3.13 we have that (G,T) is a cotilting module over Γ with $\operatorname{id}_\Gamma(G,T) \leq 2$. Moreover, ${}^\perp T = \operatorname{mod}\Lambda$ is dual to ${}^\perp(G,T)$ by the functor $\operatorname{Hom}_\Lambda(\ ,T)$. If the Γ-module (G,T) is injective, then $\operatorname{mod}\Lambda$ and $\operatorname{mod}\Gamma$ are dual. Since the number of projectives over Λ and Γ are different, this is impossible. So, assume that $\operatorname{id}_\Gamma(G,T) = 1$. Let $G = P \oplus M$, where M has no nonzero projective summands. Then $(G,T) = (P,T) \oplus (M,T)$. Let $P_1 \xrightarrow{f} P_0 \to M \to 0$ be a standard projective resolution of M in $\operatorname{mod}\Lambda$. This gives rise to the exact sequence $0 \to (M,T) \to (P_0,T) \xrightarrow{(f,T)} (P_1,T)$, where (P_i,T) is in $\operatorname{add}{}_\Gamma T$ for $i = 0, 1$. The module ${}_\Gamma T$ is injective by Theorem 3.13. If $\operatorname{id}_\Gamma(G,T) = 1$, then the image $\operatorname{Im}(f,T)$ is injective. This implies that $\operatorname{Im}(f,T)$ is a direct summand of (P_1,T). Therefore $\operatorname{Im}(f,T) \simeq (Q,T)$ for some direct summand Q of P_1 by Lemma 3.3 (b). Then we have an exact sequence $0 \to (M,T) \to (P_0,T) \to (Q,T) \to 0$, which by Proposition 3.7 is induced from an F-exact sequence $0 \to Q \to P_0 \to M \to 0$. Since M is in $\mathcal{P}(F)$, this sequence must split and M is in $\mathcal{P}(\Lambda)$. This is a contradiction, so we have shown that $\operatorname{id}_\Gamma(G,T) = 2$. □

This result shows that the upper theoretical bound for $\operatorname{id}_\Gamma(G,T)$ is sharp when T is an F-cotilting module with $\operatorname{id}_F T = 0$.

Example II.. Let $\mathcal{X}$ be a resolving contravariantly finite subcategory of $\operatorname{mod}\Lambda$ with $\widehat{\mathcal{X}} = \operatorname{mod}\Lambda$. Let $\mathcal{W}_{\text{std}} = \mathcal{X} \cap \mathcal{X}^\perp_{\text{std}}$, which by [**3**, Theorem 5.4 (a)] is $\operatorname{add} T_{\text{std}}$ for some cotilting module T_{std} in $\operatorname{mod}\Lambda$. We want to construct a relative cotilting module which has T_{std} as a summand.

Let $\mathcal{P}(F)$ be a finite subcategory of $\mathcal{X}$ containing $\mathcal{P}(\Lambda)$. Then $F = F_{\mathcal{P}(F)}$ has enough projective and injectives and the projectives are as indicated by the notation. Since $\mathcal{P}(F)$ is a subcategory of $\mathcal{X}$, the subcategory $\mathcal{X}$ is F-resolving contravariantly finite in $\operatorname{mod}\Lambda$. For each simple Λ-module S denote by n_S the minimal number such that there is an exact sequence $0 \to X_{n_S} \to \cdots X_1 \to X_0 \to S \to 0$ with X_i in $\mathcal{X}$. Let n denote the maximum of the numbers n_S. Then it follows that $\mathcal{X}^{\perp}_{\text{std}}$ is contained in $\mathcal{I}^n(\Lambda)$. Let $\cdots \to X_1 \xrightarrow{f_1} X_0 \to C_0 \to 0$ is an exact sequence where $X_i \to \operatorname{Im} f_i$ is a minimal right $\mathcal{X}$-approximation for all i. Then $\operatorname{Im} f_{n+1}$ is in $\mathcal{X}$, since $\mathcal{X} = {}^{\perp}(\mathcal{X}^{\perp}_{\text{std}})$. Hence we have that $\mathcal{X}\text{-resdim}_F(\operatorname{mod}\Lambda) \leq n$. Since $\mathcal{P}(F)$ is of finite type, we have that $\mathcal{X} \cap \mathcal{X}^{\perp}$, where $\mathcal{X}^{\perp}$ is taken relatively, is $\operatorname{add} T$ for some F-cotilting module T. Since $\mathcal{X}^{\perp} = \mathcal{Y}_F(\mathcal{X})$ and $\mathcal{X}^{\perp}_{\text{std}} = \mathcal{Y}_{\text{std}}(\mathcal{X})$, it follows that $\operatorname{add} T_{\text{std}}$ is contained in $\operatorname{add} T$ and therefore T_{std} is a direct summand of T (if we only consider basic cotilting modules).

Example III.. Let Λ be the path algebra over a field k of the following quiver

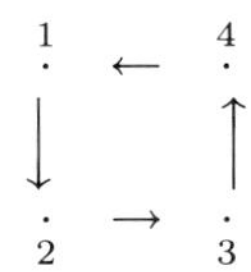

where the cube of the radical is zero. Let $P_1 = \begin{smallmatrix}1&1\\0&1\end{smallmatrix}$, $P_2 = \begin{smallmatrix}1&1\\1&0\end{smallmatrix}$, $P_3 = \begin{smallmatrix}1&1\\1&0\end{smallmatrix}$ and $P_4 = \begin{smallmatrix}0&1\\1&1\end{smallmatrix}$, which are all the indecomposable projective Λ-modules. Then let $M_i = P_i/\operatorname{soc} P_i$ and $S_i = P_i/\underline{\mathrm{r}}P_i$ for $i = 1, \ldots, 4$, where $\underline{\mathrm{r}}$ is the radical of Λ. Let $\mathcal{I}(F) = \mathcal{I}(\Lambda) \cup \operatorname{add} S_1$ and $F = F^{\mathcal{I}(F)}$, which has enough projectives and injectives. The projectives of the subfunctor are $\mathcal{P}(F) = \mathcal{P}(\Lambda) \cup \operatorname{add} S_4$ and the injectives are as indicated by the notation. Since Λ is a selfinjective algebra, any F-cotilting module must have Λ has a direct summand. Let $T = \Lambda \oplus M_2$. Since $0 \to M_2 \to P_1 \to S_1 \to 0$ is F-exact, it follows that $\operatorname{id}_F T = 1$ and that $\mathcal{I}(F)$ is contained in $\operatorname{add} T$. The relative injective dimension of T is one, so that $\operatorname{Ext}^i_F(T,T) = 0$ for all $i > 1$. Using the relative Auslander-Reiten isomorphism, we have that $\operatorname{Ext}^1_F(T,T) = D\,\underline{\operatorname{Hom}}_{\mathcal{P}(F)}(\operatorname{Tr} D(M_2), M_2)$, where $\operatorname{Tr} D(M_2) = M_1$. It is easy to see that this group is zero, so that T is an F-cotilting module and T is **not** relative projective or injective. Therefore selfinjective algebras may have nontrivial relative cotilting modules. If we let $F' = F^{\operatorname{add} T}$, then T is also an F'-cotilting module. Thus a module T can be a relative cotilting module with respect to different relative theories. In general, if a relative cotilting module T contains all the indecomposable injectives, then T is a relative cotilting module with respect to the subfunctor $F = F^{\operatorname{add} T}$.

4. Relative cotilting and global dimensions

Throughout this section let F be an additive subfunctor of $\operatorname{Ext}^1_\Lambda(\ ,\ )$ with enough projective and injectives. Let T be a fixed F-cotilting module and let $\Gamma = \operatorname{End}_\Lambda(T)$. This section is devoted to comparing the global dimensions of Λ and Γ.

If T is a standard cotilting module, then the global dimensions of Λ and Γ satisfy the formula $\operatorname{gl.dim}\Lambda - \operatorname{id}_\Lambda T \leq \operatorname{gl.dim}\Gamma \leq \operatorname{id}_\Lambda T + \operatorname{gl.dim}\Lambda$. We show by giving an example of a relative cotilting module that Λ can have infinite global dimension, but Γ can still have finite global dimension. So the above formula is not valid in the relative case. Denote by $\operatorname{gl.dim}_F \Lambda$ the relative global dimension. Then we show that $\operatorname{gl.dim}_F \Lambda$ and $\operatorname{gl.dim}\Gamma$ satisfy a similar formula but not the same as in the standard case, namely $\operatorname{gl.dim}_F \Lambda - \operatorname{id}_F T \leq \operatorname{gl.dim}\Gamma \leq \operatorname{id}_F T + 2 + \operatorname{gl.dim}_F \Lambda$. Since $\operatorname{gl.dim}\Gamma$ can be finite with $\operatorname{gl.dim}\Lambda$ being infinite, one might conclude that there is no connection between $\operatorname{gl.dim}\Lambda$

and $\mathrm{gl.dim}\,\Gamma$. But we show that when $\mathrm{gl.dim}\,\Gamma$ is finite, then $\mathrm{gl.dim}\,\Lambda$ is finite if and only if $\mathrm{id}_\Lambda T$ or $\mathrm{pd}_\Lambda T$ is finite. Moreover, $\mathrm{gl.dim}\,\Lambda \leq \min\{\mathrm{id}_\Lambda T + \mathrm{gl.dim}\,\Gamma + 1, \mathrm{pd}_\Lambda T + \mathrm{id}_F T\}$.

PROPOSITION 4.1. *Let T and Γ be as above. Then*

$$\mathrm{gl.dim}_F \Lambda - \mathrm{id}_F T \leq \mathrm{gl.dim}\,\Gamma \leq \mathrm{id}_F T + 2 + \mathrm{gl.dim}_F \Lambda.$$

PROOF. We first prove that $\mathrm{gl.dim}\,\Gamma \leq \mathrm{id}_F T + 2 + \mathrm{gl.dim}_F \Lambda$. If $\mathrm{gl.dim}_F \Lambda$ is infinite, there is nothing to prove, so assume that it is finite. Then for all X in $\mathcal{X}_T$ there is an F-exact sequence

$$0 \to P_m \to P_{m-1} \to \cdots \to P_0 \to X \to 0$$

with P_i in $\mathcal{P}(F)$ and $m \leq \mathrm{gl.dim}_F \Lambda$. When we apply $\mathrm{Hom}_\Lambda(\ ,T)$ to the sequence above it remains exact,

$$0 \to (X,T) \to (P_0,T) \to \cdots \to (P_{m-1},T) \to (P_m,T) \to 0.$$

By Theorem 3.13 $\mathrm{id}_\Gamma(\mathcal{P}(F),T) \leq \mathrm{id}_F T + 2$, hence $\mathrm{id}_\Gamma(\mathcal{X}_T,T) \leq \mathrm{id}_F T + 2 + \mathrm{gl.dim}_F \Lambda$. By Proposition 3.11 the module $\Omega_\Gamma^{\mathrm{id}_F T+2}(B)$ is in $(\mathcal{X}_T,T)$ for all modules B in $\mathrm{mod}\,\Gamma$, therefore $\mathrm{id}_\Gamma B \leq \mathrm{id}_\Gamma(\mathcal{X}_T,T) \leq \mathrm{id}_F T + 2 + \mathrm{gl.dim}_F \Lambda$. Hence, we have proved that $\mathrm{gl.dim}\,\Gamma \leq \mathrm{id}_F T + 2 + \mathrm{gl.dim}_F \Lambda$.

Now, we prove that $\mathrm{gl.dim}_F \Lambda \leq \mathrm{id}_F T + \mathrm{gl.dim}\,\Gamma$. If $\mathrm{gl.dim}\,\Gamma$ is infinite, there is nothing to prove, so assume that it is finite. By Proposition 3.7 we have that $\mathrm{Ext}_F^i(C,A) \simeq \mathrm{Ext}_\Gamma^i((A,T),(C,T))$ for all A and C in $\mathcal{X}_T$. Then it follows that $\mathrm{Ext}_F^i(C,\mathcal{X}_T) = 0$ for all $i > \mathrm{gl.dim}\,\Gamma$. If $\mathrm{Ext}_F^i(B,\mathcal{X}_T) = 0$ for $i > 0$, then it is easy to see that $\mathrm{Ext}_F^i(B,\ ) = 0$ for $i > 0$ and therefore B is in $\mathcal{P}(F)$. Hence, $\Omega_F^{\mathrm{gl.dim}\,\Gamma}(C)$ is in $\mathcal{P}(F)$ for all C in $\mathcal{X}_T$. Since $\mathcal{X}_T\text{-resdim}_F(\mathrm{mod}\,\Lambda) \leq \mathrm{id}_F T$ and $\mathcal{P}(F)$ is contained in $\mathcal{X}_T$, we have that $\Omega_F^{\mathrm{id}_F T}(M)$ is in $\mathcal{X}_T$ for all M in $\mathrm{mod}\,\Lambda$. From this it follows that $\mathrm{gl.dim}_F \Lambda \leq \mathrm{id}_F T + \mathrm{gl.dim}\,\Gamma$. □

As already remarked the standard global dimension $\mathrm{gl.dim}\,\Lambda$ can be infinite, but $\mathrm{gl.dim}\,\Gamma$ is finite. The following examples illustrate this behavior.

Let Λ be of finite type with infinite global dimension and let all the nonisomorphic indecomposable Λ-modules be $\{M_i\}_{i=1}^n$. Then if $F = 0$, the module $T = \oplus_{i=1}^n M_i$ is an F-cotilting module where $\Gamma = \mathrm{End}_\Lambda(T)$ has finite global dimension.

A less trivial example is the following. Let Λ be the path algebra of the following quiver

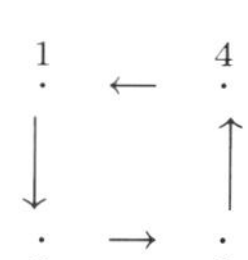

where the cube of the radical is zero. We considered this selfinjective artin algebra in section 4 and we will use the notation introduced there.

Let $\mathcal{P}(F) = \{P_1,P_2,P_3,P_4,S_1,S_4,M_3\}$. Then we have that $\mathcal{I}(F) = \{P_1,P_2,P_3,P_4,S_1,S_2,M_4\}$. It is easy to find F-exact sequences for the modules M_1, M_2, M_4, S_2 and S_3 to show that $\mathrm{gl.dim}_F \Lambda \leq 3$. By Proposition 4.1 we have that $\mathrm{gl.dim}\,\Gamma \leq \mathrm{id}_F T + 2 + 3$ for any F-cotilting module T where $\Gamma = \mathrm{End}_\Lambda(T)$. In particular, for the F-cotilting module

$$T = P_1 \oplus P_2 \oplus P_3 \oplus P_4 \oplus S_1 \oplus S_2 \oplus M_4$$

we have that $\mathrm{gl.dim}\,\Gamma \leq 5$, but $\mathrm{gl.dim}\,\Lambda = \infty$.

After having seen the above examples, it is tempting to conclude that there is no connection between $\mathrm{gl.dim}\,\Lambda$ and $\mathrm{gl.dim}\,\Gamma$. But if $\mathrm{gl.dim}\,\Gamma$ is finite, every Λ-module has a finite coresolution in $\widehat{\operatorname{add} T}$, because we have the following. Every module X in $\mathcal{X}_T$ has an F-exact coresolution in $\operatorname{add} T$ of the form $0 \to X \to T_0 \xrightarrow{f_0} T_1 \xrightarrow{f_1} T_2 \to \cdots$ with T_i in $\operatorname{add} T$ and $\operatorname{Im} f_i$ in $\mathcal{X}_T$. This sequence induces a Γ-projective resolution of (X, T) by applying $\operatorname{Hom}_\Lambda(\ , T)$. Since $\mathrm{gl.dim}\,\Gamma$ is finite, $\operatorname{Im} f_i$ is in $\operatorname{add} T$ for some $i \leq \mathrm{gl.dim}\,\Gamma - 1$. The subcategory $\widehat{\operatorname{add} T}$ is F-covariantly finite by Theorem 3.2 (b). Moreover, for every module C in $\operatorname{mod}\Lambda$ there is an F-exact sequence $0 \to C \to \omega^C \to X^C \to 0$ with ω^C in $\widehat{\operatorname{add} T}$ and X^C in $\mathcal{X}_T$. Using this we have the following connection between $\mathrm{gl.dim}\,\Lambda$ and $\mathrm{gl.dim}\,\Gamma$ when $\mathrm{gl.dim}\,\Gamma$ is finite.

PROPOSITION 4.2. *Let T and Γ be as above and assume that* $\mathrm{gl.dim}\,\Gamma$ *is finite. Then we have the following.*

(a) $\mathrm{gl.dim}\,\Lambda$ *is finite if and only if* $\mathrm{id}_\Lambda T$ *is finite.*
(b) $\mathrm{gl.dim}\,\Lambda$ *is finite if and only if* $\mathrm{pd}_\Lambda T$ *is finite.*
(c) *If* $\mathrm{id}_\Lambda T$ *is finite, then* $\mathrm{gl.dim}\,\Lambda \leq \mathrm{id}_\Lambda T + \mathrm{gl.dim}\,\Gamma + 1$.
(d) *If* $\mathrm{pd}_\Lambda T$ *is finite, then* $\mathrm{gl.dim}\,\Lambda \leq \mathrm{pd}_\Lambda T + \mathrm{id}_F T$.

PROOF. Clearly, if $\mathrm{gl.dim}\,\Lambda$ is finite, then both $\mathrm{id}_\Lambda T$ and $\mathrm{pd}_\Lambda T$ are finite.

Assume that $\mathrm{id}_\Lambda T$ is finite. Since $\mathrm{gl.dim}\,\Gamma$ is finite, there exists for all X in $\mathcal{X}_T$ an F-exact sequence

$$0 \to X \to T_0 \to T_1 \to \cdots \to T_{m-1} \to T_m \to 0 \quad (1)$$

with T_i in $\operatorname{add} T$ and $m \leq \mathrm{gl.dim}\,\Gamma$. Therefore for all C in $\operatorname{mod}\Lambda$ there exists an F-exact sequence

$$0 \to C \to \omega^C \to T_0 \to \cdots \to T_{m-1} \to T_m \to 0 \quad (2),$$

where the first short exact sequence $0 \to C \to \omega^C \to X^C \to 0$ is the minimal left F-$\widehat{\operatorname{add} T}$-approximation, the modules T_i is in $\operatorname{add} T$ and $m \leq \mathrm{gl.dim}\,\Gamma$. For all ω in $\widehat{\operatorname{add} T}$ there is an F-exact sequence

$$0 \to T'_s \to T'_{s-1} \to \cdots \to T'_0 \to \omega \to 0$$

with T'_i in $\operatorname{add} T$ and $s \leq \mathrm{id}_F T$, so that $\mathrm{id}_\Lambda \omega \leq \mathrm{id}_\Lambda T$. Then from the sequence (2) above we have that $\mathrm{id}_\Lambda C \leq \mathrm{id}_\Lambda T + m + 1 \leq \mathrm{id}_\Lambda T + \mathrm{gl.dim}\,\Gamma + 1$. Hence $\mathrm{gl.dim}\,\Lambda \leq \mathrm{id}_\Lambda T + \mathrm{gl.dim}\,\Gamma + 1$.

Assume that $\mathrm{pd}_\Lambda T$ is finite. Then it follows from the sequence (1) that $\mathrm{pd}_\Lambda X \leq \mathrm{pd}_\Lambda T$ for all X in $\mathcal{X}_T$. Since $\operatorname{Ext}^i_F(C, T) = 0$ for $i > \mathrm{id}_F T$ and all C in $\operatorname{mod}\Lambda$ and $\mathcal{P}(F) \subset \mathcal{X}_T$, we have that $\mathrm{pd}_\Lambda C \leq \mathrm{pd}_\Lambda T + \mathrm{id}_F T$. Therefore $\mathrm{gl.dim}\,\Lambda \leq \mathrm{pd}_\Lambda T + \mathrm{id}_F T$. □

5. Subcategory correspondence

Throughout this section let F be an additive subfunctor of $\operatorname{Ext}^1_\Lambda(\ ,\ )$ with enough projectives and injectives. Fix an F-cotilting module T in $\operatorname{mod}\Lambda$ and let $\Gamma = \operatorname{End}_\Lambda(T)$. This section is devoted to proving that $\widehat{\operatorname{add} T}$ is dual to $\mathcal{I}^\infty(\Gamma) \cap D(T, \mathcal{I}(F))^\perp$ by the functor $D\operatorname{Hom}_\Lambda(T,\ )\colon \operatorname{mod}\Lambda \to \operatorname{mod}\Gamma$, where $\mathcal{I}^\infty(\Gamma)$ is the subcategory of $\operatorname{mod}\Gamma$ consisting of the modules with finite injective dimension.

Let T, F and Γ be as above. We have the following functors

$$G = \operatorname{Hom}_\Lambda(T,\ )\colon \operatorname{mod}\Lambda \to \operatorname{mod}\Gamma^{\mathrm{op}}$$

and

$$H = T \otimes_{\Gamma^{\mathrm{op}}} -\colon \operatorname{mod}\Gamma^{\mathrm{op}} \to \operatorname{mod}\Lambda.$$

By the adjointness isomorphism we have that

$$\operatorname{Hom}_\Lambda(T \otimes_{\Gamma^{\mathrm{op}}} B, C) \simeq \operatorname{Hom}_{\Gamma^{\mathrm{op}}}(B, \operatorname{Hom}_\Lambda(T, C)),$$

for all B in $\operatorname{mod}\Gamma^{\mathrm{op}}$ and C in $\operatorname{mod}\Lambda$.

The following lemma is the key result in proving the claim above.

LEMMA 5.1. *For ω in $\widehat{\operatorname{add} T}$ the natural homomorphism*

$$\psi_\omega \colon T \otimes_{\Gamma^{\mathrm{op}}} (T, \omega) \to \omega$$

is an isomorphism functorial in ω.

PROOF. Clearly, $\psi_T \colon T \otimes_{\Gamma^{\mathrm{op}}} (T, T) \to T$ is an isomorphism. Therefore by additivity $\psi_{T'} \colon T \otimes_{\Gamma^{\mathrm{op}}} (T, T') \to T'$ is an isomorphism for all T' in $\operatorname{add} T$. Let ω be in $\widehat{\operatorname{add} T}$, then we have an F-exact sequence $T_1 \to T_0 \to \omega \to 0$ with T_i in $\operatorname{add} T$. This sequence remains exact when we apply $\operatorname{Hom}_\Lambda(T, \ )$, so that $(T, T_1) \to (T, T_0) \to (T, \omega) \to 0$ is exact. Applying the functor $T \otimes_{\Gamma^{\mathrm{op}}} -$ to this sequence, we get the following commutative diagram

$$\begin{array}{ccccccc}
T \otimes_{\Gamma^{\mathrm{op}}} (T, T_1) & \to & T \otimes_{\Gamma^{\mathrm{op}}} (T, T_0) & \to & T \otimes_{\Gamma^{\mathrm{op}}} (T, \omega) & \to & 0 \\
\downarrow \wr & & \downarrow \wr & & \downarrow & & \\
T_1 & \to & T_0 & \to & \omega & \to & 0.
\end{array}$$

Hence it follows that $\psi_\omega \colon T \otimes_{\Gamma^{\mathrm{op}}} (T, \omega) \to \omega$ is an isomorphism for all ω in $\widehat{\operatorname{add} T}$. It is easy to see that this isomorphism is functorial in ω. □

By the adjointness isomorphism and the above result the following result is immediate.

COROLLARY 5.2. *For all ω in $\widehat{\operatorname{add} T}$ and all C in $\operatorname{mod}\Lambda$ the homomorphism*

$$\operatorname{Hom}_\Lambda(T, \ ) \colon \operatorname{Hom}_\Lambda(\omega, C) \to \operatorname{Hom}_{\Gamma^{\mathrm{op}}}((T, \omega), (T, C))$$

is an isomorphism functorial in both variables.

This isomorphism on the homomorphism groups also extends to an isomorphism on the Ext-groups, which we show next.

COROLLARY 5.3. *For all ω in $\widehat{\operatorname{add} T}$ and any C in ${}_\Lambda T^\perp$ the homomorphism*

$$\operatorname{Hom}_\Lambda(T, \ ) \colon \operatorname{Ext}^i_F(\omega, C) \to \operatorname{Ext}^i_{\Gamma^{\mathrm{op}}}((T, \omega), (T, C))$$

is an isomorphism functorial in both variables.

PROOF. Let ω be in $\widehat{\operatorname{add} T}$, then there is an F-exact sequence $0 \to T_n \to T_{n-1} \to \cdots \to T_0 \to \omega \to 0$ with T_i in $\operatorname{add} T$ for all i, which remains exact when applying $\operatorname{Hom}_\Lambda(T, \ )$. Therefore we get the following commutative diagram by Corollary 5.2

$$\begin{array}{ccccccccc}
0 & \to & ((T, \omega), (T, C)) & \to & ((T, T_0), (T, C)) & \to & ((T, T_1), (T, C)) & \to \cdots \\
 & & \downarrow \wr & & \downarrow \wr & & \downarrow \wr & \\
0 & \to & (\omega, C) & \to & (T_0, C) & \to & (T_1, C) & \to \cdots.
\end{array}$$

The cohomology of the upper row is $\operatorname{Ext}^i_{\Gamma^{\mathrm{op}}}((T, \omega), (T, C))$ and the cohomology of the lower row is $\operatorname{Ext}^i_F(\omega, C)$ if C is in ${}_\Lambda T^\perp$, hence our desired result. Since the isomorphism is induced from the functorial isomorphism of the homomorphism groups, it follows that the isomorphism is functorial in both variables. □

Then our promised result is a consequence of the following.

PROPOSITION 5.4. *The subcategory* $\widehat{\operatorname{add} T}$ *is equivalent to* $\mathcal{P}^\infty(\Gamma^{\text{op}}) \cap {}^\perp(T, \mathcal{I}(F))$, *where the inverse equivalences are given by the functors* $\operatorname{Hom}_\Lambda(T, \ )$ *and* $T \otimes_{\Gamma^{\text{op}}} -$.

PROOF. It is clear that $\operatorname{Hom}_\Lambda(T, \widehat{\operatorname{add} T}) \subset \mathcal{P}^\infty(\Gamma^{\text{op}}) \cap {}^\perp(T, \mathcal{I}(F))$ by Corollary 5.3. Moreover, by Lemma 5.1 we have that $T \otimes_{\Gamma^{\text{op}}} (T, \omega) \simeq \omega$ for all ω in $\widehat{\operatorname{add} T}$.

Assume that C is in $\mathcal{P}^\infty(\Gamma^{\text{op}}) \cap {}^\perp(T, \mathcal{I}(F))$. To complete the proof of the proposition it is enough to show that C is isomorphic to (T, ω) for some ω in $\widehat{\operatorname{add} T}$. Let $\eta\colon 0 \to (T, T_n) \to \cdots \to (T, T_0) \to C \to 0$ be projective resolution of C. This projective resolution is induced from a complex in $\operatorname{mod} \Lambda$, namely $\theta\colon 0 \to C_{n+1} \to T_n \to \cdots \to T_2 \xrightarrow{f_2} T_1 \xrightarrow{f_1} T_0 \to C_0 \to 0$. Let C_i denote $\operatorname{Im} f_i$ for $i = 1, 2, \ldots, n$. When applying $\operatorname{Hom}_{\Gamma^{\text{op}}}(\ , (T, \mathcal{I}(F)))$ to the sequence η it remains exact by assumption and we obtain the following commutative exact diagram by Corollary 5.2

$$\begin{array}{ccccccc}
0 \to & (C, (T, \mathcal{I}(F))) & \to & ((T, T_0), (T, \mathcal{I}(F))) & \to & ((T, T_1), (T, \mathcal{I}(F))) & \cdots \\
& \downarrow \wr & & \downarrow \wr & & \downarrow \wr & \cdots \\
0 \to & (C_0, \mathcal{I}(F)) & \to & (T_0, \mathcal{I}(F)) & \to & (T_1, \mathcal{I}(F)) & \cdots .
\end{array}$$

Since $0 \to (C_i, \mathcal{I}(F)) \to (T_{i+1}, \mathcal{I}(F)) \to (C_{i+1}, \mathcal{I}(F)) \to 0$ is exact, it follows that $0 \to C_{i+1} \to T_{i+1} \to C_i \to 0$ is F-exact for $i = 0, \ldots, n$. We also have that $C_{n+1} = 0$ and therefore the sequence $\theta\colon 0 \to T_n \to \cdots \to T_1 \xrightarrow{f_1} T_0 \to C_0 \to 0$ is F-exact and C_0 is in $\widehat{\operatorname{add} T}$. Then it follows easily by applying $\operatorname{Hom}_\Lambda(T, \ )$ to θ, that C is isomorphic to (T, C_0), hence C is in $(T, \widehat{\operatorname{add} T})$. □

The duality $D\colon \operatorname{mod} \Gamma^{\text{op}} \to \operatorname{mod} \Gamma$ induces dualities between $\mathcal{P}^\infty(\Gamma^{\text{op}}) \to \mathcal{I}^\infty(\Gamma)$ and ${}^\perp\mathcal{Y} \to D(Y)^\perp$. Then our desired result is an immediate consequence of this observation and Proposition 5.4.

COROLLARY 5.5. *The subcategory* $\widehat{\operatorname{add} T}$ *is dual to* $\mathcal{I}^\infty(\Gamma) \cap D(T, \mathcal{I}(F))^\perp$ *by the functor* $D \operatorname{Hom}_\Lambda(T, \ )\colon \operatorname{mod} \Lambda \to \operatorname{mod} \Gamma$.

If we specializes Proposition 5.4 and Corollary 5.5 to the case when ${}_\Lambda T$ is a standard cotilting module these results read as follows.

COROLLARY 5.6. *Let T be a standard cotilting module in* $\operatorname{mod} \Lambda$.

(a) *The subcategory* $\widehat{\operatorname{add} T}$ *is equivalent to* $\mathcal{P}^\infty(\Gamma^{\text{op}}) \cap {}^\perp D(T)$ *by the functor* $\operatorname{Hom}_\Lambda(T, \ )\colon \operatorname{mod} \Lambda \to \operatorname{mod} \Gamma^{\text{op}}$.

(b) *The subcategory* $\widehat{\operatorname{add} T}$ *is dual to* $\mathcal{I}^\infty(\Gamma) \cap {}_\Gamma T^\perp$ *by the functor*

$$D \operatorname{Hom}_\Lambda(T, \ )\colon \operatorname{mod} \Lambda \to \operatorname{mod} \Gamma.$$

The notions of strong tilting and strong cotilting modules were introduced in [**2**]. A Λ-module T is called a *strong F-cotilting module* if $\widehat{\operatorname{add} T} = \mathcal{I}^\infty(F)$. It is shown in [**2**, Proposition 6.6] that if ${}_\Lambda T$ and ${}_\Gamma T$ are strong standard cotilting modules, then the functor $\operatorname{Hom}_\Lambda(T, \ )\colon \operatorname{mod} \Lambda \to \operatorname{mod} \Gamma^{\text{op}}$ induces an equivalence between $\mathcal{I}^\infty(\Lambda)$ and $\mathcal{P}^\infty(\Gamma^{\text{op}})$. This also follows from Corollary 5.6. If ${}_\Lambda T$ is a strong cotilting module, then $\widehat{\operatorname{add} T} = \mathcal{I}^\infty(\Lambda)$ is dual to $\mathcal{I}^\infty(\Gamma) \cap {}_\Gamma T^\perp$. The subcategory ${}_\Gamma T^\perp$ always contains $\widehat{\operatorname{add} {}_\Gamma T}$, so that if ${}_\Gamma T$ is a strong cotilting module, then $\mathcal{I}^\infty(\Gamma) \cap {}_\Gamma T^\perp = \mathcal{I}^\infty(\Gamma)$. It follows that $\operatorname{Hom}_\Lambda(T, \ )\colon \operatorname{mod} \Lambda \to \operatorname{mod} \Gamma$ induces an equivalence between $\mathcal{I}^\infty(\Gamma)$ and $\mathcal{P}^\infty(\Gamma^{\text{op}})$ in this case.

Let T be a standard cotilting module in $\operatorname{mod} \Lambda$. Denote $\operatorname{End}_\Lambda(T)$ by Γ. Let $\mathcal{E}^i_\Lambda = \{X \in \operatorname{mod} \Lambda \mid \operatorname{Ext}^j_\Lambda(X, T) = 0, j \neq i\}$ and $\mathcal{E}^i_\Gamma = \{X \in \operatorname{mod} \Gamma \mid \operatorname{Ext}^j_\Gamma(X, {}_\Gamma T) = 0, j \neq i\}$.

It is shown in **[10]** that $\mathrm{Ext}^i_\Lambda(\ ,T)\colon \mathcal{E}^i_\Lambda \to \mathcal{E}^i_\Gamma$ and $\mathrm{Ext}^i_\Gamma(\ ,{}_\Gamma T)\colon \mathcal{E}^i_\Gamma \to \mathcal{E}^i_\Lambda$ are inverse dualities for all $i \geq 0$. The analogous result for relative cotilting theory is not true. By this we mean the following. Let T be an F-cotilting module in $\mathrm{mod}\,\Lambda$. Denote $\mathrm{End}_\Lambda(T)$ by Γ. Let $\mathcal{E}^i_F = \{X \in \mathrm{mod}\,\Lambda \mid \mathrm{Ext}^j_F(X,T) = 0, j \neq i\}$ and $\mathcal{E}^i_\Gamma = \{X \in \mathrm{mod}\,\Gamma \mid \mathrm{Ext}^j_\Gamma(X,(\mathcal{P}(F),T)) = 0, j \neq i\}$. By Corollary 3.6 (a) and Proposition 3.8 (b) the subcategories $\mathcal{E}^0_F$ and $\mathcal{E}^0_\Gamma$ are dual via the functors $\mathrm{Ext}^0_F(\ ,T)$ and $\mathrm{Ext}^0_\Gamma(\ ,{}_\Gamma T)$. One can show that there is an isomorphism $M \simeq \mathrm{Ext}^i_\Gamma(\mathrm{Ext}^i_F(M,T),{}_\Gamma T)$ for all modules M in $\mathcal{E}^i_F$ for all i. But there is not necessarily an isomorphism $N \simeq \mathrm{Ext}^i_F(\mathrm{Ext}^i_\Gamma(N,T),{}_\Lambda T)$ for a module N in $\mathcal{E}^i_\Gamma$. The following example shows this.

Let $\Lambda = k[x]/(x^2)$ for a field k and let $T = \Lambda \oplus S$, where S is the simple Λ-module. Denote $F^{\mathrm{add}\,T}$ by F. Then T is an F-cotilting module. Let $\Gamma = \mathrm{End}_\Lambda(T)$. Since $\mathrm{add}\,T = \mathcal{I}(F)$, it is clear that $\mathcal{E}^i_F = \{0\}$ for $i > 0$. It is not hard to see that $\mathcal{E}^2_\Gamma$ contains a nonzero module. Therefore $\mathcal{E}^i_F$ and $\mathcal{E}^i_\Gamma$ are not dual for all i in general.

6. Finitistic dimension conjecture

Let Λ be an artin algebra and let $\mathcal{C}$ be a subcategory of $\mathrm{mod}\,\Lambda$ of finite type containing $\mathcal{P}(\Lambda)$. Then $F = F_\mathcal{C}$ has enough projectives and injectives, where $\mathcal{P}(F) = \mathcal{C}$ and $\mathcal{I}(F) = \mathcal{I}(\Lambda) \cup D\,\mathrm{Tr}\,\mathcal{C}$. We define the *$F$-finitistic dimension* $\mathrm{fin.dim}_F\,\Lambda$ of Λ to be $\sup\{\mathrm{pd}_F\,A \mid \mathrm{pd}_F\,A < \infty, A \in \mathrm{mod}\,\Lambda\}$. The *relative finitistic dimension* $\mathrm{rel.fin.dim}\,\Lambda$ of Λ is defined to be $\sup_F \mathrm{fin.dim}_F\,\Lambda$, where F ranges over all subfunctors $F = F_\mathcal{C}$ for all subcategories $\mathcal{C}$ of finite type of $\mathrm{mod}\,\Lambda$ containing $\mathcal{P}(\Lambda)$. Then an artin algebra Λ is said to satisfy the *relative finitistic dimension conjecture* if $\mathrm{rel.fin.dim}\,\Lambda$ is finite. This section is devoted to showing that the relative finitistic dimension conjecture is true for all artin algebras if and only if the standard finitistic dimension conjecture is true for all artin algebras.

Clearly, if an artin algebra Λ satisfies the relative finitistic dimension conjecture, then Λ also satisfies the standard finitistic dimension conjecture. So, if the relative finitistic dimension conjecture is true for all artin algebras, then the standard finitistic dimension conjecture is true for all artin algebras. Our aim is to prove the converse, namely the following result.

PROPOSITION 6.1. *The standard finitistic dimension conjecture is true for all artin algebras if and only if the relative finitistic dimension conjecture is true for all artin algebras.*

The proof of this result is an easy consequence of the next result. Let $F = F_\mathcal{C}$ for a finite subcategory $\mathcal{C}$ of $\mathrm{mod}\,\Lambda$ containing $\mathcal{P}(\Lambda)$. Let T be an F-cotilting module in $\mathrm{mod}\,\Lambda$ and denote by $\Gamma = \mathrm{End}_\Lambda(T)$. We denote by $\mathrm{fin.dim}\,\Gamma$ the standard finitistic dimension of Γ.

LEMMA 6.2. *The dimension* $\mathrm{fin.dim}_F\,\Lambda$ *is finite if and only if* $\mathrm{fin.dim}\,\Gamma^{\mathrm{op}}$ *is finite.*

PROOF. Let $\mathrm{id}_F\,T = r$ and $\mathrm{add}\,G = \mathcal{P}(F)$ for a module G in $\mathrm{mod}\,\Lambda$. Then by Theorem 3.13 the subcategory $(\mathcal{X}_T,T)$ is a resolving functorially finite subcategory of $\mathrm{mod}\,\Gamma$ with $(\mathcal{X}_T,T)\text{-resdim}(\mathrm{mod}\,\Gamma) \leq r+2$ and $(\mathcal{X}_T,T)^\perp = \widehat{(\mathcal{P}(F),T)}$. Moreover, (G,T) is a cotilting module with $\mathrm{id}_\Gamma(G,T) \leq r+2$ and $\mathrm{add}\,\widehat{(\mathcal{P}(F),T)} = \widehat{\mathrm{add}(G,T)} \subset \mathcal{I}^{r+2}(\Gamma)$.

Let B be in $\mathrm{mod}\,\Gamma$, then we have a minimal right $(\mathcal{X}_T,T)$-approximation

$$0 \to \omega_B \to (X_B,T) \to B \to 0$$

with ω_B in $\widehat{\mathrm{add}(P,T)}$ and X_B in $\mathcal{X}_T$. Since $\widehat{\mathrm{add}(P,T)} \subset \mathcal{I}^{r+2}(\Gamma)$, we have that $\mathrm{id}_\Gamma B$ is finite if and only if $\mathrm{id}_\Gamma(X_B,T)$ is finite. Hence, it follows that $\sup\{\mathrm{id}_\Gamma B \mid \mathrm{id}_\Gamma B < \infty, B \in \mathrm{mod}\,\Gamma\}$ is finite if and only if $\sup\{\mathrm{id}_\Gamma(X,T) \mid \mathrm{id}_\Gamma(X,T) < \infty, X \in \mathcal{X}_T\}$ is finite.

Clearly, if $\sup\{\mathrm{pd}_F A \mid \mathrm{pd}_F A < \infty, A \in \mathrm{mod}\,\Lambda\}$ is finite, then $\sup\{\mathrm{pd}_F C \mid \mathrm{pd}_F C < \infty, C \in \mathcal{X}_T\}$ is finite. Since $\mathcal{X}_T$-$\mathrm{resdim}_F(\mathrm{mod}\,\Lambda) \leq r$, we also have the converse. So, to complete the proof we only need to show that $\sup\{\mathrm{pd}_F C \mid \mathrm{pd}_F C < \infty, C \in \mathcal{X}_T\}$ is finite if and only if $\sup\{\mathrm{id}_\Gamma(X,T) \mid \mathrm{id}_\Gamma(X,T) < \infty, X \in \mathcal{X}_T\}$ is finite, since $\mathrm{fin.dim}\,\Gamma^{\mathrm{op}} = \sup\{\mathrm{id}_\Gamma B \mid \mathrm{id}_\Gamma B < \infty\}$.

By Proposition 3.7 we have that $\mathrm{Ext}^i_F(C,A) \simeq \mathrm{Ext}^i_\Gamma((A,T),(C,T))$ for all A and C in $\mathcal{X}_T$ and all $i > 0$. If $\mathrm{id}_\Gamma(X,T) = n$ is finite for X in $\mathcal{X}_T$, then $\mathrm{Ext}^i_F(X,A) = 0$ for all $i > n$ and all A in $\mathcal{X}_T$. Since $\mathcal{X}_T$-$\mathrm{resdim}_F(\mathrm{mod}\,\Lambda) \leq r$, it follows that $\mathrm{Ext}^i_F(X,\ ) = 0$ for all $i > n$ and therefore $\mathrm{pd}_F X \leq n$. Conversely, if $\mathrm{pd}_F X = n$ is finite, then $\mathrm{Ext}^i_\Gamma((\mathcal{X}_T,T),(X,T)) = 0$ for all $i > n$. Therefore $\Omega_\Gamma^{-n}(X,T)$ is in $\widehat{\mathrm{add}(P,T)}$ and consequently $\mathrm{id}_\Gamma(X,T) \leq n + r + 2$. Hence, we have shown that $\sup\{\mathrm{id}_\Gamma(X,T) \mid \mathrm{id}_\Gamma(X,T) < \infty, X \in \mathcal{X}_T\}$ is finite if and only if $\sup\{\mathrm{pd}_F C \mid \mathrm{pd}_F C < \infty, C \in \mathcal{X}_T\}$ is finite and we are done. □

References

[1] M. Auslander, R.-O. Buchweitz, *The homological theory of maximal Cohen-Macaulay approximations*, Soc. Math. de France, Mem. No. 38 (1989) 5–37.

[2] M. Auslander, I. Reiten, *Applications of homologically finite subcategories*, Adv. in Math., 86 (1991) 111–152.

[3] M. Auslander, I. Reiten, *Homologically finite subcategories*, Proceedings of the ICRA V, Tsukuba, Japan.

[4] M. Auslander, S. O. Smalø, *Preprojective modules over artin algebras*, J. Algebra 66 (1980) 61–122.

[5] M. Auslander, Ø. Solberg, *Relative homology and representation theory* I, *Relative homology and homologically finite subcategories.*

[6] M. C. R. Butler, G. Horrocks, *Classes of extensions and resolutions*, Phil. Trans. Royal Soc., London, Ser. A, 254 (1961) 155–222.

[7] D. Happel, C. M. Ringel, Tilted Algebras, Trans. Amer. Math. Soc. 274 (2) (1982) 399–443.

[8] G. Hochschild, Relative homological algebra, Trans. Amer. Math. Soc. 82 (1956) 246–269.

[9] J. Rickard, A. Scofield, *Cocovers and tilting modules*, Math. Proc. Cambridge Philos. Soc. No. 1, 106 (1989) 1–5.

[10] T. Miyashita, *Tilting modules of finite projective dimension*, Math. Z. vol. 193 (1986) 113–146.

Department of Mathematics, Brandeis University, Waltham, Mass. 02254–9110, USA

Department of Mathematics, Brandeis University, Waltham, Mass. 02254–9110, USA

Journal of Pure and Applied Algebra 92 (1994) 1–27
North-Holland

k-Gorenstein algebras and syzygy modules

Maurice Auslander*
Department of Mathematics, Brandeis University, Waltham, MA 02254, USA

Idun Reiten**
Department of Mathematics, University of Trondheim, AVH, 7055 Dragvoll, Norway

Communicated by K.W. Gruenberg
Received 5 October 1992

Abstract

Auslander, M. and I. Reiten, k-Gorenstein algebras and syzygy modules, Journal of Pure and Applied Algebra 92 (1994) 1–27.

We investigate for an artin algebra Λ when categories of dth syzygy modules are closed under extensions. In this case they are functorially finite resolving, and have an associated cotilting module. We show that this holds for all d for algebras Λ which are k-Gorenstein for all k, that is, in a minimal injective resolution $0 \to \Lambda \to I_0 \to I_1 \to \cdots \to I_j \to \cdots$ we have $\mathrm{pd}_\Lambda I_j \leq j$ for all j. From this we get a correspondence between indecomposable projective and indecomposable injective modules over these algebras, which can be applied to prove that if Λ is k-Gorenstein for all k and $\mathrm{id}_\Lambda \Lambda < \infty$, then $\mathrm{id}\, \Lambda_\Lambda < \infty$.

Introduction

Motivated by the theory of commutative noetherian Gorenstein rings, Auslander introduced the notion of a two-sided noetherian ring being k-Gorenstein as follows (see [6]). Let $0 \to \Lambda \to I_0 \to \cdots \to I_j \to \cdots$ be a minimal injective resolution for Λ viewed as a left Λ-module, where Λ is a two-sided noetherian ring. Then Λ is said to be left k-Gorenstein if $\mathrm{flatdim}_\Lambda I_i \leq i$ for all $i = 0, \ldots, k-1$. Since Λ being left k-Gorenstein is equivalent to Λ being right k-Gorenstein (see [6]), we say Λ is k-Gorenstein if it is either left or right k-Gorenstein. If Λ is k-Gorenstein for all k, then it is known in the literature as an Auslander ring. Although these notions have proven to be of considerable interest in the context of two-sided noetherian rings, except for

Correspondence to: M. Auslander, Department of Mathematics, Brandeis University, Waltham, MA 02254, USA.

* Partially supported by NSF grant DMS8904594 and NAVF (Norwegian Research Council for Science and Humanities).
** Partially supported by NAVF.

SSDI 0022-4049(93)E0020-5

the theory of dominant dimension, they have received very little attention in the context of artin algebras. One of the aims of this paper is to initiate a study of artin algebras which are k-Gorenstein for arbitrary k and all k.

Throughout this paper Λ is an artin algebra and all Λ-modules are finitely generated left Λ-modules. Let $0 \to \Lambda \to I_0 \to \cdots \to I_j \to \cdots$ be a minimal injective resolution of Λ. Since each I_i is a finitely generated Λ-module, we have that $\operatorname{flatdim} I_i = \operatorname{pd}_\Lambda I_i$. So Λ is k-Gorenstein if $\operatorname{pd}_\Lambda I_i \leq i$ for all $i = 0, \ldots, k-1$, and Λ is k-Gorenstein for all k if $\operatorname{pd} I_i \leq i$ for all $i \geq 0$. Recall that Λ is said to be of dominant dimension greater than or equal to k if $\operatorname{pd}_\Lambda I_i = 0$ for $i = 0, \ldots, k-1$. Hence if $\operatorname{dom.dim} \Lambda \geq k$, then Λ is k-Gorenstein. Selfinjective algebras Λ are obviously examples of artin algebras which are k-Gorenstein for all k. Starting from these well-known examples and using the easily established result that the two by two lower triangular matrix ring over an artin algebra Λ is k-Gorenstein if and only if Λ is k-Gorenstein (see [6]), it is possible to generate new artin algebras which are k-Gorenstein for all k.

Part of the motivation for studying artin algebras which are k-Gorenstein for some k or all k, is its connection with the well-known Nakayama conjecture, which states that if Λ is of dominant dimension k for all k, then Λ is self-injective. This is a special case of the following conjecture. If Λ is k-Gorenstein for all k, then Λ is Gorenstein, i.e. the injective dimension of Λ is finite when viewed both as a left Λ-module and left Λ^{op}-module. In Section 5 we show that if Λ is k-Gorenstein for all k and $\operatorname{id}_\Lambda \Lambda < \infty$, then $\operatorname{id}_{\Lambda^{\mathrm{op}}} \Lambda < \infty$, and so Λ is Gorenstein. This gives some support for the conjecture. This result is based on the following property of artin algebras which are k-Gorenstein for all k. The syzygy functors Ω^d and cosyzygy functors Ω^{-d} give a bijection between the isomorphism classes of indecomposable projective modules of injective dimension d and the isomorphism classes of indecomposable injective modules of projective dimension d for all $d \geq 0$. In fact, our study of k-Gorenstein artin algebras is mainly concerned with the following subcategories of $\operatorname{mod} \Lambda$, the category of finitely generated Λ-modules. We denote by $\Omega^d(\operatorname{mod} \Lambda)$ the full subcategory of $\operatorname{mod} \Lambda$ consisting of the modules of the form $\Omega^d(C) \amalg P$ for all C in $\operatorname{mod} \Lambda$ and all projective Λ-modules P. Similarly we denote by $\Omega^{-d}(\operatorname{mod} \Lambda)$ the full subcategory of $\operatorname{mod} \Lambda$ consisting of all $\Omega^{-d}(C) \amalg I$ for all C in $\operatorname{mod} \Lambda$ and all injective modules I. Amongst other things, we show that if Λ is k-Gorenstein, then $\Omega^d(\operatorname{mod} \Lambda)$ and $\Omega^{-d}(\operatorname{mod} \Lambda)$ are closed under summands and extensions as well as being functorially finite in $\operatorname{mod} \Lambda$ for all $d \leq k$. Since some of these properties hold for more general algebras than k-Gorenstein algebras, we study the problem of when these subcategories have these properties in general and then specialize to k-Gorenstein algebras as we now explain.

Let Λ be an arbitrary artin algebra and let d be a nonnegative integer. Although $\Omega^d(\operatorname{mod} \Lambda)$ is closed under finite sums it is in general not closed under summands. We denote $\operatorname{add} \Omega^d(\operatorname{mod} \Lambda)$ by $\mathscr{X}_d$ and $\operatorname{add} \Omega^{-d}(\operatorname{mod} \Lambda)$ by $\mathscr{X}_{-d}$. We know from [4] that $\mathscr{X}_d$ is covariantly finite in $\operatorname{mod} \Lambda$. In [3] it was shown that $\mathscr{X}_d$ is also contravariantly finite in $\operatorname{mod} \Lambda$ when Λ is an algebra over a field k. Smalø generalized this result to arbitrary artin algebras and we present here a modified version of his proof. Next we discuss the question of when $\mathscr{X}_d$ is closed under extensions. We show that $\mathscr{X}_d$ is closed under extensions for all $d \leq k$ if and only if a minimal injective resolution $0 \to \Lambda \to I_0 \to \cdots \to I_i \to \cdots$ of Λ as a right module has the property $\operatorname{pd}_\Lambda I_i \leq i+1$ for all $i \leq k$. The importance of $\mathscr{X}_d$ being closed under extensions lies in the fact that when $\mathscr{X}_d$ is closed under extensions then $\mathscr{X}_d$ is a resolving subcategory of $\operatorname{mod}\Lambda$ (see [4]) with the additional property that $\hat{\mathscr{X}}_d = \operatorname{mod}\Lambda$, i.e. there is an exact sequence $0 \to X_n \to \cdots \to X_0 \to C \to 0$ with the X_i in $\mathscr{X}_d$ for all C in $\operatorname{mod}\Lambda$. From this it follows that $\mathscr{X}_d$ has almost split sequences [5] and determines a unique, up to isomorphism, basic cotilting module [2]. It would be interesting to see what more can be said about these almost split sequences as well as the properties of the uniquely determined cotilting module.

1. Contravariantly finite subcategories of complexes

Let Λ be an artin algebra. By a Λ-module we always mean a module in $\operatorname{mod}\Lambda$, the category of finitely generated Λ-modules. For all integers $d > 0$ we denote by $(\operatorname{mod}\Lambda)_d$ the category consisting of all complexes $A^{\cdot}: 0 \to A_d \to A_{d-1} \to \cdots \to A_1 \to 0$ of length d with the usual maps of sequences. It is well known that $(\operatorname{mod}\Lambda)_d$ is canonically equivalent to $\operatorname{mod}(T_d(\Lambda)/\underline{a}^2)$ where $T_d(\Lambda)$ is the lower $d \times d$ matrix ring over Λ and $\underline{a}$ is the ideal in $T_d(\Lambda)$ consisting of all matrices with only zeros on the diagonal. Therefore all the usual notions for modules over artin algebras apply to the category $(\operatorname{mod}\Lambda)_d$. In particular, it makes sense to ask if a full subcategory of $(\operatorname{mod}\Lambda)_d$ is contravariantly finite in $(\operatorname{mod}\Lambda)_d$. The first part of this section is devoted to showing that if $\mathscr{P}$ is a contravariantly finite subcategory of $\operatorname{mod}\Lambda$, then the subcategory $\mathscr{P}_d$ of $(\operatorname{mod}\Lambda)_d$ consisting of all complexes $0 \to A_d \to \cdots \to A_1 \to 0$ with the A_i in $\mathscr{P}$ is contravariantly finite in $(\operatorname{mod}\Lambda)_d$. Restricting $\mathscr{P}$ to be the full subcategory of projective Λ-modules, we apply this result to show that $\mathscr{X}_d = \operatorname{add}\Omega^d(\operatorname{mod}\Lambda)$ is contravariantly finite in $\operatorname{mod}\Lambda$ for all integers $d \geq 1$.

Let $\mathscr{P} \subset \operatorname{mod}\Lambda$ be a contravariantly finite subcategory. For $A^{\cdot}: 0 \to A_d \xrightarrow{g_{d-1}} A_{d-1} \to \cdots \to A_2 \xrightarrow{g_1} A_1 \to 0$ in $(\operatorname{mod}\Lambda)_d$ we want to construct a canonical map $f^{\cdot}: P^{\cdot} \to A^{\cdot}$ with $P^{\cdot}$ in $\mathscr{P}_d$. It will turn out that this map is a minimal right $\mathscr{P}_d$-approximation.

We first define $f_1: P_1 \to A_1$ to be a minimal right $\mathscr{P}$-approximation. We then take the pullback of

$$\begin{array}{ccc} & & P_1 \\ & & \downarrow{\scriptstyle f_1} \\ A_2 & \xrightarrow{g_1} & A_1 \end{array}$$

to get the commutative exact diagram

$$\begin{array}{ccccccc}
U_2 & \longrightarrow & P_1 & \longrightarrow & K_1 & \longrightarrow & 0\\
\downarrow & & \downarrow{\scriptstyle f_1} & & \| & & \\
A_2 & \xrightarrow{g_1} & A_1 & \longrightarrow & K_1 & \longrightarrow & 0
\end{array}$$

Let $P_2 \to U_2$ be a minimal right $\mathscr{P}$-approximation, and define $h_1 : P_2 \to P_1$ to be the composition $P_2 \to U_2 \to P_1$ and $f_2 : P_2 \to A_2$ to be the composition $P_2 \to U_2 \to A_2$. Having defined $P_i \xrightarrow{h_{i-1}} P_{i-1} \to \cdots \to P_2 \xrightarrow{h_1} P_1$ and $f_j : P_j \to A_j$ for $1 \leq j \leq i$ for $i \geq 2$, we consider the induced map $\operatorname{Ker} h_{i-1} \to \operatorname{Ker} g_{i-1}$. We take the pullback of

$$\begin{array}{ccc}
 & & \operatorname{Ker} h_{i-1}\\
 & & \downarrow\\
A_{i+1} & \longrightarrow & \operatorname{Ker} g_{i-1}
\end{array}$$

to get the exact commutative diagram

$$\begin{array}{ccccccc}
U_{i+1} & \longrightarrow & \operatorname{Ker} h_{i-1} & \longrightarrow & K_i & \longrightarrow & 0\\
\downarrow & & \downarrow & & \| & & \\
A_{i+1} & \longrightarrow & \operatorname{Ker} g_{i-1} & \longrightarrow & K_i & \longrightarrow & 0
\end{array}$$

We let $P_{i+1} \to U_{i+1}$ be a minimal right $\mathscr{P}$-approximation, and define $h_i : P_{i+1} \to P_i$ to be the composition $P_{i+1} \to U_{i+1} \to \operatorname{Ker} h_{i-1} \to P_i$ and $f_{i+1} : P_{i+1} \to A_{i+1}$ to be the composition $P_{i+1} \to U_{i+1} \to A_{i+1}$. Hence we get the complex $P^{\cdot}$ together with a map $f^{\cdot} : P^{\cdot} \to A^{\cdot}$. We have the following result.

Theorem 1.1. *Let* $A^{\cdot}$ *be in* $(\operatorname{mod} \Lambda)_d$, $\mathscr{P} \subset \operatorname{mod} \Lambda$ *contravariantly finite, and let* $f^{\cdot} : P^{\cdot} \to A^{\cdot}$ *be the map defined above with* $P^{\cdot}$ *in* $\mathscr{P}_d$. *Then we have the following.*

(a) $f^{\cdot} : P^{\cdot} \to A^{\cdot}$ *is a minimal right* $\mathscr{P}_d$*-approximation, which is an epimorphism if* $\mathscr{P}$ *contains the projective* Λ*-modules.*

(b) $\mathscr{P}_d$ *is contravariantly finite in* $(\operatorname{mod} \Lambda)_d$.

(c) *If* $\mathscr{P}_d$ *contains the projective modules, then* $H_i(P^{\cdot}) \simeq H_i(A^{\cdot})$ *for* $1 \leq i < d$, *and we have an epimorphism* $H_d(P^{\cdot}) \to H_d(A^{\cdot})$.

Proof. To show that $f^{\cdot}$ is a right $\mathscr{P}_d$-approximation, let $t^{\cdot}: Q^{\cdot} \to A^{\cdot}$ be a map with $Q^{\cdot}$ in $\mathscr{P}_d$. Since $f_1: P_1 \to A_1$ is a right $\mathscr{P}$-approximation and Q_1 is in $\mathscr{P}$, there is a map $u_1: Q_1 \to P_1$ with $f_1 u_1 = t_1$. From the diagram

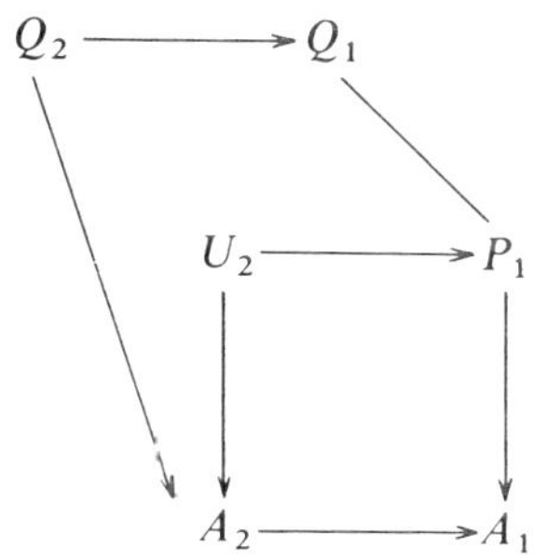

we use the property of the pullback to get a map $Q_2 \to U_2$ such that the diagram commutes. Hence we get a map $u_2: Q_2 \to P_2$ since $P_2 \to U_2$ is a right $\mathscr{P}_d$-approximation. Continuing this way we get a map $u^{\cdot}: Q^{\cdot} \to P^{\cdot}$ with $f^{\cdot}u^{\cdot} = t^{\cdot}$. This shows that $f^{\cdot}$ is a right $\mathscr{P}_d$-approximation.

We now show by induction that the map $f^{\cdot}: P^{\cdot} \to A^{\cdot}$ is a minimal right $\mathscr{P}_d$-approximation. For $d = 1, f_1: P_1 \to A_1$ is minimal by our choice. Assume by induction that

(*)

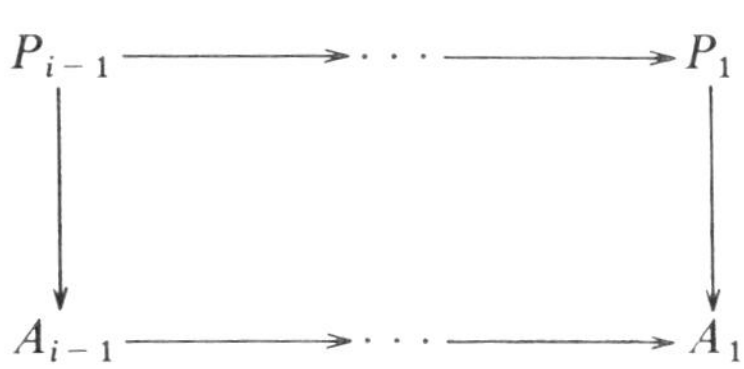

is a minimal right $\mathscr{P}_{i-1}$-approximation in $(\operatorname{mod} \Lambda)_{i-1}$ where $i > 1$. If

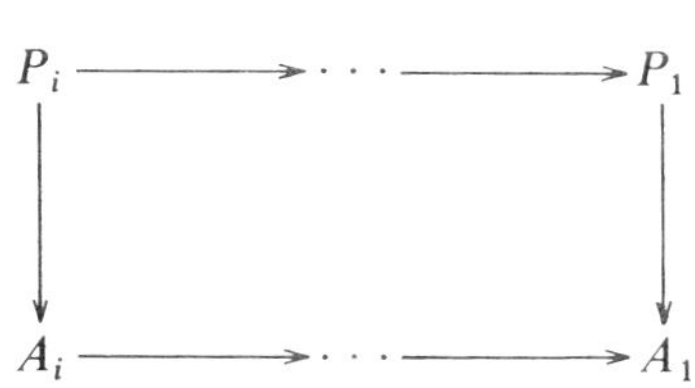

was not a minimal right $\mathscr{P}_i$-approximation in $(\operatorname{mod} \Lambda)_i$, it would, since (*) is a minimal right $\mathscr{P}_{i-1}$-approximation, have a summand of the form $P_i' \to 0 \to \cdots \to 0$ mapping to zero. It is easy to see that this is impossible since P_i is a minimal right $\mathscr{P}$-approximation of a pullback.

Assume now that $\mathscr{P}$ contains the projective modules. Then the minimal right $\mathscr{P}$-approximation $f_1: P_1 \to A_1$ is an epimorphism. Hence we have, in the previous notation, that the map $U_2 \to A_2$ and consequently $f_2: P_2 \to A_2$ is an epimorphism. Then the map $\operatorname{Ker} h_1 \to \operatorname{Ker} g_1$ is an epimorphism. Continuing this way we see that $f^{\cdot}: P^{\cdot} \to A^{\cdot}$ is an epimorphism. Since the minimal right $\mathscr{P}$-approximations $P_i \to U_i$ are epimorphisms, we have exact commutative diagrams

$$\begin{array}{ccccccc}
P_{i+1} & \longrightarrow & \operatorname{Ker} h_{i-1} & \longrightarrow & K_i & \longrightarrow & 0 \\
\downarrow & & \downarrow & & \| & & \\
A_{i+1} & \longrightarrow & \operatorname{Ker} g_{i-1} & \longrightarrow & K_i & \longrightarrow & 0
\end{array}$$

for $0 \leq i < d$. Hence we get $H_i(P^{\cdot}) \simeq H_i(A^{\cdot})$ for $1 \leq i < d$, and the induced map $H_d(P^{\cdot}) \simeq \operatorname{Ker} h_{d-1} \to H_d(A^{\cdot}) \simeq \operatorname{Ker} g_{d-1}$ is an epimorphism. $\square$

From now on we assume that $\mathscr{P}$ is the full subcategory of projective Λ-modules. We now use our construction of minimal right $\mathscr{P}_d$-approximations to construct particular right $\mathscr{X}_d$-approximations of modules in $\operatorname{mod} \Lambda$. Before doing this it is convenient to make certain conventions and notations.

For each module C choose a fixed minimal projective resolution $P_i(C) \to \cdots \to P_1(C) \xrightarrow{\delta_1} P_0(C) \xrightarrow{\delta_0} C \to 0$ and a fixed minimal injective resolution $0 \to C \xrightarrow{\varepsilon_0} I_0(C) \xrightarrow{\varepsilon_1} I_1(C) \to \cdots \xrightarrow{\varepsilon_j} I_j(C) \xrightarrow{\varepsilon_{j+1}} \cdots$ and define $\Omega^j(C) = \operatorname{Im} \delta_j$ and $\Omega^{-j}(C) = \operatorname{Im} \varepsilon_j$. Also for each complex $A^{\cdot}$ in $(\operatorname{mod} \Lambda)_d$ choose a fixed minimal right $\mathscr{P}_d$-approximation $\mathscr{P}_d(A^{\cdot}) \to A^{\cdot}$.

Let C be in $\operatorname{mod} \Lambda$. Let $I(C)_d$ be the complex $0 \to I_0(C) \to \cdots \to I_{d-1}(C) \to 0$ which is the dth truncation of the chosen minimal injective resolution $I(C)$ of C. Since the chosen minimal right $\mathscr{P}_d$-approximation $f^{\cdot}: \mathscr{P}_d(I(C)_d) \to I(C)_d$ induces isomorphisms $H_i(\mathscr{P}_d(I(C))_d) \to H_i(I(C)_d)$ for $0 < i \leq d$ and an epimorphism $H_0(\mathscr{P}_d(I(C))_d) \to H_0(I(C)_d)$ by Theorem 1.1(c), we obtain the following exact commutative diagram:

$$\begin{array}{ccccccccccccc}
0 & \longrightarrow & \operatorname{Ker} h_1 & \longrightarrow & P_0 & \xrightarrow{h_1} & P_1 & \longrightarrow \cdots \xrightarrow{h_{d-1}} & P_{d-1} & \longrightarrow & \Omega^{-d} C & \longrightarrow & 0 \\
\downarrow & & \downarrow{\scriptstyle t} & & \downarrow{\scriptstyle f_0} & & \downarrow{\scriptstyle f_1} & & \downarrow{\scriptstyle f_{n-1}} & & \| & & \\
0 & \longrightarrow & C & \longrightarrow & I_0 & \xrightarrow{g_1} & I_1 & \longrightarrow \cdots \xrightarrow{g_{d-1}} & I_{d-1} & \longrightarrow & \Omega^{-d} C & \longrightarrow & 0
\end{array}$$

Hence $\operatorname{Ker} h_1$ is in $\mathscr{X}_d$ since it is isomorphic to $\Omega^d(\Omega^{-d}C) \amalg Q$ with Q a projective module. The importance of this map $t: \Omega^d(\Omega^{-d}C) \amalg Q \to C$ lies in the following result.

Theorem 1.2. *The map $t: \operatorname{Ker} h_1 \to C$ constructed above is a right $\mathscr{X}_d$-approximation.*

Proof. Let $s: X \to C$ be a map, with X in $\mathscr{X}_d$. Then there is an exact sequence $0 \to X \to Q_0 \to Q_1 \to \cdots \to Q_{d-1}$, where the Q_i are projective. Since the I_j are injective, there is a map $u: Q^{\cdot} \to I^{\cdot}$, where $Q^{\cdot}$ is $0 \to Q_0 \to \cdots \to Q_{d-1} \to 0$, which factors through $f^{\cdot}: P^{\cdot} \to I^{\cdot}$ because this is a (minimal) right $\mathscr{P}_d$-approximation. It follows that $s: X \to C$ factors through $t: \operatorname{Ker} h_1 \to C$ by considering the induced map on kernels. □

This uniquely determined $\mathscr{X}_d$-approximation $t: \operatorname{Ker} h_1 \to C$ of C will be called the augmented minimal right $\mathscr{X}_d$-approximation of C. In general this is not a minimal right $\mathscr{X}_d$-approximation of C. Exactly when it is minimal will be discussed in Section 4.

Our proof of Theorem 1.1 was inspired by Smalø's work on contravariantly finite subcategories of mod Λ when Λ is a triangular matrix ring [9]. Our proof of Theorem 1.2 is a modified version of Smalø's proof for artin algebras of $\mathscr{X}_d$ being contravariantly finite, a proof quite different from our earlier proof for finite-dimensional algebras in [3] (oral communication).

We end this section with the following remark on derived categories. Consider the derived category $D^{\mathrm{b}}(\Lambda)$ whose objects are bounded complexes. Such an object is isomorphic, in the category $D^{-,\mathrm{b}}(\Lambda)$ of complexes bounded to the right, of bounded homology, to a complex of projective modules. The construction of this associated complex of projective modules can be done in the same way as our construction above. We start with a complex $0 \to A_d \to \cdots \to A_1 \to 0$. Rather than stopping with $P_d \to \cdots \to P_1 \to 0$, we continue the construction infinitely to the left. This way we get the same homology at each stage, and hence we have an isomorphism in the derived category. Conversely, let $f^{\cdot}: P^{\cdot} \to I^{\cdot}$ be a map where $P^{\cdot}$ is a projective complex, which is an isomorphism in the derived category, and where $I^{\cdot}$ is the complex $0 \to I_0 \to \cdots \to I_{d-1} \to 0$ coming from a minimal injective resolution of an object C. Then an $\mathscr{X}_d$-approximation is got by cutting off the complex $P^{\cdot}$ and taking the kernel at the appropriate stage.

2. When are syzygy modules closed under extensions?

This section is devoted to proving the following necessary and sufficient condition for each of the subcategories $\mathscr{X}_i = \operatorname{add} \Omega^i(\operatorname{mod} \Lambda)$ to be closed under extensions for all $i \leq d$ for some fixed integer d.

Theorem 2.1. *Let $d \geq 1$ be a fixed integer. Then each $\mathscr{X}_i$ is closed under extensions for $i \leq d$ if and only if $\operatorname{id}_{\Lambda} P_i(D\Lambda) \leq i + 1$ for $i < d$, where $D\Lambda$ is viewed as a left Λ-module.*

We start out with some preliminary results. The first one provides a criterion for deciding when $\mathscr{X}_i$ is closed under extensions.

Proposition 2.2. *$\mathscr{X}_i$ is closed under extensions if and only if in a minimal right $\mathscr{X}_i$-approximation sequence $0 \to Y_C \to X_C \to C \to 0$ we have $\mathrm{id}_\Lambda Y_C \leq i$.*

Proof. We define $\mathscr{Y}_i = \{Y \mid \mathrm{Ext}^1_\Lambda(\mathscr{X}_i, Y) = 0\}$. We then see that $\mathscr{Y}_i = \{Y \mid \mathrm{id}_\Lambda Y \leq i\}$. Since $\mathscr{X}_i$ is contravariantly finite and contains the projective modules, we know from [4] that $\mathscr{X}_i$ is closed under extensions if and only if Y_C is in $\mathscr{Y}_i$. □

Lemma 2.3. *If $\mathrm{id}_\Lambda C < i$ we have the following.*
(a) *$P_0(C) \to C$ is a minimal right $\mathscr{X}_i$-approximation.*
(b) *If $\mathscr{X}_i$ is closed under extensions, then $\mathrm{id}_\Lambda \Omega^1 C \leq i$ and $\mathrm{id}_\Lambda P_0(C) \leq i$.*

Proof. (a) Since $\Omega^i(\Omega^{-i}C) = (0)$ because $\mathrm{id}_\Lambda C < i$, it follows that the augmented minimal right $\mathscr{X}_i$-approximation of C is projective, and consequently so is the minimal one. This must then be the projective cover of C.

(b) Consider the minimal right $\mathscr{X}_i$-approximation sequence $0 \to \Omega^1 C \to P_0(C) \to C \to 0$. Since $\mathscr{X}_i$ is assumed to be closed under extensions, it follows from Proposition 2.2 that $\mathrm{id}_\Lambda \Omega^1 C \leq i$. Since $\mathrm{id}_\Lambda C \leq i$, we conclude that $\mathrm{id}_\Lambda P_0(C) \leq i$. □

The next lemma finishes the proof of Theorem 2.1 in one direction.

Lemma 2.4. *If $\mathscr{X}_i$ is closed under extensions for $i \leq d$, then $\mathrm{id}_\Lambda P_i(D\Lambda) \leq i + 1$ for all $i < d$.*

Proof. If $d = 0$ there is nothing to prove. Assume then $d \geq 1$, and consider the exact sequence $0 \to \Omega^1(D\Lambda) \to P_0(D\Lambda) \to D\Lambda \to 0$. We then have $\mathrm{id}_\Lambda \Omega^1(D\Lambda) \leq 1$ and $\mathrm{id}_\Lambda P_0(D\Lambda) \leq 1$ by Lemma 2.3. Assume that we have proved $\mathrm{id}_\Lambda \Omega^j(D\Lambda) \leq j$ and $\mathrm{id}_\Lambda P_{j-1}(D\Lambda) \leq j$ for $j < d - 1$. Considering the exact sequence $0 \to \Omega^{j+1}(D\Lambda) \to P_j(D\Lambda) \to \Omega^j(D\Lambda) \to 0$, it follows from Lemma 2.3 that since $\mathscr{X}_{j+1}$ is closed under extensions, $\mathrm{id}_\Lambda \Omega^{j+1}(D\Lambda) \leq j + 1$ and $\mathrm{id}_\Lambda P_j(D\Lambda) \leq j + 1$. Hence it follows that $\mathrm{id}_\Lambda P_i(D\Lambda) \leq i + 1$ for all $i < d$. □

We continue with some lemmas leading to the proof of Theorem 2.1 in the other direction.

Lemma 2.5. *Let $0 \to A \to B \to C \to 0$ be an exact sequence. Then $P_i(A) \mid P_i(B) \amalg P_{i+1}(C)$ and $P_i(C) \mid P_i(B) \amalg P_{i-1}(A)$, where $P_{-1}(A) = 0$.*

Proof. Consider the exact commutative diagram

$$
\begin{array}{ccccccccccc}
 & & & & 0 & & 0 & & & & \\
 & & & & \downarrow & & \downarrow & & & & \\
 & & 0 & \longrightarrow & \Omega^1 B & \longrightarrow & \Omega^1 C \amalg Q & \longrightarrow & A & \longrightarrow & 0 \\
 & & & & \downarrow & & \downarrow & & & & \\
 & & & & P_0(B) & \xrightarrow{\sim} & P_0(B) & & & & \\
 & & & & \downarrow & & \downarrow & & & & \\
0 & \longrightarrow & A & \longrightarrow & B & \longrightarrow & C & \longrightarrow & 0 & & \\
 & & & & \downarrow & & \downarrow & & & & \\
 & & & & 0 & & 0 & & & &
\end{array}
$$

where Q is a projective Λ-module. We have $P_0(C)|P_0(B)$, and the exact sequence $0 \to \Omega^1 B \to \Omega^1 C \amalg Q \to A \to 0$ shows that $P_i(C)|P_i(B) \amalg P_{i-1}(A)$ for $i \geq 1$.

Repeating the process we get an exact sequence $0 \to \Omega^2 C \to \Omega^1 A \amalg Q' \to \Omega^1 B \to 0$ with Q' projective. This shows that $P_i(A)|P_i(B) \amalg P_{i+1}(C)$ for $i \geq 1$. Further $Q|P_0(B)$ and $P_0(A)|P_1(C) \amalg Q$, and hence $P_0(A)|P_0(B) \amalg P_1(C)$. □

We give the next results in a slightly greater generality than is necessary to prove Theorem 2.1.

Lemma 2.6. (a) *Let* $\{n_0, n_1, \ldots, n_{d-1}\}$ *be nonnegative integers such that in a minimal projective resolution* $\cdots \to P_{d-1} \to \cdots \to P_1 \to P_0 \to D\Lambda \to 0$ *we have* $\mathrm{id}_\Lambda P_i \leq n_i$ *for* $0 \leq i \leq d-1$.

If $\mathrm{id}_\Lambda A = t$ *and* $\cdots \to P_j(A) \to \cdots \to P_1(A) \to P_0(A) \to A \to 0$ *is a minimal projective resolution for* A, *then* $\mathrm{id}_\Lambda P_i(A) \leq \max(n_i, \ldots, n_{i+t})$ *for* $i < d-t$.

(b) *If we have the sequence* $\{1, 2, \ldots, d\}$, *then* $\mathrm{id}_\Lambda P_i(A) \leq i+t+1$ *for* $i < d-t$.

Proof. We prove this by induction on $\mathrm{id}_\Lambda A = t$. If $t = 0$, we are done by assumption. So assume that $t > 0$ and consider the exact sequence $0 \to A \to I_0(A) \to C \to 0$, where $\mathrm{id}_\Lambda C = t-1$. Then $P_i(A)|P_i(I_0(A)) \amalg P_{i+1}(C)$ by Lemma 2.5. By induction we have $\mathrm{id}_\Lambda P_{i+1}(C) \leq \max(n_{i+1}, \ldots, n_{i+t})$ for $i+1 < d-t+1$. Hence we get that $\mathrm{id}_\Lambda P_i(A) \leq \max(n_i, n_{i+1}, \ldots, n_{i+t})$ for $i < d-t$. This finishes the proof of (a), and (b) is a direct consequence of (a). □

Lemma 2.7. *Let C be a Λ-module and $I^\cdot$ the dth truncation complex of a minimal injective resolution of C for some $d > 0$. Let $0 \to K^\cdot \to P^\cdot \xrightarrow{f^\cdot} I^\cdot$ be exact where $f^\cdot : P^\cdot \to I^\cdot$ is a minimal right $\mathscr{P}_d$-approximation. Let $0 \to K \to X \to C \to 0$ be the exact sequence induced by taking kernels. Then we have the following.*

(a) *If $\operatorname{id}_\Lambda P_i(D\Lambda) \le i + 1$ for $0 \le i < d$, then $\operatorname{id}_\Lambda P_i \le d - i$ and $\operatorname{id}_\Lambda K \le d$.*

(b) *If $\operatorname{id}_\Lambda P_i(D\Lambda) \le i$ for $0 \le i < d$, then $\operatorname{id}_\Lambda P_i \le d - i - 1$ for $0 \le i < d - 1$.*

Proof. Assume that $\operatorname{id}_\Lambda P_i(D\Lambda) \le i + 1$ for $0 \le i < d$. We have the following exact commutative diagram

$$\begin{array}{ccccccccccccccc}
 & & 0 & & 0 & & 0 & & & & 0 & & 0 & & \\
 & & \downarrow & & \downarrow & & \downarrow & & & & \downarrow & & \downarrow & & \\
0 & \to & K & \to & K_0 & \to & K_1 & \to & \cdots & \to & K_{d-2} & \to & K_{d-1} & \to & 0 \\
 & & \downarrow & & \downarrow & & \downarrow & & & & \downarrow & & \downarrow & & \\
0 & \to & X & \to & P_0 & \to & P_1 & \to & \cdots & \to & P_{d-2} & \to & P_{d-1} & \to & \Omega^{-d}C \to 0 \\
 & & \downarrow{\scriptstyle f} & & \downarrow & & \downarrow & & & & \downarrow & & \downarrow & & \| \\
0 & \to & C & \to & I_0 & \to & I_1 & \to & \cdots & \to & I_{d-2} & \to & I_{d-1} & \to & \Omega^{-d}C \to 0 \\
 & & \downarrow & & \downarrow & & \downarrow & & & & \downarrow & & \downarrow & & \\
 & & 0 & & 0 & & 0 & & & & 0 & & 0 & &
\end{array}$$

The first long exact sequence in the diagram induces short exact sequences $0 \to K'_i \to K_i \to K'_{i+1} \to 0$ for $0 \le i \le d - 1$, where $K'_d = 0$ and $K'_0 = K$. Since P_{d-1} is by construction a projective cover of $I_{d-1}(C)$, we have $\operatorname{id}_\Lambda P_{d-1} \le 1$ and hence $\operatorname{id}_\Lambda K_{d-1} \le 1$. Since $K'_{d-1} \simeq K_{d-1}$, we have $\operatorname{id}_\Lambda K'_{d-1} \le 1$. For each i with $0 < i \le d - 1$ we have an exact sequence $0 \to K'_i \to U_{i-1} \to I_{i-1}(C) \to 0$, where U_{i-1} is defined as in Section 1 and P_{i-1} is a projective cover of U_{i-1}. Hence we have $P_{i-1} | P_0(K'_i) \amalg P_0(I_{i-1}(C))$.

Since $\operatorname{id}_\Lambda K_i \le d - i$ if and only if $\operatorname{id}_\Lambda P_i \le d - i$, it is sufficient to prove $\operatorname{id}_\Lambda K_i \le d - i$ and $\operatorname{id}_\Lambda K'_i \le d - i$ for $0 \le i \le d - 1$ in order to prove (a). We shall do this by induction. We have already proved the claim for $i = d - 1$. So we assume the claim has been proved for an i with $0 < i \le d - 1$, and we want to prove it for $i - 1$. Since $\operatorname{id}_\Lambda K'_i \le d - i$ by the induction assumption, we have $\operatorname{id}_\Lambda P_0(K'_i) \le d - i + 1$ by Lemma 2.6. Further $\operatorname{id}_\Lambda P_0(I_{i-1}(C)) \le 1$ by our assumption, and hence $\operatorname{id}_\Lambda K_{i-1} \le d - (i - 1)$ is a consequence of $P_{i-1} | P_0(K'_i) \amalg P_0(I_{i-1}(C))$. Then $\operatorname{id}_\Lambda K_{i-1} \le d - (i - 1)$ follows from the exact sequence $0 \to K'_{i-1} \to K_{i-1} \to K'_i \to 0$, and we are done.

Assume now that $\operatorname{id}_\Lambda P_i(D\Lambda) \le i$ for $0 \le i < d$ and $d > 1$. We want to prove that $\operatorname{id}_\Lambda K_i \le d - i - 1$ and $\operatorname{id}_\Lambda K'_i \le d - i$ for $0 \le i \le d - 2$ by induction. We have $\operatorname{id}_\Lambda P_0(K'_{d-1}) \le 1$ by Lemma 2.6, since $\operatorname{id}_\Lambda K'_{d-1} \le 1$ and $\operatorname{id}_\Lambda P_0(D\Lambda) = 0$ and

$\mathrm{id}_\Lambda P_1(D\Lambda) \le 1$. Since $\mathrm{id}_\Lambda P_0(I_{d-2}(C)) = 0$, it follows from $P_{d-2} | P_0(K'_{d-1}) \amalg P_0(I_{d-2}(C))$ that $\mathrm{id}_\Lambda P_{d-2} \le 1$, and consequently $\mathrm{id}_\Lambda K_{d-2} \le 1 = d - (d-2) - 1$. Then $\mathrm{id}_\Lambda K'_{d-2} \le 2$ follows from the exact sequence $0 \to K'_{d-2} \to K_{d-2} \to K'_{d-1} \to 0$.

Assume now that the claim has been proved for an i with $0 < i \le d-2$. Since $\mathrm{id}_\Lambda K'_i \le d - i$ by the induction assumption, we have $\mathrm{id}_\Lambda P_0(K'_i) \le d-1$ by Lemma 2.6 and the assumption on Λ. Since $\mathrm{id}_\Lambda P_0(I_{i-1}(C)) = 0$, we get $\mathrm{id}_\Lambda P_{i-1} \le d - i$ by using $P_{i-1} | P_0(K'_i) \amalg P_0(I_{i-1}(C))$. Hence we have $\mathrm{id}_\Lambda K_{i-1} \le d - i = d - (i-1) - 1$, and then $\mathrm{id}_\Lambda K'_{i-1} \le d - i + 1$ follows from the exact sequence $0 \to K'_{i-1} \to K_{i-1} \to K'_i \to 0$. This finishes the induction step, and we have proved that $\mathrm{id}_\Lambda P_i \le d - i - 1$ for $0 \le i \le d-2$. □

Theorem 2.1 now follows from the preliminary results. One implication is given by Lemma 2.4. For the other implication, let $\cdots P_i \to \cdots \to P_1 \to P_0 \to D\Lambda \to 0$ be a minimal projective resolution of $D\Lambda$ and assume $\mathrm{id}_\Lambda P_i \le i+1$ for $i < d$. For C in $\mathrm{mod}\,\Lambda$ we have the exact sequence $0 \to K \to X \xrightarrow{h} C \to 0$ with $\mathrm{id}_\Lambda K \le d$ from Lemma 2.7, where $h: X \to C$ is a right $\mathscr{X}_d$-approximation. It then follows from Proposition 2.2 that $\mathscr{X}_d$ is closed under extensions. □

Using [2] we have the following consequence of Theorem 2.1.

Corollary 2.8. *The following are equivalent for an artin algebra Λ, where $d \ge 0$ is an integer.*

(a) $\mathrm{id}_\Lambda P_i(D\Lambda) \le i+1$ *for $i < d$.*

(b) $\mathscr{X}_i$ *is closed under extensions for $i \le d$.*

(c) $\mathscr{Y}_i = \{C \mid \mathrm{id}_\Lambda C \le i\}$ *is covariantly finite and* $\mathscr{X}_i = \{Z \mid \mathrm{Ext}^1_\Lambda(Z, \mathscr{Y}_i) = 0\}$ *for $i \le d$.* □

We point out that Hoshino has given an example where $\mathrm{pd}\, I_0({}_\Lambda\Lambda) \le 1$ but $\mathrm{pd}\, I_0(\Lambda_\Lambda) \ge 2$. Hence the torsionless left modules may be closed under extensions without the torsionless right modules being closed under extensions.

3. k-Gorenstein algebras

Recall that an artin algebra Λ is said to be k-Gorenstein if in a minimal injective resolution

$$0 \to \Lambda \to I_0 \to \cdots \to I_j \to \cdots$$

of Λ we have $\mathrm{pd}_\Lambda I_i \le i$ for all $i < k$. These are the types of algebras which are the main concern of the rest of this paper. This section is devoted to giving some properties of the subcategories $\Omega^i(\mathrm{mod}\,\Lambda)$ for $i \le k$ when Λ is k-Gorenstein, which hold in somewhat broader contexts. For example, as a special case of Theorem 2.1, we have that the subcategories $\mathscr{X}_i$ are extension closed for all $i \leq k$ when Λ is k-Gorenstein.

We note that self-injective algebras are k-Gorenstein for all k. But algebras of finite global dimension, even hereditary algebras, are usually not 1-Gorenstein. For example the path algebra of the quiver $\bullet \leftarrow \bullet \rightarrow \bullet$ over a field has no nonzero projective injective modules and is hence not 1-Gorenstein.

A basic property of k-Gorenstein algebras is that if Λ is k-Gorenstein, then so is Λ^{op} (see [6] for a proof of this result due to Auslander). Using this we get the following consequence of the results from Section 2.

Corollary 3.1. *For a k-Gorenstein algebra Λ we have the following.*

(a) $\mathscr{X}_i = \Omega^i(\operatorname{mod}\Lambda)$ *is closed under extensions for* $i \leq k$.

(b) $\mathscr{Y}_i = \{C \mid \operatorname{id}_\Lambda C \leq i\}$ *is covariantly finite in* $\operatorname{mod}\Lambda$ *for* $i \leq k$.

(c) $\mathscr{Z}_i = \{C \mid \operatorname{pd}_\Lambda C \leq i\}$ *is contravariantly finite in* $\operatorname{mod}\Lambda$ *for* $i \leq k$. □

We recall the following characterization of k-Gorenstein, which was useful in proving left–right symmetry (see [6]).

Proposition 3.2. *An artin algebra Λ is k-Gorenstein if and only if for each submodule X of $\operatorname{Ext}^i_\Lambda(C, \Lambda)$, with C in $\operatorname{mod}\Lambda$ and $i \leq k$, we have $\operatorname{grade} X \geq i$, i.e. $\operatorname{Ext}^j_\Lambda(X, \Lambda) = 0$ for all j with $0 \leq j < i$.* □

At first sight it looks complicated to test whether an algebra is k-Gorenstein by using this criterion. We shall however simplify this test. For this it is useful to define, for a Λ-module C, the strong grade $\operatorname{s.grade} C \geq i$ if $\operatorname{grade} C' \geq i$ for all submodules C' of C. We have the following preliminary result, where $\mathscr{G}_i$ denotes the category of Λ-modules having s.grade at least i. This category has some properties not in general satisfied by the category of modules of grade at least i.

Lemma 3.3. *Let Λ be an artin algebra.*

(a) *$\mathscr{G}_i$ is closed under submodules, factor modules and extensions.*

(b) *The modules in $\mathscr{G}_i$ are those whose simple composition factors have grade at least i.*

Proof. (a) $\mathscr{G}_i$ is clearly closed under submodules. Consider then the exact sequence $0 \to A \xrightarrow{g} B \xrightarrow{f} C \to 0$. Assume that B is in $\mathscr{G}_i$, and let C' be a submodule of C. Then writing $B' = f^{-1}(C')$, the exact sequence $0 \to A \to B' \to C' \to 0$ gives rise to the exact sequence $\cdots \to \operatorname{Ext}^{j-1}(A, \Lambda) \to \operatorname{Ext}^j(C', \Lambda) \to \operatorname{Ext}^j(B', \Lambda) \to \cdots$. It follows that $\operatorname{grade} C' \geq i$, so that C is in $\mathscr{G}_i$.

Assume now that A and C are in $\mathscr{G}_i$, and let B' be a submodule of B. We have the exact sequence $0 \to A \cap B' \to B' \to \operatorname{Im} f \to 0$, where $A \cap B'$ and $\operatorname{Im} f$ have grade at least i, and consequently $\operatorname{grade} B' \geq i$, so that $\operatorname{s.grade} B \geq i$.

(b) Since clearly s.grade S = grade S when S is a simple Λ-module, the claim is a direct consequence of (a). □

We deduce the following consequence of Lemma 3.3 and Proposition 3.2.

Proposition 3.4. *The following are equivalent for an artin algebra Λ.*

(a) *Λ is k-Gorenstein.*

(b) *If $i \leq k$ and S is a simple Λ-module, then all simple composition factors of* $\mathrm{Ext}^i_\Lambda(S, \Lambda)$ *have grade at least i.* □

Related to the grade conditions characterizing k-Gorenstein algebras, we have the following weaker conditions which are important in the study of syzygies and contravariantly and covariantly finite subcategories for classes of algebras somewhat larger than k-Gorenstein algebras. The conditions are the following, where (a_k) is equivalent to k-Gorenstein.

(a_k) s.grade $\mathrm{Ext}^i_\Lambda(C, \Lambda) \geq i$ for C in $\mathrm{mod}\,\Lambda$ and $i \leq k$.
(b_k) grade $\mathrm{Ext}^i_\Lambda(C, \Lambda) \geq i$ for C in $\mathrm{mod}\,\Lambda$ and $i \leq k$.
(c_k) s.grade $\mathrm{Ext}^{i+1}_\Lambda(C, \Lambda) \geq i$ for C in $\mathrm{mod}\,\Lambda$ and $i \leq k$.
(d_k) grade $\mathrm{Ext}^{i+1}_\Lambda(C, \Lambda) \geq i$ for C in $\mathrm{mod}\,\Lambda$ and $i \leq k$.

We clearly have the implications $(a_k) \Rightarrow (b_k) \Rightarrow (d_k)$ and $(a_k) \Rightarrow (c_k) \Rightarrow (d_k)$, but the implications cannot be reversed. For example we have seen that a hereditary algebra is not necessarily 1-Gorenstein, but it is not hard to see that any algebra of global dimension at most 2 satisfies the condition (d_1).

We recall a useful description of the category $\Omega^i(\mathrm{mod}\,\Lambda)$ for $i \leq k$ when Λ is k-Gorenstein, or more generally, when Λ satisfies (d_{k-1}).

For any artin algebra Λ a Λ-module C is said to be *d-torsionfree* if $\mathrm{Ext}^i_\Lambda(\mathrm{Tr}\,C, \Lambda) = 0$ for $i = 1, \ldots, d$. Note that because of the exact sequence $0 \to \mathrm{Ext}^1_\Lambda(\mathrm{Tr}\,C, \Lambda) \to C \to C^{**} \to \mathrm{Ext}^2_\Lambda(\mathrm{Tr}\,C, \Lambda) \to 0$, C is 1-torsionfree if and only if C is torsionless, or equivalently C is in $\mathscr{X}_1$. Further C is 2-torsionfree if and only if C is reflexive. In general it is easy to see that C is k-torsionfree if and only if there is an exact sequence $0 \to C \to P_0 \to P_1 \to \cdots \to P_{k-1}$ with the P_i projective, such that $P^*_{k-1} \to \cdots \to P^*_1 \to P^*_0 \to C^* \to 0$ is exact [1].

We can now state the following [1].

Proposition 3.5. *Let Λ be an artin algebra satisfying the condition* (d_k).

(a) *$\Omega^d(\mathrm{mod}\,\Lambda)$ is the category of d-torsionfree modules for $d \leq k$.*

(b) *$\Omega^d(\mathrm{mod}\,\Lambda)$ is closed under direct summands for $d \leq k$.* □

We note that (b) is a direct consequence of (a) since the d-torsionfree modules are obviously always closed under direct summands. However, $\Omega^d(\mathrm{mod}\,\Lambda)$ is not in general closed under direct summands. For example if $\Lambda = k[x, y]/(x, y)^2$ and S denotes the simple Λ-module, it is easy to see that $S \amalg S$ is in $\Omega^2(\mathrm{mod}\,\Lambda)$, but S is not.

But the category $\Omega^1(\operatorname{mod} \Lambda)$, consisting of the torsionless modules, is always closed under direct summands.

We also state the following results on covariantly and contravariantly subcategories [1].

Proposition 3.6. (a) *If Λ satisfies* (b_k), *then* $\{C \mid \operatorname{pd}_\Lambda C \leq k\}$ *is covariantly finite in* $\operatorname{mod} \Lambda$.

(b) *If Λ satisfies* (b_k), *then* $\{C \mid \operatorname{Ext}^i_\Lambda(C, \Lambda) = 0, 1 \leq i \leq k\}$ *is contravariantly finite in* $\operatorname{mod} \Lambda$. □

We end this section by giving some sufficient conditions for $\mathscr{X}_i$ to be closed under extensions different from those in the previous section.

Proposition 3.7. *Let Λ be an artin algebra satisfying* (c_k). *Then $\mathscr{X}_i$ is closed under extensions for all $i \leq k$.*

Proof. We note that when Λ satisfies (c_k), it satisfies (d_k), so that $\mathscr{X}_i$ consists of the i-torsionfree modules for $i \leq k$. Assume first that $i = 1 \leq k$ and let $0 \to A \to B \to C \to 0$ be an exact sequence with A and C in $\mathscr{X}_1$, that is with A and C torsionless. We then have the exact sequence $0 \to C^* \to B^* \to A^* \to \operatorname{Ext}^1_\Lambda(C, \Lambda)$, giving rise to short exact sequences $0 \to C^* \to B^* \to W \to 0$ and $0 \to W \to A^* \to Y \to 0$. We have $C \simeq \Omega^1 E \amalg Q$ for some E in $\operatorname{mod} \Lambda$ and some projective Λ-module Q, so that $\operatorname{Ext}^1_\Lambda(C, \Lambda) \simeq \operatorname{Ext}^2_\Lambda(E, \Lambda)$. Since Y is a submodule of $\operatorname{Ext}^1_\Lambda(C, \Lambda)$, we then have by assumption grade $Y \geq 1$, that is, $Y^* = 0$. Hence we have a monomorphism $0 \to A^{**} \to W^*$, and an exact sequence $0 \to W^* \to B^{**} \to C^{**}$. Consider then the commutative diagram

$$\begin{array}{ccccccccc} 0 & \longrightarrow & A & \longrightarrow & B & \longrightarrow & C & \longrightarrow & 0 \\ & & \downarrow & & \downarrow & & \downarrow & & \\ 0 & \longrightarrow & A^{**} & \longrightarrow & B^{**} & \longrightarrow & C^{**} & & \end{array}$$

Since $A \to A^{**}$ and $C \to C^{**}$ are monomorphisms by assumption, it follows from the above that $B \to B^{**}$ is also a monomorphism, that is, B is in $\mathscr{X}_1$.

Since (c_k) implies (d_k) we have by Proposition 3.5 that C is in $\mathscr{X}_i$ for $i \leq k$ if and only if $\operatorname{Ext}^j_\Lambda(\operatorname{Tr} C, \Lambda) = 0$ for $j = 1, \ldots, i$. When $2 \leq i \leq k$, this is equivalent to C being reflexive and $\operatorname{Ext}^j_\Lambda(C^*, \Lambda) = 0$ for $j = 1, \ldots, i-2$. Here we use that $C^* \simeq \Omega^2(\operatorname{Tr} C)$.

Assume now $i = 2 \leq k$, and let $0 \to A \to B \to C \to 0$ be an exact sequence with A and C in $\Omega^2(\operatorname{mod} \Lambda)$, that is A and C reflexive. Consider the exact sequence $0 \to C^* \to B^* \to A^* \to \operatorname{Ext}^1_\Lambda(C, \Lambda)$. This induces exact sequences $0 \to C^* \to B^* \to Z \to 0$, $0 \to Z \to A^* \to Y \to 0$ and $0 \to Y \to \operatorname{Ext}^1_\Lambda(C, \Lambda)$. Since C is in $\Omega^2(\operatorname{mod} \Lambda)$, we have $\mathrm{Ext}^1_\Lambda(C, \Lambda) \simeq \mathrm{Ext}^3_\Lambda(E, \Lambda)$ for some E, so that grade $Y \geq 2$ by assumption. This means that $Y^* = 0$ and $\mathrm{Ext}^1_\Lambda(Y, \Lambda) = 0$, so we get an isomorphism $A^{**} \xrightarrow{\sim} Z^*$. Consider then the commutative diagram

$$\begin{array}{ccccccccc}
0 & \longrightarrow & A & \longrightarrow & B & \xrightarrow{f} & C & \longrightarrow & 0 \\
 & & \downarrow & & \downarrow & & \downarrow\wr & & \\
0 & \longrightarrow & A^{**} \simeq Z^* & \longrightarrow & B^{**} & \xrightarrow{f^{**}} & C^{**} & &
\end{array}$$

Since f is onto, f^{**} must be onto, so that $B \to B^{**}$ is an isomorphism.

Let now $3 \leq i \leq k$, and let $0 \to A \to B \to C \to 0$ be an exact sequence with A and C in $\mathscr{X}_i$. With Z and Y as above, we have the exact sequence $\mathrm{Ext}^j_\Lambda(A^*, \Lambda) \to \mathrm{Ext}^j_\Lambda(Z, \Lambda) \to \mathrm{Ext}^{j+1}_\Lambda(Y, \Lambda)$. Here $\mathrm{Ext}^j_\Lambda(A^*, \Lambda) = 0$ for $j = 1, \ldots, i-2$, and grade $Y \geq i$ since $Y \subset \mathrm{Ext}^1_\Lambda(C, \Lambda) \simeq \mathrm{Ext}^{i+1}_\Lambda(E, \Lambda)$ for some E in mod Λ. Hence $\mathrm{Ext}^{j+1}_\Lambda(Y, \Lambda) = 0$ for $0 \leq j \leq i-2$, so that $\mathrm{Ext}^j_\Lambda(Z, \Lambda) = 0$ for $j = 1, \ldots, i-2$. Since $\mathrm{Ext}^j_\Lambda(C, \Lambda) = 0$ for $j = 1, \ldots, i-2$, it follows from the exact sequence $\mathrm{Ext}^j_\Lambda(Z, \Lambda) \to \mathrm{Ext}^j_\Lambda(B^*, \Lambda) \to \mathrm{Ext}^j_\Lambda(C^*, \Lambda)$ that $\mathrm{Ext}^j_\Lambda(B^*, \Lambda) = 0$ for $j = 1, \ldots, i-2$. Since we have already seen that B is reflexive, we have that B is in $\mathscr{X}_i$. □

Combining with our results from Section 2 we get the following.

Corollary 3.8. *If Λ satisfies* (c_k) *then we have* pd $I_i(\Lambda_\Lambda) \leq i + 1$ *for all $i < k$.* □

It would be interesting to know if the converse holds. In that case we would get a description of (c_k) in terms of the projective dimension of the injective modules in a minimal injective resolution of Λ_Λ.

We note that (d_k) could not replace (c_k) in Corollary 3.8. For assume $\Lambda = P_1 \amalg P_2 \amalg P_3$ is a direct sum of three indecomposable projective modules, such that $\underline{r}P_1 = S_2 \amalg S_3$, $\underline{r}P_2 = S_3$ and $P_3 = S_3$. Here S_i denotes the simple module $P_i/\underline{r}P_i$. Then gl.dim $\Lambda = 2$, and it is easy to see that Λ satisfies (d_1). However P_1 is projective cover of the injective envelope I_2 of S_2, and $\mathrm{id}_\Lambda P_1 = 2$.

It would be interesting to know if $\mathscr{X}_i$ being closed under extensions implies that it is closed under summands.

4. When are the augmented minimal approximations minimal?

In this section we show that the augmented minimal right $\mathscr{X}_i$-approximation is minimal for all $i \leq k$ if and only if Λ is k-Gorenstein. In general not even the nonprojective part of the augmented minimal right $\mathscr{X}_i$-approximation is minimal.

We first show that if the augmented minimal right $\mathscr{X}_i$-approximations are minimal for all $i \leq k$, then Λ is k-Gorenstein.

Proposition 4.1. *Let k be a positive integer, and assume that for all $i \leq k$, the augmented minimal right $\mathscr{X}_i$-approximation is an isomorphism for all projective Λ-modules. Then Λ is a k-Gorenstein algebra.*

Proof. For $k = 1$ we have the commutative diagram

$$\begin{array}{ccccccccc}
0 & \longrightarrow & X & \longrightarrow & P_0 & \longrightarrow & \Omega^{-1}(P) & \longrightarrow & 0 \\
\downarrow & & \downarrow{\scriptstyle f} & & \downarrow{\scriptstyle g} & & \| & & \\
0 & \longrightarrow & P & \longrightarrow & I_0(P) & \longrightarrow & \Omega^{-1}(P) & \longrightarrow & 0 \\
 & & & & \downarrow & & & & \\
 & & & & 0 & & & &
\end{array}$$

where P_0 is projective, and by assumption the augmented minimal right $\mathscr{X}_1$-approximation $f: X \to P$ is an isomorphism. Then $g: P_0 \to I_0(P)$ is an isomorphism, so that $I_0(P)$ is projective, and consequently Λ is 1-Gorenstein.

Assume now that $k > 1$ and that the augmented minimal right $\mathscr{X}_i$-approximation is an isomorphism for $i \leq k$ for all projective Λ-modules. By induction we can assume that Λ is $(k-1)$-Gorenstein. Consider then for a projective Λ-module P the commutative diagram

$$\begin{array}{ccccccccccccccccc}
 & & & & 0 & & 0 & & & & 0 & & 0 & & & & \\
 & & & & \downarrow & & \downarrow & & & & \downarrow & & \downarrow & & & & \\
 & & 0 & \longrightarrow & K_0 & \longrightarrow & K_1 & \longrightarrow & \cdots & \longrightarrow & K_{k-2} & \longrightarrow & K_{k-1} & \longrightarrow & 0 & & \\
 & & \downarrow & & \downarrow & & \downarrow & & & & \downarrow & & \downarrow & & & & \\
0 & \longrightarrow & X & \longrightarrow & P_0 & \longrightarrow & P_1 & \longrightarrow & \cdots & \longrightarrow & P_{k-2} & \longrightarrow & P_{k-1} & \longrightarrow & \Omega^{-k}(P) & \longrightarrow & 0 \\
 & & \downarrow{\scriptstyle \wr} & & \downarrow & & \downarrow & & & & \downarrow & & \downarrow & & \| & & \\
0 & \longrightarrow & P & \longrightarrow & I_0(P) & \longrightarrow & I_1(P) & \longrightarrow & \cdots & \longrightarrow & I_{k-2}(P) & \longrightarrow & I_{k-1}(P) & \longrightarrow & \Omega^{-k}(P) & \longrightarrow & 0 \\
 & & & & \downarrow & & \downarrow & & & & \downarrow & & \downarrow & & & & \\
 & & & & 0 & & 0 & & & & 0 & & 0 & & & &
\end{array}$$

where $P^{\cdot} \to I(P)_d$ is a minimal right $\mathscr{P}_d$-approximation. We have $\operatorname{pd}_\Lambda K_j \leq j-1$ for $1 \leq j \leq k-2$ and $\operatorname{pd}_\Lambda K_0 = 0$ since Λ is $(k-1)$-Gorenstein. It follows from the first long exact sequence that $\mathrm{pd}_\Lambda K_{k-1} \le k-2$, and hence $\mathrm{pd}_\Lambda I_{k-1}(P) \le k-1$. Hence we get that Λ is k-Gorenstein. □

We now turn to the question of proving the converse. The main difficulty associated with this is having a criterion for deciding whether a given approximation is actually minimal. To clarify what is going on we formulate the first preliminary result in a more general context.

For an artin algebra Λ, assume that we have contravariantly finite resolving subcategories $\mathscr{X}_1 \supset \mathscr{X}_2 \supset \cdots \supset \mathscr{X}_d$ and let $\mathscr{Y}_1 \subset \mathscr{Y}_2 \subset \cdots \subset \mathscr{Y}_d$ be the associated covariantly finite subcategories defined by $\mathscr{Y}_i = \{C \mid \mathrm{Ext}^1_\Lambda(\mathscr{X}_i, C) = 0\}$. We want to show that in some cases we can use a minimal left $\mathscr{Y}_{d-1}$-approximation to get a criterion for whether a given right $\mathscr{X}_d$-approximation is minimal.

Proposition 4.2. *For an artin algebra Λ, let $\{\mathscr{X}_i\}_{1 \le i \le d}$ and $\{\mathscr{Y}_i\}_{1 \le i \le d}$ be families of subcategories as introduced above satisfying the following properties.*

(i) *If Y is in $\mathscr{Y}_{i-1}$, then $\Omega^1 Y$ is in $\mathscr{Y}_i$.*

(ii) *$\mathscr{Y}_i$ is closed under projective covers.*

(iii) *If $f\colon Y \to Y'$ is a monomorphism with Y in $\mathscr{Y}_i$ and Y' in $\mathscr{Y}_{i-1}$, then* Coker f *is in $\mathscr{Y}_{i-1}$.*

If for C in mod Λ, *the exact sequences $0 \to Y_i(C) \to X_i(C) \to C \to 0$ and $0 \to C \to Y^{i-1}(C) \to X^{i-1}(C) \to 0$ are minimal right $\mathscr{X}_i$-approximation and minimal left $\mathscr{Y}_{i-1}$-approximation sequences, then $Y_i(C) \simeq \Omega^1(Y^{i-1}(C))$.*

Proof. From the minimal right $\mathscr{X}_i$-approximation sequence $0 \to Y_i(C) \to X_i(C) \to C \to 0$ we obtain by taking pushout the following commutative diagram

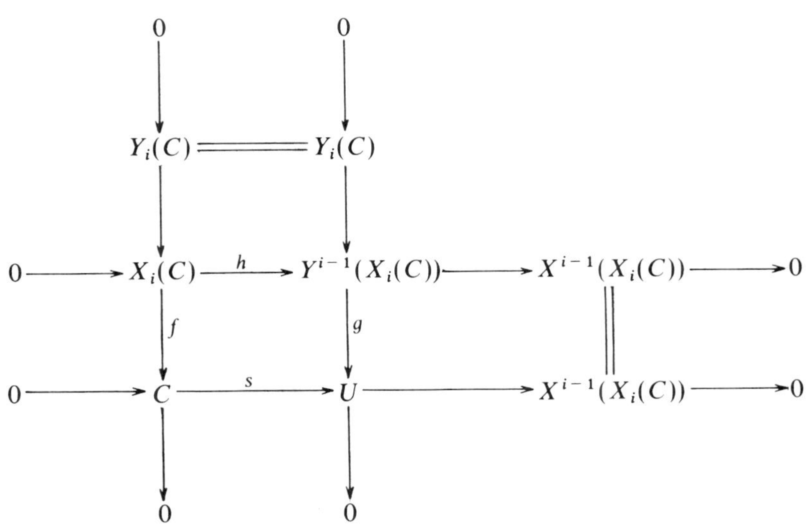

We claim that $Y^{i-1}(X_i(C))$ is a projective module. Let X be in $\mathscr{X}_i$ and Y in $\mathscr{Y}_{i-1}$. By assumption $\Omega^1 Y$ is in $\mathscr{Y}_i$, so that $\operatorname{Ext}^1_\Lambda(X, \Omega^1(Y)) = 0$, and consequently $(X, P_0(Y)) \to (X, Y) \to 0$ is exact. This shows that any map $X \to Y$ factors through $P_0(Y)$, which by assumption is in $\mathscr{Y}_{i-1}$. It then follows that $Y^{i-1}(X)$ is projective, and hence in particular $Y^{i-1}(X_i(C)) = P$ is a projective module. Since $f: X_i(C) \to C$ is right minimal, it follows easily that $g: Y^{i-1}(X_i(C)) \to U$ is right minimal. Since g is onto, it is then a projective cover. By assumption (iii), U is in $\mathscr{Y}_{i-1}$, so that $s: C \to U$ is a left $\mathscr{Y}_{i-1}$-approximation because $\operatorname{Coker} s$ is in $\mathscr{X}_{i-1}$. We want to show that $s: C \to U$ is left minimal.

Assume to the contrary that $U = U_1 \amalg U_2$ and $X^{i-1}(X_i(C)) = V_1 \amalg V_2$, with $U_2 \neq 0$, and the induced map $s: C \to U_1$ left minimal and the induced map $U_2 \to V_2$ an isomorphism. Then the exact sequence $0 \to Y_i(C) \to Y^{i-1}(X_i(C)) \to U \to 0$ can be written as a direct sum of the sequences $0 \to Y_1 \to P_0(U_1) \to U_1 \to 0$ and $0 \to Y_2 \to P_0(U_2) \to U_2 \to 0$. Since the induced map $P_0(U_2) \to V_2$ is then also a projective cover, we have that the sequence $0 \to X_i(C) \to P \to V_1 \amalg V_2 \to 0$ is a direct sum of the sequences $0 \to Y_2 \to P_2 \to V_2 \to 0$ and $0 \to Z \to P_1 \to V_1 \to 0$. Then the isomorphism $Y_2 \xrightarrow{\sim} Y_2$ is a summand of the map $Y_i(C) \to X_i(C)$, so that $Y_2 = (0)$ by the minimality of $f: X_i(C) \to C$. Then $P_0(U_2) \to V_2$ is an isomorphism, contradicting the minimality of $h: X_i(C) \to P$.

We can now conclude that $U \simeq Y^{i-1}(C)$, so that $Y_i(C) \simeq \Omega^1(Y^{i-1}(C))$. □

The following result on k-Gorenstein algebras shows that Proposition 4.2 applies to our situation. The claims are obvious or have already been proved.

Proposition 4.3. *Let Λ be a k-Gorenstein algebra. Then we have the following for $i \leq k$.*

(a) $\mathscr{X}_i = \operatorname{add} \Omega^i(\operatorname{mod} \Lambda) = \Omega^i(\operatorname{mod} \Lambda)$ *are contravariantly finite subcategories of* $\operatorname{mod} \Lambda$, *and* $\mathscr{Y}_i = \{C \mid \operatorname{id}_\Lambda C \leq i\}$ *are the associated covariantly finite subcategories.*

(b) *If $f: Y \to Y'$ is a monomorphism, with $\operatorname{id}_\Lambda Y \leq i$ and $\operatorname{id}_\Lambda Y' \leq i - 1$, then* $\operatorname{id}_\Lambda(\operatorname{Coker} f) \leq i - 1$.

(c) *If $\operatorname{id}_\Lambda C \leq i$, then $\operatorname{id}_\Lambda P_0(C) \leq i$.*

(d) *If $\operatorname{id}_\Lambda C \leq i - 1$, then $\operatorname{id}_\Lambda \Omega^1 C \leq i$.* □

We now have the following.

Theorem 4.4. *If Λ is k-Gorenstein, then for any C in* $\operatorname{mod} \Lambda$ *the augmented minimal right $\mathscr{X}_d$-approximation $X_d \to C$ is minimal for all $d \leq k$.*

Proof. Let $d \leq k$, and consider for C in $\operatorname{mod} \Lambda$ the exact sequence $0 \to C \to I_0(C) \to \Omega^{-1} C \to 0$. Assume by induction that the augmented minimal right $\mathscr{X}_{d-1}$-approximation of $\Omega^{-1} C$ is minimal. We recall that the augmented minimal right

$\mathscr{X}_d$-approximation of C is then constructed via the following two commutative diagrams:

$$
\begin{array}{ccccccccc}
 & & & & 0 & & 0 & & \\
 & & & & \downarrow & & \downarrow & & \\
 & & & & Y_{d-1}(\Omega^{-1}C) & = & Y_{d-1}(\Omega^{-1}C) & & \\
 & & & & \downarrow & & \downarrow & & \\
0 & \longrightarrow & C & \xrightarrow{h} & E & \longrightarrow & X_{d-1}(\Omega^{-1}C) & \longrightarrow & 0 \\
 & & \| & & \downarrow & & \downarrow & & \\
0 & \longrightarrow & C & \longrightarrow & I_0(C) & \longrightarrow & \Omega^{-1}(C) & \longrightarrow & 0 \\
 & & & & \downarrow & & \downarrow & & \\
 & & & & 0 & & 0 & &
\end{array}
$$

and

$$
\begin{array}{ccccccccc}
 & & 0 & & 0 & & & & \\
 & & \downarrow & & \downarrow & & & & \\
 & & Z & = & Z & & & & \\
 & & \downarrow & & \downarrow & & & & \\
0 & \longrightarrow & X & \longrightarrow & P_0(E) & \longrightarrow & X_{d-1}(\Omega^{-1}C) & \longrightarrow & 0 \\
 & & \downarrow f & & \downarrow & & \| & & \\
0 & \longrightarrow & C & \xrightarrow{h} & E & \longrightarrow & X_{d-1}(\Omega^{-1}C) & \longrightarrow & 0 \\
 & & \downarrow & & \downarrow & & & & \\
 & & 0 & & 0 & & & &
\end{array}
$$

where $\mathrm{id}_\Lambda E \leq d-1$, so that $h\colon C \to E$ is a left $\mathscr{Y}_{d-1}$-approximation.

To show that $f: X \to C$ is right minimal, consider the exact sequence $0 \to C \to Y^{d-1}(C) \to X^{d-1}(C) \to 0$, and the commutative diagram

$$\begin{array}{ccccccccc}
 & & & & 0 & & 0 & & \\
 & & & & \downarrow & & \downarrow & & \\
 & & & & K & = & K & & \\
 & & & & \downarrow & & \downarrow & & \\
0 & \to & C & \to & P \amalg Y^{d-1}(C) & \to & P \amalg X^{d-1}(C) & \to & 0 \\
 & & \| & & \downarrow{\scriptstyle t} & & \downarrow{\scriptstyle s} & & \\
0 & \to & C & \to & I_0(C) & \to & \Omega^{-1}(C) & \to & 0 \\
 & & & & \downarrow & & \downarrow & & \\
 & & & & 0 & & 0 & &
\end{array}$$

where P projective is chosen minimal to make t surjective. Since Λ is k-Gorenstein, P is then injective, and hence $\mathrm{id}_\Lambda K \leq d-1$. This shows that $s: P \amalg X^{d-1}(C) \to \Omega^{-1}(C)$ is a right $\mathscr{X}_{d-1}$-approximation.

The exact sequence $0 \to C \to P \amalg Y^{d-1}(C) \to P \amalg X^{d-1}(C) \to 0$ is obtained as the pullback of

$$\begin{array}{ccccc}
 & & P \amalg X^{d-1}(C) & & \\
 & & \downarrow & & \\
I_0(C) & \to & \Omega^{-1}(C) & \to & 0 \\
 & & \downarrow & & \\
 & & 0 & &
\end{array}$$

Consider then

$$
\begin{array}{ccccccccc}
 & & 0 & & 0 & & & & \\
 & & \downarrow & & \downarrow & & & & \\
0 & \longrightarrow & \Omega^1(Y^{d-1}C) & = & \Omega^1(Y^{d-1}(C)) & & & & \\
 & & \downarrow & & \downarrow & & & & \\
0 & \longrightarrow & \tilde{X} & \longrightarrow & P \amalg P_0(Y^{d-1}(C)) & \longrightarrow & P \amalg X^{d-1}(C) & \longrightarrow & 0 \\
 & & \downarrow{\scriptstyle g} & & \downarrow & & \| & & \\
0 & \longrightarrow & C & \longrightarrow & P \amalg Y^{d-1}(C) & \longrightarrow & P \amalg X^{d-1}(C) & \longrightarrow & 0 \\
 & & \downarrow & & \downarrow & & & & \\
 & & 0 & & 0 & & & &
\end{array}
$$

Since $\tilde{X}$ is in $\mathscr{X}_d$ and $\Omega^1(Y^{d-1}C)$ is in $\mathscr{Y}_d$, $g: \tilde{X} \to C$ is a right $\mathscr{X}_d$-approximation, which is minimal since we know from Proposition 4.2 that $Y_{d-1}(C) \simeq \Omega^1(Y^{d-1}C)$.

Since one right $\mathscr{X}_{d-1}$-approximation of $\Omega^{-1}C$ gives rise to a minimal right $\mathscr{X}_d$-approximation $g: \tilde{X} \to C$ by the above procedure, it is easy to see that the minimal right $\mathscr{X}_{d-1}$-approximation $X_{d-1}(\Omega^{-1}C) \to \Omega^{-1}C$ gives rise to a minimal right $\mathscr{X}_d$-approximation $f: X \to C$. □

5. Properties of k-Gorenstein algebras

Recall that an artin algebra is said to be Gorenstein if both the left and right injective dimensions of Λ are finite, in which case they are equal. Probably the most important question concerning algebras which are k-Gorenstein for all $k \geq 1$ is whether such algebras are Gorenstein or not. Amongst other things, an affirmative answer to this question would also give an affirmative answer to the Nakayama conjecture that an artin algebra of infinite dominant dimension is selfinjective. As a step in this direction we prove that if Λ is k-Gorenstein for all k and $\mathrm{id}_\Lambda \Lambda < \infty$, then Λ is Gorenstein. This result follows from a one-to-one correspondence we establish between indecomposable projective modules of injective dimension d and indecomposable injective modules of projective dimension d for all $d \leq k$ when Λ is k-Gorenstein. We start with a description of the $\Omega^d(\mathrm{mod}\,\Lambda)$ for all $d \leq k$ when Λ is k-Gorenstein in terms of the homological properties of injective resolutions of modules.

Proposition 5.1. *Let Λ be a k-Gorenstein artin algebra, and let $n_i \leq i$ for $0 \leq i < k$ be nonnegative integers satisfying $n_0 \leq n_1 \leq \cdots \leq n_{k-1}$ and $\operatorname{pd}_\Lambda I_i(\Lambda) \leq n_i$. Then for any $d \leq k$ we have $\mathscr{X}_d = \{C \mid \operatorname{pd}_\Lambda I_i(C) \leq i,\ 0 \leq i < d\} = \{C \mid \operatorname{pd}_\Lambda I_i(C) \leq n_i,\ 0 \leq i < d\}$.*

Proof. We first prove $\mathscr{X}_d \subseteq \{C \mid \operatorname{pd}_\Lambda I_i(C) \leq n_i, 0 \leq i < d\}$ for all $d \leq k$ by induction. It is obvious for $d = 1$. Assume that $d > 1$ and that the claim has been proved for $d - 1$. Consider for C in $\operatorname{mod} \Lambda$ the exact sequence $0 \to \Omega^d C \to P_0(\Omega^{d-1}C) \to \Omega^{d-1}C \to 0$. Then $I_i(\Omega^d C) | I_i(P_0(\Omega^{d-1}C)) \amalg I_{i-1}(\Omega^{d-1}C)$ by Lemma 2.5, where $I_{-1}(\Omega^{d-1}C) = 0$. Then $\operatorname{pd}_\Lambda I_i(P_0(\Omega^{d-1}C)) \leq n_i$ for $i < k$ and $\operatorname{pd}_\Lambda I_{i-1}(\Omega^{d-1}C) \leq n_{i-1}$ for $0 \leq i - 1 < d - 1$ by assumption. Since $n_{i-1} \leq n_i$ it follows that $\operatorname{pd}_\Lambda I_i(\Omega^d C) \leq n_i$ for $0 \leq i < d$.

We next want to prove $\{C \mid \operatorname{pd}_\Lambda I_i(C) \leq i, 0 \leq i < d\} \subseteq \mathscr{X}_d$. So assume for C in $\operatorname{mod} \Lambda$ that $\operatorname{pd}_\Lambda I_i(C) \leq i$ for $0 \leq i < d$, and we want to prove that the augmented minimal right $\mathscr{X}_d$-approximation $X \to C$ is an isomorphism. Consider then the commutative diagram

$$\begin{array}{ccccccccccccccc}
 & & 0 & & 0 & & 0 & & & & 0 & & & & \\
 & & \downarrow & & \downarrow & & \downarrow & & & & \downarrow & & & & \\
0 & \to & K & \to & K_0 & \to & K_1 & \to & \cdots & \to & K_{d-1} & \to & 0 & & \\
 & & \downarrow & & \downarrow & & \downarrow & & & & \downarrow & & & & \\
0 & \to & X & \to & P_0 & \to & P_1 & \to & \cdots & \to & P_{d-1} & \to & \Omega^{-d}C & \to & 0 \\
 & & \downarrow & & \downarrow & & \downarrow & & & & \downarrow & & \| & & \\
0 & \to & C & \to & I_0(C) & \to & I_1(C) & \to & \cdots & \to & I_{d-1}(C) & \to & \Omega^{-d}(C) & \to & 0 \\
 & & \downarrow & & \downarrow & & \downarrow & & & & \downarrow & & & & \\
 & & 0 & & 0 & & 0 & & & & 0 & & & &
\end{array}$$

and

$$\begin{array}{ccccccccc}
 & & 0 & & 0 & & 0 & & \\
 & & \downarrow & & \downarrow & & \downarrow & & \\
0 & \to & K & \to & K_0 & \to & K'_1 & \to & 0 \\
 & & \downarrow & & \downarrow & & \downarrow & & \\
0 & \to & X & \to & P_0 & \to & B & \to & 0 \\
 & & \downarrow & & \downarrow & & \downarrow & & \\
0 & \to & C & \to & I_0(C) & \to & \Omega^{-1}(C) & \to & 0 \\
 & & \downarrow & & \downarrow & & \downarrow & & \\
 & & 0 & & 0 & & 0 & &
\end{array}$$

Since $\operatorname{pd}_\Lambda K_j \le j-1$ for $1 \le j < d$, we conclude that $\operatorname{pd}_\Lambda K'_1 = 0$. Hence in the pullback diagram

$$
\begin{array}{ccccccccc}
 & & & & 0 & & 0 & & \\
 & & & & \downarrow & & \downarrow & & \\
 & & & & K'_1 & = & K'_1 & & \\
 & & & & \downarrow & & \downarrow & & \\
0 & \to & C & \to & U & \to & B & \to & 0 \\
 & & \| & & \downarrow & & \downarrow & & \\
0 & \to & C & \to & I_0(C) & \to & \Omega^{-1}(C) & \to & 0 \\
 & & & & \downarrow & & \downarrow & & \\
 & & & & 0 & & 0 & &
\end{array}
$$

U is projective, so that $C \xrightarrow{\sim} C$ is the augmented minimal right $\mathscr{X}_d$-approximation, and hence C is in $\mathscr{X}_d$. The rest of the claim is trivial. □

Note that Proposition 5.1 gives a description of $\Omega^d(\operatorname{mod} \Lambda)$ making it apparent that the category is closed under extensions.

Our result specializes to the following result of Martinez [8].

Corollary 5.2. *Let* dom.dim $\Lambda = k$. *Then* $\mathscr{X}_d = \{C \mid I_i(C)$ *is projective for* $i < d\}$ *for all* $d \le k$. □

We note that Proposition 5.1 cannot be generalized to the algebras where the syzygy modules are closed under extensions. For if Λ is hereditary and hence $\operatorname{pd} I_0 \le 1$, and not 1-Gorenstein, then $\mathscr{X}_1$ is the category of projective modules and $\{C \mid \operatorname{pd} I_0(C) \le 1\} = \operatorname{mod} \Lambda$.

We now consider for an algebra Λ which is k-Gorenstein for all k, the subcategory $\tilde{\mathscr{X}}_d = \tilde{\Omega}^d(\operatorname{mod} \Lambda)$ of $\mathscr{X}_d = \Omega^d(\operatorname{mod} \Lambda)$ whose objects are of the form $\Omega^d(C)$ for some C in $\operatorname{mod} \Lambda$. Let $\tilde{\mathscr{X}}_{-d} = \tilde{\Omega}^{-d}(\operatorname{mod} \Lambda)$ be defined similarly. The relationship between $\mathscr{X}_d$ and $\tilde{\mathscr{X}}_d$ is given by the following.

Proposition 5.3. *Let Λ be k-Gorenstein and $d \le k$.*

(a) *An object C in $\mathscr{X}_d$ is in $\tilde{\mathscr{X}}_d$ if and only if every indecomposable summand of C is nonprojective or is a projective module P with* $\operatorname{id}_\Lambda P \ge d$.

(b) *An object C in $\mathscr{X}_{-d}$ is in $\tilde{\mathscr{X}}_{-d}$ if and only if every indecomposable summand of C is noninjective or is an injective module I with* $\operatorname{pd}_\Lambda I \ge d$.

Proof. (a) We already know that $\Omega^d(\text{mod}\,\Lambda) = \mathscr{X}_d$, so it is clear that modules in $\mathscr{X}_d$ with no nonzero projective summands are in $\tilde{\mathscr{X}}_d$. Let A be an indecomposable Λ-module with $\text{id}_\Lambda A < d$, and assume $A = \Omega^d C$. Then $\text{Ext}^d_\Lambda(C, A) \simeq \text{Ext}^1_\Lambda(\Omega^{d-1}C, A) = 0$, so that $0 \to \Omega^d C \to P_0(\Omega^{d-1}C) \to \Omega^{d-1}C \to 0$ splits. Then $\Omega^{d-1}C$ is projective and hence $\Omega^d C = (0)$, which is a contradiction. In particular, if P is indecomposable projective with $\text{id}_\Lambda P < d$, then P is not in $\tilde{\mathscr{X}}_d$. If $\text{id}_\Lambda P \geq d$, then $\Omega^{-d}P \neq 0$. By the construction of the augmented minimal right $\mathscr{X}_d$-approximation we see that $P \simeq \Omega^d(\Omega^{-d}P)\amalg Q$, where Q is projective. By the dual of the above, we have $\text{pd}_\Lambda(\Omega^{-d}P) \geq d$, and hence $\Omega^d(\Omega^{-d}P) \neq 0$. Since P is indecomposable, we must then have $P \simeq \Omega^d(\Omega^{-d}P)$, so that P is in $\tilde{\mathscr{X}}_d$.

(b) follows from (a) by duality. □

Proposition 5.4. *Let Λ be k-Gorenstein and $d \leq k$.*

(a) *Ω^{-d} and Ω^d induce inverse one–one correspondences between the isomorphism classes of indecomposable objects in $\tilde{\Omega}^d(\text{mod}\,\Lambda)$ and $\tilde{\Omega}^{-d}(\text{mod}\,\Lambda)$.*

(b) *There is an induced correspondence between the isomorphism classes of indecomposable projective modules P with $\text{id}_\Lambda P = d$ and the indecomposable injective modules I with $\text{pd}_\Lambda I = d$.*

Proof. (a) If X is indecomposable in $\tilde{\Omega}^d(\text{mod}\,\Lambda)$, we have $X \simeq \Omega^d(\Omega^{-d}X)\amalg Q$. If X is not projective, then $X \simeq \Omega^d(\Omega^{-d}X)$. If X is projective, then $\text{id}_\Lambda X \geq d$ by Proposition 5.3, so that $\Omega^{-d}X \neq 0$. Since $\Omega^{-d}X$ is in $\tilde{\mathscr{X}}_{-d}$, we have $\text{pd}_\Lambda \Omega^{-d}X \geq d$ by Proposition 5.3, so that $\Omega^d(\Omega^{-d}X) \neq (0)$. Hence $X \simeq \Omega^d(\Omega^{-d}X)$ in this case also.

(b) If P is indecomposable projective with $\text{id}_\Lambda P = d$, then $\Omega^{-d}P$ is injective. If I is indecomposable injective with $\text{pd}_\Lambda I = d$, then $\Omega^d I$ is projective. □

We give various applications of this correspondence between indecomposable projective and indecomposable injective modules. It is an important problem whether Λ being k-Gorenstein for all k or $\text{id}_\Lambda \Lambda < \infty$ is sufficient for Λ to be Gorenstein. We get the following information on this problem as a direct consequence of Proposition 5.4.

Corollary 5.5. *Let Λ be k-Gorenstein for all k.*

(a) *The number of indecomposable projective modules P with $\text{id}_\Lambda P = \infty$ is equal to the number of indecomposable injective modules I with $\text{pd}_\Lambda I = \infty$.*

(b) *If $\text{id}_\Lambda \Lambda < \infty$, then Λ is Gorenstein.* □

We also get the following information on the minimal injective resolution of Λ.

Corollary 5.6. *Let Λ be k-Gorenstein.*

(a) *If I is an indecomposable injective module with $\text{pd}_\Lambda I = d \leq k$, then $I \mid I_d(\Lambda)$.*

(b) *If P is an indecomposable projective module with $\text{id}_\Lambda P = d \leq k$, then $P \mid P_d(D\Lambda)$.*

Proof. Since $\mathrm{pd}_\Lambda I = d \leq k$, there is by Proposition 5.4 an indecomposable projective module P with $I = \Omega^{-d}(P)$, and hence $I \mid I_d(\Lambda)$. The second claim follows by duality. $\square$

For any artin algebra Λ we have that $\mathscr{X}_d$ is contravariantly finite, and we clearly have $\hat{\mathscr{X}}_d = \mathrm{mod}\,\Lambda$, that is, for each C in $\mathrm{mod}\,\Lambda$ there is an exact sequence $0 \to X_n \to \cdots \to X_1 \to X_0 \to C \to 0$ with the X_i in $\mathscr{X}_d$. When $\mathscr{X}_d$ is closed under extensions, it is automatically resolving, that is, it also contains the projective modules and is closed under kernels of epimorphisms [4]. Hence the Ext-injective modules in $\mathscr{X}_d$ give rise to a cotilting module [2]. As another application of Proposition 5.4 we get an explicit description of this cotilting module when Λ is k-Gorenstein and $d \leq k$.

Proposition 5.7. *The cotilting module T associated with $\mathscr{X}_d$ for a k-Gorenstein algebra Λ with $d \leq k$ has as indecomposable summands the indecomposable projective modules P with $\mathrm{id}_\Lambda P \leq d$, and $\Omega^d I$ where I is indecomposable injective with $\mathrm{pd}_\Lambda I > d$.*

Proof. We need to compute $\mathscr{X}_d \cap \mathscr{Y}_d$ where $\mathscr{Y}_d = \{C \mid \mathrm{id}_\Lambda C \leq d\}$, that is, to find the indecomposable objects of $\mathscr{X}_d$ with injective dimension at most d. Let I be indecomposable injective. If $\mathrm{pd}_\Lambda I \geq d$, then $\Omega^{-d}(\Omega^d I) \simeq I$ by Proposition 5.4. If $\Omega^{-d}(\Omega^d I) \simeq I$, we have $\mathrm{id}_\Lambda \Omega^d T \leq d$, so that $\mathrm{id}_\Lambda \Omega^d T = d$. If $\mathrm{id}_\Lambda C = d$, then $\Omega^{-d} C$ is injective and $C \simeq \Omega^d(\Omega^{-d} C)$. Further clearly the indecomposable projective modules P with $\mathrm{id}_\Lambda P < d$ are in $\mathscr{X}_d \cap \mathscr{Y}_d$, and we are done. $\square$

We note that for k-Gorenstein algebras the categories $\mathscr{Y}_d = \{C \mid \mathrm{id}_\Lambda C \leq d\}$ and $\mathscr{X}_d = \{C \mid \mathrm{pd}_\Lambda C \leq d\}$ for $d \leq k$ also have nice properties.

Proposition 5.8. *Let Λ be a k-Gorenstein artin algebra. For $d \leq k$ the categories $\{C \mid \mathrm{id}_\Lambda C \leq d\}$ and $\{C \mid \mathrm{pd}_\Lambda C \leq d\}$ are functorially finite and closed under extensions, and hence have almost split sequences.*

Proof. We have seen that $\mathscr{Y}_d = \{C \mid \mathrm{id}_\Lambda C \leq d\}$ is covariantly finite in $\mathrm{mod}\,\Lambda$ since $\mathscr{X}_d$ is closed under extensions, and $\mathscr{Y}_d$ is clearly closed under extensions. Since Λ^{op} is also k-Gorenstein, we have that $\{C \mid \mathrm{pd}_\Lambda C \leq d\}$ is contravariantly finite. By Proposition 3.6 $\{C \mid \mathrm{pd}_\Lambda C \leq d\}$ is also covariantly finite, and hence also $\{C \mid \mathrm{id}_\Lambda C \leq d\}$ is contravariantly finite. That both subcategories have almost split sequences follows from [5]. $\square$

When Λ is k-Gorenstein for all k, it would be interesting to know if $\{C \mid \mathrm{pd}_\Lambda C < \infty\}$ is contravariantly finite. In view of Proposition 5.8, this is equivalent to Λ having finite finitistic projective dimension. Note that $\{C \mid \mathrm{pd}_\Lambda C < \infty\}$ is not always contravariantly finite for an artin algebra [7], but we do not know if there is always an almost split sequence with any indecomposable nonprojective module in the category on the right.

We also point out the following result about the finitistic injective dimension which applies to algebras which are k-Gorenstein for all k.

Proposition 5.9. *Assume that* $\mathscr{X}_d$ *is closed under extensions for all* $d \geq 0$. *Then* $\operatorname{fin.id}(\Lambda) < \infty$ *if and only if there is some integer* t *with* $\mathscr{X}_t = \mathscr{X}_{t+i}$ *for all* $i > 0$.

Proof. Assume first that $\operatorname{fin.id}(\Lambda) = t < \infty$. Then $\mathscr{Y}_t = \{C \mid \operatorname{id}_\Lambda C \leq t\} = \mathscr{Y}_{t+i}$ for all $i > 0$. Since $\mathscr{X}_d$ is contravariantly finite and resolving, we have $\mathscr{X}_d = \{C \mid \operatorname{Ext}^1_\Lambda(C, \mathscr{Y}_d) = 0\}$ by [2]. Hence we get $\mathscr{X}_t = \mathscr{X}_{t+i}$ for all $i > 0$.

If conversely there is some t with $\mathscr{X}_t = \mathscr{X}_{t+i}$ for all $i > 0$, then $\mathscr{Y}_t = \mathscr{Y}_{t+i}$ for all $i > 0$, and consequently $\operatorname{fin.id}(\Lambda) < \infty$. □

For a selfinjective algebra Λ we have that Ω^{-d} and Ω^d induce inverse equivalences between $\underline{\operatorname{mod}}\,\Lambda$ and $\underline{\operatorname{mod}}\,\Lambda$, the category $\operatorname{mod}\Lambda$ modulo projectives. We have the following generalization to k-Gorenstein algebras, containing as a special case the correspondence between indecomposable objects induced by Ω^{-d} and Ω^d from Proposition 5.4.

Let Λ be k-Gorenstein and $d \leq k$. Let $f: B \to C$ be a map in $\mathscr{X}_d$ and consider an induced exact commutative diagram

$$\begin{array}{ccccccccccccccc}
0 & \to & B & \to & I_0(B) & \to & I_1(B) & \to \cdots \to & I_{d-1}(B) & \to & \Omega^{-d}(B) & \to & 0 \\
 & & \downarrow{\scriptstyle f} & & \downarrow & & \downarrow & & \downarrow & & \downarrow{\scriptstyle g} & & \\
0 & \to & C & \to & I_0(C) & \to & I_1(C) & \to \cdots \to & I_{d-1}(C) & \to & \Omega^{-d}(C) & \to & 0
\end{array}$$

If by choosing a different lifting of maps between the indecomposable injective modules we get an induced map $g': \Omega^{-d}B \to \Omega^{-d}C$, it is easy to see that $g - g'$ factors through $I_{d-1}(C)$. Since C is in $\mathscr{X}_d$, it follows from Proposition 5.1 that $\operatorname{pd}_\Lambda I_{d-1}(C) \leq d-1$. Hence there is induced a functor $\Omega^{-d}: \mathscr{X}_d \to \mathscr{X}_{-d}/\mathscr{Z}_d$ where $\mathscr{Z}_d = \{C \mid \operatorname{pd}_\Lambda C < d\}$. If $f: B \to C$ factors through an object in $\mathscr{Y}_d = \{C \mid \operatorname{id}_\Lambda C < d\}$, we see that $\Omega^{-d}(f) = 0$, so that there is induced a functor $\Omega^{-d}: \mathscr{X}_d/\mathscr{Y}_d \to \mathscr{X}_{-d}/\mathscr{Z}_d$. Dually we have an induced functor $\Omega^d: \mathscr{X}_{-d}/\mathscr{Z}_d \to \mathscr{X}_d/\mathscr{Y}_d$ by using the fixed minimal projective resolutions of objects in $\mathscr{X}_{-d}$.

Similarly we can define functors $\tilde{\Omega}^d: \mathscr{X}_{-d}/\mathscr{Z}_d \to \mathscr{X}_d/\mathscr{Y}_d$ and $\tilde{\Omega}^{-d}: \mathscr{X}_d/\mathscr{Y}_d \to \mathscr{X}_{-d}/\mathscr{Z}_d$ by using other particular injective and projective resolutions of objects in $\tilde{\mathscr{X}}_d$ and $\tilde{\mathscr{X}}_{-d}$ respectively, as follows.

For A in $\tilde{\mathscr{X}}_{-d}$, choose C in $\tilde{\mathscr{X}}_d$ such that $\Omega^{-d}C \simeq A$. Choose a fixed minimal augmented $\mathscr{P}_d$-approximation $f: \mathscr{P}(I^{\cdot}(C)_d) \to I^{\cdot}(C)_d$. This gives rise to an exact sequence $0 \to \Omega^d A \to P_{d-1} \to \cdots \to P_1 \to P_0 \to A \to 0$. By Lemma 2.7 we have that $\operatorname{id}_\Lambda P_{d-1} < d$. Hence we can use the projective resolutions of objects in $\tilde{\mathscr{X}}_{-d}$ obtained this way to define a functor $\tilde{\Omega}^d: \mathscr{X}_{-d}/\mathscr{Z}_d \to \mathscr{X}_d/\mathscr{Y}_d$ with $\tilde{\Omega}^d(A) \simeq \Omega^d(A)$ for A in $\tilde{\mathscr{X}}_{-d}$. It is easy to see that $\tilde{\Omega}^d$ and Ω^d are isomorphic functors. Similarly we define a functor

$\tilde{\Omega}^{-d}: \mathscr{X}_d/\mathscr{Y}_d \to \mathscr{X}_{-d}/\mathscr{Z}_d$ which is isomorphic to Ω^{-d}. Given a map $h: B \to C$ in $\tilde{\mathscr{X}}_d$, we have a commutative diagram

$$\begin{array}{ccc} \mathscr{P}(I(B)_d) & \longrightarrow & \mathscr{P}(I(C)_d) \\ \downarrow & & \downarrow \\ I(B)_d & \longrightarrow & I(C)_d \end{array}$$

using that we have right $\mathscr{P}_d$-approximations. Taking kernels induces a commutative diagram

$$\begin{array}{ccc} \tilde{\Omega}^d \Omega^{-d} B & \xrightarrow{\tilde{\Omega}^d \Omega^{-d}(h)} & \tilde{\Omega}^d \Omega^{-d} C \\ \wr \downarrow \alpha_B & & \wr \downarrow \alpha_C \\ B & \xrightarrow{h} & C \end{array}$$

in $\mathscr{X}_d/\mathscr{Y}_d$. Here α_B and α_C are isomorphisms by Theorem 4.4. It is easy to see that $\tilde{\Omega}^d$ and Ω^d are isomorphic functors, and hence we have that $\Omega^d\Omega^{-d}$ is isomorphic to the identity functor on $\mathscr{X}_d/\mathscr{Y}_d$. Similarly $\Omega^{-d}\Omega^d$ is isomorphic to the identity on $\mathscr{X}_{-d}/\mathscr{Z}_d$. Hence $\Omega^{-d}: \mathscr{X}_d/\mathscr{Y}_d \to \mathscr{X}_{-d}/\mathscr{Z}_d$ and $\Omega^d: \mathscr{X}_{-d}/\mathscr{Z}_d \to \mathscr{X}_d/\mathscr{Y}_d$ are inverse equivalences of categories so that we have proved the following.

Theorem 5.10. *Let Λ be a k-Gorenstein algebra and assume $d \leq k$. Then Ω^{-d} induces a functor $\Omega^{-d}: \mathscr{X}_d/\mathscr{Y}_d \to \mathscr{X}_{-d}/\mathscr{Z}_d$ and Ω^d induces a functor $\Omega^d: \mathscr{X}_{-d}/\mathscr{Z}_d \to \mathscr{X}_d/\mathscr{Y}_d$, and these functors are inverse equivalences of categories.* □

Note added in proof. We have recently shown that the converse to Corollary 3.8 holds.

References

[1] M. Auslander and M. Bridger, Stable Module Theory, Memoirs of the American Mathematical Society, Vol. 94 (American Mathematical Society, Providence, RI, 1969).

[2] M. Auslander and I. Reiten, Applications of contravariantly finite subcategories, Adv. in Math. 86 (1) (1991) 111–152.

[3] M. Auslander and I. Reiten, On a theorem of E. Green on the dual of the transpose, Canad. Math. Soc., Conference Proceedings 11 (1991) 53–65.

[4] M. Auslander and I. Reiten, Homologically finite subcategories, in: Proceedings of the Tsukuba Conference, 1990, Representations of Algebras, London Mathematical Society Lecture Note Series, Vol. 168 (Cambridge University Press, Cambridge, 1992) 1–42.

[5] M. Auslander and S.O. Smalø, Almost split sequences in subcategories, J. Algebra 69 (1981) 426–454.

[6] R. Fossum, Ph, Griffith and I. Reiten, Trivial Extensions of Abelian Categories, Lecture Notes in Mathematics, Vol. 456 (Springer, Berlin, 1975).

[7] K. Igusa, S.O. Smalø and G. Todorov, Finite projectivity and contravariant finiteness, Proc. Amer. Math. Soc. 109(4) (1990) 937–941.

[8] R. Martinez, Modules of dominant and codominant dimensions, Preprint, 1991.

[9] S.O. Smalø, Functorial finite subcategories over triangular matrix rings, Proc. Amer. Math. Soc. 111 (1991) 651–656.

JOURNAL OF ALGEBRA **183**, 167–185 (1996)
ARTICLE NO. 0212

Syzygy Modules for Noetherian Rings

Maurice Auslander* and Idun Reiten†

Institutt for Matematikk og Statistikk, NTNU, 7055 Dragvoll, Norway

Communicated by Kent R. Fuller

Received March 17, 1995

INTRODUCTION

Throughout this paper Λ is a twosided noetherian ring and $\operatorname{mod}\Lambda$ the category of finitely generated Λ-modules. For each integer $k > 0$ we denote by $\Omega^k(\operatorname{mod}\Lambda)$ the full subcategory of $\operatorname{mod}\Lambda$ whose objects are the C in $\operatorname{mod}\Lambda$ which are the kth syzygies of modules in $\operatorname{mod}\Lambda$, that is, for which there is an exact sequence $0 \to C \to P_{k-1} \to \cdots \to P_0 \to B \to 0$ in $\operatorname{mod}\Lambda$ with the P_i projective modules. While projective resolutions of modules have been studied extensively, little seems to be known about the subcategories $\Omega^k(\operatorname{mod}\Lambda)$ of $\operatorname{mod}\Lambda$. One of the few general properties known about these categories is that they are covariantly finite in $\operatorname{mod}\Lambda$ [AR2]. This means that each module M in $\operatorname{mod}\Lambda$ has a left $\Omega^k(\operatorname{mod}\Lambda)$-approximation, that is, we have a morphism $f\colon M \to L$ with L in $\Omega^k(\operatorname{mod}\Lambda)$ with the property that if $h\colon M \to L'$ is any morphism with L' in $\Omega^k(\operatorname{mod}\Lambda)$, then there is some $g\colon L \to L'$ such that $gf = h$. While in general covariantly finite subcategories of $\operatorname{mod}\Lambda$ have proven to be of interest in various situations, for instance, in the representation theory of artin algebras and commutative ring theory, they are particularly interesting when they are extension closed, i.e., if $0 \to A \to B \to C \to 0$ is exact with A and C in $\Omega^k(\operatorname{mod}\Lambda)$, then B is in $\Omega^k(\operatorname{mod}\Lambda)$, in connection with Wakamatsu's lemma type results (see [AR1, AR2, AR3]). The main motivation for this paper was to describe when the subcategories $\Omega^k(\operatorname{mod}\Lambda)$ are extension closed. However, before stating our main result, it is convenient to recall some definitions [AB].

* The first author died on November 18, 1994.
† E-mail: idun.reiten@avh.unit.no.

0021-8693/96 $18.00

Let C be in mod Λ. We say that the grade of C, written grade C, is greater than or equal to an integer $i \geq 0$ if $\operatorname{Ext}_\Lambda^j(C, \Lambda) = 0$ for all $j < i$. We say the strong grade of C, written s.grade C, is greater than or equal to an integer $i \geq 0$ if grade $X \geq i$ for all submodules X of C. Finally, we say that a ring Λ is a noetherian R-algebra if Λ is an R-algebra where R is a commutative noetherian ring and Λ is a finitely generated R-module. Our main result is the following.

THEOREM 0.1. *Let Λ be a noetherian R-algebra. Then the following conditions are equivalent for an integer $k > 0$.*

(a) $\Omega^i(\text{mod}\,\Lambda)$ *is extension closed for* $1 \leq i \leq k$.

(b) *If* $0 \to \Lambda \to I_0 \to I_1 \to \cdots \to I_i \to \cdots$ *is a minimal injective resolution of* Λ *as a right* Λ*-module, then* flatdim $I_i \leq i + 1$ *for* $i < k$.

(c) grade $\operatorname{Ext}_\Lambda^i(C, \Lambda) \geq i$ *for all* C *in* mod Λ^{op} *and* $i \leq k$.

(d) s.grade $\operatorname{Ext}_\Lambda^{i+1}(B, \Lambda) \geq i$ *for all* B *in* mod Λ *and* $i \leq k$.

Actually, the only place the hypothesis that Λ is a noetherian R-algebra is used is in showing that (a) implies (c) when $k > 1$. The proofs of all the other implications only depend on the condition that Λ be a twosided noetherian ring.

We now make some other remarks about this theorem. First, in the course of proving (a) implies (c), it is shown that (a) implies that the modules in $\Omega^i(\text{mod}\,\Lambda)$ are i-torsionfree for $1 \leq i \leq k$ (see Section 1 for the definition). This implies for $1 \leq i \leq k$ that the $\Omega^i(\text{mod}\,\Lambda)$ are also closed under summands. It is a measure of our ignorance of the categories $\Omega^i(\text{mod}\,\Lambda)$ that we do not even know precisely when they are closed under summands.

Second, twosided noetherian rings satisfying (c) were studied in [AB], culminating in what was called the approximation theorem. However, except for the case when Λ is commutative, no criterion for when a twosided noetherian ring satisfied (c) was known. The theorem remedies this situation.

Third, the equivalence of (a) and (b) was proven in [AR3] for artin algebras and it was shown that (d) implies (a) for twosided noetherian rings. All the other results are new, in particular for artin algebras.

1. EXTENSION CLOSURE AND STRONG GRADE

The main aim of this section is to show for a twosided noetherian ring Λ that for an integer $k \geq 1$ the category $\Omega^i(\text{mod}\,\Lambda)$ is extension closed for $1 \leq i \leq k$ if and only if s.grade $\operatorname{Ext}_\Lambda^{i+1}(B, \Lambda) \geq i$ for $1 \leq i \leq k$.

Throughout this section Λ is an arbitrary twosided noetherian ring, all Λ-modules are assumed to be finitely generated, and $\operatorname{mod}\Lambda$ and $\operatorname{mod}\Lambda^{\mathrm{op}}$ denote the categories of finitely generated left and finitely generated right modules, respectively. For each integer $i \geq 0$ and each C in $\operatorname{mod}\Lambda$, we consider the abelian group $\operatorname{Ext}^i_\Lambda(C, \Lambda)$ a right Λ-module by means of the action induced by right multiplication in Λ. We recall that a module C is k-torsionfree for an integer $k > 0$ if $\operatorname{Ext}^i_\Lambda(\operatorname{Tr} C, \Lambda) = 0$ for $1 \leq i \leq k$, where $\operatorname{Tr} C$ is the transpose of C [AB].

Our first aim in this section is to show how the fact that the categories $\Omega^i(\operatorname{mod}\Lambda)$ are extension closed for $1 \leq i \leq k$ is connected with the fact that the modules in $\Omega^i(\operatorname{mod}\Lambda)$ are the i-torsionfree modules for $1 \leq i \leq k$. The results along these lines are based on the following.

THEOREM 1.1. *Suppose C is k-torsionfree with $k \geq 1$. Then the following are equivalent.*

(a) $\operatorname{s.grade} \operatorname{Ext}^1_\Lambda(C, \Lambda) \geq k$.

(b) *If* $0 \to A \to B \to C \to 0$ *is exact with A k-torsionfree, then B is k-torsionfree.*

(c) *If* $0 \to \Lambda^n \to E \to C \to 0$ *is exact, then E is k-torsionfree.*

The proof of this result proceeds by induction on $k \geq 1$ as follows. The cases $k = 1$ and $k = 2$ are proven separately with the induction starting with $k \geq 3$.

Recall that C being 1-torsionfree is the same as C being torsionless, i.e., C is in $\Omega^1(\operatorname{mod}\Lambda)$.

LEMMA 1.2. *The following are equivalent for C in $\Omega^1(\operatorname{mod}\Lambda)$.*

(a) $\operatorname{s.grade} \operatorname{Ext}^1_\Lambda(C, \Lambda) \geq 1$.

(b) *If* $0 \to A \to B \to C \to 0$ *is exact with A in $\Omega^1(\operatorname{mod}\Lambda)$, then B is in $\Omega^1(\operatorname{mod}\Lambda)$.*

(c) *If* $0 \to \Lambda^n \to E \to C \to 0$ *is exact, then E is in $\Omega^1(\operatorname{mod}\Lambda)$.*

Proof. (a) $\Rightarrow$ (b) Applying the functor $X \mapsto X^{**}$, where $X^{**} = \operatorname{Hom}_\Lambda(\operatorname{Hom}_\Lambda(X, \Lambda), \Lambda)$, to the exact sequence $0 \to A \to^f B \to^g C \to 0$ we obtain the commutative diagram

$$\begin{array}{ccccccccc}
0 & \longrightarrow & A & \xrightarrow{f} & B & \xrightarrow{g} & C & \longrightarrow & 0 \\
 & & \downarrow{\scriptstyle \alpha_A} & & \downarrow{\scriptstyle \alpha_B} & & \downarrow{\scriptstyle \alpha_C} & & \\
 & & A^{**} & \xrightarrow{f^{**}} & B^{**} & \xrightarrow{g^{**}} & C^{**} & &
\end{array}$$

where $\alpha_X\colon X \to X^{**}$ is the usual morphism. The fact that A and C are in $\Omega^1(\operatorname{mod}\Lambda)$ means that α_A and α_C are monomorphisms. It is then easily checked that α_B is also a monomorphism if $f^{**}\colon A^{**} \to B^{**}$ is a monomorphism. But f^{**} can be seen to be a monomorphism as follows.

The exact sequence $0 \to A \to^{f} B \to^{g} C \to 0$ gives rise to the exact sequence $0 \to C^* \to^{g^*} B^* \to^{f^*} A^* \to \operatorname{Ext}^1_\Lambda(C, \Lambda)$. Since s.grade $\operatorname{Ext}^1_\Lambda(C, \Lambda) \geq 1$, we have that $L^* = 0$, where $L = \operatorname{Coker} f^*$. This implies that $f^{**}\colon A^{**} \to B^{**}$ is a monomorphism, which finishes the proof.

(b) $\Rightarrow$ (c) This is trivial.

(c) $\Rightarrow$ (a) Let L be a submodule of $\operatorname{Ext}^1_\Lambda(C, \Lambda)$. We want to show that $L^* = 0$. Since $\operatorname{Ext}^1_\Lambda(C, \Lambda)$ is a finitely generated Λ^{op}-module, L is also a finitely generated submodule of $\operatorname{Ext}^1_\Lambda(C, \Lambda)$. From this it follows that there is an exact sequence $0 \to \Lambda^n \to E \to^{g} C \to 0$ such that the induced exact sequence $E^* \to (\Lambda^n)^* \to \operatorname{Ext}^1_\Lambda(C, \Lambda)$ has the property that $E^* \to (\Lambda^n)^* \to L \to 0$ is exact. We therefore have the exact commutative diagram

$$\begin{array}{ccccccccc}
 & & & & 0 & & 0 & & \\
 & & & & \downarrow & & \downarrow & & \\
 & & 0 & \longrightarrow & \Lambda^n & \longrightarrow & E & \longrightarrow & 0 \\
 & & & & \downarrow & & \downarrow & & \\
0 & \longrightarrow & L^* & \longrightarrow & (\Lambda^n)^{**} & \longrightarrow & E^{**} & & \\
 & & & & \downarrow & & & & \\
 & & & & 0 & & & &
\end{array}$$

from which it follows that $L^* = 0$. This shows that s.grade $\operatorname{Ext}^1_\Lambda(C, \Lambda) \geq 1$, completing the proof of the lemma. ▮

Having proven Theorem 1.1 when $k = 1$, we now turn our attention to proving it when $k = 2$. Remember, a module C is 2-torsionfree if and only if $\alpha_C\colon C \to C^{**}$ is an isomorphism, i.e., C is reflexive.

LEMMA 1.3. *The following are equivalent for a reflexive Λ-module C.*

(a) s.grade $\operatorname{Ext}^1_\Lambda(C, \Lambda) \geq 2$.

(b) *If $0 \to A \to B \to C \to 0$ is exact with A reflexive, then B is reflexive.*

(c) *If $0 \to \Lambda^n \to E \to C \to 0$ is exact, then E is reflexive.*

Proof. (a) $\Rightarrow$ (b) We have the exact sequence $0 \to C^* \to B^* \to A^* \to L \to 0$ with grade $L \geq 2$ since $L \subset \operatorname{Ext}^1_\Lambda(C, \Lambda)$ and s.grade $\operatorname{Ext}^1_\Lambda(C, \Lambda) \geq 2$. Then the exact sequence $0 \to K \to A^* \to L \to 0$ gives rise to the exact sequence $0 \to L^* \to A^{**} \to K^* \to \operatorname{Ext}^1_\Lambda(L, \Lambda)$, which shows that the map $A^{**} \to K^*$ is an isomorphism. This shows that $0 \to A^{**} \to B^{**} \to C^{**}$ is exact and so we have the exact commutative diagram

$$\begin{array}{ccccccccc}
0 & \longrightarrow & A & \longrightarrow & B & \longrightarrow & C & \longrightarrow & 0 \\
 & & {\scriptstyle \alpha_A}\downarrow\wr & & {\scriptstyle \alpha_B}\downarrow & & {\scriptstyle \alpha_C}\downarrow\wr & & \\
0 & \longrightarrow & A^{**} & \longrightarrow & B^{**} & \longrightarrow & C^{**} & &
\end{array}$$

which shows that B is reflexive.

(b) $\Rightarrow$ (c) This is trivial.

(c) $\Rightarrow$ (a) Let L be a submodule of $\operatorname{Ext}^1_\Lambda(C, \Lambda)$. Then $L^* = 0$ since C is in $\Omega^1(\operatorname{mod} \Lambda)$ and so $\operatorname{s.grade} \operatorname{Ext}^1_\Lambda(C, \Lambda) \geq 1$. We know there is an exact sequence $0 \to \Lambda^n \to E \to C \to 0$ such that $\operatorname{Im}((\Lambda^n)^* \to \operatorname{Ext}^1_\Lambda(C, \Lambda)) = L$. Thus we have the exact sequences $0 \to C^* \to E^* \to K \to 0$ and $0 \to K \to (\Lambda^n)^* \to L \to 0$. Then we get the exact sequence $0 \to L^* \to (\Lambda^n)^{**} \to K^* \to \operatorname{Ext}^1_\Lambda(L, \Lambda) \to 0$. So we have to show that $0 \to (\Lambda^n)^{**} \to K^* \to 0$ is exact. We have the commutative exact diagram

$$\begin{array}{ccccccccc}
0 & \longrightarrow & \Lambda^n & \longrightarrow & E & \longrightarrow & C & \longrightarrow & 0 \\
 & & \downarrow & & \downarrow\wr & & \downarrow\wr & & \\
0 & \longrightarrow & K^* & \longrightarrow & E^{**} & \longrightarrow & C^{**}, & &
\end{array}$$

from which it follows that $\Lambda^n \to K^*$ is an isomorphism, giving our desired result. ∎

Having established Theorem 1.1 when $k = 1$ and 2, we now prove it for $k \geq 3$. This proof is based on the following.

LEMMA 1.4. *Let $0 \to A \to^f B \to^g C \to 0$ be exact with C k-torsionfree, with $k \geq 3$ and B reflexive. Define the module K by the exact sequence $0 \to C^* \to^{g^*} B^* \to K \to 0$. Then B is k-torsionfree if and only if* $\operatorname{Ext}^i_\Lambda(K, \Lambda) = 0$ *for* $1 \leq i \leq k - 2$.

Proof. Since C is k-torsionfree with $k \geq 3$, we have that $\operatorname{Ext}^i_\Lambda(C^*, \Lambda) = 0$ for $i = 1, \ldots, k-2$. We have the long exact sequence $0 \to K^* \to B^{**} \to C^{**} \to \operatorname{Ext}^1_\Lambda(K, \Lambda) \to \operatorname{Ext}^1_\Lambda(B^*, \Lambda) \to \operatorname{Ext}^1_\Lambda(C^*, \Lambda) \to \operatorname{Ext}^2_\Lambda(K, \Lambda) \to \operatorname{Ext}^2_\Lambda(B^*, \Lambda) \to \operatorname{Ext}^2_\Lambda(C^*, \Lambda) \to \cdots$. Now $B^{**} \to C^{**} \to 0$ is exact since B and C are reflexive. Hence we have isomorphisms $\operatorname{Ext}^i_\Lambda(K, \Lambda) \xrightarrow{\sim} \operatorname{Ext}^i_\Lambda(B^*, \Lambda)$ for $i = 1, \ldots, k-2$. Therefore B is k-torsionfree if and only if $\operatorname{Ext}^i_\Lambda(K, \Lambda) = 0$ for $i = 1, \ldots, k-2$, since a reflexive module B is k-torsionfree for $k \geq 3$ if and only if $\operatorname{Ext}^i_\Lambda(B^*, \Lambda) = 0$ for $i = 1, \ldots, k-2$. ∎

We now prove Theorem 1.1 by induction on k. Clearly we can assume that $k \geq 3$.

(a) $\Rightarrow$ (b) Let $0 \to A \to B \to C \to 0$ be exact with A k-torsionfree, where $k \geq 3$. Since $\operatorname{s.grade} \operatorname{Ext}^1_\Lambda(C, \Lambda) \geq k \geq 3$ we have by Lemma 1.3 that B is reflexive. Therefore to show that B is k-torsionfree, it suffices to show that $\operatorname{Ext}^i_\Lambda(K, \Lambda) = 0$ for $i = 1, \dots, k-2$, where $K = \operatorname{Coker}(C^* \to B^*)$. Since $0 \to K \to A^* \to L \to 0$ is exact with $L \subset \operatorname{Ext}^1_\Lambda(C, \Lambda)$, we know that $\operatorname{Ext}^i_\Lambda(L, \Lambda) = 0$ for $i = 0, \dots, k-1$, where $k - 1 \geq 2$. Hence we have isomorphisms $\operatorname{Ext}^i_\Lambda(K, \Lambda) \xrightarrow{\sim} \operatorname{Ext}^i_\Lambda(A^*, \Lambda)$ for $1 \leq i \leq k-2$. Therefore $\operatorname{Ext}^i_\Lambda(K, \Lambda) = 0$ for $1 \leq i \leq k-2$ since the fact that A is k-torsionfree implies $\operatorname{Ext}^i_\Lambda(A^*, \Lambda) = 0$ for $i = 1, \dots, k-2$ when $k \geq 3$.

(b) $\Rightarrow$ (c) This is trivial.

(c) $\Rightarrow$ (a) Let $L \subset \operatorname{Ext}^1_\Lambda(C, \Lambda)$. Then there is an exact sequence $0 \to \Lambda^n \to E \to C \to 0$ such that $\operatorname{Im}((\Lambda^n)^* \to \operatorname{Ext}^1_\Lambda(C, \Lambda)) = L$. Thus we have exact sequences $0 \to C^* \to E^* \to K \to 0$ and $0 \to K \to (\Lambda^n)^* \to L \to 0$. Since E is k-torsionfree, we know that $\operatorname{Ext}^i_\Lambda(E^*, \Lambda) = 0$ for $i = 1, \dots, k-2$. So by Lemma 1.4 we have that $\operatorname{Ext}^i_\Lambda(K, \Lambda) = 0$ for $i = 1, \dots, k-2$. Hence we have $\operatorname{Ext}^{i+1}_\Lambda(L, \Lambda) \simeq \operatorname{Ext}^i_\Lambda(K, \Lambda) = 0$ for $i = 1, \dots, k-2$ so $\operatorname{Ext}^j_\Lambda(L, \Lambda) = 0$ for $j = 2, \dots, k-1$. But since E is reflexive, we know that $\operatorname{Ext}^i_\Lambda(L, \Lambda) = 0$ for $i = 0, 1$, and so grade $L \geq k$. This finishes the proof of Theorem 1.1.

We will be using Theorem 1.1 in the following form, which is an immediate consequence of the theorem.

COROLLARY 1.5. *The following are equivalent for an integer $k \geq 1$.*

(a) *For each i with $1 \leq i \leq k$, we have that the subcategory of i-torsionfree modules is extension closed.*

(b) *If $1 \leq i \leq k$ and C is i-torsionfree, then* $\operatorname{s.grade} \operatorname{Ext}^1_\Lambda(C, \Lambda) \geq i$.

We now turn our attention to applying our results about the i-torsionfree modules being extension closed to the main question of this paper, when the categories $\Omega^i(\operatorname{mod} \Lambda)$ are extension closed for $1 \leq i \leq k$ for some integer $k \geq 1$. The following is the first result along these lines. We denote by $\mathscr{X}_i = \operatorname{add} \Omega^i(\operatorname{mod} \Lambda)$ the subcategory of mod Λ whose objects are those modules which are summands of ith syzygies.

PROPOSITION 1.6. *Let Λ be a twosided noetherian ring. Then we have the following for an integer $k \geq 1$.*

(a) $\operatorname{grade} \operatorname{Ext}^{i+1}_\Lambda(B, \Lambda) \geq i$ *for $1 \leq i < k$ and all modules B if and only if $\Omega^i(\operatorname{mod} \Lambda)$ coincides with the i-torsionfree modules for all $1 \leq i \leq k$.*

(b) *If* $\Omega^i(\text{mod}\,\Lambda)$ *is extension closed for* $1 \le i < k$, *then* $\Omega^i(\text{mod}\,\Lambda)$ *consists of the i-torsionfree modules for* $1 \le i \le k$, *and hence* $\mathscr{X}_i = \Omega^i(\text{mod}\,\Lambda)$ *for* $1 \le i \le k$.

(c) *If* $\mathscr{X}_i$ *is extension closed for each* $1 \le i \le k$, *then* $\mathscr{X}_i = \Omega^i(\text{mod}\,\Lambda)$ *for each* $1 \le i \le k$.

Proof. (a) This is Proposition 2.26 in [AB].

(b) Proceed by induction on k. Since $\Omega^1(\text{mod}\,\Lambda)$ consists of the 1-torsionfree modules, or equivalently, the torsionless modules, we have our desired result for $k = 1$.

Suppose $k > 1$. Then by assumption each $\Omega^i(\text{mod}\,\Lambda)$ consists of the i-torsionfree modules for $i < k$ and is also extension closed. Therefore the i-torsionfree modules are extension closed for $1 \le i < k$. Hence by Corollary 1.5 we have that $\text{s.grade}\,\text{Ext}^1_\Lambda(C, \Lambda) \ge i$ for $1 \le i < k$ and all i-torsionfree modules C. Since $\Omega^i(\text{mod}\,\Lambda)$ consists of the i-torsionfree modules for $1 \le i < k$, we have that $\text{grade}\,\text{Ext}^{i+1}_\Lambda(B, \Lambda) \ge i$ for $1 \le i < k$ and all Λ-modules B. Then by part (a) we have that $\Omega^k(\text{mod}\,\Lambda)$ consists of the k-torsionfree modules. It is then obvious that $\mathscr{X}_k = \Omega^k(\text{mod}\,\Lambda)$.

(c) Proceed by induction on k. Since $\Omega^1(\text{mod}\,\Lambda) = \mathscr{X}_1$, we have our result for $k = 1$. Let $k > 1$. Then $\mathscr{X}_i = \Omega^i(\text{mod}\,\Lambda)$ is extension closed for $1 \le i < k$ by the induction hypothesis. By part (b) we have that $\Omega^k(\text{mod}\,\Lambda)$ consists of the k-torsionfree modules, and is hence closed under summands. It follows that $\Omega^k(\text{mod}\,\Lambda) = \mathscr{X}_k$. ∎

We now summarize our results so far as follows.

THEOREM 1.7. *Suppose* Λ *is a twosided noetherian ring. Then the following are equivalent for an integer* $k \ge 1$.

(a) $\text{s.grade}\,\text{Ext}^{i+1}_\Lambda(B, \Lambda) \ge i$ *for all* $1 \le i \le k$ *and all* Λ*-modules* B.

(b) $\Omega^i(\text{mod}\,\Lambda)$ *is extension closed for each* $1 \le i \le k$.

(c) $\Omega^i(\text{mod}\,\Lambda)$ *is extension closed and consists of the i-torsionfree modules for* $1 \le i \le k$.

(d) $\mathscr{X}_i$ *is extension closed for each* $1 \le i \le k$.

Proof. (b) ⇔ (c) Clearly (c) implies (b). That (b) implies (c) is Proposition 1.6(b).

(c) ⇔ (d) That (d) implies (c) follows from Proposition 1.6(c). That (c) implies (d) is trivial.

(c) ⇒ (a) Since $\Omega^i(\text{mod}\,\Lambda) = \mathscr{X}_i$ is extension closed, it follows from Corollary 1.5 that $\text{s.grade}\,\text{Ext}^1(C, \Lambda) \ge i$ for all C in $\Omega^i(\text{mod}\,\Lambda)$ with

$1 \leq i \leq k$. Hence s.grade $\operatorname{Ext}^{i+1}(B, \Lambda) \geq i$ for $1 \leq i \leq k$ and for all B in mod Λ.

(a) $\Rightarrow$ (b) By Corollary 1.5 we have that the subcategories of i-torsionfree modules are each extension closed for $1 \leq i \leq k$. But also by Proposition 1.6(a) we have that $\Omega^i(\text{mod}\,\Lambda)$ is the subcategory consisting of the i-torsionfree modules for $1 \leq i \leq k$. Hence $\Omega^i(\text{mod}\,\Lambda)$ is extension closed for each $1 \leq i \leq k$. ∎

2. STRONG GRADE AND FLAT DIMENSION

Let $0 \to \Lambda \to I_0 \to I_1 \to \cdots \to I_i \to \cdots$ be a minimal injective resolution of Λ as a right Λ-module. In this section we describe the property that s.grade $\operatorname{Ext}_\Lambda^{i+1}(C, \Lambda) \geq i$ for all C in mod Λ and $i \leq k$ in terms of the flat dimension of the first k injective modules in this resolution. The main idea is to express flat dimension in terms of maps from modules of the form $\operatorname{Ext}_\Lambda^i(C, \Lambda)$ to injective modules.

LEMMA 2.1. *Let I be an injective Λ^{op}-module and n a positive integer. Then* flatdim $I < n$ *if and only if* $\operatorname{Hom}_\Lambda(\operatorname{Ext}_\Lambda^n(C, \Lambda), I) = 0$ *for all C in* mod Λ.

Proof. From [AB, Th. 2.8] we have the exact sequence $0 \to \operatorname{Ext}_\Lambda^1(\operatorname{Tr}\Omega^n C, X) \to \operatorname{Tor}_n(X, C) \to \operatorname{Hom}_\Lambda(\operatorname{Ext}_\Lambda^n(C, \Lambda), X) \to \operatorname{Ext}_\Lambda^2(\operatorname{Tr}\Omega^n C, X)$ for C in mod Λ and X in mod Λ^{op}.

Letting $X = I$ we get $\operatorname{Tor}_n(I, C) \simeq \operatorname{Hom}_\Lambda(\operatorname{Ext}_\Lambda^n(C, \Lambda), I)$ for all C in mod Λ. This also follows from [CE, p. 120]. Then we use that flatdim $I < n$ if and only if $\operatorname{Tor}_n(I, C) = 0$ for all C in mod Λ. ∎

We now give the main result of this section.

PROPOSITION 2.2. *Let $0 \to \Lambda \to I_0 \to I_1 \to \cdots \to I_i \to \cdots$ be a minimal injective resolution of Λ as a right Λ-module and let k be a positive integer. Then* flatdim $I_i \leq i + 1$ *for $i < k$ if and only if* s.grade $\operatorname{Ext}_\Lambda^{i+1}(C, \Lambda) \geq i$ *for all C in* mod Λ *and $i \leq k$.*

Proof. We first treat the case $k = 1$. So if $I = I_0$ we want to show that flatdim $I \leq 1$ if and only if s.grade $\operatorname{Ext}_\Lambda^2(C, \Lambda) \geq 1$ for all C in mod Λ. We have by Lemma 2.1 that flatdim $I \leq 1$ if and only if $\operatorname{Hom}_\Lambda(\operatorname{Ext}_\Lambda^2(C, \Lambda), I) = 0$ for all C in mod Λ. Now $f\colon \operatorname{Ext}_\Lambda^2(C, \Lambda) \to I$ is not zero if and only if $\operatorname{Im} f \cap \Lambda \neq 0$, which is the case if and only if there is a submodule A of $\operatorname{Ext}_\Lambda^2(C, \Lambda)$ such that $\operatorname{Hom}_\Lambda(A, \Lambda) \neq 0$. Hence we are done.

Assume now that $k > 1$, and that s.grade $\operatorname{Ext}_\Lambda^{i+1}(C, \Lambda) \geq i$ for $i \leq k$ and all C in mod Λ. By the induction assumption we have flatdim $I_i \leq i + 1$ for $i < k - 1$. To show flatdim $I_{k-1} \leq k$, it suffices to show that

$\operatorname{Hom}_\Lambda(\operatorname{Ext}_\Lambda^{k+1}(C, \Lambda), I_{k-1}) = 0$. Consider the short exact sequences $0 \to K_{k-2} \to I_{k-2} \to K_{k-1} \to 0$ obtained from the injective resolution of Λ. Then $\operatorname{Hom}_\Lambda(\operatorname{Ext}_\Lambda^{k+1}(C, \Lambda), I_{k-1}) \neq 0$ if and only if there is some submodule A of $\operatorname{Ext}_\Lambda^{k+1}(C, \Lambda)$ with $\operatorname{Hom}_\Lambda(A, K_{k-1}) \neq 0$.

Consider for $A \subset \operatorname{Ext}_\Lambda^{k+1}(C, \Lambda)$ the exact sequence

$$0 \to \operatorname{Hom}_\Lambda(A, K_{k-2}) \to \operatorname{Hom}_\Lambda(A, I_{k-2}) \to \operatorname{Hom}_\Lambda(A, K_{k-1}) \to \operatorname{Ext}_\Lambda^1(A, K_{k-2}) \to 0. \qquad (*)$$

We have $\operatorname{Hom}_\Lambda(A, I_{k-2}) = 0$, since flatdim $I_{k-2} \leq k - 1$, and $\operatorname{Ext}_\Lambda^1(A, K_{k-2}) \simeq \operatorname{Ext}_\Lambda^{k-1}(A, \Lambda) = 0$ since grade $A \geq k$ by assumption. Hence we get $\operatorname{Hom}_\Lambda(A, K_{k-1}) = 0$, which shows $\operatorname{Hom}_\Lambda(\operatorname{Ext}_\Lambda^{k+1}(C, \Lambda), I_{k-1}) = 0$.

Assume now that flatdim $I_i \leq i + 1$ for $i < k$. We have s.grade $\operatorname{Ext}_\Lambda^{k+1}(C, \Lambda) \geq k - 1$ by the induction assumption. Consider the sequence $(*)$, where $\operatorname{Hom}_\Lambda(A, K_{k-1}) = 0$ since $\operatorname{Hom}_\Lambda(\operatorname{Ext}_\Lambda^{k+1}(C, \Lambda), I_{k-1}) = 0$. Hence $\operatorname{Ext}_\Lambda^1(A, K_{k-2}) \simeq \operatorname{Ext}_\Lambda^{k-1}(A, \Lambda) = 0$, so that grade $A \geq k$, and we are done. ∎

3. ADJOINTNESS

Let Λ be a twosided noetherian ring. For each C in mod Λ we choose a fixed epimorphism $f: P_C \to C$, where P_C is projective. We define $\Omega^1(C)$ to be Ker f, and we define $\Omega^n(C)$ for each $n \geq 1$ by induction. Let $\underline{\operatorname{mod}}\,\Lambda$ be the category mod Λ modulo projectives. There are induced functors $G = \Omega^n$: $\underline{\operatorname{mod}}\,\Lambda \to \underline{\operatorname{mod}}\,\Lambda$ and $F = \operatorname{Tr}\Omega^n\operatorname{Tr}$: $\underline{\operatorname{mod}}\,\Lambda \to \underline{\operatorname{mod}}\,\Lambda$ for each $n \geq 1$, and we know from [AR2] that they are adjoint functors. In this section we shall need a more explicit description of why Ω^k and $\operatorname{Tr}\Omega^k\operatorname{Tr}$ are adjoint functors. This is obtained through studying maps between complexes. We give some applications, including establishing an exact sequence which we shall need later.

Let J be the integers in an interval $\langle -\infty, \infty\rangle, [i, \infty\rangle, \langle -\infty, i]$ or $[i, j]$, where $i < j$ are integers. Let

$$P^{\cdot} \qquad \cdots \to P_i \xrightarrow{p_i} P_{i-1} \xrightarrow{p_{i-1}} P_{i-2} \to \cdots$$

and

$$Q^{\cdot} \qquad \cdots \to Q_i \xrightarrow{q_i} Q_{i-1} \xrightarrow{q_{i-1}} Q_{i-2} \to \cdots$$

be complexes of projective modules, where $i \in J$. If $I \subset J$ and I is the integers in an interval, we denote by $P_I^{\cdot}$ and $Q_I^{\cdot}$ the corresponding restrictions of the complexes, and for a map $f\colon P^{\cdot} \to Q^{\cdot}$ of complexes we have the restriction map $f_I\colon P_I^{\cdot} \to Q_I^{\cdot}$. For $i \in J$ let $I = J_{\leq i}$ be the integers $j \in J$ with $j \leq i$ and let $I = J_{\geq i}$ be the integers $j \in J$ with $j \geq i$. Consider the subgroup $U(P^{\cdot}, Q^{\cdot})$ of $\operatorname{Hom}(P^{\cdot}, Q^{\cdot})$ consisting of the maps $f\colon P^{\cdot} \to Q^{\cdot}$ for which there are maps $h_i\colon P_{i-1} \to Q_i$ such that $q_i h_i p_i = f_{i-1} p_i$ for $\{i-1, i\} \subset J$. We then have the following easily verified result.

LEMMA 3.1. *Let $P^{\cdot}$ and $Q^{\cdot}$ be complexes of projective modules on some J as above.*

(a) *Assume that $Q^{\cdot}$ is exact, and let $I = J_{\geq i}$ for some $i \in J$.*
  (i) *Given $g\colon P_I^{\cdot} \to Q_I^{\cdot}$ there is some $\tilde{g}\colon P^{\cdot} \to Q^{\cdot}$ with $\tilde{g}_I = g$.*
  (ii) *If $g \in U(P_I^{\cdot}, Q_I^{\cdot})$, then $\tilde{g} \in U(P^{\cdot}, Q^{\cdot})$.*

(b) *Assume that $P^{\cdot *}$ is exact and let $I = J_{\leq i}$ for some integer i.*
  (i) *Given $g\colon P_I^{\cdot} \to Q_I^{\cdot}$ there is some $\tilde{g}\colon P^{\cdot} \to Q^{\cdot}$ with $\tilde{g}_I = g$.*
  (ii) *If $g \in U(P_I^{\cdot}, Q_I^{\cdot})$, then $\tilde{g} \in U(P^{\cdot}, Q^{\cdot})$.*

Writing $\underset{\sim}{\operatorname{Hom}}(P^{\cdot}, Q^{\cdot}) = \operatorname{Hom}(P^{\cdot}, Q^{\cdot})/U(P^{\cdot}, Q^{\cdot})$, we have the following direct consequence.

PROPOSITION 3.2. *Let $P^{\cdot}$ and $Q^{\cdot}$ be complexes of projective modules on an interval J, where $P^{\cdot *}$ and $Q^{\cdot}$ are exact, and let $I = J_{\geq i} \cap J_{\leq j}$, where i and j are in J and $i < j$. Then the natural map $\underset{\sim}{\operatorname{Hom}}(P^{\cdot}, Q^{\cdot}) \to \underset{\sim}{\operatorname{Hom}}(P_I^{\cdot}, Q_I^{\cdot})$ is an isomorphism.*

Note that if $P^{\cdot}$ is the complex $P_i \xrightarrow{p_i} P_{i-1}$ and $Q^{\cdot}$ the complex $Q_i \xrightarrow{q_i} Q_{i-1}$, then it is not hard to see that we have a natural isomorphism $\underset{\sim}{\operatorname{Hom}}(P^{\cdot}, Q^{\cdot}) \xrightarrow{\sim} \underline{\operatorname{Hom}}(\operatorname{Coker} p_i, \operatorname{Coker} q_i)$. Hence we get the following consequence.

COROLLARY 3.3. *Let $P^{\cdot}$ and $Q^{\cdot}$ be complexes of projective modules indexed by some J, where $P^{\cdot *}$ and $Q^{\cdot}$ are exact. For each integer i with $\{i-1, i\} \subset J$ we have a natural isomorphism $\underset{\sim}{\operatorname{Hom}}(P^{\cdot}, Q^{\cdot}) \to \underline{\operatorname{Hom}}(\operatorname{Coker} p_i, \operatorname{Coker} q_i)$, and hence a natural isomorphism $\underline{\operatorname{Hom}}(\operatorname{Coker} p_i, \operatorname{Coker} q_i) \to \underline{\operatorname{Hom}}(\operatorname{Coker} p_j, \operatorname{Coker} q_j)$ when $\{j-1, j\} \subset J$.*

Now let M and N be in $\operatorname{mod} \Lambda$, and consider the projective resolutions

$$\cdots \to Q_2 \xrightarrow{q_2} Q_1 \xrightarrow{q_1} Q_0 \to N \to 0$$

and

$$\cdots \to L_2 \xrightarrow{l_2} L_1 \xrightarrow{l_1} L_0 \to \operatorname{Tr} M \to 0.$$

Denote by $Q^{\cdot}$ and $L^{\cdot}$ the corresponding complexes of projective modules, indexed by the set J of nonnegative integers. For the dual complex $L^{\cdot *}$

and a given positive integer k, let $P^{\cdot}$ be the shifted complex where $P_i = (L_{k+1-i})^*$. Let $I = J_{\leq k+1}$. Then $P_I^{\cdot}$ and $Q_I^{\cdot}$ are projective complexes on I such that $P_I^{\cdot *}$ and $Q_I^{\cdot}$ are exact. Hence we have the complexes

$$\begin{array}{ccccccccc} L_0^* & \xrightarrow{q_1^* = p_{k+1}} & L_1^* & \to & \cdots & \to & L_k^* & \to & L_{k+1}^* \\ \| & & \| & & & & \| & & \| \\ P_{k+1} & & P_k & & & & P_1 & & P_0 \end{array}$$

and

$$Q_{k+1} \to Q_k \to \cdots \to Q_1 \to Q_0.$$

Then we have the following consequence of Corollary 3.3.

COROLLARY 3.4. *Let M and N be in* $\operatorname{mod}\Lambda$. *Then we have natural isomorphisms*

$$\begin{aligned} \underline{\operatorname{Hom}}(M, \Omega^k N) &\tilde{\to} \underline{\operatorname{Hom}}(\operatorname{Tr}\Omega^1 \operatorname{Tr} M, \Omega^{k-1}N) \\ &\tilde{\to} \cdots \tilde{\to} \underline{\operatorname{Hom}}(\operatorname{Tr}\Omega^k \operatorname{Tr} M, N). \end{aligned}$$

In particular the induced isomorphism $\underline{\operatorname{Hom}}(M, \Omega^k N) \tilde{\to} \underline{\operatorname{Hom}}(\operatorname{Tr}\Omega^k \operatorname{Tr} M, N)$ *gives an adjointness isomorphism between the functors* Ω^k *and* $\operatorname{Tr}\Omega^k\operatorname{Tr}$ *on* $\underline{\operatorname{mod}}\,\Lambda$.

Write, as in [AB], $J_k(C) = \operatorname{Tr}\Omega^k C$ and $D_k(C) = \Omega^k \operatorname{Tr} C$. Letting $M = \Omega^k N$ in Corollary 3.4 we obtain associated with the identity on $\Omega^k N$ a map $t: J_k^2(N) \to N$ in $\underline{\operatorname{mod}}\,\Lambda$, which we know is (and is easily seen to be) a right $J_k(\operatorname{mod}\Lambda^{\mathrm{op}})$-approximation [AR3]. Similarly we get $s: M \to D_k^2(M)$ in $\underline{\operatorname{mod}}\,\Lambda$ by setting $N = \operatorname{Tr}\Omega^k \operatorname{Tr} M$. Note that it follows from [AB, Appendix] that the maps t and s coincide with those defined in [AB].

We give the following application of these considerations, which we shall need in the next section. We denote by $\alpha_{M,N}$ the adjointness isomorphism $\underline{\operatorname{Hom}}(M, \Omega^k N) \tilde{\to} \underline{\operatorname{Hom}}(\operatorname{Tr}\Omega^k \operatorname{Tr} M, N)$.

PROPOSITION 3.5. *Let M and N be in* $\operatorname{mod}\Lambda$. *Then we have the commutative diagram*

$$\begin{array}{ccc} \underline{\operatorname{Hom}}(M, \Omega^k N) & \xrightarrow[\sim]{\alpha_{M,N}} & \underline{\operatorname{Hom}}(\operatorname{Tr}\Omega^k \operatorname{Tr} M, N) \\ \downarrow \wr\, \operatorname{Tr} & & \downarrow \wr\, \operatorname{Tr} \\ \underline{\operatorname{Hom}}(\operatorname{Tr}\Omega^k N, \operatorname{Tr} M) & \xrightarrow[\sim]{(\alpha_{\operatorname{Tr} N, \operatorname{Tr} M})^{-1}} & \underline{\operatorname{Hom}}(\operatorname{Tr} N, \Omega^k \operatorname{Tr} M) \end{array}$$

Proof. Consider the complexes $P_I^{\cdot}$ and $Q_I^{\cdot}$ for I the integers in $[0, k+1]$ as above. Then we have a natural isomorphism $\operatorname{Hom}(P_I^{\cdot},$

$Q_I^{\cdot}) \xrightarrow{\sim} \operatorname{Hom}(Q_I^{\cdot *}, P_I^{\cdot *})$, which induces a natural isomorphism $\underline{\operatorname{Hom}}(P_I^{\cdot}, Q_I^{\cdot}) \xrightarrow{\sim} \underline{\operatorname{Hom}}(Q_I^{\cdot *}, P_I^{\cdot *})$. Using Corollary 3.3 we obtain our desired result. ∎

Consider again the adjointness morphism $s\colon M \to D_k^2(M) = \Omega^k \operatorname{Tr} \Omega^k \operatorname{Tr}(M)$ in $\overline{\operatorname{mod} \Lambda}$. When $n = 2$ we have $\Omega^2 \operatorname{Tr} M \simeq M^*$ in mod Λ, and it follows from [AB, p. 143] that if $f\colon M \to M^{**}$ is the natural map, then $\underline{f} = s$. In this case we have an exact sequence $0 \to \operatorname{Ext}_\Lambda^1(\operatorname{Tr} M, \Lambda) \to M \to^f M^{**} \to \operatorname{Ext}_\Lambda^2(\operatorname{Tr} M, \Lambda) \to 0$, and in particular, if M is torsionless, an exact sequence $0 \to M \to^f M^{**} \to \operatorname{Ext}_\Lambda^2(\operatorname{Tr} M, \Lambda) \to 0$. The following generalization of this will be useful.

PROPOSITION 3.6. *If $k \geq 2$ and M in* mod Λ *is $(k-1)$-torsionfree, we have an exact sequence* $0 \to M \to^f \Omega^k \operatorname{Tr} \Omega^k \operatorname{Tr} M \amalg Q \to \Omega^{k-2} \operatorname{Ext}_\Lambda^k(\operatorname{Tr} M, \Lambda) \to 0$, *where Q is projective and $\underline{f}\colon M \to \Omega^k \operatorname{Tr} \Omega^k \operatorname{Tr} M$ is the adjointness morphism.*

Proof. We already know the result for $k = 2$, so we can assume $k \geq 3$. Let M in mod Λ be k-torsionfree and let $N = \operatorname{Tr} \Omega^k \operatorname{Tr} M$. Consider the projective resolutions

$$\cdots \to Q_1 \to Q_0 \to \operatorname{Tr} \Omega^k \operatorname{Tr} M \to 0$$

and

$$\cdots \to L_1 \to L_0 \to \operatorname{Tr} M \to 0.$$

Then we have the commutative diagram

$$\begin{array}{ccccccccccccc}
L_0^* & \longrightarrow & L_1^* & \longrightarrow & L_2^* & \longrightarrow & \cdots & \longrightarrow & L_k^* & \to & L_{k+1}^* & \longrightarrow & \operatorname{Tr} \Omega^k \operatorname{Tr} M \longrightarrow 0 \\
\downarrow & & \downarrow & & \downarrow & & & & \downarrow\wr & & \downarrow\wr & & \downarrow\| \\
Q_{k+1} & \longrightarrow & Q_k & \longrightarrow & Q_{k-1} & \longrightarrow & \cdots & \longrightarrow & Q_1 & \longrightarrow & Q_0 & \longrightarrow & \operatorname{Tr} \Omega^k \operatorname{Tr} M \longrightarrow 0,
\end{array}$$

where we can assume that $L_k^* \to Q_1$ and $L_{k+1}^* \to Q_0$ are isomorphisms.

Since $\operatorname{Ext}_\Lambda^i(\operatorname{Tr} M, \Lambda) = 0$ for $i = 1, \ldots, k-1$, the top sequence is exact except possibly at L_k^*. Hence we get an exact commutative diagram

$$\begin{array}{ccccccccccccc}
0 \longrightarrow & M & \longrightarrow & L_2^* & \longrightarrow & L_3^* & \longrightarrow & \cdots & \longrightarrow & L_{k-1}^* & \longrightarrow & X & \longrightarrow 0 \\
& \downarrow f & & \downarrow & & \downarrow & & & & \downarrow & & \downarrow s & \\
0 \longrightarrow & \Omega^k \operatorname{Tr} \Omega^k \operatorname{Tr} M & \longrightarrow & Q_{k-1} & \longrightarrow & Q_{k-2} & \longrightarrow & \cdots & \longrightarrow & Q_2 & \longrightarrow & Y & \longrightarrow 0
\end{array}$$

and an exact sequence $0 \to X \to^s Y \to \operatorname{Ext}_\Lambda^k(\operatorname{Tr} M, \Lambda) \to 0$. Using the mapping cone we get an exact sequence $0 \to M \to^{(f, f')} \Omega^k \operatorname{Tr} \Omega^k \operatorname{Tr} M \amalg L_2^* \to Q_{k-1} \amalg L_3^* \to \cdots \to Q_3 \amalg P_{n-1}^* \to Q_2 \to \operatorname{Ext}_\Lambda^k(\operatorname{Tr} M,$

$\Lambda) \to 0$ and consequently an exact sequence $0 \to M \to^{(f, f')} \Omega^k \operatorname{Tr} \Omega^k \operatorname{Tr} M \amalg Q \to \Omega^{k-2} \operatorname{Ext}_\Lambda^k(\operatorname{Tr} M, \Lambda) \to 0$ with Q projective. ∎

We shall show in the next section that $\underline{f}\colon M \to D_k^2(M)$ in $\underline{\operatorname{mod}} \Lambda$ is sometimes left minimal, that is, if

$$\begin{array}{ccc} M & \xrightarrow{\underline{f}} & D_k^2(M) \\ & \searrow^{\underline{f}} & \downarrow{\scriptstyle \underline{g}} \\ & & D_k^2(M) \end{array}$$

is commutative, then $\underline{g}\colon D_k^2(M) \to D_k^2(M)$ is an isomorphism.

4. GRADE AND EXTENSION CLOSURE

The aim of this section is to show that if Λ is a noetherian R-algebra, that is, R is commutative noetherian and Λ is a finitely generated R-module, then $\Omega^i(\operatorname{mod} \Lambda)$ is closed under extensions for $i \leq k$ if and only if grade $\operatorname{Ext}_\Lambda^i(B, \Lambda) \geq i$ for all B in mod Λ^{op} and $i \leq k$.

We first give a part of the result which is valid for arbitrary twosided noetherian rings. We start out with the following preliminary result.

LEMMA 4.1. *Assume that grade $Z \geq k$ for some Z in* mod Λ *and some integer $k \geq 1$. Then we have $\operatorname{Ext}_\Lambda^{k-1}(Z, \mathscr{X}_k) = 0$, that is, $\operatorname{Ext}_\Lambda^{k-1}(Z, X) = 0$ for all X in $\mathscr{X}_k =$ add $\Omega^k(\operatorname{mod} \Lambda)$.*

Proof. If $k = 1$, then $Z^* = 0$ clearly implies $\operatorname{Hom}_\Lambda(Z, \mathscr{X}_1) = 0$. For $k > 1$ consider for C in mod Λ the exact sequence $0 \to \Omega^k C \to P \to \Omega^{k-1} C \to 0$ with P projective and the induced exact sequence $\operatorname{Ext}_\Lambda^{k-2}(Z, \Omega^{k-1} C) \to \operatorname{Ext}_\Lambda^{k-1}(Z, \Omega^k C) \to 0$. Since grade $Z \geq k - 1$, we have $\operatorname{Ext}_\Lambda^{k-2}(Z, \Omega^{k-1} C) = 0$ by the induction assumption, and consequently $\operatorname{Ext}_\Lambda^{k-1}(Z, \Omega^k C) = 0$. ∎

PROPOSITION 4.2. *Let Λ be a twosided noetherian ring.*

(a) $\Omega^1(\operatorname{mod} \Lambda)$ *is closed under extensions if and only if* $\operatorname{Ext}_\Lambda^1(B, \Lambda)^* = 0$ *for all B in* mod Λ^{op}.

(b) *Let $k \geq 1$ be an integer. If* grade $\operatorname{Ext}_\Lambda^i(C, \Lambda) \geq i$ *for all C in* mod Λ^{op} *and $i \leq k$, then* $\Omega^i(\operatorname{mod} \Lambda)$ *is closed under extensions for $i \leq k$.*

Proof. (a) Assume first that $\mathscr{X}_1 = \Omega^1(\operatorname{mod} \Lambda)$ is closed under extensions, and consider for C in mod Λ the exact sequence $0 \to t(C) \to C \to C/t(C) \to 0$, where $t(C)$ is given by the exact sequence $0 \to t(C) \to C \to$

C^{**}. We know that $t(C) \simeq \operatorname{Ext}^1_\Lambda(\operatorname{Tr} C, \Lambda)$ (see [AB, Prop. 2.6]). Assume that $t(C)^* \neq 0$. Then we have $t(t(C)) \neq t(C)$, and hence $V = t(C)/t(t(C)) \neq 0$. Consider the pushout diagram

$$\begin{array}{ccccccccc}
0 & \longrightarrow & t(C) & \xrightarrow{f} & C & \xrightarrow{g} & C/t(C) & \longrightarrow & 0 \\
 & & \downarrow{\scriptstyle p} & & \downarrow{\scriptstyle q} & \swarrow{\scriptstyle h} & \| & & \\
0 & \longrightarrow & V & \xrightarrow{s} & W & \longrightarrow & C/t(C) & \longrightarrow & 0. \\
 & & \downarrow & & \downarrow & & & & \\
 & & 0 & & 0 & & & &
\end{array}$$

Since V and $C/t(C)$ are in $\mathscr{X}_1$ and $\mathscr{X}_1$ is closed under extensions, we have that W is in $\mathscr{X}_1$. Then there is some $h: C/t(C) \to W$ such that $hg = q$. But then $sp(t(C)) = qf(t(C)) = hgf(t(C)) = 0$, so that $s(V) \simeq V$ is 0, which is a contradiction. Hence $t(C)^* = 0$, and consequently $\operatorname{grade} \operatorname{Ext}^1_\Lambda(\operatorname{Tr} C, \Lambda) \geq 1$ for all C in mod Λ. It follows that $\operatorname{grade} \operatorname{Ext}^1_\Lambda(B, \Lambda) \geq 1$ for all B in mod Λ^{op}.

Assume now that $t(C)^* = 0$ for all C in mod Λ, and let C be in the extension closure of $\mathscr{X}_1$. Then we have a chain of submodules $0 = C_n \subset C_{n-1} \subset \cdots \subset C_1 \subset C_0 = C$ such that C_i/C_{i+1} is in $\mathscr{X}_1$ for $i = 0, \ldots, n-1$. Let X be a nonzero submodule of C and let i be largest possible with $X \subset C_i$. Then there is a nonzero map $X \to C_i/C_{i+1}$, so that X^* is not zero. Hence we conclude that $t(C)$ is zero, so that C is in $\mathscr{X}_1$.

(b) For $k = 1$ the claim follows from (a). Assume that $k > 1$ and that the claim has been proved for $k - 1$. Assume that $\operatorname{grade} \operatorname{Ext}^i_\Lambda(C, \Lambda) \geq i$ for all C in mod Λ^{op} and $i \leq k$. By Theorem 1.7 it is sufficient to show that $\mathscr{X}_i$ is closed under extensions for $1 \leq i \leq k$.

Consider now the exact sequence $0 \to L \to M \to N \to 0$, where L and N are in $\mathscr{X}_k$. Since $\mathscr{X}_{k-1}$ is closed under extensions, M is in $\mathscr{X}_{k-1}$ and is $(k-1)$-torsionfree by Proposition 1.6. Then we have an exact sequence

$$0 \to M \overset{(f,f')}{\to} GF(M) \amalg Q \to \Omega^{k-2} \operatorname{Ext}^k_\Lambda(\operatorname{Tr} M, \Lambda) \to 0 \qquad (*)$$

by Proposition 3.6, where $G = \Omega^k$ and $F = \operatorname{Tr} \Omega^k \operatorname{Tr}$. Since $\operatorname{grade} \operatorname{Ext}^k_\Lambda(\operatorname{Tr} M, \Lambda) \geq k$, we have $\operatorname{Ext}^{k-1}_\Lambda(\operatorname{Ext}^k_\Lambda(\operatorname{Tr} M, \Lambda), \mathscr{X}_k) = 0$ by Lemma 4.1, and hence $\operatorname{Ext}^1_\Lambda(\Omega^{k-2} \operatorname{Ext}^k_\Lambda(\operatorname{Tr} M, \Lambda), \mathscr{X}_k) = 0$. It then follows that $\operatorname{Ext}^1_\Lambda(\Omega^{k-2} \operatorname{Ext}^k_\Lambda(\operatorname{Tr} M, \Lambda), M) = 0$, so that $(*)$ splits. Since $GF(M) \amalg Q$ is in $\mathscr{X}_k$, we conclude that M is in $\mathscr{X}_k$. This completes the proof. ∎

In order to prove the converse of Proposition 4.2(b) the following result will be useful for the induction step.

LEMMA 4.3. *Let* Λ *be a twosided noetherian ring and assume that* $\Omega^i(\text{mod}\,\Lambda)$ *is closed under extensions for* $i \leq k$ *and that* $\operatorname{grade}\operatorname{Ext}^i(C, \Lambda) \geq i$ *for each* C *in* $\text{mod}\,\Lambda^{\text{op}}$ *and* $i < k$. *Then we have the following.*

(a) *The modules in* $\Omega^i(\text{mod}\,\Lambda)$ *and* $\Omega^i(\text{mod}\,\Lambda^{\text{op}})$ *are* i*-torsionfree for* $i \leq k$.

(b) *If* M *in* $\text{mod}\,\Lambda$ *is* $(k-1)$*-torsionfree, then we have the following.*

(i) *The adjointness morphism* t: $J_k^2(\operatorname{Tr} M) \to \operatorname{Tr} M$ *in* $\underline{\text{mod}\,\Lambda^{\text{op}}}$ *is right minimal.*

(ii) *The adjointness morphism* s: $M \to D_k^2(M)$ *in* $\underline{\text{mod}\,\Lambda}$ *is left minimal.*

Proof. (a) $\Omega^i(\text{mod}\,\Lambda)$ is extension closed for $i \leq k$ by assumption, and $\Omega^i(\text{mod}\,\Lambda^{\text{op}})$ is extension closed for $i \leq k$ by Proposition 1.6 since $\operatorname{grade}\operatorname{Ext}^i(C, \Lambda) \geq i$ for each C in $\text{mod}\,\Lambda^{\text{op}}$ and $i < k$. It then follows from Proposition 1.6 that the modules in $\Omega^i(\text{mod}\,\Lambda)$ and $\Omega^i(\text{mod}\,\Lambda^{\text{op}})$ are i-torsionfree.

(b)(i) Let M in $\text{mod}\,\Lambda$ be k-torsionfree and let $N = \operatorname{Tr} M$. Since by (a), $\Omega^k N$ is k-torsionfree in $\text{mod}\,\Lambda^{\text{op}}$, it follows from [AB, Prop. 2.21] that the map $g = t$: $J_k^2(N) \to N$ has the property that the induced map $\Omega^k(\underline{g})$: $\Omega^k J_k^2(N) \to \Omega^k N$ is an isomorphism.

Consider then the commutative diagram

$$\begin{array}{ccc} J_k^2(N) & \xrightarrow{\;\;g\;\;} & N \\ \Big\downarrow{\scriptstyle \underline{h}} & \nearrow{\scriptstyle \underline{g}} & \\ J_k^2(N) & & \end{array}$$

which induces the commutative diagram

$$\begin{array}{ccc} \Omega^k J_k^2(N) & \xrightarrow{\;\;\sim\;\;} & \Omega^k N \\ \Big\downarrow{\scriptstyle \Omega^k(\underline{h})} & \nearrow{\scriptstyle \sim} & \\ \Omega^k J_k^2(N) & & \end{array}$$

Then $\Omega^k(\underline{h})$ is an isomorphism. Since $\Omega^k \operatorname{Tr} \Omega^k N$ is in $\Omega^k(\text{mod}\,\Lambda)$ and is hence k-torsionfree by (a), we have that $\operatorname{Ext}^i_\Lambda(J_k^2(N), \Lambda) = 0$ for $i = 1, \ldots, k$. Then it follows that the map $\underline{\operatorname{End}}(J_k^2(N)) \to \underline{\operatorname{End}}(\Omega^k J_k^2(N))$ is an isomorphism, and consequently $\underline{h}\colon J_k^2(N) \to J_k^2(N)$ is an isomorphism. This shows that $\underline{g}\colon J_k^2(N) \to N$ is right minimal.

(ii) Let $f\colon M \to D_k^2(M)$ be the natural map. Then it follows that $\operatorname{Tr} \underline{f}\colon \operatorname{Tr} D_k^2(M) \to \operatorname{Tr} M$ is isomorphic to $\underline{g}\colon J_k^2(\operatorname{Tr} M) \to \operatorname{Tr} M$, and hence $\underline{f}$ is left minimal. ∎

For artin algebras Λ there is the following lemma by Wakamatsu (see [AR1]). Let $\mathscr{B}$ be a covariantly finite subcategory of $\operatorname{mod} \Lambda$ closed under extensions and let $0 \to A \to^f B \to^g C \to 0$ be an exact sequence such that $f\colon A \to B$ is a minimal left $\mathscr{B}$-approximation. Then we have $\operatorname{Ext}_\Lambda^1(C, B) = 0$. We shall need an analogue for noetherian R-algebras, and for this the following lemma is useful.

LEMMA 4.4. *Let Λ be a noetherian R-algebra and let*

$$\begin{array}{ccccccccc}
0 & \longrightarrow & A & \xrightarrow{f} & B & \xrightarrow{g} & C & \longrightarrow & 0 \\
 & & \Big\downarrow \| & & \Big\downarrow u & & \Big\downarrow v & & \\
0 & \longrightarrow & A & \xrightarrow{f} & B & \xrightarrow{g} & C & \longrightarrow & 0
\end{array}$$

be a commutative diagram in $\operatorname{mod} \Lambda$, *and assume that $\underline{u}\colon B \to B$ is an isomorphism in* $\underline{\operatorname{mod}}\,\Lambda$. *Then $\underline{v}\colon C \to C$ is an isomorphism in* $\underline{\operatorname{mod}}\,\Lambda$.

Proof. Consider the exact commutative diagram

$$\begin{array}{ccccccccccc}
(C,\ ) & \longrightarrow & (B,\ ) & \longrightarrow & (A,\ ) & \xrightarrow{\alpha} & \operatorname{Ext}_\Lambda^1(C,\ ) & \longrightarrow & \operatorname{Ext}_\Lambda^1(B,\ ) & \longrightarrow & \operatorname{Ext}_\Lambda^1(A,\ ) \\
\Big\downarrow (v,\ ) & & \Big\downarrow (u,\ ) & & \Big\downarrow \| & & \Big\downarrow \operatorname{Ext}^1(v,\ ) & & \Big\downarrow \operatorname{Ext}^1(u,\ ) & & \Big\downarrow \| \\
(C,\ ) & \longrightarrow & (B,\ ) & \longrightarrow & (A,\ ) & \xrightarrow{\alpha} & \operatorname{Ext}_\Lambda^1(C,\ ) & \longrightarrow & \operatorname{Ext}_\Lambda^1(B,\ ) & \longrightarrow & \operatorname{Ext}_\Lambda^1(A,\ )
\end{array}$$

Since $\underline{u}\colon B \to B$ is assumed to be an isomorphism, it follows that $\operatorname{Ext}^1(u,\ )\colon \operatorname{Ext}_\Lambda^1(B,\ ) \to \operatorname{Ext}_\Lambda^1(B,\ )$ is an isomorphism. Then diagram chasing shows that $\operatorname{Ext}^1(v,\ )\colon \operatorname{Ext}_\Lambda^1(C,\ ) \to \operatorname{Ext}_\Lambda^1(C,\ )$ is an epimorphism. Since $\operatorname{End}_\Lambda(B)$ is noetherian, it follows that $\operatorname{Ext}^1(v,\ )$ is an isomorphism. By [HR, Theorem 2.4] there is a split exact sequence $0 \to K \to P \amalg C \to^{(s, v)} C \to 0$, where P is projective such that $0 \to \operatorname{Ext}_\Lambda^1(C,\ ) \to^{\operatorname{Ext}^1(v,\ )} \operatorname{Ext}_\Lambda^1(C,\ ) \to \operatorname{Ext}_\Lambda^1(K,\ ) \to 0$ is exact. Then $\operatorname{Ext}_\Lambda^1(K,\ ) = 0$ so that K is projective, and hence $\underline{v}\colon C \to C$ is an isomorphism in $\underline{\operatorname{mod}}\,\Lambda$. ∎

Now we can prove the announced analogue of Wakamatsu's lemma.

LEMMA 4.5. *Let $\mathscr{C}$ be a covariantly finite subcategory of* $\operatorname{mod} \Lambda$ *closed under extensions. Assume that we have an exact sequence $0 \to M \to^f C_M \to Y_M \to 0$, where $f\colon M \to C_M$ is a left $\mathscr{C}$-approximation such that $\underline{f}\colon M \to C_M$ is left minimal in* $\underline{\operatorname{mod}}\,\Lambda$. *Then we have* $\operatorname{Ext}_\Lambda^1(Y_M, \mathscr{C}) = 0$.

Proof. Assume that we have an exact sequence $0 \to C \to U \to Y_M \to 0$ with C in $\mathscr{C}$. Consider the pullback diagram

$$
\begin{array}{ccccccccc}
 & & & & 0 & & 0 & & \\
 & & & & \downarrow & & \downarrow & & \\
 & & & & C & = & C & & \\
 & & & & \downarrow & & \downarrow & & \\
0 & \longrightarrow & M & \longrightarrow & V & \longrightarrow & U & \longrightarrow & 0 \\
 & & \| & & \downarrow & & \downarrow & & \\
0 & \longrightarrow & M & \longrightarrow & C_M & \longrightarrow & Y_M & \longrightarrow & 0 \\
 & & & & \downarrow & & \downarrow & & \\
 & & & & 0 & & 0 & &
\end{array}
$$

Since $\mathscr{C}$ is closed under extensions, V is in $\mathscr{C}$. Using that $f\colon M \to C_M$ is a left $\mathscr{C}$-approximation, we get the commutative diagram with exact rows

$$
\begin{array}{ccccccccc}
0 & \longrightarrow & M & \xrightarrow{f} & C_M & \longrightarrow & Y_M & \longrightarrow & 0 \\
 & & \| & & \downarrow s & & \downarrow u & & \\
0 & \longrightarrow & M & \longrightarrow & V & \longrightarrow & U & \longrightarrow & 0 \\
 & & \| & & \downarrow t & & \downarrow v & & \\
0 & \longrightarrow & M & \xrightarrow{f} & C_M & \longrightarrow & Y_M & \longrightarrow & 0
\end{array}
$$

Since $\underline{f}\colon M \to C_M$ is left minimal in $\underline{\operatorname{mod}\Lambda}$, $\underline{ts}\colon C_M \to C_M$ is an isomorphism in $\underline{\operatorname{mod}\Lambda}$. Hence it follows from Lemma 4.4 that $\overline{vu}\colon Y_M \to Y_M$ is an isomorphism.

Consider now the exact sequence $0 \to C \to U \xrightarrow{v} Y_M \to 0$, and the induced exact sequence of functors $0 \to (Y_M, \ ) \to (U, \ ) \to (C, \ ) \to \operatorname{Ext}^1_\Lambda(Y_M, \ ) \xrightarrow{\operatorname{Ext}^1(v, \ )} \operatorname{Ext}^1_\Lambda(U, \ ) \to \operatorname{Ext}^1_\Lambda(C, \ )$. Since $\overline{vu}\colon Y_M \to Y_M$ is an isomorphism, $\operatorname{Ext}^1(v, \ )\colon \operatorname{Ext}^1_\Lambda(Y_M, \ ) \to \operatorname{Ext}^1_\Lambda(U, \ )$ is a split monomorphism. Hence $(U, \ ) \to (C, \ )$ is an epimorphism so that the sequence $0 \to C \to U \to Y_M \to 0$ splits. Hence we have $\operatorname{Ext}^1_\Lambda(Y_M, \mathscr{C}) = 0$. ∎

We are now ready to prove the converse of Proposition 4.2(b).

PROPOSITION 4.6. *Let Λ be a noetherian R-algebra. For a positive integer k we have that if $\Omega^i(\text{mod}\,\Lambda)$ is closed under extensions for $i \leq k$, then* $\text{grade}\,\text{Ext}^i_\Lambda(C, \Lambda) \geq i$ *for all C in* $\text{mod}\,\Lambda^{\text{op}}$ *and $i \leq k$.*

Proof. We have proved the claim for $k = 1$ in Proposition 4.2(a). Assume that $k > 1$ and that the claim has been proved for $k - 1$. So we assume that $\mathscr{X}_i$ is closed under extensions for $1 \leq i \leq k$, and that $\text{grade}\,\text{Ext}^i(C, \Lambda) \geq i$ for all C in $\text{mod}\,\Lambda^{\text{op}}$ and $i < k$. Let M be $(k-1)$-torsionfree and consider the exact sequence $0 \to M \to^{(f, f')} GF(M) \amalg Q \to \Omega^{k-2}\,\text{Ext}^k(\text{Tr}\,M, \Lambda) \to 0$ from Proposition 3.6, where $f\colon M \to GF(M)$ is left minimal by Lemma 4.3, and $G = \Omega^n$ and $F = \bar{\text{Tr}}\,\Omega^n\,\text{Tr}$. Then it follows by Lemma 4.5 that $\text{Ext}^1_\Lambda(\Omega^{k-2}\,\text{Ext}^k_\Lambda(\text{Tr}\,M, \Lambda), \Lambda) \simeq \text{Ext}^{k-1}_\Lambda(\text{Ext}^k_\Lambda(\text{Tr}\,M, \Lambda), \Lambda) = 0$. Since $\text{grade}\,\text{Ext}^k(\text{Tr}\,M, \Lambda) \geq k - 1$ by assumption, we now conclude that $\text{grade}\,\text{Ext}^k_\Lambda(\text{Tr}\,M, \Lambda) \geq k$. Now let N be in $\text{mod}\,\Lambda$. Since the modules in $\mathscr{X}_{k-1}$ are $(k-1)$-torsionfree by Proposition 1.6, it follows from [AB, Prop. 2.21] that $\text{Ext}^k_\Lambda(\text{Tr}\,N, \Lambda) \simeq \text{Ext}^k_\Lambda(\text{Tr}\,M, \Lambda)$, where M is $(k-1)$-torsionfree, and consequently $\text{grade}\,\text{Ext}^k_\Lambda(\text{Tr}\,N, \Lambda) \geq k$. ∎

Putting the results from all sections together we have the following.

THEOREM 4.7. *For a noetherian R-algebra Λ the following conditions are equivalent for a positive integer k.*

(a) *$\Omega^i(\text{mod}\,\Lambda)$ is closed under extensions for $i \leq k$.*

(b) *$\mathscr{X}_i$ is closed under extensions for $i \leq k$.*

(c) *If $0 \to \Lambda \to I_0 \to I_1 \to \cdots \to I_i \to \cdots$ is a minimal injective resolution of Λ as a right Λ-module, then* $\text{flatdim}\,I_i \leq i + 1$ *for $i < k$.*

(d) $\text{grade}\,\text{Ext}^i_\Lambda(C, \Lambda) \geq i$ *for all C in* $\text{mod}\,\Lambda^{\text{op}}$ *and $i \leq k$.*

(e) $\text{s.grade}\,\text{Ext}^{i+1}_\Lambda(B, \Lambda) \geq i$ *for all B in* $\text{mod}\,\Lambda$ *and $i \leq k$.*

As already pointed out in [AR3], contrary to the condition that Λ be k-Gorenstein, these conditions are not left–right symmetric.

REFERENCES

[AB] M. Auslander and M. Bridger, Stable module theory, *Mem. Amer. Math. Soc.* **94** (1969).

[AR1] M. Auslander and I. Reiten, Applications of contravariantly finite subcategories, *Adv. in Math.* **86**, No. 1 (1991), 111–152.

[AR2] M. Auslander and I. Reiten, Homologically finite subcategories, *in* "Proceedings, Tsukuba Workshop on Representations of Algebras," London Math. Soc. Lecture Notes, Vol. 168, Cambridge Univ. Press, Cambridge, UK.

[AR3] M. Auslander and I. Reiten, *k*-Gorenstein algebras and syzygy modules, *J. Pure Appl. Algebra* **92** (1994), 1–27.

[CE] H. Cartan and S. Eilenberg, "Homological Algebra," Princeton Univ. Press, Princeton, NJ, 1956.

[HR] P. J. Hilton and D. Rees, Natural maps of extension functors and a theorem of R. G. Swan, *Proc. Cambridge Philos. Soc.* (1961), 489–502.

Acknowledgments

The American Mathematical Society gratefully acknowledges the kindness of these institutions in granting the following permissions:

Academic Press, Inc.

(with I. Reiten) *Stable equivalence of dualizing R-varieties*, Adv. in Math. 12 (1974) No. 3, 306–366.

Large modules over artin algebras, Algebra, Topology and Category theory, (A collection of papers in honor of Samuel Eilenberg) 1–17 Academic Press, New York, 1976.

(with S. O. Smalø) *Preprojective modules over artin algebras*, J. Algebra 66, (1980) No. 1, 61–122.

(with S. O. Smalø) *Almost split sequences in subcategories*, J. Algebra 69, (1981) No. 2, 426–454, with Addendum J. Algebra, 71 (1981) No. 2 ,592–594.

(with J. F. Carlson) *Almost split sequences and group rings*, J. Algebra 103 (1986) No. 1, 122–140.

(with I. Reiten) *The Cohen-Macaulay type of Cohen-Macaulay rings*, Adv. in Math., 73 (1989), No. 1, 1–23.

(with I. Reiten) *Applications of contravariantly finite subcategories*, Adv. in Math. 86, No. 1 (1991) 111–152.

(with S. Ding and Ø. Solberg) *Liftings and weak liftings of modules*, J. Algebra 156 (1993) No. 2, 273–317.

(with M. Kleiner) *Adjoint functors and an extension of the Green correspondance for group representations*, Adv. in Math. 104, (1994) No. 2, 297–314.

(with I. Reiten) *Syzygy modules for noetherian rings*, J. Algebra 183 (1996) No. 1, 167–185.

Cambridge University Press

A survey of existence theorems for almost split sequences, Representations of algebras (Durham, 1985), 81–89, London Math. Soc. Lecture Note Ser. 116, Cambridge Univ. Press, Cambridge-New York, 1986. Reprinted with the permission of Cambridge University Press.

Canadian Mathematical Society

(with R. Bautista, M. I. Platzeck, I. Reiten and S. O. Smalø) *Almost split sequences whose middle term has at most two indecomposable summands*, Canad. J. Math. 31, (1979) No 5, 942–960. Reprinted with the permission of the Canadian Mathematical Society.

Elsevier Science–NL

(with I. Reiten) *Grothendieck groups of algebras and orders*, J. Pure Appl. Algebra 39 (1986) No. 1–2, 1–51. © 1986 with kind permission from Elsevier Science–NL, Sara Burgerhartstraat 25, 1055 KV, Amsterdam, The Netherlands.

(with I. Reiten) *k-Gorenstein algebras and syzygy modules*, J. Pure Appl. Algebra 92 (1994) 1–27. © 1994 with kind permission from Elsevier Science–NL, Sara Burgerhartstraat 25, 1055 KV, Amsterdam, The Netherlands.

Illinois Math Journal

Modules over unramified regular local rings, Illinois J. Math. 5 (1961) 631–647.

(with D. S. Rim) *Ramification index and multiplicity*, Illinois J. Math. 7 (1963) 566–581.

(with E. L. Green and I. Reiten) *Modules with waists*, Illinois J. Math. 19 (1975) 467–477.

(with I. Reiten) *Modules determined by their composition factors*, Illinois J. Math. 29 (1985) No. 2, 280–301.

Institut Mittag-Leffler
Modules over unramified regular local rings, Proc. International Congr. Mathematicians, Stockholm (1962) 230–233, Inst. Mittag-Leffler, Djursholm, 1963.

Koninklijke Nederlandse Akademie van Wetenschappen

(with A. Brumer) *Brauer groups of discrete valuation rings*, Nederl. Akad. Wetensch. Proc. Ser. A, 71, no. 3 (1968) 286–296.

Marcel Dekker, Inc.

Representation theory of artin algebras I, Comm. Algebra 1 (1974) 177–268. Marcel Dekker, Inc., NY, 1974.

Representation theory of artin algebras II, Comm. in Algebra 1 (1974) 269–310. Marcel Dekker, Inc., NY, 1974.

(with I. Reiten) *Representation theory of artin algebras* III*: Almost split sequences*, Comm. Algebra 3 (1975) 239–294. Marcel Dekker, Inc., NY, 1975.

(with I. Reiten) *Representation theory of artin algebras* IV*: Invariants given by almost split sequences*, Comm. Algebra 5 (1977) 443–518. Marcel Dekker, Inc., NY, 1977.

(with I. Reiten) *Representation theory of artin algebras* V*: Methods for computing almost split sequences and irreducible morphisms*, Comm. Algebra 5 (1977) No. 5, 519–554. Marcel Dekker, Inc., NY, 1977.

(with I. Reiten) *Representation theory of artin algebras* VI*: A functorial approach to almost split sequences*, Comm. Algebra 6 (1978) No. 3, 257–300. Marcel Dekker, Inc., NY, 1978.

(with M. I. Platzeck) *Representation theory of hereditary artin algebras*, Representation theory of algebras, (Proc. Conf. Temple Univ., 1976) 389–424. Lecture Notes in Pure Appl. Math. 37, Marcel Dekker, Inc., NY, 1978.

Functors and morphisms determined by objects, Representation theory of algebras (Proc. Conf. Temple Univ., 1976), 1–244. Lecture Notes in Pure Appl. Math. 37, Marcel Dekker, Inc., NY, 1978.

Applications of morphisms determined by modules, Representation theory of algebras (Proc. Conf. Temple Univ. 1976), 245–327. Lecture Notes in Pure Appl. Math. 37, Marcel Dekker, Inc., NY, 1978.

(with Ø. Solberg) *Relative homology and representation theory* I*, Relative homology and homologically finite subcategories*, Comm. Algebra 21 (9) (1993) 2995–3031. Marcel Dekker, Inc., NY, 1993.

(with Ø. Solberg) *Relative homology and representation theory* II*, Relative cotilting theory*, Comm. Algebra 21 (9) (1993) 3033–3079. Marcel Dekker, Inc., NY, 1993.

Mathematica Scandinavica

(with I. Reiten and S. O. Smalø) *Galois actions on rings and finite Galois coverings*, Math. Scand. 65 (1989) No. 1, 5–32.

Polish Scientific Publishers PWN

(with I. Reiten) *Graded modules and their completions*, Topics in Algebra, Part 1 (Warsaw 1989) 181–192, Banach Center Publ. 26, Part 1, PWN, Warsaw 1990.

Princeton University Press

(with D. Buchsbaum) *Codimension and multiplicity*, Ann. of Math. 68 (1958) 625–657. Correction, Ann. of Math. (2) 70 (1959) 395–397. Copyright © 1958, 1959, by Princeton University Press. Reprinted by permission of Princeton University Press.

Queen Mary and Westfield College, University of London

Representation dimension of artin algebras, Queen Mary College, Mathematics Notes, University of London (1971).

Societe Mathematique de France

(with R. O. Buchweitz) *The homological theory of maximal Cohen-Macaulay approximations*, Colloque en l'honneur de Pierre Samuel (Orsay 1987), Mém. Soc. Math. France, (N.S.) No. 38 (1989) 5–37.

Springer-Verlag New York, Inc.

Coherent functors, Proc. Conf. Categorial Algebra (La Jolla, Calif. 1965) 189–231, Springer, New York, 1966.

(with K. W. Roggenkamp) *A characterization of orders of finite lattice type*, Invent. Math. 17 (1972) 79–84.

(with I. Reiten) *Stable equivalence of artin algebras*, Proc. Conf. on Orders, Group rings and related topics (Ohio 1972) 8–71, Lecture Notes in Math. 353, Springer, Berlin 1973.

A functorial approach to representation theory, Representations of algebras. (ICRA III, Puebla 1980) 105–179. Lecture Notes in Math. 944, Springer, Berlin-New York, 1982.

Isolated singularities and existence of almost split sequences, Representation theory II (ICRA IV, Ottawa, 1984) 194–241, Lecture Notes in Math. 1178, Springer, Berlin-New York, 1986.

(with I. Reiten) *Almost split sequences for Cohen-Macaulay modules*, Math. Ann. 277 (1987) No. 2, 345–349.

(with I. Reiten) *Cohen-Macaulay modules for graded Cohen-Macaulay rings and their completions*, Commutative Algebra, (Berkeley, CA, 1987) 21–31, Math. Sci. Res. Inst. Publ. 15 Springer, Berlin-New York, 1989.

The Johns Hopkins University Press

(with R. C. Lyndon) *Commutator subgroups of free groups*, Amer. J. Math. 77 (1955) 929–931. ©The Johns Hopkins University Press.

(with D. Buchsbaum) *On ramification theory in noetherian rings*, Amer. J. Math. 81 (1959) 749–765. ©The Johns Hopkins University Press.

On the purity of the branch locus, Amer. J. Math. 84 (1962) 116–125. ©The Johns Hopkins University Press.

A remark on a paper of M. Hironaka, Amer. J. Math. 84 (1962) 8–10. ©The Johns Hopkins University Press.

University of Nagoya Press

On the dimension of modules and algebras (III), *Global dimension*, Nagoya Math. J. 9 (1955) 67-77.

On the dimension of modules and algebras (VI), *Comparison of global and algebra dimension*, Nagoya Math. J. 11 (1957) 61–65.

ISBN 0-8218-1000-6

9 780821 810002